Complete Solutions Manual
for
MULTIVARIABLE CALCULUS
SIXTH EDITION

DAN CLEGG
Palomar College

BARBARA FRANK
Cape Fear Community College

THOMSON
™
BROOKS/COLE

Australia ▫ Brazil ▫ Canada ▫ Mexico ▫ Singapore ▫ Spain ▫ United Kingdom ▫ United States

Printed in the United States of America

1 2 3 4 5 6 7 11 10 09 08 07

Printer: Thomson/West
Cover Image: © M. Neugebauer/zefa/Corbis

ISBN-13: 978-0-495-01229-0
ISBN-10: 0-495-01229-7

Thomson Higher Education
10 Davis Drive
Belmont, CA 94002-3098
USA

For more information about our products, contact us at:
Thomson Learning Academic Resource Center
1-800-423-0563

For permission to use material from this text or product, submit a request online at
http://www.thomsonrights.com.

Any additional questions about permissions can be submitted by email to **thomsonrights@thomson.com.**

☐ PREFACE

This *Complete Solutions Manual* contains detailed solutions to all exercises in the texts *Multivariable Calculus,* Sixth Edition and *Multivariable Calculus: Early Transcendentals,* Sixth Edition (Chapters 11–18 of *Calculus,* Sixth Edition and Chapters 10–17 of *Calculus: Early Transcendentals,* Sixth Edition) by James Stewart. A *Student Solutions Manual* is also available, which contains solutions to the odd-numbered exercises in each chapter section, review section, True-False Quiz, and Problems Plus section as well as all solutions to the Concept Check questions. (It does not, however, include solutions to any of the projects.)

The *Early Transcendentals* version of the text uses different chapter and page numbers; consequently, all section numbers and references are given in a dual format. Users of the *Early Transcendentals* text should use the references denoted by "ET."

While we have extended every effort to ensure the accuracy of the solutions presented, we would appreciate correspondence regarding any errors that may exist. Other suggestions or comments are also welcome, and can be sent to dan clegg at dclegg@palomar.edu or in care of the publisher: Thomson Brooks/Cole, 10 Davis Drive, Belmont CA 94002-3098.

We would like to thank James Stewart for entrusting us with the writing of this manual and offering suggestions, Kathi Townes and Stephanie Kuhns of TECH-arts for typesetting and producing this manual, and Brian Betsill of TECH-arts for creating the illustrations. We also thank Bob Pirtle and Stacy Green of Brooks/Cole for their trust, assistance, and patience.

DAN CLEGG
Palomar College

BARBARA FRANK
Cape Fear Community College

ABREVIATIONS AND SYMBOLS

CD	concave downward
CU	concave upward
D	the domain of f
FDT	First Derivative Test
HA	horizontal asymptote(s)
I	interval of convergence
I/D	Increasing/Decreasing Test
IP	inflection point(s)
R	radius of convergence
VA	vertical asymptote(s)

$\overset{CAS}{=}$ indicates the use of a computer algebra system.

$\overset{H}{=}$ indicates the use of l'Hospital's Rule.

$\overset{j}{=}$ indicates the use of Formula j in the Table of Integrals in the back endpapers.

$\overset{s}{=}$ indicates the use of the substitution $\{u = \sin x, du = \cos x\, dx\}$.

$\overset{c}{=}$ indicates the use of the substitution $\{u = \cos x, du = -\sin x\, dx\}$.

CONTENTS

11 □ PARAMETRIC EQUATIONS AND POLAR COORDINATES □ ET 10

11.1 Curves Defined by Parametric Equations

1. $x = 1 + \sqrt{t}, \quad y = t^2 - 4t, \quad 0 \le t \le 5$

t	0	1	2	3	4	5
x	1	2	$1 + \sqrt{2}$	$1 + \sqrt{3}$	3	$1 + \sqrt{5}$
			2.41	2.73		3.24
y	0	−3	−4	−3	0	5

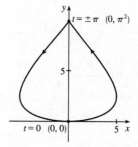

2. $x = 2 \cos t, \quad y = t - \cos t, \quad 0 \le t \le 2\pi$

t	0	$\pi/2$	π	$3\pi/2$	2π
x	2	0	−2	0	2
y	−1	$\pi/2$	$\pi + 1$	$3\pi/2$	$2\pi - 1$
		1.57	4.14	4.71	5.28

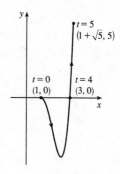

3. $x = 5 \sin t, \quad y = t^2, \quad -\pi \le t \le \pi$

t	$-\pi$	$-\pi/2$	0	$\pi/2$	π
x	0	−5	0	5	0
y	π^2	$\pi^2/4$	0	$\pi^2/4$	π^2
	9.87	2.47		2.47	9.87

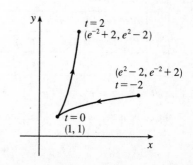

4. $x = e^{-t} + t, \quad y = e^t - t, \quad -2 \le t \le 2$

t	−2	−1	0	1	2
x	$e^2 - 2$	$e - 1$	1	$e^{-1} + 1$	$e^{-2} + 2$
	5.39	1.72		1.37	2.14
y	$e^{-2} + 2$	$e^{-1} + 1$	1	$e - 1$	$e^2 - 2$
	2.14	1.37		1.72	5.39

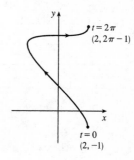

5. $x = 3t - 5$, $y = 2t + 1$

(a)

t	-2	-1	0	1	2	3	4
x	-11	-8	-5	-2	1	4	7
y	-3	-1	1	3	5	7	9

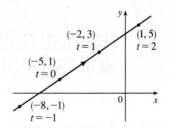

(b) $x = 3t - 5 \Rightarrow 3t = x + 5 \Rightarrow t = \frac{1}{3}(x + 5) \Rightarrow$

$y = 2 \cdot \frac{1}{3}(x + 5) + 1$, so $y = \frac{2}{3}x + \frac{13}{3}$.

6. $x = 1 + t$, $y = 5 - 2t$, $-2 \le t \le 3$

(a)

t	-2	-1	0	1	2	3
x	-1	0	1	2	3	4
y	9	7	5	3	1	-1

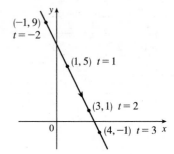

(b) $x = 1 + t \Rightarrow t = x - 1 \Rightarrow y = 5 - 2(x - 1)$,

so $y = -2x + 7$, $-1 \le x \le 4$.

7. $x = t^2 - 2$, $y = 5 - 2t$, $-3 \le t \le 4$

(a)

t	-3	-2	-1	0	1	2	3	4
x	7	2	-1	-2	-1	2	7	14
y	11	9	7	5	3	1	-1	-3

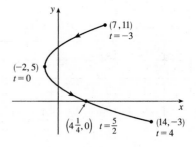

(b) $y = 5 - 2t \Rightarrow 2t = 5 - y \Rightarrow t = \frac{1}{2}(5 - y) \Rightarrow$

$x = \left[\frac{1}{2}(5 - y)\right]^2 - 2$, so $x = \frac{1}{4}(5 - y)^2 - 2$, $-3 \le y \le 11$.

8. $x = 1 + 3t$, $y = 2 - t^2$

(a)

t	-3	-2	-1	0	1	2	3
x	-8	-5	-2	1	4	7	10
y	-7	-2	1	2	1	-2	-7

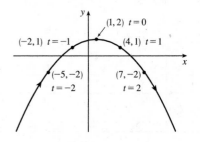

(b) $x = 1 + 3t \Rightarrow t = \frac{1}{3}(x - 1) \Rightarrow y = 2 - \left[\frac{1}{3}(x - 1)\right]^2$,

so $y = -\frac{1}{9}(x - 1)^2 + 2$.

9. $x = \sqrt{t}$, $y = 1 - t$

(a)

t	0	1	2	3	4
x	0	1	1.414	1.732	2
y	1	0	-1	-2	-3

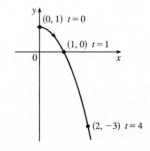

(b) $x = \sqrt{t} \Rightarrow t = x^2 \Rightarrow y = 1 - t = 1 - x^2$. Since $t \ge 0$, $x \ge 0$.

So the curve is the right half of the parabola $y = 1 - x^2$.

10. $x = t^2$, $y = t^3$

(a)

t	-2	-1	0	1	2
x	4	1	0	1	4
y	-8	-1	0	1	8

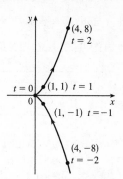

(b) $y = t^3$ $\Rightarrow$ $t = \sqrt[3]{y}$ $\Rightarrow$ $x = t^2 = \left(\sqrt[3]{y}\right)^2 = y^{2/3}$. $t \in \mathbb{R}$, $y \in \mathbb{R}$, $x \geq 0$.

11. (a) $x = \sin\theta$, $y = \cos\theta$, $0 \leq \theta \leq \pi$.

$x^2 + y^2 = \sin^2\theta + \cos^2\theta = 1$. Since $0 \leq \theta \leq \pi$, we have $\sin\theta \geq 0$, so $x \geq 0$. Thus, the curve is the right half of the circle $x^2 + y^2 = 1$.

(b)
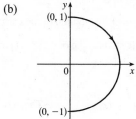

12. (a) $x = 4\cos\theta$, $y = 5\sin\theta$, $-\pi/2 \leq \theta \leq \pi/2$.

$\left(\frac{x}{4}\right)^2 + \left(\frac{y}{5}\right)^2 = \cos^2\theta + \sin^2\theta = 1$, which is an ellipse with x-intercepts $(\pm 4, 0)$ and y-intercepts $(0, \pm 5)$. We obtain the portion of the ellipse with $x \geq 0$ since $4\cos\theta \geq 0$ for $-\pi/2 \leq \theta \leq \pi/2$.

(b)
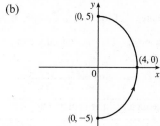

13. (a) $x = \sin t$, $y = \csc t$, $0 < t < \frac{\pi}{2}$.

$y = \csc t = \dfrac{1}{\sin t} = \dfrac{1}{x}$. For $0 < t < \frac{\pi}{2}$, we have $0 < x < 1$ and $y > 1$. Thus, the curve is the portion of the hyperbola $y = 1/x$ with $y > 1$.

(b)
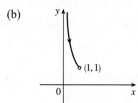

14. (a) $x = e^t - 1$, $y = e^{2t}$. $y = (e^t)^2 = (x+1)^2$ and since $x > -1$, we have the right side of the parabola $y = (x+1)^2$.

(b)
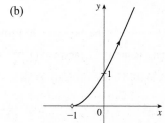

15. (a) $x = e^{2t}$ $\Rightarrow$ $2t = \ln x$ $\Rightarrow$ $t = \frac{1}{2}\ln x$.

$y = t + 1 = \frac{1}{2}\ln x + 1$.

(b)
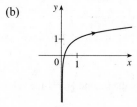

16. (a) $x = \ln t$, $y = \sqrt{t}$, $t \geq 1$.

$x = \ln t$ $\Rightarrow$ $t = e^x$ $\Rightarrow$ $y = \sqrt{t} = e^{x/2}$, $x \geq 0$.

(b)
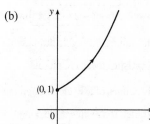

17. (a) $x = \sinh t$, $y = \cosh t$ $\Rightarrow$ $y^2 - x^2 = \cosh^2 t - \sinh^2 t = 1$. Since

(b)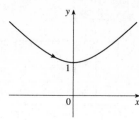

$y = \cosh t \geq 1$, we have the upper branch of the hyperbola $y^2 - x^2 = 1$.

18. (a) $x = 2\cosh t$, $y = 5\sinh t$ $\Rightarrow$ $\dfrac{x}{2} = \cosh t$, $\dfrac{y}{5} = \sinh t$ $\Rightarrow$

(b)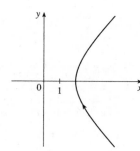

$\left(\dfrac{x}{2}\right)^2 = \cosh^2 t$, $\left(\dfrac{y}{5}\right)^2 = \sinh^2 t$. Since $\cosh^2 t - \sinh^2 t = 1$, we have

$\dfrac{x^2}{4} - \dfrac{y^2}{25} = 1$, a hyperbola. Because $x \geq 2$, we have the right branch of the

hyperbola.

19. $x = 3 + 2\cos t$, $y = 1 + 2\sin t$, $\pi/2 \leq t \leq 3\pi/2$. By Example 4 with $r = 2$, $h = 3$, and $k = 1$, the motion of the particle

takes place on a circle centered at $(3, 1)$ with a radius of 2. As t goes from $\frac{\pi}{2}$ to $\frac{3\pi}{2}$, the particle starts at the point $(3, 3)$ and

moves counterclockwise to $(3, -1)$ [one-half of a circle].

20. $x = 2\sin t$, $y = 4 + \cos t$ $\Rightarrow$ $\sin t = \dfrac{x}{2}$, $\cos t = y - 4$. $\sin^2 t + \cos^2 t = 1$ $\Rightarrow$ $\left(\dfrac{x}{2}\right)^2 + (y - 4)^2 = 1$. The motion

of the particle takes place on an ellipse centered at $(0, 4)$. As t goes from 0 to $\frac{3\pi}{2}$, the particle starts at the point $(0, 5)$ and

moves clockwise to $(-2, 4)$ [three-quarters of an ellipse].

21. $x = 5\sin t$, $y = 2\cos t$ $\Rightarrow$ $\sin t = \dfrac{x}{5}$, $\cos t = \dfrac{y}{2}$. $\sin^2 t + \cos^2 t = 1$ $\Rightarrow$ $\left(\dfrac{x}{5}\right)^2 + \left(\dfrac{y}{2}\right)^2 = 1$. The motion of the

particle takes place on an ellipse centered at $(0, 0)$. As t goes from $-\pi$ to 5π, the particle starts at the point $(0, -2)$ and moves

clockwise around the ellipse 3 times.

22. $y = \cos^2 t = 1 - \sin^2 t = 1 - x^2$. The motion of the particle takes place on the parabola $y = 1 - x^2$. As t goes from -2π to

$-\pi$, the particle starts at the point $(0, 1)$, moves to $(1, 0)$, and goes back to $(0, 1)$. As t goes from $-\pi$ to 0, the particle moves

to $(-1, 0)$ and goes back to $(0, 1)$. The particle repeats this motion as t goes from 0 to 2π.

23. We must have $1 \leq x \leq 4$ and $2 \leq y \leq 3$. So the graph of the curve must be contained in the rectangle $[1, 4]$ by $[2, 3]$.

24. (a) From the first graph, we have $1 \leq x \leq 2$. From the second graph, we have $-1 \leq y \leq 1$. The only choice that satisfies

either of those conditions is III.

(b) From the first graph, the values of x cycle through the values from -2 to 2 four times. From the second graph, the values

of y cycle through the values from -2 to 2 six times. Choice I satisfies these conditions.

(c) From the first graph, the values of x cycle through the values from -2 to 2 three times. From the second graph, we have

$0 \leq y \leq 2$. Choice IV satisfies these conditions.

(d) From the first graph, the values of x cycle through the values from -2 to 2 two times. From the second graph, the values of

y do the same thing. Choice II satisfies these conditions.

25. When $t = -1$, $(x, y) = (0, -1)$. As t increases to 0, x decreases to -1 and y

increases to 0. As t increases from 0 to 1, x increases to 0 and y increases to 1.

As t increases beyond 1, both x and y increase. For $t < -1$, x is positive and

decreasing and y is negative and increasing. We could achieve greater accuracy

by estimating x- and y-values for selected values of t from the given graphs and

plotting the corresponding points.

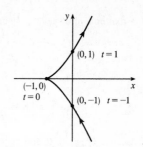

26. For $t < -1$, x is positive and decreasing, while y is negative and increasing (these

points are in Quadrant IV). When $t = -1$, $(x, y) = (0, 0)$ and, as t increases from

-1 to 0, x becomes negative and y increases from 0 to 1. At $t = 0$, $(x, y) = (0, 1)$

and, as t increases from 0 to 1, y decreases from 1 to 0 and x is positive. At

$t = 1$, $(x, y) = (0, 0)$ again, so the loop is completed. For $t > 1$, x and y both

become large negative. This enables us to draw a rough sketch. We could achieve greater accuracy by estimating x- and

y-values for selected values of t from the given graphs and plotting the corresponding points.

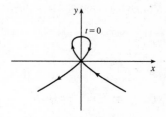

27. When $t = 0$ we see that $x = 0$ and $y = 0$, so the curve starts at the origin. As t

increases from 0 to $\frac{1}{2}$, the graphs show that y increases from 0 to 1 while x

increases from 0 to 1, decreases to 0 and to -1, then increases back to 0, so we

arrive at the point $(0, 1)$. Similarly, as t increases from $\frac{1}{2}$ to 1, y decreases from 1

to 0 while x repeats its pattern, and we arrive back at the origin. We could achieve greater accuracy by estimating x- and

y-values for selected values of t from the given graphs and plotting the corresponding points.

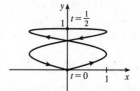

28. (a) $x = t^4 - t + 1 = (t^4 + 1) - t > 0$ [think of the graphs of $y = t^4 + 1$ and $y = t$] and $y = t^2 \geq 0$, so these equations

are matched with graph V.

(b) $y = \sqrt{t} \geq 0$. $x = t^2 - 2t = t(t - 2)$ is negative for $0 < t < 2$, so these equations are matched with graph I.

(c) $x = \sin 2t$ has period $2\pi/2 = \pi$. Note that

$y(t + 2\pi) = \sin[t + 2\pi + \sin 2(t + 2\pi)] = \sin(t + 2\pi + \sin 2t) = \sin(t + \sin 2t) = y(t)$, so y has period 2π.

These equations match graph II since x cycles through the values -1 to 1 twice as y cycles through those values once.

(d) $x = \cos 5t$ has period $2\pi/5$ and $y = \sin 2t$ has period π, so x will take on the values -1 to 1, and then 1 to -1, before y

takes on the values -1 to 1. Note that when $t = 0$, $(x, y) = (1, 0)$. These equations are matched with graph VI.

(e) $x = t + \sin 4t$, $y = t^2 + \cos 3t$.   As t becomes large, t and t^2 become the dominant terms in the expressions for x and

y, so the graph will look like the graph of $y = x^2$, but with oscillations. These equations are matched with graph IV.

(f) $x = \dfrac{\sin 2t}{4 + t^2}$, $y = \dfrac{\cos 2t}{4 + t^2}$.   As $t \to \infty$, x and y both approach 0. These equations are matched with graph III.

29. As in Example 6, we let $y = t$ and $x = t - 3t^3 + t^5$ and use a t-interval of $[-3, 3]$.

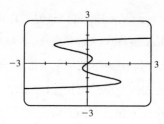

30. We use $x_1 = t$, $y_1 = t^5$ and $x_2 = t(t-1)^2$, $y_2 = t$ with $-3 \le t \le 3$.

There are 3 points of intersection; $(0, 0)$ is fairly obvious. The point in quadrant III

is approximately $(-0.8, -0.4)$ and the point in quadrant I is approximately

$(1.1, 1.8)$.

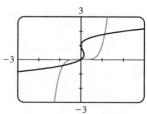

31. (a) $x = x_1 + (x_2 - x_1)t$, $y = y_1 + (y_2 - y_1)t$, $0 \le t \le 1$. Clearly the curve passes through $P_1(x_1, y_1)$ when $t = 0$ and

through $P_2(x_2, y_2)$ when $t = 1$. For $0 < t < 1$, x is strictly between x_1 and x_2 and y is strictly between y_1 and y_2. For

every value of t, x and y satisfy the relation $y - y_1 = \dfrac{y_2 - y_1}{x_2 - x_1}(x - x_1)$, which is the equation of the line through

$P_1(x_1, y_1)$ and $P_2(x_2, y_2)$.

Finally, any point (x, y) on that line satisfies $\dfrac{y - y_1}{y_2 - y_1} = \dfrac{x - x_1}{x_2 - x_1}$; if we call that common value t, then the given

parametric equations yield the point (x, y); and any (x, y) on the line between $P_1(x_1, y_1)$ and $P_2(x_2, y_2)$ yields a value of

t in $[0, 1]$. So the given parametric equations exactly specify the line segment from $P_1(x_1, y_1)$ to $P_2(x_2, y_2)$.

(b) $x = -2 + [3 - (-2)]t = -2 + 5t$ and $y = 7 + (-1 - 7)t = 7 - 8t$ for $0 \le t \le 1$.

32. For the side of the triangle from A to B, use $(x_1, y_1) = (1, 1)$ and $(x_2, y_2) = (4, 2)$.

Hence, the equations are

$$x = x_1 + (x_2 - x_1)\,t = 1 + (4 - 1)\,t = 1 + 3t,$$
$$y = y_1 + (y_2 - y_1)\,t = 1 + (2 - 1)\,t = 1 + t.$$

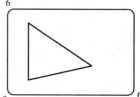

Graphing $x = 1 + 3t$ and $y = 1 + t$ with $0 \le t \le 1$ gives us the side of the

triangle from A to B. Similarly, for the side BC we use $x = 4 - 3t$ and $y = 2 + 3t$, and for the side AC we use $x = 1$

and $y = 1 + 4t$.

33. The circle $x^2 + (y - 1)^2 = 4$ has center $(0, 1)$ and radius 2, so by Example 4 it can be represented by $x = 2\cos t$,

$y = 1 + 2\sin t$, $0 \le t \le 2\pi$. This representation gives us the circle with a counterclockwise orientation starting at $(2, 1)$.

(a) To get a clockwise orientation, we could change the equations to $x = 2\cos t$, $y = 1 - 2\sin t$, $0 \le t \le 2\pi$.

(b) To get three times around in the counterclockwise direction, we use the original equations $x = 2\cos t$, $y = 1 + 2\sin t$ with

the domain expanded to $0 \le t \le 6\pi$.

(c) To start at $(0, 3)$ using the original equations, we must have $x_1 = 0$; that is, $2\cos t = 0$. Hence, $t = \frac{\pi}{2}$. So we use

$x = 2\cos t$, $y = 1 + 2\sin t$, $\frac{\pi}{2} \le t \le \frac{3\pi}{2}$.

Alternatively, if we want t to start at 0, we could change the equations of the curve. For example, we could use

$x = -2\sin t$, $y = 1 + 2\cos t$, $0 \le t \le \pi$.

34. (a) Let $x^2/a^2 = \sin^2 t$ and $y^2/b^2 = \cos^2 t$ to obtain $x = a \sin t$ and

$y = b \cos t$ with $0 \le t \le 2\pi$ as possible parametric equations for the ellipse

$x^2/a^2 + y^2/b^2 = 1$.

(b) The equations are $x = 3 \sin t$ and $y = b \cos t$ for $b \in \{1, 2, 4, 8\}$.

(c) As b increases, the ellipse stretches vertically.

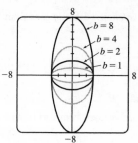

35. *Big circle:* It's centered at $(2, 2)$ with a radius of 2, so by Example 4, parametric equations are

$$x = 2 + 2\cos t, \qquad y = 2 + 2\sin t, \qquad 0 \le t \le 2\pi$$

Small circles: They are centered at $(1, 3)$ and $(3, 3)$ with a radius of 0.1. By Example 4, parametric equations are

$\qquad\qquad$ *(left)* $\qquad x = 1 + 0.1\cos t, \qquad y = 3 + 0.1\sin t, \qquad 0 \le t \le 2\pi$

and $\qquad\qquad$ *(right)* $\qquad x = 3 + 0.1\cos t, \qquad y = 3 + 0.1\sin t, \qquad 0 \le t \le 2\pi$

Semicircle: It's the lower half of a circle centered at $(2, 2)$ with radius 1. By Example 4, parametric equations are

$$x = 2 + 1\cos t, \qquad y = 2 + 1\sin t, \qquad \pi \le t \le 2\pi$$

To get all four graphs on the same screen with a typical graphing calculator, we need to change the last t-interval to $[0, 2\pi]$ in order to match the others. We can do this by changing t to $0.5t$. This change gives us the upper half. There are several ways to get the lower half—one is to change the "+" to a "−" in the y-assignment, giving us

$$x = 2 + 1\cos(0.5t), \qquad y = 2 - 1\sin(0.5t), \qquad 0 \le t \le 2\pi$$

36. If you are using a calculator or computer that can overlay graphs (using multiple t-intervals), the following is appropriate.

Left side: $x = 1$ and y goes from 1.5 to 4, so use

$$x = 1, \qquad y = t, \qquad 1.5 \le t \le 4$$

Right side: $x = 10$ and y goes from 1.5 to 4, so use

$$x = 10, \qquad y = t, \qquad 1.5 \le t \le 4$$

Bottom: x goes from 1 to 10 and $y = 1.5$, so use

$$x = t, \qquad y = 1.5, \qquad 1 \le t \le 10$$

Handle: It starts at $(10, 4)$ and ends at $(13, 7)$, so use

$$x = 10 + t, \qquad y = 4 + t, \qquad 0 \le t \le 3$$

Left wheel: It's centered at $(3, 1)$, has a radius of 1, and appears to go about $30°$ above the horizontal, so use

$$x = 3 + 1\cos t, \qquad y = 1 + 1\sin t, \qquad \tfrac{5\pi}{6} \le t \le \tfrac{13\pi}{6}$$

Right wheel: Similar to the left wheel with center $(8, 1)$, so use

$$x = 8 + 1\cos t, \qquad y = 1 + 1\sin t, \qquad \tfrac{5\pi}{6} \le t \le \tfrac{13\pi}{6}$$

$\quad$ If you are using a calculator or computer that cannot overlay graphs (using one t-interval), the following is appropriate. We'll start by picking the t-interval $[0, 2.5]$ since it easily matches the t-values for the two sides. We now need to find parametric equations for all graphs with $0 \le t \le 2.5$.

Left side: $x = 1$ and y goes from 1.5 to 4, so use

$$x = 1, \qquad y = 1.5 + t, \qquad 0 \le t \le 2.5$$

Right side: $x = 10$ and y goes from 1.5 to 4, so use

$$x = 10, \qquad y = 1.5 + t, \qquad 0 \le t \le 2.5$$

Bottom: x goes from 1 to 10 and $y = 1.5$, so use

$$x = 1 + 3.6t, \qquad y = 1.5, \qquad 0 \le t \le 2.5$$

To get the x-assignment, think of creating a linear function such that when $t = 0$, $x = 1$ and when $t = 2.5$, $x = 10$. We can use the point-slope form of a line with $(t_1, x_1) = (0, 1)$ and $(t_2, x_2) = (2.5, 10)$.

$$x - 1 = \frac{10 - 1}{2.5 - 0}(t - 0) \quad \Rightarrow \quad x = 1 + 3.6t.$$

Handle: It starts at $(10, 4)$ and ends at $(13, 7)$, so use

$$x = 10 + 1.2t, \qquad y = 4 + 1.2t, \qquad 0 \le t \le 2.5$$

$(t_1, x_1) = (0, 10)$ and $(t_2, x_2) = (2.5, 13)$ gives us $x - 10 = \dfrac{13 - 10}{2.5 - 0}(t - 0) \quad \Rightarrow \quad x = 10 + 1.2t.$

$(t_1, y_1) = (0, 4)$ and $(t_2, y_2) = (2.5, 7)$ gives us $y - 4 = \dfrac{7 - 4}{2.5 - 0}(t - 0) \quad \Rightarrow \quad y = 4 + 1.2t.$

Left wheel: It's centered at $(3, 1)$, has a radius of 1, and appears to go about $30°$ above the horizontal, so use

$$x = 3 + 1\cos\left(\tfrac{8\pi}{15}t + \tfrac{5\pi}{6}\right), \qquad y = 1 + 1\sin\left(\tfrac{8\pi}{15}t + \tfrac{5\pi}{6}\right), \qquad 0 \le t \le 2.5$$

$(t_1, \theta_1) = \left(0, \tfrac{5\pi}{6}\right)$ and $(t_2, \theta_2) = \left(\tfrac{5}{2}, \tfrac{13\pi}{6}\right)$ gives us $\theta - \tfrac{5\pi}{6} = \dfrac{\frac{13\pi}{6} - \frac{5\pi}{6}}{\frac{5}{2} - 0}(t - 0) \quad \Rightarrow \quad \theta = \tfrac{5\pi}{6} + \tfrac{8\pi}{15}t.$

Right wheel: Similar to the left wheel with center $(8, 1)$, so use

$$x = 8 + 1\cos\left(\tfrac{8\pi}{15}t + \tfrac{5\pi}{6}\right), \qquad y = 1 + 1\sin\left(\tfrac{8\pi}{15}t + \tfrac{5\pi}{6}\right), \qquad 0 \le t \le 2.5$$

37. (a) $x = t^3 \quad \Rightarrow \quad t = x^{1/3}$, so $y = t^2 = x^{2/3}$.

We get the entire curve $y = x^{2/3}$ traversed in a left to right direction.

(b) $x = t^6 \quad \Rightarrow \quad t = x^{1/6}$, so $y = t^4 = x^{4/6} = x^{2/3}$.

Since $x = t^6 \ge 0$, we only get the right half of the curve $y = x^{2/3}$.

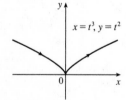

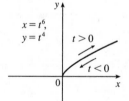

(c) $x = e^{-3t} = (e^{-t})^3 \quad$ [so $e^{-t} = x^{1/3}$],

$y = e^{-2t} = (e^{-t})^2 = (x^{1/3})^2 = x^{2/3}$.

If $t < 0$, then x and y are both larger than 1. If $t > 0$, then x and y are between 0 and 1. Since $x > 0$ and $y > 0$, the curve never quite reaches the origin.

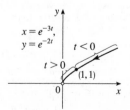

38. (a) $x = t$, so $y = t^{-2} = x^{-2}$. We get the entire curve $y = 1/x^2$ traversed in a

left-to-right direction.

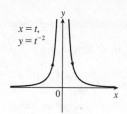

(b) $x = \cos t$, $y = \sec^2 t = \dfrac{1}{\cos^2 t} = \dfrac{1}{x^2}$. Since $\sec t \geq 1$, we only get the

parts of the curve $y = 1/x^2$ with $y \geq 1$. We get the first quadrant portion of

the curve when $x > 0$, that is, $\cos t > 0$, and we get the second quadrant

portion of the curve when $x < 0$, that is, $\cos t < 0$.

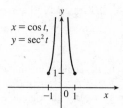

(c) $x = e^t$, $y = e^{-2t} = (e^t)^{-2} = x^{-2}$. Since e^t and e^{-2t} are both positive, we

only get the first quadrant portion of the curve $y = 1/x^2$.

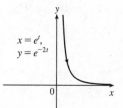

39. The case $\frac{\pi}{2} < \theta < \pi$ is illustrated. C has coordinates $(r\theta, r)$ as in Example 7,

and Q has coordinates $(r\theta, r + r\cos(\pi - \theta)) = (r\theta, r(1 - \cos\theta))$

[since $\cos(\pi - \alpha) = \cos\pi\cos\alpha + \sin\pi\sin\alpha = -\cos\alpha$], so P has coordinates

$(r\theta - r\sin(\pi - \theta), r(1 - \cos\theta)) = (r(\theta - \sin\theta), r(1 - \cos\theta))$

[since $\sin(\pi - \alpha) = \sin\pi\cos\alpha - \cos\pi\sin\alpha = \sin\alpha$]. Again we have the

parametric equations $x = r(\theta - \sin\theta)$, $y = r(1 - \cos\theta)$.

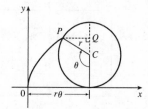

40. The first two diagrams depict the case $\pi < \theta < \frac{3\pi}{2}$, $d < r$. As in Example 7, C has coordinates $(r\theta, r)$. Now Q (in the second

diagram) has coordinates $(r\theta, r + d\cos(\theta - \pi)) = (r\theta, r - d\cos\theta)$, so a typical point P of the trochoid has coordinates

$(r\theta + d\sin(\theta - \pi), r - d\cos\theta)$. That is, P has coordinates (x, y), where $x = r\theta - d\sin\theta$ and $y = r - d\cos\theta$. When

$d = r$, these equations agree with those of the cycloid.

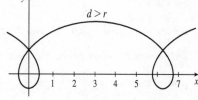

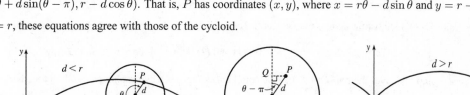

41. It is apparent that $x = |OQ|$ and $y = |QP| = |ST|$. From the diagram,

$x = |OQ| = a\cos\theta$ and $y = |ST| = b\sin\theta$. Thus, the parametric equations are

$x = a\cos\theta$ and $y = b\sin\theta$. To eliminate θ we rearrange: $\sin\theta = y/b \;\Rightarrow\;$

$\sin^2\theta = (y/b)^2$ and $\cos\theta = x/a \;\Rightarrow\; \cos^2\theta = (x/a)^2$. Adding the two

equations: $\sin^2\theta + \cos^2\theta = 1 = x^2/a^2 + y^2/b^2$. Thus, we have an ellipse.

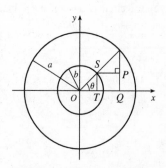

42. A has coordinates $(a \cos \theta, a \sin \theta)$. Since OA is perpendicular to AB, $\triangle OAB$ is a right triangle and B has coordinates

$(a \sec \theta, 0)$. It follows that P has coordinates $(a \sec \theta, b \sin \theta)$. Thus, the parametric equations are $x = a \sec \theta$, $y = b \sin \theta$.

43. $C = (2a \cot \theta, 2a)$, so the x-coordinate of P is $x = 2a \cot \theta$. Let $B = (0, 2a)$.

Then $\angle OAB$ is a right angle and $\angle OBA = \theta$, so $|OA| = 2a \sin \theta$ and

$A = ((2a \sin \theta) \cos \theta, (2a \sin \theta) \sin \theta)$. Thus, the y-coordinate of P

is $y = 2a \sin^2 \theta$.

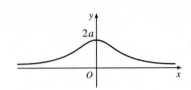

44. (a) Let θ be the angle of inclination of segment OP. Then $|OB| = \dfrac{2a}{\cos \theta}$. Let $C = (2a, 0)$.   (b)

Then by use of right triangle OAC we see that $|OA| = 2a \cos \theta$. Now

$$|OP| = |AB| = |OB| - |OA|$$

$$= 2a \left(\frac{1}{\cos \theta} - \cos \theta \right) = 2a \frac{1 - \cos^2 \theta}{\cos \theta} = 2a \frac{\sin^2 \theta}{\cos \theta} = 2a \sin \theta \tan \theta$$

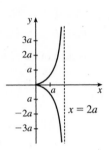

So P has coordinates $x = 2a \sin \theta \tan \theta \cdot \cos \theta = 2a \sin^2 \theta$ and

$y = 2a \sin \theta \tan \theta \cdot \sin \theta = 2a \sin^2 \theta \tan \theta$.

45. (a)

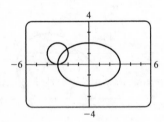

There are 2 points of intersection:

$(-3, 0)$ and approximately $(-2.1, 1.4)$.

(b) A collision point occurs when $x_1 = x_2$ and $y_1 = y_2$ for the same t. So solve the equations:

$$3 \sin t = -3 + \cos t \quad \textbf{(1)}$$

$$2 \cos t = 1 + \sin t \quad \textbf{(2)}$$

From **(2)**, $\sin t = 2 \cos t - 1$. Substituting into **(1)**, we get $3(2 \cos t - 1) = -3 + \cos t \Rightarrow 5 \cos t = 0 \ (\star) \Rightarrow$

$\cos t = 0 \Rightarrow t = \frac{\pi}{2}$ or $\frac{3\pi}{2}$. We check that $t = \frac{3\pi}{2}$ satisfies **(1)** and **(2)** but $t = \frac{\pi}{2}$ does not. So the only collision point

occurs when $t = \frac{3\pi}{2}$, and this gives the point $(-3, 0)$. [We could check our work by graphing x_1 and x_2 together as

functions of t and, on another plot, y_1 and y_2 as functions of t. If we do so, we see that the only value of t for which *both*

pairs of graphs intersect is $t = \frac{3\pi}{2}$.]

(c) The circle is centered at $(3, 1)$ instead of $(-3, 1)$. There are still 2 intersection points: $(3, 0)$ and $(2.1, 1.4)$, but there are

no collision points, since $(\star)$ in part (b) becomes $5 \cos t = 6 \Rightarrow \cos t = \frac{6}{5} > 1$.

46. (a) If $\alpha = 30°$ and $v_0 = 500$ m/s, then the equations become $x = (500 \cos 30°)t = 250 \sqrt{3}t$ and

$y = (500 \sin 30°)t - \frac{1}{2}(9.8)t^2 = 250t - 4.9t^2$. $y = 0$ when $t = 0$ (when the gun is fired) and again when

$t = \frac{250}{4.9} \approx 51$ s. Then $x = \left(250\sqrt{3}\right)\left(\frac{250}{4.9}\right) \approx 22{,}092$ m, so the bullet hits the ground about 22 km from the gun.

The formula for y is quadratic in t. To find the maximum y-value, we will complete the square:

$$y = -4.9\left(t^2 - \frac{250}{4.9}t\right) = -4.9\left[t^2 - \frac{250}{4.9}t + \left(\frac{125}{4.9}\right)^2\right] + \frac{125^2}{4.9} = -4.9\left(t - \frac{125}{4.9}\right)^2 + \frac{125^2}{4.9} \leq \frac{125^2}{4.9}$$

with equality when $t = \frac{125}{4.9}$ s, so the maximum height attained is $\frac{125^2}{4.9} \approx 3189$ m.

(b)

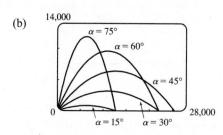

As α ($0° < \alpha < 90°$) increases up to $45°$, the projectile attains a greater height and a greater range. As α increases past $45°$, the projectile attains a greater height, but its range decreases.

(c) $x = (v_0 \cos\alpha)t \quad \Rightarrow \quad t = \dfrac{x}{v_0 \cos\alpha}$.

$$y = (v_0 \sin\alpha)t - \tfrac{1}{2}gt^2 \quad \Rightarrow \quad y = (v_0 \sin\alpha)\frac{x}{v_0\cos\alpha} - \frac{g}{2}\left(\frac{x}{v_0\cos\alpha}\right)^2 = (\tan\alpha)x - \left(\frac{g}{2v_0^2\cos^2\alpha}\right)x^2,$$

which is the equation of a parabola (quadratic in x).

47. $x = t^2, y = t^3 - ct$. We use a graphing device to produce the graphs for various values of c with $-\pi \leq t \leq \pi$. Note that all the members of the family are symmetric about the x-axis. For $c < 0$, the graph does not cross itself, but for $c = 0$ it has a cusp at $(0,0)$ and for $c > 0$ the graph crosses itself at $x = c$, so the loop grows larger as c increases.

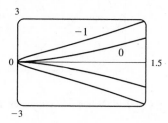
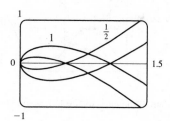

48. $x = 2ct - 4t^3, y = -ct^2 + 3t^4$. We use a graphing device to produce the graphs for various values of c with $-\pi \leq t \leq \pi$. Note that all the members of the family are symmetric about the y-axis. When $c < 0$, the graph resembles that of a polynomial of even degree, but when $c = 0$ there is a corner at the origin, and when $c > 0$, the graph crosses itself at the origin, and has two cusps below the x-axis. The size of the "swallowtail" increases as c increases.

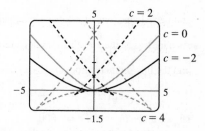

49. Note that all the Lissajous figures are symmetric about the x-axis. The parameters a and b simply stretch the graph in the x- and y-directions respectively. For $a = b = n = 1$ the graph is simply a circle with radius 1. For $n = 2$ the graph crosses itself at the origin and there are loops above and below the x-axis. In general, the figures have $n - 1$ points of intersection, all of which are on the y-axis, and a total of n closed loops.

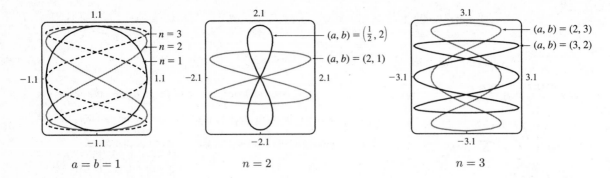

50. $x = \cos t$, $y = \sin t - \sin ct$. If $c = 1$, then $y = 0$, and the curve is simply the line segment from $(-1, 0)$ to $(1, 0)$. The graphs are shown for $c = 2, 3, 4$ and 5.

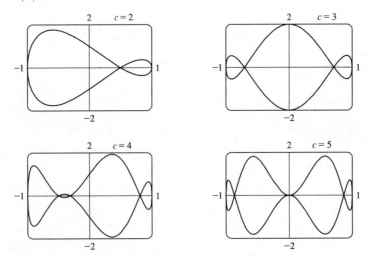

It is easy to see that all the curves lie in the rectangle $[-1, 1]$ by $[-2, 2]$. When c is an integer, $x(t + 2\pi) = x(t)$ and $y(t + 2\pi) = y(t)$, so the curve is closed. When c is a positive integer greater than 1, the curve intersects the x-axis $c + 1$ times and has c loops (one of which degenerates to a tangency at the origin when c is an odd integer of the form $4k + 1$).

As c increases, the curve's loops become thinner, but stay in the region bounded by the semicircles $y = \pm\left(1 + \sqrt{1 - x^2}\right)$ and the line segments from $(-1, -1)$ to $(-1, 1)$ and from $(1, -1)$ to $(1, 1)$. This is true because

$|y| = |\sin t - \sin ct| \leq |\sin t| + |\sin ct| \leq \sqrt{1 - x^2} + 1$. This curve appears to fill the entire region when c is very large, as shown in the figure for $c = 1000$.

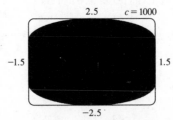

When c is a fraction, we get a variety of shapes with multiple loops, but always within the same region. For some fractional values, such as $c = 2.359$, the curve again appears to fill the region.

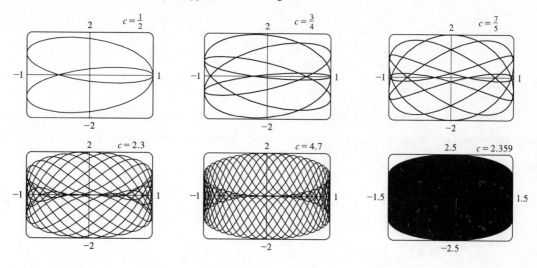

LABORATORY PROJECT   Running Circles Around Circles

1. The center Q of the smaller circle has coordinates $((a - b)\cos\theta, (a - b)\sin\theta)$. Arc PS on circle C has length $a\theta$ since it is equal in length to arc AS (the smaller circle rolls without slipping against the larger.)

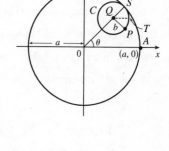

Thus, $\angle PQS = \dfrac{a}{b}\theta$ and $\angle PQT = \dfrac{a}{b}\theta - \theta$, so P has coordinates

$$x = (a - b)\cos\theta + b\cos(\angle PQT) = (a - b)\cos\theta + b\cos\left(\frac{a - b}{b}\theta\right)$$

and $y = (a - b)\sin\theta - b\sin(\angle PQT) = (a - b)\sin\theta - b\sin\left(\dfrac{a - b}{b}\theta\right)$.

2. With $b = 1$ and a a positive integer greater than 2, we obtain a hypocycloid of a cusps. Shown in the figure is the graph for $a = 4$. Let $a = 4$ and $b = 1$. Using the sum identities to expand $\cos 3\theta$ and $\sin 3\theta$, we obtain

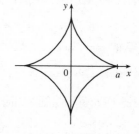

$$x = 3\cos\theta + \cos 3\theta = 3\cos\theta + \left(4\cos^3\theta - 3\cos\theta\right) = 4\cos^3\theta$$

and $y = 3\sin\theta - \sin 3\theta = 3\sin\theta - \left(3\sin\theta - 4\sin^3\theta\right) = 4\sin^3\theta$.

3. The graphs at the right are obtained with $b = 1$ and $a = \frac{1}{2}, \frac{1}{3}, \frac{1}{4}$, and $\frac{1}{10}$ with $-2\pi \le \theta \le 2\pi$. We conclude that as the denominator d increases, the graph gets smaller, but maintains the basic shape shown.

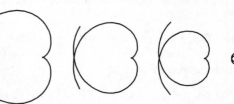

Letting $d = 2$ and $n = 3$, 5, and 7 with $-2\pi \leq \theta \leq 2\pi$ gives us the following:

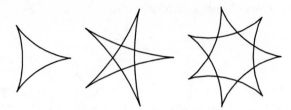

So if d is held constant and n varies, we get a graph with n cusps (assuming n/d is in lowest form). When $n = d + 1$, we obtain a hypocycloid of n cusps. As n increases, we must expand the range of θ in order to get a closed curve. The following graphs have $a = \frac{3}{2}, \frac{5}{4}$, and $\frac{11}{10}$.

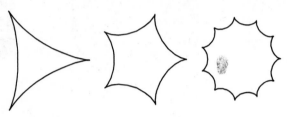

4. If $b = 1$, the equations for the hypocycloid are

$$x = (a - 1)\cos\theta + \cos((a-1)\theta) \qquad y = (a-1)\sin\theta - \sin((a-1)\theta)$$

which is a hypocycloid of a cusps (from Problem 2). In general, if $a > 1$, we get a figure with cusps on the "outside ring" and if $a < 1$, the cusps are on the "inside ring". In any case, as the values of θ get larger, we get a figure that looks more and more like a washer. If we were to graph the hypocycloid for all values of θ, every point on the washer would eventually be arbitrarily close to a point on the curve.

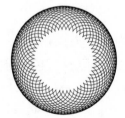

$$a = \sqrt{2}, \quad -10\pi \leq \theta \leq 10\pi \qquad\qquad a = e - 2, \quad 0 \leq \theta \leq 446$$

5. The center Q of the smaller circle has coordinates $((a + b)\cos\theta, (a + b)\sin\theta)$.

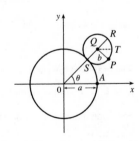

Arc PS has length $a\theta$ (as in Problem 1), so that $\angle PQS = \dfrac{a\theta}{b}$, $\angle PQR = \pi - \dfrac{a\theta}{b}$,

and $\angle PQT = \pi - \dfrac{a\theta}{b} - \theta = \pi - \left(\dfrac{a+b}{b}\right)\theta$ since $\angle RQT = \theta$.

Thus, the coordinates of P are

$$x = (a+b)\cos\theta + b\cos\left(\pi - \frac{a+b}{b}\theta\right) = (a+b)\cos\theta - b\cos\left(\frac{a+b}{b}\theta\right)$$

and $\quad y = (a+b)\sin\theta - b\sin\left(\pi - \dfrac{a+b}{b}\theta\right) = (a+b)\sin\theta - b\sin\left(\dfrac{a+b}{b}\theta\right)$.

6. Let $b = 1$ and the equations become

$$x = (a + 1)\cos\theta - \cos((a + 1)\theta) \qquad y = (a + 1)\sin\theta - \sin((a + 1)\theta)$$

If $a = 1$, we have a cardioid. If a is a positive integer greater than 1, we get the graph of an "a-leafed clover", with cusps that are a units from the origin. (Some of the pairs of figures are not to scale.)

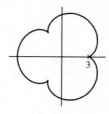

$a = 3, -2\pi \leq \theta \leq 2\pi$

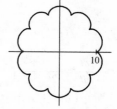

$a = 10, -2\pi \leq \theta \leq 2\pi$

If $a = n/d$ with $n = 1$, we obtain a figure that does not increase in size and requires $-d\pi \leq \theta \leq d\pi$ to be a closed curve traced exactly once.

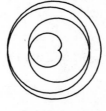

$a = \frac{1}{4}, -4\pi \leq \theta \leq 4\pi$

$a = \frac{1}{7}, -7\pi \leq \theta \leq 7\pi$

Next, we keep d constant and let n vary. As n increases, so does the size of the figure. There is an n-pointed star in the middle.

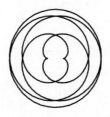

$a = \frac{2}{5}, -5\pi \leq \theta \leq 5\pi$

$a = \frac{7}{5}, -5\pi \leq \theta \leq 5\pi$

Now if $n = d + 1$ we obtain figures similar to the previous ones, but the size of the figure does not increase.

$a = \frac{4}{3}, -3\pi \leq \theta \leq 3\pi$

$a = \frac{7}{6}, -6\pi \leq \theta \leq 6\pi$

If a is irrational, we get washers that increase in size as a increases.

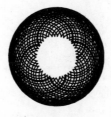

$a = \sqrt{2}, 0 \leq \theta \leq 200$

$a = e - 2, 0 \leq \theta \leq 446$

11.2 Calculus with Parametric Curves

1. $x = t \sin t$, $y = t^2 + t$ $\Rightarrow$ $\dfrac{dy}{dt} = 2t + 1$, $\dfrac{dx}{dt} = t \cos t + \sin t$, and $\dfrac{dy}{dx} = \dfrac{dy/dt}{dx/dt} = \dfrac{2t+1}{t \cos t + \sin t}$.

2. $x = \dfrac{1}{t}$, $y = \sqrt{t} \, e^{-t}$ $\Rightarrow$ $\dfrac{dy}{dt} = t^{1/2}(-e^{-t}) + e^{-t}\left(\frac{1}{2} t^{-1/2}\right) = \frac{1}{2} t^{-1/2} e^{-t}(-2t+1) = \dfrac{-2t+1}{2t^{1/2}e^t}$, $\dfrac{dx}{dt} = -\dfrac{1}{t^2}$, and

$\dfrac{dy}{dx} = \dfrac{dy/dt}{dx/dt} = \dfrac{-2t+1}{2t^{1/2}e^t}\left(-\dfrac{t^2}{1}\right) = \dfrac{(2t-1)t^{3/2}}{2e^t}$.

3. $x = t^4 + 1$, $y = t^3 + t$; $t = -1$. $\dfrac{dy}{dt} = 3t^2 + 1$, $\dfrac{dx}{dt} = 4t^3$, and $\dfrac{dy}{dx} = \dfrac{dy/dt}{dx/dt} = \dfrac{3t^2+1}{4t^3}$. When $t = -1$,

$(x, y) = (2, -2)$ and $dy/dx = \frac{4}{-4} = -1$, so an equation of the tangent to the curve at the point corresponding to $t = -1$

is $y - (-2) = (-1)(x - 2)$, or $y = -x$.

4. $x = t - t^{-1}$, $y = 1 + t^2$; $t = 1$. $\dfrac{dy}{dt} = 2t$, $\dfrac{dx}{dt} = 1 + t^{-2} = \dfrac{t^2+1}{t^2}$, and $\dfrac{dy}{dx} = \dfrac{dy/dt}{dx/dt} = 2t\left(\dfrac{t^2}{t^2+1}\right) = \dfrac{2t^3}{t^2+1}$.

When $t = 1$, $(x, y) = (0, 2)$ and $dy/dx = \frac{2}{2} = 1$, so an equation of the tangent to the curve at the point corresponding to

$t = 1$ is $y - 2 = 1(x - 0)$, or $y = x + 2$.

5. $x = e^{\sqrt{t}}$, $y = t - \ln t^2$; $t = 1$. $\dfrac{dy}{dt} = 1 - \dfrac{2t}{t^2} = 1 - \dfrac{2}{t}$, $\dfrac{dx}{dt} = \dfrac{e^{\sqrt{t}}}{2\sqrt{t}}$, and $\dfrac{dy}{dx} = \dfrac{dy/dt}{dx/dt} = \dfrac{1 - 2/t}{e^{\sqrt{t}}/(2\sqrt{t})} \cdot \dfrac{2t}{2t} = \dfrac{2t-4}{\sqrt{t}e^{\sqrt{t}}}$.

When $t = 1$, $(x, y) = (e, 1)$ and $\dfrac{dy}{dx} = -\dfrac{2}{e}$, so an equation of the tangent line is $y - 1 = -\dfrac{2}{e}(x - e)$, or $y = -\dfrac{2}{e}x + 3$.

6. $x = \cos\theta + \sin 2\theta$, $y = \sin\theta + \cos 2\theta$; $\theta = 0$. $\dfrac{dy}{dx} = \dfrac{dy/d\theta}{dx/d\theta} = \dfrac{\cos\theta - 2\sin 2\theta}{-\sin\theta + 2\cos 2\theta}$. When $\theta = 0$, $(x, y) = (1, 1)$ and

$dy/dx = \frac{1}{2}$, so an equation of the tangent to the curve is $y - 1 = \frac{1}{2}(x - 1)$, or $y = \frac{1}{2}x + \frac{1}{2}$.

7. (a) $x = 1 + \ln t$, $y = t^2 + 2$; $(1, 3)$. $\dfrac{dy}{dt} = 2t$, $\dfrac{dx}{dt} = \dfrac{1}{t}$, and $\dfrac{dy}{dx} = \dfrac{dy/dt}{dx/dt} = \dfrac{2t}{1/t} = 2t^2$.

At $(1, 3)$, $x = 1 + \ln t = 1$ $\Rightarrow$ $\ln t = 0$ $\Rightarrow$ $t = 1$ and $\dfrac{dy}{dx} = 2$, so an equation of the tangent is $y - 3 = 2(x - 1)$,

or $y = 2x + 1$.

(b) $x = 1 + \ln t$ $\Rightarrow$ $x - 1 = \ln t$ $\Rightarrow$ $t = e^{x-1}$, so $y = (e^{x-1})^2 + 2 = e^{2x-2} + 2$ and $\dfrac{dy}{dx} = 2e^{2x-2}$.

When $x = 1$, $\dfrac{dy}{dx} = 2e^0 = 2$, so an equation of the tangent is $y = 2x + 1$, as in part (a).

8. (a) $x = \tan\theta$, $y = \sec\theta$; $(1, \sqrt{2})$. $\dfrac{dy}{dx} = \dfrac{dy/d\theta}{dx/d\theta} = \dfrac{\sec\theta \tan\theta}{\sec^2\theta} = \dfrac{\tan\theta}{\sec\theta} = \sin\theta$.

When $(x, y) = (1, \sqrt{2})$, $\theta = \frac{\pi}{4}$ (or $\frac{\pi}{4} + 2\pi n$ for some integer n), so $dy/dx = \sin\frac{\pi}{4} = \sqrt{2}/2$.

Thus, an equation of the tangent to the curve is $y - \sqrt{2} = (\sqrt{2}/2)(x - 1)$, or $y = (\sqrt{2}/2)x + (\sqrt{2}/2)$.

(b) $\tan^2\theta + 1 = \sec^2\theta \;\Rightarrow\; x^2 + 1 = y^2$, so $\dfrac{d}{dx}(x^2 + 1) = \dfrac{d}{dx}(y^2) \;\Rightarrow\; 2x = 2y\,\dfrac{dy}{dx}$. When $(x, y) = (1, \sqrt{2}\,)$,

$\dfrac{dy}{dx} = \dfrac{x}{y} = \dfrac{1}{\sqrt{2}} = \dfrac{\sqrt{2}}{2}$, so an equation of the tangent is $y - \sqrt{2} = \left(\sqrt{2}/2\right)(x - 1)$, as in part (a).

9. $x = 6\sin t,\ \ y = t^2 + t;\ \ (0, 0)$.

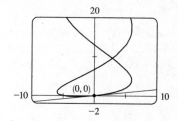

$\dfrac{dy}{dx} = \dfrac{dy/dt}{dx/dt} = \dfrac{2t + 1}{6\cos t}$. The point $(0, 0)$ corresponds to $t = 0$, so the

slope of the tangent at that point is $\frac{1}{6}$. An equation of the tangent is therefore

$y - 0 = \frac{1}{6}(x - 0)$, or $y = \frac{1}{6}x$.

10. $x = \cos t + \cos 2t,\ \ y = \sin t + \sin 2t;\ \ (-1, 1)$.

$\dfrac{dy}{dx} = \dfrac{dy/dt}{dx/dt} = \dfrac{\cos t + 2\cos 2t}{-\sin t - 2\sin 2t}$. To find the value of t corresponding to

the point $(-1, 1)$, solve $x = -1 \;\Rightarrow\; \cos t + \cos 2t = -1 \;\Rightarrow$

$\cos t + 2\cos^2 t - 1 = -1 \;\Rightarrow\; \cos t\,(1 + 2\cos t) = 0 \;\Rightarrow\; \cos t = 0$ or

$\cos t = -\frac{1}{2}$. The interval $[0, 2\pi]$ gives the complete curve, so we need only find

the values of t in this interval. Thus, $t = \frac{\pi}{2}$ or $t = \frac{2\pi}{3}$ or $t = \frac{4\pi}{3}$. Checking $t = \frac{\pi}{2}, \frac{3\pi}{2}, \frac{2\pi}{3}$, and $\frac{4\pi}{3}$ in the equation for y,

we find that $t = \frac{\pi}{2}$ corresponds to $(-1, 1)$. The slope of the tangent at $(-1, 1)$ with $t = \frac{\pi}{2}$ is $\dfrac{0 - 2}{-1 - 0} = 2$. An equation

of the tangent is therefore $y - 1 = 2(x + 1)$, or $y = 2x + 3$.

11. $x = 4 + t^2,\ \ y = t^2 + t^3 \;\Rightarrow\; \dfrac{dy}{dx} = \dfrac{dy/dt}{dx/dt} = \dfrac{2t + 3t^2}{2t} = 1 + \dfrac{3}{2}t \;\Rightarrow$

$\dfrac{d^2y}{dx^2} = \dfrac{d}{dx}\left(\dfrac{dy}{dx}\right) = \dfrac{d(dy/dx)/dt}{dx/dt} = \dfrac{(d/dt)\left(1 + \frac{3}{2}t\right)}{2t} = \dfrac{3/2}{2t} = \dfrac{3}{4t}$.

The curve is CU when $\dfrac{d^2y}{dx^2} > 0$, that is, when $t > 0$.

12. $x = t^3 - 12t,\ \ y = t^2 - 1 \;\Rightarrow\; \dfrac{dy}{dx} = \dfrac{dy/dt}{dx/dt} = \dfrac{2t}{3t^2 - 12} \;\Rightarrow$

$\dfrac{d^2y}{dx^2} = \dfrac{\dfrac{d}{dt}\left(\dfrac{dy}{dx}\right)}{dx/dt} = \dfrac{\dfrac{(3t^2 - 12)\cdot 2 - 2t(6t)}{(3t^2 - 12)^2}}{3t^2 - 12} = \dfrac{-6t^2 - 24}{(3t^2 - 12)^3} = \dfrac{-6(t^2 + 4)}{3^3(t^2 - 4)^3} = \dfrac{-2(t^2 + 4)}{9(t^2 - 4)^3}$.

Thus, the curve is CU when $t^2 - 4 < 0 \;\Rightarrow\; |t| < 2 \;\Rightarrow\; -2 < t < 2$.

13. $x = t - e^t,\ \ y = t + e^{-t} \;\Rightarrow$

$\dfrac{dy}{dx} = \dfrac{dy/dt}{dx/dt} = \dfrac{1 - e^{-t}}{1 - e^t} = \dfrac{1 - \dfrac{1}{e^t}}{1 - e^t} = \dfrac{\dfrac{e^t - 1}{e^t}}{1 - e^t} = -e^{-t} \;\Rightarrow\; \dfrac{d^2y}{dx^2} = \dfrac{\dfrac{d}{dt}\left(\dfrac{dy}{dx}\right)}{dx/dt} = \dfrac{\dfrac{d}{dt}(-e^{-t})}{dx/dt} = \dfrac{e^{-t}}{1 - e^t}$.

The curve is CU when $e^t < 1$ [since $e^{-t} > 0$] $\;\Rightarrow\; t < 0$.

14. $x = t + \ln t$, $y = t - \ln t$ $\Rightarrow$ $\dfrac{dy}{dx} = \dfrac{dy/dt}{dx/dt} = \dfrac{1 - 1/t}{1 + 1/t} = \dfrac{t-1}{t+1} = 1 - \dfrac{2}{t+1}$ $\Rightarrow$

$\dfrac{d^2y}{dx^2} = \dfrac{\dfrac{d}{dt}\left(\dfrac{dy}{dx}\right)}{dx/dt} = \dfrac{\dfrac{d}{dt}\left(1 - \dfrac{2}{t+1}\right)}{1 + 1/t} = \dfrac{2/(t+1)^2}{(t+1)/t} = \dfrac{2t}{(t+1)^3}$, so the curve is CU for all t in its domain,

that is, $t > 0$ [$t < -1$ not in domain].

15. $x = 2\sin t$, $y = 3\cos t$, $0 < t < 2\pi$.

$\dfrac{dy}{dx} = \dfrac{dy/dt}{dx/dt} = \dfrac{-3\sin t}{2\cos t} = -\dfrac{3}{2}\tan t$, so $\dfrac{d^2y}{dx^2} = \dfrac{\dfrac{d}{dt}\left(\dfrac{dy}{dx}\right)}{dx/dt} = \dfrac{-\frac{3}{2}\sec^2 t}{2\cos t} = -\dfrac{3}{4}\sec^3 t$.

The curve is CU when $\sec^3 t < 0$ $\Rightarrow$ $\sec t < 0$ $\Rightarrow$ $\cos t < 0$ $\Rightarrow$ $\frac{\pi}{2} < t < \frac{3\pi}{2}$.

16. $x = \cos 2t$, $y = \cos t$, $0 < t < \pi$.

$\dfrac{dy}{dx} = \dfrac{dy/dt}{dx/dt} = \dfrac{-\sin t}{-2\sin 2t} = \dfrac{\sin t}{2\cdot 2\sin t\cos t} = \dfrac{1}{4\cos t} = \dfrac{1}{4}\sec t$, so $\dfrac{d^2y}{dx^2} = \dfrac{\dfrac{d}{dt}\left(\dfrac{dy}{dx}\right)}{dx/dt} = \dfrac{\frac{1}{4}\sec t\tan t}{-4\sin t\cos t} = -\dfrac{1}{16}\sec^3 t$.

The curve is CU when $\sec^3 t < 0$ $\Rightarrow$ $\sec t < 0$ $\Rightarrow$ $\cos t < 0$ $\Rightarrow$ $\frac{\pi}{2} < t < \pi$.

17. $x = 10 - t^2$, $y = t^3 - 12t$.

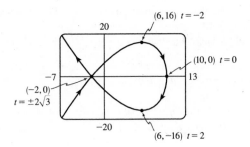

$\dfrac{dy}{dt} = 3t^2 - 12 = 3(t+2)(t-2)$, so $\dfrac{dy}{dt} = 0$ $\Leftrightarrow$

$t = \pm 2$ $\Leftrightarrow$ $(x, y) = (6, \mp 16)$.

$\dfrac{dx}{dt} = -2t$, so $\dfrac{dx}{dt} = 0$ $\Leftrightarrow$ $t = 0$ $\Leftrightarrow$ $(x, y) = (10, 0)$.

The curve has horizontal tangents at $(6, \pm 16)$ and a vertical

tangent at $(10, 0)$.

18. $x = 2t^3 + 3t^2 - 12t$, $y = 2t^3 + 3t^2 + 1$.

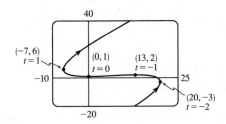

$\dfrac{dy}{dt} = 6t^2 + 6t = 6t(t+1)$, so $\dfrac{dy}{dt} = 0$ $\Leftrightarrow$

$t = 0$ or -1 $\Leftrightarrow$ $(x, y) = (0, 1)$ or $(13, 2)$.

$\dfrac{dx}{dt} = 6t^2 + 6t - 12 = 6(t+2)(t-1)$, so $\dfrac{dx}{dt} = 0$ $\Leftrightarrow$

$t = -2$ or 1 $\Leftrightarrow$ $(x, y) = (20, -3)$ or $(-7, 6)$.

The curve has horizontal tangents at $(0, 1)$ and $(13, 2)$, and vertical tangents at $(20, -3)$ and $(-7, 6)$.

19. $x = 2\cos\theta$, $y = \sin 2\theta$.

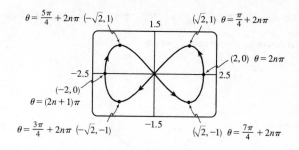

$\dfrac{dy}{d\theta} = 2\cos 2\theta$, so $\dfrac{dy}{d\theta} = 0 \iff 2\theta = \frac{\pi}{2} + n\pi$

[n an integer] $\iff \theta = \frac{\pi}{4} + \frac{\pi}{2}n \iff$

$(x, y) = (\pm\sqrt{2}, \pm 1)$. Also, $\dfrac{dx}{d\theta} = -2\sin\theta$, so

$\dfrac{dx}{d\theta} = 0 \iff \theta = n\pi \iff (x, y) = (\pm 2, 0)$.

The curve has horizontal tangents at $\left(\pm\sqrt{2}, \pm 1\right)$ (four points), and vertical tangents at $(\pm 2, 0)$.

20. $x = \cos 3\theta$, $y = 2\sin\theta$. $dy/d\theta = 2\cos\theta$, so $dy/d\theta = 0 \iff \theta = \frac{\pi}{2} + n\pi$ (n an integer) $\iff (x, y) = (0, \pm 2)$.

Also, $dx/d\theta = -3\sin 3\theta$, so $dx/d\theta = 0 \iff 3\theta = n\pi \iff \theta = \frac{\pi}{3}n \iff (x, y) = (\pm 1, 0)$ or $\left(\pm 1, \pm\sqrt{3}\right)$.

The curve has horizontal tangents at $(0, \pm 2)$, and vertical tangents at $(\pm 1, 0)$, $\left(\pm 1, -\sqrt{3}\right)$ and $\left(\pm 1, \sqrt{3}\right)$.

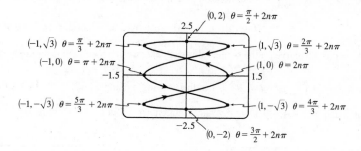

21. From the graph, it appears that the rightmost point on the curve $x = t - t^6$, $y = e^t$

is about $(0.6, 2)$. To find the exact coordinates, we find the value of t for which the

graph has a vertical tangent, that is, $0 = dx/dt = 1 - 6t^5 \iff t = 1/\sqrt[5]{6}$.

Hence, the rightmost point is

$$\left(1/\sqrt[5]{6} - 1/\left(6\sqrt[5]{6}\right),\, e^{1/\sqrt[5]{6}}\right) = \left(5 \cdot 6^{-6/5},\, e^{6^{-1/5}}\right) \approx (0.58, 2.01).$$

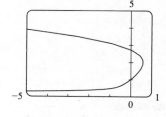

22. From the graph, it appears that the lowest point and the leftmost point on the curve

$x = t^4 - 2t$, $y = t + t^4$ are $(1.5, -0.5)$ and $(-1.2, 1.2)$, respectively. To find the

exact coordinates, we solve $dy/dt = 0$ (horizontal tangents) and $dx/dt = 0$

(vertical tangents).

$\dfrac{dy}{dt} = 0 \iff 1 + 4t^3 = 0 \iff t = -\dfrac{1}{\sqrt[3]{4}}$, so the lowest point is

$$\left(\dfrac{1}{\sqrt[3]{256}} + \dfrac{2}{\sqrt[3]{4}}, -\dfrac{1}{\sqrt[3]{4}} + \dfrac{1}{\sqrt[3]{256}}\right) = \left(\dfrac{9}{\sqrt[3]{256}}, -\dfrac{3}{\sqrt[3]{256}}\right) \approx (1.42, -0.47).$$

$\dfrac{dx}{dt} = 0 \iff 4t^3 - 2 = 0 \iff t = \dfrac{1}{\sqrt[3]{2}}$, so the leftmost point is

$$\left(\dfrac{1}{\sqrt[3]{16}} - \dfrac{2}{\sqrt[3]{2}}, \dfrac{1}{\sqrt[3]{2}} + \dfrac{1}{\sqrt[3]{16}}\right) = \left(-\dfrac{3}{\sqrt[3]{16}}, \dfrac{3}{\sqrt[3]{16}}\right) \approx (-1.19, 1.19).$$

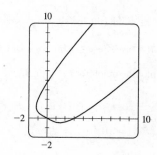

23. We graph the curve $x = t^4 - 2t^3 - 2t^2$, $y = t^3 - t$ in the viewing rectangle $[-2, 1.1]$ by $[-0.5, 0.5]$. This rectangle

corresponds approximately to $t \in [-1, 0.8]$.

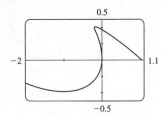

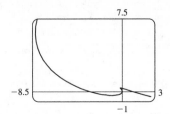

We estimate that the curve has horizontal tangents at about $(-1, -0.4)$ and $(-0.17, 0.39)$ and vertical tangents at

about $(0, 0)$ and $(-0.19, 0.37)$. We calculate $\dfrac{dy}{dx} = \dfrac{dy/dt}{dx/dt} = \dfrac{3t^2 - 1}{4t^3 - 6t^2 - 4t}$. The horizontal tangents occur when

$dy/dt = 3t^2 - 1 = 0 \iff t = \pm\frac{1}{\sqrt{3}}$, so both horizontal tangents are shown in our graph. The vertical tangents occur when

$dx/dt = 2t(2t^2 - 3t - 2) = 0 \iff 2t(2t + 1)(t - 2) = 0 \iff t = 0, -\frac{1}{2}$ or 2. It seems that we have missed one vertical

tangent, and indeed if we plot the curve on the t-interval $[-1.2, 2.2]$ we see that there is another vertical tangent at $(-8, 6)$.

24. We graph the curve $x = t^4 + 4t^3 - 8t^2$, $y = 2t^2 - t$ in the viewing rectangle $[-3.7, 0.2]$ by $[-0.2, 1.4]$. It appears that there

is a horizontal tangent at about $(-0.4, -0.1)$, and vertical tangents at about $(-3, 1)$ and $(0, 0)$.

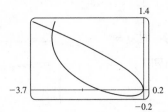

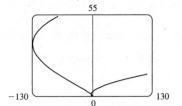

We calculate $\dfrac{dy}{dx} = \dfrac{dy/dt}{dx/dt} = \dfrac{4t - 1}{4t^3 + 12t^2 - 16t}$, so there is a horizontal tangent where $dy/dt = 4t - 1 = 0 \iff t = \frac{1}{4}$.

This point (the lowest point) is shown in the first graph. There are vertical tangents where $dx/dt = 4t^3 + 12t^2 - 16t = 0 \iff$

$4t(t^2 + 3t - 4) = 0 \iff 4t(t + 4)(t - 1) = 0$. We have missed one vertical tangent corresponding to $t = -4$, and if we

plot the graph for $t \in [-5, 3]$, we see that the curve has another vertical tangent line at approximately $(-128, 36)$.

25. $x = \cos t$, $y = \sin t \cos t$. $dx/dt = -\sin t$, $dy/dt = -\sin^2 t + \cos^2 t = \cos 2t$.

$(x, y) = (0, 0) \iff \cos t = 0 \iff t$ is an odd multiple of $\frac{\pi}{2}$. When $t = \frac{\pi}{2}$,

$dx/dt = -1$ and $dy/dt = -1$, so $dy/dx = 1$. When $t = \frac{3\pi}{2}$, $dx/dt = 1$ and

$dy/dt = -1$. So $dy/dx = -1$. Thus, $y = x$ and $y = -x$ are both tangent to the

curve at $(0, 0)$.

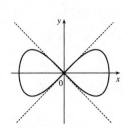

26.

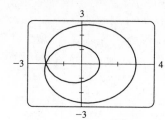

From the graph, we discover that the graph of the curve $x = \cos t + 2\cos 2t$, $y = \sin t + 2\sin 2t$ crosses itself at the point $(-2, 0)$. To find t at $(-2, 0)$, solve $y = 0 \iff \sin t + 2\sin 2t = 0 \iff \sin t + 4\sin t \cos t = 0 \iff \sin t (1 + 4\cos t) = 0 \iff \sin t = 0$ or $\cos t = -\frac{1}{4}$. We find that $t = \pm \arccos\left(-\frac{1}{4}\right)$ corresponds to $(-2, 0)$.

Now $\dfrac{dy}{dx} = \dfrac{dy/dt}{dx/dt} = \dfrac{\cos t + 4\cos 2t}{-\sin t - 4\sin 2t} = -\dfrac{\cos t + 8\cos^2 t - 4}{\sin t + 8\sin t \cos t}$. When $t = \arccos\left(-\frac{1}{4}\right)$, $\cos t = -\frac{1}{4}$, $\sin t = \dfrac{\sqrt{15}}{4}$,

and $\dfrac{dy}{dx} = -\dfrac{-\frac{1}{4} + \frac{1}{2} - 4}{\frac{\sqrt{15}}{4} - \frac{\sqrt{15}}{2}} = -\dfrac{-\frac{15}{4}}{-\frac{\sqrt{15}}{4}} = -\sqrt{15}$. By symmetry, $t = -\arccos\left(-\frac{1}{4}\right) \implies \dfrac{dy}{dx} = \sqrt{15}$.

The tangent lines are $y - 0 = \pm\sqrt{15}\,(x + 2)$, or $y = \sqrt{15}\,x + 2\sqrt{15}$ and $y = -\sqrt{15}\,x - 2\sqrt{15}$.

27. $x = r\theta - d\sin\theta, \ y = r - d\cos\theta$.

(a) $\dfrac{dx}{d\theta} = r - d\cos\theta, \dfrac{dy}{d\theta} = d\sin\theta$, so $\dfrac{dy}{dx} = \dfrac{d\sin\theta}{r - d\cos\theta}$.

(b) If $0 < d < r$, then $|d\cos\theta| \le d < r$, so $r - d\cos\theta \ge r - d > 0$. This shows that $dx/d\theta$ never vanishes, so the trochoid can have no vertical tangent if $d < r$.

28. $x = a\cos^3\theta, \ y = a\sin^3\theta$.

(a) $\dfrac{dx}{d\theta} = -3a\cos^2\theta \sin\theta, \dfrac{dy}{d\theta} = 3a\sin^2\theta\cos\theta$, so $\dfrac{dy}{dx} = -\dfrac{\sin\theta}{\cos\theta} = -\tan\theta$.

(b) The tangent is horizontal $\iff dy/dx = 0 \iff \tan\theta = 0 \iff \theta = n\pi \iff (x, y) = (\pm a, 0)$.

The tangent is vertical $\iff \cos\theta = 0 \iff \theta$ is an odd multiple of $\frac{\pi}{2} \iff (x, y) = (0, \pm a)$.

(c) $dy/dx = \pm 1 \iff \tan\theta = \pm 1 \iff \theta$ is an odd multiple of $\frac{\pi}{4} \iff (x, y) = \left(\pm\frac{\sqrt{2}}{4}a, \pm\frac{\sqrt{2}}{4}a\right)$

[All sign choices are valid.]

29. $x = 2t^3, \ y = 1 + 4t - t^2 \implies \dfrac{dy}{dx} = \dfrac{dy/dt}{dx/dt} = \dfrac{4 - 2t}{6t^2}$. Now solve $\dfrac{dy}{dx} = 1 \iff \dfrac{4 - 2t}{6t^2} = 1 \iff$

$6t^2 + 2t - 4 = 0 \iff 2(3t - 2)(t + 1) = 0 \iff t = \frac{2}{3}$ or $t = -1$. If $t = \frac{2}{3}$, the point is $\left(\frac{16}{27}, \frac{29}{9}\right)$, and if $t = -1$, the point is $(-2, -4)$.

30. $x = 3t^2 + 1, y = 2t^3 + 1, \dfrac{dx}{dt} = 6t, \dfrac{dy}{dt} = 6t^2$, so $\dfrac{dy}{dx} = \dfrac{6t^2}{6t} = t$ [even where $t = 0$].

So at the point corresponding to parameter value t, an equation of the tangent line is $y - (2t^3 + 1) = t[x - (3t^2 + 1)]$.

If this line is to pass through $(4, 3)$, we must have $3 - (2t^3 + 1) = t[4 - (3t^2 + 1)] \iff 2t^3 - 2 = 3t^3 - 3t \iff$

$t^3 - 3t + 2 = 0 \iff (t - 1)^2(t + 2) = 0 \iff t = 1$ or -2. Hence, the desired equations are $y - 3 = x - 4$, or

$y = x - 1$, tangent to the curve at $(4, 3)$, and $y - (-15) = -2(x - 13)$, or $y = -2x + 11$, tangent to the curve at $(13, -15)$.

31. By symmetry of the ellipse about the x- and y-axes,

$$A = 4\int_0^a y\,dx = 4\int_{\pi/2}^0 b\sin\theta\,(-a\sin\theta)\,d\theta = 4ab\int_0^{\pi/2}\sin^2\theta\,d\theta = 4ab\int_0^{\pi/2}\tfrac{1}{2}(1-\cos 2\theta)\,d\theta$$

$$= 2ab\big[\theta - \tfrac{1}{2}\sin 2\theta\big]_0^{\pi/2} = 2ab\big(\tfrac{\pi}{2}\big) = \pi ab$$

32. The curve $x = t^2 - 2t = t(t-2)$, $y = \sqrt{t}$ intersects the y-axis when $x = 0$,

that is, when $t = 0$ and $t = 2$. The corresponding values of y are 0 and $\sqrt{2}$.

The shaded area is given by

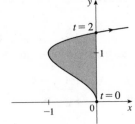

$$\int_{y=0}^{y=\sqrt{2}} (x_R - x_L)\,dy = \int_{t=0}^{t=2} [0 - x(t)]\,y'(t)\,dt = -\int_0^2 (t^2 - 2t)\left(\frac{1}{2\sqrt{t}}\,dt\right)$$

$$= -\int_0^2\left(\tfrac{1}{2}t^{3/2} - t^{1/2}\right)dt = -\left[\tfrac{1}{5}t^{5/2} - \tfrac{2}{3}t^{3/2}\right]_0^2$$

$$= -\left(\tfrac{1}{5}\cdot 2^{5/2} - \tfrac{2}{3}\cdot 2^{3/2}\right) = -2^{1/2}\big(\tfrac{4}{5} - \tfrac{4}{3}\big) = -\sqrt{2}\big(-\tfrac{8}{15}\big) = \tfrac{8}{15}\sqrt{2}$$

33. The curve $x = 1 + e^t$, $y = t - t^2 = t(1-t)$ intersects the x-axis when $y = 0$,

that is, when $t = 0$ and $t = 1$. The corresponding values of x are 2 and $1 + e$.

The shaded area is given by

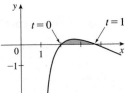

$$\int_{x=2}^{x=1+e} (y_T - y_B)\,dx = \int_{t=0}^{t=1} [y(t) - 0]\,x'(t)\,dt = \int_0^1 (t - t^2)e^t\,dt$$

$$= \int_0^1 te^t\,dt - \int_0^1 t^2 e^t\,dt = \int_0^1 te^t\,dt - \big[t^2 e^t\big]_0^1 + 2\int_0^1 te^t\,dt \qquad \text{[Formula 97 or parts]}$$

$$= 3\int_0^1 te^t\,dt - (e - 0) = 3\big[(t-1)e^t\big]_0^1 - e \qquad \text{[Formula 96 or parts]}$$

$$= 3[0 - (-1)] - e = 3 - e$$

34. By symmetry, $A = 4\int_0^a y\,dx = 4\int_{\pi/2}^0 a\sin^3\theta(-3a\cos^2\theta\,\sin\theta)\,d\theta = 12a^2\int_0^{\pi/2}\sin^4\theta\,\cos^2\theta\,d\theta$. Now

$$\int \sin^4\theta\,\cos^2\theta\,d\theta = \int \sin^2\theta\big(\tfrac{1}{4}\sin^2 2\theta\big)\,d\theta = \tfrac{1}{8}\int(1-\cos 2\theta)\sin^2 2\theta\,d\theta$$

$$= \tfrac{1}{8}\int\big[\tfrac{1}{2}(1-\cos 4\theta) - \sin^2 2\theta\,\cos 2\theta\big]\,d\theta = \tfrac{1}{16}\theta - \tfrac{1}{64}\sin 4\theta - \tfrac{1}{48}\sin^3 2\theta + C$$

so $\int_0^{\pi/2}\sin^4\theta\,\cos^2\theta\,d\theta = \big[\tfrac{1}{16}\theta - \tfrac{1}{64}\sin 4\theta - \tfrac{1}{48}\sin^3 2\theta\big]_0^{\pi/2} = \tfrac{\pi}{32}$. Thus, $A = 12a^2\big(\tfrac{\pi}{32}\big) = \tfrac{3}{8}\pi a^2$.

35. $x = r\theta - d\sin\theta$, $y = r - d\cos\theta$.

$$A = \int_0^{2\pi r} y\,dx = \int_0^{2\pi}(r - d\cos\theta)(r - d\cos\theta)\,d\theta = \int_0^{2\pi}(r^2 - 2dr\cos\theta + d^2\cos^2\theta)\,d\theta$$

$$= \big[r^2\theta - 2dr\sin\theta + \tfrac{1}{2}d^2\big(\theta + \tfrac{1}{2}\sin 2\theta\big)\big]_0^{2\pi} = 2\pi r^2 + \pi d^2$$

36. (a) By symmetry, the area of $\mathcal{R}$ is twice the area inside $\mathcal{R}$ above the x-axis. The top half of the loop is described by

$x = t^2$, $y = t^3 - 3t$, $-\sqrt{3} \le t \le 0$, so, using the Substitution Rule with $y = t^3 - 3t$ and $dx = 2t\,dt$, we find that

$$\text{area} = 2\int_0^3 y\,dx = 2\int_0^{-\sqrt{3}}(t^3 - 3t)2t\,dt = 2\int_0^{-\sqrt{3}}(2t^4 - 6t^2)\,dt = 2\big[\tfrac{2}{5}t^5 - 2t^3\big]_0^{-\sqrt{3}}$$

$$= 2\big[\tfrac{2}{5}(-3^{1/2})^5 - 2(-3^{1/2})^3\big] = 2\big[\tfrac{2}{5}(-9\sqrt{3}) - 2(-3\sqrt{3})\big] = \tfrac{24}{5}\sqrt{3}$$

(b) Here we use the formula for disks and use the Substitution Rule as in part (a):

$$\text{volume} = \pi \int_0^3 y^2 \, dx = \pi \int_0^{-\sqrt{3}} (t^3 - 3t)^2 2t \, dt = 2\pi \int_0^{-\sqrt{3}} (t^6 - 6t^4 + 9t^2) t \, dt = 2\pi \left[\tfrac{1}{8} t^8 - t^6 + \tfrac{9}{4} t^4 \right]_0^{-\sqrt{3}}$$

$$= 2\pi \left[\tfrac{1}{8} (-3^{1/2})^8 - (-3^{1/2})^6 + \tfrac{9}{4} (-3^{1/2})^4 \right] = 2\pi \left[\tfrac{81}{8} - 27 + \tfrac{81}{4} \right] = \tfrac{27}{4} \pi$$

(c) By symmetry, the y-coordinate of the centroid is 0. To find the x-coordinate, we note that it is the same as the x-coordinate of the centroid of the top half of $\mathcal{R}$, the area of which is $\frac{1}{2} \cdot \frac{24}{5} \sqrt{3} = \frac{12}{5} \sqrt{3}$. So, using Formula 9.3.8 [ET 8.3.8] with $A = \frac{12}{5} \sqrt{3}$, we get

$$\bar{x} = \tfrac{5}{12\sqrt{3}} \int_0^3 xy \, dx = \tfrac{5}{12\sqrt{3}} \int_0^{-\sqrt{3}} t^2 (t^3 - 3t) 2t \, dt = \tfrac{5}{6\sqrt{3}} \left[\tfrac{1}{7} t^7 - \tfrac{3}{5} t^5 \right]_0^{-\sqrt{3}}$$

$$= \tfrac{5}{6\sqrt{3}} \left[\tfrac{1}{7} (-3^{1/2})^7 - \tfrac{3}{5} (-3^{1/2})^5 \right] = \tfrac{5}{6\sqrt{3}} \left[-\tfrac{27}{7} \sqrt{3} + \tfrac{27}{5} \sqrt{3} \right] = \tfrac{9}{7}$$

So the coordinates of the centroid of $\mathcal{R}$ are $(x, y) = \left(\tfrac{9}{7}, 0 \right)$.

37. $x = t - t^2$, $y = \tfrac{4}{3} t^{3/2}$, $1 \le t \le 2$. $dx/dt = 1 - 2t$ and $dy/dt = 2t^{1/2}$, so

$(dx/dt)^2 + (dy/dt)^2 = (1 - 2t)^2 + (2t^{1/2})^2 = 1 - 4t + 4t^2 + 4t = 1 + 4t^2$.

Thus, $L = \int_a^b \sqrt{(dx/dt)^2 + (dy/dt)^2} \, dt = \int_1^2 \sqrt{1 + 4t^2} \, dt \approx 3.1678$.

38. $x = 1 + e^t$, $y = t^2$, $-3 \le t \le 3$. $dx/dt = e^t$ and $dy/dt = 2t$, so $(dx/dt)^2 + (dy/dt)^2 = e^{2t} + 4t^2$.

Thus, $L = \int_a^b \sqrt{(dx/dt)^2 + (dy/dt)^2} \, dt = \int_{-3}^3 \sqrt{e^{2t} + 4t^2} \, dt \approx 30.5281$.

39. $x = t + \cos t$, $y = t - \sin t$, $0 \le t \le 2\pi$. $dx/dt = 1 - \sin t$ and $dy/dt = 1 - \cos t$, so

$\left(\tfrac{dx}{dt} \right)^2 + \left(\tfrac{dy}{dt} \right)^2 = (1 - \sin t)^2 + (1 - \cos t)^2 = (1 - 2\sin t + \sin^2 t) + (1 - 2\cos t + \cos^2 t) = 3 - 2\sin t - 2\cos t$.

Thus, $L = \int_a^b \sqrt{(dx/dt)^2 + (dy/dt)^2} \, dt = \int_0^{2\pi} \sqrt{3 - 2\sin t - 2\cos t} \, dt \approx 10.0367$.

40. $x = \ln t$, $y = \sqrt{t + 1}$, $1 \le t \le 5$. $\dfrac{dx}{dt} = \dfrac{1}{t}$ and $\dfrac{dy}{dt} = \dfrac{1}{2\sqrt{t+1}}$, so $\left(\dfrac{dx}{dt} \right)^2 + \left(\dfrac{dy}{dt} \right)^2 = \dfrac{1}{t^2} + \dfrac{1}{4(t+1)} = \dfrac{t^2 + 4t + 4}{4t^2(t+1)}$.

Thus, $L = \displaystyle\int_a^b \sqrt{\left(\dfrac{dx}{dt} \right)^2 + \left(\dfrac{dy}{dt} \right)^2} \, dt = \int_1^5 \sqrt{\dfrac{t^2 + 4t + 4}{4t^2(t+1)}} \, dt = \int_1^5 \sqrt{\dfrac{(t+2)^2}{(2t)^2(t+1)}} \, dt = \int_1^5 \dfrac{t + 2}{2t\sqrt{t+1}} \, dt \approx 1.9310$.

41.

$x = 1 + 3t^2$, $y = 4 + 2t^3$, $0 \le t \le 1$.

$dx/dt = 6t$ and $dy/dt = 6t^2$, so $(dx/dt)^2 + (dy/dt)^2 = 36t^2 + 36t^4$.

Thus, $L = \int_0^1 \sqrt{36t^2 + 36t^4} \, dt = \int_0^1 6t \sqrt{1 + t^2} \, dt$

$= 6 \int_1^2 \sqrt{u} \left(\tfrac{1}{2} du \right)$ $[u = 1 + t^2, du = 2t \, dt]$

$= 3 \left[\tfrac{2}{3} u^{3/2} \right]_1^2 = 2(2^{3/2} - 1) = 2(2\sqrt{2} - 1)$

42.

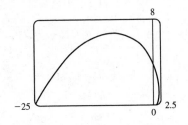

$x = e^t + e^{-t}$, $y = 5 - 2t$, $0 \le t \le 3$.

$dx/dt = e^t - e^{-t}$ and $dy/dt = -2$, so

$\left(\frac{dx}{dt}\right)^2 + \left(\frac{dy}{dt}\right)^2 = e^{2t} - 2 + e^{-2t} + 4 = e^{2t} + 2 + e^{-2t} = (e^t + e^{-t})^2$ and

$L = \int_0^3 (e^t + e^{-t})\, dt = \left[e^t - e^{-t}\right]_0^3 = e^3 - e^{-3} - (1 - 1) = e^3 - e^{-3}$.

43. $x = \dfrac{t}{1+t}$, $y = \ln(1+t)$, $0 \le t \le 2$. $\dfrac{dx}{dt} = \dfrac{(1+t)\cdot 1 - t \cdot 1}{(1+t)^2} = \dfrac{1}{(1+t)^2}$ and $\dfrac{dy}{dt} = \dfrac{1}{1+t}$,

so $\left(\dfrac{dx}{dt}\right)^2 + \left(\dfrac{dy}{dt}\right)^2 = \dfrac{1}{(1+t)^4} + \dfrac{1}{(1+t)^2} = \dfrac{1}{(1+t)^4}\left[1 + (1+t)^2\right] = \dfrac{t^2 + 2t + 2}{(1+t)^4}$. Thus,

$$L = \int_0^2 \frac{\sqrt{t^2 + 2t + 2}}{(1+t)^2}\, dt = \int_1^3 \frac{\sqrt{u^2 + 1}}{u^2}\, du \quad \begin{bmatrix} u = t+1, \\ du = dt \end{bmatrix} \overset{24}{=} \left[-\frac{\sqrt{u^2+1}}{u} + \ln\left(u + \sqrt{u^2+1}\right)\right]_1^3$$

$$= -\frac{\sqrt{10}}{3} + \ln\left(3 + \sqrt{10}\right) + \sqrt{2} - \ln\left(1 + \sqrt{2}\right)$$

44. $x = 3\cos t - \cos 3t$, $y = 3\sin t - \sin 3t$, $0 \le t \le \pi$. $\frac{dx}{dt} = -3\sin t + 3\sin 3t$ and $\frac{dy}{dt} = 3\cos t - 3\cos 3t$, so

$$\left(\tfrac{dx}{dt}\right)^2 + \left(\tfrac{dy}{dt}\right)^2 = 9\sin^2 t - 18\sin t \sin 3t + 9\sin^2(3t) + 9\cos^2 t - 18\cos t \cos 3t + 9\cos^2(3t)$$

$$= 9(\cos^2 t + \sin^2 t) - 18(\cos t \cos 3t + \sin t \sin 3t) + 9[\cos^2(3t) + \sin^2(3t)]$$

$$= 9(1) - 18\cos(t - 3t) + 9(1) = 18 - 18\cos(-2t) = 18(1 - \cos 2t)$$

$$= 18[1 - (1 - 2\sin^2 t)] = 36\sin^2 t.$$

Thus, $L = \int_0^\pi \sqrt{36\sin^2 t}\, dt = 6\int_0^\pi |\sin t|\, dt = 6\int_0^\pi \sin t\, dt = -6\left[\cos t\right]_0^\pi = -6(-1 - 1) = 12$.

45. $x = e^t \cos t$, $y = e^t \sin t$, $0 \le t \le \pi$.

$$\left(\tfrac{dx}{dt}\right)^2 + \left(\tfrac{dy}{dt}\right)^2 = [e^t(\cos t - \sin t)]^2 + [e^t(\sin t + \cos t)]^2$$

$$= (e^t)^2(\cos^2 t - 2\cos t \sin t + \sin^2 t)$$

$$\qquad + (e^t)^2(\sin^2 t + 2\sin t \cos t + \cos^2 t)$$

$$= e^{2t}(2\cos^2 t + 2\sin^2 t) = 2e^{2t}$$

Thus, $L = \int_0^\pi \sqrt{2e^{2t}}\, dt = \int_0^\pi \sqrt{2}\, e^t\, dt = \sqrt{2}\left[e^t\right]_0^\pi = \sqrt{2}\,(e^\pi - 1)$.

46. $x = \cos t + \ln(\tan \tfrac{1}{2} t)$, $\quad y = \sin t$, $\quad \pi/4 \le t \le 3\pi/4$.

$$\frac{dx}{dt} = -\sin t + \frac{\tfrac{1}{2}\sec^2(t/2)}{\tan(t/2)} = -\sin t + \frac{1}{2\sin(t/2)\cos(t/2)} = -\sin t + \frac{1}{\sin t} \text{ and } \frac{dy}{dt} = \cos t, \text{ so}$$

$$\left(\frac{dx}{dt}\right)^2 + \left(\frac{dy}{dt}\right)^2 = \sin^2 t - 2 + \frac{1}{\sin^2 t} + \cos^2 t = 1 - 2 + \csc^2 t = \cot^2 t. \text{ Thus,}$$

$$L = \int_{\pi/4}^{3\pi/4} |\cot t|\, dt = 2\int_{\pi/4}^{\pi/2} \cot t\, dt$$

$$= 2\Big[\ln|\sin t|\Big]_{\pi/4}^{\pi/2} = 2\Big(\ln 1 - \ln \frac{1}{\sqrt{2}}\Big)$$

$$= 2\big(0 + \ln\sqrt{2}\big) = 2\big(\tfrac{1}{2}\ln 2\big) = \ln 2.$$

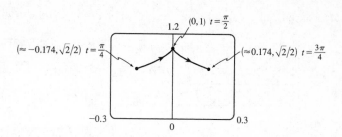

47.

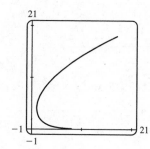

$$x = e^t - t,\; y = 4e^{t/2},\; -8 \le t \le 3$$

$$\Big(\frac{dx}{dt}\Big)^2 + \Big(\frac{dy}{dt}\Big)^2 = (e^t - 1)^2 + (2e^{t/2})^2 = e^{2t} - 2e^t + 1 + 4e^t$$

$$= e^{2t} + 2e^t + 1 = (e^t + 1)^2$$

$$L = \int_{-8}^{3} \sqrt{(e^t + 1)^2}\, dt = \int_{-8}^{3}(e^t + 1)\, dt = \big[e^t + t\big]_{-8}^{3t}$$

$$= (e^3 + 3) - (e^{-8} - 8) = e^3 - e^{-8} + 11$$

48. $x = 3t - t^3, y = 3t^2.$ $dx/dt = 3 - 3t^2$ and $dy/dt = 6t$, so

$$\Big(\frac{dx}{dt}\Big)^2 + \Big(\frac{dy}{dt}\Big)^2 = (3 - 3t^2)^2 + (6t)^2 = (3 + 3t^2)^2$$

and the length of the loop is given by

$$L = \int_{-\sqrt{3}}^{\sqrt{3}}(3 + 3t^2)\, dt = 2\int_0^{\sqrt{3}}(3 + 3t^2)\, dt = 2\big[3t + t^3\big]_0^{\sqrt{3}}$$

$$= 2\big(3\sqrt{3} + 3\sqrt{3}\big) = 12\sqrt{3}.$$

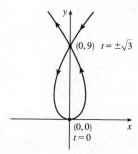

49. $x = t - e^t,\; y = t + e^t,\; -6 \le t \le 6.$

$\Big(\frac{dx}{dt}\Big)^2 + \Big(\frac{dy}{dt}\Big)^2 = (1 - e^t)^2 + (1 + e^t)^2 = (1 - 2e^t + e^{2t}) + (1 + 2e^t + e^{2t}) = 2 + 2e^{2t}$, so $L = \int_{-6}^{6} \sqrt{2 + 2e^{2t}}\, dt.$

Set $f(t) = \sqrt{2 + 2e^{2t}}$. Then by Simpson's Rule with $n = 6$ and $\Delta t = \frac{6 - (-6)}{6} = 2$, we get

$L \approx \frac{2}{3}[f(-6) + 4f(-4) + 2f(-2) + 4f(0) + 2f(2) + 4f(4) + f(6)] \approx 612.3053.$

50. $x = 2a\cot\theta \;\Rightarrow\; dx/dt = -2a\csc^2\theta$ and $y = 2a\sin^2\theta \;\Rightarrow\; dy/dt = 4a\sin\theta\cos\theta = 2a\sin 2\theta.$

So $L = \int_{\pi/4}^{\pi/2} \sqrt{4a^2\csc^4\theta + 4a^2\sin^2 2\theta}\, d\theta = 2a\int_{\pi/4}^{\pi/2} \sqrt{\csc^4\theta + \sin^2 2\theta}\, d\theta.$ Using Simpson's Rule with

$n = 4,\; \Delta\theta = \frac{\pi/2 - \pi/4}{4} = \frac{\pi}{16},$ and $f(\theta) = \sqrt{\csc^4\theta + \sin^2 2\theta}$, we get

$L \approx 2a \cdot S_4 = (2a)\frac{\pi}{16\cdot 3}\big[f\big(\frac{\pi}{4}\big) + 4f\big(\frac{5\pi}{16}\big) + 2f\big(\frac{3\pi}{8}\big) + 4f\big(\frac{7\pi}{16}\big) + f\big(\frac{\pi}{2}\big)\big] \approx 2.2605a.$

51. $x = \sin^2 t,\; y = \cos^2 t,\; 0 \le t \le 3\pi.$

$(dx/dt)^2 + (dy/dt)^2 = (2\sin t\cos t)^2 + (-2\cos t\sin t)^2 = 8\sin^2 t\cos^2 t = 2\sin^2 2t \;\Rightarrow$

Distance $= \int_0^{3\pi} \sqrt{2}\,|\sin 2t|\, dt = 6\sqrt{2}\int_0^{\pi/2}\sin 2t\, dt$ [by symmetry] $= -3\sqrt{2}\Big[\cos 2t\Big]_0^{\pi/2} = -3\sqrt{2}(-1 - 1) = 6\sqrt{2}.$

The full curve is traversed as t goes from 0 to $\frac{\pi}{2}$, because the curve is the segment of $x + y = 1$ that lies in the first quadrant

(since $x, y \ge 0$), and this segment is completely traversed as t goes from 0 to $\frac{\pi}{2}$. Thus, $L = \int_0^{\pi/2}\sin 2t\, dt = \sqrt{2}$, as above.

52. $x = \cos^2 t$, $y = \cos t$, $0 \leq t \leq 4\pi$. $\left(\frac{dx}{dt}\right)^2 + \left(\frac{dy}{dt}\right)^2 = (-2\cos t \sin t)^2 + (-\sin t)^2 = \sin^2 t\,(4\cos^2 t + 1)$

$$\text{Distance} = \int_0^{4\pi} |\sin t|\,\sqrt{4\cos^2 t + 1}\,dt = 4\int_0^\pi \sin t\,\sqrt{4\cos^2 t + 1}\,dt$$

$$= -4\int_1^{-1} \sqrt{4u^2 + 1}\,du \quad [u = \cos t,\ du = -\sin t\,dt] \ = 4\int_{-1}^1 \sqrt{4u^2 + 1}\,du$$

$$= 8\int_0^1 \sqrt{4u^2 + 1}\,du = 8\int_0^{\tan^{-1} 2} \sec\theta \cdot \tfrac{1}{2}\sec^2\theta\,d\theta \qquad [2u = \tan\theta,\ 2\,du = \sec^2\theta\,d\theta]$$

$$= 4\int_0^{\tan^{-1} 2} \sec^3\theta\,d\theta \overset{71}{=} \left[2\sec\theta\tan\theta + 2\ln|\sec\theta + \tan\theta|\right]_0^{\tan^{-1} 2} = 4\sqrt{5} + 2\ln\!\left(\sqrt{5} + 2\right)$$

Thus, $L = \int_0^\pi |\sin t|\,\sqrt{4\cos^2 t + 1}\,dt = \sqrt{5} + \tfrac{1}{2}\ln\!\left(\sqrt{5} + 2\right)$.

53. $x = a\sin\theta$, $y = b\cos\theta$, $0 \leq \theta \leq 2\pi$.

$$\left(\tfrac{dx}{dt}\right)^2 + \left(\tfrac{dy}{dt}\right)^2 = (a\cos\theta)^2 + (-b\sin\theta)^2 = a^2\cos^2\theta + b^2\sin^2\theta = a^2(1 - \sin^2\theta) + b^2\sin^2\theta$$

$$= a^2 - (a^2 - b^2)\sin^2\theta = a^2 - c^2\sin^2\theta = a^2\left(1 - \frac{c^2}{a^2}\sin^2\theta\right) = a^2(1 - e^2\sin^2\theta)$$

So $L = 4\int_0^{\pi/2} \sqrt{a^2\left(1 - e^2\sin^2\theta\right)}\,d\theta \quad$ [by symmetry] $\ = 4a\int_0^{\pi/2} \sqrt{1 - e^2\sin^2\theta}\,d\theta$.

54. $x = a\cos^3\theta$, $y = a\sin^3\theta$.

$$\left(\tfrac{dx}{dt}\right)^2 + \left(\tfrac{dy}{dt}\right)^2 = (-3a\cos^2\theta\,\sin\theta)^2 + (3a\sin^2\theta\,\cos\theta)^2$$

$$= 9a^2\cos^4\theta\,\sin^2\theta + 9a^2\sin^4\theta\,\cos^2\theta$$

$$= 9a^2\sin^2\theta\,\cos^2\theta(\cos^2\theta + \sin^2\theta) = 9a^2\sin^2\theta\,\cos^2\theta.$$

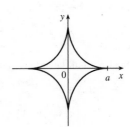

The graph has four-fold symmetry and the curve in the first quadrant corresponds to $0 \leq \theta \leq \pi/2$. Thus,

$$L = 4\int_0^{\pi/2} 3a\sin\theta\,\cos\theta\,d\theta \qquad [\text{since } a > 0 \text{ and } \sin\theta \text{ and } \cos\theta \text{ are positive for } 0 \leq \theta \leq \pi/2]$$

$$= 12a\left[\tfrac{1}{2}\sin^2\theta\right]_0^{\pi/2} = 12a\left(\tfrac{1}{2} - 0\right) = 6a$$

55. (a) $x = 11\cos t - 4\cos(11t/2)$, $y = 11\sin t - 4\sin(11t/2)$.

Notice that $0 \leq t \leq 2\pi$ does not give the complete curve because $x(0) \neq x(2\pi)$. In fact, we must take $t \in [0, 4\pi]$ in order to obtain the complete curve, since the first term in each of the parametric equations has period 2π and the second has period $\frac{2\pi}{11/2} = \frac{4\pi}{11}$, and the least common integer multiple of these two numbers is 4π.

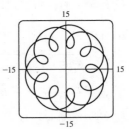

(b) We use the CAS to find the derivatives dx/dt and dy/dt, and then use Theorem 6 to find the arc length. Recent versions of Maple express the integral $\int_0^{4\pi} \sqrt{(dx/dt)^2 + (dy/dt)^2}\,dt$ as $88E\!\left(2\sqrt{2}\,i\right)$, where $E(x)$ is the elliptic integral

$$\int_0^1 \frac{\sqrt{1 - x^2 t^2}}{\sqrt{1 - t^2}}\,dt \text{ and } i \text{ is the imaginary number } \sqrt{-1}.$$

Some earlier versions of Maple (as well as Mathematica) cannot do the integral exactly, so we use the command `evalf(Int(sqrt(diff(x,t)^2+diff(y,t)^2),t=0..4*Pi));` to estimate the length, and find that the arc length is approximately 294.03. Derive's `Para_arc_length` function in the utility file `Int_apps` simplifies the integral to $11 \int_0^{4\pi} \sqrt{-4\cos t \cos\left(\frac{11t}{2}\right) - 4\sin t \sin\left(\frac{11t}{2}\right) + 5}\, dt$.

56. (a) It appears that as $t \to \infty$, $(x, y) \to \left(\frac{1}{2}, \frac{1}{2}\right)$, and as $t \to -\infty$, $(x, y) \to \left(-\frac{1}{2}, -\frac{1}{2}\right)$.

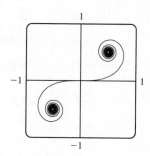

(b) By the Fundamental Theorem of Calculus, $dx/dt = \cos\left(\frac{\pi}{2}t^2\right)$ and $dy/dt = \sin\left(\frac{\pi}{2}t^2\right)$, so by Formula 4, the length of the curve from the origin to the point with parameter value t is

$$L = \int_0^t \sqrt{\left(\frac{dx}{du}\right)^2 + \left(\frac{dy}{du}\right)^2}\, du = \int_0^t \sqrt{\cos^2\left(\frac{\pi}{2}u^2\right) + \sin^2\left(\frac{\pi}{2}u^2\right)}\, du$$
$$= \int_0^t 1\, du = t \qquad [\text{or } -t \text{ if } t < 0]$$

We have used u as the dummy variable so as not to confuse it with the upper limit of integration.

57. $x = 1 + te^t$, $y = (t^2 + 1)e^t$, $0 \le t \le 1$.

$$\left(\frac{dx}{dt}\right)^2 + \left(\frac{dy}{dt}\right)^2 = (te^t + e^t)^2 + [(t^2+1)e^t + e^t(2t)]^2 = [e^t(t+1)]^2 + [e^t(t^2+2t+1)]^2$$
$$= e^{2t}(t+1)^2 + e^{2t}(t+1)^4 = e^{2t}(t+1)^2[1 + (t+1)^2], \quad \text{so}$$

$$S = \int 2\pi y\, ds = \int_0^1 2\pi(t^2+1)e^t \sqrt{e^{2t}(t+1)^2(t^2+2t+2)}\, dt = \int_0^1 2\pi(t^2+1)e^{2t}(t+1)\sqrt{t^2+2t+2}\, dt \approx 103.5999$$

58. $x = \sin^2 t$, $y = \sin 3t$, $0 \le t \le \frac{\pi}{3}$. $dx/dt = 2\sin t \cos t = \sin 2t$ and $dy/dt = 3\cos 3t$, so

$$(dx/dt)^2 + (dy/dt)^2 = \sin^2 2t + 9\cos^2 3t \text{ and } S = \int 2\pi y\, ds = \int_0^{\pi/3} 2\pi \sin 3t \sqrt{\sin^2 2t + 9\cos^2 3t}\, dt \approx 7.4775.$$

59. $x = t^3$, $y = t^2$, $0 \le t \le 1$. $\left(\frac{dx}{dt}\right)^2 + \left(\frac{dy}{dt}\right)^2 = (3t^2)^2 + (2t)^2 = 9t^4 + 4t^2$.

$$S = \int_0^1 2\pi y \sqrt{\left(\frac{dx}{dt}\right)^2 + \left(\frac{dy}{dt}\right)^2}\, dt = \int_0^1 2\pi t^2 \sqrt{9t^4 + 4t^2}\, dt = 2\pi \int_0^1 t^2 \sqrt{t^2(9t^2 + 4)}\, dt$$

$$= 2\pi \int_4^{13} \left(\frac{u-4}{9}\right) \sqrt{u}\left(\frac{1}{18}\, du\right) \quad \begin{bmatrix} u = 9t^2 + 4, \ t^2 = (u-4)/9, \\ du = 18t\, dt, \text{ so } t\, dt = \frac{1}{18}\, du \end{bmatrix} = \frac{2\pi}{9 \cdot 18} \int_4^{13} (u^{3/2} - 4u^{1/2})\, du$$

$$= \frac{\pi}{81}\left[\frac{2}{5}u^{5/2} - \frac{8}{3}u^{3/2}\right]_4^{13} = \frac{\pi}{81} \cdot \frac{2}{15}\left[3u^{5/2} - 20u^{3/2}\right]_4^{13}$$

$$= \frac{2\pi}{1215}\left[(3 \cdot 13^2\sqrt{13} - 20 \cdot 13\sqrt{13}) - (3 \cdot 32 - 20 \cdot 8)\right] = \frac{2\pi}{1215}\left(247\sqrt{13} + 64\right)$$

60. $x = 3t - t^3$, $y = 3t^2$, $0 \le t \le 1$. $\left(\frac{dx}{dt}\right)^2 + \left(\frac{dy}{dt}\right)^2 = (3 - 3t^2)^2 + (6t)^2 = 9(1 + 2t^2 + t^4) = [3(1 + t^2)]^2$.

$$S = \int_0^1 2\pi \cdot 3t^2 \cdot 3(1 + t^2)\, dt = 18\pi \int_0^1 (t^2 + t^4)\, dt = 18\pi\left[\frac{1}{3}t^3 + \frac{1}{5}t^5\right]_0^1 = \frac{48}{5}\pi$$

61. $x = a\cos^3\theta$, $y = a\sin^3\theta$, $0 \le \theta \le \frac{\pi}{2}$. $\left(\frac{dx}{d\theta}\right)^2 + \left(\frac{dy}{d\theta}\right)^2 = (-3a\cos^2\theta \sin\theta)^2 + (3a\sin^2\theta \cos\theta)^2 = 9a^2\sin^2\theta \cos^2\theta$.

$$S = \int_0^{\pi/2} 2\pi \cdot a\sin^3\theta \cdot 3a\sin\theta \cos\theta\, d\theta = 6\pi a^2 \int_0^{\pi/2} \sin^4\theta \cos\theta\, d\theta = \frac{6}{5}\pi a^2\left[\sin^5\theta\right]_0^{\pi/2} = \frac{6}{5}\pi a^2$$

62. $\left(\frac{dx}{d\theta}\right)^2 + \left(\frac{dy}{d\theta}\right)^2 = (-2\sin\theta + 2\sin 2\theta)^2 + (2\cos\theta - 2\cos 2\theta)^2$

$$= 4[(\sin^2\theta - 2\sin\theta\sin 2\theta + \sin^2 2\theta) + (\cos^2\theta - 2\cos\theta\cos 2\theta + \cos^2 2\theta)]$$

$$= 4[1 + 1 - 2(\cos 2\theta\cos\theta + \sin 2\theta\sin\theta)] = 8[1 - \cos(2\theta - \theta)] = 8(1 - \cos\theta)$$

We plot the graph with parameter interval $[0, 2\pi]$, and see that we should only

integrate between 0 and π. (If the interval $[0, 2\pi]$ were taken, the surface of

revolution would be generated twice.) Also note that

$y = 2\sin\theta - \sin 2\theta = 2\sin\theta(1 - \cos\theta)$. So

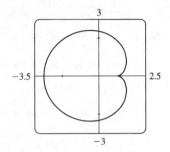

$S = \int_0^\pi 2\pi \cdot 2\sin\theta(1 - \cos\theta) 2\sqrt{2}\sqrt{1 - \cos\theta}\, d\theta$

$= 8\sqrt{2}\pi \int_0^\pi (1 - \cos\theta)^{3/2}\sin\theta\, d\theta = 8\sqrt{2}\pi \int_0^2 \sqrt{u^3}\, du \quad \begin{bmatrix} u = 1 - \cos\theta, \\ du = \sin\theta\, d\theta \end{bmatrix}$

$= 8\sqrt{2}\pi \left[\left(\frac{2}{5}\right)u^{5/2}\right]_0^2 = \frac{16}{5}\sqrt{2}\pi(2^{5/2}) = \frac{128}{5}\pi$

63. $x = t + t^3,\ y = t - \frac{1}{t^2},\ 1 \le t \le 2.$ $\quad \frac{dx}{dt} = 1 + 3t^2$ and $\frac{dy}{dt} = 1 + \frac{2}{t^3}$, so $\left(\frac{dx}{dt}\right)^2 + \left(\frac{dy}{dt}\right)^2 = (1 + 3t^2)^2 + \left(1 + \frac{2}{t^3}\right)^2$

and $S = \int 2\pi y\, ds = \int_1^2 2\pi\left(t - \frac{1}{t^2}\right)\sqrt{(1 + 3t^2)^2 + \left(1 + \frac{2}{t^3}\right)^2}\, dt \approx 59.101.$

64. $S = \int_{\pi/4}^{\pi/2} 2\pi \cdot 2a\sin^2\theta\sqrt{\csc^4\theta + \sin^2 2\theta}\, dt = 4\pi a \int_{\pi/4}^{\pi/2}\sin^2\theta\sqrt{\csc^4\theta + \sin^2 2\theta}\, d\theta.$

Using Simpson's Rule with $n = 4$, $\Delta\theta = \frac{\pi/2 - \pi/4}{4} = \frac{\pi}{16}$, and $f(\theta) = \sin^2\theta\sqrt{\csc^4\theta + \sin^2 2\theta}$, we get

$S \approx (4\pi a)\frac{\pi}{16 \cdot 3}\left[f\left(\frac{\pi}{4}\right) + 4f\left(\frac{5\pi}{16}\right) + 2f\left(\frac{3\pi}{8}\right) + 4f\left(\frac{7\pi}{16}\right) + f\left(\frac{\pi}{2}\right)\right] \approx 11.0893a.$

65. $x = 3t^2,\ y = 2t^3,\ 0 \le t \le 5 \quad \Rightarrow \quad \left(\frac{dx}{dt}\right)^2 + \left(\frac{dy}{dt}\right)^2 = (6t)^2 + (6t^2)^2 = 36t^2(1 + t^2) \quad \Rightarrow$

$S = \int_0^5 2\pi x \sqrt{(dx/dt)^2 + (dy/dt)^2}\, dt = \int_0^5 2\pi(3t^2)6t\sqrt{1 + t^2}\, dt = 18\pi\int_0^5 t^2\sqrt{1 + t^2}\, 2t\, dt$

$= 18\pi\int_1^{26}(u - 1)\sqrt{u}\, du \quad \begin{bmatrix} u = 1 + t^2, \\ du = 2t\, dt \end{bmatrix} = 18\pi\int_1^{26}(u^{3/2} - u^{1/2})\, du = 18\pi\left[\frac{2}{5}u^{5/2} - \frac{2}{3}u^{3/2}\right]_1^{26}$

$= 18\pi\left[\left(\frac{2}{5}\cdot 676\sqrt{26} - \frac{2}{3}\cdot 26\sqrt{26}\right) - \left(\frac{2}{5} - \frac{2}{3}\right)\right] = \frac{24}{5}\pi(949\sqrt{26} + 1)$

66. $x = e^t - t,\ y = 4e^{t/2},\ 0 \le t \le 1.$ $\left(\frac{dx}{dt}\right)^2 + \left(\frac{dy}{dt}\right)^2 = (e^t - 1)^2 + (2e^{t/2})^2 = e^{2t} + 2e^t + 1 = (e^t + 1)^2.$

$S = \int_0^1 2\pi(e^t - t)\sqrt{(e^t - 1)^2 + (2e^{t/2})^2}\, dt = \int_0^1 2\pi(e^t - t)(e^t + 1)d$

$= 2\pi\left[\frac{1}{2}e^{2t} + e^t - (t - 1)e^t - \frac{1}{2}t^2\right]_0^1 = \pi(e^2 + 2e - 6)$

67. If f' is continuous and $f'(t) \ne 0$ for $a \le t \le b$, then either $f'(t) > 0$ for all t in $[a, b]$ or $f'(t) < 0$ for all t in $[a, b]$. Thus, f

is monotonic (in fact, strictly increasing or strictly decreasing) on $[a, b]$. It follows that f has an inverse. Set $F = g \circ f^{-1}$,

that is, define F by $F(x) = g(f^{-1}(x))$. Then $x = f(t) \quad \Rightarrow \quad f^{-1}(x) = t$, so $y = g(t) = g(f^{-1}(x)) = F(x)$.

68. By Formula 9.2.5 [ET 8.2.5] with $y = F(x)$, $S = \int_a^b 2\pi F(x)\sqrt{1 + [F'(x)]^2}\,dx$. But by Formula 11.2.2 [ET 10.2.2],

$$1 + [F'(x)]^2 = 1 + \left(\frac{dy}{dx}\right)^2 = 1 + \left(\frac{dy/dt}{dx/dt}\right)^2 = \frac{(dx/dt)^2 + (dy/dt)^2}{(dx/dt)^2}.$$ Using the Substitution Rule with $x = x(t)$,

where $a = x(\alpha)$ and $b = x(\beta)$, we have $\left[\text{since } dx = \frac{dx}{dt}\,dt\right]$

$$S = \int_\alpha^\beta 2\pi\, F(x(t))\sqrt{\frac{(dx/dt)^2 + (dy/dt)^2}{(dx/dt)^2}}\,\frac{dx}{dt}\,dt = \int_\alpha^\beta 2\pi y\sqrt{\left(\frac{dx}{dt}\right)^2 + \left(\frac{dy}{dt}\right)^2}\,dt,$$ which is Formula 11.2.7

[ET 10.2.7].

69. (a) $\phi = \tan^{-1}\left(\dfrac{dy}{dx}\right) \;\Rightarrow\; \dfrac{d\phi}{dt} = \dfrac{d}{dt}\tan^{-1}\left(\dfrac{dy}{dx}\right) = \dfrac{1}{1 + (dy/dx)^2}\left[\dfrac{d}{dt}\left(\dfrac{dy}{dx}\right)\right]$. But $\dfrac{dy}{dx} = \dfrac{dy/dt}{dx/dt} = \dfrac{\dot{y}}{\dot{x}} \;\Rightarrow\;$

$\dfrac{d}{dt}\left(\dfrac{dy}{dx}\right) = \dfrac{d}{dt}\left(\dfrac{\dot{y}}{\dot{x}}\right) = \dfrac{\ddot{y}\dot{x} - \ddot{x}\dot{y}}{\dot{x}^2} \;\Rightarrow\; \dfrac{d\phi}{dt} = \dfrac{1}{1 + (\dot{y}/\dot{x})^2}\left(\dfrac{\ddot{y}\dot{x} - \ddot{x}\dot{y}}{\dot{x}^2}\right) = \dfrac{\dot{x}\ddot{y} - \ddot{x}\dot{y}}{\dot{x}^2 + \dot{y}^2}$. Using the Chain Rule, and the

fact that $s = \displaystyle\int_0^t \sqrt{\left(\frac{dx}{dt}\right)^2 + \left(\frac{dy}{dt}\right)^2}\,dt \;\Rightarrow\; \dfrac{ds}{dt} = \sqrt{\left(\frac{dx}{dt}\right)^2 + \left(\frac{dy}{dt}\right)^2} = (\dot{x}^2 + \dot{y}^2)^{1/2}$, we have that

$\dfrac{d\phi}{ds} = \dfrac{d\phi/dt}{ds/dt} = \left(\dfrac{\dot{x}\ddot{y} - \ddot{x}\dot{y}}{\dot{x}^2 + \dot{y}^2}\right)\dfrac{1}{(\dot{x}^2 + \dot{y}^2)^{1/2}} = \dfrac{\dot{x}\ddot{y} - \ddot{x}\dot{y}}{(\dot{x}^2 + \dot{y}^2)^{3/2}}$. So $\kappa = \left|\dfrac{d\phi}{ds}\right| = \left|\dfrac{\dot{x}\ddot{y} - \ddot{x}\dot{y}}{(\dot{x}^2 + \dot{y}^2)^{3/2}}\right| = \dfrac{|\dot{x}\ddot{y} - \ddot{x}\dot{y}|}{(\dot{x}^2 + \dot{y}^2)^{3/2}}$.

(b) $x = x$ and $y = f(x) \;\Rightarrow\; \dot{x} = 1,\, \ddot{x} = 0$ and $\dot{y} = \dfrac{dy}{dx},\, \ddot{y} = \dfrac{d^2y}{dx^2}$.

So $\kappa = \dfrac{|1 \cdot (d^2y/dx^2) - 0 \cdot (dy/dx)|}{[1 + (dy/dx)^2]^{3/2}} = \dfrac{|d^2y/dx^2|}{[1 + (dy/dx)^2]^{3/2}}$.

70. (a) $y = x^2 \;\Rightarrow\; \dfrac{dy}{dx} = 2x \;\Rightarrow\; \dfrac{d^2y}{dx^2} = 2$. So $\kappa = \dfrac{|d^2y/dx^2|}{[1 + (dy/dx)^2]^{3/2}} = \dfrac{2}{(1 + 4x^2)^{3/2}}$, and at $(1, 1)$,

$\kappa = \dfrac{2}{5^{3/2}} = \dfrac{2}{5\sqrt{5}}$.

(b) $\kappa' = \dfrac{d\kappa}{dx} = -3(1 + 4x^2)^{-5/2}(8x) = 0 \;\Leftrightarrow\; x = 0 \;\Rightarrow\; y = 0$. This is a maximum since $\kappa' > 0$ for $x < 0$ and

$\kappa' < 0$ for $x > 0$. So the parabola $y = x^2$ has maximum curvature at the origin.

71. $x = \theta - \sin\theta \;\Rightarrow\; \dot{x} = 1 - \cos\theta \;\Rightarrow\; \ddot{x} = \sin\theta$, and $y = 1 - \cos\theta \;\Rightarrow\; \dot{y} = \sin\theta \;\Rightarrow\; \ddot{y} = \cos\theta$. Therefore,

$\kappa = \dfrac{|\cos\theta - \cos^2\theta - \sin^2\theta|}{[(1 - \cos\theta)^2 + \sin^2\theta]^{3/2}} = \dfrac{|\cos\theta - (\cos^2\theta + \sin^2\theta)|}{(1 - 2\cos\theta + \cos^2\theta + \sin^2\theta)^{3/2}} = \dfrac{|\cos\theta - 1|}{(2 - 2\cos\theta)^{3/2}}$. The top of the arch is

characterized by a horizontal tangent, and from Example 2(b) in Section 11.2 [ET 10.2], the tangent is horizontal when

$\theta = (2n - 1)\pi$, so take $n = 1$ and substitute $\theta = \pi$ into the expression for κ: $\kappa = \dfrac{|\cos\pi - 1|}{(2 - 2\cos\pi)^{3/2}} = \dfrac{|-1 - 1|}{[2 - 2(-1)]^{3/2}} = \dfrac{1}{4}$.

72. (a) Every straight line has parametrizations of the form $x = a + vt$, $y = b + wt$, where a, b are arbitrary and $v, w \neq 0$.

For example, a straight line passing through distinct points (a, b) and (c, d) can be described as the parametrized curve

$x = a + (c - a)t$, $y = b + (d - b)t$. Starting with $x = a + vt$, $y = b + wt$, we compute $\dot{x} = v$, $\dot{y} = w$, $\ddot{x} = \ddot{y} = 0$,

and $\kappa = \dfrac{|v \cdot 0 - w \cdot 0|}{(v^2 + w^2)^{3/2}} = 0$.

(b) Parametric equations for a circle of radius r are $x = r\cos\theta$ and $y = r\sin\theta$. We can take the center to be the origin.

So $\dot{x} = -r\sin\theta \;\Rightarrow\; \ddot{x} = -r\cos\theta$ and $\dot{y} = r\cos\theta \;\Rightarrow\; \ddot{y} = -r\sin\theta$. Therefore,

$$\kappa = \frac{\left| r^2\sin^2\theta + r^2\cos^2\theta \right|}{(r^2\sin^2\theta + r^2\cos^2\theta)^{3/2}} = \frac{r^2}{r^3} = \frac{1}{r}. \text{ And so for any } \theta \text{ (and thus any point), } \kappa = \frac{1}{r}.$$

73. The coordinates of T are $(r\cos\theta, r\sin\theta)$. Since TP was unwound from

arc TA, TP has length $r\theta$. Also $\angle PTQ = \angle PTR - \angle QTR = \frac{1}{2}\pi - \theta$,

so P has coordinates $x = r\cos\theta + r\theta\cos\left(\frac{1}{2}\pi - \theta\right) = r(\cos\theta + \theta\sin\theta)$,

$y = r\sin\theta - r\theta\sin\left(\frac{1}{2}\pi - \theta\right) = r(\sin\theta - \theta\cos\theta)$.

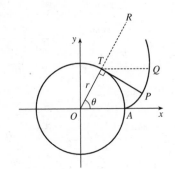

74. If the cow walks with the rope taut, it traces out the portion of the
involute in Exercise 73 corresponding to the range $0 \le \theta \le \pi$, arriving at
the point $(-r, \pi r)$ when $\theta = \pi$. With the rope now fully extended, the
cow walks in a semicircle of radius πr, arriving at $(-r, -\pi r)$. Finally,
the cow traces out another portion of the involute, namely the reflection
about the x-axis of the initial involute path. (This corresponds to the
range $-\pi \le \theta \le 0$.) Referring to the figure, we see that the total grazing

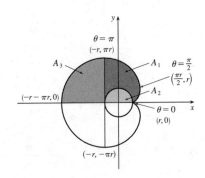

area is $2(A_1 + A_3)$. A_3 is one-quarter of the area of a circle of radius πr, so $A_3 = \frac{1}{4}\pi(\pi r)^2 = \frac{1}{4}\pi^3 r^2$. We will compute

$A_1 + A_2$ and then subtract $A_2 = \frac{1}{2}\pi r^2$ to obtain A_1.

To find $A_1 + A_2$, first note that the rightmost point of the involute is $\left(\frac{\pi}{2}r, r\right)$. [To see this, note that $dx/d\theta = 0$ when

$\theta = 0$ or $\frac{\pi}{2}$. $\theta = 0$ corresponds to the cusp at $(r, 0)$ and $\theta = \frac{\pi}{2}$ corresponds to $\left(\frac{\pi}{2}r, r\right)$.] The leftmost point of the involute is

$(-r, \pi r)$. Thus, $A_1 + A_2 = \int_{\theta=\pi}^{\pi/2} y\, dx - \int_{\theta=0}^{\pi/2} y\, dx = \int_{\theta=\pi}^{0} y\, dx$.

Now $y\, dx = r(\sin\theta - \theta\cos\theta)\, r\theta\cos\theta\, d\theta = r^2(\theta\sin\theta\cos\theta - \theta^2\cos^2\theta)d\theta$. Integrate:

$(1/r^2)\int y\, dx = -\theta\cos^2\theta - \frac{1}{2}(\theta^2 - 1)\sin\theta\,\cos\theta - \frac{1}{6}\theta^3 + \frac{1}{2}\theta + C$. This enables us to compute

$$A_1 + A_2 = r^2\left[-\theta\cos^2\theta - \tfrac{1}{2}(\theta^2 - 1)\sin\theta\,\cos\theta - \tfrac{1}{6}\theta^3 + \tfrac{1}{2}\theta\right]_\pi^0 = r^2\left[0 - \left(-\pi - \frac{\pi^3}{6} + \frac{\pi}{2}\right)\right] = r^2\left(\frac{\pi}{2} + \frac{\pi^3}{6}\right)$$

Therefore, $A_1 = (A_1 + A_2) - A_2 = \frac{1}{6}\pi^3 r^2$, so the grazing area is $2(A_1 + A_3) = 2\left(\frac{1}{6}\pi^3 r^2 + \frac{1}{4}\pi^3 r^2\right) = \frac{5}{6}\pi^3 r^2$.

LABORATORY PROJECT   Bézier Curves

1. The parametric equations for a cubic Bézier curve are

$$x = x_0(1-t)^3 + 3x_1 t(1-t)^2 + 3x_2 t^2(1-t) + x_3 t^3$$

$$y = y_0(1-t)^3 + 3y_1 t(1-t)^2 + 3y_2 t^2(1-t) + y_3 t^3$$

where $0 \le t \le 1$. We are given the points $P_0(x_0, y_0) = (4, 1)$, $P_1(x_1, y_1) = (28, 48)$, $P_2(x_2, y_2) = (50, 42)$, and $P_3(x_3, y_3) = (40, 5)$. The curve is then given by

$$x(t) = 4(1-t)^3 + 3 \cdot 28t(1-t)^2 + 3 \cdot 50t^2(1-t) + 40t^3$$

$$y(t) = 1(1-t)^3 + 3 \cdot 48t(1-t)^2 + 3 \cdot 42t^2(1-t) + 5t^3$$

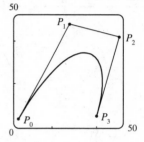

where $0 \le t \le 1$. The line segments are of the form $x = x_0 + (x_1 - x_0)t$, $y = y_0 + (y_1 - y_0)t$:

$P_0 P_1$	$x = 4 + 24t,$	$y = 1 + 47t$
$P_1 P_2$	$x = 28 + 22t,$	$y = 48 - 6t$
$P_2 P_3$	$x = 50 - 10t,$	$y = 42 - 37t$

2. It suffices to show that the slope of the tangent at P_0 is the same as that of line segment $P_0 P_1$, namely $\dfrac{y_1 - y_0}{x_1 - x_0}$. We calculate the slope of the tangent to the Bézier curve:

$$\frac{dy/dt}{dx/dt} = \frac{-3y_0(1-t)^2 + 3y_1\left[-2t(1-t) + (1-t)^2\right] + 3y_2\left[-t^2 + (2t)(1-t)\right] + 3y_3 t^2}{-3x_0^2(1-t) + 3x_1\left[-2t(1-t) + (1-t)^2\right] + 3x_2\left[-t^2 + (2t)(1-t)\right] + 3x_3 t^2}$$

At point P_0, $t = 0$, so the slope of the tangent is $\dfrac{-3y_0 + 3y_1}{-3x_0 + 3x_1} = \dfrac{y_1 - y_0}{x_1 - x_0}$. So the tangent to the curve at P_0 passes through P_1. Similarly, the slope of the tangent at point P_3 [where $t = 1$] is $\dfrac{-3y_2 + 3y_3}{-3x_2 + 3x_3} = \dfrac{y_3 - y_2}{x_3 - x_2}$, which is also the slope of line $P_2 P_3$.

3. It seems that if P_1 were to the right of P_2, a loop would appear.

We try setting $P_1 = (110, 30)$, and the resulting curve does indeed have a loop.

4. Based on the behavior of the Bézier curve in Problems 1–3, we suspect that the four control points should be in an exaggerated C shape. We try $P_0(10, 12)$, $P_1(4, 15)$, $P_2(4, 5)$, and $P_3(10, 8)$, and these produce a decent C. If you are using a CAS, it may be necessary to instruct it to make the x- and y-scales the same so as not to distort the figure (this is called a "constrained projection" in Maple.)

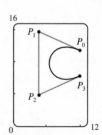

5. We use the same P_0 and P_1 as in Problem 4, and use part of our C as the top of an S. To prevent the center line from slanting up too much, we move P_2 up to $(4, 6)$ and P_3 down and to the left, to $(8, 7)$. In order to have a smooth joint between the top and bottom halves of the S (and a symmetric S), we determine points P_4, P_5, and P_6 by rotating points P_2, P_1, and P_0 about the center of the letter (point P_3). The points are therefore $P_4(12, 8)$, $P_5(12, -1)$, and $P_6(6, 2)$.

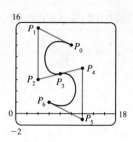

11.3 Polar Coordinates

1. (a) $\left(2, \frac{\pi}{3}\right)$

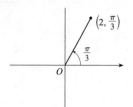

By adding 2π to $\frac{\pi}{3}$, we obtain the point $\left(2, \frac{7\pi}{3}\right)$. The direction opposite $\frac{\pi}{3}$ is $\frac{4\pi}{3}$, so $\left(-2, \frac{4\pi}{3}\right)$ is a point that satisfies the $r < 0$ requirement.

(b) $\left(1, -\frac{3\pi}{4}\right)$

$r > 0$: $\left(1, -\frac{3\pi}{4} + 2\pi\right) = \left(1, \frac{5\pi}{4}\right)$

$r < 0$: $\left(-1, -\frac{3\pi}{4} + \pi\right) = \left(-1, \frac{\pi}{4}\right)$

(c) $\left(-1, \frac{\pi}{2}\right)$

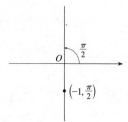

$r > 0$: $\left(-(-1), \frac{\pi}{2} + \pi\right) = \left(1, \frac{3\pi}{2}\right)$

$r < 0$: $\left(-1, \frac{\pi}{2} + 2\pi\right) = \left(-1, \frac{5\pi}{2}\right)$

2. (a) $\left(1, \frac{7\pi}{4}\right)$

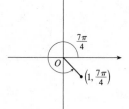

$r > 0$: $\left(1, \frac{7\pi}{4} - 2\pi\right) = \left(1, -\frac{\pi}{4}\right)$

$r < 0$: $\left(-1, \frac{7\pi}{4} - \pi\right) = \left(-1, \frac{3\pi}{4}\right)$

(b) $\left(-3, \frac{\pi}{6}\right)$

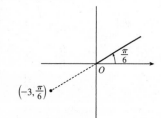

$r > 0$: $\left(-(-3), \frac{\pi}{6} + \pi\right) = \left(3, \frac{7\pi}{6}\right)$

$r < 0$: $\left(-3, \frac{\pi}{6} + 2\pi\right) = \left(-3, \frac{13\pi}{6}\right)$

(c) $(1, -1)$

$\theta = -1$ radian $\approx -57.3°$

$\boldsymbol{r > 0:}\ (1, -1 + 2\pi)$

$\boldsymbol{r < 0:}\ (-1, -1 + \pi)$

3. (a)

$x = 1\cos\pi = 1(-1) = -1$ and

$y = 1\sin\pi = 1(0) = 0$ give us

the Cartesian coordinates $(-1, 0)$.

(b)

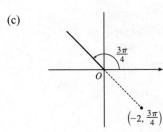

$x = 2\cos\left(-\frac{2\pi}{3}\right) = 2\left(-\frac{1}{2}\right) = -1$ and

$y = 2\sin\left(-\frac{2\pi}{3}\right) = 2\left(-\frac{\sqrt{3}}{2}\right) = -\sqrt{3}$

give us $\left(-1, -\sqrt{3}\right)$.

(c)

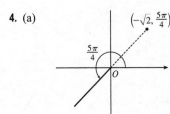

$x = -2\cos\frac{3\pi}{4} = -2\left(-\frac{\sqrt{2}}{2}\right) = \sqrt{2}$ and

$y = -2\sin\frac{3\pi}{4} = -2\left(\frac{\sqrt{2}}{2}\right) = -\sqrt{2}$

gives us $\left(\sqrt{2}, -\sqrt{2}\right)$.

4. (a)

$x = -\sqrt{2}\cos\frac{5\pi}{4} = -\sqrt{2}\left(-\frac{\sqrt{2}}{2}\right) = 1$ and

$y = -\sqrt{2}\sin\frac{5\pi}{4} = -\sqrt{2}\left(-\frac{\sqrt{2}}{2}\right) = 1$

gives us $(1, 1)$.

(b)

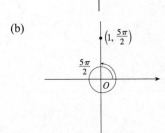

$x = 1\cos\frac{5\pi}{2} = 1(0) = 0$ and

$y = 1\sin\frac{5\pi}{2} = 1(1) = 1$

gives us $(0, 1)$.

(c)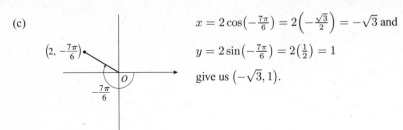

$x = 2\cos\left(-\frac{7\pi}{6}\right) = 2\left(-\frac{\sqrt{3}}{2}\right) = -\sqrt{3}$ and

$y = 2\sin\left(-\frac{7\pi}{6}\right) = 2\left(\frac{1}{2}\right) = 1$

give us $\left(-\sqrt{3}, 1\right)$.

5. (a) $x = 2$ and $y = -2$ $\Rightarrow$ $r = \sqrt{2^2 + (-2)^2} = 2\sqrt{2}$ and $\theta = \tan^{-1}\left(\frac{-2}{2}\right) = -\frac{\pi}{4}$. Since $(2, -2)$ is in the fourth

quadrant, the polar coordinates are (i) $\left(2\sqrt{2}, \frac{7\pi}{4}\right)$ and (ii) $\left(-2\sqrt{2}, \frac{3\pi}{4}\right)$.

(b) $x = -1$ and $y = \sqrt{3}$ $\Rightarrow$ $r = \sqrt{(-1)^2 + \left(\sqrt{3}\right)^2} = 2$ and $\theta = \tan^{-1}\left(\frac{\sqrt{3}}{-1}\right) = \frac{2\pi}{3}$. Since $\left(-1, \sqrt{3}\right)$ is in the second

quadrant, the polar coordinates are (i) $\left(2, \frac{2\pi}{3}\right)$ and (ii) $\left(-2, \frac{5\pi}{3}\right)$.

6. (a) $x = 3\sqrt{3}$ and $y = 3$ $\Rightarrow$ $r = \sqrt{\left(3\sqrt{3}\right)^2 + 3^2} = \sqrt{27 + 9} = 6$ and $\theta = \tan^{-1}\left(\frac{3}{3\sqrt{3}}\right) = \tan^{-1}\left(\frac{1}{\sqrt{3}}\right) = \frac{\pi}{6}$. Since

$\left(3\sqrt{3}, 3\right)$ is in the first quadrant, the polar coordinates are (i) $\left(6, \frac{\pi}{6}\right)$ and (ii) $\left(-6, \frac{7\pi}{6}\right)$.

(b) $x = 1$ and $y = -2$ $\Rightarrow$ $r = \sqrt{1^2 + (-2)^2} = \sqrt{5}$ and $\theta = \tan^{-1}\left(\frac{-2}{1}\right) = -\tan^{-1} 2$. Since $(1, -2)$ is in the fourth

quadrant, the polar coordinates are (i) $\left(\sqrt{5}, 2\pi - \tan^{-1} 2\right)$ and (ii) $\left(-\sqrt{5}, \pi - \tan^{-1} 2\right)$.

7. The curves $r = 1$ and $r = 2$ represent circles with center O and radii 1 and 2. The region in the plane satisfying $1 \le r \le 2$ consists of both circles and the shaded region between them in the figure.

8. $r \ge 0$, $\pi/3 \le \theta \le 2\pi/3$

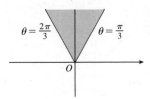

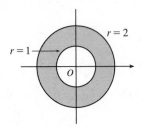

9. The region satisfying $0 \le r < 4$ and $-\pi/2 \le \theta < \pi/6$ does not include the circle $r = 4$ nor the line $\theta = \frac{\pi}{6}$.

10. $2 < r \le 5$, $3\pi/4 < \theta < 5\pi/4$

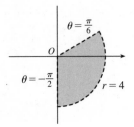

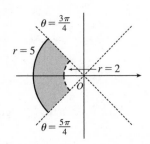

11. $2 < r < 3, \quad \frac{5\pi}{3} \le \theta \le \frac{7\pi}{3}$

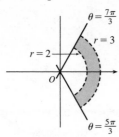

12. $r \ge 1, \pi \le \theta \le 2\pi$

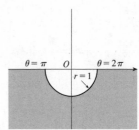

13. Converting the polar coordinates $(2, \pi/3)$ and $(4, 2\pi/3)$ to Cartesian coordinates gives us $\left(2\cos\frac{\pi}{3}, 2\sin\frac{\pi}{3}\right) = \left(1, \sqrt{3}\right)$ and $\left(4\cos\frac{2\pi}{3}, 4\sin\frac{2\pi}{3}\right) = \left(-2, 2\sqrt{3}\right)$. Now use the distance formula.

$$d = \sqrt{(x_2 - x_1)^2 + (y_2 - y_1)^2} = \sqrt{(-2-1)^2 + \left(2\sqrt{3} - \sqrt{3}\right)^2} = \sqrt{9+3} = \sqrt{12} = 2\sqrt{3}$$

14. The points (r_1, θ_1) and (r_2, θ_2) in Cartesian coordinates are $(r_1\cos\theta_1, r_1\sin\theta_1)$ and $(r_2\cos\theta_2, r_2\sin\theta_2)$, respectively. The *square* of the distance between them is

$$(r_2\cos\theta_2 - r_1\cos\theta_1)^2 + (r_2\sin\theta_2 - r_1\sin\theta_1)^2$$
$$= \left(r_2^2\cos^2\theta_2 - 2r_1r_2\cos\theta_1\cos\theta_2 + r_1^2\cos^2\theta_1\right) + \left(r_2^2\sin^2\theta_2 - 2r_1r_2\sin\theta_1\sin\theta_2 + r_1^2\sin^2\theta_1\right)$$
$$= r_1^2\left(\sin^2\theta_1 + \cos^2\theta_1\right) + r_2^2\left(\sin^2\theta_2 + \cos^2\theta_2\right) - 2r_1r_2(\cos\theta_1\cos\theta_2 + \sin\theta_1\sin\theta_2)$$
$$= r_1^2 - 2r_1r_2\cos(\theta_1 - \theta_2) + r_2^2,$$

so the distance between them is $\sqrt{r_1^2 - 2r_1r_2\cos(\theta_1 - \theta_2) + r_2^2}$.

15. $r = 2 \iff \sqrt{x^2 + y^2} = 2 \iff x^2 + y^2 = 4$, a circle of radius 2 centered at the origin.

16. $r\cos\theta = 1 \iff x = 1$, a vertical line.

17. $r = 3\sin\theta \Rightarrow r^2 = 3r\sin\theta \iff x^2 + y^2 = 3y \iff x^2 + \left(y - \frac{3}{2}\right)^2 = \left(\frac{3}{2}\right)^2$, a circle of radius $\frac{3}{2}$ centered at $\left(0, \frac{3}{2}\right)$. The first two equations are actually equivalent since $r^2 = 3r\sin\theta \Rightarrow r(r - 3\sin\theta) = 0 \Rightarrow r = 0$ or $r = 3\sin\theta$. But $r = 3\sin\theta$ gives the point $r = 0$ (the pole) when $\theta = 0$. Thus, the single equation $r = 3\sin\theta$ is equivalent to the compound condition $(r = 0$ or $r = 3\sin\theta)$.

18. $r = 2\sin\theta + 2\cos\theta \Rightarrow r^2 = 2r\sin\theta + 2r\cos\theta \iff x^2 + y^2 = 2y + 2x \iff$ $(x^2 - 2x + 1) + (y^2 - 2y + 1) = 2 \iff (x - 1)^2 + (y - 1)^2 = 2$. The first implication is reversible since $r^2 = 2r\sin\theta + 2r\cos\theta \Rightarrow r = 0$ or $r = 2\sin\theta + 2\cos\theta$, but the curve $r = 2\sin\theta + 2\cos\theta$ passes through the pole $(r = 0)$ when $\theta = -\frac{\pi}{4}$, so $r = 2\sin\theta + 2\cos\theta$ includes the single point of $r = 0$. The curve is a circle of radius $\sqrt{2}$, centered at $(1, 1)$.

19. $r = \csc\theta \iff r = \dfrac{1}{\sin\theta} \iff r\sin\theta = 1 \iff y = 1$, a horizontal line 1 unit above the x-axis.

20. $r = \tan\theta\sec\theta = \dfrac{\sin\theta}{\cos^2\theta} \Rightarrow r\cos^2\theta = \sin\theta \iff (r\cos\theta)^2 = r\sin\theta \iff x^2 = y$, a parabola with vertex at the origin opening upward. The first implication is reversible since $\cos\theta = 0$ would imply $\sin\theta = r\cos^2\theta = 0$, contradicting the fact that $\cos^2\theta + \sin^2\theta = 1$.

21. $x = 3 \Leftrightarrow r\cos\theta = 3 \Leftrightarrow r = 3/\cos\theta \Leftrightarrow r = 3\sec\theta.$

22. $x^2 + y^2 = 9 \Leftrightarrow r^2 = 9 \Leftrightarrow r = 3.$ $[r = -3$ gives the same curve.$]$

23. $x = -y^2 \Leftrightarrow r\cos\theta = -r^2\sin^2\theta \Leftrightarrow \cos\theta = -r\sin^2\theta \Leftrightarrow r = -\dfrac{\cos\theta}{\sin^2\theta} = -\cot\theta\csc\theta.$

24. $x + y = 9 \Leftrightarrow r\cos\theta + r\sin\theta = 9 \Leftrightarrow r = 9/(\cos\theta + \sin\theta).$

25. $x^2 + y^2 = 2cx \Leftrightarrow r^2 = 2cr\cos\theta \Leftrightarrow r^2 - 2cr\cos\theta = 0 \Leftrightarrow r(r - 2c\cos\theta) = 0 \Leftrightarrow r = 0$ or $r = 2c\cos\theta.$
$r = 0$ is included in $r = 2c\cos\theta$ when $\theta = \frac{\pi}{2} + n\pi$, so the curve is represented by the single equation $r = 2c\cos\theta.$

26. $xy = 4 \Leftrightarrow (r\cos\theta)(r\sin\theta) = 4 \Leftrightarrow r^2\left(\frac{1}{2}\cdot 2\sin\theta\cos\theta\right) = 4 \Leftrightarrow r^2\sin 2\theta = 8 \Rightarrow r^2 = 8\csc 2\theta$

27. (a) The description leads immediately to the polar equation $\theta = \frac{\pi}{6}$, and the Cartesian equation $y = \tan\left(\frac{\pi}{6}\right)x = \frac{1}{\sqrt{3}}x$ is slightly more difficult to derive.

(b) The easier description here is the Cartesian equation $x = 3.$

28. (a) Because its center is not at the origin, it is more easily described by its Cartesian equation, $(x - 2)^2 + (y - 3)^2 = 5^2.$

(b) This circle is more easily given in polar coordinates: $r = 4.$ The Cartesian equation is also simple: $x^2 + y^2 = 16.$

29. $\theta = -\pi/6$

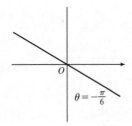

30. $r^2 - 3r + 2 = 0 \Leftrightarrow (r - 1)(r - 2) = 0 \Leftrightarrow$
$\quad\quad r = 1$ or $r = 2$

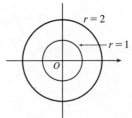

31. $r = \sin\theta \Leftrightarrow r^2 = r\sin\theta \Leftrightarrow x^2 + y^2 = y \Leftrightarrow$
$x^2 + \left(y - \frac{1}{2}\right)^2 = \left(\frac{1}{2}\right)^2.$ The reasoning here is the same
as in Exercise 17. This is a circle of radius $\frac{1}{2}$ centered
at $\left(0, \frac{1}{2}\right).$

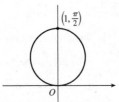

32. $r = -3\cos\theta \Leftrightarrow r^2 = -3r\cos\theta \Leftrightarrow$
$x^2 + y^2 = -3x \Leftrightarrow \left(x + \frac{3}{2}\right)^2 + y^2 = \left(\frac{3}{2}\right)^2.$
This curve is a circle of radius $\frac{3}{2}$ centered at $\left(-\frac{3}{2}, 0\right).$

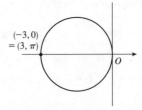

33. $r = 2(1 - \sin\theta).$ This curve is a cardioid.

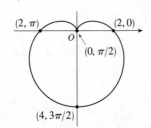

34. $r = 1 - 3\cos\theta.$ This is a limaçon.

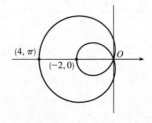

35. $r = \theta, \quad \theta \geq 0$

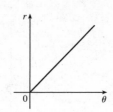

36. $r = \ln \theta, \; \theta \geq 1$

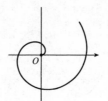

37. $r = 4 \sin 3\theta$

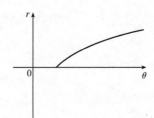

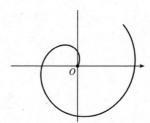

38. $r = \cos 5\theta$

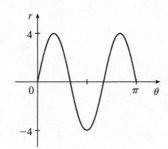

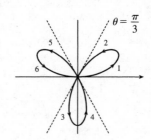

39. $r = 2 \cos 4\theta$

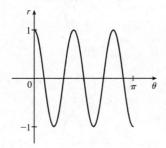

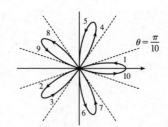

40. $r = 3 \cos 6\theta$

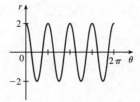

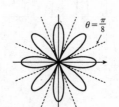

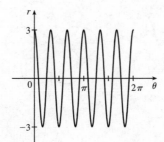

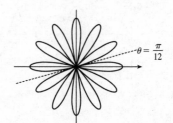

41. $r = 1 - 2\sin\theta$

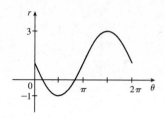 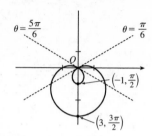

42. $r = 2 + \sin\theta$

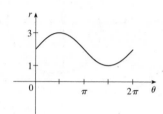 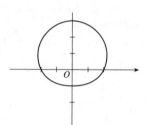

43. $r^2 = 9\sin 2\theta$

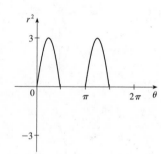 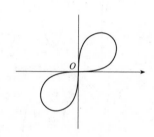

44. $r^2 = \cos 4\theta$

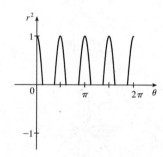 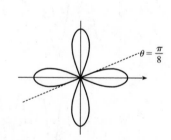

45. $r = 2\cos\left(\frac{3}{2}\theta\right)$

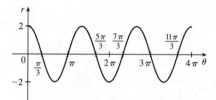 

46. $r^2\theta = 1 \quad \Leftrightarrow \quad r = \pm 1/\sqrt{\theta}$ for $\theta > 0$

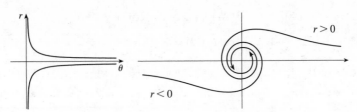

47. $r = 1 + 2\cos 2\theta$

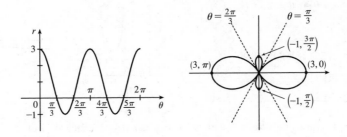

48. $r = 1 + 2\cos(\theta/2)$

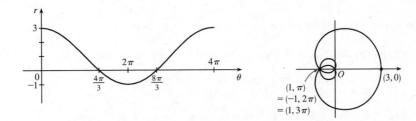

49. For $\theta = 0$, π, and 2π, r has its minimum value of about 0.5. For $\theta = \frac{\pi}{2}$ and $\frac{3\pi}{2}$, r attains its maximum value of 2.

We see that the graph has a similar shape for $0 \le \theta \le \pi$ and $\pi \le \theta \le 2\pi$.

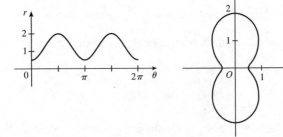

50.

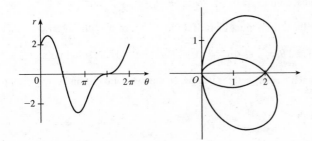

51. $x = (r)\cos\theta = (4 + 2\sec\theta)\cos\theta = 4\cos\theta + 2$. Now, $r \to \infty \Rightarrow$

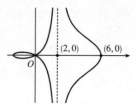

$(4 + 2\sec\theta) \to \infty \Rightarrow \theta \to \left(\frac{\pi}{2}\right)^- $ or $\theta \to \left(\frac{3\pi}{2}\right)^+$ [since we need only

consider $0 \le \theta < 2\pi$], so $\lim\limits_{r\to\infty} x = \lim\limits_{\theta\to\pi/2^-} (4\cos\theta + 2) = 2$. Also,

$r \to -\infty \Rightarrow (4 + 2\sec\theta) \to -\infty \Rightarrow \theta \to \left(\frac{\pi}{2}\right)^+ $ or $\theta \to \left(\frac{3\pi}{2}\right)^-$, so

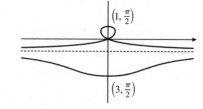

$\lim\limits_{r\to-\infty} x = \lim\limits_{\theta\to\pi/2^+} (4\cos\theta + 2) = 2$. Therefore, $\lim\limits_{r\to\pm\infty} x = 2 \Rightarrow x = 2$ is a vertical asymptote.

52. $y = r\sin\theta = 2\sin\theta - \csc\theta\sin\theta = 2\sin\theta - 1$.

$r \to \infty \Rightarrow (2 - \csc\theta) \to \infty \Rightarrow$

$\csc\theta \to -\infty \Rightarrow \theta \to \pi^+$ [since we need

only consider $0 \le \theta < 2\pi$] and so

$\lim\limits_{r\to\infty} y = \lim\limits_{\theta\to\pi^+} 2\sin\theta - 1 = -1$.

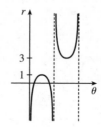

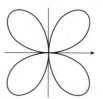

Also $r \to -\infty \Rightarrow (2 - \csc\theta) \to -\infty \Rightarrow \csc\theta \to \infty \Rightarrow \theta \to \pi^-$ and so $\lim\limits_{r\to-\infty} x = \lim\limits_{\theta\to\pi^-} 2\sin\theta - 1 = -1$.

Therefore $\lim\limits_{r\to\pm\infty} y = -1 \Rightarrow y = -1$ is a horizontal asymptote.

53. To show that $x = 1$ is an asymptote we must prove $\lim\limits_{r\to\pm\infty} x = 1$.

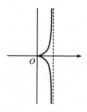

$x = (r)\cos\theta = (\sin\theta\tan\theta)\cos\theta = \sin^2\theta$. Now, $r \to \infty \Rightarrow \sin\theta\tan\theta \to \infty \Rightarrow$

$\theta \to \left(\frac{\pi}{2}\right)^-$, so $\lim\limits_{r\to\infty} x = \lim\limits_{\theta\to\pi/2^-} \sin^2\theta = 1$. Also, $r \to -\infty \Rightarrow \sin\theta\tan\theta \to -\infty \Rightarrow$

$\theta \to \left(\frac{\pi}{2}\right)^+$, so $\lim\limits_{r\to-\infty} x = \lim\limits_{\theta\to\pi/2^+} \sin^2\theta = 1$. Therefore, $\lim\limits_{r\to\pm\infty} x = 1 \Rightarrow x = 1$ is

a vertical asymptote. Also notice that $x = \sin^2\theta \ge 0$ for all θ, and $x = \sin^2\theta \le 1$ for all θ. And $x \ne 1$, since the curve is not

defined at odd multiples of $\frac{\pi}{2}$. Therefore, the curve lies entirely within the vertical strip $0 \le x < 1$.

54. The equation is $(x^2 + y^2)^3 = 4x^2y^2$, but using polar coordinates we know that

$x^2 + y^2 = r^2$ and $x = r\cos\theta$ and $y = r\sin\theta$. Substituting into the given

equation: $r^6 = 4r^2\cos^2\theta\, r^2\sin^2\theta \Rightarrow r^2 = 4\cos^2\theta\sin^2\theta \Rightarrow$

$r = \pm 2\cos\theta\sin\theta = \pm\sin 2\theta$. $r = \pm\sin 2\theta$ is sketched at right.

55. (a) We see that the curve $r = 1 + c\sin\theta$ crosses itself at the origin, where $r = 0$ (in fact the inner loop corresponds to

negative r-values,) so we solve the equation of the limaçon for $r = 0 \Leftrightarrow c\sin\theta = -1 \Leftrightarrow \sin\theta = -1/c$. Now if

$|c| < 1$, then this equation has no solution and hence there is no inner loop. But if $c < -1$, then on the interval $(0, 2\pi)$

the equation has the two solutions $\theta = \sin^{-1}(-1/c)$ and $\theta = \pi - \sin^{-1}(-1/c)$, and if $c > 1$, the solutions are

$\theta = \pi + \sin^{-1}(1/c)$ and $\theta = 2\pi - \sin^{-1}(1/c)$. In each case, $r < 0$ for θ between the two solutions, indicating a loop.

(b) For $0 < c < 1$, the dimple (if it exists) is characterized by the fact that y has a local maximum at $\theta = \frac{3\pi}{2}$. So we determine

for what c-values $\dfrac{d^2y}{d\theta^2}$ is negative at $\theta = \frac{3\pi}{2}$, since by the Second Derivative Test this indicates a maximum:

$$y = r\sin\theta = \sin\theta + c\sin^2\theta \;\Rightarrow\; \frac{dy}{d\theta} = \cos\theta + 2c\sin\theta\cos\theta = \cos\theta + c\sin 2\theta \;\Rightarrow\; \frac{d^2y}{d\theta^2} = -\sin\theta + 2c\cos 2\theta.$$

At $\theta = \frac{3\pi}{2}$, this is equal to $-(-1) + 2c(-1) = 1 - 2c$, which is negative only for $c > \frac{1}{2}$. A similar argument shows that

for $-1 < c < 0$, y only has a local minimum at $\theta = \frac{\pi}{2}$ (indicating a dimple) for $c < -\frac{1}{2}$.

56. (a) $r = \sqrt{\theta}$, $0 \le \theta \le 16\pi$. r increases as θ increases and there are eight full revolutions. The graph must be either II or V.

When $\theta = 2\pi$, $r = \sqrt{2\pi} \approx 2.5$ and when $\theta = 16\pi$, $r = \sqrt{16\pi} \approx 7$, so the last revolution intersects the polar axis at

approximately 3 times the distance that the first revolution intersects the polar axis, which is depicted in graph V.

(b) $r = \theta^2$, $0 \le \theta \le 16\pi$. See part (a). This is graph II.

(c) $r = \cos(\theta/3)$. $0 \le \frac{\theta}{3} \le 2\pi \;\Rightarrow\; 0 \le \theta \le 6\pi$, so this curve will repeat itself every 6π radians.

$\cos\left(\frac{\theta}{3}\right) = 0 \;\Rightarrow\; \frac{\theta}{3} = \frac{\pi}{2} + \pi n \;\Rightarrow\; \theta = \frac{3\pi}{2} + 3\pi n$, so there will be two "pole" values, $\frac{3\pi}{2}$ and $\frac{9\pi}{2}$.

This is graph VI.

(d) $r = 1 + 2\cos\theta$ is a limaçon [see Exercise 55(a)] with $c = 2$. This is graph III.

(e) Since $-1 \le \sin 3\theta \le 1$, $1 \le 2 + \sin 3\theta \le 3$, so $r = 2 + \sin 3\theta$ is never 0; that is, the curve never intersects the pole.

This is graph I.

(f) $r = 1 + 2\sin 3\theta$. Solving $r = 0$ will give us many "pole" values, so this is graph IV.

57. $r = 2\sin\theta \;\Rightarrow\; x = r\cos\theta = 2\sin\theta\cos\theta = \sin 2\theta$, $y = r\sin\theta = 2\sin^2\theta \;\Rightarrow\;$

$$\frac{dy}{dx} = \frac{dy/d\theta}{dx/d\theta} = \frac{2 \cdot 2\sin\theta\cos\theta}{\cos 2\theta \cdot 2} = \frac{\sin 2\theta}{\cos 2\theta} = \tan 2\theta$$

When $\theta = \frac{\pi}{6}$, $\dfrac{dy}{dx} = \tan\left(2 \cdot \frac{\pi}{6}\right) = \tan\frac{\pi}{3} = \sqrt{3}$. [*Another method:* Use Equation 3.]

58. $r = 2 - \sin\theta \;\Rightarrow\; x = r\cos\theta = (2 - \sin\theta)\cos\theta$, $y = r\sin\theta = (2 - \sin\theta)\sin\theta \;\Rightarrow\;$

$$\frac{dy}{dx} = \frac{dy/d\theta}{dx/d\theta} = \frac{(2 - \sin\theta)\cos\theta + \sin\theta(-\cos\theta)}{(2 - \sin\theta)(-\sin\theta) + \cos\theta(-\cos\theta)} = \frac{2\cos\theta - 2\sin\theta\cos\theta}{-2\sin\theta + \sin^2\theta - \cos^2\theta} = \frac{2\cos\theta - \sin 2\theta}{-2\sin\theta - \cos 2\theta}$$

When $\theta = \frac{\pi}{3}$, $\dfrac{dy}{dx} = \dfrac{2(1/2) - (\sqrt{3}/2)}{-2(\sqrt{3}/2) - (-1/2)} = \dfrac{1 - \sqrt{3}/2}{-\sqrt{3} + 1/2} \cdot \dfrac{2}{2} = \dfrac{2 - \sqrt{3}}{1 - 2\sqrt{3}}$.

59. $r = 1/\theta \;\Rightarrow\; x = r\cos\theta = (\cos\theta)/\theta$, $y = r\sin\theta = (\sin\theta)/\theta \;\Rightarrow\;$

$$\frac{dy}{dx} = \frac{dy/d\theta}{dx/d\theta} = \frac{\sin\theta(-1/\theta^2) + (1/\theta)\cos\theta}{\cos\theta(-1/\theta^2) - (1/\theta)\sin\theta} \cdot \frac{\theta^2}{\theta^2} = \frac{-\sin\theta + \theta\cos\theta}{-\cos\theta - \theta\sin\theta}$$

When $\theta = \pi$, $\dfrac{dy}{dx} = \dfrac{-0 + \pi(-1)}{-(-1) - \pi(0)} = \dfrac{-\pi}{1} = -\pi$.

60. $r = \cos(\theta/3) \Rightarrow x = r\cos\theta = \cos(\theta/3)\cos\theta, y = r\sin\theta = \cos(\theta/3)\sin\theta \Rightarrow$

$$\frac{dy}{dx} = \frac{dy/d\theta}{dx/d\theta} = \frac{\cos(\theta/3)\cos\theta + \sin\theta \left(-\frac{1}{3}\sin(\theta/3)\right)}{\cos(\theta/3)(-\sin\theta) + \cos\theta \left(-\frac{1}{3}\sin(\theta/3)\right)}$$

When $\theta = \pi$, $\dfrac{dy}{dx} = \dfrac{\frac{1}{2}(-1) + (0)(-\sqrt{3}/6)}{\frac{1}{2}(0) + (-1)(-\sqrt{3}/6)} = \dfrac{-1/2}{\sqrt{3}/6} = -\dfrac{3}{\sqrt{3}} = -\sqrt{3}$.

61. $r = \cos 2\theta \Rightarrow x = r\cos\theta = \cos 2\theta \cos\theta, y = r\sin\theta = \cos 2\theta \sin\theta \Rightarrow$

$$\frac{dy}{dx} = \frac{dy/d\theta}{dx/d\theta} = \frac{\cos 2\theta \cos\theta + \sin\theta(-2\sin 2\theta)}{\cos 2\theta(-\sin\theta) + \cos\theta(-2\sin 2\theta)}$$

When $\theta = \dfrac{\pi}{4}$, $\dfrac{dy}{dx} = \dfrac{0(\sqrt{2}/2) + (\sqrt{2}/2)(-2)}{0(-\sqrt{2}/2) + (\sqrt{2}/2)(-2)} = \dfrac{-\sqrt{2}}{-\sqrt{2}} = 1$.

62. $r = 1 + 2\cos\theta \Rightarrow x = r\cos\theta = (1 + 2\cos\theta)\cos\theta, y = r\sin\theta = (1 + 2\cos\theta)\sin\theta \Rightarrow$

$$\frac{dy}{dx} = \frac{dy/d\theta}{dx/d\theta} = \frac{(1 + 2\cos\theta)\cos\theta + \sin\theta(-2\sin\theta)}{(1 + 2\cos\theta)(-\sin\theta) + \cos\theta(-2\sin\theta)}$$

When $\theta = \dfrac{\pi}{3}$, $\dfrac{dy}{dx} = \dfrac{2(\frac{1}{2}) + (\sqrt{3}/2)(-\sqrt{3})}{2(-\sqrt{3}/2) + \frac{1}{2}(-\sqrt{3})} \cdot \dfrac{2}{2} = \dfrac{2 - 3}{-2\sqrt{3} - \sqrt{3}} = \dfrac{-1}{-3\sqrt{3}} = \dfrac{\sqrt{3}}{9}$.

63. $r = 3\cos\theta \Rightarrow x = r\cos\theta = 3\cos\theta\cos\theta, y = r\sin\theta = 3\cos\theta\sin\theta \Rightarrow$

$\frac{dy}{d\theta} = -3\sin^2\theta + 3\cos^2\theta = 3\cos 2\theta = 0 \Rightarrow 2\theta = \frac{\pi}{2}$ or $\frac{3\pi}{2} \Leftrightarrow \theta = \frac{\pi}{4}$ or $\frac{3\pi}{4}$.

So the tangent is horizontal at $\left(\frac{3}{\sqrt{2}}, \frac{\pi}{4}\right)$ and $\left(-\frac{3}{\sqrt{2}}, \frac{3\pi}{4}\right)$ $\left[\text{same as } \left(\frac{3}{\sqrt{2}}, -\frac{\pi}{4}\right)\right]$.

$\frac{dx}{d\theta} = -6\sin\theta\cos\theta = -3\sin 2\theta = 0 \Rightarrow 2\theta = 0$ or $\pi \Leftrightarrow \theta = 0$ or $\frac{\pi}{2}$. So the tangent is vertical at $(3, 0)$ and $\left(0, \frac{\pi}{2}\right)$.

64. $r = 1 - \sin\theta \Rightarrow x = r\cos\theta = \cos\theta(1 - \sin\theta), y = r\sin\theta = \sin\theta(1 - \sin\theta) \Rightarrow$

$\frac{dy}{d\theta} = \sin\theta(-\cos\theta) + (1 - \sin\theta)\cos\theta = \cos\theta(1 - 2\sin\theta) = 0 \Rightarrow \cos\theta = 0$ or $\sin\theta = \frac{1}{2} \Rightarrow$

$\theta = \frac{\pi}{6}, \frac{\pi}{2}, \frac{5\pi}{6}$, or $\frac{3\pi}{2} \Rightarrow$ horizontal tangent at $\left(\frac{1}{2}, \frac{\pi}{6}\right), \left(\frac{1}{2}, \frac{5\pi}{6}\right)$, and $\left(2, \frac{3\pi}{2}\right)$.

$\frac{dx}{d\theta} = \cos\theta(-\cos\theta) + (1 - \sin\theta)(-\sin\theta) = -\cos^2\theta - \sin\theta + \sin^2\theta = 2\sin^2\theta - \sin\theta - 1$

$\quad = (2\sin\theta + 1)(\sin\theta - 1) = 0 \Rightarrow$

$\sin\theta = -\frac{1}{2}$ or $1 \Rightarrow \theta = \frac{7\pi}{6}, \frac{11\pi}{6}$, or $\frac{\pi}{2} \Rightarrow$ vertical tangent at $\left(\frac{3}{2}, \frac{7\pi}{6}\right), \left(\frac{3}{2}, \frac{11\pi}{6}\right)$, and $\left(0, \frac{\pi}{2}\right)$.

Note that the tangent is vertical, not horizontal, when $\theta = \frac{\pi}{2}$, since

$$\lim_{\theta \to (\pi/2)^-} \frac{dy/d\theta}{dx/d\theta} = \lim_{\theta \to (\pi/2)^-} \frac{\cos\theta(1 - 2\sin\theta)}{(2\sin\theta + 1)(\sin\theta - 1)} = \infty \text{ and } \lim_{\theta \to (\pi/2)^+} \frac{dy/d\theta}{dx/d\theta} = -\infty.$$

65. $r = 1 + \cos\theta \Rightarrow x = r\cos\theta = \cos\theta(1 + \cos\theta), y = r\sin\theta = \sin\theta(1 + \cos\theta) \Rightarrow$

$\frac{dy}{d\theta} = (1 + \cos\theta)\cos\theta - \sin^2\theta = 2\cos^2\theta + \cos\theta - 1 = (2\cos\theta - 1)(\cos\theta + 1) = 0 \Rightarrow \cos\theta = \frac{1}{2}$ or $-1 \Rightarrow$

$\theta = \frac{\pi}{3}, \pi$, or $\frac{5\pi}{3} \Rightarrow$ horizontal tangent at $\left(\frac{3}{2}, \frac{\pi}{3}\right), (0, \pi)$, and $\left(\frac{3}{2}, \frac{5\pi}{3}\right)$.

$\frac{dx}{d\theta} = -(1 + \cos\theta)\sin\theta - \cos\theta\sin\theta = -\sin\theta(1 + 2\cos\theta) = 0 \Rightarrow \sin\theta = 0$ or $\cos\theta = -\frac{1}{2} \Rightarrow$

$\theta = 0, \pi, \frac{2\pi}{3}$, or $\frac{4\pi}{3} \Rightarrow$ vertical tangent at $(2, 0), \left(\frac{1}{2}, \frac{2\pi}{3}\right)$, and $\left(\frac{1}{2}, \frac{4\pi}{3}\right)$.

Note that the tangent is horizontal, not vertical when $\theta = \pi$, since $\lim\limits_{\theta \to \pi} \dfrac{dy/d\theta}{dx/d\theta} = 0$.

66. $r = e^\theta \Rightarrow x = r \cos\theta = e^\theta \cos\theta, \; y = r \sin\theta = e^\theta \sin\theta \Rightarrow$

$\frac{dy}{d\theta} = e^\theta \sin\theta + e^\theta \cos\theta = e^\theta(\sin\theta + \cos\theta) = 0 \Rightarrow \sin\theta = -\cos\theta \Rightarrow \tan\theta = -1 \Rightarrow$

$\theta = -\frac{1}{4}\pi + n\pi$ [n any integer] $\Rightarrow$ horizontal tangents at $\left(e^{\pi(n-1/4)}, \pi\left(n - \frac{1}{4}\right)\right)$.

$\frac{dx}{d\theta} = e^\theta \cos\theta - e^\theta \sin\theta = e^\theta(\cos\theta - \sin\theta) = 0 \Rightarrow \sin\theta = \cos\theta \Rightarrow \tan\theta = 1 \Rightarrow$

$\theta = \frac{1}{4}\pi + n\pi$ [n any integer] $\Rightarrow$ vertical tangents at $\left(e^{\pi(n+1/4)}, \pi\left(n + \frac{1}{4}\right)\right)$.

67. $r = 2 + \sin\theta \Rightarrow x = r \cos\theta = (2 + \sin\theta)\cos\theta, \; y = r \sin\theta = (2 + \sin\theta)\sin\theta \Rightarrow$

$\frac{dy}{d\theta} = (2 + \sin\theta)\cos\theta + \sin\theta \cos\theta = \cos\theta \cdot 2(1 + \sin\theta) = 0 \Rightarrow \cos\theta = 0$ or $\sin\theta = -1 \Rightarrow$

$\theta = \frac{\pi}{2}$ or $\frac{3\pi}{2} \Rightarrow$ horizontal tangent at $\left(3, \frac{\pi}{2}\right)$ and $\left(1, \frac{3\pi}{2}\right)$.

$\frac{dx}{d\theta} = (2 + \sin\theta)(-\sin\theta) + \cos\theta \cos\theta = -2\sin\theta - \sin^2\theta + 1 - \sin^2\theta = -2\sin^2\theta - 2\sin\theta + 1 \Rightarrow$

$\sin\theta = \dfrac{2 \pm \sqrt{4 + 8}}{-4} = \dfrac{2 \pm 2\sqrt{3}}{-4} = \dfrac{1 - \sqrt{3}}{-2} \quad \left[\dfrac{1 + \sqrt{3}}{-2} < -1\right] \Rightarrow$

$\theta_1 = \sin^{-1}\left(-\frac{1}{2} + \frac{1}{2}\sqrt{3}\right)$ and $\theta_2 = \pi - \theta_1 \Rightarrow$ vertical tangent at $\left(\frac{3}{2} + \frac{1}{2}\sqrt{3}, \theta_1\right)$ and $\left(\frac{3}{2} + \frac{1}{2}\sqrt{3}, \theta_2\right)$.

Note that $r(\theta_1) = 2 + \sin\left[\sin^{-1}\left(-\frac{1}{2} + \frac{1}{2}\sqrt{3}\right)\right] = 2 - \frac{1}{2} + \frac{1}{2}\sqrt{3} = \frac{3}{2} + \frac{1}{2}\sqrt{3}$.

68. By differentiating implicitly, $r^2 = \sin 2\theta \Rightarrow 2r(dr/d\theta) = 2\cos 2\theta \Rightarrow$

$dr/d\theta = (1/r)\cos 2\theta$, so $dy/d\theta = (dr/d\theta)\sin\theta + r\cos\theta \Rightarrow$

$$\frac{dy}{d\theta} = \frac{1}{r}\cos 2\theta \sin\theta + r\cos\theta = \frac{1}{r}\left(\cos 2\theta \sin\theta + r^2 \cos\theta\right)$$

$$= \frac{1}{r}\left(\cos 2\theta \sin\theta + \sin 2\theta \cos\theta\right) = \frac{1}{r}\sin 3\theta$$

This is 0 when $\sin 3\theta = 0 \Rightarrow \theta = 0, \frac{\pi}{3}$ or $\frac{4\pi}{3}$ (restricting θ to the domain of the lemniscate), so there are horizontal

tangents at $\left(\sqrt[4]{\frac{3}{4}}, \frac{\pi}{3}\right), \left(\sqrt[4]{\frac{3}{4}}, \frac{4\pi}{3}\right)$ and $(0, 0)$. Similarly, $dx/d\theta = (1/r)\cos 3\theta = 0$ when $\theta = \frac{\pi}{6}$ or $\frac{7\pi}{6}$, so there are vertical

tangents at $\left(\sqrt[4]{\frac{3}{4}}, \frac{\pi}{6}\right)$ and $\left(\sqrt[4]{\frac{3}{4}}, \frac{7\pi}{6}\right)$ [and $(0, 0)$].

69. $r = a\sin\theta + b\cos\theta \Rightarrow r^2 = ar\sin\theta + br\cos\theta \Rightarrow x^2 + y^2 = ay + bx \Rightarrow$

$x^2 - bx + \left(\frac{1}{2}b\right)^2 + y^2 - ay + \left(\frac{1}{2}a\right)^2 = \left(\frac{1}{2}b\right)^2 + \left(\frac{1}{2}a\right)^2 \Rightarrow \left(x - \frac{1}{2}b\right)^2 + \left(y - \frac{1}{2}a\right)^2 = \frac{1}{4}(a^2 + b^2)$, and this is a circle

with center $\left(\frac{1}{2}b, \frac{1}{2}a\right)$ and radius $\frac{1}{2}\sqrt{a^2 + b^2}$.

70. These curves are circles which intersect at the origin and at $\left(\frac{1}{\sqrt{2}}a, \frac{\pi}{4}\right)$. At the origin, the first circle has a horizontal

tangent and the second a vertical one, so the tangents are perpendicular here. For the first circle [$r = a\sin\theta$],

$dy/d\theta = a\cos\theta \sin\theta + a\sin\theta \cos\theta = a\sin 2\theta = a$ at $\theta = \frac{\pi}{4}$ and $dx/d\theta = a\cos^2\theta - a\sin^2\theta = a\cos 2\theta = 0$

at $\theta = \frac{\pi}{4}$, so the tangent here is vertical. Similarly, for the second circle [$r = a\cos\theta$], $dy/d\theta = a\cos 2\theta = 0$ and

$dx/d\theta = -a\sin 2\theta = -a$ at $\theta = \frac{\pi}{4}$, so the tangent is horizontal, and again the tangents are perpendicular.

Note for Exercises 71–76: Maple is able to plot polar curves using the `polarplot` command, or using the `coords=polar` option in a regular `plot` command. In Mathematica, use `PolarPlot`. In Derive, change to `Polar` under `Options State`. If your graphing device cannot plot polar equations, you must convert to parametric equations. For example, in Exercise 71, $x = r \cos\theta = [1 + 2\sin(\theta/2)]\cos\theta$, $y = r\sin\theta = [1 + 2\sin(\theta/2)]\sin\theta$.

71. $r = 1 + 2\sin(\theta/2)$. The parameter interval is $[0, 4\pi]$.

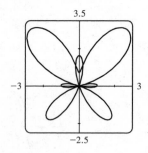

72. $r = \sqrt{1 - 0.8\sin^2\theta}$. The parameter interval is $[0, 2\pi]$.

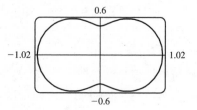

73. $r = e^{\sin\theta} - 2\cos(4\theta)$. The parameter interval is $[0, 2\pi]$.

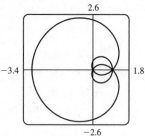

74. $r = \sin^2(4\theta) + \cos(4\theta)$. The parameter interval is $[0, 2\pi]$.

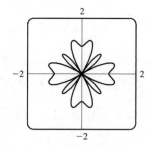

75. $r = 2 - 5\sin(\theta/6)$. The parameter interval is $[-6\pi, 6\pi]$.

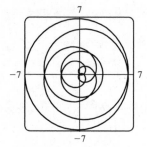

76. $r = \cos(\theta/2) + \cos(\theta/3)$. The parameter interval is $[-6\pi, 6\pi]$.

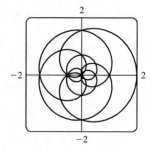

77. It appears that the graph of $r = 1 + \sin\left(\theta - \frac{\pi}{6}\right)$ is the same shape as the graph of $r = 1 + \sin\theta$, but rotated counterclockwise about the origin by $\frac{\pi}{6}$. Similarly, the graph of $r = 1 + \sin\left(\theta - \frac{\pi}{3}\right)$ is rotated by $\frac{\pi}{3}$. In general, the graph of $r = f(\theta - \alpha)$ is the same shape as that of $r = f(\theta)$, but rotated counterclockwise through α about the origin. That is, for any point (r_0, θ_0) on the curve $r = f(\theta)$, the point $(r_0, \theta_0 + \alpha)$ is on the curve $r = f(\theta - \alpha)$, since $r_0 = f(\theta_0) = f((\theta_0 + \alpha) - \alpha)$.

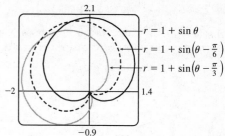

78.

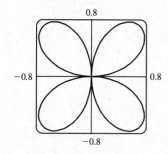

From the graph, the highest points seem to have $y \approx 0.77$. To find the exact value, we solve $dy/d\theta = 0$. $y = r\sin\theta = \sin\theta \sin2\theta \Rightarrow$

$$dy/d\theta = 2\sin\theta \cos2\theta + \cos\theta \sin2\theta$$
$$= 2\sin\theta \left(2\cos^2\theta - 1\right) + \cos\theta \left(2\sin\theta \cos\theta\right)$$
$$= 2\sin\theta \left(3\cos^2\theta - 1\right)$$

In the first quadrant, this is 0 when $\cos\theta = \frac{1}{\sqrt{3}} \Leftrightarrow \sin\theta = \sqrt{\frac{2}{3}} \Leftrightarrow$
$y = 2\sin^2\theta \cos\theta = 2 \cdot \frac{2}{3} \cdot \frac{1}{\sqrt{3}} = \frac{4}{9}\sqrt{3} \approx 0.77$.

79. (a) $r = \sin n\theta$.

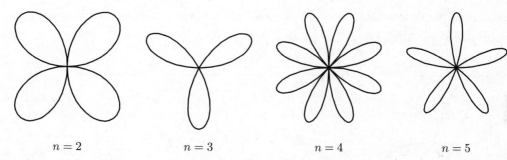

$n = 2$ $n = 3$ $n = 4$ $n = 5$

From the graphs, it seems that when n is even, the number of loops in the curve (called a rose) is $2n$, and when n is odd, the number of loops is simply n. This is because in the case of n odd, every point on the graph is traversed twice, due to the fact that

$$r(\theta + \pi) = \sin[n(\theta + \pi)] = \sin n\theta \cos n\pi + \cos n\theta \sin n\pi = \begin{cases} \sin n\theta & \text{if } n \text{ is even} \\ -\sin n\theta & \text{if } n \text{ is odd} \end{cases}$$

(b) The graph of $r = |\sin n\theta|$ has $2n$ loops whether n is odd or even, since $r(\theta + \pi) = r(\theta)$.

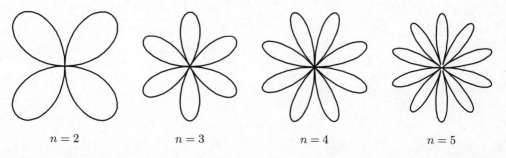

$n = 2$ $n = 3$ $n = 4$ $n = 5$

80. $r = 1 + c\sin n\theta$. We vary n while keeping c constant at 2. As n changes, the curves change in the same way as those in Exercise 79: the number of loops increases. Note that if n is even, the smaller loops are outside the larger ones; if n is odd, they are inside.

$$c = 2$$

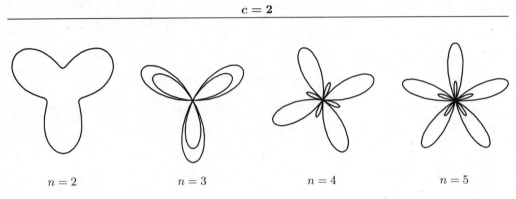

$n = 2$ $n = 3$ $n = 4$ $n = 5$

Now we vary c while keeping $n = 3$. As c increases toward 0, the entire graph gets smaller (the graphs below are not to scale) and the smaller loops shrink in relation to the large ones. At $c = -1$, the small loops disappear entirely, and for $-1 < c < 1$, the graph is a simple, closed curve (at $c = 0$ it is a circle). As c continues to increase, the same changes are seen, but in reverse order, since $1 + (-c)\sin n\theta = 1 + c\sin n(\theta + \pi)$, so the graph for $c = c_0$ is the same as that for $c = -c_0$, with a rotation through π. As $c \to \infty$, the smaller loops get relatively closer in size to the large ones. Note that the distance between the outermost points of corresponding inner and outer loops is always 2. Maple's `animate` command (or Mathematica's `Animate`) is very useful for seeing the changes that occur as c varies.

$$n = 3$$

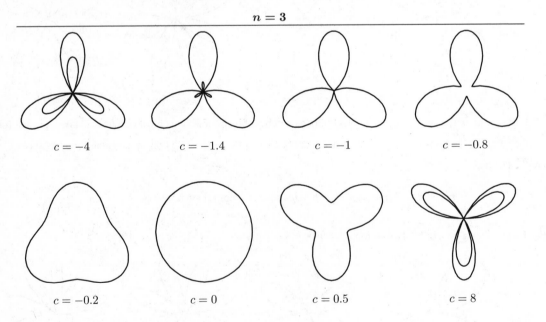

$c = -4$ $c = -1.4$ $c = -1$ $c = -0.8$

$c = -0.2$ $c = 0$ $c = 0.5$ $c = 8$

81. $r = \dfrac{1 - a\cos\theta}{1 + a\cos\theta}$. We start with $a = 0$, since in this case the curve is simply the circle $r = 1$.

As a increases, the graph moves to the left, and its right side becomes flattened. As a increases through about 0.4, the right side seems to grow a dimple, which upon closer investigation (with narrower θ-ranges) seems to appear at $a \approx 0.42$ [the actual value is $\sqrt{2} - 1$]. As $a \to 1$, this dimple becomes more pronounced, and the curve begins to stretch out horizontally, until at $a = 1$ the denominator vanishes at $\theta = \pi$, and the dimple becomes an actual cusp. For $a > 1$ we must choose our parameter interval carefully, since $r \to \infty$ as $1 + a\cos\theta \to 0 \iff \theta \to \pm\cos^{-1}(-1/a)$. As a increases from 1, the curve splits into two parts. The left part has a loop, which grows larger as a increases, and the right part grows broader vertically, and its left tip develops a dimple when $a \approx 2.42$ [actually, $\sqrt{2} + 1$]. As a increases, the dimple grows more and more pronounced. If $a < 0$, we get the same graph as we do for the corresponding positive a-value, but with a rotation through π about the pole, as happened when c was replaced with $-c$ in Exercise 80.

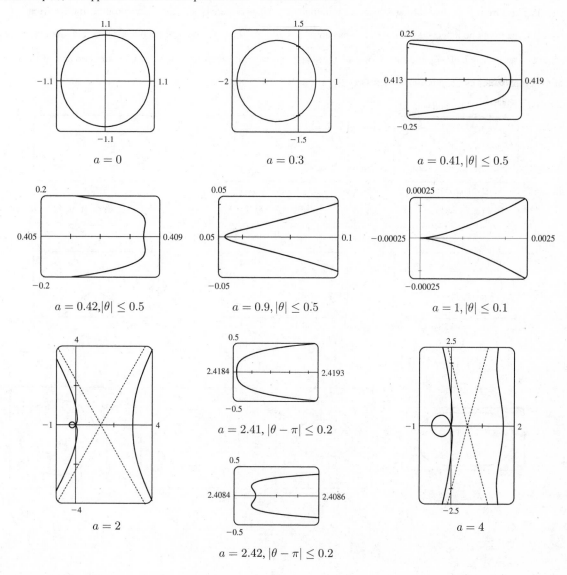

82. Most graphing devices cannot plot implicit polar equations, so we must first find an explicit expression (or expressions) for r in terms of θ, a, and c. We note that the given equation, $r^4 - 2c^2r^2\cos 2\theta + c^4 - a^4 = 0$, is a quadratic in r^2, so we use the quadratic formula and find that

$$r^2 = \frac{2c^2\cos 2\theta \pm \sqrt{4c^4\cos^2 2\theta - 4(c^4 - a^4)}}{2} = c^2\cos 2\theta \pm \sqrt{a^4 - c^4\sin^2 2\theta}$$

so $r = \pm\sqrt{c^2\cos 2\theta \pm \sqrt{a^4 - c^4\sin^2 2\theta}}$. So for each graph, we must plot four curves to be sure of plotting all the points which satisfy the given equation. Note that all four functions have period π.

We start with the case $a = c = 1$, and the resulting curve resembles the symbol for infinity. If we let a decrease, the curve splits into two symmetric parts, and as a decreases further, the parts become smaller, further apart, and rounder. If instead we let a increase from 1, the two lobes of the curve join together, and as a increases further they continue to merge, until at $a \approx 1.4$, the graph no longer has dimples, and has an oval shape. As $a \to \infty$, the oval becomes larger and rounder, since the c^2 and c^4 terms lose their significance. Note that the shape of the graph seems to depend only on the ratio c/a, while the size of the graph varies as c and a jointly increase.

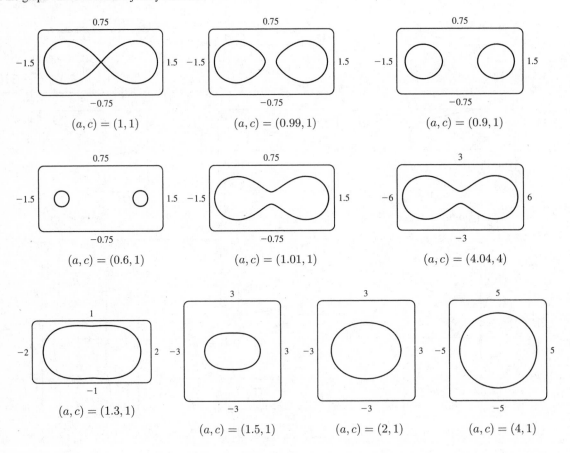

$(a, c) = (1, 1)$

$(a, c) = (0.99, 1)$

$(a, c) = (0.9, 1)$

$(a, c) = (0.6, 1)$

$(a, c) = (1.01, 1)$

$(a, c) = (4.04, 4)$

$(a, c) = (1.3, 1)$

$(a, c) = (1.5, 1)$

$(a, c) = (2, 1)$

$(a, c) = (4, 1)$

83. $\tan\psi = \tan(\phi - \theta) = \dfrac{\tan\phi - \tan\theta}{1 + \tan\phi\,\tan\theta} = \dfrac{\dfrac{dy}{dx} - \tan\theta}{1 + \dfrac{dy}{dx}\tan\theta} = \dfrac{\dfrac{dy/d\theta}{dx/d\theta} - \tan\theta}{1 + \dfrac{dy/d\theta}{dx/d\theta}\tan\theta}$

$= \dfrac{\dfrac{dy}{d\theta} - \dfrac{dx}{d\theta}\tan\theta}{\dfrac{dx}{d\theta} + \dfrac{dy}{d\theta}\tan\theta} = \dfrac{\left(\dfrac{dr}{d\theta}\sin\theta + r\cos\theta\right) - \tan\theta\left(\dfrac{dr}{d\theta}\cos\theta - r\sin\theta\right)}{\left(\dfrac{dr}{d\theta}\cos\theta - r\sin\theta\right) + \tan\theta\left(\dfrac{dr}{d\theta}\sin\theta + r\cos\theta\right)} = \dfrac{r\cos\theta + r\cdot\dfrac{\sin^2\theta}{\cos\theta}}{\dfrac{dr}{d\theta}\cos\theta + \dfrac{dr}{d\theta}\cdot\dfrac{\sin^2\theta}{\cos\theta}}$

$= \dfrac{r\cos^2\theta + r\sin^2\theta}{\dfrac{dr}{d\theta}\cos^2\theta + \dfrac{dr}{d\theta}\sin^2\theta} = \dfrac{r}{dr/d\theta}$

84. (a) $r = e^\theta \ \Rightarrow\ dr/d\theta = e^\theta$, so by Exercise 83, $\tan\psi = r/e^\theta = 1 \ \Rightarrow$

$\psi = \arctan 1 = \frac{\pi}{4}$.

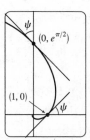

(b) The Cartesian equation of the tangent line at $(1,0)$ is $y = x - 1$, and that of

the tangent line at $(0, e^{\pi/2})$ is $y = e^{\pi/2} - x$.

(c) Let a be the tangent of the angle between the tangent and radial lines, that

is, $a = \tan\psi$. Then, by Exercise 83, $a = \dfrac{r}{dr/d\theta} \ \Rightarrow\ \dfrac{dr}{d\theta} = \dfrac{1}{a}r \ \Rightarrow$

$r = Ce^{\theta/a}$ (by Theorem 10.4.2 [ET 9.4.2]).

11.4 Areas and Lengths in Polar Coordinates ET 10.4

1. $r = \theta^2,\ 0 \le \theta \le \frac{\pi}{4}$. $A = \displaystyle\int_0^{\pi/4}\frac{1}{2}r^2\,d\theta = \int_0^{\pi/4}\frac{1}{2}(\theta^2)^2\,d\theta = \int_0^{\pi/4}\frac{1}{2}\theta^4\,d\theta = \left[\frac{1}{10}\theta^5\right]_0^{\pi/4} = \frac{1}{10}\left(\frac{\pi}{4}\right)^5 = \frac{1}{10{,}240}\pi^5$

2. $r = e^{\theta/2},\ \pi \le \theta \le 2\pi$. $A = \displaystyle\int_\pi^{2\pi}\frac{1}{2}(e^{\theta/2})^2\,d\theta = \int_\pi^{2\pi}\frac{1}{2}e^\theta\,d\theta = \frac{1}{2}\left[e^\theta\right]_\pi^{2\pi} = \frac{1}{2}(e^{2\pi} - e^\pi)$

3. $r = \sin\theta,\ \frac{\pi}{3} \le \theta \le \frac{2\pi}{3}$.

$A = \displaystyle\int_{\pi/3}^{2\pi/3}\frac{1}{2}\sin^2\theta\,d\theta = \frac{1}{4}\int_{\pi/3}^{2\pi/3}(1 - \cos 2\theta)\,d\theta = \frac{1}{4}\left[\theta - \frac{1}{2}\sin 2\theta\right]_{\pi/3}^{2\pi/3} = \frac{1}{4}\left[\frac{2\pi}{3} - \frac{1}{2}\sin\frac{4\pi}{3} - \frac{\pi}{3} + \frac{1}{2}\sin\frac{2\pi}{3}\right]$

$= \frac{1}{4}\left[\frac{2\pi}{3} - \frac{1}{2}\left(-\frac{\sqrt{3}}{2}\right) - \frac{\pi}{3} + \frac{1}{2}\left(\frac{\sqrt{3}}{2}\right)\right] = \frac{1}{4}\left(\frac{\pi}{3} + \frac{\sqrt{3}}{2}\right) = \frac{\pi}{12} + \frac{\sqrt{3}}{8}$

4. $r = \sqrt{\sin\theta},\ 0 \le \theta \le \pi$. $A = \displaystyle\int_0^\pi\frac{1}{2}\left(\sqrt{\sin\theta}\right)^2 d\theta = \int_0^\pi\frac{1}{2}\sin\theta\,d\theta = \left[-\frac{1}{2}\cos\theta\right]_0^\pi = \frac{1}{2} + \frac{1}{2} = 1$

5. $r = \sqrt{\theta},\ 0 \le \theta \le 2\pi$. $A = \displaystyle\int_0^{2\pi}\frac{1}{2}r^2\,d\theta = \int_0^{2\pi}\frac{1}{2}\left(\sqrt{\theta}\right)^2 d\theta = \int_0^{2\pi}\frac{1}{2}\theta\,d\theta = \left[\frac{1}{4}\theta^2\right]_0^{2\pi} = \pi^2$

6. $r = 1 + \cos\theta,\ 0 \le \theta \le \pi$.

$A = \displaystyle\int_0^\pi\frac{1}{2}(1 + \cos\theta)^2\,d\theta = \frac{1}{2}\int_0^\pi(1 + 2\cos\theta + \cos^2\theta)\,d\theta = \frac{1}{2}\int_0^\pi\left[1 + 2\cos\theta + \frac{1}{2}(1 + \cos 2\theta)\right]d\theta$

$= \frac{1}{2}\int_0^\pi\left(\frac{3}{2} + 2\cos\theta + \frac{1}{2}\cos 2\theta\right)d\theta = \frac{1}{2}\left[\frac{3}{2}\theta + 2\sin\theta + \frac{1}{4}\sin 2\theta\right]_0^\pi = \frac{1}{2}\left(\frac{3}{2}\pi + 0 + 0\right) - \frac{1}{2}(0) = \frac{3\pi}{4}$

7. $r = 4 + 3\sin\theta$, $-\frac{\pi}{2} \le \theta \le \frac{\pi}{2}$.

$$A = \int_{-\pi/2}^{\pi/2} \frac{1}{2}((4 + 3\sin\theta)^2 \, d\theta = \frac{1}{2} \int_{-\pi/2}^{\pi/2} (16 + 24\sin\theta + 9\sin^2\theta) \, d\theta$$

$$= \frac{1}{2} \int_{-\pi/2}^{\pi/2} (16 + 9\sin^2\theta) \, d\theta \qquad \text{[by Theorem 5.5.6(a) [ET 5.5.7(a)]]}$$

$$= \frac{1}{2} \cdot 2 \int_{0}^{\pi/2} \left[16 + 9 \cdot \frac{1}{2}(1 - \cos 2\theta)\right] d\theta \qquad \text{[by Theorem 5.5.6(a) [ET 5.5.7(a)]]}$$

$$= \int_{0}^{\pi/2} \left(\frac{41}{2} - \frac{9}{2}\cos 2\theta\right) d\theta = \left[\frac{41}{2}\theta - \frac{9}{4}\sin 2\theta\right]_{0}^{\pi/2} = \left(\frac{41\pi}{4} - 0\right) - (0 - 0) = \frac{41\pi}{4}$$

8. $r = \sin 2\theta$, $0 \le \theta \le \frac{\pi}{2}$.

$$A = \int_{0}^{\pi/2} \frac{1}{2}\sin^2 2\theta \, d\theta = \frac{1}{2} \int_{0}^{\pi/2} \frac{1}{2}(1 - \cos 4\theta) \, d\theta = \frac{1}{4}\left[\theta - \frac{1}{4}\sin 4\theta\right]_{0}^{\pi/2} = \frac{1}{4}\left(\frac{\pi}{2}\right) = \frac{\pi}{8}$$

9. The area above the polar axis is bounded by $r = 3\cos\theta$ for $\theta = 0$

to $\theta = \pi/2$ [*not* π]. By symmetry,

$$A = 2 \int_{0}^{\pi/2} \frac{1}{2}r^2 \, d\theta = \int_{0}^{\pi/2} (3\cos\theta)^2 \, d\theta = 3^2 \int_{0}^{\pi/2} \cos^2\theta \, d\theta$$

$$= 9 \int_{0}^{\pi/2} \frac{1}{2}(1 + \cos 2\theta) \, d\theta = \frac{9}{2}\left[\theta + \frac{1}{2}\sin 2\theta\right]_{0}^{\pi/2} = \frac{9}{2}\left[\left(\frac{\pi}{2} + 0\right) - (0 + 0)\right] = \frac{9\pi}{4}$$

Also, note that this is a circle with radius $\frac{3}{2}$, so its area is $\pi\left(\frac{3}{2}\right)^2 = \frac{9\pi}{4}$.

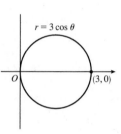

10. $A = \int_{0}^{2\pi} \frac{1}{2}r^2 \, d\theta = \int_{0}^{2\pi} \frac{1}{2}[3(1 + \cos\theta)]^2 \, d\theta$

$$= \frac{9}{2} \int_{0}^{2\pi} (1 + 2\cos\theta + \cos^2\theta) \, d\theta$$

$$= \frac{9}{2} \int_{0}^{2\pi} \left[1 + 2\cos\theta + \frac{1}{2}(1 + \cos 2\theta)\right] d\theta$$

$$= \frac{9}{2}\left[\frac{3}{2}\theta + 2\sin\theta + \frac{1}{4}\sin 2\theta\right]_{0}^{2\pi} = \frac{27}{2}\pi$$

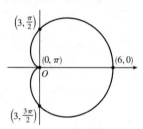

11. The curve goes through the pole when $\theta = \pi/4$, so we'll find the area for

$0 \le \theta \le \pi/4$ and multiply it by 4.

$$A = 4 \int_{0}^{\pi/4} \frac{1}{2}r^2 \, d\theta = 2 \int_{0}^{\pi/4} (4\cos 2\theta) \, d\theta$$

$$= 8 \int_{0}^{\pi/4} \cos 2\theta \, d\theta = 4\left[\sin 2\theta\right]_{0}^{\pi/4} = 4$$

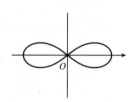

12. To find the area that the curve encloses, we'll double the area to the left of the

vertical axis.

$$A = 2 \int_{\pi/2}^{3\pi/2} \frac{1}{2}(2 - \sin\theta)^2 \, d\theta = \int_{\pi/2}^{3\pi/2} (4 - 4\sin\theta + \sin^2\theta) \, d\theta$$

$$= \int_{\pi/2}^{3\pi/2} \left[4 - 4\sin\theta + \frac{1}{2}(1 - \cos 2\theta)\right] d\theta = \int_{\pi/2}^{3\pi/2} \left(\frac{9}{2} - 4\sin\theta - \frac{1}{2}\cos 2\theta\right) d\theta$$

$$= \left[\frac{9}{2}\theta + 4\cos\theta - \frac{1}{4}\sin 2\theta\right]_{\pi/2}^{3\pi/2} = \left(\frac{27\pi}{4}\right) - \left(\frac{9\pi}{4}\right) = \frac{9\pi}{2}$$

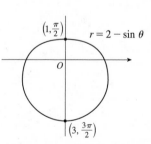

Or: We could have doubled the area to the right of the vertical axis and integrated from $-\pi/2$ to $\pi/2$.

Or: We could have integrated from 0 to 2π [simpler arithmetic].

13. One-sixth of the area lies above the polar axis and is bounded by the curve

$r = 2\cos 3\theta$ for $\theta = 0$ to $\theta = \pi/6$.

$$A = 6 \int_0^{\pi/6} \tfrac{1}{2}(2\cos 3\theta)^2 \, d\theta = 12 \int_0^{\pi/6} \cos^2 3\theta \, d\theta$$

$$= \tfrac{12}{2} \int_0^{\pi/6}(1 + \cos 6\theta) \, d\theta$$

$$= 6\big[\theta + \tfrac{1}{6}\sin 6\theta\big]_0^{\pi/6} = 6\big(\tfrac{\pi}{6}\big) = \pi$$

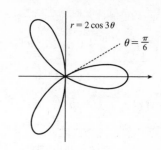

$r = 2\cos 3\theta$

$\theta = \frac{\pi}{6}$

14. $A = \int_0^{2\pi} \tfrac{1}{2}(2 + \cos 2\theta)^2 \, d\theta = \tfrac{1}{2}\int_0^{2\pi}(4 + 4\cos 2\theta + \cos^2 2\theta) \, d\theta$

$= \tfrac{1}{2}\int_0^{2\pi}\big(4 + 4\cos 2\theta + \tfrac{1}{2} + \tfrac{1}{2}\cos 4\theta\big) \, d\theta$

$= \tfrac{1}{2}\big[\tfrac{9}{2}\theta + 2\sin 2\theta + \tfrac{1}{8}\sin 4\theta\big]_0^{2\pi} = \tfrac{1}{2}(9\pi) = \tfrac{9\pi}{2}$

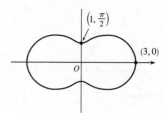

$\big(1, \tfrac{\pi}{2}\big)$

$(3, 0)$

O

15. $A = \int_0^{2\pi} \tfrac{1}{2}(1 + 2\sin 6\theta)^2 \, d\theta = \tfrac{1}{2}\int_0^{2\pi}(1 + 4\sin 6\theta + 4\sin^2 6\theta) \, d\theta$

$= \tfrac{1}{2}\int_0^{2\pi}\big[1 + 4\sin 6\theta + 4 \cdot \tfrac{1}{2}(1 - \cos 12\theta)\big] \, d\theta$

$= \tfrac{1}{2}\int_0^{2\pi}(3 + 4\sin 6\theta - 2\cos 12\theta) \, d\theta$

$= \tfrac{1}{2}\big[3\theta - \tfrac{2}{3}\cos 6\theta - \tfrac{1}{6}\sin 12\theta\big]_0^{2\pi}$

$= \tfrac{1}{2}\big[(6\pi - \tfrac{2}{3} - 0) - (0 - \tfrac{2}{3} - 0)\big] = 3\pi$

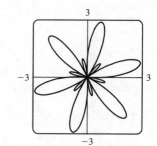

3

−3   3

−3

16. $A = \int_0^{\pi} \tfrac{1}{2}(2\sin\theta + 3\sin 9\theta)^2 \, d\theta = 2\int_0^{\pi/2} \tfrac{1}{2}(2\sin\theta + 3\sin 9\theta)^2 \, d\theta$

$= \int_0^{\pi/2}(4\sin^2\theta + 12\sin\theta\sin 9\theta + 9\sin^2 9\theta) \, d\theta$

$= \int_0^{\pi/2}\big[2(1 - \cos 2\theta) + 12 \cdot \tfrac{1}{2}(\cos(\theta - 9\theta) - \cos(\theta + 9\theta)) + \tfrac{9}{2}(1 - \cos 18\theta)\big] \, d\theta$

[integration by parts could be used for $\int \sin\theta \sin 9\theta \, d\theta$]

$= \int_0^{\pi/2}\big(2 - 2\cos 2\theta + 6\cos 8\theta - 6\cos 10\theta + \tfrac{9}{2} - \tfrac{9}{2}\cos 18\theta\big) \, d\theta$

$= \big[\tfrac{13}{2}\theta - \sin 2\theta + \tfrac{3}{4}\sin 8\theta - \tfrac{3}{5}\sin 10\theta - \tfrac{1}{4}\sin 18\theta\big]_0^{\pi/2} = \tfrac{13\pi}{4}$

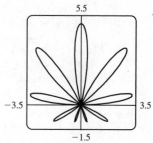

5.5

−3.5   3.5

−1.5

17. The shaded loop is traced out from $\theta = 0$ to $\theta = \pi/2$.

$$A = \int_0^{\pi/2} \tfrac{1}{2}r^2 \, d\theta = \tfrac{1}{2}\int_0^{\pi/2}\sin^2 2\theta \, d\theta$$

$$= \tfrac{1}{2}\int_0^{\pi/2} \tfrac{1}{2}(1 - \cos 4\theta) \, d\theta = \tfrac{1}{4}\big[\theta - \tfrac{1}{4}\sin 4\theta\big]_0^{\pi/2}$$

$$= \tfrac{1}{4}\big(\tfrac{\pi}{2}\big) = \tfrac{\pi}{8}$$

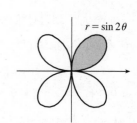

$r = \sin 2\theta$

18. $A = \int_0^{\pi/3} \tfrac{1}{2}(4\sin 3\theta)^2 \, d\theta = 8\int_0^{\pi/3}\sin^2 3\theta \, d\theta$

$= 4\int_0^{\pi/3}(1 - \cos 6\theta) \, d\theta$

$= 4\big[\theta - \tfrac{1}{6}\sin 6\theta\big]_0^{\pi/3} = \tfrac{4\pi}{3}$

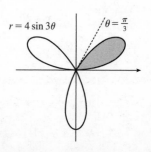

$r = 4\sin 3\theta$

$\theta = \frac{\pi}{3}$

19. $r = 0 \implies 3\cos 5\theta = 0 \implies 5\theta = \frac{\pi}{2} \implies \theta = \frac{\pi}{10}$.

$A = \int_{-\pi/10}^{\pi/10} \frac{1}{2}(3\cos 5\theta)^2 \, d\theta = \int_0^{\pi/10} 9\cos^2 5\theta \, d\theta = \frac{9}{2} \int_0^{\pi/10} (1 + \cos 10\theta) \, d\theta = \frac{9}{2}\left[\theta + \frac{1}{10}\sin 10\theta\right]_0^{\pi/10} = \frac{9\pi}{20}$

20. $r = 0 \implies 2\sin 6\theta = 0 \implies 6\theta = 0 \text{ or } \pi \implies \theta = 0 \text{ or } \frac{\pi}{6}$.

$A = \int_0^{\pi/6} \frac{1}{2}(2\sin 6\theta)^2 \, d\theta = \int_0^{\pi/6} 2\sin^2 6\theta \, d\theta = 2\int_0^{\pi/6} \frac{1}{2}(1 - \cos 12\theta) \, d\theta = \left[\theta - \frac{1}{12}\sin 12\theta\right]_0^{\pi/6} = \frac{\pi}{6}$

21.

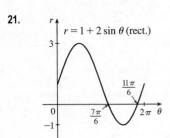

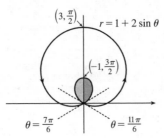

This is a limaçon, with inner loop traced out between $\theta = \frac{7\pi}{6}$ and $\frac{11\pi}{6}$ [found by solving $r = 0$].

$A = 2\int_{7\pi/6}^{3\pi/2} \frac{1}{2}(1 + 2\sin\theta)^2 \, d\theta = \int_{7\pi/6}^{3\pi/2} \left(1 + 4\sin\theta + 4\sin^2\theta\right) \, d\theta = \int_{7\pi/6}^{3\pi/2} \left[1 + 4\sin\theta + 4\cdot\frac{1}{2}(1 - \cos 2\theta)\right] \, d\theta$

$= \left[\theta - 4\cos\theta + 2\theta - \sin 2\theta\right]_{7\pi/6}^{3\pi/2} = \left(\frac{9\pi}{2}\right) - \left(\frac{7\pi}{2} + 2\sqrt{3} - \frac{\sqrt{3}}{2}\right) = \pi - \frac{3\sqrt{3}}{2}$

22. To determine when the strophoid $r = 2\cos\theta - \sec\theta$ passes through the pole, we solve

$r = 0 \implies 2\cos\theta - \frac{1}{\cos\theta} = 0 \implies 2\cos^2\theta - 1 = 0 \implies \cos^2\theta = \frac{1}{2} \implies$

$\cos\theta = \pm\frac{1}{\sqrt{2}} \implies \theta = \frac{\pi}{4} \text{ or } \theta = \frac{3\pi}{4} \text{ for } 0 \le \theta \le \pi \text{ with } \theta \ne \frac{\pi}{2}$.

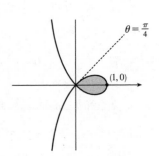

$A = 2\int_0^{\pi/4} \frac{1}{2}(2\cos\theta - \sec\theta)^2 \, d\theta = \int_0^{\pi/4} (4\cos^2\theta - 4 + \sec^2\theta) \, d\theta$

$= \int_0^{\pi/4} \left[4\cdot\frac{1}{2}(1 + \cos 2\theta) - 4 + \sec^2\theta\right] \, d\theta = \int_0^{\pi/4} (-2 + 2\cos 2\theta + \sec^2\theta) \, d\theta$

$= \left[-2\theta + \sin 2\theta + \tan\theta\right]_0^{\pi/4} = \left(-\frac{\pi}{2} + 1 + 1\right) - 0 = 2 - \frac{\pi}{2}$

23. $2\cos\theta = 1 \implies \cos\theta = \frac{1}{2} \implies \theta = \frac{\pi}{3} \text{ or } \frac{5\pi}{3}$.

$A = 2\int_0^{\pi/3} \frac{1}{2}\left[(2\cos\theta)^2 - 1^2\right] \, d\theta = \int_0^{\pi/3} (4\cos^2\theta - 1) \, d\theta$

$= \int_0^{\pi/3} \left\{4\left[\frac{1}{2}(1 + \cos 2\theta)\right] - 1\right\} \, d\theta = \int_0^{\pi/3} (1 + 2\cos 2\theta) \, d\theta$

$= \left[\theta + \sin 2\theta\right]_0^{\pi/3} = \frac{\pi}{3} + \frac{\sqrt{3}}{2}$

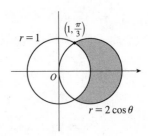

24. $1 - \sin\theta = 1 \implies \sin\theta = 0 \implies \theta = 0 \text{ or } \pi \implies$

$A = \int_\pi^{2\pi} \frac{1}{2}\left[(1 - \sin\theta)^2 - 1\right] \, d\theta = \frac{1}{2}\int_\pi^{2\pi} (\sin^2\theta - 2\sin\theta) \, d\theta$

$= \frac{1}{4}\int_\pi^{2\pi} (1 - \cos 2\theta - 4\sin\theta) \, d\theta = \frac{1}{4}\left[\theta - \frac{1}{2}\sin 2\theta + 4\cos\theta\right]_\pi^{2\pi}$

$= \frac{1}{4}\pi + 2$

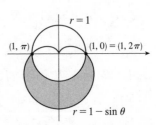

25. To find the area inside the leminiscate $r^2 = 8\cos 2\theta$ and outside the circle $r = 2$,

we first note that the two curves intersect when $r^2 = 8\cos 2\theta$ and $r = 2$,

that is, when $\cos 2\theta = \frac{1}{2}$. For $-\pi < \theta \le \pi$, $\cos 2\theta = \frac{1}{2}$ ⇔ $2\theta = \pm\pi/3$

or $\pm 5\pi/3$ ⇔ $\theta = \pm\pi/6$ or $\pm 5\pi/6$. The figure shows that the desired area is

4 times the area between the curves from 0 to $\pi/6$. Thus,

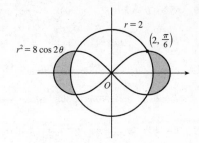

$$A = 4\int_0^{\pi/6}\left[\tfrac{1}{2}(8\cos 2\theta) - \tfrac{1}{2}(2)^2\right]d\theta = 8\int_0^{\pi/6}(2\cos 2\theta - 1)\,d\theta$$

$$= 8\left[\sin 2\theta - \theta\right]_0^{\pi/6} = 8(\sqrt{3}/2 - \pi/6) = 4\sqrt{3} - 4\pi/3$$

26. To find the shaded area A, we'll find the area A_1 inside the curve $r = 2 + \sin\theta$

and subtract $\pi\left(\frac{3}{2}\right)^2$ since $r = 3\sin\theta$ is a circle with radius $\frac{3}{2}$.

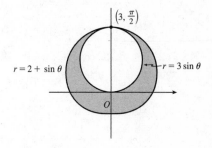

$$A_1 = \int_0^{2\pi}\tfrac{1}{2}(2 + \sin\theta)^2\,d\theta = \tfrac{1}{2}\int_0^{2\pi}(4 + 4\sin\theta + \sin^2\theta)\,d\theta$$

$$= \tfrac{1}{2}\int_0^{2\pi}\left[4 + 4\sin\theta + \tfrac{1}{2}\cdot(1 - \cos 2\theta)\right]d\theta$$

$$= \tfrac{1}{2}\int_0^{2\pi}\left(\tfrac{9}{2} + 4\sin\theta - \tfrac{1}{2}\cos 2\theta\right)d\theta$$

$$= \tfrac{1}{2}\left[\tfrac{9}{2}\theta - 4\cos\theta - \tfrac{1}{4}\sin 2\theta\right]_0^{2\pi} = \tfrac{1}{2}[(9\pi - 4) - (-4)] = \tfrac{9\pi}{2}$$

So $A = A_1 - \frac{9\pi}{4} = \frac{9\pi}{2} - \frac{9\pi}{4} = \frac{9\pi}{4}$.

27. $3\cos\theta = 1 + \cos\theta$ ⇔ $\cos\theta = \frac{1}{2}$ ⇒ $\theta = \frac{\pi}{3}$ or $-\frac{\pi}{3}$.

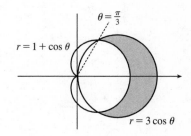

$$A = 2\int_0^{\pi/3}\tfrac{1}{2}[(3\cos\theta)^2 - (1 + \cos\theta)^2]\,d\theta$$

$$= \int_0^{\pi/3}(8\cos^2\theta - 2\cos\theta - 1)\,d\theta = \int_0^{\pi/3}[4(1 + \cos 2\theta) - 2\cos\theta - 1]\,d\theta$$

$$= \int_0^{\pi/3}(3 + 4\cos 2\theta - 2\cos\theta)\,d\theta = \left[3\theta + 2\sin 2\theta - 2\sin\theta\right]_0^{\pi/3}$$

$$= \pi + \sqrt{3} - \sqrt{3} = \pi$$

28. $3\sin\theta = 2 - \sin\theta$ ⇒ $4\sin\theta = 2$ ⇒ $\sin\theta = \frac{1}{2}$ ⇒ $\theta = \frac{\pi}{6}$ or $\frac{5\pi}{6}$.

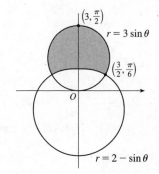

$$A = 2\int_{\pi/6}^{\pi/2}\tfrac{1}{2}[(3\sin\theta)^2 - (2 - \sin\theta)^2]\,d\theta$$

$$= \int_{\pi/6}^{\pi/2}(9\sin^2\theta - 4 + 4\sin\theta - \sin^2\theta]\,d\theta$$

$$= \int_{\pi/6}^{\pi/2}(8\sin^2\theta + 4\sin\theta - 4)\,d\theta$$

$$= 4\int_{\pi/6}^{\pi/2}\left[2\cdot\tfrac{1}{2}(1 - \cos 2\theta) + \sin\theta - 1\right]d\theta$$

$$= 4\int_{\pi/6}^{\pi/2}(\sin\theta - \cos 2\theta)\,d\theta = 4\left[-\cos\theta - \tfrac{1}{2}\sin 2\theta\right]_{\pi/6}^{\pi/2}$$

$$= 4\left[(0 - 0) - \left(-\tfrac{\sqrt{3}}{2} - \tfrac{\sqrt{3}}{4}\right)\right] = 4\left(\tfrac{3\sqrt{3}}{4}\right) = 3\sqrt{3}$$

29. $\sqrt{3}\cos\theta = \sin\theta \;\Rightarrow\; \sqrt{3} = \dfrac{\sin\theta}{\cos\theta} \;\Rightarrow\; \tan\theta = \sqrt{3} \;\Rightarrow\; \theta = \frac{\pi}{3}.$

$A = \int_0^{\pi/3} \frac{1}{2}(\sin\theta)^2\, d\theta + \int_{\pi/3}^{\pi/2} \frac{1}{2}\left(\sqrt{3}\cos\theta\right)^2 d\theta$

$\quad = \int_0^{\pi/3} \frac{1}{2}\cdot\frac{1}{2}(1 - \cos 2\theta)\, d\theta + \int_{\pi/3}^{\pi/2} \frac{1}{2}\cdot 3\cdot\frac{1}{2}(1 + \cos 2\theta)\, d\theta$

$\quad = \frac{1}{4}\left[\theta - \frac{1}{2}\sin 2\theta\right]_0^{\pi/3} + \frac{3}{4}\left[\theta + \frac{1}{2}\sin 2\theta\right]_{\pi/3}^{\pi/2}$

$\quad = \frac{1}{4}\left[\left(\frac{\pi}{3} - \frac{\sqrt{3}}{4}\right) - 0\right] + \frac{3}{4}\left[\left(\frac{\pi}{2} + 0\right) - \left(\frac{\pi}{3} + \frac{\sqrt{3}}{4}\right)\right]$

$\quad = \frac{\pi}{12} - \frac{\sqrt{3}}{16} + \frac{\pi}{8} - \frac{3\sqrt{3}}{16} = \frac{5\pi}{24} - \frac{\sqrt{3}}{4}$

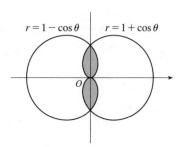

30. $A = 4\int_0^{\pi/2} \frac{1}{2}(1 - \cos\theta)^2\, d\theta = 2\int_0^{\pi/2}(1 - 2\cos\theta + \cos^2\theta)\, d\theta$

$\quad = 2\int_0^{\pi/2}\left[1 - 2\cos\theta + \frac{1}{2}(1 + \cos 2\theta)\right] d\theta$

$\quad = 2\int_0^{\pi/2}\left(\frac{3}{2} - 2\cos\theta + \frac{1}{2}\cos 2\theta\right) d\theta = \int_0^{\pi/2}(3 - 4\cos\theta + \cos 2\theta)\, d\theta$

$\quad = \left[3\theta - 4\sin\theta + \frac{1}{2}\sin 2\theta\right]_0^{\pi/2} = \frac{3\pi}{2} - 4$

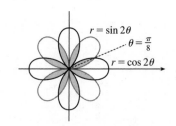

31. $\sin 2\theta = \cos 2\theta \;\Rightarrow\; \dfrac{\sin 2\theta}{\cos 2\theta} = 1 \;\Rightarrow\; \tan 2\theta = 1 \;\Rightarrow\; 2\theta = \frac{\pi}{4} \;\Rightarrow\;$
$\theta = \frac{\pi}{8} \;\Rightarrow\;$

$\quad\quad A = 8\cdot 2\int_0^{\pi/8} \frac{1}{2}\sin^2 2\theta\, d\theta = 8\int_0^{\pi/8} \frac{1}{2}(1 - \cos 4\theta)\, d\theta$

$\quad\quad\quad = 4\left[\theta - \frac{1}{4}\sin 4\theta\right]_0^{\pi/8} = 4\left(\frac{\pi}{8} - \frac{1}{4}\cdot 1\right) = \frac{\pi}{2} - 1$

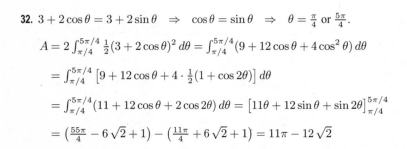

32. $3 + 2\cos\theta = 3 + 2\sin\theta \;\Rightarrow\; \cos\theta = \sin\theta \;\Rightarrow\; \theta = \frac{\pi}{4}$ or $\frac{5\pi}{4}.$

$A = 2\int_{\pi/4}^{5\pi/4} \frac{1}{2}(3 + 2\cos\theta)^2\, d\theta = \int_{\pi/4}^{5\pi/4}(9 + 12\cos\theta + 4\cos^2\theta)\, d\theta$

$\quad = \int_{\pi/4}^{5\pi/4}\left[9 + 12\cos\theta + 4\cdot\frac{1}{2}(1 + \cos 2\theta)\right] d\theta$

$\quad = \int_{\pi/4}^{5\pi/4}(11 + 12\cos\theta + 2\cos 2\theta)\, d\theta = \left[11\theta + 12\sin\theta + \sin 2\theta\right]_{\pi/4}^{5\pi/4}$

$\quad = \left(\frac{55\pi}{4} - 6\sqrt{2} + 1\right) - \left(\frac{11\pi}{4} + 6\sqrt{2} + 1\right) = 11\pi - 12\sqrt{2}$

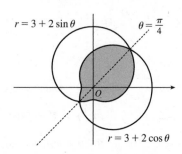

33. $\sin 2\theta = \cos 2\theta \;\Rightarrow\; \tan 2\theta = 1 \;\Rightarrow\; 2\theta = \frac{\pi}{4} \;\Rightarrow\; \theta = \frac{\pi}{8}$

$\quad\quad A = 4\int_0^{\pi/8} \frac{1}{2}\sin 2\theta\, d\theta \quad [\text{since } r^2 = \sin 2\theta]$

$\quad\quad\quad = \int_0^{\pi/8} 2\sin 2\theta\, d\theta = \left[-\cos 2\theta\right]_0^{\pi/8}$

$\quad\quad\quad = -\frac{1}{2}\sqrt{2} - (-1) = 1 - \frac{1}{2}\sqrt{2}$

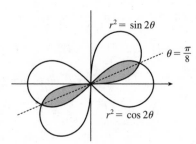

34. Let $\alpha = \tan^{-1}(b/a)$. Then

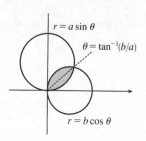

$r = a \sin \theta$

$\theta = \tan^{-1}(b/a)$

$r = b \cos \theta$

$$A = \int_0^\alpha \tfrac{1}{2}(a \sin \theta)^2 \, d\theta + \int_\alpha^{\pi/2} \tfrac{1}{2}(b \cos \theta)^2 \, d\theta$$

$$= \tfrac{1}{4}a^2 \left[\theta - \tfrac{1}{2}\sin 2\theta\right]_0^\alpha + \tfrac{1}{4}b^2 \left[\theta + \tfrac{1}{2}\sin 2\theta\right]_\alpha^{\pi/2}$$

$$= \tfrac{1}{4}\alpha(a^2 - b^2) + \tfrac{1}{8}\pi b^2 - \tfrac{1}{4}(a^2 + b^2)(\sin\alpha \, \cos\alpha)$$

$$= \tfrac{1}{4}(a^2 - b^2)\tan^{-1}(b/a) + \tfrac{1}{8}\pi b^2 - \tfrac{1}{4}ab$$

35. The darker shaded region (from $\theta = 0$ to $\theta = 2\pi/3$) represents $\tfrac{1}{2}$ of the desired area plus $\tfrac{1}{2}$ of the area of the inner loop. From this area, we'll subtract $\tfrac{1}{2}$ of the area of the inner loop (the lighter shaded region from $\theta = 2\pi/3$ to $\theta = \pi$), and then double that difference to obtain the desired area.

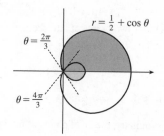

$r = \tfrac{1}{2} + \cos\theta$

$\theta = \dfrac{2\pi}{3}$

$\theta = \dfrac{4\pi}{3}$

$$A = 2\left[\int_0^{2\pi/3} \tfrac{1}{2}\left(\tfrac{1}{2} + \cos\theta\right)^2 d\theta - \int_{2\pi/3}^{\pi} \tfrac{1}{2}\left(\tfrac{1}{2} + \cos\theta\right)^2 d\theta\right]$$

$$= \int_0^{2\pi/3}\left(\tfrac{1}{4} + \cos\theta + \cos^2\theta\right) d\theta - \int_{2\pi/3}^{\pi}\left(\tfrac{1}{4} + \cos\theta + \cos^2\theta\right) d\theta$$

$$= \int_0^{2\pi/3}\left[\tfrac{1}{4} + \cos\theta + \tfrac{1}{2}(1 + \cos 2\theta)\right] d\theta$$

$$\qquad\qquad - \int_{2\pi/3}^{\pi}\left[\tfrac{1}{4} + \cos\theta + \tfrac{1}{2}(1 + \cos 2\theta)\right] d\theta$$

$$= \left[\frac{\theta}{4} + \sin\theta + \frac{\theta}{2} + \frac{\sin 2\theta}{4}\right]_0^{2\pi/3} - \left[\frac{\theta}{4} + \sin\theta + \frac{\theta}{2} + \frac{\sin 2\theta}{4}\right]_{2\pi/3}^{\pi}$$

$$= \left(\frac{\pi}{6} + \frac{\sqrt{3}}{2} + \frac{\pi}{3} - \frac{\sqrt{3}}{8}\right) - \left(\frac{\pi}{4} + \frac{\pi}{2}\right) + \left(\frac{\pi}{6} + \frac{\sqrt{3}}{2} + \frac{\pi}{3} - \frac{\sqrt{3}}{8}\right)$$

$$= \frac{\pi}{4} + \frac{3}{4}\sqrt{3} = \frac{1}{4}\left(\pi + 3\sqrt{3}\right)$$

36. $r = 0 \;\Rightarrow\; 1 + 2\cos 3\theta = 0 \;\Rightarrow\; \cos 3\theta = -\tfrac{1}{2} \;\Rightarrow\; 3\theta = \frac{2\pi}{3}, \frac{4\pi}{3}$ [for $0 \le 3\theta \le 2\pi$] $\;\Rightarrow\; \theta = \frac{2\pi}{9}, \frac{4\pi}{9}$. The darker shaded region (from $\theta = 0$ to $\theta = 2\pi/9$) represents $\tfrac{1}{2}$ of the desired area plus $\tfrac{1}{2}$ of the area of the inner loop. From this area, we'll subtract $\tfrac{1}{2}$ of the area of the inner loop (the lighter shaded region from $\theta = 2\pi/9$ to $\theta = \pi/3$), and then double that difference to obtain the desired area.

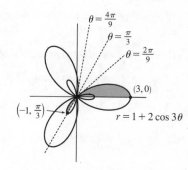

$\theta = \dfrac{4\pi}{9}$

$\theta = \dfrac{\pi}{3}$

$\theta = \dfrac{2\pi}{9}$

$(3, 0)$

$\left(-1, \dfrac{\pi}{3}\right)$

$r = 1 + 2\cos 3\theta$

$$A = 2\left[\int_0^{2\pi/9} \tfrac{1}{2}(1 + 2\cos 3\theta)^2 \, d\theta - \int_{2\pi/9}^{\pi/3} \tfrac{1}{2}(1 + 2\cos 3\theta)^2 \, d\theta\right]$$

Now

$$r^2 = (1 + 2\cos 3\theta)^2 = 1 + 4\cos 3\theta + 4\cos^2 3\theta = 1 + 4\cos 3\theta + 4 \cdot \tfrac{1}{2}(1 + \cos 6\theta)$$

$$= 1 + 4\cos 3\theta + 2 + 2\cos 6\theta = 3 + 4\cos 3\theta + 2\cos 6\theta$$

and $\int r^2 \, d\theta = 3\theta + \tfrac{4}{3}\sin 3\theta + \tfrac{1}{3}\sin 6\theta + C$, so

$$A = \left[3\theta + \tfrac{4}{3}\sin 3\theta + \tfrac{1}{3}\sin 6\theta\right]_0^{2\pi/9} - \left[3\theta + \tfrac{4}{3}\sin 3\theta + \tfrac{1}{3}\sin 6\theta\right]_{2\pi/9}^{\pi/3}$$

$$= \left[\left(\frac{2\pi}{3} + \frac{4}{3} \cdot \frac{\sqrt{3}}{2} + \frac{1}{3} \cdot \frac{-\sqrt{3}}{2}\right) - 0\right] - \left[(\pi + 0 + 0) - \left(\frac{2\pi}{3} + \frac{4}{3} \cdot \frac{\sqrt{3}}{2} + \frac{1}{3} \cdot \frac{-\sqrt{3}}{2}\right)\right]$$

$$= \frac{4\pi}{3} + \frac{4}{3}\sqrt{3} - \frac{1}{3}\sqrt{3} - \pi = \frac{\pi}{3} + \sqrt{3}$$

37. The pole is a point of intersection.

$$1 + \sin\theta = 3\sin\theta \quad\Rightarrow\quad 1 = 2\sin\theta \quad\Rightarrow\quad \sin\theta = \tfrac{1}{2} \quad\Rightarrow$$

$\theta = \tfrac{\pi}{6}$ or $\tfrac{5\pi}{6}$.

The other two points of intersection are $\left(\tfrac{3}{2}, \tfrac{\pi}{6}\right)$ and $\left(\tfrac{3}{2}, \tfrac{5\pi}{6}\right)$.

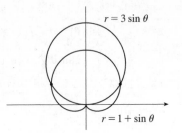

38. The pole is a point of intersection.

$$1 - \cos\theta = 1 + \sin\theta \quad\Rightarrow\quad -\cos\theta = \sin\theta \quad\Rightarrow\quad -1 = \tan\theta \quad\Rightarrow$$

$\theta = \tfrac{3\pi}{4}$ or $\tfrac{7\pi}{4}$.

The other two points of intersection are $\left(1 + \tfrac{\sqrt{2}}{2}, \tfrac{3\pi}{4}\right)$ and $\left(1 - \tfrac{\sqrt{2}}{2}, \tfrac{7\pi}{4}\right)$.

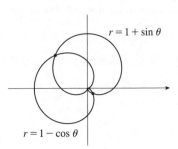

39. $2\sin 2\theta = 1 \quad\Rightarrow\quad \sin 2\theta = \tfrac{1}{2} \quad\Rightarrow\quad 2\theta = \tfrac{\pi}{6}, \tfrac{5\pi}{6}, \tfrac{13\pi}{6}, \text{ or } \tfrac{17\pi}{6}.$

By symmetry, the eight points of intersection are given by

$(1, \theta)$, where $\theta = \tfrac{\pi}{12}, \tfrac{5\pi}{12}, \tfrac{13\pi}{12}, \text{ and } \tfrac{17\pi}{12}$, and

$(-1, \theta)$, where $\theta = \tfrac{7\pi}{12}, \tfrac{11\pi}{12}, \tfrac{19\pi}{12}, \text{ and } \tfrac{23\pi}{12}$.

[There are many ways to describe these points.]

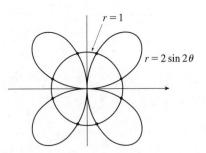

40. Clearly the pole lies on both curves. $\sin 3\theta = \cos 3\theta \quad\Rightarrow\quad \tan 3\theta = 1 \quad\Rightarrow$

$3\theta = \tfrac{\pi}{4} + n\pi$ [n any integer] $\quad\Rightarrow\quad \theta = \tfrac{\pi}{12} + \tfrac{\pi}{3}n \quad\Rightarrow$

$\theta = \tfrac{\pi}{12}, \tfrac{5\pi}{12}, \text{ or } \tfrac{3\pi}{4}$, so the three remaining intersection points are $\left(\tfrac{1}{\sqrt{2}}, \tfrac{\pi}{12}\right)$,

$\left(-\tfrac{1}{\sqrt{2}}, \tfrac{5\pi}{12}\right), \text{ and } \left(\tfrac{1}{\sqrt{2}}, \tfrac{3\pi}{4}\right).$

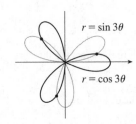

41. The pole is a point of intersection. $\sin\theta = \sin 2\theta = 2\sin\theta\cos\theta \quad\Leftrightarrow$

$\sin\theta \, (1 - 2\cos\theta) = 0 \quad\Leftrightarrow\quad \sin\theta = 0 \text{ or } \cos\theta = \tfrac{1}{2} \quad\Rightarrow$

$\theta = 0, \pi, \tfrac{\pi}{3}, \text{ or } -\tfrac{\pi}{3} \quad\Rightarrow\quad$ the other intersection points are $\left(\tfrac{\sqrt{3}}{2}, \tfrac{\pi}{3}\right)$

and $\left(\tfrac{\sqrt{3}}{2}, \tfrac{2\pi}{3}\right)$ [by symmetry].

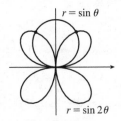

42. Clearly the pole is a point of intersection. $\sin 2\theta = \cos 2\theta \quad\Rightarrow$

$\tan 2\theta = 1 \quad\Rightarrow\quad 2\theta = \tfrac{\pi}{4} + 2n\pi$ [since $\sin 2\theta$ and $\cos 2\theta$ must be

positive in the equations] $\quad\Rightarrow\quad \theta = \tfrac{\pi}{8} + n\pi \quad\Rightarrow\quad \theta = \tfrac{\pi}{8} \text{ or } \tfrac{9\pi}{8}.$

So the curves also intersect at $\left(\tfrac{1}{\sqrt[4]{2}}, \tfrac{\pi}{8}\right)$ and $\left(\tfrac{1}{\sqrt[4]{2}}, \tfrac{9\pi}{8}\right).$

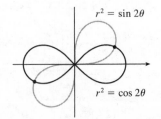

43.

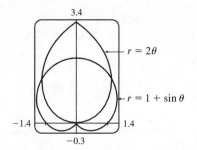
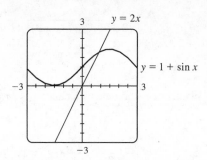

From the first graph, we see that the pole is one point of intersection. By zooming in or using the cursor, we find the θ-values of the intersection points to be $\alpha \approx 0.88786 \approx 0.89$ and $\pi - \alpha \approx 2.25$. (The first of these values may be more easily estimated by plotting $y = 1 + \sin x$ and $y = 2x$ in rectangular coordinates; see the second graph.) By symmetry, the total area contained is twice the area contained in the first quadrant, that is,

$$A = 2\int_0^\alpha \tfrac{1}{2}(2\theta)^2\, d\theta + 2\int_\alpha^{\pi/2} \tfrac{1}{2}(1 + \sin\theta)^2\, d\theta = \int_0^\alpha 4\theta^2\, d\theta + \int_\alpha^{\pi/2} \left[1 + 2\sin\theta + \tfrac{1}{2}(1 - \cos 2\theta)\right] d\theta$$

$$= \left[\tfrac{4}{3}\theta^3\right]_0^\alpha + \left[\theta - 2\cos\theta + \left(\tfrac{1}{2}\theta - \tfrac{1}{4}\sin 2\theta\right)\right]_\alpha^{\pi/2} = \tfrac{4}{3}\alpha^3 + \left[\left(\tfrac{\pi}{2} + \tfrac{\pi}{4}\right) - \left(\alpha - 2\cos\alpha + \tfrac{1}{2}\alpha - \tfrac{1}{4}\sin 2\alpha\right)\right] \approx 3.4645$$

44.

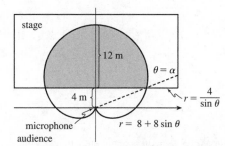

We need to find the shaded area A in the figure. The horizontal line representing the front of the stage has equation $y = 4 \iff r\sin\theta = 4 \implies r = 4/\sin\theta$. This line intersects the curve $r = 8 + 8\sin\theta$ when $8 + 8\sin\theta = \dfrac{4}{\sin\theta} \implies$

$8\sin\theta + 8\sin^2\theta = 4 \implies 2\sin^2\theta + 2\sin\theta - 1 = 0 \implies$

$\sin\theta = \dfrac{-2 \pm \sqrt{4 + 8}}{4} = \dfrac{-2 \pm 2\sqrt{3}}{4} = \dfrac{-1 + \sqrt{3}}{2}$ [the other value is less than -1] $\implies \theta = \sin^{-1}\left(\dfrac{\sqrt{3} - 1}{2}\right)$.

This angle is about $21.5°$ and is denoted by α in the figure.

$A = 2\int_\alpha^{\pi/2} \tfrac{1}{2}(8 + 8\sin\theta)^2\, d\theta - 2\int_\alpha^{\pi/2} \tfrac{1}{2}(4\csc\theta)^2\, d\theta = 64\int_\alpha^{\pi/2}(1 + 2\sin\theta + \sin^2\theta)\, d\theta - 16\int_\alpha^{\pi/2} \csc^2\theta\, d\theta$

$= 64\int_\alpha^{\pi/2}\left(1 + 2\sin\theta + \tfrac{1}{2} - \tfrac{1}{2}\cos 2\theta\right) d\theta + 16\int_\alpha^{\pi/2}(-\csc^2\theta)\, d\theta = 64\left[\tfrac{3}{2}\theta - 2\cos\theta - \tfrac{1}{4}\sin 2\theta\right]_\alpha^{\pi/2} + 16\left[\cot\theta\right]_\alpha^{\pi/2}$

$= 16\left[6\theta - 8\cos\theta - \sin 2\theta + \cot\theta\right]_\alpha^{\pi/a} = 16[(3\pi - 0 - 0 + 0) - (6\alpha - 8\cos\alpha - \sin 2\alpha + \cot\alpha)]$

$= 48\pi - 96\alpha + 128\cos\alpha + 16\sin 2\alpha - 16\cot\alpha$

From the figure, $x^2 + (\sqrt{3} - 1)^2 = 2^2 \implies x^2 = 4 - (3 - 2\sqrt{3} + 1) \implies$

$x^2 = 2\sqrt{3} = \sqrt{12}$, so $x = \sqrt{2\sqrt{3}} = \sqrt[4]{12}$. Using the trigonometric relationships for a right triangle and the identity $\sin 2\alpha = 2\sin\alpha\cos\alpha$, we continue:

$$A = 48\pi - 96\alpha + 128 \cdot \frac{\sqrt[4]{12}}{2} + 16 \cdot 2 \cdot \frac{\sqrt{3} - 1}{2} \cdot \frac{\sqrt[4]{12}}{2} - 16 \cdot \frac{\sqrt[4]{12}}{\sqrt{3} - 1} \cdot \frac{\sqrt{3} + 1}{\sqrt{3} + 1}$$

$$= 48\pi - 96\alpha + 64\sqrt[4]{12} + 8\sqrt[4]{12}\left(\sqrt{3} - 1\right) - 8\sqrt[4]{12}\left(\sqrt{3} + 1\right) = 48\pi + 48\sqrt[4]{12} - 96\sin^{-1}\left(\frac{\sqrt{3} - 1}{2}\right)$$

$$\approx 204.16 \text{ m}^2$$

45. $L = \int_a^b \sqrt{r^2 + (dr/d\theta)^2}\, d\theta = \int_0^{\pi/3} \sqrt{(3\sin\theta)^2 + (3\cos\theta)^2}\, d\theta = \int_0^{\pi/3} \sqrt{9(\sin^2\theta + \cos^2\theta)}\, d\theta$

$= 3\int_0^{\pi/3} d\theta = 3\big[\theta\big]_0^{\pi/3} = 3\big(\frac{\pi}{3}\big) = \pi.$

As a check, note that the circumference of a circle with radius $\frac{3}{2}$ is $2\pi\big(\frac{3}{2}\big) = 3\pi$, and since $\theta = 0$ to $\pi = \frac{\pi}{3}$ traces out $\frac{1}{3}$ of the

circle (from $\theta = 0$ to $\theta = \pi$), $\frac{1}{3}(3\pi) = \pi.$

46. $L = \int_a^b \sqrt{r^2 + (dr/d\theta)^2}\, d\theta = \int_0^{2\pi} \sqrt{(e^{2\theta})^2 + (2e^{2\theta})^2}\, d\theta = \int_0^{2\pi} \sqrt{e^{4\theta} + 4e^{4\theta}}\, d\theta = \int_0^{2\pi} \sqrt{5e^{4\theta}}\, d\theta$

$= \sqrt{5}\int_0^{2\pi} e^{2\theta}\, d\theta = \frac{\sqrt{5}}{2}\big[e^{2\theta}\big]_0^{2\pi} = \frac{\sqrt{5}}{2}(e^{4\pi} - 1)$

47. $L = \int_a^b \sqrt{r^2 + (dr/d\theta)^2}\, d\theta = \int_0^{2\pi} \sqrt{(\theta^2)^2 + (2\theta)^2}\, d\theta = \int_0^{2\pi} \sqrt{\theta^4 + 4\theta^2}\, d\theta$

$= \int_0^{2\pi} \sqrt{\theta^2(\theta^2 + 4)}\, d\theta = \int_0^{2\pi} \theta\sqrt{\theta^2 + 4}\, d\theta$

Now let $u = \theta^2 + 4$, so that $du = 2\theta\, d\theta$ $\big[\theta\, d\theta = \frac{1}{2}\, du\big]$ and

$\int_0^{2\pi} \theta\sqrt{\theta^2 + 4}\, d\theta = \int_4^{4\pi^2 + 4} \frac{1}{2}\sqrt{u}\, du = \frac{1}{2}\cdot\frac{2}{3}\big[u^{3/2}\big]_4^{4(\pi^2+1)} = \frac{1}{3}\big[4^{3/2}(\pi^2 + 1)^{3/2} - 4^{3/2}\big] = \frac{8}{3}\big[(\pi^2 + 1)^{3/2} - 1\big]$

48. $L = \int_a^b \sqrt{r^2 + (dr/d\theta)^2}\, d\theta = \int_0^{2\pi} \sqrt{\theta^2 + 1}\, d\theta \overset{21}{=} \Big[\frac{\theta}{2}\sqrt{\theta^2 + 1} + \frac{1}{2}\ln\big(\theta + \sqrt{\theta^2 + 1}\big)\Big]_0^{2\pi}$

$= \pi\sqrt{4\pi^2 + 1} + \frac{1}{2}\ln\big(2\pi + \sqrt{4\pi^2 + 1}\big)$

49. The curve $r = 3\sin 2\theta$ is completely traced with

$0 \le \theta \le 2\pi.$

$r^2 + \big(\frac{dr}{d\theta}\big)^2 = (3\sin 2\theta)^2 + (6\cos 2\theta)^2 \quad\Rightarrow$

$L = \int_0^{2\pi} \sqrt{9\sin^2 2\theta + 36\cos^2 2\theta}\, d\theta \approx 29.0653$

50. The curve $r = 4\sin 3\theta$ is completely traced with

$0 \le \theta \le \pi.$

$r^2 + \big(\frac{dr}{d\theta}\big)^2 = (4\sin 3\theta)^2 + (12\cos 3\theta)^2 \quad\Rightarrow$

$L = \int_0^{\pi} \sqrt{16\sin^2 3\theta + 144\cos^2 3\theta}\, d\theta \approx 26.7298$

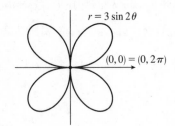

$r = 3\sin 2\theta$

$(0, 0) = (0, 2\pi)$

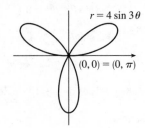

$r = 4\sin 3\theta$

$(0, 0) = (0, \pi)$

51. The curve $r = \sin\left(\frac{\theta}{2}\right)$ is completely traced
with $0 \le \theta \le 4\pi$.

$r^2 + \left(\frac{dr}{d\theta}\right)^2 = \sin^2\left(\frac{\theta}{2}\right) + \left[\frac{1}{2}\cos\left(\frac{\theta}{2}\right)\right]^2 \Rightarrow$

$L = \displaystyle\int_0^{4\pi} \sqrt{\sin^2\left(\frac{\theta}{2}\right) + \frac{1}{4}\cos^2\left(\frac{\theta}{2}\right)}\, d\theta$

≈ 9.6884

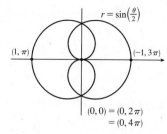

52. The curve $r = 1 + \cos\left(\frac{\theta}{3}\right)$ is completely traced
with $0 \le \theta \le 6\pi$.

$r^2 + \left(\frac{dr}{d\theta}\right)^2 = \left[1 + \cos\left(\frac{\theta}{3}\right)\right]^2 + \left[-\frac{1}{3}\sin\left(\frac{\theta}{3}\right)\right]^2 \Rightarrow$

$L = \displaystyle\int_0^{6\pi} \sqrt{\left[1 + \cos\left(\frac{\theta}{3}\right)\right]^2 + \frac{1}{9}\sin^2\left(\frac{\theta}{3}\right)}\, d\theta$

≈ 19.6676

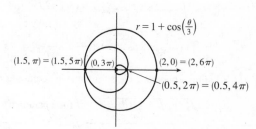

53. The curve $r = \cos^4(\theta/4)$ is completely traced with $0 \le \theta \le 4\pi$.

$r^2 + (dr/d\theta)^2 = [\cos^4(\theta/4)]^2 + \left[4\cos^3(\theta/4) \cdot (-\sin(\theta/4)) \cdot \frac{1}{4}\right]^2$

$\qquad = \cos^8(\theta/4) + \cos^6(\theta/4)\sin^2(\theta/4)$

$\qquad = \cos^6(\theta/4)[\cos^2(\theta/4) + \sin^2(\theta/4)] = \cos^6(\theta/4)$

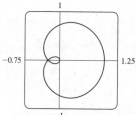

$L = \int_0^{4\pi} \sqrt{\cos^6(\theta/4)}\, d\theta = \int_0^{4\pi} \left|\cos^3(\theta/4)\right|\, d\theta$

$= 2\int_0^{2\pi} \cos^3(\theta/4)\, d\theta \quad [\text{since } \cos^3(\theta/4) \ge 0 \text{ for } 0 \le \theta \le 2\pi] \quad = 8\int_0^{\pi/2} \cos^3 u\, du \quad \left[u = \frac{1}{4}\theta\right]$

$\overset{68}{=} 8\left[\frac{1}{3}(2 + \cos^2 u)\sin u\right]_0^{\pi/2} = \frac{8}{3}[(2 \cdot 1) - (3 \cdot 0)] = \frac{16}{3}$

54. The curve $r = \cos^2(\theta/2)$ is completely traced with $0 \le \theta \le 2\pi$.

$r^2 + (dr/d\theta)^2 = [\cos^2(\theta/2)]^2 + \left[2\cos(\theta/2) \cdot (-\sin(\theta/2)) \cdot \frac{1}{2}\right]^2$

$\qquad = \cos^4(\theta/2) + \cos^2(\theta/2)\sin^2(\theta/2)$

$\qquad = \cos^2(\theta/2)[\cos^2(\theta/2) + \sin^2(\theta/2)]$

$\qquad = \cos^2(\theta/2)$

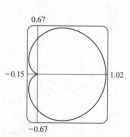

$L = \int_0^{2\pi} \sqrt{\cos^2(\theta/2)}\, d\theta = \int_0^{2\pi} \left|\cos(\theta/2)\right|\, d\theta = 2\int_0^{\pi} \cos(\theta/2)\, d\theta \quad [\text{since } \cos(\theta/2) \ge 0 \text{ for } 0 \le \theta \le \pi]$

$= 4\int_0^{\pi/2} \cos u\, du \quad \left[u = \frac{1}{2}\theta\right] \quad = 4\left[\sin u\right]_0^{\pi/2} = 4(1 - 0) = 4$

55. (a) From (11.2.7) [ET (10.2.7)],

$\qquad S = \int_a^b 2\pi y\sqrt{(dx/d\theta)^2 + (dy/d\theta)^2}\, d\theta$

$\qquad = \int_a^b 2\pi y\sqrt{r^2 + (dr/d\theta)^2}\, d\theta \qquad [\text{from the derivation of Equation 11.4.5 [ET 10.4.5]}]$

$\qquad = \int_a^b 2\pi r\sin\theta\sqrt{r^2 + (dr/d\theta)^2}\, d\theta$

(b) The curve $r^2 = \cos 2\theta$ goes through the pole when $\cos 2\theta = 0 \Rightarrow$

$2\theta = \frac{\pi}{2} \Rightarrow \theta = \frac{\pi}{4}$. We'll rotate the curve from $\theta = 0$ to $\theta = \frac{\pi}{4}$ and double

this value to obtain the total surface area generated.

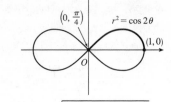

$r^2 = \cos 2\theta \Rightarrow 2r\frac{dr}{d\theta} = -2\sin 2\theta \Rightarrow \left(\frac{dr}{d\theta}\right)^2 = \frac{\sin^2 2\theta}{r^2} = \frac{\sin^2 2\theta}{\cos 2\theta}.$

$S = 2\int_0^{\pi/4} 2\pi\sqrt{\cos 2\theta}\sin\theta\sqrt{\cos 2\theta + (\sin^2 2\theta)/\cos 2\theta}\,d\theta = 4\pi\int_0^{\pi/4}\sqrt{\cos 2\theta}\sin\theta\sqrt{\frac{\cos^2 2\theta + \sin^2 2\theta}{\cos 2\theta}}\,d\theta$

$= 4\pi\int_0^{\pi/4}\sqrt{\cos 2\theta}\sin\theta\frac{1}{\sqrt{\cos 2\theta}}\,d\theta = 4\pi\int_0^{\pi/4}\sin\theta\,d\theta = 4\pi\big[-\cos\theta\big]_0^{\pi/4} = -4\pi\left(\frac{\sqrt{2}}{2} - 1\right) = 2\pi\left(2 - \sqrt{2}\right)$

56. (a) Rotation around $\theta = \frac{\pi}{2}$ is the same as rotation around the y-axis, that is, $S = \int_a^b 2\pi x\,ds$ where

$ds = \sqrt{(dx/dt)^2 + (dy/dt)^2}\,dt$ for a parametric equation, and for the special case of a polar equation, $x = r\cos\theta$ and

$ds = \sqrt{(dx/d\theta)^2 + (dy/d\theta)^2}\,d\theta = \sqrt{r^2 + (dr/d\theta)^2}\,d\theta$ [see the derivation of Equation 11.4.5 [ET 10.4.5]]. Therefore,

for a polar equation rotated around $\theta = \frac{\pi}{2}$, $S = \int_a^b 2\pi r\cos\theta\sqrt{r^2 + (dr/d\theta)^2}\,d\theta$.

(b) As in the solution for Exercise 55(b), we can double the surface area generated by rotating the curve from $\theta = 0$ to $\theta = \frac{\pi}{4}$

to obtain the total surface area.

$S = 2\int_0^{\pi/4} 2\pi\sqrt{\cos 2\theta}\cos\theta\sqrt{\cos 2\theta + (\sin^2 2\theta)/\cos 2\theta}\,d\theta = 4\pi\int_0^{\pi/4}\sqrt{\cos 2\theta}\cos\theta\sqrt{\frac{\cos^2 2\theta + \sin^2 2\theta}{\cos 2\theta}}\,d\theta$

$= 4\pi\int_0^{\pi/4}\sqrt{\cos 2\theta}\cos\theta\frac{1}{\sqrt{\cos 2\theta}}\,d\theta = 4\pi\int_0^{\pi/4}\cos\theta\,d\theta = 4\pi\big[\sin\theta\big]_0^{\pi/4} = 4\pi\left(\frac{\sqrt{2}}{2} - 0\right) = 2\sqrt{2}\,\pi$

11.5 Conic Sections   ET 10.5

1. $x = 2y^2 \Rightarrow y^2 = \frac{1}{2}x$. $4p = \frac{1}{2}$, so $p = \frac{1}{8}$. The vertex

is $(0,0)$, the focus is $\left(\frac{1}{8}, 0\right)$, and the directrix is $x = -\frac{1}{8}$.

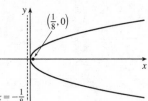

2. $4y + x^2 = 0 \Rightarrow x^2 = -4y$. $4p = -4$, so $p = -1$.

The vertex is $(0,0)$, the focus is $(0,-1)$, and the directrix

is $y = 1$.

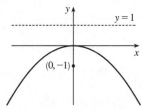

3. $4x^2 = -y \Rightarrow x^2 = -\frac{1}{4}y$. $4p = -\frac{1}{4}$, so $p = -\frac{1}{16}$.

The vertex is $(0,0)$, the focus is $\left(0, -\frac{1}{16}\right)$, and the

directrix is $y = \frac{1}{16}$.

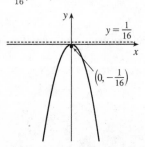

4. $y^2 = 12x$. $4p = 12$, so $p = 3$. The vertex is $(0,0)$, the

focus is $(3,0)$, and the directrix is $x = -3$.

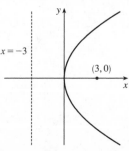

5. $(x + 2)^2 = 8(y - 3)$. $4p = 8$, so $p = 2$. The vertex is $(-2, 3)$, the focus is $(-2, 5)$, and the directrix is $y = 1$.

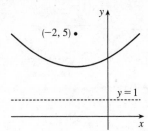

6. $x - 1 = (y + 5)^2$. $4p = 1$, so $p = \frac{1}{4}$. The vertex is $(1, -5)$, the focus is $\left(\frac{5}{4}, -5\right)$, and the directrix is $x = \frac{3}{4}$.

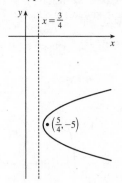

7. $y^2 + 2y + 12x + 25 = 0 \Rightarrow$
$y^2 + 2y + 1 = -12x - 24 \Rightarrow$
$(y + 1)^2 = -12(x + 2)$. $4p = -12$, so $p = -3$.
The vertex is $(-2, -1)$, the focus is $(-5, -1)$, and the directrix is $x = 1$.

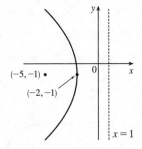

8. $y + 12x - 2x^2 = 16 \Rightarrow 2x^2 - 12x = y - 16 \Rightarrow$
$2(x^2 - 6x + 9) = y - 16 + 18 \Rightarrow$
$2(x - 3)^2 = y + 2 \Rightarrow (x - 3)^2 = \frac{1}{2}(y + 2)$.
$4p = \frac{1}{2}$, so $p = \frac{1}{8}$. The vertex is $(3, -2)$, the focus is $\left(3, -\frac{15}{8}\right)$, and the directrix is $y = -\frac{17}{8}$.

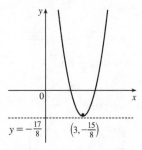

9. The equation has the form $y^2 = 4px$, where $p < 0$. Since the parabola passes through $(-1, 1)$, we have $1^2 = 4p(-1)$, so $4p = -1$ and an equation is $y^2 = -x$ or $x = -y^2$. $4p = -1$, so $p = -\frac{1}{4}$ and the focus is $\left(-\frac{1}{4}, 0\right)$ while the directrix is $x = \frac{1}{4}$.

10. The vertex is $(2, -2)$, so the equation is of the form $(x - 2)^2 = 4p(y + 2)$, where $p > 0$. The point $(0, 0)$ is on the parabola, so $4 = 4p(2)$ and $4p = 2$. Thus, an equation is $(x - 2)^2 = 2(y + 2)$. $4p = 2$, so $p = \frac{1}{2}$ and the focus is $\left(2, -\frac{3}{2}\right)$ while the directrix is $y = -\frac{5}{2}$.

11. $\dfrac{x^2}{9} + \dfrac{y^2}{5} = 1 \Rightarrow a = \sqrt{9} = 3, b = \sqrt{5}$,
$c = \sqrt{a^2 - b^2} = \sqrt{9 - 5} = 2$. The ellipse is centered at $(0, 0)$, with vertices at $(\pm 3, 0)$. The foci are $(\pm 2, 0)$.

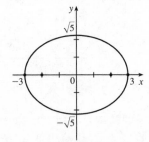

12. $\dfrac{x^2}{64} + \dfrac{y^2}{100} = 1 \Rightarrow a = \sqrt{100} = 10, b = \sqrt{64} = 8$,
$c = \sqrt{a^2 - b^2} = \sqrt{100 - 64} = 6$. The ellipse is centered at $(0, 0)$, with vertices at $(0, \pm 10)$. The foci are $(0, \pm 6)$.

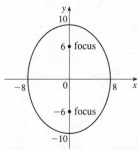

13. $4x^2 + y^2 = 16 \;\Rightarrow\; \dfrac{x^2}{4} + \dfrac{y^2}{16} = 1 \;\Rightarrow$

$a = \sqrt{16} = 4, b = \sqrt{4} = 2,$

$c = \sqrt{a^2 - b^2} = \sqrt{16 - 4} = 2\sqrt{3}$. The ellipse is

centered at $(0, 0)$, with vertices at $(0, \pm 4)$. The foci

are $\left(0, \pm 2\sqrt{3}\right)$.

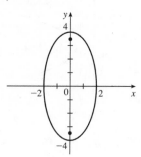

14. $4x^2 + 25y^2 = 25 \;\Rightarrow\; \dfrac{x^2}{25/4} + \dfrac{y^2}{1} = 1 \;\Rightarrow$

$a = \sqrt{\dfrac{25}{4}} = \dfrac{5}{2}, b = \sqrt{1} = 1,$

$c = \sqrt{a^2 - b^2} = \sqrt{\dfrac{25}{4} - 1} = \sqrt{\dfrac{21}{4}} = \dfrac{\sqrt{21}}{2}$. The ellipse

is centered at $(0, 0)$, with vertices at $\left(\pm\dfrac{5}{2}, 0\right)$. The foci

are $\left(\pm\dfrac{\sqrt{21}}{2}, 0\right)$.

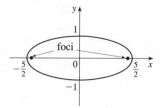

15. $9x^2 - 18x + 4y^2 = 27 \;\Leftrightarrow$

$9(x^2 - 2x + 1) + 4y^2 = 27 + 9 \;\Leftrightarrow$

$9(x - 1)^2 + 4y^2 = 36 \;\Leftrightarrow\; \dfrac{(x-1)^2}{4} + \dfrac{y^2}{9} = 1 \;\Rightarrow$

$a = 3, b = 2, c = \sqrt{5} \;\Rightarrow\;$ center $(1, 0)$,

vertices $(1, \pm 3)$, foci $\left(1, \pm\sqrt{5}\right)$

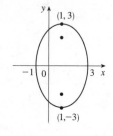

16. $x^2 + 3y^2 + 2x - 12y + 10 = 0 \;\Leftrightarrow$

$x^2 + 2x + 1 + 3(y^2 - 4y + 4) = -10 + 1 + 12 \;\Leftrightarrow$

$(x + 1)^2 + 3(y - 2)^2 = 3 \;\Leftrightarrow$

$\dfrac{(x+1)^2}{3} + \dfrac{(y-2)^2}{1} = 1 \;\Rightarrow\; a = \sqrt{3}, b = 1,$

$c = \sqrt{2} \;\Rightarrow\;$ center $(-1, 2)$, vertices $\left(-1 \pm \sqrt{3}, 2\right)$,

foci $\left(-1 \pm \sqrt{2}, 2\right)$

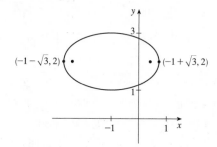

17. The center is $(0, 0)$, $a = 3$, and $b = 2$, so an equation is $\dfrac{x^2}{4} + \dfrac{y^2}{9} = 1$. $c = \sqrt{a^2 - b^2} = \sqrt{5}$, so the foci are $\left(0, \pm\sqrt{5}\right)$.

18. The ellipse is centered at $(2, 1)$, with $a = 3$ and $b = 2$. An equation is $\dfrac{(x-2)^2}{9} + \dfrac{(y-1)^2}{4} = 1$. $c = \sqrt{a^2 - b^2} = \sqrt{5}$, so the foci are $\left(2 \pm \sqrt{5}, 1\right)$.

19. $\dfrac{x^2}{144} - \dfrac{y^2}{25} = 1 \;\Rightarrow\; a = 12, b = 5, c = \sqrt{144 + 25} = 13 \;\Rightarrow$

center $(0, 0)$, vertices $(\pm 12, 0)$, foci $(\pm 13, 0)$, asymptotes $y = \pm\dfrac{5}{12}x$.

Note: It is helpful to draw a $2a$-by-$2b$ rectangle whose center is the center

of the hyperbola. The asymptotes are the extended diagonals of the

rectangle.

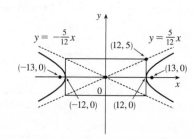

20. $\dfrac{y^2}{16} - \dfrac{x^2}{36} = 1 \ \Rightarrow \ a = 4, b = 6,$

$c = \sqrt{a^2 + b^2} = \sqrt{16 + 36} = \sqrt{52} = 2\sqrt{13}.$ The center is $(0,0),$ the

vertices are $(0, \pm 4),$ the foci are $\left(0, \pm 2\sqrt{13}\right),$ and the asymptotes are the

lines $y = \pm \frac{a}{b} x = \pm \frac{2}{3} x.$

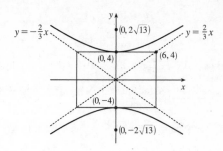

21. $y^2 - x^2 = 4 \ \Leftrightarrow \ \dfrac{y^2}{4} - \dfrac{x^2}{4} = 1 \ \Rightarrow \ a = \sqrt{4} = 2 = b,$

$c = \sqrt{4 + 4} = 2\sqrt{2} \ \Rightarrow \ \text{center } (0,0), \text{ vertices } (0, \pm 2),$

foci $\left(0, \pm 2\sqrt{2}\right),$ asymptotes $y = \pm x$

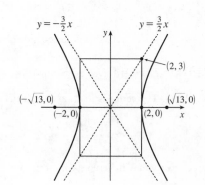

22. $9x^2 - 4y^2 = 36 \ \Leftrightarrow \ \dfrac{x^2}{4} - \dfrac{y^2}{9} = 1 \ \Rightarrow \ a = \sqrt{4} = 2, b = \sqrt{9} = 3,$

$c = \sqrt{4 + 9} = \sqrt{13} \ \Rightarrow \ \text{center } (0,0), \text{ vertices } (\pm 2, 0),$

foci $\left(\pm\sqrt{13}, 0\right),$ asymptotes $y = \pm \frac{3}{2} x$

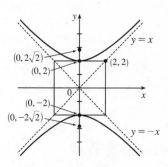

23. $4x^2 - y^2 - 24x - 4y + 28 = 0 \ \Leftrightarrow$

$4(x^2 - 6x + 9) - (y^2 + 4y + 4) = -28 + 36 - 4 \ \Leftrightarrow$

$4(x - 3)^2 - (y + 2)^2 = 4 \ \Leftrightarrow \ \dfrac{(x-3)^2}{1} - \dfrac{(y+2)^2}{4} = 1 \ \Rightarrow$

$a = \sqrt{1} = 1, b = \sqrt{4} = 2, c = \sqrt{1 + 4} = \sqrt{5} \ \Rightarrow$

center $(3, -2),$ vertices $(4, -2)$ and $(2, -2),$ foci $\left(3 \pm \sqrt{5}, -2\right),$

asymptotes $y + 2 = \pm 2(x - 3).$

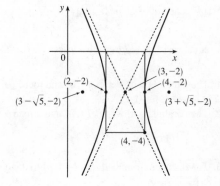

24. $y^2 - 4x^2 - 2y + 16x = 31$ $\Leftrightarrow$

$(y^2 - 2y + 1) - 4(x^2 - 4x + 4) = 31 + 1 - 16$ $\Leftrightarrow$

$(y - 1)^2 - 4(x - 2)^2 = 16$ $\Leftrightarrow$

$\dfrac{(y-1)^2}{16} - \dfrac{(x-2)^2}{4} = 1$ $\Rightarrow$ $a = \sqrt{16} = 4, b = \sqrt{4} = 2,$

$c = \sqrt{16 + 4} = \sqrt{20}$ $\Rightarrow$ center $(2, 1)$, vertices $(2, 1 \pm 4)$,

foci $\left(2, 1 \pm \sqrt{20}\right)$, asymptotes $y - 1 = \pm 2(x - 2)$.

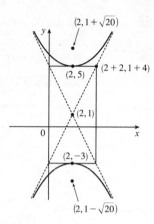

25. $x^2 = y + 1$ $\Leftrightarrow$ $x^2 = 1(y + 1)$. This is an equation of a *parabola* with $4p = 1$, so $p = \frac{1}{4}$. The vertex is $(0, -1)$ and the

focus is $\left(0, -\frac{3}{4}\right)$.

26. $x^2 = y^2 + 1$ $\Leftrightarrow$ $x^2 - y^2 = 1$. This is an equation of a *hyperbola* with vertices $(\pm 1, 0)$. The foci are at

$\left(\pm\sqrt{1 + 1}, 0\right) = \left(\pm\sqrt{2}, 0\right)$.

27. $x^2 = 4y - 2y^2$ $\Leftrightarrow$ $x^2 + 2y^2 - 4y = 0$ $\Leftrightarrow$ $x^2 + 2(y^2 - 2y + 1) = 2$ $\Leftrightarrow$ $x^2 + 2(y - 1)^2 = 2$ $\Leftrightarrow$

$\dfrac{x^2}{2} + \dfrac{(y-1)^2}{1} = 1$. This is an equation of an *ellipse* with vertices at $\left(\pm\sqrt{2}, 1\right)$. The foci are at $\left(\pm\sqrt{2 - 1}, 1\right) = (\pm 1, 1)$.

28. $y^2 - 8y = 6x - 16$ $\Leftrightarrow$ $y^2 - 8y + 16 = 6x$ $\Leftrightarrow$ $(y - 4)^2 = 6x$. This is an equation of a *parabola* with $4p = 6$,

so $p = \frac{3}{2}$. The vertex is $(0, 4)$ and the focus is $\left(\frac{3}{2}, 4\right)$.

29. $y^2 + 2y = 4x^2 + 3$ $\Leftrightarrow$ $y^2 + 2y + 1 = 4x^2 + 4$ $\Leftrightarrow$ $(y + 1)^2 - 4x^2 = 4$ $\Leftrightarrow$ $\dfrac{(y+1)^2}{4} - x^2 = 1$. This is an equation

of a *hyperbola* with vertices $(0, -1 \pm 2) = (0, 1)$ and $(0, -3)$. The foci are at $\left(0, -1 \pm \sqrt{4 + 1}\right) = \left(0, -1 \pm \sqrt{5}\right)$.

30. $4x^2 + 4x + y^2 = 0$ $\Leftrightarrow$ $4\left(x^2 + x + \frac{1}{4}\right) + y^2 = 1$ $\Leftrightarrow$ $4\left(x + \frac{1}{2}\right)^2 + y^2 = 1$ $\Leftrightarrow$ $\dfrac{\left(x + \frac{1}{2}\right)^2}{1/4} + y^2 = 1$. This is an

equation of an *ellipse* with vertices $\left(-\frac{1}{2}, 0 \pm 1\right) = \left(-\frac{1}{2}, \pm 1\right)$. The foci are at $\left(-\frac{1}{2}, 0 \pm \sqrt{1 - \frac{1}{4}}\right) = \left(-\frac{1}{2}, \pm\sqrt{3}/2\right)$.

31. The parabola with vertex $(0, 0)$ and focus $(0, -2)$ opens downward and has $p = -2$, so its equation is $x^2 = 4py = -8y$.

32. The parabola with vertex $(1, 0)$ and directrix $x = -5$ opens to the right and has $p = 6$, so its equation is

$y^2 = 4p(x - 1) = 24(x - 1)$.

33. The distance from the focus $(-4, 0)$ to the directrix $x = 2$ is $2 - (-4) = 6$, so the distance from the focus to the vertex is

$\frac{1}{2}(6) = 3$ and the vertex is $(-1, 0)$. Since the focus is to the left of the vertex, $p = -3$. An equation is $y^2 = 4p(x + 1)$ $\Rightarrow$

$y^2 = -12(x + 1)$.

34. The distance from the focus $(3, 6)$ to the vertex $(3, 2)$ is $6 - 2 = 4$. Since the focus is above the vertex, $p = 4$.

An equation is $(x - 3)^2 = 4p(y - 2)$ $\Rightarrow$ $(x - 3)^2 = 16(y - 2)$.

35. A parabola with vertical axis and vertex $(2, 3)$ has equation $y - 3 = a(x - 2)^2$. Since it passes through $(1, 5)$, we have

$5 - 3 = a(1 - 2)^2 \quad \Rightarrow \quad a = 2$, so an equation is $y - 3 = 2(x - 2)^2$.

36. A parabola with horizontal axis has equation $x = ay^2 + by + c$. Since the parabola passes through the point $(-1, 0)$,

substitute -1 for x and 0 for y: $-1 = 0 + 0 + c$. Now with $c = -1$, substitute 1 for x and -1 for y: $1 = a - b - 1$ **(1)**;

and then 3 for x and 1 for y: $3 = a + b - 1$ **(2)**. Add **(1)** and **(2)** to get $4 = 2a - 2 \quad \Rightarrow \quad a = 3$ and then $b = 1$.

Thus, the equation is $x = 3y^2 + y - 1$.

37. The ellipse with foci $(\pm 2, 0)$ and vertices $(\pm 5, 0)$ has center $(0, 0)$ and a horizontal major axis, with $a = 5$ and $c = 2$,

so $b^2 = a^2 - c^2 = 25 - 4 = 21$. An equation is $\dfrac{x^2}{25} + \dfrac{y^2}{21} = 1$.

38. The ellipse with foci $(0, \pm 5)$ and vertices $(0, \pm 13)$ has center $(0, 0)$ and a vertical major axis, with $c = 5$ and $a = 13$,

so $b = \sqrt{a^2 - c^2} = 12$. An equation is $\dfrac{x^2}{144} + \dfrac{y^2}{169} = 1$.

39. Since the vertices are $(0, 0)$ and $(0, 8)$, the ellipse has center $(0, 4)$ with a vertical axis and $a = 4$. The foci at $(0, 2)$ and $(0, 6)$

are 2 units from the center, so $c = 2$ and $b = \sqrt{a^2 - c^2} = \sqrt{4^2 - 2^2} = \sqrt{12}$. An equation is $\dfrac{(x - 0)^2}{b^2} + \dfrac{(y - 4)^2}{a^2} = 1 \quad \Rightarrow$

$\dfrac{x^2}{12} + \dfrac{(y - 4)^2}{16} = 1$.

40. Since the foci are $(0, -1)$ and $(8, -1)$, the ellipse has center $(4, -1)$ with a horizontal axis and $c = 4$.

The vertex $(9, -1)$ is 5 units from the center, so $a = 5$ and $b = \sqrt{a^2 - c^2} = \sqrt{5^2 - 4^2} = \sqrt{9}$. An equation is

$\dfrac{(x - 4)^2}{a^2} + \dfrac{(y + 1)^2}{b^2} = 1 \quad \Rightarrow \quad \dfrac{(x - 4)^2}{25} + \dfrac{(y + 1)^2}{9} = 1$.

41. An equation of an ellipse with center $(-1, 4)$ and vertex $(-1, 0)$ is $\dfrac{(x + 1)^2}{b^2} + \dfrac{(y - 4)^2}{4^2} = 1$. The focus $(-1, 6)$ is 2 units

from the center, so $c = 2$. Thus, $b^2 + 2^2 = 4^2 \quad \Rightarrow \quad b^2 = 12$, and the equation is $\dfrac{(x + 1)^2}{12} + \dfrac{(y - 4)^2}{16} = 1$.

42. Foci $F_1(-4, 0)$ and $F_2(4, 0) \quad \Rightarrow \quad c = 4$ and an equation is $\dfrac{x^2}{a^2} + \dfrac{y^2}{b^2} = 1$. The ellipse passes through $P(-4, 1.8)$, so

$2a = |PF_1| + |PF_2| \quad \Rightarrow \quad 2a = 1.8 + \sqrt{8^2 + (1.8)^2} \quad \Rightarrow \quad 2a = 1.8 + 8.2 \quad \Rightarrow \quad a = 5$.

$b^2 = a^2 - c^2 = 25 - 16 = 9$ and the equation is $\dfrac{x^2}{25} + \dfrac{y^2}{9} = 1$.

43. An equation of a hyperbola with vertices $(\pm 3, 0)$ is $\dfrac{x^2}{3^2} - \dfrac{y^2}{b^2} = 1$. Foci $(\pm 5, 0) \quad \Rightarrow \quad c = 5$ and $3^2 + b^2 = 5^2 \quad \Rightarrow$

$b^2 = 25 - 9 = 16$, so the equation is $\dfrac{x^2}{9} - \dfrac{y^2}{16} = 1$.

44. An equation of a hyperbola with vertices $(0, \pm 2)$ is $\dfrac{y^2}{2^2} - \dfrac{x^2}{b^2} = 1$. Foci $(0, \pm 5) \quad \Rightarrow \quad c = 5$ and $2^2 + b^2 = 5^2 \quad \Rightarrow$

$b^2 = 25 - 4 = 21$, so the equation is $\dfrac{y^2}{4} - \dfrac{x^2}{21} = 1$.

45. The center of a hyperbola with vertices $(-3, -4)$ and $(-3, 6)$ is $(-3, 1)$, so $a = 5$ and an equation is

$\dfrac{(y-1)^2}{5^2} - \dfrac{(x+3)^2}{b^2} = 1$. Foci $(-3, -7)$ and $(-3, 9)$ $\Rightarrow$ $c = 8$, so $5^2 + b^2 = 8^2$ $\Rightarrow$ $b^2 = 64 - 25 = 39$ and the

equation is $\dfrac{(y-1)^2}{25} - \dfrac{(x+3)^2}{39} = 1$.

46. The center of a hyperbola with vertices $(-1, 2)$ and $(7, 2)$ is $(3, 2)$, so $a = 4$ and an equation is $\dfrac{(x-3)^2}{4^2} - \dfrac{(y-2)^2}{b^2} = 1$.

Foci $(-2, 2)$ and $(8, 2)$ $\Rightarrow$ $c = 5$, so $4^2 + b^2 = 5^2$ $\Rightarrow$ $b^2 = 25 - 16 = 9$ and the equation is

$\dfrac{(x-3)^2}{16} - \dfrac{(y-2)^2}{9} = 1$.

47. The center of a hyperbola with vertices $(\pm 3, 0)$ is $(0, 0)$, so $a = 3$ and an equation is $\dfrac{x^2}{3^2} - \dfrac{y^2}{b^2} = 1$.

Asymptotes $y = \pm 2x$ $\Rightarrow$ $\dfrac{b}{a} = 2$ $\Rightarrow$ $b = 2(3) = 6$ and the equation is $\dfrac{x^2}{9} - \dfrac{y^2}{36} = 1$.

48. The center of a hyperbola with foci $(2, 0)$ and $(2, 8)$ is $(2, 4)$, so $c = 4$ and an equation is $\dfrac{(y-4)^2}{a^2} - \dfrac{(x-2)^2}{b^2} = 1$.

The asymptote $y = 3 + \frac{1}{2}x$ has slope $\frac{1}{2}$, so $\dfrac{a}{b} = \dfrac{1}{2}$ $\Rightarrow$ $b = 2a$ and $a^2 + b^2 = c^2$ $\Rightarrow$ $a^2 + (2a)^2 = 4^2$ $\Rightarrow$

$5a^2 = 16$ $\Rightarrow$ $a^2 = \frac{16}{5}$ and so $b^2 = 16 - \frac{16}{5} = \frac{64}{5}$. Thus, an equation is $\dfrac{(y-4)^2}{16/5} - \dfrac{(x-2)^2}{64/5} = 1$.

49. In Figure 8, we see that the point on the ellipse closest to a focus is the closer vertex (which is a distance

$a - c$ from it) while the farthest point is the other vertex (at a distance of $a + c$). So for this lunar orbit,

$(a - c) + (a + c) = 2a = (1728 + 110) + (1728 + 314)$, or $a = 1940$; and $(a + c) - (a - c) = 2c = 314 - 110$,

or $c = 102$. Thus, $b^2 = a^2 - c^2 = 3{,}753{,}196$, and the equation is $\dfrac{x^2}{3{,}763{,}600} + \dfrac{y^2}{3{,}753{,}196} = 1$.

50. (a) Choose V to be the origin, with x-axis through V and F. Then F is $(p, 0)$, A is $(p, 5)$, so substituting A into the

equation $y^2 = 4px$ gives $25 = 4p^2$ so $p = \frac{5}{2}$ and $y^2 = 10x$.

(b) $x = 11$ $\Rightarrow$ $y = \sqrt{110}$ $\Rightarrow$ $|CD| = 2\sqrt{110}$

51. (a) Set up the coordinate system so that A is $(-200, 0)$ and B is $(200, 0)$.

$|PA| - |PB| = (1200)(980) = 1{,}176{,}000 \text{ ft} = \frac{2450}{11} \text{ mi} = 2a$ $\Rightarrow$ $a = \frac{1225}{11}$, and $c = 200$ so

$b^2 = c^2 - a^2 = \dfrac{3{,}339{,}375}{121}$ $\Rightarrow$ $\dfrac{121x^2}{1{,}500{,}625} - \dfrac{121y^2}{3{,}339{,}375} = 1$.

(b) Due north of B $\Rightarrow$ $x = 200$ $\Rightarrow$ $\dfrac{(121)(200)^2}{1{,}500{,}625} - \dfrac{121y^2}{3{,}339{,}375} = 1$ $\Rightarrow$ $y = \dfrac{133{,}575}{539} \approx 248 \text{ mi}$

52. $|PF_1| - |PF_2| = \pm 2a$ $\Leftrightarrow$ $\sqrt{(x+c)^2 + y^2} - \sqrt{(x-c)^2 + y^2} = \pm 2a$ $\Leftrightarrow$

$\sqrt{(x+c)^2 + y^2} = \sqrt{(x-c)^2 + y^2} \pm 2a$ $\Leftrightarrow$ $(x+c)^2 + y^2 = (x-c)^2 + y^2 + 4a^2 \pm 4a\sqrt{(x-c)^2 + y^2}$ $\Leftrightarrow$

$$4cx - 4a^2 = \pm 4a\sqrt{(x-c)^2 + y^2} \quad \Leftrightarrow \quad c^2x^2 - 2a^2cx + a^4 = a^2(x^2 - 2cx + c^2 + y^2) \quad \Leftrightarrow$$

$$(c^2 - a^2)x^2 - a^2y^2 = a^2(c^2 - a^2) \quad \Leftrightarrow \quad b^2x^2 - a^2y^2 = a^2b^2 \text{ [where } b^2 = c^2 - a^2] \quad \Leftrightarrow \quad \frac{x^2}{a^2} - \frac{y^2}{b^2} = 1$$

53. The function whose graph is the upper branch of this hyperbola is concave upward. The function is

$$y = f(x) = a\sqrt{1 + \frac{x^2}{b^2}} = \frac{a}{b}\sqrt{b^2 + x^2}, \text{ so } y' = \frac{a}{b}x(b^2 + x^2)^{-1/2} \text{ and}$$

$$y'' = \frac{a}{b}\left[(b^2 + x^2)^{-1/2} - x^2(b^2 + x^2)^{-3/2}\right] = ab(b^2 + x^2)^{-3/2} > 0 \text{ for all } x, \text{ and so } f \text{ is concave upward.}$$

54. We can follow exactly the same sequence of steps as in the derivation of Formula 4, except we use the points $(1, 1)$ and

$(-1, -1)$ in the distance formula (first equation of that derivation) so $\sqrt{(x-1)^2 + (y-1)^2} + \sqrt{(x+1)^2 + (y+1)^2} = 4$

will lead (after moving the second term to the right, squaring, and simplifying) to $2\sqrt{(x+1)^2 + (y+1)^2} = x + y + 4$,

which, after squaring and simplifying again, leads to $3x^2 - 2xy + 3y^2 = 8$.

55. (a) If $k > 16$, then $k - 16 > 0$, and $\dfrac{x^2}{k} + \dfrac{y^2}{k - 16} = 1$ is an *ellipse* since it is the sum of two squares on the left side.

(b) If $0 < k < 16$, then $k - 16 < 0$, and $\dfrac{x^2}{k} + \dfrac{y^2}{k - 16} = 1$ is a *hyperbola* since it is the difference of two squares on the

left side.

(c) If $k < 0$, then $k - 16 < 0$, and there is *no curve* since the left side is the sum of two negative terms, which cannot equal 1.

(d) In case (a), $a^2 = k$, $b^2 = k - 16$, and $c^2 = a^2 - b^2 = 16$, so the foci are at $(\pm 4, 0)$. In case (b), $k - 16 < 0$, so $a^2 = k$,

$b^2 = 16 - k$, and $c^2 = a^2 + b^2 = 16$, and so again the foci are at $(\pm 4, 0)$.

56. (a) $y^2 = 4px \quad \Rightarrow \quad 2yy' = 4p \quad \Rightarrow \quad y' = \dfrac{2p}{y}$, so the tangent line is

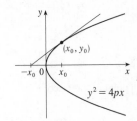

$$y - y_0 = \frac{2p}{y_0}(x - x_0) \quad \Rightarrow \quad yy_0 - y_0^2 = 2p(x - x_0) \quad \Leftrightarrow$$

$$yy_0 - 4px_0 = 2px - 2px_0 \quad \Rightarrow \quad yy_0 = 2p(x + x_0).$$

(b) The x-intercept is $-x_0$.

57. $x^2 = 4py \quad \Rightarrow \quad 2x = 4py' \quad \Rightarrow \quad y' = \dfrac{x}{2p}$, so the tangent line at (x_0, y_0) is

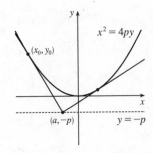

$y - \dfrac{x_0^2}{4p} = \dfrac{x_0}{2p}(x - x_0)$. This line passes through the point $(a, -p)$ on the

directrix, so $-p - \dfrac{x_0^2}{4p} = \dfrac{x_0}{2p}(a - x_0) \quad \Rightarrow \quad -4p^2 - x_0^2 = 2ax_0 - 2x_0^2 \quad \Leftrightarrow$

$x_0^2 - 2ax_0 - 4p^2 = 0 \quad \Leftrightarrow \quad x_0^2 - 2ax_0 + a^2 = a^2 + 4p^2 \quad \Leftrightarrow$

$(x_0 - a)^2 = a^2 + 4p^2$ ⇔ $x_0 = a \pm \sqrt{a^2 + 4p^2}$. The slopes of the tangent lines at $x = a \pm \sqrt{a^2 + 4p^2}$

are $\dfrac{a \pm \sqrt{a^2 + 4p^2}}{2p}$, so the product of the two slopes is

$$\frac{a + \sqrt{a^2 + 4p^2}}{2p} \cdot \frac{a - \sqrt{a^2 + 4p^2}}{2p} = \frac{a^2 - (a^2 + 4p^2)}{4p^2} = \frac{-4p^2}{4p^2} = -1,$$

showing that the tangent lines are perpendicular.

58. Without a loss of generality, let the ellipse, hyperbola, and foci be as shown in the figure.

The curves intersect (eliminate y^2)   ⇒

$B^2\left(\dfrac{x^2}{A^2} - \dfrac{y^2}{B^2}\right) + b^2\left(\dfrac{x^2}{a^2} + \dfrac{y^2}{b^2}\right) = B^2 + b^2$ ⇒

$\dfrac{B^2 x^2}{A^2} + \dfrac{b^2 x^2}{a^2} = B^2 + b^2$ ⇒ $x^2\left(\dfrac{B^2}{A^2} + \dfrac{b^2}{a^2}\right) = B^2 + b^2$ ⇒

$x^2 = \dfrac{B^2 + b^2}{\dfrac{a^2 B^2 + b^2 A^2}{A^2 a^2}} = \dfrac{A^2 a^2 (B^2 + b^2)}{a^2 B^2 + b^2 A^2}.$

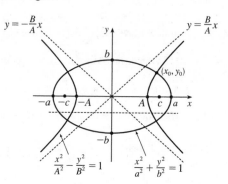

Similarly, $y^2 = \dfrac{B^2 b^2 (a^2 - A^2)}{b^2 A^2 + a^2 B^2}.$

Next we find the slopes of the tangent lines of the curves: $\dfrac{x^2}{a^2} + \dfrac{y^2}{b^2} = 1$ ⇒ $\dfrac{2x}{a^2} + \dfrac{2yy'}{b^2} = 0$ ⇒ $\dfrac{yy'}{b^2} = -\dfrac{x}{a^2}$ ⇒

$y'_E = -\dfrac{b^2}{a^2}\dfrac{x}{y}$ and $\dfrac{x^2}{A^2} - \dfrac{y^2}{B^2} = 1$ ⇒ $\dfrac{2x}{A^2} - \dfrac{2yy'}{B^2} = 0$ ⇒ $\dfrac{yy'}{B^2} = \dfrac{x}{A^2}$ ⇒ $y'_H = \dfrac{B^2}{A^2}\dfrac{x}{y}$. The product of the slopes

at (x_0, y_0) is $y'_E\, y'_H = -\dfrac{b^2 B^2 x_0^2}{a^2 A^2 y_0^2} = -\dfrac{b^2 B^2 \left[\dfrac{A^2 a^2 (B^2 + b^2)}{a^2 B^2 + b^2 A^2}\right]}{a^2 A^2 \left[\dfrac{B^2 b^2 (a^2 - A^2)}{b^2 A^2 + a^2 B^2}\right]} = -\dfrac{B^2 + b^2}{a^2 - A^2}$. Since $a^2 - b^2 = c^2$ and $A^2 + B^2 = c^2$,

we have $a^2 - b^2 = A^2 + B^2$ ⇒ $a^2 - A^2 = b^2 + B^2$, so the product of the slopes is -1, and hence, the tangent lines at

each point of intersection are perpendicular.

59. For $x^2 + 4y^2 = 4$, or $x^2/4 + y^2 = 1$, use the parametrization $x = 2\cos t$, $y = \sin t$, $0 \le t \le 2\pi$ to get

$$L = 4\int_0^{\pi/2} \sqrt{(dx/dt)^2 + (dy/dt)^2}\, dt = 4\int_0^{\pi/2} \sqrt{4\sin^2 t + \cos^2 t}\, dt = 4\int_0^{\pi/2} \sqrt{3\sin^2 t + 1}\, dt$$

Using Simpson's Rule with $n = 10$, $\Delta t = \dfrac{\pi/2 - 0}{10} = \dfrac{\pi}{20}$, and $f(t) = \sqrt{3\sin^2 t + 1}$, we get

$$L \approx \tfrac{4}{3}\left(\tfrac{\pi}{20}\right)\left[f(0) + 4f\left(\tfrac{\pi}{20}\right) + 2f\left(\tfrac{2\pi}{20}\right) + \cdots + 2f\left(\tfrac{8\pi}{20}\right) + 4f\left(\tfrac{9\pi}{20}\right) + f\left(\tfrac{\pi}{2}\right)\right] \approx 9.69$$

60. The length of the major axis is $2a$, so $a = \tfrac{1}{2}(1.18 \times 10^{10}) = 5.9 \times 10^9$. The length of the minor axis is $2b$, so

$b = \tfrac{1}{2}(1.14 \times 10^{10}) = 5.7 \times 10^9$. An equation of the ellipse is $\dfrac{x^2}{a^2} + \dfrac{y^2}{b^2} = 1$, or converting into parametric equations,

$x = a\cos\theta$ and $y = b\sin\theta$. So

$$L = 4\int_0^{\pi/2} \sqrt{(dx/d\theta)^2 + (dy/d\theta)^2}\, d\theta = 4\int_0^{\pi/2} \sqrt{a^2 \sin^2\theta + b^2 \cos^2\theta}\, d\theta$$

Using Simpson's Rule with $n = 10$, $\Delta\theta = \frac{\pi/2 - 0}{10} = \frac{\pi}{20}$, and $f(\theta) = \sqrt{a^2 \sin^2\theta + b^2 \cos^2\theta}$, we get

$$L \approx 4 \cdot S_{10} = 4 \cdot \tfrac{\pi}{20 \cdot 3}\left[f(0) + 4f\left(\tfrac{\pi}{20}\right) + 2f\left(\tfrac{2\pi}{20}\right) + \cdots + 2f\left(\tfrac{8\pi}{20}\right) + 4f\left(\tfrac{9\pi}{20}\right) + f\left(\tfrac{\pi}{2}\right)\right] \approx 3.64 \times 10^{10} \text{ km}$$

61. $\dfrac{x^2}{a^2} - \dfrac{y^2}{b^2} = 1 \;\Rightarrow\; \dfrac{y^2}{b^2} = \dfrac{x^2 - a^2}{a^2} \;\Rightarrow\; y = \pm\dfrac{b}{a}\sqrt{x^2 - a^2}$.

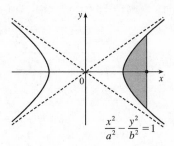

$A = 2\displaystyle\int_a^c \frac{b}{a}\sqrt{x^2 - a^2}\,dx \overset{39}{=} \frac{2b}{a}\left[\frac{x}{2}\sqrt{x^2 - a^2} - \frac{a^2}{2}\ln\left|x + \sqrt{x^2 - a^2}\right|\right]_a^c$

$= \dfrac{b}{a}\left[c\sqrt{c^2 - a^2} - a^2\ln\left|c + \sqrt{c^2 - a^2}\right| + a^2\ln|a|\right]$

Since $a^2 + b^2 = c^2$, $c^2 - a^2 = b^2$, and $\sqrt{c^2 - a^2} = b$.

$= \dfrac{b}{a}\left[cb - a^2\ln(c + b) + a^2\ln a\right] = \dfrac{b}{a}\left[cb + a^2(\ln a - \ln(b + c))\right]$

$= b^2 c/a + ab\ln[a/(b + c)]$, where $c^2 = a^2 + b^2$.

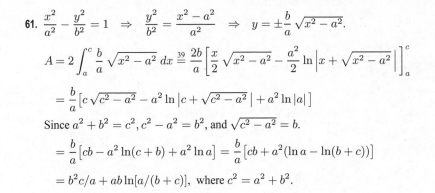

62. (a) $\dfrac{x^2}{a^2} + \dfrac{y^2}{b^2} = 1 \;\Rightarrow\; \dfrac{y^2}{b^2} = \dfrac{a^2 - x^2}{a^2} \;\Rightarrow\; y = \pm\dfrac{b}{a}\sqrt{a^2 - x^2}$.

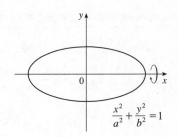

$V = \displaystyle\int_{-a}^a \pi\left(\frac{b}{a}\sqrt{a^2 - x^2}\right)^2 dx = 2\pi\frac{b^2}{a^2}\int_0^a (a^2 - x^2)\,dx$

$= \dfrac{2\pi b^2}{a^2}\left[a^2 x - \tfrac{1}{3}x^3\right]_0^a = \dfrac{2\pi b^2}{a^2}\left(\dfrac{2a^3}{3}\right) = \dfrac{4}{3}\pi b^2 a$

(b) $\dfrac{x^2}{a^2} + \dfrac{y^2}{b^2} = 1 \;\Rightarrow\; \dfrac{x^2}{a^2} = \dfrac{b^2 - y^2}{b^2} \;\Rightarrow\; x = \pm\dfrac{a}{b}\sqrt{b^2 - y^2}$.

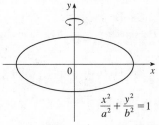

$V = \displaystyle\int_{-b}^b \pi\left(\frac{a}{b}\sqrt{b^2 - y^2}\right)^2 dy = 2\pi\frac{a^2}{b^2}\int_0^b (b^2 - y^2)\,dy$

$= \dfrac{2\pi a^2}{b^2}\left[b^2 y - \tfrac{1}{3}y^3\right]_0^b = \dfrac{2\pi a^2}{b^2}\left(\dfrac{2b^3}{3}\right) = \dfrac{4}{3}\pi a^2 b$

63. Differentiating implicitly, $\dfrac{x^2}{a^2} + \dfrac{y^2}{b^2} = 1 \;\Rightarrow\; \dfrac{2x}{a^2} + \dfrac{2yy'}{b^2} = 0 \;\Rightarrow\; y' = -\dfrac{b^2 x}{a^2 y}$ $[y \neq 0]$. Thus, the slope of the tangent

line at P is $-\dfrac{b^2 x_1}{a^2 y_1}$. The slope of $F_1 P$ is $\dfrac{y_1}{x_1 + c}$ and of $F_2 P$ is $\dfrac{y_1}{x_1 - c}$. By the formula from Problems Plus, we have

$$\tan\alpha = \frac{\dfrac{y_1}{x_1 + c} + \dfrac{b^2 x_1}{a^2 y_1}}{1 - \dfrac{b^2 x_1 y_1}{a^2 y_1(x_1 + c)}} = \frac{a^2 y_1^2 + b^2 x_1(x_1 + c)}{a^2 y_1(x_1 + c) - b^2 x_1 y_1} = \frac{a^2 b^2 + b^2 c x_1}{c^2 x_1 y_1 + a^2 c y_1} \quad \left[\begin{array}{l}\text{using } b^2 x_1^2 + a^2 y_1^2 = a^2 b^2, \\ \text{and } a^2 - b^2 = c^2\end{array}\right]$$

$$= \frac{b^2(cx_1 + a^2)}{cy_1(cx_1 + a^2)} = \frac{b^2}{cy_1}$$

and $\tan\beta = \dfrac{-\dfrac{b^2 x_1}{a^2 y_1} - \dfrac{y_1}{x_1 - c}}{1 - \dfrac{b^2 x_1 y_1}{a^2 y_1(x_1 - c)}} = \dfrac{-a^2 y_1^2 - b^2 x_1(x_1 - c)}{a^2 y_1(x_1 - c) - b^2 x_1 y_1} = \dfrac{-a^2 b^2 + b^2 c x_1}{c^2 x_1 y_1 - a^2 c y_1} = \dfrac{b^2(cx_1 - a^2)}{cy_1(cx_1 - a^2)} = \dfrac{b^2}{cy_1}$

Thus, $\alpha = \beta$.

64. The slopes of the line segments F_1P and F_2P are $\dfrac{y_1}{x_1 + c}$ and $\dfrac{y_1}{x_1 - c}$, where P is (x_1, y_1). Differentiating implicitly,

$$\frac{2x}{a^2} - \frac{2yy'}{b^2} = 0 \quad \Rightarrow \quad y' = \frac{b^2 x}{a^2 y} \quad \Rightarrow \quad \text{the slope of the tangent at } P \text{ is } \frac{b^2 x_1}{a^2 y_1}, \text{ so by the formula in Problem 17 on text}$$

page 268 [ET Problem 15 on text page 202],

$$\tan\alpha = \frac{\dfrac{b^2 x_1}{a^2 y_1} - \dfrac{y_1}{x_1 + c}}{1 + \dfrac{b^2 x_1 y_1}{a^2 y_1 (x_1 + c)}} = \frac{b^2 x_1 (x_1 + c) - a^2 y_1^2}{a^2 y_1 (x_1 + c) + b^2 x_1 y_1} = \frac{b^2(cx_1 + a^2)}{cy_1(cx_1 + a^2)} \quad \left[\begin{array}{l} \text{using } x_1^2/a^2 - y_1^2/b^2 = 1, \\ \text{and } a^2 + b^2 = c^2 \end{array}\right] = \frac{b^2}{cy_1}$$

and

$$\tan\beta = \frac{-\dfrac{b^2 x_1}{a^2 y_1} + \dfrac{y_1}{x_1 - c}}{1 + \dfrac{b^2 x_1 y_1}{a^2 y_1 (x_1 - c)}} = \frac{-b^2 x_1 (x_1 - c) + a^2 y_1^2}{a^2 y_1 (x_1 - c) + b^2 x_1 y_1} = \frac{b^2(cx_1 - a^2)}{cy_1(cx_1 - a^2)} = \frac{b^2}{cy_1}$$

So $\alpha = \beta$.

11.6   Conic Sections in Polar Coordinates   ET 10.6

1. The directrix $y = 6$ is above the focus at the origin, so we use the form with "$+ e\sin\theta$" in the denominator. [See Theorem 6

and Figure 2(c).] $\quad r = \dfrac{ed}{1 + e\sin\theta} = \dfrac{\frac{7}{4} \cdot 6}{1 + \frac{7}{4}\sin\theta} = \dfrac{42}{4 + 7\sin\theta}$

2. The directrix $x = 4$ is to the right of the focus at the origin, so we use the form with "$+ e\cos\theta$" in the denominator.

$e = 1$ for a parabola, so an equation is $r = \dfrac{ed}{1 + e\cos\theta} = \dfrac{1 \cdot 4}{1 + 1\cos\theta} = \dfrac{4}{1 + \cos\theta}$

3. The directrix $x = -5$ is to the left of the focus at the origin, so we use the form with "$- e\cos\theta$" in the denominator.

$r = \dfrac{ed}{1 - e\cos\theta} = \dfrac{\frac{3}{4} \cdot 5}{1 - \frac{3}{4}\cos\theta} = \dfrac{15}{4 - 3\cos\theta}$

4. The directrix $y = -2$ is below the focus at the origin, so we use the form with "$- e\sin\theta$" in the denominator.

$r = \dfrac{ed}{1 - e\sin\theta} = \dfrac{2 \cdot 2}{1 - 2\sin\theta} = \dfrac{4}{1 - 2\sin\theta}$

5. The vertex $(4, 3\pi/2)$ is 4 units below the focus at the origin, so the directrix is 8 units below the focus $(d = 8)$, and we use the

form with "$-e\sin\theta$" in the denominator. $e = 1$ for a parabola, so an equation is $r = \dfrac{ed}{1 - e\sin\theta} = \dfrac{1(8)}{1 - 1\sin\theta} = \dfrac{8}{1 - \sin\theta}$.

6. The vertex $P(1, \pi/2)$ is 1 unit above the focus F at the origin, so $|PF| = 1$ and we use the form with "$+e\sin\theta$" in the

denominator. The distance from the focus to the directrix l is d, so

$e = \dfrac{|PF|}{|Pl|} \quad \Rightarrow \quad 0.8 = \dfrac{1}{d - 1} \quad \Rightarrow \quad 0.8d - 0.8 = 1 \quad \Rightarrow \quad 0.8d = 1.8 \quad \Rightarrow \quad d = 2.25.$

An equation is $r = \dfrac{ed}{1 + e\sin\theta} = \dfrac{0.8(2.25)}{1 + 0.8\sin\theta} \cdot \dfrac{5}{5} = \dfrac{9}{5 + 4\sin\theta}.$

7. The directrix $r = 4\sec\theta$ (equivalent to $r\cos\theta = 4$ or $x = 4$) is to the right of the focus at the origin, so we will use the form

with "$+e\cos\theta$" in the denominator. The distance from the focus to the directrix is $d = 4$, so an equation is

$r = \dfrac{ed}{1 + e\cos\theta} = \dfrac{\frac{1}{2}(4)}{1 + \frac{1}{2}\cos\theta} \cdot \dfrac{2}{2} = \dfrac{4}{2 + \cos\theta}.$

8. The directrix $r = -6 \csc \theta$ (equivalent to $r \sin \theta = -6$ or $y = -6$) is below the focus at the origin, so we will use the form

with "$-e \sin \theta$" in the denominator. The distance from the focus to the directrix is $d = 6$, so an equation is

$$r = \frac{ed}{1 - e \sin \theta} = \frac{3(6)}{1 - 3 \sin \theta} = \frac{18}{1 - 3 \sin \theta}.$$

9. $r = \dfrac{1}{1 + \sin \theta} = \dfrac{ed}{1 + e \sin \theta}$, where $d = e = 1$.

(a) Eccentricity $= e = 1$

(b) Since $e = 1$, the conic is a parabola.

(c) Since "$+e \sin \theta$" appears in the denominator, the directrix is above the

focus at the origin. $d = |Fl| = 1$, so an equation of the directrix is $y = 1$.

(d) The vertex is at $\left(\frac{1}{2}, \frac{\pi}{2}\right)$, midway between the focus and the directrix.

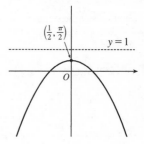

10. $r = \dfrac{12}{3 - 10 \cos \theta} \cdot \dfrac{1/3}{1/3} = \dfrac{4}{1 - \frac{10}{3} \cos \theta}$, where $e = \frac{10}{3}$ and $ed = 4 \;\Rightarrow\; d = 4\left(\frac{3}{10}\right) = \frac{6}{5}$.

(a) Eccentricity $= e = \frac{10}{3}$

(b) Since $e = \frac{10}{3} > 1$, the conic is a hyperbola.

(c) Since "$-e \cos \theta$" appears in the denominator, the directrix is to the left of the

focus at the origin. $d = |Fl| = \frac{6}{5}$, so an equation of the directrix is $x = -\frac{6}{5}$.

(d) The vertices are $\left(-\frac{12}{7}, 0\right)$ and $\left(\frac{12}{13}, \pi\right)$, so the center is midway between them,

that is, $\left(\frac{120}{91}, \pi\right)$.

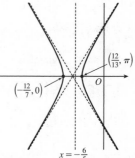

11. $r = \dfrac{12}{4 - \sin \theta} \cdot \dfrac{1/4}{1/4} = \dfrac{3}{1 - \frac{1}{4} \sin \theta}$, where $e = \frac{1}{4}$ and $ed = 3 \;\Rightarrow\; d = 12$.

(a) Eccentricity $= e = \frac{1}{4}$

(b) Since $e = \frac{1}{4} < 1$, the conic is an ellipse.

(c) Since "$-e \sin \theta$" appears in the denominator, the directrix is below the focus

at the origin. $d = |Fl| = 12$, so an equation of the directrix is $y = -12$.

(d) The vertices are $\left(4, \frac{\pi}{2}\right)$ and $\left(\frac{12}{5}, \frac{3\pi}{2}\right)$, so the center is midway between them,

that is, $\left(\frac{4}{5}, \frac{\pi}{2}\right)$.

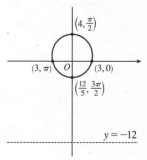

12. $r = \dfrac{3}{2 + 2 \cos \theta} \cdot \dfrac{1/2}{1/2} = \dfrac{3/2}{1 + 1 \cos \theta}$, where $e = 1$ and $ed = \frac{3}{2} \;\Rightarrow\; d = \frac{3}{2}$.

(a) Eccentricity $= e = 1$

(b) Since $e = 1$, the conic is a parabola.

(c) Since "$+e \cos \theta$" appears in the denominator, the directrix is to the right of

the focus at the origin. $d = |Fl| = \frac{3}{2}$, so an equation of the directrix is

$x = \frac{3}{2}$.

(d) The vertex is at $\left(\frac{3}{4}, 0\right)$, midway between the focus and directrix.

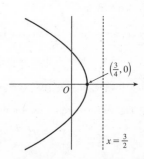

13. $r = \dfrac{9}{6 + 2\cos\theta} \cdot \dfrac{1/6}{1/6} = \dfrac{3/2}{1 + \frac{1}{3}\cos\theta}$, where $e = \frac{1}{3}$ and $ed = \frac{3}{2}$ $\Rightarrow$ $d = \frac{9}{2}$.

(a) Eccentricity $= e = \frac{1}{3}$

(b) Since $e = \frac{1}{3} < 1$, the conic is an ellipse.

(c) Since "$+e\cos\theta$" appears in the denominator, the directrix is to the right of

the focus at the origin. $d = |Fl| = \frac{9}{2}$, so an equation of the directrix is

$x = \frac{9}{2}$.

(d) The vertices are $\left(\frac{9}{8}, 0\right)$ and $\left(\frac{9}{4}, \pi\right)$, so the center is midway between them,

that is, $\left(\frac{9}{16}, \pi\right)$.

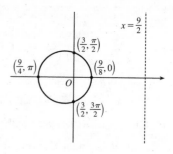

14. $r = \dfrac{8}{4 + 5\sin\theta} \cdot \dfrac{1/4}{1/4} = \dfrac{2}{1 + \frac{5}{4}\sin\theta}$, where $e = \frac{5}{4}$ and $ed = 2$ $\Rightarrow$ $d = 2\left(\frac{4}{5}\right) = \frac{8}{5}$.

(a) Eccentricity $= e = \frac{5}{4}$

(b) Since $e = \frac{5}{4} > 1$, the conic is a hyperbola.

(c) Since "$+e\sin\theta$" appears in the denominator, the directrix is above the

focus at the origin. $d = |Fl| = \frac{8}{5}$, so an equation of the directrix is $y = \frac{8}{5}$.

(d) The vertices are $\left(\frac{8}{9}, \frac{\pi}{2}\right)$ and $\left(-8, \frac{3\pi}{2}\right)$, so the center is midway between them,

that is, $\left(\frac{40}{9}, \frac{\pi}{2}\right)$.

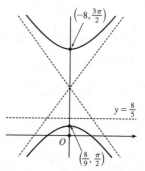

15. $r = \dfrac{3}{4 - 8\cos\theta} \cdot \dfrac{1/4}{1/4} = \dfrac{3/4}{1 - 2\cos\theta}$, where $e = 2$ and $ed = \frac{3}{4}$ $\Rightarrow$ $d = \frac{3}{8}$.

(a) Eccentricity $= e = 2$

(b) Since $e = 2 > 1$, the conic is a hyperbola.

(c) Since "$-e\cos\theta$" appears in the denominator, the directrix is to the left of

the focus at the origin. $d = |Fl| = \frac{3}{8}$, so an equation of the directrix is

$x = -\frac{3}{8}$.

(d) The vertices are $\left(-\frac{3}{4}, 0\right)$ and $\left(\frac{1}{4}, \pi\right)$, so the center is midway between them,

that is, $\left(\frac{1}{2}, \pi\right)$.

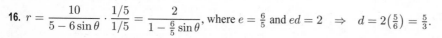

16. $r = \dfrac{10}{5 - 6\sin\theta} \cdot \dfrac{1/5}{1/5} = \dfrac{2}{1 - \frac{6}{5}\sin\theta}$, where $e = \frac{6}{5}$ and $ed = 2$ $\Rightarrow$ $d = 2\left(\frac{5}{6}\right) = \frac{5}{3}$.

(a) Eccentricity $= e = \frac{6}{5}$

(b) Since $e = \frac{6}{5} > 1$, the conic is a hyperbola.

(c) Since "$-e\sin\theta$" appears in the denominator, the directrix is below the focus

at the origin. $d = |Fl| = \frac{5}{3}$, so an equation of the directrix is $y = -\frac{5}{3}$.

(d) The vertices are $\left(-10, \frac{\pi}{2}\right)$ and $\left(\frac{10}{11}, \frac{3\pi}{2}\right)$, so the center is midway between them,

that is, $\left(\frac{60}{11}, \frac{3\pi}{2}\right)$.

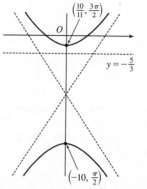

17. (a) $r = \dfrac{1}{1 - 2\sin\theta}$, where $e = 2$ and $ed = 1 \Rightarrow d = \frac{1}{2}$. The eccentricity

$e = 2 > 1$, so the conic is a hyperbola. Since "$-e\sin\theta$" appears in the

denominator, the directrix is below the focus at the origin. $d = |Fl| = \frac{1}{2}$,

so an equation of the directrix is $y = -\frac{1}{2}$. The vertices are $\left(-1, \frac{\pi}{2}\right)$ and

$\left(\frac{1}{3}, \frac{3\pi}{2}\right)$, so the center is midway between them, that is, $\left(\frac{2}{3}, \frac{3\pi}{2}\right)$.

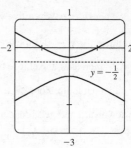

(b) By the discussion that precedes Example 4, the equation

is $r = \dfrac{1}{1 - 2\sin\left(\theta - \frac{3\pi}{4}\right)}$.

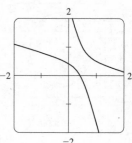

18. $r = \dfrac{4}{5 + 6\cos\theta} = \dfrac{4/5}{1 + \frac{6}{5}\cos\theta}$, so $e = \frac{6}{5}$ and $ed = \frac{4}{5} \Rightarrow d = \frac{2}{3}$.

An equation of the directrix is $x = \frac{2}{3} \Rightarrow r\cos\theta = \frac{2}{3} \Rightarrow r = \dfrac{2}{3\cos\theta}$.

If the hyperbola is rotated about its focus (the origin) through an angle $\pi/3$,

its equation is the same as that of the original, with θ replaced by $\theta - \frac{\pi}{3}$

(see Example 4), so $r = \dfrac{4}{5 + 6\cos\left(\theta - \frac{\pi}{3}\right)}$.

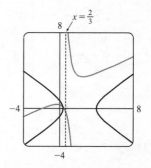

19. For $e < 1$ the curve is an ellipse. It is nearly circular when e is close to 0. As e

increases, the graph is stretched out to the right, and grows larger (that is, its

right-hand focus moves to the right while its left-hand focus remains at the

origin.) At $e = 1$, the curve becomes a parabola with focus at the origin.

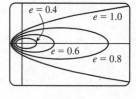

20. (a) The value of d does not seem to affect the shape of the conic (a parabola) at

all, just its size, position, and orientation (for $d < 0$ it opens upward, for

$d > 0$ it opens downward).

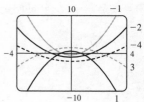

(b) We consider only positive values of e. When $0 < e < 1$, the conic is an ellipse. As $e \to 0^+$, the graph approaches perfect roundness and zero size. As e increases, the ellipse becomes more elongated, until at $e = 1$ it turns into a parabola. For $e > 1$, the conic is a hyperbola, which moves downward and gets broader as e continues to increase.

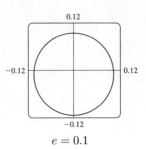

$e = 0.1$

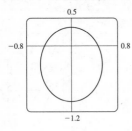

$e = 0.5$

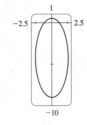

$e = 0.9$

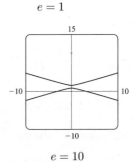

$e = 1$

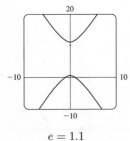

$e = 1.1$

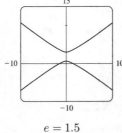

$e = 1.5$

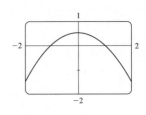

$e = 10$

21. $|PF| = e |Pl| \quad \Rightarrow \quad r = e[d - r \cos(\pi - \theta)] = e(d + r \cos \theta) \quad \Rightarrow$

$r(1 - e \cos \theta) = ed \quad \Rightarrow \quad r = \dfrac{ed}{1 - e \cos \theta}$

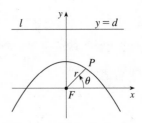

22. $|PF| = e |Pl| \quad \Rightarrow \quad r = e[d - r \sin \theta] \quad \Rightarrow \quad r(1 + e \sin \theta) = ed \quad \Rightarrow$

$r = \dfrac{ed}{1 + e \sin \theta}$

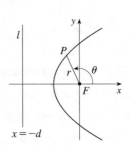

23. $|PF| = e |Pl| \quad \Rightarrow \quad r = e[d - r \sin(\theta - \pi)] = e(d + r \sin \theta) \quad \Rightarrow$

$r(1 - e \sin \theta) = ed \quad \Rightarrow \quad r = \dfrac{ed}{1 - e \sin \theta}$

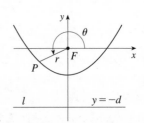

24. The parabolas intersect at the two points where $\dfrac{c}{1+\cos\theta} = \dfrac{d}{1-\cos\theta} \Rightarrow \cos\theta = \dfrac{c-d}{c+d} \Rightarrow r = \dfrac{c+d}{2}$.

For the first parabola, $\dfrac{dr}{d\theta} = \dfrac{c\sin\theta}{(1+\cos\theta)^2}$, so

$$\frac{dy}{dx} = \frac{(dr/d\theta)\sin\theta + r\cos\theta}{(dr/d\theta)\cos\theta - r\sin\theta} = \frac{c\sin^2\theta + c\cos\theta(1+\cos\theta)}{c\sin\theta\cos\theta - c\sin\theta(1+\cos\theta)} = \frac{1+\cos\theta}{-\sin\theta}$$

and similarly for the second, $\dfrac{dy}{dx} = \dfrac{1-\cos\theta}{\sin\theta} = \dfrac{\sin\theta}{1+\cos\theta}$. Since the product of these slopes is -1, the parabolas intersect

at right angles.

25. We are given $e = 0.093$ and $a = 2.28 \times 10^8$. By (7), we have

$$r = \frac{a(1-e^2)}{1+e\cos\theta} = \frac{2.28 \times 10^8[1-(0.093)^2]}{1+0.093\cos\theta} \approx \frac{2.26 \times 10^8}{1+0.093\cos\theta}$$

26. We are given $e = 0.048$ and $2a = 1.56 \times 10^9 \Rightarrow a = 7.8 \times 10^8$. By (7), we have

$$r = \frac{a(1-e^2)}{1+e\cos\theta} = \frac{7.8 \times 10^8[1-(0.048)^2]}{1+0.048\cos\theta} \approx \frac{7.78 \times 10^8}{1+0.048\cos\theta}$$

27. Here $2a = $ length of major axis $= 36.18$ AU $\Rightarrow a = 18.09$ AU and $e = 0.97$. By (7), the equation of the orbit is

$r = \dfrac{18.09[1-(0.97)^2]}{1-0.97\cos\theta} \approx \dfrac{1.07}{1-0.97\cos\theta}$. By (8), the maximum distance from the comet to the sun is

$18.09(1+0.97) \approx 35.64$ AU or about 3.314 billion miles.

28. Here $2a = $ length of major axis $= 356.5$ AU $\Rightarrow a = 178.25$ AU and $e = 0.9951$. By (7), the equation of the orbit

is $r = \dfrac{178.25[1-(0.9951)^2]}{1-0.9951\cos\theta} \approx \dfrac{1.7426}{1-0.9951\cos\theta}$. By (8), the minimum distance from the comet to the sun is

$178.25(1-0.9951) \approx 0.8734$ AU or about 81 million miles.

29. The minimum distance is at perihelion, where $4.6 \times 10^7 = r = a(1-e) = a(1-0.206) = a(0.794) \Rightarrow$

$a = 4.6 \times 10^7/0.794$. So the maximum distance, which is at aphelion, is

$r = a(1+e) = (4.6 \times 10^7/0.794)(1.206) \approx 7.0 \times 10^7$ km.

30. At perihelion, $r = a(1-e) = 4.43 \times 10^9$, and at aphelion, $r = a(1+e) = 7.37 \times 10^9$. Adding, we get $2a = 11.80 \times 10^9$,

so $a = 5.90 \times 10^9$ km. Therefore $1 + e = a(1+e)/a = \frac{7.37}{5.90} \approx 1.249$ and $e \approx 0.249$.

31. From Exercise 29, we have $e = 0.206$ and $a(1-e) = 4.6 \times 10^7$ km. Thus, $a = 4.6 \times 10^7/0.794$. From (7), we can write the

equation of Mercury's orbit as $r = a\dfrac{1-e^2}{1-e\cos\theta}$. So since

$$\frac{dr}{d\theta} = \frac{-a(1-e^2)e\sin\theta}{(1-e\cos\theta)^2} \Rightarrow$$

$$r^2 + \left(\frac{dr}{d\theta}\right)^2 = \frac{a^2(1-e^2)^2}{(1-e\cos\theta)^2} + \frac{a^2(1-e^2)^2 e^2\sin^2\theta}{(1-e\cos\theta)^4} = \frac{a^2(1-e^2)^2}{(1-e\cos\theta)^4}(1-2e\cos\theta + e^2)$$

the length of the orbit is

$$L = \int_0^{2\pi} \sqrt{r^2 + (dr/d\theta)^2}\, d\theta = a(1-e^2)\int_0^{2\pi} \frac{\sqrt{1+e^2-2e\cos\theta}}{(1-e\cos\theta)^2}\, d\theta \approx 3.6 \times 10^8 \text{ km}$$

This seems reasonable, since Mercury's orbit is nearly circular, and the circumference of a circle of radius a

is $2\pi a \approx 3.6 \times 10^8$ km.

11 Review

ET 10

CONCEPT CHECK

1. (a) A parametric curve is a set of points of the form $(x, y) = (f(t), g(t))$, where f and g are continuous functions of a variable t.

   (b) Sketching a parametric curve, like sketching the graph of a function, is difficult to do in general. We can plot points on the curve by finding $f(t)$ and $g(t)$ for various values of t, either by hand or with a calculator or computer. Sometimes, when f and g are given by formulas, we can eliminate t from the equations $x = f(t)$ and $y = g(t)$ to get a Cartesian equation relating x and y. It may be easier to graph that equation than to work with the original formulas for x and y in terms of t.

2. (a) You can find $\dfrac{dy}{dx}$ as a function of t by calculating $\dfrac{dy}{dx} = \dfrac{dy/dt}{dx/dt}$ [if $dx/dt \neq 0$].

   (b) Calculate the area as $\int_a^b y\, dx = \int_\alpha^\beta g(t) f'(t) dt$ [or $\int_\beta^\alpha g(t) f'(t) dt$ if the leftmost point is $(f(\beta), g(\beta))$ rather than $(f(\alpha), g(\alpha))$].

3. (a) $L = \int_\alpha^\beta \sqrt{(dx/dt)^2 + (dy/dt)^2}\, dt = \int_\alpha^\beta \sqrt{[f'(t)]^2 + [g'(t)]^2}\, dt$

   (b) $S = \int_\alpha^\beta 2\pi y \sqrt{(dx/dt)^2 + (dy/dt)^2}\, dt = \int_\alpha^\beta 2\pi g(t) \sqrt{[f'(t)]^2 + [g'(t)]^2}\, dt$

4. (a) See Figure 5 in Section 11.3 [ET 10.3].

   (b) $x = r\cos\theta,\ y = r\sin\theta$

   (c) To find a polar representation (r, θ) with $r \geq 0$ and $0 \leq \theta < 2\pi$, first calculate $r = \sqrt{x^2 + y^2}$. Then θ is specified by $\cos\theta = x/r$ and $\sin\theta = y/r$.

5. (a) Calculate $\dfrac{dy}{dx} = \dfrac{\dfrac{dy}{d\theta}}{\dfrac{dx}{d\theta}} = \dfrac{\dfrac{d}{d\theta}(y)}{\dfrac{d}{d\theta}(x)} = \dfrac{\dfrac{d}{d\theta}(r\sin\theta)}{\dfrac{d}{d\theta}(r\cos\theta)} = \dfrac{\left(\dfrac{dr}{d\theta}\right)\sin\theta + r\cos\theta}{\left(\dfrac{dr}{d\theta}\right)\cos\theta - r\sin\theta}$, where $r = f(\theta)$.

   (b) Calculate $A = \int_a^b \frac{1}{2} r^2\, d\theta = \int_a^b \frac{1}{2}[f(\theta)]^2\, d\theta$

   (c) $L = \int_a^b \sqrt{(dx/d\theta)^2 + (dy/d\theta)^2}\, d\theta = \int_a^b \sqrt{r^2 + (dr/d\theta)^2}\, d\theta = \int_a^b \sqrt{[f(\theta)]^2 + [f'(\theta)]^2}\, d\theta$

6. (a) A parabola is a set of points in a plane whose distances from a fixed point F (the focus) and a fixed line l (the directrix) are equal.

   (b) $x^2 = 4py;\ y^2 = 4px$

7. (a) An ellipse is a set of points in a plane the sum of whose distances from two fixed points (the foci) is a constant.

   (b) $\dfrac{x^2}{a^2} + \dfrac{y^2}{a^2 - c^2} = 1.$

8. (a) A hyperbola is a set of points in a plane the difference of whose distances from two fixed points (the foci) is a constant. This difference should be interpreted as the larger distance minus the smaller distance.

   (b) $\dfrac{x^2}{a^2} - \dfrac{y^2}{c^2 - a^2} = 1$

   (c) $y = \pm\dfrac{\sqrt{c^2 - a^2}}{a}x$

9. (a) If a conic section has focus F and corresponding directrix l, then the eccentricity e is the fixed ratio $|PF|\,/\,|Pl|$ for points P of the conic section.

(b) $e < 1$ for an ellipse; $e > 1$ for a hyperbola; $e = 1$ for a parabola.

(c) $x = d: r = \dfrac{ed}{1 + e \cos \theta}$. $x = -d: r = \dfrac{ed}{1 - e \cos \theta}$. $y = d: r = \dfrac{ed}{1 + e \sin \theta}$. $y = -d: r = \dfrac{ed}{1 - e \sin \theta}$.

TRUE-FALSE QUIZ

1. False. Consider the curve defined by $x = f(t) = (t - 1)^3$ and $y = g(t) = (t - 1)^2$. Then $g'(t) = 2(t - 1)$, so $g'(1) = 0$, but its graph has a *vertical* tangent when $t = 1$. *Note:* The statement is true if $f'(1) \neq 0$ when $g'(1) = 0$.

2. False. If $x = f(t)$ and $y = g(t)$ are twice differentiable, then $\dfrac{d^2 y}{dx^2} = \dfrac{d}{dx}\left(\dfrac{dy}{dx}\right) = \dfrac{\dfrac{d}{dt}\left(\dfrac{dy}{dx}\right)}{\dfrac{dx}{dt}}$.

3. False. For example, if $f(t) = \cos t$ and $g(t) = \sin t$ for $0 \leq t \leq 4\pi$, then the curve is a circle of radius 1, hence its length is 2π, but $\int_0^{4\pi} \sqrt{[f'(t)]^2 + [g'(t)]^2}\, dt = \int_0^{4\pi} \sqrt{(-\sin t)^2 + (\cos t)^2}\, dt = \int_0^{4\pi} 1\, dt = 4\pi$, since as t increases from 0 to 4π, the circle is traversed twice.

4. False. If $(r, \theta) = (1, \pi)$, then $(x, y) = (-1, 0)$, so $\tan^{-1}(y/x) = \tan^{-1} 0 = 0 \neq \theta$. The statement is true for points in quadrants I and IV.

5. True. The curve $r = 1 - \sin 2\theta$ is unchanged if we rotate it through $180°$ about O because $1 - \sin 2(\theta + \pi) = 1 - \sin(2\theta + 2\pi) = 1 - \sin 2\theta$. So it's unchanged if we replace r by $-r$. (See the discussion after Example 8 in Section 11.3 [ET 10.3].) In other words, it's the same curve as $r = -(1 - \sin 2\theta) = \sin 2\theta - 1$.

6. True. The polar equation $r = 2$, the Cartesian equation $x^2 + y^2 = 4$, and the parametric equations $x = 2\sin 3t$, $y = 2\cos 3t$ $[0 \leq t \leq 2\pi]$ all describe the circle of radius 2 centered at the origin.

7. False. The first pair of equations gives the portion of the parabola $y = x^2$ with $x \geq 0$, whereas the second pair of equations traces out the whole parabola $y = x^2$.

8. True. $y^2 = 2y + 3x$ $\Leftrightarrow$ $(y - 1)^2 = 3x + 1 = 3\left(x + \frac{1}{3}\right) = 4\left(\frac{3}{4}\right)\left(x + \frac{1}{3}\right)$, which is the equation of a parabola with vertex $\left(-\frac{1}{3}, 1\right)$ and focus $\left(-\frac{1}{3} + \frac{3}{4}, 1\right)$, opening to the right.

9. True. By rotating and translating the parabola, we can assume it has an equation of the form $y = cx^2$, where $c > 0$. The tangent at the point (a, ca^2) is the line $y - ca^2 = 2ca(x - a)$; i.e., $y = 2cax - ca^2$. This tangent meets the parabola at the points (x, cx^2) where $cx^2 = 2cax - ca^2$. This equation is equivalent to $x^2 = 2ax - a^2$ [since $c > 0$]. But $x^2 = 2ax - a^2$ $\Leftrightarrow$ $x^2 - 2ax + a^2 = 0$ $\Leftrightarrow$ $(x - a)^2 = 0$ $\Leftrightarrow$ $x = a$ $\Leftrightarrow$ $(x, cx^2) = (a, ca^2)$. This shows that each tangent meets the parabola at exactly one point.

10. True. Consider a hyperbola with focus at the origin, oriented so that its polar equation is $r = \dfrac{ed}{1 + e \cos \theta}$, where $e > 1$.

The directrix is $x = d$, but along the hyperbola we have $x = r \cos \theta = \dfrac{ed \cos \theta}{1 + e \cos \theta} = d\left(\dfrac{e \cos \theta}{1 + e \cos \theta}\right) \neq d$.

EXERCISES

1. $x = t^2 + 4t$, $y = 2 - t$, $-4 \le t \le 1$. $t = 2 - y$, so

$x = (2 - y)^2 + 4(2 - y) = 4 - 4y + y^2 + 8 - 4y = y^2 - 8y + 12 \quad \Leftrightarrow$

$x + 4 = y^2 - 8y + 16 = (y - 4)^2$. This is part of a parabola with vertex

$(-4, 4)$, opening to the right.

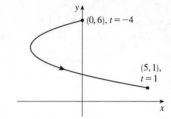

2. $x = 1 + e^{2t}$, $y = e^t$.

$x = 1 + e^{2t} = 1 + (e^t)^2 = 1 + y^2$, $y > 0$.

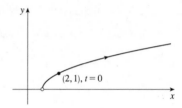

3. $y = \sec\theta = \dfrac{1}{\cos\theta} = \dfrac{1}{x}$. Since $0 \le \theta \le \pi/2$, $0 < x \le 1$ and $y \ge 1$.

This is part of the hyperbola $y = 1/x$.

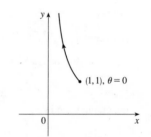

4. $x = 2\cos\theta$, $y = 1 + \sin\theta$, $\cos^2\theta + \sin^2\theta = 1 \quad \Rightarrow$

$\left(\dfrac{x}{2}\right)^2 + (y - 1)^2 = 1 \quad \Rightarrow \quad \dfrac{x^2}{4} + (y - 1)^2 = 1$. This is an ellipse,

centered at $(0, 1)$, with semimajor axis of length 2 and semiminor axis of

length 1.

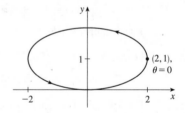

5. Three different sets of parametric equations for the curve $y = \sqrt{x}$ are

(i) $x = t$, $y = \sqrt{t}$

(ii) $x = t^4$, $y = t^2$

(iii) $x = \tan^2 t$, $y = \tan t$, $0 \le t < \pi/2$

There are many other sets of equations that also give this curve.

6. For $t < -1$, $x > 0$ and $y < 0$ with x decreasing and y increasing. When

$t = -1$, $(x, y) = (0, 0)$. When $-1 < t < 0$, we have $-1 < x < 0$ and

$0 < y < 1/2$. When $t = 0$, $(x, y) = (-1, 0)$. When $0 < t < 1$,

$-1 < x < 0$ and $-\frac{1}{2} < y < 0$. When $t = 1$, $(x, y) = (0, 0)$ again.

When $t > 1$, both x and y are positive and increasing.

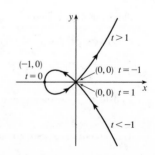

7. (a) 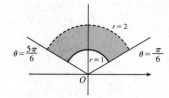 The Cartesian coordinates are $x = 4\cos\frac{2\pi}{3} = 4\left(-\frac{1}{2}\right) = -2$ and

$y = 4\sin\frac{2\pi}{3} = 4\left(\frac{\sqrt{3}}{2}\right) = 2\sqrt{3}$, that is, the point $\left(-2, 2\sqrt{3}\right)$.

(b) Given $x = -3$ and $y = 3$, we have $r = \sqrt{(-3)^2 + 3^2} = \sqrt{18} = 3\sqrt{2}$. Also, $\tan\theta = \frac{y}{x} \Rightarrow \tan\theta = \frac{3}{-3}$, and since

$(-3, 3)$ is in the second quadrant, $\theta = \frac{3\pi}{4}$. Thus, one set of polar coordinates for $(-3, 3)$ is $\left(3\sqrt{2}, \frac{3\pi}{4}\right)$, and two others are

$\left(3\sqrt{2}, \frac{11\pi}{4}\right)$ and $\left(-3\sqrt{2}, \frac{7\pi}{4}\right)$.

8. $1 \le r < 2$, $\frac{\pi}{6} \le \theta \le \frac{5\pi}{6}$

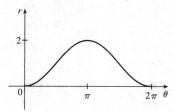

9. $r = 1 - \cos\theta$. This cardioid is symmetric about the polar axis.

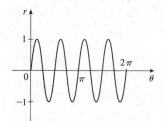
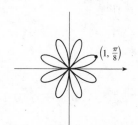

10. $r = \sin 4\theta$. This is an eight-leaved rose.

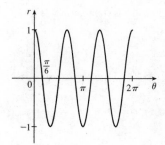
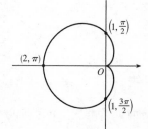

11. $r = \cos 3\theta$. This is a three-leaved rose. The curve is traced twice.

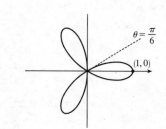

12. $r = 3 + \cos 3\theta$. The curve is symmetric about the horizontal axis.

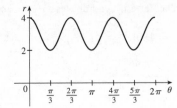

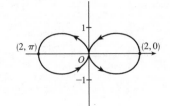

13. $r = 1 + \cos 2\theta$. The curve is symmetric about the pole and both the horizontal and vertical axes.

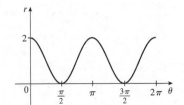

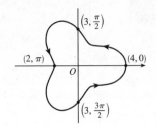

14. $r = 2\cos(\theta/2)$. The curve is symmetric about the pole and both the horizontal and vertical axes.

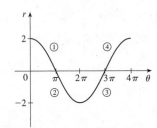

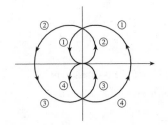

15. $r = \dfrac{3}{1 + 2\sin\theta}$ $\Rightarrow$ $e = 2 > 1$, so the conic is a hyperbola. $de = 3$ $\Rightarrow$
$d = \frac{3}{2}$ and the form "$+2\sin\theta$" imply that the directrix is above the focus at
the origin and has equation $y = \frac{3}{2}$. The vertices are $\left(1, \frac{\pi}{2}\right)$ and $\left(-3, \frac{3\pi}{2}\right)$.

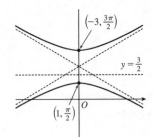

16. $r = \dfrac{3}{2 - 2\cos\theta} \cdot \dfrac{1/2}{1/2} = \dfrac{3/2}{1 - 1\cos\theta}$ $\Rightarrow$ $e = 1$, so the conic is a
parabola. $de = \frac{3}{2}$ $\Rightarrow$ $d = \frac{3}{2}$ and the form "$-2\cos\theta$" imply that the
directrix is to the left of the focus at the origin and has equation $x = -\frac{3}{2}$.
The vertex is $\left(\frac{3}{4}, \pi\right)$.

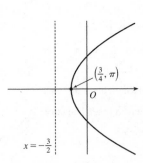

17. $x + y = 2$ $\Leftrightarrow$ $r\cos\theta + r\sin\theta = 2$ $\Leftrightarrow$ $r(\cos\theta + \sin\theta) = 2$ $\Leftrightarrow$ $r = \dfrac{2}{\cos\theta + \sin\theta}$

18. $x^2 + y^2 = 2$ $\Rightarrow$ $r^2 = 2$ $\Rightarrow$ $r = \sqrt{2}$. [$r = -\sqrt{2}$ gives the same curve.]

19. $r = (\sin\theta)/\theta$. As $\theta \to \pm\infty, r \to 0$.
As $\theta \to 0, r \to 1$. In the first figure,
there are an infinite number of
x-intercepts at $x = \pi n$, n a nonzero
integer. These correspond to pole
points in the second figure.

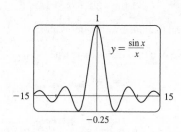

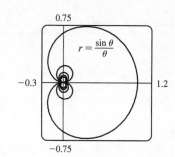

20. $r = \dfrac{2}{4 - 3\cos\theta} = \dfrac{1/2}{1 - \frac{3}{4}\cos\theta}$ $\Rightarrow$ $e = \frac{3}{4}$ and $d = \frac{2}{3}$. The equation of

the directrix is $x = -\frac{2}{3}$ $\Rightarrow$ $r = -2/(3\cos\theta)$. To obtain the equation

of the rotated ellipse, we replace θ in the original equation with $\theta - \frac{2\pi}{3}$,

and get $r = \dfrac{2}{4 - 3\cos\left(\theta - \frac{2\pi}{3}\right)}$.

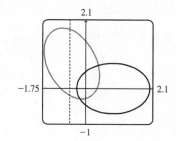

21. $x = \ln t, y = 1 + t^2; t = 1$. $\dfrac{dy}{dt} = 2t$ and $\dfrac{dx}{dt} = \dfrac{1}{t}$, so $\dfrac{dy}{dx} = \dfrac{dy/dt}{dx/dt} = \dfrac{2t}{1/t} = 2t^2$.

When $t = 1$, $(x, y) = (0, 2)$ and $dy/dx = 2$.

22. $x = t^3 + 6t + 1$, $y = 2t - t^2$; $t = -1$. $\dfrac{dy}{dx} = \dfrac{dy/dt}{dx/dt} = \dfrac{2 - 2t}{3t^2 + 6}$. When $t = -1$, $(x, y) = (-6, -3)$ and $\dfrac{dy}{dx} = \dfrac{4}{9}$.

23. $r = e^{-\theta}$ $\Rightarrow$ $y = r\sin\theta = e^{-\theta}\sin\theta$ and $x = r\cos\theta = e^{-\theta}\cos\theta$ $\Rightarrow$

$$\frac{dy}{dx} = \frac{dy/d\theta}{dx/d\theta} = \frac{\frac{dr}{d\theta}\sin\theta + r\cos\theta}{\frac{dr}{d\theta}\cos\theta - r\sin\theta} = \frac{-e^{-\theta}\sin\theta + e^{-\theta}\cos\theta}{-e^{-\theta}\cos\theta - e^{-\theta}\sin\theta} \cdot \frac{-e^{\theta}}{-e^{\theta}} = \frac{\sin\theta - \cos\theta}{\cos\theta + \sin\theta}.$$

When $\theta = \pi$, $\dfrac{dy}{dx} = \dfrac{0 - (-1)}{-1 + 0} = \dfrac{1}{-1} = -1$.

24. $r = 3 + \cos 3\theta$ $\Rightarrow$ $\dfrac{dy}{dx} = \dfrac{dy/d\theta}{dx/d\theta} = \dfrac{\frac{dr}{d\theta}\sin\theta + r\cos\theta}{\frac{dr}{d\theta}\cos\theta - r\sin\theta} = \dfrac{-3\sin 3\theta\sin\theta + (3 + \cos 3\theta)\cos\theta}{-3\sin 3\theta\cos\theta - (3 + \cos 3\theta)\sin\theta}$.

When $\theta = \pi/2$, $\dfrac{dy}{dx} = \dfrac{(-3)(-1)(1) + (3 + 0)\cdot 0}{(-3)(-1)(0) - (3 + 0)\cdot 1} = \dfrac{3}{-3} = -1$.

25. $x = t + \sin t$, $y = t - \cos t$ $\Rightarrow$ $\dfrac{dy}{dx} = \dfrac{dy/dt}{dx/dt} = \dfrac{1 + \sin t}{1 + \cos t}$ $\Rightarrow$

$$\frac{d^2y}{dx^2} = \frac{\frac{d}{dt}\left(\frac{dy}{dx}\right)}{dx/dt} = \frac{\dfrac{(1 + \cos t)\cos t - (1 + \sin t)(-\sin t)}{(1 + \cos t)^2}}{1 + \cos t} = \frac{\cos t + \cos^2 t + \sin t + \sin^2 t}{(1 + \cos t)^3} = \frac{1 + \cos t + \sin t}{(1 + \cos t)^3}$$

26. $x = 1 + t^2$, $y = t - t^3$. $\dfrac{dy}{dt} = 1 - 3t^2$ and $\dfrac{dx}{dt} = 2t$, so $\dfrac{dy}{dx} = \dfrac{dy/dt}{dx/dt} = \dfrac{1 - 3t^2}{2t} = \frac{1}{2}t^{-1} - \frac{3}{2}t$.

$$\frac{d^2y}{dx^2} = \frac{d(dy/dx)/dt}{dx/dt} = \frac{-\frac{1}{2}t^{-2} - \frac{3}{2}}{2t} = -\frac{1}{4}t^{-3} - \frac{3}{4}t^{-1} = -\frac{1}{4t^3}(1 + 3t^2) = -\frac{3t^2 + 1}{4t^3}.$$

27. We graph the curve $x = t^3 - 3t$, $y = t^2 + t + 1$ for $-2.2 \le t \le 1.2$.

By zooming in or using a cursor, we find that the lowest point is about

$(1.4, 0.75)$. To find the exact values, we find the t-value at which

$dy/dt = 2t + 1 = 0 \iff t = -\frac{1}{2} \iff (x, y) = \left(\frac{11}{8}, \frac{3}{4}\right)$.

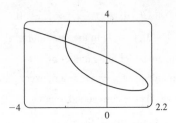

28. We estimate the coordinates of the point of intersection to be $(-2, 3)$. In fact this is exact, since both $t = -2$ and $t = 1$ give

the point $(-2, 3)$. So the area enclosed by the loop is

$$\int_{t=-2}^{t=1} y \, dx = \int_{-2}^{1} (t^2 + t + 1)(3t^2 - 3) \, dt = \int_{-2}^{1} (3t^4 + 3t^3 - 3t - 3) \, dt$$

$$= \left[\tfrac{3}{5}t^5 + \tfrac{3}{4}t^4 - \tfrac{3}{2}t^2 - 3t\right]_{-2}^{1} = \left(\tfrac{3}{5} + \tfrac{3}{4} - \tfrac{3}{2} - 3\right) - \left[-\tfrac{96}{5} + 12 - 6 - (-6)\right] = \tfrac{81}{20}$$

29. $x = 2a \cos t - a \cos 2t \Rightarrow \dfrac{dx}{dt} = -2a \sin t + 2a \sin 2t = 2a \sin t (2 \cos t - 1) = 0 \iff$

$\sin t = 0$ or $\cos t = \tfrac{1}{2} \Rightarrow t = 0, \tfrac{\pi}{3}, \pi$, or $\tfrac{5\pi}{3}$.

$y = 2a \sin t - a \sin 2t \Rightarrow \dfrac{dy}{dt} = 2a \cos t - 2a \cos 2t = 2a(1 + \cos t - 2\cos^2 t) = 2a(1 - \cos t)(1 + 2\cos t) = 0 \Rightarrow$

$t = 0, \tfrac{2\pi}{3}$, or $\tfrac{4\pi}{3}$.

Thus the graph has vertical tangents where

$t = \tfrac{\pi}{3}, \pi$ and $\tfrac{5\pi}{3}$, and horizontal tangents where

$t = \tfrac{2\pi}{3}$ and $\tfrac{4\pi}{3}$. To determine what the slope is

where $t = 0$, we use l'Hospital's Rule to evaluate

$\lim\limits_{t \to 0} \dfrac{dy/dt}{dx/dt} = 0$, so there is a horizontal tangent

there.

t	x	y
0	a	0
$\tfrac{\pi}{3}$	$\tfrac{3}{2}a$	$\tfrac{\sqrt{3}}{2}a$
$\tfrac{2\pi}{3}$	$-\tfrac{1}{2}a$	$\tfrac{3\sqrt{3}}{2}a$
π	$-3a$	0
$\tfrac{4\pi}{3}$	$-\tfrac{1}{2}a$	$-\tfrac{3\sqrt{3}}{2}a$
$\tfrac{5\pi}{3}$	$\tfrac{3}{2}a$	$-\tfrac{\sqrt{3}}{2}a$

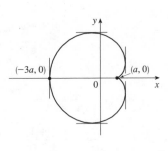

30. From Exercise 29, $x = 2a \cos t - a \cos 2t$, $y = 2a \sin t - a \sin 2t \Rightarrow$

$$A = 2 \int_{\pi}^{0} (2a \sin t - a \sin 2t)(-2a \sin t + 2a \sin 2t) \, dt = 4a^2 \int_{0}^{\pi} (2\sin^2 t + \sin^2 2t - 3 \sin t \sin 2t) \, dt$$

$$= 4a^2 \int_{0}^{\pi} \left[(1 - \cos 2t) + \tfrac{1}{2}(1 - \cos 4t) - 6\sin^2 t \, \cos t\right] dt = 4a^2 \left[t - \tfrac{1}{2}\sin 2t + \tfrac{1}{2}t - \tfrac{1}{8}\sin 4t - 2\sin^3 t\right]_{0}^{\pi}$$

$$= 4a^2 \left(\tfrac{3}{2}\right)\pi = 6\pi a^2$$

31. The curve $r^2 = 9 \cos 5\theta$ has 10 "petals." For instance, for $-\tfrac{\pi}{10} \le \theta \le \tfrac{\pi}{10}$, there are two petals, one with $r > 0$ and one

with $r < 0$.

$$A = 10 \int_{-\pi/10}^{\pi/10} \tfrac{1}{2}r^2 \, d\theta = 5 \int_{-\pi/10}^{\pi/10} 9 \cos 5\theta \, d\theta = 5 \cdot 9 \cdot 2 \int_{0}^{\pi/10} \cos 5\theta \, d\theta = 18 \left[\sin 5\theta\right]_{0}^{\pi/10} = 18$$

32. $r = 1 - 3 \sin \theta$. The inner loop is traced out as θ goes from $\alpha = \sin^{-1}\left(\tfrac{1}{3}\right)$ to $\pi - \alpha$, so

$$A = \int_{\alpha}^{\pi - \alpha} \tfrac{1}{2}r^2 \, d\theta = \int_{\alpha}^{\pi/2} (1 - 3\sin\theta)^2 \, d\theta = \int_{\alpha}^{\pi/2} \left[1 - 6\sin\theta + \tfrac{9}{2}(1 - \cos 2\theta)\right] d\theta$$

$$= \left[\tfrac{11}{2}\theta + 6\cos\theta - \tfrac{9}{4}\sin 2\theta\right]_{\alpha}^{\pi/2} = \tfrac{11}{4}\pi - \tfrac{11}{2}\sin^{-1}\left(\tfrac{1}{3}\right) - 3\sqrt{2}$$

33. The curves intersect when $4\cos\theta = 2 \Rightarrow \cos\theta = \frac{1}{2} \Rightarrow \theta = \pm\frac{\pi}{3}$

for $-\pi \le \theta \le \pi$. The points of intersection are $\left(2, \frac{\pi}{3}\right)$ and $\left(2, -\frac{\pi}{3}\right)$.

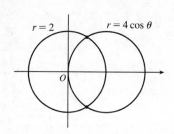

34. The two curves clearly both contain the pole. For other points of intersection, $\cot\theta = 2\cos(\theta + 2n\pi)$ or

$-2\cos(\theta + \pi + 2n\pi)$, both of which reduce to $\cot\theta = 2\cos\theta \Leftrightarrow \cos\theta = 2\sin\theta\cos\theta \Leftrightarrow \cos\theta(1 - 2\sin\theta) = 0 \Rightarrow$

$\cos\theta = 0$ or $\sin\theta = \frac{1}{2} \Rightarrow \theta = \frac{\pi}{6}, \frac{\pi}{2}, \frac{5\pi}{6}$ or $\frac{3\pi}{2} \Rightarrow$ intersection points are $\left(0, \frac{\pi}{2}\right)$, $\left(\sqrt{3}, \frac{\pi}{6}\right)$, and $\left(\sqrt{3}, \frac{11\pi}{6}\right)$.

35. The curves intersect where $2\sin\theta = \sin\theta + \cos\theta \Rightarrow$

$\sin\theta = \cos\theta \Rightarrow \theta = \frac{\pi}{4}$, and also at the origin (at which $\theta = \frac{3\pi}{4}$

on the second curve).

$$A = \int_0^{\pi/4} \tfrac{1}{2}(2\sin\theta)^2\,d\theta + \int_{\pi/4}^{3\pi/4} \tfrac{1}{2}(\sin\theta + \cos\theta)^2\,d\theta$$

$$= \int_0^{\pi/4}(1 - \cos 2\theta)\,d\theta + \tfrac{1}{2}\int_{\pi/4}^{3\pi/4}(1 + \sin 2\theta)\,d\theta$$

$$= \left[\theta - \tfrac{1}{2}\sin 2\theta\right]_0^{\pi/4} + \left[\tfrac{1}{2}\theta - \tfrac{1}{4}\cos 2\theta\right]_{\pi/4}^{3\pi/4} = \tfrac{1}{2}(\pi - 1)$$

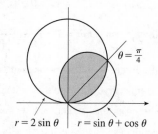

36. $A = 2\int_{-\pi/2}^{\pi/6} \tfrac{1}{2}\left[(2 + \cos 2\theta)^2 - (2 + \sin\theta)^2\right]d\theta$

$$= \int_{-\pi/2}^{\pi/6}\left[4\cos 2\theta + \cos^2 2\theta - 4\sin\theta - \sin^2\theta\right]d\theta$$

$$= \left[2\sin 2\theta + \tfrac{1}{2}\theta + \tfrac{1}{8}\sin 4\theta + 4\cos\theta - \tfrac{1}{2}\theta + \tfrac{1}{4}\sin 2\theta\right]_{-\pi/2}^{\pi/6}$$

$$= \tfrac{51}{16}\sqrt{3}$$

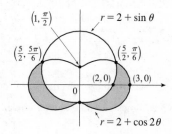

37. $x = 3t^2$, $y = 2t^3$.

$$L = \int_0^2 \sqrt{(dx/dt)^2 + (dy/dt)^2}\,dt = \int_0^2 \sqrt{(6t)^2 + (6t^2)^2}\,dt = \int_0^2 \sqrt{36t^2 + 36t^4}\,dt = \int_0^2 \sqrt{36t^2}\sqrt{1 + t^2}\,dt$$

$$= \int_0^2 6\,|t|\,\sqrt{1 + t^2}\,dt = 6\int_0^2 t\sqrt{1 + t^2}\,dt = 6\int_1^5 u^{1/2}\left(\tfrac{1}{2}du\right) \qquad \left[u = 1 + t^2,\, du = 2t\,dt\right]$$

$$= 6\cdot\tfrac{1}{2}\cdot\tfrac{2}{3}\left[u^{3/2}\right]_1^5 = 2(5^{3/2} - 1) = 2\left(5\sqrt{5} - 1\right)$$

38. $x = 2 + 3t$, $y = \cosh 3t \Rightarrow (dx/dt)^2 + (dy/dt)^2 = 3^2 + (3\sinh 3t)^2 = 9(1 + \sinh^2 3t) = 9\cosh^2 3t$, so

$$L = \int_0^1 \sqrt{9\cosh^2 3t}\,dt = \int_0^1 |3\cosh 3t|\,dt = \int_0^1 3\cosh 3t\,dt = \left[\sinh 3t\right]_0^1 = \sinh 3 - \sinh 0 = \sinh 3.$$

39. $L = \int_\pi^{2\pi} \sqrt{r^2 + (dr/d\theta)^2}\,d\theta = \int_\pi^{2\pi} \sqrt{(1/\theta)^2 + (-1/\theta^2)^2}\,d\theta = \int_\pi^{2\pi} \dfrac{\sqrt{\theta^2 + 1}}{\theta^2}\,d\theta$

$$\overset{24}{=} \left[-\dfrac{\sqrt{\theta^2 + 1}}{\theta} + \ln\left(\theta + \sqrt{\theta^2 + 1}\right)\right]_\pi^{2\pi} = \dfrac{\sqrt{\pi^2 + 1}}{\pi} - \dfrac{\sqrt{4\pi^2 + 1}}{2\pi} + \ln\left(\dfrac{2\pi + \sqrt{4\pi^2 + 1}}{\pi + \sqrt{\pi^2 + 1}}\right)$$

$$= \dfrac{2\sqrt{\pi^2 + 1} - \sqrt{4\pi^2 + 1}}{2\pi} + \ln\left(\dfrac{2\pi + \sqrt{4\pi^2 + 1}}{\pi + \sqrt{\pi^2 + 1}}\right)$$

40. $L = \int_0^\pi \sqrt{r^2 + (dr/d\theta)^2}\, d\theta = \int_0^\pi \sqrt{\sin^6\left(\frac{1}{3}\theta\right) + \sin^4\left(\frac{1}{3}\theta\right)\cos^2\left(\frac{1}{3}\theta\right)}\, d\theta$

$= \int_0^\pi \sin^2\left(\frac{1}{3}\theta\right) d\theta = \left[\frac{1}{2}\left(\theta - \frac{3}{2}\sin\left(\frac{2}{3}\theta\right)\right)\right]_0^\pi = \frac{1}{2}\pi - \frac{3}{8}\sqrt{3}$

41. $x = 4\sqrt{t}, \; y = \dfrac{t^3}{3} + \dfrac{1}{2t^2}, \; 1 \le t \le 4 \;\Rightarrow$

$S = \int_1^4 2\pi y \sqrt{(dx/dt)^2 + (dy/dt)^2}\, dt = \int_1^4 2\pi\left(\frac{1}{3}t^3 + \frac{1}{2}t^{-2}\right)\sqrt{\left(2/\sqrt{t}\right)^2 + (t^2 - t^{-3})^2}\, dt$

$= 2\pi \int_1^4 \left(\frac{1}{3}t^3 + \frac{1}{2}t^{-2}\right)\sqrt{(t^2 + t^{-3})^2}\, dt = 2\pi \int_1^4 \left(\frac{1}{3}t^5 + \frac{5}{6} + \frac{1}{2}t^{-5}\right) dt = 2\pi\left[\frac{1}{18}t^6 + \frac{5}{6}t - \frac{1}{8}t^{-4}\right]_1^4 = \frac{471{,}295}{1024}\pi$

42. $x = 2 + 3t, \quad y = \cosh 3t \;\Rightarrow\; (dx/dt)^2 + (dy/dt)^2 = 3^2 + (3\sinh 3t)^2 = 9(1 + \sinh^2 3t) = 9\cosh^2 3t$, so

$S = \int_0^1 2\pi y\, ds = \int_0^1 2\pi \cosh 3t \sqrt{9\cosh^2 3t}\, dt = \int_0^1 2\pi \cosh 3t\, |3\cosh 3t|\, dt = \int_0^1 2\pi \cosh 3t \cdot 3\cosh 3t\, dt$

$= 6\pi \int_0^1 \cosh^2 3t\, dt = 6\pi \int_0^1 \frac{1}{2}(1 + \cosh 6t)\, dt = 3\pi\left[t + \frac{1}{6}\sinh 6t\right]_0^1 = 3\pi\left(1 + \frac{1}{6}\sinh 6\right) = 3\pi + \frac{\pi}{2}\sinh 6$

43. For all c except -1, the curve is asymptotic to the line $x = 1$. For $c < -1$, the curve bulges to the right near $y = 0$. As c increases, the bulge becomes smaller, until at $c = -1$ the curve is the straight line $x = 1$. As c continues to increase, the curve bulges to the left, until at $c = 0$ there is a cusp at the origin. For $c > 0$, there is a loop to the left of the origin, whose size and roundness increase as c increases. Note that the x-intercept of the curve is always $-c$.

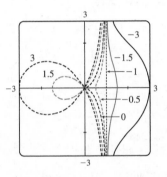

44. For a close to 0, the graph consists of four thin petals. As a increases, the petals get wider, until as $a \to \infty$, each petal occupies almost its entire quarter-circle.

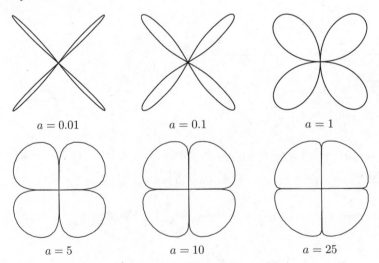

$a = 0.01 \qquad\qquad a = 0.1 \qquad\qquad a = 1$

$a = 5 \qquad\qquad a = 10 \qquad\qquad a = 25$

45. $\dfrac{x^2}{9} + \dfrac{y^2}{8} = 1$ is an ellipse with center $(0,0)$.

$a = 3, b = 2\sqrt{2}, c = 1 \ \Rightarrow$
foci $(\pm 1, 0)$, vertices $(\pm 3, 0)$.

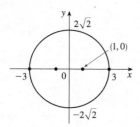

46. $4x^2 - y^2 = 16 \ \Leftrightarrow \ \dfrac{x^2}{4} - \dfrac{y^2}{16} = 1$ is a hyperbola

with center $(0,0)$, vertices $(\pm 2, 0)$, $a = 2, b = 4$,

$c = \sqrt{16 + 4} = 2\sqrt{5}$, foci $\left(\pm 2\sqrt{5}, 0\right)$ and

asymptotes $y = \pm 2x$.

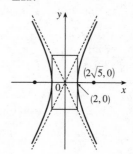

47. $6y^2 + x - 36y + 55 = 0 \ \Leftrightarrow$

$6(y^2 - 6y + 9) = -(x + 1) \ \Leftrightarrow$

$(y - 3)^2 = -\frac{1}{6}(x + 1)$, a parabola with vertex $(-1, 3)$,

opening to the left, $p = -\frac{1}{24} \ \Rightarrow \ $ focus $\left(-\frac{25}{24}, 3\right)$ and

directrix $x = -\frac{23}{24}$.

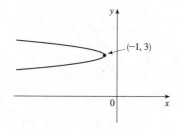

48. $25x^2 + 4y^2 + 50x - 16y = 59 \ \Leftrightarrow$

$25(x + 1)^2 + 4(y - 2)^2 = 100 \ \Leftrightarrow$

$\frac{1}{4}(x + 1)^2 + \frac{1}{25}(y - 2)^2 = 1$ is an ellipse centered at

$(-1, 2)$ with foci on the line $x = -1$, vertices $(-1, 7)$

and $(-1, -3)$; $a = 5, b = 2 \ \Rightarrow \ c = \sqrt{21} \ \Rightarrow$

foci $\left(-1, 2 \pm \sqrt{21}\right)$.

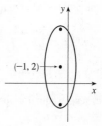

49. The ellipse with foci $(\pm 4, 0)$ and vertices $(\pm 5, 0)$ has center $(0,0)$ and a horizontal major axis, with $a = 5$ and $c = 4$,

so $b^2 = a^2 - c^2 = 5^2 - 4^2 = 9$. An equation is $\dfrac{x^2}{25} + \dfrac{y^2}{9} = 1$.

50. The distance from the focus $(2, 1)$ to the directrix $x = -4$ is $2 - (-4) = 6$, so the distance from the focus to the vertex

is $\frac{1}{2}(6) = 3$ and the vertex is $(-1, 1)$. Since the focus is to the right of the vertex, $p = 3$. An equation is

$(y - 1)^2 = 4 \cdot 3[x - (-1)]$, or $(y - 1)^2 = 12(x + 1)$.

51. The center of a hyperbola with foci $(0, \pm 4)$ is $(0, 0)$, so $c = 4$ and an equation is $\dfrac{y^2}{a^2} - \dfrac{x^2}{b^2} = 1$.

The asymptote $y = 3x$ has slope 3, so $\dfrac{a}{b} = \dfrac{3}{1} \ \Rightarrow \ a = 3b$ and $a^2 + b^2 = c^2 \ \Rightarrow \ (3b)^2 + b^2 = 4^2 \ \Rightarrow$

$10b^2 = 16 \ \Rightarrow \ b^2 = \frac{8}{5}$ and so $a^2 = 16 - \frac{8}{5} = \frac{72}{5}$. Thus, an equation is $\dfrac{y^2}{72/5} - \dfrac{x^2}{8/5} = 1$, or $\dfrac{5y^2}{72} - \dfrac{5x^2}{8} = 1$.

52. Center is $(3, 0)$, and $a = \frac{8}{2} = 4, c = 2 \ \Leftrightarrow \ b = \sqrt{4^2 - 2^2} = \sqrt{12} \ \Rightarrow$

an equation of the ellipse is $\dfrac{(x - 3)^2}{12} + \dfrac{y^2}{16} = 1$.

53. $x^2 = -(y - 100)$ has its vertex at $(0, 100)$, so one of the vertices of the ellipse is $(0, 100)$. Another form of the equation of a parabola is $x^2 = 4p(y - 100)$ so $4p(y - 100) = -(y - 100)$ $\Rightarrow$ $4p = -1$ $\Rightarrow$ $p = -\frac{1}{4}$. Therefore the shared focus is found at $\left(0, \frac{399}{4}\right)$ so $2c = \frac{399}{4} - 0$ $\Rightarrow$ $c = \frac{399}{8}$ and the center of the ellipse is $\left(0, \frac{399}{8}\right)$. So $a = 100 - \frac{399}{8} = \frac{401}{8}$ and

$b^2 = a^2 - c^2 = \dfrac{401^2 - 399^2}{8^2} = 25$. So the equation of the ellipse is $\dfrac{x^2}{b^2} + \dfrac{\left(y - \frac{399}{8}\right)^2}{a^2} = 1$ $\Rightarrow$ $\dfrac{x^2}{25} + \dfrac{\left(y - \frac{399}{8}\right)^2}{\left(\frac{401}{8}\right)^2} = 1$,

or $\dfrac{x^2}{25} + \dfrac{(8y - 399)^2}{160{,}801} = 1$.

54. $\dfrac{x^2}{a^2} + \dfrac{y^2}{b^2} = 1$ $\Rightarrow$ $\dfrac{2x}{a^2} + \dfrac{2y}{b^2} \dfrac{dy}{dx} = 0$ $\Rightarrow$ $\dfrac{dy}{dx} = -\dfrac{b^2}{a^2} \dfrac{x}{y}$. Therefore $\dfrac{dy}{dx} = m$ $\Leftrightarrow$ $y = -\dfrac{b^2}{a^2} \dfrac{x}{m}$. Combining this

condition with $\dfrac{x^2}{a^2} + \dfrac{y^2}{b^2} = 1$, we find that $x = \pm\dfrac{a^2 m}{\sqrt{a^2 m^2 + b^2}}$. In other words, the two points on the ellipse where the

tangent has slope m are $\left(\pm\dfrac{a^2 m}{\sqrt{a^2 m^2 + b^2}}, \mp\dfrac{b^2}{\sqrt{a^2 m^2 + b^2}}\right)$. The tangent lines at these points have the equations

$y \pm \dfrac{b^2}{\sqrt{a^2 m^2 + b^2}} = m\left(x \mp \dfrac{a^2 m}{\sqrt{a^2 m^2 + b^2}}\right)$ or $y = mx \mp \dfrac{a^2 m^2}{\sqrt{a^2 m^2 + b^2}} \mp \dfrac{b^2}{\sqrt{a^2 m^2 + b^2}} = mx \mp \sqrt{a^2 m^2 + b^2}$.

55. Directrix $x = 4$ $\Rightarrow$ $d = 4$, so $e = \frac{1}{3}$ $\Rightarrow$ $r = \dfrac{ed}{1 + e\cos\theta} = \dfrac{4}{3 + \cos\theta}$.

56. See the end of the proof of Theorem 11.6.1 [ET 10.6.1]. If $e > 1$, then $1 - e^2 < 0$ and Equations 11.6.4 [ET 10.6.4] become

$a^2 = \dfrac{e^2 d^2}{(e^2 - 1)^2}$ and $b^2 = \dfrac{e^2 d^2}{e^2 - 1}$, so $\dfrac{b^2}{a^2} = e^2 - 1$. The asymptotes $y = \pm\dfrac{b}{a}x$ have slopes $\pm\dfrac{b}{a} = \pm\sqrt{e^2 - 1}$, so the angles

they make with the polar axis are $\pm\tan^{-1}\left[\sqrt{e^2 - 1}\right] = \cos^{-1}(\pm 1/e)$.

57. In polar coordinates, an equation for the circle is $r = 2a\sin\theta$. Thus, the coordinates of Q are $x = r\cos\theta = 2a\sin\theta\cos\theta$ and $y = r\sin\theta = 2a\sin^2\theta$. The coordinates of R are $x = 2a\cot\theta$ and $y = 2a$. Since P is the midpoint of QR, we use the midpoint formula to get $x = a(\sin\theta\cos\theta + \cot\theta)$ and $y = a(1 + \sin^2\theta)$.

☐ PROBLEMS PLUS

1. $x = \int_1^t \dfrac{\cos u}{u}\, du,\ y = \int_1^t \dfrac{\sin u}{u}\, du$, so by FTC1, we have $\dfrac{dx}{dt} = \dfrac{\cos t}{t}$ and $\dfrac{dy}{dt} = \dfrac{\sin t}{t}$. Vertical tangent lines occur when

$\dfrac{dx}{dt} = 0 \quad\Leftrightarrow\quad \cos t = 0$. The parameter value corresponding to $(x, y) = (0, 0)$ is $t = 1$, so the nearest vertical tangent

occurs when $t = \frac{\pi}{2}$. Therefore, the arc length between these points is

$$L = \int_1^{\pi/2} \sqrt{\left(\dfrac{dx}{dt}\right)^2 + \left(\dfrac{dy}{dt}\right)^2}\, dt = \int_1^{\pi/2} \sqrt{\dfrac{\cos^2 t}{t^2} + \dfrac{\sin^2 t}{t^2}}\, dt = \int_1^{\pi/2} \dfrac{dt}{t} = \big[\ln t\big]_1^{\pi/2} = \ln \tfrac{\pi}{2}$$

2. (a) The curve $x^4 + y^4 = x^2 + y^2$ is symmetric about both axes and about the line $y = x$ (since interchanging x

and y does not change the equation) so we need only consider $y \ge x \ge 0$ to begin with. Implicit differentiation gives

$4x^3 + 4y^3 y' = 2x + 2yy' \quad\Rightarrow\quad y' = \dfrac{x(1 - 2x^2)}{y(2y^2 - 1)} \quad\Rightarrow\quad y' = 0$ when $x = 0$ and when $x = \pm\frac{1}{\sqrt{2}}$. If $x = 0$, then

$y^4 = y^2 \quad\Rightarrow\quad y^2(y^2 - 1) = 0 \quad\Rightarrow\quad y = 0$ or ± 1. The point $(0, 0)$ can't be a highest or lowest point because it is

isolated. [If $-1 < x < 1$ and $-1 < y < 1$, then $x^4 < x^2$ and $y^4 < y^2 \quad\Rightarrow\quad x^4 + y^4 < x^2 + y^2$, except for $(0, 0)$.]

If $x = \frac{1}{\sqrt{2}}$, then $x^2 = \frac{1}{2},\ x^4 = \frac{1}{4}$, so $\frac{1}{4} + y^4 = \frac{1}{2} + y^2 \quad\Rightarrow\quad 4y^4 - 4y^2 - 1 = 0 \quad\Rightarrow\quad y^2 = \dfrac{4 \pm \sqrt{16+16}}{8} = \dfrac{1 \pm \sqrt{2}}{2}$.

But $y^2 > 0$, so $y^2 = \dfrac{1 + \sqrt{2}}{2} \quad\Rightarrow\quad y = \pm\sqrt{\frac{1}{2}(1 + \sqrt{2})}$. Near the point $(0, 1)$, the denominator of y' is positive and the

numerator changes from negative to positive as x increases through 0, so $(0, 1)$ is a local minimum point. At

$\left(\frac{1}{\sqrt{2}}, \sqrt{\dfrac{1 + \sqrt{2}}{2}}\right),\ y'$ changes from positive to negative, so that point gives a maximum. By symmetry, the highest points

on the curve are $\left(\pm\frac{1}{\sqrt{2}}, \sqrt{\dfrac{1 + \sqrt{2}}{2}}\right)$ and the lowest points are $\left(\pm\frac{1}{\sqrt{2}}, -\sqrt{\dfrac{1 + \sqrt{2}}{2}}\right)$.

(b) We use the information from part (a), together with symmetry with respect to the
axes and the lines $y = \pm x$, to sketch the curve.

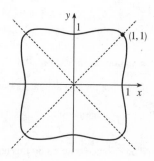

(c) In polar coordinates, $x^4 + y^4 = x^2 + y^2$ becomes $r^4 \cos^4 \theta + r^4 \sin^4 \theta = r^2$ or

$r^2 = \dfrac{1}{\cos^4 \theta + \sin^4 \theta}$. By the symmetry shown in part (b), the area enclosed by

the curve is $A = 8 \displaystyle\int_0^{\pi/4} \frac{1}{2} r^2\, d\theta = 4 \int_0^{\pi/4} \dfrac{d\theta}{\cos^4 \theta + \sin^4 \theta} \overset{\text{CAS}}{=} \sqrt{2}\, \pi$.

3. In terms of x and y, we have $x = r\cos\theta = (1 + c\sin\theta)\cos\theta = \cos\theta + c\sin\theta\cos\theta = \cos\theta + \frac{1}{2}c\sin 2\theta$ and

$y = r\sin\theta = (1 + c\sin\theta)\sin\theta = \sin\theta + c\sin^2\theta$. Now $-1 \le \sin\theta \le 1 \quad\Rightarrow\quad -1 \le \sin\theta + c\sin^2\theta \le 1 + c \le 2$, so

$-1 \le y \le 2$. Furthermore, $y = 2$ when $c = 1$ and $\theta = \frac{\pi}{2}$, while $y = -1$ for $c = 0$ and $\theta = \frac{3\pi}{2}$. Therefore, we need a viewing

rectangle with $-1 \le y \le 2$.

To find the x-values, look at the equation $x = \cos\theta + \frac{1}{2}c\sin 2\theta$ and use the fact that $\sin 2\theta \ge 0$ for $0 \le \theta \le \frac{\pi}{2}$ and

$\sin 2\theta \le 0$ for $-\frac{\pi}{2} \le \theta \le 0$. [Because $r = 1 + c\sin\theta$ is symmetric about the y-axis, we only need to consider

$-\frac{\pi}{2} \le \theta \le \frac{\pi}{2}$.] So for $-\frac{\pi}{2} \le \theta \le 0$, x has a maximum value when $c = 0$ and then $x = \cos\theta$ has a maximum value

of 1 at $\theta = 0$. Thus, the maximum value of x must occur on $\left[0, \frac{\pi}{2}\right]$ with $c = 1$. Then $x = \cos\theta + \frac{1}{2}\sin 2\theta \Rightarrow$

$\frac{dx}{d\theta} = -\sin\theta + \cos 2\theta = -\sin\theta + 1 - 2\sin^2\theta \Rightarrow \frac{dx}{d\theta} = -(2\sin\theta - 1)(\sin\theta + 1) = 0$ when $\sin\theta = -1$ or $\frac{1}{2}$

[but $\sin\theta \ne -1$ for $0 \le \theta \le \frac{\pi}{2}$]. If $\sin\theta = \frac{1}{2}$, then $\theta = \frac{\pi}{6}$ and

$x = \cos\frac{\pi}{6} + \frac{1}{2}\sin\frac{\pi}{3} = \frac{3}{4}\sqrt{3}$. Thus, the maximum value of x is $\frac{3}{4}\sqrt{3}$, and,

by symmetry, the minimum value is $-\frac{3}{4}\sqrt{3}$. Therefore, the smallest

viewing rectangle that contains every member of the family of polar curves

$r = 1 + c\sin\theta$, where $0 \le c \le 1$, is $\left[-\frac{3}{4}\sqrt{3}, \frac{3}{4}\sqrt{3}\right] \times [-1, 2]$.

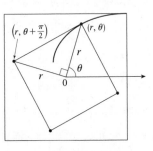

4. (a) Let us find the polar equation of the path of the bug that starts in the upper

right corner of the square. If the polar coordinates of this bug, at a

particular moment, are (r, θ), then the polar coordinates of the bug that it is

crawling toward must be $\left(r, \theta + \frac{\pi}{2}\right)$. (The next bug must be the same

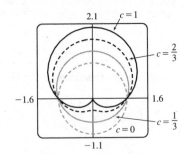

distance from the origin and the angle between the lines joining the bugs to

the pole must be $\frac{\pi}{2}$.) The Cartesian coordinates of the first bug are

$(r\cos\theta, r\sin\theta)$ and for the second bug we have

$x = r\cos\left(\theta + \frac{\pi}{2}\right) = -r\sin\theta$, $y = r\sin\left(\theta + \frac{\pi}{2}\right) = r\cos\theta$. So the slope of the line joining the bugs is

$\dfrac{r\cos\theta - r\sin\theta}{-r\sin\theta - r\cos\theta} = \dfrac{\sin\theta - \cos\theta}{\sin\theta + \cos\theta}$. This must be equal to the slope of the tangent line at (r, θ), so by

Equation 11.3.3 [ET 10.3.3] we have $\dfrac{(dr/d\theta)\sin\theta + r\cos\theta}{(dr/d\theta)\cos\theta - r\sin\theta} = \dfrac{\sin\theta - \cos\theta}{\sin\theta + \cos\theta}$. Solving for $\dfrac{dr}{d\theta}$, we get

$\dfrac{dr}{d\theta}\sin^2\theta + \dfrac{dr}{d\theta}\sin\theta\cos\theta + r\sin\theta\cos\theta + r\cos^2\theta = \dfrac{dr}{d\theta}\sin\theta\cos\theta - \dfrac{dr}{d\theta}\cos^2\theta - r\sin^2\theta + r\sin\theta\cos\theta \Rightarrow$

$\dfrac{dr}{d\theta}(\sin^2\theta + \cos^2\theta) + r(\cos^2\theta + \sin^2\theta) = 0 \Rightarrow \dfrac{dr}{d\theta} = -r$. Solving this differential equation as a separable

equation (as in Section 10.3 [ET 9.3]), or using Theorem 10.4.2 [ET 9.4.2] with $k = -1$, we get $r = Ce^{-\theta}$. To determine

C we use the fact that, at its starting position, $\theta = \frac{\pi}{4}$ and $r = \frac{1}{\sqrt 2}a$, so $\frac{1}{\sqrt 2}a = Ce^{-\pi/4} \Rightarrow C = \frac{1}{\sqrt 2}ae^{\pi/4}$. Therefore,

a polar equation of the bug's path is $r = \frac{1}{\sqrt 2}ae^{\pi/4}e^{-\theta}$ or $r = \frac{1}{\sqrt 2}ae^{(\pi/4)-\theta}$.

(b) The distance traveled by this bug is $L = \int_{\pi/4}^{\infty}\sqrt{r^2 + (dr/d\theta)^2}\,d\theta$, where $\dfrac{dr}{d\theta} = \dfrac{a}{\sqrt 2}e^{\pi/4}(-e^{-\theta})$ and so

$r^2 + (dr/d\theta)^2 = \frac{1}{2}a^2e^{\pi/2}e^{-2\theta} + \frac{1}{2}a^2e^{\pi/2}e^{-2\theta} = a^2e^{\pi/2}e^{-2\theta}$. Thus

$$L = \int_{\pi/4}^{\infty}ae^{\pi/4}e^{-\theta}\,d\theta = ae^{\pi/4}\lim_{t\to\infty}\int_{\pi/4}^{t}e^{-\theta}\,d\theta = ae^{\pi/4}\lim_{t\to\infty}\left[-e^{-\theta}\right]_{\pi/4}^{t}$$

$$= ae^{\pi/4}\lim_{t\to\infty}\left[e^{-\pi/4} - e^{-t}\right] = ae^{\pi/4}e^{-\pi/4} = a$$

5. (a) If (a, b) lies on the curve, then there is some parameter value t_1 such that $\dfrac{3t_1}{1+t_1^3} = a$ and $\dfrac{3t_1^2}{1+t_1^3} = b$. If $t_1 = 0$,

the point is $(0, 0)$, which lies on the line $y = x$. If $t_1 \neq 0$, then the point corresponding to $t = \dfrac{1}{t_1}$ is given by

$$x = \frac{3(1/t_1)}{1+(1/t_1)^3} = \frac{3t_1^2}{t_1^3+1} = b, \quad y = \frac{3(1/t_1)^2}{1+(1/t_1)^3} = \frac{3t_1}{t_1^3+1} = a. \text{ So } (b, a) \text{ also lies on the curve. [Another way to see}$$

this is to do part (e) first; the result is immediate.] The curve intersects the line $y = x$ when $\dfrac{3t}{1+t^3} = \dfrac{3t^2}{1+t^3} \;\Rightarrow\;$

$t = t^2 \;\Rightarrow\; t = 0$ or 1, so the points are $(0, 0)$ and $\left(\frac{3}{2}, \frac{3}{2}\right)$.

(b) $\dfrac{dy}{dt} = \dfrac{(1+t^3)(6t) - 3t^2(3t^2)}{(1+t^3)^2} = \dfrac{6t - 3t^4}{(1+t^3)^2} = 0$ when $6t - 3t^4 = 3t(2 - t^3) = 0 \;\Rightarrow\; t = 0$ or $t = \sqrt[3]{2}$, so there are

horizontal tangents at $(0, 0)$ and $\left(\sqrt[3]{2}, \sqrt[3]{4}\right)$. Using the symmetry from part (a), we see that there are vertical tangents at

$(0, 0)$ and $\left(\sqrt[3]{4}, \sqrt[3]{2}\right)$.

(c) Notice that as $t \to -1^+$, we have $x \to -\infty$ and $y \to \infty$. As $t \to -1^-$, we have $x \to \infty$ and $y \to -\infty$. Also

$$y - (-x - 1) = y + x + 1 = \frac{3t + 3t^2 + (1+t^3)}{1+t^3} = \frac{(t+1)^3}{1+t^3} = \frac{(t+1)^2}{t^2 - t + 1} \to 0 \text{ as } t \to -1. \text{ So } y = -x - 1 \text{ is a}$$

slant asymptote.

(d) $\dfrac{dx}{dt} = \dfrac{(1+t^3)(3) - 3t(3t^2)}{(1+t^3)^2} = \dfrac{3 - 6t^3}{(1+t^3)^2}$ and from part (b) we have $\dfrac{dy}{dt} = \dfrac{6t - 3t^4}{(1+t^3)^2}$. So $\dfrac{dy}{dx} = \dfrac{dy/dt}{dx/dt} = \dfrac{t(2 - t^3)}{1 - 2t^3}$.

Also $\dfrac{d^2y}{dx^2} = \dfrac{\dfrac{d}{dt}\left(\dfrac{dy}{dx}\right)}{dx/dt} = \dfrac{2(1+t^3)^4}{3(1 - 2t^3)^3} > 0 \;\Leftrightarrow\; t < \dfrac{1}{\sqrt[3]{2}}$.

So the curve is concave upward there and has a minimum point at $(0, 0)$

and a maximum point at $\left(\sqrt[3]{2}, \sqrt[3]{4}\right)$. Using this together with the

information from parts (a), (b), and (c), we sketch the curve.

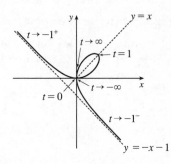

(e) $x^3 + y^3 = \left(\dfrac{3t}{1+t^3}\right)^3 + \left(\dfrac{3t^2}{1+t^3}\right)^3 = \dfrac{27t^3 + 27t^6}{(1+t^3)^3} = \dfrac{27t^3(1+t^3)}{(1+t^3)^3} = \dfrac{27t^3}{(1+t^3)^2}$ and

$3xy = 3\left(\dfrac{3t}{1+t^3}\right)\left(\dfrac{3t^2}{1+t^3}\right) = \dfrac{27t^3}{(1+t^3)^2}$, so $x^3 + y^3 = 3xy$.

(f) We start with the equation from part (e) and substitute $x = r\cos\theta$, $y = r\sin\theta$. Then $x^3 + y^3 = 3xy \;\Rightarrow\;$

$r^3\cos^3\theta + r^3\sin^3\theta = 3r^2\cos\theta\sin\theta$. For $r \neq 0$, this gives $r = \dfrac{3\cos\theta\sin\theta}{\cos^3\theta + \sin^3\theta}$. Dividing numerator and denominator

by $\cos^3\theta$, we obtain $r = \dfrac{3\left(\dfrac{1}{\cos\theta}\right)\dfrac{\sin\theta}{\cos\theta}}{1 + \dfrac{\sin^3\theta}{\cos^3\theta}} = \dfrac{3\sec\theta\tan\theta}{1 + \tan^3\theta}$.

(g) The loop corresponds to $\theta \in \left(0, \frac{\pi}{2}\right)$, so its area is

$$A = \int_0^{\pi/2} \frac{r^2}{2}\,d\theta = \frac{1}{2}\int_0^{\pi/2}\left(\frac{3\sec\theta\tan\theta}{1+\tan^3\theta}\right)^2 d\theta = \frac{9}{2}\int_0^{\pi/2}\frac{\sec^2\theta\tan^2\theta}{(1+\tan^3\theta)^2}\,d\theta = \frac{9}{2}\int_0^\infty \frac{u^2\,du}{(1+u^3)^2} \quad [\text{let } u = \tan\theta]$$

$$= \lim_{b\to\infty}\frac{9}{2}\left[-\frac{1}{3}(1+u^3)^{-1}\right]_0^b = \frac{3}{2}$$

(h) By symmetry, the area between the folium and the line $y = -x - 1$ is equal to the enclosed area in the third quadrant, plus twice the enclosed area in the fourth quadrant. The area in the third quadrant is $\frac{1}{2}$, and since $y = -x - 1 \implies$

$r\sin\theta = -r\cos\theta - 1 \implies r = -\dfrac{1}{\sin\theta + \cos\theta}$, the area in the fourth quadrant is

$\dfrac{1}{2}\displaystyle\int_{-\pi/2}^{-\pi/4}\left[\left(-\dfrac{1}{\sin\theta + \cos\theta}\right)^2 - \left(\dfrac{3\sec\theta\tan\theta}{1+\tan^3\theta}\right)^2\right]d\theta \overset{\text{CAS}}{=} \dfrac{1}{2}$. Therefore, the total area is $\frac{1}{2} + 2\left(\frac{1}{2}\right) = \frac{3}{2}$.

6. (a) Since the smaller circle rolls without slipping around C, the amount of arc traversed on C ($2r\theta$ in the figure) must equal the amount of arc of the smaller circle that has been in contact with C. Since the smaller circle has radius r, it must have turned through an angle of $2r\theta/r = 2\theta$. In addition to turning through an angle 2θ, the little circle has rolled through an angle θ against C. Thus, P has turned through an angle of 3θ as shown in the figure. (If the little circle had turned through an angle of 2θ with its center pinned to the x-axis,

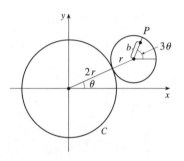

then P would have turned only 2θ instead of 3θ. The movement of the little circle around C adds θ to the angle.) From the figure, we see that the center of the small circle has coordinates $(3r\cos\theta, 3r\sin\theta)$. Thus, P has coordinates (x, y), where $x = b\cos 3\theta + 3r\cos\theta$ and $y = b\sin 3\theta + 3r\sin\theta$.

(b)

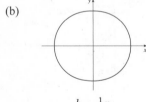

$b = \frac{1}{5}r$

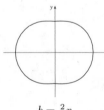

$b = \frac{2}{5}r$

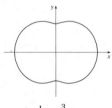

$b = \frac{3}{5}r$

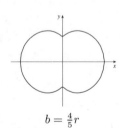

$b = \frac{4}{5}r$

(c) The diagram gives an alternate description of point P on the epitrochoid. Q moves around a circle of radius b, and P rotates one-third as fast with respect to Q at a distance of $3r$. Place an equilateral triangle with sides of length $3\sqrt{3}\,r$ so that its centroid is at Q and one vertex is at P. (The distance from the centroid to a vertex is $\frac{1}{\sqrt{3}}$ times the length of a side of the equilateral triangle.)

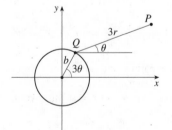

As θ increases by $\frac{2\pi}{3}$, the point Q travels once around the circle of radius b, returning to its original position. At the same time, P (and the rest of the triangle) rotate through an angle of $\frac{2\pi}{3}$ about Q, so P's position is occupied by another vertex. In this way, we see that the epitrochoid traced out by P is simultaneously traced out by the other two vertices as well. The whole equilateral triangle sits inside the epitrochoid (touching it only with its vertices) and each vertex traces out the curve once while the centroid moves around the circle three times.

(d) We view the epitrochoid as being traced out in the same way as in part (c), by a rotor for which the distance from its center to each vertex is $3r$, so it has radius $6r$. To show that the rotor fits inside the epitrochoid, it suffices to show that for any position of the tracing point P, there are no points on the opposite side of the rotor which are outside the epitrochoid. But the most likely case of intersection is when P is on the y-axis, so as long as the diameter of the rotor $\left(\text{which is } 3\sqrt{3}\,r\right)$ is less than the distance between the y-intercepts, the rotor will fit. The y-intercepts occur when $\theta = \frac{\pi}{2}$ or $\theta = \frac{3\pi}{2}$ $\Rightarrow$ $y = -b + 3r$ or $y = b - 3r$, so the distance between the intercepts is $(-b + 3r) - (b - 3r) = 6r - 2b$, and the rotor will fit if $3\sqrt{3}\,r \le 6r - 2b$ $\Leftrightarrow$ $2b \le 6r - 3\sqrt{3}\,r$ $\Leftrightarrow$ $b \le \frac{3}{2}\left(2 - \sqrt{3}\right)r$.

12 □ INFINITE SEQUENCES AND SERIES □ ET 11

12.1 Sequences

1. (a) A sequence is an ordered list of numbers. It can also be defined as a function whose domain is the set of positive integers.

(b) The terms a_n approach 8 as n becomes large. In fact, we can make a_n as close to 8 as we like by taking n sufficiently large.

(c) The terms a_n become large as n becomes large. In fact, we can make a_n as large as we like by taking n sufficiently large.

2. (a) From Definition 1, a convergent sequence is a sequence for which $\lim\limits_{n \to \infty} a_n$ exists. Examples: $\{1/n\}$, $\{1/2^n\}$

(b) A divergent sequence is a sequence for which $\lim\limits_{n \to \infty} a_n$ *does not* exist. Examples: $\{n\}$, $\{\sin n\}$

3. $a_n = 1 - (0.2)^n$, so the sequence is $\{0.8, 0.96, 0.992, 0.9984, 0.99968, \ldots\}$.

4. $a_n = \dfrac{n+1}{3n-1}$, so the sequence is $\left\{\dfrac{2}{2}, \dfrac{3}{5}, \dfrac{4}{8}, \dfrac{5}{11}, \dfrac{6}{14}, \ldots\right\} = \left\{1, \dfrac{3}{5}, \dfrac{1}{2}, \dfrac{5}{11}, \dfrac{3}{7}, \ldots\right\}$.

5. $a_n = \dfrac{3(-1)^n}{n!}$, so the sequence is $\left\{\dfrac{-3}{1}, \dfrac{3}{2}, \dfrac{-3}{6}, \dfrac{3}{24}, \dfrac{-3}{120}, \ldots\right\} = \left\{-3, \dfrac{3}{2}, -\dfrac{1}{2}, \dfrac{1}{8}, -\dfrac{1}{40}, \ldots\right\}$.

6. $a_n = 2 \cdot 4 \cdot 6 \cdots (2n)$, so the sequence is $\{2, 2 \cdot 4, 2 \cdot 4 \cdot 6, 2 \cdot 4 \cdot 6 \cdot 8, 2 \cdot 4 \cdot 6 \cdot 8 \cdot 10, \ldots\} = \{2, 8, 48, 384, 3840, \ldots\}$.

7. $a_1 = 3$, $a_{n+1} = 2a_n - 1$. Each term is defined in terms of the preceding term.
$a_2 = 2a_1 - 1 = 2(3) - 1 = 5$. $a_3 = 2a_2 - 1 = 2(5) - 1 = 9$. $a_4 = 2a_3 - 1 = 2(9) - 1 = 17$.
$a_5 = 2a_4 - 1 = 2(17) - 1 = 33$. The sequence is $\{3, 5, 9, 17, 33, \ldots\}$.

8. $a_1 = 4$, $a_{n+1} = \dfrac{a_n}{a_n - 1}$. Each term is defined in terms of the preceding term.

$a_2 = \dfrac{a_1}{a_1 - 1} = \dfrac{4}{4-1} = \dfrac{4}{3}$. $a_3 = \dfrac{a_2}{a_2 - 1} = \dfrac{4/3}{\frac{4}{3} - 1} = \dfrac{4/3}{1/3} = 4$. Since $a_3 = a_1$, we can see that the terms of the sequence

will alternately equal 4 and 4/3, so the sequence is $\left\{4, \dfrac{4}{3}, 4, \dfrac{4}{3}, 4, \ldots\right\}$.

9. $\left\{1, \dfrac{1}{3}, \dfrac{1}{5}, \dfrac{1}{7}, \dfrac{1}{9}, \ldots\right\}$. The denominator of the nth term is the nth positive odd integer, so $a_n = \dfrac{1}{2n-1}$.

10. $\left\{1, \dfrac{1}{3}, \dfrac{1}{9}, \dfrac{1}{27}, \dfrac{1}{81}, \ldots\right\}$. The denominator of the nth term is the $(n-1)$st power of 3, so $a_n = \dfrac{1}{3^{n-1}}$.

11. $\{2, 7, 12, 17, \ldots\}$. Each term is larger than the preceding one by 5, so $a_n = a_1 + d(n-1) = 2 + 5(n-1) = 5n - 3$.

12. $\left\{-\dfrac{1}{4}, \dfrac{2}{9}, -\dfrac{3}{16}, \dfrac{4}{25}, \ldots\right\}$. The numerator of the nth term is n and its denominator is $(n+1)^2$. Including the alternating signs,

we get $a_n = (-1)^n \dfrac{n}{(n+1)^2}$.

13. $\left\{1, -\dfrac{2}{3}, \dfrac{4}{9}, -\dfrac{8}{27}, \ldots\right\}$. Each term is $-\dfrac{2}{3}$ times the preceding one, so $a_n = \left(-\dfrac{2}{3}\right)^{n-1}$.

14. $\{5, 1, 5, 1, 5, 1, \ldots\}$. The average of 5 and 1 is 3, so we can think of the sequence as alternately adding 2 and -2 to 3.

Thus, $a_n = 3 + (-1)^{n+1} \cdot 2$.

15. The first six terms of $a_n = \dfrac{n}{2n+1}$ are $\dfrac{1}{3}, \dfrac{2}{5}, \dfrac{3}{7}, \dfrac{4}{9}, \dfrac{5}{11}, \dfrac{6}{13}$. It appears that the sequence is approaching $\dfrac{1}{2}$.

$$\lim_{n \to \infty} \frac{n}{2n+1} = \lim_{n \to \infty} \frac{1}{2 + 1/n} = \frac{1}{2}$$

16. $\{\cos(n\pi/3)\}_{n=1}^{9} = \{\frac{1}{2}, -\frac{1}{2}, -1, -\frac{1}{2}, \frac{1}{2}, 1, \frac{1}{2}, -\frac{1}{2}, -1\}$. The sequence does not appear to have a limit. The values will cycle through the first six numbers in the sequence—never approaching a particular number.

17. $a_n = 1 - (0.2)^n$, so $\lim\limits_{n \to \infty} a_n = 1 - 0 = 1$ by (9). Converges

18. $a_n = \dfrac{n^3}{n^3 + 1} = \dfrac{n^3/n^3}{(n^3+1)/n^3} = \dfrac{1}{1 + 1/n^3}$, so $a_n \to \dfrac{1}{1+0} = 1$ as $n \to \infty$. Converges

19. $a_n = \dfrac{3 + 5n^2}{n + n^2} = \dfrac{(3+5n^2)/n^2}{(n+n^2)/n^2} = \dfrac{5 + 3/n^2}{1 + 1/n}$, so $a_n \to \dfrac{5+0}{1+0} = 5$ as $n \to \infty$. Converges

20. $a_n = \dfrac{n^3}{n+1} = \dfrac{n^3/n}{(n+1)/n} = \dfrac{n^2}{1 + 1/n^2}$, so $a_n \to \infty$ as $n \to \infty$ since $\lim\limits_{n \to \infty} n^2 = \infty$ and $\lim\limits_{n \to \infty}(1 + 1/n^2) = 1$. Diverges

21. Because the natural exponential function is continuous at 0, Theorem 7 enables us to write

$$\lim_{n \to \infty} a_n = \lim_{n \to \infty} e^{1/n} = e^{\lim_{n \to \infty}(1/n)} = e^0 = 1. \quad \text{Converges}$$

22. $a_n = \dfrac{3^{n+2}}{5^n} = \dfrac{3^2 3^n}{5^n} = 9\left(\frac{3}{5}\right)^n$, so $\lim\limits_{n \to \infty} a_n = 9 \lim\limits_{n \to \infty} \left(\frac{3}{5}\right)^n = 9 \cdot 0 = 0$ by (9) with $r = \frac{3}{5}$. Converges

23. If $b_n = \dfrac{2n\pi}{1 + 8n}$, then $\lim\limits_{n \to \infty} b_n = \lim\limits_{n \to \infty} \dfrac{(2n\pi)/n}{(1+8n)/n} = \lim\limits_{n \to \infty} \dfrac{2\pi}{1/n + 8} = \dfrac{2\pi}{8} = \dfrac{\pi}{4}$. Since $\tan$ is continuous at $\frac{\pi}{4}$, by

Theorem 7, $\lim\limits_{n \to \infty} \tan\left(\dfrac{2n\pi}{1 + 8n}\right) = \tan\left(\lim\limits_{n \to \infty} \dfrac{2n\pi}{1 + 8n}\right) = \tan\dfrac{\pi}{4} = 1$. Converges

24. Using the last limit law for sequences and the continuity of the square root function,

$$\lim_{n \to \infty} a_n = \lim_{n \to \infty} \sqrt{\frac{n+1}{9n+1}} = \sqrt{\lim_{n \to \infty} \frac{n+1}{9n+1}} = \sqrt{\lim_{n \to \infty} \frac{1 + 1/n}{9 + 1/n}} = \sqrt{\frac{1}{9}} = \frac{1}{3}. \quad \text{Converges}$$

25. $a_n = \dfrac{(-1)^{n-1} n}{n^2 + 1} = \dfrac{(-1)^{n-1}}{n + 1/n}$, so $0 \leq |a_n| = \dfrac{1}{n + 1/n} \leq \dfrac{1}{n} \to 0$ as $n \to \infty$, so $a_n \to 0$ by the Squeeze Theorem and

Theorem 6. Converges

26. $a_n = \dfrac{(-1)^n n^3}{n^3 + 2n^2 + 1}$. Now $|a_n| = \dfrac{n^3}{n^3 + 2n^2 + 1} = \dfrac{1}{1 + \frac{2}{n} + \frac{1}{n^3}} \to 1$ as $n \to \infty$, but the terms of the sequence $\{a_n\}$

alternate in sign, so the sequence $a_1, a_3, a_5, \ldots$ converges to -1 and the sequence $a_2, a_4, a_6, \ldots$ converges to $+1$.

This shows that the given sequence diverges since its terms don't approach a single real number.

27. $a_n = \cos(n/2)$.   This sequence diverges since the terms don't approach any particular real number as $n \to \infty$. The terms take on values between -1 and 1.

28. $a_n = \cos(2/n)$.   As $n \to \infty$, $2/n \to 0$, so $\cos(2/n) \to \cos 0 = 1$ because cos is continuous.   Converges

29. $a_n = \dfrac{(2n-1)!}{(2n+1)!} = \dfrac{(2n-1)!}{(2n+1)(2n)(2n-1)!} = \dfrac{1}{(2n+1)(2n)} \to 0$ as $n \to \infty$.   Converges

30. $2n \to \infty$ as $n \to \infty$, so since $\lim\limits_{x \to \infty} \arctan x = \dfrac{\pi}{2}$, we have $\lim\limits_{n \to \infty} \arctan 2n = \dfrac{\pi}{2}$.   Converges

31. $a_n = \dfrac{e^n + e^{-n}}{e^{2n} - 1} \cdot \dfrac{e^{-n}}{e^{-n}} = \dfrac{1 + e^{-2n}}{e^n - e^{-n}} \to 0$ as $n \to \infty$ because $1 + e^{-2n} \to 1$ and $e^n - e^{-n} \to \infty$.   Converges

32. $a_n = \dfrac{\ln n}{\ln 2n} = \dfrac{\ln n}{\ln 2 + \ln n} = \dfrac{1}{\frac{\ln 2}{\ln n} + 1} \to \dfrac{1}{0 + 1} = 1$ as $n \to \infty$.   Converges

33. $a_n = n^2 e^{-n} = \dfrac{n^2}{e^n}$. Since $\lim\limits_{x \to \infty} \dfrac{x^2}{e^x} \overset{\mathrm{H}}{=} \lim\limits_{x \to \infty} \dfrac{2x}{e^x} \overset{\mathrm{H}}{=} \lim\limits_{x \to \infty} \dfrac{2}{e^x} = 0$, it follows from Theorem 3 that $\lim\limits_{n \to \infty} a_n = 0$.   Converges

34. $a_n = n \cos n\pi = n(-1)^n$.   Since $|a_n| = n \to \infty$ as $n \to \infty$, the given sequence diverges.

35. $0 \le \dfrac{\cos^2 n}{2^n} \le \dfrac{1}{2^n}$ [since $0 \le \cos^2 n \le 1$],   so since $\lim\limits_{n \to \infty} \dfrac{1}{2^n} = 0$, $\left\{ \dfrac{\cos^2 n}{2^n} \right\}$ converges to 0 by the Squeeze Theorem.

36. $a_n = \ln(n+1) - \ln n = \ln\left(\dfrac{n+1}{n}\right) = \ln\left(1 + \dfrac{1}{n}\right) \to \ln(1) = 0$ as $n \to \infty$ because ln is continuous.   Converges

37. $a_n = n \sin(1/n) = \dfrac{\sin(1/n)}{1/n}$.   Since $\lim\limits_{x \to \infty} \dfrac{\sin(1/x)}{1/x} = \lim\limits_{t \to 0^+} \dfrac{\sin t}{t}$ [where $t = 1/x$] $= 1$, it follows from Theorem 3 that $\{a_n\}$ converges to 1.

38. $a_n = \sqrt[n]{2^{1+3n}} = (2^{1+3n})^{1/n} = (2^1 2^{3n})^{1/n} = 2^{1/n} 2^3 = 8 \cdot 2^{1/n}$, so

$\lim\limits_{n \to \infty} a_n = 8 \lim\limits_{n \to \infty} 2^{1/n} = 8 \cdot 2^{\lim_{n \to \infty}(1/n)} = 8 \cdot 2^0 = 8$ by Theorem 7, since the function $f(x) = 2^x$ is continuous at 0. Convergent

39. $y = \left(1 + \dfrac{2}{x}\right)^x \;\Rightarrow\; \ln y = x \ln\left(1 + \dfrac{2}{x}\right)$, so

$\lim\limits_{x \to \infty} \ln y = \lim\limits_{x \to \infty} \dfrac{\ln(1 + 2/x)}{1/x} \overset{\mathrm{H}}{=} \lim\limits_{x \to \infty} \dfrac{\left(\dfrac{1}{1 + 2/x}\right)\left(-\dfrac{2}{x^2}\right)}{-1/x^2} = \lim\limits_{x \to \infty} \dfrac{2}{1 + 2/x} = 2 \;\Rightarrow$

$\lim\limits_{x \to \infty} \left(1 + \dfrac{2}{x}\right)^x = \lim\limits_{x \to \infty} e^{\ln y} = e^2$, so by Theorem 3, $\lim\limits_{n \to \infty} \left(1 + \dfrac{2}{n}\right)^n = e^2$.   Convergent

40. $a_n = \dfrac{\sin 2n}{1 + \sqrt{n}}$. $|a_n| \le \dfrac{1}{1 + \sqrt{n}}$ and $\lim\limits_{n \to \infty} \dfrac{1}{1 + \sqrt{n}} = 0$, so $\dfrac{-1}{1 + \sqrt{n}} \le a_n \le \dfrac{1}{1 + \sqrt{n}} \;\Rightarrow\; \lim\limits_{n \to \infty} a_n = 0$ by the Squeeze Theorem.   Converges

41. $a_n = \ln(2n^2 + 1) - \ln(n^2 + 1) = \ln\left(\dfrac{2n^2 + 1}{n^2 + 1}\right) = \ln\left(\dfrac{2 + 1/n^2}{1 + 1/n^2}\right) \to \ln 2$ as $n \to \infty$.   Convergent

42. $\displaystyle\lim_{x\to\infty}\frac{(\ln x)^2}{x}\overset{\text{H}}{=}\lim_{x\to\infty}\frac{2(\ln x)(1/x)}{1}=2\lim_{x\to\infty}\frac{\ln x}{x}\overset{\text{H}}{=}2\lim_{x\to\infty}\frac{1/x}{1}=0$, so by Theorem 3, $\displaystyle\lim_{n\to\infty}\frac{(\ln n)^2}{n}=0$.   Convergent

43. $\{0,1,0,0,1,0,0,0,1,\ldots\}$ diverges since the sequence takes on only two values, 0 and 1, and never stays arbitrarily close to either one (or any other value) for n sufficiently large.

44. $\left\{\frac{1}{1},\frac{1}{3},\frac{1}{2},\frac{1}{4},\frac{1}{3},\frac{1}{5},\frac{1}{4},\frac{1}{6},\ldots\right\}$. $a_{2n-1}=\dfrac{1}{n}$ and $a_{2n}=\dfrac{1}{n+2}$ for all positive integers n. $\displaystyle\lim_{n\to\infty}a_n=0$ since

$\displaystyle\lim_{n\to\infty}a_{2n-1}=\lim_{n\to\infty}\frac{1}{n}=0$ and $\displaystyle\lim_{n\to\infty}a_{2n}=\lim_{n\to\infty}\frac{1}{n+2}=0$. For n sufficiently large, a_n can be made as close to 0

as we like.   Converges

45. $a_n=\dfrac{n!}{2^n}=\dfrac{1}{2}\cdot\dfrac{2}{2}\cdot\dfrac{3}{2}\cdot\ \cdots\ \cdot\dfrac{(n-1)}{2}\cdot\dfrac{n}{2}\geq\dfrac{1}{2}\cdot\dfrac{n}{2}$ [for $n>1$] $=\dfrac{n}{4}\to\infty$ as $n\to\infty$, so $\{a_n\}$ diverges.

46. $0<|a_n|=\dfrac{3^n}{n!}=\dfrac{3}{1}\cdot\dfrac{3}{2}\cdot\dfrac{3}{3}\cdot\ \cdots\ \cdot\dfrac{3}{(n-1)}\cdot\dfrac{3}{n}\leq\dfrac{3}{1}\cdot\dfrac{3}{2}\cdot\dfrac{3}{n}$ [for $n>2$] $=\dfrac{27}{2n}\to0$ as $n\to\infty$, so by the Squeeze

Theorem and Theorem 6, $\{(-3)^n/n!\}$ converges to 0.

47.

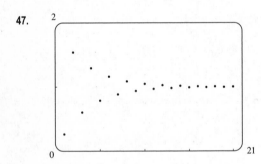

From the graph, it appears that the sequence converges to 1.

$\{(-2/e)^n\}$ converges to 0 by (9), and hence $\{1+(-2/e)^n\}$ converges to $1+0=1$.

48.

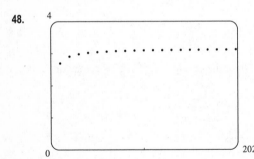

From the graph, it appears that the sequence converges to a number greater than 3.

$\displaystyle\lim_{n\to\infty}a_n=\lim_{n\to\infty}\sqrt{n}\sin\left(\frac{\pi}{\sqrt{n}}\right)=\lim_{n\to\infty}\frac{\sin\left(\pi/\sqrt{n}\right)}{\pi/\sqrt{n}}\cdot\pi$

$\displaystyle=\lim_{x\to0^+}\frac{\sin x}{x}\cdot\pi\quad\left[x=\pi/\sqrt{n}\right]\ =1\cdot\pi=\pi.$

49.

From the graph, it appears that the sequence converges to $\frac{1}{2}$.

As $n\to\infty$,

$a_n=\sqrt{\dfrac{3+2n^2}{8n^2+n}}=\sqrt{\dfrac{3/n^2+2}{8+1/n}}\ \Rightarrow\ \sqrt{\dfrac{0+2}{8+0}}=\sqrt{\dfrac{1}{4}}=\dfrac{1}{2},$

so $\displaystyle\lim_{n\to\infty}a_n=\frac{1}{2}.$

50.

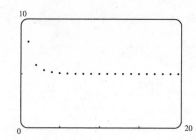

From the graph, it appears that the sequence converges to 5.

$$5 = \sqrt[n]{5^n} \le \sqrt[n]{3^n + 5^n} \le \sqrt[n]{5^n + 5^n} = \sqrt[n]{2}\,\sqrt[n]{5^n}$$

$$= \sqrt[n]{2} \cdot 5 \to 5 \text{ as } n \to \infty \quad \left[\lim_{n \to \infty} 2^{1/n} = 2^0 = 1 \right]$$

Hence, $a_n \to 5$ by the Squeeze Theorem.

Alternate solution: Let $y = (3^x + 5^x)^{1/x}$. Then

$$\lim_{x \to \infty} \ln y = \lim_{x \to \infty} \frac{\ln(3^x + 5^x)}{x} \overset{\text{H}}{=} \lim_{x \to \infty} \frac{3^x \ln 3 + 5^x \ln 5}{3^x + 5^x} = \lim_{x \to \infty} \frac{\left(\frac{3}{5}\right)^x \ln 3 + \ln 5}{\left(\frac{3}{5}\right)^x + 1} = \ln 5,$$

so $\lim\limits_{x \to \infty} y = e^{\ln 5} = 5$, and so $\left\{ \sqrt[n]{3^n + 5^n} \right\}$ converges to 5.

51.

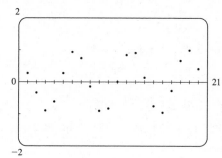

From the graph, it appears that the sequence $\{a_n\} = \left\{ \dfrac{n^2 \cos n}{1 + n^2} \right\}$ is

divergent, since it oscillates between 1 and -1 (approximately). To

prove this, suppose that $\{a_n\}$ converges to L. If $b_n = \dfrac{n^2}{1 + n^2}$, then

$\{b_n\}$ converges to 1, and $\lim\limits_{n \to \infty} \dfrac{a_n}{b_n} = \dfrac{L}{1} = L$. But $\dfrac{a_n}{b_n} = \cos n$, so

$\lim\limits_{n \to \infty} \dfrac{a_n}{b_n}$ does not exist. This contradiction shows that $\{a_n\}$ diverges.

52.

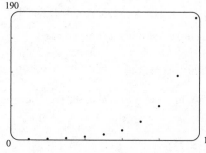

From the graphs, it seems that the sequence diverges. $a_n = \dfrac{1 \cdot 3 \cdot 5 \cdot \cdots \cdot (2n-1)}{n!}$. We first prove by induction that

$a_n \ge \left(\dfrac{3}{2}\right)^{n-1}$ for all n. This is clearly true for $n = 1$, so let $P(n)$ be the statement that the above is true for n. We must

show it is then true for $n + 1$. $a_{n+1} = a_n \cdot \dfrac{2n+1}{n+1} \ge \left(\dfrac{3}{2}\right)^{n-1} \cdot \dfrac{2n+1}{n+1}$ (induction hypothesis). But $\dfrac{2n+1}{n+1} \ge \dfrac{3}{2}$

[since $2(2n+1) \ge 3(n+1) \iff 4n+2 \ge 3n+3 \iff n \ge 1$], and so we get that $a_{n+1} \ge \left(\frac{3}{2}\right)^{n-1} \cdot \frac{3}{2} = \left(\frac{3}{2}\right)^n$ which

is $P(n+1)$. Thus, we have proved our first assertion, so since $\left\{ \left(\frac{3}{2}\right)^{n-1} \right\}$ diverges [by (9)], so does the given sequence $\{a_n\}$.

53.

From the graph, it appears that the sequence approaches 0.

$$0 < a_n = \frac{1 \cdot 3 \cdot 5 \cdot \cdots \cdot (2n-1)}{(2n)^n} = \frac{1}{2n} \cdot \frac{3}{2n} \cdot \frac{5}{2n} \cdot \cdots \cdot \frac{2n-1}{2n}$$

$$\leq \frac{1}{2n} \cdot (1) \cdot (1) \cdot \cdots \cdot (1) = \frac{1}{2n} \to 0 \text{ as } n \to \infty$$

So by the Squeeze Theorem, $\left\{ \dfrac{1 \cdot 3 \cdot 5 \cdot \cdots \cdot (2n-1)}{(2n)^n} \right\}$ converges to 0.

54. (a) $a_1 = 1$, $a_{n+1} = 4 - a_n$ for $n \geq 1$. $a_1 = 1$, $a_2 = 4 - a_1 = 4 - 1 = 3$, $a_3 = 4 - a_2 = 4 - 3 = 1$,

$a_4 = 4 - a_3 = 4 - 1 = 3$, $a_5 = 4 - a_4 = 4 - 3 = 1$. Since the terms of the sequence alternate between 1 and 3,

the sequence is divergent.

(b) $a_1 = 2$, $a_2 = 4 - a_1 = 4 - 2 = 2$, $a_3 = 4 - a_2 = 4 - 2 = 2$. Since all of the terms are 2, $\lim\limits_{n \to \infty} a_n = 2$ and hence, the

sequence is convergent.

55. (a) $a_n = 1000(1.06)^n$ $\Rightarrow$ $a_1 = 1060$, $a_2 = 1123.60$, $a_3 = 1191.02$, $a_4 = 1262.48$, and $a_5 = 1338.23$.

(b) $\lim\limits_{n \to \infty} a_n = 1000 \lim\limits_{n \to \infty} (1.06)^n$, so the sequence diverges by (9) with $r = 1.06 > 1$.

56. $a_{n+1} = \begin{cases} \frac{1}{2} a_n & \text{if } a_n \text{ is an even number} \\ 3a_n + 1 & \text{if } a_n \text{ is an odd number} \end{cases}$ When $a_1 = 11$, the first 40 terms are 11, 34, 17, 52, 26, 13, 40, 20, 10, 5,

16, 8, 4, 2, 1, 4, 2, 1, 4, 2, 1, 4, 2, 1, 4, 2, 1, 4, 2, 1, 4, 2, 1, 4, 2, 1, 4, 2, 1, 4. When $a_1 = 25$, the first 40 terms are 25, 76, 38,

19, 58, 29, 88, 44, 22, 11, 34, 17, 52, 26, 13, 40, 20, 10, 5, 16, 8, 4, 2, 1, 4, 2, 1, 4, 2, 1, 4, 2, 1, 4, 2, 1, 4, 2, 1, 4.

The famous Collatz conjecture is that this sequence always reaches 1, regardless of the starting point a_1.

57. If $|r| \geq 1$, then $\{r^n\}$ diverges by (9), so $\{nr^n\}$ diverges also, since $|nr^n| = n |r^n| \geq |r^n|$. If $|r| < 1$ then

$$\lim_{x \to \infty} x r^x = \lim_{x \to \infty} \frac{x}{r^{-x}} \overset{\text{H}}{=} \lim_{x \to \infty} \frac{1}{(-\ln r) r^{-x}} = \lim_{x \to \infty} \frac{r^x}{-\ln r} = 0, \text{ so } \lim_{n \to \infty} nr^n = 0, \text{ and hence } \{nr^n\} \text{ converges}$$

whenever $|r| < 1$.

58. (a) Let $\lim\limits_{n \to \infty} a_n = L$. By Definition 2, this means that for every $\varepsilon > 0$ there is an integer N such that $|a_n - L| < \varepsilon$

whenever $n > N$. Thus, $|a_{n+1} - L| < \varepsilon$ whenever $n + 1 > N$ $\Leftrightarrow$ $n > N - 1$. It follows that $\lim\limits_{n \to \infty} a_{n+1} = L$ and so

$\lim\limits_{n \to \infty} a_n = \lim\limits_{n \to \infty} a_{n+1}$.

(b) If $L = \lim\limits_{n \to \infty} a_n$ then $\lim\limits_{n \to \infty} a_{n+1} = L$ also, so L must satisfy $L = 1/(1 + L)$ $\Rightarrow$ $L^2 + L - 1 = 0$ $\Rightarrow$ $L = \frac{-1 + \sqrt{5}}{2}$

(since L has to be nonnegative if it exists).

59. Since $\{a_n\}$ is a decreasing sequence, $a_n > a_{n+1}$ for all $n \geq 1$. Because all of its terms lie between 5 and 8, $\{a_n\}$ is a

bounded sequence. By the Monotonic Sequence Theorem, $\{a_n\}$ is convergent; that is, $\{a_n\}$ has a limit L. L must be less than

8 since $\{a_n\}$ is decreasing, so $5 \leq L < 8$.

60. The terms of $a_n = (-2)^{n+1}$ alternate in sign, so the sequence is not monotonic. The first five terms are $4, -8, 16, -32$, and 64. Since $\lim\limits_{n \to \infty} |a_n| = \lim\limits_{n \to \infty} 2^{n+1} = \infty$, the sequence is not bounded.

61. $a_n = \dfrac{1}{2n+3}$ is decreasing since $a_{n+1} = \dfrac{1}{2(n+1)+3} = \dfrac{1}{2n+5} < \dfrac{1}{2n+3} = a_n$ for each $n \geq 1$. The sequence is bounded since $0 < a_n \leq \frac{1}{5}$ for all $n \geq 1$. Note that $a_1 = \frac{1}{5}$.

62. $a_n = \dfrac{2n-3}{3n+4}$ defines an increasing sequence since for $f(x) = \dfrac{2x-3}{3x+4}$,

$f'(x) = \dfrac{(3x+4)(2) - (2x-3)(3)}{(3x+4)^2} = \dfrac{17}{(3x+4)^2} > 0$. The sequence is bounded since $a_n \geq a_1 = -\frac{1}{7}$ for $n \geq 1$,

and $a_n < \dfrac{2n-3}{3n} < \dfrac{2n}{3n} = \dfrac{2}{3}$ for $n \geq 1$.

63. The terms of $a_n = n(-1)^n$ alternate in sign, so the sequence is not monotonic. The first five terms are $-1, 2, -3, 4$, and -5. Since $\lim\limits_{n \to \infty} |a_n| = \lim\limits_{n \to \infty} n = \infty$, the sequence is not bounded.

64. $a_n = ne^{-n}$ defines a positive decreasing sequence since the function $f(x) = xe^{-x}$ is decreasing for $x > 1$.

$[f'(x) = e^{-x} - xe^{-x} = e^{-x}(1-x) < 0$ for $x > 1$.] The sequence is bounded above by $a_1 = \frac{1}{e}$ and below by 0.

65. $a_n = \dfrac{n}{n^2+1}$ defines a decreasing sequence since for $f(x) = \dfrac{x}{x^2+1}$, $f'(x) = \dfrac{(x^2+1)(1) - x(2x)}{(x^2+1)^2} = \dfrac{1-x^2}{(x^2+1)^2} \leq 0$

for $x \geq 1$. The sequence is bounded since $0 < a_n \leq \frac{1}{2}$ for all $n \geq 1$.

66. $a_n = n + \dfrac{1}{n}$ defines an increasing sequence since the function $g(x) = x + \dfrac{1}{x}$ is increasing for $x > 1$. $[g'(x) = 1 - 1/x^2 > 0$

for $x > 1$.] The sequence is unbounded since $a_n \to \infty$ as $n \to \infty$. (It is, however, bounded below by $a_1 = 2$.)

67. For $\left\{ \sqrt{2}, \sqrt{2\sqrt{2}}, \sqrt{2\sqrt{2\sqrt{2}}}, \ldots \right\}$, $a_1 = 2^{1/2}$, $a_2 = 2^{3/4}$, $a_3 = 2^{7/8}, \ldots$, so $a_n = 2^{(2^n - 1)/2^n} = 2^{1-(1/2^n)}$.

$\lim\limits_{n \to \infty} a_n = \lim\limits_{n \to \infty} 2^{1-(1/2^n)} = 2^1 = 2$.

Alternate solution: Let $L = \lim\limits_{n \to \infty} a_n$. (We could show the limit exists by showing that $\{a_n\}$ is bounded and increasing.)

Then L must satisfy $L = \sqrt{2 \cdot L} \Rightarrow L^2 = 2L \Rightarrow L(L-2) = 0$. $L \neq 0$ since the sequence increases, so $L = 2$.

68. (a) Let P_n be the statement that $a_{n+1} \geq a_n$ and $a_n \leq 3$. P_1 is obviously true. We will assume that P_n is true and then show that as a consequence P_{n+1} must also be true. $a_{n+2} \geq a_{n+1} \Leftrightarrow \sqrt{2 + a_{n+1}} \geq \sqrt{2 + a_n} \Leftrightarrow 2 + a_{n+1} \geq 2 + a_n \Leftrightarrow a_{n+1} \geq a_n$, which is the induction hypothesis. $a_{n+1} \leq 3 \Leftrightarrow \sqrt{2 + a_n} \leq 3 \Leftrightarrow 2 + a_n \leq 9 \Leftrightarrow a_n \leq 7$, which is certainly true because we are assuming that $a_n \leq 3$. So P_n is true for all n, and so $a_1 \leq a_n \leq 3$ (showing that the sequence is bounded), and hence by the Monotonic Sequence Theorem, $\lim\limits_{n \to \infty} a_n$ exists.

(b) If $L = \lim\limits_{n \to \infty} a_n$, then $\lim\limits_{n \to \infty} a_{n+1} = L$ also, so $L = \sqrt{2 + L} \Rightarrow L^2 = 2 + L \Leftrightarrow L^2 - L - 2 = 0 \Leftrightarrow (L+1)(L-2) = 0 \Leftrightarrow L = 2$ [since L can't be negative].

69. $a_1 = 1$, $a_{n+1} = 3 - \dfrac{1}{a_n}$. We show by induction that $\{a_n\}$ is increasing and bounded above by 3. Let P_n be the proposition

that $a_{n+1} > a_n$ and $0 < a_n < 3$. Clearly P_1 is true. Assume that P_n is true. Then $a_{n+1} > a_n \;\Rightarrow\; \dfrac{1}{a_{n+1}} < \dfrac{1}{a_n} \;\Rightarrow$

$-\dfrac{1}{a_{n+1}} > -\dfrac{1}{a_n}$. Now $a_{n+2} = 3 - \dfrac{1}{a_{n+1}} > 3 - \dfrac{1}{a_n} = a_{n+1} \;\Leftrightarrow\; P_{n+1}$. This proves that $\{a_n\}$ is increasing and bounded

above by 3, so $1 = a_1 < a_n < 3$, that is, $\{a_n\}$ is bounded, and hence convergent by the Monotonic Sequence Theorem.

If $L = \lim\limits_{n\to\infty} a_n$, then $\lim\limits_{n\to\infty} a_{n+1} = L$ also, so L must satisfy $L = 3 - 1/L \;\Rightarrow\; L^2 - 3L + 1 = 0 \;\Rightarrow\; L = \frac{3\pm\sqrt{5}}{2}$.

But $L > 1$, so $L = \frac{3+\sqrt{5}}{2}$.

70. $a_1 = 2$, $a_{n+1} = \dfrac{1}{3 - a_n}$. We use induction. Let P_n be the statement that $0 < a_{n+1} \le a_n \le 2$. Clearly P_1 is true, since

$a_2 = 1/(3 - 2) = 1$. Now assume that P_n is true. Then $a_{n+1} \le a_n \;\Rightarrow\; -a_{n+1} \ge -a_n \;\Rightarrow\; 3 - a_{n+1} \ge 3 - a_n \;\Rightarrow$

$a_{n+2} = \dfrac{1}{3 - a_{n+1}} \le \dfrac{1}{3 - a_n} = a_{n+1}$. Also $a_{n+2} > 0$ [since $3 - a_{n+1}$ is positive] and $a_{n+1} \le 2$ by the induction

hypothesis, so P_{n+1} is true. To find the limit, we use the fact that $\lim\limits_{n\to\infty} a_n = \lim\limits_{n\to\infty} a_{n+1} \;\Rightarrow\; L = \frac{1}{3-L} \;\Rightarrow$

$L^2 - 3L + 1 = 0 \;\Rightarrow\; L = \frac{3\pm\sqrt{5}}{2}$. But $L \le 2$, so we must have $L = \frac{3-\sqrt{5}}{2}$.

71. (a) Let a_n be the number of rabbit pairs in the nth month. Clearly $a_1 = 1 = a_2$. In the nth month, each pair that is

2 or more months old (that is, a_{n-2} pairs) will produce a new pair to add to the a_{n-1} pairs already present. Thus,

$a_n = a_{n-1} + a_{n-2}$, so that $\{a_n\} = \{f_n\}$, the Fibonacci sequence.

(b) $a_n = \dfrac{f_{n+1}}{f_n} \;\Rightarrow\; a_{n-1} = \dfrac{f_n}{f_{n-1}} = \dfrac{f_{n-1} + f_{n-2}}{f_{n-1}} = 1 + \dfrac{f_{n-2}}{f_{n-1}} = 1 + \dfrac{1}{f_{n-1}/f_{n-2}} = 1 + \dfrac{1}{a_{n-2}}$. If $L = \lim\limits_{n\to\infty} a_n$,

then $L = \lim\limits_{n\to\infty} a_{n-1}$ and $L = \lim\limits_{n\to\infty} a_{n-2}$, so L must satisfy $L = 1 + \dfrac{1}{L} \;\Rightarrow\; L^2 - L - 1 = 0 \;\Rightarrow\; L = \frac{1+\sqrt{5}}{2}$

[since L must be positive].

72. (a) If f is continuous, then $f(L) = f\left(\lim\limits_{n\to\infty} a_n\right) = \lim\limits_{n\to\infty} f(a_n) = \lim\limits_{n\to\infty} a_{n+1} = \lim\limits_{n\to\infty} a_n = L$ by Exercise 58(a).

(b) By repeatedly pressing the cosine key on the calculator (that is, taking cosine of the previous answer) until the displayed

value stabilizes, we see that $L \approx 0.73909$.

73. (a)

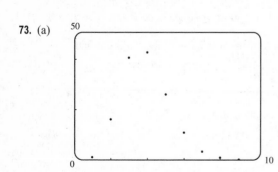

From the graph, it appears that the sequence $\left\{\dfrac{n^5}{n!}\right\}$

converges to 0, that is, $\lim\limits_{n\to\infty} \dfrac{n^5}{n!} = 0$.

(b)

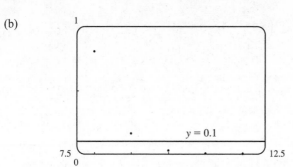 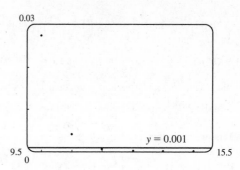

From the first graph, it seems that the smallest possible value of N corresponding to $\varepsilon = 0.1$ is 9, since $n^5/n! < 0.1$ whenever $n \geq 10$, but $9^5/9! > 0.1$. From the second graph, it seems that for $\varepsilon = 0.001$, the smallest possible value for N is 11 since $n^5/n! < 0.001$ whenever $n \geq 12$.

74. Let $\varepsilon > 0$ and let N be any positive integer larger than $\ln(\varepsilon)/\ln|r|$. If $n > N$, then $n > \ln(\varepsilon)/\ln|r|$ $\Rightarrow$ $n\ln|r| < \ln\varepsilon$ [since $|r| < 1 \Rightarrow \ln|r| < 0$] $\Rightarrow$ $\ln(|r|^n) < \ln\varepsilon$ $\Rightarrow$ $|r|^n < \varepsilon$ $\Rightarrow$ $|r^n - 0| < \varepsilon$, and so by Definition 2, $\lim\limits_{n\to\infty} r^n = 0$.

75. Theorem 6: If $\lim\limits_{n\to\infty} |a_n| = 0$ then $\lim\limits_{n\to\infty} -|a_n| = 0$, and since $-|a_n| \leq a_n \leq |a_n|$, we have that $\lim\limits_{n\to\infty} a_n = 0$ by the Squeeze Theorem.

76. Theorem 7: If $\lim\limits_{n\to\infty} a_n = L$ and the function f is continuous at L, then $\lim\limits_{n\to\infty} f(a_n) = f(L)$.

Proof: We must show that, given a number $\varepsilon > 0$, there is an integer N such that $|f(a_n) - f(L)| < \varepsilon$ whenever $n > N$. Suppose $\varepsilon > 0$. Since f is continuous at L, there is a number $\delta > 0$ such that $|f(x) - f(L)| < \varepsilon$ if $|x - L| < \delta$. Since $\lim\limits_{n\to\infty} a_n = L$, there is an integer N such that $|a_n - L| < \delta$ if $n > N$. Suppose $n > N$. Then $0 < |a_n - L| < \delta$, so $|f(a_n) - f(L)| < \varepsilon$.

77. To Prove: If $\lim\limits_{n\to\infty} a_n = 0$ and $\{b_n\}$ is bounded, then $\lim\limits_{n\to\infty} (a_n b_n) = 0$.

Proof: Since $\{b_n\}$ is bounded, there is a positive number M such that $|b_n| \leq M$ and hence, $|a_n|\,|b_n| \leq |a_n|\,M$ for all $n \geq 1$. Let $\varepsilon > 0$ be given. Since $\lim\limits_{n\to\infty} a_n = 0$, there is an integer N such that $|a_n - 0| < \dfrac{\varepsilon}{M}$ if $n > N$. Then

$$|a_n b_n - 0| = |a_n b_n| = |a_n|\,|b_n| \leq |a_n|\,M = |a_n - 0|\,M < \frac{\varepsilon}{M} \cdot M = \varepsilon \text{ for all } n > N. \text{ Since } \varepsilon \text{ was arbitrary,}$$

$\lim\limits_{n\to\infty} (a_n b_n) = 0$.

78. (a)
$$\frac{b^{n+1} - a^{n+1}}{b - a} = b^n + b^{n-1}a + b^{n-2}a^2 + b^{n-3}a^3 + \cdots + ba^{n-1} + a^n$$
$$< b^n + b^{n-1}b + b^{n-2}b^2 + b^{n-3}b^3 + \cdots + bb^{n-1} + b^n = (n+1)b^n$$

(b) Since $b - a > 0$, we have $b^{n+1} - a^{n+1} < (n+1)b^n(b - a)$ $\Rightarrow$ $b^{n+1} - (n+1)b^n(b-a) < a^{n+1}$ $\Rightarrow$ $b^n[(n+1)a - nb] < a^{n+1}$.

(c) With this substitution, $(n+1)a - nb = 1$, and so $b^n = \left(1 + \dfrac{1}{n}\right)^n < a^{n+1} = \left(1 + \dfrac{1}{n+1}\right)^{n+1}$.

(d) With this substitution, we get $\left(1+\frac{1}{2n}\right)^n\left(\frac{1}{2}\right) < 1 \;\Rightarrow\; \left(1+\frac{1}{2n}\right)^n < 2 \;\Rightarrow\; \left(1+\frac{1}{2n}\right)^{2n} < 4$.

(e) $a_n < a_{2n}$ since $\{a_n\}$ is increasing, so $a_n < a_{2n} < 4$.

(f) Since $\{a_n\}$ is increasing and bounded above by 4, $a_1 \le a_n \le 4$, and so $\{a_n\}$ is bounded and monotonic, and hence has a limit by the Monotonic Sequence Theorem.

79. (a) First we show that $a > a_1 > b_1 > b$.

$a_1 - b_1 = \frac{a+b}{2} - \sqrt{ab} = \frac{1}{2}\left(a - 2\sqrt{ab} + b\right) = \frac{1}{2}\left(\sqrt{a} - \sqrt{b}\right)^2 > 0$ [since $a > b$] $\Rightarrow$ $a_1 > b_1$. Also

$a - a_1 = a - \frac{1}{2}(a+b) = \frac{1}{2}(a-b) > 0$ and $b - b_1 = b - \sqrt{ab} = \sqrt{b}\left(\sqrt{b} - \sqrt{a}\right) < 0$, so $a > a_1 > b_1 > b$. In the same

way we can show that $a_1 > a_2 > b_2 > b_1$ and so the given assertion is true for $n = 1$. Suppose it is true for $n = k$, that is,

$a_k > a_{k+1} > b_{k+1} > b_k$. Then

$a_{k+2} - b_{k+2} = \frac{1}{2}(a_{k+1} + b_{k+1}) - \sqrt{a_{k+1}b_{k+1}} = \frac{1}{2}\left(a_{k+1} - 2\sqrt{a_{k+1}b_{k+1}} + b_{k+1}\right) = \frac{1}{2}\left(\sqrt{a_{k+1}} - \sqrt{b_{k+1}}\right)^2 > 0$,

$a_{k+1} - a_{k+2} = a_{k+1} - \frac{1}{2}(a_{k+1} + b_{k+1}) = \frac{1}{2}(a_{k+1} - b_{k+1}) > 0$, and

$b_{k+1} - b_{k+2} = b_{k+1} - \sqrt{a_{k+1}b_{k+1}} = \sqrt{b_{k+1}}\left(\sqrt{b_{k+1}} - \sqrt{a_{k+1}}\right) < 0 \;\Rightarrow\; a_{k+1} > a_{k+2} > b_{k+2} > b_{k+1}$,

so the assertion is true for $n = k + 1$. Thus, it is true for all n by mathematical induction.

(b) From part (a) we have $a > a_n > a_{n+1} > b_{n+1} > b_n > b$, which shows that both sequences, $\{a_n\}$ and $\{b_n\}$, are monotonic and bounded. So they are both convergent by the Monotonic Sequence Theorem.

(c) Let $\lim\limits_{n\to\infty} a_n = \alpha$ and $\lim\limits_{n\to\infty} b_n = \beta$. Then $\lim\limits_{n\to\infty} a_{n+1} = \lim\limits_{n\to\infty} \dfrac{a_n + b_n}{2} \;\Rightarrow\; \alpha = \dfrac{\alpha + \beta}{2} \;\Rightarrow$

$2\alpha = \alpha + \beta \;\Rightarrow\; \alpha = \beta$.

80. (a) Let $\varepsilon > 0$. Since $\lim\limits_{n\to\infty} a_{2n} = L$, there exists N_1 such that $|a_{2n} - L| < \varepsilon$ for $n > N_1$. Since $\lim\limits_{n\to\infty} a_{2n+1} = L$, there

exists N_2 such that $|a_{2n+1} - L| < \varepsilon$ for $n > N_2$. Let $N = \max\{2N_1, 2N_2 + 1\}$ and let $n > N$. If n is even, then

$n = 2m$ where $m > N_1$, so $|a_n - L| = |a_{2m} - L| < \varepsilon$. If n is odd, then $n = 2m + 1$, where $m > N_2$, so

$|a_n - L| = |a_{2m+1} - L| < \varepsilon$. Therefore $\lim\limits_{n\to\infty} a_n = L$.

(b) $a_1 = 1$, $a_2 = 1 + \frac{1}{1+1} = \frac{3}{2} = 1.5$, $a_3 = 1 + \frac{1}{5/2} = \frac{7}{5} = 1.4$, $a_4 = 1 + \frac{1}{12/5} = \frac{17}{12} = 1.41\overline{6}$,

$a_5 = 1 + \frac{1}{29/12} = \frac{41}{29} \approx 1.413793$, $a_6 = 1 + \frac{1}{70/29} = \frac{99}{70} \approx 1.414286$, $a_7 = 1 + \frac{1}{169/70} = \frac{239}{169} \approx 1.414201$,

$a_8 = 1 + \frac{1}{408/169} = \frac{577}{408} \approx 1.414216$. Notice that $a_1 < a_3 < a_5 < a_7$ and $a_2 > a_4 > a_6 > a_8$. It appears that the

odd terms are increasing and the even terms are decreasing. Let's prove that $a_{2n-2} > a_{2n}$ and $a_{2n-1} < a_{2n+1}$ by

mathematical induction. Suppose that $a_{2k-2} > a_{2k}$. Then $1 + a_{2k-2} > 1 + a_{2k} \;\Rightarrow\; \dfrac{1}{1 + a_{2k-2}} < \dfrac{1}{1 + a_{2k}} \;\Rightarrow$

$1 + \dfrac{1}{1 + a_{2k-2}} < 1 + \dfrac{1}{1 + a_{2k}} \;\Rightarrow\; a_{2k-1} < a_{2k+1} \;\Rightarrow\; 1 + a_{2k-1} < 1 + a_{2k+1} \;\Rightarrow$

$\dfrac{1}{1 + a_{2k-1}} > \dfrac{1}{1 + a_{2k+1}} \;\Rightarrow\; 1 + \dfrac{1}{1 + a_{2k-1}} > 1 + \dfrac{1}{1 + a_{2k+1}} \;\Rightarrow\; a_{2k} > a_{2k+2}$. We have thus shown, by induction, that the odd terms are increasing and the even terms are decreasing. Also all terms lie between 1 and 2, so both $\{a_n\}$ and $\{b_n\}$ are bounded monotonic sequences and are therefore convergent by the Monotonic Sequence Theorem. Let $\lim\limits_{n\to\infty} a_{2n} = L$. Then $\lim\limits_{n\to\infty} a_{2n+2} = L$ also. We have

$$a_{n+2} = 1 + \cfrac{1}{1+1+1/(1+a_n)} = 1 + \cfrac{1}{(3+2a_n)/(1+a_n)} = \frac{4+3a_n}{3+2a_n}$$

so $a_{2n+2} = \dfrac{4+3a_{2n}}{3+2a_{2n}}$. Taking limits of both sides, we get $L = \dfrac{4+3L}{3+2L} \;\Rightarrow\; 3L + 2L^2 = 4 + 3L \;\Rightarrow\; L^2 = 2 \;\Rightarrow$

$L = \sqrt{2}$ [since $L > 0$]. Thus, $\lim\limits_{n\to\infty} a_{2n} = \sqrt{2}$. Similarly we find that $\lim\limits_{n\to\infty} a_{2n+1} = \sqrt{2}$. So, by part (a),

$\lim\limits_{n\to\infty} a_n = \sqrt{2}$.

81. (a) Suppose $\{p_n\}$ converges to p. Then $p_{n+1} = \dfrac{bp_n}{a+p_n} \;\Rightarrow\; \lim\limits_{n\to\infty} p_{n+1} = \dfrac{b \lim\limits_{n\to\infty} p_n}{a + \lim\limits_{n\to\infty} p_n} \;\Rightarrow\; p = \dfrac{bp}{a+p} \;\Rightarrow$

$p^2 + ap = bp \;\Rightarrow\; p(p+a-b) = 0 \;\Rightarrow\; p = 0 \text{ or } p = b - a.$

(b) $p_{n+1} = \dfrac{bp_n}{a+p_n} = \dfrac{\left(\dfrac{b}{a}\right)p_n}{1+\dfrac{p_n}{a}} < \left(\dfrac{b}{a}\right)p_n$ since $1 + \dfrac{p_n}{a} > 1$.

(c) By part (b), $p_1 < \left(\dfrac{b}{a}\right)p_0$, $p_2 < \left(\dfrac{b}{a}\right)p_1 < \left(\dfrac{b}{a}\right)^2 p_0$, $p_3 < \left(\dfrac{b}{a}\right)p_2 < \left(\dfrac{b}{a}\right)^3 p_0$, etc. In general, $p_n < \left(\dfrac{b}{a}\right)^n p_0$,

so $\lim\limits_{n\to\infty} p_n \le \lim\limits_{n\to\infty} \left(\dfrac{b}{a}\right)^n \cdot p_0 = 0$ since $b < a$. $\left[\text{By result 9, } \lim\limits_{n\to\infty} r^n = 0 \text{ if } -1 < r < 1. \text{ Here } r = \dfrac{b}{a} \in (0,1).\right]$

(d) Let $a < b$. We first show, by induction, that if $p_0 < b - a$, then $p_n < b - a$ and $p_{n+1} > p_n$.

For $n = 0$, we have $p_1 - p_0 = \dfrac{bp_0}{a+p_0} - p_0 = \dfrac{p_0(b-a-p_0)}{a+p_0} > 0$ since $p_0 < b - a$. So $p_1 > p_0$.

Now we suppose the assertion is true for $n = k$, that is, $p_k < b - a$ and $p_{k+1} > p_k$. Then

$b - a - p_{k+1} = b - a - \dfrac{bp_k}{a+p_k} = \dfrac{a(b-a) + bp_k - ap_k - bp_k}{a+p_k} = \dfrac{a(b-a-p_k)}{a+p_k} > 0$ because $p_k < b - a$. So

$p_{k+1} < b - a$. And $p_{k+2} - p_{k+1} = \dfrac{bp_{k+1}}{a+p_{k+1}} - p_{k+1} = \dfrac{p_{k+1}(b-a-p_{k+1})}{a+p_{k+1}} > 0$ since $p_{k+1} < b - a$. Therefore,

$p_{k+2} > p_{k+1}$. Thus, the assertion is true for $n = k + 1$. It is therefore true for all n by mathematical induction.

A similar proof by induction shows that if $p_0 > b - a$, then $p_n > b - a$ and $\{p_n\}$ is decreasing.

In either case the sequence $\{p_n\}$ is bounded and monotonic, so it is convergent by the Monotonic Sequence Theorem.

It then follows from part (a) that $\lim\limits_{n\to\infty} p_n = b - a$.

LABORATORY PROJECT Logistic Sequences

1. To write such a program in Maple it is best to calculate all the points first and then graph them. One possible sequence of commands [taking $p_0 = \frac{1}{2}$ and $k = 1.5$ for the difference equation] is

```
t:='t';p(0):=1/2;k:=1.5;

for j from 1 to 20  do p(j):=k*p(j-1)*(1-p(j-1)) od;

plot([seq([t,p(t)] t=0..20)],t=0..20,p=0..0.5,style=point);
```

In Mathematica, we can use the following program:

```
p[0]=1/2

k=1.5

p[j_]:=k*p[j-1]*(1-p[j-1])

P=Table[p[t],{t,20}]

ListPlot[P]
```

With $p_0 = \frac{1}{2}$ and $k = 1.5$:

n	p_n	n	p_n	n	p_n
0	0.5	7	0.3338465076	14	0.3333373303
1	0.375	8	0.3335895255	15	0.3333353318
2	0.3515625	9	0.3334613309	16	0.3333343326
3	0.3419494629	10	0.3333973076	17	0.3333338329
4	0.3375300416	11	0.3333653143	18	0.3333335831
5	0.3354052689	12	0.3333493223	19	0.3333334582
6	0.3343628617	13	0.3333413274	20	0.3333333958

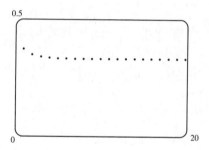

With $p_0 = \frac{1}{2}$ and $k = 2.5$:

n	p_n	n	p_n	n	p_n
0	0.5	7	0.6004164790	14	0.5999967417
1	0.625	8	0.5997913269	15	0.6000016291
2	0.5859375	9	0.6001042277	16	0.5999991854
3	0.6065368651	10	0.5999478590	17	0.6000004073
4	0.5966247409	11	0.6000260637	18	0.5999997964
5	0.6016591486	12	0.5999869664	19	0.6000001018
6	0.5991635437	13	0.6000065164	20	0.5999999491

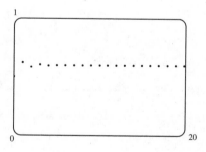

Both of these sequences seem to converge (the first to about $\frac{1}{3}$, the second to about 0.60).

With $p_0 = \frac{7}{8}$ and $k = 1.5$:

n	p_n	n	p_n	n	p_n
0	0.875	7	0.3239166554	14	0.3332554829
1	0.1640625	8	0.3284919837	15	0.3332943990
2	0.2057189941	9	0.3308775005	16	0.3333138639
3	0.2450980344	10	0.3320963702	17	0.3333235980
4	0.2775374819	11	0.3327125567	18	0.3333284655
5	0.3007656421	12	0.3330223670	19	0.3333308994
6	0.3154585059	13	0.3331777051	20	0.3333321164

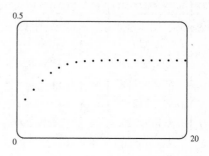

With $p_0 = \frac{7}{8}$ and $k = 2.5$:

n	p_n	n	p_n	n	p_n
0	0.875	7	0.6016572368	14	0.5999869815
1	0.2734375	8	0.5991645155	15	0.6000065088
2	0.4966735840	9	0.6004159972	16	0.5999967455
3	0.6249723374	10	0.5997915688	17	0.6000016272
4	0.5859547872	11	0.6001041070	18	0.5999991864
5	0.6065294364	12	0.5999479194	19	0.6000004068
6	0.5966286980	13	0.6000260335	20	0.5999997966

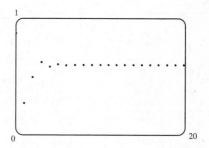

The limit of the sequence seems to depend on k, but not on p_0.

2. With $p_0 = \frac{7}{8}$ and $k = 3.2$:

n	p_n	n	p_n	n	p_n
0	0.875	7	0.5830728495	14	0.7990633827
1	0.35	8	0.7779164854	15	0.5137954979
2	0.728	9	0.5528397669	16	0.7993909896
3	0.6336512	10	0.7910654689	17	0.5131681132
4	0.7428395416	11	0.5288988570	18	0.7994451225
5	0.6112926626	12	0.7973275394	19	0.5130643795
6	0.7603646184	13	0.5171082698	20	0.7994538304

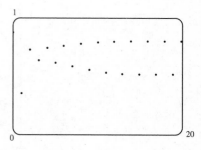

It seems that eventually the terms fluctuate between two values (about 0.5 and 0.8 in this case).

3. With $p_0 = \frac{7}{8}$ and $k = 3.42$:

n	p_n	n	p_n	n	p_n
0	0.875	7	0.4523028596	14	0.8442074951
1	0.3740625	8	0.8472194412	15	0.4498025048
2	0.8007579316	9	0.4426802161	16	0.8463823232
3	0.5456427596	10	0.8437633929	17	0.4446659586
4	0.8478752457	11	0.4508474156	18	0.8445284520
5	0.4411212220	12	0.8467373602	19	0.4490464985
6	0.8431438501	13	0.4438243545	20	0.8461207931

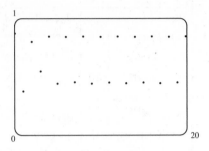

With $p_0 = \frac{7}{8}$ and $k = 3.45$:

n	p_n	n	p_n	n	p_n
0	0.875	7	0.4670259170	14	0.8403376122
1	0.37734375	8	0.8587488490	15	0.4628875685
2	0.8105962830	9	0.4184824586	16	0.8577482026
3	0.5296783241	10	0.8395743720	17	0.4209559716
4	0.8594612299	11	0.4646778983	18	0.8409445432
5	0.4167173034	12	0.8581956045	19	0.4614610237
6	0.8385707740	13	0.4198508858	20	0.8573758782

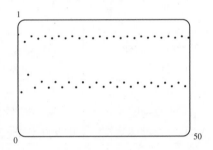

From the graphs above, it seems that for k between 3.4 and 3.5, the terms eventually fluctuate between four values. In the graph below, the pattern followed by the terms is $0.395, 0.832, 0.487, 0.869, 0.395, \ldots$. Note that even for $k = 3.42$ (as in the first graph), there are four distinct "branches"; even after 1000 terms, the first and third terms in the pattern differ by about 2×10^{-9}, while the first and fifth terms differ by only 2×10^{-10}. With $p_0 = \frac{7}{8}$ and $k = 3.48$:

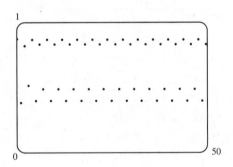

4.

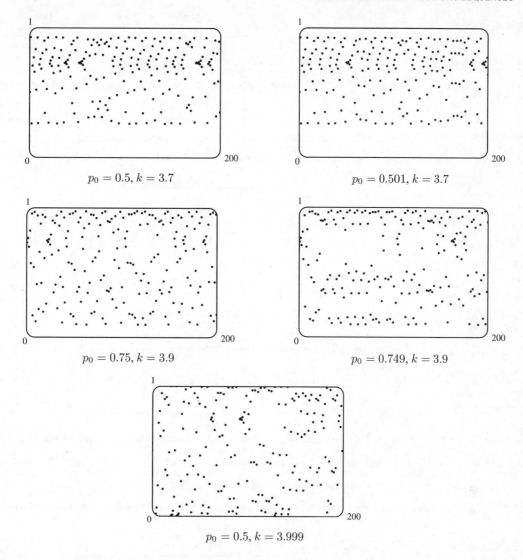

$p_0 = 0.5, k = 3.7$

$p_0 = 0.501, k = 3.7$

$p_0 = 0.75, k = 3.9$

$p_0 = 0.749, k = 3.9$

$p_0 = 0.5, k = 3.999$

From the graphs, it seems that if p_0 is changed by 0.001, the whole graph changes completely. (Note, however, that this might be partially due to accumulated round-off error in the CAS. These graphs were generated by Maple with 100-digit accuracy, and different degrees of accuracy give different graphs.) There seem to be some some fleeting patterns in these graphs, but on the whole they are certainly very chaotic. As k increases, the graph spreads out vertically, with more extreme values close to 0 or 1.

12.2 Series

1. (a) A sequence is an ordered list of numbers whereas a series is the *sum* of a list of numbers.

(b) A series is convergent if the sequence of partial sums is a convergent sequence. A series is divergent if it is not convergent.

2. $\sum_{n=1}^{\infty} a_n = 5$ means that by adding sufficiently many terms of the series we can get as close as we like to the number 5.

In other words, it means that $\lim_{n\to\infty} s_n = 5$, where s_n is the nth partial sum, that is, $\sum_{i=1}^{n} a_i$.

3.

n	s_n
1	-2.40000
2	-1.92000
3	-2.01600
4	-1.99680
5	-2.00064
6	-1.99987
7	-2.00003
8	-1.99999
9	-2.00000
10	-2.00000

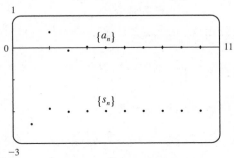

From the graph and the table, it seems that the series converges to -2. In fact, it is a geometric

series with $a = -2.4$ and $r = -\frac{1}{5}$, so its sum is $\sum_{n=1}^{\infty} \dfrac{12}{(-5)^n} = \dfrac{-2.4}{1-\left(-\frac{1}{5}\right)} = \dfrac{-2.4}{1.2} = -2.$

Note that the dot corresponding to $n = 1$ is part of both $\{a_n\}$ and $\{s_n\}$.

TI-86 Note: To graph $\{a_n\}$ and $\{s_n\}$, set your calculator to Param mode and DrawDot mode. (DrawDot is under

GRAPH, MORE, FORMT (F3).) Now under E(t) = make the assignments: xt1=t, yt1=12/(-5)^t, xt2=t,

yt2=sum seq(yt1,t,1,t,1). (sum and seq are under LIST, OPS (F5), MORE.) Under WIND use

1,10,1,0,10,1,-3,1,1 to obtain a graph similar to the one above. Then use TRACE (F4) to see the values.

4.

n	s_n
1	0.50000
2	1.90000
3	3.60000
4	5.42353
5	7.30814
6	9.22706
7	11.16706
8	13.12091
9	15.08432
10	17.05462

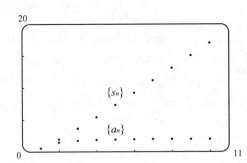

The series $\sum_{n=1}^{\infty} \dfrac{2n^2 - 1}{n^2 + 1}$ diverges, since its terms do not approach 0.

5.

n	s_n
1	1.55741
2	−0.62763
3	−0.77018
4	0.38764
5	−2.99287
6	−3.28388
7	−2.41243
8	−9.21214
9	−9.66446
10	−9.01610

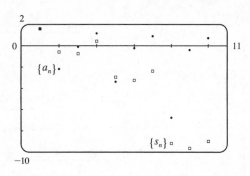

The series $\displaystyle\sum_{n=1}^{\infty} \tan n$ diverges, since its terms do not approach 0.

6.

n	s_n
1	1.00000
2	1.60000
3	1.96000
4	2.17600
5	2.30560
6	2.38336
7	2.43002
8	2.45801
9	2.47481
10	2.48488

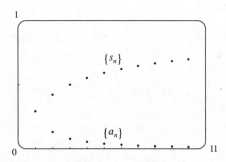

From the graph and the table, it seems that the series converges to 2.5. In fact, it is a geometric series with $a = 1$ and $r = 0.6$, so its sum is $\displaystyle\sum_{n=1}^{\infty} (0.6)^{n-1} = \frac{1}{1 - 0.6} = \frac{1}{2/5} = 2.5$.

7.

n	s_n
1	0.29289
2	0.42265
3	0.50000
4	0.55279
5	0.59175
6	0.62204
7	0.64645
8	0.66667
9	0.68377
10	0.69849

From the graph and the table, it seems that the series converges.

$$\sum_{n=1}^{k} \left(\frac{1}{\sqrt{n}} - \frac{1}{\sqrt{n+1}} \right) = \left(\frac{1}{\sqrt{1}} - \frac{1}{\sqrt{2}} \right) + \left(\frac{1}{\sqrt{2}} - \frac{1}{\sqrt{3}} \right) + \cdots + \left(\frac{1}{\sqrt{k}} - \frac{1}{\sqrt{k+1}} \right)$$
$$= 1 - \frac{1}{\sqrt{k+1}},$$

so $\displaystyle\sum_{n=1}^{\infty} \left(\frac{1}{\sqrt{n}} - \frac{1}{\sqrt{n+1}} \right) = \lim_{k \to \infty} \left(1 - \frac{1}{\sqrt{k+1}} \right) = 1.$

8.

n	s_n
2	0.12500
3	0.19167
4	0.23333
5	0.26190
6	0.28274
7	0.29861
8	0.31111
9	0.32121
10	0.32955
11	0.33654

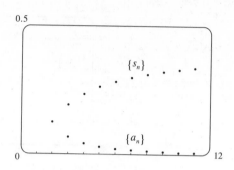

From the graph and the table, it seems that the series converges.

$$\frac{1}{n(n+2)} = \frac{1/2}{n} - \frac{1/2}{n+2} = \frac{1}{2}\left(\frac{1}{n} - \frac{1}{n+2}\right), \text{ so}$$

$$\sum_{n=2}^{k} \frac{1}{n(n+2)} = \frac{1}{2}\left(\frac{1}{2} - \frac{1}{4}\right) + \frac{1}{2}\left(\frac{1}{3} - \frac{1}{5}\right) + \frac{1}{2}\left(\frac{1}{4} - \frac{1}{6}\right) + \cdots + \frac{1}{2}\left(\frac{1}{k} - \frac{1}{k+2}\right)$$

$$= \frac{1}{2}\left(\frac{1}{2} + \frac{1}{3} - \frac{1}{k+1} - \frac{1}{k+2}\right).$$

As $k \to \infty$, this sum approaches $\frac{1}{2}\left(\frac{5}{6} - 0\right) = \frac{5}{12}$.

9. (a) $\lim\limits_{n\to\infty} a_n = \lim\limits_{n\to\infty} \frac{2n}{3n+1} = \frac{2}{3}$, so the *sequence* $\{a_n\}$ is convergent by (12.1.1) [ET (11.1.1)].

(b) Since $\lim\limits_{n\to\infty} a_n = \frac{2}{3} \neq 0$, the *series* $\sum\limits_{n=1}^{\infty} a_n$ is divergent by the Test for Divergence.

10. (a) Both $\sum\limits_{i=1}^{n} a_i$ and $\sum\limits_{j=1}^{n} a_j$ represent the sum of the first n terms of the sequence $\{a_n\}$, that is, the nth partial sum.

(b) $\sum\limits_{i=1}^{n} a_j = \underbrace{a_j + a_j + \cdots + a_j}_{n \text{ terms}} = na_j$, which, in general, is not the same as $\sum\limits_{i=1}^{n} a_i = a_1 + a_2 + \cdots + a_n$.

11. $3 + 2 + \frac{4}{3} + \frac{8}{9} + \cdots$ is a geometric series with first term $a = 3$ and common ratio $r = \frac{2}{3}$. Since $|r| = \frac{2}{3} < 1$, the series converges to $\frac{a}{1-r} = \frac{3}{1-2/3} = \frac{3}{1/3} = 9$.

12. $\frac{1}{8} - \frac{1}{4} + \frac{1}{2} - 1 + \cdots$ is a geometric series with ratio $r = -2$. Since $|r| = 2 > 1$, the series diverges.

13. $3 - 4 + \frac{16}{3} - \frac{64}{9} + \cdots$ is a geometric series with ratio $r = -\frac{4}{3}$. Since $|r| = \frac{4}{3} > 1$, the series diverges.

14. $1 + 0.4 + 0.16 + 0.064 + \cdots$ is a geometric series with ratio $r = 0.4 = \frac{2}{5}$. Since $|r| = \frac{2}{5} < 1$, the series converges to

$$\frac{a}{1-r} = \frac{1}{1-2/5} = \frac{5}{3}.$$

15. $\sum\limits_{n=1}^{\infty} 6(0.9)^{n-1}$ is a geometric series with first term $a = 6$ and ratio $r = 0.9$. Since $|r| = 0.9 < 1$, the series converges to

$$\frac{a}{1-r} = \frac{6}{1-0.9} = \frac{6}{0.1} = 60.$$

16. $\sum_{n=1}^{\infty} \frac{10^n}{(-9)^{n-1}} = \sum_{n=1}^{\infty} \frac{10(10)^{n-1}}{(-9)^{n-1}} = 10 \sum_{n=1}^{\infty} \left(-\frac{10}{9}\right)^{n-1}$. The latter series is geometric with $a = 10$ and ratio $r = -\frac{10}{9}$.

Since $|r| = \frac{10}{9} > 1$, the series diverges.

17. $\sum_{n=1}^{\infty} \frac{(-3)^{n-1}}{4^n} = \frac{1}{4} \sum_{n=1}^{\infty} \left(-\frac{3}{4}\right)^{n-1}$. The latter series is geometric with $a = 1$ and ratio $r = -\frac{3}{4}$. Since $|r| = \frac{3}{4} < 1$, it

converges to $\dfrac{1}{1-(-3/4)} = \frac{4}{7}$. Thus, the given series converges to $\left(\frac{1}{4}\right)\left(\frac{4}{7}\right) = \frac{1}{7}$.

18. $\sum_{n=0}^{\infty} \frac{1}{(\sqrt{2})^n}$ is a geometric series with ratio $r = \dfrac{1}{\sqrt{2}}$. Since $|r| = \dfrac{1}{\sqrt{2}} < 1$, the series converges. Its sum is

$$\frac{1}{1-1/\sqrt{2}} = \frac{\sqrt{2}}{\sqrt{2}-1} = \frac{\sqrt{2}}{\sqrt{2}-1} \cdot \frac{\sqrt{2}+1}{\sqrt{2}+1} = \sqrt{2}\left(\sqrt{2}+1\right) = 2 + \sqrt{2}.$$

19. $\sum_{n=0}^{\infty} \frac{\pi^n}{3^{n+1}} = \frac{1}{3} \sum_{n=0}^{\infty} \left(\frac{\pi}{3}\right)^n$ is a geometric series with ratio $r = \dfrac{\pi}{3}$. Since $|r| > 1$, the series diverges.

20. $\sum_{n=1}^{\infty} \frac{e^n}{3^{n-1}} = 3 \sum_{n=1}^{\infty} \left(\frac{e}{3}\right)^n$ is a geometric series with first term $3(e/3) = e$ and ratio $r = \dfrac{e}{3}$. Since $|r| < 1$, the series

converges. Its sum is $\dfrac{e}{1-e/3} = \dfrac{3e}{3-e}$.

21. $\sum_{n=1}^{\infty} \frac{1}{2n} = \frac{1}{2} \sum_{n=1}^{\infty} \frac{1}{n}$ diverges since each of its partial sums is $\frac{1}{2}$ times the corresponding partial sum of the harmonic series

$\sum_{n=1}^{\infty} \frac{1}{n}$, which diverges. $\left[$ If $\sum_{n=1}^{\infty} \frac{1}{2n}$ were to converge, then $\sum_{n=1}^{\infty} \frac{1}{n}$ would also have to converge by Theorem 8(i). $\right]$

In general, constant multiples of divergent series are divergent.

22. $\sum_{n=1}^{\infty} \frac{n+1}{2n-3}$ diverges by the Test for Divergence since $\lim_{n\to\infty} a_n = \lim_{n\to\infty} \frac{n+1}{2n-3} = \frac{1}{2} \neq 0$.

23. $\sum_{k=2}^{\infty} \frac{k^2}{k^2-1}$ diverges by the Test for Divergence since $\lim_{k\to\infty} a_k = \lim_{k\to\infty} \frac{k^2}{k^2-1} = 1 \neq 0$.

24. $\sum_{k=1}^{\infty} \frac{k(k+2)}{(k+3)^2}$ diverges by the Test for Divergence since $\lim_{k\to\infty} a_k = \lim_{k\to\infty} \frac{k(k+2)}{(k+3)^2} = \lim_{k\to\infty} \frac{1 \cdot (1+2/k)}{(1+3/k)^2} = 1 \neq 0$.

25. Converges.

$$\sum_{n=1}^{\infty} \frac{1+2^n}{3^n} = \sum_{n=1}^{\infty} \left(\frac{1}{3^n} + \frac{2^n}{3^n}\right) = \sum_{n=1}^{\infty} \left[\left(\frac{1}{3}\right)^n + \left(\frac{2}{3}\right)^n\right] \qquad \text{[sum of two convergent geometric series]}$$

$$= \frac{1/3}{1-1/3} + \frac{2/3}{1-2/3} = \frac{1}{2} + 2 = \frac{5}{2}$$

26. $\sum_{n=1}^{\infty} \frac{1+3^n}{2^n} = \sum_{n=1}^{\infty} \left(\frac{1}{2^n} + \frac{3^n}{2^n}\right) = \sum_{n=1}^{\infty} \left[\left(\frac{1}{2}\right)^n + \left(\frac{3}{2}\right)^n\right] = \sum_{n=1}^{\infty} \left(\frac{1}{2}\right)^n + \sum_{n=1}^{\infty} \left(\frac{3}{2}\right)^n$. The first series is a convergent

geometric series ($|r| = \frac{1}{2} < 1$), but the second series is a divergent geometric series ($|r| = \frac{3}{2} \geq 1$), so the original series

is divergent.

27. $\sum\limits_{n=1}^{\infty} \sqrt[n]{2} = 2 + \sqrt{2} + \sqrt[3]{2} + \sqrt[4]{2} + \cdots$ diverges by the Test for Divergence since

$$\lim_{n\to\infty} a_n = \lim_{n\to\infty} \sqrt[n]{2} = \lim_{n\to\infty} 2^{1/n} = 2^0 = 1 \neq 0.$$

28. $\sum\limits_{n=1}^{\infty} \left[(0.8)^{n-1} - (0.3)^n\right] = \sum\limits_{n=1}^{\infty} (0.8)^{n-1} - \sum\limits_{n=1}^{\infty} (0.3)^n$ [difference of two convergent geometric series]

$$= \frac{1}{1 - 0.8} - \frac{0.3}{1 - 0.3} = 5 - \frac{3}{7} = \frac{32}{7}$$

29. $\sum\limits_{n=1}^{\infty} \ln\left(\dfrac{n^2 + 1}{2n^2 + 1}\right)$ diverges by the Test for Divergence since

$$\lim_{n\to\infty} a_n = \lim_{n\to\infty} \ln\left(\frac{n^2 + 1}{2n^2 + 1}\right) = \ln\left(\lim_{n\to\infty} \frac{n^2 + 1}{2n^2 + 1}\right) = \ln\tfrac{1}{2} \neq 0.$$

30. $\sum\limits_{k=1}^{\infty} (\cos 1)^k$ is a geometric series with ratio $r = \cos 1 \approx 0.540302$. It converges because $|r| < 1$. Its sum is

$$\frac{\cos 1}{1 - \cos 1} \approx 1.175343.$$

31. $\sum\limits_{n=1}^{\infty} \arctan n$ diverges by the Test for Divergence since $\lim\limits_{n\to\infty} a_n = \lim\limits_{n\to\infty} \arctan n = \frac{\pi}{2} \neq 0.$

32. $\sum\limits_{n=1}^{\infty} \left(\dfrac{3}{5^n} + \dfrac{2}{n}\right)$ diverges because $\sum\limits_{n=1}^{\infty} \dfrac{2}{n} = 2 \sum\limits_{n=1}^{\infty} \dfrac{1}{n}$ diverges. (If it converged, then $\dfrac{1}{2} \cdot 2 \sum\limits_{n=1}^{\infty} \dfrac{1}{n}$ would also converge by

Theorem 8(i), but we know from Example 7 that the harmonic series $\sum\limits_{n=1}^{\infty} \dfrac{1}{n}$ diverges.) If the given series converges, then the

difference $\sum\limits_{n=1}^{\infty} \left(\dfrac{3}{5^n} + \dfrac{2}{n}\right) - \sum\limits_{n=1}^{\infty} \dfrac{3}{5^n}$ must converge (since $\sum\limits_{n=1}^{\infty} \dfrac{3}{5^n}$ is a convergent geometric series) and equal $\sum\limits_{n=1}^{\infty} \dfrac{2}{n}$, but we

have just seen that $\sum\limits_{n=1}^{\infty} \dfrac{2}{n}$ diverges, so the given series must also diverge.

33. $\sum\limits_{n=1}^{\infty} \dfrac{1}{e^n} = \sum\limits_{n=1}^{\infty} \left(\dfrac{1}{e}\right)^n$ is a geometric series with first term $a = \dfrac{1}{e}$ and ratio $r = \dfrac{1}{e}$. Since $|r| = \dfrac{1}{e} < 1$, the series converges

to $\dfrac{1/e}{1 - 1/e} = \dfrac{1/e}{1 - 1/e} \cdot \dfrac{e}{e} = \dfrac{1}{e - 1}$. By Example 6, $\sum\limits_{n=1}^{\infty} \dfrac{1}{n(n + 1)} = 1$. Thus, by Theorem 8(ii),

$$\sum\limits_{n=1}^{\infty} \left(\frac{1}{e^n} + \frac{1}{n(n + 1)}\right) = \sum\limits_{n=1}^{\infty} \frac{1}{e^n} + \sum\limits_{n=1}^{\infty} \frac{1}{n(n + 1)} = \frac{1}{e - 1} + 1 = \frac{1}{e - 1} + \frac{e - 1}{e - 1} = \frac{e}{e - 1}.$$

34. $\sum\limits_{n=1}^{\infty} \dfrac{e^n}{n^2}$ diverges by the Test for Divergence since $\lim\limits_{n\to\infty} a_n = \lim\limits_{n\to\infty} \dfrac{e^n}{n^2} = \lim\limits_{x\to\infty} \dfrac{e^x}{x^2} \overset{\text{H}}{=} \lim\limits_{x\to\infty} \dfrac{e^x}{2x} \overset{\text{H}}{=} \lim\limits_{x\to\infty} \dfrac{e^x}{2} = \infty \neq 0.$

35. Using partial fractions, the partial sums of the series $\sum\limits_{n=2}^{\infty} \dfrac{2}{n^2 - 1}$ are

$$s_n = \sum_{i=2}^{n} \frac{2}{(i - 1)(i + 1)} = \sum_{i=2}^{n} \left(\frac{1}{i - 1} - \frac{1}{i + 1}\right)$$

$$= \left(1 - \frac{1}{3}\right) + \left(\frac{1}{2} - \frac{1}{4}\right) + \left(\frac{1}{3} - \frac{1}{5}\right) + \cdots + \left(\frac{1}{n - 3} - \frac{1}{n - 1}\right) + \left(\frac{1}{n - 2} - \frac{1}{n}\right)$$

This sum is a telescoping series and $s_n = 1 + \frac{1}{2} - \frac{1}{n-1} - \frac{1}{n}$.

Thus, $\displaystyle\sum_{n=2}^{\infty} \frac{2}{n^2-1} = \lim_{n\to\infty} s_n = \lim_{n\to\infty}\left(1 + \frac{1}{2} - \frac{1}{n-1} - \frac{1}{n}\right) = \frac{3}{2}$.

36. For the series $\displaystyle\sum_{n=1}^{\infty} \frac{2}{n^2+4n+3}$, $s_n = \displaystyle\sum_{i=1}^{n}\frac{2}{i^2+4i+3} = \sum_{i=1}^{n}\left(\frac{1}{i+1} - \frac{1}{i+3}\right)$ [using partial fractions]. The latter sum is

$\left(\frac{1}{2} - \frac{1}{4}\right) + \left(\frac{1}{3} - \frac{1}{5}\right) + \left(\frac{1}{4} - \frac{1}{6}\right) + \left(\frac{1}{5} - \frac{1}{7}\right) + \cdots + \left(\frac{1}{n} - \frac{1}{n+2}\right) + \left(\frac{1}{n+1} - \frac{1}{n+3}\right) = \frac{1}{2} + \frac{1}{3} - \frac{1}{n+2} - \frac{1}{n+3}$

[telescoping series]

Thus, $\displaystyle\sum_{n=1}^{\infty} \frac{2}{n^2+4n+3} = \lim_{n\to\infty} s_n = \lim_{n\to\infty}\left(\frac{1}{2} + \frac{1}{3} - \frac{1}{n+2} - \frac{1}{n+3}\right) = \frac{1}{2} + \frac{1}{3} = \frac{5}{6}$. Converges

37. For the series $\displaystyle\sum_{n=1}^{\infty} \frac{3}{n(n+3)}$, $s_n = \displaystyle\sum_{i=1}^{n}\frac{3}{i(i+3)} = \sum_{i=1}^{n}\left(\frac{1}{i} - \frac{1}{i+3}\right)$ [using partial fractions]. The latter sum is

$\left(1 - \frac{1}{4}\right) + \left(\frac{1}{2} - \frac{1}{5}\right) + \left(\frac{1}{3} - \frac{1}{6}\right) + \left(\frac{1}{4} - \frac{1}{7}\right) + \cdots + \left(\frac{1}{n-3} - \frac{1}{n}\right) + \left(\frac{1}{n-2} - \frac{1}{n+1}\right) + \left(\frac{1}{n-1} - \frac{1}{n+2}\right) + \left(\frac{1}{n} - \frac{1}{n+3}\right)$

$= 1 + \frac{1}{2} + \frac{1}{3} - \frac{1}{n+1} - \frac{1}{n+2} - \frac{1}{n+3}$ [telescoping series]

Thus, $\displaystyle\sum_{n=1}^{\infty} \frac{3}{n(n+3)} = \lim_{n\to\infty} s_n = \lim_{n\to\infty}\left(1 + \frac{1}{2} + \frac{1}{3} - \frac{1}{n+1} - \frac{1}{n+2} - \frac{1}{n+3}\right) = 1 + \frac{1}{2} + \frac{1}{3} = \frac{11}{6}$. Converges

38. For the series $\displaystyle\sum_{n=1}^{\infty} \ln\frac{n}{n+1}$,

$s_n = (\ln 1 - \ln 2) + (\ln 2 - \ln 3) + (\ln 3 - \ln 4) + \cdots + [\ln n - \ln(n+1)] = \ln 1 - \ln(n+1) = -\ln(n+1)$

[telescoping series]

Thus, $\displaystyle\lim_{n\to\infty} s_n = -\infty$, so the series is divergent.

39. For the series $\displaystyle\sum_{n=1}^{\infty}\left(e^{1/n} - e^{1/(n+1)}\right)$,

$s_n = \displaystyle\sum_{i=1}^{n}\left(e^{1/i} - e^{1/(i+1)}\right) = \left(e^1 - e^{1/2}\right) + \left(e^{1/2} - e^{1/3}\right) + \cdots + \left(e^{1/n} - e^{1/(n+1)}\right) = e - e^{1/(n+1)}$

[telescoping series]

Thus, $\displaystyle\sum_{n=1}^{\infty}\left(e^{1/n} - e^{1/(n+1)}\right) = \lim_{n\to\infty} s_n = \lim_{n\to\infty}\left(e - e^{1/(n+1)}\right) = e - e^0 = e - 1$. Converges

40. For the series $\displaystyle\sum_{n=1}^{\infty}\left(\cos\frac{1}{n^2} - \cos\frac{1}{(n+1)^2}\right)$,

$s_n = \displaystyle\sum_{i=1}^{n}\left(\cos\frac{1}{i^2} - \cos\frac{1}{(i+1)^2}\right) = \left(\cos 1 - \cos\frac{1}{4}\right) + \left(\cos\frac{1}{4} - \cos\frac{1}{9}\right) + \cdots + \left(\cos\frac{1}{n^2} - \cos\frac{1}{(n+1)^2}\right)$

$= \cos 1 - \cos\frac{1}{(n+1)^2}$ [telescoping series]

Thus, $\displaystyle\sum_{n=1}^{\infty}\left(\cos\frac{1}{n^2} - \cos\frac{1}{(n+1)^2}\right) = \lim_{n\to\infty} s_n = \lim_{n\to\infty}\left(\cos 1 - \cos\frac{1}{(n+1)^2}\right) = \cos 1 - \cos 0 = \cos 1 - 1$.
Converges

41. $0.\overline{2} = \dfrac{2}{10} + \dfrac{2}{10^2} + \cdots$ is a geometric series with $a = \dfrac{2}{10}$ and $r = \dfrac{1}{10}$. It converges to $\dfrac{a}{1-r} = \dfrac{2/10}{1-1/10} = \dfrac{2}{9}$.

42. $0.\overline{73} = \dfrac{73}{10^2} + \dfrac{73}{10^4} + \cdots = \dfrac{73/10^2}{1-1/10^2} = \dfrac{73/100}{99/100} = \dfrac{73}{99}$

43. $3.\overline{417} = 3 + \dfrac{417}{10^3} + \dfrac{417}{10^6} + \cdots$. Now $\dfrac{417}{10^3} + \dfrac{417}{10^6} + \cdots$ is a geometric series with $a = \dfrac{417}{10^3}$ and $r = \dfrac{1}{10^3}$.

It converges to $\dfrac{a}{1-r} = \dfrac{417/10^3}{1 - 1/10^3} = \dfrac{417/10^3}{999/10^3} = \dfrac{417}{999}$. Thus, $3.\overline{417} = 3 + \dfrac{417}{999} = \dfrac{3414}{999} = \dfrac{1138}{333}$.

44. $6.2\overline{54} = 6.2 + \dfrac{54}{10^3} + \dfrac{54}{10^5} + \cdots = 6.2 + \dfrac{54/10^3}{1 - 1/10^2} = \dfrac{62}{10} + \dfrac{54}{990} = \dfrac{6192}{990} = \dfrac{344}{55}$

45. $1.53\overline{42} = 1.53 + \dfrac{42}{10^4} + \dfrac{42}{10^6} + \cdots$. Now $\dfrac{42}{10^4} + \dfrac{42}{10^6} + \cdots$ is a geometric series with $a = \dfrac{42}{10^4}$ and $r = \dfrac{1}{10^2}$.

It converges to $\dfrac{a}{1-r} = \dfrac{42/10^4}{1 - 1/10^2} = \dfrac{42/10^4}{99/10^2} = \dfrac{42}{9900}$.

Thus, $1.53\overline{42} = 1.53 + \dfrac{42}{9900} = \dfrac{153}{100} + \dfrac{42}{9900} = \dfrac{15{,}147}{9900} + \dfrac{42}{9900} = \dfrac{15{,}189}{9900}$ or $\dfrac{5063}{3300}$.

46. $7.\overline{12345} = 7 + \dfrac{12{,}345}{10^5} + \dfrac{12{,}345}{10^{10}} + \cdots$. Now $\dfrac{12{,}345}{10^5} + \dfrac{12{,}345}{10^{10}} + \cdots$ is a geometric series with $a = \dfrac{12{,}345}{10^5}$ and $r = \dfrac{1}{10^5}$.

It converges to $\dfrac{a}{1-r} = \dfrac{12{,}345/10^5}{1 - 1/10^5} = \dfrac{12{,}345/10^5}{99{,}999/10^5} = \dfrac{12{,}345}{99{,}999}$.

Thus, $7.\overline{12345} = 7 + \dfrac{12{,}345}{99{,}999} = \dfrac{699{,}993}{99{,}999} + \dfrac{12{,}345}{99{,}999} = \dfrac{712{,}338}{99{,}999}$ or $\dfrac{237{,}446}{33{,}333}$.

47. $\displaystyle\sum_{n=1}^{\infty} \dfrac{x^n}{3^n} = \sum_{n=1}^{\infty} \left(\dfrac{x}{3}\right)^n$ is a geometric series with $r = \dfrac{x}{3}$, so the series converges $\Leftrightarrow$ $|r| < 1$ $\Leftrightarrow$ $\dfrac{|x|}{3} < 1$ $\Leftrightarrow$ $|x| < 3$;

that is, $-3 < x < 3$. In that case, the sum of the series is $\dfrac{a}{1-r} = \dfrac{x/3}{1 - x/3} = \dfrac{x/3}{1 - x/3} \cdot \dfrac{3}{3} = \dfrac{x}{3 - x}$.

48. $\displaystyle\sum_{n=1}^{\infty} (x-4)^n$ is a geometric series with $r = x - 4$, so the series converges $\Leftrightarrow$ $|r| < 1$ $\Leftrightarrow$ $|x - 4| < 1$ $\Leftrightarrow$

$3 < x < 5$. In that case, the sum of the series is $\dfrac{x-4}{1 - (x-4)} = \dfrac{x-4}{5 - x}$.

49. $\displaystyle\sum_{n=0}^{\infty} 4^n x^n = \sum_{n=0}^{\infty} (4x)^n$ is a geometric series with $r = 4x$, so the series converges $\Leftrightarrow$ $|r| < 1$ $\Leftrightarrow$ $4\,|x| < 1$ $\Leftrightarrow$

$|x| < \frac{1}{4}$. In that case, the sum of the series is $\dfrac{1}{1 - 4x}$.

50. $\displaystyle\sum_{n=0}^{\infty} \dfrac{(x+3)^n}{2^n}$ is a geometric series with $r = \dfrac{x+3}{2}$, so the series converges $\Leftrightarrow$ $|r| < 1$ $\Leftrightarrow$ $\dfrac{|x+3|}{2} < 1$ $\Leftrightarrow$

$|x + 3| < 2$ $\Leftrightarrow$ $-5 < x < -1$. For these values of x, the sum of the series is $\dfrac{1}{1 - (x+3)/2} = \dfrac{2}{2 - (x+3)} = -\dfrac{2}{x+1}$.

51. $\displaystyle\sum_{n=0}^{\infty} \dfrac{\cos^n x}{2^n}$ is a geometric series with first term 1 and ratio $r = \dfrac{\cos x}{2}$, so it converges $\Leftrightarrow$ $|r| < 1$. But $|r| = \dfrac{|\cos x|}{2} \leq \dfrac{1}{2}$

for all x. Thus, the series converges for all real values of x and the sum of the series is $\dfrac{1}{1 - (\cos x)/2} = \dfrac{2}{2 - \cos x}$.

52. Because $\dfrac{1}{n} \to 0$ and ln is continuous, we have $\displaystyle\lim_{n\to\infty} \ln\left(1 + \dfrac{1}{n}\right) = \ln 1 = 0$.

We now show that the series $\displaystyle\sum_{n=1}^{\infty} \ln\left(1 + \dfrac{1}{n}\right) = \sum_{n=1}^{\infty} \ln\left(\dfrac{n+1}{n}\right) = \sum_{n=1}^{\infty} [\ln(n+1) - \ln n]$ diverges.

$s_n = (\ln 2 - \ln 1) + (\ln 3 - \ln 2) + \cdots + (\ln(n+1) - \ln n) = \ln(n+1) - \ln 1 = \ln(n+1)$.

As $n \to \infty$, $s_n = \ln(n+1) \to \infty$, so the series diverges.

53. After defining f, We use `convert(f,parfrac);` in Maple, `Apart` in Mathematica, or `Expand Rational` and

`Simplify` in Derive to find that the general term is $\dfrac{3n^2 + 3n + 1}{(n^2 + n)^3} = \dfrac{1}{n^3} - \dfrac{1}{(n+1)^3}$. So the nth partial sum is

$$s_n = \sum_{k=1}^{n}\left(\frac{1}{k^3} - \frac{1}{(k+1)^3}\right) = \left(1 - \frac{1}{2^3}\right) + \left(\frac{1}{2^3} - \frac{1}{3^3}\right) + \cdots + \left(\frac{1}{n^3} - \frac{1}{(n+1)^3}\right) = 1 - \frac{1}{(n+1)^3}$$

The series converges to $\lim\limits_{n\to\infty} s_n = 1$. This can be confirmed by directly computing the sum using `sum(f,1..infinity);`

(in Maple), `Sum[f,{n,1,Infinity}]` (in Mathematica), or `Calculus Sum` (from 1 to ∞) and `Simplify` (in Derive).

54. See Exercise 53 for specific CAS commands. $\dfrac{1}{n^3 - n} = \dfrac{1/2}{n-1} - \dfrac{1}{n} + \dfrac{1/2}{n+1}$. So the nth partial sum is

$$s_n = \sum_{k=2}^{n}\left(\frac{1/2}{k-1} - \frac{1}{k} + \frac{1/2}{k+1}\right)$$

$$= \left(\frac{1/2}{1} - \frac{1}{2} + \frac{1/2}{3}\right) + \left(\frac{1/2}{2} - \frac{1}{3} + \frac{1/2}{4}\right) + \left(\frac{1/2}{3} - \frac{1}{4} + \frac{1/2}{5}\right)$$

$$\quad + \left(\frac{1/2}{4} - \frac{1}{5} + \frac{1/2}{6}\right) + \cdots + \left(\frac{1/2}{n-2} - \frac{1}{n-1} + \frac{1/2}{n}\right) + \left(\frac{1/2}{n-1} - \frac{1}{n} + \frac{1/2}{n+1}\right)$$

$$= \frac{1/2}{1} + \left(-\frac{1}{2} + \frac{1/2}{2}\right) + \left(\frac{1/2}{3} - \frac{1}{3} + \frac{1/2}{3}\right) + \left(\frac{1/2}{4} - \frac{1}{4} + \frac{1/2}{4}\right) + \cdots + \left(\frac{1/2}{n} - \frac{1}{n} + \frac{1/2}{n+1}\right)$$

$$= \frac{1}{2} + \left(-\frac{1}{4}\right) + 0 + 0 + \cdots + \frac{1/2}{n} - \frac{1}{n} + \frac{1/2}{n+1}$$

The series converges to $\lim\limits_{n\to\infty} s_n = \frac{1}{4}$.

55. For $n = 1$, $a_1 = 0$ since $s_1 = 0$. For $n > 1$,

$$a_n = s_n - s_{n-1} = \frac{n-1}{n+1} - \frac{(n-1)-1}{(n-1)+1} = \frac{(n-1)n - (n+1)(n-2)}{(n+1)n} = \frac{2}{n(n+1)}$$

Also, $\displaystyle\sum_{n=1}^{\infty} a_n = \lim\limits_{n\to\infty} s_n = \lim\limits_{n\to\infty}\dfrac{1 - 1/n}{1 + 1/n} = 1$.

56. $a_1 = s_1 = 3 - \frac{1}{2} = \frac{5}{2}$. For $n \neq 1$,

$$a_n = s_n - s_{n-1} = \left(3 - n2^{-n}\right) - \left[3 - (n-1)2^{-(n-1)}\right] = -\frac{n}{2^n} + \frac{n-1}{2^{n-1}}\cdot\frac{2}{2} = \frac{2(n-1)}{2^n} - \frac{n}{2^n} = \frac{n-2}{2^n}$$

Also, $\displaystyle\sum_{n=1}^{\infty} a_n = \lim\limits_{n\to\infty} s_n = \lim\limits_{n\to\infty}\left(3 - \dfrac{n}{2^n}\right) = 3$ because $\lim\limits_{x\to\infty}\dfrac{x}{2^x} \overset{\text{H}}{=} \lim\limits_{x\to\infty}\dfrac{1}{2^x \ln 2} = 0$.

57. (a) The first step in the chain occurs when the local government spends D dollars. The people who receive it spend a fraction c

of those D dollars, that is, Dc dollars. Those who receive the Dc dollars spend a fraction c of it, that is, Dc^2 dollars.

Continuing in this way, we see that the total spending after n transactions is

$$S_n = D + Dc + Dc^2 + \cdots + Dc^{n-1} = \frac{D(1 - c^n)}{1 - c} \text{ by (3).}$$

(b) $\lim\limits_{n\to\infty} S_n = \lim\limits_{n\to\infty}\dfrac{D(1 - c^n)}{1 - c} = \dfrac{D}{1 - c}\lim\limits_{n\to\infty}(1 - c^n) = \dfrac{D}{1 - c}\ \left[\text{since } 0 < c < 1 \quad \Rightarrow \quad \lim\limits_{n\to\infty} c^n = 0\right]$

$$= \frac{D}{s}\ [\text{since } c + s = 1] = kD\ [\text{since } k = 1/s]$$

If $c = 0.8$, then $s = 1 - c = 0.2$ and the multiplier is $k = 1/s = 5$.

58. (a) Initially, the ball falls a distance H, then rebounds a distance rH, falls rH, rebounds r^2H, falls r^2H, etc. The total distance it travels is

$$H + 2rH + 2r^2H + 2r^3H + \cdots = H\left(1 + 2r + 2r^2 + 2r^3 + \cdots\right) = H\left[1 + 2r\left(1 + r + r^2 + \cdots\right)\right]$$

$$= H\left[1 + 2r\left(\frac{1}{1-r}\right)\right] = H\left(\frac{1+r}{1-r}\right) \text{ meters}$$

(b) From Example 3 in Section 2.1, we know that a ball falls $\frac{1}{2}gt^2$ meters in t seconds, where g is the gravitational acceleration. Thus, a ball falls h meters in $t = \sqrt{2h/g}$ seconds. The total travel time in seconds is

$$\sqrt{\frac{2H}{g}} + 2\sqrt{\frac{2H}{g}}r + 2\sqrt{\frac{2H}{g}}r^2 + 2\sqrt{\frac{2H}{g}}r^3 + \cdots = \sqrt{\frac{2H}{g}}\left[1 + 2\sqrt{r} + 2\sqrt{r}^2 + 2\sqrt{r}^3 + \cdots\right]$$

$$= \sqrt{\frac{2H}{g}}\left(1 + 2\sqrt{r}\left[1 + \sqrt{r} + \sqrt{r}^2 + \cdots\right]\right)$$

$$= \sqrt{\frac{2H}{g}}\left[1 + 2\sqrt{r}\left(\frac{1}{1-\sqrt{r}}\right)\right] = \sqrt{\frac{2H}{g}}\frac{1+\sqrt{r}}{1-\sqrt{r}}$$

(c) It will help to make a chart of the time for each descent and each rebound of the ball, together with the velocity just before and just after each bounce. Recall that the time in seconds needed to fall h meters is $\sqrt{2h/g}$. The ball hits the ground with velocity $-g\sqrt{2h/g} = -\sqrt{2hg}$ (taking the upward direction to be positive) and rebounds with velocity $kg\sqrt{2h/g} = k\sqrt{2hg}$, taking time $k\sqrt{2h/g}$ to reach the top of its bounce, where its velocity is 0. At that point, its height is k^2h. All these results follow from the formulas for vertical motion with gravitational acceleration $-g$:

$$\frac{d^2y}{dt^2} = -g \quad \Rightarrow \quad v = \frac{dy}{dt} = v_0 - gt \quad \Rightarrow \quad y = y_0 + v_0t - \frac{1}{2}gt^2.$$

number of descent	time of descent	speed before bounce	speed after bounce	time of ascent	peak height
1	$\sqrt{2H/g}$	$\sqrt{2Hg}$	$k\sqrt{2Hg}$	$k\sqrt{2H/g}$	k^2H
2	$\sqrt{2k^2H/g}$	$\sqrt{2k^2Hg}$	$k\sqrt{2k^2Hg}$	$k\sqrt{2k^2H/g}$	k^4H
3	$\sqrt{2k^4H/g}$	$\sqrt{2k^4Hg}$	$k\sqrt{2k^4Hg}$	$k\sqrt{2k^4H/g}$	k^6H
$\cdots$	$\cdots$	$\cdots$	$\cdots$	$\cdots$	$\cdots$

The total travel time in seconds is

$$\sqrt{\frac{2H}{g}} + k\sqrt{\frac{2H}{g}} + k\sqrt{\frac{2H}{g}} + k^2\sqrt{\frac{2H}{g}} + k^2\sqrt{\frac{2H}{g}} + \cdots = \sqrt{\frac{2H}{g}}\left(1 + 2k + 2k^2 + 2k^3 + \cdots\right)$$

$$= \sqrt{\frac{2H}{g}}\left[1 + 2k(1 + k + k^2 + \cdots)\right]$$

$$= \sqrt{\frac{2H}{g}}\left[1 + 2k\left(\frac{1}{1-k}\right)\right] = \sqrt{\frac{2H}{g}}\frac{1+k}{1-k}$$

Another method: We could use part (b). At the top of the bounce, the height is $k^2h = rh$, so $\sqrt{r} = k$ and the result follows from part (b).

59. $\sum_{n=2}^{\infty} (1+c)^{-n}$ is a geometric series with $a = (1+c)^{-2}$ and $r = (1+c)^{-1}$, so the series converges when

$\left|(1+c)^{-1}\right| < 1 \iff |1+c| > 1 \iff 1+c > 1$ or $1+c < -1 \iff c > 0$ or $c < -2$. We calculate the sum of the

series and set it equal to 2: $\dfrac{(1+c)^{-2}}{1-(1+c)^{-1}} = 2 \iff \left(\dfrac{1}{1+c}\right)^2 = 2 - 2\left(\dfrac{1}{1+c}\right) \iff 1 = 2(1+c)^2 - 2(1+c) \iff$

$2c^2 + 2c - 1 = 0 \iff c = \dfrac{-2 \pm \sqrt{12}}{4} = \dfrac{\pm\sqrt{3}-1}{2}$. However, the negative root is inadmissible because $-2 < \dfrac{-\sqrt{3}-1}{2} < 0$.

So $c = \dfrac{\sqrt{3}-1}{2}$.

60. $\sum_{n=0}^{\infty} e^{nc} = \sum_{n=0}^{\infty} (e^c)^n$ is a geometric series with $a = (e^c)^0 = 1$ and $r = e^c$. If $e^c < 1$, it has sum $\dfrac{1}{1-e^c}$, so $\dfrac{1}{1-e^c} = 10 \implies$

$\dfrac{1}{10} = 1 - e^c \implies e^c = \dfrac{9}{10} \implies c = \ln\dfrac{9}{10}$.

61. $e^{s_n} = e^{1 + \frac{1}{2} + \frac{1}{3} + \cdots + \frac{1}{n}} = e^1 e^{1/2} e^{1/3} \cdots e^{1/n} > (1+1)\left(1+\tfrac{1}{2}\right)\left(1+\tfrac{1}{3}\right)\cdots\left(1+\tfrac{1}{n}\right)$ $[e^x > 1 + x]$

$= \dfrac{2}{1}\dfrac{3}{2}\dfrac{4}{3}\cdots\dfrac{n+1}{n} = n+1$

Thus, $e^{s_n} > n+1$ and $\lim_{n\to\infty} e^{s_n} = \infty$. Since $\{s_n\}$ is increasing, $\lim_{n\to\infty} s_n = \infty$, implying that the harmonic series is

divergent.

62. The area between $y = x^{n-1}$ and $y = x^n$ for $0 \le x \le 1$ is

$$\int_0^1 (x^{n-1} - x^n)\,dx = \left[\dfrac{x^n}{n} - \dfrac{x^{n+1}}{n+1}\right]_0^1 = \dfrac{1}{n} - \dfrac{1}{n+1}$$

$$= \dfrac{(n+1)-n}{n(n+1)} = \dfrac{1}{n(n+1)}$$

We can see from the diagram that as $n \to \infty$, the sum of the areas

between the successive curves approaches the area of the unit square, that

is, 1. So $\sum_{n=1}^{\infty} \dfrac{1}{n(n+1)} = 1$.

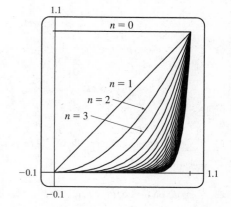

63. Let d_n be the diameter of C_n. We draw lines from the centers of the C_i to

the center of D (or C), and using the Pythagorean Theorem, we can write

$1^2 + \left(1 - \tfrac{1}{2}d_1\right)^2 = \left(1 + \tfrac{1}{2}d_1\right)^2 \iff$

$1 = \left(1 + \tfrac{1}{2}d_1\right)^2 - \left(1 - \tfrac{1}{2}d_1\right)^2 = 2d_1$ [difference of squares] $\implies d_1 = \tfrac{1}{2}$.

Similarly,

$1 = \left(1 + \tfrac{1}{2}d_2\right)^2 - \left(1 - d_1 - \tfrac{1}{2}d_2\right)^2 = 2d_2 + 2d_1 - d_1^2 - d_1 d_2$

$= (2 - d_1)(d_1 + d_2) \iff$

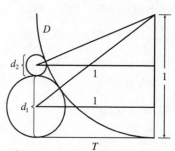

$d_2 = \dfrac{1}{2-d_1} - d_1 = \dfrac{(1-d_1)^2}{2-d_1}$, $1 = \left(1 + \tfrac{1}{2}d_3\right)^2 - \left(1 - d_1 - d_2 - \tfrac{1}{2}d_3\right)^2 \iff d_3 = \dfrac{[1-(d_1+d_2)]^2}{2-(d_1+d_2)}$, and in general,

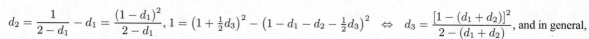

$d_{n+1} = \dfrac{\left(1 - \sum_{i=1}^{n} d_i\right)^2}{2 - \sum_{i=1}^{n} d_i}$. If we actually calculate d_2 and d_3 from the formulas above, we find that they are $\dfrac{1}{6} = \dfrac{1}{2 \cdot 3}$ and

$\dfrac{1}{12} = \dfrac{1}{3 \cdot 4}$ respectively, so we suspect that in general, $d_n = \dfrac{1}{n(n+1)}$. To prove this, we use induction: Assume that for all

$k \leq n$, $d_k = \dfrac{1}{k(k+1)} = \dfrac{1}{k} - \dfrac{1}{k+1}$. Then $\sum_{i=1}^{n} d_i = 1 - \dfrac{1}{n+1} = \dfrac{n}{n+1}$ [telescoping sum]. Substituting this into our

formula for d_{n+1}, we get $d_{n+1} = \dfrac{\left[1 - \dfrac{n}{n+1}\right]^2}{2 - \left(\dfrac{n}{n+1}\right)} = \dfrac{\dfrac{1}{(n+1)^2}}{\dfrac{n+2}{n+1}} = \dfrac{1}{(n+1)(n+2)}$, and the induction is complete.

Now, we observe that the partial sums $\sum_{i=1}^{n} d_i$ of the diameters of the circles approach 1 as $n \to \infty$; that is,

$\sum_{n=1}^{\infty} a_n = \sum_{n=1}^{\infty} \dfrac{1}{n(n+1)} = 1$, which is what we wanted to prove.

64. $|CD| = b \sin \theta$, $|DE| = |CD| \sin \theta = b \sin^2 \theta$, $|EF| = |DE| \sin \theta = b \sin^3 \theta$, ... Therefore,

$|CD| + |DE| + |EF| + |FG| + \cdots = b \sum_{n=1}^{\infty} \sin^n \theta = b \left(\dfrac{\sin \theta}{1 - \sin \theta}\right)$ since this is a geometric series with $r = \sin \theta$

and $|\sin \theta| < 1$ $\left[\text{because } 0 < \theta < \frac{\pi}{2}\right]$.

65. The series $1 - 1 + 1 - 1 + 1 - 1 + \cdots$ diverges (geometric series with $r = -1$) so we cannot say that

$0 = 1 - 1 + 1 - 1 + 1 - 1 + \cdots$.

66. If $\sum_{n=1}^{\infty} a_n$ is convergent, then $\lim_{n \to \infty} a_n = 0$ by Theorem 6, so $\lim_{n \to \infty} \dfrac{1}{a_n} \neq 0$, and so $\sum_{n=1}^{\infty} \dfrac{1}{a_n}$ is divergent by the Test for

Divergence.

67. $\sum_{n=1}^{\infty} ca_n = \lim_{n \to \infty} \sum_{i=1}^{n} ca_i = \lim_{n \to \infty} c \sum_{i=1}^{n} a_i = c \lim_{n \to \infty} \sum_{i=1}^{n} a_i = c \sum_{n=1}^{\infty} a_n$, which exists by hypothesis.

68. If $\sum ca_n$ were convergent, then $\sum (1/c)(ca_n) = \sum a_n$ would be also, by Theorem 8(i). But this is not the case, so $\sum ca_n$

must diverge.

69. Suppose on the contrary that $\sum (a_n + b_n)$ converges. Then $\sum (a_n + b_n)$ and $\sum a_n$ are convergent series. So by

Theorem 8(iii), $\sum [(a_n + b_n) - a_n]$ would also be convergent. But $\sum [(a_n + b_n) - a_n] = \sum b_n$, a contradiction, since

$\sum b_n$ is given to be divergent.

70. No. For example, take $\sum a_n = \sum n$ and $\sum b_n = \sum (-n)$, which both diverge, yet $\sum (a_n + b_n) = \sum 0$, which converges

with sum 0.

71. The partial sums $\{s_n\}$ form an increasing sequence, since $s_n - s_{n-1} = a_n > 0$ for all n. Also, the sequence $\{s_n\}$ is bounded

since $s_n \leq 1000$ for all n. So by the Monotonic Sequence Theorem, the sequence of partial sums converges, that is, the series

$\sum a_n$ is convergent.

72. (a) RHS $= \dfrac{1}{f_{n-1} f_n} - \dfrac{1}{f_n f_{n+1}} = \dfrac{f_n f_{n+1} - f_n f_{n-1}}{f_n^2 f_{n-1} f_{n+1}} = \dfrac{f_{n+1} - f_{n-1}}{f_n f_{n-1} f_{n+1}} = \dfrac{(f_{n-1} + f_n) - f_{n-1}}{f_n f_{n-1} f_{n+1}} = \dfrac{1}{f_{n-1} f_{n+1}} =$ LHS

(b) $\displaystyle\sum_{n=2}^{\infty} \frac{1}{f_{n-1}f_{n+1}} = \sum_{n=2}^{\infty}\left(\frac{1}{f_{n-1}f_n} - \frac{1}{f_n f_{n+1}}\right)$ [from part (a)]

$\displaystyle = \lim_{n\to\infty}\left[\left(\frac{1}{f_1 f_2} - \frac{1}{f_2 f_3}\right) + \left(\frac{1}{f_2 f_3} - \frac{1}{f_3 f_4}\right) + \left(\frac{1}{f_3 f_4} - \frac{1}{f_4 f_5}\right) + \cdots + \left(\frac{1}{f_{n-1}f_n} - \frac{1}{f_n f_{n+1}}\right)\right]$

$\displaystyle = \lim_{n\to\infty}\left(\frac{1}{f_1 f_2} - \frac{1}{f_n f_{n+1}}\right) = \frac{1}{f_1 f_2} - 0 = \frac{1}{1\cdot 1} = 1 \quad \text{because } f_n \to \infty \text{ as } n \to \infty.$

(c) $\displaystyle\sum_{n=2}^{\infty} \frac{f_n}{f_{n-1}f_{n+1}} = \sum_{n=2}^{\infty}\left(\frac{f_n}{f_{n-1}f_n} - \frac{f_n}{f_n f_{n+1}}\right)$ [as above]

$\displaystyle = \sum_{n=2}^{\infty}\left(\frac{1}{f_{n-1}} - \frac{1}{f_{n+1}}\right)$

$\displaystyle = \lim_{n\to\infty}\left[\left(\frac{1}{f_1} - \frac{1}{f_3}\right) + \left(\frac{1}{f_2} - \frac{1}{f_4}\right) + \left(\frac{1}{f_3} - \frac{1}{f_5}\right) + \left(\frac{1}{f_4} - \frac{1}{f_6}\right) + \cdots + \left(\frac{1}{f_{n-1}} - \frac{1}{f_{n+1}}\right)\right]$

$\displaystyle = \lim_{n\to\infty}\left(\frac{1}{f_1} + \frac{1}{f_2} - \frac{1}{f_n} - \frac{1}{f_{n+1}}\right) = 1 + 1 - 0 - 0 = 2 \quad \text{because } f_n \to \infty \text{ as } n \to \infty.$

73. (a) At the first step, only the interval $\left(\frac{1}{3}, \frac{2}{3}\right)$ (length $\frac{1}{3}$) is removed. At the second step, we remove the intervals $\left(\frac{1}{9}, \frac{2}{9}\right)$ and $\left(\frac{7}{9}, \frac{8}{9}\right)$, which have a total length of $2 \cdot \left(\frac{1}{3}\right)^2$. At the third step, we remove 2^2 intervals, each of length $\left(\frac{1}{3}\right)^3$. In general, at the nth step we remove 2^{n-1} intervals, each of length $\left(\frac{1}{3}\right)^n$, for a length of $2^{n-1}\cdot \left(\frac{1}{3}\right)^n = \frac{1}{3}\left(\frac{2}{3}\right)^{n-1}$. Thus, the total length of all removed intervals is $\displaystyle\sum_{n=1}^{\infty} \frac{1}{3}\left(\frac{2}{3}\right)^{n-1} = \frac{1/3}{1-2/3} = 1$ [geometric series with $a = \frac{1}{3}$ and $r = \frac{2}{3}$]. Notice that at the nth step, the leftmost interval that is removed is $\left(\left(\frac{1}{3}\right)^n, \left(\frac{2}{3}\right)^n\right)$, so we never remove 0, and 0 is in the Cantor set. Also, the rightmost interval removed is $\left(1 - \left(\frac{2}{3}\right)^n, 1 - \left(\frac{1}{3}\right)^n\right)$, so 1 is never removed. Some other numbers in the Cantor set are $\frac{1}{3}, \frac{2}{3}, \frac{1}{9}, \frac{2}{9}, \frac{7}{9}$, and $\frac{8}{9}$.

(b) The area removed at the first step is $\frac{1}{9}$; at the second step, $8 \cdot \left(\frac{1}{9}\right)^2$; at the third step, $(8)^2 \cdot \left(\frac{1}{9}\right)^3$. In general, the area removed at the nth step is $(8)^{n-1}\left(\frac{1}{9}\right)^n = \frac{1}{9}\left(\frac{8}{9}\right)^{n-1}$, so the total area of all removed squares is

$$\sum_{n=1}^{\infty} \frac{1}{9}\left(\frac{8}{9}\right)^{n-1} = \frac{1/9}{1 - 8/9} = 1.$$

74. (a)

a_1	1	2	4	1	1	1000
a_2	2	3	1	4	1000	1
a_3	1.5	2.5	2.5	2.5	500.5	500.5
a_4	1.75	2.75	1.75	3.25	750.25	250.75
a_5	1.625	2.625	2.125	2.875	625.375	375.625
a_6	1.6875	2.6875	1.9375	3.0625	687.813	313.188
a_7	1.65625	2.65625	2.03125	2.96875	656.594	344.406
a_8	1.67188	2.67188	1.98438	3.01563	672.203	328.797
a_9	1.66406	2.66406	2.00781	2.99219	664.398	336.602
a_{10}	1.66797	2.66797	1.99609	3.00391	668.301	332.699
a_{11}	1.66602	2.66602	2.00195	2.99805	666.350	334.650
a_{12}	1.66699	2.66699	1.99902	3.00098	667.325	333.675

The limits seem to be $\frac{5}{3}, \frac{8}{3}$, 2, 3, 667, and 334. Note that the limits appear to be "weighted" more toward a_2. In general, we guess that the limit is $\dfrac{a_1 + 2a_2}{3}$.

(b) $a_{n+1} - a_n = \frac{1}{2}(a_n + a_{n-1}) - a_n = -\frac{1}{2}(a_n - a_{n-1}) = -\frac{1}{2}\left[\frac{1}{2}(a_{n-1} + a_{n-2}) - a_{n-1}\right]$

$= -\frac{1}{2}\left[-\frac{1}{2}(a_{n-1} - a_{n-2})\right] = \cdots = \left(-\frac{1}{2}\right)^{n-1}(a_2 - a_1)$

Note that we have used the formula $a_k = \frac{1}{2}(a_{k-1} + a_{k-2})$ a total of $n-1$ times in this calculation, once for each k

between 3 and $n+1$. Now we can write

$$a_n = a_1 + (a_2 - a_1) + (a_3 - a_2) + \cdots + (a_{n-1} - a_{n-2}) + (a_n - a_{n-1})$$

$$= a_1 + \sum_{k=1}^{n-1}(a_{k+1} - a_k) = a_1 + \sum_{k=1}^{n-1}\left(-\frac{1}{2}\right)^{k-1}(a_2 - a_1)$$

and so

$$\lim_{n\to\infty} a_n = a_1 + (a_2 - a_1)\sum_{k=1}^{\infty}\left(-\frac{1}{2}\right)^{k-1} = a_1 + (a_2 - a_1)\left[\frac{1}{1-(-1/2)}\right] = a_1 + \frac{2}{3}(a_2 - a_1) = \frac{a_1 + 2a_2}{3}.$$

75. (a) For $\sum_{n=1}^{\infty} \frac{n}{(n+1)!}$, $s_1 = \frac{1}{1\cdot 2} = \frac{1}{2}$, $s_2 = \frac{1}{2} + \frac{2}{1\cdot 2\cdot 3} = \frac{5}{6}$, $s_3 = \frac{5}{6} + \frac{3}{1\cdot 2\cdot 3\cdot 4} = \frac{23}{24}$,

$s_4 = \frac{23}{24} + \frac{4}{1\cdot 2\cdot 3\cdot 4\cdot 5} = \frac{119}{120}$. The denominators are $(n+1)!$, so a guess would be $s_n = \frac{(n+1)! - 1}{(n+1)!}$.

(b) For $n = 1$, $s_1 = \frac{1}{2} = \frac{2! - 1}{2!}$, so the formula holds for $n = 1$. Assume $s_k = \frac{(k+1)! - 1}{(k+1)!}$. Then

$$s_{k+1} = \frac{(k+1)! - 1}{(k+1)!} + \frac{k+1}{(k+2)!} = \frac{(k+1)! - 1}{(k+1)!} + \frac{k+1}{(k+1)!(k+2)} = \frac{(k+2)! - (k+2) + k + 1}{(k+2)!}$$

$$= \frac{(k+2)! - 1}{(k+2)!}$$

Thus, the formula is true for $n = k + 1$. So by induction, the guess is correct.

(c) $\lim_{n\to\infty} s_n = \lim_{n\to\infty} \frac{(n+1)! - 1}{(n+1)!} = \lim_{n\to\infty}\left[1 - \frac{1}{(n+1)!}\right] = 1$ and so $\sum_{n=1}^{\infty}\frac{n}{(n+1)!} = 1$.

76.

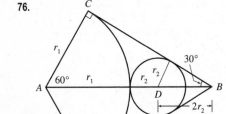

Let $r_1 =$ radius of the large circle, $r_2 =$ radius of next circle, and so on.

From the figure we have $\angle BAC = 60°$ and $\cos 60° = r_1/|AB|$, so

$|AB| = 2r_1$ and $|DB| = 2r_2$. Therefore, $2r_1 = r_1 + r_2 + 2r_2$ $\Rightarrow$

$r_1 = 3r_2$. In general, we have $r_{n+1} = \frac{1}{3}r_n$, so the total area is

$$A = \pi r_1^2 + 3\pi r_2^2 + 3\pi r_3^2 + \cdots = \pi r_1^2 + 3\pi r_2^2\left(1 + \frac{1}{3^2} + \frac{1}{3^4} + \frac{1}{3^6} + \cdots\right)$$

$$= \pi r_1^2 + 3\pi r_2^2 \cdot \frac{1}{1 - 1/9} = \pi r_1^2 + \frac{27}{8}\pi r_2^2$$

Since the sides of the triangle have length 1, $|BC| = \frac{1}{2}$ and $\tan 30° = \frac{r_1}{1/2}$. Thus, $r_1 = \frac{\tan 30°}{2} = \frac{1}{2\sqrt{3}}$ $\Rightarrow$ $r_2 = \frac{1}{6\sqrt{3}}$,

so $A = \pi\left(\frac{1}{2\sqrt{3}}\right)^2 + \frac{27\pi}{8}\left(\frac{1}{6\sqrt{3}}\right)^2 = \frac{\pi}{12} + \frac{\pi}{32} = \frac{11\pi}{96}$. The area of the triangle is $\frac{\sqrt{3}}{4}$, so the circles occupy about 83.1%

of the area of the triangle.

12.3 The Integral Test and Estimates of Sums

ET 11.3

1. The picture shows that $a_2 = \dfrac{1}{2^{1.3}} < \displaystyle\int_1^2 \dfrac{1}{x^{1.3}}\,dx$,

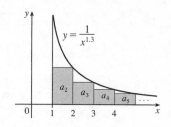

$a_3 = \dfrac{1}{3^{1.3}} < \displaystyle\int_2^3 \dfrac{1}{x^{1.3}}\,dx$, and so on, so $\displaystyle\sum_{n=2}^{\infty} \dfrac{1}{n^{1.3}} < \int_1^{\infty} \dfrac{1}{x^{1.3}}\,dx$. The

integral converges by (8.8.2) [ET (7.8.2)] with $p = 1.3 > 1$, so the series
converges.

2. From the first figure, we see that

$\displaystyle\int_1^6 f(x)\,dx < \sum_{i=1}^{5} a_i$. From the second figure,

we see that $\displaystyle\sum_{i=2}^{6} a_i < \int_1^6 f(x)\,dx$. Thus, we

have $\displaystyle\sum_{i=2}^{6} a_i < \int_1^6 f(x)\,dx < \sum_{i=1}^{5} a_i$.

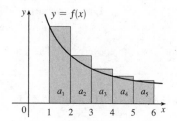

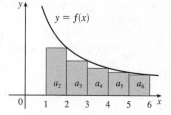

3. The function $f(x) = 1/\sqrt[5]{x} = x^{-1/5}$ is continuous, positive, and decreasing on $[1, \infty)$, so the Integral Test applies.

$\displaystyle\int_1^{\infty} x^{-1/5}\,dx = \lim_{t\to\infty} \int_1^t x^{-1/5}\,dx = \lim_{t\to\infty}\left[\tfrac{5}{4}x^{4/5}\right]_1^t = \lim_{t\to\infty}\left(\tfrac{5}{4}t^{4/5} - \tfrac{5}{4}\right) = \infty$, so $\displaystyle\sum_{n=1}^{\infty} 1/\sqrt[5]{n}$ diverges.

4. The function $f(x) = 1/x^5$ is continuous, positive, and decreasing on $[1, \infty)$, so the Integral Test applies.

$\displaystyle\int_1^{\infty} \frac{1}{x^5}\,dx = \lim_{t\to\infty} \int_1^t x^{-5}\,dx = \lim_{t\to\infty}\left[\frac{x^{-4}}{-4}\right]_1^t = \lim_{t\to\infty}\left(-\frac{1}{4t^4} + \frac{1}{4}\right) = \frac{1}{4}$.

Since this improper integral is convergent, the series $\displaystyle\sum_{n=1}^{\infty} \frac{1}{n^5}$ is also convergent by the Integral Test.

5. The function $f(x) = \dfrac{1}{(2x+1)^3}$ is continuous, positive, and decreasing on $[1, \infty)$, so the Integral Test applies.

$\displaystyle\int_1^{\infty} \frac{1}{(2x+1)^3}\,dx = \lim_{t\to\infty} \int_1^t \frac{1}{(2x+1)^3}\,dx = \lim_{t\to\infty}\left[-\frac{1}{4}\frac{1}{(2x+1)^2}\right]_1^t = \lim_{t\to\infty}\left(-\frac{1}{4(2t+1)^2} + \frac{1}{36}\right) = \frac{1}{36}$.

Since this improper integral is convergent, the series $\displaystyle\sum_{n=1}^{\infty} \frac{1}{(2n+1)^3}$ is also convergent by the Integral Test.

6. The function $f(x) = 1/\sqrt{x+4} = (x+4)^{-1/2}$ is continuous, positive, and decreasing on $[1, \infty)$, so the Integral Test applies.

$\displaystyle\int_1^{\infty} (x+4)^{-1/2}\,dx = \lim_{t\to\infty} \int_1^t (x+4)^{-1/2}\,dx = \lim_{t\to\infty}\left[2(x+4)^{1/2}\right]_1^t = \lim_{t\to\infty}\left(2\sqrt{t+4} - 2\sqrt{5}\right) = \infty$, so the series

$\displaystyle\sum_{n=1}^{\infty} 1/\sqrt{n+4}$ diverges.

7. $f(x) = xe^{-x}$ is continuous and positive on $[1, \infty)$. $f'(x) = -xe^{-x} + e^{-x} = e^{-x}(1 - x) < 0$ for $x > 1$, so f is decreasing

on $[1, \infty)$. Thus, the Integral Test applies.

$\displaystyle\int_1^{\infty} xe^{-x}\,dx = \lim_{b\to\infty} \int_1^b xe^{-x}\,dx = \lim_{b\to\infty}\left[-xe^{-x} - e^{-x}\right]_1^b \quad \text{[by parts]} \; = \lim_{b\to\infty}\left[-be^{-b} - e^{-b} + e^{-1} + e^{-1}\right] = 2/e$

since $\displaystyle\lim_{b\to\infty} be^{-b} = \lim_{b\to\infty}(b/e^b) \overset{\text{H}}{=} \lim_{b\to\infty}(1/e^b) = 0$ and $\displaystyle\lim_{b\to\infty} e^{-b} = 0$. Thus, $\sum_{n=1}^{\infty} ne^{-n}$ converges.

8. The function $f(x) = \dfrac{x+2}{x+1} = 1 + \dfrac{1}{x+1}$ is continuous, positive, and decreasing on $[1, \infty)$, so the Integral Test applies.

$$\int_1^\infty f(x)\,dx = \lim_{t \to \infty} \int_1^t \left(1 + \frac{1}{x+1}\right) dx = \lim_{t \to \infty} \left[x + \ln(x+1)\right]_1^t = \lim_{t \to \infty} \left(t + \ln(t+1) - 1 - \ln 2\right) = \infty, \text{ so}$$

$\displaystyle\int_1^\infty \dfrac{x+2}{x+1}\,dx$ is divergent and the series $\displaystyle\sum_{n=1}^\infty \dfrac{n+2}{n+1}$ is divergent.

Note: $\displaystyle\lim_{n \to \infty} \dfrac{n+2}{n+1} = 1$, so the given series diverges by the Test for Divergence.

9. The series $\displaystyle\sum_{n=1}^\infty \dfrac{1}{n^{0.85}}$ is a p-series with $p = 0.85 \le 1$, so it diverges by (1). Therefore, the series $\displaystyle\sum_{n=1}^\infty \dfrac{2}{n^{0.85}}$ must also diverge,

for if it converged, then $\displaystyle\sum_{n=1}^\infty \dfrac{1}{n^{0.85}}$ would have to converge [by Theorem 8(i) in Section 12.2 [ET 11.2]].

10. $\displaystyle\sum_{n=1}^\infty n^{-1.4}$ and $\displaystyle\sum_{n=1}^\infty n^{-1.2}$ are p-series with $p > 1$, so they converge by (1). Thus, $\displaystyle\sum_{n=1}^\infty 3n^{-1.2}$ converges by Theorem 8(i) in

Section 12.2 [ET 11.2]. It follows from Theorem 8(ii) that the given series $\displaystyle\sum_{n=1}^\infty \left(n^{-1.4} + 3n^{-1.2}\right)$ also converges.

11. $1 + \dfrac{1}{8} + \dfrac{1}{27} + \dfrac{1}{64} + \dfrac{1}{125} + \cdots = \displaystyle\sum_{n=1}^\infty \dfrac{1}{n^3}$. This is a p-series with $p = 3 > 1$, so it converges by (1).

12. $1 + \dfrac{1}{2\sqrt{2}} + \dfrac{1}{3\sqrt{3}} + \dfrac{1}{4\sqrt{4}} + \dfrac{1}{5\sqrt{5}} + \cdots = \displaystyle\sum_{n=1}^\infty \dfrac{1}{n\sqrt{n}} = \displaystyle\sum_{n=1}^\infty \dfrac{1}{n^{3/2}}$. This is a p-series with $p = \frac{3}{2} > 1$, so it converges by (1).

13. $1 + \dfrac{1}{3} + \dfrac{1}{5} + \dfrac{1}{7} + \dfrac{1}{9} + \cdots = \displaystyle\sum_{n=1}^\infty \dfrac{1}{2n-1}$. The function $f(x) = \dfrac{1}{2x-1}$ is

continuous, positive, and decreasing on $[1, \infty)$, so the Integral Test applies.

$$\int_1^\infty \frac{1}{2x-1}\,dx = \lim_{t \to \infty} \int_1^t \frac{1}{2x-1}\,dx = \lim_{t \to \infty} \left[\tfrac{1}{2} \ln|2x-1|\right]_1^t = \tfrac{1}{2} \lim_{t \to \infty} \left(\ln(2t-1) - 0\right) = \infty, \text{ so the series } \sum_{n=1}^\infty \frac{1}{2n-1}$$

diverges.

14. $\dfrac{1}{5} + \dfrac{1}{8} + \dfrac{1}{11} + \dfrac{1}{14} + \dfrac{1}{17} + \cdots = \displaystyle\sum_{n=1}^\infty \dfrac{1}{3n+2}$. The function $f(x) = \dfrac{1}{3x+2}$ is continuous, positive, and decreasing on

$[1, \infty)$, so the Integral Test applies.

$$\int_1^\infty \frac{1}{3x+2}\,dx = \lim_{t \to \infty} \int_1^t \frac{1}{3x+2}\,dx = \lim_{t \to \infty} \left[\tfrac{1}{3} \ln|3x+2|\right]_1^t = \tfrac{1}{3} \lim_{t \to \infty} \left(\ln(3t+2) - \ln 5\right) = \infty, \text{ so the series}$$

$\displaystyle\sum_{n=1}^\infty \dfrac{1}{3n+2}$ diverges.

15. $\displaystyle\sum_{n=1}^\infty \dfrac{5 - 2\sqrt{n}}{n^3} = 5 \displaystyle\sum_{n=1}^\infty \dfrac{1}{n^3} - 2 \displaystyle\sum_{n=1}^\infty \dfrac{1}{n^{5/2}}$ by Theorem 12.2.8 [ET 11.2.8], since $\displaystyle\sum_{n=1}^\infty \dfrac{1}{n^3}$ and $\displaystyle\sum_{n=1}^\infty \dfrac{1}{n^{5/2}}$ both converge by (1)

$\left[\text{with } p = 3 > 1 \text{ and } p = \frac{5}{2} > 1\right]$. Thus, $\displaystyle\sum_{n=1}^\infty \dfrac{5 - 2\sqrt{n}}{n^3}$ converges.

16. $f(x) = \dfrac{x^2}{x^3 + 1}$ is continuous and positive on $[2, \infty)$, and also decreasing since $f'(x) = \dfrac{x(2 - x^3)}{(x^3 + 1)^2} < 0$ for $x \geq 2$,

so we can use the Integral Test [note that f is *not* decreasing on $[1, \infty)$].

$\displaystyle\int_2^\infty \dfrac{x^2}{x^3 + 1}\, dx = \lim_{t \to \infty} \left[\tfrac{1}{3}\ln(x^3 + 1)\right]_2^t = \tfrac{1}{3}\lim_{t \to \infty}\left[\ln(t^3 + 1) - \ln 9\right] = \infty$, so the series $\displaystyle\sum_{n=2}^{\infty} \dfrac{n^2}{n^3 + 1}$ diverges, and so does

the given series, $\displaystyle\sum_{n=1}^{\infty} \dfrac{n^2}{n^3 + 1}$.

17. The function $f(x) = \dfrac{1}{x^2 + 4}$ is continuous, positive, and decreasing on $[1, \infty)$, so we can apply the Integral Test.

$$\int_1^\infty \dfrac{1}{x^2 + 4}\, dx = \lim_{t \to \infty}\int_1^t \dfrac{1}{x^2 + 4}\, dx = \lim_{t \to \infty}\left[\dfrac{1}{2}\tan^{-1}\dfrac{x}{2}\right]_1^t = \dfrac{1}{2}\lim_{t \to \infty}\left[\tan^{-1}\left(\dfrac{t}{2}\right) - \tan^{-1}\left(\dfrac{1}{2}\right)\right]$$

$$= \dfrac{1}{2}\left[\dfrac{\pi}{2} - \tan^{-1}\left(\dfrac{1}{2}\right)\right]$$

Therefore, the series $\displaystyle\sum_{n=1}^{\infty} \dfrac{1}{n^2 + 4}$ converges.

18. The function $f(x) = \dfrac{3x + 2}{x(x + 1)} = \dfrac{2}{x} + \dfrac{1}{x + 1}$ [by partial fractions] is continuous, positive, and decreasing on $[1, \infty)$ since it

is the sum of two such functions. Thus, we can apply the Integral Test.

$$\int_1^\infty \dfrac{3x + 2}{x(x + 1)}\, dx = \lim_{t \to \infty}\int_1^t \left[\dfrac{2}{x} + \dfrac{1}{x + 1}\right] dx = \lim_{t \to \infty}\left[2\ln x + \ln(x + 1)\right]_1^t = \lim_{t \to \infty}\left[2\ln t + \ln(t + 1) - \ln 2\right] = \infty.$$

Thus, the series $\displaystyle\sum_{n=1}^{\infty} \dfrac{3n + 2}{n(n + 1)}$ diverges.

19. $\displaystyle\sum_{n=1}^{\infty} \dfrac{\ln n}{n^3} = \sum_{n=2}^{\infty} \dfrac{\ln n}{n^3}$ since $\dfrac{\ln 1}{1} = 0$. The function $f(x) = \dfrac{\ln x}{x^3}$ is continuous and positive on $[2, \infty)$.

$$f'(x) = \dfrac{x^3(1/x) - (\ln x)(3x^2)}{(x^3)^2} = \dfrac{x^2 - 3x^2 \ln x}{x^6} = \dfrac{1 - 3\ln x}{x^4} < 0 \iff 1 - 3\ln x < 0 \iff \ln x > \tfrac{1}{3} \iff$$

$x > e^{1/3} \approx 1.4$, so f is decreasing on $[2, \infty)$, and the Integral Test applies.

$\displaystyle\int_2^\infty \dfrac{\ln x}{x^3}\, dx = \lim_{t \to \infty}\int_2^t \dfrac{\ln x}{x^3}\, dx \overset{(\star)}{=} \lim_{t \to \infty}\left[-\dfrac{\ln x}{2x^2} - \dfrac{1}{4x^2}\right]_1^t = \lim_{t \to \infty}\left[-\dfrac{1}{4t^2}(2\ln t + 1) + \dfrac{1}{4}\right] \overset{(\star\star)}{=} \dfrac{1}{4}$, so the series $\displaystyle\sum_{n=2}^{\infty} \dfrac{\ln n}{n^3}$

converges.

$(\star)$: $u = \ln x$, $dv = x^{-3}\, dx$ $\Rightarrow$ $du = (1/x)\, dx$, $v = -\tfrac{1}{2}x^{-2}$, so

$$\int \dfrac{\ln x}{x^3}\, dx = -\tfrac{1}{2}x^{-2}\ln x - \int -\tfrac{1}{2}x^{-2}(1/x)\, dx = -\tfrac{1}{2}x^{-2}\ln x + \tfrac{1}{2}\int x^{-3}\, dx = -\tfrac{1}{2}x^{-2}\ln x - \tfrac{1}{4}x^{-2} + C.$$

$(\star\star)$: $\displaystyle\lim_{t \to \infty}\left(-\dfrac{2\ln t + 1}{4t^2}\right) \overset{\text{H}}{=} -\lim_{t \to \infty}\dfrac{2/t}{8t} = -\tfrac{1}{4}\lim_{t \to \infty}\dfrac{1}{t^2} = 0.$

20. The function $f(x) = \dfrac{1}{x^2 - 4x + 5} = \dfrac{1}{(x-2)^2 + 1}$ is continuous, positive, and decreasing on $[2, \infty)$, so the Integral Test

applies.

$$\int_2^\infty f(x)\,dx = \lim_{t \to \infty} \int_2^t f(x)\,dx = \lim_{t \to \infty} \int_2^t \frac{1}{(x-2)^2 + 1}\,dx = \lim_{t \to \infty} \left[\tan^{-1}(x-2) \right]_2^t = \lim_{t \to \infty} [\tan^{-1}(t-2) - \tan^{-1} 0]$$

$$= \frac{\pi}{2} - 0 = \frac{\pi}{2}$$

so the series $\displaystyle\sum_{n=2}^\infty \frac{1}{n^2 - 4n + 5}$ converges. Of course, this means that $\displaystyle\sum_{n=1}^\infty \frac{1}{n^2 - 4n + 5}$ converges too.

21. $f(x) = \dfrac{1}{x \ln x}$ is continuous and positive on $[2, \infty)$, and also decreasing since $f'(x) = -\dfrac{1 + \ln x}{x^2 (\ln x)^2} < 0$ for $x > 2$, so we can

use the Integral Test. $\displaystyle\int_2^\infty \frac{1}{x \ln x}\,dx = \lim_{t \to \infty} [\ln(\ln x)]_2^t = \lim_{t \to \infty} [\ln(\ln t) - \ln(\ln 2)] = \infty$, so the series $\displaystyle\sum_{n=2}^\infty \frac{1}{n \ln n}$ diverges.

22. The function $f(x) = \dfrac{1}{x(\ln x)^2}$ is continuous, positive, and decreasing on $[2, \infty)$, so the Integral Test applies.

$$\int_2^\infty f(x)\,dx = \lim_{t \to \infty} \int_2^t \frac{1}{x(\ln x)^2}\,dx = \lim_{t \to \infty} \left[\frac{-1}{\ln x} \right]_2^t \quad \text{[by substitution with } u = \ln x\text{]} \quad = -\lim_{t \to \infty} \left(\frac{1}{\ln t} - \frac{1}{\ln 2} \right) = \frac{1}{\ln 2},$$

so the series $\displaystyle\sum_{n=2}^\infty \frac{1}{n(\ln n)^2}$ converges.

23. The function $f(x) = e^{1/x}/x^2$ is continuous, positive, and decreasing on $[1, \infty)$, so the Integral Test applies.

$[g(x) = e^{1/x}$ is decreasing and dividing by x^2 doesn't change that fact.]

$$\int_1^\infty f(x)\,dx = \lim_{t \to \infty} \int_1^t \frac{e^{1/x}}{x^2}\,dx = \lim_{t \to \infty} \left[-e^{1/x} \right]_1^t = -\lim_{t \to \infty} (e^{1/t} - e) = -(1 - e) = e - 1, \text{ so the series } \sum_{n=1}^\infty \frac{e^{1/n}}{n^2}$$

converges.

24. $f(x) = \dfrac{x^2}{e^x} \quad \Rightarrow \quad f'(x) = \dfrac{e^x(2x) - x^2 e^x}{(e^x)^2} = \dfrac{xe^x(2 - x)}{(e^x)^2} = \dfrac{x(2 - x)}{e^x} < 0$ for $x > 2$, so f is continuous, positive, and

decreasing on $[3, \infty)$ and so the Integral Test applies.

$$\int_3^\infty f(x)\,dx = \lim_{t \to \infty} \int_3^t \frac{x^2}{e^x}\,dx \overset{(*)}{=} \lim_{t \to \infty} \left[-e^{-x}(x^2 + 2x + 2) \right]_3^t = -\lim_{t \to \infty} \left[e^{-t}(t^2 + 2t + 2) - e^{-3}(17) \right] \overset{(**)}{=} \frac{17}{e^3},$$

so the series $\displaystyle\sum_{n=3}^\infty \frac{n^2}{e^n}$ converges.

$(*)$: $\int x^2 e^{-x}\,dx \overset{97}{=} -x^2 e^{-x} + 2 \int xe^{-x}\,dx \overset{97}{=} -x^2 e^{-x} + 2\left(-xe^{-x} + \int e^{-x}\,dx \right)$

$\qquad = -x^2 e^{-x} - 2xe^{-x} - 2e^{-x} + C = -e^{-x}(x^2 + 2x + 2) + C.$

$(**)$: $\displaystyle\lim_{t \to \infty} \frac{t^2 + 2t + 2}{e^t} \overset{\text{H}}{=} \lim_{t \to \infty} \frac{2t + 2}{e^t} \overset{\text{H}}{=} \lim_{t \to \infty} \frac{2}{e^t} = 0.$

25. The function $f(x) = \dfrac{1}{x^3 + x}$ is continuous, positive, and decreasing on $[1, \infty)$, so the Integral Test applies. We use partial

fractions to evaluate the integral:

$$\int_1^\infty \frac{1}{x^3 + x}\,dx = \lim_{t \to \infty} \int_1^t \left[\frac{1}{x} - \frac{x}{1 + x^2} \right]dx = \lim_{t \to \infty} \left[\ln x - \frac{1}{2}\ln(1 + x^2) \right]_1^t = \lim_{t \to \infty} \left[\ln \frac{x}{\sqrt{1 + x^2}} \right]_1^t$$

$$= \lim_{t \to \infty} \left(\ln \frac{t}{\sqrt{1 + t^2}} - \ln \frac{1}{\sqrt{2}} \right) = \lim_{t \to \infty} \left(\ln \frac{1}{\sqrt{1 + 1/t^2}} + \frac{1}{2}\ln 2 \right) = \frac{1}{2}\ln 2$$

so the series $\displaystyle\sum_{n=1}^\infty \frac{1}{n^3 + n}$ converges.

26. The function $f(x) = \dfrac{x}{x^4 + 1}$ is positive, continuous, and decreasing on $[1, \infty)$. [Note that

$$f'(x) = \frac{x^4 + 1 - 4x^4}{(x^4 + 1)^2} = \frac{1 - 3x^4}{(x^4 + 1)^2} < 0 \text{ on } [1, \infty).] \text{ Thus, we can apply the Integral Test.}$$

$$\int_1^\infty \frac{x}{x^4 + 1}\,dx = \lim_{t \to \infty} \int_1^t \frac{\frac{1}{2}(2x)}{1 + (x^2)^2}\,dx = \lim_{t \to \infty} \left[\frac{1}{2}\tan^{-1}(x^2) \right]_1^t = \frac{1}{2}\lim_{t \to \infty}[\tan^{-1}(t^2) - \tan^{-1} 1] = \frac{1}{2}\left(\frac{\pi}{2} - \frac{\pi}{4} \right) = \frac{\pi}{8}$$

so the series $\displaystyle\sum_{n=1}^\infty \frac{n}{n^4 + 1}$ converges.

27. We have already shown (in Exercise 21) that when $p = 1$ the series $\displaystyle\sum_{n=2}^\infty \frac{1}{n(\ln n)^p}$ diverges, so assume that $p \neq 1$.

$f(x) = \dfrac{1}{x(\ln x)^p}$ is continuous and positive on $[2, \infty)$, and $f'(x) = -\dfrac{p + \ln x}{x^2(\ln x)^{p+1}} < 0$ if $x > e^{-p}$, so that f is eventually

decreasing and we can use the Integral Test.

$$\int_2^\infty \frac{1}{x(\ln x)^p}\,dx = \lim_{t \to \infty} \left[\frac{(\ln x)^{1-p}}{1 - p} \right]_2^t \quad \text{[for } p \neq 1] = \lim_{t \to \infty} \left[\frac{(\ln t)^{1-p}}{1 - p} - \frac{(\ln 2)^{1-p}}{1 - p} \right]$$

This limit exists whenever $1 - p < 0 \iff p > 1$, so the series converges for $p > 1$.

28. $f(x) = \dfrac{1}{x \ln x \,[\ln(\ln x)]^p}$ is positive and continuous on $[3, \infty)$. For $p \geq 0$, f clearly decreases on $[3, \infty)$; and for $p < 0$,

it can be verified that f is ultimately decreasing. Thus, we can apply the Integral Test.

$$I = \int_3^\infty \frac{dx}{x \ln x \,[\ln(\ln x)]^p} = \lim_{t \to \infty} \int_3^t \frac{[\ln(\ln x)]^{-p}}{x \ln x}\,dx = \lim_{t \to \infty} \left[\frac{[\ln(\ln x)]^{-p+1}}{-p + 1} \right]_3^t \quad \text{[for } p \neq 1]$$

$$= \lim_{t \to \infty} \left[\frac{[\ln(\ln t)]^{-p+1}}{-p + 1} - \frac{[\ln(\ln 3)]^{-p+1}}{-p + 1} \right],$$

which exists whenever $-p + 1 < 0 \iff p > 1$. If $p = 1$, then $I = \lim_{t \to \infty} \left[\ln(\ln(\ln x)) \right]_3^t = \infty$. Therefore,

$\displaystyle\sum_{n=3}^\infty \frac{1}{n \ln n \,[\ln(\ln n)]^p}$ converges for $p > 1$.

29. Clearly the series cannot converge if $p \geq -\frac{1}{2}$, because then $\lim\limits_{n \to \infty} n(1 + n^2)^p \neq 0$. So assume $p < -\frac{1}{2}$. Then

$f(x) = x(1 + x^2)^p$ is continuous, positive, and eventually decreasing on $[1, \infty)$, and we can use the Integral Test.

$$\int_1^\infty x(1 + x^2)^p \, dx = \lim_{t \to \infty} \left[\frac{1}{2} \cdot \frac{(1 + x^2)^{p+1}}{p+1} \right]_1^t = \frac{1}{2(p+1)} \lim_{t \to \infty} \left[(1 + t^2)^{p+1} - 2^{p+1} \right].$$

This limit exists and is finite $\Leftrightarrow p + 1 < 0 \Leftrightarrow p < -1$, so the series converges whenever $p < -1$.

30. If $p \leq 0$, $\lim\limits_{n \to \infty} \dfrac{\ln n}{n^p} = \infty$ and the series diverges, so assume $p > 0$. $f(x) = \dfrac{\ln x}{x^p}$ is positive and continuous and $f'(x) < 0$

for $x > e^{1/p}$, so f is eventually decreasing and we can use the Integral Test. Integration by parts gives

$$\int_1^\infty \frac{\ln x}{x^p} \, dx = \lim_{t \to \infty} \left[\frac{x^{1-p}\left[(1-p)\ln x - 1\right]}{(1-p)^2} \right]_1^t \quad \text{(for } p \neq 1\text{)} = \frac{1}{(1-p)^2} \left[\lim_{t \to \infty} t^{1-p}\left[(1-p)\ln t - 1\right] + 1 \right], \text{ which exists}$$

whenever $1 - p < 0 \Leftrightarrow p > 1$. Thus, $\sum\limits_{n=1}^\infty \dfrac{\ln n}{n^p}$ converges $\Leftrightarrow p > 1$.

31. Since this is a p-series with $p = x$, $\zeta(x)$ is defined when $x > 1$. Unless specified otherwise, the domain of a function f is the set of real numbers x such that the expression for $f(x)$ makes sense and defines a real number. So, in the case of a series, it's the set of real numbers x such that the series is convergent.

32. (a) $f(x) = 1/x^4$ is positive and continuous and $f'(x) = -4/x^5$ is negative for $x > 0$, and so the Integral Test applies.

$$\sum_{n=1}^\infty \frac{1}{n^4} \approx s_{10} = \frac{1}{1^4} + \frac{1}{2^4} + \frac{1}{3^4} + \cdots + \frac{1}{10^4} \approx 1.082037.$$

$$R_{10} \leq \int_{10}^\infty \frac{1}{x^4} \, dx = \lim_{t \to \infty} \left[\frac{1}{-3x^3} \right]_{10}^t = \lim_{t \to \infty} \left(-\frac{1}{3t^3} + \frac{1}{3\,(10)^3} \right) = \frac{1}{3000}, \text{ so the error is at most } 0.000\overline{3}.$$

(b) $s_{10} + \displaystyle\int_{11}^\infty \frac{1}{x^4} \, dx \leq s \leq s_{10} + \int_{10}^\infty \frac{1}{x^4} \, dx \Rightarrow s_{10} + \frac{1}{3(11)^3} \leq s \leq s_{10} + \frac{1}{3(10)^3} \Rightarrow$

$1.082037 + 0.000250 = 1.082287 \leq s \leq 1.082037 + 0.000333 = 1.082370$, so we get $s \approx 1.08233$ with error ≤ 0.00005.

(c) $R_n \leq \displaystyle\int_n^\infty \frac{1}{x^4} \, dx = \frac{1}{3n^3}$. So $R_n < 0.00001 \Rightarrow \dfrac{1}{3n^3} < \dfrac{1}{10^5} \Rightarrow 3n^3 > 10^5 \Rightarrow n > \sqrt[3]{(10)^5/3} \approx 32.2$,

that is, for $n > 32$.

33. (a) $f(x) = \dfrac{1}{x^2}$ is positive and continuous and $f'(x) = -\dfrac{2}{x^3}$ is negative for $x > 0$, and so the Integral Test applies.

$$\sum_{n=1}^\infty \frac{1}{n^2} \approx s_{10} = \frac{1}{1^2} + \frac{1}{2^2} + \frac{1}{3^2} + \cdots + \frac{1}{10^2} \approx 1.549768.$$

$$R_{10} \leq \int_{10}^\infty \frac{1}{x^2} \, dx = \lim_{t \to \infty} \left[\frac{-1}{x} \right]_{10}^t = \lim_{t \to \infty} \left(-\frac{1}{t} + \frac{1}{10} \right) = \frac{1}{10}, \text{ so the error is at most } 0.1.$$

(b) $s_{10} + \displaystyle\int_{11}^\infty \frac{1}{x^2} \, dx \leq s \leq s_{10} + \int_{10}^\infty \frac{1}{x^2} \, dx \Rightarrow s_{10} + \frac{1}{11} \leq s \leq s_{10} + \frac{1}{10} \Rightarrow$

$1.549768 + 0.090909 = 1.640677 \leq s \leq 1.549768 + 0.1 = 1.649768$, so we get $s \approx 1.64522$ (the average of 1.640677 and 1.649768) with error ≤ 0.005 (the maximum of $1.649768 - 1.64522$ and $1.64522 - 1.640677$, rounded up).

(c) $R_n \leq \displaystyle\int_n^\infty \frac{1}{x^2} \, dx = \frac{1}{n}$. So $R_n < 0.001$ if $\dfrac{1}{n} < \dfrac{1}{1000} \Leftrightarrow n > 1000$.

34. $f(x) = 1/x^5$ is positive and continuous and $f'(x) = -5/x^6$ is negative for $x > 0$, and so the Integral Test applies. Using (2),

$$R_n \le \int_n^\infty x^{-5}\, dx = \lim_{t\to\infty}\left[\frac{-1}{4x^4}\right]_n^t = \frac{1}{4n^4}.$$ If we take $n = 5$, then $s_5 \approx 1.036662$ and $R_5 \le 0.0004$. So $s \approx s_5 \approx 1.037$.

35. $f(x) = 1/(2x + 1)^6$ is continuous, positive, and decreasing on $[1, \infty)$, so the Integral Test applies. Using (2),

$$R_n \le \int_n^\infty (2x + 1)^{-6}\, dx = \lim_{t\to\infty}\left[\frac{-1}{10(2x+1)^5}\right]_n^t = \frac{1}{10(2n+1)^5}.$$ To be correct to five decimal places, we want

$$\frac{1}{10(2n+1)^5} \le \frac{5}{10^6} \quad\Leftrightarrow\quad (2n+1)^5 \ge 20{,}000 \quad\Leftrightarrow\quad n \ge \tfrac{1}{2}\left(\sqrt[5]{20{,}000} - 1\right) \approx 3.12,\ \text{so use } n = 4.$$

$$s_4 = \sum_{n=1}^4 \frac{1}{(2n+1)^6} = \frac{1}{3^6} + \frac{1}{5^6} + \frac{1}{7^6} + \frac{1}{9^6} \approx 0.001\,446 \approx 0.00145.$$

36. $f(x) = \dfrac{1}{x(\ln x)^2}$ is positive and continuous and $f'(x) = -\dfrac{\ln x + 2}{x^2(\ln x)^3}$ is negative for $x > 1$, so the Integral Test applies.

Using (2), we need $0.01 > \displaystyle\int_n^\infty \frac{dx}{x(\ln x)^2} = \lim_{t\to\infty}\left[\frac{-1}{\ln x}\right]_n^t = \frac{1}{\ln n}$. This is true for $n > e^{100}$, so we would have to take this

many terms, which would be problematic because $e^{100} \approx 2.7 \times 10^{43}$.

37. $\displaystyle\sum_{n=1}^\infty n^{-1.001} = \sum_{n=1}^\infty \frac{1}{n^{1.001}}$ is a convergent p-series with $p = 1.001 > 1$. Using (2), we get

$$R_n \le \int_n^\infty x^{-1.001}\, dx = \lim_{t\to\infty}\left[\frac{x^{-0.001}}{-0.001}\right]_n^t = -1000 \lim_{t\to\infty}\left[\frac{1}{x^{0.001}}\right]_n^t = -1000\left(-\frac{1}{n^{0.001}}\right) = \frac{1000}{n^{0.001}}.$$

We want $R_n < 0.000\,000\,005 \quad\Leftrightarrow\quad \dfrac{1000}{n^{0.001}} < 5 \times 10^{-9} \quad\Leftrightarrow\quad n^{0.001} > \dfrac{1000}{5 \times 10^{-9}} \quad\Leftrightarrow$

$$n > \left(2 \times 10^{11}\right)^{1000} = 2^{1000} \times 10^{11{,}000} \approx 1.07 \times 10^{301} \times 10^{11{,}000} = 1.07 \times 10^{11{,}301}.$$

38. (a) $f(x) = \left(\dfrac{\ln x}{x}\right)^2$ is continuous and positive for $x > 1$, and since $f'(x) = \dfrac{2\ln x\,(1 - \ln x)}{x^3} < 0$ for $x > e$, we can apply

the Integral Test. Using a CAS, we get $\displaystyle\int_1^\infty \left(\frac{\ln x}{x}\right)^2 dx = 2$, so the series also converges.

(b) Since the Integral Test applies, the error in $s \approx s_n$ is $R_n \le \displaystyle\int_n^\infty \left(\frac{\ln x}{x}\right)^2 dx = \frac{(\ln n)^2 + 2\ln n + 2}{n}$.

(c) By graphing the functions $y_1 = \dfrac{(\ln x)^2 + 2\ln x + 2}{x}$ and $y_2 = 0.05$, we see that $y_1 < y_2$ for $n \ge 1373$.

(d) Using the CAS to sum the first 1373 terms, we get $s_{1373} \approx 1.94$.

39. (a) From the figure, $a_2 + a_3 + \cdots + a_n \le \int_1^n f(x)\, dx$, so with

$$f(x) = \frac{1}{x},\ \frac{1}{2} + \frac{1}{3} + \frac{1}{4} + \cdots + \frac{1}{n} \le \int_1^n \frac{1}{x}\, dx = \ln n.$$

Thus, $s_n = 1 + \dfrac{1}{2} + \dfrac{1}{3} + \dfrac{1}{4} + \cdots + \dfrac{1}{n} \le 1 + \ln n$.

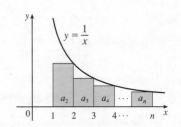

(b) By part (a), $s_{10^6} \le 1 + \ln 10^6 \approx 14.82 < 15$ and

$s_{10^9} \le 1 + \ln 10^9 \approx 21.72 < 22$.

40. (a) The sum of the areas of the n rectangles in the graph to the right is

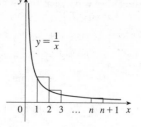

$1 + \dfrac{1}{2} + \dfrac{1}{3} + \cdots + \dfrac{1}{n}$. Now $\displaystyle\int_1^{n+1} \dfrac{dx}{x}$ is less than this sum because

the rectangles extend above the curve $y = 1/x$, so

$\displaystyle\int_1^{n+1} \dfrac{1}{x}\, dx = \ln(n+1) < 1 + \dfrac{1}{2} + \dfrac{1}{3} + \cdots + \dfrac{1}{n}$, and since

$\ln n < \ln(n+1)$, $0 < 1 + \dfrac{1}{2} + \dfrac{1}{3} + \cdots + \dfrac{1}{n} - \ln n = t_n$.

(b) The area under $f(x) = 1/x$ between $x = n$ and $x = n+1$ is

$\displaystyle\int_n^{n+1} \dfrac{dx}{x} = \ln(n+1) - \ln n$, and this is clearly greater than the area of

the inscribed rectangle in the figure to the right $\left[\text{which is } \dfrac{1}{n+1}\right]$, so

$t_n - t_{n+1} = [\ln(n+1) - \ln n] - \dfrac{1}{n+1} > 0$, and so $t_n > t_{n+1}$, so $\{t_n\}$ is a decreasing sequence.

(c) We have shown that $\{t_n\}$ is decreasing and that $t_n > 0$ for all n. Thus, $0 < t_n \le t_1 = 1$, so $\{t_n\}$ is a bounded monotonic sequence, and hence converges by the Monotonic Sequence Theorem.

41. $b^{\ln n} = \left(e^{\ln b}\right)^{\ln n} = \left(e^{\ln n}\right)^{\ln b} = n^{\ln b} = \dfrac{1}{n^{-\ln b}}$. This is a p-series, which converges for all b such that $-\ln b > 1$ $\Leftrightarrow$

$\ln b < -1$ $\Leftrightarrow$ $b < e^{-1}$ $\Leftrightarrow$ $b < 1/e$ [with $b > 0$].

42. For the series $\displaystyle\sum_{n=1}^{\infty} \left(\dfrac{c}{n} - \dfrac{1}{n+1}\right)$,

$$s_n = \sum_{i=1}^{n} \left(\dfrac{c}{i} - \dfrac{1}{i+1}\right) = \left(\dfrac{c}{1} - \dfrac{1}{2}\right) + \left(\dfrac{c}{2} - \dfrac{1}{3}\right) + \left(\dfrac{c}{3} - \dfrac{1}{4}\right) + \cdots + \left(\dfrac{c}{n} - \dfrac{1}{n+1}\right)$$

$$= \dfrac{c}{1} + \dfrac{c-1}{2} + \dfrac{c-1}{3} + \dfrac{c-1}{4} + \cdots + \dfrac{c-1}{n} - \dfrac{1}{n+1} = c + (c-1)\left(\dfrac{1}{2} + \dfrac{1}{3} + \dfrac{1}{4} + \cdots + \dfrac{1}{n}\right) - \dfrac{1}{n+1}$$

Thus, $\displaystyle\sum_{n=1}^{\infty} \left(\dfrac{c}{n} - \dfrac{1}{n+1}\right) = \lim_{n\to\infty} s_n = \lim_{n\to\infty}\left[c + (c-1)\sum_{i=2}^{n}\dfrac{1}{i} - \dfrac{1}{n+1}\right]$. Since a constant multiple of a divergent series

is divergent, the last limit exists only if $c - 1 = 0$, so the original series converges only if $c = 1$.

12.4 The Comparison Tests

1. (a) We cannot say anything about $\sum a_n$. If $a_n > b_n$ for all n and $\sum b_n$ is convergent, then $\sum a_n$ could be convergent or divergent. (See the note after Example 2.)

(b) If $a_n < b_n$ for all n, then $\sum a_n$ is convergent. [This is part (i) of the Comparison Test.]

2. (a) If $a_n > b_n$ for all n, then $\sum a_n$ is divergent. [This is part (ii) of the Comparison Test.]

(b) We cannot say anything about $\sum a_n$. If $a_n < b_n$ for all n and $\sum b_n$ is divergent, then $\sum a_n$ could be convergent or divergent.

3. $\dfrac{n}{2n^3+1} < \dfrac{n}{2n^3} = \dfrac{1}{2n^2} < \dfrac{1}{n^2}$ for all $n \geq 1$, so $\displaystyle\sum_{n=1}^{\infty} \dfrac{n}{2n^3+1}$ converges by comparison with $\displaystyle\sum_{n=1}^{\infty} \dfrac{1}{n^2}$, which converges

because it is a p-series with $p = 2 > 1$.

4. $\dfrac{n^3}{n^4-1} > \dfrac{n^3}{n^4} = \dfrac{1}{n}$ for all $n \geq 2$, so $\displaystyle\sum_{n=2}^{\infty} \dfrac{n^3}{n^4-1}$ diverges by comparison with $\displaystyle\sum_{n=2}^{\infty} \dfrac{1}{n}$, which diverges because it is a p-series

with $p = 1 \leq 1$ (the harmonic series).

5. $\dfrac{n+1}{n\sqrt{n}} > \dfrac{n}{n\sqrt{n}} = \dfrac{1}{\sqrt{n}}$ for all $n \geq 1$, so $\displaystyle\sum_{n=1}^{\infty} \dfrac{n+1}{n\sqrt{n}}$ diverges by comparison with $\displaystyle\sum_{n=1}^{\infty} \dfrac{1}{\sqrt{n}}$, which diverges because it is a

p-series with $p = \frac{1}{2} \leq 1$.

6. $\dfrac{n-1}{n^2\sqrt{n}} < \dfrac{n}{n^2\,n^{1/2}} = \dfrac{1}{n^{3/2}}$ for all $n \geq 1$, so $\displaystyle\sum_{n=1}^{\infty} \dfrac{n-1}{n^2\sqrt{n}}$ converges by comparison with $\displaystyle\sum_{n=1}^{\infty} \dfrac{1}{n^{3/2}}$, which converges because

it is a p-series with $p = \frac{3}{2} > 1$.

7. $\dfrac{9^n}{3+10^n} < \dfrac{9^n}{10^n} = \left(\dfrac{9}{10}\right)^n$ for all $n \geq 1$. $\displaystyle\sum_{n=1}^{\infty} \left(\frac{9}{10}\right)^n$ is a convergent geometric series $\left(|r| = \frac{9}{10} < 1\right)$, so $\displaystyle\sum_{n=1}^{\infty} \dfrac{9^n}{3+10^n}$

converges by the Comparison Test.

8. $\dfrac{4+3^n}{2^n} > \dfrac{3^n}{2^n} = \left(\dfrac{3}{2}\right)^n$ for all $n \geq 1$, so $\displaystyle\sum_{n=1}^{\infty} \dfrac{4+3^n}{2^n}$ diverges by comparison with the divergent geometric series $\displaystyle\sum_{n=1}^{\infty} \left(\frac{3}{2}\right)^n$.

9. $\dfrac{\cos^2 n}{n^2+1} \leq \dfrac{1}{n^2+1} < \dfrac{1}{n^2}$, so the series $\displaystyle\sum_{n=1}^{\infty} \dfrac{\cos^2 n}{n^2+1}$ converges by comparison with the p-series $\displaystyle\sum_{n=1}^{\infty} \dfrac{1}{n^2}$ $[p = 2 > 1]$.

10. $\dfrac{n^2-1}{3n^4+1} < \dfrac{n^2}{3n^4+1} < \dfrac{n^2}{3n^4} = \dfrac{1}{3}\dfrac{1}{n^2}$. $\displaystyle\sum_{n=1}^{\infty} \dfrac{n^2-1}{3n^4+1}$ converges by comparison with $\displaystyle\sum_{n=1}^{\infty} \dfrac{1}{3n^2}$, which converges because it is

a constant multiple of a convergent p-series $[p = 2 > 1]$. The terms of the given series are positive for $n > 1$, which is good

enough.

11. $\dfrac{n-1}{n\,4^n}$ is positive for $n > 1$ and $\dfrac{n-1}{n\,4^n} < \dfrac{n}{n\,4^n} = \dfrac{1}{4^n} = \left(\dfrac{1}{4}\right)^n$, so $\displaystyle\sum_{n=1}^{\infty} \dfrac{n-1}{n\,4^n}$ converges by comparison with the convergent

geometric series $\displaystyle\sum_{n=1}^{\infty} \left(\frac{1}{4}\right)^n$.

12. $\dfrac{1+\sin n}{10^n} \leq \dfrac{2}{10^n}$ and $\displaystyle\sum_{n=0}^{\infty} \dfrac{2}{10^n} = 2\displaystyle\sum_{n=0}^{\infty} \left(\dfrac{1}{10}\right)^n$, so the given series converges by comparison with a constant multiple of a

convergent geometric series.

13. $\dfrac{\arctan n}{n^{1.2}} < \dfrac{\pi/2}{n^{1.2}}$ for all $n \geq 1$, so $\displaystyle\sum_{n=1}^{\infty} \dfrac{\arctan n}{n^{1.2}}$ converges by comparison with $\dfrac{\pi}{2}\displaystyle\sum_{n=1}^{\infty} \dfrac{1}{n^{1.2}}$, which converges because it is a

constant times a p-series with $p = 1.2 > 1$.

14. $\dfrac{\sqrt{n}}{n-1} > \dfrac{\sqrt{n}}{n} = \dfrac{1}{\sqrt{n}}$, so $\displaystyle\sum_{n=2}^{\infty} \dfrac{\sqrt{n}}{n-1}$ diverges by comparison with the divergent (partial) p-series $\displaystyle\sum_{n=2}^{\infty} \dfrac{1}{\sqrt{n}}$ $\left[p = \frac{1}{2} \leq 1\right]$.

15. $\dfrac{2+(-1)^n}{n\sqrt{n}} \le \dfrac{3}{n\sqrt{n}}$, and $\displaystyle\sum_{n=1}^{\infty} \dfrac{3}{n\sqrt{n}}$ converges because it is a constant multiple of the convergent p-series $\displaystyle\sum_{n=1}^{\infty} \dfrac{1}{n\sqrt{n}}$

$\left[p = \frac{3}{2} > 1\right]$, so the given series converges by the Comparison Test.

16. $\dfrac{1}{\sqrt{n^3+1}} < \dfrac{1}{\sqrt{n^3}} = \dfrac{1}{n^{3/2}}$, so $\displaystyle\sum_{n=1}^{\infty} \dfrac{1}{\sqrt{n^3+1}}$ converges by comparison with the convergent p-series

$\displaystyle\sum_{n=1}^{\infty} \dfrac{1}{n^{3/2}}$ $\left[p = \frac{3}{2} > 1\right]$.

17. Use the Limit Comparison Test with $a_n = \dfrac{1}{\sqrt{n^2+1}}$ and $b_n = \dfrac{1}{n}$:

$\displaystyle\lim_{n\to\infty} \dfrac{a_n}{b_n} = \lim_{n\to\infty} \dfrac{n}{\sqrt{n^2+1}} = \lim_{n\to\infty} \dfrac{1}{\sqrt{1+(1/n^2)}} = 1 > 0$. Since the harmonic series $\displaystyle\sum_{n=1}^{\infty} \dfrac{1}{n}$ diverges, so does

$\displaystyle\sum_{n=1}^{\infty} \dfrac{1}{\sqrt{n^2+1}}$.

18. Use the Limit Comparison Test with $a_n = \dfrac{1}{2n+3}$ and $b_n = \dfrac{1}{n}$: $\displaystyle\lim_{n\to\infty} \dfrac{a_n}{b_n} = \lim_{n\to\infty} \dfrac{n}{2n+3} = \lim_{n\to\infty} \dfrac{1}{2+(3/n)} = \dfrac{1}{2} > 0$.

Since the harmonic series $\displaystyle\sum_{n=1}^{\infty} \dfrac{1}{n}$ diverges, so does $\displaystyle\sum_{n=1}^{\infty} \dfrac{1}{2n+3}$.

19. Use the Limit Comparison Test with $a_n = \dfrac{1+4^n}{1+3^n}$ and $b_n = \dfrac{4^n}{3^n}$:

$\displaystyle\lim_{n\to\infty} \dfrac{a_n}{b_n} = \lim_{n\to\infty} \dfrac{\dfrac{1+4^n}{1+3^n}}{\dfrac{4^n}{3^n}} = \lim_{n\to\infty} \dfrac{1+4^n}{1+3^n} \cdot \dfrac{3^n}{4^n} = \lim_{n\to\infty} \dfrac{1+4^n}{4^n} \cdot \dfrac{3^n}{1+3^n} = \lim_{n\to\infty} \left(\dfrac{1}{4^n}+1\right) \cdot \dfrac{1}{\dfrac{1}{3^n}+1} = 1 > 0$

Since the geometric series $\sum b_n = \sum \left(\frac{4}{3}\right)^n$ diverges, so does $\displaystyle\sum_{n=1}^{\infty} \dfrac{1+4^n}{1+3^n}$. Alternatively, use the Comparison Test with

$\dfrac{1+4^n}{1+3^n} > \dfrac{1+4^n}{3^n+3^n} > \dfrac{4^n}{2(3^n)} = \dfrac{1}{2}\left(\dfrac{4}{3}\right)^n$ or use the Test for Divergence.

20. $4^n > n$ for all $n \ge 1$ since the function $f(x) = 4^x - x$ satisfies $f(1) = 3$ and $f'(x) = 4^x \ln 4 - 1 > 0$ for $x \ge 1$, so

$\dfrac{n+4^n}{n+6^n} < \dfrac{4^n+4^n}{n+6^n} < \dfrac{2\cdot 4^n}{6^n} = 2\left(\frac{4}{6}\right)^n$, so the series $\displaystyle\sum_{n=1}^{\infty} \dfrac{n+4^n}{n+6^n}$ converges by comparison with $2\displaystyle\sum_{n=1}^{\infty} \left(\frac{2}{3}\right)^n$, which is a

constant multiple of a convergent geometric series $\left[|r| = \frac{2}{3} < 1\right]$.

Or: Use the Limit Comparison Test with $a_n = \dfrac{n+4^n}{n+6^n}$ and $b_n = \left(\frac{2}{3}\right)^n$.

21. Use the Limit Comparison Test with $a_n = \dfrac{\sqrt{n+2}}{2n^2+n+1}$ and $b_n = \dfrac{1}{n^{3/2}}$:

$\displaystyle\lim_{n\to\infty} \dfrac{a_n}{b_n} = \lim_{n\to\infty} \dfrac{n^{3/2}\sqrt{n+2}}{2n^2+n+1} = \lim_{n\to\infty} \dfrac{(n^{3/2}\sqrt{n+2})/(n^{3/2}\sqrt{n})}{(2n^2+n+1)/n^2} = \lim_{n\to\infty} \dfrac{\sqrt{1+2/n}}{2+1/n+1/n^2} = \dfrac{\sqrt{1}}{2} = \dfrac{1}{2} > 0$.

Since $\displaystyle\sum_{n=1}^{\infty} \dfrac{1}{n^{3/2}}$ is a convergent p-series $\left[p = \frac{3}{2} > 1\right]$, the series $\displaystyle\sum_{n=1}^{\infty} \dfrac{\sqrt{n+2}}{2n^2+n+1}$ also converges.

22. Use the Limit Comparison Test with $a_n = \dfrac{n+2}{(n+1)^3}$ and $b_n = \dfrac{1}{n^2}$:

$$\lim_{n\to\infty}\frac{a_n}{b_n} = \lim_{n\to\infty}\frac{n^2(n+2)}{(n+1)^3} = \lim_{n\to\infty}\frac{1+\frac{2}{n}}{\left(1+\frac{1}{n}\right)^3} = 1 > 0. \text{ Since } \sum_{n=3}^{\infty}\frac{1}{n^2} \text{ is a convergent (partial) } p\text{-series } [p=2>1],$$

the series $\displaystyle\sum_{n=3}^{\infty}\frac{n+2}{(n+1)^3}$ also converges.

23. Use the Limit Comparison Test with $a_n = \dfrac{5+2n}{(1+n^2)^2}$ and $b_n = \dfrac{1}{n^3}$:

$$\lim_{n\to\infty}\frac{a_n}{b_n} = \lim_{n\to\infty}\frac{n^3(5+2n)}{(1+n^2)^2} = \lim_{n\to\infty}\frac{5n^3+2n^4}{(1+n^2)^2}\cdot\frac{1/n^4}{1/(n^2)^2} = \lim_{n\to\infty}\frac{\frac{5}{n}+2}{\left(\frac{1}{n^2}+1\right)^2} = 2 > 0. \text{ Since } \sum_{n=1}^{\infty}\frac{1}{n^3} \text{ is a convergent}$$

p-series $[p=3>1]$, the series $\displaystyle\sum_{n=1}^{\infty}\frac{5+2n}{(1+n^2)^2}$ also converges.

24. If $a_n = \dfrac{n^2-5n}{n^3+n+1}$ and $b_n = \dfrac{1}{n}$, then $\displaystyle\lim_{n\to\infty}\frac{a_n}{b_n} = \lim_{n\to\infty}\frac{n^3-5n^2}{n^3+n+1} = \lim_{n\to\infty}\frac{1-5/n}{1+1/n^2+1/n^3} = 1 > 0,$

so $\displaystyle\sum_{n=1}^{\infty}\frac{n^2-5n}{n^3+n+1}$ diverges by the Limit Comparison Test with the divergent harmonic series $\displaystyle\sum_{n=1}^{\infty}\frac{1}{n}$.

(Note that $a_n > 0$ for $n \ge 6$.)

25. If $a_n = \dfrac{1+n+n^2}{\sqrt{1+n^2+n^6}}$ and $b_n = \dfrac{1}{n}$, then $\displaystyle\lim_{n\to\infty}\frac{a_n}{b_n} = \lim_{n\to\infty}\frac{n+n^2+n^3}{\sqrt{1+n^2+n^6}} = \lim_{n\to\infty}\frac{1/n^2+1/n+1}{\sqrt{1/n^6+1/n^4+1}} = 1 > 0,$

so $\displaystyle\sum_{n=1}^{\infty}\frac{1+n+n^2}{\sqrt{1+n^2+n^6}}$ diverges by the Limit Comparison Test with the divergent harmonic series $\displaystyle\sum_{n=1}^{\infty}\frac{1}{n}$.

26. If $a_n = \dfrac{n+5}{\sqrt[3]{n^7+n^2}}$ and $b_n = \dfrac{n}{\sqrt[3]{n^7}} = \dfrac{n}{n^{7/3}} = \dfrac{1}{n^{4/3}}$, then

$$\lim_{n\to\infty}\frac{a_n}{b_n} = \lim_{n\to\infty}\frac{n^{7/3}+5n^{4/3}}{(n^7+n^2)^{1/3}}\cdot\frac{n^{-7/3}}{n^{-7/3}} = \lim_{n\to\infty}\frac{1+5/n}{[(n^7+n^2)/n^7]^{1/3}} = \lim_{n\to\infty}\frac{1+5/n}{(1+1/n^5)^{1/3}} = \frac{1+0}{(1+0)^{1/3}} = 1 > 0,$$

so $\displaystyle\sum_{n=1}^{\infty}\frac{n+5}{\sqrt[3]{n^7+n^2}}$ converges by the Limit Comparison Test with the convergent p-series $\displaystyle\sum_{n=1}^{\infty}\frac{1}{n^{4/3}}$.

27. Use the Limit Comparison Test with $a_n = \left(1+\dfrac{1}{n}\right)^2 e^{-n}$ and $b_n = e^{-n}$: $\displaystyle\lim_{n\to\infty}\frac{a_n}{b_n} = \lim_{n\to\infty}\left(1+\frac{1}{n}\right)^2 = 1 > 0.$ Since

$\displaystyle\sum_{n=1}^{\infty}e^{-n} = \sum_{n=1}^{\infty}\frac{1}{e^n}$ is a convergent geometric series $\left[|r|=\frac{1}{e}<1\right]$, the series $\displaystyle\sum_{n=1}^{\infty}\left(1+\frac{1}{n}\right)^2 e^{-n}$ also converges.

28. $\dfrac{e^{1/n}}{n} > \dfrac{1}{n}$ for all $n \ge 1$, so $\displaystyle\sum_{n=1}^{\infty}\frac{e^{1/n}}{n}$ diverges by comparison with the harmonic series $\displaystyle\sum_{n=1}^{\infty}\frac{1}{n}$.

29. Clearly $n! = n(n-1)(n-2)\cdots(3)(2) \ge 2\cdot2\cdot2\cdots\cdots2\cdot2 = 2^{n-1}$, so $\dfrac{1}{n!} \le \dfrac{1}{2^{n-1}}$. $\displaystyle\sum_{n=1}^{\infty}\frac{1}{2^{n-1}}$ is a convergent geometric

series $\left[|r|=\frac{1}{2}<1\right]$, so $\displaystyle\sum_{n=1}^{\infty}\frac{1}{n!}$ converges by the Comparison Test.

30. $\dfrac{n!}{n^n} = \dfrac{1 \cdot 2 \cdot 3 \cdot \cdots \cdot (n-1)n}{n \cdot n \cdot n \cdot \cdots \cdot n \cdot n} \le \dfrac{1}{n} \cdot \dfrac{2}{n} \cdot 1 \cdot 1 \cdot \cdots \cdot 1$ for $n \ge 2$, so since $\displaystyle\sum_{n=1}^{\infty} \dfrac{2}{n^2}$ converges $[p = 2 > 1]$, $\displaystyle\sum_{n=1}^{\infty} \dfrac{n!}{n^n}$ converges also by the Comparison Test.

31. Use the Limit Comparison Test with $a_n = \sin\left(\dfrac{1}{n}\right)$ and $b_n = \dfrac{1}{n}$. Then $\sum a_n$ and $\sum b_n$ are series with positive terms and

$$\lim_{n \to \infty} \dfrac{a_n}{b_n} = \lim_{n \to \infty} \dfrac{\sin(1/n)}{1/n} = \lim_{\theta \to 0} \dfrac{\sin\theta}{\theta} = 1 > 0. \text{ Since } \sum_{n=1}^{\infty} b_n \text{ is the divergent harmonic series,}$$

$\displaystyle\sum_{n=1}^{\infty} \sin(1/n)$ also diverges.   [Note that we could also use l'Hospital's Rule to evaluate the limit:

$$\lim_{x \to \infty} \dfrac{\sin(1/x)}{1/x} \overset{\text{H}}{=} \lim_{x \to \infty} \dfrac{\cos(1/x) \cdot (-1/x^2)}{-1/x^2} = \lim_{x \to \infty} \cos\dfrac{1}{x} = \cos 0 = 1.]$$

32. Use the Limit Comparison Test with $a_n = \dfrac{1}{n^{1+1/n}}$ and $b_n = \dfrac{1}{n}$. $\displaystyle\lim_{n \to \infty} \dfrac{a_n}{b_n} = \lim_{n \to \infty} \dfrac{n}{n^{1+1/n}} = \lim_{n \to \infty} \dfrac{1}{n^{1/n}} = 1$

$\left[\text{since } \displaystyle\lim_{x \to \infty} x^{1/x} = 1 \text{ by l'Hospital's Rule}\right]$, so $\displaystyle\sum_{n=1}^{\infty} \dfrac{1}{n}$ diverges   [harmonic series] $\Rightarrow$ $\displaystyle\sum_{n=1}^{\infty} \dfrac{1}{n^{1+1/n}}$ diverges.

33. $\displaystyle\sum_{n=1}^{10} \dfrac{1}{\sqrt{n^4+1}} = \dfrac{1}{\sqrt{2}} + \dfrac{1}{\sqrt{17}} + \dfrac{1}{\sqrt{82}} + \cdots + \dfrac{1}{\sqrt{10,001}} \approx 1.24856$. Now $\dfrac{1}{\sqrt{n^4+1}} < \dfrac{1}{\sqrt{n^4}} = \dfrac{1}{n^2}$, so the error is

$$R_{10} \le T_{10} \le \int_{10}^{\infty} \dfrac{1}{x^2}\, dx = \lim_{t \to \infty} \left[-\dfrac{1}{x}\right]_{10}^{t} = \lim_{t \to \infty} \left(-\dfrac{1}{t} + \dfrac{1}{10}\right) = \dfrac{1}{10} = 0.1.$$

34. $\displaystyle\sum_{n=1}^{10} \dfrac{\sin^2 n}{n^3} = \dfrac{\sin^2 1}{1} + \dfrac{\sin^2 2}{8} + \dfrac{\sin^2 3}{27} + \cdots + \dfrac{\sin^2 10}{1000} \approx 0.83253$. Now $\dfrac{\sin^2 n}{n^3} \le \dfrac{1}{n^3}$, so the error is

$$R_{10} \le T_{10} \le \int_{10}^{\infty} \dfrac{1}{x^3}\, dx = \lim_{t \to \infty} \left[-\dfrac{1}{2x^2}\right]_{10}^{t} = \lim_{t \to \infty} \left(-\dfrac{1}{2t^2} + \dfrac{1}{200}\right) = \dfrac{1}{200} = 0.005.$$

35. $\displaystyle\sum_{n=1}^{10} \dfrac{1}{1+2^n} = \dfrac{1}{3} + \dfrac{1}{5} + \dfrac{1}{9} + \cdots + \dfrac{1}{1025} \approx 0.76352$. Now $\dfrac{1}{1+2^n} < \dfrac{1}{2^n}$, so the error is

$$R_{10} \le T_{10} = \sum_{n=11}^{\infty} \dfrac{1}{2^n} = \dfrac{1/2^{11}}{1 - 1/2} \quad \text{[geometric series]} \approx 0.00098.$$

36. $\displaystyle\sum_{n=1}^{10} \dfrac{n}{(n+1)3^n} = \dfrac{1}{6} + \dfrac{2}{27} + \dfrac{3}{108} + \cdots + \dfrac{10}{649,539} \approx 0.283597$. Now $\dfrac{n}{(n+1)3^n} < \dfrac{n}{n \cdot 3^n} = \dfrac{1}{3^n}$, so the error is

$$R_{10} \le T_{10} = \sum_{n=11}^{\infty} \dfrac{1}{3^n} = \dfrac{1/3^{11}}{1 - 1/3} \approx 0.0000085.$$

37. Since $\dfrac{d_n}{10^n} \le \dfrac{9}{10^n}$ for each n, and since $\displaystyle\sum_{n=1}^{\infty} \dfrac{9}{10^n}$ is a convergent geometric series $\left(|r| = \dfrac{1}{10} < 1\right)$, $0.d_1 d_2 d_3 \ldots = \displaystyle\sum_{n=1}^{\infty} \dfrac{d_n}{10^n}$ will always converge by the Comparison Test.

38. Clearly, if $p < 0$ then the series diverges, since $\displaystyle\lim_{n \to \infty} \dfrac{1}{n^p \ln n} = \infty$. If $0 \le p \le 1$, then $n^p \ln n \le n \ln n$ $\Rightarrow$

$\dfrac{1}{n^p \ln n} \ge \dfrac{1}{n \ln n}$ and $\displaystyle\sum_{n=2}^{\infty} \dfrac{1}{n \ln n}$ diverges (Exercise 12.3.21 [ET 11.3.21]), so $\displaystyle\sum_{n=2}^{\infty} \dfrac{1}{n^p \ln n}$ diverges. If $p > 1$, use the Limit

Comparison Test with $a_n = \dfrac{1}{n^p \ln n}$ and $b_n = \dfrac{1}{n^p}$. $\displaystyle\sum_{n=2}^{\infty} b_n$ converges, and $\displaystyle\lim_{n\to\infty} \frac{a_n}{b_n} = \lim_{n\to\infty} \frac{1}{\ln n} = 0$, so $\displaystyle\sum_{n=2}^{\infty} \frac{1}{n^p \ln n}$ also

converges. (Or use the Comparison Test, since $n^p \ln n > n^p$ for $n > e$.) In summary, the series converges if and only if $p > 1$.

39. Since $\sum a_n$ converges, $\displaystyle\lim_{n\to\infty} a_n = 0$, so there exists N such that $|a_n - 0| < 1$ for all $n > N \;\;\Rightarrow\;\; 0 \le a_n < 1$ for

all $n > N \;\;\Rightarrow\;\; 0 \le a_n^2 \le a_n$. Since $\sum a_n$ converges, so does $\sum a_n^2$ by the Comparison Test.

40. (a) Since $\displaystyle\lim_{n\to\infty} (a_n/b_n) = 0$, there is a number $N > 0$ such that $|a_n/b_n - 0| < 1$ for all $n > N$, and so $a_n < b_n$ since a_n

and b_n are positive. Thus, since $\sum b_n$ converges, so does $\sum a_n$ by the Comparison Test.

(b) (i) If $a_n = \dfrac{\ln n}{n^3}$ and $b_n = \dfrac{1}{n^2}$, then $\displaystyle\lim_{n\to\infty} \frac{a_n}{b_n} = \lim_{n\to\infty} \frac{\ln n}{n} = \lim_{x\to\infty} \frac{\ln x}{x} \overset{\text{H}}{=} \lim_{x\to\infty} \frac{1/x}{1} = 0$, so $\displaystyle\sum_{n=1}^{\infty} \frac{\ln n}{n^3}$ converges by

part (a).

(ii) If $a_n = \dfrac{\ln n}{\sqrt{n}\, e^n}$ and $b_n = \dfrac{1}{e^n}$, then $\displaystyle\lim_{n\to\infty} \frac{a_n}{b_n} = \lim_{n\to\infty} \frac{\ln n}{\sqrt{n}} = \lim_{x\to\infty} \frac{\ln x}{\sqrt{x}} \overset{\text{H}}{=} \lim_{x\to\infty} \frac{1/x}{1/(2\sqrt{x})} = \lim_{x\to\infty} \frac{2}{\sqrt{x}} = 0$. Now

$\sum b_n$ is a convergent geometric series with ratio $r = 1/e$ $[\,|r| < 1\,]$, so $\sum a_n$ converges by part (a).

41. (a) Since $\displaystyle\lim_{n\to\infty} \frac{a_n}{b_n} = \infty$, there is an integer N such that $\dfrac{a_n}{b_n} > 1$ whenever $n > N$. (Take $M = 1$ in Definition 12.1.5

[ET 11.1.5].) Then $a_n > b_n$ whenever $n > N$ and since $\sum b_n$ is divergent, $\sum a_n$ is also divergent by the Comparison

Test.

(b) (i) If $a_n = \dfrac{1}{\ln n}$ and $b_n = \dfrac{1}{n}$ for $n \ge 2$, then $\displaystyle\lim_{n\to\infty} \frac{a_n}{b_n} = \lim_{n\to\infty} \frac{n}{\ln n} = \lim_{x\to\infty} \frac{x}{\ln x} \overset{\text{H}}{=} \lim_{x\to\infty} \frac{1}{1/x} = \lim_{x\to\infty} x = \infty$,

so by part (a), $\displaystyle\sum_{n=2}^{\infty} \frac{1}{\ln n}$ is divergent.

(ii) If $a_n = \dfrac{\ln n}{n}$ and $b_n = \dfrac{1}{n}$, then $\displaystyle\sum_{n=1}^{\infty} b_n$ is the divergent harmonic series and $\displaystyle\lim_{n\to\infty} \frac{a_n}{b_n} = \lim_{n\to\infty} \ln n = \lim_{x\to\infty} \ln x = \infty$,

so $\displaystyle\sum_{n=1}^{\infty} a_n$ diverges by part (a).

42. Let $a_n = \dfrac{1}{n^2}$ and $b_n = \dfrac{1}{n}$. Then $\displaystyle\lim_{n\to\infty} \frac{a_n}{b_n} = \lim_{n\to\infty} \frac{1}{n} = 0$, but $\sum b_n$ diverges while $\sum a_n$ converges.

43. $\displaystyle\lim_{n\to\infty} n a_n = \lim_{n\to\infty} \frac{a_n}{1/n}$, so we apply the Limit Comparison Test with $b_n = \dfrac{1}{n}$. Since $\displaystyle\lim_{n\to\infty} n a_n > 0$ we know that either both

series converge or both series diverge, and we also know that $\displaystyle\sum_{n=1}^{\infty} \frac{1}{n}$ diverges [p-series with $p = 1$]. Therefore, $\sum a_n$ must be

divergent.

44. First we observe that, by l'Hospital's Rule, $\displaystyle\lim_{x\to0} \frac{\ln(1+x)}{x} = \lim_{x\to0} \frac{1}{1+x} = 1$. Also, if $\sum a_n$ converges, then $\displaystyle\lim_{n\to\infty} a_n = 0$ by

Theorem 12.2.6 [ET 11.2.6]. Therefore, $\displaystyle\lim_{n\to\infty} \frac{\ln(1+a_n)}{a_n} = \lim_{x\to0} \frac{\ln(1+x)}{x} = 1 > 0$. We are given that $\sum a_n$ is convergent

and $a_n > 0$. Thus, $\sum \ln(1+a_n)$ is convergent by the Limit Comparison Test.

45. Yes. Since $\sum a_n$ is a convergent series with positive terms, $\lim_{n \to \infty} a_n = 0$ by Theorem 12.2.6 [ET 11.2.6], and

$\sum b_n = \sum \sin(a_n)$ is a series with positive terms (for large enough n). We have $\lim_{n \to \infty} \dfrac{b_n}{a_n} = \lim_{n \to \infty} \dfrac{\sin(a_n)}{a_n} = 1 > 0$ by

Theorem 3.4.2 [ET 3.3.2]. Thus, $\sum b_n$ is also convergent by the Limit Comparison Test.

46. Yes. Since $\sum a_n$ converges, its terms approach 0 as $n \to \infty$, so for some integer N, $a_n \leq 1$ for all $n \geq N$. But then

$\sum_{n=1}^{\infty} a_n b_n = \sum_{n=1}^{N-1} a_n b_n + \sum_{n=N}^{\infty} a_n b_n \leq \sum_{n=1}^{N-1} a_n b_n + \sum_{n=N}^{\infty} b_n$. The first term is a finite sum, and the second term

converges since $\sum_{n=1}^{\infty} b_n$ converges. So $\sum a_n b_n$ converges by the Comparison Test.

12.5 Alternating Series

1. (a) An alternating series is a series whose terms are alternately positive and negative.

(b) An alternating series $\sum_{n=1}^{\infty} (-1)^{n-1} b_n$ converges if $0 < b_{n+1} \leq b_n$ for all n and $\lim_{n \to \infty} b_n = 0$. (This is the Alternating

Series Test.)

(c) The error involved in using the partial sum s_n as an approximation to the total sum s is the remainder $R_n = s - s_n$ and the

size of the error is smaller than b_{n+1}; that is, $|R_n| \leq b_{n+1}$. (This is the Alternating Series Estimation Theorem.)

2. $-\dfrac{1}{3} + \dfrac{2}{4} - \dfrac{3}{5} + \dfrac{4}{6} - \dfrac{5}{7} + \cdots = \sum_{n=1}^{\infty} (-1)^n \dfrac{n}{n+2}$. Here $a_n = (-1)^n \dfrac{n}{n+2}$. Since $\lim_{n \to \infty} a_n \neq 0$ (in fact the limit does not

exist), the series diverges by the Test for Divergence.

3. $\dfrac{4}{7} - \dfrac{4}{8} + \dfrac{4}{9} - \dfrac{4}{10} + \dfrac{4}{11} - \cdots = \sum_{n=1}^{\infty} (-1)^{n-1} \dfrac{4}{n+6}$. Now $b_n = \dfrac{4}{n+6} > 0$, $\{b_n\}$ is decreasing, and $\lim_{n \to \infty} b_n = 0$, so the

series converges by the Alternating Series Test.

4. $\dfrac{1}{\sqrt{2}} - \dfrac{1}{\sqrt{3}} + \dfrac{1}{\sqrt{4}} - \dfrac{1}{\sqrt{5}} + \dfrac{1}{\sqrt{6}} - \cdots = \sum_{n=1}^{\infty} (-1)^{n-1} \dfrac{1}{\sqrt{n+1}}$. Now $b_n = \dfrac{1}{\sqrt{n+1}} > 0$, $\{b_n\}$ is decreasing, and

$\lim_{n \to \infty} b_n = 0$, so the series converges by the Alternating Series Test.

5. $\sum_{n=1}^{\infty} a_n = \sum_{n=1}^{\infty} (-1)^{n-1} \dfrac{1}{2n+1} = \sum_{n=1}^{\infty} (-1)^{n-1} b_n$. Now $b_n = \dfrac{1}{2n+1} > 0$, $\{b_n\}$ is decreasing, and $\lim_{n \to \infty} b_n = 0$, so the

series converges by the Alternating Series Test.

6. $\sum_{n=1}^{\infty} a_n = \sum_{n=1}^{\infty} (-1)^{n-1} \dfrac{1}{\ln(n+4)} = \sum_{n=1}^{\infty} (-1)^{n-1} b_n$. Now $b_n = \dfrac{1}{\ln(n+4)} > 0$, $\{b_n\}$ is decreasing, and $\lim_{n \to \infty} b_n = 0$, so

the series converges by the Alternating Series Test.

7. $\sum_{n=1}^{\infty} a_n = \sum_{n=1}^{\infty} (-1)^n \dfrac{3n-1}{2n+1} = \sum_{n=1}^{\infty} (-1)^n b_n$. Now $\lim_{n \to \infty} b_n = \lim_{n \to \infty} \dfrac{3 - 1/n}{2 + 1/n} = \dfrac{3}{2} \neq 0$. Since $\lim_{n \to \infty} a_n \neq 0$

(in fact the limit does not exist), the series diverges by the Test for Divergence.

8. $b_n = \dfrac{n}{\sqrt{n^3 + 2}} > 0$ for $n \geq 1$. {b_n} is decreasing for $n \geq 2$ since

$$\left(\frac{x}{\sqrt{x^3+2}}\right)' = \frac{(x^3+2)^{1/2}(1) - x \cdot \frac{1}{2}(x^3+2)^{-1/2}(3x^2)}{\left(\sqrt{x^3+2}\right)^2} = \frac{\frac{1}{2}(x^3+2)^{-1/2}[2(x^3+2) - 3x^3]}{(x^3+2)^1} = \frac{4 - x^3}{2(x^3+2)^{3/2}} < 0 \text{ for}$$

$x > \sqrt[3]{4} \approx 1.6$. Also, $\displaystyle\lim_{n\to\infty} b_n = \lim_{n\to\infty} \frac{n/n}{\sqrt{n^3+2}/\sqrt{n^2}} = \lim_{n\to\infty} \frac{1}{\sqrt{n + 2/n^2}} = 0$. Thus, the series $\displaystyle\sum_{n=1}^{\infty} (-1)^n \frac{n}{\sqrt{n^3+2}}$

converges by the Alternating Series Test.

9. $b_n = \dfrac{n}{10^n} > 0$ for $n \geq 1$. {b_n} is decreasing for $n \geq 1$ since

$$\left(\frac{x}{10^x}\right)' = \frac{10^x(1) - x \cdot 10^x \ln 10}{(10^x)^2} = \frac{10^x(1 - x \ln 10)}{(10^x)^2} = \frac{1 - x \ln 10}{10^x} < 0 \text{ for } 1 - x\ln 10 < 0 \;\Rightarrow\; x \ln 10 > 1 \;\Rightarrow\;$$

$x > \dfrac{1}{\ln 10} \approx 0.4$. Also, $\displaystyle\lim_{n\to\infty} b_n = \lim_{n\to\infty} \frac{n}{10^n} = \lim_{x\to\infty} \frac{x}{10^x} \overset{\text{H}}{=} \lim_{x\to\infty} \frac{x}{10^x \ln 10} = 0$. Thus, the series $\displaystyle\sum_{n=1}^{\infty} (-1)^n \frac{n}{10^n}$

converges by the Alternating Series Test.

10. $\displaystyle\sum_{n=1}^{\infty} a_n = \sum_{n=1}^{\infty} (-1)^n \frac{\sqrt{n}}{1 + 2\sqrt{n}} = \sum_{n=1}^{\infty} (-1)^n b_n$. Now $\displaystyle\lim_{n\to\infty} b_n = \lim_{n\to\infty} \frac{1}{2 + 1/\sqrt{n}} = \frac{1}{2} \neq 0$. Since $\displaystyle\lim_{n\to\infty} a_n \neq 0$

(in fact the limit does not exist), the series diverges by the Test for Divergence.

11. $b_n = \dfrac{n^2}{n^3 + 4} > 0$ for $n \geq 1$. {b_n} is decreasing for $n \geq 2$ since

$$\left(\frac{x^2}{x^3+4}\right)' = \frac{(x^3+4)(2x) - x^2(3x^2)}{(x^3+4)^2} = \frac{x(2x^3 + 8 - 3x^3)}{(x^3+4)^2} = \frac{x(8 - x^3)}{(x^3+4)^2} < 0 \text{ for } x > 2. \text{ Also,}$$

$\displaystyle\lim_{n\to\infty} b_n = \lim_{n\to\infty} \frac{1/n}{1 + 4/n^3} = 0$. Thus, the series $\displaystyle\sum_{n=1}^{\infty} (-1)^{n+1} \frac{n^2}{n^3 + 4}$ converges by the Alternating Series Test.

12. $b_n = \dfrac{e^{1/n}}{n} > 0$ for $n \geq 1$. {b_n} is decreasing since $\left(\dfrac{e^{1/x}}{x}\right)' = \dfrac{x \cdot e^{1/x}(-1/x^2) - e^{1/x} \cdot 1}{x^2} = \dfrac{-e^{1/x}(1 + x)}{x^3} < 0$ for

$x > 0$. Also, $\displaystyle\lim_{n\to\infty} b_n = 0$ since $\displaystyle\lim_{n\to\infty} e^{1/n} = 1$. Thus, the series $\displaystyle\sum_{n=1}^{\infty} (-1)^{n-1} \frac{e^{1/n}}{n}$ converges by the Alternating Series Test.

13. $\displaystyle\sum_{n=2}^{\infty} (-1)^n \frac{n}{\ln n}$. $\displaystyle\lim_{n\to\infty} \frac{n}{\ln n} = \lim_{x\to\infty} \frac{x}{\ln x} \overset{\text{H}}{=} \lim_{x\to\infty} \frac{1}{1/x} = \infty$, so the series diverges by the Test for Divergence.

14. $\displaystyle\sum_{n=1}^{\infty} (-1)^{n-1} \left(\frac{\ln n}{n}\right) = 0 + \sum_{n=2}^{\infty} (-1)^{n-1} \left(\frac{\ln n}{n}\right)$. $b_n = \dfrac{\ln n}{n} > 0$ for $n \geq 2$, and if $f(x) = \dfrac{\ln x}{x}$, then

$f'(x) = \dfrac{1 - \ln x}{x^2} < 0$ for $x > e$, so {b_n} is eventually decreasing. Also,

$\displaystyle\lim_{n\to\infty} b_n = \lim_{n\to\infty} \frac{\ln n}{n} = \lim_{x\to\infty} \frac{\ln x}{x} \overset{\text{H}}{=} \lim_{x\to\infty} \frac{1/x}{1} = 0$, so the series converges by the Alternating Series Test.

15. $\displaystyle\sum_{n=1}^{\infty} \frac{\cos n\pi}{n^{3/4}} = \sum_{n=1}^{\infty} \frac{(-1)^n}{n^{3/4}}$. $b_n = \dfrac{1}{n^{3/4}}$ is decreasing and positive and $\displaystyle\lim_{n\to\infty} \frac{1}{n^{3/4}} = 0$, so the series converges by the

Alternating Series Test.

16. $\sin\left(\dfrac{n\pi}{2}\right) = 0$ if n is even and $(-1)^k$ if $n = 2k+1$, so the series $\displaystyle\sum_{n=1}^{\infty} \dfrac{\sin(n\pi/2)}{n!} = \sum_{n=0}^{\infty} \dfrac{(-1)^n}{(2n+1)!}$.

$b_n = \dfrac{1}{(2n+1)!} > 0$, $\{b_n\}$ is decreasing, and $\displaystyle\lim_{n\to\infty} \dfrac{1}{(2n+1)!} = 0$, so the series converges by the Alternating Series Test.

17. $\displaystyle\sum_{n=1}^{\infty} (-1)^n \sin\left(\dfrac{\pi}{n}\right)$. $b_n = \sin\left(\dfrac{\pi}{n}\right) > 0$ for $n \geq 2$ and $\sin\left(\dfrac{\pi}{n}\right) \geq \sin\left(\dfrac{\pi}{n+1}\right)$, and $\displaystyle\lim_{n\to\infty} \sin\left(\dfrac{\pi}{n}\right) = \sin 0 = 0$, so the series

converges by the Alternating Series Test.

18. $\displaystyle\sum_{n=1}^{\infty} (-1)^n \cos\left(\dfrac{\pi}{n}\right)$. $\displaystyle\lim_{n\to\infty} \cos\left(\dfrac{\pi}{n}\right) = \cos(0) = 1$, so $\displaystyle\lim_{n\to\infty} (-1)^n \cos\left(\dfrac{\pi}{n}\right)$ does not exist and the series diverges by the Test

for Divergence.

19. $\dfrac{n^n}{n!} = \dfrac{n \cdot n \cdots\cdots n}{1 \cdot 2 \cdots\cdots n} \geq n \;\;\Rightarrow\;\; \displaystyle\lim_{n\to\infty} \dfrac{n^n}{n!} = \infty \;\;\Rightarrow\;\; \displaystyle\lim_{n\to\infty} \dfrac{(-1)^n n^n}{n!}$ does not exist. So the series diverges by the Test for

Divergence.

20. $\displaystyle\sum_{n=1}^{\infty} \left(-\dfrac{n}{5}\right)^n$ diverges by the Test for Divergence since $\displaystyle\lim_{n\to\infty} \left(\dfrac{n}{5}\right)^n = \infty \;\;\Rightarrow\;\; \displaystyle\lim_{n\to\infty} \left(-\dfrac{n}{5}\right)^n$ does not exist.

21.

n	a_n	s_n
1	1	1
2	-0.35355	0.64645
3	0.19245	0.83890
4	-0.125	0.71390
5	0.08944	0.80334
6	-0.06804	0.73530
7	0.05399	0.78929
8	-0.04419	0.74510
9	0.03704	0.78214
10	-0.03162	0.75051

By the Alternating Series Estimation Theorem, the error in the approximation

$\displaystyle\sum_{n=1}^{\infty} \dfrac{(-1)^{n-1}}{n^{3/2}} \approx 0.75051$ is $|s - s_{10}| \leq b_{11} = 1/(11)^{3/2} \approx 0.0275$ (to four

decimal places, rounded up).

22.

n	a_n	s_n
1	1	1
2	-0.125	0.875
3	0.03704	0.91204
4	-0.01563	0.89641
5	0.008	0.90441
6	-0.00463	0.89978
7	0.00292	0.90270
8	-0.00195	0.90074
9	0.00137	0.90212
10	-0.001	0.90112

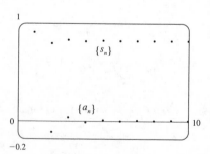

By the Alternating Series Estimation Theorem, the error in the approximation

$\displaystyle\sum_{n=1}^{\infty} \dfrac{(-1)^{n-1}}{n^3} \approx 0.90112$ is $|s - s_{10}| \leq b_{11} = 1/11^3 \approx 0.0007513$.

23. The series $\sum\limits_{n=1}^{\infty} \dfrac{(-1)^{n+1}}{n^6}$ satisfies (i) of the Alternating Series Test because $\dfrac{1}{(n+1)^6} < \dfrac{1}{n^6}$ and (ii) $\lim\limits_{n \to \infty} \dfrac{1}{n^6} = 0$, so the

series is convergent. Now $b_5 = \dfrac{1}{5^6} = 0.000064 > 0.00005$ and $b_6 = \dfrac{1}{6^6} \approx 0.00002 < 0.00005$, so by the Alternating Series

Estimation Theorem, $n = 5$. (That is, since the 6th term is less than the desired error, we need to add the first 5 terms to get the

sum to the desired accuracy.)

24. The series $\sum\limits_{n=1}^{\infty} \dfrac{(-1)^{n}}{n\,5^n}$ satisfies (i) of the Alternating Series Test because $\dfrac{1}{(n+1)5^{n+1}} < \dfrac{1}{n\,5^n}$ and (ii) $\lim\limits_{n \to \infty} \dfrac{1}{n\,5^n} = 0$, so

the series is convergent. Now $b_4 = \dfrac{1}{4 \cdot 5^4} = 0.0004 > 0.0001$ and $b_5 = \dfrac{1}{5 \cdot 5^5} = 0.000064 < 0.0001$, so by the Alternating

Series Estimation Theorem, $n = 4$. (That is, since the 5th term is less than the desired error, we need to add the first 4 terms to

get the sum to the desired accuracy.)

25. The series $\sum\limits_{n=0}^{\infty} \dfrac{(-1)^{n}}{10^n\, n!}$ satisfies (i) of the Alternating Series Test because $\dfrac{1}{10^{n+1}(n+1)!} < \dfrac{1}{10^n\, n!}$ and (ii) $\lim\limits_{n \to \infty} \dfrac{1}{10^n\, n!} = 0$,

so the series is convergent. Now $b_3 = \dfrac{1}{10^3\,3!} \approx 0.000\,167 > 0.000\,005$ and $b_4 = \dfrac{1}{10^4\,4!} = 0.000\,004 < 0.000\,005$, so by

the Alternating Series Estimation Theorem, $n = 4$ (since the series starts with $n = 0$, not $n = 1$). (That is, since the 5th term

is less than the desired error, we need to add the first 4 terms to get the sum to the desired accuracy.)

26. The series $\sum\limits_{n=1}^{\infty} (-1)^{n-1} n e^{-n} = \sum\limits_{n=1}^{\infty} (-1)^{n-1} \dfrac{n}{e^n}$ satisfies (i) of the Alternating Series Test because

$\left(\dfrac{x}{e^x}\right)' = \dfrac{e^x(1) - xe^x}{(e^x)^2} = \dfrac{e^x(1-x)}{(e^x)^2} = \dfrac{1-x}{e^x} < 0$ for $x > 1$ and (ii) $\lim\limits_{n \to \infty} \dfrac{n}{e^n} = \lim\limits_{x \to \infty} \dfrac{x}{e^x} \overset{\text{H}}{=} \lim\limits_{x \to \infty} \dfrac{1}{e^x} = 0$, so the series is

convergent. Now $b_6 = 6/e^6 \approx 0.015 > 0.01$ and $b_7 = 7/e^7 \approx 0.006 < 0.01$, so by the Alternating Series Estimation

Theorem, $n = 6$. (That is, since the 7th term is less than the desired error, we need to add the first 6 terms to get the sum to the

desired accuracy.)

27. $b_7 = \dfrac{1}{7^5} = \dfrac{1}{16{,}807} \approx 0.000\,059\,5$, so

$\sum\limits_{n=1}^{\infty} \dfrac{(-1)^{n+1}}{n^5} \approx s_6 = \sum\limits_{n=1}^{6} \dfrac{(-1)^{n+1}}{n^5} = 1 - \tfrac{1}{32} + \tfrac{1}{243} - \tfrac{1}{1024} + \tfrac{1}{3125} - \tfrac{1}{7776} \approx 0.972\,080$. Adding b_7 to s_6 does not change

the fourth decimal place of s_6, so the sum of the series, correct to four decimal places, is 0.9721.

28. $b_6 = \dfrac{6}{8^6} = \dfrac{6}{262{,}144} \approx 0.000\,023$, so $\sum\limits_{n=1}^{\infty} \dfrac{(-1)^{n} n}{8^n} \approx s_5 = \sum\limits_{n=1}^{5} \dfrac{(-1)^{n} n}{8^n} = -\tfrac{1}{8} + \tfrac{2}{64} - \tfrac{3}{512} + \tfrac{4}{4096} - \tfrac{5}{32{,}768} \approx -0.098\,785$.

Adding b_6 to s_5 does not change the fourth decimal place of s_5, so the sum of the series, correct to four decimal places,

is -0.0988.

29. $b_7 = \dfrac{7^2}{10^7} = 0.000\,004\,9$, so

$\sum\limits_{n=1}^{\infty} \dfrac{(-1)^{n-1} n^2}{10^n} \approx s_6 = \sum\limits_{n=1}^{6} \dfrac{(-1)^{n-1} n^2}{10^n} = \tfrac{1}{10} - \tfrac{4}{100} + \tfrac{9}{1000} - \tfrac{16}{10{,}000} + \tfrac{25}{100{,}000} - \tfrac{36}{1{,}000{,}000} = 0.067\,614$. Adding b_7 to s_6

does not change the fourth decimal place of s_6, so the sum of the series, correct to four decimal places, is 0.0676.

30. $b_6 = \dfrac{1}{3^6 \cdot 6!} = \dfrac{1}{524{,}880} \approx 0.000\,001\,9$, so

$$\sum_{n=1}^{\infty} \frac{(-1)^n}{3^n n!} \approx s_5 = \sum_{n=1}^{5} \frac{(-1)^n}{3^n n!} = -\frac{1}{3} + \frac{1}{18} - \frac{1}{162} + \frac{1}{1944} - \frac{1}{29{,}160} \approx -0.283\,471.$$ Adding b_6 to s_5 does not change the

fourth decimal place of s_5, so the sum of the series, correct to four decimal places, is -0.2835.

31. $\displaystyle\sum_{n=1}^{\infty} \frac{(-1)^{n-1}}{n} = 1 - \frac{1}{2} + \frac{1}{3} - \frac{1}{4} + \cdots + \frac{1}{49} - \frac{1}{50} + \frac{1}{51} - \frac{1}{52} + \cdots$. The 50th partial sum of this series is an

underestimate, since $\displaystyle\sum_{n=1}^{\infty} \frac{(-1)^{n-1}}{n} = s_{50} + \left(\frac{1}{51} - \frac{1}{52} \right) + \left(\frac{1}{53} - \frac{1}{54} \right) + \cdots$, and the terms in parentheses are all positive.
The result can be seen geometrically in Figure 1.

32. If $p > 0$, $\dfrac{1}{(n+1)^p} \leq \dfrac{1}{n^p}$ ($\{1/n^p\}$ is decreasing) and $\displaystyle\lim_{n \to \infty} \frac{1}{n^p} = 0$, so the series converges by the Alternating Series Test.

If $p \leq 0$, $\displaystyle\lim_{n \to \infty} \frac{(-1)^{n-1}}{n^p}$ does not exist, so the series diverges by the Test for Divergence. Thus, $\displaystyle\sum_{n=1}^{\infty} \frac{(-1)^{n-1}}{n^p}$

converges $\Leftrightarrow$ $p > 0$.

33. Clearly $b_n = \dfrac{1}{n+p}$ is decreasing and eventually positive and $\displaystyle\lim_{n \to \infty} b_n = 0$ for any p. So the series converges (by the

Alternating Series Test) for any p for which every b_n is defined, that is, $n + p \neq 0$ for $n \geq 1$, or p is not a negative integer.

34. Let $f(x) = \dfrac{(\ln x)^p}{x}$. Then $f'(x) = \dfrac{(\ln x)^{p-1}(p - \ln x)}{x^2} < 0$ if $x > e^p$ so f is eventually decreasing for every p. Clearly

$\displaystyle\lim_{n \to \infty} \frac{(\ln n)^p}{n} = 0$ if $p \leq 0$, and if $p > 0$ we can apply l'Hospital's Rule $[\![p + 1]\!]$ times to get a limit of 0 as well. So the series

converges for all p (by the Alternating Series Test).

35. $\sum b_{2n} = \sum 1/(2n)^2$ clearly converges (by comparison with the p-series for $p = 2$). So suppose that $\sum (-1)^{n-1} b_n$
converges. Then by Theorem 12.2.8(ii) [ET 11.2.8(ii)], so does

$$\sum \left[(-1)^{n-1} b_n + b_n \right] = 2 \left(1 + \frac{1}{3} + \frac{1}{5} + \cdots \right) = 2 \sum \frac{1}{2n-1}.$$ But this diverges by comparison with the harmonic series,

a contradiction. Therefore, $\sum (-1)^{n-1} b_n$ must diverge. The Alternating Series Test does not apply since $\{b_n\}$ is not
decreasing.

36. (a) We will prove this by induction. Let $P(n)$ be the proposition that $s_{2n} = h_{2n} - h_n$. $P(1)$ is the statement $s_2 = h_2 - h_1$,

which is true since $1 - \frac{1}{2} = \left(1 + \frac{1}{2} \right) - 1$. So suppose that $P(n)$ is true. We will show that $P(n + 1)$ must be true as a
consequence.

$$h_{2n+2} - h_{n+1} = \left(h_{2n} + \frac{1}{2n+1} + \frac{1}{2n+2} \right) - \left(h_n + \frac{1}{n+1} \right) = (h_{2n} - h_n) + \frac{1}{2n+1} - \frac{1}{2n+2}$$

$$= s_{2n} + \frac{1}{2n+1} - \frac{1}{2n+2} = s_{2n+2}$$

which is $P(n + 1)$, and proves that $s_{2n} = h_{2n} - h_n$ for all n.

(b) We know that $h_{2n} - \ln(2n) \to \gamma$ and $h_n - \ln n \to \gamma$ as $n \to \infty$. So

$$s_{2n} = h_{2n} - h_n = [h_{2n} - \ln(2n)] - (h_n - \ln n) + [\ln(2n) - \ln n], \text{ and}$$

$$\lim_{n \to \infty} s_{2n} = \gamma - \gamma + \lim_{n \to \infty} [\ln(2n) - \ln n] = \lim_{n \to \infty} (\ln 2 + \ln n - \ln n) = \ln 2.$$

12.6   Absolute Convergence and the Ratio and Root Tests   ET 11.6

1. (a) Since $\lim\limits_{n\to\infty}\left|\dfrac{a_{n+1}}{a_n}\right| = 8 > 1$, part (b) of the Ratio Test tells us that the series $\sum a_n$ is divergent.

   (b) Since $\lim\limits_{n\to\infty}\left|\dfrac{a_{n+1}}{a_n}\right| = 0.8 < 1$, part (a) of the Ratio Test tells us that the series $\sum a_n$ is absolutely convergent (and

   therefore convergent).

   (c) Since $\lim\limits_{n\to\infty}\left|\dfrac{a_{n+1}}{a_n}\right| = 1$, the Ratio Test fails and the series $\sum a_n$ might converge or it might diverge.

2. The series $\sum\limits_{n=1}^{\infty}\dfrac{n^2}{2^n}$ has positive terms and $\lim\limits_{n\to\infty}\dfrac{a_{n+1}}{a_n} = \lim\limits_{n\to\infty}\left[\dfrac{(n+1)^2}{2^{n+1}}\cdot\dfrac{2^n}{n^2}\right] = \lim\limits_{n\to\infty}\left(1+\dfrac{1}{n}\right)^2\cdot\dfrac{1}{2} = \dfrac{1}{2} < 1$, so the

   series is absolutely convergent by the Ratio Test.

3. $\sum\limits_{n=0}^{\infty}\dfrac{(-10)^n}{n!}$. Using the Ratio Test, $\lim\limits_{n\to\infty}\left|\dfrac{a_{n+1}}{a_n}\right| = \lim\limits_{n\to\infty}\left|\dfrac{(-10)^{n+1}}{(n+1)!}\cdot\dfrac{n!}{(-10)^n}\right| = \lim\limits_{n\to\infty}\left|\dfrac{-10}{n+1}\right| = 0 < 1$, so the series is

   absolutely convergent.

4. $\sum\limits_{n=1}^{\infty}(-1)^{n-1}\dfrac{2^n}{n^4}$ diverges by the Test for Divergence. $\lim\limits_{n\to\infty}\dfrac{2^n}{n^4} = \infty$, so $\lim\limits_{n\to\infty}(-1)^{n-1}\dfrac{2^n}{n^4}$ does not exist.

5. $\sum\limits_{n=1}^{\infty}\dfrac{(-1)^{n+1}}{\sqrt[4]{n}}$ converges by the Alternating Series Test, but $\sum\limits_{n=1}^{\infty}\dfrac{1}{\sqrt[4]{n}}$ is a divergent p-series $\left(p=\frac{1}{4}\le 1\right)$, so the given series

   is conditionally convergent.

6. $\sum\limits_{n=1}^{\infty}\dfrac{1}{n^4}$ is a convergent p-series $(p=4>1)$, so $\sum\limits_{n=1}^{\infty}\dfrac{(-1)^n}{n^4}$ is absolutely convergent.

7. $\lim\limits_{k\to\infty}\left|\dfrac{a_{k+1}}{a_k}\right| = \lim\limits_{k\to\infty}\left[\dfrac{(k+1)\left(\frac{2}{3}\right)^{k+1}}{k\left(\frac{2}{3}\right)^k}\right] = \lim\limits_{k\to\infty}\dfrac{k+1}{k}\left(\dfrac{2}{3}\right)^1 = \dfrac{2}{3}\lim\limits_{k\to\infty}\left(1+\dfrac{1}{k}\right) = \dfrac{2}{3}(1) = \dfrac{2}{3} < 1$, so the series

   $\sum\limits_{n=1}^{\infty}k\left(\frac{2}{3}\right)^k$ is absolutely convergent by the Ratio Test. Since the terms of this series are positive, absolute convergence is the

   same as convergence.

8. $\lim\limits_{n\to\infty}\left|\dfrac{a_{n+1}}{a_n}\right| = \lim\limits_{n\to\infty}\left[\dfrac{(n+1)!}{100^{n+1}}\cdot\dfrac{100^n}{n!}\right] = \lim\limits_{n\to\infty}\dfrac{n+1}{100} = \infty$, so the series $\sum\limits_{n=1}^{\infty}\dfrac{n!}{100^n}$ diverges by the Ratio Test.

9. $\lim\limits_{n\to\infty}\left|\dfrac{a_{n+1}}{a_n}\right| = \lim\limits_{n\to\infty}\left[\dfrac{(1.1)^{n+1}}{(n+1)^4}\cdot\dfrac{n^4}{(1.1)^n}\right] = \lim\limits_{n\to\infty}\dfrac{(1.1)n^4}{(n+1)^4} = (1.1)\lim\limits_{n\to\infty}\dfrac{1}{\dfrac{(n+1)^4}{n^4}} = (1.1)\lim\limits_{n\to\infty}\dfrac{1}{(1+1/n)^4}$

   $= (1.1)(1) = 1.1 > 1$,

   so the series $\sum\limits_{n=1}^{\infty}(-1)^n\dfrac{(1.1)^n}{n^4}$ diverges by the Ratio Test.

10. $\sum_{n=1}^{\infty} (-1)^n \dfrac{n}{\sqrt{n^3 + 2}}$ converges by the Alternating Series Test (see Exercise 12.5.8 [ET 11.5.8]). Let $a_n = \dfrac{1}{\sqrt{n}}$ with

$b_n = \dfrac{n}{\sqrt{n^3 + 2}}$. Then $\lim\limits_{n\to\infty} \dfrac{a_n}{b_n} = \lim\limits_{n\to\infty} \left(\dfrac{1}{\sqrt{n}} \cdot \dfrac{\sqrt{n^3 + 2}}{n} \right) = \lim\limits_{n\to\infty} \dfrac{\sqrt{n^3 + 2}}{\sqrt{n^3}} = \lim\limits_{n\to\infty} \sqrt{1 + \dfrac{2}{n^3}} = 1 > 0$, so

$\sum_{n=1}^{\infty} \dfrac{n}{\sqrt{n^3 + 2}}$ diverges by limit comparision with the divergent p-series $\sum_{n=1}^{\infty} \dfrac{1}{\sqrt{n}}$ $\left[p = \tfrac{1}{2} \le 1 \right]$. Thus, $\sum_{n=1}^{\infty} (-1)^n \dfrac{n}{\sqrt{n^3 + 2}}$

is conditionally convergent.

11. Since $0 \le \dfrac{e^{1/n}}{n^3} \le \dfrac{e}{n^3} = e \left(\dfrac{1}{n^3} \right)$ and $\sum_{n=1}^{\infty} \dfrac{1}{n^3}$ is a convergent p-series $[p = 3 > 1]$, $\sum_{n=1}^{\infty} \dfrac{e^{1/n}}{n^3}$ converges, and so

$\sum_{n=1}^{\infty} \dfrac{(-1)^n e^{1/n}}{n^3}$ is absolutely convergent.

12. $\left| \dfrac{\sin 4n}{4^n} \right| \le \dfrac{1}{4^n}$, so $\sum_{n=1}^{\infty} \left| \dfrac{\sin 4n}{4^n} \right|$ converges by comparison with the convergent geometric series $\sum_{n=1}^{\infty} \dfrac{1}{4^n}$ $\left[|r| = \tfrac{1}{4} < 1 \right]$.

Thus, $\sum_{n=1}^{\infty} \dfrac{\sin 4n}{4^n}$ is absolutely convergent.

13. $\lim\limits_{n\to\infty} \left| \dfrac{a_{n+1}}{a_n} \right| = \lim\limits_{n\to\infty} \left[\dfrac{10^{n+1}}{(n+2)\,4^{2n+3}} \cdot \dfrac{(n+1)\,4^{2n+1}}{10^n} \right] = \lim\limits_{n\to\infty} \left(\dfrac{10}{4^2} \cdot \dfrac{n+1}{n+2} \right) = \dfrac{5}{8} < 1$, so the series $\sum_{n=1}^{\infty} \dfrac{10^n}{(n+1)4^{2n+1}}$

is absolutely convergent by the Ratio Test. Since the terms of this series are positive, absolute convergence is the same as

convergence.

14. $\lim\limits_{n\to\infty} \left| \dfrac{a_{n+1}}{a_n} \right| = \lim\limits_{n\to\infty} \left[\dfrac{(n+1)^2\,2^{n+1}}{(n+1)!} \cdot \dfrac{n!}{n^2 2^n} \right] = \lim\limits_{n\to\infty} \left[\left(1 + \dfrac{1}{n} \right)^2 \cdot \dfrac{2}{n+1} \right] = 0$, so the series $\sum_{n=1}^{\infty} (-1)^{n+1} \dfrac{n^2 2^n}{n!}$ is

absolutely convergent by the Ratio Test.

15. $\left| \dfrac{(-1)^n \arctan n}{n^2} \right| < \dfrac{\pi/2}{n^2}$, so since $\sum_{n=1}^{\infty} \dfrac{\pi/2}{n^2} = \dfrac{\pi}{2} \sum_{n=1}^{\infty} \dfrac{1}{n^2}$ converges $(p = 2 > 1)$, the given series $\sum_{n=1}^{\infty} \dfrac{(-1)^n \arctan n}{n^2}$

converges absolutely by the Comparison Test.

16. $n^{2/3} - 2 > 0$ for $n \ge 3$, so $\dfrac{3 - \cos n}{n^{2/3} - 2} > \dfrac{1}{n^{2/3} - 2} > \dfrac{1}{n^{2/3}}$ for $n \ge 3$. Since $\sum_{n=1}^{\infty} \dfrac{1}{n^{2/3}}$ diverges $\left[p = \tfrac{2}{3} \le 1 \right]$, so does

$\sum_{n=1}^{\infty} \dfrac{3 - \cos n}{n^{2/3} - 2}$ by the Comparison Test.

17. $\sum_{n=2}^{\infty} \dfrac{(-1)^n}{\ln n}$ converges by the Alternating Series Test since $\lim\limits_{n\to\infty} \dfrac{1}{\ln n} = 0$ and $\left\{ \dfrac{1}{\ln n} \right\}$ is decreasing. Now $\ln n < n$, so

$\dfrac{1}{\ln n} > \dfrac{1}{n}$, and since $\sum_{n=2}^{\infty} \dfrac{1}{n}$ is the divergent (partial) harmonic series, $\sum_{n=2}^{\infty} \dfrac{1}{\ln n}$ diverges by the Comparison Test. Thus,

$\sum_{n=2}^{\infty} \dfrac{(-1)^n}{\ln n}$ is conditionally convergent.

18. $\lim\limits_{n\to\infty} \left| \dfrac{a_{n+1}}{a_n} \right| = \lim\limits_{n\to\infty} \dfrac{(n+1)! / (n+1)^{n+1}}{n! / n^n} = \lim\limits_{n\to\infty} \dfrac{n^n}{(n+1)^n} = \lim\limits_{n\to\infty} \dfrac{1}{(1 + 1/n)^n} = \dfrac{1}{e} < 1$, so the series $\sum_{n=1}^{\infty} \dfrac{n!}{n^n}$

converges absolutely by the Ratio Test.

19. $\dfrac{|\cos(n\pi/3)|}{n!} \le \dfrac{1}{n!}$ and $\displaystyle\sum_{n=1}^{\infty} \dfrac{1}{n!}$ converges (use the Ratio Test), so the series $\displaystyle\sum_{n=1}^{\infty} \dfrac{\cos(n\pi/3)}{n!}$ converges absolutely by the

Comparison Test.

20. $\displaystyle\lim_{n\to\infty} \sqrt[n]{|a_n|} = \lim_{n\to\infty} \sqrt[n]{\left|\dfrac{(-2)^n}{n^n}\right|} = \lim_{n\to\infty} \dfrac{2}{n} = 0 < 1$, so the series $\displaystyle\sum_{n=1}^{\infty} \dfrac{(-2)^n}{n^n}$ is absolutely convergent by the Root Test.

21. $\displaystyle\lim_{n\to\infty} \sqrt[n]{|a_n|} = \lim_{n\to\infty} \dfrac{n^2+1}{2n^2+1} = \lim_{n\to\infty} \dfrac{1+1/n^2}{2+1/n^2} = \dfrac{1}{2} < 1$, so the series $\displaystyle\sum_{n=1}^{\infty} \left(\dfrac{n^2+1}{2n^2+1}\right)^n$ is absolutely convergent by the

Root Test.

22. $\displaystyle\lim_{n\to\infty} \sqrt[n]{|a_n|} = \lim_{n\to\infty} \sqrt[n]{\left|\left(\dfrac{-2n}{n+1}\right)^{5n}\right|} = \lim_{n\to\infty} \dfrac{2^5 n^5}{(n+1)^5} = 32 \lim_{n\to\infty} \dfrac{1}{\left(\dfrac{n+1}{n}\right)^5} = 32 \lim_{n\to\infty} \dfrac{1}{(1+1/n)^5}$

$$= 32(1) = 32 > 1,$$

so the series $\displaystyle\sum_{n=2}^{\infty} \left(\dfrac{-2n}{n+1}\right)^{5n}$ diverges by the Root Test.

23. $\displaystyle\lim_{n\to\infty} \sqrt[n]{|a_n|} = \lim_{n\to\infty} \sqrt[n]{\left(1+\dfrac{1}{n}\right)^{n^2}} = \lim_{n\to\infty} \left(1+\dfrac{1}{n}\right)^n = e > 1$ [by Equation 7.4.9 (or 7.4*.9) [ET 3.6.6]],

so the series $\displaystyle\sum_{n=1}^{\infty} \left(1+\dfrac{1}{n}\right)^{n^2}$ diverges by the Root Test.

24. $\displaystyle\lim_{n\to\infty} \sqrt[n]{|a_n|} = \lim_{n\to\infty} \sqrt[n]{\dfrac{n}{(\ln n)^n}} = \lim_{n\to\infty} \dfrac{n^{1/n}}{\ln n} \overset{(\star)}{=} 0 < 1$, so the series $\displaystyle\sum_{n=2}^{\infty} \dfrac{n}{(\ln n)^n}$ is absolutely convergent by the

Root Test.

$(\star)$ Let $y = x^{1/x}$. Then $\ln y = \dfrac{1}{x}\ln x$, so $\displaystyle\lim_{x\to\infty} \ln y = \lim_{x\to\infty} \dfrac{\ln x}{x} \overset{\text{H}}{=} \lim_{x\to\infty} \dfrac{1/x}{1} = 0 \ \Rightarrow \ \lim_{x\to\infty} y = e^0 = 1.$

25. Use the Ratio Test with the series

$$1 - \dfrac{1\cdot 3}{3!} + \dfrac{1\cdot 3\cdot 5}{5!} - \dfrac{1\cdot 3\cdot 5\cdot 7}{7!} + \cdots + (-1)^{n-1}\dfrac{1\cdot 3\cdot 5\cdots(2n-1)}{(2n-1)!} + \cdots = \sum_{n=1}^{\infty} (-1)^{n-1}\dfrac{1\cdot 3\cdot 5\cdots(2n-1)}{(2n-1)!}.$$

$$\lim_{n\to\infty}\left|\dfrac{a_{n+1}}{a_n}\right| = \lim_{n\to\infty}\left|\dfrac{(-1)^n\cdot 1\cdot 3\cdot 5\cdots(2n-1)[2(n+1)-1]}{[2(n+1)-1]!}\cdot\dfrac{(2n-1)!}{(-1)^{n-1}\cdot 1\cdot 3\cdot 5\cdots(2n-1)}\right|$$

$$= \lim_{n\to\infty}\left|\dfrac{(-1)(2n+1)(2n-1)!}{(2n+1)(2n)(2n-1)!}\right| = \lim_{n\to\infty}\dfrac{1}{2n} = 0 < 1,$$

so the given series is absolutely convergent and therefore convergent.

26. Use the Ratio Test with the series $\dfrac{2}{5} + \dfrac{2\cdot 6}{5\cdot 8} + \dfrac{2\cdot 6\cdot 10}{5\cdot 8\cdot 11} + \dfrac{2\cdot 6\cdot 10\cdot 14}{5\cdot 8\cdot 11\cdot 14} + \cdots = \displaystyle\sum_{n=1}^{\infty} \dfrac{2\cdot 6\cdot 10\cdot 14\cdots(4n-2)}{5\cdot 8\cdot 11\cdot 14\cdots(3n+2)}.$

$$\lim_{n\to\infty}\left|\dfrac{a_{n+1}}{a_n}\right| = \lim_{n\to\infty}\left|\dfrac{2\cdot 6\cdot 10\cdots(4n-2)[4(n+1)-2]}{5\cdot 8\cdot 11\cdots(3n+2)[3(n+1)+2]}\cdot\dfrac{5\cdot 8\cdot 11\cdots(3n+2)}{2\cdot 6\cdot 10\cdots(4n-2)}\right| = \lim_{n\to\infty}\dfrac{4n+2}{3n+5} = \dfrac{4}{3} > 1,$$

so the given series is divergent.

27. $\displaystyle\sum_{n=1}^{\infty} \dfrac{2\cdot 4\cdot 6\cdots(2n)}{n!} = \sum_{n=1}^{\infty} \dfrac{(2\cdot 1)\cdot(2\cdot 2)\cdot(2\cdot 3)\cdots(2\cdot n)}{n!} = \sum_{n=1}^{\infty} \dfrac{2^n n!}{n!} = \sum_{n=1}^{\infty} 2^n$, which diverges by the Test for

Divergence since $\displaystyle\lim_{n\to\infty} 2^n = \infty.$

28. $\lim\limits_{n\to\infty}\left|\dfrac{a_{n+1}}{a_n}\right| = \lim\limits_{n\to\infty}\left|\dfrac{\dfrac{2^{n+1}\,(n+1)!}{5\cdot 8\cdot 11\cdot\cdots\cdot(3n+5)}}{\dfrac{2^n n!}{5\cdot 8\cdot 11\cdot\cdots\cdot(3n+2)}}\right| = \lim\limits_{n\to\infty}\dfrac{2(n+1)}{3n+5} = \dfrac{2}{3} < 1$, so the series converges absolutely by the

Ratio Test.

29. By the recursive definition, $\lim\limits_{n\to\infty}\left|\dfrac{a_{n+1}}{a_n}\right| = \lim\limits_{n\to\infty}\left|\dfrac{5n+1}{4n+3}\right| = \dfrac{5}{4} > 1$, so the series diverges by the Ratio Test.

30. By the recursive definition, $\lim\limits_{n\to\infty}\left|\dfrac{a_{n+1}}{a_n}\right| = \lim\limits_{n\to\infty}\left|\dfrac{2+\cos n}{\sqrt{n}}\right| = 0 < 1$, so the series converges absolutely by the Ratio Test.

31. (a) $\lim\limits_{n\to\infty}\left|\dfrac{1/(n+1)^3}{1/n^3}\right| = \lim\limits_{n\to\infty}\dfrac{n^3}{(n+1)^3} = \lim\limits_{n\to\infty}\dfrac{1}{(1+1/n)^3} = 1$. Inconclusive

(b) $\lim\limits_{n\to\infty}\left|\dfrac{(n+1)}{2^{n+1}}\cdot\dfrac{2^n}{n}\right| = \lim\limits_{n\to\infty}\dfrac{n+1}{2n} = \lim\limits_{n\to\infty}\left(\dfrac{1}{2}+\dfrac{1}{2n}\right) = \dfrac{1}{2}$. Conclusive (convergent)

(c) $\lim\limits_{n\to\infty}\left|\dfrac{(-3)^n}{\sqrt{n+1}}\cdot\dfrac{\sqrt{n}}{(-3)^{n-1}}\right| = 3\lim\limits_{n\to\infty}\sqrt{\dfrac{n}{n+1}} = 3\lim\limits_{n\to\infty}\sqrt{\dfrac{1}{1+1/n}} = 3$. Conclusive (divergent)

(d) $\lim\limits_{n\to\infty}\left|\dfrac{\sqrt{n+1}}{1+(n+1)^2}\cdot\dfrac{1+n^2}{\sqrt{n}}\right| = \lim\limits_{n\to\infty}\left[\sqrt{1+\dfrac{1}{n}}\cdot\dfrac{1/n^2+1}{1/n^2+(1+1/n)^2}\right] = 1$. Inconclusive

32. We use the Ratio Test:

$$\lim\limits_{n\to\infty}\left|\dfrac{a_{n+1}}{a_n}\right| = \lim\limits_{n\to\infty}\left|\dfrac{[(n+1)!]^2/[k(n+1)]!}{(n!)^2/(kn)!}\right| = \lim\limits_{n\to\infty}\left|\dfrac{(n+1)^2}{[k(n+1)]\,[k(n+1)-1]\cdots[kn+1]}\right|$$

Now if $k = 1$, then this is equal to $\lim\limits_{n\to\infty}\left|\dfrac{(n+1)^2}{(n+1)}\right| = \infty$, so the series diverges; if $k = 2$, the limit is

$\lim\limits_{n\to\infty}\left|\dfrac{(n+1)^2}{(2n+2)(2n+1)}\right| = \dfrac{1}{4} < 1$, so the series converges, and if $k > 2$, then the highest power of n in the denominator is

larger than 2, and so the limit is 0, indicating convergence. So the series converges for $k \geq 2$.

33. (a) $\lim\limits_{n\to\infty}\left|\dfrac{a_{n+1}}{a_n}\right| = \lim\limits_{n\to\infty}\left|\dfrac{x^{n+1}}{(n+1)!}\cdot\dfrac{n!}{x^n}\right| = \lim\limits_{n\to\infty}\left|\dfrac{x}{n+1}\right| = |x|\lim\limits_{n\to\infty}\dfrac{1}{n+1} = |x|\cdot 0 = 0 < 1$, so by the Ratio Test the

series $\sum\limits_{n=0}^{\infty}\dfrac{x^n}{n!}$ converges for all x.

(b) Since the series of part (a) always converges, we must have $\lim\limits_{n\to\infty}\dfrac{x^n}{n!} = 0$ by Theorem 12.2.6 [ET 11.2.6].

34. (a) $R_n = a_{n+1} + a_{n+2} + a_{n+3} + a_{n+4} + \cdots = a_{n+1}\left(1 + \dfrac{a_{n+2}}{a_{n+1}} + \dfrac{a_{n+3}}{a_{n+1}} + \dfrac{a_{n+4}}{a_{n+1}} + \cdots\right)$

$= a_{n+1}\left(1 + \dfrac{a_{n+2}}{a_{n+1}} + \dfrac{a_{n+3}}{a_{n+2}}\dfrac{a_{n+2}}{a_{n+1}} + \dfrac{a_{n+4}}{a_{n+3}}\dfrac{a_{n+3}}{a_{n+2}}\dfrac{a_{n+2}}{a_{n+1}} + \cdots\right)$

$= a_{n+1}(1 + r_{n+1} + r_{n+2}r_{n+1} + r_{n+3}r_{n+2}r_{n+1} + \cdots)$ $\quad(\star)$

$\leq a_{n+1}\left(1 + r_{n+1} + r_{n+1}^2 + r_{n+1}^3 + \cdots\right)$ [since $\{r_n\}$ is decreasing] $= \dfrac{a_{n+1}}{1-r_{n+1}}$

(b) Note that since $\{r_n\}$ is increasing and $r_n \to L$ as $n \to \infty$, we have $r_n < L$ for all n. So, starting with equation $(\star)$,

$R_n = a_{n+1}(1 + r_{n+1} + r_{n+2}r_{n+1} + r_{n+3}r_{n+2}r_{n+1} + \cdots) \leq a_{n+1}\left(1 + L + L^2 + L^3 + \cdots\right) = \dfrac{a_{n+1}}{1-L}$.

35. (a) $s_5 = \sum_{n=1}^{5} \frac{1}{n2^n} = \frac{1}{2} + \frac{1}{8} + \frac{1}{24} + \frac{1}{64} + \frac{1}{160} = \frac{661}{960} \approx 0.68854$. Now the ratios

$$r_n = \frac{a_{n+1}}{a_n} = \frac{n2^n}{(n+1)2^{n+1}} = \frac{n}{2(n+1)} \quad \text{form an increasing sequence, since}$$

$$r_{n+1} - r_n = \frac{n+1}{2(n+2)} - \frac{n}{2(n+1)} = \frac{(n+1)^2 - n(n+2)}{2(n+1)(n+2)} = \frac{1}{2(n+1)(n+2)} > 0. \text{ So by Exercise 34(b), the error}$$

in using s_5 is $R_5 \le \dfrac{a_6}{1 - \lim\limits_{n\to\infty} r_n} = \dfrac{1/(6\cdot 2^6)}{1 - 1/2} = \dfrac{1}{192} \approx 0.00521$.

(b) The error in using s_n as an approximation to the sum is $R_n = \dfrac{a_{n+1}}{1 - \frac{1}{2}} = \dfrac{2}{(n+1)2^{n+1}}$. We want $R_n < 0.00005 \quad \Leftrightarrow$

$$\frac{1}{(n+1)2^n} < 0.00005 \quad \Leftrightarrow \quad (n+1)2^n > 20,000. \text{ To find such an } n \text{ we can use trial and error or a graph. We calculate}$$

$(11+1)2^{11} = 24,576$, so $s_{11} = \sum_{n=1}^{11} \frac{1}{n2^n} \approx 0.693109$ is within 0.00005 of the actual sum.

36. $s_{10} = \sum_{n=1}^{10} \frac{n}{2^n} = \frac{1}{2} + \frac{2}{4} + \frac{3}{8} + \cdots + \frac{10}{1024} \approx 1.988$. The ratios $r_n = \dfrac{a_{n+1}}{a_n} = \dfrac{n+1}{2^{n+1}} \cdot \dfrac{2^n}{n} = \dfrac{n+1}{2n} = \dfrac{1}{2}\left(1 + \dfrac{1}{n}\right)$ form a

decreasing sequence, and $r_{11} = \dfrac{11+1}{2(11)} = \dfrac{12}{22} = \dfrac{6}{11} < 1$, so by Exercise 34(a), the error in using s_{10} to approximate the sum

of the series $\sum_{n=1}^{\infty} \frac{n}{2^n}$ is $R_{10} \le \dfrac{a_{11}}{1 - r_{11}} = \dfrac{\frac{11}{2048}}{1 - \frac{6}{11}} = \dfrac{121}{10,240} \approx 0.0118$.

37. (i) Following the hint, we get that $|a_n| < r^n$ for $n \ge N$, and so since the geometric series $\sum_{n=1}^{\infty} r^n$ converges $[0 < r < 1]$,

the series $\sum_{n=N}^{\infty} |a_n|$ converges as well by the Comparison Test, and hence so does $\sum_{n=1}^{\infty} |a_n|$, so $\sum_{n=1}^{\infty} a_n$ is absolutely

convergent.

(ii) If $\lim\limits_{n\to\infty} \sqrt[n]{|a_n|} = L > 1$, then there is an integer N such that $\sqrt[n]{|a_n|} > 1$ for all $n \ge N$, so $|a_n| > 1$ for $n \ge N$. Thus,

$\lim\limits_{n\to\infty} a_n \ne 0$, so $\sum_{n=1}^{\infty} a_n$ diverges by the Test for Divergence.

(iii) Consider $\sum_{n=1}^{\infty} \frac{1}{n}$ [diverges] and $\sum_{n=1}^{\infty} \frac{1}{n^2}$ [converges]. For each sum, $\lim\limits_{n\to\infty} \sqrt[n]{|a_n|} = 1$, so the Root Test is inconclusive.

38. (a) $\lim\limits_{n\to\infty} \left|\dfrac{a_{n+1}}{a_n}\right| = \lim\limits_{n\to\infty} \left|\dfrac{[4(n+1)]!\,[1103 + 26{,}390(n+1)]}{[(n+1)!]^4\, 396^{4(n+1)}} \cdot \dfrac{(n!)^4\, 396^{4n}}{(4n)!\,(1103 + 26{,}390n)}\right|$

$$= \lim\limits_{n\to\infty} \frac{(4n+4)(4n+3)(4n+2)(4n+1)(26{,}390n + 27{,}493)}{(n+1)^4\, 396^4\,(26{,}390n + 1103)} = \frac{4^4}{396^4} = \frac{1}{99^4} < 1,$$

so by the Ratio Test, the series $\sum_{n=0}^{\infty} \dfrac{(4n)!\,(1103 + 26{,}390n)}{(n!)^4\, 396^{4n}}$ converges.

(b) $\dfrac{1}{\pi} = \dfrac{2\sqrt{2}}{9801} \sum_{n=0}^{\infty} \dfrac{(4n)!\,(1103 + 26{,}390n)}{(n!)^4\, 396^{4n}}$

With the first term ($n = 0$), $\dfrac{1}{\pi} \approx \dfrac{2\sqrt{2}}{9801} \cdot \dfrac{1103}{1} \quad \Rightarrow \quad \pi \approx 3.141\,592\,73$, so we get 6 correct decimal places of π, which is

$3.141\,592\,653\,589\,793\,238$ to 18 decimal places.

With the second term ($n = 1$), $\dfrac{1}{\pi} \approx \dfrac{2\sqrt{2}}{9801}\left(\dfrac{1103}{1} + \dfrac{4!\,(1103 + 26{,}390)}{396^4}\right)$ $\Rightarrow$ $\pi \approx 3.141\,592\,653\,589\,793\,878$, so

we get 15 correct decimal places of π.

39. (a) Since $\sum a_n$ is absolutely convergent, and since $\left|a_n^+\right| \le \left|a_n\right|$ and $\left|a_n^-\right| \le \left|a_n\right|$ (because a_n^+ and a_n^- each equal

either a_n or 0), we conclude by the Comparison Test that both $\sum a_n^+$ and $\sum a_n^-$ must be absolutely convergent.

Or: Use Theorem 12.2.8 [ET 11.2.8].

(b) We will show by contradiction that both $\sum a_n^+$ and $\sum a_n^-$ must diverge. For suppose that $\sum a_n^+$ converged. Then so

would $\sum\left(a_n^+ - \frac{1}{2}a_n\right)$ by Theorem 12.2.8 [ET 11.2.8]. But $\sum\left(a_n^+ - \frac{1}{2}a_n\right) = \sum\left[\frac{1}{2}\left(a_n + |a_n|\right) - \frac{1}{2}a_n\right] = \frac{1}{2}\sum|a_n|$,

which diverges because $\sum a_n$ is only conditionally convergent. Hence, $\sum a_n^+$ can't converge. Similarly, neither

can $\sum a_n^-$.

40. Let $\sum b_n$ be the rearranged series constructed in the hint. [This series can be constructed by virtue of the result of

Exercise 39(b).] This series will have partial sums s_n that oscillate in value back and forth across r. Since $\lim\limits_{n\to\infty} a_n = 0$

(by Theorem 12.2.6 [ET 11.2.6]), and since the size of the oscillations $|s_n - r|$ is always less than $|a_n|$ because of the way

$\sum b_n$ was constructed, we have that $\sum b_n = \lim\limits_{n\to\infty} s_n = r$.

12.7 Strategy for Testing Series

ET 11.7

1. $\dfrac{1}{n + 3^n} < \dfrac{1}{3^n} = \left(\dfrac{1}{3}\right)^n$ for all $n \ge 1$. $\displaystyle\sum_{n=1}^{\infty}\left(\dfrac{1}{3}\right)^n$ is a convergent geometric series $\left[|r| = \frac{1}{3} < 1\right]$, so $\displaystyle\sum_{n=1}^{\infty}\dfrac{1}{n + 3^n}$

converges by the Comparison Test.

2. $\lim\limits_{n\to\infty}\sqrt[n]{|a_n|} = \lim\limits_{n\to\infty}\sqrt[n]{\left|\dfrac{(2n+1)^n}{n^{2n}}\right|} = \lim\limits_{n\to\infty}\dfrac{2n+1}{n^2} = \lim\limits_{n\to\infty}\left(\dfrac{2}{n} + \dfrac{1}{n^2}\right) = 0 < 1$, so the series $\displaystyle\sum_{n=1}^{\infty}\dfrac{(2n+1)^n}{n^{2n}}$

converges by the Root Test.

3. $\lim\limits_{n\to\infty}|a_n| = \lim\limits_{n\to\infty}\dfrac{n}{n+2} = 1$, so $\lim\limits_{n\to\infty}a_n = \lim\limits_{n\to\infty}(-1)^n\dfrac{n}{n+2}$ does not exist. Thus, the series $\displaystyle\sum_{n=1}^{\infty}(-1)^n\dfrac{n}{n+2}$ diverges by

the Test for Divergence.

4. $b_n = \dfrac{n}{n^2 + 2} > 0$ for $n \ge 1$. $\{b_n\}$ is decreasing for $n \ge 2$ since $\left(\dfrac{x}{x^2 + 2}\right)' = \dfrac{(x^2 + 2)(1) - x(2x)}{(x^2 + 2)^2} = \dfrac{2 - x^2}{(x^2 + 2)^2} < 0$

for $x \ge \sqrt{2}$. Also, $\lim\limits_{n\to\infty}b_n = \lim\limits_{n\to\infty}\dfrac{n}{n^2 + 2} = \lim\limits_{n\to\infty}\dfrac{1/n}{1 + 2/n^2} = 0$. Thus, the series $\displaystyle\sum_{n=1}^{\infty}(-1)^n\dfrac{n}{n^2 + 2}$ converges by the

Alternating Series Test.

5. $\lim\limits_{n\to\infty}\left|\dfrac{a_{n+1}}{a_n}\right| = \lim\limits_{n\to\infty}\left|\dfrac{(n+1)^2\,2^n}{(-5)^{n+1}} \cdot \dfrac{(-5)^n}{n^2\,2^{n-1}}\right| = \lim\limits_{n\to\infty}\dfrac{2(n+1)^2}{5n^2} = \dfrac{2}{5}\lim\limits_{n\to\infty}\left(1 + \dfrac{1}{n}\right)^2 = \dfrac{2}{5}(1) = \dfrac{2}{5} < 1$, so the series

$\displaystyle\sum_{n=1}^{\infty}\dfrac{n^2\,2^{n-1}}{(-5)^n}$ converges by the Ratio Test.

6. Use the Limit Comparison Test with $a_n = \dfrac{1}{2n+1}$ and $b_n = \dfrac{1}{n}$: $\displaystyle\lim_{n\to\infty} \frac{a_n}{b_n} = \lim_{n\to\infty} \frac{n}{2n+1} = \lim_{n\to\infty} \frac{1}{2 + (1/n)} = \frac{1}{2} > 0$.

Since the harmonic series $\displaystyle\sum_{n=1}^{\infty} \frac{1}{n}$ diverges, so does $\displaystyle\sum_{n=1}^{\infty} \frac{1}{2n+1}$.   [*Or:* Use the Integral Test.]

7. Let $f(x) = \dfrac{1}{x\sqrt{\ln x}}$. Then f is positive, continuous, and decreasing on $[2, \infty)$, so we can apply the Integral Test.

Since $\displaystyle\int \frac{1}{x\sqrt{\ln x}}\, dx \begin{bmatrix} u = \ln x, \\ du = dx/x \end{bmatrix} = \int u^{-1/2}\, du = 2u^{1/2} + C = 2\sqrt{\ln x} + C$, we find

$\displaystyle\int_2^{\infty} \frac{dx}{x\sqrt{\ln x}} = \lim_{t\to\infty} \int_2^t \frac{dx}{x\sqrt{\ln x}} = \lim_{t\to\infty} \left[2\sqrt{\ln x} \right]_2^t = \lim_{t\to\infty} \left(2\sqrt{\ln t} - 2\sqrt{\ln 2} \right) = \infty$. Since the integral diverges, the

given series $\displaystyle\sum_{n=2}^{\infty} \frac{1}{n\sqrt{\ln n}}$ diverges.

8. $\displaystyle\sum_{k=1}^{\infty} \frac{2^k k!}{(k+2)!} = \sum_{k=1}^{\infty} \frac{2^k}{(k+1)(k+2)}$. Using the Ratio Test, we get

$\displaystyle\lim_{k\to\infty} \left| \frac{a_{k+1}}{a_k} \right| = \lim_{k\to\infty} \left| \frac{2^{k+1}}{(k+2)(k+3)} \cdot \frac{(k+1)(k+2)}{2^k} \right| = \lim_{k\to\infty} \left(2 \cdot \frac{k+1}{k+3} \right) = 2 > 1$, so the series diverges.

Or: Use the Test for Divergence.

9. $\displaystyle\sum_{k=1}^{\infty} k^2 e^{-k} = \sum_{k=1}^{\infty} \frac{k^2}{e^k}$. Using the Ratio Test, we get

$\displaystyle\lim_{k\to\infty} \left| \frac{a_{k+1}}{a_k} \right| = \lim_{k\to\infty} \left| \frac{(k+1)^2}{e^{k+1}} \cdot \frac{e^k}{k^2} \right| = \lim_{k\to\infty} \left[\left(\frac{k+1}{k} \right)^2 \cdot \frac{1}{e} \right] = 1^2 \cdot \frac{1}{e} = \frac{1}{e} < 1$, so the series converges.

10. Let $f(x) = x^2 e^{-x^3}$. Then f is continuous and positive on $[1, \infty)$, and $f'(x) = \dfrac{x(2 - 3x^3)}{e^{x^3}} < 0$ for $x \geq 1$, so f is

decreasing on $[1, \infty)$ as well, and we can apply the Integral Test. $\displaystyle\int_1^{\infty} x^2 e^{-x^3}\, dx = \lim_{t\to\infty} \left[-\frac{1}{3} e^{-x^3} \right]_1^t = \frac{1}{3e}$, so the integral

converges, and hence, the series converges.

11. $b_n = \dfrac{1}{n \ln n} > 0$ for $n \geq 2$, $\{b_n\}$ is decreasing, and $\displaystyle\lim_{n\to\infty} b_n = 0$, so the given series $\displaystyle\sum_{n=2}^{\infty} \frac{(-1)^{n+1}}{n \ln n}$ converges by the

Alternating Series Test.

12. The series $\displaystyle\sum_{n=1}^{\infty} \sin n$ diverges by the Test for Divergence since $\displaystyle\lim_{n\to\infty} \sin n$ does not exist.

13. $\displaystyle\lim_{n\to\infty} \left| \frac{a_{n+1}}{a_n} \right| = \lim_{n\to\infty} \left| \frac{3^{n+1}(n+1)^2}{(n+1)!} \cdot \frac{n!}{3^n n^2} \right| = \lim_{n\to\infty} \frac{3(n+1)^2}{(n+1)n^2} = 3 \lim_{n\to\infty} \frac{n+1}{n^2} = 0 < 1$, so the series $\displaystyle\sum_{n=1}^{\infty} \frac{3^n n^2}{n!}$

converges by the Ratio Test.

14. $\left| \dfrac{\sin 2n}{1 + 2^n} \right| \leq \dfrac{1}{1 + 2^n} < \dfrac{1}{2^n} = \left(\dfrac{1}{2} \right)^n$, so the series $\displaystyle\sum_{n=1}^{\infty} \left| \frac{\sin 2n}{1 + 2^n} \right|$ converges by comparison with the geometric series

$\displaystyle\sum_{n=1}^{\infty} \left(\frac{1}{2} \right)^n$ with $|r| = \frac{1}{2} < 1$. Thus, the series $\displaystyle\sum_{n=1}^{\infty} \frac{\sin 2n}{1 + 2^n}$ converges absolutely, implying convergence.

15. $\lim\limits_{n\to\infty} \left|\dfrac{a_{n+1}}{a_n}\right| = \lim\limits_{n\to\infty} \left|\dfrac{(n+1)!}{2\cdot 5\cdot 8\cdot\,\cdots\,\cdot(3n+2)[3(n+1)+2]} \cdot \dfrac{2\cdot 5\cdot 8\cdot\,\cdots\,\cdot(3n+2)}{n!}\right| = \lim\limits_{n\to\infty} \dfrac{n+1}{3n+5} = \dfrac{1}{3} < 1,$

so the series $\displaystyle\sum_{n=0}^{\infty} \dfrac{n!}{2\cdot 5\cdot 8\cdot\,\cdots\,\cdot(3n+2)}$ converges by the Ratio Test.

16. Using the Limit Comparison Test with $a_n = \dfrac{n^2+1}{n^3+1}$ and $b_n = \dfrac{1}{n}$, we have

$\lim\limits_{n\to\infty}\dfrac{a_n}{b_n} = \lim\limits_{n\to\infty}\left(\dfrac{n^2+1}{n^3+1}\cdot\dfrac{n}{1}\right) = \lim\limits_{n\to\infty}\dfrac{n^3+n}{n^3+1} = \lim\limits_{n\to\infty}\dfrac{1+1/n^2}{1+1/n^3} = 1 > 0.$ Since $\displaystyle\sum_{n=1}^{\infty} b_n$ is the divergent harmonic

series, $\displaystyle\sum_{n=1}^{\infty} a_n$ is also divergent.

17. $\lim\limits_{n\to\infty} 2^{1/n} = 2^0 = 1$, so $\lim\limits_{n\to\infty} (-1)^n\, 2^{1/n}$ does not exist and the series $\displaystyle\sum_{n=1}^{\infty} (-1)^n 2^{1/n}$ diverges by the Test for Divergence.

18. $b_n = \dfrac{1}{\sqrt{n-1}}$ for $n \geq 2$. $\{b_n\}$ is a decreasing sequence of positive numbers and $\lim\limits_{n\to\infty} b_n = 0$, so $\displaystyle\sum_{n=2}^{\infty} \dfrac{(-1)^{n-1}}{\sqrt{n-1}}$ converges by

the Alternating Series Test.

19. Let $f(x) = \dfrac{\ln x}{\sqrt{x}}$. Then $f'(x) = \dfrac{2-\ln x}{2x^{3/2}} < 0$ when $\ln x > 2$ or $x > e^2$, so $\dfrac{\ln n}{\sqrt{n}}$ is decreasing for $n > e^2$.

By l'Hospital's Rule, $\lim\limits_{n\to\infty}\dfrac{\ln n}{\sqrt{n}} = \lim\limits_{n\to\infty}\dfrac{1/n}{1/\left(2\sqrt{n}\right)} = \lim\limits_{n\to\infty}\dfrac{2}{\sqrt{n}} = 0$, so the series $\displaystyle\sum_{n=1}^{\infty} (-1)^n\dfrac{\ln n}{\sqrt{n}}$ converges by the

Alternating Series Test.

20. $\lim\limits_{k\to\infty} \left|\dfrac{a_{k+1}}{a_k}\right| = \lim\limits_{k\to\infty}\left|\dfrac{k+6}{5^{k+1}}\cdot\dfrac{5^k}{k+5}\right| = \dfrac{1}{5}\lim\limits_{k\to\infty}\dfrac{k+6}{k+5} = \dfrac{1}{5} < 1$, so the series $\displaystyle\sum_{k=1}^{\infty}\dfrac{k+5}{5^k}$ converges by the Ratio Test.

21. $\displaystyle\sum_{n=1}^{\infty}\dfrac{(-2)^{2n}}{n^n} = \sum_{n=1}^{\infty}\left(\dfrac{4}{n}\right)^n$. $\lim\limits_{n\to\infty}\sqrt[n]{|a_n|} = \lim\limits_{n\to\infty}\dfrac{4}{n} = 0 < 1$, so the given series is absolutely convergent by the Root Test.

22. $\dfrac{\sqrt{n^2-1}}{n^3+2n^2+5} < \dfrac{n}{n^3+2n^2+5} < \dfrac{n}{n^3} = \dfrac{1}{n^2}$ for $n \geq 1$, so $\displaystyle\sum_{n=1}^{\infty}\dfrac{\sqrt{n^2-1}}{n^3+2n^2+5}$ converges by the Comparison Test with the

convergent p-series $\displaystyle\sum_{n=1}^{\infty} 1/n^2$ $[p = 2 > 1]$.

23. Using the Limit Comparison Test with $a_n = \tan\left(\dfrac{1}{n}\right)$ and $b_n = \dfrac{1}{n}$, we have

$\lim\limits_{n\to\infty}\dfrac{a_n}{b_n} = \lim\limits_{n\to\infty}\dfrac{\tan(1/n)}{1/n} = \lim\limits_{x\to\infty}\dfrac{\tan(1/x)}{1/x} \overset{\text{H}}{=} \lim\limits_{x\to\infty}\dfrac{\sec^2(1/x)\cdot(-1/x^2)}{-1/x^2} = \lim\limits_{x\to\infty}\sec^2(1/x) = 1^2 = 1 > 0.$ Since

$\displaystyle\sum_{n=1}^{\infty} b_n$ is the divergent harmonic series, $\displaystyle\sum_{n=1}^{\infty} a_n$ is also divergent.

24. $\lim\limits_{n\to\infty} a_n = \lim\limits_{n\to\infty}\left(n\sin\dfrac{1}{n}\right) = \lim\limits_{n\to\infty}\dfrac{\sin(1/n)}{1/n} = \lim\limits_{x\to 0^+}\dfrac{\sin x}{x} = 1 \neq 0$, so the series $\displaystyle\sum_{n=1}^{\infty} n\sin(1/n)$ diverges by the

Test for Divergence.

25. Use the Ratio Test. $\lim\limits_{n \to \infty} \left| \dfrac{a_{n+1}}{a_n} \right| = \lim\limits_{n \to \infty} \left| \dfrac{(n+1)!}{e^{(n+1)^2}} \cdot \dfrac{e^{n^2}}{n!} \right| = \lim\limits_{n \to \infty} \dfrac{(n+1)n! \cdot e^{n^2}}{e^{n^2+2n+1}n!} = \lim\limits_{n \to \infty} \dfrac{n+1}{e^{2n+1}} = 0 < 1$, so $\sum\limits_{n=1}^{\infty} \dfrac{n!}{e^{n^2}}$

converges.

26. $\lim\limits_{n \to \infty} \left| \dfrac{a_{n+1}}{a_n} \right| = \lim\limits_{n \to \infty} \dfrac{a_{n+1}}{a_n} = \lim\limits_{n \to \infty} \left(\dfrac{n^2+2n+2}{5^{n+1}} \cdot \dfrac{5^n}{n^2+1} \right) = \lim\limits_{n \to \infty} \left(\dfrac{1+2/n+2/n^2}{1+1/n^2} \cdot \dfrac{1}{5} \right) = \dfrac{1}{5} < 1$, so $\sum\limits_{n=1}^{\infty} \dfrac{n^2+1}{5^n}$

converges by the Ratio Test.

27. $\displaystyle\int_{2}^{\infty} \dfrac{\ln x}{x^2}\,dx = \lim\limits_{t \to \infty} \left[-\dfrac{\ln x}{x} - \dfrac{1}{x} \right]_{1}^{t}$ [using integration by parts] $\overset{\text{H}}{=} 1$. So $\sum\limits_{n=1}^{\infty} \dfrac{\ln n}{n^2}$ converges by the Integral Test, and since

$\dfrac{k \ln k}{(k+1)^3} < \dfrac{k \ln k}{k^3} = \dfrac{\ln k}{k^2}$, the given series $\sum\limits_{k=1}^{\infty} \dfrac{k \ln k}{(k+1)^3}$ converges by the Comparison Test.

28. Since $\left\{ \dfrac{1}{n} \right\}$ is a decreasing sequence, $e^{1/n} \le e^{1/1} = e$ for all $n \ge 1$, and $\sum\limits_{n=1}^{\infty} \dfrac{e}{n^2}$ converges ($p = 2 > 1$), so $\sum\limits_{n=1}^{\infty} \dfrac{e^{1/n}}{n^2}$

converges by the Comparison Test. (Or use the Integral Test.)

29. $\sum\limits_{n=1}^{\infty} a_n = \sum\limits_{n=1}^{\infty} (-1)^n \dfrac{1}{\cosh n} = \sum\limits_{n=1}^{\infty} (-1)^n b_n$. Now $b_n = \dfrac{1}{\cosh n} > 0$, $\{b_n\}$ is decreasing, and $\lim\limits_{n \to \infty} b_n = 0$, so the series

converges by the Alternating Series Test.

Or: Write $\dfrac{1}{\cosh n} = \dfrac{2}{e^n + e^{-n}} < \dfrac{2}{e^n}$ and $\sum\limits_{n=1}^{\infty} \dfrac{1}{e^n}$ is a convergent geometric series, so $\sum\limits_{n=1}^{\infty} \dfrac{1}{\cosh n}$ is convergent by the

Comparison Test. So $\sum\limits_{n=1}^{\infty} (-1)^n \dfrac{1}{\cosh n}$ is absolutely convergent and therefore convergent.

30. Let $f(x) = \dfrac{\sqrt{x}}{x+5}$. Then $f(x)$ is continuous and positive on $[1, \infty)$, and since $f'(x) = \dfrac{5-x}{2\sqrt{x}\,(x+5)^2} < 0$ for $x > 5$, $f(x)$ is

eventually decreasing, so we can use the Alternating Series Test. $\lim\limits_{n \to \infty} \dfrac{\sqrt{n}}{n+5} = \lim\limits_{n \to \infty} \dfrac{1}{n^{1/2} + 5n^{-1/2}} = 0$, so the series

$\sum\limits_{j=1}^{\infty} (-1)^j \dfrac{\sqrt{j}}{j+5}$ converges.

31. $\lim\limits_{k \to \infty} a_k = \lim\limits_{k \to \infty} \dfrac{5^k}{3^k + 4^k} = $ [divide by 4^k] $\lim\limits_{k \to \infty} \dfrac{(5/4)^k}{(3/4)^k + 1} = \infty$ since $\lim\limits_{k \to \infty} \left(\dfrac{3}{4} \right)^k = 0$ and $\lim\limits_{k \to \infty} \left(\dfrac{5}{4} \right)^k = \infty$.

Thus, $\sum\limits_{k=1}^{\infty} \dfrac{5^k}{3^k + 4^k}$ diverges by the Test for Divergence.

32. $\lim\limits_{n \to \infty} \sqrt[n]{|a_n|} = \lim\limits_{n \to \infty} \sqrt[n]{\dfrac{(n!)^n}{n^{4n}}} = \lim\limits_{n \to \infty} \dfrac{n!}{n^4} = \lim\limits_{n \to \infty} \left[\dfrac{n}{n} \cdot \dfrac{n-1}{n} \cdot \dfrac{n-2}{n} \cdot \dfrac{n-3}{n} \cdot (n-4)! \right]$

$= \lim\limits_{n \to \infty} \left[\left(1 - \dfrac{1}{n} \right) \left(1 - \dfrac{2}{n} \right) \left(1 - \dfrac{3}{n} \right) (n-4)! \right] = \infty$,

so the series $\sum\limits_{n=1}^{\infty} \dfrac{(n!)^n}{n^{4n}}$ diverges by the Root Test.

33. Let $a_n = \dfrac{\sin(1/n)}{\sqrt{n}}$ and $b_n = \dfrac{1}{n\sqrt{n}}$. Then $\lim\limits_{n \to \infty} \dfrac{a_n}{b_n} = \lim\limits_{n \to \infty} \dfrac{\sin(1/n)}{1/n} = 1 > 0$, so $\sum\limits_{n=1}^{\infty} \dfrac{\sin(1/n)}{\sqrt{n}}$ converges by limit

comparison with the convergent p-series $\sum\limits_{n=1}^{\infty} \dfrac{1}{n^{3/2}}$ $[p = 3/2 > 1]$.

34. $0 \le n \cos^2 n \le n$, so $\dfrac{1}{n + n \cos^2 n} \ge \dfrac{1}{n+n} = \dfrac{1}{2n}$. Thus, $\displaystyle\sum_{n=1}^{\infty} \dfrac{1}{n + n \cos^2 n}$ diverges by comparison with $\displaystyle\sum_{n=1}^{\infty} \dfrac{1}{2n}$, which is

a constant multiple of the (divergent) harmonic series.

35. $\displaystyle\lim_{n\to\infty} \sqrt[n]{|a_n|} = \lim_{n\to\infty} \left(\dfrac{n}{n+1}\right)^{n^2/n} = \lim_{n\to\infty} \dfrac{1}{[(n+1)/n]^n} = \dfrac{1}{\displaystyle\lim_{n\to\infty} (1 + 1/n)^n} = \dfrac{1}{e} < 1$, so the series $\displaystyle\sum_{n=1}^{\infty} \left(\dfrac{n}{n+1}\right)^{n^2}$

converges by the Root Test.

36. Note that $(\ln n)^{\ln n} = \left(e^{\ln \ln n}\right)^{\ln n} = \left(e^{\ln n}\right)^{\ln \ln n} = n^{\ln \ln n}$ and $\ln \ln n \to \infty$ as $n \to \infty$, so $\ln \ln n > 2$ for sufficiently

large n. For these n we have $(\ln n)^{\ln n} > n^2$, so $\dfrac{1}{(\ln n)^{\ln n}} < \dfrac{1}{n^2}$. Since $\displaystyle\sum_{n=2}^{\infty} \dfrac{1}{n^2}$ converges $[p = 2 > 1]$, so does

$\displaystyle\sum_{n=2}^{\infty} \dfrac{1}{(\ln n)^{\ln n}}$ by the Comparison Test.

37. $\displaystyle\lim_{n\to\infty} \sqrt[n]{|a_n|} = \lim_{n\to\infty} (2^{1/n} - 1) = 1 - 1 = 0 < 1$, so the series $\displaystyle\sum_{n=1}^{\infty} \left(\sqrt[n]{2} - 1\right)^n$ converges by the Root Test.

38. Use the Limit Comparison Test with $a_n = \sqrt[n]{2} - 1$ and $b_n = 1/n$. Then

$$\lim_{n\to\infty} \frac{a_n}{b_n} = \lim_{n\to\infty} \frac{2^{1/n} - 1}{1/n} = \lim_{x\to\infty} \frac{2^{1/x} - 1}{1/x} \overset{\text{H}}{=} \lim_{x\to\infty} \frac{2^{1/x} \cdot \ln 2 \cdot (-1/x^2)}{-1/x^2} = \lim_{x\to\infty} (2^{1/x} \cdot \ln 2) = 1 \cdot \ln 2 = \ln 2 > 0.$$

So since $\displaystyle\sum_{n=1}^{\infty} b_n$ diverges (harmonic series), so does $\displaystyle\sum_{n=1}^{\infty} \left(\sqrt[n]{2} - 1\right)$.

Alternate solution: $\sqrt[n]{2} - 1 = \dfrac{1}{2^{(n-1)/n} + 2^{(n-2)/n} + 2^{(n-3)/n} + \cdots + 2^{1/n} + 1}$ [rationalize the numerator] $\ge \dfrac{1}{2n}$,

and since $\displaystyle\sum_{n=1}^{\infty} \dfrac{1}{2n} = \dfrac{1}{2} \sum_{n=1}^{\infty} \dfrac{1}{n}$ diverges (harmonic series), so does $\displaystyle\sum_{n=1}^{\infty} \left(\sqrt[n]{2} - 1\right)$ by the Comparison Test.

12.8 Power Series

ET 11.8

1. A power series is a series of the form $\sum_{n=0}^{\infty} c_n x^n = c_0 + c_1 x + c_2 x^2 + c_3 x^3 + \cdots$, where x is a variable and the c_n's are

constants called the coefficients of the series.

More generally, a series of the form $\sum_{n=0}^{\infty} c_n (x - a)^n = c_0 + c_1 (x - a) + c_2 (x - a)^2 + \cdots$ is called a power series in

$(x - a)$ or a power series centered at a or a power series about a, where a is a constant.

2. (a) Given the power series $\sum_{n=0}^{\infty} c_n (x - a)^n$, the radius of convergence is:

  (i) 0 if the series converges only when $x = a$

  (ii) ∞ if the series converges for all x, or

  (iii) a positive number R such that the series converges if $|x - a| < R$ and diverges if $|x - a| > R$.

  In most cases, R can be found by using the Ratio Test.

 (b) The interval of convergence of a power series is the interval that consists of all values of x for which the series converges.

  Corresponding to the cases in part (a), the interval of convergence is: (i) the single point $\{a\}$, (ii) all real numbers; that is,

  the real number line $(-\infty, \infty)$, or (iii) an interval with endpoints $a - R$ and $a + R$ which can contain neither, either, or

  both of the endpoints. In this case, we must test the series for convergence at each endpoint to determine the interval of

  convergence.

3. If $a_n = \dfrac{x^n}{\sqrt{n}}$, then $\lim\limits_{n\to\infty}\left|\dfrac{a_{n+1}}{a_n}\right| = \lim\limits_{n\to\infty}\left|\dfrac{x^{n+1}}{\sqrt{n+1}}\cdot\dfrac{\sqrt{n}}{x^n}\right| = \lim\limits_{n\to\infty}\left|\dfrac{x}{\sqrt{n+1}/\sqrt{n}}\right| = \lim\limits_{n\to\infty}\dfrac{|x|}{\sqrt{1+1/n}} = |x|$.

By the Ratio Test, the series $\sum\limits_{n=1}^{\infty}\dfrac{x^n}{\sqrt{n}}$ converges when $|x| < 1$, so the radius of convergence $R = 1$. Now we'll check the

endpoints, that is, $x = \pm 1$. When $x = 1$, the series $\sum\limits_{n=1}^{\infty}\dfrac{1}{\sqrt{n}}$ diverges because it is a p-series with $p = \frac{1}{2} \le 1$. When $x = -1$,

the series $\sum\limits_{n=1}^{\infty}\dfrac{(-1)^n}{\sqrt{n}}$ converges by the Alternating Series Test. Thus, the interval of convergence is $I = [-1, 1)$.

4. If $a_n = \dfrac{(-1)^n x^n}{n+1}$, then $\lim\limits_{n\to\infty}\left|\dfrac{a_{n+1}}{a_n}\right| = \lim\limits_{n\to\infty}\left|\dfrac{x^{n+1}}{n+2}\cdot\dfrac{n+1}{x^n}\right| = \lim\limits_{n\to\infty}\dfrac{|x|}{1+1/(n+1)} = |x|$.

By the Ratio Test, the series $\sum\limits_{n=0}^{\infty}\dfrac{(-1)^n x^n}{n+1}$ converges when $|x| < 1$, so $R = 1$. When $x = -1$, the series diverges because it

is the harmonic series; when $x = 1$, it is the alternating harmonic series, which converges by the Alternating Series Test.

Thus, $I = (-1, 1]$.

5. If $a_n = \dfrac{(-1)^{n-1} x^n}{n^3}$, then

$\lim\limits_{n\to\infty}\left|\dfrac{a_{n+1}}{a_n}\right| = \lim\limits_{n\to\infty}\left|\dfrac{(-1)^n x^{n+1}}{(n+1)^3}\cdot\dfrac{n^3}{(-1)^{n-1}x^n}\right| = \lim\limits_{n\to\infty}\left|\dfrac{(-1)xn^3}{(n+1)^3}\right| = \lim\limits_{n\to\infty}\left[\left(\dfrac{n}{n+1}\right)^3|x|\right] = 1^3\cdot|x| = |x|$. By the

Ratio Test, the series $\sum\limits_{n=1}^{\infty}\dfrac{(-1)^{n-1} x^n}{n^3}$ converges when $|x| < 1$, so the radius of convergence $R = 1$. Now we'll check the

endpoints, that is, $x = \pm 1$. When $x = 1$, the series $\sum\limits_{n=1}^{\infty}\dfrac{(-1)^{n-1}}{n^3}$ converges by the Alternating Series Test. When $x = -1$,

the series $\sum\limits_{n=1}^{\infty}\dfrac{(-1)^{n-1}(-1)^n}{n^3} = -\sum\limits_{n=1}^{\infty}\dfrac{1}{n^3}$ converges because it is a constant multiple of a convergent p-series $[p = 3 > 1]$.

Thus, the interval of convergence is $I = [-1, 1]$.

6. $a_n = \sqrt{n}\, x^n$, so we need $\lim\limits_{n\to\infty}\left|\dfrac{a_{n+1}}{a_n}\right| = \lim\limits_{n\to\infty}\dfrac{\sqrt{n+1}\,|x|^{n+1}}{\sqrt{n}\,|x|^n} = \lim\limits_{n\to\infty}\sqrt{1+\dfrac{1}{n}}\,|x| = |x| < 1$ for convergence (by the

Ratio Test), so $R = 1$. When $x = \pm 1$, $\lim\limits_{n\to\infty}|a_n| = \lim\limits_{n\to\infty}\sqrt{n} = \infty$, so the series diverges by the Test for Divergence.

Thus, $I = (-1, 1)$.

7. If $a_n = \dfrac{x^n}{n!}$, then $\lim\limits_{n\to\infty}\left|\dfrac{a_{n+1}}{a_n}\right| = \lim\limits_{n\to\infty}\left|\dfrac{x^{n+1}}{(n+1)!}\cdot\dfrac{n!}{x^n}\right| = \lim\limits_{n\to\infty}\left|\dfrac{x}{n+1}\right| = |x|\lim\limits_{n\to\infty}\dfrac{1}{n+1} = |x|\cdot 0 = 0 < 1$ for *all* real x.

So, by the Ratio Test, $R = \infty$ and $I = (-\infty, \infty)$.

8. Here the Root Test is easier. If $a_n = n^n x^n$ then $\lim\limits_{n\to\infty}\sqrt[n]{|a_n|} = \lim\limits_{n\to\infty}n\,|x| = \infty$ if $x \ne 0$, so $R = 0$ and $I = \{0\}$.

9. If $a_n = (-1)^n\dfrac{n^2 x^n}{2^n}$, then

$\lim\limits_{n\to\infty}\left|\dfrac{a_{n+1}}{a_n}\right| = \lim\limits_{n\to\infty}\left|\dfrac{(n+1)^2\,x^{n+1}}{2^{n+1}}\cdot\dfrac{2^n}{n^2\,x^n}\right| = \lim\limits_{n\to\infty}\left|\dfrac{x(n+1)^2}{2n^2}\right| = \lim\limits_{n\to\infty}\left[\dfrac{|x|}{2}\left(1+\dfrac{1}{n}\right)^2\right] = \dfrac{|x|}{2}(1)^2 = \dfrac{1}{2}|x|$. By the

Ratio Test, the series $\sum\limits_{n=1}^{\infty} (-1)^n \dfrac{n^2 x^n}{2^n}$ converges when $\frac{1}{2}|x| < 1$ $\Leftrightarrow$ $|x| < 2$, so the radius of convergence is $R = 2$.

When $x = \pm 2$, both series $\sum\limits_{n=1}^{\infty} (-1)^n \dfrac{n^2 (\pm 2)^n}{2^n} = \sum\limits_{n=1}^{\infty} (\mp 1)^n n^2$ diverge by the Test for Divergence since

$\lim\limits_{n \to \infty} \left| (\mp 1)^n n^2 \right| = \infty$. Thus, the interval of convergence is $I = (-2, 2)$.

10. If $a_n = \dfrac{10^n x^n}{n^3}$, then

$$\lim_{n \to \infty} \left| \frac{a_{n+1}}{a_n} \right| = \lim_{n \to \infty} \left| \frac{10^{n+1} x^{n+1}}{(n+1)^3} \cdot \frac{n^3}{10^n x^n} \right| = \lim_{n \to \infty} \left| \frac{10x\, n^3}{(n+1)^3} \right| = \lim_{n \to \infty} \frac{10\,|x|}{(1+1/n)^3} = \frac{10\,|x|}{1^3} = 10\,|x|$$

By the Ratio Test, the series $\sum\limits_{n=1}^{\infty} \dfrac{10^n x^n}{n^3}$ converges when $10\,|x| < 1$ $\Leftrightarrow$ $|x| < \frac{1}{10}$, so the radius of convergence is $R = \frac{1}{10}$.

When $x = -\frac{1}{10}$, the series converges by the Alternating Series Test; when $x = \frac{1}{10}$, the series converges because it is a p-series

with $p = 3 > 1$. Thus, the interval of convergence is $I = \left[-\frac{1}{10}, \frac{1}{10} \right]$.

11. $a_n = \dfrac{(-2)^n x^n}{\sqrt[4]{n}}$, so $\lim\limits_{n \to \infty} \left| \dfrac{a_{n+1}}{a_n} \right| = \lim\limits_{n \to \infty} \dfrac{2^{n+1} |x|^{n+1}}{\sqrt[4]{n+1}} \cdot \dfrac{\sqrt[4]{n}}{2^n |x|^n} = \lim\limits_{n \to \infty} 2\,|x|\, \sqrt[4]{\dfrac{n}{n+1}} = 2\,|x|$, so by the Ratio Test, the

series converges when $2\,|x| < 1$ $\Leftrightarrow$ $|x| < \frac{1}{2}$, so $R = \frac{1}{2}$. When $x = -\frac{1}{2}$, we get the divergent p-series $\sum\limits_{n=1}^{\infty} \dfrac{1}{\sqrt[4]{n}}$

$\left[p = \frac{1}{4} \leq 1 \right]$. When $x = \frac{1}{2}$, we get the series $\sum\limits_{n=1}^{\infty} \dfrac{(-1)^n}{\sqrt[4]{n}}$, which converges by the Alternating Series Test.

Thus, $I = \left(-\frac{1}{2}, \frac{1}{2} \right]$.

12. $a_n = \dfrac{x^n}{5^n n^5}$, so $\lim\limits_{n \to \infty} \left| \dfrac{a_{n+1}}{a_n} \right| = \lim\limits_{n \to \infty} \left| \dfrac{x^{n+1}}{5^{n+1}(n+1)^5} \cdot \dfrac{5^n n^5}{x^n} \right| = \lim\limits_{n \to \infty} \dfrac{|x|}{5} \left(\dfrac{n}{n+1} \right)^5 = \dfrac{|x|}{5}$. By the Ratio Test, the series

$\sum\limits_{n=0}^{\infty} \dfrac{x^n}{5^n n^5}$ converges when $\dfrac{|x|}{5} < 1$ $\Leftrightarrow$ $|x| < 5$, so $R = 5$. When $x = -5$, we get the series $\sum\limits_{n=1}^{\infty} \dfrac{(-1)^n}{n^5}$, which

converges by the Alternating Series Test. When $x = 5$, we get the convergent p-series $\sum\limits_{n=1}^{\infty} \dfrac{1}{n^5}$ $\;[p = 5 > 1]$.

Thus, $I = [-5, 5]$.

13. If $a_n = (-1)^n \dfrac{x^n}{4^n \ln n}$, then $\lim\limits_{n \to \infty} \left| \dfrac{a_{n+1}}{a_n} \right| = \lim\limits_{n \to \infty} \left| \dfrac{x^{n+1}}{4^{n+1} \ln(n+1)} \cdot \dfrac{4^n \ln n}{x^n} \right| = \dfrac{|x|}{4} \lim\limits_{n \to \infty} \dfrac{\ln n}{\ln(n+1)} = \dfrac{|x|}{4} \cdot 1$

[by l'Hospital's Rule] $= \dfrac{|x|}{4}$. By the Ratio Test, the series converges when $\dfrac{|x|}{4} < 1$ $\Leftrightarrow$ $|x| < 4$, so $R = 4$. When

$x = -4$, $\sum\limits_{n=2}^{\infty} (-1)^n \dfrac{x^n}{4^n \ln n} = \sum\limits_{n=2}^{\infty} \dfrac{[(-1)(-4)]^n}{4^n \ln n} = \sum\limits_{n=2}^{\infty} \dfrac{1}{\ln n}$. Since $\ln n < n$ for $n \geq 2$, $\dfrac{1}{\ln n} > \dfrac{1}{n}$ and $\sum\limits_{n=2}^{\infty} \dfrac{1}{n}$ is the

divergent harmonic series (without the $n = 1$ term), $\sum\limits_{n=2}^{\infty} \dfrac{1}{\ln n}$ is divergent by the Comparison Test. When $x = 4$,

$\sum\limits_{n=2}^{\infty} (-1)^n \dfrac{x^n}{4^n \ln n} = \sum\limits_{n=2}^{\infty} (-1)^n \dfrac{1}{\ln n}$, which converges by the Alternating Series Test. Thus, $I = (-4, 4]$.

14. $a_n = (-1)^n \dfrac{x^{2n}}{(2n)!}$, so $\lim\limits_{n \to \infty} \left| \dfrac{a_{n+1}}{a_n} \right| = \lim\limits_{n \to \infty} \dfrac{|x|^{2n+2}}{(2n+2)!} \cdot \dfrac{(2n)!}{|x|^{2n}} = \lim\limits_{n \to \infty} \dfrac{|x|^2}{(2n+1)(2n+2)} = 0 < 1$. Thus, by the Ratio

Test, the series converges for *all* real x and we have $R = \infty$ and $I = (-\infty, \infty)$.

15. If $a_n = \dfrac{(x-2)^n}{n^2+1}$, then $\lim\limits_{n\to\infty}\left|\dfrac{a_{n+1}}{a_n}\right| = \lim\limits_{n\to\infty}\left|\dfrac{(x-2)^{n+1}}{(n+1)^2+1}\cdot\dfrac{n^2+1}{(x-2)^n}\right| = |x-2|\lim\limits_{n\to\infty}\dfrac{n^2+1}{(n+1)^2+1} = |x-2|$. By the

Ratio Test, the series $\sum\limits_{n=0}^{\infty}\dfrac{(x-2)^n}{n^2+1}$ converges when $|x-2|<1$ $[R=1]$ $\Leftrightarrow$ $-1 < x-2 < 1$ $\Leftrightarrow$ $1 < x < 3$. When

$x=1$, the series $\sum\limits_{n=0}^{\infty}(-1)^n\dfrac{1}{n^2+1}$ converges by the Alternating Series Test; when $x=3$, the series $\sum\limits_{n=0}^{\infty}\dfrac{1}{n^2+1}$ converges by

comparison with the p-series $\sum\limits_{n=1}^{\infty}\dfrac{1}{n^2}$ $[p=2>1]$. Thus, the interval of convergence is $I = [1,3]$.

16. If $a_n = (-1)^n\dfrac{(x-3)^n}{2n+1}$, then $\lim\limits_{n\to\infty}\left|\dfrac{a_{n+1}}{a_n}\right| = \lim\limits_{n\to\infty}\left|\dfrac{(x-3)^{n+1}}{2n+3}\cdot\dfrac{2n+1}{(x-3)^n}\right| = |x-3|\lim\limits_{n\to\infty}\dfrac{2n+1}{2n+3} = |x-3|$. By the

Ratio Test, the series $\sum\limits_{n=0}^{\infty}(-1)^n\dfrac{(x-3)^n}{2n+1}$ converges when $|x-3|<1$ $[R=1]$ $\Leftrightarrow$ $-1 < x-3 < 1$ $\Leftrightarrow$ $2 < x < 4$.

When $x=2$, the series $\sum\limits_{n=0}^{\infty}\dfrac{1}{2n+1}$ diverges by limit comparison with the harmonic series (or by the Integral Test); when

$x=4$, the series $\sum\limits_{n=0}^{\infty}(-1)^n\dfrac{1}{2n+1}$ converges by the Alternating Series Test. Thus, the interval of convergence is $I=(2,4]$.

17. If $a_n = \dfrac{3^n(x+4)^n}{\sqrt{n}}$, then $\lim\limits_{n\to\infty}\left|\dfrac{a_{n+1}}{a_n}\right| = \lim\limits_{n\to\infty}\left|\dfrac{3^{n+1}(x+4)^{n+1}}{\sqrt{n+1}}\cdot\dfrac{\sqrt{n}}{3^n(x+4)^n}\right| = 3|x+4|\lim\limits_{n\to\infty}\dfrac{\sqrt{n}}{\sqrt{n+1}} = 3|x+4|$.

By the Ratio Test, the series $\sum\limits_{n=1}^{\infty}\dfrac{3^n(x+4)^n}{\sqrt{n}}$ converges when $3|x+4|<1$ $\Leftrightarrow$ $|x+4|<\frac{1}{3}$ $\left[R=\frac{1}{3}\right]$ $\Leftrightarrow$

$-\frac{1}{3} < x+4 < \frac{1}{3}$ $\Leftrightarrow$ $-\frac{13}{3} < x < -\frac{11}{3}$. When $x=-\frac{13}{3}$, the series $\sum\limits_{n=1}^{\infty}(-1)^n\dfrac{1}{\sqrt{n}}$ converges by the Alternating Series

Test; when $x=-\frac{11}{3}$, the series $\sum\limits_{n=1}^{\infty}\dfrac{1}{\sqrt{n}}$ diverges $\left[p=\frac{1}{2}\le 1\right]$. Thus, the interval of convergence is $I=\left[-\frac{13}{3},-\frac{11}{3}\right)$.

18. If $a_n = \dfrac{n}{4^n}(x+1)^n$, then $\lim\limits_{n\to\infty}\left|\dfrac{a_{n+1}}{a_n}\right| = \lim\limits_{n\to\infty}\left|\dfrac{(n+1)(x+1)^{n+1}}{4^{n+1}}\cdot\dfrac{4^n}{n(x+1)^n}\right| = \dfrac{|x+1|}{4}\lim\limits_{n\to\infty}\dfrac{n+1}{n} = \dfrac{|x+1|}{4}$.

By the Ratio Test, the series $\sum\limits_{n=1}^{\infty}\dfrac{n}{4^n}(x+1)^n$ converges when $\dfrac{|x+1|}{4}<1$ $\Leftrightarrow$ $|x+1|<4$ $[R=4]$ $\Leftrightarrow$

$-4 < x+1 < 4$ $\Leftrightarrow$ $-5 < x < 3$. When $x=-5$ or 3, both series $\sum\limits_{n=1}^{\infty}(\mp1)^n n$ diverge by the Test for Divergence since

$\lim\limits_{n\to\infty}|(\mp1)^n n| = \infty$. Thus, the interval of convergence is $I=(-5,3)$.

19. If $a_n = \dfrac{(x-2)^n}{n^n}$, then $\lim\limits_{n\to\infty}\sqrt[n]{|a_n|} = \lim\limits_{n\to\infty}\dfrac{|x-2|}{n} = 0$, so the series converges for all x (by the Root Test).

$R=\infty$ and $I=(-\infty,\infty)$.

20. $\lim\limits_{n\to\infty}\left|\dfrac{a_{n+1}}{a_n}\right| = \lim\limits_{n\to\infty}\left|\dfrac{(3x-2)^{n+1}}{(n+1)3^{n+1}}\cdot\dfrac{n3^n}{(3x-2)^n}\right| = \lim\limits_{n\to\infty}\left(\dfrac{|3x-2|}{3}\cdot\dfrac{1}{1+1/n}\right) = \dfrac{|3x-2|}{3} = \left|x-\frac{2}{3}\right|$, so by the Ratio

Test, the series converges when $\left|x-\frac{2}{3}\right|<1$ $\Leftrightarrow$ $-\frac{1}{3} < x < \frac{5}{3}$. $R=1$. When $x=-\frac{1}{3}$, the series is $\sum\limits_{n=1}^{\infty}\dfrac{(-1)^n}{n}$, the

convergent alternating harmonic series. When $x=\frac{5}{3}$, the series becomes the divergent harmonic series. Thus, $I=\left[-\frac{1}{3},\frac{5}{3}\right)$.

21. $a_n = \dfrac{n}{b^n}(x-a)^n$, where $b > 0$.

$$\lim_{n\to\infty}\left|\frac{a_{n+1}}{a_n}\right| = \lim_{n\to\infty}\frac{(n+1)\,|x-a|^{n+1}}{b^{n+1}}\cdot\frac{b^n}{n\,|x-a|^n} = \lim_{n\to\infty}\left(1+\frac{1}{n}\right)\frac{|x-a|}{b} = \frac{|x-a|}{b}.$$

By the Ratio Test, the series converges when $\dfrac{|x-a|}{b} < 1$ $\Leftrightarrow$ $|x-a| < b$ [so $R = b$] $\Leftrightarrow$ $-b < x-a < b$ $\Leftrightarrow$

$a-b < x < a+b$. When $|x-a| = b$, $\lim\limits_{n\to\infty}|a_n| = \lim\limits_{n\to\infty} n = \infty$, so the series diverges. Thus, $I = (a-b, a+b)$.

22. $a_n = \dfrac{n(x-4)^n}{n^3+1}$, so

$$\lim_{n\to\infty}\left|\frac{a_{n+1}}{a_n}\right| = \lim_{n\to\infty}\frac{(n+1)\,|x-4|^{n+1}}{(n+1)^3+1}\cdot\frac{n^3+1}{n\,|x-4|^n} = \lim_{n\to\infty}\left(1+\frac{1}{n}\right)\frac{n^3+1}{n^3+3n^2+3n+2}\,|x-4| = |x-4|.$$

By the Ratio Test, the series converges when $|x-4| < 1$ [so $R = 1$] $\Leftrightarrow$ $-1 < x-4 < 1$ $\Leftrightarrow$ $3 < x < 5$. When

$|x-4| = 1$, $\sum\limits_{n=1}^{\infty}|a_n| = \sum\limits_{n=1}^{\infty}\dfrac{n}{n^3+1}$, which converges by comparison with the convergent p-series $\sum\limits_{n=1}^{\infty}\dfrac{1}{n^2}$ [$p = 2 > 1$].

Thus, $I = [3, 5]$.

23. If $a_n = n!\,(2x-1)^n$, then $\lim\limits_{n\to\infty}\left|\dfrac{a_{n+1}}{a_n}\right| = \lim\limits_{n\to\infty}\left|\dfrac{(n+1)!\,(2x-1)^{n+1}}{n!(2x-1)^n}\right| = \lim\limits_{n\to\infty}(n+1)\,|2x-1| \to \infty$ as $n\to\infty$

for all $x \neq \frac{1}{2}$. Since the series diverges for all $x \neq \frac{1}{2}$, $R = 0$ and $I = \left\{\frac{1}{2}\right\}$.

24. $a_n = \dfrac{n^2 x^n}{2\cdot 4\cdot 6\cdots(2n)} = \dfrac{n^2 x^n}{2^n n!} = \dfrac{n x^n}{2^n(n-1)!}$, so

$$\lim_{n\to\infty}\left|\frac{a_{n+1}}{a_n}\right| = \lim_{n\to\infty}\frac{(n+1)\,|x|^{n+1}}{2^{n+1}n!}\cdot\frac{2^n(n-1)!}{n\,|x|^n} = \lim_{n\to\infty}\frac{n+1}{n^2}\frac{|x|}{2} = 0.$$ Thus, by the Ratio Test, the series converges for

all real x and we have $R = \infty$ and $I = (-\infty, \infty)$.

25. $\lim\limits_{n\to\infty}\left|\dfrac{a_{n+1}}{a_n}\right| = \lim\limits_{n\to\infty}\left[\dfrac{|4x+1|^{n+1}}{(n+1)^2}\cdot\dfrac{n^2}{|4x+1|^n}\right] = \lim\limits_{n\to\infty}\dfrac{|4x+1|}{(1+1/n)^2} = |4x+1|$, so by the Ratio Test, the series

converges when $|4x+1| < 1$ $\Leftrightarrow$ $-1 < 4x+1 < 1$ $\Leftrightarrow$ $-2 < 4x < 0$ $\Leftrightarrow$ $-\frac{1}{2} < x < 0$, so $R = \frac{1}{4}$. When $x = -\frac{1}{2}$,

the series becomes $\sum\limits_{n=1}^{\infty}\dfrac{(-1)^n}{n^2}$, which converges by the Alternating Series Test. When $x = 0$, the series becomes $\sum\limits_{n=1}^{\infty}\dfrac{1}{n^2}$,

a convergent p-series [$p = 2 > 1$]. $I = \left[-\frac{1}{2}, 0\right]$.

26. If $a_n = \dfrac{x^{2n}}{n\,(\ln n)^2}$, then $\lim\limits_{n\to\infty}\left|\dfrac{a_{n+1}}{a_n}\right| = \lim\limits_{n\to\infty}\left|\dfrac{x^{2n+2}}{(n+1)[\ln(n+1)]^2}\cdot\dfrac{n\,(\ln n)^2}{x^{2n}}\right| = |x^2|\lim\limits_{n\to\infty}\dfrac{n\,(\ln n)^2}{(n+1)[\ln(n+1)]^2} = x^2.$

By the Ratio Test, the series $\sum\limits_{n=2}^{\infty}\dfrac{x^{2n}}{n\,(\ln n)^2}$ converges when $x^2 < 1$ $\Leftrightarrow$ $|x| < 1$, so $R = 1$. When $x = \pm 1$, $x^{2n} = 1$, the

series $\sum\limits_{n=2}^{\infty}\dfrac{1}{n\,(\ln n)^2}$ converges by the Integral Test (see Exercise 12.3.22 [ET 11.3.22]). Thus, the interval of convergence

is $I = [-1, 1]$.

27. If $a_n = \dfrac{x^n}{1 \cdot 3 \cdot 5 \cdot \ldots \cdot (2n-1)}$, then

$$\lim_{n\to\infty} \left| \frac{a_{n+1}}{a_n} \right| = \lim_{n\to\infty} \left| \frac{x^{n+1}}{1 \cdot 3 \cdot 5 \cdot \ldots \cdot (2n-1)(2n+1)} \cdot \frac{1 \cdot 3 \cdot 5 \cdot \ldots \cdot (2n-1)}{x^n} \right| = \lim_{n\to\infty} \frac{|x|}{2n+1} = 0 < 1. \text{ Thus, by the}$$

Ratio Test, the series $\displaystyle\sum_{n=1}^{\infty} \dfrac{x^n}{1 \cdot 3 \cdot 5 \cdot \ldots \cdot (2n-1)}$ converges for *all* real x and we have $R = \infty$ and $I = (-\infty, \infty)$.

28. If $a_n = \dfrac{n!\, x^n}{1 \cdot 3 \cdot 5 \cdot \ldots \cdot (2n-1)}$, then

$$\lim_{n\to\infty} \left| \frac{a_{n+1}}{a_n} \right| = \lim_{n\to\infty} \left| \frac{(n+1)!\, x^{n+1}}{1 \cdot 3 \cdot 5 \cdot \ldots \cdot (2n-1)(2n+1)} \cdot \frac{1 \cdot 3 \cdot 5 \cdot \ldots \cdot (2n-1)}{n!\, x^n} \right| = \lim_{n\to\infty} \frac{(n+1)\,|x|}{2n+1} = \tfrac{1}{2}\,|x|.$$

By the Ratio Test, the series $\displaystyle\sum_{n=1}^{\infty} a_n$ converges when $\tfrac{1}{2}|x| < 1 \;\Rightarrow\; |x| < 2$, so $R = 2$. When $x = \pm 2$,

$$|a_n| = \frac{n!\, 2^n}{1 \cdot 3 \cdot 5 \cdot \ldots \cdot (2n-1)} = \frac{[1 \cdot 2 \cdot 3 \cdot \ldots \cdot n]\, 2^n}{[1 \cdot 3 \cdot 5 \cdot \ldots \cdot (2n-1)]} = \frac{2 \cdot 4 \cdot 6 \cdot \ldots \cdot 2n}{1 \cdot 3 \cdot 5 \cdot \ldots \cdot (2n-1)} > 1, \text{ so both endpoint series}$$

diverge by the Test for Divergence. Thus, the interval of convergence is $I = (-2, 2)$.

29. (a) We are given that the power series $\sum_{n=0}^{\infty} c_n x^n$ is convergent for $x = 4$. So by Theorem 3, it must converge for at least

$-4 < x \le 4$. In particular, it converges when $x = -2$; that is, $\sum_{n=0}^{\infty} c_n (-2)^n$ is convergent.

(b) It does not follow that $\sum_{n=0}^{\infty} c_n (-4)^n$ is necessarily convergent. [See the comments after Theorem 3 about convergence at

the endpoint of an interval. An example is $c_n = (-1)^n / (n 4^n)$.]

30. We are given that the power series $\sum_{n=0}^{\infty} c_n x^n$ is convergent for $x = -4$ and divergent when $x = 6$. So by Theorem 3 it

converges for at least $-4 \le x < 4$ and diverges for at least $x \ge 6$ and $x < -6$. Therefore:

(a) It converges when $x = 1$; that is, $\sum c_n$ is convergent.

(b) It diverges when $x = 8$; that is, $\sum c_n 8^n$ is divergent.

(c) It converges when $x = -3$; that is, $\sum c_n (-3^n)$ is convergent.

(d) It diverges when $x = -9$; that is, $\sum c_n (-9)^n = \sum (-1)^n c_n 9^n$ is divergent.

31. If $a_n = \dfrac{(n!)^k}{(kn)!}\, x^n$, then

$$\lim_{n\to\infty} \left| \frac{a_{n+1}}{a_n} \right| = \lim_{n\to\infty} \frac{[(n+1)!]^k\, (kn)!}{(n!)^k\, [k(n+1)]!}\, |x| = \lim_{n\to\infty} \frac{(n+1)^k}{(kn+k)(kn+k-1)\cdots(kn+2)(kn+1)}\, |x|$$

$$= \lim_{n\to\infty} \left[\frac{(n+1)}{(kn+1)} \frac{(n+1)}{(kn+2)} \cdots \frac{(n+1)}{(kn+k)} \right] |x|$$

$$= \lim_{n\to\infty} \left[\frac{n+1}{kn+1} \right] \lim_{n\to\infty} \left[\frac{n+1}{kn+2} \right] \cdots \lim_{n\to\infty} \left[\frac{n+1}{kn+k} \right] |x|$$

$$= \left(\frac{1}{k} \right)^k |x| < 1 \quad\Leftrightarrow\quad |x| < k^k \text{ for convergence, and the radius of convergence is } R = k^k.$$

32. (a) Note that the four intervals in parts (a)–(d) have midpoint $m = \frac{1}{2}(p+q)$ and radius of convergence $r = \frac{1}{2}(q-p)$. We also

know that the power series $\sum\limits_{n=0}^{\infty} x^n$ has interval of convergence $(-1, 1)$. To change the radius of convergence to r, we can

change x^n to $\left(\dfrac{x}{r}\right)^n$. To shift the midpoint of the interval of convergence, we can replace x with $x - m$. Thus, a power

series whose interval of convergence is (p, q) is $\sum\limits_{n=0}^{\infty} \left(\dfrac{x-m}{r}\right)^n$, where $m = \frac{1}{2}(p+q)$ and $r = \frac{1}{2}(q-p)$.

(b) Similar to Example 2, we know that $\sum\limits_{n=1}^{\infty} \dfrac{x^n}{n}$ has interval of convergence $[-1, 1)$. By introducing the factor $(-1)^n$

in a_n, the interval of convergence changes to $(-1, 1]$. Now change the midpoint and radius as in part (a) to get

$\sum\limits_{n=1}^{\infty} (-1)^n \dfrac{1}{n} \left(\dfrac{x-m}{r}\right)^n$ as a power series whose interval of convergence is $(p, q]$.

(c) As in part (b), $\sum\limits_{n=1}^{\infty} \dfrac{1}{n} \left(\dfrac{x-m}{r}\right)^n$ is a power series whose interval of convergence is $[p, q)$.

(d) If we increase the exponent on n (to say, $n = 2$), in the power series in part (c), then when $x = q$, the power series

$\sum\limits_{n=1}^{\infty} \dfrac{1}{n^2} \left(\dfrac{x-m}{r}\right)^n$ will converge by comparison to the p-series with $p = 2 > 1$, and the interval of convergence will

be $[p, q]$.

33. No. If a power series is centered at a, its interval of convergence is symmetric about a. If a power series has an infinite radius

of convergence, then its interval of convergence must be $(-\infty, \infty)$, not $[0, \infty)$.

34. The partial sums of the series $\sum_{n=0}^{\infty} x^n$ definitely do not converge

to $f(x) = 1/(1-x)$ for $x \geq 1$, since f is undefined at $x = 1$ and

negative on $(1, \infty)$, while all the partial sums are positive on this

interval. The partial sums also fail to converge to f for $x \leq -1$,

since $0 < f(x) < 1$ on this interval, while the partial sums are

either larger than 1 or less than 0. The partial sums seem to

converge to f on $(-1, 1)$. This graphical evidence is consistent

with what we know about geometric series: convergence for

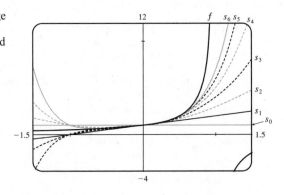

$|x| < 1$, divergence for $|x| \geq 1$ (see Examples 1 and 5 in Section 12.2 [ET 11.2]).

35. (a) If $a_n = \dfrac{(-1)^n x^{2n+1}}{n!(n+1)! \, 2^{2n+1}}$, then

$$\lim_{n\to\infty} \left| \frac{a_{n+1}}{a_n} \right| = \lim_{n\to\infty} \left| \frac{x^{2n+3}}{(n+1)!(n+2)! \, 2^{2n+3}} \cdot \frac{n!(n+1)! \, 2^{2n+1}}{x^{2n+1}} \right| = \left(\frac{x}{2}\right)^2 \lim_{n\to\infty} \frac{1}{(n+1)(n+2)} = 0 \text{ for all } x.$$

So $J_1(x)$ converges for all x and its domain is $(-\infty, \infty)$.

(b), (c) The initial terms of $J_1(x)$ up to $n = 5$ are $a_0 = \dfrac{x}{2}$,

$$a_1 = -\frac{x^3}{16}, a_2 = \frac{x^5}{384}, a_3 = -\frac{x^7}{18{,}432}, a_4 = \frac{x^9}{1{,}474{,}560},$$

and $a_5 = -\dfrac{x^{11}}{176{,}947{,}200}$. The partial sums seem to

approximate $J_1(x)$ well near the origin, but as $|x|$ increases,
we need to take a large number of terms to get a good
approximation.

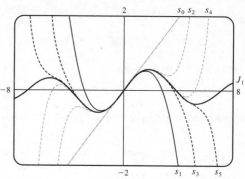

36. (a) $A(x) = 1 + \displaystyle\sum_{n=1}^{\infty} a_n$, where $a_n = \dfrac{x^{3n}}{2 \cdot 3 \cdot 5 \cdot 6 \cdots (3n-1)(3n)}$, so $\displaystyle\lim_{n \to \infty} \left| \frac{a_{n+1}}{a_n} \right| = |x|^3 \lim_{n \to \infty} \dfrac{1}{(3n+2)(3n+3)} = 0$

for all x, so the domain is $\mathbb{R}$.

(b), (c)

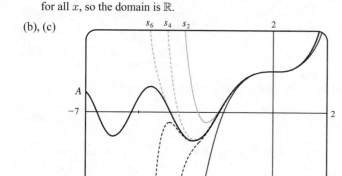

$s_0 = 1$ has been omitted from the graph. The
partial sums seem to approximate $A(x)$ well
near the origin, but as $|x|$ increases, we need to
take a large number of terms to get a good
approximation.

To plot A, we must first define $A(x)$ for the CAS. Note that for $n \geq 1$, the denominator of a_n is

$$2 \cdot 3 \cdot 5 \cdot 6 \cdots (3n-1) \cdot 3n = \frac{(3n)!}{1 \cdot 4 \cdot 7 \cdots (3n-2)} = \frac{(3n)!}{\prod_{k=1}^{n}(3k-2)}, \text{ so } a_n = \frac{\prod_{k=1}^{n}(3k-2)}{(3n)!}x^{3n} \text{ and thus}$$

$A(x) = 1 + \displaystyle\sum_{n=1}^{\infty} \frac{\prod_{k=1}^{n}(3k-2)}{(3n)!}x^{3n}$. Both Maple and Mathematica are able to plot A if we define it this way, and Derive

is able to produce a similar graph using a suitable partial sum of $A(x)$.

Derive, Maple and Mathematica all have two initially known Airy functions, called `AI·SERIES(z,m)` and
`BI·SERIES(z,m)` from `BESSEL.MTH` in Derive and `AiryAi` and `AiryBi` in Maple and Mathematica (just `Ai` and
`Bi` in older versions of Maple). However, it is very difficult to solve for A in terms of the CAS's Airy functions, although

in fact $A(x) = \dfrac{\sqrt{3}\,\texttt{AiryAi}(x) + \texttt{AiryBi}(x)}{\sqrt{3}\,\texttt{AiryAi}(0) + \texttt{AiryBi}(0)}$.

37. $s_{2n-1} = 1 + 2x + x^2 + 2x^3 + x^4 + 2x^5 + \cdots + x^{2n-2} + 2x^{2n-1}$

$= 1(1+2x) + x^2(1+2x) + x^4(1+2x) + \cdots + x^{2n-2}(1+2x) = (1+2x)(1 + x^2 + x^4 + \cdots + x^{2n-2})$

$= (1+2x)\dfrac{1 - x^{2n}}{1 - x^2}$ [by (12.2.3) [ET (11.2.3)] with $r = x^2$] $\to \dfrac{1 + 2x}{1 - x^2}$ as $n \to \infty$

[by (12.2.4) [ET (11.2.4)]], when $|x| < 1$.

Also $s_{2n} = s_{2n-1} + x^{2n} \to \dfrac{1+2x}{1-x^2}$ since $x^{2n} \to 0$ for $|x| < 1$. Therefore, $s_n \to \dfrac{1+2x}{1-x^2}$ since s_{2n} and s_{2n-1} both

approach $\dfrac{1+2x}{1-x^2}$ as $n \to \infty$. Thus, the interval of convergence is $(-1, 1)$ and $f(x) = \dfrac{1+2x}{1-x^2}$.

38. $s_{4n-1} = c_0 + c_1 x + c_2 x^2 + c_3 x^3 + c_0 x^4 + c_1 x^5 + c_2 x^6 + c_3 x^7 + \cdots + c_3 x^{4n-1}$

$= (c_0 + c_1 x + c_2 x^2 + c_3 x^3)(1 + x^4 + x^8 + \cdots + x^{4n-4}) \to \dfrac{c_0 + c_1 x + c_2 x^2 + c_3 x^3}{1 - x^4}$ as $n \to \infty$

[by (12.2.4) [ET (11.2.4)] with $r = x^4$] for $|x^4| < 1 \quad \Leftrightarrow \quad |x| < 1$. Also s_{4n}, s_{4n+1}, s_{4n+2} have the same limits

(for example, $s_{4n} = s_{4n-1} + c_0 x^{4n}$ and $x^{4n} \to 0$ for $|x| < 1$). So if at least one of c_0, c_1, c_2, and c_3 is nonzero, then the

interval of convergence is $(-1, 1)$ and $f(x) = \dfrac{c_0 + c_1 x + c_2 x^2 + c_3 x^3}{1 - x^4}$.

39. We use the Root Test on the series $\sum c_n x^n$. We need $\lim\limits_{n \to \infty} \sqrt[n]{|c_n x^n|} = |x| \lim\limits_{n \to \infty} \sqrt[n]{|c_n|} = c|x| < 1$ for convergence, or

$|x| < 1/c$, so $R = 1/c$.

40. Suppose $c_n \neq 0$. Applying the Ratio Test to the series $\sum c_n (x - a)^n$, we find that

$L = \lim\limits_{n \to \infty} \left| \dfrac{a_{n+1}}{a_n} \right| = \lim\limits_{n \to \infty} \left| \dfrac{c_{n+1}(x-a)^{n+1}}{c_n(x-a)^n} \right| = \lim\limits_{n \to \infty} \dfrac{|x-a|}{|c_n/c_{n+1}|} (*) = \dfrac{|x-a|}{\lim\limits_{n \to \infty} |c_n/c_{n+1}|}$ (if $\lim\limits_{n \to \infty} |c_n/c_{n+1}| \neq 0$), so the

series converges when $\dfrac{|x-a|}{\lim\limits_{n \to \infty} |c_n/c_{n+1}|} < 1 \quad \Leftrightarrow \quad |x-a| < \lim\limits_{n \to \infty} \left| \dfrac{c_n}{c_{n+1}} \right|$. Thus, $R = \lim\limits_{n \to \infty} \left| \dfrac{c_n}{c_{n+1}} \right|$. If $\lim\limits_{n \to \infty} \left| \dfrac{c_n}{c_{n+1}} \right| = 0$

and $|x - a| \neq 0$, then $(*)$ shows that $L = \infty$ and so the series diverges, and hence, $R = 0$. Thus, in all cases,

$R = \lim\limits_{n \to \infty} \left| \dfrac{c_n}{c_{n+1}} \right|$.

41. For $2 < x < 3$, $\sum c_n x^n$ diverges and $\sum d_n x^n$ converges. By Exercise 12.2.69 [ET 11.2.69], $\sum (c_n + d_n) x^n$ diverges. Since

both series converge for $|x| < 2$, the radius of convergence of $\sum (c_n + d_n) x^n$ is 2.

42. Since $\sum c_n x^n$ converges whenever $|x| < R$, $\sum c_n x^{2n} = \sum c_n (x^2)^n$ converges whenever $|x^2| < R \quad \Leftrightarrow \quad |x| < \sqrt{R}$, so the

second series has radius of convergence $\sqrt{R}$.

12.9 Representations of Functions as Power Series ET 11.9

1. If $f(x) = \sum\limits_{n=0}^{\infty} c_n x^n$ has radius of convergence 10, then $f'(x) = \sum\limits_{n=1}^{\infty} n c_n x^{n-1}$ also has radius of convergence 10 by

Theorem 2.

2. If $f(x) = \sum\limits_{n=0}^{\infty} b_n x^n$ converges on $(-2, 2)$, then $\int f(x)\,dx = C + \sum\limits_{n=0}^{\infty} \dfrac{b_n}{n+1} x^{n+1}$ has the same radius of convergence

(by Theorem 2), but may not have the same interval of convergence—it may happen that the integrated series converges at an

endpoint (or both endpoints).

3. Our goal is to write the function in the form $\dfrac{1}{1 - r}$, and then use Equation (1) to represent the function as a sum of a power

series. $f(x) = \dfrac{1}{1 + x} = \dfrac{1}{1 - (-x)} = \sum\limits_{n=0}^{\infty} (-x)^n = \sum\limits_{n=0}^{\infty} (-1)^n x^n$ with $|-x| < 1 \quad \Leftrightarrow \quad |x| < 1$, so $R = 1$ and $I = (-1, 1)$.

4. $f(x) = \dfrac{3}{1 - x^4} = 3\left(\dfrac{1}{1 - x^4}\right) = 3(1 + x^4 + x^8 + x^{12} + \cdots) = 3\sum\limits_{n=0}^{\infty} (x^4)^n = \sum\limits_{n=0}^{\infty} 3x^{4n}$

with $|x^4| < 1 \quad \Leftrightarrow \quad |x| < 1$, so $R = 1$ and $I = (-1, 1)$.

$\left[\text{Note that } 3\sum\limits_{n=0}^{\infty} (x^4)^n \text{ converges} \quad \Leftrightarrow \quad \sum\limits_{n=0}^{\infty} (x^4)^n \text{ converges, so the appropriate condition [from equation (1)] is } |x^4| < 1. \right]$

5. $f(x) = \dfrac{2}{3-x} = \dfrac{2}{3}\left(\dfrac{1}{1-x/3}\right) = \dfrac{2}{3}\sum\limits_{n=0}^{\infty}\left(\dfrac{x}{3}\right)^n$ or, equivalently, $2\sum\limits_{n=0}^{\infty}\dfrac{1}{3^{n+1}}\,x^n$. The series converges when $\left|\dfrac{x}{3}\right| < 1$,

that is, when $|x| < 3$, so $R = 3$ and $I = (-3, 3)$.

6. $f(x) = \dfrac{1}{x+10} = \dfrac{1}{10}\left(\dfrac{1}{1-(-x/10)}\right) = \dfrac{1}{10}\sum\limits_{n=0}^{\infty}\left(-\dfrac{x}{10}\right)^n$ or, equivalently, $\sum\limits_{n=0}^{\infty}(-1)^n\dfrac{1}{10^{n+1}}\,x^n$. The series converges

when $\left|\dfrac{x}{10}\right| < 1$, that is, when $|x| < 10$, so $R = 10$ and $I = (-10, 10)$.

7. $f(x) = \dfrac{x}{9+x^2} = \dfrac{x}{9}\left[\dfrac{1}{1+(x/3)^2}\right] = \dfrac{x}{9}\left[\dfrac{1}{1-\{-(x/3)^2\}}\right] = \dfrac{x}{9}\sum\limits_{n=0}^{\infty}\left[-\left(\dfrac{x}{3}\right)^2\right]^n = \dfrac{x}{9}\sum\limits_{n=0}^{\infty}(-1)^n\dfrac{x^{2n}}{9^n} = \sum\limits_{n=0}^{\infty}(-1)^n\dfrac{x^{2n+1}}{9^{n+1}}$

The geometric series $\sum\limits_{n=0}^{\infty}\left[-\left(\dfrac{x}{3}\right)^2\right]^n$ converges when $\left|-\left(\dfrac{x}{3}\right)^2\right| < 1$ $\Leftrightarrow$ $\dfrac{|x^2|}{9} < 1$ $\Leftrightarrow$ $|x|^2 < 9$ $\Leftrightarrow$ $|x| < 3$, so

$R = 3$ and $I = (-3, 3)$.

8. $f(x) = \dfrac{x}{2x^2+1} = x\left(\dfrac{1}{1-(-2x^2)}\right) = x\sum\limits_{n=0}^{\infty}(-2x^2)^n$ or, equivalently, $\sum\limits_{n=0}^{\infty}(-1)^n 2^n x^{2n+1}$. The series converges when

$|-2x^2| < 1$ $\Rightarrow$ $|x^2| < \frac{1}{2}$ $\Rightarrow$ $|x| < \dfrac{1}{\sqrt{2}}$, so $R = \dfrac{1}{\sqrt{2}}$ and $I = \left(-\dfrac{1}{\sqrt{2}}, \dfrac{1}{\sqrt{2}}\right)$.

9. $f(x) = \dfrac{1+x}{1-x} = (1+x)\left(\dfrac{1}{1-x}\right) = (1+x)\sum\limits_{n=0}^{\infty}x^n = \sum\limits_{n=0}^{\infty}x^n + \sum\limits_{n=0}^{\infty}x^{n+1} = 1 + \sum\limits_{n=1}^{\infty}x^n + \sum\limits_{n=1}^{\infty}x^n = 1 + 2\sum\limits_{n=1}^{\infty}x^n$.

The series converges when $|x| < 1$, so $R = 1$ and $I = (-1, 1)$.

A second approach: $f(x) = \dfrac{1+x}{1-x} = \dfrac{-(1-x)+2}{1-x} = -1 + 2\left(\dfrac{1}{1-x}\right) = -1 + 2\sum\limits_{n=0}^{\infty}x^n = 1 + 2\sum\limits_{n=1}^{\infty}x^n$.

A third approach:

$f(x) = \dfrac{1+x}{1-x} = (1+x)\left(\dfrac{1}{1-x}\right) = (1+x)(1 + x + x^2 + x^3 + \cdots)$

$= (1 + x + x^2 + x^3 + \cdots) + (x + x^2 + x^3 + x^4 + \cdots) = 1 + 2x + 2x^2 + 2x^3 + \cdots = 1 + 2\sum\limits_{n=1}^{\infty}x^n$.

10. $f(x) = \dfrac{x^2}{a^3-x^3} = \dfrac{x^2}{a^3}\cdot\dfrac{1}{1-x^3/a^3} = \dfrac{x^2}{a^3}\sum\limits_{n=0}^{\infty}\left(\dfrac{x^3}{a^3}\right)^n = \sum\limits_{n=0}^{\infty}\dfrac{x^{3n+2}}{a^{3n+3}}$. The series converges when $|x^3/a^3| < 1$ $\Leftrightarrow$

$|x^3| < |a^3|$ $\Leftrightarrow$ $|x| < |a|$, so $R = |a|$ and $I = (-|a|, |a|)$.

11. $f(x) = \dfrac{3}{x^2-x-2} = \dfrac{3}{(x-2)(x+1)} = \dfrac{A}{x-2} + \dfrac{B}{x+1}$ $\Rightarrow$ $3 = A(x+1) + B(x-2)$. Let $x = 2$ to get $A = 1$ and

$x = -1$ to get $B = -1$. Thus

$\dfrac{3}{x^2-x-2} = \dfrac{1}{x-2} - \dfrac{1}{x+1} = \dfrac{1}{-2}\left(\dfrac{1}{1-(x/2)}\right) - \dfrac{1}{1-(-x)} = -\dfrac{1}{2}\sum\limits_{n=0}^{\infty}\left(\dfrac{x}{2}\right)^n - \sum\limits_{n=0}^{\infty}(-x)^n$

$= \sum\limits_{n=0}^{\infty}\left[-\dfrac{1}{2}\left(\dfrac{1}{2}\right)^n - 1(-1)^n\right]x^n = \sum\limits_{n=0}^{\infty}\left[(-1)^{n+1} - \dfrac{1}{2^{n+1}}\right]x^n$

We represented f as the sum of two geometric series; the first converges for $x \in (-2, 2)$ and the second converges for $(-1, 1)$.

Thus, the sum converges for $x \in (-1, 1) = I$.

12. $f(x) = \dfrac{x+2}{2x^2 - x - 1} = \dfrac{x+2}{(2x+1)(x-1)} = \dfrac{A}{2x+1} + \dfrac{B}{x-1} \Rightarrow x + 2 = A(x-1) + B(2x+1)$. Let $x = 1$ to get

$3 = 3B \Rightarrow B = 1$ and $x = -\frac{1}{2}$ to get $\frac{3}{2} = -\frac{3}{2}A \Rightarrow A = -1$. Thus,

$$\dfrac{x+2}{2x^2 - x - 1} = \dfrac{-1}{2x+1} + \dfrac{1}{x-1} = -1\left(\dfrac{1}{1-(-2x)}\right) - 1\left(\dfrac{1}{1-x}\right) = -\sum_{n=0}^{\infty}(-2x)^n - \sum_{n=0}^{\infty} x^n$$

$$= -\sum_{n=0}^{\infty}\left[(-2)^n + 1\right]x^n$$

We represented f as the sum of two geometric series; the first converges for $x \in \left(-\frac{1}{2}, \frac{1}{2}\right)$ and the second converges for

$(-1, 1)$. Thus, the sum converges for $x \in \left(-\frac{1}{2}, \frac{1}{2}\right) = I$.

13. (a) $f(x) = \dfrac{1}{(1+x)^2} = \dfrac{d}{dx}\left(\dfrac{-1}{1+x}\right) = -\dfrac{d}{dx}\left[\sum_{n=0}^{\infty}(-1)^n x^n\right]$ [from Exercise 3]

$= \sum_{n=1}^{\infty}(-1)^{n+1}nx^{n-1}$ [from Theorem 2(i)] $= \sum_{n=0}^{\infty}(-1)^n(n+1)x^n$ with $R = 1$.

In the last step, note that we *decreased* the initial value of the summation variable n by 1, and then *increased* each

occurrence of n in the term by 1 [also note that $(-1)^{n+2} = (-1)^n$].

(b) $f(x) = \dfrac{1}{(1+x)^3} = -\dfrac{1}{2}\dfrac{d}{dx}\left[\dfrac{1}{(1+x)^2}\right] = -\dfrac{1}{2}\dfrac{d}{dx}\left[\sum_{n=0}^{\infty}(-1)^n(n+1)x^n\right]$ [from part (a)]

$= -\dfrac{1}{2}\sum_{n=1}^{\infty}(-1)^n(n+1)nx^{n-1} = \dfrac{1}{2}\sum_{n=0}^{\infty}(-1)^n(n+2)(n+1)x^n$ with $R = 1$.

(c) $f(x) = \dfrac{x^2}{(1+x)^3} = x^2 \cdot \dfrac{1}{(1+x)^3} = x^2 \cdot \dfrac{1}{2}\sum_{n=0}^{\infty}(-1)^n(n+2)(n+1)x^n$ [from part (b)]

$= \dfrac{1}{2}\sum_{n=0}^{\infty}(-1)^n(n+2)(n+1)x^{n+2}$

To write the power series with x^n rather than x^{n+2}, we will *decrease* each occurrence of n in the term by 2 and *increase*

the initial value of the summation variable by 2. This gives us $\dfrac{1}{2}\sum_{n=2}^{\infty}(-1)^n(n)(n-1)x^n$ with $R = 1$.

14. (a) $\dfrac{1}{1+x} = \dfrac{1}{1-(-x)} = \sum_{n=0}^{\infty}(-1)^n x^n$ [geometric series with $R = 1$], so

$f(x) = \ln(1+x) = \displaystyle\int \dfrac{dx}{1+x} = \int\left[\sum_{n=0}^{\infty}(-1)^n x^n\right]dx = C + \sum_{n=0}^{\infty}(-1)^n\dfrac{x^{n+1}}{n+1} = \sum_{n=1}^{\infty}\dfrac{(-1)^{n-1}x^n}{n}$

$[C = 0$ since $f(0) = \ln 1 = 0]$, with $R = 1$

(b) $f(x) = x\ln(1+x) = x\left[\sum_{n=1}^{\infty}\dfrac{(-1)^{n-1}x^n}{n}\right]$ [by part (a)] $= \sum_{n=1}^{\infty}\dfrac{(-1)^{n-1}x^{n+1}}{n} = \sum_{n=2}^{\infty}\dfrac{(-1)^n x^n}{n-1}$ with $R = 1$.

(c) $f(x) = \ln(x^2 + 1) = \sum_{n=1}^{\infty}\dfrac{(-1)^{n-1}(x^2)^n}{n}$ [by part (a)] $= \sum_{n=1}^{\infty}\dfrac{(-1)^{n-1}x^{2n}}{n}$ with $R = 1$.

15. $f(x) = \ln(5-x) = -\displaystyle\int \dfrac{dx}{5-x} = -\dfrac{1}{5}\int\dfrac{dx}{1-x/5} = -\dfrac{1}{5}\int\left[\sum_{n=0}^{\infty}\left(\dfrac{x}{5}\right)^n\right]dx = C - \dfrac{1}{5}\sum_{n=0}^{\infty}\dfrac{x^{n+1}}{5^n(n+1)} = C - \sum_{n=1}^{\infty}\dfrac{x^n}{n\,5^n}$

Putting $x = 0$, we get $C = \ln 5$. The series converges for $|x/5| < 1 \Leftrightarrow |x| < 5$, so $R = 5$.

16. We know that $\dfrac{1}{1-2x} = \sum\limits_{n=0}^{\infty} (2x)^n$. Differentiating, we get $\dfrac{2}{(1-2x)^2} = \sum\limits_{n=1}^{\infty} 2^n n x^{n-1} = \sum\limits_{n=0}^{\infty} 2^{n+1}(n+1)x^n$, so

$$f(x) = \frac{x^2}{(1-2x)^2} = \frac{x^2}{2}\cdot\frac{2}{(1-2x)^2} = \frac{x^2}{2}\sum\limits_{n=0}^{\infty} 2^{n+1}(n+1)x^n = \sum\limits_{n=0}^{\infty} 2^n(n+1)x^{n+2} \text{ or } \sum\limits_{n=2}^{\infty} 2^{n-2}(n-1)x^n,$$

with $R = \frac{1}{2}$.

17. $\dfrac{1}{2-x} = \dfrac{1}{2(1-x/2)} = \dfrac{1}{2}\sum\limits_{n=0}^{\infty}\left(\dfrac{x}{2}\right)^n = \sum\limits_{n=0}^{\infty}\dfrac{1}{2^{n+1}}x^n$ for $\left|\dfrac{x}{2}\right| < 1 \ \Leftrightarrow \ |x| < 2$. Now

$$\frac{1}{(x-2)^2} = \frac{d}{dx}\left(\frac{1}{2-x}\right) = \frac{d}{dx}\left(\sum\limits_{n=0}^{\infty}\frac{1}{2^{n+1}}x^n\right) = \sum\limits_{n=1}^{\infty}\frac{n}{2^{n+1}}x^{n-1} = \sum\limits_{n=0}^{\infty}\frac{n+1}{2^{n+2}}x^n. \text{ So}$$

$$f(x) = \frac{x^3}{(x-2)^2} = x^3\sum\limits_{n=0}^{\infty}\frac{n+1}{2^{n+2}}x^n = \sum\limits_{n=0}^{\infty}\frac{n+1}{2^{n+2}}x^{n+3} \text{ or } \sum\limits_{n=3}^{\infty}\frac{n-2}{2^{n-1}}x^n \text{ for } |x| < 2. \text{ Thus, } R = 2 \text{ and } I = (-2,2).$$

18. From Example 7, $g(x) = \arctan x = \sum\limits_{n=0}^{\infty}(-1)^n\dfrac{x^{2n+1}}{2n+1}$. Thus,

$$f(x) = \arctan(x/3) = \sum\limits_{n=0}^{\infty}(-1)^n\frac{(x/3)^{2n+1}}{2n+1} = \sum\limits_{n=0}^{\infty}(-1)^n\frac{1}{3^{2n+1}(2n+1)}x^{2n+1} \text{ for } \left|\frac{x}{3}\right| < 1 \ \Leftrightarrow \ |x| < 3, \text{ so } R = 3.$$

19. $f(x) = \dfrac{x}{x^2+16} = \dfrac{x}{16}\left(\dfrac{1}{1-(-x^2/16)}\right) = \dfrac{x}{16}\sum\limits_{n=0}^{\infty}\left(-\dfrac{x^2}{16}\right)^n = \dfrac{x}{16}\sum\limits_{n=0}^{\infty}(-1)^n\dfrac{1}{16^n}x^{2n} = \sum\limits_{n=0}^{\infty}(-1)^n\dfrac{1}{16^{n+1}}x^{2n+1}.$

The series converges when $\left|-x^2/16\right| < 1 \ \Leftrightarrow \ x^2 < 16 \ \Leftrightarrow \ |x| < 4$, so $R = 4$. The partial sums are $s_1 = \dfrac{x}{16}$,

$s_2 = s_1 - \dfrac{x^3}{16^2}$, $s_3 = s_2 + \dfrac{x^5}{16^3}$, $s_4 = s_3 - \dfrac{x^7}{16^4}$, $s_5 = s_4 + \dfrac{x^9}{16^5}$, ... Note that s_1 corresponds to the first term of the infinite

sum, regardless of the value of the summation variable and the value of the exponent.

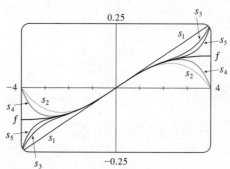

As n increases, $s_n(x)$ approximates f better on the interval of convergence, which is $(-4,4)$.

20. $f(x) = \ln(x^2+4) \ \Rightarrow \ f'(x) = \dfrac{2x}{x^2+4} = \dfrac{2x}{4}\left(\dfrac{1}{1-(-x^2/4)}\right) = \dfrac{x}{2}\sum\limits_{n=0}^{\infty}\left(-\dfrac{x^2}{4}\right)^n = \sum\limits_{n=0}^{\infty}(-1)^n\dfrac{x^{2n+1}}{2^{2n+1}}$,

so $f(x) = \displaystyle\int\sum\limits_{n=0}^{\infty}(-1)^n\frac{x^{2n+1}}{2^{2n+1}}\,dx = C + \sum\limits_{n=0}^{\infty}(-1)^n\frac{x^{2n+2}}{2^{2n+1}(2n+2)} = \ln 4 + \sum\limits_{n=0}^{\infty}(-1)^n\frac{x^{2n+2}}{(n+1)2^{2n+2}}$

$[f(0) = \ln 4, \text{ so } C = \ln 4]$. The series converges when $\left|-x^2/4\right| < 1 \ \Leftrightarrow \ x^2 < 4 \ \Leftrightarrow \ |x| < 2$, so $R = 2$. If

$x = \pm 2$, then $f(x) = \ln 4 + \sum_{n=0}^{\infty} (-1)^n \dfrac{1}{n+1}$, which converges by the Alternating Series Test. The partial sums

are $s_0 = \ln 4 \; [\approx 1.39]$, $s_1 = s_0 + \dfrac{x^2}{4}$, $s_2 = s_1 - \dfrac{x^4}{2 \cdot 2^4}$, $s_3 = s_2 + \dfrac{x^6}{3 \cdot 2^6}$, $s_4 = s_3 - \dfrac{x^8}{4 \cdot 2^8}$, $\ldots$.

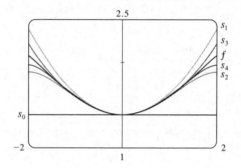

As n increases, $s_n(x)$ approximates f better on the interval of convergence, which is $[-2, 2]$.

21. $f(x) = \ln\left(\dfrac{1+x}{1-x}\right) = \ln(1+x) - \ln(1-x) = \displaystyle\int \dfrac{dx}{1+x} + \int \dfrac{dx}{1-x} = \int \dfrac{dx}{1-(-x)} + \int \dfrac{dx}{1-x}$

$= \displaystyle\int \left[\sum_{n=0}^{\infty} (-1)^n x^n + \sum_{n=0}^{\infty} x^n\right] dx = \int [(1 - x + x^2 - x^3 + x^4 - \cdots) + (1 + x + x^2 + x^3 + x^4 + \cdots)]\, dx$

$= \displaystyle\int (2 + 2x^2 + 2x^4 + \cdots)\, dx = \int \sum_{n=0}^{\infty} 2x^{2n}\, dx = C + \sum_{n=0}^{\infty} \dfrac{2x^{2n+1}}{2n+1}$

But $f(0) = \ln \frac{1}{1} = 0$, so $C = 0$ and we have $f(x) = \displaystyle\sum_{n=0}^{\infty} \dfrac{2x^{2n+1}}{2n+1}$ with $R = 1$. If $x = \pm 1$, then $f(x) = \pm 2 \displaystyle\sum_{n=0}^{\infty} \dfrac{1}{2n+1}$,

which both diverge by the Limit Comparison Test with $b_n = \dfrac{1}{n}$. The partial sums are $s_1 = \dfrac{2x}{1}$, $s_2 = s_1 + \dfrac{2x^3}{3}$,

$s_3 = s_2 + \dfrac{2x^5}{5}$, $\ldots$.

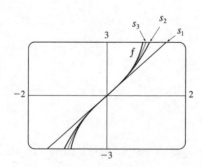

As n increases, $s_n(x)$ approximates f better on the interval of convergence, which is $(-1, 1)$.

22. $f(x) = \tan^{-1}(2x) = 2 \displaystyle\int \dfrac{dx}{1+4x^2} = 2 \int \sum_{n=0}^{\infty} (-1)^n \left(4x^2\right)^n\, dx = 2 \int \sum_{n=0}^{\infty} (-1)^n 4^n x^{2n}\, dx$

$= C + 2 \displaystyle\sum_{n=0}^{\infty} \dfrac{(-1)^n 4^n x^{2n+1}}{2n+1} = \sum_{n=0}^{\infty} \dfrac{(-1)^n 2^{2n+1} x^{2n+1}}{2n+1} \qquad [f(0) = \tan^{-1} 0 = 0, \text{ so } C = 0]$

The series converges when $\left|4x^2\right| < 1 \;\Leftrightarrow\; |x| < \frac{1}{2}$, so $R = \frac{1}{2}$. If $x = \pm\frac{1}{2}$, then $f(x) = \displaystyle\sum_{n=0}^{\infty} (-1)^n \dfrac{1}{2n+1}$ and

$f(x) = \sum_{n=0}^{\infty} (-1)^{n+1} \frac{1}{2n+1}$, respectively. Both series converge by the Alternating Series Test. The partial sums are

$s_1 = \frac{2x}{1}, \ s_2 = s_1 - \frac{2^3 x^3}{3}, \ s_3 = s_2 + \frac{2^5 x^5}{5}, \dots$

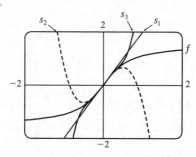

As n increases, $s_n(x)$ approximates f better on the interval of convergence, which is $\left[-\frac{1}{2}, \frac{1}{2}\right]$.

23. $\frac{t}{1 - t^8} = t \cdot \frac{1}{1 - t^8} = t \sum_{n=0}^{\infty} (t^8)^n = \sum_{n=0}^{\infty} t^{8n+1} \quad \Rightarrow \quad \int \frac{t}{1 - t^8} \, dt = C + \sum_{n=0}^{\infty} \frac{t^{8n+2}}{8n + 2}$. The series for $\frac{1}{1 - t^8}$ converges

when $|t^8| < 1 \quad \Leftrightarrow \quad |t| < 1$, so $R = 1$ for that series and also the series for $t/(1 - t^8)$. By Theorem 2, the series for

$\int \frac{t}{1 - t^8} \, dt$ also has $R = 1$.

24. By Example 6, $\ln(1 - t) = -\sum_{n=1}^{\infty} \frac{t^n}{n}$ for $|t| < 1$, so $\frac{\ln(1 - t)}{t} = -\sum_{n=1}^{\infty} \frac{t^{n-1}}{n}$ and $\int \frac{\ln(1 - t)}{t} \, dt = C - \sum_{n=1}^{\infty} \frac{t^n}{n^2}$.

By Theorem 2, $R = 1$.

25. By Example 7, $\tan^{-1} x = \sum_{n=0}^{\infty} (-1)^n \frac{x^{2n+1}}{2n + 1}$ with $R = 1$, so

$x - \tan^{-1} x = x - \left(x - \frac{x^3}{3} + \frac{x^5}{5} - \frac{x^7}{7} + \cdots \right) = \frac{x^3}{3} - \frac{x^5}{5} + \frac{x^7}{7} - \cdots = \sum_{n=1}^{\infty} (-1)^{n+1} \frac{x^{2n+1}}{2n + 1}$ and

$\frac{x - \tan^{-1} x}{x^3} = \sum_{n=1}^{\infty} (-1)^{n+1} \frac{x^{2n-2}}{2n + 1}$, so

$\int \frac{x - \tan^{-1} x}{x^3} \, dx = C + \sum_{n=1}^{\infty} (-1)^{n+1} \frac{x^{2n-1}}{(2n + 1)(2n - 1)} = C + \sum_{n=1}^{\infty} (-1)^{n+1} \frac{x^{2n-1}}{4n^2 - 1}$. By Theorem 2, $R = 1$.

26. By Example 7, $\int \tan^{-1}(x^2) \, dx = \int \sum_{n=0}^{\infty} (-1)^n \frac{(x^2)^{2n+1}}{2n + 1} \, dx = C + \sum_{n=0}^{\infty} (-1)^n \frac{x^{4n+3}}{(2n + 1)(4n + 3)}$ with $R = 1$.

27. $\frac{1}{1 + x^5} = \frac{1}{1 - (-x^5)} = \sum_{n=0}^{\infty} (-x^5)^n = \sum_{n=0}^{\infty} (-1)^n x^{5n} \quad \Rightarrow$

$\int \frac{1}{1 + x^5} \, dx = \int \sum_{n=0}^{\infty} (-1)^n x^{5n} \, dx = C + \sum_{n=0}^{\infty} (-1)^n \frac{x^{5n+1}}{5n + 1}$. Thus,

$I = \int_0^{0.2} \frac{1}{1 + x^5} \, dx = \left[x - \frac{x^6}{6} + \frac{x^{11}}{11} - \cdots \right]_0^{0.2} = 0.2 - \frac{(0.2)^6}{6} + \frac{(0.2)^{11}}{11} - \cdots$. The series is alternating, so if we use

the first two terms, the error is at most $(0.2)^{11}/11 \approx 1.9 \times 10^{-9}$. So $I \approx 0.2 - (0.2)^6/6 \approx 0.199989$ to six decimal places.

28. From Example 6, we know $\ln(1-x) = -\sum\limits_{n=1}^{\infty} \dfrac{x^n}{n}$, so

$$\ln(1+x^4) = \ln\left[1-(-x^4)\right] = -\sum_{n=1}^{\infty} \frac{(-x^4)^n}{n} = \sum_{n=1}^{\infty}(-1)^{n+1}\frac{x^{4n}}{n} \quad \Rightarrow$$

$$\int \ln(1+x^4)\,dx = \int \sum_{n=1}^{\infty}(-1)^{n+1}\frac{x^{4n}}{n}\,dx = C + \sum_{n=1}^{\infty}(-1)^{n+1}\frac{x^{4n+1}}{n(4n+1)}. \text{ Thus,}$$

$$I = \int_0^{0.4} \ln(1+x^4)\,dx = \left[\frac{x^5}{5} - \frac{x^9}{18} + \frac{x^{13}}{39} - \frac{x^{17}}{68} + \cdots\right]_0^{0.4} = \frac{(0.4)^5}{5} - \frac{(0.4)^9}{18} + \frac{(0.4)^{13}}{39} - \frac{(0.4)^{17}}{68} + \cdots.$$

The series is alternating, so if we use the first three terms, the error is at most $(0.4)^{17}/68 \approx 2.5 \times 10^{-9}$.

So $I \approx (0.4)^5/5 - (0.4)^9/18 + (0.9)^{13}/39 \approx 0.002034$ to six decimal places.

29. We substitute $3x$ for x in Example 7, and find that

$$\int x\arctan(3x)\,dx = \int x\sum_{n=0}^{\infty}(-1)^n\frac{(3x)^{2n+1}}{2n+1}\,dx = \int \sum_{n=0}^{\infty}(-1)^n\frac{3^{2n+1}x^{2n+2}}{2n+1}\,dx = C + \sum_{n=0}^{\infty}(-1)^n\frac{3^{2n+1}x^{2n+3}}{(2n+1)(2n+3)}$$

So

$$\int_0^{0.1} x\arctan(3x)\,dx = \left[\frac{3x^3}{1\cdot 3} - \frac{3^3 x^5}{3\cdot 5} + \frac{3^5 x^7}{5\cdot 7} - \frac{3^7 x^9}{7\cdot 9} + \cdots\right]_0^{0.1}$$

$$= \frac{1}{10^3} - \frac{9}{5\times 10^5} + \frac{243}{35\times 10^7} - \frac{2187}{63\times 10^9} + \cdots.$$

The series is alternating, so if we use three terms, the error is at most $\dfrac{2187}{63\times 10^9} \approx 3.5\times 10^{-8}$. So

$$\int_0^{0.1} x\arctan(3x)\,dx \approx \frac{1}{10^3} - \frac{9}{5\times 10^5} + \frac{243}{35\times 10^7} \approx 0.000\,983 \text{ to six decimal places.}$$

30. $\displaystyle\int_0^{0.3}\frac{x^2}{1+x^4}\,dx = \int_0^{0.3} x^2\sum_{n=0}^{\infty}(-1)^n x^{4n}\,dx = \sum_{n=0}^{\infty}\left[\frac{(-1)^n x^{4n+3}}{4n+3}\right]_0^{0.3} = \sum_{n=0}^{\infty}\frac{(-1)^n 3^{4n+3}}{(4n+3)10^{4n+3}}$

$$= \frac{3^3}{3\times 10^3} - \frac{3^7}{7\times 10^7} + \frac{3^{11}}{11\times 10^{11}} - \cdots$$

The series is alternating, so if we use only two terms, the error is at most $\dfrac{3^{11}}{11\times 10^{11}} \approx 0.000\,000\,16$. So, to six decimal

places, $\displaystyle\int_0^{0.3}\frac{x^2}{1+x^4}\,dx \approx \frac{3^3}{3\times 10^3} - \frac{3^7}{7\times 10^7} \approx 0.008\,969.$

31. Using the result of Example 6, $\ln(1-x) = -\sum\limits_{n=1}^{\infty}\dfrac{x^n}{n}$, with $x = -0.1$, we have

$$\ln 1.1 = \ln[1-(-0.1)] = 0.1 - \frac{0.01}{2} + \frac{0.001}{3} - \frac{0.0001}{4} + \frac{0.00001}{5} - \cdots. \text{ The series is alternating, so if we use only}$$

the first four terms, the error is at most $\dfrac{0.00001}{5} = 0.000002$. So $\ln 1.1 \approx 0.1 - \dfrac{0.01}{2} + \dfrac{0.001}{3} - \dfrac{0.0001}{4} \approx 0.09531.$

32. $f(x) = \sum\limits_{n=0}^{\infty}\dfrac{(-1)^n x^{2n}}{(2n)!} \quad \Rightarrow \quad f'(x) = \sum\limits_{n=1}^{\infty}\dfrac{(-1)^n 2nx^{2n-1}}{(2n)!}$ [the first term disappears], so

$$f''(x) = \sum_{n=1}^{\infty}\frac{(-1)^n(2n)(2n-1)x^{2n-2}}{(2n)!} = \sum_{n=1}^{\infty}\frac{(-1)^n x^{2(n-1)}}{[2(n-1)]!} = \sum_{n=0}^{\infty}\frac{(-1)^{n+1}x^{2n}}{(2n)!} \quad \text{[substituting } n+1 \text{ for } n\text{]}$$

$$= -\sum_{n=0}^{\infty}\frac{(-1)^n x^{2n}}{(2n)!} = -f(x) \quad \Rightarrow \quad f''(x) + f(x) = 0.$$

33. (a) $J_0(x) = \sum\limits_{n=0}^{\infty} \dfrac{(-1)^n \, x^{2n}}{2^{2n}(n!)^2}$, $J_0'(x) = \sum\limits_{n=1}^{\infty} \dfrac{(-1)^n \, 2nx^{2n-1}}{2^{2n}(n!)^2}$, and $J_0''(x) = \sum\limits_{n=1}^{\infty} \dfrac{(-1)^n \, 2n(2n-1)x^{2n-2}}{2^{2n}(n!)^2}$, so

$$x^2 J_0''(x) + x J_0'(x) + x^2 J_0(x) = \sum_{n=1}^{\infty} \frac{(-1)^n \, 2n(2n-1)x^{2n}}{2^{2n}(n!)^2} + \sum_{n=1}^{\infty} \frac{(-1)^n \, 2nx^{2n}}{2^{2n}(n!)^2} + \sum_{n=0}^{\infty} \frac{(-1)^n \, x^{2n+2}}{2^{2n}(n!)^2}$$

$$= \sum_{n=1}^{\infty} \frac{(-1)^n \, 2n(2n-1)x^{2n}}{2^{2n}(n!)^2} + \sum_{n=1}^{\infty} \frac{(-1)^n \, 2nx^{2n}}{2^{2n}(n!)^2} + \sum_{n=1}^{\infty} \frac{(-1)^{n-1} \, x^{2n}}{2^{2n-2}\,[(n-1)!]^2}$$

$$= \sum_{n=1}^{\infty} \frac{(-1)^n \, 2n(2n-1)x^{2n}}{2^{2n}(n!)^2} + \sum_{n=1}^{\infty} \frac{(-1)^n \, 2nx^{2n}}{2^{2n}(n!)^2} + \sum_{n=1}^{\infty} \frac{(-1)^n(-1)^{-1}2^2 n^2 x^{2n}}{2^{2n}(n!)^2}$$

$$= \sum_{n=1}^{\infty} (-1)^n \left[\frac{2n(2n-1) + 2n - 2^2 n^2}{2^{2n}(n!)^2}\right] x^{2n}$$

$$= \sum_{n=1}^{\infty} (-1)^n \left[\frac{4n^2 - 2n + 2n - 4n^2}{2^{2n}(n!)^2}\right] x^{2n} = 0$$

(b) $\displaystyle \int_0^1 J_0(x)\,dx = \int_0^1 \left[\sum_{n=0}^{\infty} \frac{(-1)^n \, x^{2n}}{2^{2n}(n!)^2}\right] dx = \int_0^1 \left(1 - \frac{x^2}{4} + \frac{x^4}{64} - \frac{x^6}{2304} + \cdots\right) dx$

$$= \left[x - \frac{x^3}{3\cdot 4} + \frac{x^5}{5\cdot 64} - \frac{x^7}{7\cdot 2304} + \cdots\right]_0^1 = 1 - \frac{1}{12} + \frac{1}{320} - \frac{1}{16{,}128} + \cdots$$

Since $\frac{1}{16{,}128} \approx 0.000062$, it follows from The Alternating Series Estimation Theorem that, correct to three decimal places,

$\int_0^1 J_0(x)\,dx \approx 1 - \frac{1}{12} + \frac{1}{320} \approx 0.920$.

34. (a) $J_1(x) = \sum\limits_{n=0}^{\infty} \dfrac{(-1)^n x^{2n+1}}{n!\,(n+1)!\,2^{2n+1}}$, $J_1'(x) = \sum\limits_{n=0}^{\infty} \dfrac{(-1)^n \, (2n+1)\,x^{2n}}{n!\,(n+1)!\,2^{2n+1}}$, and $J_1''(x) = \sum\limits_{n=1}^{\infty} \dfrac{(-1)^n \, (2n+1)\,(2n)\,x^{2n-1}}{n!\,(n+1)!\,2^{2n+1}}$.

$x^2 J_1''(x) + x J_1'(x) + (x^2 - 1) J_1(x)$

$$= \sum_{n=1}^{\infty} \frac{(-1)^n \, (2n+1)(2n)x^{2n+1}}{n!\,(n+1)!\,2^{2n+1}} + \sum_{n=0}^{\infty} \frac{(-1)^n \, (2n+1)x^{2n+1}}{n!\,(n+1)!\,2^{2n+1}}$$

$$+ \sum_{n=0}^{\infty} \frac{(-1)^n \, x^{2n+3}}{n!\,(n+1)!\,2^{2n+1}} - \sum_{n=0}^{\infty} \frac{(-1)^n \, x^{2n+1}}{n!\,(n+1)!\,2^{2n+1}}$$

$$= \sum_{n=1}^{\infty} \frac{(-1)^n \, (2n+1)(2n)x^{2n+1}}{n!\,(n+1)!\,2^{2n+1}} + \sum_{n=0}^{\infty} \frac{(-1)^n \, (2n+1)x^{2n+1}}{n!\,(n+1)!\,2^{2n+1}}$$

$$- \sum_{n=1}^{\infty} \frac{(-1)^n \, x^{2n+1}}{(n-1)!\,n!\,2^{2n-1}} - \sum_{n=0}^{\infty} \frac{(-1)^n \, x^{2n+1}}{n!\,(n+1)!\,2^{2n+1}} \qquad \left[\begin{array}{l} \text{Replace } n \text{ with } n-1 \\ \text{in the third term} \end{array}\right]$$

$$= \frac{x}{2} - \frac{x}{2} + \sum_{n=1}^{\infty} (-1)^n \left[\frac{(2n+1)(2n) + (2n+1) - (n)(n+1)2^2 - 1}{n!\,(n+1)!\,2^{2n+1}}\right] x^{2n+1} = 0$$

(b) $J_0(x) = \sum\limits_{n=0}^{\infty} \dfrac{(-1)^n \, x^{2n}}{2^{2n}(n!)^2} \quad \Rightarrow$

$$J_0'(x) = \sum_{n=1}^{\infty} \frac{(-1)^n \, (2n)x^{2n-1}}{2^{2n}(n!)^2} = \sum_{n=0}^{\infty} \frac{(-1)^{n+1} \, 2(n+1)x^{2n+1}}{2^{2n+2}\,[(n+1)!]^2} \qquad [\text{Replace } n \text{ with } n+1]$$

$$= -\sum_{n=0}^{\infty} \frac{(-1)^n \, x^{2n+1}}{2^{2n+1}(n+1)!\,n!} \qquad [\text{cancel 2 and } n+1;\ \text{take } -1 \text{ outside sum}] \quad = -J_1(x)$$

35. (a) $f(x) = \sum_{n=0}^{\infty} \dfrac{x^n}{n!}$ $\Rightarrow$ $f'(x) = \sum_{n=1}^{\infty} \dfrac{nx^{n-1}}{n!} = \sum_{n=1}^{\infty} \dfrac{x^{n-1}}{(n-1)!} = \sum_{n=0}^{\infty} \dfrac{x^n}{n!} = f(x)$

(b) By Theorem 10.4.2 [ET 9.4.2], the only solution to the differential equation $df(x)/dx = f(x)$ is $f(x) = Ke^x$, but

$f(0) = 1$, so $K = 1$ and $f(x) = e^x$.

Or: We could solve the equation $df(x)/dx = f(x)$ as a separable differential equation.

36. $\dfrac{|\sin nx|}{n^2} \le \dfrac{1}{n^2}$, so $\sum_{n=1}^{\infty} \dfrac{\sin nx}{n^2}$ converges by the Comparison Test. $\dfrac{d}{dx}\left(\dfrac{\sin nx}{n^2}\right) = \dfrac{\cos nx}{n}$, so when $x = 2k\pi$

[k an integer], $\sum_{n=1}^{\infty} f_n'(x) = \sum_{n=1}^{\infty} \dfrac{\cos(2kn\pi)}{n} = \sum_{n=1}^{\infty} \dfrac{1}{n}$, which diverges [harmonic series]. $f_n''(x) = -\sin nx$, so

$\sum_{n=1}^{\infty} f_n''(x) = -\sum_{n=1}^{\infty} \sin nx$, which converges only if $\sin nx = 0$, or $x = k\pi$ [k an integer].

37. If $a_n = \dfrac{x^n}{n^2}$, then by the Ratio Test, $\lim_{n\to\infty}\left|\dfrac{a_{n+1}}{a_n}\right| = \lim_{n\to\infty}\left|\dfrac{x^{n+1}}{(n+1)^2}\cdot\dfrac{n^2}{x^n}\right| = |x|\lim_{n\to\infty}\left(\dfrac{n}{n+1}\right)^2 = |x| < 1$ for

convergence, so $R = 1$. When $x = \pm 1$, $\sum_{n=1}^{\infty}\left|\dfrac{x^n}{n^2}\right| = \sum_{n=1}^{\infty} \dfrac{1}{n^2}$ which is a convergent p-series ($p = 2 > 1$), so the interval of

convergence for f is $[-1, 1]$. By Theorem 2, the radii of convergence of f' and f'' are both 1, so we need only check the

endpoints. $f(x) = \sum_{n=1}^{\infty} \dfrac{x^n}{n^2}$ $\Rightarrow$ $f'(x) = \sum_{n=1}^{\infty} \dfrac{nx^{n-1}}{n^2} = \sum_{n=0}^{\infty} \dfrac{x^n}{n+1}$, and this series diverges for $x = 1$ (harmonic series)

and converges for $x = -1$ (Alternating Series Test), so the interval of convergence is $[-1, 1)$. $f''(x) = \sum_{n=1}^{\infty} \dfrac{nx^{n-1}}{n+1}$ diverges

at both 1 and -1 (Test for Divergence) since $\lim_{n\to\infty} \dfrac{n}{n+1} = 1 \ne 0$, so its interval of convergence is $(-1, 1)$.

38. (a) $\sum_{n=1}^{\infty} nx^{n-1} = \sum_{n=0}^{\infty} \dfrac{d}{dx} x^n = \dfrac{d}{dx}\left[\sum_{n=0}^{\infty} x^n\right] = \dfrac{d}{dx}\left[\dfrac{1}{1-x}\right] = -\dfrac{1}{(1-x)^2}(-1) = \dfrac{1}{(1-x)^2}$, $|x| < 1$.

(b) (i) $\sum_{n=1}^{\infty} nx^n = x\sum_{n=1}^{\infty} nx^{n-1} = x\left[\dfrac{1}{(1-x)^2}\right]$ [from part (a)] $= \dfrac{x}{(1-x)^2}$ for $|x| < 1$.

(ii) Put $x = \frac{1}{2}$ in (i): $\sum_{n=1}^{\infty} \dfrac{n}{2^n} = \sum_{n=1}^{\infty} n\left(\tfrac{1}{2}\right)^n = \dfrac{1/2}{(1-1/2)^2} = 2$.

(c) (i) $\sum_{n=2}^{\infty} n(n-1)x^n = x^2\sum_{n=2}^{\infty} n(n-1)x^{n-2} = x^2\dfrac{d}{dx}\left[\sum_{n=1}^{\infty} nx^{n-1}\right] = x^2\dfrac{d}{dx}\dfrac{1}{(1-x)^2}$

$= x^2\dfrac{2}{(1-x)^3} = \dfrac{2x^2}{(1-x)^3}$ for $|x| < 1$.

(ii) Put $x = \frac{1}{2}$ in (i): $\sum_{n=2}^{\infty} \dfrac{n^2 - n}{2^n} = \sum_{n=2}^{\infty} n(n-1)\left(\tfrac{1}{2}\right)^n = \dfrac{2(1/2)^2}{(1-1/2)^3} = 4$.

(iii) From (b)(ii) and (c)(ii), we have $\sum_{n=1}^{\infty} \dfrac{n^2}{2^n} = \sum_{n=1}^{\infty} \dfrac{n^2-n}{2^n} + \sum_{n=1}^{\infty} \dfrac{n}{2^n} = 4 + 2 = 6$.

39. By Example 7, $\tan^{-1} x = \sum\limits_{n=0}^{\infty} (-1)^n \dfrac{x^{2n+1}}{2n+1}$ for $|x| < 1$. In particular, for $x = \dfrac{1}{\sqrt{3}}$, we

have $\dfrac{\pi}{6} = \tan^{-1}\left(\dfrac{1}{\sqrt{3}}\right) = \sum\limits_{n=0}^{\infty} (-1)^n \dfrac{(1/\sqrt{3})^{2n+1}}{2n+1} = \sum\limits_{n=0}^{\infty} (-1)^n \left(\dfrac{1}{3}\right)^n \dfrac{1}{\sqrt{3}} \dfrac{1}{2n+1}$, so

$\pi = \dfrac{6}{\sqrt{3}} \sum\limits_{n=0}^{\infty} \dfrac{(-1)^n}{(2n+1)3^n} = 2\sqrt{3} \sum\limits_{n=0}^{\infty} \dfrac{(-1)^n}{(2n+1)3^n}$.

40. (a) $\displaystyle\int_0^{1/2} \dfrac{dx}{x^2 - x + 1} = \int_0^{1/2} \dfrac{dx}{(x-1/2)^2 + 3/4}$ $\left[x - \dfrac{1}{2} = \dfrac{\sqrt{3}}{2}u,\ u = \dfrac{2}{\sqrt{3}}\left(x - \dfrac{1}{2}\right),\ dx = \dfrac{\sqrt{3}}{2}\,du\right]$

$= \displaystyle\int_{-1/\sqrt{3}}^{0} \dfrac{(\sqrt{3}/2)\,du}{(3/4)(u^2+1)} = \dfrac{2\sqrt{3}}{3}\Big[\tan^{-1} u\Big]_{-1/\sqrt{3}}^{0} = \dfrac{2}{\sqrt{3}}\left[0 - \left(-\dfrac{\pi}{6}\right)\right] = \dfrac{\pi}{3\sqrt{3}}$

(b) $\dfrac{1}{x^3+1} = \dfrac{1}{(x+1)(x^2-x+1)} \quad \Rightarrow$

$\dfrac{1}{x^2 - x + 1} = (x+1)\left(\dfrac{1}{1+x^3}\right) = (x+1)\dfrac{1}{1-(-x^3)} = (x+1)\sum\limits_{n=0}^{\infty}(-1)^n x^{3n}$

$= \sum\limits_{n=0}^{\infty}(-1)^n x^{3n+1} + \sum\limits_{n=0}^{\infty}(-1)^n x^{3n}$ for $|x| < 1 \quad \Rightarrow$

$\displaystyle\int \dfrac{dx}{x^2 - x + 1} = C + \sum\limits_{n=0}^{\infty}(-1)^n \dfrac{x^{3n+2}}{3n+2} + \sum\limits_{n=0}^{\infty}(-1)^n \dfrac{x^{3n+1}}{3n+1}$ for $|x| < 1 \quad \Rightarrow$

$\displaystyle\int_0^{1/2} \dfrac{dx}{x^2-x+1} = \sum\limits_{n=0}^{\infty}(-1)^n \left[\dfrac{1}{4 \cdot 8^n(3n+2)} + \dfrac{1}{2 \cdot 8^n(3n+1)}\right] = \dfrac{1}{4}\sum\limits_{n=0}^{\infty}\dfrac{(-1)^n}{8^n}\left(\dfrac{2}{3n+1} + \dfrac{1}{3n+2}\right)$.

By part (a), this equals $\dfrac{\pi}{3\sqrt{3}}$, so $\pi = \dfrac{3\sqrt{3}}{4}\sum\limits_{n=0}^{\infty}\dfrac{(-1)^n}{8^n}\left(\dfrac{2}{3n+1} + \dfrac{1}{3n+2}\right)$.

12.10 Taylor and Maclaurin Series   ET 11.10

1. Using Theorem 5 with $\sum\limits_{n=0}^{\infty} b_n(x-5)^n$, $b_n = \dfrac{f^{(n)}(a)}{n!}$, so $b_8 = \dfrac{f^{(8)}(5)}{8!}$.

2. (a) Using Equation 6, a power series expansion of f at 1 must have the form $f(1) + f'(1)(x-1) + \cdots$. Comparing to the

given series, $1.6 - 0.8(x-1) + \cdots$, we must have $f'(1) = -0.8$. But from the graph, $f'(1)$ is positive. Hence, the given

series is *not* the Taylor series of f centered at 1.

(b) A power series expansion of f at 2 must have the form $f(2) + f'(2)(x-2) + \frac{1}{2}f''(2)(x-2)^2 + \cdots$. Comparing to the

given series, $2.8 + 0.5(x-2) + 1.5(x-2)^2 - 0.1(x-2)^3 + \cdots$, we must have $\frac{1}{2}f''(2) = 1.5$; that is, $f''(2)$ is positive.

But from the graph, f is concave downward near $x = 2$, so $f''(2)$ must be negative. Hence, the given series is *not* the

Taylor series of f centered at 2.

3. Since $f^{(n)}(0) = (n + 1)!$, Equation 7 gives the Maclaurin series

$$\sum_{n=0}^{\infty} \frac{f^{(n)}(0)}{n!} x^n = \sum_{n=0}^{\infty} \frac{(n+1)!}{n!} x^n = \sum_{n=0}^{\infty} (n+1)x^n. \text{ Applying the Ratio Test with } a_n = (n+1)x^n \text{ gives us}$$

$$\lim_{n \to \infty} \left| \frac{a_{n+1}}{a_n} \right| = \lim_{n \to \infty} \left| \frac{(n+2)x^{n+1}}{(n+1)x^n} \right| = |x| \lim_{n \to \infty} \frac{n+2}{n+1} = |x| \cdot 1 = |x|. \text{ For convergence, we must have } |x| < 1, \text{ so the}$$

radius of convergence $R = 1$.

4. Since $f^{(n)}(4) = \dfrac{(-1)^n \, n!}{3^n(n+1)}$, Equation 6 gives the Taylor series

$$\sum_{n=0}^{\infty} \frac{f^{(n)}(4)}{n!} (x-4)^n = \sum_{n=0}^{\infty} \frac{(-1)^n \, n!}{3^n(n+1)\,n!} (x-4)^n = \sum_{n=0}^{\infty} \frac{(-1)^n}{3^n(n+1)} (x-4)^n, \text{ which is the Taylor series for } f \text{ centered}$$

at 4. Apply the Ratio Test to find the radius of convergence R.

$$\lim_{n \to \infty} \left| \frac{a_{n+1}}{a_n} \right| = \lim_{n \to \infty} \left| \frac{(-1)^{n+1}(x-4)^{n+1}}{3^{n+1}(n+2)} \cdot \frac{3^n(n+1)}{(-1)^n(x-4)^n} \right| = \lim_{n \to \infty} \left| \frac{(-1)(x-4)(n+1)}{3(n+2)} \right|$$

$$= \frac{1}{3} |x-4| \lim_{n \to \infty} \frac{n+1}{n+2} = \frac{1}{3} |x-4|$$

For convergence, $\frac{1}{3} |x-4| < 1 \iff |x-4| < 3$, so $R = 3$.

5.

n	$f^{(n)}(x)$	$f^{(n)}(0)$
0	$(1-x)^{-2}$	1
1	$2(1-x)^{-3}$	2
2	$6(1-x)^{-4}$	6
3	$24(1-x)^{-5}$	24
4	$120(1-x)^{-6}$	120
⋮	⋮	⋮

$$(1-x)^{-2} = f(0) + f'(0)x + \frac{f''(0)}{2!} x^2 + \frac{f'''(0)}{3!} x^3 + \frac{f^{(4)}(0)}{4!} x^4 + \cdots$$

$$= 1 + 2x + \frac{6}{2} x^2 + \frac{24}{6} x^3 + \frac{120}{24} x^4 + \cdots$$

$$= 1 + 2x + 3x^2 + 4x^3 + 5x^4 + \cdots = \sum_{n=0}^{\infty} (n+1)x^n$$

$$\lim_{n \to \infty} \left| \frac{a_{n+1}}{a_n} \right| = \lim_{n \to \infty} \left| \frac{(n+2)x^{n+1}}{(n+1)x^n} \right| = |x| \lim_{n \to \infty} \frac{n+2}{n+1} = |x|\,(1) = |x| < 1$$

for convergence, so $R = 1$.

6.

n	$f^{(n)}(x)$	$f^{(n)}(0)$
0	$\ln(1+x)$	0
1	$(1+x)^{-1}$	1
2	$-(1+x)^{-2}$	-1
3	$2(1+x)^{-3}$	2
4	$-6(1+x)^{-4}$	-6
5	$24(1+x)^{-5}$	24
⋮	⋮	⋮

$$\ln(1+x) = f(0) + f'(0)x + \frac{f''(0)}{2!} x^2$$

$$+ \frac{f'''(0)}{3!} x^3 + \frac{f^{(4)}(0)}{4!} x^4 + \frac{f^{(5)}(0)}{5!} x^5 + \cdots$$

$$= 0 + x - \frac{1}{2} x^2 + \frac{2}{6} x^3 - \frac{6}{24} x^4 + \frac{24}{120} x^5 - \cdots$$

$$= x - \frac{x^2}{2} + \frac{x^3}{3} - \frac{x^4}{4} + \frac{x^5}{5} - \cdots = \sum_{n=1}^{\infty} \frac{(-1)^{n-1}}{n} x^n$$

$$\lim_{n \to \infty} \left| \frac{a_{n+1}}{a_n} \right| = \lim_{n \to \infty} \left| \frac{x^{n+1}}{n+1} \cdot \frac{n}{x^n} \right| = \lim_{n \to \infty} \frac{|x|}{1 + 1/n} = |x| < 1 \text{ for convergence,}$$

so $R = 1$.

7.

n	$f^{(n)}(x)$	$f^{(n)}(0)$
0	$\sin \pi x$	0
1	$\pi \cos \pi x$	π
2	$-\pi^2 \sin \pi x$	0
3	$-\pi^3 \cos \pi x$	$-\pi^3$
4	$\pi^4 \sin \pi x$	0
5	$\pi^5 \cos \pi x$	π^5
$\vdots$	$\vdots$	$\vdots$

$$\sin \pi x = f(0) + f'(0)x + \frac{f''(0)}{2!}x^2 + \frac{f'''(0)}{3!}x^3$$
$$+ \frac{f^{(4)}(0)}{4!}x^4 + \frac{f^{(5)}(0)}{5!}x^5 + \cdots$$
$$= 0 + \pi x + 0 - \frac{\pi^3}{3!}x^3 + 0 + \frac{\pi^5}{5!}x^5 + \cdots$$
$$= \pi x - \frac{\pi^3}{3!}x^3 + \frac{\pi^5}{5!}x^5 - \frac{\pi^7}{7!}x^7 + \cdots$$
$$= \sum_{n=0}^{\infty} (-1)^n \frac{\pi^{2n+1}}{(2n+1)!} x^{2n+1}$$

$$\lim_{n \to \infty} \left| \frac{a_{n+1}}{a_n} \right| = \lim_{n \to \infty} \left| \frac{\pi^{2n+3} x^{2n+3}}{(2n+3)!} \cdot \frac{(2n+1)!}{\pi^{2n+1} x^{2n+1}} \right| = \lim_{n \to \infty} \frac{\pi^2 x^2}{(2n+3)(2n+2)}$$
$$= 0 < 1 \quad \text{for all } x, \text{ so } R = \infty.$$

8.

n	$f^{(n)}(x)$	$f^{(n)}(0)$
0	$\cos 3x$	1
1	$-3 \sin 3x$	0
2	$-3^2 \cos 3x$	-3^2
3	$3^3 \sin 3x$	0
4	$3^4 \cos 3x$	3^4
$\vdots$	$\vdots$	$\vdots$

$$\cos 3x = \sum_{n=0}^{\infty} \frac{f^{(n)}(0)}{n!} x^n = \sum_{n=0}^{\infty} (-1)^n \frac{3^{2n}}{(2n)!} x^{2n}$$

$$\lim_{n \to \infty} \left| \frac{a_{n+1}}{a_n} \right| = \lim_{n \to \infty} \left| \frac{3^{2n+2} x^{2n+2}}{(2n+2)!} \cdot \frac{(2n)!}{3^{2n} x^{2n}} \right| = \lim_{n \to \infty} \frac{3^2 x^2}{(2n+2)(2n+1)}$$
$$= 0 < 1 \quad \text{for all } x, \text{ so } R = \infty.$$

9.

n	$f^{(n)}(x)$	$f^{(n)}(0)$
0	e^{5x}	1
1	$5e^{5x}$	5
2	$5^2 e^{5x}$	25
3	$5^3 e^{5x}$	125
4	$5^4 e^{5x}$	625
$\vdots$	$\vdots$	$\vdots$

$$e^{5x} = \sum_{n=0}^{\infty} \frac{f^{(n)}(0)}{n!} x^n = \sum_{n=0}^{\infty} \frac{5^n}{n!} x^n.$$

$$\lim_{n \to \infty} \left| \frac{a_{n+1}}{a_n} \right| = \lim_{n \to \infty} \left[\frac{5^{n+1} |x|^{n+1}}{(n+1)!} \cdot \frac{n!}{5^n |x|^n} \right] = \lim_{n \to \infty} \frac{5|x|}{n+1} = 0 < 1$$
for all x, so $R = \infty$.

10.

n	$f^{(n)}(x)$	$f^{(n)}(0)$
0	xe^x	0
1	$(x+1)e^x$	1
2	$(x+2)e^x$	2
3	$(x+3)e^x$	3
$\vdots$	$\vdots$	$\vdots$

$$xe^x = \sum_{n=0}^{\infty} \frac{f^{(n)}(0)}{n!} x^n = \sum_{n=0}^{\infty} \frac{n}{n!} x^n = \sum_{n=1}^{\infty} \frac{n}{n!} x^n = \sum_{n=1}^{\infty} \frac{x^n}{(n-1)!}.$$

$$\lim_{n \to \infty} \left| \frac{a_{n+1}}{a_n} \right| = \lim_{n \to \infty} \left[\frac{|x|^{n+1}}{n!} \cdot \frac{(n-1)!}{|x|^n} \right] = \lim_{n \to \infty} \frac{|x|}{n} = 0 < 1 \text{ for all } x,$$
so $R = \infty$.

11.

n	$f^{(n)}(x)$	$f^{(n)}(0)$
0	$\sinh x$	0
1	$\cosh x$	1
2	$\sinh x$	0
3	$\cosh x$	1
4	$\sinh x$	0
$\vdots$	$\vdots$	$\vdots$

$f^{(n)}(0) = \begin{cases} 0 & \text{if } n \text{ is even} \\ 1 & \text{if } n \text{ is odd} \end{cases}$ so $\sinh x = \sum_{n=0}^{\infty} \dfrac{x^{2n+1}}{(2n+1)!}$.

Use the Ratio Test to find R. If $a_n = \dfrac{x^{2n+1}}{(2n+1)!}$, then

$$\lim_{n\to\infty} \left| \frac{a_{n+1}}{a_n} \right| = \lim_{n\to\infty} \left| \frac{x^{2n+3}}{(2n+3)!} \cdot \frac{(2n+1)!}{x^{2n+1}} \right| = x^2 \cdot \lim_{n\to\infty} \frac{1}{(2n+3)(2n+2)}$$

$$= 0 < 1 \quad \text{for all } x, \text{ so } R = \infty.$$

12.

n	$f^{(n)}(x)$	$f^{(n)}(0)$
0	$\cosh x$	1
1	$\sinh x$	0
2	$\cosh x$	1
3	$\sinh x$	0
$\vdots$	$\vdots$	$\vdots$

$f^{(n)}(0) = \begin{cases} 1 & \text{if } n \text{ is even} \\ 0 & \text{if } n \text{ is odd} \end{cases}$ so $\cosh x = \sum_{n=0}^{\infty} \dfrac{x^{2n}}{(2n)!}$.

Use the Ratio Test to find R. If $a_n = \dfrac{x^{2n}}{(2n)!}$, then

$$\lim_{n\to\infty} \left| \frac{a_{n+1}}{a_n} \right| = \lim_{n\to\infty} \left| \frac{x^{2n+2}}{(2n+2)!} \cdot \frac{(2n)!}{x^{2n}} \right| = x^2 \cdot \lim_{n\to\infty} \frac{1}{(2n+2)(2n+1)}$$

$$= 0 < 1 \quad \text{for all } x, \text{ so } R = \infty.$$

13.

n	$f^{(n)}(x)$	$f^{(n)}(1)$
0	$x^4 - 3x^2 + 1$	-1
1	$4x^3 - 6x$	-2
2	$12x^2 - 6$	6
3	$24x$	24
4	24	24
5	0	0
6	0	0
$\vdots$	$\vdots$	$\vdots$

$f^{(n)}(x) = 0$ for $n \geq 5$, so f has a finite series expansion about $a = 1$.

$$f(x) = x^4 - 3x^2 + 1 = \sum_{n=0}^{4} \frac{f^{(n)}(1)}{n!} (x-1)^n$$

$$= \frac{-1}{0!} (x-1)^0 + \frac{-2}{1!} (x-1)^1 + \frac{6}{2!} (x-1)^2 + \frac{24}{3!} (x-1)^3 + \frac{24}{4!} (x-1)^4$$

$$= -1 - 2(x-1) + 3(x-1)^2 + 4(x-1)^3 + (x-1)^4$$

A finite series converges for all x, so $R = \infty$.

14.

n	$f^{(n)}(x)$	$f^{(n)}(-2)$
0	$x - x^3$	6
1	$1 - 3x^2$	-11
2	$-6x$	12
3	-6	-6
4	0	0
5	0	0
$\vdots$	$\vdots$	$\vdots$

$f^{(n)}(x) = 0$ for $n \geq 4$, so f has a finite series expansion about $a = -2$.

$$f(x) = x - x^3 = \sum_{n=0}^{3} \frac{f^{(n)}(-2)}{n!} (x+2)^n$$

$$= \frac{6}{0!} (x+2)^0 + \frac{-11}{1!} (x+2)^1 + \frac{12}{2!} (x+2)^2 + \frac{-6}{3!} (x+2)^3$$

$$= 6 - 11(x+2) + 6(x+2)^2 - (x+2)^3$$

A finite series converges for all x, so $R = \infty$.

15. $f(x) = e^x \Rightarrow f^{(n)}(x) = e^x$, so $f^{(n)}(3) = e^3$ and $e^x = \sum_{n=0}^{\infty} \dfrac{e^3}{n!} (x-3)^n$. If $a_n = \dfrac{e^3}{n!} (x-3)^n$, then

$$\lim_{n\to\infty} \left| \frac{a_{n+1}}{a_n} \right| = \lim_{n\to\infty} \left| \frac{e^3 (x-3)^{n+1}}{(n+1)!} \cdot \frac{n!}{e^3 (x-3)^n} \right| = \lim_{n\to\infty} \frac{|x-3|}{n+1} = 0 < 1 \text{ for all } x, \text{ so } R = \infty.$$

16.

n	$f^{(n)}(x)$	$f^{(n)}(-3)$
0	$1/x$	$-1/3$
1	$-1/x^2$	$-1/3^2$
2	$2/x^3$	$-2/3^3$
3	$-6/x^4$	$-6/3^4$
4	$24/x^5$	$-24/3^5$
$\vdots$	$\vdots$	$\vdots$

$$f(x) = \frac{1}{x} = \sum_{n=0}^{\infty} \frac{f^{(n)}(-3)}{n!}(x+3)^n$$

$$= \frac{-1/3}{0!}(x+3)^0 + \frac{-1/3^2}{1!}(x+3)^1 + \frac{-2/3^3}{2!}(x+3)^2$$

$$+ \frac{-6/3^4}{3!}(x+3)^3 + \frac{-24/3^5}{4!}(x+3)^4 + \cdots$$

$$= \sum_{n=0}^{\infty} \frac{-n!/3^{n+1}}{n!}(x+3)^n = -\sum_{n=0}^{\infty} \frac{(x+3)^n}{3^{n+1}}$$

$$\lim_{n\to\infty}\left|\frac{a_{n+1}}{a_n}\right| = \lim_{n\to\infty}\left|\frac{(x+3)^{n+1}}{3^{n+2}} \cdot \frac{3^{n+1}}{(x+3)^n}\right| = \lim_{n\to\infty}\frac{|x+3|}{3} = \frac{|x+3|}{3} < 1$$

for convergence, so $|x+3| < 3$ and $R = 3$.

17.

n	$f^{(n)}(x)$	$f^{(n)}(\pi)$
0	$\cos x$	-1
1	$-\sin x$	0
2	$-\cos x$	1
3	$\sin x$	0
4	$\cos x$	-1
$\vdots$	$\vdots$	$\vdots$

$$\cos x = \sum_{k=0}^{\infty} \frac{f^{(k)}(\pi)}{k!}(x-\pi)^k = -1 + \frac{(x-\pi)^2}{2!} - \frac{(x-\pi)^4}{4!} + \frac{(x-\pi)^6}{6!} - \cdots$$

$$= \sum_{n=0}^{\infty}(-1)^{n+1}\frac{(x-\pi)^{2n}}{(2n)!}$$

$$\lim_{n\to\infty}\left|\frac{a_{n+1}}{a_n}\right| = \lim_{n\to\infty}\left[\frac{|x-\pi|^{2n+2}}{(2n+2)!} \cdot \frac{(2n)!}{|x-\pi|^{2n}}\right] = \lim_{n\to\infty}\frac{|x-\pi|^2}{(2n+2)(2n+1)} = 0 < 1$$

for all x, so $R = \infty$.

18.

n	$f^{(n)}(x)$	$f^{(n)}(\pi/2)$
0	$\sin x$	1
1	$\cos x$	0
2	$-\sin x$	-1
3	$-\cos x$	0
4	$\sin x$	1
$\vdots$	$\vdots$	$\vdots$

$$\sin x = \sum_{k=0}^{\infty}\frac{f^{(k)}(\pi/2)}{k!}\left(x - \frac{\pi}{2}\right)^k$$

$$= 1 - \frac{(x-\pi/2)^2}{2!} + \frac{(x-\pi/2)^4}{4!} - \frac{(x-\pi/2)^6}{6!} + \cdots$$

$$= \sum_{n=0}^{\infty}(-1)^n\frac{(x-\pi/2)^{2n}}{(2n)!}$$

$$\lim_{n\to\infty}\left|\frac{a_{n+1}}{a_n}\right| = \lim_{n\to\infty}\left[\frac{|x-\pi/2|^{2n+2}}{(2n+2)!} \cdot \frac{(2n)!}{|x-\pi/2|^{2n}}\right]$$

$$= \lim_{n\to\infty}\frac{|x-\pi/2|^2}{(2n+2)(2n+1)} = 0 < 1 \quad \text{for all } x \text{, so } R = \infty.$$

19.

n	$f^{(n)}(x)$	$f^{(n)}(9)$
0	$x^{-1/2}$	$\frac{1}{3}$
1	$-\frac{1}{2}x^{-3/2}$	$-\frac{1}{2}\cdot\frac{1}{3^3}$
2	$\frac{3}{4}x^{-5/2}$	$-\frac{1}{2}\cdot\left(-\frac{3}{2}\right)\cdot\frac{1}{3^5}$
3	$-\frac{15}{8}x^{-7/2}$	$-\frac{1}{2}\cdot\left(-\frac{3}{2}\right)\cdot\left(-\frac{5}{2}\right)\cdot\frac{1}{3^7}$
$\vdots$	$\vdots$	$\vdots$

$$\frac{1}{\sqrt{x}} = \frac{1}{3} - \frac{1}{2\cdot3^3}(x-9) + \frac{3}{2^2\cdot3^5}\frac{(x-9)^2}{2!}$$

$$- \frac{3\cdot5}{2^3\cdot3^7}\frac{(x-9)^3}{3!} + \cdots$$

$$= \frac{1}{3} + \sum_{n=1}^{\infty}(-1)^n\frac{1\cdot3\cdot5\cdot\cdots\cdot(2n-1)}{2^n\cdot3^{2n+1}\cdot n!}(x-9)^n.$$

[continued]

$$\lim_{n\to\infty}\left|\frac{a_{n+1}}{a_n}\right| = \lim_{n\to\infty}\left[\frac{1\cdot3\cdot5\cdot\,\cdots\,\cdot(2n-1)[2(n+1)-1]\,|x-9|^{n+1}}{2^{n+1}\cdot3^{[2(n+1)+1]}\cdot(n+1)!}\cdot\frac{2^n\cdot3^{2n+1}\cdot n!}{1\cdot3\cdot5\cdot\,\cdots\,\cdot(2n-1)\,|x-9|^n}\right]$$

$$= \lim_{n\to\infty}\left[\frac{(2n+1)\,|x-9|}{2\cdot3^2(n+1)}\right] = \frac{1}{9}\,|x-9| < 1$$

for convergence, so $|x-9| < 9$ and $R = 9$.

20.

n	$f^{(n)}(x)$	$f^{(n)}(1)$
0	x^{-2}	1
1	$-2x^{-3}$	-2
2	$6x^{-4}$	6
3	$-24x^{-5}$	-24
4	$120x^{-6}$	120
⋮	⋮	⋮

$$x^{-2} = 1 - 2(x-1) + 6\cdot\frac{(x-1)^2}{2!} - 24\cdot\frac{(x-1)^3}{3!} + 120\cdot\frac{(x-1)^4}{4!} - \cdots$$

$$= 1 - 2(x-1) + 3(x-1)^2 - 4(x-1)^3 + 5(x-1)^4 - \cdots$$

$$= \sum_{n=0}^{\infty}(-1)^n(n+1)(x-1)^n.$$

$$\lim_{n\to\infty}\left|\frac{a_{n+1}}{a_n}\right| = \lim_{n\to\infty}\frac{(n+2)\,|x-1|^{n+1}}{(n+1)\,|x-1|^n} = \lim_{n\to\infty}\left[\frac{n+2}{n+1}\cdot|x-1|\right]$$

$$= |x-1| < 1 \quad \text{for convergence, so } R = 1.$$

21. If $f(x) = \sin\pi x$, then $f^{(n+1)}(x) = \pm\pi^{n+1}\sin\pi x$ or $\pm\pi^{n+1}\cos\pi x$. In each case, $\left|f^{(n+1)}(x)\right| \le \pi^{n+1}$, so by Formula 9

with $a = 0$ and $M = \pi^{n+1}$, $|R_n(x)| \le \dfrac{\pi^{n+1}}{(n+1)!}\,|x|^{n+1} = \dfrac{|\pi x|^{n+1}}{(n+1)!}$. Thus, $|R_n(x)| \to 0$ as $n \to \infty$ by Equation 10.

So $\lim\limits_{n\to\infty} R_n(x) = 0$ and, by Theorem 8, the series in Exercise 7 represents $\sin\pi x$ for all x.

22. If $f(x) = \sin x$, then $f^{(n+1)}(x) = \pm\sin x$ or $\pm\cos x$. In each case, $\left|f^{(n+1)}(x)\right| \le 1$, so by Formula 9 with $a = 0$ and

$M = 1$, $|R_n(x)| \le \dfrac{1}{(n+1)!}\left|x - \dfrac{\pi}{2}\right|^{n+1}$. Thus, $|R_n(x)| \to 0$ as $n \to \infty$ by Equation 10. So $\lim\limits_{n\to\infty} R_n(x) = 0$ and, by

Theorem 8, the series in Exercise 18 represents $\sin x$ for all x.

23. If $f(x) = \sinh x$, then for all n, $f^{(n+1)}(x) = \cosh x$ or $\sinh x$. Since $|\sinh x| < |\cosh x| = \cosh x$ for all x, we have

$\left|f^{(n+1)}(x)\right| \le \cosh x$ for all n. If d is any positive number and $|x| \le d$, then $\left|f^{(n+1)}(x)\right| \le \cosh x \le \cosh d$, so by

Formula 9 with $a = 0$ and $M = \cosh d$, we have $|R_n(x)| \le \dfrac{\cosh d}{(n+1)!}\,|x|^{n+1}$. It follows that $|R_n(x)| \to 0$ as $n \to \infty$ for

$|x| \le d$ (by Equation 10). But d was an arbitrary positive number. So by Theorem 8, the series represents $\sinh x$ for all x.

24. If $f(x) = \cosh x$, then for all n, $f^{(n+1)}(x) = \cosh x$ or $\sinh x$. Since $|\sinh x| < |\cosh x| = \cosh x$ for all x, we have

$\left|f^{(n+1)}(x)\right| \le \cosh x$ for all n. If d is any positive number and $|x| \le d$, then $\left|f^{(n+1)}(x)\right| \le \cosh x \le \cosh d$, so by

Formula 9 with $a = 0$ and $M = \cosh d$, we have $|R_n(x)| \le \dfrac{\cosh d}{(n+1)!}\,|x|^{n+1}$. It follows that $|R_n(x)| \to 0$ as $n \to \infty$ for

$|x| \le d$ (by Equation 10). But d was an arbitrary positive number. So by Theorem 8, the series represents $\cosh x$ for all x.

25. The general binomial series in (17) is

$$(1+x)^k = \sum_{n=0}^{\infty} \binom{k}{n} x^n = 1 + kx + \frac{k(k-1)}{2!}x^2 + \frac{k(k-1)(k-2)}{3!}x^3 + \cdots.$$

$$(1+x)^{1/2} = \sum_{n=0}^{\infty} \binom{\frac{1}{2}}{n} x^n = 1 + \left(\tfrac{1}{2}\right)x + \frac{\left(\frac{1}{2}\right)\left(-\frac{1}{2}\right)}{2!}x^2 + \frac{\left(\frac{1}{2}\right)\left(-\frac{1}{2}\right)\left(-\frac{3}{2}\right)}{3!}x^3 + \cdots$$

$$= 1 + \frac{x}{2} - \frac{x^2}{2^2 \cdot 2!} + \frac{1 \cdot 3 \cdot x^3}{2^3 \cdot 3!} - \frac{1 \cdot 3 \cdot 5 \cdot x^4}{2^4 \cdot 4!} + \cdots$$

$$= 1 + \frac{x}{2} + \sum_{n=2}^{\infty} \frac{(-1)^{n-1} 1 \cdot 3 \cdot 5 \cdot \cdots \cdot (2n-3)x^n}{2^n \cdot n!} \quad \text{for } |x| < 1, \quad \text{so } R = 1.$$

26. $\dfrac{1}{(1+x)^4} = (1+x)^{-4} = \displaystyle\sum_{n=0}^{\infty} \binom{-4}{n} x^n$. The binomial coefficient is

$$\binom{-4}{n} = \frac{(-4)(-5)(-6) \cdots \cdots (-4-n+1)}{n!} = \frac{(-4)(-5)(-6) \cdots \cdots [-(n+3)]}{n!}$$

$$= \frac{(-1)^n \cdot 2 \cdot 3 \cdot 4 \cdot 5 \cdot 6 \cdots \cdots (n+1)(n+2)(n+3)}{2 \cdot 3 \cdot n!} = \frac{(-1)^n (n+1)(n+2)(n+3)}{6}$$

Thus, $\dfrac{1}{(1+x)^4} = \displaystyle\sum_{n=0}^{\infty} \frac{(-1)^n (n+1)(n+2)(n+3)}{6} x^n$ for $|x| < 1, \quad \text{so } R = 1$.

27. $\dfrac{1}{(2+x)^3} = \dfrac{1}{[2(1+x/2)]^3} = \dfrac{1}{8}\left(1+\dfrac{x}{2}\right)^{-3} = \dfrac{1}{8}\displaystyle\sum_{n=0}^{\infty} \binom{-3}{n}\left(\dfrac{x}{2}\right)^n$. The binomial coefficient is

$$\binom{-3}{n} = \frac{(-3)(-4)(-5) \cdots \cdots (-3-n+1)}{n!} = \frac{(-3)(-4)(-5) \cdots \cdots [-(n+2)]}{n!}$$

$$= \frac{(-1)^n \cdot 2 \cdot 3 \cdot 4 \cdot 5 \cdots \cdots (n+1)(n+2)}{2 \cdot n!} = \frac{(-1)^n (n+1)(n+2)}{2}$$

Thus, $\dfrac{1}{(2+x)^3} = \dfrac{1}{8}\displaystyle\sum_{n=0}^{\infty} \frac{(-1)^n (n+1)(n+2)}{2} \frac{x^n}{2^n} = \displaystyle\sum_{n=0}^{\infty} \frac{(-1)^n (n+1)(n+2)x^n}{2^{n+4}}$ for $\left|\dfrac{x}{2}\right| < 1 \Leftrightarrow |x| < 2$, so $R = 2$.

28. $(1-x)^{2/3} = \displaystyle\sum_{n=0}^{\infty} \binom{\frac{2}{3}}{n} (-x)^n = 1 + \tfrac{2}{3}(-x) + \frac{\frac{2}{3}\left(-\frac{1}{3}\right)}{2!}(-x)^2 + \frac{\frac{2}{3}\left(-\frac{1}{3}\right)\left(-\frac{4}{3}\right)}{3!}(-x)^3 + \cdots$

$$= 1 - \tfrac{2}{3}x + \sum_{n=2}^{\infty} \frac{(-1)^{n-1}(-1)^n \cdot 2 \cdot [1 \cdot 4 \cdot 7 \cdots \cdots (3n-5)]}{3^n \cdot n!} x^n$$

$$= 1 - \tfrac{2}{3}x - 2\sum_{n=2}^{\infty} \frac{1 \cdot 4 \cdot 7 \cdots \cdots (3n-5)}{3^n \cdot n!} x^n$$

and $|-x| < 1 \Leftrightarrow |x| < 1$, so $R = 1$.

29. $\sin x = \displaystyle\sum_{n=0}^{\infty} (-1)^n \frac{x^{2n+1}}{(2n+1)!} \Rightarrow f(x) = \sin(\pi x) = \displaystyle\sum_{n=0}^{\infty} (-1)^n \frac{(\pi x)^{2n+1}}{(2n+1)!} = \displaystyle\sum_{n=0}^{\infty} (-1)^n \frac{\pi^{2n+1}}{(2n+1)!} x^{2n+1}, R = \infty.$

30. $\cos x = \sum\limits_{n=0}^{\infty} (-1)^n \dfrac{x^{2n}}{(2n)!}$ $\Rightarrow$ $f(x) = \cos(\pi x/2) = \sum\limits_{n=0}^{\infty} (-1)^n \dfrac{(\pi x/2)^{2n}}{(2n)!} = \sum\limits_{n=0}^{\infty} (-1)^n \dfrac{\pi^{2n}}{2^{2n}(2n)!} x^{2n}$, $R = \infty$.

31. $e^x = \sum\limits_{n=0}^{\infty} \dfrac{x^n}{n!}$ $\Rightarrow$ $e^{2x} = \sum\limits_{n=0}^{\infty} \dfrac{(2x)^n}{n!} = \sum\limits_{n=0}^{\infty} \dfrac{2^n x^n}{n!}$, so $f(x) = e^x + e^{2x} = \sum\limits_{n=0}^{\infty} \dfrac{1}{n!} x^n + \sum\limits_{n=0}^{\infty} \dfrac{2^n}{n!} x^n = \sum\limits_{n=0}^{\infty} \dfrac{2^n+1}{n!} x^n$,

$R = \infty$.

32. $e^x = \sum\limits_{n=0}^{\infty} \dfrac{x^n}{n!}$ $\Rightarrow$ $2e^{-x} = 2\sum\limits_{n=0}^{\infty} \dfrac{(-x)^n}{n!} = 2\sum\limits_{n=0}^{\infty} \dfrac{(-1)^n x^n}{n!}$, so $f(x) = e^x + 2e^{-x} = \sum\limits_{n=0}^{\infty} \dfrac{[1+2(-1)^n]}{n!} x^n$, $R = \infty$.

33. $\cos x = \sum\limits_{n=0}^{\infty} (-1)^n \dfrac{x^{2n}}{(2n)!}$ $\Rightarrow$ $\cos(\tfrac{1}{2}x^2) = \sum\limits_{n=0}^{\infty} (-1)^n \dfrac{\left(\tfrac{1}{2}x^2\right)^{2n}}{(2n)!} = \sum\limits_{n=0}^{\infty} (-1)^n \dfrac{x^{4n}}{2^{2n}(2n)!}$, so

$f(x) = x\cos(\tfrac{1}{2}x^2) = \sum\limits_{n=0}^{\infty} (-1)^n \dfrac{1}{2^{2n}(2n)!} x^{4n+1}$, $R = \infty$.

34. $\tan^{-1} x = \sum\limits_{n=0}^{\infty} (-1)^n \dfrac{x^{2n+1}}{2n+1}$ $\Rightarrow$ $\tan^{-1}(x^3) = \sum\limits_{n=0}^{\infty} (-1)^n \dfrac{(x^3)^{2n+1}}{2n+1} = \sum\limits_{n=0}^{\infty} (-1)^n \dfrac{x^{6n+3}}{2n+1}$, so

$x^2 \tan^{-1}(x^3) = \sum\limits_{n=0}^{\infty} (-1)^n \dfrac{1}{2n+1} x^{6n+5}$; $|x^3| < 1$ $\Leftrightarrow$ $|x| < 1$, so $R = 1$.

35. We must write the binomial in the form $(1+$ expression$)$, so we'll factor out a 4.

$$\dfrac{x}{\sqrt{4+x^2}} = \dfrac{x}{\sqrt{4(1+x^2/4)}} = \dfrac{x}{2\sqrt{1+x^2/4}} = \dfrac{x}{2}\left(1+\dfrac{x^2}{4}\right)^{-1/2} = \dfrac{x}{2}\sum\limits_{n=0}^{\infty}\binom{-\tfrac{1}{2}}{n}\left(\dfrac{x^2}{4}\right)^n$$

$$= \dfrac{x}{2}\left[1 + \left(-\tfrac{1}{2}\right)\dfrac{x^2}{4} + \dfrac{\left(-\tfrac{1}{2}\right)\left(-\tfrac{3}{2}\right)}{2!}\left(\dfrac{x^2}{4}\right)^2 + \dfrac{\left(-\tfrac{1}{2}\right)\left(-\tfrac{3}{2}\right)\left(-\tfrac{5}{2}\right)}{3!}\left(\dfrac{x^2}{4}\right)^3 + \cdots\right]$$

$$= \dfrac{x}{2} + \dfrac{x}{2}\sum\limits_{n=1}^{\infty}(-1)^n \dfrac{1\cdot 3\cdot 5 \cdot\cdots\cdot (2n-1)}{2^n\cdot 4^n\cdot n!} x^{2n}$$

$$= \dfrac{x}{2} + \sum\limits_{n=1}^{\infty}(-1)^n \dfrac{1\cdot 3\cdot 5 \cdot\cdots\cdot (2n-1)}{n!\, 2^{3n+1}} x^{2n+1} \text{ and } \dfrac{x^2}{4} < 1 \Leftrightarrow \dfrac{|x|}{2} < 1 \Leftrightarrow |x| < 2, \text{ so } R = 2.$$

36. $$\dfrac{x^2}{\sqrt{2+x}} = \dfrac{x^2}{\sqrt{2(1+x/2)}} = \dfrac{x^2}{\sqrt{2}}\left(1+\dfrac{x}{2}\right)^{-1/2} = \dfrac{x^2}{\sqrt{2}}\sum\limits_{n=0}^{\infty}\binom{-\tfrac{1}{2}}{n}\left(\dfrac{x}{2}\right)^n$$

$$= \dfrac{x^2}{\sqrt{2}}\left[1 + \left(-\tfrac{1}{2}\right)\left(\dfrac{x}{2}\right) + \dfrac{\left(-\tfrac{1}{2}\right)\left(-\tfrac{3}{2}\right)}{2!}\left(\dfrac{x}{2}\right)^2 + \dfrac{\left(-\tfrac{1}{2}\right)\left(-\tfrac{3}{2}\right)\left(-\tfrac{5}{2}\right)}{3!}\left(\dfrac{x}{2}\right)^3 + \cdots\right]$$

$$= \dfrac{x^2}{\sqrt{2}} + \dfrac{x^2}{\sqrt{2}}\sum\limits_{n=1}^{\infty}(-1)^n \dfrac{1\cdot 3\cdot 5 \cdot\cdots\cdot (2n-1)}{n!\, 2^{2n}} x^n$$

$$= \dfrac{x^2}{\sqrt{2}} + \sum\limits_{n=1}^{\infty}(-1)^n \dfrac{1\cdot 3\cdot 5 \cdot\cdots\cdot (2n-1)}{n!\, 2^{2n+1/2}} x^{n+2} \text{ and } \left|\dfrac{x}{2}\right| < 1 \Leftrightarrow |x| < 2, \text{ so } R = 2.$$

37. $\sin^2 x = \dfrac{1}{2}(1-\cos 2x) = \dfrac{1}{2}\left[1 - \sum\limits_{n=0}^{\infty}\dfrac{(-1)^n(2x)^{2n}}{(2n)!}\right] = \dfrac{1}{2}\left[1 - 1 - \sum\limits_{n=1}^{\infty}\dfrac{(-1)^n(2x)^{2n}}{(2n)!}\right] = \sum\limits_{n=1}^{\infty}\dfrac{(-1)^{n+1}2^{2n-1}x^{2n}}{(2n)!}$,

$R = \infty$

38. $\dfrac{x - \sin x}{x^3} = \dfrac{1}{x^3}\left[x - \displaystyle\sum_{n=0}^{\infty} \dfrac{(-1)^n x^{2n+1}}{(2n+1)!} \right] = \dfrac{1}{x^3}\left[x - x - \displaystyle\sum_{n=1}^{\infty} \dfrac{(-1)^n x^{2n+1}}{(2n+1)!} \right] = \dfrac{1}{x^3}\left[-\displaystyle\sum_{n=0}^{\infty} \dfrac{(-1)^{n+1} x^{2n+3}}{(2n+3)!} \right]$

$= \dfrac{1}{x^3} \displaystyle\sum_{n=0}^{\infty} \dfrac{(-1)^n x^{2n+3}}{(2n+3)!} = \displaystyle\sum_{n=0}^{\infty} \dfrac{(-1)^n x^{2n}}{(2n+3)!}$

and this series also gives the required value at $x = 0$ (namely $1/6$); $R = \infty$.

39. $\cos x = \displaystyle\sum_{n=0}^{\infty} (-1)^n \dfrac{x^{2n}}{(2n)!} \quad \Rightarrow \quad f(x) = \cos(x^2) = \displaystyle\sum_{n=0}^{\infty} \dfrac{(-1)^n \left(x^2\right)^{2n}}{(2n)!} = \displaystyle\sum_{n=0}^{\infty} \dfrac{(-1)^n x^{4n}}{(2n)!}, R = \infty$

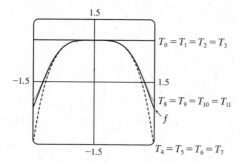

Notice that, as n increases, $T_n(x)$

becomes a better approximation to $f(x)$.

40. $e^x \overset{(11)}{=} \displaystyle\sum_{n=0}^{\infty} \dfrac{x^n}{n!}$, so $e^{-x^2} = \displaystyle\sum_{n=0}^{\infty} \dfrac{(-x^2)^n}{n!} = \displaystyle\sum_{n=0}^{\infty} (-1)^n \dfrac{x^{2n}}{n!}$.

Also, $\cos x \overset{(16)}{=} \displaystyle\sum_{n=0}^{\infty} (-1)^n \dfrac{x^{2n}}{(2n)!}$, so

$$f(x) = e^{-x^2} + \cos x = \sum_{n=0}^{\infty} (-1)^n \left(\dfrac{1}{n!} + \dfrac{1}{(2n)!} \right) x^{2n}$$

$$= 2 - \dfrac{3}{2}x^2 + \dfrac{13}{24}x^4 - \dfrac{121}{720}x^6 + \cdots .$$

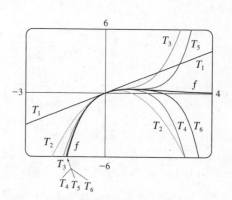

The series for e^x and $\cos x$ converge for all x, so the same is true of the series

for $f(x)$; that is, $R = \infty$. From the graphs of f and the first few Taylor

polynomials, we see that $T_n(x)$ provides a closer fit to $f(x)$ near 0 as n increases.

41. $e^x \overset{(11)}{=} \displaystyle\sum_{n=0}^{\infty} \dfrac{x^n}{n!}$, so $e^{-x} = \displaystyle\sum_{n=0}^{\infty} \dfrac{(-x)^n}{n!} = \displaystyle\sum_{n=0}^{\infty} (-1)^n \dfrac{x^n}{n!}$, so

$$f(x) = xe^{-x} = \sum_{n=0}^{\infty} (-1)^n \dfrac{1}{n!} x^{n+1}$$

$$= x - x^2 + \tfrac{1}{2}x^3 - \tfrac{1}{6}x^4 + \tfrac{1}{24}x^5 - \tfrac{1}{120}x^6 + \cdots$$

$$= \sum_{n=1}^{\infty} (-1)^{n-1} \dfrac{x^n}{(n-1)!}$$

The series for e^x converges for all x, so the same is true of the series

for $f(x)$; that is, $R = \infty$. From the graphs of f and the first few Taylor

polynomials, we see that $T_n(x)$ provides a closer fit to $f(x)$ near 0 as n increases.

42. From Example 6 in Section 12.9 [ET 11.9],

$$\ln(1-x) = -\sum_{n=1}^{\infty} \frac{x^n}{n} \text{ for } |x| < 1.$$

$$\ln(1+x^2) = \ln[1-(-x^2)] = -\sum_{n=1}^{\infty} \frac{(-x^2)^n}{n} = \sum_{n=1}^{\infty} (-1)^{n+1} \frac{1}{n} x^{2n},$$

so $f(x) = \ln(1+x^2) = x^2 - \frac{1}{2}x^4 + \frac{1}{3}x^6 - \frac{1}{4}x^8 + \frac{1}{5}x^{10} - \cdots$. This

series converges for $|x^2| < 1 \Leftrightarrow |x| < 1$, so $R = 1$. From the graphs

of f and the first few Taylor polynomials, we see that $T_n(x)$ provides a

closer fit to $f(x)$ near 0 as n increases.

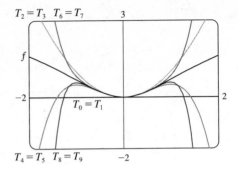

43. $e^x = \sum_{n=0}^{\infty} \frac{x^n}{n!}$, so $e^{-0.2} = \sum_{n=0}^{\infty} \frac{(-0.2)^n}{n!} = 1 - 0.2 + \frac{1}{2!}(0.2)^2 - \frac{1}{3!}(0.2)^3 + \frac{1}{4!}(0.2)^4 - \frac{1}{5!}(0.2)^5 + \frac{1}{6!}(0.2)^6 - \cdots$.

But $\frac{1}{6!}(0.2)^6 = 8.\overline{8} \times 10^{-8}$, so by the Alternating Series Estimation Theorem, $e^{-0.2} \approx \sum_{n=0}^{5} \frac{(-0.2)^n}{n!} \approx 0.81873$, correct to

five decimal places.

44. $3° = \frac{\pi}{60}$ radians and $\sin x = \sum_{n=0}^{\infty} \frac{(-1)^n x^{2n+1}}{(2n+1)!}$, so

$$\sin \frac{\pi}{60} = \frac{\pi}{60} - \frac{\left(\frac{\pi}{60}\right)^3}{3!} + \frac{\left(\frac{\pi}{60}\right)^5}{5!} - \cdots = \frac{\pi}{60} - \frac{\pi^3}{1,296,000} + \frac{\pi^5}{93,312,000,000} - \cdots. \text{ But } \frac{\pi^5}{93,312,000,000} < 10^{-8}, \text{ so by}$$

the Alternating Series Estimation Theorem, $\sin \frac{\pi}{60} \approx \frac{\pi}{60} - \frac{\pi^3}{1,296,000} \approx 0.05234$.

45. (a) $1/\sqrt{1-x^2} = [1+(-x^2)]^{-1/2} = 1 + \left(-\frac{1}{2}\right)\left(-x^2\right) + \frac{\left(-\frac{1}{2}\right)\left(-\frac{3}{2}\right)}{2!}\left(-x^2\right)^2 + \frac{\left(-\frac{1}{2}\right)\left(-\frac{3}{2}\right)\left(-\frac{5}{2}\right)}{3!}\left(-x^2\right)^3 + \cdots$

$$= 1 + \sum_{n=1}^{\infty} \frac{1 \cdot 3 \cdot 5 \cdots (2n-1)}{2^n \cdot n!} x^{2n}$$

(b) $\sin^{-1} x = \int \frac{1}{\sqrt{1-x^2}} \, dx = C + x + \sum_{n=1}^{\infty} \frac{1 \cdot 3 \cdot 5 \cdots (2n-1)}{(2n+1)2^n \cdot n!} x^{2n+1}$

$$= x + \sum_{n=1}^{\infty} \frac{1 \cdot 3 \cdot 5 \cdots (2n-1)}{(2n+1)2^n \cdot n!} x^{2n+1} \quad \text{since } 0 = \sin^{-1} 0 = C.$$

46. (a) $1/\sqrt[4]{1+x} = (1+x)^{-1/4} = \sum_{n=0}^{\infty} \binom{-\frac{1}{4}}{n} x^n = 1 - \frac{1}{4}x + \frac{\left(-\frac{1}{4}\right)\left(-\frac{5}{4}\right)}{2!}x^2 + \frac{\left(-\frac{1}{4}\right)\left(-\frac{5}{4}\right)\left(-\frac{9}{4}\right)}{3!}x^3 + \cdots$

$$= 1 - \frac{1}{4}x + \sum_{n=2}^{\infty} (-1)^n \frac{1 \cdot 5 \cdot 9 \cdots (4n-3)}{4^n \cdot n!} x^n$$

(b) $1/\sqrt[4]{1+x} = 1 - \frac{1}{4}x + \frac{5}{32}x^2 - \frac{15}{128}x^3 + \frac{195}{2048}x^4 - \cdots$. $1/\sqrt[4]{1.1} = 1/\sqrt[4]{1+0.1}$, so let $x = 0.1$. The sum of the first four

terms is then $1 - \frac{1}{4}(0.1) + \frac{5}{32}(0.1)^2 - \frac{15}{128}(0.1)^3 \approx 0.976$. The fifth term is $\frac{195}{2048}(0.1)^4 \approx 0.000\,009\,5$, which does not

affect the third decimal place of the sum, so we have $1/\sqrt[4]{1.1} \approx 0.976$. (Note that the third decimal place of the sum of the

first three terms is affected by the fourth term, so we need to use more than three terms for the sum.)

47. $\cos x \overset{(16)}{=} \sum\limits_{n=0}^{\infty} (-1)^n \dfrac{x^{2n}}{(2n)!} \;\Rightarrow\; \cos(x^3) = \sum\limits_{n=0}^{\infty} (-1)^n \dfrac{(x^3)^{2n}}{(2n)!} = \sum\limits_{n=0}^{\infty} (-1)^n \dfrac{x^{6n}}{(2n)!} \;\Rightarrow$

$x\cos(x^3) = \sum\limits_{n=0}^{\infty} (-1)^n \dfrac{x^{6n+1}}{(2n)!} \;\Rightarrow\; \int x\cos(x^3)\,dx = C + \sum\limits_{n=0}^{\infty} (-1)^n \dfrac{x^{6n+2}}{(6n+2)(2n)!}$, with $R = \infty$.

48. $e^x \overset{(11)}{=} \sum\limits_{n=0}^{\infty} \dfrac{x^n}{n!} \;\Rightarrow\; e^x - 1 = \sum\limits_{n=1}^{\infty} \dfrac{x^n}{n!} \;\Rightarrow\; \dfrac{e^x - 1}{x} = \sum\limits_{n=1}^{\infty} \dfrac{x^{n-1}}{n!} \;\Rightarrow\; \int \dfrac{e^x - 1}{x}\,dx = C + \sum\limits_{n=1}^{\infty} \dfrac{x^n}{n \cdot n!}$,

with $R = \infty$.

49. $\cos x \overset{(16)}{=} \sum\limits_{n=0}^{\infty} (-1)^n \dfrac{x^{2n}}{(2n)!} \;\Rightarrow\; \cos x - 1 = \sum\limits_{n=1}^{\infty} (-1)^n \dfrac{x^{2n}}{(2n)!} \;\Rightarrow\; \dfrac{\cos x - 1}{x} = \sum\limits_{n=1}^{\infty} (-1)^n \dfrac{x^{2n-1}}{(2n)!} \;\Rightarrow$

$\int \dfrac{\cos x - 1}{x}\,dx = C + \sum\limits_{n=1}^{\infty} (-1)^n \dfrac{x^{2n}}{2n \cdot (2n)!}$, with $R = \infty$.

50. $\arctan x = \sum\limits_{n=0}^{\infty} (-1)^n \dfrac{x^{2n+1}}{2n+1} \;\Rightarrow\; \arctan(x^2) = \sum\limits_{n=0}^{\infty} (-1)^n \dfrac{(x^2)^{2n+1}}{2n+1} = \sum\limits_{n=0}^{\infty} (-1)^n \dfrac{x^{4n+2}}{2n+1} \;\Rightarrow$

$\int \arctan(x^2)\,dx = C + \sum\limits_{n=0}^{\infty} (-1)^n \dfrac{x^{4n+3}}{(2n+1)(4n+3)}$, with $R = 1$.

51. By Exercise 47, $\int x\cos(x^3)\,dx = C + \sum\limits_{n=0}^{\infty} (-1)^n \dfrac{x^{6n+2}}{(6n+2)(2n)!}$, so

$\int_0^1 x\cos(x^3)\,dx = \left[\sum\limits_{n=0}^{\infty} (-1)^n \dfrac{x^{6n+2}}{(6n+2)(2n)!} \right]_0^1 = \sum\limits_{n=0}^{\infty} \dfrac{(-1)^n}{(6n+2)(2n)!} = \dfrac{1}{2} - \dfrac{1}{8 \cdot 2!} + \dfrac{1}{14 \cdot 4!} - \dfrac{1}{20 \cdot 6!} + \cdots$, but

$\dfrac{1}{20 \cdot 6!} = \dfrac{1}{14{,}400} \approx 0.000\,069$, so $\int_0^1 x\cos(x^3)\,dx \approx \dfrac{1}{2} - \dfrac{1}{16} + \dfrac{1}{336} \approx 0.440$ (correct to three decimal places) by the

Alternating Series Estimation Theorem.

52. From the table of Maclaurin series in this section, we see that

$\tan^{-1} x = \sum\limits_{n=0}^{\infty} (-1)^n \dfrac{x^{2n+1}}{2n+1}$ for x in $[-1, 1]$ and $\sin x = \sum\limits_{n=0}^{\infty} (-1)^n \dfrac{x^{2n+1}}{(2n+1)!}$ for all real numbers x, so

$\tan^{-1}(x^3) + \sin(x^3) = \sum\limits_{n=0}^{\infty} (-1)^n \dfrac{x^{6n+3}}{2n+1} + \sum\limits_{n=0}^{\infty} (-1)^n \dfrac{x^{6n+3}}{(2n+1)!}$ for x^3 in $[-1, 1] \;\Leftrightarrow\; x$ in $[-1, 1]$. Thus,

$I = \displaystyle\int_0^{0.2} [\tan^{-1}(x^3) + \sin(x^3)]\,dx = \int_0^{0.2} \sum\limits_{n=0}^{\infty} (-1)^n x^{6n+3} \left(\dfrac{1}{2n+1} + \dfrac{1}{(2n+1)!} \right) dx$

$= \left[\sum\limits_{n=0}^{\infty} (-1)^n \dfrac{x^{6n+4}}{6n+4} \left(\dfrac{1}{2n+1} + \dfrac{1}{(2n+1)!} \right) \right]_0^{0.2} = \sum\limits_{n=0}^{\infty} (-1)^n \dfrac{(0.2)^{6n+4}}{6n+4} \left(\dfrac{1}{2n+1} + \dfrac{1}{(2n+1)!} \right)$

$= \dfrac{(0.2)^4}{4} (1 + 1) - \dfrac{(0.2)^{10}}{10} \left(\dfrac{1}{3} + \dfrac{1}{3!} \right) + \cdots$

But $\dfrac{(0.2)^{10}}{10} \left(\dfrac{1}{3} + \dfrac{1}{3!} \right) = \dfrac{(0.2)^{10}}{20} = 5.12 \times 10^{-9}$, so by the Alternating Series Estimation Theorem,

$I \approx \dfrac{(0.2)^4}{2} = 0.000\,80$ (correct to five decimal places). [Actually, the value is $0.000\,800\,0$, correct to seven decimal places.]

53. $\sqrt{1+x^4} = (1+x^4)^{1/2} = \sum\limits_{n=0}^{\infty} \binom{1/2}{n}(x^4)^n$, so $\int \sqrt{1+x^4}\,dx = C + \sum\limits_{n=0}^{\infty} \binom{1/2}{n}\dfrac{x^{4n+1}}{4n+1}$ and hence, since $0.4 < 1$,

we have

$$I = \int_0^{0.4} \sqrt{1+x^4}\,dx = \sum_{n=0}^{\infty} \binom{1/2}{n}\frac{(0.4)^{4n+1}}{4n+1}$$

$$= (1)\frac{(0.4)^1}{0!} + \frac{\frac{1}{2}}{1!}(0.4)^5 + \frac{\frac{1}{2}\left(-\frac{1}{2}\right)}{2!}\frac{(0.4)^9}{9} + \frac{\frac{1}{2}\left(-\frac{1}{2}\right)\left(-\frac{3}{2}\right)}{3!}\frac{(0.4)^{13}}{13} + \frac{\frac{1}{2}\left(-\frac{1}{2}\right)\left(-\frac{3}{2}\right)\left(-\frac{5}{2}\right)}{4!}\frac{(0.4)^{17}}{17} + \cdots$$

$$= 0.4 + \frac{(0.4)^5}{10} - \frac{(0.4)^9}{72} + \frac{(0.4)^{13}}{208} - \frac{5(0.4)^{17}}{2176} + \cdots$$

Now $\dfrac{(0.4)^9}{72} \approx 3.6 \times 10^{-6} < 5 \times 10^{-6}$, so by the Alternating Series Estimation Theorem, $I \approx 0.4 + \dfrac{(0.4)^5}{10} \approx 0.40102$

(correct to five decimal places).

54. $\int_0^{0.5} x^2 e^{-x^2}\,dx = \int_0^{0.5} \sum\limits_{n=0}^{\infty} \dfrac{(-1)^n x^{2n+2}}{n!}\,dx = \sum\limits_{n=0}^{\infty}\left[\dfrac{(-1)^n x^{2n+3}}{n!(2n+3)}\right]_0^{1/2} = \sum\limits_{n=0}^{\infty}\dfrac{(-1)^n}{n!(2n+3)2^{2n+3}}$ and since the term

with $n=2$ is $\dfrac{1}{1792} < 0.001$, we use $\sum\limits_{n=0}^{1}\dfrac{(-1)^n}{n!(2n+3)2^{2n+3}} = \dfrac{1}{24} - \dfrac{1}{160} \approx 0.0354.$

55. $\lim\limits_{x\to 0}\dfrac{x - \tan^{-1}x}{x^3} = \lim\limits_{x\to 0}\dfrac{x - \left(x - \frac{1}{3}x^3 + \frac{1}{5}x^5 - \frac{1}{7}x^7 + \cdots\right)}{x^3} = \lim\limits_{x\to 0}\dfrac{\frac{1}{3}x^3 - \frac{1}{5}x^5 + \frac{1}{7}x^7 - \cdots}{x^3}$

$$= \lim\limits_{x\to 0}\left(\tfrac{1}{3} - \tfrac{1}{5}x^2 + \tfrac{1}{7}x^4 - \cdots\right) = \tfrac{1}{3}$$

since power series are continuous functions.

56. $\lim\limits_{x\to 0}\dfrac{1 - \cos x}{1 + x - e^x} = \lim\limits_{x\to 0}\dfrac{1 - \left(1 - \frac{1}{2!}x^2 + \frac{1}{4!}x^4 - \frac{1}{6!}x^6 + \cdots\right)}{1 + x - \left(1 + x + \frac{1}{2!}x^2 + \frac{1}{3!}x^3 + \frac{1}{4!}x^4 + \frac{1}{5!}x^5 + \frac{1}{6!}x^6 + \cdots\right)}$

$$= \lim\limits_{x\to 0}\dfrac{\frac{1}{2!}x^2 - \frac{1}{4!}x^4 + \frac{1}{6!}x^6 - \cdots}{-\frac{1}{2!}x^2 - \frac{1}{3!}x^3 - \frac{1}{4!}x^4 - \frac{1}{5!}x^5 - \frac{1}{6!}x^6 - \cdots}$$

$$= \lim\limits_{x\to 0}\dfrac{\frac{1}{2!} - \frac{1}{4!}x^2 + \frac{1}{6!}x^4 - \cdots}{-\frac{1}{2!} - \frac{1}{3!}x - \frac{1}{4!}x^2 - \frac{1}{5!}x^3 - \frac{1}{6!}x^4 - \cdots} = \dfrac{\frac{1}{2} - 0}{-\frac{1}{2} - 0} = -1$$

since power series are continuous functions.

57. $\lim\limits_{x\to 0}\dfrac{\sin x - x + \frac{1}{6}x^3}{x^5} = \lim\limits_{x\to 0}\dfrac{\left(x - \frac{1}{3!}x^3 + \frac{1}{5!}x^5 - \frac{1}{7!}x^7 + \cdots\right) - x + \frac{1}{6}x^3}{x^5}$

$$= \lim\limits_{x\to 0}\dfrac{\frac{1}{5!}x^5 - \frac{1}{7!}x^7 + \cdots}{x^5} = \lim\limits_{x\to 0}\left(\dfrac{1}{5!} - \dfrac{x^2}{7!} + \dfrac{x^4}{9!} - \cdots\right) = \dfrac{1}{5!} = \dfrac{1}{120}$$

since power series are continuous functions.

58. $\lim\limits_{x\to 0}\dfrac{\tan x - x}{x^3} = \lim\limits_{x\to 0}\dfrac{\left(x + \frac{1}{3}x^3 + \frac{2}{15}x^5 + \cdots\right) - x}{x^3} = \lim\limits_{x\to 0}\dfrac{\frac{1}{3}x^3 + \frac{2}{15}x^5 + \cdots}{x^3} = \lim\limits_{x\to 0}\left(\tfrac{1}{3} + \tfrac{2}{15}x^2 + \cdots\right) = \tfrac{1}{3}$

since power series are continuous functions.

59. From Equation 11, we have $e^{-x^2} = 1 - \frac{x^2}{1!} + \frac{x^4}{2!} - \frac{x^6}{3!} + \cdots$ and we know that $\cos x = 1 - \frac{x^2}{2!} + \frac{x^4}{4!} - \cdots$ from

Equation 16. Therefore, $e^{-x^2}\cos x = \left(1 - x^2 + \frac{1}{2}x^4 - \cdots\right)\left(1 - \frac{1}{2}x^2 + \frac{1}{24}x^4 - \cdots\right)$. Writing only the terms with

degree ≤ 4, we get $e^{-x^2}\cos x = 1 - \frac{1}{2}x^2 + \frac{1}{24}x^4 - x^2 + \frac{1}{2}x^4 + \frac{1}{2}x^4 + \cdots = 1 - \frac{3}{2}x^2 + \frac{25}{24}x^4 + \cdots$.

60. $\sec x = \frac{1}{\cos x} \overset{(16)}{=} \frac{1}{1 - \frac{1}{2}x^2 + \frac{1}{24}x^4 - \cdots}$.

$$
\begin{array}{r}
1 + \frac{1}{2}x^2 + \frac{5}{24}x^4 + \cdots \\
\hline
1 - \frac{1}{2}x^2 + \frac{1}{24}x^4 - \cdots \ \Big|\ 1 \\
1 - \frac{1}{2}x^2 + \frac{1}{24}x^4 - \cdots \\
\hline
\frac{1}{2}x^2 - \frac{1}{24}x^4 + \cdots \\
\frac{1}{2}x^2 - \frac{1}{4}x^4 + \cdots \\
\hline
\frac{5}{24}x^4 + \cdots \\
\frac{5}{24}x^4 + \cdots \\
\hline
\cdots
\end{array}
$$

From the long division above, $\sec x = 1 + \frac{1}{2}x^2 + \frac{5}{24}x^4 + \cdots$.

61. $\frac{x}{\sin x} \overset{(15)}{=} \frac{x}{x - \frac{1}{6}x^3 + \frac{1}{120}x^5 - \cdots}$.

$$
\begin{array}{r}
1 + \frac{1}{6}x^2 + \frac{7}{360}x^4 + \cdots \\
\hline
x - \frac{1}{6}x^3 + \frac{1}{120}x^5 - \cdots \ \Big|\ x \\
x - \frac{1}{6}x^3 + \frac{1}{120}x^5 - \cdots \\
\hline
\frac{1}{6}x^3 - \frac{1}{120}x^5 + \cdots \\
\frac{1}{6}x^3 - \frac{1}{36}x^5 + \cdots \\
\hline
\frac{7}{360}x^5 + \cdots \\
\frac{7}{360}x^5 + \cdots \\
\hline
\cdots
\end{array}
$$

From the long division above, $\dfrac{x}{\sin x} = 1 + \frac{1}{6}x^2 + \frac{7}{360}x^4 + \cdots$.

62. From Example 6 in Section 12.9 [ET 11.9], we have $\ln(1-x) = -x - \frac{1}{2}x^2 - \frac{1}{3}x^3 - \cdots$, $|x| < 1$. Therefore,

$$e^x \ln(1-x) = \left(1 + x + \frac{1}{2}x^2 + \cdots\right)\left(-x - \frac{1}{2}x^2 - \frac{1}{3}x^3 - \cdots\right)$$

$$= -x - \frac{1}{2}x^2 - \frac{1}{3}x^3 - x^2 - \frac{1}{2}x^3 - \frac{1}{2}x^3 - \cdots = -x - \frac{3}{2}x^2 - \frac{4}{3}x^3 - \cdots, \ |x| < 1$$

63. $\displaystyle\sum_{n=0}^{\infty} (-1)^n \frac{x^{4n}}{n!} = \sum_{n=0}^{\infty} \frac{\left(-x^4\right)^n}{n!} = e^{-x^4}$, by (11).

64. $\displaystyle\sum_{n=0}^{\infty} \frac{(-1)^n\, \pi^{2n}}{6^{2n}(2n)!} = \sum_{n=0}^{\infty} (-1)^n \frac{\left(\frac{\pi}{6}\right)^{2n}}{(2n)!} = \cos\frac{\pi}{6} = \frac{\sqrt{3}}{2}$, by (16).

65. $\displaystyle\sum_{n=0}^{\infty} \frac{(-1)^n \pi^{2n+1}}{4^{2n+1}(2n+1)!} = \sum_{n=0}^{\infty} \frac{(-1)^n \left(\frac{\pi}{4}\right)^{2n+1}}{(2n+1)!} = \sin\frac{\pi}{4} = \frac{1}{\sqrt{2}}$, by (15).

66. $\displaystyle\sum_{n=0}^{\infty} \frac{3^n}{5^n\, n!} = \sum_{n=0}^{\infty} \frac{(3/5)^n}{n!} = e^{3/5}$, by (11).

67. $3 + \dfrac{9}{2!} + \dfrac{27}{3!} + \dfrac{81}{4!} + \cdots = \dfrac{3^1}{1!} + \dfrac{3^2}{2!} + \dfrac{3^3}{3!} + \dfrac{3^4}{4!} + \cdots = \displaystyle\sum_{n=1}^{\infty} \dfrac{3^n}{n!} = \sum_{n=0}^{\infty} \dfrac{3^n}{n!} - 1 = e^3 - 1$, by (11).

68. $1 - \ln 2 + \dfrac{(\ln 2)^2}{2!} - \dfrac{(\ln 2)^3}{3!} + \cdots = \displaystyle\sum_{n=0}^{\infty} \dfrac{(-\ln 2)^n}{n!} = e^{-\ln 2} = \left(e^{\ln 2}\right)^{-1} = 2^{-1} = \frac{1}{2}$, by (11).

69. Assume that $|f'''(x)| \le M$, so $f'''(x) \le M$ for $a \le x \le a+d$. Now $\int_a^x f'''(t)\,dt \le \int_a^x M\,dt \quad\Rightarrow$

$f''(x) - f''(a) \le M(x-a) \quad\Rightarrow\quad f''(x) \le f''(a) + M(x-a)$. Thus, $\int_a^x f''(t)\,dt \le \int_a^x [f''(a) + M(t-a)]\,dt \quad\Rightarrow$

$f'(x) - f'(a) \le f''(a)(x-a) + \frac{1}{2}M(x-a)^2 \quad\Rightarrow\quad f'(x) \le f'(a) + f''(a)(x-a) + \frac{1}{2}M(x-a)^2 \quad\Rightarrow$

$\int_a^x f'(t)\,dt \le \int_a^x \left[f'(a) + f''(a)(t-a) + \frac{1}{2}M(t-a)^2\right]dt \quad\Rightarrow$

$f(x) - f(a) \le f'(a)(x-a) + \frac{1}{2}f''(a)(x-a)^2 + \frac{1}{6}M(x-a)^3$. So

$f(x) - f(a) - f'(a)(x-a) - \frac{1}{2}f''(a)(x-a)^2 \le \frac{1}{6}M(x-a)^3$. But

$R_2(x) = f(x) - T_2(x) = f(x) - f(a) - f'(a)(x-a) - \frac{1}{2}f''(a)(x-a)^2$, so $R_2(x) \le \frac{1}{6}M(x-a)^3$.

A similar argument using $f'''(x) \ge -M$ shows that $R_2(x) \ge -\frac{1}{6}M(x-a)^3$. So $|R_2(x_2)| \le \frac{1}{6}M\,|x-a|^3$.

Although we have assumed that $x > a$, a similar calculation shows that this inequality is also true if $x < a$.

70. (a) $f(x) = \begin{cases} e^{-1/x^2} & \text{if } x \ne 0 \\ 0 & \text{if } x = 0 \end{cases}$ so $f'(0) = \displaystyle\lim_{x\to 0} \frac{f(x) - f(0)}{x - 0} = \lim_{x\to 0} \frac{e^{-1/x^2}}{x} = \lim_{x\to 0} \frac{1/x}{e^{1/x^2}} = \lim_{x\to 0} \frac{x}{2e^{1/x^2}} = 0$

(using l'Hospital's Rule and simplifying in the penultimate step). Similarly, we can use the definition of the derivative and

l'Hospital's Rule to show that $f''(0) = 0$, $f^{(3)}(0) = 0$, ..., $f^{(n)}(0) = 0$, so that the Maclaurin series for f consists

entirely of zero terms. But since $f(x) \ne 0$ except for $x = 0$, we see that f cannot equal its Maclaurin series except

at $x = 0$.

(b)

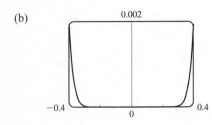

From the graph, it seems that the function is extremely flat at the origin.
In fact, it could be said to be "infinitely flat" at $x = 0$, since all of its
derivatives are 0 there.

71. (a) $g(x) = \sum_{n=0}^{\infty} \binom{k}{n} x^n \Rightarrow g'(x) = \sum_{n=1}^{\infty} \binom{k}{n} nx^{n-1}$, so

$$(1+x)g'(x) = (1+x)\sum_{n=1}^{\infty} \binom{k}{n} nx^{n-1} = \sum_{n=1}^{\infty} \binom{k}{n} nx^{n-1} + \sum_{n=1}^{\infty} \binom{k}{n} nx^n$$

$$= \sum_{n=0}^{\infty} \binom{k}{n+1} (n+1)x^n + \sum_{n=0}^{\infty} \binom{k}{n} nx^n \qquad \begin{bmatrix} \text{Replace } n \text{ with } n+1 \\ \text{in the first series} \end{bmatrix}$$

$$= \sum_{n=0}^{\infty} (n+1)\frac{k(k-1)(k-2)\cdots(k-n+1)(k-n)}{(n+1)!} x^n + \sum_{n=0}^{\infty} \left[(n)\frac{k(k-1)(k-2)\cdots(k-n+1)}{n!}\right] x^n$$

$$= \sum_{n=0}^{\infty} \frac{(n+1)k(k-1)(k-2)\cdots(k-n+1)}{(n+1)!} [(k-n)+n] x^n$$

$$= k\sum_{n=0}^{\infty} \frac{k(k-1)(k-2)\cdots(k-n+1)}{n!} x^n = k\sum_{n=0}^{\infty} \binom{k}{n} x^n = kg(x)$$

Thus, $g'(x) = \dfrac{kg(x)}{1+x}$.

(b) $h(x) = (1+x)^{-k} g(x) \Rightarrow$

$$h'(x) = -k(1+x)^{-k-1}g(x) + (1+x)^{-k} g'(x) \qquad \text{[Product Rule]}$$

$$= -k(1+x)^{-k-1}g(x) + (1+x)^{-k} \frac{kg(x)}{1+x} \qquad \text{[from part (a)]}$$

$$= -k(1+x)^{-k-1}g(x) + k(1+x)^{-k-1}g(x) = 0$$

(c) From part (b) we see that $h(x)$ must be constant for $x \in (-1, 1)$, so $h(x) = h(0) = 1$ for $x \in (-1, 1)$.

Thus, $h(x) = 1 = (1+x)^{-k} g(x) \Leftrightarrow g(x) = (1+x)^k$ for $x \in (-1, 1)$.

72. By Exercise 25, $\sqrt{1+x} = (1+x)^{1/2} = 1 + \dfrac{x}{2} + \sum_{n=2}^{\infty} \dfrac{(-1)^{n-1} 1 \cdot 3 \cdot 5 \cdots (2n-3)x^n}{2^n \cdot n!}$, so

$$\left(1-x^2\right)^{1/2} = 1 - \frac{1}{2}x^2 - \sum_{n=2}^{\infty} \frac{1 \cdot 3 \cdot 5 \cdots (2n-3)}{2^n \cdot n!} x^{2n} \text{ and}$$

$$\sqrt{1 - e^2 \sin^2 \theta} = 1 - \frac{1}{2}e^2 \sin^2 \theta - \sum_{n=2}^{\infty} \frac{1 \cdot 3 \cdot 5 \cdots (2n-3)}{2^n \cdot n!} e^{2n} \sin^{2n} \theta. \text{ Thus,}$$

$$L = 4a \int_0^{\pi/2} \sqrt{1 - e^2 \sin^2 \theta}\, d\theta = 4a \int_0^{\pi/2} \left(1 - \frac{1}{2}e^2 \sin^2 \theta - \sum_{n=2}^{\infty} \frac{1 \cdot 3 \cdot 5 \cdots (2n-3)}{2^n \cdot n!} e^{2n} \sin^{2n} \theta \right) d\theta$$

$$= 4a \left[\frac{\pi}{2} - \frac{e^2}{2}S_1 - \sum_{n=2}^{\infty} \frac{1 \cdot 3 \cdot 5 \cdots (2n-3)}{n!} \left(\frac{e^2}{2}\right)^n S_n \right]$$

where $S_n = \displaystyle\int_0^{\pi/2} \sin^{2n}\theta\, d\theta = \dfrac{1 \cdot 3 \cdot 5 \cdots (2n-1)}{2 \cdot 4 \cdot 6 \cdots 2n} \dfrac{\pi}{2}$ by Exercise 46 of Section 8.1 [ET 7.1].

[continued]

$$L = 4a\left(\frac{\pi}{2}\right)\left[1 - \frac{e^2}{2}\cdot\frac{1}{2} - \sum_{n=2}^{\infty}\frac{1\cdot3\cdot5\cdot\cdots\cdot(2n-3)}{n!}\left(\frac{e^2}{2}\right)^n\frac{1\cdot3\cdot5\cdot\cdots\cdot(2n-1)}{2\cdot4\cdot6\cdot\cdots\cdot2n}\right]$$

$$= 2\pi a\left[1 - \frac{e^2}{4} - \sum_{n=2}^{\infty}\frac{e^{2n}}{2^n}\cdot\frac{1^2\cdot3^2\cdot5^2\cdot\cdots\cdot(2n-3)^2(2n-1)}{n!\cdot2^n\cdot n!}\right]$$

$$= 2\pi a\left[1 - \frac{e^2}{4} - \sum_{n=2}^{\infty}\frac{e^{2n}}{4^n}\left(\frac{1\cdot3\cdot\cdots\cdot(2n-3)}{n!}\right)^2(2n-1)\right]$$

$$= 2\pi a\left[1 - \frac{e^2}{4} - \frac{3e^4}{64} - \frac{5e^6}{256} - \cdots\right] = \frac{\pi a}{128}(256 - 64e^2 - 12e^4 - 5e^6 - \cdots)$$

LABORATORY PROJECT   An Elusive Limit

1. $f(x) = \dfrac{n(x)}{d(x)} = \dfrac{\sin(\tan x) - \tan(\sin x)}{\arcsin(\arctan x) - \arctan(\arcsin x)}$

x	$f(x)$
1	1.1838
0.1	0.9821
0.01	2.0000
0.001	3.3333
0.0001	3.3333

The table of function values were obtained using Maple with 10 digits of precision. The results of this project will vary depending on the CAS and precision level. It appears that as $x \to 0^+$, $f(x) \to \frac{10}{3}$. Since f is an even function, we have $f(x) \to \frac{10}{3}$ as $x \to 0$.

2. The graph is inconclusive about the limit of f as $x \to 0$.

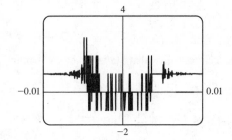

3. The limit has the indeterminate form $\frac{0}{0}$. Applying l'Hospital's Rule, we obtain the form $\frac{0}{0}$ six times. Finally, on the seventh application we obtain $\lim\limits_{x\to0}\dfrac{n^{(7)}(x)}{d^{(7)}(x)} = \dfrac{-168}{-168} = 1$.

4. $\lim\limits_{x\to0}f(x) = \lim\limits_{x\to0}\dfrac{n(x)}{d(x)} \overset{\text{CAS}}{=} \lim\limits_{x\to0}\dfrac{-\frac{1}{30}x^7 - \frac{29}{756}x^9 + \cdots}{-\frac{1}{30}x^7 + \frac{13}{756}x^9 + \cdots}$

$= \lim\limits_{x\to0}\dfrac{\left(-\frac{1}{30}x^7 - \frac{29}{756}x^9 + \cdots\right)/x^7}{\left(-\frac{1}{30}x^7 + \frac{13}{756}x^9 + \cdots\right)/x^7} = \lim\limits_{x\to0}\dfrac{-\frac{1}{30} - \frac{29}{756}x^2 + \cdots}{-\frac{1}{30} + \frac{13}{756}x^2 + \cdots} = \dfrac{-\frac{1}{30}}{-\frac{1}{30}} = 1$

Note that $n^{(7)}(x) = d^{(7)}(x) = -\frac{7!}{30} = -\frac{5040}{30} = -168$, which agrees with the result in Problem 3.

5. The limit command gives the result that $\lim\limits_{x\to0}f(x) = 1$.

6. The strange results (with only 10 digits of precision) must be due to the fact that the terms being subtracted in the numerator and denominator are very close in value when $|x|$ is small. Thus, the differences are imprecise (have few correct digits).

12.11 Applications of Taylor Polynomials

1. (a)

n	$f^{(n)}(x)$	$f^{(n)}(0)$	$T_n(x)$
0	$\cos x$	1	1
1	$-\sin x$	0	1
2	$-\cos x$	−1	$1 - \frac{1}{2}x^2$
3	$\sin x$	0	$1 - \frac{1}{2}x^2$
4	$\cos x$	1	$1 - \frac{1}{2}x^2 + \frac{1}{24}x^4$
5	$-\sin x$	0	$1 - \frac{1}{2}x^2 + \frac{1}{24}x^4$
6	$-\cos x$	−1	$1 - \frac{1}{2}x^2 + \frac{1}{24}x^4 - \frac{1}{720}x^6$

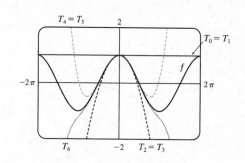

(b)

x	f	$T_0 = T_1$	$T_2 = T_3$	$T_4 = T_5$	T_6
$\frac{\pi}{4}$	0.7071	1	0.6916	0.7074	0.7071
$\frac{\pi}{2}$	0	1	−0.2337	0.0200	−0.0009
π	−1	1	−3.9348	0.1239	−1.2114

(c) As n increases, $T_n(x)$ is a good approximation to $f(x)$ on a larger and larger interval.

2. (a)

n	$f^{(n)}(x)$	$f^{(n)}(1)$	$T_n(x)$
0	x^{-1}	1	1
1	$-x^{-2}$	−1	$1 - (x-1) = 2 - x$
2	$2x^{-3}$	2	$1 - (x-1) + (x-1)^2 = x^2 - 3x + 3$
3	$-6x^{-4}$	−6	$1 - (x-1) + (x-1)^2 - (x-1)^3 = -x^3 + 4x^2 - 6x + 4$

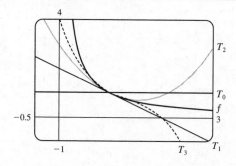

(b)

x	f	T_0	T_1	T_2	T_3
0.9	$1.\overline{1}$	1	1.1	1.11	1.111
1.3	0.7692	1	0.7	0.79	0.763

(c) As n increases, $T_n(x)$ is a good approximation to $f(x)$ on a larger and larger interval.

3.

n	$f^{(n)}(x)$	$f^{(n)}(2)$
0	$1/x$	$\frac{1}{2}$
1	$-1/x^2$	$-\frac{1}{4}$
2	$2/x^3$	$\frac{1}{4}$
3	$-6/x^4$	$-\frac{3}{8}$

$T_3(x) = \displaystyle\sum_{n=0}^{3} \frac{f^{(n)}(2)}{n!} (x-2)^n$

$= \dfrac{\frac{1}{2}}{0!} - \dfrac{\frac{1}{4}}{1!}(x-2) + \dfrac{\frac{1}{4}}{2!}(x-2)^2 - \dfrac{\frac{3}{8}}{3!}(x-2)^3$

$= \frac{1}{2} - \frac{1}{4}(x-2) + \frac{1}{8}(x-2)^2 - \frac{1}{16}(x-2)^3$

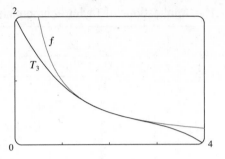

4.

n	$f^{(n)}(x)$	$f^{(n)}(0)$
0	$x + e^{-x}$	1
1	$1 - e^{-x}$	0
2	e^{-x}	1
3	$-e^{-x}$	-1

$T_3(x) = \displaystyle\sum_{n=0}^{3} \frac{f^{(n)}(0)}{n!} (x-0)^n$

$= \dfrac{1}{0!} + \dfrac{0}{1!}x + \dfrac{1}{2!}x^2 - \dfrac{1}{3!}x^3 = 1 + \frac{1}{2}x^2 - \frac{1}{6}x^3$

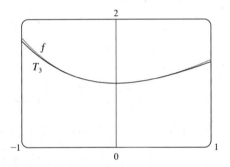

5.

n	$f^{(n)}(x)$	$f^{(n)}(\pi/2)$
0	$\cos x$	0
1	$-\sin x$	-1
2	$-\cos x$	0
3	$\sin x$	1

$T_3(x) = \displaystyle\sum_{n=0}^{3} \frac{f^{(n)}(\pi/2)}{n!} \left(x - \frac{\pi}{2}\right)^n$

$= -\left(x - \frac{\pi}{2}\right) + \frac{1}{6}\left(x - \frac{\pi}{2}\right)^3$

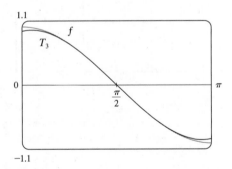

6.

n	$f^{(n)}(x)$	$f^{(n)}(0)$
0	$e^{-x}\sin x$	0
1	$e^{-x}(\cos x - \sin x)$	1
2	$-2e^{-x}\cos x$	-2
3	$2e^{-x}(\cos x + \sin x)$	2

$T_3(x) = \displaystyle\sum_{n=0}^{3} \frac{f^{(n)}(0)}{n!} x^n = x - x^2 + \frac{1}{3}x^3$

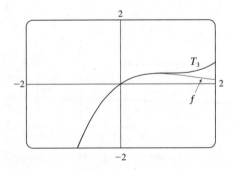

7.

n	$f^{(n)}(x)$	$f^{(n)}(0)$
0	$\arcsin x$	0
1	$\dfrac{1}{\sqrt{1-x^2}}$	1
2	$\dfrac{x}{(1-x^2)^{3/2}}$	0
3	$\dfrac{2x^2+1}{(1-x^2)^{5/2}}$	1

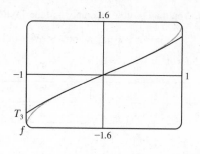

$$T_3(x) = \sum_{n=0}^{3} \frac{f^{(n)}(0)}{n!}\, x^n = x + \frac{x^3}{6}$$

8.

n	$f^{(n)}(x)$	$f^{(n)}(1)$
0	$\dfrac{\ln x}{x}$	0
1	$\dfrac{1-\ln x}{x^2}$	1
2	$\dfrac{-3+2\ln x}{x^3}$	-3
3	$\dfrac{11-6\ln x}{x^4}$	11

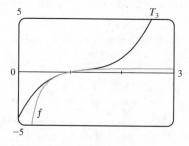

$$T_3(x) = \sum_{n=0}^{3} \frac{f^{(n)}(1)}{n!}\,(x-1)^n = (x-1) - \tfrac{3}{2}(x-1)^2 + \tfrac{11}{6}(x-1)^3$$

9.

n	$f^{(n)}(x)$	$f^{(n)}(0)$
0	xe^{-2x}	0
1	$(1-2x)e^{-2x}$	1
2	$4(x-1)e^{-2x}$	-4
3	$4(3-2x)e^{-2x}$	12

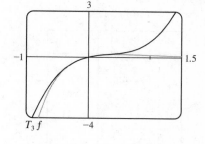

$$T_3(x) = \sum_{n=0}^{3} \frac{f^{(n)}(0)}{n!}\, x^n = \tfrac{0}{1}\cdot 1 + \tfrac{1}{1}x^1 + \tfrac{-4}{2}x^2 + \tfrac{12}{6}x^3 = x - 2x^2 + 2x^3$$

10.

n	$f^{(n)}(x)$	$f^{(n)}(1)$
0	$\tan^{-1} x$	$\dfrac{\pi}{4}$
1	$\dfrac{1}{1+x^2}$	$\dfrac{1}{2}$
2	$\dfrac{-2x}{(1+x^2)^2}$	$-\dfrac{1}{2}$
3	$\dfrac{6x^2-2}{(1+x^2)^3}$	$\dfrac{1}{2}$

$$T_3(x) = \sum_{n=0}^{3} \frac{f^{(n)}(1)}{n!}\,(x-1)^n = \frac{\pi}{4} + \frac{1/2}{1}(x-1)^1 + \frac{-1/2}{2}(x-1)^2 + \frac{1/2}{6}(x-1)^3$$

$$= \tfrac{\pi}{4} + \tfrac{1}{2}(x-1) - \tfrac{1}{4}(x-1)^2 + \tfrac{1}{12}(x-1)^3$$

11. You may be able to simply find the Taylor polynomials for

$f(x) = \cot x$ using your CAS. We will list the values of $f^{(n)}(\pi/4)$

for $n = 0$ to $n = 5$.

n	0	1	2	3	4	5
$f^{(n)}(\pi/4)$	1	-2	4	-16	80	-512

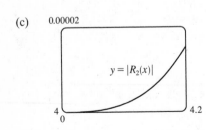

$$T_5(x) = \sum_{n=0}^{5} \frac{f^{(n)}(\pi/4)}{n!} \left(x - \tfrac{\pi}{4}\right)^n$$

$$= 1 - 2\left(x - \tfrac{\pi}{4}\right) + 2\left(x - \tfrac{\pi}{4}\right)^2 - \tfrac{8}{3}\left(x - \tfrac{\pi}{4}\right)^3 + \tfrac{10}{3}\left(x - \tfrac{\pi}{4}\right)^4 - \tfrac{64}{15}\left(x - \tfrac{\pi}{4}\right)^5$$

For $n = 2$ to $n = 5$, $T_n(x)$ is the polynomial consisting of all the terms up to and including the $\left(x - \tfrac{\pi}{4}\right)^n$ term.

12. You may be able to simply find the Taylor polynomials for

$f(x) = \sqrt[3]{1 + x^2}$ using your CAS. We will list the values of $f^{(n)}(0)$

for $n = 0$ to $n = 5$.

n	0	1	2	3	4	5
$f^{(n)}(0)$	1	0	$\tfrac{2}{3}$	0	$-\tfrac{8}{3}$	0

$$T_5(x) = \sum_{n=0}^{5} \frac{f^{(n)}(0)}{n!} x^n = 1 + \tfrac{1}{3}x^2 - \tfrac{1}{9}x^4$$

For $n = 2$ to $n = 5$, $T_n(x)$ is the polynomial consisting of all the terms up to and including the x^n term. Note that $T_2 = T_3$ and $T_4 = T_5$.

13.

n	$f^{(n)}(x)$	$f^{(n)}(4)$
0	$\sqrt{x}$	2
1	$\tfrac{1}{2}x^{-1/2}$	$\tfrac{1}{4}$
2	$-\tfrac{1}{4}x^{-3/2}$	$-\tfrac{1}{32}$
3	$\tfrac{3}{8}x^{-5/2}$	

(a) $f(x) = \sqrt{x} \approx T_2(x) = 2 + \dfrac{1}{4}(x - 4) - \dfrac{1/32}{2!}(x - 4)^2$

$$= 2 + \tfrac{1}{4}(x - 4) - \tfrac{1}{64}(x - 4)^2$$

(b) $|R_2(x)| \le \dfrac{M}{3!} |x - 4|^3$, where $|f'''(x)| \le M$. Now $4 \le x \le 4.2 \Rightarrow$

$|x - 4| \le 0.2 \Rightarrow |x - 4|^3 \le 0.008$. Since $f'''(x)$ is decreasing

on $[4, 4.2]$, we can take $M = |f'''(4)| = \tfrac{3}{8}4^{-5/2} = \tfrac{3}{256}$, so

$$|R_2(x)| \le \frac{3/256}{6}(0.008) = \frac{0.008}{512} = 0.000\,015\,625.$$

(c) 0.00002

$y = |R_2(x)|$

4 ⌞____⌟ 4.2
0

From the graph of $|R_2(x)| = |\sqrt{x} - T_2(x)|$, it seems that the

error is less than 1.52×10^{-5} on $[4, 4.2]$.

14.

n	$f^{(n)}(x)$	$f^{(n)}(1)$
0	x^{-2}	1
1	$-2x^{-3}$	-2
2	$6x^{-4}$	6
3	$-24x^{-5}$	

(a) $f(x) = x^{-2} \approx T_2(x) = 1 - 2(x-1) + \dfrac{6}{2!}(x-1)^2 = 1 - 2(x-1) + 3(x-1)^2$

(b) $|R_2(x)| \le \dfrac{M}{3!}|x-1|^3$, where $|f'''(x)| \le M$. Now $0.9 \le x \le 1.1 \Rightarrow$

$|x-1| \le 0.1 \Rightarrow |x-1|^3 \le 0.001$. Since $f'''(x)$ is decreasing on

$[0.9, 1.1]$, we can take $M = |f'''(0.9)| = \dfrac{24}{(0.9)^5}$, so

$|R_2(x)| \le \dfrac{24/(0.9)^5}{6}(0.001) = \dfrac{0.004}{0.59049} \approx 0.006\,774\,04.$

(c)

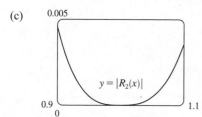

From the graph of $|R_2(x)| = |x^{-2} - T_2(x)|$, it seems that

the error is less than 0.0046 on $[0.9, 1.1]$.

15.

n	$f^{(n)}(x)$	$f^{(n)}(1)$
0	$x^{2/3}$	1
1	$\frac{2}{3}x^{-1/3}$	$\frac{2}{3}$
2	$-\frac{2}{9}x^{-4/3}$	$-\frac{2}{9}$
3	$\frac{8}{27}x^{-7/3}$	$\frac{8}{27}$
4	$-\frac{56}{81}x^{-10/3}$	

(a) $f(x) = x^{2/3} \approx T_3(x) = 1 + \dfrac{2}{3}(x-1) - \dfrac{2/9}{2!}(x-1)^2 + \dfrac{8/27}{3!}(x-1)^3$

$\qquad = 1 + \dfrac{2}{3}(x-1) - \dfrac{1}{9}(x-1)^2 + \dfrac{4}{81}(x-1)^3$

(b) $|R_3(x)| \le \dfrac{M}{4!}|x-1|^4$, where $\left|f^{(4)}(x)\right| \le M$. Now $0.8 \le x \le 1.2 \Rightarrow$

$|x-1| \le 0.2 \Rightarrow |x-1|^4 \le 0.0016$. Since $\left|f^{(4)}(x)\right|$ is decreasing

on $[0.8, 1.2]$, we can take $M = \left|f^{(4)}(0.8)\right| = \frac{56}{81}(0.8)^{-10/3}$, so

$|R_3(x)| \le \dfrac{\frac{56}{81}(0.8)^{-10/3}}{24}(0.0016) \approx 0.000\,096\,97.$

(c)

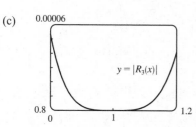

From the graph of $|R_3(x)| = \left|x^{2/3} - T_3(x)\right|$, it seems that the

error is less than $0.000\,053\,3$ on $[0.8, 1.2]$.

16.

n	$f^{(n)}(x)$	$f^{(n)}(\pi/6)$
0	$\sin x$	$1/2$
1	$\cos x$	$\sqrt{3}/2$
2	$-\sin x$	$-1/2$
3	$-\cos x$	$-\sqrt{3}/2$
4	$\sin x$	$1/2$
5	$\cos x$	

(a) $f(x) = \sin x \approx T_4(x)$

$\qquad = \frac{1}{2} + \frac{\sqrt{3}}{2}\left(x - \frac{\pi}{6}\right) - \frac{1}{4}\left(x - \frac{\pi}{6}\right)^2 - \frac{\sqrt{3}}{12}\left(x - \frac{\pi}{6}\right)^3 + \frac{1}{48}\left(x - \frac{\pi}{6}\right)^4$

(b) $|R_4(x)| \le \dfrac{M}{5!}\left|x - \frac{\pi}{6}\right|^5$, where $\left|f^{(5)}(x)\right| \le M$. Now $0 \le x \le \frac{\pi}{3} \Rightarrow$

$-\frac{\pi}{6} \le x - \frac{\pi}{6} \le \frac{\pi}{6} \Rightarrow \left|x - \frac{\pi}{6}\right| \le \frac{\pi}{6} \Rightarrow \left|x - \frac{\pi}{6}\right|^5 \le \left(\frac{\pi}{6}\right)^5$. Since

$\left|f^{(5)}(x)\right|$ is decreasing on $\left[0, \frac{\pi}{3}\right]$, we can take $M = \left|f^{(5)}(0)\right| = \cos 0 = 1$,

so $|R_4(x)| \le \dfrac{1}{5!}\left(\dfrac{\pi}{6}\right)^5 \approx 0.000\,328.$

(c)

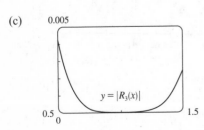

From the graph of $|R_4(x)| = |\sin x - T_4(x)|$, it seems that the error is less than 0.000 297 on $\left[0, \frac{\pi}{3}\right]$.

17.

(a) $f(x) = \sec x \approx T_2(x) = 1 + \frac{1}{2}x^2$

n	$f^{(n)}(x)$	$f^{(n)}(0)$
0	$\sec x$	1
1	$\sec x \tan x$	0
2	$\sec x \left(2\sec^2 x - 1\right)$	1
3	$\sec x \tan x \left(6\sec^2 x - 1\right)$	

(b) $|R_2(x)| \le \dfrac{M}{3!}\,|x|^3$, where $\left|f^{(3)}(x)\right| \le M$. Now $-0.2 \le x \le 0.2 \;\Rightarrow\; |x| \le 0.2 \;\Rightarrow\; |x|^3 \le (0.2)^3$.

$f^{(3)}(x)$ is an odd function and it is increasing on $[0, 0.2]$ since $\sec x$ and $\tan x$ are increasing on $[0, 0.2]$,

so $\left|f^{(3)}(x)\right| \le f^{(3)}(0.2) \approx 1.085\,158\,892$. Thus, $|R_2(x)| \le \dfrac{f^{(3)}(0.2)}{3!}\,(0.2)^3 \approx 0.001\,447$.

(c)

From the graph of $|R_2(x)| = |\sec x - T_2(x)|$, it seems that the error is less than 0.000 339 on $[-0.2, 0.2]$.

18.

(a) $f(x) = \ln(1 + 2x) \approx T_3(x)$

$$= \ln 3 + \tfrac{2}{3}(x - 1) - \frac{4/9}{2!}(x - 1)^2 + \frac{16/27}{3!}(x - 1)^3$$

n	$f^{(n)}(x)$	$f^{(n)}(1)$
0	$\ln(1 + 2x)$	$\ln 3$
1	$2/(1 + 2x)$	$\frac{2}{3}$
2	$-4/(1 + 2x)^2$	$-\frac{4}{9}$
3	$16/(1 + 2x)^3$	$\frac{16}{27}$
4	$-96/(1 + 2x)^4$	

(b) $|R_3(x)| \le \dfrac{M}{4!}\,|x - 1|^4$, where $\left|f^{(4)}(x)\right| \le M$. Now $0.5 \le x \le 1.5 \;\Rightarrow$

$-0.5 \le x - 1 \le 0.5 \;\Rightarrow\; |x - 1| \le 0.5 \;\Rightarrow\; |x - 1|^4 \le \frac{1}{16}$, and

letting $x = 0.5$ gives $M = 6$, so $|R_3(x)| \le \dfrac{6}{4!} \cdot \dfrac{1}{16} = \dfrac{1}{64} = 0.015\,625$.

(c)

From the graph of $|R_3(x)| = |\ln(1 + 2x) - T_3(x)|$, it seems that the error is less than 0.005 on $[0.5, 1.5]$.

19.

n	$f^{(n)}(x)$	$f^{(n)}(0)$
0	e^{x^2}	1
1	$e^{x^2}(2x)$	0
2	$e^{x^2}(2 + 4x^2)$	2
3	$e^{x^2}(12x + 8x^3)$	0
4	$e^{x^2}(12 + 48x^2 + 16x^4)$	

(a) $f(x) = e^{x^2} \approx T_3(x) = 1 + \dfrac{2}{2!}x^2 = 1 + x^2$

(b) $|R_3(x)| \le \dfrac{M}{4!}|x|^4$, where $\left|f^{(4)}(x)\right| \le M$. Now $0 \le x \le 0.1 \;\Rightarrow$

$x^4 \le (0.1)^4$, and letting $x = 0.1$ gives

$$|R_3(x)| \le \frac{e^{0.01}(12 + 0.48 + 0.0016)}{24}(0.1)^4 \approx 0.00006.$$

(c)   0.00008

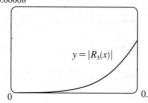

$y = |R_3(x)|$

0   0.1

From the graph of $|R_3(x)| = \left|e^{x^2} - T_3(x)\right|$, it appears that the

error is less than $0.000\,051$ on $[0, 0.1]$.

20.

n	$f^{(n)}(x)$	$f^{(n)}(1)$
0	$x \ln x$	0
1	$\ln x + 1$	1
2	$1/x$	1
3	$-1/x^2$	-1
4	$2/x^3$	

(a) $f(x) = x \ln x \approx T_3(x) = (x-1) + \frac{1}{2}(x-1)^2 - \frac{1}{6}(x-1)^3$

(b) $|R_3(x)| \le \dfrac{M}{4!}|x-1|^4$, where $\left|f^{(4)}(x)\right| \le M$. Now $0.5 \le x \le 1.5 \;\Rightarrow$

$|x-1| \le \frac{1}{2} \;\Rightarrow\; |x-1|^4 \le \frac{1}{16}$. Since $\left|f^{(4)}(x)\right|$ is decreasing on

$[0.5, 1.5]$, we can take $M = \left|f^{(4)}(0.5)\right| = 2/(0.5)^3 = 16$, so

$|R_3(x)| \le \frac{16}{24}(1/16) = \frac{1}{24} = 0.041\overline{6}$.

(c)   0.008

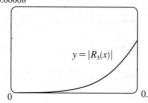

$y = |R_3(x)|$

0.5   1.5
            1

From the graph of $|R_3(x)| = |x \ln x - T_3(x)|$, it seems that the

error is less than 0.0076 on $[0.5, 1.5]$.

21.

n	$f^{(n)}(x)$	$f^{(n)}(0)$
0	$x \sin x$	0
1	$\sin x + x \cos x$	0
2	$2\cos x - x \sin x$	2
3	$-3\sin x - x \cos x$	0
4	$-4\cos x + x \sin x$	-4
5	$5\sin x + x \cos x$	

(a) $f(x) = x \sin x \approx T_4(x) = \dfrac{2}{2!}(x-0)^2 + \dfrac{-4}{4!}(x-0)^4 = x^2 - \dfrac{1}{6}x^4$

(b) $|R_4(x)| \le \dfrac{M}{5!}|x|^5$, where $\left|f^{(5)}(x)\right| \le M$. Now $-1 \le x \le 1 \;\Rightarrow$

$|x| \le 1$, and a graph of $f^{(5)}(x)$ shows that $\left|f^{(5)}(x)\right| \le 5$ for $-1 \le x \le 1$.

Thus, we can take $M = 5$ and get $|R_4(x)| \le \dfrac{5}{5!}\cdot 1^5 = \dfrac{1}{24} = 0.041\overline{6}$.

(c)

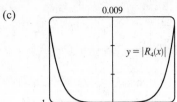

From the graph of $|R_4(x)| = |x \sin x - T_4(x)|$, it seems that the error is less than 0.0082 on $[-1, 1]$.

22.

n	$f^{(n)}(x)$	$f^{(n)}(0)$
0	$\sinh 2x$	0
1	$2 \cosh 2x$	2
2	$4 \sinh 2x$	0
3	$8 \cosh 2x$	8
4	$16 \sinh 2x$	0
5	$32 \cosh 2x$	32
6	$64 \sinh 2x$	

(a) $f(x) = \sinh 2x \approx T_5(x) = 2x + \frac{8}{3!}x^3 + \frac{32}{5!}x^5 = 2x + \frac{4}{3}x^3 + \frac{4}{15}x^5$

(b) $|R_5(x)| \le \frac{M}{6!} |x|^6$, where $\left| f^{(6)}(x) \right| \le M$. For x in $[-1, 1]$, we have $|x| \le 1$. Since $f^{(6)}(x)$ is an increasing odd function on $[-1, 1]$, we see that $\left| f^{(6)}(x) \right| \le f^{(6)}(1) = 64 \sinh 2 = 32(e^2 - e^{-2}) \approx 232.119$, so we can take $M = 232.12$ and get $|R_5(x)| \le \frac{232.12}{720} \cdot 1^6 \approx 0.3224$.

(c)

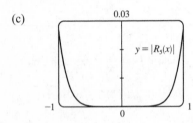

From the graph of $|R_5(x)| = |\sinh 2x - T_5(x)|$, it seems that the error is less than 0.027 on $[-1, 1]$.

23. From Exercise 5, $\cos x = -\left(x - \frac{\pi}{2}\right) + \frac{1}{6}\left(x - \frac{\pi}{2}\right)^3 + R_3(x)$, where $|R_3(x)| \le \frac{M}{4!} \left| x - \frac{\pi}{2} \right|^4$ with $\left| f^{(4)}(x) \right| = |\cos x| \le M = 1$. Now $x = 80° = (90° - 10°) = \left(\frac{\pi}{2} - \frac{\pi}{18}\right) = \frac{4\pi}{9}$ radians, so the error is $\left| R_3\left(\frac{4\pi}{9}\right) \right| \le \frac{1}{24}\left(\frac{\pi}{18}\right)^4 \approx 0.000\,039$, which means our estimate would *not* be accurate to five decimal places. However, $T_3 = T_4$, so we can use $\left| R_4\left(\frac{4\pi}{9}\right) \right| \le \frac{1}{120}\left(\frac{\pi}{18}\right)^5 \approx 0.000\,001$. Therefore, to five decimal places, $\cos 80° \approx -\left(-\frac{\pi}{18}\right) + \frac{1}{6}\left(-\frac{\pi}{18}\right)^3 \approx 0.17365$.

24. From Exercise 16, $\sin x = \frac{1}{2} + \frac{\sqrt{3}}{2}\left(x - \frac{\pi}{6}\right) - \frac{1}{4}\left(x - \frac{\pi}{6}\right)^2 - \frac{\sqrt{3}}{12}\left(x - \frac{\pi}{6}\right)^3 + \frac{1}{48}\left(x - \frac{\pi}{6}\right)^4 + R_4(x)$, where $|R_4(x)| \le \frac{M}{5!} \left| x - \frac{\pi}{6} \right|^5$ with $\left| f^{(5)}(x) \right| = |\cos x| \le M = 1$. Now $x = 38° = (30° + 8°) = \left(\frac{\pi}{6} + \frac{2\pi}{45}\right)$ radians, so the error is $\left| R_4\left(\frac{38\pi}{180}\right) \right| \le \frac{1}{120}\left(\frac{2\pi}{45}\right)^5 \approx 0.000\,000\,44$, which means our estimate will be accurate to five decimal places. Therefore, to five decimal places, $\sin 38° = \frac{1}{2} + \frac{\sqrt{3}}{2}\left(\frac{2\pi}{45}\right) - \frac{1}{4}\left(\frac{2\pi}{45}\right)^2 - \frac{\sqrt{3}}{12}\left(\frac{2\pi}{45}\right)^3 + \frac{1}{48}\left(\frac{2\pi}{45}\right)^4 \approx 0.61566$.

25. All derivatives of e^x are e^x, so $|R_n(x)| \le \frac{e^x}{(n+1)!} |x|^{n+1}$, where $0 < x < 0.1$. Letting $x = 0.1$, $R_n(0.1) \le \frac{e^{0.1}}{(n+1)!} (0.1)^{n+1} < 0.00001$, and by trial and error we find that $n = 3$ satisfies this inequality since $R_3(0.1) < 0.0000046$. Thus, by adding the four terms of the Maclaurin series for e^x corresponding to $n = 0, 1, 2,$ and 3, we can estimate $e^{0.1}$ to within 0.00001. (In fact, this sum is $1.1051\overline{6}$ and $e^{0.1} \approx 1.10517$.)

26. Example 6 in Section 12.9 [ET 11.9] gives the Maclaurin series for $\ln(1-x)$ as $-\sum\limits_{n=1}^{\infty} \dfrac{x^n}{n}$ for $|x| < 1$. Thus,

$\ln 1.4 = \ln[1-(-0.4)] = -\sum\limits_{n=1}^{\infty} \dfrac{(-0.4)^n}{n} = \sum\limits_{n=1}^{\infty} (-1)^{n+1}\dfrac{(0.4)^n}{n}$. Since this is an alternating series, the error is less than the

first neglected term by the Alternating Series Estimation Theorem, and we find that $|a_6| = (0.4)^6/6 \approx 0.0007 < 0.001$. So

we need the first five (nonzero) terms of the Maclaurin series for the desired accuracy. (In fact, this sum is approximately

0.33698 and $\ln 1.4 \approx 0.33647$.)

27. $\sin x = x - \dfrac{1}{3!}x^3 + \dfrac{1}{5!}x^5 - \cdots$. By the Alternating Series

Estimation Theorem, the error in the approximation

$\sin x = x - \dfrac{1}{3!}x^3$ is less than $\left|\dfrac{1}{5!}x^5\right| < 0.01 \quad \Leftrightarrow$

$|x^5| < 120(0.01) \quad \Leftrightarrow \quad |x| < (1.2)^{1/5} \approx 1.037$. The curves

$y = x - \frac{1}{6}x^3$ and $y = \sin x - 0.01$ intersect at $x \approx 1.043$, so

the graph confirms our estimate. Since both the sine function

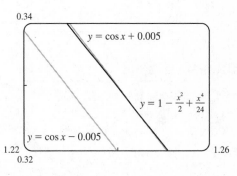

and the given approximation are odd functions, we need to check the estimate only for $x > 0$. Thus, the desired range of

values for x is $-1.037 < x < 1.037$.

28. $\cos x = 1 - \dfrac{1}{2!}x^2 + \dfrac{1}{4!}x^4 - \dfrac{1}{6!}x^6 + \cdots$. By the Alternating Series

Estimation Theorem, the error is less than $\left|-\dfrac{1}{6!}x^6\right| < 0.005 \quad \Leftrightarrow$

$x^6 < 720(0.005) \quad \Leftrightarrow \quad |x| < (3.6)^{1/6} \approx 1.238$. The curves

$y = 1 - \frac{1}{2}x^2 + \frac{1}{24}x^4$ and $y = \cos x + 0.005$ intersect at $x \approx 1.244$,

so the graph confirms our estimate. Since both the cosine function

and the given approximation are even functions, we need to check

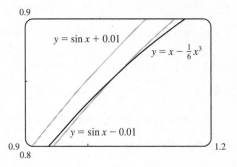

the estimate only for $x > 0$. Thus, the desired range of values for x is $-1.238 < x < 1.238$.

29. $\arctan x = x - \dfrac{x^3}{3} + \dfrac{x^5}{5} - \dfrac{x^7}{7} + \cdots$. By the Alternating Series

Estimation Theorem, the error is less than $\left|-\frac{1}{7}x^7\right| < 0.05 \quad \Leftrightarrow$

$|x^7| < 0.35 \quad \Leftrightarrow \quad |x| < (0.35)^{1/7} \approx 0.8607$. The curves

$y = x - \frac{1}{3}x^3 + \frac{1}{5}x^5$ and $y = \arctan x + 0.05$ intersect at

$x \approx 0.9245$, so the graph confirms our estimate. Since both the

arctangent function and the given approximation are odd functions,

we need to check the estimate only for $x > 0$. Thus, the desired

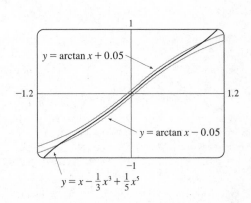

range of values for x is $-0.86 < x < 0.86$.

30. $f(x) = \sum_{n=0}^{\infty} \dfrac{f^{(n)}(4)}{n!}(x-4)^n = \sum_{n=0}^{\infty} \dfrac{(-1)^n\, n!}{3^n(n+1)\,n!}(x-4)^n = \sum_{n=0}^{\infty} \dfrac{(-1)^n}{3^n(n+1)}(x-4)^n$. Now

$f(5) = \sum_{n=0}^{\infty} \dfrac{(-1)^n}{3^n(n+1)} = \sum_{n=0}^{\infty}(-1)^n b_n$ is the sum of an alternating series that satisfies (i) $b_{n+1} \le b_n$ and

(ii) $\lim\limits_{n\to\infty} b_n = 0$, so by the Alternating Series Estimation Theorem, $|R_5(5)| = |f(5) - T_5(5)| \le b_6$, and

$b_6 = \dfrac{1}{3^6(7)} = \dfrac{1}{5103} \approx 0.000196 < 0.0002$; that is, the fifth-degree Taylor polynomial approximates $f(5)$ with error less

than 0.0002.

31. Let $s(t)$ be the position function of the car, and for convenience set $s(0) = 0$. The velocity of the car is $v(t) = s'(t)$ and the

acceleration is $a(t) = s''(t)$, so the second degree Taylor polynomial is $T_2(t) = s(0) + v(0)t + \dfrac{a(0)}{2}t^2 = 20t + t^2$. We

estimate the distance traveled during the next second to be $s(1) \approx T_2(1) = 20 + 1 = 21$ m. The function $T_2(t)$ would not be

accurate over a full minute, since the car could not possibly maintain an acceleration of 2 m/s² for that long (if it did, its final

speed would be 140 m/s ≈ 313 mi/h!).

32. (a)

n	$\rho^{(n)}(t)$	$\rho^{(n)}(20)$
0	$\rho_{20}e^{\alpha(t-20)}$	ρ_{20}
1	$\alpha\rho_{20}e^{\alpha(t-20)}$	$\alpha\rho_{20}$
2	$\alpha^2\rho_{20}e^{\alpha(t-20)}$	$\alpha^2\rho_{20}$

The linear approximation is

$$T_1(t) = \rho(20) + \rho'(20)(t-20) = \rho_{20}[1 + \alpha(t-20)]$$

The quadratic approximation is

$$T_2(t) = \rho(20) + \rho'(20)(t-20) + \dfrac{\rho''(20)}{2}(t-20)^2$$
$$= \rho_{20}\left[1 + \alpha(t-20) + \tfrac{1}{2}\alpha^2(t-20)^2\right]$$

(b)

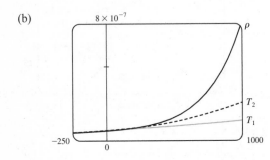

(c)

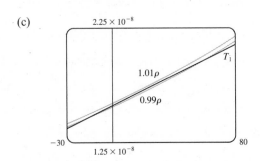

From the graph, it seems that $T_1(t)$ is within 1% of $\rho(t)$, that

is, $0.99\rho(t) \le T_1(t) \le 1.01\rho(t)$, for $-14°\mathrm{C} \le t \le 58°\mathrm{C}$.

33. $E = \dfrac{q}{D^2} - \dfrac{q}{(D+d)^2} = \dfrac{q}{D^2} - \dfrac{q}{D^2(1+d/D)^2} = \dfrac{q}{D^2}\left[1 - \left(1 + \dfrac{d}{D}\right)^{-2}\right]$.

We use the Binomial Series to expand $(1 + d/D)^{-2}$:

$E = \dfrac{q}{D^2}\left[1 - \left(1 - 2\left(\dfrac{d}{D}\right) + \dfrac{2\cdot 3}{2!}\left(\dfrac{d}{D}\right)^2 - \dfrac{2\cdot 3\cdot 4}{3!}\left(\dfrac{d}{D}\right)^3 + \cdots\right)\right] = \dfrac{q}{D^2}\left[2\left(\dfrac{d}{D}\right) - 3\left(\dfrac{d}{D}\right)^2 + 4\left(\dfrac{d}{D}\right)^3 - \cdots\right]$

$\approx \dfrac{q}{D^2}\cdot 2\left(\dfrac{d}{D}\right) = 2qd\cdot\dfrac{1}{D^3}$

when D is much larger than d; that is, when P is far away from the dipole.

34. (a) $\dfrac{n_1}{\ell_o} + \dfrac{n_2}{\ell_i} = \dfrac{1}{R}\left(\dfrac{n_2 s_i}{\ell_i} - \dfrac{n_1 s_o}{\ell_o}\right)$ [Equation 1] where

$$\ell_o = \sqrt{R^2 + (s_o + R)^2 - 2R(s_o + R)\cos\phi} \quad \text{and} \quad \ell_i = \sqrt{R^2 + (s_i - R)^2 + 2R(s_i - R)\cos\phi} \quad \textbf{(2)}$$

Using $\cos\phi \approx 1$ gives

$$\ell_o = \sqrt{R^2 + (s_o + R)^2 - 2R(s_o + R)} = \sqrt{R^2 + s_o^2 + 2Rs_o + R^2 - 2Rs_o - 2R^2} = \sqrt{s_o^2} = s_o \ .$$

and similarly, $\ell_i = s_i$. Thus, Equation 1 becomes $\dfrac{n_1}{s_o} + \dfrac{n_2}{s_i} = \dfrac{1}{R}\left(\dfrac{n_2 s_i}{s_i} - \dfrac{n_1 s_o}{s_o}\right) \ \Rightarrow \ \dfrac{n_1}{s_o} + \dfrac{n_2}{s_i} = \dfrac{n_2 - n_1}{R}$.

(b) Using $\cos\phi \approx 1 - \tfrac{1}{2}\phi^2$ in **(2)** gives us

$$\ell_o = \sqrt{R^2 + (s_o + R)^2 - 2R(s_o + R)\left(1 - \tfrac{1}{2}\phi^2\right)}$$

$$= \sqrt{R^2 + s_o^2 + 2Rs_o + R^2 - 2Rs_o + Rs_o\phi^2 - 2R^2 + R^2\phi^2} = \sqrt{s_o^2 + Rs_o\phi^2 + R^2\phi^2}$$

Anticipating that we will use the binomial series expansion $(1+x)^k \approx 1 + kx$, we can write the last expression for ℓ_o as

$s_o\sqrt{1 + \phi^2\left(\dfrac{R}{s_o} + \dfrac{R^2}{s_o^2}\right)}$ and similarly, $\ell_i = s_i\sqrt{1 - \phi^2\left(\dfrac{R}{s_i} - \dfrac{R^2}{s_i^2}\right)}$. Thus, from Equation 1,

$$\dfrac{n_1}{\ell_o} + \dfrac{n_2}{\ell_i} = \dfrac{1}{R}\left(\dfrac{n_2 s_i}{\ell_i} - \dfrac{n_1 s_o}{\ell_o}\right) \ \Leftrightarrow \ n_1\ell_o^{-1} + n_2\ell_i^{-1} = \dfrac{n_2}{R}\cdot\dfrac{s_i}{\ell_i} - \dfrac{n_1}{R}\cdot\dfrac{s_o}{\ell_o} \ \Leftrightarrow$$

$$\dfrac{n_1}{s_o}\left[1 + \phi^2\left(\dfrac{R}{s_o} + \dfrac{R^2}{s_o^2}\right)\right]^{-1/2} + \dfrac{n_2}{s_i}\left[1 - \phi^2\left(\dfrac{R}{s_i} - \dfrac{R^2}{s_i^2}\right)\right]^{-1/2}$$

$$= \dfrac{n_2}{R}\left[1 - \phi^2\left(\dfrac{R}{s_i} - \dfrac{R^2}{s_i^2}\right)\right]^{-1/2} - \dfrac{n_1}{R}\left[1 + \phi^2\left(\dfrac{R}{s_o} + \dfrac{R^2}{s_o^2}\right)\right]^{-1/2}$$

Approximating the expressions for ℓ_o^{-1} and ℓ_i^{-1} by the first two terms in their binomial series, we get

$$\dfrac{n_1}{s_o}\left[1 - \tfrac{1}{2}\phi^2\left(\dfrac{R}{s_o} + \dfrac{R^2}{s_o^2}\right)\right] + \dfrac{n_2}{s_i}\left[1 + \tfrac{1}{2}\phi^2\left(\dfrac{R}{s_i} - \dfrac{R^2}{s_i^2}\right)\right]$$

$$= \dfrac{n_2}{R}\left[1 + \tfrac{1}{2}\phi^2\left(\dfrac{R}{s_i} - \dfrac{R^2}{s_i^2}\right)\right] - \dfrac{n_1}{R}\left[1 - \tfrac{1}{2}\phi^2\left(\dfrac{R}{s_o} + \dfrac{R^2}{s_o^2}\right)\right] \ \Leftrightarrow$$

$$\dfrac{n_1}{s_o} - \dfrac{n_1\phi^2}{2s_o}\left(\dfrac{R}{s_o} + \dfrac{R^2}{s_o^2}\right) + \dfrac{n_2}{s_i} + \dfrac{n_2\phi^2}{2s_i}\left(\dfrac{R}{s_i} - \dfrac{R^2}{s_i^2}\right) = \dfrac{n_2}{R} + \dfrac{n_2\phi^2}{2R}\left(\dfrac{R}{s_i} - \dfrac{R^2}{s_i^2}\right) - \dfrac{n_1}{R} + \dfrac{n_1\phi^2}{2R}\left(\dfrac{R}{s_o} + \dfrac{R^2}{s_o^2}\right) \ \Leftrightarrow$$

$$\dfrac{n_1}{s_o} + \dfrac{n_2}{s_i} = \dfrac{n_2}{R} - \dfrac{n_1}{R} + \dfrac{n_1\phi^2}{2s_o}\left(\dfrac{R}{s_o} + \dfrac{R^2}{s_o^2}\right) + \dfrac{n_1\phi^2}{2R}\left(\dfrac{R}{s_o} + \dfrac{R^2}{s_o^2}\right) + \dfrac{n_2\phi^2}{2R}\left(\dfrac{R}{s_i} - \dfrac{R^2}{s_i^2}\right) - \dfrac{n_2\phi^2}{2s_i}\left(\dfrac{R}{s_i} - \dfrac{R^2}{s_i^2}\right)$$

$$= \dfrac{n_2 - n_1}{R} + \dfrac{n_1\phi^2}{2}\left(\dfrac{R}{s_o} + \dfrac{R^2}{s_o^2}\right)\left(\dfrac{1}{s_o} + \dfrac{1}{R}\right) + \dfrac{n_2\phi^2}{2}\left(\dfrac{R}{s_i} - \dfrac{R^2}{s_i^2}\right)\left(\dfrac{1}{R} - \dfrac{1}{s_i}\right)$$

$$= \dfrac{n_2 - n_1}{R} + \dfrac{n_1\phi^2 R^2}{2s_o}\left(\dfrac{1}{R} + \dfrac{1}{s_o}\right)\left(\dfrac{1}{R} + \dfrac{1}{s_o}\right) + \dfrac{n_2\phi^2 R^2}{2s_i}\left(\dfrac{1}{R} - \dfrac{1}{s_i}\right)\left(\dfrac{1}{R} - \dfrac{1}{s_i}\right)$$

$$= \dfrac{n_2 - n_1}{R} + \phi^2 R^2\left[\dfrac{n_1}{2s_o}\left(\dfrac{1}{R} + \dfrac{1}{s_o}\right)^2 + \dfrac{n_2}{2s_i}\left(\dfrac{1}{R} - \dfrac{1}{s_i}\right)^2\right]$$

From Figure 8, we see that $\sin\phi = h/R$. So if we approximate $\sin\phi$ with ϕ, we get $h = R\phi$ and $h^2 = \phi^2 R^2$ and hence, Equation 4, as desired.

35. (a) If the water is deep, then $2\pi d/L$ is large, and we know that $\tanh x \to 1$ as $x \to \infty$. So we can approximate

$\tanh(2\pi d/L) \approx 1$, and so $v^2 \approx gL/(2\pi) \quad \Leftrightarrow \quad v \approx \sqrt{gL/(2\pi)}$.

(b) From the table, the first term in the Maclaurin series of

$\tanh x$ is x, so if the water is shallow, we can approximate

$\tanh \dfrac{2\pi d}{L} \approx \dfrac{2\pi d}{L}$, and so $v^2 \approx \dfrac{gL}{2\pi} \cdot \dfrac{2\pi d}{L} \quad \Leftrightarrow \quad v \approx \sqrt{gd}$.

n	$f^{(n)}(x)$	$f^{(n)}(0)$
0	$\tanh x$	0
1	$\operatorname{sech}^2 x$	1
2	$-2\operatorname{sech}^2 x \tanh x$	0
3	$2\operatorname{sech}^2 x\,(3\tanh^2 x - 1)$	-2

(c) Since $\tanh x$ is an odd function, its Maclaurin series is alternating, so the error in the approximation

$\tanh \dfrac{2\pi d}{L} \approx \dfrac{2\pi d}{L}$ is less than the first neglected term, which is $\dfrac{|f'''(0)|}{3!}\left(\dfrac{2\pi d}{L}\right)^3 = \dfrac{1}{3}\left(\dfrac{2\pi d}{L}\right)^3$.

If $L > 10d$, then $\dfrac{1}{3}\left(\dfrac{2\pi d}{L}\right)^3 < \dfrac{1}{3}\left(2\pi \cdot \dfrac{1}{10}\right)^3 = \dfrac{\pi^3}{375}$, so the error in the approximation $v^2 = gd$ is less

than $\dfrac{gL}{2\pi} \cdot \dfrac{\pi^3}{375} \approx 0.0132 gL$.

36. (a) $4\sqrt{\dfrac{L}{g}} \displaystyle\int_0^{\pi/2} \dfrac{dx}{\sqrt{1 - k^2 \sin^2 x}} = 4\sqrt{\dfrac{L}{g}} \int_0^{\pi/2} \left[1 + (-k^2 \sin^2 x)\right]^{-1/2} dx$

$= 4\sqrt{\dfrac{L}{g}} \displaystyle\int_0^{\pi/2} \left[1 - \dfrac{1}{2}(-k^2 \sin^2 x) + \dfrac{\frac{1}{2} \cdot \frac{3}{2}}{2!}(-k^2 \sin^2 x)^2 - \dfrac{\frac{1}{2} \cdot \frac{3}{2} \cdot \frac{5}{2}}{3!}(-k^2 \sin^2 x)^3 + \cdots\right] dx$

$= 4\sqrt{\dfrac{L}{g}} \displaystyle\int_0^{\pi/2} \left[1 + \left(\dfrac{1}{2}\right)k^2 \sin^2 x + \left(\dfrac{1 \cdot 3}{2 \cdot 4}\right)k^4 \sin^4 x + \left(\dfrac{1 \cdot 3 \cdot 5}{2 \cdot 4 \cdot 6}\right)k^6 \sin^6 x + \cdots\right] dx$

$= 4\sqrt{\dfrac{L}{g}} \left[\dfrac{\pi}{2} + \left(\dfrac{1}{2}\right)\left(\dfrac{1}{2} \cdot \dfrac{\pi}{2}\right)k^2 + \left(\dfrac{1 \cdot 3}{2 \cdot 4}\right)\left(\dfrac{1 \cdot 3}{2 \cdot 4} \cdot \dfrac{\pi}{2}\right)k^4 + \left(\dfrac{1 \cdot 3 \cdot 5}{2 \cdot 4 \cdot 6}\right)\left(\dfrac{1 \cdot 3 \cdot 5}{2 \cdot 4 \cdot 6} \cdot \dfrac{\pi}{2}\right)k^6 + \cdots\right]$

[split up the integral and use the result from Exercise 7.1.46]

$= 2\pi\sqrt{\dfrac{L}{g}} \left[1 + \dfrac{1^2}{2^2}k^2 + \dfrac{1^2 \cdot 3^2}{2^2 \cdot 4^2}k^4 + \dfrac{1^2 \cdot 3^2 \cdot 5^2}{2^2 \cdot 4^2 \cdot 6^2}k^6 + \cdots\right]$

(b) The first of the two inequalities is true because all of the terms in the series are positive. For the second,

$$T = 2\pi\sqrt{\dfrac{L}{g}} \left[1 + \dfrac{1^2}{2^2}k^2 + \dfrac{1^2 \cdot 3^2}{2^2 \cdot 4^2}k^4 + \dfrac{1^2 \cdot 3^2 \cdot 5^2}{2^2 \cdot 4^2 \cdot 6^2}k^6 + \dfrac{1^2 \cdot 3^2 \cdot 5^2 \cdot 7^2}{2^2 \cdot 4^2 \cdot 6^2 \cdot 8^2}k^8 + \cdots\right]$$

$$\leq 2\pi\sqrt{\dfrac{L}{g}} \left[1 + \tfrac{1}{4}k^2 + \tfrac{1}{4}k^4 + \tfrac{1}{4}k^6 + \tfrac{1}{4}k^8 + \cdots\right]$$

The terms in brackets (after the first) form a geometric series with $a = \tfrac{1}{4}k^2$ and $r = k^2 = \sin^2\left(\tfrac{1}{2}\theta_0\right) < 1$.

So $T \leq 2\pi\sqrt{\dfrac{L}{g}} \left[1 + \dfrac{k^2/4}{1 - k^2}\right] = 2\pi\sqrt{\dfrac{L}{g}} \dfrac{4 - 3k^2}{4 - 4k^2}$.

(c) We substitute $L = 1$, $g = 9.8$, and $k = \sin(10°/2) \approx 0.08716$, and the inequality from part (b) becomes

$2.01090 \leq T \leq 2.01093$, so $T \approx 2.0109$. The estimate $T \approx 2\pi\sqrt{L/g} \approx 2.0071$ differs by about 0.2%.

If $\theta_0 = 42°$, then $k \approx 0.35837$ and the inequality becomes $2.07153 \leq T \leq 2.08103$, so $T \approx 2.0763$.

The one-term estimate is the same, and the discrepancy between the two estimates increases to about 3.4%.

37. (a) L is the length of the arc subtended by the angle θ, so $L = R\theta$ $\Rightarrow$

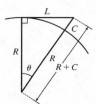

$\theta = L/R$. Now $\sec\theta = (R + C)/R$ $\Rightarrow$ $R\sec\theta = R + C$ $\Rightarrow$

$C = R\sec\theta - R = R\sec(L/R) - R.$

(b) Extending the result in Exercise 17, we have $f^{(4)}(x) = \sec x\,(18\sec^2 x \tan^2 x + 6\sec^4 x - \sec^2 x - \tan^2 x)$,

so $f^{(4)}(0) = 5$, and $\sec x \approx T_4(x) = 1 + \frac{1}{2}x^2 + \frac{5}{24}x^4$. By part (a),

$$C \approx R\left[1 + \frac{1}{2}\left(\frac{L}{R}\right)^2 + \frac{5}{24}\left(\frac{L}{R}\right)^4\right] - R = R + \frac{1}{2}R\cdot\frac{L^2}{R^2} + \frac{5}{24}R\cdot\frac{L^4}{R^4} - R = \frac{L^2}{2R} + \frac{5L^4}{24R^3}.$$

(c) Taking $L = 100$ km and $R = 6370$ km, the formula in part (a) says that

$C = R\sec(L/R) - R = 6370\sec(100/6370) - 6370 \approx 0.785\,009\,965\,44$ km.

The formula in part (b) says that $C \approx \dfrac{L^2}{2R} + \dfrac{5L^4}{24R^3} = \dfrac{100^2}{2\cdot 6370} + \dfrac{5\cdot 100^4}{24\cdot 6370^3} \approx 0.785\,009\,957\,36$ km.

The difference between these two results is only $0.000\,000\,008\,08$ km, or $0.000\,008\,08$ m!

38. $T_n(x) = f(a) + \dfrac{f'(a)}{1!}(x - a) + \dfrac{f''(a)}{2!}(x - a)^2 + \cdots + \dfrac{f^{(n)}(a)}{n!}(x - a)^n$. Let $0 \leq m \leq n$. Then

$$T_n^{(m)}(x) = m!\,\frac{f^{(m)}(a)}{m!}(x - a)^0 + (m + 1)(m)\cdots(2)\frac{f^{(m+1)}(a)}{(m+1)!}(x - a)^1 + \cdots$$

$$+ n(n - 1)\cdots(n - m + 1)\frac{f^{(n)}(a)}{n!}(x - a)^{n-m}$$

For $x = a$, all terms in this sum except the first one are 0, so $T_n^{(m)}(a) = \dfrac{m!\,f^{(m)}(a)}{m!} = f^{(m)}(a).$

39. Using $f(x) = T_n(x) + R_n(x)$ with $n = 1$ and $x = r$, we have $f(r) = T_1(r) + R_1(r)$, where T_1 is the first-degree Taylor

polynomial of f at a. Because $a = x_n$, $f(r) = f(x_n) + f'(x_n)(r - x_n) + R_1(r)$. But r is a root of f, so $f(r) = 0$

and we have $0 = f(x_n) + f'(x_n)(r - x_n) + R_1(r)$. Taking the first two terms to the left side gives us

$f'(x_n)(x_n - r) - f(x_n) = R_1(r)$. Dividing by $f'(x_n)$, we get $x_n - r - \dfrac{f(x_n)}{f'(x_n)} = \dfrac{R_1(r)}{f'(x_n)}$. By the formula for Newton's

method, the left side of the preceding equation is $x_{n+1} - r$, so $|x_{n+1} - r| = \left|\dfrac{R_1(r)}{f'(x_n)}\right|$. Taylor's Inequality gives us

$|R_1(r)| \leq \dfrac{|f''(r)|}{2!}|r - x_n|^2$. Combining this inequality with the facts $|f''(x)| \leq M$ and $|f'(x)| \geq K$ gives us

$|x_{n+1} - r| \leq \dfrac{M}{2K}|x_n - r|^2.$

APPLIED PROJECT   Radiation from the Stars

1. If we write $f(\lambda) = \dfrac{8\pi hc\lambda^{-5}}{e^{hc/(\lambda kT)} - 1} = \dfrac{a\lambda^{-5}}{e^{b/(\lambda T)} - 1}$, then as $\lambda \to 0^+$, it is of the form ∞/∞, and as $\lambda \to \infty$ it is of the form

$0/0$, so in either case we can use l'Hospital's Rule. First of all,

$$\lim_{\lambda \to \infty} f(\lambda) \overset{\mathrm{H}}{=} \lim_{\lambda \to \infty} \frac{a\left(-5\lambda^{-6}\right)}{-\dfrac{bT}{(\lambda T)^2}e^{b/(\lambda T)}} = 5\,\frac{aT}{b}\lim_{\lambda \to \infty}\frac{\lambda^2\lambda^{-6}}{e^{b/(\lambda T)}} = 5\,\frac{aT}{b}\lim_{\lambda \to \infty}\frac{\lambda^{-4}}{e^{b/(\lambda T)}} = 0$$

Also, $\displaystyle \lim_{\lambda \to 0^+} f(\lambda) \overset{\mathrm{H}}{=} 5\,\frac{aT}{b}\lim_{\lambda \to 0^+}\frac{\lambda^{-4}}{e^{b/(\lambda T)}} \overset{\mathrm{H}}{=} 5\,\frac{aT}{b}\lim_{\lambda \to 0^+}\frac{-4\lambda^{-5}}{-\dfrac{bT}{(\lambda T)^2}e^{b/(\lambda T)}} = 20\,\frac{aT^2}{b^2}\lim_{\lambda \to 0^+}\frac{\lambda^{-3}}{e^{b/(\lambda T)}}$

This is still indeterminate, but note that each time we use l'Hospital's Rule, we gain a factor of λ in the numerator, as well as a constant factor, and the denominator is unchanged. So if we use l'Hospital's Rule three more times, the exponent of λ in the numerator will become 0. That is, for some $\{k_i\}$, all constant,

$$\lim_{\lambda \to 0^+} f(\lambda) \overset{\mathrm{H}}{=} k_1\lim_{\lambda \to 0^+}\frac{\lambda^{-3}}{e^{b/(\lambda T)}} \overset{\mathrm{H}}{=} k_2\lim_{\lambda \to 0^+}\frac{\lambda^{-2}}{e^{b/(\lambda T)}} \overset{\mathrm{H}}{=} k_3\lim_{\lambda \to 0^+}\frac{\lambda^{-1}}{e^{b/(\lambda T)}} \overset{\mathrm{H}}{=} k_4\lim_{\lambda \to 0^+}\frac{1}{e^{b/(\lambda T)}} = 0$$

2. We expand the denominator of Planck's Law using the Taylor series $e^x = 1 + x + \dfrac{x^2}{2!} + \dfrac{x^3}{3!} + \cdots$ with $x = \dfrac{hc}{\lambda kT}$, and use

the fact that if λ is large, then all subsequent terms in the Taylor expansion are very small compared to the first one, so we can

approximate using the Taylor polynomial T_1:

$$f(\lambda) = \frac{8\pi hc\lambda^{-5}}{e^{hc/(\lambda kT)} - 1} = \frac{8\pi hc\lambda^{-5}}{\left[1 + \dfrac{hc}{\lambda kT} + \dfrac{1}{2!}\left(\dfrac{hc}{\lambda kT}\right)^2 + \dfrac{1}{3!}\left(\dfrac{hc}{\lambda kT}\right)^3 + \cdots\right] - 1} \approx \frac{8\pi hc\lambda^{-5}}{\left(1 + \dfrac{hc}{\lambda kT}\right) - 1} = \frac{8\pi kT}{\lambda^4}$$

which is the Rayleigh-Jeans Law.

3. To convert to μm, we substitute $\lambda/10^6$ for λ in both laws. The first figure shows that the two laws are similar for large λ. The

second figure shows that the two laws are very different for short wavelengths (Planck's Law gives a maximum at

$\lambda \approx 0.51\ \mu$m; the Rayleigh-Jeans Law gives no minimum or maximum.).

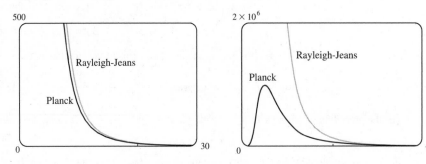

4. From the graph in Problem 3, $f(\lambda)$ has a maximum under Planck's Law at $\lambda \approx 0.51\ \mu$m.

5.

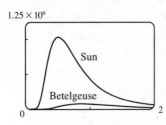
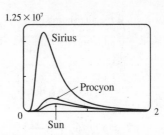

As T gets larger, the total area under the curve increases, as we would expect: the hotter the star, the more energy it emits. Also, as T increases, the λ-value of the maximum decreases, so the higher the temperature, the shorter the peak wavelength (and consequently the average wavelength) of light emitted. This is why Sirius is a blue star and Betelgeuse is a red star: most of Sirius's light is of a fairly short wavelength; that is, a higher frequency, toward the blue end of the spectrum, whereas most of Betelgeuse's light is of a lower frequency, toward the red end of the spectrum.

12 Review

ET 11

CONCEPT CHECK

1. (a) See Definition 12.1.1 [ET 11.1.1].

   (b) See Definition 12.2.2 [ET 11.2.2].

   (c) The terms of the sequence $\{a_n\}$ approach 3 as n becomes large.

   (d) By adding sufficiently many terms of the series, we can make the partial sums as close to 3 as we like.

2. (a) See Definition 12.1.11 [ET 11.1.11].

   (b) A sequence is monotonic if it is either increasing or decreasing.

   (c) By Theorem 12.1.12 [ET 11.1.12], every bounded, monotonic sequence is convergent.

3. (a) See (4) in Section 12.2 [ET 11.2].

   (b) The p-series $\sum_{n=1}^{\infty} \dfrac{1}{n^p}$ is convergent if $p > 1$.

4. If $\sum a_n = 3$, then $\lim\limits_{n \to \infty} a_n = 0$ and $\lim\limits_{n \to \infty} s_n = 3$.

5. (a) *Test for Divergence:* If $\lim\limits_{n \to \infty} a_n$ does not exist or if $\lim\limits_{n \to \infty} a_n \neq 0$, then the series $\sum_{n=1}^{\infty} a_n$ is divergent.

   (b) *Integral Test:* Suppose f is a continuous, positive, decreasing function on $[1, \infty)$ and let $a_n = f(n)$. Then the series $\sum_{n=1}^{\infty} a_n$ is convergent if and only if the improper integral $\int_1^{\infty} f(x)\,dx$ is convergent. In other words:

     (i) If $\int_1^{\infty} f(x)\,dx$ is convergent, then $\sum_{n=1}^{\infty} a_n$ is convergent.

     (ii) If $\int_1^{\infty} f(x)\,dx$ is divergent, then $\sum_{n=1}^{\infty} a_n$ is divergent.

   (c) *Comparison Test:* Suppose that $\sum a_n$ and $\sum b_n$ are series with positive terms.

     (i) If $\sum b_n$ is convergent and $a_n \leq b_n$ for all n, then $\sum a_n$ is also convergent.

     (ii) If $\sum b_n$ is divergent and $a_n \geq b_n$ for all n, then $\sum a_n$ is also divergent.

   (d) *Limit Comparison Test:* Suppose that $\sum a_n$ and $\sum b_n$ are series with positive terms. If $\lim\limits_{n \to \infty} (a_n/b_n) = c$, where c is a finite number and $c > 0$, then either both series converge or both diverge.

(e) *Alternating Series Test:* If the alternating series $\sum_{n=1}^{\infty}(-1)^{n-1}b_n = b_1 - b_2 + b_3 - b_4 + b_5 - b_6 + \cdots$ $[b_n > 0]$ satisfies (i) $b_{n+1} \le b_n$ for all n and (ii) $\lim_{n \to \infty} b_n = 0$, then the series is convergent.

(f) *Ratio Test:*

    (i) If $\lim_{n \to \infty} \left| \dfrac{a_{n+1}}{a_n} \right| = L < 1$, then the series $\sum_{n=1}^{\infty} a_n$ is absolutely convergent (and therefore convergent).

    (ii) If $\lim_{n \to \infty} \left| \dfrac{a_{n+1}}{a_n} \right| = L > 1$ or $\lim_{n \to \infty} \left| \dfrac{a_{n+1}}{a_n} \right| = \infty$, then the series $\sum_{n=1}^{\infty} a_n$ is divergent.

    (iii) If $\lim_{n \to \infty} \left| \dfrac{a_{n+1}}{a_n} \right| = 1$, the Ratio Test is inconclusive; that is, no conclusion can be drawn about the convergence or divergence of $\sum a_n$.

(g) *Root Test:*

    (i) If $\lim_{n \to \infty} \sqrt[n]{|a_n|} = L < 1$, then the series $\sum_{n=1}^{\infty} a_n$ is absolutely convergent (and therefore convergent).

    (ii) If $\lim_{n \to \infty} \sqrt[n]{|a_n|} = L > 1$ or $\lim_{n \to \infty} \sqrt[n]{|a_n|} = \infty$, then the series $\sum_{n=1}^{\infty} a_n$ is divergent.

    (iii) If $\lim_{n \to \infty} \sqrt[n]{|a_n|} = 1$, the Root Test is inconclusive.

6. (a) A series $\sum a_n$ is called *absolutely convergent* if the series of absolute values $\sum |a_n|$ is convergent.

  (b) If a series $\sum a_n$ is absolutely convergent, then it is convergent.

  (c) A series $\sum a_n$ is called *conditionally convergent* if it is convergent but not absolutely convergent.

7. (a) Use (3) in Section 12.3 [ET 11.3].

  (b) See Example 5 in Section 12.4 [ET 11.4].

  (c) By adding terms until you reach the desired accuracy given by the Alternating Series Estimation Theorem on page 712 [ET 748].

8. (a) $\sum_{n=0}^{\infty} c_n(x-a)^n$

  (b) Given the power series $\sum_{n=0}^{\infty} c_n(x-a)^n$, the radius of convergence is:

    (i) 0 if the series converges only when $x = a$

    (ii) ∞ if the series converges for all x, or

    (iii) a positive number R such that the series converges if $|x - a| < R$ and diverges if $|x - a| > R$.

  (c) The interval of convergence of a power series is the interval that consists of all values of x for which the series converges. Corresponding to the cases in part (b), the interval of convergence is: (i) the single point $\{a\}$, (ii) all real numbers, that is, the real number line $(-\infty, \infty)$, or (iii) an interval with endpoints $a - R$ and $a + R$ which can contain neither, either, or both of the endpoints. In this case, we must test the series for convergence at each endpoint to determine the interval of convergence.

9. (a), (b) See Theorem 12.9.2 [ET 11.9.2].

10. (a) $T_n(x) = \sum_{i=0}^{n} \dfrac{f^{(i)}(a)}{i!}(x-a)^i$

  (b) $\sum_{n=0}^{\infty} \dfrac{f^{(n)}(a)}{n!}(x-a)^n$

(c) $\sum_{n=0}^{\infty} \dfrac{f^{(n)}(0)}{n!} x^n$ $[a = 0$ in part (b)]

(d) See Theorem 12.10.8 [ET 11.10.8].

(e) See Taylor's Inequality (12.10.9) [ET (11.10.9)].

11. (a)–(e) See Table 1 on page 779 [ET 743].

12. See the binomial series (12.10.17) [ET (11.10.17)] for the expansion. The radius of convergence for the binomial series is 1.

TRUE-FALSE QUIZ

1. False.   See Note 2 after Theorem 12.2.6 [ET 11.2.6].

2. False.   The series $\sum_{n=1}^{\infty} n^{-\sin 1} = \sum_{n=1}^{\infty} \dfrac{1}{n^{\sin 1}}$ is a p-series with $p = \sin 1 \approx 0.84 \le 1$, so the series diverges.

3. True.   If $\lim_{n\to\infty} a_n = L$, then given any $\varepsilon > 0$, we can find a positive integer N such that $|a_n - L| < \varepsilon$ whenever $n > N$.

If $n > N$, then $2n + 1 > N$ and $|a_{2n+1} - L| < \varepsilon$. Thus, $\lim_{n\to\infty} a_{2n+1} = L$.

4. True by Theorem 12.8.3 [ET 11.8.3].

Or: Use the Comparison Test to show that $\sum c_n(-2)^n$ converges absolutely.

5. False.   For example, take $c_n = (-1)^n/(n6^n)$.

6. True by Theorem 12.8.3 [ET 11.8.3].

7. False, since $\lim_{n\to\infty} \left| \dfrac{a_{n+1}}{a_n} \right| = \lim_{n\to\infty} \left| \dfrac{1}{(n+1)^3} \cdot \dfrac{n^3}{1} \right| = \lim_{n\to\infty} \left| \dfrac{n^3}{(n+1)^3} \cdot \dfrac{1/n^3}{1/n^3} \right| = \lim_{n\to\infty} \dfrac{1}{(1+1/n)^3} = 1.$

8. True, since $\lim_{n\to\infty} \left| \dfrac{a_{n+1}}{a_n} \right| = \lim_{n\to\infty} \left| \dfrac{1}{(n+1)!} \cdot \dfrac{n!}{1} \right| = \lim_{n\to\infty} \dfrac{1}{n+1} = 0 < 1.$

9. False.   See the note after Example 2 in Section 12.4 [ET 11.4].

10. True, since $\dfrac{1}{e} = e^{-1}$ and $e^x = \sum_{n=0}^{\infty} \dfrac{x^n}{n!}$, so $e^{-1} = \sum_{n=0}^{\infty} \dfrac{(-1)^n}{n!}.$

11. True.   See (9) in Section 12.1 [ET 11.1].

12. True, because if $\sum |a_n|$ is convergent, then so is $\sum a_n$ by Theorem 12.6.3 [ET 11.6.3].

13. True.   By Theorem 12.10.5 [ET 11.10.5] the coefficient of x^3 is $\dfrac{f'''(0)}{3!} = \dfrac{1}{3} \Rightarrow f'''(0) = 2.$

Or: Use Theorem 12.9.2 [ET 11.9.2] to differentiate f three times.

14. False.   Let $a_n = n$ and $b_n = -n$. Then $\{a_n\}$ and $\{b_n\}$ are divergent, but $a_n + b_n = 0$, so $\{a_n + b_n\}$ is convergent.

15. False.   For example, let $a_n = b_n = (-1)^n$. Then $\{a_n\}$ and $\{b_n\}$ are divergent, but $a_n b_n = 1$, so $\{a_n b_n\}$ is convergent.

16. True by the Monotonic Sequence Theorem, since $\{a_n\}$ is decreasing and $0 < a_n \le a_1$ for all $n \Rightarrow \{a_n\}$ is bounded.

17. True by Theorem 12.6.3 [ET 11.6.3]. $\left[\sum (-1)^n a_n \text{ is absolutely convergent and hence convergent.} \right]$

18. True. $\displaystyle\lim_{n\to\infty} \frac{a_{n+1}}{a_n} < 1 \Rightarrow \sum a_n$ converges (Ratio Test) $\Rightarrow \displaystyle\lim_{n\to\infty} a_n = 0$ [Theorem 12.2.6 [ET 11.2.6]].

19. True. $0.99999\ldots = 0.9 + 0.9(0.1)^1 + 0.9(0.1)^2 + 0.9(0.1)^3 + \cdots = \displaystyle\sum_{n=1}^{\infty} (0.9)(0.1)^{n-1} = \frac{0.9}{1-0.1} = 1$ by the formula

for the sum of a geometric series $[S = a_1/(1-r)]$ with ratio r satisfying $|r| < 1$.

20. False. Let $a_n = (0.1)^n$ and $b_n = (0.2)^n$. Then $\displaystyle\sum_{n=1}^{\infty} a_n = \sum_{n=1}^{\infty} (0.1)^n = \frac{0.1}{1-0.1} = \frac{1}{9} = A$,

$\displaystyle\sum_{n=1}^{\infty} b_n = \sum_{n=1}^{\infty} (0.2)^n = \frac{0.2}{1-0.2} = \frac{1}{4} = B$, and $\displaystyle\sum_{n=1}^{\infty} a_n b_n = \sum_{n=1}^{\infty} (0.02)^n = \frac{0.02}{1-0.02} = \frac{1}{49}$, but

$AB = \frac{1}{9} \cdot \frac{1}{4} = \frac{1}{36}$.

EXERCISES

1. $\left\{ \dfrac{2+n^3}{1+2n^3} \right\}$ converges since $\displaystyle\lim_{n\to\infty} \frac{2+n^3}{1+2n^3} = \lim_{n\to\infty} \frac{2/n^3+1}{1/n^3+2} = \frac{1}{2}$.

2. $a_n = \dfrac{9^{n+1}}{10^n} = 9 \cdot \left(\frac{9}{10}\right)^n$, so $\displaystyle\lim_{n\to\infty} a_n = 9 \lim_{n\to\infty} \left(\frac{9}{10}\right)^n = 9 \cdot 0 = 0$ by (12.1.9) [ET (11.1.9)].

3. $\displaystyle\lim_{n\to\infty} a_n = \lim_{n\to\infty} \frac{n^3}{1+n^2} = \lim_{n\to\infty} \frac{n}{1/n^2 + 1} = \infty$, so the sequence diverges.

4. $a_n = \cos(n\pi/2)$, so $a_n = 0$ if n is odd and $a_n = \pm 1$ if n is even. As n increases, a_n keeps cycling through the values

0, 1, 0, −1, so the sequence $\{a_n\}$ is divergent.

5. $|a_n| = \left| \dfrac{n \sin n}{n^2+1} \right| \leq \dfrac{n}{n^2+1} < \dfrac{1}{n}$, so $|a_n| \to 0$ as $n \to \infty$. Thus, $\displaystyle\lim_{n\to\infty} a_n = 0$. The sequence $\{a_n\}$ is convergent.

6. $a_n = \dfrac{\ln n}{\sqrt{n}}$. Let $f(x) = \dfrac{\ln x}{\sqrt{x}}$ for $x > 0$. Then $\displaystyle\lim_{x\to\infty} f(x) = \lim_{x\to\infty} \frac{\ln x}{\sqrt{x}} \overset{H}{=} \lim_{x\to\infty} \frac{1/x}{1/(2\sqrt{x})} = \lim_{x\to\infty} \frac{2}{\sqrt{x}} = 0$.

Thus, by Theorem 3 in Section 12.1 [ET 11.1], $\{a_n\}$ converges and $\displaystyle\lim_{n\to\infty} a_n = 0$.

7. $\left\{ \left(1+\dfrac{3}{n}\right)^{4n} \right\}$ is convergent. Let $y = \left(1+\dfrac{3}{x}\right)^{4x}$. Then

$\displaystyle\lim_{x\to\infty} \ln y = \lim_{x\to\infty} 4x \ln(1+3/x) = \lim_{x\to\infty} \frac{\ln(1+3/x)}{1/(4x)} \overset{H}{=} \lim_{x\to\infty} \frac{\dfrac{1}{1+3/x}\left(-\dfrac{3}{x^2}\right)}{-1/(4x^2)} = \lim_{x\to\infty} \frac{12}{1+3/x} = 12$, so

$\displaystyle\lim_{x\to\infty} y = \lim_{n\to\infty} \left(1+\frac{3}{n}\right)^{4n} = e^{12}$.

8. $\left\{ \dfrac{(-10)^n}{n!} \right\}$ converges, since $\dfrac{10^n}{n!} = \dfrac{10 \cdot 10 \cdot 10 \cdots \cdots 10}{1 \cdot 2 \cdot 3 \cdots \cdots 10} \cdot \dfrac{10 \cdot 10 \cdots \cdots 10}{11 \cdot 12 \cdots \cdots n} \leq 10^{10} \left(\dfrac{10}{11}\right)^{n-10} \to 0$ as $n \to \infty$, so

$\displaystyle\lim_{n\to\infty} \frac{(-10)^n}{n!} = 0$ [Squeeze Theorem]. *Or:* Use (12.10.10) [ET (11.10.10)].

9. We use induction, hypothesizing that $a_{n-1} < a_n < 2$. Note first that $1 < a_2 = \frac{1}{3}(1+4) = \frac{5}{3} < 2$, so the hypothesis holds for $n = 2$. Now assume that $a_{k-1} < a_k < 2$. Then $a_k = \frac{1}{3}(a_{k-1}+4) < \frac{1}{3}(a_k+4) < \frac{1}{3}(2+4) = 2$. So $a_k < a_{k+1} < 2$, and the induction is complete. To find the limit of the sequence, we note that $L = \lim\limits_{n\to\infty} a_n = \lim\limits_{n\to\infty} a_{n+1} \Rightarrow$

$L = \frac{1}{3}(L+4) \Rightarrow L = 2$.

10. $\lim\limits_{x\to\infty} \dfrac{x^4}{e^x} \overset{H}{=} \lim\limits_{x\to\infty} \dfrac{4x^3}{e^x} \overset{H}{=} \lim\limits_{x\to\infty} \dfrac{12x^2}{e^x} \overset{H}{=} \lim\limits_{x\to\infty} \dfrac{24x}{e^x} \overset{H}{=} \lim\limits_{x\to\infty} \dfrac{24}{e^x} = 0$

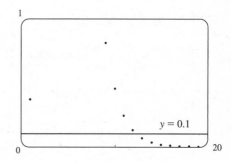

Then we conclude from Theorem 12.1.3 [ET 11.1.3] that

$\lim\limits_{n\to\infty} n^4 e^{-n} = 0$. From the graph, it seems that $12^4 e^{-12} > 0.1$,

but $n^4 e^{-n} < 0.1$ whenever $n > 12$. So the smallest value of N

corresponding to $\varepsilon = 0.1$ in the definition of the limit is $N = 12$.

11. $\dfrac{n}{n^3+1} < \dfrac{n}{n^3} = \dfrac{1}{n^2}$, so $\sum\limits_{n=1}^{\infty} \dfrac{n}{n^3+1}$ converges by the Comparison Test with the convergent p-series $\sum\limits_{n=1}^{\infty} \dfrac{1}{n^2}$ $[p = 2 > 1]$.

12. Let $a_n = \dfrac{n^2+1}{n^3+1}$ and $b_n = \dfrac{1}{n}$, so $\lim\limits_{n\to\infty} \dfrac{a_n}{b_n} = \lim\limits_{n\to\infty} \dfrac{n^3+n}{n^3+1} = \lim\limits_{n\to\infty} \dfrac{1+1/n^2}{1+1/n^3} = 1 > 0$.

Since $\sum\limits_{n=1}^{\infty} b_n$ is the divergent harmonic series, $\sum\limits_{n=1}^{\infty} a_n$ also diverges by the Limit Comparison Test.

13. $\lim\limits_{n\to\infty} \left| \dfrac{a_{n+1}}{a_n} \right| = \lim\limits_{n\to\infty} \left[\dfrac{(n+1)^3}{5^{n+1}} \cdot \dfrac{5^n}{n^3} \right] = \lim\limits_{n\to\infty} \left(1 + \dfrac{1}{n}\right)^3 \cdot \dfrac{1}{5} = \dfrac{1}{5} < 1$, so $\sum\limits_{n=1}^{\infty} \dfrac{n^3}{5^n}$ converges by the Ratio Test.

14. Let $b_n = \dfrac{1}{\sqrt{n+1}}$. Then b_n is positive for $n \geq 1$, the sequence $\{b_n\}$ is decreasing, and $\lim\limits_{n\to\infty} b_n = 0$, so the series

$\sum\limits_{n=1}^{\infty} \dfrac{(-1)^n}{\sqrt{n+1}}$ converges by the Alternating Series Test.

15. Let $f(x) = \dfrac{1}{x\sqrt{\ln x}}$. Then f is continuous, positive, and decreasing on $[2, \infty)$, so the Integral Test applies.

$$\int_2^{\infty} f(x)\, dx = \lim\limits_{t\to\infty} \int_2^t \dfrac{1}{x\sqrt{\ln x}}\, dx \;\left[u = \ln x,\, du = \dfrac{1}{x}\, dx \right] = \lim\limits_{t\to\infty} \int_{\ln 2}^{\ln t} u^{-1/2}\, du = \lim\limits_{t\to\infty} \left[2\sqrt{u} \right]_{\ln 2}^{\ln t}$$

$$= \lim\limits_{t\to\infty} \left(2\sqrt{\ln t} - 2\sqrt{\ln 2} \right) = \infty,$$

so the series $\sum\limits_{n=2}^{\infty} \dfrac{1}{n\sqrt{\ln n}}$ diverges.

16. $\lim\limits_{n\to\infty} \dfrac{n}{3n+1} = \dfrac{1}{3}$, so $\lim\limits_{n\to\infty} \ln\left(\dfrac{n}{3n+1} \right) = \ln\dfrac{1}{3} \neq 0$. Thus, the series $\sum\limits_{n=1}^{\infty} \ln\left(\dfrac{n}{3n+1} \right)$ diverges by the Test for Divergence.

17. $|a_n| = \left| \dfrac{\cos 3n}{1+(1.2)^n} \right| \leq \dfrac{1}{1+(1.2)^n} < \dfrac{1}{(1.2)^n} = \left(\dfrac{5}{6} \right)^n$, so $\sum\limits_{n=1}^{\infty} |a_n|$ converges by comparison with the convergent geometric

series $\sum\limits_{n=1}^{\infty} \left(\dfrac{5}{6} \right)^n$ $\left[r = \dfrac{5}{6} < 1 \right]$. It follows that $\sum\limits_{n=1}^{\infty} a_n$ converges (by Theorem 3 in Section 12.6 [ET 11.6]).

18. $\lim\limits_{n\to\infty} \sqrt[n]{|a_n|} = \lim\limits_{n\to\infty} \sqrt[n]{\left|\dfrac{n^{2n}}{(1+2n^2)^n}\right|} = \lim\limits_{n\to\infty} \dfrac{n^2}{1+2n^2} = \lim\limits_{n\to\infty} \dfrac{1}{1/n^2+2} = \dfrac{1}{2} < 1$, so $\sum\limits_{n=1}^{\infty} \dfrac{n^{2n}}{(1+2n^2)^n}$ converges by the

Root Test.

19. $\lim\limits_{n\to\infty} \left|\dfrac{a_{n+1}}{a_n}\right| = \lim\limits_{n\to\infty} \dfrac{1 \cdot 3 \cdot 5 \cdots (2n-1)(2n+1)}{5^{n+1}(n+1)!} \cdot \dfrac{5^n\, n!}{1 \cdot 3 \cdot 5 \cdots (2n-1)} = \lim\limits_{n\to\infty} \dfrac{2n+1}{5(n+1)} = \dfrac{2}{5} < 1$, so the series

converges by the Ratio Test.

20. $\sum\limits_{n=1}^{\infty} \dfrac{(-5)^{2n}}{n^2\, 9^n} = \sum\limits_{n=1}^{\infty} \dfrac{1}{n^2}\left(\dfrac{25}{9}\right)^n$. Now $\lim\limits_{n\to\infty} \left|\dfrac{a_{n+1}}{a_n}\right| = \lim\limits_{n\to\infty} \dfrac{25^{n+1}}{(n+1)^2 \cdot 9^{n+1}} \cdot \dfrac{n^2 \cdot 9^n}{25^n} = \lim\limits_{n\to\infty} \dfrac{25n^2}{9(n+1)^2} = \dfrac{25}{9} > 1$,

so the series diverges by the Ratio Test.

21. $b_n = \dfrac{\sqrt{n}}{n+1} > 0$, $\{b_n\}$ is decreasing, and $\lim\limits_{n\to\infty} b_n = 0$, so the series $\sum\limits_{n=1}^{\infty} (-1)^{n-1} \dfrac{\sqrt{n}}{n+1}$ converges by the Alternating

Series Test.

22. Use the Limit Comparison Test with $a_n = \dfrac{\sqrt{n+1}-\sqrt{n-1}}{n} = \dfrac{2}{n(\sqrt{n+1}+\sqrt{n-1})}$ (rationalizing the numerator) and

$b_n = \dfrac{1}{n^{3/2}}$. $\lim\limits_{n\to\infty} \dfrac{a_n}{b_n} = \lim\limits_{n\to\infty} \dfrac{2\sqrt{n}}{\sqrt{n+1}+\sqrt{n-1}} = 1$, so since $\sum\limits_{n=1}^{\infty} b_n$ converges $\left[p = \frac{3}{2} > 1\right]$, $\sum\limits_{n=1}^{\infty} a_n$ converges also.

23. Consider the series of absolute values: $\sum\limits_{n=1}^{\infty} n^{-1/3}$ is a p-series with $p = \frac{1}{3} \le 1$ and is therefore divergent. But if we apply the

Alternating Series Test, we see that $b_n = \dfrac{1}{\sqrt[3]{n}} > 0$, $\{b_n\}$ is decreasing, and $\lim\limits_{n\to\infty} b_n = 0$, so the series $\sum\limits_{n=1}^{\infty} (-1)^{n-1}\, n^{-1/3}$

converges. Thus, $\sum\limits_{n=1}^{\infty} (-1)^{n-1}\, n^{-1/3}$ is conditionally convergent.

24. $\sum\limits_{n=1}^{\infty} |(-1)^{n-1}\, n^{-3}| = \sum\limits_{n=1}^{\infty} n^{-3}$ is a convergent p-series $[p = 3 > 1]$. Therefore, $\sum\limits_{n=1}^{\infty} (-1)^{n-1}\, n^{-3}$ is absolutely convergent.

25. $\left|\dfrac{a_{n+1}}{a_n}\right| = \left|\dfrac{(-1)^{n+1}(n+2)3^{n+1}}{2^{2n+3}} \cdot \dfrac{2^{2n+1}}{(-1)^n(n+1)3^n}\right| = \dfrac{n+2}{n+1} \cdot \dfrac{3}{4} = \dfrac{1+(2/n)}{1+(1/n)} \cdot \dfrac{3}{4} \to \dfrac{3}{4} < 1$ as $n \to \infty$, so by the Ratio

Test, $\sum\limits_{n=1}^{\infty} \dfrac{(-1)^n(n+1)3^n}{2^{2n+1}}$ is absolutely convergent.

26. $\lim\limits_{x\to\infty} \dfrac{\sqrt{x}}{\ln x} \overset{\text{H}}{=} \lim\limits_{x\to\infty} \dfrac{1/(2\sqrt{x})}{1/x} = \lim\limits_{x\to\infty} \dfrac{\sqrt{x}}{2} = \infty$. Therefore, $\lim\limits_{n\to\infty} \dfrac{(-1)^n \sqrt{n}}{\ln n} \ne 0$, so the given series is divergent by the

Test for Divergence.

27. $\sum\limits_{n=1}^{\infty} \dfrac{(-3)^{n-1}}{2^{3n}} = \sum\limits_{n=1}^{\infty} \dfrac{(-3)^{n-1}}{(2^3)^n} = \sum\limits_{n=1}^{\infty} \dfrac{(-3)^{n-1}}{8^n} = \dfrac{1}{8}\sum\limits_{n=1}^{\infty} \dfrac{(-3)^{n-1}}{8^{n-1}} = \dfrac{1}{8}\sum\limits_{n=1}^{\infty} \left(-\dfrac{3}{8}\right)^{n-1} = \dfrac{1}{8}\left(\dfrac{1}{1-(-3/8)}\right)$

$\qquad = \dfrac{1}{8} \cdot \dfrac{8}{11} = \dfrac{1}{11}$

28. $\displaystyle\sum_{n=1}^{\infty} \frac{1}{n(n+3)} = \sum_{n=1}^{\infty} \left[\frac{1}{3n} - \frac{1}{3(n+3)}\right]$ [partial fractions].

$s_n = \displaystyle\sum_{i=1}^{n} \left[\frac{1}{3i} - \frac{1}{3(i+3)}\right] = \frac{1}{3} + \frac{1}{6} + \frac{1}{9} - \frac{1}{3(n+1)} - \frac{1}{3(n+2)} - \frac{1}{3(n+3)}$ (telescoping sum), so

$\displaystyle\sum_{n=1}^{\infty} \frac{1}{n(n+3)} = \lim_{n\to\infty} s_n = \frac{1}{3} + \frac{1}{6} + \frac{1}{9} = \frac{11}{18}$.

29. $\displaystyle\sum_{n=1}^{\infty} [\tan^{-1}(n+1) - \tan^{-1} n] = \lim_{n\to\infty} s_n$

$= \displaystyle\lim_{n\to\infty} [(\tan^{-1} 2 - \tan^{-1} 1) + (\tan^{-1} 3 - \tan^{-1} 2) + \cdots + (\tan^{-1}(n+1) - \tan^{-1} n)]$

$= \displaystyle\lim_{n\to\infty} [\tan^{-1}(n+1) - \tan^{-1} 1] = \frac{\pi}{2} - \frac{\pi}{4} = \frac{\pi}{4}$

30. $\displaystyle\sum_{n=0}^{\infty} \frac{(-1)^n \pi^n}{3^{2n}\,(2n)!} = \sum_{n=0}^{\infty} (-1)^n \frac{1}{(2n)!} \cdot \frac{\pi^n}{3^{2n}} = \sum_{n=0}^{\infty} (-1)^n \frac{1}{(2n)!} \cdot \left(\frac{\sqrt{\pi}}{3}\right)^{2n} = \cos\left(\frac{\sqrt{\pi}}{3}\right)$ since $\cos x = \displaystyle\sum_{n=0}^{\infty} (-1)^n \frac{x^{2n}}{(2n)!}$

for all x.

31. $1 - e + \dfrac{e^2}{2!} - \dfrac{e^3}{3!} + \dfrac{e^4}{4!} - \cdots = \displaystyle\sum_{n=0}^{\infty} (-1)^n \frac{e^n}{n!} = \sum_{n=0}^{\infty} \frac{(-e)^n}{n!} = e^{-e}$ since $e^x = \displaystyle\sum_{n=0}^{\infty} \frac{x^n}{n!}$ for all x.

32. $4.17\overline{326} = 4.17 + \dfrac{326}{10^5} + \dfrac{326}{10^8} + \cdots = 4.17 + \dfrac{326/10^5}{1 - 1/10^3} = \dfrac{417}{100} + \dfrac{326}{99{,}900} = \dfrac{416{,}909}{99{,}900}$

33. $\cosh x = \dfrac{1}{2}(e^x + e^{-x}) = \dfrac{1}{2}\left(\displaystyle\sum_{n=0}^{\infty} \frac{x^n}{n!} + \sum_{n=0}^{\infty} \frac{(-x)^n}{n!}\right)$

$= \dfrac{1}{2}\left[\left(1 + x + \dfrac{x^2}{2!} + \dfrac{x^3}{3!} + \dfrac{x^4}{4!} + \cdots\right) + \left(1 - x + \dfrac{x^2}{2!} - \dfrac{x^3}{3!} + \dfrac{x^4}{4!} - \cdots\right)\right]$

$= \dfrac{1}{2}\left(2 + 2 \cdot \dfrac{x^2}{2!} + 2 \cdot \dfrac{x^4}{4!} + \cdots\right) = 1 + \dfrac{1}{2}x^2 + \displaystyle\sum_{n=2}^{\infty} \frac{x^{2n}}{(2n)!} \geq 1 + \dfrac{1}{2}x^2$ for all x

34. $\displaystyle\sum_{n=1}^{\infty} (\ln x)^n$ is a geometric series which converges whenever $|\ln x| < 1 \;\Rightarrow\; -1 < \ln x < 1 \;\Rightarrow\; e^{-1} < x < e$.

35. $\displaystyle\sum_{n=1}^{\infty} \frac{(-1)^{n+1}}{n^5} = 1 - \frac{1}{32} + \frac{1}{243} - \frac{1}{1024} + \frac{1}{3125} - \frac{1}{7776} + \frac{1}{16{,}807} - \frac{1}{32{,}768} + \cdots$.

Since $b_8 = \dfrac{1}{8^5} = \dfrac{1}{32{,}768} < 0.000031$, $\displaystyle\sum_{n=1}^{\infty} \frac{(-1)^{n+1}}{n^5} \approx \sum_{n=1}^{7} \frac{(-1)^{n+1}}{n^5} \approx 0.9721$.

36. (a) $s_5 = \displaystyle\sum_{n=1}^{5} \frac{1}{n^6} = 1 + \frac{1}{2^6} + \cdots + \frac{1}{5^6} \approx 1.017305$. The series $\displaystyle\sum_{n=1}^{\infty} \frac{1}{n^6}$ converges by the Integral Test, so we estimate the

remainder R_5 with (12.3.2) [ET (11.3.2)]: $R_5 \leq \displaystyle\int_5^{\infty} \frac{dx}{x^6} = \left[-\frac{x^{-5}}{5}\right]_5^{\infty} = \frac{5^{-5}}{5} = 0.000064$. So the error is at

most 0.000064.

(b) In general, $R_n \leq \displaystyle\int_n^{\infty} \frac{dx}{x^6} = \frac{1}{5n^5}$. If we take $n = 9$, then $s_9 \approx 1.01734$ and $R_9 \leq \dfrac{1}{5 \cdot 9^5} \approx 3.4 \times 10^{-6}$.

So to five decimal places, $\displaystyle\sum_{n=1}^{\infty} \frac{1}{n^5} \approx \sum_{n=1}^{9} \frac{1}{n^5} \approx 1.01734$.

Another method: Use (12.3.3) [ET (11.3.3)] instead of (12.3.2) [ET (11.3.2)].

37. $\displaystyle\sum_{n=1}^{\infty} \frac{1}{2+5^n} \approx \sum_{n=1}^{8} \frac{1}{2+5^n} \approx 0.18976224$. To estimate the error, note that $\dfrac{1}{2+5^n} < \dfrac{1}{5^n}$, so the remainder term is

$$R_8 = \sum_{n=9}^{\infty} \frac{1}{2+5^n} < \sum_{n=9}^{\infty} \frac{1}{5^n} = \frac{1/5^9}{1-1/5} = 6.4 \times 10^{-7} \quad [\text{geometric series with } a = \tfrac{1}{5^9} \text{ and } r = \tfrac{1}{5}].$$

38. (a) $\displaystyle\lim_{n\to\infty} \left| \frac{a_{n+1}}{a_n} \right| = \lim_{n\to\infty} \left| \frac{(n+1)^{n+1}}{[2(n+1)]!} \cdot \frac{(2n)!}{n^n} \right| = \lim_{n\to\infty} \frac{(n+1)^n (n+1)^1}{(2n+2)(2n+1)n^n} = \lim_{n\to\infty} \left(\frac{n+1}{n} \right)^n \frac{1}{2(2n+1)}$

$$= \lim_{n\to\infty} \left(1 + \frac{1}{n} \right)^n \frac{1}{2(2n+1)} = e \cdot 0 = 0 < 1$$

so the series converges by the Ratio Test.

(b) The series in part (a) is convergent, so $\displaystyle\lim_{n\to\infty} a_n = \lim_{n\to\infty} \frac{n^n}{(2n)!} = 0$ by Theorem 12.2.6 [ET 11.2.6].

39. Use the Limit Comparison Test. $\displaystyle\lim_{n\to\infty} \left| \frac{\left(\frac{n+1}{n}\right)a_n}{a_n} \right| = \lim_{n\to\infty} \frac{n+1}{n} = \lim_{n\to\infty} \left(1 + \frac{1}{n} \right) = 1 > 0.$

Since $\sum |a_n|$ is convergent, so is $\displaystyle\sum \left| \left(\frac{n+1}{n} \right) a_n \right|$, by the Limit Comparison Test.

40. $\displaystyle\lim_{n\to\infty} \left| \frac{a_{n+1}}{a_n} \right| = \lim_{n\to\infty} \left| \frac{x^{n+1}}{(n+1)^2 5^{n+1}} \cdot \frac{n^2 5^n}{x^n} \right| = \lim_{n\to\infty} \frac{1}{(1+1/n)^2} \frac{|x|}{5} = \frac{|x|}{5}$, so by the Ratio Test, $\displaystyle\sum_{n=1}^{\infty} (-1)^n \frac{x^n}{n^2 5^n}$

converges when $\dfrac{|x|}{5} < 1 \iff |x| < 5$, so $R = 5$. When $x = -5$, the series becomes the convergent p-series $\displaystyle\sum_{n=1}^{\infty} \frac{1}{n^2}$ with

$p = 2 > 1$. When $x = 5$, the series becomes $\displaystyle\sum_{n=1}^{\infty} \frac{(-1)^n}{n^2}$, which converges by the Alternating Series Test. Thus, $I = [-5, 5]$.

41. $\displaystyle\lim_{n\to\infty} \left| \frac{a_{n+1}}{a_n} \right| = \lim_{n\to\infty} \left[\frac{|x+2|^{n+1}}{(n+1)4^{n+1}} \cdot \frac{n\,4^n}{|x+2|^n} \right] = \lim_{n\to\infty} \left[\frac{n}{n+1} \frac{|x+2|}{4} \right] = \frac{|x+2|}{4} < 1 \iff |x+2| < 4$, so $R = 4$.

$|x+2| < 4 \iff -4 < x+2 < 4 \iff -6 < x < 2$. If $x = -6$, then the series $\displaystyle\sum_{n=1}^{\infty} \frac{(x+2)^n}{n\,4^n}$ becomes

$\displaystyle\sum_{n=1}^{\infty} \frac{(-4)^n}{n4^n} = \sum_{n=1}^{\infty} \frac{(-1)^n}{n}$, the alternating harmonic series, which converges by the Alternating Series Test. When $x = 2$, the

series becomes the harmonic series $\displaystyle\sum_{n=1}^{\infty} \frac{1}{n}$, which diverges. Thus, $I = [-6, 2)$.

42. $\displaystyle\lim_{n\to\infty} \left| \frac{a_{n+1}}{a_n} \right| = \lim_{n\to\infty} \left| \frac{2^{n+1} (x-2)^{n+1}}{(n+3)!} \cdot \frac{(n+2)!}{2^n(x-2)^n} \right| = \lim_{n\to\infty} \frac{2}{n+3} |x-2| = 0 < 1$, so the series $\displaystyle\sum_{n=1}^{\infty} \frac{2^n (x-2)^n}{(n+2)!}$

converges for all x. $R = \infty$ and $I = (-\infty, \infty)$.

43. $\displaystyle\lim_{n\to\infty} \left| \frac{a_{n+1}}{a_n} \right| = \lim_{n\to\infty} \left| \frac{2^{n+1}(x-3)^{n+1}}{\sqrt{n+4}} \cdot \frac{\sqrt{n+3}}{2^n(x-3)^n} \right| = 2|x-3| \lim_{n\to\infty} \sqrt{\frac{n+3}{n+4}} = 2|x-3| < 1 \iff |x-3| < \tfrac{1}{2}$,

so $R = \tfrac{1}{2}$. $|x-3| < \tfrac{1}{2} \iff -\tfrac{1}{2} < x-3 < \tfrac{1}{2} \iff \tfrac{5}{2} < x < \tfrac{7}{2}$. For $x = \tfrac{7}{2}$, the series $\displaystyle\sum_{n=1}^{\infty} \frac{2^n(x-3)^n}{\sqrt{n+3}}$ becomes

$\displaystyle\sum_{n=0}^{\infty} \frac{1}{\sqrt{n+3}} = \sum_{n=3}^{\infty} \frac{1}{n^{1/2}}$, which diverges $\left[p = \tfrac{1}{2} \le 1 \right]$, but for $x = \tfrac{5}{2}$, we get $\displaystyle\sum_{n=0}^{\infty} \frac{(-1)^n}{\sqrt{n+3}}$, which is a convergent

alternating series, so $I = \left[\tfrac{5}{2}, \tfrac{7}{2} \right)$.

44. $\lim\limits_{n\to\infty}\left|\dfrac{a_{n+1}}{a_n}\right| = \lim\limits_{n\to\infty}\left|\dfrac{(2n+2)!\,x^{n+1}}{[(n+1)!]^2}\cdot\dfrac{(n!)^2}{(2n)!\,x^n}\right| = \lim\limits_{n\to\infty}\dfrac{(2n+2)(2n+1)}{(n+1)(n+1)}\,|x| = 4\,|x|.$

To converge, we must have $4\,|x| < 1 \quad\Leftrightarrow\quad |x| < \frac14$, so $R = \frac14$.

45.

n	$f^{(n)}(x)$	$f^{(n)}\!\left(\frac{\pi}{6}\right)$
0	$\sin x$	$\frac12$
1	$\cos x$	$\frac{\sqrt3}{2}$
2	$-\sin x$	$-\frac12$
3	$-\cos x$	$-\frac{\sqrt3}{2}$
4	$\sin x$	$\frac12$
$\vdots$	$\vdots$	$\vdots$

$\sin x = f\!\left(\dfrac{\pi}{6}\right) + f'\!\left(\dfrac{\pi}{6}\right)\!\left(x - \dfrac{\pi}{6}\right) + \dfrac{f''\!\left(\frac{\pi}{6}\right)}{2!}\!\left(x - \dfrac{\pi}{6}\right)^2 + \dfrac{f^{(3)}\!\left(\frac{\pi}{6}\right)}{3!}\!\left(x - \dfrac{\pi}{6}\right)^3 + \dfrac{f^{(4)}\!\left(\frac{\pi}{6}\right)}{4!}\!\left(x - \dfrac{\pi}{6}\right)^4 + \cdots$

$= \dfrac12\left[1 - \dfrac{1}{2!}\!\left(x - \dfrac{\pi}{6}\right)^2 + \dfrac{1}{4!}\!\left(x - \dfrac{\pi}{6}\right)^4 - \cdots\right] + \dfrac{\sqrt3}{2}\left[\left(x - \dfrac{\pi}{6}\right) - \dfrac{1}{3!}\!\left(x - \dfrac{\pi}{6}\right)^3 + \cdots\right]$

$= \dfrac12\displaystyle\sum_{n=0}^{\infty}(-1)^n\dfrac{1}{(2n)!}\!\left(x - \dfrac{\pi}{6}\right)^{2n} + \dfrac{\sqrt3}{2}\sum_{n=0}^{\infty}(-1)^n\dfrac{1}{(2n+1)!}\!\left(x - \dfrac{\pi}{6}\right)^{2n+1}$

46.

n	$f^{(n)}(x)$	$f^{(n)}\!\left(\frac{\pi}{3}\right)$
0	$\cos x$	$\frac12$
1	$-\sin x$	$-\frac{\sqrt3}{2}$
2	$-\cos x$	$-\frac12$
3	$\sin x$	$\frac{\sqrt3}{2}$
4	$\cos x$	$\frac12$
$\vdots$	$\vdots$	$\vdots$

$\cos x = f\!\left(\dfrac{\pi}{3}\right) + f'\!\left(\dfrac{\pi}{3}\right)\!\left(x - \dfrac{\pi}{3}\right) + \dfrac{f''\!\left(\frac{\pi}{3}\right)}{2!}\!\left(x - \dfrac{\pi}{3}\right)^2 + \dfrac{f^{(3)}\!\left(\frac{\pi}{3}\right)}{3!}\!\left(x - \dfrac{\pi}{3}\right)^3 + \dfrac{f^{(4)}\!\left(\frac{\pi}{3}\right)}{4!}\!\left(x - \dfrac{\pi}{3}\right)^4 + \cdots$

$= \dfrac12\left[1 - \dfrac{1}{2!}\!\left(x - \dfrac{\pi}{3}\right)^2 + \dfrac{1}{4!}\!\left(x - \dfrac{\pi}{3}\right)^4 - \cdots\right] + \dfrac{\sqrt3}{2}\left[-\left(x - \dfrac{\pi}{3}\right) + \dfrac{1}{3!}\!\left(x - \dfrac{\pi}{3}\right)^3 - \cdots\right]$

$= \dfrac12\displaystyle\sum_{n=0}^{\infty}(-1)^n\dfrac{1}{(2n)!}\!\left(x - \dfrac{\pi}{3}\right)^{2n} + \dfrac{\sqrt3}{2}\sum_{n=0}^{\infty}(-1)^{n+1}\dfrac{1}{(2n+1)!}\!\left(x - \dfrac{\pi}{3}\right)^{2n+1}$

47. $\dfrac{1}{1+x} = \dfrac{1}{1-(-x)} = \displaystyle\sum_{n=0}^{\infty}(-x)^n = \sum_{n=0}^{\infty}(-1)^n x^n$ for $|x| < 1 \quad\Rightarrow\quad \dfrac{x^2}{1+x} = \displaystyle\sum_{n=0}^{\infty}(-1)^n x^{n+2}$ with $R = 1$.

48. $\tan^{-1}x = \displaystyle\sum_{n=0}^{\infty}(-1)^n\dfrac{x^{2n+1}}{2n+1}$ with interval of convergence $[-1, 1]$, so

$\tan^{-1}(x^2) = \displaystyle\sum_{n=0}^{\infty}(-1)^n\dfrac{(x^2)^{2n+1}}{2n+1} = \sum_{n=0}^{\infty}(-1)^n\dfrac{x^{4n+2}}{2n+1}$, which converges when $x^2 \in [-1, 1] \quad\Leftrightarrow\quad x \in [-1, 1]$.

Therefore, $R = 1$.

49. $\dfrac{1}{1-x} = \sum\limits_{n=0}^{\infty} x^n$ for $|x| < 1$ $\Rightarrow$ $\ln(1-x) = -\displaystyle\int \dfrac{dx}{1-x} = -\int \sum\limits_{n=0}^{\infty} x^n \, dx = C - \sum\limits_{n=0}^{\infty} \dfrac{x^{n+1}}{n+1}$.

$\ln(1-0) = C - 0$ $\Rightarrow$ $C = 0$ $\Rightarrow$ $\ln(1-x) = -\sum\limits_{n=0}^{\infty} \dfrac{x^{n+1}}{n+1} = \sum\limits_{n=1}^{\infty} \dfrac{-x^n}{n}$ with $R = 1$.

50. $e^x = \sum\limits_{n=0}^{\infty} \dfrac{x^n}{n!}$ $\Rightarrow$ $e^{2x} = \sum\limits_{n=0}^{\infty} \dfrac{(2x)^n}{n!}$ $\Rightarrow$ $xe^{2x} = x\sum\limits_{n=0}^{\infty} \dfrac{2^n x^n}{n!} = \sum\limits_{n=0}^{\infty} \dfrac{2^n x^{n+1}}{n!}$, $R = \infty$

51. $\sin x = \sum\limits_{n=0}^{\infty} \dfrac{(-1)^n x^{2n+1}}{(2n+1)!}$ $\Rightarrow$ $\sin(x^4) = \sum\limits_{n=0}^{\infty} \dfrac{(-1)^n (x^4)^{2n+1}}{(2n+1)!} = \sum\limits_{n=0}^{\infty} \dfrac{(-1)^n x^{8n+4}}{(2n+1)!}$ for all x, so the radius of

convergence is ∞.

52. $e^x = \sum\limits_{n=0}^{\infty} \dfrac{x^n}{n!}$ $\Rightarrow$ $10^x = e^{(\ln 10)x} = \sum\limits_{n=0}^{\infty} \dfrac{[(\ln 10)x]^n}{n!} = \sum\limits_{n=0}^{\infty} \dfrac{(\ln 10)^n x^n}{n!}$, $R = \infty$

53. $f(x) = \dfrac{1}{\sqrt[4]{16-x}} = \dfrac{1}{\sqrt[4]{16(1-x/16)}} = \dfrac{1}{\sqrt[4]{16}\left(1-\frac{1}{16}x\right)^{1/4}} = \frac{1}{2}\left(1-\frac{1}{16}x\right)^{-1/4}$

$= \dfrac{1}{2}\left[1 + \left(-\dfrac{1}{4}\right)\left(-\dfrac{x}{16}\right) + \dfrac{\left(-\frac{1}{4}\right)\left(-\frac{5}{4}\right)}{2!}\left(-\dfrac{x}{16}\right)^2 + \dfrac{\left(-\frac{1}{4}\right)\left(-\frac{5}{4}\right)\left(-\frac{9}{4}\right)}{3!}\left(-\dfrac{x}{16}\right)^3 + \cdots \right]$

$= \dfrac{1}{2} + \sum\limits_{n=1}^{\infty} \dfrac{1 \cdot 5 \cdot 9 \cdot \,\cdots\, \cdot (4n-3)}{2 \cdot 4^n \cdot n! \cdot 16^n} x^n = \dfrac{1}{2} + \sum\limits_{n=1}^{\infty} \dfrac{1 \cdot 5 \cdot 9 \cdot \,\cdots\, \cdot (4n-3)}{2^{6n+1}\, n!} x^n$

for $\left|-\dfrac{x}{16}\right| < 1$ $\Leftrightarrow$ $|x| < 16$, so $R = 16$.

54. $(1-3x)^{-5} = \sum\limits_{n=0}^{\infty} \binom{-5}{n}(-3x)^n = 1 + (-5)(-3x) + \dfrac{(-5)(-6)}{2!}(-3x)^2 + \dfrac{(-5)(-6)(-7)}{3!}(-3x)^3 + \cdots$

$= 1 + \sum\limits_{n=1}^{\infty} \dfrac{5 \cdot 6 \cdot 7 \cdot \,\cdots\, \cdot (n+4) \cdot 3^n x^n}{n!}$ for $|-3x| < 1$ $\Leftrightarrow$ $|x| < \frac{1}{3}$, so $R = \frac{1}{3}$.

55. $e^x = \sum\limits_{n=0}^{\infty} \dfrac{x^n}{n!}$, so $\dfrac{e^x}{x} = \dfrac{1}{x}\sum\limits_{n=0}^{\infty} \dfrac{x^n}{n!} = \sum\limits_{n=0}^{\infty} \dfrac{x^{n-1}}{n!} = x^{-1} + \sum\limits_{n=1}^{\infty} \dfrac{x^{n-1}}{n!} = \dfrac{1}{x} + \sum\limits_{n=1}^{\infty} \dfrac{x^{n-1}}{n!}$ and

$\displaystyle\int \dfrac{e^x}{x}\, dx = C + \ln|x| + \sum\limits_{n=1}^{\infty} \dfrac{x^n}{n \cdot n!}$.

56. $(1+x^4)^{1/2} = \sum\limits_{n=0}^{\infty} \binom{\frac{1}{2}}{n}(x^4)^n = 1 + \left(\frac{1}{2}\right)x^4 + \dfrac{\left(\frac{1}{2}\right)\left(-\frac{1}{2}\right)}{2!}(x^4)^2 + \dfrac{\left(\frac{1}{2}\right)\left(-\frac{1}{2}\right)\left(-\frac{3}{2}\right)}{3!}(x^4)^3 + \cdots$

$= 1 + \frac{1}{2}x^4 - \frac{1}{8}x^8 + \frac{1}{16}x^{12} - \cdots$

so $\int_0^1 (1+x^4)^{1/2}\, dx = \left[x + \frac{1}{10}x^5 - \frac{1}{72}x^9 + \frac{1}{208}x^{13} - \cdots\right]_0^1 = 1 + \frac{1}{10} - \frac{1}{72} + \frac{1}{208} - \cdots$.

This is an alternating series, so by the Alternating Series Test, the error in the approximation

$\int_0^1 (1+x^4)^{1/2}\, dx \approx 1 + \frac{1}{10} - \frac{1}{72} \approx 1.086$ is less than $\frac{1}{208}$, sufficient for the desired accuracy.

Thus, correct to two decimal places, $\int_0^1 (1+x^4)^{1/2}\, dx \approx 1.09$.

57. (a)

n	$f^{(n)}(x)$	$f^{(n)}(1)$
0	$x^{1/2}$	1
1	$\frac{1}{2}x^{-1/2}$	$\frac{1}{2}$
2	$-\frac{1}{4}x^{-3/2}$	$-\frac{1}{4}$
3	$\frac{3}{8}x^{-5/2}$	$\frac{3}{8}$
4	$-\frac{15}{16}x^{-7/2}$	$-\frac{15}{16}$
⋮	⋮	⋮

$$\sqrt{x} \approx T_3(x) = 1 + \frac{1/2}{1!}(x-1) - \frac{1/4}{2!}(x-1)^2 + \frac{3/8}{3!}(x-1)^3$$
$$= 1 + \tfrac{1}{2}(x-1) - \tfrac{1}{8}(x-1)^2 + \tfrac{1}{16}(x-1)^3$$

(b)

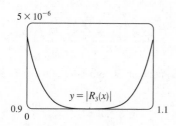

(c) $|R_3(x)| \le \dfrac{M}{4!}|x-1|^4$, where $\left|f^{(4)}(x)\right| \le M$ with

$f^{(4)}(x) = -\frac{15}{16}x^{-7/2}$. Now $0.9 \le x \le 1.1 \Rightarrow$

$-0.1 \le x-1 \le 0.1 \Rightarrow (x-1)^4 \le (0.1)^4$,

and letting $x = 0.9$ gives $M = \dfrac{15}{16(0.9)^{7/2}}$, so

$$|R_3(x)| \le \frac{15}{16(0.9)^{7/2}\, 4!}(0.1)^4 \approx 0.000\,005\,648$$
$$\approx 0.000\,006 = 6 \times 10^{-6}$$

(d)

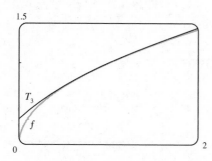

From the graph of $|R_3(x)| = |\sqrt{x} - T_3(x)|$, it appears

that the error is less than 5×10^{-6} on $[0.9, 1.1]$.

58. (a)

n	$f^{(n)}(x)$	$f^{(n)}(0)$
0	$\sec x$	1
1	$\sec x \tan x$	0
2	$\sec x \tan^2 x + \sec^3 x$	1
3	$\sec x \tan^3 x + 5\sec^3 x \tan x$	0
⋮	⋮	⋮

$$\sec x \approx T_2(x) = 1 + \tfrac{1}{2}x^2$$

(b)

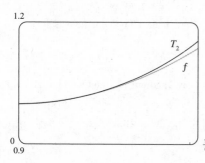

(c) $|R_2(x)| \le \dfrac{M}{3!}|x|^3$, where $\left|f^{(3)}(x)\right| \le M$ with

$f^{(3)}(x) = \sec x \tan^3 x + 5\sec^3 x \tan x$.

Now $0 \le x \le \frac{\pi}{6} \Rightarrow x^3 \le \left(\frac{\pi}{6}\right)^3$, and letting $x = \frac{\pi}{6}$ gives

$M = \frac{14}{3}$, so $|R_2(x)| \le \dfrac{14}{3 \cdot 6}\left(\dfrac{\pi}{6}\right)^3 \approx 0.111648$.

(d)

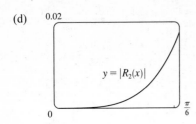

From the graph of $|R_2(x)| = |\sec x - T_2(x)|$, it appears that the error is less than 0.02 on $\left[0, \frac{\pi}{6}\right]$.

59. $\sin x = \sum\limits_{n=0}^{\infty} (-1)^n \dfrac{x^{2n+1}}{(2n+1)!} = x - \dfrac{x^3}{3!} + \dfrac{x^5}{5!} - \dfrac{x^7}{7!} + \cdots$, so $\sin x - x = -\dfrac{x^3}{3!} + \dfrac{x^5}{5!} - \dfrac{x^7}{7!} + \cdots$ and

$\dfrac{\sin x - x}{x^3} = -\dfrac{1}{3!} + \dfrac{x^2}{5!} - \dfrac{x^4}{7!} + \cdots$. Thus, $\lim\limits_{x\to 0} \dfrac{\sin x - x}{x^3} = \lim\limits_{x\to 0} \left(-\dfrac{1}{6} + \dfrac{x^2}{120} - \dfrac{x^4}{5040} + \cdots\right) = -\dfrac{1}{6}$.

60. (a) $F = \dfrac{mgR^2}{(R+h)^2} = \dfrac{mg}{(1+h/R)^2} = mg \sum\limits_{n=0}^{\infty} \binom{-2}{n} \left(\dfrac{h}{R}\right)^n$ [binomial series]

(b) We expand $F = mg \left[1 - 2\,(h/R) + 3\,(h/R)^2 - \cdots\right]$.

This is an alternating series, so by the Alternating Series Estimation Theorem, the error in the approximation $F = mg$ is less than $2mgh/R$, so for accuracy within 1% we want

$\left|\dfrac{2mgh/R}{mgR^2/(R+h)^2}\right| < 0.01 \quad\Leftrightarrow\quad \dfrac{2h(R+h)^2}{R^3} < 0.01.$

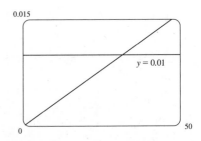

This inequality would be difficult to solve for h, so we substitute $R = 6{,}400$ km and plot both sides of the inequality. It appears that the approximation is accurate to within 1% for $h < 31$ km.

61. $f(x) = \sum\limits_{n=0}^{\infty} c_n x^n \quad\Rightarrow\quad f(-x) = \sum\limits_{n=0}^{\infty} c_n (-x)^n = \sum\limits_{n=0}^{\infty} (-1)^n c_n x^n$

(a) If f is an odd function, then $f(-x) = -f(x) \quad\Rightarrow\quad \sum\limits_{n=0}^{\infty} (-1)^n c_n x^n = \sum\limits_{n=0}^{\infty} -c_n x^n$. The coefficients of any power series are uniquely determined (by Theorem 12.10.5 [ET 11.10.5]), so $(-1)^n c_n = -c_n$.

If n is even, then $(-1)^n = 1$, so $c_n = -c_n \quad\Rightarrow\quad 2c_n = 0 \quad\Rightarrow\quad c_n = 0$. Thus, all even coefficients are 0, that is, $c_0 = c_2 = c_4 = \cdots = 0$.

(b) If f is even, then $f(-x) = f(x) \quad\Rightarrow\quad \sum\limits_{n=0}^{\infty} (-1)^n c_n x^n = \sum\limits_{n=0}^{\infty} c_n x^n \quad\Rightarrow\quad (-1)^n c_n = c_n$.

If n is odd, then $(-1)^n = -1$, so $-c_n = c_n \quad\Rightarrow\quad 2c_n = 0 \quad\Rightarrow\quad c_n = 0$. Thus, all odd coefficients are 0, that is, $c_1 = c_3 = c_5 = \cdots = 0$.

62. $e^x = \sum\limits_{n=0}^{\infty} \dfrac{x^n}{n!} \quad\Rightarrow\quad f(x) = e^{x^2} = \sum\limits_{n=0}^{\infty} \dfrac{(x^2)^n}{n!} = \sum\limits_{n=0}^{\infty} \dfrac{x^{2n}}{n!} = \sum\limits_{n=0}^{\infty} \dfrac{1}{n!} x^{2n}$. By Theorem 12.10.6 [ET 11.10.6] with $a = 0$,

we also have $f(x) = \sum\limits_{k=0}^{\infty} \dfrac{f^{(k)}(0)}{k!} x^k$. Comparing coefficients for $k = 2n$, we have $\dfrac{f^{(2n)}(0)}{(2n)!} = \dfrac{1}{n!} \quad\Rightarrow\quad f^{(2n)}(0) = \dfrac{(2n)!}{n!}$.

⬜ PROBLEMS PLUS

1. It would be far too much work to compute 15 derivatives of f. The key idea is to remember that $f^{(n)}(0)$ occurs in the coefficient of x^n in the Maclaurin series of f. We start with the Maclaurin series for sin: $\sin x = x - \dfrac{x^3}{3!} + \dfrac{x^5}{5!} - \cdots$.

Then $\sin(x^3) = x^3 - \dfrac{x^9}{3!} + \dfrac{x^{15}}{5!} - \cdots$, and so the coefficient of x^{15} is $\dfrac{f^{(15)}(0)}{15!} = \dfrac{1}{5!}$. Therefore,

$$f^{(15)}(0) = \frac{15!}{5!} = 6 \cdot 7 \cdot 8 \cdot 9 \cdot 10 \cdot 11 \cdot 12 \cdot 13 \cdot 14 \cdot 15 = 10{,}897{,}286{,}400.$$

2. We use the problem-solving strategy of taking cases:

Case (i):   If $|x| < 1$, then $0 \le x^2 < 1$, so $\lim\limits_{n \to \infty} x^{2n} = 0$ [see Example 10 in Section 12.1 [ET 11.1]]

       and $f(x) = \lim\limits_{n \to \infty} \dfrac{x^{2n} - 1}{x^{2n} + 1} = \dfrac{0 - 1}{0 + 1} = -1$.

Case (ii):   If $|x| = 1$, that is, $x = \pm 1$, then $x^2 = 1$, so $f(x) = \lim\limits_{n \to \infty} \dfrac{x^{2n} - 1}{x^{2n} + 1} = \lim\limits_{n \to \infty} \dfrac{1 - 1}{1 + 1} = 0$.

Case (iii):   If $|x| > 1$, then $x^2 > 1$, so $\lim\limits_{n \to \infty} x^{2n} = \infty$ and $f(x) = \lim\limits_{n \to \infty} \dfrac{x^{2n} - 1}{x^{2n} + 1} = \lim\limits_{n \to \infty} \dfrac{1 - (1/x^{2n})}{1 + (1/x^{2n})} = \dfrac{1 - 0}{1 + 0} = 1$.

Thus, $f(x) = \begin{cases} 1 & \text{if } x < -1 \\ 0 & \text{if } x = -1 \\ -1 & \text{if } -1 < x < 1 \\ 0 & \text{if } x = 1 \\ 1 & \text{if } x > 1 \end{cases}$

The graph shows that f is continuous everywhere except at $x = \pm 1$.

3. (a) From Formula 14a in Appendix D, with $x = y = \theta$, we get $\tan 2\theta = \dfrac{2\tan\theta}{1 - \tan^2\theta}$, so $\cot 2\theta = \dfrac{1 - \tan^2\theta}{2\tan\theta}$ $\Rightarrow$

$$2\cot 2\theta = \frac{1 - \tan^2\theta}{\tan\theta} = \cot\theta - \tan\theta. \text{ Replacing } \theta \text{ by } \tfrac{1}{2}x, \text{ we get } 2\cot x = \cot\tfrac{1}{2}x - \tan\tfrac{1}{2}x, \text{ or}$$

$\tan\tfrac{1}{2}x = \cot\tfrac{1}{2}x - 2\cot x$.

(b) From part (a) with $\dfrac{x}{2^{n-1}}$ in place of x, $\tan\dfrac{x}{2^n} = \cot\dfrac{x}{2^n} - 2\cot\dfrac{x}{2^{n-1}}$, so the nth partial sum of $\displaystyle\sum_{n=1}^{\infty} \dfrac{1}{2^n} \tan\dfrac{x}{2^n}$ is

$$\begin{aligned}
s_n &= \frac{\tan(x/2)}{2} + \frac{\tan(x/4)}{4} + \frac{\tan(x/8)}{8} + \cdots + \frac{\tan(x/2^n)}{2^n} \\
&= \left[\frac{\cot(x/2)}{2} - \cot x\right] + \left[\frac{\cot(x/4)}{4} - \frac{\cot(x/2)}{2}\right] + \left[\frac{\cot(x/8)}{8} - \frac{\cot(x/4)}{4}\right] + \cdots \\
&\quad + \left[\frac{\cot(x/2^n)}{2^n} - \frac{\cot(x/2^{n-1})}{2^{n-1}}\right] = -\cot x + \frac{\cot(x/2^n)}{2^n} \quad \text{[telescoping sum]}
\end{aligned}$$

Now $\dfrac{\cot(x/2^n)}{2^n} = \dfrac{\cos(x/2^n)}{2^n \sin(x/2^n)} = \dfrac{\cos(x/2^n)}{x} \cdot \dfrac{x/2^n}{\sin(x/2^n)} \to \dfrac{1}{x} \cdot 1 = \dfrac{1}{x}$ as $n \to \infty$ since $x/2^n \to 0$

for $x \neq 0$. Therefore, if $x \neq 0$ and $x \neq k\pi$ where k is any integer, then

$$\sum_{n=1}^{\infty} \frac{1}{2^n} \tan \frac{x}{2^n} = \lim_{n \to \infty} s_n = \lim_{n \to \infty} \left(-\cot x + \frac{1}{2^n} \cot \frac{x}{2^n} \right) = -\cot x + \frac{1}{x}$$

If $x = 0$, then all terms in the series are 0, so the sum is 0.

4. $|AP_2|^2 = 2$, $|AP_3|^2 = 2 + 2^2$, $|AP_4|^2 = 2 + 2^2 + \left(2^2\right)^2$, $|AP_5|^2 = 2 + 2^2 + \left(2^2\right)^2 + \left(2^3\right)^2, \ldots$,

$|AP_n|^2 = 2 + 2^2 + \left(2^2\right)^2 + \cdots + (2^{n-2})^2$ [for $n \geq 3$] $= 2 + (4 + 4^2 + 4^3 + \cdots + 4^{n-2})$

$\qquad = 2 + \dfrac{4(4^{n-2} - 1)}{4 - 1}$ [finite geometric sum with $a = 4, r = 4$] $= \dfrac{6}{3} + \dfrac{4^{n-1} - 4}{3} = \dfrac{2}{3} + \dfrac{4^{n-1}}{3}$

So $\tan \angle P_n A P_{n+1} = \dfrac{|P_n P_{n+1}|}{|AP_n|} = \dfrac{2^{n-1}}{\sqrt{\dfrac{2}{3} + \dfrac{4^{n-1}}{3}}} = \dfrac{\sqrt{4^{n-1}}}{\sqrt{\dfrac{2}{3} + \dfrac{4^{n-1}}{3}}} = \dfrac{1}{\sqrt{\dfrac{2}{3 \cdot 4^{n-1}} + \dfrac{1}{3}}} \to \sqrt{3}$ as $n \to \infty$.

Thus, $\angle P_n A P_{n+1} \to \frac{\pi}{3}$ as $n \to \infty$.

5. (a) At each stage, each side is replaced by four shorter sides, each of length $\frac{1}{3}$ of the side length at the preceding stage. Writing s_0 and ℓ_0 for the number of sides and the length of the side of the initial triangle, we generate the table at right. In general, we have $s_n = 3 \cdot 4^n$ and $\ell_n = \left(\frac{1}{3}\right)^n$, so the length of the perimeter at the nth stage of construction is $p_n = s_n \ell_n = 3 \cdot 4^n \cdot \left(\frac{1}{3}\right)^n = 3 \cdot \left(\frac{4}{3}\right)^n$.

$s_0 = 3$	$\ell_0 = 1$
$s_1 = 3 \cdot 4$	$\ell_1 = 1/3$
$s_2 = 3 \cdot 4^2$	$\ell_2 = 1/3^2$
$s_3 = 3 \cdot 4^3$	$\ell_3 = 1/3^3$
$\vdots$	$\vdots$

(b) $p_n = \dfrac{4^n}{3^{n-1}} = 4\left(\dfrac{4}{3}\right)^{n-1}$. Since $\frac{4}{3} > 1$, $p_n \to \infty$ as $n \to \infty$.

(c) The area of each of the small triangles added at a given stage is one-ninth of the area of the triangle added at the preceding stage. Let a be the area of the original triangle. Then the area a_n of each of the small triangles added at stage n is

$a_n = a \cdot \dfrac{1}{9^n} = \dfrac{a}{9^n}$. Since a small triangle is added to each side at every stage, it follows that the total area A_n added to the

figure at the nth stage is $A_n = s_{n-1} \cdot a_n = 3 \cdot 4^{n-1} \cdot \dfrac{a}{9^n} = a \cdot \dfrac{4^{n-1}}{3^{2n-1}}$. Then the total area enclosed by the snowflake

curve is $A = a + A_1 + A_2 + A_3 + \cdots = a + a \cdot \dfrac{1}{3} + a \cdot \dfrac{4}{3^3} + a \cdot \dfrac{4^2}{3^5} + a \cdot \dfrac{4^3}{3^7} + \cdots$. After the first term, this is a

geometric series with common ratio $\dfrac{4}{9}$, so $A = a + \dfrac{a/3}{1 - \frac{4}{9}} = a + \dfrac{a}{3} \cdot \dfrac{9}{5} = \dfrac{8a}{5}$. But the area of the original equilateral

triangle with side 1 is $a = \dfrac{1}{2} \cdot 1 \cdot \sin \dfrac{\pi}{3} = \dfrac{\sqrt{3}}{4}$. So the area enclosed by the snowflake curve is $\dfrac{8}{5} \cdot \dfrac{\sqrt{3}}{4} = \dfrac{2\sqrt{3}}{5}$.

6. Let the series $S = 1 + \frac{1}{2} + \frac{1}{3} + \frac{1}{4} + \frac{1}{6} + \frac{1}{8} + \frac{1}{9} + \frac{1}{12} + \cdots$. Then every term in S is of the form $\dfrac{1}{2^m 3^n}$, $m, n \geq 0$, and furthermore each term occurs only once. So we can write

$$S = \sum_{m=0}^{\infty} \sum_{n=0}^{\infty} \frac{1}{2^m 3^n} = \sum_{m=0}^{\infty} \sum_{n=0}^{\infty} \frac{1}{2^m} \frac{1}{3^n} = \sum_{m=0}^{\infty} \frac{1}{2^m} \sum_{n=0}^{\infty} \frac{1}{3^n} = \frac{1}{1 - \frac{1}{2}} \cdot \frac{1}{1 - \frac{1}{3}} = 2 \cdot \frac{3}{2} = 3$$

7. (a) Let $a = \arctan x$ and $b = \arctan y$. Then, from Formula 14b in Appendix D,

$$\tan(a-b) = \frac{\tan a - \tan b}{1 + \tan a \, \tan b} = \frac{\tan(\arctan x) - \tan(\arctan y)}{1 + \tan(\arctan x)\tan(\arctan y)} = \frac{x-y}{1+xy}$$

Now $\arctan x - \arctan y = a - b = \arctan(\tan(a-b)) = \arctan \dfrac{x-y}{1+xy}$ since $-\frac{\pi}{2} < a - b < \frac{\pi}{2}$.

(b) From part (a) we have

$$\arctan \tfrac{120}{119} - \arctan \tfrac{1}{239} = \arctan \frac{\frac{120}{119} - \frac{1}{239}}{1 + \frac{120}{119}\cdot\frac{1}{239}} = \arctan \frac{\frac{28,561}{28,441}}{\frac{28,561}{28,441}} = \arctan 1 = \tfrac{\pi}{4}$$

(c) Replacing y by $-y$ in the formula of part (a), we get $\arctan x + \arctan y = \arctan \dfrac{x+y}{1-xy}$. So

$$4\arctan \tfrac{1}{5} = 2\left(\arctan \tfrac{1}{5} + \arctan \tfrac{1}{5}\right) = 2\arctan \frac{\frac{1}{5} + \frac{1}{5}}{1 - \frac{1}{5}\cdot\frac{1}{5}} = 2\arctan \tfrac{5}{12} = \arctan \tfrac{5}{12} + \arctan \tfrac{5}{12}$$

$$= \arctan \frac{\frac{5}{12} + \frac{5}{12}}{1 - \frac{5}{12}\cdot\frac{5}{12}} = \arctan \tfrac{120}{119}$$

Thus, from part (b), we have $4\arctan \tfrac{1}{5} - \arctan \tfrac{1}{239} = \arctan \tfrac{120}{119} - \arctan \tfrac{1}{239} = \tfrac{\pi}{4}$.

(d) From Example 7 in Section 12.9 [ET 11.9] we have $\arctan x = x - \dfrac{x^3}{3} + \dfrac{x^5}{5} - \dfrac{x^7}{7} + \dfrac{x^9}{9} - \dfrac{x^{11}}{11} + \cdots$, so

$$\arctan \frac{1}{5} = \frac{1}{5} - \frac{1}{3\cdot 5^3} + \frac{1}{5\cdot 5^5} - \frac{1}{7\cdot 5^7} + \frac{1}{9\cdot 5^9} - \frac{1}{11\cdot 5^{11}} + \cdots$$

This is an alternating series and the size of the terms decreases to 0, so by the Alternating Series Estimation Theorem, the sum lies between s_5 and s_6, that is, $0.197395560 < \arctan \tfrac{1}{5} < 0.197395562$.

(e) From the series in part (d) we get $\arctan \dfrac{1}{239} = \dfrac{1}{239} - \dfrac{1}{3\cdot 239^3} + \dfrac{1}{5\cdot 239^5} - \cdots$. The third term is less than 2.6×10^{-13}, so by the Alternating Series Estimation Theorem, we have, to nine decimal places, $\arctan \tfrac{1}{239} \approx s_2 \approx 0.004184076$. Thus, $0.004184075 < \arctan \tfrac{1}{239} < 0.004184077$.

(f) From part (c) we have $\pi = 16\arctan \tfrac{1}{5} - 4\arctan \tfrac{1}{239}$, so from parts (d) and (e) we have

$16(0.197395560) - 4(0.004184077) < \pi < 16(0.197395562) - 4(0.004184075) \quad \Rightarrow$

$3.141592652 < \pi < 3.141592692$. So, to 7 decimal places, $\pi \approx 3.1415927$.

8. (a) Let $a = \operatorname{arccot} x$ and $b = \operatorname{arccot} y$ where $0 < a - b < \pi$. Then

$$\cot(a-b) = \frac{1}{\tan(a-b)} = \frac{1 + \tan a \, \tan b}{\tan a - \tan b} = \frac{\frac{1}{\cot a}\cdot\frac{1}{\cot b} + 1}{\frac{1}{\cot a} - \frac{1}{\cot b}} \cdot \frac{\cot a \, \cot b}{\cot a \, \cot b}$$

$$= \frac{1 + \cot a \, \cot b}{\cot b - \cot a} = \frac{1 + \cot(\operatorname{arccot} x)\cot(\operatorname{arccot} y)}{\cot(\operatorname{arccot} y) - \cot(\operatorname{arccot} x)} = \frac{1 + xy}{y - x}$$

Now $\operatorname{arccot} x - \operatorname{arccot} y = a - b = \operatorname{arccot}(\cot(a-b)) = \operatorname{arccot} \dfrac{1 + xy}{y - x}$ since $0 < a - b < \pi$.

(b) Applying the identity in part (a) with $x = n$ and $y = n + 1$, we have

$$\operatorname{arccot}(n^2 + n + 1) = \operatorname{arccot}(1 + n(n + 1)) = \operatorname{arccot} \frac{1 + n(n + 1)}{(n + 1) - n} = \operatorname{arccot} n - \operatorname{arccot}(n + 1)$$

Thus, we have a telescoping series with nth partial sum

$$s_n = [\operatorname{arccot} 0 - \operatorname{arccot} 1] + [\operatorname{arccot} 1 - \operatorname{arccot} 2] + \cdots + [\operatorname{arccot} n - \operatorname{arccot}(n + 1)] = \operatorname{arccot} 0 - \operatorname{arccot}(n + 1).$$

Thus, $\displaystyle\sum_{n=0}^{\infty} \operatorname{arccot}(n^2 + n + 1) = \lim_{n \to \infty} s_n = \lim_{n \to \infty} [\operatorname{arccot} 0 - \operatorname{arccot}(n + 1)] = \frac{\pi}{2} - 0 = \frac{\pi}{2}$.

9. We start with the geometric series $\displaystyle\sum_{n=0}^{\infty} x^n = \frac{1}{1 - x}$, $|x| < 1$, and differentiate:

$$\sum_{n=1}^{\infty} nx^{n-1} = \frac{d}{dx}\left(\sum_{n=0}^{\infty} x^n \right) = \frac{d}{dx}\left(\frac{1}{1 - x} \right) = \frac{1}{(1 - x)^2} \text{ for } |x| < 1 \quad \Rightarrow \quad \sum_{n=1}^{\infty} nx^n = x \sum_{n=1}^{\infty} nx^{n-1} = \frac{x}{(1 - x)^2}$$

for $|x| < 1$. Differentiate again:

$$\sum_{n=1}^{\infty} n^2 x^{n-1} = \frac{d}{dx} \frac{x}{(1 - x)^2} = \frac{(1 - x)^2 - x \cdot 2(1 - x)(-1)}{(1 - x)^4} = \frac{x + 1}{(1 - x)^3} \quad \Rightarrow \quad \sum_{n=1}^{\infty} n^2 x^n = \frac{x^2 + x}{(1 - x)^3} \quad \Rightarrow$$

$$\sum_{n=1}^{\infty} n^3 x^{n-1} = \frac{d}{dx} \frac{x^2 + x}{(1 - x)^3} = \frac{(1 - x)^3(2x + 1) - (x^2 + x)3(1 - x)^2(-1)}{(1 - x)^6} = \frac{x^2 + 4x + 1}{(1 - x)^4} \quad \Rightarrow$$

$$\sum_{n=1}^{\infty} n^3 x^n = \frac{x^3 + 4x^2 + x}{(1 - x)^4}, |x| < 1.$$ The radius of convergence is 1 because that is the radius of convergence for the

geometric series we started with. If $x = \pm 1$, the series is $\sum n^3(\pm 1)^n$, which diverges by the Test For Divergence, so the

interval of convergence is $(-1, 1)$.

10. Let's first try the case $k = 1$: $a_0 + a_1 = 0 \quad \Rightarrow \quad a_1 = -a_0 \quad \Rightarrow$

$$\lim_{n \to \infty} \left(a_0 \sqrt{n} + a_1 \sqrt{n + 1} \right) = \lim_{n \to \infty} \left(a_0 \sqrt{n} - a_0 \sqrt{n + 1} \right) = a_0 \lim_{n \to \infty} \left(\sqrt{n} - \sqrt{n + 1} \right) \frac{\sqrt{n} + \sqrt{n + 1}}{\sqrt{n} + \sqrt{n + 1}}$$

$$= a_0 \lim_{n \to \infty} \frac{-1}{\sqrt{n} + \sqrt{n + 1}} = 0$$

In general we have $a_0 + a_1 + \cdots + a_k = 0 \quad \Rightarrow \quad a_k = -a_0 - a_1 - \cdots - a_{k-1} \quad \Rightarrow$

$$\lim_{n \to \infty} \left(a_0 \sqrt{n} + a_1 \sqrt{n + 1} + a_2 \sqrt{n + 2} + \cdots + a_k \sqrt{n + k} \right)$$

$$= \lim_{n \to \infty} \left(a_0 \sqrt{n} + a_1 \sqrt{n + 1} + \cdots + a_{k-1} \sqrt{n + k - 1} - a_0 \sqrt{n + k} - a_1 \sqrt{n + k} - \cdots - a_{k-1} \sqrt{n + k} \right)$$

$$= a_0 \lim_{n \to \infty} \left(\sqrt{n} - \sqrt{n + k} \right) + a_1 \lim_{n \to \infty} \left(\sqrt{n + 1} - \sqrt{n + k} \right) + \cdots + a_{k-1} \lim_{n \to \infty} \left(\sqrt{n + k - 1} - \sqrt{n + k} \right)$$

Each of these limits is 0 by the same type of simplification as in the case $k = 1$. So we have

$$\lim_{n \to \infty} \left(a_0 \sqrt{n} + a_1 \sqrt{n + 1} + a_2 \sqrt{n + 2} + \cdots + a_k \sqrt{n + k} \right) = a_0(0) + a_1(0) + \cdots + a_{k-1}(0) = 0$$

11. $\ln\left(1 - \frac{1}{n^2} \right) = \ln\left(\frac{n^2 - 1}{n^2} \right) = \ln \frac{(n + 1)(n - 1)}{n^2} = \ln[(n + 1)(n - 1)] - \ln n^2$

$$= \ln(n + 1) + \ln(n - 1) - 2\ln n = \ln(n - 1) - \ln n - \ln n + \ln(n + 1)$$

$$= \ln \frac{n - 1}{n} - [\ln n - \ln(n + 1)] = \ln \frac{n - 1}{n} - \ln \frac{n}{n + 1}.$$

Let $s_k = \sum_{n=2}^{k} \ln\left(1 - \frac{1}{n^2}\right) = \sum_{n=2}^{k} \left(\ln\frac{n-1}{n} - \ln\frac{n}{n+1}\right)$ for $k \geq 2$. Then

$s_k = \left(\ln\frac{1}{2} - \ln\frac{2}{3}\right) + \left(\ln\frac{2}{3} - \ln\frac{3}{4}\right) + \cdots + \left(\ln\frac{k-1}{k} - \ln\frac{k}{k+1}\right) = \ln\frac{1}{2} - \ln\frac{k}{k+1}$, so

$\sum_{n=2}^{\infty} \ln\left(1 - \frac{1}{n^2}\right) = \lim_{k\to\infty} s_k = \lim_{k\to\infty}\left(\ln\frac{1}{2} - \ln\frac{k}{k+1}\right) = \ln\frac{1}{2} - \ln 1 = \ln 1 - \ln 2 - \ln 1 = -\ln 2$.

12. Place the y-axis as shown and let the length of each book be L. We want to show that the center of mass of the system of n books lies above the table, that is, $\overline{x} < L$. The x-coordinates of the centers of mass of the books are

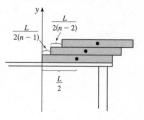

$x_1 = \frac{L}{2}, x_2 = \frac{L}{2(n-1)} + \frac{L}{2}, x_3 = \frac{L}{2(n-1)} + \frac{L}{2(n-2)} + \frac{L}{2}$, and so on.

Each book has the same mass m, so if there are n books, then

$$\overline{x} = \frac{mx_1 + mx_2 + \cdots + mx_n}{mn} = \frac{x_1 + x_2 + \cdots + x_n}{n}$$

$$= \frac{1}{n}\left[\frac{L}{2} + \left(\frac{L}{2(n-1)} + \frac{L}{2}\right) + \left(\frac{L}{2(n-1)} + \frac{L}{2(n-2)} + \frac{L}{2}\right) + \cdots\right.$$

$$\left. + \left(\frac{L}{2(n-1)} + \frac{L}{2(n-2)} + \cdots + \frac{L}{4} + \frac{L}{2} + \frac{L}{2}\right)\right]$$

$$= \frac{L}{n}\left[\frac{n-1}{2(n-1)} + \frac{n-2}{2(n-2)} + \cdots + \frac{2}{4} + \frac{1}{2} + \frac{n}{2}\right] = \frac{L}{n}\left[(n-1)\frac{1}{2} + \frac{n}{2}\right] = \frac{2n-1}{2n}L < L$$

This shows that, no matter how many books are added according to the given scheme, the center of mass lies above the table. It remains to observe that the series $\frac{1}{2} + \frac{1}{4} + \frac{1}{6} + \frac{1}{8} + \cdots = \frac{1}{2}\sum(1/n)$ is divergent (harmonic series), so we can make the top book extend as far as we like beyond the edge of the table if we add enough books.

13. (a)

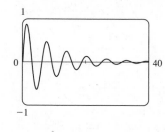

The x-intercepts of the curve occur where $\sin x = 0 \iff x = n\pi$, n an integer. So using the formula for disks (and either a CAS or $\sin^2 x = \frac{1}{2}(1 - \cos 2x)$ and Formula 99 to evaluate the integral), the volume of the nth bead is

$V_n = \pi \int_{(n-1)\pi}^{n\pi} \left(e^{-x/10}\sin x\right)^2 dx = \pi \int_{(n-1)\pi}^{n\pi} e^{-x/5}\sin^2 x \, dx$

$= \frac{250\pi}{101}\left(e^{-(n-1)\pi/5} - e^{-n\pi/5}\right)$

(b) The total volume is

$\pi \int_0^{\infty} e^{-x/5}\sin^2 x \, dx = \sum_{n=1}^{\infty} V_n = \frac{250\pi}{101}\sum_{n=1}^{\infty}\left[e^{-(n-1)\pi/5} - e^{-n\pi/5}\right] = \frac{250\pi}{101}$ [telescoping sum].

Another method: If the volume in part (a) has been written as $V_n = \frac{250\pi}{101}e^{-n\pi/5}(e^{\pi/5} - 1)$, then we recognize $\sum_{n=1}^{\infty} V_n$ as a geometric series with $a = \frac{250\pi}{101}(1 - e^{-\pi/5})$ and $r = e^{-\pi/5}$.

14. First notice that both series are absolutely convergent (p-series with $p > 1$.) Let the given expression be called x. Then

$$x = \frac{1 + \dfrac{1}{2^p} + \dfrac{1}{3^p} + \dfrac{1}{4^p} + \cdots}{1 - \dfrac{1}{2^p} + \dfrac{1}{3^p} - \dfrac{1}{4^p} + \cdots} = \frac{1 + \left(2 \cdot \dfrac{1}{2^p} - \dfrac{1}{2^p}\right) + \dfrac{1}{3^p} + \left(2 \cdot \dfrac{1}{4^p} - \dfrac{1}{4^p}\right) + \cdots}{1 - \dfrac{1}{2^p} + \dfrac{1}{3^p} - \dfrac{1}{4^p} + \cdots}$$

$$= \frac{\left(1 - \dfrac{1}{2^p} + \dfrac{1}{3^p} - \dfrac{1}{4^p} + \cdots\right) + \left(2 \cdot \dfrac{1}{2^p} + 2 \cdot \dfrac{1}{4^p} + 2 \cdot \dfrac{1}{6^p} + \cdots\right)}{1 - \dfrac{1}{2^p} + \dfrac{1}{3^p} - \dfrac{1}{4^p} + \cdots}$$

$$= 1 + \frac{2\left(\dfrac{1}{2^p} + \dfrac{1}{4^p} + \dfrac{1}{6^p} + \dfrac{1}{8^p} + \cdots\right)}{1 - \dfrac{1}{2^p} + \dfrac{1}{3^p} - \dfrac{1}{4^p} + \cdots} = 1 + \frac{\dfrac{1}{2^{p-1}}\left(1 + \dfrac{1}{2^p} + \dfrac{1}{3^p} + \dfrac{1}{4^p} + \cdots\right)}{1 - \dfrac{1}{2^p} + \dfrac{1}{3^p} - \dfrac{1}{4^p} + \cdots} = 1 + 2^{1-p}x$$

Therefore, $x = 1 + 2^{1-p}x \iff x - 2^{1-p}x = 1 \iff x(1 - 2^{1-p}) = 1 \iff x = \dfrac{1}{1 - 2^{1-p}}$.

15. If L is the length of a side of the equilateral triangle, then the area is $A = \frac{1}{2}L \cdot \frac{\sqrt{3}}{2}L = \frac{\sqrt{3}}{4}L^2$ and so $L^2 = \frac{4}{\sqrt{3}}A$.

Let r be the radius of one of the circles. When there are n rows of circles, the figure shows that

$$L = \sqrt{3}\,r + r + (n-2)(2r) + r + \sqrt{3}\,r = r\left(2n - 2 + 2\sqrt{3}\right), \text{ so } r = \frac{L}{2(n + \sqrt{3} - 1)}.$$

The number of circles is $1 + 2 + \cdots + n = \dfrac{n(n+1)}{2}$, and so the total area of the circles is

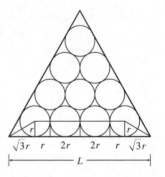

$$A_n = \frac{n(n+1)}{2}\pi r^2 = \frac{n(n+1)}{2}\,\pi\,\frac{L^2}{4(n + \sqrt{3} - 1)^2}$$

$$= \frac{n(n+1)}{2}\,\pi\,\frac{4A/\sqrt{3}}{4(n + \sqrt{3} - 1)^2} = \frac{n(n+1)}{(n + \sqrt{3} - 1)^2}\,\frac{\pi A}{2\sqrt{3}} \Rightarrow$$

$$\frac{A_n}{A} = \frac{n(n+1)}{(n + \sqrt{3} - 1)^2}\,\frac{\pi}{2\sqrt{3}}$$

$$= \frac{1 + 1/n}{\left[1 + (\sqrt{3} - 1)/n\right]^2}\,\frac{\pi}{2\sqrt{3}} \to \frac{\pi}{2\sqrt{3}} \text{ as } n \to \infty$$

16. Given $a_0 = a_1 = 1$ and $a_n = \dfrac{(n-1)(n-2)a_{n-1} - (n-3)a_{n-2}}{n(n-1)}$, we calculate the next few terms of the sequence:

$$a_2 = \frac{1 \cdot 0 \cdot a_1 - (-1)a_0}{2 \cdot 1} = \frac{1}{2}, a_3 = \frac{2 \cdot 1 \cdot a_2 - 0 \cdot a_1}{3 \cdot 2} = \frac{1}{6}, a_4 = \frac{3 \cdot 2 \cdot a_3 - 1 \cdot a_2}{4 \cdot 3} = \frac{1}{24}.$$

It seems that $a_n = \dfrac{1}{n!}$, so we try to prove this by induction. The first step is done, so assume $a_k = \dfrac{1}{k!}$ and $a_{k-1} = \dfrac{1}{(k-1)!}$.

Then $a_{k+1} = \dfrac{k(k-1)a_k - (k-2)a_{k-1}}{(k+1)k} = \dfrac{\dfrac{k(k-1)}{k!} - \dfrac{k-2}{(k-1)!}}{(k+1)k} = \dfrac{(k-1) - (k-2)}{[(k+1)(k)](k-1)!} = \dfrac{1}{(k+1)!}$ and the induction

is complete. Therefore, $\displaystyle\sum_{n=0}^{\infty} a_n = \sum_{n=0}^{\infty} \frac{1}{n!} = e$.

17. As in Section 12.9 [ET 11.9] we have to integrate the function x^x by integrating series. Writing $x^x = (e^{\ln x})^x = e^{x \ln x}$ and

using the Maclaurin series for e^x, we have $x^x = (e^{\ln x})^x = e^{x \ln x} = \sum_{n=0}^{\infty} \frac{(x \ln x)^n}{n!} = \sum_{n=0}^{\infty} \frac{x^n (\ln x)^n}{n!}$. As with power series,

we can integrate this series term-by-term: $\int_0^1 x^x \, dx = \sum_{n=0}^{\infty} \int_0^1 \frac{x^n (\ln x)^n}{n!} \, dx = \sum_{n=0}^{\infty} \frac{1}{n!} \int_0^1 x^n (\ln x)^n \, dx$. We integrate by

parts with $u = (\ln x)^n$, $dv = x^n \, dx$, so $du = \frac{n(\ln x)^{n-1}}{x} \, dx$ and $v = \frac{x^{n+1}}{n+1}$:

$$\int_0^1 x^n (\ln x)^n \, dx = \lim_{t \to 0^+} \int_t^1 x^n (\ln x)^n \, dx = \lim_{t \to 0^+} \left[\frac{x^{n+1}}{n+1} (\ln x)^n \right]_t^1 - \lim_{t \to 0^+} \int_t^1 \frac{n}{n+1} x^n (\ln x)^{n-1} \, dx$$

$$= 0 - \frac{n}{n+1} \int_0^1 x^n (\ln x)^{n-1} \, dx$$

(where l'Hospital's Rule was used to help evaluate the first limit). Further integration by parts gives

$$\int_0^1 x^n (\ln x)^k \, dx = -\frac{k}{n+1} \int_0^1 x^n (\ln x)^{k-1} \, dx \text{ and, combining these steps, we get}$$

$$\int_0^1 x^n (\ln x)^n \, dx = \frac{(-1)^n \, n!}{(n+1)^n} \int_0^1 x^n \, dx = \frac{(-1)^n \, n!}{(n+1)^{n+1}} \quad \Rightarrow$$

$$\int_0^1 x^x \, dx = \sum_{n=0}^{\infty} \frac{1}{n!} \int_0^1 x^n (\ln x)^n \, dx = \sum_{n=0}^{\infty} \frac{1}{n!} \frac{(-1)^n \, n!}{(n+1)^{n+1}} = \sum_{n=0}^{\infty} \frac{(-1)^n}{(n+1)^{n+1}} = \sum_{n=1}^{\infty} \frac{(-1)^{n-1}}{n^n}.$$

18. (a) Since P_n is defined as the midpoint of $P_{n-4}P_{n-3}$, $x_n = \frac{1}{2}(x_{n-4} + x_{n-3})$ for $n \geq 5$. So we prove by induction that

$\frac{1}{2}x_n + x_{n+1} + x_{n+2} + x_{n+3} = 2$. The case $n = 1$ is immediate, since $\frac{1}{2} \cdot 0 + 1 + 1 + 0 = 2$. Assume that the result

holds for $n = k - 1$, that is, $\frac{1}{2}x_{k-1} + x_k + x_{k+1} + x_{k+2} = 2$. Then for $n = k$,

$$\frac{1}{2}x_k + x_{k+1} + x_{k+2} + x_{k+3} = \frac{1}{2}x_k + x_{k+1} + x_{k+2} + \frac{1}{2}(x_{k+3-4} + x_{k+3-3}) \quad \text{[by above]}$$

$$= \frac{1}{2}x_{k-1} + x_k + x_{k+1} + x_{k+2} = 2 \quad \text{[by the induction hypothesis]}$$

Similarly, for $n \geq 5$, $y_n = \frac{1}{2}(y_{n-4} + y_{n-3})$, so the same argument as above holds for y, with 2 replaced by

$\frac{1}{2}y_1 + y_2 + y_3 + y_4 = \frac{1}{2} \cdot 1 + 1 + 0 + 0 = \frac{3}{2}$. So $\frac{1}{2}y_n + y_{n+1} + y_{n+2} + y_{n+3} = \frac{3}{2}$ for all n.

(b) $\lim_{n \to \infty} \left(\frac{1}{2}x_n + x_{n+1} + x_{n+2} + x_{n+3} \right) = \frac{1}{2} \lim_{n \to \infty} x_n + \lim_{n \to \infty} x_{n+1} + \lim_{n \to \infty} x_{n+2} + \lim_{n \to \infty} x_{n+3} = 2$. Since all

the limits on the left hand side are the same, we get $\frac{7}{2} \lim_{n \to \infty} x_n = 2 \quad \Rightarrow \quad \lim_{n \to \infty} x_n = \frac{4}{7}$. In the same way,

$\frac{7}{2} \lim_{n \to \infty} y_n = \frac{3}{2} \quad \Rightarrow \quad \lim_{n \to \infty} y_n = \frac{3}{7}$, so $P = \left(\frac{4}{7}, \frac{3}{7} \right)$.

19. Let $f(x) = \sum_{m=0}^{\infty} c_m x^m$ and $g(x) = e^{f(x)} = \sum_{n=0}^{\infty} d_n x^n$. Then $g'(x) = \sum_{n=0}^{\infty} n d_n x^{n-1}$, so $n d_n$ occurs as the coefficient

of x^{n-1}. But also

$$g'(x) = e^{f(x)} f'(x) = \left(\sum_{n=0}^{\infty} d_n x^n \right) \left(\sum_{m=1}^{\infty} m c_m x^{m-1} \right)$$

$$= \left(d_0 + d_1 x + d_2 x^2 + \cdots + d_{n-1} x^{n-1} + \cdots \right) \left(c_1 + 2c_2 x + 3c_3 x^2 + \cdots + n c_n x^{n-1} + \cdots \right)$$

so the coefficient of x^{n-1} is $c_1 d_{n-1} + 2c_2 d_{n-2} + 3c_3 d_{n-3} + \cdots + n c_n d_0 = \sum_{i=1}^{n} i c_i d_{n-i}$. Therefore, $n d_n = \sum_{i=1}^{n} i c_i d_{n-i}$.

20. Suppose the base of the first right triangle has length a. Then by repeated use of the Pythagorean theorem, we find that the base of the second right triangle has length $\sqrt{1+a^2}$, the base of the third right triangle has length $\sqrt{2+a^2}$, and in general, the nth right triangle has base of length $\sqrt{n-1+a^2}$ and hypotenuse of length $\sqrt{n+a^2}$. Thus, $\theta_n = \tan^{-1}\left(1/\sqrt{n-1+a^2}\right)$ and

$$\sum_{n=1}^{\infty} \theta_n = \sum_{n=1}^{\infty} \tan^{-1}\left(\frac{1}{\sqrt{n-1+a^2}}\right) = \sum_{n=0}^{\infty} \tan^{-1}\left(\frac{1}{\sqrt{n+a^2}}\right).$$ We wish to show that this series diverges.

First notice that the series $\displaystyle\sum_{n=1}^{\infty} \frac{1}{\sqrt{n+a^2}}$ diverges by the Limit Comparison Test with the divergent p-series $\displaystyle\sum_{n=1}^{\infty} \frac{1}{\sqrt{n}}$

$\left[p = \frac{1}{2} \le 1\right]$ since $\displaystyle\lim_{n\to\infty} \frac{1/\sqrt{n+a^2}}{1/\sqrt{n}} = \lim_{n\to\infty} \frac{\sqrt{n}}{\sqrt{n+a^2}} = \lim_{n\to\infty} \sqrt{\frac{n}{n+a^2}} = \lim_{n\to\infty} \sqrt{\frac{1}{1+a^2/n}} = 1 > 0$. Thus,

$\displaystyle\sum_{n=0}^{\infty} \frac{1}{\sqrt{n+a^2}}$ also diverges. Now $\displaystyle\sum_{n=0}^{\infty} \tan^{-1}\left(\frac{1}{\sqrt{n+a^2}}\right)$ diverges by the Limit Comparison Test with $\displaystyle\sum_{n=0}^{\infty} \frac{1}{\sqrt{n+a^2}}$ since

$$\lim_{n\to\infty} \frac{\tan^{-1}\left(1/\sqrt{n+a^2}\right)}{1/\sqrt{n+a^2}} = \lim_{x\to\infty} \frac{\tan^{-1}\left(1/\sqrt{x+a^2}\right)}{1/\sqrt{x+a^2}} = \lim_{y\to\infty} \frac{\tan^{-1}(1/y)}{1/y} \quad \left[y = \sqrt{x+a^2}\,\right]$$

$$= \lim_{z\to0^+} \frac{\tan^{-1} z}{z} \quad \left[z = 1/y\right] \quad \overset{\mathrm{H}}{=} \lim_{z\to0^+} \frac{1/(1+z^2)}{1} = 1 > 0$$

Thus, $\displaystyle\sum_{n=1}^{\infty} \theta_n$ is a divergent series.

21. Call the series S. We group the terms according to the number of digits in their denominators:

$$S = \underbrace{\left(\tfrac{1}{1} + \tfrac{1}{2} + \cdots + \tfrac{1}{8} + \tfrac{1}{9}\right)}_{g_1} + \underbrace{\left(\tfrac{1}{11} + \cdots + \tfrac{1}{99}\right)}_{g_2} + \underbrace{\left(\tfrac{1}{111} + \cdots + \tfrac{1}{999}\right)}_{g_3} + \cdots$$

Now in the group g_n, since we have 9 choices for each of the n digits in the denominator, there are 9^n terms.

Furthermore, each term in g_n is less than $\frac{1}{10^{n-1}}$ [except for the first term in g_1]. So $g_n < 9^n \cdot \frac{1}{10^{n-1}} = 9\left(\frac{9}{10}\right)^{n-1}$.

Now $\displaystyle\sum_{n=1}^{\infty} 9\left(\frac{9}{10}\right)^{n-1}$ is a geometric series with $a = 9$ and $r = \frac{9}{10} < 1$. Therefore, by the Comparison Test,

$$S = \sum_{n=1}^{\infty} g_n < \sum_{n=1}^{\infty} 9\left(\tfrac{9}{10}\right)^{n-1} = \frac{9}{1-9/10} = 90.$$

22. (a) Let $f(x) = \dfrac{x}{1-x-x^2} = \displaystyle\sum_{n=0}^{\infty} c_n x^n = c_0 + c_1 x + c_2 x^2 + c_3 x^3 + \cdots$. Then

$$x = (1 - x - x^2)(c_0 + c_1 x + c_2 x^2 + c_3 x^3 + \cdots)$$

$$x = c_0 + c_1 x + c_2 x^2 + c_3 x^3 + c_4 x^4 + c_5 x^5 + \cdots$$
$$- c_0 x - c_1 x^2 - c_2 x^3 - c_3 x^4 - c_4 x^5 - \cdots$$
$$- c_0 x^2 - c_1 x^3 - c_2 x^4 - c_3 x^5 - \cdots$$

$$x = c_0 + (c_1 - c_0)x + (c_2 - c_1 - c_0)x^2 + (c_3 - c_2 - c_1)x^3 + \cdots$$

Comparing coefficients of powers of x gives us $c_0 = 0$ and

$$c_1 - c_0 = 1 \quad \Rightarrow \quad c_1 = c_0 + 1 = 1$$

$$c_2 - c_1 - c_0 = 0 \quad \Rightarrow \quad c_2 = c_1 + c_0 = 1 + 0 = 1$$

$$c_3 - c_2 - c_1 = 0 \quad \Rightarrow \quad c_3 = c_2 + c_1 = 1 + 1 = 2$$

In general, we have $c_n = c_{n-1} + c_{n-2}$ for $n \geq 3$. Each c_n is equal to the nth Fibonacci number, that is,

$$\sum_{n=0}^{\infty} c_n x^n = \sum_{n=1}^{\infty} c_n x^n = \sum_{n=1}^{\infty} f_n x^n$$

(b) Completing the square on $x^2 + x - 1$ gives us

$$\left(x^2 + x + \frac{1}{4} \right) - 1 - \frac{1}{4} = \left(x + \frac{1}{2} \right)^2 - \frac{5}{4} = \left(x + \frac{1}{2} \right)^2 - \left(\frac{\sqrt{5}}{2} \right)^2$$

$$= \left(x + \frac{1}{2} + \frac{\sqrt{5}}{2} \right) \left(x + \frac{1}{2} - \frac{\sqrt{5}}{2} \right) = \left(x + \frac{1+\sqrt{5}}{2} \right) \left(x + \frac{1-\sqrt{5}}{2} \right)$$

So $\dfrac{x}{1 - x - x^2} = \dfrac{-x}{x^2 + x - 1} = \dfrac{-x}{\left(x + \frac{1+\sqrt{5}}{2} \right) \left(x + \frac{1-\sqrt{5}}{2} \right)}$. The factors in the denominator are linear,

so the partial fraction decomposition is

$$\frac{-x}{\left(x + \frac{1+\sqrt{5}}{2} \right) \left(x + \frac{1-\sqrt{5}}{2} \right)} = \frac{A}{x + \frac{1+\sqrt{5}}{2}} + \frac{B}{x + \frac{1-\sqrt{5}}{2}} \quad - x = A\left(x + \frac{1-\sqrt{5}}{2} \right) + B\left(x + \frac{1+\sqrt{5}}{2} \right)$$

If $x = \frac{-1+\sqrt{5}}{2}$, then $-\frac{-1+\sqrt{5}}{2} = B\sqrt{5} \;\Rightarrow\; B = \frac{1-\sqrt{5}}{2\sqrt{5}}$.

If $x = \frac{-1-\sqrt{5}}{2}$, then $-\frac{-1-\sqrt{5}}{2} = A\left(-\sqrt{5}\right) \;\Rightarrow\; A = \frac{1+\sqrt{5}}{-2\sqrt{5}}$. Thus,

$$\frac{x}{1 - x - x^2} = \frac{\frac{1+\sqrt{5}}{-2\sqrt{5}}}{x + \frac{1+\sqrt{5}}{2}} + \frac{\frac{1-\sqrt{5}}{2\sqrt{5}}}{x + \frac{1-\sqrt{5}}{2}} = \frac{\frac{1+\sqrt{5}}{-2\sqrt{5}}}{x + \frac{1+\sqrt{5}}{2}} \cdot \frac{\frac{2}{1+\sqrt{5}}}{\frac{2}{1+\sqrt{5}}} + \frac{\frac{1-\sqrt{5}}{2\sqrt{5}}}{x + \frac{1-\sqrt{5}}{2}} \cdot \frac{\frac{2}{1-\sqrt{5}}}{\frac{2}{1-\sqrt{5}}}$$

$$= \frac{-1/\sqrt{5}}{1 + \frac{2}{1+\sqrt{5}}x} + \frac{1/\sqrt{5}}{1 + \frac{2}{1-\sqrt{5}}x} = -\frac{1}{\sqrt{5}} \sum_{n=0}^{\infty} \left(-\frac{2}{1+\sqrt{5}}x \right)^n + \frac{1}{\sqrt{5}} \sum_{n=0}^{\infty} \left(-\frac{2}{1-\sqrt{5}}x \right)^n$$

$$= \frac{1}{\sqrt{5}} \sum_{n=0}^{\infty} \left[\left(\frac{-2}{1-\sqrt{5}} \right)^n - \left(\frac{-2}{1+\sqrt{5}} \right)^n \right] x^n$$

$$= \frac{1}{\sqrt{5}} \sum_{n=1}^{\infty} \left[\frac{(-2)^n \left(1+\sqrt{5} \right)^n - (-2)^n \left(1-\sqrt{5} \right)^n}{\left(1-\sqrt{5} \right)^n \left(1+\sqrt{5} \right)^n} \right] x^n \qquad \text{[the } n = 0 \text{ term is 0]}$$

$$= \frac{1}{\sqrt{5}} \sum_{n=1}^{\infty} \left[\frac{(-2)^n \left(\left(1+\sqrt{5} \right)^n - \left(1-\sqrt{5} \right)^n \right)}{(1-5)^n} \right] x^n$$

$$= \frac{1}{\sqrt{5}} \sum_{n=1}^{\infty} \left[\frac{\left(1+\sqrt{5} \right)^n - \left(1-\sqrt{5} \right)^n}{2^n} \right] x^n \qquad [(-4)^n = (-2)^n \cdot 2^n]$$

From part (a), this series must equal $\displaystyle\sum_{n=1}^{\infty} f_n x^n$, so $f_n = \dfrac{\left(1+\sqrt{5} \right)^n - \left(1-\sqrt{5} \right)^n}{2^n \sqrt{5}}$, which is an explicit formula for

the nth Fibonacci number.

23. $u = 1 + \dfrac{x^3}{3!} + \dfrac{x^6}{6!} + \dfrac{x^9}{9!} + \cdots$, $v = x + \dfrac{x^4}{4!} + \dfrac{x^7}{7!} + \dfrac{x^{10}}{10!} + \cdots$, $w = \dfrac{x^2}{2!} + \dfrac{x^5}{5!} + \dfrac{x^8}{8!} + \cdots$.

Use the Ratio Test to show that the series for u, v, and w have positive radii of convergence (∞ in each case), so

Theorem 12.9.2 [ET 11.9.2] applies, and hence, we may differentiate each of these series:

$$\frac{du}{dx} = \frac{3x^2}{3!} + \frac{6x^5}{6!} + \frac{9x^8}{9!} + \cdots = \frac{x^2}{2!} + \frac{x^5}{5!} + \frac{x^8}{8!} + \cdots = w$$

Similarly, $\dfrac{dv}{dx} = 1 + \dfrac{x^3}{3!} + \dfrac{x^6}{6!} + \dfrac{x^9}{9!} + \cdots = u$, and $\dfrac{dw}{dx} = x + \dfrac{x^4}{4!} + \dfrac{x^7}{7!} + \dfrac{x^{10}}{10!} + \cdots = v$.

So $u' = w$, $v' = u$, and $w' = v$. Now differentiate the left hand side of the desired equation:

$$\frac{d}{dx}(u^3 + v^3 + w^3 - 3uvw) = 3u^2 u' + 3v^2 v' + 3w^2 w' - 3(u'vw + uv'w + uvw')$$

$$= 3u^2 w + 3v^2 u + 3w^2 v - 3(vw^2 + u^2 w + uv^2) = 0 \quad \Rightarrow$$

$u^3 + v^3 + w^3 - 3uvw = C$. To find the value of the constant C, we put $x = 0$ in the last equation and get

$1^3 + 0^3 + 0^3 - 3(1 \cdot 0 \cdot 0) = C \quad \Rightarrow \quad C = 1$, so $u^3 + v^3 + w^3 - 3uvw = 1$.

24. To prove: If $n > 1$, then the nth partial sum $s_n = \displaystyle\sum_{i=1}^{n} \frac{1}{i}$ of the harmonic series is not an integer.

Proof: Let 2^k be the largest power of 2 that is less than or equal to n and let M be the product of all the odd positive integers

that are less than or equal to n. Suppose that $s_n = m$, an integer. Then $M2^k s_n = M2^k m$. Since $n \geq 2$, we have $k \geq 1$, and

hence, $M2^k m$ is an even integer. We will show that $M2^k s_n$ is an odd integer, contradicting the equality $M2^k s_n = M2^k m$

and showing that the supposition that s_n is an integer must have been wrong.

$M2^k s_n = M2^k \displaystyle\sum_{i=1}^{n} \frac{1}{i} = \sum_{i=1}^{n} \frac{M2^k}{i}$. If $1 \leq i \leq n$ and i is odd, then $\dfrac{M}{i}$ is an odd integer since i is one of the odd integers

that were multiplied together to form M. Thus, $\dfrac{M2^k}{i}$ is an even integer in this case. If $1 \leq i \leq n$ and i is even, then we can

write $i = 2^r l$, where 2^r is the largest power of 2 dividing i and l is odd. If $r < k$, then $\dfrac{M2^k}{i} = \dfrac{2^k}{2^r} \cdot \dfrac{M}{l} = 2^{k-r} \dfrac{M}{l}$, which is

an even integer, the product of the even integer 2^{k-r} and the odd integer $\dfrac{M}{l}$. If $r = k$, then $l = 1$, since $l > 1 = l \geq 2 \quad \Rightarrow$

$i = 2^k l \geq 2^k \cdot 2 = 2^{k+1}$, contrary to the choice of 2^k as the largest power of 2 that is less than or equal to n. This shows that

$r = k$ only when $i = 2^k$. In that case, $\dfrac{M2^k}{i} = M$, an *odd* integer. Since $\dfrac{M2^k}{i}$ is an even integer for every i except 2^k and

$\dfrac{M2^k}{i}$ is an odd integer when $i = 2^k$, we see that $M2^k s_n$ is an odd integer. This concludes the proof.

13 ☐ VECTORS AND THE GEOMETRY OF SPACE ☐ ET 12

13.1 Three-Dimensional Coordinate Systems

1. We start at the origin, which has coordinates $(0, 0, 0)$. First we move 4 units along the positive x-axis, affecting only the x-coordinate, bringing us to the point $(4, 0, 0)$. We then move 3 units straight downward, in the negative z-direction. Thus only the z-coordinate is affected, and we arrive at $(4, 0, -3)$.

2.

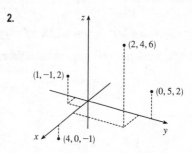

3. The distance from a point to the xz-plane is the absolute value of the y-coordinate of the point. $Q(-5, -1, 4)$ has the y-coordinate with the smallest absolute value, so Q is the point closest to the xz-plane. $R(0, 3, 8)$ must lie in the yz-plane since the distance from R to the yz-plane, given by the x-coordinate of R, is 0.

4. The projection of $(2, 3, 5)$ on the xy-plane is $(2, 3, 0)$; on the yz-plane, $(0, 3, 5)$; on the xz-plane, $(2, 0, 5)$.

The length of the diagonal of the box is the distance between the origin and $(2, 3, 5)$, given by

$$\sqrt{(2-0)^2 + (3-0)^2 + (5-0)^2} = \sqrt{38} \approx 6.16$$

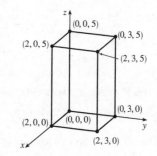

5. The equation $x + y = 2$ represents the set of all points in $\mathbb{R}^3$ whose x- and y-coordinates have a sum of 2, or equivalently where $y = 2 - x$. This is the set $\{(x, 2 - x, z) \mid x \in \mathbb{R}, z \in \mathbb{R}\}$ which is a vertical plane that intersects the xy-plane in the line $y = 2 - x$, $z = 0$.

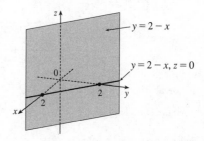

6. (a) In $\mathbb{R}^2$, the equation $x = 4$ represents a line parallel to the y-axis. In $\mathbb{R}^3$, the equation $x = 4$ represents the set $\{(x, y, z) \mid x = 4\}$, the set of all points whose x-coordinate is 4. This is the vertical plane that is parallel to the yz-plane and 4 units in front of it.

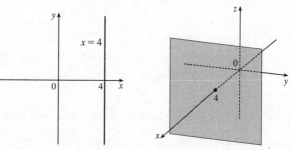

(b) In $\mathbb{R}^3$, the equation $y = 3$ represents a vertical plane that is parallel to the xz-plane and 3 units to the right of it. The equation $z = 5$ represents a horizontal plane parallel to the xy-plane and 5 units above it. The pair of equations $y = 3$, $z = 5$ represents the set of points that are simultaneously on both planes, or in other words, the line of intersection of the planes $y = 3$, $z = 5$. This line can also be described as the set $\{(x, 3, 5) \mid x \in \mathbb{R}\}$, which is the set of all points in $\mathbb{R}^3$ whose x-coordinate may vary but whose y- and z-coordinates are fixed at 3 and 5, respectively. Thus the line is parallel to the x-axis and intersects the yz-plane in the point $(0, 3, 5)$.

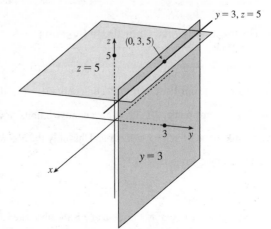

7. We can find the lengths of the sides of the triangle by using the distance formula between pairs of vertices:

$$|PQ| = \sqrt{(7-3)^2 + [0-(-2)]^2 + [1-(-3)]^2} = \sqrt{16 + 4 + 16} = 6$$

$$|QR| = \sqrt{(1-7)^2 + (2-0)^2 + (1-1)^2} = \sqrt{36 + 4 + 0} = \sqrt{40} = 2\sqrt{10}$$

$$|RP| = \sqrt{(3-1)^2 + (-2-2)^2 + (-3-1)^2} = \sqrt{4 + 16 + 16} = 6$$

The longest side is QR, but the Pythagorean Theorem is not satisfied: $|PQ|^2 + |RP|^2 \neq |QR|^2$. Thus PQR is not a right triangle. PQR is isosceles, as two sides have the same length.

8. Compute the lengths of the sides of the triangle by using the distance formula between pairs of vertices:

$$|PQ| = \sqrt{(4-2)^2 + [1-(-1)]^2 + (1-0)^2} = \sqrt{4 + 4 + 1} = 3$$

$$|QR| = \sqrt{(4-4)^2 + (-5-1)^2 + (4-1)^2} = \sqrt{0 + 36 + 9} = \sqrt{45} = 3\sqrt{5}$$

$$|RP| = \sqrt{(2-4)^2 + [-1-(-5)]^2 + (0-4)^2} = \sqrt{4 + 16 + 16} = 6$$

Since the Pythagorean Theorem is satisfied by $|PQ|^2 + |RP|^2 = |QR|^2$, PQR is a right triangle. PQR is not isosceles, as no two sides have the same length.

9. (a) First we find the distances between points:

$$|AB| = \sqrt{(3-2)^2 + (7-4)^2 + (-2-2)^2} = \sqrt{26}$$

$$|BC| = \sqrt{(1-3)^2 + (3-7)^2 + [3-(-2)]^2} = \sqrt{45} = 3\sqrt{5}$$

$$|AC| = \sqrt{(1-2)^2 + (3-4)^2 + (3-2)^2} = \sqrt{3}$$

In order for the points to lie on a straight line, the sum of the two shortest distances must be equal to the longest distance. Since $\sqrt{26} + \sqrt{3} \neq 3\sqrt{5}$, the three points do not lie on a straight line.

(b) First we find the distances between points:

$$|DE| = \sqrt{(1-0)^2 + [-2-(-5)]^2 + (4-5)^2} = \sqrt{11}$$

$$|EF| = \sqrt{(3-1)^2 + [4-(-2)]^2 + (2-4)^2} = \sqrt{44} = 2\sqrt{11}$$

$$|DF| = \sqrt{(3-0)^2 + [4-(-5)]^2 + (2-5)^2} = \sqrt{99} = 3\sqrt{11}$$

Since $|DE| + |EF| = |DF|$, the three points lie on a straight line.

10. (a) The distance from a point to the xy-plane is the absolute value of the z-coordinate of the point. Thus, the distance
is $|-5| = 5$.

(b) Similarly, the distance to the yz-plane is the absolute value of the x-coordinate of the point: $|3| = 3$.

(c) The distance to the xz-plane is the absolute value of the y-coordinate of the point: $|7| = 7$.

(d) The point on the x-axis closest to $(3, 7, -5)$ is the point $(3, 0, 0)$. (Approach the x-axis perpendicularly.)

The distance from $(3, 7, -5)$ to the x-axis is the distance between these two points:

$$\sqrt{(3-3)^2 + (7-0)^2 + (-5-0)^2} = \sqrt{74} \approx 8.60.$$

(e) The point on the y-axis closest to $(3, 7, -5)$ is $(0, 7, 0)$. The distance between these points is

$$\sqrt{(3-0)^2 + (7-7)^2 + (-5-0)^2} = \sqrt{34} \approx 5.83.$$

(f) The point on the z-axis closest to $(3, 7, -5)$ is $(0, 0, -5)$. The distance between these points is

$$\sqrt{(3-0)^2 + (7-0)^2 + [-5-(-5)]^2} = \sqrt{58} \approx 7.62.$$

11. An equation of the sphere with center $(1, -4, 3)$ and radius 5 is $(x-1)^2 + [y-(-4)]^2 + (z-3)^2 = 5^2$ or
$(x-1)^2 + (y+4)^2 + (z-3)^2 = 25$. The intersection of this sphere with the xz-plane is the set of points on the sphere
whose y-coordinate is 0. Putting $y = 0$ into the equation, we have $(x-1)^2 + 4^2 + (z-3)^2 = 25$, $y = 0$ or
$(x-1)^2 + (z-3)^2 = 9$, $y = 0$, which represents a circle in the xz-plane with center $(1, 0, 3)$ and radius 3.

12. An equation of the sphere with center $(2, -6, 4)$ and radius 5 is $(x-2)^2 + [y-(-6)]^2 + (z-4)^2 = 5^2$ or
$(x-2)^2 + (y+6)^2 + (z-4)^2 = 25$. The intersection of this sphere with the xy-plane is the set of points on the sphere
whose z-coordinate is 0. Putting $z = 0$ into the equation, we have $(x-2)^2 + (y+6)^2 = 9$, $z = 0$ which represents a circle
in the xy-plane with center $(2, -6, 0)$ and radius 3. To find the intersection with the xz-plane, we set $y = 0$:
$(x-2)^2 + (z-4)^2 = -11$. Since no points satisfy this equation, the sphere does not intersect the xz-plane. (Also note that
the distance from the center of the sphere to the xz-plane is greater than the radius of the sphere.) To find the intersection with
the yz-plane, we set $x = 0$: $(y+6)^2 + (z-4)^2 = 21$, $x = 0$, a circle in the yz-plane with center $(0, -6, 4)$ and radius $\sqrt{21}$.

13. The radius of the sphere is the distance between $(4, 3, -1)$ and $(3, 8, 1)$: $r = \sqrt{(3-4)^2 + (8-3)^2 + [1-(-1)]^2} = \sqrt{30}$.

Thus, an equation of the sphere is $(x-3)^2 + (y-8)^2 + (z-1)^2 = 30$.

14. If the sphere passes through the origin, the radius of the sphere must be the distance from the origin to the point $(1, 2, 3)$:

$r = \sqrt{(1-0)^2 + (2-0)^2 + (3-0)^2} = \sqrt{14}$. Then an equation of the sphere is $(x-1)^2 + (y-2)^2 + (z-3)^2 = 14$.

15. Completing squares in the equation $x^2 + y^2 + z^2 - 6x + 4y - 2z = 11$ gives

$(x^2 - 6x + 9) + (y^2 + 4y + 4) + (z^2 - 2z + 1) = 11 + 9 + 4 + 1 \quad\Rightarrow\quad (x-3)^2 + (y+2)^2 + (z-1)^2 = 25$, which we

recognize as an equation of a sphere with center $(3, -2, 1)$ and radius 5.

16. Completing squares in the equation gives $(x^2 + 8x + 16) + (y^2 - 6y + 9) + (z^2 + 2z + 1) = -17 + 16 + 9 + 1 \quad\Rightarrow$

$(x+4)^2 + (y-3)^2 + (z+1)^2 = 9$, which we recognize as an equation of a sphere with center $(-4, 3, -1)$ and radius 3.

17. Completing squares in the equation $2x^2 - 8x + 2y^2 + 2z^2 + 24z = 1$ gives

$2(x^2 - 4x + 4) + 2y^2 + 2(z^2 + 12z + 36) = 1 + 8 + 72 \quad\Rightarrow\quad 2(x-2)^2 + 2y^2 + 2(z+6)^2 = 81 \quad\Rightarrow$

$(x-2)^2 + y^2 + (z+6)^2 = \frac{81}{2}$, which we recognize as an equation of a sphere with center $(2, 0, -6)$ and radius

$\sqrt{\frac{81}{2}} = 9/\sqrt{2}$.

18. Completing squares in the equation gives $4(x^2 - 2x + 1) + 4(y^2 + 4y + 4) + 4z^2 = 1 + 4 + 16 \quad\Rightarrow$

$4(x-1)^2 + 4(y+2)^2 + 4z^2 = 21 \quad\Rightarrow\quad (x-1)^2 + (y+2)^2 + z^2 = \frac{21}{4}$, which we recognize as an equation of a sphere

with center $(1, -2, 0)$ and radius $\sqrt{\frac{21}{4}} = \frac{\sqrt{21}}{2}$.

19. (a) If the midpoint of the line segment from $P_1(x_1, y_1, z_1)$ to $P_2(x_2, y_2, z_2)$ is $Q = \left(\dfrac{x_1 + x_2}{2}, \dfrac{y_1 + y_2}{2}, \dfrac{z_1 + z_2}{2}\right)$,

then the distances $|P_1Q|$ and $|QP_2|$ are equal, and each is half of $|P_1P_2|$. We verify that this is the case:

$$|P_1P_2| = \sqrt{(x_2 - x_1)^2 + (y_2 - y_1)^2 + (z_2 - z_1)^2}$$

$$|P_1Q| = \sqrt{\left[\tfrac{1}{2}(x_1 + x_2) - x_1\right]^2 + \left[\tfrac{1}{2}(y_1 + y_2) - y_1\right]^2 + \left[\tfrac{1}{2}(z_1 + z_2) - z_1\right]^2}$$

$$= \sqrt{\left(\tfrac{1}{2}x_2 - \tfrac{1}{2}x_1\right)^2 + \left(\tfrac{1}{2}y_2 - \tfrac{1}{2}y_1\right)^2 + \left(\tfrac{1}{2}z_2 - \tfrac{1}{2}z_1\right)^2}$$

$$= \sqrt{\left(\tfrac{1}{2}\right)^2 \left[(x_2 - x_1)^2 + (y_2 - y_1)^2 + (z_2 - z_1)^2\right]} = \tfrac{1}{2}\sqrt{(x_2 - x_1)^2 + (y_2 - y_1)^2 + (z_2 - z_1)^2}$$

$$= \tfrac{1}{2}|P_1P_2|$$

$$|QP_2| = \sqrt{\left[x_2 - \tfrac{1}{2}(x_1 + x_2)\right]^2 + \left[y_2 - \tfrac{1}{2}(y_1 + y_2)\right]^2 + \left[z_2 - \tfrac{1}{2}(z_1 + z_2)\right]^2}$$

$$= \sqrt{\left(\tfrac{1}{2}x_2 - \tfrac{1}{2}x_1\right)^2 + \left(\tfrac{1}{2}y_2 - \tfrac{1}{2}y_1\right)^2 + \left(\tfrac{1}{2}z_2 - \tfrac{1}{2}z_1\right)^2} = \sqrt{\left(\tfrac{1}{2}\right)^2 \left[(x_2 - x_1)^2 + (y_2 - y_1)^2 + (z_2 - z_1)^2\right]}$$

$$= \tfrac{1}{2}\sqrt{(x_2 - x_1)^2 + (y_2 - y_1)^2 + (z_2 - z_1)^2} = \tfrac{1}{2}|P_1P_2|$$

So Q is indeed the midpoint of P_1P_2.

(b) By part (a), the midpoints of sides AB, BC and CA are $P_1\left(-\frac{1}{2}, 1, 4\right)$, $P_2\left(1, \frac{1}{2}, 5\right)$ and $P_3\left(\frac{5}{2}, \frac{3}{2}, 4\right)$. (Recall that a median of a triangle is a line segment from a vertex to the midpoint of the opposite side.) Then the lengths of the medians are:

$$|AP_2| = \sqrt{0^2 + \left(\tfrac{1}{2} - 2\right)^2 + (5 - 3)^2} = \sqrt{\tfrac{9}{4} + 4} = \sqrt{\tfrac{25}{4}} = \tfrac{5}{2}$$

$$|BP_3| = \sqrt{\left(\tfrac{5}{2} + 2\right)^2 + \left(\tfrac{3}{2}\right)^2 + (4 - 5)^2} = \sqrt{\tfrac{81}{4} + \tfrac{9}{4} + 1} = \sqrt{\tfrac{94}{4}} = \tfrac{1}{2}\sqrt{94}$$

$$|CP_1| = \sqrt{\left(-\tfrac{1}{2} - 4\right)^2 + (1 - 1)^2 + (4 - 5)^2} = \sqrt{\tfrac{81}{4} + 1} = \tfrac{1}{2}\sqrt{85}$$

20. By Exercise 19(a), the midpoint of the diameter (and thus the center of the sphere) is $C(3, 2, 7)$. The radius is half the diameter, so $r = \frac{1}{2}\sqrt{(4 - 2)^2 + (3 - 1)^2 + (10 - 4)^2} = \frac{1}{2}\sqrt{44} = \sqrt{11}$. Therefore an equation of the sphere is $(x - 3)^2 + (y - 2)^2 + (z - 7)^2 = 11$.

21. (a) Since the sphere touches the xy-plane, its radius is the distance from its center, $(2, -3, 6)$, to the xy-plane, namely 6. Therefore $r = 6$ and an equation of the sphere is $(x - 2)^2 + (y + 3)^2 + (z - 6)^2 = 6^2 = 36$.

(b) The radius of this sphere is the distance from its center $(2, -3, 6)$ to the yz-plane, which is 2. Therefore, an equation is $(x - 2)^2 + (y + 3)^2 + (z - 6)^2 = 4$.

(c) Here the radius is the distance from the center $(2, -3, 6)$ to the xz-plane, which is 3. Therefore, an equation is $(x - 2)^2 + (y + 3)^2 + (z - 6)^2 = 9$.

22. The largest sphere contained in the first octant must have a radius equal to the minimum distance from the center $(5, 4, 9)$ to any of the three coordinate planes. The shortest such distance is to the xz-plane, a distance of 4. Thus an equation of the sphere is $(x - 5)^2 + (y - 4)^2 + (z - 9)^2 = 16$.

23. The equation $y = -4$ represents a plane parallel to the xz-plane and 4 units to the left of it.

24. The equation $x = 10$ represents a plane parallel to the yz-plane and 10 units in front of it.

25. The inequality $x > 3$ represents a half-space consisting of all points in front of the plane $x = 3$.

26. The inequality $y \geq 0$ represents a half-space consisting of all points on or to the right of the xz-plane.

27. The inequality $0 \leq z \leq 6$ represents all points on or between the horizontal planes $z = 0$ (the xy-plane) and $z = 6$.

28. The equation $z^2 = 1$ ⇔ $z = \pm 1$ represents two horizontal planes; $z = 1$ is parallel to the xy-plane, one unit above it, and $z = -1$ is one unit below it.

29. The inequality $x^2 + y^2 + z^2 \leq 3$ is equivalent to $\sqrt{x^2 + y^2 + z^2} \leq \sqrt{3}$, so the region consists of those points whose distance from the origin is at most $\sqrt{3}$. This is the set of all points on or inside the sphere with radius $\sqrt{3}$ and center $(0, 0, 0)$.

30. The equation $x = z$ represents a plane perpendicular to the xz-plane and intersecting the xz-plane in the line $x = z$, $y = 0$.

31. Here $x^2 + z^2 \leq 9$ or equivalently $\sqrt{x^2 + z^2} \leq 3$ which describes the set of all points in $\mathbb{R}^3$ whose distance from the y-axis is at most 3. Thus, the inequality represents the region consisting of all points on or inside a circular cylinder of radius 3 with axis the y-axis.

32. The inequality $x^2 + y^2 + z^2 > 2z$ $\Leftrightarrow$ $x^2 + y^2 + (z-1)^2 > 1$ is equivalent to $\sqrt{x^2 + y^2 + (z-1)^2} > 1$, so the region consists of those points whose distance from the point $(0,0,1)$ is greater than 1. This is the set of all points outside the sphere with radius 1 and center $(0,0,1)$.

33. This describes all points whose x-coordinate is between 0 and 5, that is, $0 < x < 5$.

34. For any point on or above the disk in the xy-plane with center the origin and radius 2 we have $x^2 + y^2 \leq 4$. Also each point lies on or between the planes $z = 0$ and $z = 8$, so the region is described by $x^2 + y^2 \leq 4$, $0 \leq z \leq 8$.

35. This describes a region all of whose points have a distance to the origin which is greater than r, but smaller than R. So inequalities describing the region are $r < \sqrt{x^2 + y^2 + z^2} < R$, or $r^2 < x^2 + y^2 + z^2 < R^2$.

36. The solid sphere itself is represented by $\sqrt{x^2 + y^2 + z^2} \leq 2$. Since we want only the upper hemisphere, we restrict the z-coordinate to nonnegative values. Then inequalities describing the region are $\sqrt{x^2 + y^2 + z^2} \leq 2$, $z \geq 0$, or $x^2 + y^2 + z^2 \leq 4$, $z \geq 0$.

37. (a) To find the x- and y-coordinates of the point P, we project it onto L_2 and project the resulting point Q onto the x- and y-axes. To find the z-coordinate, we project P onto either the xz-plane or the yz-plane (using our knowledge of its x- or y-coordinate) and then project the resulting point onto the z-axis. (Or, we could draw a line parallel to QO from P to the z-axis.) The coordinates of P are $(2, 1, 4)$.

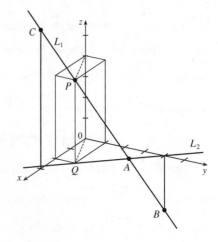

(b) A is the intersection of L_1 and L_2, B is directly below the y-intercept of L_2, and C is directly above the x-intercept of L_2.

38. Let $P = (x, y, z)$. Then $2\,|PB| = |PA|$ $\Leftrightarrow$ $4\,|PB|^2 = |PA|^2$ $\Leftrightarrow$

$4\left((x-6)^2 + (y-2)^2 + (z+2)^2\right) = (x+1)^2 + (y-5)^2 + (z-3)^2$ $\Leftrightarrow$

$4\left(x^2 - 12x + 36\right) - x^2 - 2x + 4\left(y^2 - 4y + 4\right) - y^2 + 10y + 4\left(z^2 + 4z + 4\right) - z^2 + 6z = 35$ $\Leftrightarrow$

$3x^2 - 50x + 3y^2 - 6y + 3z^2 + 22z = 35 - 144 - 16 - 16$ $\Leftrightarrow$ $x^2 - \frac{50}{3}x + y^2 - 2y + z^2 + \frac{22}{3}z = -\frac{141}{3}$.

By completing the square three times we get $\left(x - \frac{25}{3}\right)^2 + (y-1)^2 + \left(z + \frac{11}{3}\right)^2 = \frac{332}{9}$, which is an equation of a sphere with center $\left(\frac{25}{3}, 1, -\frac{11}{3}\right)$ and radius $\frac{\sqrt{332}}{3}$.

39. We need to find a set of points $\{P(x, y, z) \mid |AP| = |BP|\}$.

$$\sqrt{(x+1)^2 + (y-5)^2 + (z-3)^2} = \sqrt{(x-6)^2 + (y-2)^2 + (z+2)^2} \quad \Rightarrow$$

$$(x+1)^2 + (y-5) + (z-3)^2 = (x-6)^2 + (y-2)^2 + (z+2)^2 \quad \Rightarrow$$

$$x^2 + 2x + 1 + y^2 - 10y + 25 + z^2 - 6z + 9 = x^2 - 12x + 36 + y^2 - 4y + 4 + z^2 + 4z + 4 \quad \Rightarrow \quad 14x - 6y - 10z = 9.$$

Thus the set of points is a plane perpendicular to the line segment joining A and B (since this plane must contain the perpendicular bisector of the line segment AB).

40. Completing the square three times in the first equation gives $(x+2)^2 + (y-1)^2 + (z+2)^2 = 2^2$, a sphere with center $(-2, 1, 2)$ and radius 2. The second equation is that of a sphere with center $(0, 0, 0)$ and radius 2. The distance between the centers of the spheres is $\sqrt{(-2-0)^2 + (1-0)^2 + (-2-0)^2} = \sqrt{4+1+4} = 3$. Since the spheres have the same radius, the volume inside both spheres is symmetrical about the plane containing the circle of intersection of the spheres. The distance from this plane to the center of the circles is $\frac{3}{2}$. So the region inside both spheres consists of two caps of spheres of height $h = 2 - \frac{3}{2} = \frac{1}{2}$. From Exercise 6.2.51 [ET 6.2.51], the volume of a cap of a sphere is

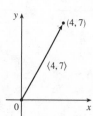

$$V = \tfrac{1}{3}\pi h^2(3r - h) = \tfrac{1}{3}\pi\left(\tfrac{1}{2}\right)^2\left(3 \cdot 2 - \tfrac{1}{2}\right) = \tfrac{11\pi}{24}.$$

So the total volume is $2 \cdot \frac{11\pi}{24} = \frac{11\pi}{12}$.

13.2 Vectors

ET 12.2

1. (a) The cost of a theater ticket is a scalar, because it has only magnitude.

(b) The current in a river is a vector, because it has both magnitude (the speed of the current) and direction at any given location.

(c) If we assume that the initial path is linear, the initial flight path from Houston to Dallas is a vector, because it has both magnitude (distance) and direction.

(d) The population of the world is a scalar, because it has only magnitude.

2. If the initial point of the vector $\langle 4, 7 \rangle$ is placed at the origin, then

$\langle 4, 7 \rangle$ is the position vector of the point $(4, 7)$.

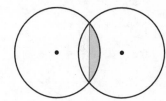

3. Vectors are equal when they share the same length and direction (but not necessarily location). Using the symmetry of the parallelogram as a guide, we see that $\overrightarrow{AB} = \overrightarrow{DC}$, $\overrightarrow{DA} = \overrightarrow{CB}$, $\overrightarrow{DE} = \overrightarrow{EB}$, and $\overrightarrow{EA} = \overrightarrow{CE}$.

4. (a) The initial point of $\overrightarrow{QR}$ is positioned at the terminal point of $\overrightarrow{PQ}$, so by the Triangle Law the sum $\overrightarrow{PQ} + \overrightarrow{QR}$ is the vector with initial point P and terminal point R, namely $\overrightarrow{PR}$.

(b) By the Triangle Law, $\overrightarrow{RP} + \overrightarrow{PS}$ is the vector with initial point R and terminal point S, namely $\overrightarrow{RS}$.

(c) First we consider $\overrightarrow{QS} - \overrightarrow{PS}$ as $\overrightarrow{QS} + \left(-\overrightarrow{PS}\right)$. Then since $-\overrightarrow{PS}$ has the same length as $\overrightarrow{PS}$ but points in the opposite

direction, we have $-\overrightarrow{PS} = \overrightarrow{SP}$ and so $\overrightarrow{QS} - \overrightarrow{PS} = \overrightarrow{QS} + \overrightarrow{SP} = \overrightarrow{QP}$.

(d) We use the Triangle Law twice: $\overrightarrow{RS} + \overrightarrow{SP} + \overrightarrow{PQ} = \left(\overrightarrow{RS} + \overrightarrow{SP}\right) + \overrightarrow{PQ} = \overrightarrow{RP} + \overrightarrow{PQ} = \overrightarrow{RQ}$

5. (a)

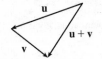

(b)

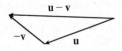

(c)

(d)

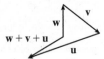

6. (a)

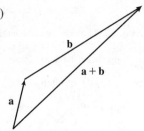

(b)

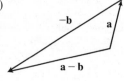

(c)

(d)

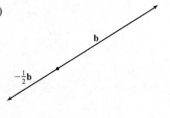

(e)

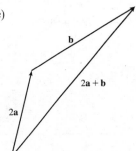

(f)

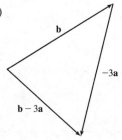

7. $\mathbf{a} = \langle -2 - 2, 1 - 3 \rangle = \langle -4, -2 \rangle$

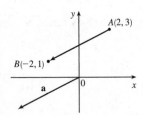

8. $\mathbf{a} = \langle 5 - (-2), 3 - (-2) \rangle = \langle 7, 5 \rangle$

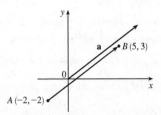

9. $\mathbf{a} = \langle 2 - (-1), 2 - 3 \rangle = \langle 3, -1 \rangle$

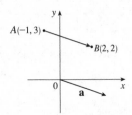

10. $\mathbf{a} = \langle 0 - 2, 6 - 1 \rangle = \langle -2, 5 \rangle$

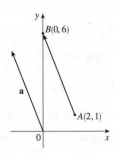

11. $\mathbf{a} = \langle 2 - 0, 3 - 3, -1 - 1 \rangle = \langle 2, 0, -2 \rangle$

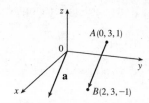

12. $\mathbf{a} = \langle 4 - 4, 2 - 0, 1 - (-2) \rangle = \langle 0, 2, 3 \rangle$

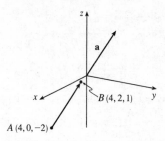

13. $\langle -1, 4 \rangle + \langle 6, -2 \rangle = \langle -1 + 6, 4 + (-2) \rangle = \langle 5, 2 \rangle$

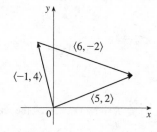

14. $\langle -2, -1 \rangle + \langle 5, 7 \rangle = \langle -2 + 5, -1 + 7 \rangle = \langle 3, 6 \rangle$

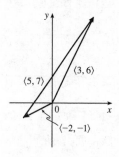

15. $\langle 0, 1, 2 \rangle + \langle 0, 0, -3 \rangle = \langle 0 + 0, 1 + 0, 2 + (-3) \rangle$

$$= \langle 0, 1, -1 \rangle$$

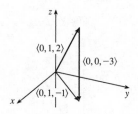

16. $\langle -1, 0, 2 \rangle + \langle 0, 4, 0 \rangle = \langle -1 + 0, 0 + 4, 2 + 0 \rangle$

$$= \langle -1, 4, 2 \rangle$$

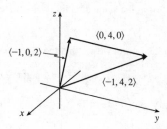

17. $\mathbf{a} + \mathbf{b} = \langle 5 + (-3), -12 + (-6) \rangle = \langle 2, -18 \rangle$

$2\mathbf{a} + 3\mathbf{b} = \langle 10, -24 \rangle + \langle -9, -18 \rangle = \langle 1, -42 \rangle$

$|\mathbf{a}| = \sqrt{5^2 + (-12)^2} = \sqrt{169} = 13$

$|\mathbf{a} - \mathbf{b}| = |\langle 5 - (-3), -12 - (-6) \rangle| = |\langle 8, -6 \rangle| = \sqrt{8^2 + (-6)^2} = \sqrt{100} = 10$

18. $\mathbf{a} + \mathbf{b} = (4\mathbf{i} + \mathbf{j}) + (\mathbf{i} - 2\mathbf{j}) = 5\mathbf{i} - \mathbf{j}$

$2\mathbf{a} + 3\mathbf{b} = 2(4\mathbf{i} + \mathbf{j}) + 3(\mathbf{i} - 2\mathbf{j}) = 8\mathbf{i} + 2\mathbf{j} + 3\mathbf{i} - 6\mathbf{j} = 11\mathbf{i} - 4\mathbf{j}$

$|\mathbf{a}| = \sqrt{4^2 + 1^2} = \sqrt{17}$

$|\mathbf{a} - \mathbf{b}| = |(4\mathbf{i} + \mathbf{j}) - (\mathbf{i} - 2\mathbf{j})| = |3\mathbf{i} + 3\mathbf{j}| = \sqrt{3^2 + 3^2} = \sqrt{18} = 3\sqrt{2}$

19. $\mathbf{a} + \mathbf{b} = (\mathbf{i} + 2\mathbf{j} - 3\mathbf{k}) + (-2\mathbf{i} - \mathbf{j} + 5\mathbf{k}) = -\mathbf{i} + \mathbf{j} + 2\mathbf{k}$

$2\mathbf{a} + 3\mathbf{b} = 2(\mathbf{i} + 2\mathbf{j} - 3\mathbf{k}) + 3(-2\mathbf{i} - \mathbf{j} + 5\mathbf{k}) = 2\mathbf{i} + 4\mathbf{j} - 6\mathbf{k} - 6\mathbf{i} - 3\mathbf{j} + 15\mathbf{k} = -4\mathbf{i} + \mathbf{j} + 9\mathbf{k}$

$|\mathbf{a}| = \sqrt{1^2 + 2^2 + (-3)^2} = \sqrt{14}$

$|\mathbf{a} - \mathbf{b}| = |(\mathbf{i} + 2\mathbf{j} - 3\mathbf{k}) - (-2\mathbf{i} - \mathbf{j} + 5\mathbf{k})| = |3\mathbf{i} + 3\mathbf{j} - 8\mathbf{k}| = \sqrt{3^2 + 3^2 + (-8)^2} = \sqrt{82}$

20. $\mathbf{a} + \mathbf{b} = (2\mathbf{i} - 4\mathbf{j} + 4\mathbf{k}) + (2\mathbf{j} - \mathbf{k}) = 2\mathbf{i} - 2\mathbf{j} + 3\mathbf{k}$

$2\mathbf{a} + 3\mathbf{b} = 2(2\mathbf{i} - 4\mathbf{j} + 4\mathbf{k}) + 3(2\mathbf{j} - \mathbf{k}) = 4\mathbf{i} - 8\mathbf{j} + 8\mathbf{k} + 6\mathbf{j} - 3\mathbf{k} = 4\mathbf{i} - 2\mathbf{j} + 5\mathbf{k}$

$|\mathbf{a}| = \sqrt{2^2 + (-4)^2 + 4^2} = \sqrt{36} = 6$

$|\mathbf{a} - \mathbf{b}| = |(2\mathbf{i} - 4\mathbf{j} + 4\mathbf{k}) - (2\mathbf{j} - \mathbf{k})| = |2\mathbf{i} - 6\mathbf{j} + 5\mathbf{k}| = \sqrt{2^2 + (-6)^2 + 5^2} = \sqrt{65}$

21. $|-3\mathbf{i} + 7\mathbf{j}| = \sqrt{(-3)^2 + 7^2} = \sqrt{58}$, so $\mathbf{u} = \dfrac{1}{\sqrt{58}}(-3\mathbf{i} + 7\mathbf{j}) = -\dfrac{3}{\sqrt{58}}\mathbf{i} + \dfrac{7}{\sqrt{58}}\mathbf{j}$.

22. $|\langle -4, 2, 4\rangle| = \sqrt{(-4)^2 + 2^2 + 4^2} = \sqrt{36} = 6$, so $\mathbf{u} = \frac{1}{6}\langle -4, 2, 4\rangle = \left\langle -\frac{2}{3}, \frac{1}{3}, \frac{2}{3}\right\rangle$.

23. The vector $8\mathbf{i} - \mathbf{j} + 4\mathbf{k}$ has length $|8\mathbf{i} - \mathbf{j} + 4\mathbf{k}| = \sqrt{8^2 + (-1)^2 + 4^2} = \sqrt{81} = 9$, so by Equation 4 the unit vector with

the same direction is $\frac{1}{9}(8\mathbf{i} - \mathbf{j} + 4\mathbf{k}) = \frac{8}{9}\mathbf{i} - \frac{1}{9}\mathbf{j} + \frac{4}{9}\mathbf{k}$.

24. $|\langle -2, 4, 2\rangle| = \sqrt{(-2)^2 + 4^2 + 2^2} = \sqrt{24} = 2\sqrt{6}$, so a unit vector in the direction of $\langle -2, 4, 2\rangle$ is $\mathbf{u} = \dfrac{1}{2\sqrt{6}}\langle -2, 4, 2\rangle$.

A vector in the same direction but with length 6 is $6\mathbf{u} = 6 \cdot \dfrac{1}{2\sqrt{6}}\langle -2, 4, 2\rangle = \left\langle -\dfrac{6}{\sqrt{6}}, \dfrac{12}{\sqrt{6}}, \dfrac{6}{\sqrt{6}}\right\rangle$ or $\langle -\sqrt{6}, 2\sqrt{6}, \sqrt{6}\rangle$.

25. From the figure, we see that the x-component of $\mathbf{v}$ is

$v_1 = |\mathbf{v}|\cos(\pi/3) = 4 \cdot \frac{1}{2} = 2$ and the y-component is

$v_2 = |\mathbf{v}|\sin(\pi/3) = 4 \cdot \frac{\sqrt{3}}{2} = 2\sqrt{3}$. Thus

$\mathbf{v} = \langle v_1, v_2\rangle = \langle 2, 2\sqrt{3}\rangle$.

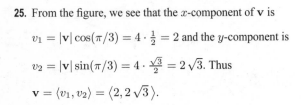

26. From the figure, we see that the horizontal component of the

force $\mathbf{F}$ is $|\mathbf{F}|\cos 38° = 50\cos 38° \approx 39.4$ N, and the

vertical component is $|\mathbf{F}|\sin 38° = 50\sin 38° \approx 30.8$ N.

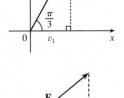

27. The velocity vector $\mathbf{v}$ makes an angle of $40°$ with the horizontal and

has magnitude equal to the speed at which the football was thrown.

From the figure, we see that the horizontal component of $\mathbf{v}$ is

$|\mathbf{v}|\cos 40° = 60\cos 40° \approx 45.96$ ft/s and the vertical component is

$|\mathbf{v}|\sin 40° = 60\sin 40° \approx 38.57$ ft/s.

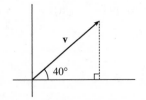

28. The given force vectors can be expressed in terms of their horizontal and vertical components as

$20\cos 45°\,\mathbf{i} + 20\sin 45°\,\mathbf{j} = 10\sqrt{2}\,\mathbf{i} + 10\sqrt{2}\,\mathbf{j}$ and $16\cos 30°\,\mathbf{i} - 16\sin 30°\,\mathbf{j} = 8\sqrt{3}\,\mathbf{i} - 8\,\mathbf{j}$. The resultant force $\mathbf{F}$ is the

sum of these two vectors: $\mathbf{F} = \left(10\sqrt{2} + 8\sqrt{3}\right)\mathbf{i} + \left(10\sqrt{2} - 8\right)\mathbf{j} \approx 28.00\,\mathbf{i} + 6.14\,\mathbf{j}$. Then we have

$|\mathbf{F}| \approx \sqrt{(28.00)^2 + (6.14)^2} \approx 28.7$ lb and, letting θ be the angle $\mathbf{F}$ makes with the positive x-axis,

$$\tan\theta = \frac{10\sqrt{2} - 8}{10\sqrt{2} + 8\sqrt{3}} \quad\Rightarrow\quad \theta = \tan^{-1}\left(\frac{10\sqrt{2} - 8}{10\sqrt{2} + 8\sqrt{3}}\right) \approx 12.4°.$$

29. The given force vectors can be expressed in terms of their horizontal and vertical components as $-300\,\mathbf{i}$ and

$200\cos 60°\,\mathbf{i} + 200\sin 60°\,\mathbf{j} = 200\left(\frac{1}{2}\right)\mathbf{i} + 200\left(\frac{\sqrt{3}}{2}\right)\mathbf{j} = 100\,\mathbf{i} + 100\sqrt{3}\,\mathbf{j}$. The resultant force $\mathbf{F}$ is the sum of these

two vectors: $\mathbf{F} = (-300 + 100)\,\mathbf{i} + \left(0 + 100\sqrt{3}\right)\mathbf{j} = -200\,\mathbf{i} + 100\sqrt{3}\,\mathbf{j}$. Then we have

$|\mathbf{F}| \approx \sqrt{(-200)^2 + \left(100\sqrt{3}\right)^2} = \sqrt{70{,}000} = 100\sqrt{7} \approx 264.6$ N. Let θ be the angle $\mathbf{F}$ makes with the positive x-axis.

Then $\tan\theta = \dfrac{100\sqrt{3}}{-200} = -\dfrac{\sqrt{3}}{2}$ and the terminal point of $\mathbf{F}$ lies in the second quadrant, so

$\theta = \tan^{-1}\left(-\dfrac{\sqrt{3}}{2}\right) + 180° \approx -40.9° + 180° = 139.1°$.

30. Set up the coordinate axes so that north is the positive y-direction, and east is the positive x-direction. The wind is blowing at

50 km/h from the direction N45°W, so that its velocity vector is 50 km/h S45°E, which can be written as

$\mathbf{v}_{\text{wind}} = 50(\cos 45°\,\mathbf{i} - \sin 45°\,\mathbf{j})$. With respect to the still air, the velocity vector of the plane is 250 km/h N 60°E, or

equivalently $\mathbf{v}_{\text{plane}} = 250(\cos 30°\,\mathbf{i} + \sin 30°\,\mathbf{j})$. The velocity of the plane relative to the ground is

$$\mathbf{v} = \mathbf{v}_{\text{wind}} + \mathbf{v}_{\text{plane}} = (50\cos 45° + 250\cos 30°)\,\mathbf{i} + (-50\sin 45° + 250\sin 30°)\,\mathbf{j}$$

$$= \left(25\sqrt{2} + 125\sqrt{3}\right)\mathbf{i} + \left(125 - 25\sqrt{2}\right)\mathbf{j} \approx 251.9\,\mathbf{i} + 89.6\,\mathbf{j}$$

The ground speed is $|\mathbf{v}| \approx \sqrt{(251.9)^2 + (89.6)^2} \approx 267$ km/h. The angle the velocity vector makes with the x-axis is

$\theta \approx \tan^{-1}\left(\frac{89.6}{251.9}\right) \approx 20°$. Therefore, the true course of the plane is about $N(90 - 20)°E = N\,70°E$.

31. With respect to the water's surface, the woman's velocity is the vector sum of the velocity of the ship with respect to the water,

and the woman's velocity with respect to the ship. If we let north be the positive y-direction, then

$\mathbf{v} = \langle 0, 22 \rangle + \langle -3, 0 \rangle = \langle -3, 22 \rangle$. The woman's speed is $|\mathbf{v}| = \sqrt{9 + 484} \approx 22.2$ mi/h. The vector $\mathbf{v}$ makes an angle θ

with the east, where $\theta = \tan^{-1}\left(\frac{22}{-3}\right) \approx 98°$. Therefore, the woman's direction is about $N(98 - 90)°W = N8°W$.

32. Call the two tensile forces $\mathbf{T}_3$ and $\mathbf{T}_5$, corresponding to the ropes of length 3 m and 5 m. In terms of vertical and horizontal

components,

$$\mathbf{T}_3 = -|\mathbf{T}_3|\cos 52°\,\mathbf{i} + |\mathbf{T}_3|\sin 52°\,\mathbf{j} \quad (1) \qquad \text{and} \qquad \mathbf{T}_5 = |\mathbf{T}_5|\cos 40°\,\mathbf{i} + |\mathbf{T}_5|\sin 40°\,\mathbf{j} \quad (2)$$

The resultant of these forces, $\mathbf{T}_3 + \mathbf{T}_5$, counterbalances the force of gravity acting on the decoration [which is

$-5g\,\mathbf{j} \approx -5(9.8)\,\mathbf{j} = -49\,\mathbf{j}$]. So $\mathbf{T}_3 + \mathbf{T}_5 = 49\,\mathbf{j}$. Hence

$\mathbf{T}_3 + \mathbf{T}_5 = (-|\mathbf{T}_3|\cos 52° + |\mathbf{T}_5|\cos 40°)\,\mathbf{i} + (|\mathbf{T}_3|\sin 52° + |\mathbf{T}_5|\sin 40°)\,\mathbf{j} = 49\,\mathbf{j}.$

Thus $-|\mathbf{T}_3|\cos 52° + |\mathbf{T}_5|\cos 40° = 0$ and $|\mathbf{T}_3|\sin 52° + |\mathbf{T}_5|\sin 40° = 49$.

From the first of these two equations $|\mathbf{T}_3| = |\mathbf{T}_5|\dfrac{\cos 40°}{\cos 52°}$. Substituting this into the second equation gives

$|\mathbf{T}_5| = \dfrac{49}{\cos 40° \tan 52° + \sin 40°} \approx 30$ N. Therefore, $|\mathbf{T}_3| = |\mathbf{T}_5|\dfrac{\cos 40°}{\cos 52°} \approx 38$ N. Finally, from **(1)** and **(2)**,

$\mathbf{T}_3 \approx -23\,\mathbf{i} + 30\,\mathbf{j}$, and $\mathbf{T}_5 \approx 23\,\mathbf{i} + 19\,\mathbf{j}$.

33. Let $\mathbf{T}_1$ and $\mathbf{T}_2$ represent the tension vectors in each side of the
clothesline as shown in the figure. $\mathbf{T}_1$ and $\mathbf{T}_2$ have equal vertical
components and opposite horizontal components, so we can write

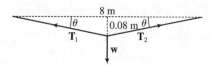

$\mathbf{T}_1 = -a\,\mathbf{i} + b\,\mathbf{j}$ and $\mathbf{T}_2 = a\,\mathbf{i} + b\,\mathbf{j}$ $[a, b > 0]$. By similar triangles, $\dfrac{b}{a} = \dfrac{0.08}{4} \Rightarrow a = 50b$. The force due to gravity

acting on the shirt has magnitude $0.8g \approx (0.8)(9.8) = 7.84$ N, hence we have $\mathbf{w} = -7.84\,\mathbf{j}$. The resultant $\mathbf{T}_1 + \mathbf{T}_2$
of the tensile forces counterbalances $\mathbf{w}$, so $\mathbf{T}_1 + \mathbf{T}_2 = -\mathbf{w} \Rightarrow (-a\,\mathbf{i} + b\,\mathbf{j}) + (a\,\mathbf{i} + b\,\mathbf{j}) = 7.84\,\mathbf{j} \Rightarrow$

$(-50b\,\mathbf{i} + b\,\mathbf{j}) + (50b\,\mathbf{i} + b\,\mathbf{j}) = 2b\,\mathbf{j} = 7.84\,\mathbf{j} \Rightarrow b = \frac{7.84}{2} = 3.92$ and $a = 50b = 196$. Thus the tensions are

$\mathbf{T}_1 = -a\,\mathbf{i} + b\,\mathbf{j} = -196\,\mathbf{i} + 3.92\,\mathbf{j}$ and $\mathbf{T}_2 = a\,\mathbf{i} + b\,\mathbf{j} = 196\,\mathbf{i} + 3.92\,\mathbf{j}$.

Alternatively, we can find the value of θ and proceed as in Example 7.

34. We can consider the weight of the chain to be concentrated at its midpoint. The
forces acting on the chain then are the tension vectors $\mathbf{T}_1$, $\mathbf{T}_2$ in each end of the
chain and the weight $\mathbf{w}$, as shown in the figure. We know $|\mathbf{T}_1| = |\mathbf{T}_2| = 25$ N
so, in terms of vertical and horizontal components, we have

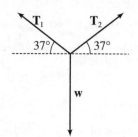

$$\mathbf{T}_1 = -25\cos 37°\,\mathbf{i} + 25\sin 37°\,\mathbf{j}$$
$$\mathbf{T}_2 = 25\cos 37°\,\mathbf{i} + 25\sin 37°\,\mathbf{j}$$

The resultant vector $\mathbf{T}_1 + \mathbf{T}_2$ of the tensions counterbalances the weight $\mathbf{w}$, giving $\mathbf{T}_1 + \mathbf{T}_2 = -\mathbf{w}$. Since $\mathbf{w} = -|\mathbf{w}|\,\mathbf{j}$,
we have $(-25\cos 37°\,\mathbf{i} + 25\sin 37°\,\mathbf{j}) + (25\cos 37°\,\mathbf{i} + 25\sin 37°\,\mathbf{j}) = |\mathbf{w}|\,\mathbf{j} \Rightarrow 50\sin 37°\,\mathbf{j} = |\mathbf{w}|\,\mathbf{j} \Rightarrow$

$|\mathbf{w}| = 50\sin 37° \approx 30.1$. So the weight is 30.1 N, and since $w = mg$, the mass is $\frac{30.1}{9.8} \approx 3.07$ kg.

35. The slope of the tangent line to the graph of $y = x^2$ at the point $(2, 4)$ is

$$\left.\frac{dy}{dx}\right|_{x=2} = 2x\Big|_{x=2} = 4$$

and a parallel vector is $\mathbf{i} + 4\,\mathbf{j}$ which has length $|\mathbf{i} + 4\,\mathbf{j}| = \sqrt{1^2 + 4^2} = \sqrt{17}$, so unit vectors parallel to the tangent line are
$\pm\frac{1}{\sqrt{17}}(\mathbf{i} + 4\,\mathbf{j})$.

36. (a) The slope of the tangent line to the graph of $y = 2\sin x$ at the point $(\pi/6, 1)$ is

$$\left.\frac{dy}{dx}\right|_{x=\pi/6} = \left.2\cos x\right|_{x=\pi/6} = 2\cdot\frac{\sqrt{3}}{2} = \sqrt{3}$$

and a parallel vector is $\mathbf{i} + \sqrt{3}\,\mathbf{j}$ which has length $\left|\mathbf{i} + \sqrt{3}\,\mathbf{j}\right| = \sqrt{1^2 + \left(\sqrt{3}\right)^2} = \sqrt{4} = 2$, so unit vectors parallel to the

tangent line are $\pm\frac{1}{2}\left(\mathbf{i} + \sqrt{3}\,\mathbf{j}\right)$.

(b) The slope of the tangent line is $\sqrt{3}$, so the slope of a line

perpendicular to the tangent line is $-\frac{1}{\sqrt{3}}$ and a vector in this direction

is $\sqrt{3}\,\mathbf{i} - \mathbf{j}$. Since $\left|\sqrt{3}\,\mathbf{i} - \mathbf{j}\right| = \sqrt{\left(\sqrt{3}\right)^2 + (-1)^2} = 2$, unit vectors

perpendicular to the tangent line are $\pm\frac{1}{2}\left(\sqrt{3}\,\mathbf{i} - \mathbf{j}\right)$.

(c)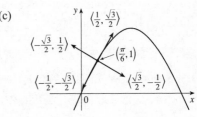

37. By the Triangle Law, $\overrightarrow{AB} + \overrightarrow{BC} = \overrightarrow{AC}$. Then $\overrightarrow{AB} + \overrightarrow{BC} + \overrightarrow{CA} = \overrightarrow{AC} + \overrightarrow{CA}$, but $\overrightarrow{AC} + \overrightarrow{CA} = \overrightarrow{AC} + \left(-\overrightarrow{AC}\right) = \mathbf{0}$. So

$\overrightarrow{AB} + \overrightarrow{BC} + \overrightarrow{CA} = \mathbf{0}$.

38. $\overrightarrow{AC} = \frac{1}{3}\overrightarrow{AB}$ and $\overrightarrow{BC} = \frac{2}{3}\overrightarrow{BA}$. $\mathbf{c} = \overrightarrow{OA} + \overrightarrow{AC} = \mathbf{a} + \frac{1}{3}\overrightarrow{AB} \;\Rightarrow\; \overrightarrow{AB} = 3\mathbf{c} - 3\mathbf{a}$. $\mathbf{c} = \overrightarrow{OB} + \overrightarrow{BC} = \overrightarrow{OA} + \frac{2}{3}\overrightarrow{BA} \;\Rightarrow\;$

$\overrightarrow{BA} = \frac{3}{2}\mathbf{c} - \frac{3}{2}\mathbf{b}$. $\overrightarrow{BA} = -\overrightarrow{AB}$, so $\frac{3}{2}\mathbf{c} - \frac{3}{2}\mathbf{b} = 3\mathbf{a} - 3\mathbf{c} \;\Leftrightarrow\; \mathbf{c} + 2\mathbf{c} = 2\mathbf{a} + \mathbf{b} \;\Leftrightarrow\; \mathbf{c} = \frac{2}{3}\mathbf{a} + \frac{1}{3}\mathbf{b}$.

39. (a), (b)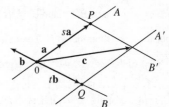

(c) From the sketch, we estimate that $s \approx 1.3$ and $t \approx 1.6$.

(d) $\mathbf{c} = s\mathbf{a} + t\mathbf{b} \;\Leftrightarrow\; 7 = 3s + 2t$ and $1 = 2s - t$.

Solving these equations gives $s = \frac{9}{7}$ and $t = \frac{11}{7}$.

40. Draw $\mathbf{a}$, $\mathbf{b}$, and $\mathbf{c}$ emanating from the origin. Extend $\mathbf{a}$ and $\mathbf{b}$ to form lines A

and B, and draw lines A' and B' parallel to these two lines through the terminal

point of $\mathbf{c}$. Since $\mathbf{a}$ and $\mathbf{b}$ are not parallel, A and B' must meet (at P), and A'

and B must also meet (at Q). Now we see that $\overrightarrow{OP} + \overrightarrow{OQ} = \mathbf{c}$, so if

$s = \dfrac{\left|\overrightarrow{OP}\right|}{|\mathbf{a}|}$ (or its negative, if $\mathbf{a}$ points in the direction opposite $\overrightarrow{OP}$) and $t = \dfrac{\left|\overrightarrow{OQ}\right|}{|\mathbf{b}|}$ (or its negative, as in the diagram), then

$\mathbf{c} = s\mathbf{a} + t\mathbf{b}$, as required.

Argument using components: Since $\mathbf{a}$, $\mathbf{b}$, and $\mathbf{c}$ all lie in the same plane, we can consider them to be vectors in two

dimensions. Let $\mathbf{a} = \langle a_1, a_2\rangle$, $\mathbf{b} = \langle b_1, b_2\rangle$, and $\mathbf{c} = \langle c_1, c_2\rangle$. We need $sa_1 + tb_1 = c_1$ and $sa_2 + tb_2 = c_2$. Multiplying the

first equation by a_2 and the second by a_1 and subtracting, we get $t = \dfrac{c_2 a_1 - c_1 a_2}{b_2 a_1 - b_1 a_2}$. Similarly $s = \dfrac{b_2 c_1 - b_1 c_2}{b_2 a_1 - b_1 a_2}$. Since $\mathbf{a} \neq \mathbf{0}$

and $\mathbf{b} \neq \mathbf{0}$ and $\mathbf{a}$ is not a scalar multiple of $\mathbf{b}$, the denominator is not zero.

41. $|\mathbf{r} - \mathbf{r}_0|$ is the distance between the points (x, y, z) and (x_0, y_0, z_0), so the set of points is a sphere with radius 1 and center (x_0, y_0, z_0).

Alternate method: $|\mathbf{r} - \mathbf{r}_0| = 1 \quad\Leftrightarrow\quad \sqrt{(x - x_0)^2 + (y - y_0)^2 + (z - z_0)^2} = 1 \quad\Leftrightarrow\quad$

$(x - x_0)^2 + (y - y_0)^2 + (z - z_0)^2 = 1$, which is the equation of a sphere with radius 1 and center (x_0, y_0, z_0).

42. Let P_1 and P_2 be the points with position vectors $\mathbf{r}_1$ and $\mathbf{r}_2$ respectively. Then $|\mathbf{r} - \mathbf{r}_1| + |\mathbf{r} - \mathbf{r}_2|$ is the sum of the distances from (x, y) to P_1 and P_2. Since this sum is constant, the set of points (x, y) represents an ellipse with foci P_1 and P_2. The condition $k > |\mathbf{r}_1 - \mathbf{r}_2|$ assures us that the ellipse is not degenerate.

43. $\mathbf{a} + (\mathbf{b} + \mathbf{c}) = \langle a_1, a_2 \rangle + (\langle b_1, b_2 \rangle + \langle c_1, c_2 \rangle) = \langle a_1, a_2 \rangle + \langle b_1 + c_1, b_2 + c_2 \rangle$

$$= \langle a_1 + b_1 + c_1, a_2 + b_2 + c_2 \rangle = \langle (a_1 + b_1) + c_1, (a_2 + b_2) + c_2 \rangle$$

$$= \langle a_1 + b_1, a_2 + b_2 \rangle + \langle c_1, c_2 \rangle = (\langle a_1, a_2 \rangle + \langle b_1, b_2 \rangle) + \langle c_1, c_2 \rangle$$

$$= (\mathbf{a} + \mathbf{b}) + \mathbf{c}$$

44. *Algebraically:*

$c(\mathbf{a} + \mathbf{b}) = c(\langle a_1, a_2, a_3 \rangle + \langle b_1, b_2, b_3 \rangle) = c\langle a_1 + b_1, a_2 + b_2, a_3 + b_3 \rangle$

$$= \langle c(a_1 + b_1), c(a_2 + b_2), c(a_3 + b_3) \rangle = \langle ca_1 + cb_1, ca_2 + cb_2, ca_3 + cb_3 \rangle$$

$$= \langle ca_1, ca_2, ca_3 \rangle + \langle cb_1, cb_2, cb_3 \rangle = c\mathbf{a} + c\mathbf{b}$$

Geometrically:

According to the Triangle Law, if $\mathbf{a} = \overrightarrow{PQ}$ and $\mathbf{b} = \overrightarrow{QR}$, then $\mathbf{a} + \mathbf{b} = \overrightarrow{PR}$. Construct triangle PST as shown so that $\overrightarrow{PS} = c\,\mathbf{a}$ and $\overrightarrow{ST} = c\,\mathbf{b}$. (We have drawn the case where $c > 1$.) By the Triangle Law, $\overrightarrow{PT} = c\,\mathbf{a} + c\,\mathbf{b}$. But triangle PQR and triangle PST are similar triangles because $c\,\mathbf{b}$ is parallel to $\mathbf{b}$. Therefore, $\overrightarrow{PR}$ and $\overrightarrow{PT}$ are parallel and, in fact, $\overrightarrow{PT} = c\overrightarrow{PR}$. Thus, $c\,\mathbf{a} + c\,\mathbf{b} = c(\mathbf{a} + \mathbf{b})$.

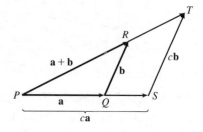

45. Consider triangle ABC, where D and E are the midpoints of AB and BC. We know that $\overrightarrow{AB} + \overrightarrow{BC} = \overrightarrow{AC}$ **(1)** and $\overrightarrow{DB} + \overrightarrow{BE} = \overrightarrow{DE}$ **(2)**. However, $\overrightarrow{DB} = \frac{1}{2}\overrightarrow{AB}$, and $\overrightarrow{BE} = \frac{1}{2}\overrightarrow{BC}$. Substituting these expressions for $\overrightarrow{DB}$ and $\overrightarrow{BE}$ into **(2)** gives $\frac{1}{2}\overrightarrow{AB} + \frac{1}{2}\overrightarrow{BC} = \overrightarrow{DE}$. Comparing this with **(1)** gives $\overrightarrow{DE} = \frac{1}{2}\overrightarrow{AC}$. Therefore $\overrightarrow{AC}$ and $\overrightarrow{DE}$ are parallel and $\left|\overrightarrow{DE}\right| = \frac{1}{2}\left|\overrightarrow{AC}\right|$.

46. The question states that the light ray strikes all three mirrors, so it is not parallel to any of them and $a_1 \neq 0$, $a_2 \neq 0$ and $a_3 \neq 0$. Let $\mathbf{b} = \langle b_1, b_2, b_3 \rangle$, as in the diagram. We can let $|\mathbf{b}| = |\mathbf{a}|$, since only its direction is important. Then

$$\frac{|b_2|}{|\mathbf{b}|} = \sin\theta = \frac{|a_2|}{|\mathbf{a}|} \quad\Rightarrow\quad |b_2| = |a_2|.$$

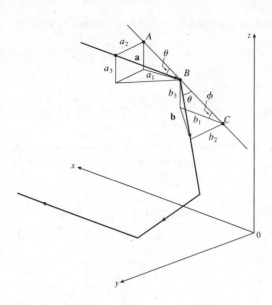

From the diagram $b_2\,\mathbf{j}$ and $a_2\,\mathbf{j}$ point in opposite directions, so $b_2 = -a_2$. $|AB| = |BC|$, so

$|b_3| = \sin\phi\,|BC| = \sin\phi\,|AB| = |a_3|$, and

$|b_1| = \cos\phi\,|BC| = \cos\phi\,|AB| = |a_1|$.

$b_3\,\mathbf{k}$ and $a_3\,\mathbf{k}$ have the same direction, as do $b_1\,\mathbf{i}$ and $a_1\,\mathbf{i}$, so

$\mathbf{b} = \langle a_1, -a_2, a_3 \rangle$. When the ray hits the other mirrors, similar arguments show that these reflections will reverse the signs of the other two coordinates, so the final reflected ray will be

$\langle -a_1, -a_2, -a_3 \rangle = -\mathbf{a}$, which is parallel to $\mathbf{a}$.

13.3 The Dot Product

ET 12.3

1. (a) $\mathbf{a}\cdot\mathbf{b}$ is a scalar, and the dot product is defined only for vectors, so $(\mathbf{a}\cdot\mathbf{b})\cdot\mathbf{c}$ has no meaning.

(b) $(\mathbf{a}\cdot\mathbf{b})\,\mathbf{c}$ is a scalar multiple of a vector, so it does have meaning.

(c) Both $|\mathbf{a}|$ and $\mathbf{b}\cdot\mathbf{c}$ are scalars, so $|\mathbf{a}|\,(\mathbf{b}\cdot\mathbf{c})$ is an ordinary product of real numbers, and has meaning.

(d) Both $\mathbf{a}$ and $\mathbf{b}+\mathbf{c}$ are vectors, so the dot product $\mathbf{a}\cdot(\mathbf{b}+\mathbf{c})$ has meaning.

(e) $\mathbf{a}\cdot\mathbf{b}$ is a scalar, but $\mathbf{c}$ is a vector, and so the two quantities cannot be added and $\mathbf{a}\cdot\mathbf{b}+\mathbf{c}$ has no meaning.

(f) $|\mathbf{a}|$ is a scalar, and the dot product is defined only for vectors, so $|\mathbf{a}|\cdot(\mathbf{b}+\mathbf{c})$ has no meaning.

2. Let the vectors be $\mathbf{a}$ and $\mathbf{b}$. Then by Theorem 3, $\mathbf{a}\cdot\mathbf{b} = |\mathbf{a}|\,|\mathbf{b}|\cos\theta = (6)\left(\tfrac{1}{3}\right)\cos\tfrac{\pi}{4} = \tfrac{6}{3\sqrt{2}} = \sqrt{2}$.

3. $\mathbf{a}\cdot\mathbf{b} = \left\langle -2, \tfrac{1}{3} \right\rangle \cdot \langle -5, 12 \rangle = (-2)(-5) + \left(\tfrac{1}{3}\right)(12) = 10 + 4 = 14$

4. $\mathbf{a}\cdot\mathbf{b} = \langle -2, 3 \rangle \cdot \langle 0.7, 1.2 \rangle = (-2)(0.7) + (3)(1.2) = 2.2$

5. $\mathbf{a}\cdot\mathbf{b} = \left\langle 4, 1, \tfrac{1}{4} \right\rangle \cdot \langle 6, -3, -8 \rangle = (4)(6) + (1)(-3) + \left(\tfrac{1}{4}\right)(-8) = 19$

6. $\mathbf{a}\cdot\mathbf{b} = \langle s, 2s, 3s \rangle \cdot \langle t, -t, 5t \rangle = (s)(t) + (2s)(-t) + (3s)(5t) = st - 2st + 15st = 14st$

7. $\mathbf{a}\cdot\mathbf{b} = (\mathbf{i} - 2\,\mathbf{j} + 3\,\mathbf{k}) \cdot (5\,\mathbf{i} + 9\,\mathbf{k}) = (1)(5) + (-2)(0) + (3)(9) = 32$

8. $\mathbf{a}\cdot\mathbf{b} = (4\,\mathbf{j} - 3\,\mathbf{k}) \cdot (2\,\mathbf{i} + 4\,\mathbf{j} + 6\,\mathbf{k}) = (0)(2) + (4)(4) + (-3)(6) = -2$

9. $\mathbf{a}\cdot\mathbf{b} = |\mathbf{a}|\,|\mathbf{b}|\cos\theta = (6)(5)\cos\tfrac{2\pi}{3} = 30\left(-\tfrac{1}{2}\right) = -15$

10. Use Theorem 3: $\mathbf{a}\cdot\mathbf{b} = |\mathbf{a}|\,|\mathbf{b}|\cos\theta = (3)\left(\sqrt{6}\right)\cos 45^\circ = 3\sqrt{6}\left(\tfrac{\sqrt{2}}{2}\right) = \tfrac{3}{2}\cdot 2\sqrt{3} = 3\sqrt{3} \approx 5.20$

11. $\mathbf{u}$, $\mathbf{v}$, and $\mathbf{w}$ are all unit vectors, so the triangle is an equilateral triangle. Thus the angle between $\mathbf{u}$ and $\mathbf{v}$ is 60° and $\mathbf{u}\cdot\mathbf{v} = |\mathbf{u}|\,|\mathbf{v}|\cos 60^\circ = (1)(1)\left(\tfrac{1}{2}\right) = \tfrac{1}{2}$. If $\mathbf{w}$ is moved so it has the same initial point as $\mathbf{u}$, we can see that the angle between them is 120° and we have $\mathbf{u}\cdot\mathbf{w} = |\mathbf{u}|\,|\mathbf{w}|\cos 120^\circ = (1)(1)\left(-\tfrac{1}{2}\right) = -\tfrac{1}{2}$.

12. $\mathbf{u}$ is a unit vector, so $\mathbf{w}$ is also a unit vector, and $|\mathbf{v}|$ can be determined by examining the right triangle formed by $\mathbf{u}$ and $\mathbf{v}$.

Since the angle between $\mathbf{u}$ and $\mathbf{v}$ is $45°$, we have $|\mathbf{v}| = |\mathbf{u}| \cos 45° = \frac{\sqrt{2}}{2}$. Then $\mathbf{u} \cdot \mathbf{v} = |\mathbf{u}| \, |\mathbf{v}| \cos 45° = (1)\left(\frac{\sqrt{2}}{2}\right)\frac{\sqrt{2}}{2} = \frac{1}{2}$.

Since $\mathbf{u}$ and $\mathbf{w}$ are orthogonal, $\mathbf{u} \cdot \mathbf{w} = 0$.

13. (a) $\mathbf{i} \cdot \mathbf{j} = \langle 1, 0, 0 \rangle \cdot \langle 0, 1, 0 \rangle = (1)(0) + (0)(1) + (0)(0) = 0$. Similarly, $\mathbf{j} \cdot \mathbf{k} = (0)(0) + (1)(0) + (0)(1) = 0$ and

$\mathbf{k} \cdot \mathbf{i} = (0)(1) + (0)(0) + (1)(0) = 0$.

Another method: Because $\mathbf{i}$, $\mathbf{j}$, and $\mathbf{k}$ are mutually perpendicular, the cosine factor in each dot product (see Theorem 3)

is $\cos \frac{\pi}{2} = 0$.

(b) By Property 1 of the dot product, $\mathbf{i} \cdot \mathbf{i} = |\mathbf{i}|^2 = 1^2 = 1$ since $\mathbf{i}$ is a unit vector. Similarly, $\mathbf{j} \cdot \mathbf{j} = |\mathbf{j}|^2 = 1$ and

$\mathbf{k} \cdot \mathbf{k} = |\mathbf{k}|^2 = 1$.

14. The dot product $\mathbf{A} \cdot \mathbf{P}$ is

$$\langle a, b, c \rangle \cdot \langle 2, 1.5, 1 \rangle = a(2) + b(1.5) + c(1)$$

$$= (\text{number of hamburgers sold})(\text{price per hamburger})$$

$$+ (\text{number of hot dogs sold})(\text{price per hot dog})$$

$$+ (\text{number of soft drinks sold})(\text{price per soft drink})$$

so it is equal to the vendor's total revenue for that day.

15. $|\mathbf{a}| = \sqrt{(-8)^2 + 6^2} = 10$, $|\mathbf{b}| = \sqrt{\left(\sqrt{7}\right)^2 + 3^2} = 4$, and $\mathbf{a} \cdot \mathbf{b} = (-8)\left(\sqrt{7}\right) + (6)(3) = 18 - 8\sqrt{7}$. From Corollary 6,

we have $\cos \theta = \dfrac{\mathbf{a} \cdot \mathbf{b}}{|\mathbf{a}| \, |\mathbf{b}|} = \dfrac{18 - 8\sqrt{7}}{10 \cdot 4} = \dfrac{9 - 4\sqrt{7}}{20}$. So the angle between $\mathbf{a}$ and $\mathbf{b}$ is $\theta = \cos^{-1}\left(\dfrac{9 - 4\sqrt{7}}{20}\right) \approx 95°$.

16. $|\mathbf{a}| = \sqrt{\left(\sqrt{3}\right)^2 + 1^2} = 2$, $|\mathbf{b}| = \sqrt{0 + 25} = 5$, and $\mathbf{a} \cdot \mathbf{b} = \left(\sqrt{3}\right)(0) + (1)(5) = 5$. Using Corollary 6, we have

$\cos \theta = \dfrac{\mathbf{a} \cdot \mathbf{b}}{|\mathbf{a}| \, |\mathbf{b}|} = \dfrac{5}{2 \cdot 5} = \dfrac{1}{2}$ and the angle between $\mathbf{a}$ and $\mathbf{b}$ is $\cos^{-1}\left(\frac{1}{2}\right) = 60°$.

17. $|\mathbf{a}| = \sqrt{3^2 + (-1)^2 + 5^2} = \sqrt{35}$, $|\mathbf{b}| = \sqrt{(-2)^2 + 4^2 + 3^2} = \sqrt{29}$, and $\mathbf{a} \cdot \mathbf{b} = (3)(-2) + (-1)(4) + (5)(3) = 5$. Then

$\cos \theta = \dfrac{\mathbf{a} \cdot \mathbf{b}}{|\mathbf{a}| \, |\mathbf{b}|} = \dfrac{5}{\sqrt{35} \cdot \sqrt{29}} = \dfrac{5}{\sqrt{1015}}$ and the angle between $\mathbf{a}$ and $\mathbf{b}$ is $\theta = \cos^{-1}\left(\dfrac{5}{\sqrt{1015}}\right) \approx 81°$.

18. $|\mathbf{a}| = \sqrt{4^2 + 0^2 + 2^2} = \sqrt{20}$, $|\mathbf{b}| = \sqrt{2^2 + (-1)^2 + 0^2} = \sqrt{5}$, and $\mathbf{a} \cdot \mathbf{b} = (4)(2) + (0)(-1) + (2)(0) = 8$.

Then $\cos \theta = \dfrac{\mathbf{a} \cdot \mathbf{b}}{|\mathbf{a}| \, |\mathbf{b}|} = \dfrac{8}{\sqrt{20} \cdot \sqrt{5}} = \dfrac{4}{5}$ and $\theta = \cos^{-1}\left(\frac{4}{5}\right) \approx 37°$.

19. $|\mathbf{a}| = \sqrt{0^2 + 1^2 + 1^2} = \sqrt{2}$, $|\mathbf{b}| = \sqrt{1^2 + 2^2 + (-3)^2} = \sqrt{14}$, and $\mathbf{a} \cdot \mathbf{b} = (0)(1) + (1)(2) + (1)(-3) = -1$.

Then $\cos \theta = \dfrac{\mathbf{a} \cdot \mathbf{b}}{|\mathbf{a}| \, |\mathbf{b}|} = \dfrac{-1}{\sqrt{2} \cdot \sqrt{14}} = \dfrac{-1}{2\sqrt{7}}$ and $\theta = \cos^{-1}\left(-\dfrac{1}{2\sqrt{7}}\right) \approx 101°$.

20. $|\mathbf{a}| = \sqrt{1^2 + 2^2 + (-2)^2} = \sqrt{9} = 3$, $|\mathbf{b}| = \sqrt{4^2 + 0^2 + (-3)^2} = \sqrt{25} = 5$, and

$\mathbf{a} \cdot \mathbf{b} = (1)(4) + (2)(0) + (-2)(-3) = 10$. Then $\cos\theta = \dfrac{\mathbf{a} \cdot \mathbf{b}}{|\mathbf{a}|\,|\mathbf{b}|} = \dfrac{10}{3 \cdot 5} = \dfrac{2}{3}$ and $\theta = \cos^{-1}\left(\frac{2}{3}\right) \approx 48°$.

21. Let a, b, and c be the angles at vertices A, B, and C respectively.

Then a is the angle between vectors $\overrightarrow{AB}$ and $\overrightarrow{AC}$, b is the angle

between vectors $\overrightarrow{BA}$ and $\overrightarrow{BC}$, and c is the angle between vectors

$\overrightarrow{CA}$ and $\overrightarrow{CB}$.

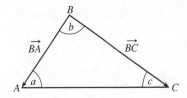

Thus $\cos a = \dfrac{\overrightarrow{AB} \cdot \overrightarrow{AC}}{\left|\overrightarrow{AB}\right|\left|\overrightarrow{AC}\right|} = \dfrac{\langle 2, 6\rangle \cdot \langle -2, 4\rangle}{\sqrt{2^2 + 6^2}\sqrt{(-2)^2 + 4^2}} = \dfrac{1}{\sqrt{40}\sqrt{20}}(-4 + 24) = \dfrac{20}{\sqrt{800}} = \dfrac{\sqrt{2}}{2}$ and

$a = \cos^{-1}\left(\dfrac{\sqrt{2}}{2}\right) = 45°$. Similarly, $\cos b = \dfrac{\overrightarrow{BA} \cdot \overrightarrow{BC}}{\left|\overrightarrow{BA}\right|\left|\overrightarrow{BC}\right|} = \dfrac{\langle -2, -6\rangle \cdot \langle -4, -2\rangle}{\sqrt{4 + 36}\sqrt{16 + 4}} = \dfrac{1}{\sqrt{40}\sqrt{20}}(8 + 12) = \dfrac{20}{\sqrt{800}} = \dfrac{\sqrt{2}}{2}$

so $b = \cos^{-1}\left(\dfrac{\sqrt{2}}{2}\right) = 45°$ and $c = 180° - (45° + 45°) = 90°$.

Alternate solution: Apply the Law of Cosines three times as follows: $\cos a = \dfrac{\left|\overrightarrow{BC}\right|^2 - \left|\overrightarrow{AB}\right|^2 - \left|\overrightarrow{AC}\right|^2}{2\left|\overrightarrow{AB}\right|\left|\overrightarrow{AC}\right|}$,

$\cos b = \dfrac{\left|\overrightarrow{AC}\right|^2 - \left|\overrightarrow{AB}\right|^2 - \left|\overrightarrow{BC}\right|^2}{2\left|\overrightarrow{AB}\right|\left|\overrightarrow{BC}\right|}$, and $\cos c = \dfrac{\left|\overrightarrow{AB}\right|^2 - \left|\overrightarrow{AC}\right|^2 - \left|\overrightarrow{BC}\right|^2}{2\left|\overrightarrow{AC}\right|\left|\overrightarrow{BC}\right|}$.

22. As in Exercise 21, let d, e, and f be the angles at vertices D, E, and F.

Then d is the angle between vectors $\overrightarrow{DE}$ and $\overrightarrow{DF}$, e is the angle between

vectors $\overrightarrow{ED}$ and $\overrightarrow{EF}$, and f is the angle between vectors $\overrightarrow{FD}$ and $\overrightarrow{FE}$.

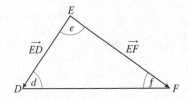

Thus $\cos d = \dfrac{\overrightarrow{DE} \cdot \overrightarrow{DF}}{\left|\overrightarrow{DE}\right|\left|\overrightarrow{DF}\right|} = \dfrac{\langle -2, 3, 2\rangle \cdot \langle 1, 1, -2\rangle}{\sqrt{(-2)^2 + 3^2 + 2^2}\sqrt{1^2 + 1^2 + (-2)^2}} = \dfrac{1}{\sqrt{17}\sqrt{6}}(-2 + 3 - 4) = -\dfrac{3}{\sqrt{102}}$

and $d = \cos^{-1}\left(-\dfrac{3}{\sqrt{102}}\right) \approx 107°$. Similarly,

$\cos e = \dfrac{\overrightarrow{ED} \cdot \overrightarrow{EF}}{\left|\overrightarrow{ED}\right|\left|\overrightarrow{EF}\right|} = \dfrac{\langle 2, -3, -2\rangle \cdot \langle 3, -2, -4\rangle}{\sqrt{4 + 9 + 4}\sqrt{9 + 4 + 16}} = \dfrac{1}{\sqrt{17}\sqrt{29}}(6 + 6 + 8) = \dfrac{20}{\sqrt{493}}$ so $e = \cos^{-1}\left(\dfrac{20}{\sqrt{493}}\right) \approx 26°$

and $f \approx 180° - (107° + 26°) = 47°$.

Alternate solution: Apply the Law of Cosines three times as follows:

$\cos d = \dfrac{\left|\overrightarrow{EF}\right|^2 - \left|\overrightarrow{DE}\right|^2 - \left|\overrightarrow{DF}\right|^2}{2\left|\overrightarrow{DE}\right|\left|\overrightarrow{DF}\right|}$ $\quad\cos e = \dfrac{\left|\overrightarrow{DF}\right|^2 - \left|\overrightarrow{DE}\right|^2 - \left|\overrightarrow{EF}\right|^2}{2\left|\overrightarrow{DE}\right|\left|\overrightarrow{EF}\right|}$ $\quad\cos f = \dfrac{\left|\overrightarrow{DE}\right|^2 - \left|\overrightarrow{DF}\right|^2 - \left|\overrightarrow{EF}\right|^2}{2\left|\overrightarrow{DF}\right|\left|\overrightarrow{EF}\right|}$

23. (a) $\mathbf{a} \cdot \mathbf{b} = (-5)(6) + (3)(-8) + (7)(2) = -40 \neq 0$, so $\mathbf{a}$ and $\mathbf{b}$ are not orthogonal. Also, since $\mathbf{a}$ is not a scalar multiple of $\mathbf{b}$, $\mathbf{a}$ and $\mathbf{b}$ are not parallel.

(b) $\mathbf{a} \cdot \mathbf{b} = (4)(-3) + (6)(2) = 0$, so $\mathbf{a}$ and $\mathbf{b}$ are orthogonal (and not parallel).

(c) $\mathbf{a} \cdot \mathbf{b} = (-1)(3) + (2)(4) + (5)(-1) = 0$, so $\mathbf{a}$ and $\mathbf{b}$ are orthogonal (and not parallel).

(d) Because $\mathbf{a} = -\frac{2}{3}\mathbf{b}$, $\mathbf{a}$ and $\mathbf{b}$ are parallel.

24. (a) Because $\mathbf{u} = -\frac{3}{4}\mathbf{v}$, $\mathbf{u}$ and $\mathbf{v}$ are parallel vectors (and thus not orthogonal).

(b) $\mathbf{u} \cdot \mathbf{v} = (1)(2) + (-1)(-1) + (2)(1) = 5 \neq 0$, so $\mathbf{u}$ and $\mathbf{v}$ are not orthogonal. Also, $\mathbf{u}$ is not a scalar multiple of $\mathbf{v}$, so $\mathbf{u}$ and $\mathbf{v}$ are not parallel.

(c) $\mathbf{u} \cdot \mathbf{v} = (a)(-b) + (b)(a) + (c)(0) = -ab + ab + 0 = 0$, so $\mathbf{u}$ and $\mathbf{v}$ are orthogonal (and not parallel).

25. $\overrightarrow{QP} = \langle -1, -3, 2 \rangle$, $\overrightarrow{QR} = \langle 4, -2, -1 \rangle$, and $\overrightarrow{QP} \cdot \overrightarrow{QR} = -4 + 6 - 2 = 0$. Thus $\overrightarrow{QP}$ and $\overrightarrow{QR}$ are orthogonal, so the angle of the triangle at vertex Q is a right angle.

26. $\langle -6, b, 2 \rangle$ and $\langle b, b^2, b \rangle$ are orthogonal when $\langle -6, b, 2 \rangle \cdot \langle b, b^2, b \rangle = 0$ $\Leftrightarrow$ $(-6)(b) + (b)(b^2) + (2)(b) = 0$ $\Leftrightarrow$ $b^3 - 4b = 0$ $\Leftrightarrow$ $b(b+2)(b-2) = 0$ $\Leftrightarrow$ $b = 0$ or $b = \pm 2$.

27. Let $\mathbf{a} = a_1\mathbf{i} + a_2\mathbf{j} + a_3\mathbf{k}$ be a vector orthogonal to both $\mathbf{i} + \mathbf{j}$ and $\mathbf{i} + \mathbf{k}$. Then $\mathbf{a} \cdot (\mathbf{i} + \mathbf{j}) = 0$ $\Leftrightarrow$ $a_1 + a_2 = 0$ and $\mathbf{a} \cdot (\mathbf{i} + \mathbf{k}) = 0$ $\Leftrightarrow$ $a_1 + a_3 = 0$, so $a_1 = -a_2 = -a_3$. Furthermore $\mathbf{a}$ is to be a unit vector, so $1 = a_1^2 + a_2^2 + a_3^2 = 3a_1^2$ implies $a_1 = \pm\frac{1}{\sqrt{3}}$. Thus $\mathbf{a} = \frac{1}{\sqrt{3}}\mathbf{i} - \frac{1}{\sqrt{3}}\mathbf{j} - \frac{1}{\sqrt{3}}\mathbf{k}$ and $\mathbf{a} = -\frac{1}{\sqrt{3}}\mathbf{i} + \frac{1}{\sqrt{3}}\mathbf{j} + \frac{1}{\sqrt{3}}\mathbf{k}$ are two such unit vectors.

28. Let $\mathbf{u} = \langle a, b \rangle$ be a unit vector. By Theorem 3 we need $\mathbf{u} \cdot \mathbf{v} = |\mathbf{u}|\,|\mathbf{v}|\cos 60°$ $\Leftrightarrow$ $3a + 4b = (1)(5)\frac{1}{2}$ $\Leftrightarrow$ $b = \frac{5}{8} - \frac{3}{4}a$. Since $\mathbf{u}$ is a unit vector, $|\mathbf{u}| = \sqrt{a^2 + b^2} = 1$ $\Leftrightarrow$ $a^2 + b^2 = 1$ $\Leftrightarrow$ $a^2 + \left(\frac{5}{8} - \frac{3}{4}a\right)^2 = 1$ $\Leftrightarrow$ $\frac{25}{16}a^2 - \frac{15}{16}a + \frac{25}{64} = 1$ $\Leftrightarrow$ $100a^2 - 60a - 39 = 0$. By the quadratic formula,

$a = \dfrac{-(-60) \pm \sqrt{(-60)^2 - 4(100)(-39)}}{2(100)} = \dfrac{60 \pm \sqrt{19{,}200}}{200} = \dfrac{3 \pm 4\sqrt{3}}{10}$. If $a = \dfrac{3 + 4\sqrt{3}}{10}$ then

$b = \dfrac{5}{8} - \dfrac{3}{4}\left(\dfrac{3 + 4\sqrt{3}}{10}\right) = \dfrac{4 - 3\sqrt{3}}{10}$, and if $a = \dfrac{3 - 4\sqrt{3}}{10}$ then $b = \dfrac{5}{8} - \dfrac{3}{4}\left(\dfrac{3 - 4\sqrt{3}}{10}\right) = \dfrac{4 + 3\sqrt{3}}{10}$. Thus the two

unit vectors are $\left\langle \dfrac{3 + 4\sqrt{3}}{10}, \dfrac{4 - 3\sqrt{3}}{10} \right\rangle \approx \langle 0.9928, -0.1196 \rangle$ and $\left\langle \dfrac{3 - 4\sqrt{3}}{10}, \dfrac{4 + 3\sqrt{3}}{10} \right\rangle \approx \langle -0.3928, 0.9196 \rangle$.

29. Since $|\langle 3, 4, 5 \rangle| = \sqrt{9 + 16 + 25} = \sqrt{50} = 5\sqrt{2}$, using Equations 8 and 9 we have $\cos\alpha = \frac{3}{5\sqrt{2}}$, $\cos\beta = \frac{4}{5\sqrt{2}}$, and $\cos\gamma = \frac{5}{5\sqrt{2}} = \frac{1}{\sqrt{2}}$. The direction angles are given by $\alpha = \cos^{-1}\left(\frac{3}{5\sqrt{2}}\right) \approx 65°$, $\beta = \cos^{-1}\left(\frac{4}{5\sqrt{2}}\right) \approx 56°$, and $\gamma = \cos^{-1}\left(\frac{1}{\sqrt{2}}\right) = 45°$.

30. Since $|\langle 1, -2, -1 \rangle| = \sqrt{1 + 4 + 1} = \sqrt{6}$, using Equations 8 and 9 we have $\cos\alpha = \frac{1}{\sqrt{6}}$, $\cos\beta = \frac{-2}{\sqrt{6}}$, and $\cos\gamma = \frac{-1}{\sqrt{6}}$. The direction angles are given by $\alpha = \cos^{-1}\left(\frac{1}{\sqrt{6}}\right) \approx 66°$, $\beta = \cos^{-1}\left(-\frac{2}{\sqrt{6}}\right) \approx 145°$, and $\gamma = \cos^{-1}\left(-\frac{1}{\sqrt{6}}\right) \approx 114°$.

31. Since $|2\mathbf{i} + 3\mathbf{j} - 6\mathbf{k}| = \sqrt{4 + 9 + 36} = \sqrt{49} = 7$, Equations 8 and 9 give $\cos\alpha = \frac{2}{7}$, $\cos\beta = \frac{3}{7}$, and $\cos\gamma = \frac{-6}{7}$, while $\alpha = \cos^{-1}\left(\frac{2}{7}\right) \approx 73°$, $\beta = \cos^{-1}\left(\frac{3}{7}\right) \approx 65°$, and $\gamma = \cos^{-1}\left(-\frac{6}{7}\right) \approx 149°$.

32. Since $|2\mathbf{i} - \mathbf{j} + 2\mathbf{k}| = \sqrt{4 + 1 + 4} = \sqrt{9} = 3$, Equations 8 and 9 give $\cos\alpha = \frac{2}{3}$, $\cos\beta = \frac{-1}{3}$, and $\cos\gamma = \frac{2}{3}$, while $\alpha = \gamma = \cos^{-1}\left(\frac{2}{3}\right) \approx 48°$ and $\beta = \cos^{-1}\left(-\frac{1}{3}\right) \approx 109°$.

33. $|\langle c, c, c\rangle| = \sqrt{c^2 + c^2 + c^2} = \sqrt{3}\,c$ [since $c > 0$], so $\cos\alpha = \cos\beta = \cos\gamma = \dfrac{c}{\sqrt{3}\,c} = \dfrac{1}{\sqrt{3}}$ and $\alpha = \beta = \gamma = \cos^{-1}\left(\dfrac{1}{\sqrt{3}}\right) \approx 55°$.

34. Since $\cos^2\alpha + \cos^2\beta + \cos^2\gamma = 1$, $\cos^2\gamma = 1 - \cos^2\alpha - \cos^2\beta = 1 - \cos^2\left(\frac{\pi}{4}\right) - \cos^2\left(\frac{\pi}{3}\right) = 1 - \left(\frac{1}{\sqrt{2}}\right)^2 - \left(\frac{1}{2}\right)^2 = \frac{1}{4}$. Thus $\cos\gamma = \pm\frac{1}{2}$ and $\gamma = \frac{\pi}{3}$ or $\gamma = \frac{2\pi}{3}$.

35. $|\mathbf{a}| = \sqrt{3^2 + (-4)^2} = 5$. The scalar projection of $\mathbf{b}$ onto $\mathbf{a}$ is $\text{comp}_{\mathbf{a}}\,\mathbf{b} = \dfrac{\mathbf{a}\cdot\mathbf{b}}{|\mathbf{a}|} = \dfrac{3\cdot 5 + (-4)\cdot 0}{5} = 3$ and the vector projection of $\mathbf{b}$ onto $\mathbf{a}$ is $\text{proj}_{\mathbf{a}}\,\mathbf{b} = \left(\dfrac{\mathbf{a}\cdot\mathbf{b}}{|\mathbf{a}|}\right)\dfrac{\mathbf{a}}{|\mathbf{a}|} = 3\cdot\frac{1}{5}\langle 3, -4\rangle = \left\langle \frac{9}{5}, -\frac{12}{5}\right\rangle$.

36. $|\mathbf{a}| = \sqrt{1^2 + 2^2} = \sqrt{5}$, so the scalar projection of $\mathbf{b}$ onto $\mathbf{a}$ is $\text{comp}_{\mathbf{a}}\,\mathbf{b} = \dfrac{\mathbf{a}\cdot\mathbf{b}}{|\mathbf{a}|} = \dfrac{1(-4) + 2\cdot 1}{\sqrt{5}} = -\dfrac{2}{\sqrt{5}}$ and the vector projection of $\mathbf{b}$ onto $\mathbf{a}$ is $\text{proj}_{\mathbf{a}}\,\mathbf{b} = \left(\dfrac{\mathbf{a}\cdot\mathbf{b}}{|\mathbf{a}|}\right)\dfrac{\mathbf{a}}{|\mathbf{a}|} = -\dfrac{2}{\sqrt{5}}\cdot\dfrac{1}{\sqrt{5}}\langle 1, 2\rangle = \left\langle -\frac{2}{5}, -\frac{4}{5}\right\rangle$.

37. $|\mathbf{a}| = \sqrt{9 + 36 + 4} = 7$ so the scalar projection of $\mathbf{b}$ onto $\mathbf{a}$ is $\text{comp}_{\mathbf{a}}\mathbf{b} = \dfrac{\mathbf{a}\cdot\mathbf{b}}{|\mathbf{a}|} = \frac{1}{7}(3 + 12 - 6) = \frac{9}{7}$. The vector projection of $\mathbf{b}$ onto $\mathbf{a}$ is $\text{proj}_{\mathbf{a}}\mathbf{b} = \dfrac{9}{7}\dfrac{\mathbf{a}}{|\mathbf{a}|} = \frac{9}{7}\cdot\frac{1}{7}\langle 3, 6, -2\rangle = \frac{9}{49}\langle 3, 6, -2\rangle = \left\langle \frac{27}{49}, \frac{54}{49}, -\frac{18}{49}\right\rangle$.

38. $|\mathbf{a}| = \sqrt{4 + 9 + 36} = 7$ so the scalar projection of $\mathbf{b}$ onto $\mathbf{a}$ is $\text{comp}_{\mathbf{a}}\,\mathbf{b} = \dfrac{\mathbf{a}\cdot\mathbf{b}}{|\mathbf{a}|} = \frac{1}{7}(-10 - 3 - 24) = -\frac{37}{7}$, while the vector projection is $\text{proj}_{\mathbf{a}}\,\mathbf{b} = -\dfrac{37}{7}\dfrac{\mathbf{a}}{|\mathbf{a}|} = -\frac{37}{7}\cdot\frac{1}{7}\langle -2, 3, -6\rangle = -\frac{37}{49}\langle -2, 3, -6\rangle = \left\langle \frac{74}{49}, -\frac{111}{49}, \frac{222}{49}\right\rangle$.

39. $|\mathbf{a}| = \sqrt{4 + 1 + 16} = \sqrt{21}$ so the scalar projection of $\mathbf{b}$ onto $\mathbf{a}$ is $\text{comp}_{\mathbf{a}}\,\mathbf{b} = \dfrac{\mathbf{a}\cdot\mathbf{b}}{|\mathbf{a}|} = \dfrac{0 - 1 + 2}{\sqrt{21}} = \dfrac{1}{\sqrt{21}}$ while the vector projection of $\mathbf{b}$ onto $\mathbf{a}$ is $\text{proj}_{\mathbf{a}}\,\mathbf{b} = \dfrac{1}{\sqrt{21}}\dfrac{\mathbf{a}}{|\mathbf{a}|} = \dfrac{1}{\sqrt{21}}\cdot\dfrac{2\mathbf{i} - \mathbf{j} + 4\mathbf{k}}{\sqrt{21}} = \frac{1}{21}(2\mathbf{i} - \mathbf{j} + 4\mathbf{k}) = \frac{2}{21}\mathbf{i} - \frac{1}{21}\mathbf{j} + \frac{4}{21}\mathbf{k}$.

40. $|\mathbf{a}| = \sqrt{1 + 1 + 1} = \sqrt{3}$, so the scalar projection of $\mathbf{b}$ onto $\mathbf{a}$ is $\text{comp}_{\mathbf{a}}\,\mathbf{b} = \dfrac{\mathbf{a}\cdot\mathbf{b}}{|\mathbf{a}|} = \dfrac{1 - 1 + 1}{\sqrt{3}} = \dfrac{1}{\sqrt{3}}$ while the vector projection of $\mathbf{b}$ onto $\mathbf{a}$ is $\text{proj}_{\mathbf{a}}\,\mathbf{b} = \dfrac{1}{\sqrt{3}}\dfrac{\mathbf{a}}{|\mathbf{a}|} = \dfrac{1}{\sqrt{3}}\cdot\dfrac{\mathbf{i} + \mathbf{j} + \mathbf{k}}{\sqrt{3}} = \frac{1}{3}(\mathbf{i} + \mathbf{j} + \mathbf{k})$.

41. $(\text{orth}_{\mathbf{a}}\,\mathbf{b})\cdot\mathbf{a} = (\mathbf{b} - \text{proj}_{\mathbf{a}}\,\mathbf{b})\cdot\mathbf{a} = \mathbf{b}\cdot\mathbf{a} - (\text{proj}_{\mathbf{a}}\,\mathbf{b})\cdot\mathbf{a} = \mathbf{b}\cdot\mathbf{a} - \dfrac{\mathbf{a}\cdot\mathbf{b}}{|\mathbf{a}|^2}\mathbf{a}\cdot\mathbf{a} = \mathbf{b}\cdot\mathbf{a} - \dfrac{\mathbf{a}\cdot\mathbf{b}}{|\mathbf{a}|^2}|\mathbf{a}|^2 = \mathbf{b}\cdot\mathbf{a} - \mathbf{a}\cdot\mathbf{b} = 0$.

So they are orthogonal by (7).

42. Using the formula in Exercise 41 and the result of Exercise 36, we have

$$\text{orth}_a \mathbf{b} = \mathbf{b} - \text{proj}_a \mathbf{b} = \langle -4, 1 \rangle - \langle -\tfrac{2}{5}, -\tfrac{4}{5} \rangle = \langle -\tfrac{18}{5}, \tfrac{9}{5} \rangle.$$

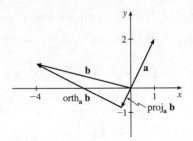

43. $\text{comp}_a \mathbf{b} = \dfrac{\mathbf{a} \cdot \mathbf{b}}{|\mathbf{a}|} = 2 \iff \mathbf{a} \cdot \mathbf{b} = 2\,|\mathbf{a}| = 2\sqrt{10}$. If $\mathbf{b} = \langle b_1, b_2, b_3 \rangle$, then we need $3b_1 + 0b_2 - 1b_3 = 2\sqrt{10}$.

One possible solution is obtained by taking $b_1 = 0$, $b_2 = 0$, $b_3 = -2\sqrt{10}$. In general, $\mathbf{b} = \langle s, t, 3s - 2\sqrt{10} \rangle$, $s, t \in \mathbb{R}$.

44. (a) $\text{comp}_a \mathbf{b} = \text{comp}_b \mathbf{a} \iff \dfrac{\mathbf{a} \cdot \mathbf{b}}{|\mathbf{a}|} = \dfrac{\mathbf{b} \cdot \mathbf{a}}{|\mathbf{b}|} \iff \dfrac{1}{|\mathbf{a}|} = \dfrac{1}{|\mathbf{b}|}$ or $\mathbf{a} \cdot \mathbf{b} = 0 \iff |\mathbf{b}| = |\mathbf{a}|$ or $\mathbf{a} \cdot \mathbf{b} = 0$.

That is, if $\mathbf{a}$ and $\mathbf{b}$ are orthogonal or if they have the same length.

(b) $\text{proj}_a \mathbf{b} = \text{proj}_b \mathbf{a} \iff \dfrac{\mathbf{a} \cdot \mathbf{b}}{|\mathbf{a}|^2}\mathbf{a} = \dfrac{\mathbf{b} \cdot \mathbf{a}}{|\mathbf{b}|^2}\mathbf{b} \iff \mathbf{a} \cdot \mathbf{b} = 0$ or $\dfrac{\mathbf{a}}{|\mathbf{a}|^2} = \dfrac{\mathbf{b}}{|\mathbf{b}|^2}$.

But $\dfrac{\mathbf{a}}{|\mathbf{a}|^2} = \dfrac{\mathbf{b}}{|\mathbf{b}|^2} \Rightarrow \dfrac{|\mathbf{a}|}{|\mathbf{a}|^2} = \dfrac{|\mathbf{b}|}{|\mathbf{b}|^2} \Rightarrow |\mathbf{a}| = |\mathbf{b}|$. Substituting this into the previous equation gives $\mathbf{a} = \mathbf{b}$.

So $\text{proj}_a \mathbf{b} = \text{proj}_b \mathbf{a} \iff \mathbf{a}$ and $\mathbf{b}$ are orthogonal, or they are equal.

45. The displacement vector is $\mathbf{D} = (6 - 0)\,\mathbf{i} + (12 - 10)\,\mathbf{j} + (20 - 8)\,\mathbf{k} = 6\,\mathbf{i} + 2\,\mathbf{j} + 12\,\mathbf{k}$ so by Equation 12 the work done is

$W = \mathbf{F} \cdot \mathbf{D} = (8\,\mathbf{i} - 6\,\mathbf{j} + 9\,\mathbf{k}) \cdot (6\,\mathbf{i} + 2\,\mathbf{j} + 12\,\mathbf{k}) = 48 - 12 + 108 = 144$ joules.

46. Here $|\mathbf{D}| = 1000$ m, $|\mathbf{F}| = 1500$ N, and $\theta = 30°$. Thus

$W = \mathbf{F} \cdot \mathbf{D} = |\mathbf{F}|\,|\mathbf{D}|\cos\theta = (1500)(1000)\left(\dfrac{\sqrt{3}}{2}\right) = 750{,}000\sqrt{3}$ joules.

47. Here $|\mathbf{D}| = 80$ ft, $|\mathbf{F}| = 30$ lb, and $\theta = 40°$. Thus

$W = \mathbf{F} \cdot \mathbf{D} = |\mathbf{F}|\,|\mathbf{D}|\cos\theta = (30)(80)\cos 40° = 2400\cos 40° \approx 1839$ ft-lb.

48. $W = \mathbf{F} \cdot \mathbf{D} = |\mathbf{F}|\,|\mathbf{D}|\cos\theta = (400)(120)\cos 36° \approx 38{,}833$ ft-lb

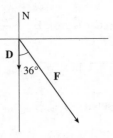

49. First note that $\mathbf{n} = \langle a, b \rangle$ is perpendicular to the line, because if $Q_1 = (a_1, b_1)$ and $Q_2 = (a_2, b_2)$ lie on the line, then

$\mathbf{n} \cdot \overrightarrow{Q_1 Q_2} = aa_2 - aa_1 + bb_2 - bb_1 = 0$, since $aa_2 + bb_2 = -c = aa_1 + bb_1$ from the equation of the line.

Let $P_2 = (x_2, y_2)$ lie on the line. Then the distance from P_1 to the line is the absolute value of the scalar projection

of $\overrightarrow{P_1 P_2}$ onto $\mathbf{n}$. $\quad \text{comp}_{\mathbf{n}}\left(\overrightarrow{P_1 P_2}\right) = \dfrac{|\mathbf{n} \cdot \langle x_2 - x_1, y_2 - y_1 \rangle|}{|\mathbf{n}|} = \dfrac{|ax_2 - ax_1 + by_2 - by_1|}{\sqrt{a^2 + b^2}} = \dfrac{|ax_1 + by_1 + c|}{\sqrt{a^2 + b^2}}$

since $ax_2 + by_2 = -c$. The required distance is $\dfrac{|3 \cdot -2 + -4 \cdot 3 + 5|}{\sqrt{3^2 + 4^2}} = \dfrac{13}{5}$.

50. $(\mathbf{r} - \mathbf{a}) \cdot (\mathbf{r} - \mathbf{b}) = 0$ implies that the vectors $\mathbf{r} - \mathbf{a}$ and $\mathbf{r} - \mathbf{b}$ are orthogonal.

From the diagram (in which A, B and R are the terminal points of the vectors), we

see that this implies that R lies on a sphere whose diameter is the line from A to B.

The center of this circle is the midpoint of AB, that is,

$\frac{1}{2}(\mathbf{a} + \mathbf{b}) = \langle \frac{1}{2}(a_1 + b_1), \frac{1}{2}(a_2 + b_2), \frac{1}{2}(a_3 + b_3) \rangle$, and its radius is

$\frac{1}{2}|\mathbf{a} - \mathbf{b}| = \frac{1}{2}\sqrt{(a_1 - b_1)^2 + (a_2 - b_2)^2 + (a_3 - b_3)^2}$.

Or: Expand the given equation, substitute $\mathbf{r} \cdot \mathbf{r} = x^2 + y^2 + z^2$ and complete the squares.

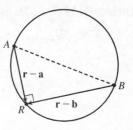

51. For convenience, consider the unit cube positioned so that its back left corner is at the origin, and its edges lie along the

coordinate axes. The diagonal of the cube that begins at the origin and ends at $(1, 1, 1)$ has vector representation $\langle 1, 1, 1 \rangle$.

The angle θ between this vector and the vector of the edge which also begins at the origin and runs along the x-axis [that is,

$\langle 1, 0, 0 \rangle$] is given by $\cos \theta = \dfrac{\langle 1, 1, 1 \rangle \cdot \langle 1, 0, 0 \rangle}{|\langle 1, 1, 1 \rangle| \, |\langle 1, 0, 0 \rangle|} = \dfrac{1}{\sqrt{3}} \quad \Rightarrow \quad \theta = \cos^{-1}\left(\frac{1}{\sqrt{3}}\right) \approx 55°$.

52. Consider a cube with sides of unit length, wholly within the first octant and with edges along each of the three coordinate axes.

$\mathbf{i} + \mathbf{j} + \mathbf{k}$ and $\mathbf{i} + \mathbf{j}$ are vector representations of a diagonal of the cube and a diagonal of one of its faces. If θ is the angle

between these diagonals, then $\cos \theta = \dfrac{(\mathbf{i} + \mathbf{j} + \mathbf{k}) \cdot (\mathbf{i} + \mathbf{j})}{|\mathbf{i} + \mathbf{j} + \mathbf{k}| \, |\mathbf{i} + \mathbf{j}|} = \dfrac{1 + 1}{\sqrt{3}\sqrt{2}} = \sqrt{\dfrac{2}{3}} \quad \Rightarrow \quad \theta = \cos^{-1}\sqrt{\dfrac{2}{3}} \approx 35°$.

53. Consider the H—C—H combination consisting of the sole carbon atom and the two hydrogen atoms that are at $(1, 0, 0)$ and

$(0, 1, 0)$ (or any H—C—H combination, for that matter). Vector representations of the line segments emanating from the

carbon atom and extending to these two hydrogen atoms are $\langle 1 - \frac{1}{2}, 0 - \frac{1}{2}, 0 - \frac{1}{2} \rangle = \langle \frac{1}{2}, -\frac{1}{2}, -\frac{1}{2} \rangle$ and

$\langle 0 - \frac{1}{2}, 1 - \frac{1}{2}, 0 - \frac{1}{2} \rangle = \langle -\frac{1}{2}, \frac{1}{2}, -\frac{1}{2} \rangle$. The bond angle, θ, is therefore given by

$\cos \theta = \dfrac{\langle \frac{1}{2}, -\frac{1}{2}, -\frac{1}{2} \rangle \cdot \langle -\frac{1}{2}, \frac{1}{2}, -\frac{1}{2} \rangle}{|\langle \frac{1}{2}, -\frac{1}{2}, -\frac{1}{2} \rangle| \, |\langle -\frac{1}{2}, \frac{1}{2}, -\frac{1}{2} \rangle|} = \dfrac{-\frac{1}{4} - \frac{1}{4} + \frac{1}{4}}{\sqrt{\frac{3}{4}}\sqrt{\frac{3}{4}}} = -\dfrac{1}{3} \quad \Rightarrow \quad \theta = \cos^{-1}\left(-\frac{1}{3}\right) \approx 109.5°$.

54. Let α be the angle between $\mathbf{a}$ and $\mathbf{c}$ and β be the angle between $\mathbf{c}$ and $\mathbf{b}$. We need to show that $\alpha = \beta$. Now

$\cos \alpha = \dfrac{\mathbf{a} \cdot \mathbf{c}}{|\mathbf{a}| \, |\mathbf{c}|} = \dfrac{\mathbf{a} \cdot |\mathbf{a}| \, \mathbf{b} + \mathbf{a} \cdot |\mathbf{b}| \, \mathbf{a}}{|\mathbf{a}| \, |\mathbf{c}|} = \dfrac{|\mathbf{a}| \, \mathbf{a} \cdot \mathbf{b} + |\mathbf{a}|^2 \, |\mathbf{b}|}{|\mathbf{a}| \, |\mathbf{c}|} = \dfrac{\mathbf{a} \cdot \mathbf{b} + |\mathbf{a}| \, |\mathbf{b}|}{|\mathbf{c}|}$. Similarly,

$\cos \beta = \dfrac{\mathbf{b} \cdot \mathbf{c}}{|\mathbf{b}| \, |\mathbf{c}|} = \dfrac{|\mathbf{a}| \, |\mathbf{b}| + \mathbf{b} \cdot \mathbf{a}}{|\mathbf{c}|}$. Thus $\cos \alpha = \cos \beta$. However $0° \le \alpha \le 180°$ and $0° \le \beta \le 180°$, so $\alpha = \beta$ and

$\mathbf{c}$ bisects the angle between $\mathbf{a}$ and $\mathbf{b}$.

55. Let $\mathbf{a} = \langle a_1, a_2, a_3 \rangle$ and $= \langle b_1, b_2, b_3 \rangle$.

Property 2: $\mathbf{a} \cdot \mathbf{b} = \langle a_1, a_2, a_3 \rangle \cdot \langle b_1, b_2, b_3 \rangle = a_1 b_1 + a_2 b_2 + a_3 b_3$

$= b_1 a_1 + b_2 a_2 + b_3 a_3 = \langle b_1, b_2, b_3 \rangle \cdot \langle a_1, a_2, a_3 \rangle = \mathbf{b} \cdot \mathbf{a}$

Property 4: $(c\,\mathbf{a}) \cdot \mathbf{b} = \langle ca_1, ca_2, ca_3 \rangle \cdot \langle b_1, b_2, b_3 \rangle = (ca_1)b_1 + (ca_2)b_2 + (ca_3)b_3$

$= c\,(a_1 b_1 + a_2 b_2 + a_3 b_3) = c\,(\mathbf{a} \cdot \mathbf{b}) = a_1(cb_1) + a_2(cb_2) + a_3(cb_3)$

$= \langle a_1, a_2, a_3 \rangle \cdot \langle cb_1, cb_2, cb_3 \rangle = \mathbf{a} \cdot (c\,\mathbf{b})$

Property 5: $\mathbf{0} \cdot \mathbf{a} = \langle 0, 0, 0 \rangle \cdot \langle a_1, a_2, a_3 \rangle = (0)(a_1) + (0)(a_2) + (0)(a_3) = 0$

56. Let the figure be called quadrilateral $ABCD$. The diagonals can be represented by $\overrightarrow{AC}$ and $\overrightarrow{BD}$. $\overrightarrow{AC} = \overrightarrow{AB} + \overrightarrow{BC}$ and

$\overrightarrow{BD} = \overrightarrow{BC} + \overrightarrow{CD} = \overrightarrow{BC} - \overrightarrow{DC} = \overrightarrow{BC} - \overrightarrow{AB}$ (Since opposite sides of the object are of the same length and parallel,

$\overrightarrow{AB} = \overrightarrow{DC}$.) Thus

$$\overrightarrow{AC} \cdot \overrightarrow{BD} = \left(\overrightarrow{AB} + \overrightarrow{BC}\right) \cdot \left(\overrightarrow{BC} - \overrightarrow{AB}\right) = \overrightarrow{AB} \cdot \left(\overrightarrow{BC} - \overrightarrow{AB}\right) + \overrightarrow{BC} \cdot \left(\overrightarrow{BC} - \overrightarrow{AB}\right)$$

$$= \overrightarrow{AB} \cdot \overrightarrow{BC} - \left|\overrightarrow{AB}\right|^2 + \left|\overrightarrow{BC}\right|^2 - \overrightarrow{AB} \cdot \overrightarrow{BC} = \left|\overrightarrow{BC}\right|^2 - \left|\overrightarrow{AB}\right|^2$$

But $\left|\overrightarrow{AB}\right|^2 = \left|\overrightarrow{BC}\right|^2$ because all sides of the quadrilateral are equal in length. Therefore $\overrightarrow{AC} \cdot \overrightarrow{BD} = 0$, and since both of

these vectors are nonzero this tells us that the diagonals of the quadrilateral are perpendicular.

57. $|\mathbf{a} \cdot \mathbf{b}| = |\, |\mathbf{a}| \, |\mathbf{b}| \cos\theta| = |\mathbf{a}| \, |\mathbf{b}| \, |\cos\theta|$. Since $|\cos\theta| \le 1$, $|\mathbf{a} \cdot \mathbf{b}| = |\mathbf{a}| \, |\mathbf{b}| \, |\cos\theta| \le |\mathbf{a}| \, |\mathbf{b}|$.

Note: We have equality in the case of $\cos\theta = \pm 1$, so $\theta = 0$ or $\theta = \pi$, thus equality when $\mathbf{a}$ and $\mathbf{b}$ are parallel.

58. (a)

The Triangle Inequality states that the length of the longest side of a triangle is less than or equal to the sum of the lengths of the two shortest sides.

(b) $|\mathbf{a} + \mathbf{b}|^2 = (\mathbf{a} + \mathbf{b}) \cdot (\mathbf{a} + \mathbf{b}) = (\mathbf{a} \cdot \mathbf{a}) + 2(\mathbf{a} \cdot \mathbf{b}) + (\mathbf{b} \cdot \mathbf{b}) = |\mathbf{a}|^2 + 2(\mathbf{a} \cdot \mathbf{b}) + |\mathbf{b}|^2$

$\qquad \le |\mathbf{a}|^2 + 2\,|\mathbf{a}|\,|\mathbf{b}| + |\mathbf{b}|^2$ [by the Cauchy-Schwartz Inequality]

$\qquad = (|\mathbf{a}| + |\mathbf{b}|)^2$

Thus, taking the square root of both sides, $|\mathbf{a} + \mathbf{b}| \le |\mathbf{a}| + |\mathbf{b}|$.

59. (a)

The Parallelogram Law states that the sum of the squares of the lengths of the diagonals of a parallelogram equals the sum of the squares of its (four) sides.

(b) $|\mathbf{a} + \mathbf{b}|^2 = (\mathbf{a} + \mathbf{b}) \cdot (\mathbf{a} + \mathbf{b}) = |\mathbf{a}|^2 + 2(\mathbf{a} \cdot \mathbf{b}) + |\mathbf{b}|^2$ and $|\mathbf{a} - \mathbf{b}|^2 = (\mathbf{a} - \mathbf{b}) \cdot (\mathbf{a} - \mathbf{b}) = |\mathbf{a}|^2 - 2(\mathbf{a} \cdot \mathbf{b}) + |\mathbf{b}|^2$.

Adding these two equations gives $|\mathbf{a} + \mathbf{b}|^2 + |\mathbf{a} - \mathbf{b}|^2 = 2\,|\mathbf{a}|^2 + 2\,|\mathbf{b}|^2$.

60. If the vectors $\mathbf{u} + \mathbf{v}$ and $\mathbf{u} - \mathbf{v}$ are orthogonal then $(\mathbf{u} + \mathbf{v}) \cdot (\mathbf{u} - \mathbf{v}) = 0$. But

$$(\mathbf{u} + \mathbf{v}) \cdot (\mathbf{u} - \mathbf{v}) = (\mathbf{u} + \mathbf{v}) \cdot \mathbf{u} - (\mathbf{u} + \mathbf{v}) \cdot \mathbf{v} \qquad \text{by Property 3 of the dot product}$$

$$= \mathbf{u} \cdot \mathbf{u} + \mathbf{v} \cdot \mathbf{u} - \mathbf{u} \cdot \mathbf{v} - \mathbf{v} \cdot \mathbf{v} \qquad \text{by Property 3}$$

$$= |\mathbf{u}|^2 + \mathbf{u} \cdot \mathbf{v} - \mathbf{u} \cdot \mathbf{v} - |\mathbf{v}|^2 \qquad \text{by Properties 1 and 2}$$

$$= |\mathbf{u}|^2 - |\mathbf{v}|^2$$

Thus $|\mathbf{u}|^2 - |\mathbf{v}|^2 = 0 \;\Rightarrow\; |\mathbf{u}|^2 = |\mathbf{v}|^2 \;\Rightarrow\; |\mathbf{u}| = |\mathbf{v}|$ [since $|\mathbf{u}|, |\mathbf{v}| \ge 0$].

13.4   The Cross Product

ET 12.4

1. $\mathbf{a} \times \mathbf{b} = \begin{vmatrix} \mathbf{i} & \mathbf{j} & \mathbf{k} \\ 6 & 0 & -2 \\ 0 & 8 & 0 \end{vmatrix} = \begin{vmatrix} 0 & -2 \\ 8 & 0 \end{vmatrix} \mathbf{i} - \begin{vmatrix} 6 & -2 \\ 0 & 0 \end{vmatrix} \mathbf{j} + \begin{vmatrix} 6 & 0 \\ 0 & 8 \end{vmatrix} \mathbf{k}$

$\qquad = [0 - (-16)]\,\mathbf{i} - (0 - 0)\,\mathbf{j} + (48 - 0)\,\mathbf{k} = 16\,\mathbf{i} + 48\,\mathbf{k}$

Now $(\mathbf{a} \times \mathbf{b}) \cdot \mathbf{a} = \langle 16, 0, 48 \rangle \cdot \langle 6, 0, -2 \rangle = 96 + 0 - 96 = 0$ and $(\mathbf{a} \times \mathbf{b}) \cdot \mathbf{b} = \langle 16, 0, 48 \rangle \cdot \langle 0, 8, 0 \rangle = 0 + 0 + 0 = 0$, so $\mathbf{a} \times \mathbf{b}$ is orthogonal to both $\mathbf{a}$ and $\mathbf{b}$.

2. $\mathbf{a} \times \mathbf{b} = \begin{vmatrix} \mathbf{i} & \mathbf{j} & \mathbf{k} \\ 1 & 1 & -1 \\ 2 & 4 & 6 \end{vmatrix} = \begin{vmatrix} 1 & -1 \\ 4 & 6 \end{vmatrix} \mathbf{i} - \begin{vmatrix} 1 & -1 \\ 2 & 6 \end{vmatrix} \mathbf{j} + \begin{vmatrix} 1 & 1 \\ 2 & 4 \end{vmatrix} \mathbf{k}$

$\qquad = [6 - (-4)]\,\mathbf{i} - [6 - (-2)]\,\mathbf{j} + (4 - 2)\,\mathbf{k} = 10\,\mathbf{i} - 8\,\mathbf{j} + 2\,\mathbf{k}$

Now $(\mathbf{a} \times \mathbf{b}) \cdot \mathbf{a} = \langle 10, -8, 2 \rangle \cdot \langle 1, 1, -1 \rangle = 10 - 8 - 2 = 0$ and $(\mathbf{a} \times \mathbf{b}) \cdot \mathbf{b} = \langle 10, -8, 2 \rangle \cdot \langle 2, 4, 6 \rangle = 20 - 32 + 12 = 0$, so $\mathbf{a} \times \mathbf{b}$ is orthogonal to both $\mathbf{a}$ and $\mathbf{b}$.

3. $\mathbf{a} \times \mathbf{b} = \begin{vmatrix} \mathbf{i} & \mathbf{j} & \mathbf{k} \\ 1 & 3 & -2 \\ -1 & 0 & 5 \end{vmatrix} = \begin{vmatrix} 3 & -2 \\ 0 & 5 \end{vmatrix} \mathbf{i} - \begin{vmatrix} 1 & -2 \\ -1 & 5 \end{vmatrix} \mathbf{j} + \begin{vmatrix} 1 & 3 \\ -1 & 0 \end{vmatrix} \mathbf{k}$

$\qquad = (15 - 0)\,\mathbf{i} - (5 - 2)\,\mathbf{j} + [0 - (-3)]\,\mathbf{k} = 15\,\mathbf{i} - 3\,\mathbf{j} + 3\,\mathbf{k}$

Since $(\mathbf{a} \times \mathbf{b}) \cdot \mathbf{a} = (15\,\mathbf{i} - 3\,\mathbf{j} + 3\,\mathbf{k}) \cdot (\mathbf{i} + 3\,\mathbf{j} - 2\,\mathbf{k}) = 15 - 9 - 6 = 0$, $\mathbf{a} \times \mathbf{b}$ is orthogonal to $\mathbf{a}$.

Since $(\mathbf{a} \times \mathbf{b}) \cdot \mathbf{b} = (15\,\mathbf{i} - 3\,\mathbf{j} + 3\,\mathbf{k}) \cdot (-\mathbf{i} + 5\,\mathbf{k}) = -15 + 0 + 15 = 0$, $\mathbf{a} \times \mathbf{b}$ is orthogonal to $\mathbf{b}$.

4. $\mathbf{a} \times \mathbf{b} = \begin{vmatrix} \mathbf{i} & \mathbf{j} & \mathbf{k} \\ 0 & 1 & 7 \\ 2 & -1 & 4 \end{vmatrix} = \begin{vmatrix} 1 & 7 \\ -1 & 4 \end{vmatrix} \mathbf{i} - \begin{vmatrix} 0 & 7 \\ 2 & 4 \end{vmatrix} \mathbf{j} + \begin{vmatrix} 0 & 1 \\ 2 & -1 \end{vmatrix} \mathbf{k}$

$\qquad = [4 - (-7)]\,\mathbf{i} - (0 - 14)\,\mathbf{j} + (0 - 2)\,\mathbf{k} = 11\,\mathbf{i} + 14\,\mathbf{j} - 2\,\mathbf{k}$

Since $(\mathbf{a} \times \mathbf{b}) \cdot \mathbf{a} = (11\,\mathbf{i} + 14\,\mathbf{j} - 2\,\mathbf{k}) \cdot (\mathbf{j} + 7\,\mathbf{k}) = 0 + 14 - 14 = 0$, $\mathbf{a} \times \mathbf{b}$ is orthogonal to $\mathbf{a}$.

Since $(\mathbf{a} \times \mathbf{b}) \cdot \mathbf{b} = (11\,\mathbf{i} + 14\,\mathbf{j} - 2\,\mathbf{k}) \cdot (2\,\mathbf{i} - \mathbf{j} + 4\,\mathbf{k}) = 22 - 14 - 8 = 0$, $\mathbf{a} \times \mathbf{b}$ is orthogonal to $\mathbf{b}$.

5. $\mathbf{a} \times \mathbf{b} = \begin{vmatrix} \mathbf{i} & \mathbf{j} & \mathbf{k} \\ 1 & -1 & -1 \\ \frac{1}{2} & 1 & \frac{1}{2} \end{vmatrix} = \begin{vmatrix} -1 & -1 \\ 1 & \frac{1}{2} \end{vmatrix} \mathbf{i} - \begin{vmatrix} 1 & -1 \\ \frac{1}{2} & \frac{1}{2} \end{vmatrix} \mathbf{j} + \begin{vmatrix} 1 & -1 \\ \frac{1}{2} & 1 \end{vmatrix} \mathbf{k}$

$\qquad = \left[-\frac{1}{2} - (-1) \right]\mathbf{i} - \left[\frac{1}{2} - \left(-\frac{1}{2} \right) \right]\mathbf{j} + \left[1 - \left(-\frac{1}{2} \right) \right]\mathbf{k} = \frac{1}{2}\,\mathbf{i} - \mathbf{j} + \frac{3}{2}\,\mathbf{k}$

Now $(\mathbf{a} \times \mathbf{b}) \cdot \mathbf{a} = \left(\frac{1}{2}\,\mathbf{i} - \mathbf{j} + \frac{3}{2}\,\mathbf{k} \right) \cdot (\mathbf{i} - \mathbf{j} - \mathbf{k}) = \frac{1}{2} + 1 - \frac{3}{2} = 0$ and

$(\mathbf{a} \times \mathbf{b}) \cdot \mathbf{b} = \left(\frac{1}{2}\,\mathbf{i} - \mathbf{j} + \frac{3}{2}\,\mathbf{k} \right) \cdot \left(\frac{1}{2}\,\mathbf{i} + \mathbf{j} + \frac{1}{2}\,\mathbf{k} \right) = \frac{1}{4} - 1 + \frac{3}{4} = 0$, so $\mathbf{a} \times \mathbf{b}$ is orthogonal to both $\mathbf{a}$ and $\mathbf{b}$.

6. $\mathbf{a} \times \mathbf{b} = \begin{vmatrix} \mathbf{i} & \mathbf{j} & \mathbf{k} \\ 1 & e^t & e^{-t} \\ 2 & e^t & -e^{-t} \end{vmatrix} = \begin{vmatrix} e^t & e^{-t} \\ e^t & -e^{-t} \end{vmatrix} \mathbf{i} - \begin{vmatrix} 1 & e^{-t} \\ 2 & -e^{-t} \end{vmatrix} \mathbf{j} + \begin{vmatrix} 1 & e^t \\ 2 & e^t \end{vmatrix} \mathbf{k}$

$\qquad = (-1 - 1)\,\mathbf{i} - (-e^{-t} - 2e^{-t})\,\mathbf{j} + (e^t - 2e^t)\,\mathbf{k} = -2\,\mathbf{i} + 3e^{-t}\,\mathbf{j} - e^t\,\mathbf{k}$

Since $(\mathbf{a} \times \mathbf{b}) \cdot \mathbf{a} = (-2\,\mathbf{i} + 3e^{-t}\,\mathbf{j} - e^t\,\mathbf{k}) \cdot (\mathbf{i} + e^t\,\mathbf{j} + e^{-t}\,\mathbf{k}) = -2 + 3 - 1 = 0$, $\mathbf{a} \times \mathbf{b}$ is orthogonal to $\mathbf{a}$.

Since $(\mathbf{a} \times \mathbf{b}) \cdot \mathbf{b} = (-2\,\mathbf{i} + 3e^{-t}\,\mathbf{j} - e^t\,\mathbf{k}) \cdot (2\,\mathbf{i} + e^t\,\mathbf{j} - e^{-t}\,\mathbf{k}) = -4 + 3 + 1 = 0$, $\mathbf{a} \times \mathbf{b}$ is orthogonal to $\mathbf{b}$.

7. $\mathbf{a} \times \mathbf{b} = \begin{vmatrix} \mathbf{i} & \mathbf{j} & \mathbf{k} \\ t & t^2 & t^3 \\ 1 & 2t & 3t^2 \end{vmatrix} = \begin{vmatrix} t^2 & t^3 \\ 2t & 3t^2 \end{vmatrix} \mathbf{i} - \begin{vmatrix} t & t^3 \\ 1 & 3t^2 \end{vmatrix} \mathbf{j} + \begin{vmatrix} t & t^2 \\ 1 & 2t \end{vmatrix} \mathbf{k}$

$\qquad = (3t^4 - 2t^4)\,\mathbf{i} - (3t^3 - t^3)\,\mathbf{j} + (2t^2 - t^2)\,\mathbf{k} = t^4\,\mathbf{i} - 2t^3\,\mathbf{j} + t^2\,\mathbf{k}$

Since $(\mathbf{a} \times \mathbf{b}) \cdot \mathbf{a} = \langle t^4, -2t^3, t^2 \rangle \cdot \langle t, t^2, t^3 \rangle = t^5 - 2t^5 + t^5 = 0$, $\mathbf{a} \times \mathbf{b}$ is orthogonal to $\mathbf{a}$.

Since $(\mathbf{a} \times \mathbf{b}) \cdot \mathbf{b} = \langle t^4, -2t^3, t^2 \rangle \cdot \langle 1, 2t, 3t^2 \rangle = t^4 - 4t^4 + 3t^4 = 0$, $\mathbf{a} \times \mathbf{b}$ is orthogonal to $\mathbf{b}$.

8. $\mathbf{a} \times \mathbf{b} = \begin{vmatrix} \mathbf{i} & \mathbf{j} & \mathbf{k} \\ 1 & 0 & -2 \\ 0 & 1 & 1 \end{vmatrix}$

$\qquad = \begin{vmatrix} 0 & -2 \\ 1 & 1 \end{vmatrix} \mathbf{i} - \begin{vmatrix} 1 & -2 \\ 0 & 1 \end{vmatrix} \mathbf{j} + \begin{vmatrix} 1 & 0 \\ 0 & 1 \end{vmatrix} \mathbf{k}$

$\qquad = 2\,\mathbf{i} - \mathbf{j} + \mathbf{k}$

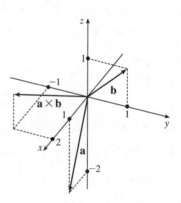

9. According to the discussion preceding Theorem 8, $\mathbf{i} \times \mathbf{j} = \mathbf{k}$, so $(\mathbf{i} \times \mathbf{j}) \times \mathbf{k} = \mathbf{k} \times \mathbf{k} = \mathbf{0}$ [by Example 2].

10. $\mathbf{k} \times (\mathbf{i} - 2\,\mathbf{j}) = \mathbf{k} \times \mathbf{i} + \mathbf{k} \times (-2\,\mathbf{j})$ $\qquad$ by Property 3 of Theorem 8

$\qquad\qquad\qquad\quad = \mathbf{k} \times \mathbf{i} + (-2)\,(\mathbf{k} \times \mathbf{j})$ $\qquad$ by Property 2 of Theorem 8

$\qquad\qquad\qquad\quad = \mathbf{j} + (-2)(-\mathbf{i}) = 2\,\mathbf{i} + \mathbf{j}$ $\qquad$ by the discussion preceding Theorem 8

11. $(\mathbf{j} - \mathbf{k}) \times (\mathbf{k} - \mathbf{i}) = (\mathbf{j} - \mathbf{k}) \times \mathbf{k} + (\mathbf{j} - \mathbf{k}) \times (-\mathbf{i})$ $\qquad$ by Property 3 of Theorem 8

$\qquad\qquad\qquad = \mathbf{j} \times \mathbf{k} + (-\mathbf{k}) \times \mathbf{k} + \mathbf{j} \times (-\mathbf{i}) + (-\mathbf{k}) \times (-\mathbf{i})$ $\qquad$ by Property 4 of Theorem 8

$\qquad\qquad\qquad = (\mathbf{j} \times \mathbf{k}) + (-1)(\mathbf{k} \times \mathbf{k}) + (-1)(\mathbf{j} \times \mathbf{i}) + (-1)^2(\mathbf{k} \times \mathbf{i})$ $\qquad$ by Property 2 of Theorem 8

$\qquad\qquad\qquad = \mathbf{i} + (-1)\,\mathbf{0} + (-1)(-\mathbf{k}) + \mathbf{j} = \mathbf{i} + \mathbf{j} + \mathbf{k}$ $\qquad$ by Example 2 and the

$\qquad\qquad\qquad\qquad\qquad\qquad\qquad\qquad\qquad\qquad\qquad\qquad\qquad\qquad$ discussion preceding Theorem 8

12. $(\mathbf{i}+\mathbf{j}) \times (\mathbf{i}-\mathbf{j}) = (\mathbf{i}+\mathbf{j}) \times \mathbf{i} + (\mathbf{i}+\mathbf{j}) \times (-\mathbf{j})$ by Property 3 of Theorem 8

$\qquad\qquad = \mathbf{i}\times\mathbf{i} + \mathbf{j}\times\mathbf{i} + \mathbf{i}\times(-\mathbf{j}) + \mathbf{j}\times(-\mathbf{j})$ by Property 4 of Theorem 8

$\qquad\qquad = (\mathbf{i}\times\mathbf{i}) + (\mathbf{j}\times\mathbf{i}) + (-1)(\mathbf{i}\times\mathbf{j}) + (-1)(\mathbf{j}\times\mathbf{j})$ by Property 2 of Theorem 8

$\qquad\qquad = \mathbf{0} + (-\mathbf{k}) + (-1)\,\mathbf{k} + (-1)\,\mathbf{0} = -2\,\mathbf{k}$ by Example 2 and the

discussion preceding Theorem 8

13. (a) Since $\mathbf{b} \times \mathbf{c}$ is a vector, the dot product $\mathbf{a} \cdot (\mathbf{b} \times \mathbf{c})$ is meaningful and is a scalar.

(b) $\mathbf{b} \cdot \mathbf{c}$ is a scalar, so $\mathbf{a} \times (\mathbf{b} \cdot \mathbf{c})$ is meaningless, as the cross product is defined only for two *vectors*.

(c) Since $\mathbf{b} \times \mathbf{c}$ is a vector, the cross product $\mathbf{a} \times (\mathbf{b} \times \mathbf{c})$ is meaningful and results in another vector.

(d) $\mathbf{a} \cdot \mathbf{b}$ is a scalar, so the cross product $(\mathbf{a} \cdot \mathbf{b}) \times \mathbf{c}$ is meaningless.

(e) Since $(\mathbf{a} \cdot \mathbf{b})$ and $(\mathbf{c} \cdot \mathbf{d})$ are both scalars, the cross product $(\mathbf{a} \cdot \mathbf{b}) \times (\mathbf{c} \cdot \mathbf{d})$ is meaningless.

(f) $\mathbf{a} \times \mathbf{b}$ and $\mathbf{c} \times \mathbf{d}$ are both vectors, so the dot product $(\mathbf{a} \times \mathbf{b}) \cdot (\mathbf{c} \times \mathbf{d})$ is meaningful and is a scalar.

14. Using Theorem 6, we have $|\mathbf{u} \times \mathbf{v}| = |\mathbf{u}|\,|\mathbf{v}| \sin\theta = (5)(10)\sin 60° = 25\sqrt{3}$. By the right-hand rule, $\mathbf{u} \times \mathbf{v}$ is directed into the page.

15. If we sketch $\mathbf{u}$ and $\mathbf{v}$ starting from the same initial point, we see that

the angle between them is $30°$. Using Theorem 6, we have

$|\mathbf{u} \times \mathbf{v}| = |\mathbf{u}|\,|\mathbf{v}| \sin 30° = (6)(8)\left(\frac{1}{2}\right) = 24$.

By the right-hand rule, $\mathbf{u} \times \mathbf{v}$ is directed into the page.

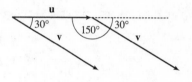

16. (a) $|\mathbf{a} \times \mathbf{b}| = |\mathbf{a}|\,|\mathbf{b}| \sin\theta = 3 \cdot 2 \cdot \sin\frac{\pi}{2} = 6$

(b) $\mathbf{a} \times \mathbf{b}$ is orthogonal to $\mathbf{k}$, so it lies in the xy-plane, and its z-coordinate is 0.

By the right-hand rule, its y-component is negative and its x-component is

positive.

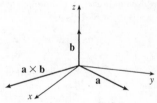

17. $\mathbf{a} \times \mathbf{b} = \begin{vmatrix} \mathbf{i} & \mathbf{j} & \mathbf{k} \\ 1 & 2 & 1 \\ 0 & 1 & 3 \end{vmatrix} = \begin{vmatrix} 2 & 1 \\ 1 & 3 \end{vmatrix}\mathbf{i} - \begin{vmatrix} 1 & 1 \\ 0 & 3 \end{vmatrix}\mathbf{j} + \begin{vmatrix} 1 & 2 \\ 0 & 1 \end{vmatrix}\mathbf{k} = (6-1)\,\mathbf{i} - (3-0)\,\mathbf{j} + (1-0)\,\mathbf{k} = 5\,\mathbf{i} - 3\,\mathbf{j} + \mathbf{k}$

$\mathbf{b} \times \mathbf{a} = \begin{vmatrix} \mathbf{i} & \mathbf{j} & \mathbf{k} \\ 0 & 1 & 3 \\ 1 & 2 & 1 \end{vmatrix} = \begin{vmatrix} 1 & 3 \\ 2 & 1 \end{vmatrix}\mathbf{i} - \begin{vmatrix} 0 & 3 \\ 1 & 1 \end{vmatrix}\mathbf{j} + \begin{vmatrix} 0 & 1 \\ 1 & 2 \end{vmatrix}\mathbf{k} = (1-6)\,\mathbf{i} - (0-3)\,\mathbf{j} + (0-1)\,\mathbf{k} = -5\,\mathbf{i} + 3\,\mathbf{j} - \mathbf{k}$

Notice $\mathbf{a} \times \mathbf{b} = -\mathbf{b} \times \mathbf{a}$ here, as we know is always true by Theorem 8.

18. $\mathbf{b} \times \mathbf{c} = \begin{vmatrix} \mathbf{i} & \mathbf{j} & \mathbf{k} \\ -1 & 1 & 0 \\ 0 & 0 & -4 \end{vmatrix} = \begin{vmatrix} 1 & 0 \\ 0 & -4 \end{vmatrix}\mathbf{i} - \begin{vmatrix} -1 & 0 \\ 0 & -4 \end{vmatrix}\mathbf{j} + \begin{vmatrix} -1 & 1 \\ 0 & 0 \end{vmatrix}\mathbf{k} = -4\,\mathbf{i} - 4\,\mathbf{j}$ so

$\mathbf{a} \times (\mathbf{b} \times \mathbf{c}) = \begin{vmatrix} \mathbf{i} & \mathbf{j} & \mathbf{k} \\ 3 & 1 & 2 \\ -4 & -4 & 0 \end{vmatrix} = \begin{vmatrix} 1 & 2 \\ -4 & 0 \end{vmatrix}\mathbf{i} - \begin{vmatrix} 3 & 2 \\ -4 & 0 \end{vmatrix}\mathbf{j} + \begin{vmatrix} 3 & 1 \\ -4 & -4 \end{vmatrix}\mathbf{k} = 8\,\mathbf{i} - 8\,\mathbf{j} - 8\,\mathbf{k}.$

$$\mathbf{a} \times \mathbf{b} = \begin{vmatrix} \mathbf{i} & \mathbf{j} & \mathbf{k} \\ 3 & 1 & 2 \\ -1 & 1 & 0 \end{vmatrix} = \begin{vmatrix} 1 & 2 \\ 1 & 0 \end{vmatrix}\mathbf{i} - \begin{vmatrix} 3 & 2 \\ -1 & 0 \end{vmatrix}\mathbf{j} + \begin{vmatrix} 3 & 1 \\ -1 & 1 \end{vmatrix}\mathbf{k} = -2\,\mathbf{i} - 2\,\mathbf{j} + 4\,\mathbf{k} \text{ so}$$

$$(\mathbf{a} \times \mathbf{b}) \times \mathbf{c} = \begin{vmatrix} \mathbf{i} & \mathbf{j} & \mathbf{k} \\ -2 & -2 & 4 \\ 0 & 0 & -4 \end{vmatrix} = \begin{vmatrix} -2 & 4 \\ 0 & -4 \end{vmatrix}\mathbf{i} - \begin{vmatrix} -2 & 4 \\ 0 & -4 \end{vmatrix}\mathbf{j} + \begin{vmatrix} -2 & -2 \\ 0 & 0 \end{vmatrix}\mathbf{k} = 8\,\mathbf{i} - 8\,\mathbf{j}.$$

Thus $\mathbf{a} \times (\mathbf{b} \times \mathbf{c}) \neq (\mathbf{a} \times \mathbf{b}) \times \mathbf{c}$.

19. We know that the cross product of two vectors is orthogonal to both. So we calculate

$$\langle 1, -1, 1 \rangle \times \langle 0, 4, 4 \rangle = \begin{vmatrix} \mathbf{i} & \mathbf{j} & \mathbf{k} \\ 1 & -1 & 1 \\ 0 & 4 & 4 \end{vmatrix} = \begin{vmatrix} -1 & 1 \\ 4 & 4 \end{vmatrix}\mathbf{i} - \begin{vmatrix} 1 & 1 \\ 0 & 4 \end{vmatrix}\mathbf{j} + \begin{vmatrix} 1 & -1 \\ 0 & 4 \end{vmatrix}\mathbf{k} = -8\,\mathbf{i} - 4\,\mathbf{j} + 4\,\mathbf{k}.$$

So two unit vectors orthogonal to both are $\pm\dfrac{\langle -8, -4, 4 \rangle}{\sqrt{64 + 16 + 16}} = \pm\dfrac{\langle -8, -4, 4 \rangle}{4\sqrt{6}}$, that is, $\left\langle -\dfrac{2}{\sqrt{6}}, -\dfrac{1}{\sqrt{6}}, \dfrac{1}{\sqrt{6}} \right\rangle$

and $\left\langle \dfrac{2}{\sqrt{6}}, \dfrac{1}{\sqrt{6}}, -\dfrac{1}{\sqrt{6}} \right\rangle$.

20. We know that the cross product of two vectors is orthogonal to both. So we calculate

$$\begin{vmatrix} \mathbf{i} & \mathbf{j} & \mathbf{k} \\ 1 & 1 & 1 \\ 2 & 0 & 1 \end{vmatrix} = \begin{vmatrix} 1 & 1 \\ 0 & 1 \end{vmatrix}\mathbf{i} - \begin{vmatrix} 1 & 1 \\ 2 & 1 \end{vmatrix}\mathbf{j} + \begin{vmatrix} 1 & 1 \\ 2 & 0 \end{vmatrix}\mathbf{k} = \mathbf{i} + \mathbf{j} - 2\,\mathbf{k}$$

Thus, two unit vectors orthogonal to both are $\pm\dfrac{1}{\sqrt{6}}\langle 1, 1, -2 \rangle$, that is, $\left\langle \dfrac{1}{\sqrt{6}}, \dfrac{1}{\sqrt{6}}, -\dfrac{2}{\sqrt{6}} \right\rangle$ and $\left\langle -\dfrac{1}{\sqrt{6}}, -\dfrac{1}{\sqrt{6}}, \dfrac{2}{\sqrt{6}} \right\rangle$.

21. Let $\mathbf{a} = \langle a_1, a_2, a_3 \rangle$. Then

$$\mathbf{0} \times \mathbf{a} = \begin{vmatrix} \mathbf{i} & \mathbf{j} & \mathbf{k} \\ 0 & 0 & 0 \\ a_1 & a_2 & a_3 \end{vmatrix} = \begin{vmatrix} 0 & 0 \\ a_2 & a_3 \end{vmatrix}\mathbf{i} - \begin{vmatrix} 0 & 0 \\ a_1 & a_3 \end{vmatrix}\mathbf{j} + \begin{vmatrix} 0 & 0 \\ a_1 & a_2 \end{vmatrix}\mathbf{k} = \mathbf{0},$$

$$\mathbf{a} \times \mathbf{0} = \begin{vmatrix} \mathbf{i} & \mathbf{j} & \mathbf{k} \\ a_1 & a_2 & a_3 \\ 0 & 0 & 0 \end{vmatrix} = \begin{vmatrix} a_2 & a_3 \\ 0 & 0 \end{vmatrix}\mathbf{i} - \begin{vmatrix} a_1 & a_3 \\ 0 & 0 \end{vmatrix}\mathbf{j} + \begin{vmatrix} a_1 & a_2 \\ 0 & 0 \end{vmatrix}\mathbf{k} = \mathbf{0}.$$

22. Let $\mathbf{a} = \langle a_1, a_2, a_3 \rangle$ and $\mathbf{b} = \langle b_1, b_2, b_3 \rangle$.

$$(\mathbf{a} \times \mathbf{b}) \cdot \mathbf{b} = \left\langle \begin{vmatrix} a_2 & a_3 \\ b_2 & b_3 \end{vmatrix}, \begin{vmatrix} a_1 & a_3 \\ b_1 & b_3 \end{vmatrix}, \begin{vmatrix} a_1 & a_2 \\ b_1 & b_2 \end{vmatrix} \right\rangle \cdot \langle b_1, b_2, b_3 \rangle = \begin{vmatrix} a_2 & a_3 \\ b_2 & b_3 \end{vmatrix} b_1 - \begin{vmatrix} a_1 & a_3 \\ b_1 & b_3 \end{vmatrix} b_2 + \begin{vmatrix} a_1 & a_2 \\ b_1 & b_2 \end{vmatrix} b_3$$

$$= (a_2 b_3 b_1 - a_3 b_2 b_1) - (a_1 b_3 b_2 - a_3 b_1 b_2) + (a_1 b_2 b_3 - a_2 b_1 b_3) = 0$$

23. $\mathbf{a} \times \mathbf{b} = \langle a_2 b_3 - a_3 b_2, a_3 b_1 - a_1 b_3, a_1 b_2 - a_2 b_1 \rangle$

$$= \langle (-1)(b_2 a_3 - b_3 a_2), (-1)(b_3 a_1 - b_1 a_3), (-1)(b_1 a_2 - b_2 a_1) \rangle$$

$$= -\langle b_2 a_3 - b_3 a_2, b_3 a_1 - b_1 a_3, b_1 a_2 - b_2 a_1 \rangle = -\mathbf{b} \times \mathbf{a}$$

24. $c\mathbf{a} = \langle ca_1, ca_2, ca_3\rangle$, so

$$(c\mathbf{a}) \times \mathbf{b} = \langle ca_2b_3 - ca_3b_2, ca_3b_1 - ca_1b_3, ca_1b_2 - ca_2b_1\rangle$$

$$= c\langle a_2b_3 - a_3b_2, a_3b_1 - a_1b_3, a_1b_2 - a_2b_1\rangle = c(\mathbf{a} \times \mathbf{b})$$

$$= \langle ca_2b_3 - ca_3b_2, ca_3b_1 - ca_1b_3, ca_1b_2 - ca_2b_1\rangle$$

$$= \langle a_2(cb_3) - a_3(cb_2), a_3(cb_1) - a_1(cb_3), a_1(cb_2) - a_2(cb_1)\rangle$$

$$= \mathbf{a} \times c\mathbf{b}$$

25. $\mathbf{a} \times (\mathbf{b} + \mathbf{c}) = \mathbf{a} \times \langle b_1 + c_1, b_2 + c_2, b_3 + c_3\rangle$

$$= \langle a_2(b_3 + c_3) - a_3(b_2 + c_2), a_3(b_1 + c_1) - a_1(b_3 + c_3), a_1(b_2 + c_2) - a_2(b_1 + c_1)\rangle$$

$$= \langle a_2b_3 + a_2c_3 - a_3b_2 - a_3c_2, a_3b_1 + a_3c_1 - a_1b_3 - a_1c_3, a_1b_2 + a_1c_2 - a_2b_1 - a_2c_1\rangle$$

$$= \langle (a_2b_3 - a_3b_2) + (a_2c_3 - a_3c_2), (a_3b_1 - a_1b_3) + (a_3c_1 - a_1c_3), (a_1b_2 - a_2b_1) + (a_1c_2 - a_2c_1)\rangle$$

$$= \langle a_2b_3 - a_3b_2, a_3b_1 - a_1b_3, a_1b_2 - a_2b_1\rangle + \langle a_2c_3 - a_3c_2, a_3c_1 - a_1c_3, a_1c_2 - a_2c_1\rangle$$

$$= (\mathbf{a} \times \mathbf{b}) + (\mathbf{a} \times \mathbf{c})$$

26. $(\mathbf{a} + \mathbf{b}) \times \mathbf{c} = -\mathbf{c} \times (\mathbf{a} + \mathbf{b})$ by Property 1 of Theorem 8

$\qquad = -(\mathbf{c} \times \mathbf{a} + \mathbf{c} \times \mathbf{b})$ by Property 3 of Theorem 8

$\qquad = -(-\mathbf{a} \times \mathbf{c} + (-\mathbf{b} \times \mathbf{c}))$ by Property 1 of Theorem 8

$\qquad = \mathbf{a} \times \mathbf{c} + \mathbf{b} \times \mathbf{c}$ by Property 2 of Theorem 8

27. By plotting the vertices, we can see that the parallelogram is determined by the

vectors $\overrightarrow{AB} = \langle 2, 3\rangle$ and $\overrightarrow{AD} = \langle 4, -2\rangle$. We know that the area of the parallelogram

determined by two vectors is equal to the length of the cross product of these vectors.

In order to compute the cross product, we consider the vector $\overrightarrow{AB}$ as the three-

dimensional vector $\langle 2, 3, 0\rangle$ (and similarly for $\overrightarrow{AD}$), and then the area of

parallelogram $ABCD$ is

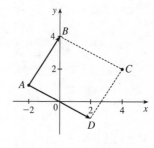

$$\left|\overrightarrow{AB} \times \overrightarrow{AD}\right| = \begin{vmatrix} \mathbf{i} & \mathbf{j} & \mathbf{k} \\ 2 & 3 & 0 \\ 4 & -2 & 0 \end{vmatrix} = |(0)\mathbf{i} - (0)\mathbf{j} + (-4 - 12)\mathbf{k}| = |-16\mathbf{k}| = 16$$

28. The parallelogram is determined by the vectors $\overrightarrow{KL} = \langle 0, 1, 3\rangle$ and $\overrightarrow{KN} = \langle 2, 5, 0\rangle$, so the area of parallelogram $KLMN$ is

$$\left|\overrightarrow{KL} \times \overrightarrow{KN}\right| = \begin{vmatrix} \mathbf{i} & \mathbf{j} & \mathbf{k} \\ 0 & 1 & 3 \\ 2 & 5 & 0 \end{vmatrix} = |(-15)\mathbf{i} - (-6)\mathbf{j} + (-2)\mathbf{k}| = |-15\mathbf{i} + 6\mathbf{j} - 2\mathbf{k}| = \sqrt{265} \approx 16.28$$

29. (a) Because the plane through P, Q, and R contains the vectors $\overrightarrow{PQ}$ and $\overrightarrow{PR}$, a vector orthogonal to both of these vectors

(such as their cross product) is also orthogonal to the plane. Here $\overrightarrow{PQ} = \langle -1, 2, 0\rangle$ and $\overrightarrow{PR} = \langle -1, 0, 3\rangle$, so

$$\overrightarrow{PQ} \times \overrightarrow{PR} = \langle (2)(3) - (0)(0), (0)(-1) - (-1)(3), (-1)(0) - (2)(-1)\rangle = \langle 6, 3, 2\rangle$$

Therefore, $\langle 6, 3, 2\rangle$ (or any scalar multiple thereof) is orthogonal to the plane through P, Q, and R.

(b) Note that the area of the triangle determined by P, Q, and R is equal to half of the area of the parallelogram determined by the three points. From part (a), the area of the parallelogram is $\left| \overrightarrow{PQ} \times \overrightarrow{PR} \right| = |\langle 6, 3, 2 \rangle| = \sqrt{36 + 9 + 4} = 7$, so the area of the triangle is $\frac{1}{2}(7) = \frac{7}{2}$.

30. (a) $\overrightarrow{PQ} = \langle -3, 2, -1 \rangle$ and $\overrightarrow{PR} = \langle 1, -1, 1 \rangle$, so a vector orthogonal to the plane through P, Q, and R is

$\overrightarrow{PQ} \times \overrightarrow{PR} = \langle (2)(1) - (-1)(-1), (-1)(1) - (-3)(1), (-3)(-1) - (2)(1) \rangle = \langle 1, 2, 1 \rangle$ (or any scalar mutiple thereof).

(b) The area of the parallelogram determined by $\overrightarrow{PQ}$ and $\overrightarrow{PR}$ is $\left| \overrightarrow{PQ} \times \overrightarrow{PR} \right| = |\langle 1, 2, 1 \rangle| = \sqrt{1^2 + 2^2 + 1^2} = \sqrt{6}$,

so the area of triangle PQR is $\frac{1}{2}\sqrt{6}$.

31. (a) $\overrightarrow{PQ} = \langle 4, 3, -2 \rangle$ and $\overrightarrow{PR} = \langle 5, 5, 1 \rangle$, so a vector orthogonal to the plane through P, Q, and R is

$\overrightarrow{PQ} \times \overrightarrow{PR} = \langle (3)(1) - (-2)(5), (-2)(5) - (4)(1), (4)(5) - (3)(5) \rangle = \langle 13, -14, 5 \rangle$ [or any scalar mutiple thereof].

(b) The area of the parallelogram determined by $\overrightarrow{PQ}$ and $\overrightarrow{PR}$ is

$\left| \overrightarrow{PQ} \times \overrightarrow{PR} \right| = |\langle 13, -14, 5 \rangle| = \sqrt{13^2 + (-14)^2 + 5^2} = \sqrt{390}$, so the area of triangle PQR is $\frac{1}{2}\sqrt{390}$.

32. (a) $\overrightarrow{PQ} = \langle 1, 2, 1 \rangle$ and $\overrightarrow{PR} = \langle 5, 0, -2 \rangle$, so a vector orthogonal to the plane through P, Q, and R is

$\overrightarrow{PQ} \times \overrightarrow{PR} = \langle (2)(-2) - (1)(0), (1)(5) - (1)(-2), (1)(0) - (2)(5) \rangle = \langle -4, 7, -10 \rangle$ [or any scalar multiple thereof].

(b) The area of the parallelogram determined by $\overrightarrow{PQ}$ and $\overrightarrow{PR}$ is $\left| \overrightarrow{PQ} \times \overrightarrow{PR} \right| = |\langle -4, 7, -10 \rangle| = \sqrt{16 + 49 + 100} = \sqrt{165}$,

so the area of triangle PQR is $\frac{1}{2}\sqrt{165}$.

33. We know that the volume of the parallelepiped determined by $\mathbf{a}$, $\mathbf{b}$, and $\mathbf{c}$ is the magnitude of their scalar triple product, which

is $\mathbf{a} \cdot (\mathbf{b} \times \mathbf{c}) = \begin{vmatrix} 6 & 3 & -1 \\ 0 & 1 & 2 \\ 4 & -2 & 5 \end{vmatrix} = 6 \begin{vmatrix} 1 & 2 \\ -2 & 5 \end{vmatrix} - 3 \begin{vmatrix} 0 & 2 \\ 4 & 5 \end{vmatrix} + (-1) \begin{vmatrix} 0 & 1 \\ 4 & -2 \end{vmatrix} = 6(5 + 4) - 3(0 - 8) - (0 - 4) = 82$.

Thus the volume of the parallelepiped is 82 cubic units.

34. $\mathbf{a} \cdot (\mathbf{b} \times \mathbf{c}) = \begin{vmatrix} 1 & 1 & -1 \\ 1 & -1 & 1 \\ -1 & 1 & 1 \end{vmatrix} = 1 \begin{vmatrix} -1 & 1 \\ 1 & 1 \end{vmatrix} - 1 \begin{vmatrix} 1 & 1 \\ -1 & 1 \end{vmatrix} + (-1) \begin{vmatrix} 1 & -1 \\ -1 & 1 \end{vmatrix} = -2 - 2 + 0 = -4$.

So the volume of the parallelepiped determined by $\mathbf{a}$, $\mathbf{b}$, and $\mathbf{c}$ is $|-4| = 4$ cubic units.

35. $\mathbf{a} = \overrightarrow{PQ} = \langle 2, 1, 1 \rangle$, $\mathbf{b} = \overrightarrow{PR} = \langle 1, -1, 2 \rangle$, and $\mathbf{c} = \overrightarrow{PS} = \langle 0, -2, 3 \rangle$.

$\mathbf{a} \cdot (\mathbf{b} \times \mathbf{c}) = \begin{vmatrix} 2 & 1 & 1 \\ 1 & -1 & 2 \\ 0 & -2 & 3 \end{vmatrix} = 2 \begin{vmatrix} -1 & 2 \\ -2 & 3 \end{vmatrix} - 1 \begin{vmatrix} 1 & 2 \\ 0 & 3 \end{vmatrix} + 1 \begin{vmatrix} 1 & -1 \\ 0 & -2 \end{vmatrix} = 2 - 3 - 2 = -3$,

so the volume of the parallelepiped is 3 cubic units.

36. $\mathbf{a} = \overrightarrow{PQ} = \langle -4, 2, 4 \rangle$, $\mathbf{b} = \overrightarrow{PR} = \langle 2, 1, -2 \rangle$ and $\mathbf{c} = \overrightarrow{PS} = \langle -3, 4, 1 \rangle$.

$$\mathbf{a} \cdot (\mathbf{b} \times \mathbf{c}) = \begin{vmatrix} -4 & 2 & 4 \\ 2 & 1 & -2 \\ -3 & 4 & 1 \end{vmatrix} = -4 \begin{vmatrix} 1 & -2 \\ 4 & 1 \end{vmatrix} - 2 \begin{vmatrix} 2 & -2 \\ -3 & 1 \end{vmatrix} + 4 \begin{vmatrix} 2 & 1 \\ -3 & 4 \end{vmatrix} = -36 + 8 + 44 = 16, \text{ so the volume of the}$$

parallelepiped is 16 cubic units.

37. $\mathbf{u} \cdot (\mathbf{v} \times \mathbf{w}) = \begin{vmatrix} 1 & 5 & -2 \\ 3 & -1 & 0 \\ 5 & 9 & -4 \end{vmatrix} = 1 \begin{vmatrix} -1 & 0 \\ 9 & -4 \end{vmatrix} - 5 \begin{vmatrix} 3 & 0 \\ 5 & -4 \end{vmatrix} + (-2) \begin{vmatrix} 3 & -1 \\ 5 & 9 \end{vmatrix} = 4 + 60 - 64 = 0$, which says that the volume

of the parallelepiped determined by $\mathbf{u}$, $\mathbf{v}$ and $\mathbf{w}$ is 0, and thus these three vectors are coplanar.

38. $\mathbf{u} = \overrightarrow{AB} = \langle 2, -4, 4 \rangle$, $\mathbf{v} = \overrightarrow{AC} = \langle 4, -1, -2 \rangle$ and $\mathbf{w} = \overrightarrow{AD} = \langle 2, 3, -6 \rangle$.

$$\mathbf{u} \cdot (\mathbf{v} \times \mathbf{w}) = \begin{vmatrix} 2 & -4 & 4 \\ 4 & -1 & -2 \\ 2 & 3 & -6 \end{vmatrix} = 2 \begin{vmatrix} -1 & -2 \\ 3 & -6 \end{vmatrix} - (-4) \begin{vmatrix} 4 & -2 \\ 2 & -6 \end{vmatrix} + 4 \begin{vmatrix} 4 & -1 \\ 2 & 3 \end{vmatrix} = 24 - 80 + 56 = 0, \text{ so the volume of the}$$

parallelepiped determined by $\mathbf{u}$, $\mathbf{v}$ and $\mathbf{w}$ is 0, which says these vectors lie in the same plane. Therefore, their initial and

terminal points A, B, C and D also lie in the same plane.

39. The magnitude of the torque is $|\boldsymbol{\tau}| = |\mathbf{r} \times \mathbf{F}| = |\mathbf{r}| \, |\mathbf{F}| \sin \theta = (0.18 \text{ m})(60 \text{ N}) \sin(70 + 10)° = 10.8 \sin 80° \approx 10.6 \text{ N·m.}$

40. $|\mathbf{r}| = \sqrt{4^2 + 4^2} = 4\sqrt{2}$ ft. A line drawn from the point P to the point of application of the force makes an angle of

$180° - (45 + 30)° = 105°$ with the force vector. Therefore,

$$|\boldsymbol{\tau}| = |\mathbf{r} \times \mathbf{F}| = |\mathbf{r}| \, |\mathbf{F}| \sin \theta = (4\sqrt{2})(36) \sin 105° \approx 197 \text{ ft-lb.}$$

41. Using the notation of the text, $\mathbf{r} = \langle 0, 0.3, 0 \rangle$ and $\mathbf{F}$ has direction $\langle 0, 3, -4 \rangle$. The angle θ between them can be determined by

$$\cos \theta = \frac{\langle 0, 0.3, 0 \rangle \cdot \langle 0, 3, -4 \rangle}{|\langle 0, 0.3, 0 \rangle| \, |\langle 0, 3, -4 \rangle|} \quad \Rightarrow \quad \cos \theta = \frac{0.9}{(0.3)(5)} \quad \Rightarrow \quad \cos \theta = 0.6 \quad \Rightarrow \quad \theta \approx 53.1°. \text{ Then } |\boldsymbol{\tau}| = |\mathbf{r}| \, |\mathbf{F}| \sin \theta \quad \Rightarrow$$

$100 = 0.3 \, |\mathbf{F}| \sin 53.1° \quad \Rightarrow \quad |\mathbf{F}| \approx 417 \text{ N.}$

42. Since $|\mathbf{u} \times \mathbf{v}| = |\mathbf{u}| \, |\mathbf{v}| \sin \theta$, $0 \le \theta \le \pi$, $|\mathbf{u} \times \mathbf{v}|$ achieves its maximum value for $\sin \theta = 1 \quad \Rightarrow \quad \theta = \frac{\pi}{2}$, in which case

$|\mathbf{u} \times \mathbf{v}| = |\mathbf{u}| \, |\mathbf{v}| = 15$. The minimum value is zero, which occurs when $\sin \theta = 0 \quad \Rightarrow \quad \theta = 0$ or π, so when $\mathbf{u}$, $\mathbf{v}$ are

parallel. Thus, when $\mathbf{u}$ points in the same direction as $\mathbf{v}$, so $\mathbf{u} = 3\mathbf{j}$, $|\mathbf{u} \times \mathbf{v}| = 0$. As $\mathbf{u}$ rotates counterclockwise, $\mathbf{u} \times \mathbf{v}$ is

directed in the negative z-direction (by the right-hand rule) and the length increases until $\theta = \frac{\pi}{2}$, in which case $\mathbf{u} = -3\mathbf{i}$ and

$|\mathbf{u} \times \mathbf{v}| = 15$. As $\mathbf{u}$ rotates to the negative y-axis, $\mathbf{u} \times \mathbf{v}$ remains pointed in the negative z-direction and the length of $\mathbf{u} \times \mathbf{v}$

decreases to 0, after which the direction of $\mathbf{u} \times \mathbf{v}$ reverses to point in the positive z-direction and $|\mathbf{u} \times \mathbf{v}|$ increases. When

$\mathbf{u} = 3\mathbf{i}$ (so $\theta = \frac{\pi}{2}$), $|\mathbf{u} \times \mathbf{v}|$ again reaches its maximum of 15, after which $|\mathbf{u} \times \mathbf{v}|$ decreases to 0 as $\mathbf{u}$ rotates to the positive

y-axis.

43. (a)

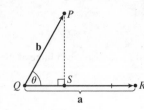

The distance between a point and a line is the length of the perpendicular from the point to the line, here $\left|\overrightarrow{PS}\right| = d$. But referring to triangle PQS,

$$d = \left|\overrightarrow{PS}\right| = \left|\overrightarrow{QP}\right| \sin\theta = |\mathbf{b}| \sin\theta. \text{ But } \theta \text{ is the angle between } \overrightarrow{QP} = \mathbf{b}$$

and $\overrightarrow{QR} = \mathbf{a}$. Thus by Theorem 6, $\sin\theta = \dfrac{|\mathbf{a} \times \mathbf{b}|}{|\mathbf{a}|\,|\mathbf{b}|}$

and so $d = |\mathbf{b}| \sin\theta = \dfrac{|\mathbf{b}|\,|\mathbf{a} \times \mathbf{b}|}{|\mathbf{a}|\,|\mathbf{b}|} = \dfrac{|\mathbf{a} \times \mathbf{b}|}{|\mathbf{a}|}$.

(b) $\mathbf{a} = \overrightarrow{QR} = \langle -1, -2, -1 \rangle$ and $\mathbf{b} = \overrightarrow{QP} = \langle 1, -5, -7 \rangle$. Then

$\mathbf{a} \times \mathbf{b} = \langle (-2)(-7) - (-1)(-5), (-1)(1) - (-1)(-7), (-1)(-5) - (-2)(1) \rangle = \langle 9, -8, 7 \rangle$.

Thus the distance is $d = \dfrac{|\mathbf{a} \times \mathbf{b}|}{|\mathbf{a}|} = \dfrac{1}{\sqrt{6}} \sqrt{81 + 64 + 49} = \sqrt{\dfrac{194}{6}} = \sqrt{\dfrac{97}{3}}$.

44. (a) The distance between a point and a plane is the length of the perpendicular from the point to the plane, here $\left|\overrightarrow{TP}\right| = d$. But $\overrightarrow{TP}$ is parallel to $\mathbf{b} \times \mathbf{a}$ (because $\mathbf{b} \times \mathbf{a}$ is perpendicular to $\mathbf{b}$ and $\mathbf{a}$) and $d = \left|\overrightarrow{TP}\right| = $ the absolute value of the scalar projection of $\mathbf{c}$ along

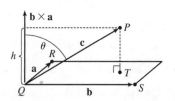

$\mathbf{b} \times \mathbf{a}$, which is $|\mathbf{c}|\,|\cos\theta|$. (Notice that this is the same setup as the development of the volume of a parallelepiped with $h = |\mathbf{c}|\,|\cos\theta|$). Thus $d = |\mathbf{c}|\,|\cos\theta| = h = V/A$ where $A = |\mathbf{a} \times \mathbf{b}|$, the area of the base. So finally

$d = \dfrac{V}{A} = \dfrac{|\mathbf{a} \cdot (\mathbf{b} \times \mathbf{c})|}{|\mathbf{a} \times \mathbf{b}|} = \dfrac{|(\mathbf{a} \times \mathbf{b}) \cdot \mathbf{c}|}{|\mathbf{a} \times \mathbf{b}|}$ by Property 5 of Theorem 8.

(b) $\mathbf{a} = \overrightarrow{QR} = \langle -1, 2, 0 \rangle$, $\mathbf{b} = \overrightarrow{QS} = \langle -1, 0, 3 \rangle$ and $\mathbf{c} = \overrightarrow{QP} = \langle 1, 1, 4 \rangle$. Then

$$(\mathbf{a} \times \mathbf{b}) \cdot \mathbf{c} = \begin{vmatrix} -1 & 2 & 0 \\ -1 & 0 & 3 \\ 1 & 1 & 4 \end{vmatrix} = (-1)\begin{vmatrix} 0 & 3 \\ 1 & 4 \end{vmatrix} - 2\begin{vmatrix} -1 & 3 \\ 1 & 4 \end{vmatrix} + 0 = 17$$

and $$\mathbf{a} \times \mathbf{b} = \begin{vmatrix} \mathbf{i} & \mathbf{j} & \mathbf{k} \\ -1 & 2 & 0 \\ -1 & 0 & 3 \end{vmatrix} = \begin{vmatrix} 2 & 0 \\ 0 & 3 \end{vmatrix}\mathbf{i} - \begin{vmatrix} -1 & 0 \\ -1 & 3 \end{vmatrix}\mathbf{j} + \begin{vmatrix} -1 & 2 \\ -1 & 0 \end{vmatrix}\mathbf{k} = 6\mathbf{i} + 3\mathbf{j} + 2\mathbf{k}$$

Thus $d = \dfrac{|(\mathbf{a} \times \mathbf{b}) \cdot \mathbf{c}|}{|\mathbf{a} \times \mathbf{b}|} = \dfrac{17}{\sqrt{36 + 9 + 4}} = \dfrac{17}{7}$.

45. $(\mathbf{a} - \mathbf{b}) \times (\mathbf{a} + \mathbf{b}) = (\mathbf{a} - \mathbf{b}) \times \mathbf{a} + (\mathbf{a} - \mathbf{b}) \times \mathbf{b}$ by Property 3 of Theorem 8

$\qquad\qquad\qquad\qquad = \mathbf{a} \times \mathbf{a} + (-\mathbf{b}) \times \mathbf{a} + \mathbf{a} \times \mathbf{b} + (-\mathbf{b}) \times \mathbf{b}$ by Property 4 of Theorem 8

$\qquad\qquad\qquad\qquad = (\mathbf{a} \times \mathbf{a}) - (\mathbf{b} \times \mathbf{a}) + (\mathbf{a} \times \mathbf{b}) - (\mathbf{b} \times \mathbf{b})$ by Property 2 of Theorem 8 (with $c = -1$)

$\qquad\qquad\qquad\qquad = \mathbf{0} - (\mathbf{b} \times \mathbf{a}) + (\mathbf{a} \times \mathbf{b}) - \mathbf{0}$ by Example 2

$\qquad\qquad\qquad\qquad = (\mathbf{a} \times \mathbf{b}) + (\mathbf{a} \times \mathbf{b})$ by Property 1 of Theorem 8

$\qquad\qquad\qquad\qquad = 2(\mathbf{a} \times \mathbf{b})$

46. Let $\mathbf{a} = \langle a_1, a_2, a_3 \rangle$, $\mathbf{b} = \langle b_1, b_2, b_3 \rangle$ and $\mathbf{c} = \langle c_1, c_2, c_3 \rangle$, so $\mathbf{b} \times \mathbf{c} = \langle b_2 c_3 - b_3 c_2, b_3 c_1 - b_1 c_3, b_1 c_2 - b_2 c_1 \rangle$ and

$$\mathbf{a} \times (\mathbf{b} \times \mathbf{c}) = \langle a_2(b_1 c_2 - b_2 c_1) - a_3(b_3 c_1 - b_1 c_3), a_3(b_2 c_3 - b_3 c_2) - a_1(b_1 c_2 - b_2 c_1),$$
$$a_1(b_3 c_1 - b_1 c_3) - a_2(b_2 c_3 - b_3 c_2) \rangle$$

$$= \langle a_2 b_1 c_2 - a_2 b_2 c_1 - a_3 b_3 c_1 + a_3 b_1 c_3, a_3 b_2 c_3 - a_3 b_3 c_2 - a_1 b_1 c_2 + a_1 b_2 c_1,$$
$$a_1 b_3 c_1 - a_1 b_1 c_3 - a_2 b_2 c_3 + a_2 b_3 c_2 \rangle$$

$$= \langle (a_2 c_2 + a_3 c_3)b_1 - (a_2 b_2 + a_3 b_3)c_1, (a_1 c_1 + a_3 c_3)b_2 - (a_1 b_1 + a_3 b_3)c_2,$$
$$(a_1 c_1 + a_2 c_2)b_3 - (a_1 b_1 + a_2 b_2)c_3 \rangle$$

$(\star) \quad = \langle (a_2 c_2 + a_3 c_3)b_1 - (a_2 b_2 + a_3 b_3)c_1 + a_1 b_1 c_1 - a_1 b_1 c_1,$
$$(a_1 c_1 + a_3 c_3)b_2 - (a_1 b_1 + a_3 b_3)c_2 + a_2 b_2 c_2 - a_2 b_2 c_2,$$
$$(a_1 c_1 + a_2 c_2)b_3 - (a_1 b_1 + a_2 b_2)c_3 + a_3 b_3 c_3 - a_3 b_3 c_3 \rangle$$

$$= \langle (a_1 c_1 + a_2 c_2 + a_3 c_3)b_1 - (a_1 b_1 + a_2 b_2 + a_3 b_3)c_1,$$
$$(a_1 c_1 + a_2 c_2 + a_3 c_3)b_2 - (a_1 b_1 + a_2 b_2 + a_3 b_3)c_2,$$
$$(a_1 c_1 + a_2 c_2 + a_3 c_3)b_3 - (a_1 b_1 + a_2 b_2 + a_3 b_3)c_3 \rangle$$

$$= (a_1 c_1 + a_2 c_2 + a_3 c_3) \langle b_1, b_2, b_3 \rangle - (a_1 b_1 + a_2 b_2 + a_3 b_3) \langle c_1, c_2, c_3 \rangle$$

$$= (\mathbf{a} \cdot \mathbf{c})\mathbf{b} - (\mathbf{a} \cdot \mathbf{b})\mathbf{c}$$

$(\star)$ Here we look ahead to see what terms are still needed to arrive at the desired equation. By adding and subtracting the same terms, we don't change the value of the component.

47. $\mathbf{a} \times (\mathbf{b} \times \mathbf{c}) + \mathbf{b} \times (\mathbf{c} \times \mathbf{a}) + \mathbf{c} \times (\mathbf{a} \times \mathbf{b})$

$$= [(\mathbf{a} \cdot \mathbf{c})\mathbf{b} - (\mathbf{a} \cdot \mathbf{b})\mathbf{c}] + [(\mathbf{b} \cdot \mathbf{a})\mathbf{c} - (\mathbf{b} \cdot \mathbf{c})\mathbf{a}] + [(\mathbf{c} \cdot \mathbf{b})\mathbf{a} - (\mathbf{c} \cdot \mathbf{a})\mathbf{b}] \qquad \text{by Exercise 46}$$

$$= (\mathbf{a} \cdot \mathbf{c})\mathbf{b} - (\mathbf{a} \cdot \mathbf{b})\mathbf{c} + (\mathbf{a} \cdot \mathbf{b})\mathbf{c} - (\mathbf{b} \cdot \mathbf{c})\mathbf{a} + (\mathbf{b} \cdot \mathbf{c})\mathbf{a} - (\mathbf{a} \cdot \mathbf{c})\mathbf{b} = 0$$

48. Let $\mathbf{c} \times \mathbf{d} = \mathbf{v}$. Then

$$(\mathbf{a} \times \mathbf{b}) \cdot (\mathbf{c} \times \mathbf{d}) = (\mathbf{a} \times \mathbf{b}) \cdot \mathbf{v} = \mathbf{a} \cdot (\mathbf{b} \times \mathbf{v}) \qquad \text{by Property 5 of Theorem 8}$$

$$= \mathbf{a} \cdot [\mathbf{b} \times (\mathbf{c} \times \mathbf{d})] = \mathbf{a} \cdot [(\mathbf{b} \cdot \mathbf{d})\mathbf{c} - (\mathbf{b} \cdot \mathbf{c})\mathbf{d}] \qquad \text{by Exercise 46}$$

$$= (\mathbf{b} \cdot \mathbf{d})(\mathbf{a} \cdot \mathbf{c}) - (\mathbf{b} \cdot \mathbf{c})(\mathbf{a} \cdot \mathbf{d}) \qquad \text{by Properties 3 and 4 of the dot product}$$

$$= \begin{vmatrix} \mathbf{a} \cdot \mathbf{c} & \mathbf{b} \cdot \mathbf{c} \\ \mathbf{a} \cdot \mathbf{d} & \mathbf{b} \cdot \mathbf{d} \end{vmatrix}$$

49. (a) No. If $\mathbf{a} \cdot \mathbf{b} = \mathbf{a} \cdot \mathbf{c}$, then $\mathbf{a} \cdot (\mathbf{b} - \mathbf{c}) = 0$, so $\mathbf{a}$ is perpendicular to $\mathbf{b} - \mathbf{c}$, which can happen if $\mathbf{b} \neq \mathbf{c}$. For example, let $\mathbf{a} = \langle 1, 1, 1 \rangle$, $\mathbf{b} = \langle 1, 0, 0 \rangle$ and $\mathbf{c} = \langle 0, 1, 0 \rangle$.

(b) No. If $\mathbf{a} \times \mathbf{b} = \mathbf{a} \times \mathbf{c}$ then $\mathbf{a} \times (\mathbf{b} - \mathbf{c}) = \mathbf{0}$, which implies that $\mathbf{a}$ is parallel to $\mathbf{b} - \mathbf{c}$, which of course can happen if $\mathbf{b} \neq \mathbf{c}$.

(c) Yes. Since $\mathbf{a} \cdot \mathbf{c} = \mathbf{a} \cdot \mathbf{b}$, $\mathbf{a}$ is perpendicular to $\mathbf{b} - \mathbf{c}$, by part (a). From part (b), $\mathbf{a}$ is also parallel to $\mathbf{b} - \mathbf{c}$. Thus since $\mathbf{a} \neq \mathbf{0}$ but is both parallel and perpendicular to $\mathbf{b} - \mathbf{c}$, we have $\mathbf{b} - \mathbf{c} = \mathbf{0}$, so $\mathbf{b} = \mathbf{c}$.

50. (a) $\mathbf{k}_i$ is perpendicular to $\mathbf{v}_i$ if $i \neq j$ by the definition of $\mathbf{k}_i$ and Theorem 5.

(b) $\mathbf{k}_1 \cdot \mathbf{v}_1 = \dfrac{\mathbf{v}_2 \times \mathbf{v}_3}{\mathbf{v}_1 \cdot (\mathbf{v}_2 \times \mathbf{v}_3)} \cdot \mathbf{v}_1 = \dfrac{\mathbf{v}_1 \cdot (\mathbf{v}_2 \times \mathbf{v}_3)}{\mathbf{v}_1 \cdot (\mathbf{v}_2 \times \mathbf{v}_3)} = 1$

$\mathbf{k}_2 \cdot \mathbf{v}_2 = \dfrac{\mathbf{v}_3 \times \mathbf{v}_1}{\mathbf{v}_1 \cdot (\mathbf{v}_2 \times \mathbf{v}_3)} \cdot \mathbf{v}_2 = \dfrac{\mathbf{v}_2 \cdot (\mathbf{v}_3 \times \mathbf{v}_1)}{\mathbf{v}_1 \cdot (\mathbf{v}_2 \times \mathbf{v}_3)} = \dfrac{(\mathbf{v}_2 \times \mathbf{v}_3) \cdot \mathbf{v}_1}{\mathbf{v}_1 \cdot (\mathbf{v}_2 \times \mathbf{v}_3)} = 1$ [by Property 5 of Theorem 8]

$\mathbf{k}_3 \cdot \mathbf{v}_3 = \dfrac{(\mathbf{v}_1 \times \mathbf{v}_2) \cdot \mathbf{v}_3}{\mathbf{v}_1 \cdot (\mathbf{v}_2 \times \mathbf{v}_3)} = \dfrac{\mathbf{v}_1 \cdot (\mathbf{v}_2 \times \mathbf{v}_3)}{\mathbf{v}_1 \cdot (\mathbf{v}_2 \times \mathbf{v}_3)} = 1$ [by Property 5 of Theorem 8]

(c) $\mathbf{k}_1 \cdot (\mathbf{k}_2 \times \mathbf{k}_3) = \mathbf{k}_1 \cdot \left(\dfrac{\mathbf{v}_3 \times \mathbf{v}_1}{\mathbf{v}_1 \cdot (\mathbf{v}_2 \times \mathbf{v}_3)} \times \dfrac{\mathbf{v}_1 \times \mathbf{v}_2}{\mathbf{v}_1 \cdot (\mathbf{v}_2 \times \mathbf{v}_3)} \right) = \dfrac{\mathbf{k}_1}{\left[\mathbf{v}_1 \cdot (\mathbf{v}_2 \times \mathbf{v}_3) \right]^2} \cdot \left[(\mathbf{v}_3 \times \mathbf{v}_1) \times (\mathbf{v}_1 \times \mathbf{v}_2) \right]$

$= \dfrac{\mathbf{k}_1}{\left[\mathbf{v}_1 \cdot (\mathbf{v}_2 \times \mathbf{v}_3) \right]^2} \cdot \left(\left[(\mathbf{v}_3 \times \mathbf{v}_1) \cdot \mathbf{v}_2 \right] \mathbf{v}_1 - \left[(\mathbf{v}_3 \times \mathbf{v}_1) \cdot \mathbf{v}_1 \right] \mathbf{v}_2 \right)$ [by Exercise 46]

But $(\mathbf{v}_3 \times \mathbf{v}_1) \cdot \mathbf{v}_1 = 0$ since $\mathbf{v}_3 \times \mathbf{v}_1$ is orthogonal to $\mathbf{v}_1$, and

$(\mathbf{v}_3 \times \mathbf{v}_1) \cdot \mathbf{v}_2 = \mathbf{v}_2 \cdot (\mathbf{v}_3 \times \mathbf{v}_1) = (\mathbf{v}_2 \times \mathbf{v}_3) \cdot \mathbf{v}_1 = \mathbf{v}_1 \cdot (\mathbf{v}_2 \times \mathbf{v}_3)$. Thus

$\mathbf{k}_1 \cdot (\mathbf{k}_2 \times \mathbf{k}_3) = \dfrac{\mathbf{k}_1}{\left[\mathbf{v}_1 \cdot (\mathbf{v}_2 \times \mathbf{v}_3) \right]^2} \cdot \left[\mathbf{v}_1 \cdot (\mathbf{v}_2 \times \mathbf{v}_3) \right] \mathbf{v}_1 = \dfrac{\mathbf{k}_1 \cdot \mathbf{v}_1}{\mathbf{v}_1 \cdot (\mathbf{v}_2 \times \mathbf{v}_3)} = \dfrac{1}{\mathbf{v}_1 \cdot (\mathbf{v}_2 \times \mathbf{v}_3)}$ [by part (b)]

DISCOVERY PROJECT The Geometry of a Tetrahedron

1. Set up a coordinate system so that vertex S is at the origin, $R = (0, y_1, 0)$, $Q = (x_2, y_2, 0)$, $P = (x_3, y_3, z_3)$.

Then $\overrightarrow{SR} = \langle 0, y_1, 0 \rangle$, $\overrightarrow{SQ} = \langle x_2, y_2, 0 \rangle$, $\overrightarrow{SP} = \langle x_3, y_3, z_3 \rangle$, $\overrightarrow{QR} = \langle -x_2, y_1 - y_2, 0 \rangle$, and $\overrightarrow{QP} = \langle x_3 - x_2, y_3 - y_2, z_3 \rangle$.

Let

$$\mathbf{v}_S = \overrightarrow{QR} \times \overrightarrow{QP} = (y_1 z_3 - y_2 z_3) \mathbf{i} + x_2 z_3 \mathbf{j} + (-x_2 y_3 - x_3 y_1 + x_3 y_2 + x_2 y_1) \mathbf{k}$$

Then $\mathbf{v}_S$ is an outward normal to the face opposite vertex S. Similarly,

$\mathbf{v}_R = \overrightarrow{SQ} \times \overrightarrow{SP} = y_2 z_3 \mathbf{i} - x_2 z_3 \mathbf{j} + (x_2 y_3 - x_3 y_2) \mathbf{k}$, $\mathbf{v}_Q = \overrightarrow{SP} \times \overrightarrow{SR} = -y_1 z_3 \mathbf{i} + x_3 y_1 \mathbf{k}$, and

$\mathbf{v}_P = \overrightarrow{SR} \times \overrightarrow{SQ} = -x_2 y_1 \mathbf{k}$ $\Rightarrow$ $\mathbf{v}_S + \mathbf{v}_R + \mathbf{v}_Q + \mathbf{v}_P = \mathbf{0}$. Now

$$|\mathbf{v}_S| = \text{area of the parallelogram determined by } \overrightarrow{QR} \text{ and } \overrightarrow{QP}$$
$$= 2 \,(\text{area of triangle } RQP) = 2|\mathbf{v}_1|$$

So $\mathbf{v}_S = 2\mathbf{v}_1$, and similarly $\mathbf{v}_R = 2\mathbf{v}_2$, $\mathbf{v}_Q = 2\mathbf{v}_3$, $\mathbf{v}_P = 2\mathbf{v}_4$. Thus $\mathbf{v}_1 + \mathbf{v}_2 + \mathbf{v}_3 + \mathbf{v}_4 = \mathbf{0}$.

2. (a) Let $S = (x_0, y_0, z_0)$, $R = (x_1, y_1, z_1)$, $Q = (x_2, y_2, z_2)$, $P = (x_3, y_3, z_3)$ be the four vertices. Then

$$\text{Volume} = \tfrac{1}{3}(\text{distance from } S \text{ to plane } RQP) \times (\text{area of triangle } RQP)$$

$$= \tfrac{1}{3} \frac{\left| \mathbf{N} \cdot \overrightarrow{SR} \right|}{|\mathbf{N}|} \cdot \tfrac{1}{2} \left| \overrightarrow{RQ} \times \overrightarrow{RP} \right|$$

where $\mathbf{N}$ is a vector which is normal to the face RQP. Thus $\mathbf{N} = \overrightarrow{RQ} \times \overrightarrow{RP}$. Therefore

$$V = \left| \tfrac{1}{6} \left(\overrightarrow{RQ} \times \overrightarrow{RP} \right) \cdot \overrightarrow{SR} \right| = \frac{1}{6} \begin{vmatrix} x_0 - x & y_0 - y_1 & z_0 - z_1 \\ x_2 - x_1 & y_2 - y_1 & z_2 - z_1 \\ x_3 - x_1 & y_3 - y_1 & z_3 - z_1 \end{vmatrix}$$

(b) Using the formula from part (a), $V = \dfrac{1}{6} \begin{vmatrix} 1-1 & 1-2 & 1-3 \\ 1-1 & 1-2 & 2-3 \\ 3-1 & -1-2 & 2-3 \end{vmatrix} = \dfrac{1}{6}\left|2(1-2)\right| = \dfrac{1}{3}.$

3. We define a vector $\mathbf{v}_1$ to have length equal to the area of the face opposite vertex P, so we can say $|\mathbf{v}_1| = A$, and direction perpendicular to the face and pointing outward, as in Problem 1. Similarly, we define $\mathbf{v}_2$, $\mathbf{v}_3$, and $\mathbf{v}_4$ so that $|\mathbf{v}_2| = B$, $|\mathbf{v}_3| = C$, and $|\mathbf{v}_4| = D$ and with the analogous directions. From Problem 1, we know $\mathbf{v}_1 + \mathbf{v}_2 + \mathbf{v}_3 + \mathbf{v}_4 = \mathbf{0}$ $\Rightarrow$

$\mathbf{v}_4 = -(\mathbf{v}_1 + \mathbf{v}_2 + \mathbf{v}_3)$ $\Rightarrow$ $|\mathbf{v}_4| = |-(\mathbf{v}_1 + \mathbf{v}_2 + \mathbf{v}_3)| = |\mathbf{v}_1 + \mathbf{v}_2 + \mathbf{v}_3|$ $\Rightarrow$ $|\mathbf{v}_4|^2 = |\mathbf{v}_1 + \mathbf{v}_2 + \mathbf{v}_3|^2$ $\Rightarrow$

$\qquad \mathbf{v}_4 \cdot \mathbf{v}_4 = (\mathbf{v}_1 + \mathbf{v}_2 + \mathbf{v}_3) \cdot (\mathbf{v}_1 + \mathbf{v}_2 + \mathbf{v}_3)$

$\qquad\qquad = \mathbf{v}_1 \cdot \mathbf{v}_1 + \mathbf{v}_1 \cdot \mathbf{v}_2 + \mathbf{v}_1 \cdot \mathbf{v}_3 + \mathbf{v}_2 \cdot \mathbf{v}_1 + \mathbf{v}_2 \cdot \mathbf{v}_2 + \mathbf{v}_2 \cdot \mathbf{v}_3 + \mathbf{v}_3 \cdot \mathbf{v}_1 + \mathbf{v}_3 \cdot \mathbf{v}_2 + \mathbf{v}_3 \cdot \mathbf{v}_3$

Since the vertex S is trirectangular, we know the three faces meeting at S are mutually perpendicular, so the vectors $\mathbf{v}_1$, $\mathbf{v}_2$, $\mathbf{v}_3$ are also mutually perpendicular. Therefore, $\mathbf{v}_i \cdot \mathbf{v}_j = 0$ for $i \neq j$ and $i, j \in \{1, 2, 3\}$. Thus we have

$\mathbf{v}_4 \cdot \mathbf{v}_4 = \mathbf{v}_1 \cdot \mathbf{v}_1 + \mathbf{v}_2 \cdot \mathbf{v}_2 + \mathbf{v}_3 \cdot \mathbf{v}_3$ $\Rightarrow$ $|\mathbf{v}_4|^2 = |\mathbf{v}_1|^2 + |\mathbf{v}_2|^2 + |\mathbf{v}_3|^2$ $\Rightarrow$ $D^2 = A^2 + B^2 + C^2$.

Another method: We introduce a coordinate system, as shown. Recall that the area of the parallelogram spanned by two vectors is equal to the length of their cross product, so since

$\mathbf{u} \times \mathbf{v} = \langle -q, r, 0 \rangle \times \langle -q, 0, p \rangle = \langle pr, pq, qr \rangle$, we have

$|\mathbf{u} \times \mathbf{v}| = \sqrt{(pr)^2 + (pq)^2 + (qr)^2}$, and therefore

$\qquad D^2 = \left(\tfrac{1}{2}|\mathbf{u} \times \mathbf{v}|\right)^2 = \tfrac{1}{4}[(pr)^2 + (pq)^2 + (qr)^2]$

$\qquad\quad = \left(\tfrac{1}{2}pr\right)^2 + \left(\tfrac{1}{2}pq\right)^2 + \left(\tfrac{1}{2}qr\right)^2 = A^2 + B^2 + C^2.$

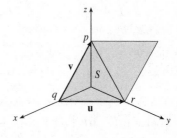

A third method: We draw a line from S perpendicular to QR, as shown.

Now $D = \tfrac{1}{2}ch$, so $D^2 = \tfrac{1}{4}c^2h^2$. Substituting $h^2 = p^2 + k^2$, we get

$D^2 = \tfrac{1}{4}c^2(p^2 + k^2) = \tfrac{1}{4}c^2p^2 + \tfrac{1}{4}c^2k^2$. But $C = \tfrac{1}{2}ck$, so

$D^2 = \tfrac{1}{4}c^2p^2 + C^2$. Now substituting $c^2 = q^2 + r^2$ gives

$D^2 = \tfrac{1}{4}p^2q^2 + \tfrac{1}{4}q^2r^2 + C^2 = A^2 + B^2 + C^2.$

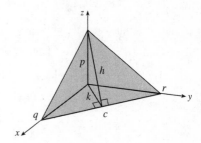

13.5   Equations of Lines and Planes

ET 12.5

1. (a) True; each of the first two lines has a direction vector parallel to the direction vector of the third line, so these vectors are each scalar multiples of the third direction vector. Then the first two direction vectors are also scalar multiples of each other, so these vectors, and hence the two lines, are parallel.

(b) False; for example, the x- and y-axes are both perpendicular to the z-axis, yet the x- and y-axes are not parallel.

(c) True; each of the first two planes has a normal vector parallel to the normal vector of the third plane, so these two normal vectors are parallel to each other and the planes are parallel.

(d) False; for example, the xy- and yz-planes are not parallel, yet they are both perpendicular to the xz-plane.

(e) False; the x- and y-axes are not parallel, yet they are both parallel to the plane $z = 1$.

(f) True; if each line is perpendicular to a plane, then the lines' direction vectors are both parallel to a normal vector for the plane. Thus, the direction vectors are parallel to each other and the lines are parallel.

(g) False; the planes $y = 1$ and $z = 1$ are not parallel, yet they are both parallel to the x-axis.

(h) True; if each plane is perpendicular to a line, then any normal vector for each plane is parallel to a direction vector for the line. Thus, the normal vectors are parallel to each other and the planes are parallel.

(i) True; see Figure 9 and the accompanying discussion.

(j) False; they can be skew, as in Example 3.

(k) True. Consider any normal vector for the plane and any direction vector for the line. If the normal vector is perpendicular to the direction vector, the line and plane are parallel. Otherwise, the vectors meet at an angle θ, $0° \le \theta < 90°$, and the line will intersect the plane at an angle $90° - \theta$.

2. For this line, we have $\mathbf{r}_0 = 6\,\mathbf{i} - 5\,\mathbf{j} + 2\,\mathbf{k}$ and $\mathbf{v} = \mathbf{i} + 3\,\mathbf{j} - \frac{2}{3}\,\mathbf{k}$, so a vector equation is

$\mathbf{r} = \mathbf{r}_0 + t\,\mathbf{v} = (6\,\mathbf{i} - 5\,\mathbf{j} + 2\,\mathbf{k}) + t\left(\mathbf{i} + 3\,\mathbf{j} - \frac{2}{3}\,\mathbf{k}\right) = (6 + t)\,\mathbf{i} + (-5 + 3t)\,\mathbf{j} + \left(2 - \frac{2}{3}t\right)\mathbf{k}$ and parametric equations are

$x = 6 + t$, $y = -5 + 3t$, $z = 2 - \frac{2}{3}t$.

3. For this line, we have $\mathbf{r}_0 = 2\,\mathbf{i} + 2.4\,\mathbf{j} + 3.5\,\mathbf{k}$ and $\mathbf{v} = 3\,\mathbf{i} + 2\,\mathbf{j} - \mathbf{k}$, so a vector equation is

$\mathbf{r} = \mathbf{r}_0 + t\,\mathbf{v} = (2\,\mathbf{i} + 2.4\,\mathbf{j} + 3.5\,\mathbf{k}) + t(3\,\mathbf{i} + 2\,\mathbf{j} - \mathbf{k}) = (2 + 3t)\,\mathbf{i} + (2.4 + 2t)\,\mathbf{j} + (3.5 - t)\,\mathbf{k}$ and parametric equations are

$x = 2 + 3t$, $y = 2.4 + 2t$, $z = 3.5 - t$.

4. This line has the same direction as the given line, $\mathbf{v} = 2\,\mathbf{i} - 3\,\mathbf{j} + 9\,\mathbf{k}$. Here $\mathbf{r}_0 = 14\,\mathbf{j} - 10\,\mathbf{k}$, so a vector equation is

$\mathbf{r} = (14\,\mathbf{j} - 10\,\mathbf{k}) + t(2\,\mathbf{i} - 3\,\mathbf{j} + 9\,\mathbf{k}) = 2t\,\mathbf{i} + (14 - 3t)\,\mathbf{j} + (-10 + 9t)\,\mathbf{k}$ and parametric equations are $x = 2t$,

$y = 14 - 3t$, $z = -10 + 9t$.

5. A line perpendicular to the given plane has the same direction as a normal vector to the plane, such as

$\mathbf{n} = \langle 1, 3, 1 \rangle$. So $\mathbf{r}_0 = \mathbf{i} + 6\,\mathbf{k}$, and we can take $\mathbf{v} = \mathbf{i} + 3\,\mathbf{j} + \mathbf{k}$. Then a vector equation is

$\mathbf{r} = (\mathbf{i} + 6\,\mathbf{k}) + t(\mathbf{i} + 3\,\mathbf{j} + \mathbf{k}) = (1 + t)\,\mathbf{i} + 3t\,\mathbf{j} + (6 + t)\,\mathbf{k}$, and parametric equations are $x = 1 + t$, $y = 3t$, $z = 6 + t$.

6. The vector $\mathbf{v} = \langle 1 - 0, 2 - 0, 3 - 0 \rangle = \langle 1, 2, 3 \rangle$ is parallel to the line. Letting $P_0 = (0, 0, 0)$, parametric equations are

$x = 0 + 1 \cdot t = t$, $y = 0 + 2 \cdot t = 2t$, $z = 0 + 3 \cdot t = 3t$, while symmetric equations are $x = \dfrac{y}{2} = \dfrac{z}{3}$.

7. The vector $\mathbf{v} = \langle -4 - 1, 3 - 3, 0 - 2 \rangle = \langle -5, 0, -2 \rangle$ is parallel to the line. Letting $P_0 = (1, 3, 2)$, parametric equations are

$x = 1 - 5t$, $y = 3 + 0t = 3$, $z = 2 - 2t$, while symmetric equations are $\dfrac{x - 1}{-5} = \dfrac{z - 2}{-2}$, $y = 3$. Notice here that the

direction number $b = 0$, so rather than writing $\dfrac{y - 3}{0}$ in the symmetric equation we must write the equation $y = 3$ separately.

8. $\mathbf{v} = \langle 2 - 6, 4 - 1, 5 - (-3) \rangle = \langle -4, 3, 8 \rangle$, and letting $P_0 = (6, 1, -3)$, parametric equations are $x = 6 - 4t$, $y = 1 + 3t$,

$z = -3 + 8t$, while symmetric equations are $\dfrac{x - 6}{-4} = \dfrac{y - 1}{3} = \dfrac{z + 3}{8}$.

9. $\mathbf{v} = \langle 2 - 0, 1 - \frac{1}{2}, -3 - 1 \rangle = \langle 2, \frac{1}{2}, -4 \rangle$, and letting $P_0 = (2, 1, -3)$, parametric equations are $x = 2 + 2t$, $y = 1 + \frac{1}{2}t$,

$z = -3 - 4t$, while symmetric equations are $\dfrac{x - 2}{2} = \dfrac{y - 1}{1/2} = \dfrac{z + 3}{-4}$ or $\dfrac{x - 2}{2} = 2y - 2 = \dfrac{z + 3}{-4}$.

10. $\mathbf{v} = (\mathbf{i} + \mathbf{j}) \times (\mathbf{j} + \mathbf{k}) = \begin{vmatrix} \mathbf{i} & \mathbf{j} & \mathbf{k} \\ 1 & 1 & 0 \\ 0 & 1 & 1 \end{vmatrix} = \mathbf{i} - \mathbf{j} + \mathbf{k}$ is the direction of the line perpendicular to both $\mathbf{i} + \mathbf{j}$ and $\mathbf{j} + \mathbf{k}$.

With $P_0 = (2, 1, 0)$, parametric equations are $x = 2 + t$, $y = 1 - t$, $z = t$ and symmetric equations are $x - 2 = \dfrac{y - 1}{-1} = z$

or $x - 2 = 1 - y = z$.

11. The line has direction $\mathbf{v} = \langle 1, 2, 1 \rangle$. Letting $P_0 = (1, -1, 1)$, parametric equations are $x = 1 + t$, $y = -1 + 2t$, $z = 1 + t$

and symmetric equations are $x - 1 = \dfrac{y + 1}{2} = z - 1$.

12. Setting $x = 0$, we see that $(0, 1, 0)$ satisfies the equations of both planes, so they do in fact have a line of intersection.

$\mathbf{v} = \mathbf{n}_1 \times \mathbf{n}_2 = \langle 1, 1, 1 \rangle \times \langle 1, 0, 1 \rangle = \langle 1, 0, -1 \rangle$ is the direction of this line. Taking the point $(0, 1, 0)$ as P_0, parametric

equations are $x = t$, $y = 1$, $z = -t$, and symmetric equations are $x = -z$, $y = 1$.

13. Direction vectors of the lines are $\mathbf{v}_1 = \langle -2 - (-4), 0 - (-6), -3 - 1 \rangle = \langle 2, 6, -4 \rangle$ and

$\mathbf{v}_2 = \langle 5 - 10, 3 - 18, 14 - 4 \rangle = \langle -5, -15, 10 \rangle$, and since $\mathbf{v}_2 = -\frac{5}{2}\mathbf{v}_1$, the direction vectors and thus the lines are parallel.

14. Direction vectors of the lines are $\mathbf{v}_1 = \langle -2, 4, 4 \rangle$ and $\mathbf{v}_2 = \langle 8, -1, 4 \rangle$. Since $\mathbf{v}_1 \cdot \mathbf{v}_2 = -16 - 4 + 16 \neq 0$, the vectors and

thus the lines are not perpendicular.

15. (a) The line passes through the point $(1, -5, 6)$ and a direction vector for the line is $\langle -1, 2, -3 \rangle$, so symmetric equations for

the line are $\dfrac{x - 1}{-1} = \dfrac{y + 5}{2} = \dfrac{z - 6}{-3}$.

(b) The line intersects the xy-plane when $z = 0$, so we need $\dfrac{x - 1}{-1} = \dfrac{y + 5}{2} = \dfrac{0 - 6}{-3}$ or $\dfrac{x - 1}{-1} = 2 \Rightarrow x = -1$,

$\dfrac{y + 5}{2} = 2 \Rightarrow y = -1$. Thus the point of intersection with the xy-plane is $(-1, -1, 0)$. Similarly for the yz-plane,

we need $x = 0 \Rightarrow 1 = \dfrac{y + 5}{2} = \dfrac{z - 6}{-3} \Rightarrow y = -3, z = 3$. Thus the line intersects the yz-plane at $(0, -3, 3)$. For

the xz-plane, we need $y = 0 \Rightarrow \dfrac{x - 1}{-1} = \dfrac{5}{2} = \dfrac{z - 6}{-3} \Rightarrow x = -\frac{3}{2}, z = -\frac{3}{2}$. So the line intersects the xz-plane

at $\left(-\frac{3}{2}, 0, -\frac{3}{2} \right)$.

16. (a) A vector normal to the plane $x - y + 3z = 7$ is $\mathbf{n} = \langle 1, -1, 3 \rangle$, and since the line is to be perpendicular to the plane, $\mathbf{n}$ is

also a direction vector for the line. Thus parametric equations of the line are $x = 2 + t$, $y = 4 - t$, $z = 6 + 3t$.

(b) On the xy-plane, $z = 0$. So $z = 6 + 3t = 0 \Rightarrow t = -2$ in the parametric equations of the line, and therefore $x = 0$

and $y = 6$, giving the point of intersection $(0, 6, 0)$. For the yz-plane, $x = 0$ so we get the same point of interesection:

$(0, 6, 0)$. For the xz-plane, $y = 0$ which implies $t = 4$, so $x = 6$ and $z = 18$ and the point of intersection is $(6, 0, 18)$.

17. From Equation 4, the line segment from $\mathbf{r}_0 = 2\mathbf{i} - \mathbf{j} + 4\mathbf{k}$ to $\mathbf{r}_1 = 4\mathbf{i} + 6\mathbf{j} + \mathbf{k}$ is

$\mathbf{r}(t) = (1 - t)\mathbf{r}_0 + t\mathbf{r}_1 = (1 - t)(2\mathbf{i} - \mathbf{j} + 4\mathbf{k}) + t(4\mathbf{i} + 6\mathbf{j} + \mathbf{k}) = (2\mathbf{i} - \mathbf{j} + 4\mathbf{k}) + t(2\mathbf{i} + 7\mathbf{j} - 3\mathbf{k})$, $0 \leq t \leq 1$.

18. From Equation 4, the line segment from $\mathbf{r}_0 = 10\mathbf{i} + 3\mathbf{j} + \mathbf{k}$ to $\mathbf{r}_1 = 5\mathbf{i} + 6\mathbf{j} - 3\mathbf{k}$ is

$$\mathbf{r}(t) = (1-t)\,\mathbf{r}_0 + t\,\mathbf{r}_1 = (1-t)(10\mathbf{i} + 3\mathbf{j} + \mathbf{k}) + t(5\mathbf{i} + 6\mathbf{j} - 3\mathbf{k})$$
$$= (10\mathbf{i} + 3\mathbf{j} + \mathbf{k}) + t(-5\mathbf{i} + 3\mathbf{j} - 4\mathbf{k}), \quad 0 \le t \le 1.$$

The corresponding parametric equations are $x = 10 - 5t$, $y = 3 + 3t$, $z = 1 - 4t$, $0 \le t \le 1$.

19. Since the direction vectors are $\mathbf{v}_1 = \langle -6, 9, -3 \rangle$ and $\mathbf{v}_2 = \langle 2, -3, 1 \rangle$, we have $\mathbf{v}_1 = -3\mathbf{v}_2$ so the lines are parallel.

20. The lines aren't parallel since the direction vectors $\langle 2, 3, -1 \rangle$ and $\langle 1, 1, 3 \rangle$ aren't parallel. For the lines to intersect we must be able to find one value of t and one value of s that produce the same point from the respective parametric equations. Thus we need to satisfy the following three equations: $1 + 2t = -1 + s$, $3t = 4 + s$, $2 - t = 1 + 3s$. Solving the first two equations we get $t = 6$, $s = 14$ and checking, we see that these values don't satisfy the third equation. Thus L_1 and L_2 aren't parallel and don't intersect, so they must be skew lines.

21. Since the direction vectors $\langle 1, 2, 3 \rangle$ and $\langle -4, -3, 2 \rangle$ are not scalar multiples of each other, the lines are not parallel, so we check to see if the lines intersect. The parametric equations of the lines are L_1: $x = t$, $y = 1 + 2t$, $z = 2 + 3t$ and L_2: $x = 3 - 4s$, $y = 2 - 3s$, $z = 1 + 2s$. For the lines to intersect, we must be able to find one value of t and one value of s that produce the same point from the respective parametric equations. Thus we need to satisfy the following three equations: $t = 3 - 4s$, $1 + 2t = 2 - 3s$, $2 + 3t = 1 + 2s$. Solving the first two equations we get $t = -1$, $s = 1$ and checking, we see that these values don't satisfy the third equation. Thus the lines aren't parallel and don't intersect, so they must be skew lines.

22. Since the direction vectors $\langle 2, 2, -1 \rangle$ and $\langle 1, -1, 3 \rangle$ aren't parallel, the lines aren't parallel. Here the parametric equations are L_1: $x = 1 + 2t$, $y = 3 + 2t$, $z = 2 - t$ and L_2: $x = 2 + s$, $y = 6 - s$, $z = -2 + 3s$. Thus, for the lines to intersect, the three equations $1 + 2t = 2 + s$, $3 + 2t = 6 - s$, and $2 - t = -2 + 3s$ must be satisfied simultaneously. Solving the first two equations gives $t = 1$, $s = 1$ and, checking, we see that these values do satisfy the third equation, so the lines intersect when $t = 1$ and $s = 1$, that is, at the point $(3, 5, 1)$.

23. Since the plane is perpendicular to the vector $\langle -2, 1, 5 \rangle$, we can take $\langle -2, 1, 5 \rangle$ as a normal vector to the plane. $(6, 3, 2)$ is a point on the plane, so setting $a = -2$, $b = 1$, $c = 5$ and $x_0 = 6$, $y_0 = 3$, $z_0 = 2$ in Equation 7 gives $-2(x - 6) + 1(y - 3) + 5(z - 2) = 0$ or $-2x + y + 5z = 1$ to be an equation of the plane.

24. $\mathbf{j} + 2\mathbf{k} = \langle 0, 1, 2 \rangle$ is a normal vector to the plane and $(4, 0, -3)$ is a point on the plane, so setting $a = 0$, $b = 1$, $c = 2$, $x_0 = 4$, $y_0 = 0$, $z_0 = -3$ in Equation 7 gives $0(x - 4) + 1(y - 0) + 2[z - (-3)] = 0$ or $y + 2z = -6$ to be an equation of the plane.

25. $\mathbf{i} + \mathbf{j} - \mathbf{k} = \langle 1, 1, -1 \rangle$ is a normal vector to the plane and $(1, -1, 1)$ is a point on the plane, so setting $a = 1$, $b = 1$, $c = -1$, $x_0 = 1$, $y_0 = -1$, $z_0 = 1$ in Equation 7 gives $1(x - 1) + 1[y - (-1)] - 1(z - 1) = 0$ or $x + y - z = -1$ to be an equation of the plane.

26. Since the line is perpendicular to the plane, its direction vector $\langle 1, 2, -3 \rangle$ is a normal vector to the plane. An equation of the plane, then, is $1[x - (-2)] + 2(y - 8) - 3(z - 10) = 0$ or $x + 2y - 3z = -16$.

27. Since the two planes are parallel, they will have the same normal vectors. So we can take $\mathbf{n} = \langle 2, -1, 3 \rangle$, and an equation of the plane is $2(x - 0) - 1(y - 0) + 3(z - 0) = 0$ or $2x - y + 3z = 0$.

28. Since the two planes are parallel, they will have the same normal vectors. So we can take $\mathbf{n} = \langle 1, 1, 1 \rangle$, and an equation of the plane is $1[x - (-1)] + 1(y - 6) + 1[z - (-5)] = 0$ or $x + y + z = 0$.

29. Since the two planes are parallel, they will have the same normal vectors. So we can take $\mathbf{n} = \langle 3, 0, -7 \rangle$, and an equation of the plane is $3(x - 4) + 0[y - (-2)] - 7(z - 3) = 0$ or $3x - 7z = -9$.

30. First, a normal vector for the plane $2x + 4y + 8z = 17$ is $\mathbf{n} = \langle 2, 4, 8 \rangle$. A direction vector for the line is $\mathbf{v} = \langle 2, 1, -1 \rangle$, and since $\mathbf{n} \cdot \mathbf{v} = 0$ we know the line is perpendicular to $\mathbf{n}$ and hence parallel to the plane. Thus, there is a parallel plane which contains the line. By putting $t = 0$, we know the point $(3, 0, 8)$ is on the line and hence the new plane. We can use the same normal vector $\mathbf{n} = \langle 2, 4, 8 \rangle$, so an equation of the plane is $2(x - 3) + 4(y - 0) + 8(z - 8) = 0$ or $x + 2y + 4z = 35$.

31. Here the vectors $\mathbf{a} = \langle 1 - 0, 0 - 1, 1 - 1 \rangle = \langle 1, -1, 0 \rangle$ and $\mathbf{b} = \langle 1 - 0, 1 - 1, 0 - 1 \rangle = \langle 1, 0, -1 \rangle$ lie in the plane, so $\mathbf{a} \times \mathbf{b}$ is a normal vector to the plane. Thus, we can take $\mathbf{n} = \mathbf{a} \times \mathbf{b} = \langle 1 - 0, 0 + 1, 0 + 1 \rangle = \langle 1, 1, 1 \rangle$. If P_0 is the point $(0, 1, 1)$, an equation of the plane is $1(x - 0) + 1(y - 1) + 1(z - 1) = 0$ or $x + y + z = 2$.

32. Here the vectors $\mathbf{a} = \langle 2, -4, 6 \rangle$ and $\mathbf{b} = \langle 5, 1, 3 \rangle$ lie in the plane, so $\mathbf{n} = \mathbf{a} \times \mathbf{b} = \langle -12 - 6, 30 - 6, 2 + 20 \rangle = \langle -18, 24, 22 \rangle$ is a normal vector to the plane and an equation of the plane is $-18(x - 0) + 24(y - 0) + 22(z - 0) = 0$ or $-18x + 24y + 22z = 0$.

33. Here the vectors $\mathbf{a} = \langle 8 - 3, 2 - (-1), 4 - 2 \rangle = \langle 5, 3, 2 \rangle$ and $\mathbf{b} = \langle -1 - 3, -2 - (-1), -3 - 2 \rangle = \langle -4, -1, -5 \rangle$ lie in the plane, so a normal vector to the plane is $\mathbf{n} = \mathbf{a} \times \mathbf{b} = \langle -15 + 2, -8 + 25, -5 + 12 \rangle = \langle -13, 17, 7 \rangle$ and an equation of the plane is $-13(x - 3) + 17[y - (-1)] + 7(z - 2) = 0$ or $-13x + 17y + 7z = -42$.

34. If we first find two nonparallel vectors in the plane, their cross product will be a normal vector to the plane. Since the given line lies in the plane, its direction vector $\mathbf{a} = \langle 3, 1, -1 \rangle$ is one vector in the plane. We can verify that the given point $(1, 2, 3)$ does not lie on this line, so to find another nonparallel vector $\mathbf{b}$ which lies in the plane, we can pick any point on the line and find a vector connecting the points. If we put $t = 0$, we see that $(0, 1, 2)$ is on the line, so $\mathbf{b} = \langle 1 - 0, 2 - 1, 3 - 2 \rangle = \langle 1, 1, 1 \rangle$ and $\mathbf{n} = \mathbf{a} \times \mathbf{b} = \langle 1 + 1, -1 - 3, 3 - 1 \rangle = \langle 2, -4, 2 \rangle$. Thus, an equation of the plane is $2(x - 1) - 4(y - 2) + 2(z - 3) = 0$ or $2x - 4y + 2z = 0$. (Equivalently, we can write $x - 2y + z = 0$.)

35. If we first find two nonparallel vectors in the plane, their cross product will be a normal vector to the plane. Since the given line lies in the plane, its direction vector $\mathbf{a} = \langle -2, 5, 4 \rangle$ is one vector in the plane. We can verify that the given point $(6, 0, -2)$ does not lie on this line, so to find another nonparallel vector $\mathbf{b}$ which lies in the plane, we can pick any point on the line and find a vector connecting the points. If we put $t = 0$, we see that $(4, 3, 7)$ is on the line, so $\mathbf{b} = \langle 6 - 4, 0 - 3, -2 - 7 \rangle = \langle 2, -3, -9 \rangle$ and $\mathbf{n} = \mathbf{a} \times \mathbf{b} = \langle -45 + 12, 8 - 18, 6 - 10 \rangle = \langle -33, -10, -4 \rangle$. Thus, an equation of the plane is $-33(x - 6) - 10(y - 0) - 4[z - (-2)] = 0$ or $33x + 10y + 4z = 190$.

36. Since the line $x = 2y = 3z$, or $x = \dfrac{y}{1/2} = \dfrac{z}{1/3}$, lies in the plane, its direction vector $\mathbf{a} = \left\langle 1, \frac{1}{2}, \frac{1}{3} \right\rangle$ is parallel to the plane.

The point $(0, 0, 0)$ is on the line (put $t = 0$), and we can verify that the given point $(1, -1, 1)$ in the plane is not on the line.

The vector connecting these two points, $\mathbf{b} = \langle 1, -1, 1 \rangle$, is therefore parallel to the plane, but not parallel to $\langle 1, 2, 3 \rangle$. Then

$\mathbf{a} \times \mathbf{b} = \left\langle \frac{1}{2} + \frac{1}{3}, \frac{1}{3} - 1, -1 - \frac{1}{2} \right\rangle = \left\langle \frac{5}{6}, -\frac{2}{3}, -\frac{3}{2} \right\rangle$ is a normal vector to the plane, and an equation of the plane is

$\frac{5}{6}(x - 0) - \frac{2}{3}(y - 0) - \frac{3}{2}(z - 0) = 0$ or $5x - 4y - 9z = 0$.

37. A direction vector for the line of intersection is $\mathbf{a} = \mathbf{n}_1 \times \mathbf{n}_2 = \langle 1, 1, -1 \rangle \times \langle 2, -1, 3 \rangle = \langle 2, -5, -3 \rangle$, and $\mathbf{a}$ is parallel to the

desired plane. Another vector parallel to the plane is the vector connecting any point on the line of intersection to the given

point $(-1, 2, 1)$ in the plane. Setting $x = 0$, the equations of the planes reduce to $y - z = 2$ and $-y + 3z = 1$ with

simultaneous solution $y = \frac{7}{2}$ and $z = \frac{3}{2}$. So a point on the line is $\left(0, \frac{7}{2}, \frac{3}{2}\right)$ and another vector parallel to the plane is

$\left\langle -1, -\frac{3}{2}, -\frac{1}{2} \right\rangle$. Then a normal vector to the plane is $\mathbf{n} = \langle 2, -5, -3 \rangle \times \left\langle -1, -\frac{3}{2}, -\frac{1}{2} \right\rangle = \langle -2, 4, -8 \rangle$ and an equation of

the plane is $-2(x + 1) + 4(y - 2) - 8(z - 1) = 0$ or $x - 2y + 4z = -1$.

38. $\mathbf{n}_1 = \langle 1, 0, -1 \rangle$ and $\mathbf{n}_2 = \langle 0, 1, 2 \rangle$. Setting $z = 0$, it is easy to see that $(1, 3, 0)$ is a point on the line of intersection of

$x - z = 1$ and $y + 2z = 3$. The direction of this line is $\mathbf{v}_1 = \mathbf{n}_1 \times \mathbf{n}_2 = \langle 1, -2, 1 \rangle$. A second vector parallel to the desired

plane is $\mathbf{v}_2 = \langle 1, 1, -2 \rangle$, since it is perpendicular to $x + y - 2z = 1$. Therefore, a normal of the plane in question is

$\mathbf{n} = \mathbf{v}_1 \times \mathbf{v}_2 = \langle 4 - 1, 1 + 2, 1 + 2 \rangle = \langle 3, 3, 3 \rangle$, or we can use $\langle 1, 1, 1 \rangle$. Taking $(x_0, y_0, z_0) = (1, 3, 0)$, the equation we are

looking for is $(x - 1) + (y - 3) + z = 0 \quad \Leftrightarrow \quad x + y + z = 4$.

39. To find the x-intercept we set $y = z = 0$ in the equation $2x + 5y + z = 10$

and obtain $2x = 10 \quad \Rightarrow \quad x = 5$ so the x-intercept is $(5, 0, 0)$. When

$x = z = 0$ we get $5y = 10 \quad \Rightarrow \quad y = 2$, so the y-intercept is $(0, 2, 0)$.

Setting $x = y = 0$ gives $z = 10$, so the z-intercept is $(0, 0, 10)$ and we

graph the portion of the plane that lies in the first octant.

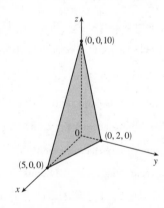

40. To find the x-intercept we set $y = z = 0$ in the equation $3x + y + 2z = 6$

and obtain $3x = 6 \quad \Rightarrow \quad x = 2$ so the x-intercept is $(2, 0, 0)$. When

$x = z = 0$ we get $y = 6$ so the y-intercept is $(0, 6, 0)$. Setting $x = y = 0$

gives $2z = 6 \quad \Rightarrow \quad z = 3$, so the z-intercept is $(0, 0, 3)$. The figure shows

the portion of the plane that lies in the first octant.

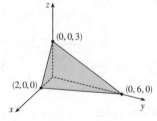

41. Setting $y = z = 0$ in the equation $6x - 3y + 4z = 6$ gives $6x = 6$ $\Rightarrow$
    $x = 1$, when $x = z = 0$ we have $-3y = 6$ $\Rightarrow$ $y = -2$, and $x = y = 0$
    implies $4z = 6$ $\Rightarrow$ $z = \frac{3}{2}$, so the intercepts are $(1, 0, 0)$, $(0, -2, 0)$, and
    $(0, 0, \frac{3}{2})$. The figure shows the portion of the plane cut off by the coordinate
    planes.

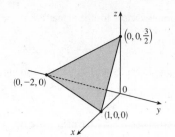

42. Setting $y = z = 0$ in the equation $6x + 5y - 3z = 15$ gives $6x = 15$ $\Rightarrow$
    $x = \frac{5}{2}$, when $x = z = 0$ we have $5y = 15$ $\Rightarrow$ $y = 3$, and $x = y = 0$
    implies $-3z = 15$ $\Rightarrow$ $z = -5$, so the intercepts are $(\frac{5}{2}, 0, 0)$, $(0, 3, 0)$,
    and $(0, 0, -5)$. The figure shows the portion of the plane cut off by the
    coordinate planes.

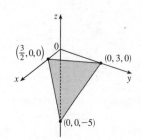

43. Substitute the parametric equations of the line into the equation of the plane: $(3 - t) - (2 + t) + 2(5t) = 9$ $\Rightarrow$
    $8t = 8$ $\Rightarrow$ $t = 1$. Therefore, the point of intersection of the line and the plane is given by $x = 3 - 1 = 2$, $y = 2 + 1 = 3$,
    and $z = 5(1) = 5$, that is, the point $(2, 3, 5)$.

44. Substitute the parametric equations of the line into the equation of the plane: $(1 + 2t) + 2(4t) - (2 - 3t) + 1 = 0$ $\Rightarrow$
    $13t = 0$ $\Rightarrow$ $t = 0$. Therefore, the point of intersection of the line and the plane is given by $x = 1 + 2(0) = 1$,
    $y = 4(0) = 0$, and $z = 2 - 3(0) = 2$, that is, the point $(1, 0, 2)$.

45. Parametric equations for the line are $x = t$, $y = 1 + t$, $z = \frac{1}{2}t$ and substituting into the equation of the plane gives
    $4(t) - (1 + t) + 3(\frac{1}{2}t) = 8$ $\Rightarrow$ $\frac{9}{2}t = 9$ $\Rightarrow$ $t = 2$. Thus $x = 2$, $y = 1 + 2 = 3$, $z = \frac{1}{2}(2) = 1$ and the point of
    intersection is $(2, 3, 1)$.

46. A direction vector for the line through $(1, 0, 1)$ and $(4, -2, 2)$ is $\mathbf{v} = \langle 3, -2, 1 \rangle$ and, taking $P_0 = (1, 0, 1)$, parametric
    equations for the line are $x = 1 + 3t$, $y = -2t$, $z = 1 + t$. Substitution of the parametric equations into the equation of the
    plane gives $1 + 3t - 2t + 1 + t = 6$ $\Rightarrow$ $t = 2$. Then $x = 1 + 3(2) = 7$, $y = -2(2) = -4$, and $z = 1 + 2 = 3$ so the point
    of intersection is $(7, -4, 3)$.

47. Setting $x = 0$, we see that $(0, 1, 0)$ satisfies the equations of both planes, so that they do in fact have a line of intersection.
    $\mathbf{v} = \mathbf{n_1} \times \mathbf{n_2} = \langle 1, 1, 1 \rangle \times \langle 1, 0, 1 \rangle = \langle 1, 0, -1 \rangle$ is the direction of this line. Therefore, direction numbers of the intersecting
    line are $1, 0, -1$.

48. The angle between the two planes is the same as the angle between their normal vectors. The normal vectors of the
    two planes are $\langle 1, 1, 1 \rangle$ and $\langle 1, 2, 3 \rangle$. The cosine of the angle θ between these two planes is

$$\cos \theta = \frac{\langle 1, 1, 1 \rangle \cdot \langle 1, 2, 3 \rangle}{|\langle 1, 1, 1 \rangle| \, |\langle 1, 2, 3 \rangle|} = \frac{1 + 2 + 3}{\sqrt{1 + 1 + 1} \, \sqrt{1 + 4 + 9}} = \frac{6}{\sqrt{42}} = \sqrt{\frac{6}{7}}.$$

49. Normal vectors for the planes are $\mathbf{n}_1 = \langle 1, 4, -3 \rangle$ and $\mathbf{n}_2 = \langle -3, 6, 7 \rangle$, so the normals (and thus the planes) aren't parallel.

But $\mathbf{n}_1 \cdot \mathbf{n}_2 = -3 + 24 - 21 = 0$, so the normals (and thus the planes) are perpendicular.

50. Normal vectors for the planes are $\mathbf{n}_1 = \langle -1, 4, -2 \rangle$ and $\mathbf{n}_2 = \langle 3, -12, 6 \rangle$. Since $\mathbf{n}_2 = -3\mathbf{n}_1$, the normals (and thus the planes) are parallel.

51. Normal vectors for the planes are $\mathbf{n}_1 = \langle 1, 1, 1 \rangle$ and $\mathbf{n}_2 = \langle 1, -1, 1 \rangle$. The normals are not parallel, so neither are the planes. Furthermore, $\mathbf{n}_1 \cdot \mathbf{n}_2 = 1 - 1 + 1 = 1 \neq 0$, so the planes aren't perpendicular. The angle between them is given by

$$\cos\theta = \frac{\mathbf{n}_1 \cdot \mathbf{n}_2}{|\mathbf{n}_1|\,|\mathbf{n}_2|} = \frac{1}{\sqrt{3}\sqrt{3}} = \frac{1}{3} \quad\Rightarrow\quad \theta = \cos^{-1}\left(\tfrac{1}{3}\right) \approx 70.5°.$$

52. The normals are $\mathbf{n}_1 = \langle 2, -3, 4 \rangle$ and $\mathbf{n}_2 = \langle 1, 6, 4 \rangle$ so the planes aren't parallel. Since $\mathbf{n}_1 \cdot \mathbf{n}_2 = 2 - 18 + 16 = 0$, the normals (and thus the planes) are perpendicular.

53. The normals are $\mathbf{n}_1 = \langle 1, -4, 2 \rangle$ and $\mathbf{n}_2 = \langle 2, -8, 4 \rangle$. Since $\mathbf{n}_2 = 2\mathbf{n}_1$, the normals (and thus the planes) are parallel.

54. The normal vectors are $\mathbf{n}_1 = \langle 1, 2, 2 \rangle$ and $\mathbf{n}_2 = \langle 2, -1, 2 \rangle$. The normals are not parallel, so neither are the planes. Furthermore, $\mathbf{n}_1 \cdot \mathbf{n}_2 = 2 - 2 + 4 = 4 \neq 0$, so the planes aren't perpendicular. The angle between them is given by

$$\cos\theta = \frac{\mathbf{n}_1 \cdot \mathbf{n}_2}{|\mathbf{n}_1|\,|\mathbf{n}_2|} = \frac{4}{\sqrt{9}\sqrt{9}} = \frac{4}{9} \quad\Rightarrow\quad \theta = \cos^{-1}\left(\tfrac{4}{9}\right) \approx 63.6°.$$

55. (a) To find a point on the line of intersection, set one of the variables equal to a constant, say $z = 0$. (This will fail if the line of intersection does not cross the xy-plane; in that case, try setting x or y equal to 0.) The equations of the two planes reduce to $x + y = 1$ and $x + 2y = 1$. Solving these two equations gives $x = 1$, $y = 0$. Thus a point on the line is $(1, 0, 0)$.

A vector $\mathbf{v}$ in the direction of this intersecting line is perpendicular to the normal vectors of both planes, so we can take $\mathbf{v} = \mathbf{n}_1 \times \mathbf{n}_2 = \langle 1, 1, 1 \rangle \times \langle 1, 2, 2 \rangle = \langle 2 - 2, 1 - 2, 2 - 1 \rangle = \langle 0, -1, 1 \rangle$. By Equations 2, parametric equations for the line are $x = 1$, $y = -t$, $z = t$.

(b) The angle between the planes satisfies $\cos\theta = \dfrac{\mathbf{n}_1 \cdot \mathbf{n}_2}{|\mathbf{n}_1|\,|\mathbf{n}_2|} = \dfrac{1 + 2 + 2}{\sqrt{3}\sqrt{9}} = \dfrac{5}{3\sqrt{3}}$. Therefore $\theta = \cos^{-1}\left(\dfrac{5}{3\sqrt{3}}\right) \approx 15.8°.$

56. (a) If we set $z = 0$ then the equations of the planes reduce to $3x - 2y = 1$ and $2x + y = 3$ and solving these two equations gives $x = 1$, $y = 1$. Thus a point on the line of intersection is $(1, 1, 0)$. A vector $\mathbf{v}$ in the direction of this intersecting line is perpendicular to the normal vectors of both planes, so let $\mathbf{v} = \mathbf{n}_1 \times \mathbf{n}_2 = \langle 3, -2, 1 \rangle \times \langle 2, 1, -3 \rangle = \langle 5, 11, 7 \rangle$. By Equations 2, parametric equations for the line are $x = 1 + 5t$, $y = 1 + 11t$, $z = 7t$.

(b) $\cos\theta = \dfrac{\mathbf{n}_1 \cdot \mathbf{n}_2}{|\mathbf{n}_1|\,|\mathbf{n}_2|} = \dfrac{6 - 2 - 3}{\sqrt{14}\sqrt{14}} = \dfrac{1}{14} \quad\Rightarrow\quad \theta = \cos^{-1}\left(\tfrac{1}{14}\right) \approx 85.9°.$

57. Setting $z = 0$, the equations of the two planes become $5x - 2y = 1$ and $4x + y = 6$. Solving these two equations gives $x = 1$, $y = 2$ so a point on the line of intersection is $(1, 2, 0)$. A vector $\mathbf{v}$ in the direction of this intersecting line is perpendicular to the normal vectors of both planes. So we can use $\mathbf{v} = \mathbf{n}_1 \times \mathbf{n}_2 = \langle 5, -2, -2 \rangle \times \langle 4, 1, 1 \rangle = \langle 0, -13, 13 \rangle$ or equivalently we can take $\mathbf{v} = \langle 0, -1, 1 \rangle$, and symmetric equations for the line are $x = 1$, $\dfrac{y - 2}{-1} = \dfrac{z}{1}$ or $x = 1$, $y - 2 = -z$.

58. If we set $z = 0$ then the equations of the planes reduce to $2x - y - 5 = 0$ and $4x + 3y - 5 = 0$ and solving these two

equations gives $x = 2$, $y = -1$. Thus a point on the line of intersection is $(2, -1, 0)$. A vector $\mathbf{v}$ in the

direction of this intersecting line is perpendicular to the normal vectors of both planes, so take

$\mathbf{v} = \mathbf{n}_1 \times \mathbf{n}_2 = \langle 2, -1, -1 \rangle \times \langle 4, 3, -1 \rangle = \langle 4, -2, 10 \rangle$ or equivalently we can take $\mathbf{v} = \langle 2, -1, 5 \rangle$. Symmetric equations for

the line are $\dfrac{x - 2}{2} = \dfrac{y + 1}{-1} = \dfrac{z}{5}$.

59. The distance from a point (x, y, z) to $(1, 0, -2)$ is $d_1 = \sqrt{(x - 1)^2 + y^2 + (z + 2)^2}$ and the distance from (x, y, z) to

$(3, 4, 0)$ is $\sqrt{(x - 3)^2 + (y - 4)^2 + z^2}$. The plane consists of all points (x, y, z) where $d_1 = d_2 \;\Rightarrow\; d_1^2 = d_2^2 \;\Leftrightarrow$

$(x - 1)^2 + y^2 + (z + 2)^2 = (x - 3)^2 + (y - 4)^2 + z^2 \;\Leftrightarrow$

$x^2 - 2x + y^2 + z^2 + 4z + 5 = x^2 - 6x + y^2 - 8y + z^2 + 25 \;\Leftrightarrow\; 4x + 8y + 4z = 20$ so an equation for the plane is

$4x + 8y + 4z = 20$ or equivalently $x + 2y + z = 5$.

Alternatively, you can argue that the segment joining points $(1, 0, -2)$ and $(3, 4, 0)$ is perpendicular to the plane and the plane

includes the midpoint of the segment.

60. The distance from a point (x, y, z) to $(2, 5, 5)$ is $d_1 = \sqrt{(x - 2)^2 + (y - 5)^2 + (z - 5)^2}$ and the distance from (x, y, z)

to $(-6, 3, 1)$ is $\sqrt{(x + 6)^2 + (y - 3)^2 + (z - 1)^2}$. The plane consists of all points (x, y, z) where $d_1 = d_2 \;\Rightarrow$

$d_1^2 = d_2^2 \;\Leftrightarrow\; (x - 2)^2 + (y - 5)^2 + (z - 5)^2 = (x + 6)^2 + (y - 3)^2 + (z - 1)^2 \;\Leftrightarrow$

$x^2 - 4x + y^2 - 10y + z^2 - 10z + 54 = x^2 + 12x + y^2 - 6y + z^2 - 2z + 46 \;\Leftrightarrow\; 16x + 4y + 8z = 8$ so an equation

for the plane is $16x + 4y + 8z = 8$ or equivalently $4x + y + 2z = 2$.

61. The plane contains the points $(a, 0, 0)$, $(0, b, 0)$ and $(0, 0, c)$. Thus the vectors $\mathbf{a} = \langle -a, b, 0 \rangle$ and $\mathbf{b} = \langle -a, 0, c \rangle$ lie in the

plane, and $\mathbf{n} = \mathbf{a} \times \mathbf{b} = \langle bc - 0, 0 + ac, 0 + ab \rangle = \langle bc, ac, ab \rangle$ is a normal vector to the plane. The equation of the plane is

therefore $bcx + acy + abz = abc + 0 + 0$ or $bcx + acy + abz = abc$. Notice that if $a \neq 0$, $b \neq 0$ and $c \neq 0$ then we can

rewrite the equation as $\dfrac{x}{a} + \dfrac{y}{b} + \dfrac{z}{c} = 1$. This is a good equation to remember!

62. (a) For the lines to intersect, we must be able to find one value of t and one value of s satisfying the three equations

$1 + t = 2 - s$, $1 - t = s$ and $2t = 2$. From the third we get $t = 1$, and putting this in the second gives $s = 0$. These values

of s and t do satisfy the first equation, so the lines intersect at the point $P_0 = (1 + 1, 1 - 1, 2(1)) = (2, 0, 2)$.

(b) The direction vectors of the lines are $\langle 1, -1, 2 \rangle$ and $\langle -1, 1, 0 \rangle$, so a normal vector for the plane is

$\langle -1, 1, 0 \rangle \times \langle 1, -1, 2 \rangle = \langle 2, 2, 0 \rangle$ and it contains the point $(2, 0, 2)$. Then the equation of the plane is

$2(x - 2) + 2(y - 0) + 0(z - 2) = 0 \;\Leftrightarrow\; x + y = 2$.

63. Two vectors which are perpendicular to the required line are the normal of the given plane, $\langle 1, 1, 1 \rangle$, and a direction vector for

the given line, $\langle 1, -1, 2 \rangle$. So a direction vector for the required line is $\langle 1, 1, 1 \rangle \times \langle 1, -1, 2 \rangle = \langle 3, -1, -2 \rangle$. Thus L is given

by $\langle x, y, z \rangle = \langle 0, 1, 2 \rangle + t \langle 3, -1, -2 \rangle$, or in parametric form, $x = 3t$, $y = 1 - t$, $z = 2 - 2t$.

64. Let L be the given line. Then $(1, 1, 0)$ is the point on L corresponding to $t = 0$. L is in the direction of $\mathbf{a} = \langle 1, -1, 2 \rangle$

and $\mathbf{b} = \langle -1, 0, 2 \rangle$ is the vector joining $(1, 1, 0)$ and $(0, 1, 2)$. Then

$$\mathbf{b} - \text{proj}_{\mathbf{a}}\,\mathbf{b} = \langle -1, 0, 2 \rangle - \frac{\langle 1, -1, 2 \rangle \cdot \langle -1, 0, 2 \rangle}{1^2 + (-1)^2 + 2^2}\langle 1, -1, 2 \rangle = \langle -1, 0, 2 \rangle - \tfrac{1}{2}\langle 1, -1, 2 \rangle = \langle -\tfrac{3}{2}, \tfrac{1}{2}, 1 \rangle \text{ is a direction vector}$$

for the required line. Thus $2\langle -\tfrac{3}{2}, \tfrac{1}{2}, 1 \rangle = \langle -3, 1, 2 \rangle$ is also a direction vector, and the line has parametric equations $x = -3t$,

$y = 1 + t$, $z = 2 + 2t$. (Notice that this is the same line as in Exercise 63.)

65. Let P_i have normal vector $\mathbf{n}_i$. Then $\mathbf{n}_1 = \langle 4, -2, 6 \rangle$, $\mathbf{n}_2 = \langle 4, -2, -2 \rangle$, $\mathbf{n}_3 = \langle -6, 3, -9 \rangle$, $\mathbf{n}_4 = \langle 2, -1, -1 \rangle$. Now

$\mathbf{n}_1 = -\tfrac{2}{3}\mathbf{n}_3$, so $\mathbf{n}_1$ and $\mathbf{n}_3$ are parallel, and hence P_1 and P_3 are parallel; similarly P_2 and P_4 are parallel because $\mathbf{n}_2 = 2\mathbf{n}_4$.

However, $\mathbf{n}_1$ and $\mathbf{n}_2$ are not parallel. $\left(0, 0, \tfrac{1}{2}\right)$ lies on P_1, but not on P_3, so they are not the same plane, but both P_2 and P_4

contain the point $(0, 0, -3)$, so these two planes are identical.

66. Let L_i have direction vector $\mathbf{v}_i$. Then $\mathbf{v}_1 = \langle 1, 1, -5 \rangle$, $\mathbf{v}_2 = \langle 1, 1, -1 \rangle$, $\mathbf{v}_3 = \langle 1, 1, -1 \rangle$, $\mathbf{v}_4 = \langle 2, 2, -10 \rangle$. $\mathbf{v}_2$ and $\mathbf{v}_3$ are

equal so they're parallel. $\mathbf{v}_4 = 2\mathbf{v}_1$, so L_4 and L_1 are parallel. L_3 contains the point $(1, 4, 1)$, but this point does not lie on L_2,

so they're not equal. $(2, 1, -3)$ lies on L_4, and on L_1, with $t = 1$. So L_1 and L_4 are identical.

67. Let $Q = (1, 3, 4)$ and $R = (2, 1, 1)$, points on the line corresponding to $t = 0$ and $t = 1$. Let

$P = (4, 1, -2)$. Then $\mathbf{a} = \overrightarrow{QR} = \langle 1, -2, -3 \rangle$, $\mathbf{b} = \overrightarrow{QP} = \langle 3, -2, -6 \rangle$. The distance is

$$d = \frac{|\mathbf{a} \times \mathbf{b}|}{|\mathbf{a}|} = \frac{|\langle 1, -2, -3 \rangle \times \langle 3, -2, -6 \rangle|}{|\langle 1, -2, -3 \rangle|} = \frac{|\langle 6, -3, 4 \rangle|}{|\langle 1, -2, -3 \rangle|} = \frac{\sqrt{6^2 + (-3)^2 + 4^2}}{\sqrt{1^2 + (-2)^2 + (-3)^2}} = \frac{\sqrt{61}}{\sqrt{14}} = \sqrt{\frac{64}{14}}.$$

68. Let $Q = (0, 6, 3)$ and $R = (2, 4, 4)$, points on the line corresponding to $t = 0$ and $t = 1$. Let

$P = (0, 1, 3)$. Then $\mathbf{a} = \overrightarrow{QR} = \langle 2, -2, 1 \rangle$ and $\mathbf{b} = \overrightarrow{QP} = \langle 0, -5, 0 \rangle$. The distance is

$$d = \frac{|\mathbf{a} \times \mathbf{b}|}{|\mathbf{a}|} = \frac{|\langle 2, -2, 1 \rangle \times \langle 0, -5, 0 \rangle|}{|\langle 2, -2, 1 \rangle|} = \frac{|\langle 5, 0, -10 \rangle|}{|\langle 2, -2, 1 \rangle|} = \frac{\sqrt{5^2 + 0^2 + (-10)^2}}{\sqrt{2^2 + (-2)^2 + 1^2}} = \frac{\sqrt{125}}{\sqrt{9}} = \frac{5\sqrt{5}}{3}.$$

69. By Equation 9, the distance is $D = \dfrac{|ax_1 + by_1 + cz_1 + d|}{\sqrt{a^2 + b^2 + c^2}} = \dfrac{|3(1) + 2(-2) + 6(4) - 5|}{\sqrt{3^2 + 2^2 + 6^2}} = \dfrac{|18|}{\sqrt{49}} = \dfrac{18}{7}$.

70. By Equation 9, the distance is $D = \dfrac{|1(-6) - 2(3) - 4(5) - 8|}{\sqrt{1^2 + (-2)^2 + (-4)^2}} = \dfrac{|-40|}{\sqrt{21}} = \dfrac{40}{\sqrt{21}}$.

71. Put $y = z = 0$ in the equation of the first plane to get the point $(2, 0, 0)$ on the plane. Because the planes are parallel, the

distance D between them is the distance from $(2, 0, 0)$ to the second plane. By Equation 9,

$$D = \frac{|4(2) - 6(0) + 2(0) - 3|}{\sqrt{4^2 + (-6)^2 + (2)^2}} = \frac{5}{\sqrt{56}} = \frac{5}{2\sqrt{14}} \text{ or } \frac{5\sqrt{14}}{28}.$$

72. Put $x = y = 0$ in the equation of the first plane to get the point $(0, 0, 0)$ on the plane. Because the planes are parallel the

distance D between them is the distance from $(0, 0, 0)$ to the second plane $3x - 6y + 9z - 1 = 0$. By Equation 9,

$$D = \frac{|3(0) - 6(0) + 9(0) - 1|}{\sqrt{3^2 + (-6)^2 + 9^2}} = \frac{1}{\sqrt{126}} = \frac{1}{3\sqrt{14}}.$$

73. The distance between two parallel planes is the same as the distance between a point on one of the planes and the other plane.

Let $P_0 = (x_0, y_0, z_0)$ be a point on the plane given by $ax + by + cz + d_1 = 0$. Then $ax_0 + by_0 + cz_0 + d_1 = 0$ and the distance between P_0 and the plane given by $ax + by + cz + d_2 = 0$ is, from Equation 9,

$$D = \frac{|ax_0 + by_0 + cz_0 + d_2|}{\sqrt{a^2 + b^2 + c^2}} = \frac{|-d_1 + d_2|}{\sqrt{a^2 + b^2 + c^2}} = \frac{|d_1 - d_2|}{\sqrt{a^2 + b^2 + c^2}}.$$

74. The planes must have parallel normal vectors, so if $ax + by + cz + d = 0$ is such a plane, then for some $t \neq 0$,

$\langle a, b, c \rangle = t \langle 1, 2, -2 \rangle = \langle t, 2t, -2t \rangle$. So this plane is given by the equation $x + 2y - 2z + k = 0$, where $k = d/t$. By

Exercise 73, the distance between the planes is $2 = \dfrac{|1 - k|}{\sqrt{1^2 + 2^2 + (-2)^2}}$ $\Leftrightarrow$ $6 = |1 - k|$ $\Leftrightarrow$ $k = 7$ or -5. So the

desired planes have equations $x + 2y - 2z = 7$ and $x + 2y - 2z = -5$.

75. L_1: $x = y = z$ $\Rightarrow$ $x = y$ **(1)**. L_2: $x + 1 = y/2 = z/3$ $\Rightarrow$ $x + 1 = y/2$ **(2)**. The solution of **(1)** and **(2)** is

$x = y = -2$. However, when $x = -2$, $x = z$ $\Rightarrow$ $z = -2$, but $x + 1 = z/3$ $\Rightarrow$ $z = -3$, a contradiction. Hence the

lines do not intersect. For L_1, $\mathbf{v}_1 = \langle 1, 1, 1 \rangle$, and for L_2, $\mathbf{v}_2 = \langle 1, 2, 3 \rangle$, so the lines are not parallel. Thus the lines are skew

lines. If two lines are skew, they can be viewed as lying in two parallel planes and so the distance between the skew lines

would be the same as the distance between these parallel planes. The common normal vector to the planes must be

perpendicular to both $\langle 1, 1, 1 \rangle$ and $\langle 1, 2, 3 \rangle$, the direction vectors of the two lines. So set

$\mathbf{n} = \langle 1, 1, 1 \rangle \times \langle 1, 2, 3 \rangle = \langle 3 - 2, -3 + 1, 2 - 1 \rangle = \langle 1, -2, 1 \rangle$. From above, we know that $(-2, -2, -2)$ and $(-2, -2, -3)$

are points of L_1 and L_2 respectively. So in the notation of Equation 8, $1(-2) - 2(-2) + 1(-2) + d_1 = 0$ $\Rightarrow$ $d_1 = 0$ and

$1(-2) - 2(-2) + 1(-3) + d_2 = 0$ $\Rightarrow$ $d_2 = 1$.

By Exercise 73, the distance between these two skew lines is $D = \dfrac{|0 - 1|}{\sqrt{1 + 4 + 1}} = \dfrac{1}{\sqrt{6}}$.

Alternate solution (without reference to planes): A vector which is perpendicular to both of the lines is

$\mathbf{n} = \langle 1, 1, 1 \rangle \times \langle 1, 2, 3 \rangle = \langle 1, -2, 1 \rangle$. Pick any point on each of the lines, say $(-2, -2, -2)$ and $(-2, -2, -3)$, and form the

vector $\mathbf{b} = \langle 0, 0, 1 \rangle$ connecting the two points. The distance between the two skew lines is the absolute value of the scalar

projection of $\mathbf{b}$ along $\mathbf{n}$, that is, $D = \dfrac{|\mathbf{n} \cdot \mathbf{b}|}{|\mathbf{n}|} = \dfrac{|1 \cdot 0 - 2 \cdot 0 + 1 \cdot 1|}{\sqrt{1 + 4 + 1}} = \dfrac{1}{\sqrt{6}}$.

76. First notice that if two lines are skew, they can be viewed as lying in two parallel planes and so the distance between the skew

lines would be the same as the distance between these parallel planes. The common normal vector to the planes must be

perpendicular to both $\mathbf{v}_1 = \langle 1, 6, 2 \rangle$ and $\mathbf{v}_2 = \langle 2, 15, 6 \rangle$, the direction vectors of the two lines respectively. Thus set

$\mathbf{n} = \mathbf{v}_1 \times \mathbf{v}_2 = \langle 36 - 30, 4 - 6, 15 - 12 \rangle = \langle 6, -2, 3 \rangle$. Setting $t = 0$ and $s = 0$ gives the points $(1, 1, 0)$ and $(1, 5, -2)$.

So in the notation of Equation 8, $6 - 2 + 0 + d_1 = 0$ $\Rightarrow$ $d_1 = -4$ and $6 - 10 - 6 + d_2 = 0$ $\Rightarrow$ $d_2 = 10$.

Then by Exercise 73, the distance between the two skew lines is given by $D = \dfrac{|-4 - 10|}{\sqrt{36 + 4 + 9}} = \dfrac{14}{7} = 2$.

Alternate solution (without reference to planes): We already know that the direction vectors of the two lines are

$\mathbf{v}_1 = \langle 1, 6, 2 \rangle$ and $\mathbf{v}_2 = \langle 2, 15, 6 \rangle$. Then $\mathbf{n} = \mathbf{v}_1 \times \mathbf{v}_2 = \langle 6, -2, 3 \rangle$ is perpendicular to both lines. Pick any point on each of the lines, say $(1, 1, 0)$ and $(1, 5, -2)$, and form the vector $\mathbf{b} = \langle 0, 4, -2 \rangle$ connecting the two points. Then the distance between the two skew lines is the absolute value of the scalar projection of $\mathbf{b}$ along $\mathbf{n}$, that is,

$$D = \frac{|\mathbf{n} \cdot \mathbf{b}|}{|\mathbf{n}|} = \frac{1}{\sqrt{36 + 4 + 9}} |0 - 8 - 6| = \frac{14}{7} = 2.$$

77. If $a \neq 0$, then $ax + by + cz + d = 0 \Rightarrow a(x + d/a) + b(y - 0) + c(z - 0) = 0$ which by (7) is the scalar equation of the plane through the point $(-d/a, 0, 0)$ with normal vector $\langle a, b, c \rangle$. Similarly, if $b \neq 0$ (or if $c \neq 0$) the equation of the plane can be rewritten as $a(x - 0) + b(y + d/b) + c(z - 0) = 0$ [or as $a(x - 0) + b(y - 0) + c(z + d/c) = 0$] which by (7) is the scalar equation of a plane through the point $(0, -d/b, 0)$ [or the point $(0, 0, -d/c)$] with normal vector $\langle a, b, c \rangle$.

78. (a) The planes $x + y + z = c$ have normal vector $\langle 1, 1, 1 \rangle$, so they are all parallel. Their x-, y-, and z-intercepts are all c. When $c > 0$ their intersection with the first octant is an equilateral triangle and when $c < 0$ their intersection with the octant diagonally opposite the first is an equilateral triangle.

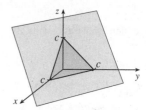

(b) The planes $x + y + cz = 1$ have x-intercept 1, y-intercept 1, and z-intercept $1/c$. The plane with $c = 0$ is parallel to the z-axis. As c gets larger, the planes get closer to the xy-plane.

(c) The planes $y \cos \theta + z \cos \theta = 1$ have normal vectors $\langle 0, \cos \theta, \sin \theta \rangle$, which are perpendicular to the x-axis, and so the planes are parallel to the x-axis. We look at their intersection with the yz-plane. These are lines that are perpendicular to $\langle \cos \theta, \sin \theta \rangle$ and pass through $(\cos \theta, \sin \theta)$, since $\cos^2 \theta + \sin^2 \theta = 1$. So these are the tangent lines to the unit circle. Thus the family consists of all planes tangent to the circular cylinder with radius 1 and axis the x-axis.

LABORATORY PROJECT   Putting 3D in Perspective

1. If we view the screen from the camera's location, the vertical clipping plane on the left passes through the points $(1000, 0, 0)$, $(0, -400, 0)$, and $(0, -400, 600)$. A vector from the first point to the second is $\mathbf{v}_1 = \langle -1000, -400, 0 \rangle$ and a vector from the first point to the third is $\mathbf{v}_2 = \langle -1000, -400, 600 \rangle$. A normal vector for the clipping plane is $\mathbf{v}_1 \times \mathbf{v}_2 = -240{,}000\,\mathbf{i} + 600{,}000\,\mathbf{j}$ or $-2\,\mathbf{i} + 5\,\mathbf{j}$, and an equation for the plane is $-2(x - 1000) + 5(y - 0) + 0(z - 0) = 0 \Rightarrow 2x - 5y = 2000$. By symmetry, the vertical clipping plane on the right is given by $2x + 5y = 2000$. The lower clipping plane is $z = 0$. The upper clipping plane passes through the points $(1000, 0, 0)$, $(0, -400, 600)$, and $(0, 400, 600)$. Vectors from the first point to the second and third points are $\mathbf{v}_1 = \langle -1000, -400, 600 \rangle$ and $\mathbf{v}_2 = \langle -1000, 400, 600 \rangle$, and a normal vector for the plane is $\mathbf{v}_1 \times \mathbf{v}_2 = -480{,}000\,\mathbf{i} - 800{,}000\,\mathbf{k}$ or $3\,\mathbf{i} + 5\,\mathbf{k}$. An equation for the plane is $3(x - 1000) + 0(y - 0) + 5(z - 0) = 0 \Rightarrow 3x + 5z = 3000$.

A direction vector for the line L is $\mathbf{v} = \langle 630, 390, 162 \rangle$ and taking $P_0 = (230, -285, 102)$, parametric equations are $x = 230 + 630t$, $y = -285 + 390t$, $z = 102 + 162t$. L intersects the left clipping plane when $2(230 + 630t) - 5(-285 + 390t) = 2000 \Rightarrow t = -\frac{1}{6}$. The corresponding point is $(125, -350, 75)$. L intersects the right clipping plane when $2(230 + 630t) + 5(-285 + 390t) = 2000$ ⟹ $t = \frac{593}{642}$. The corresponding point is

approximately $(811.9, 75.2, 251.6)$, but this point is not contained within the viewing volume. L intersects the upper clipping

plane when $3(230 + 630t) + 5(102 + 162t) = 3000$ ⟹ $t = \frac{2}{3}$, corresponding to the point $(650, -25, 210)$, and L

intersects the lower clipping plane when $z = 0$ ⟹ $102 + 162t = 0$ ⟹ $t = -\frac{17}{27}$. The corresponding point is

approximately $(-166.7, -530.6, 0)$, which is not contained within the viewing volume. Thus L should be clipped at the

points $(125, -350, 75)$ and $(650, -25, 210)$.

2. A sight line from the camera at $(1000, 0, 0)$ to the left endpoint $(125, -350, 75)$ of the clipped line has direction

$\mathbf{v} = \langle -875, -350, 75 \rangle$. Parametric equations are $x = 1000 - 875t$, $y = -350t$, $z = 75t$. This line intersects the screen

when $x = 0$ ⟹ $1000 - 875t = 0$ ⟹ $t = \frac{8}{7}$, corresponding to the point $\left(0, -400, \frac{600}{7}\right)$. Similarly, a sight line from

the camera to the right endpoint $(650, -25, 210)$ of the clipped line has direction $\langle -350, -25, 210 \rangle$ and parametric equations

are $x = 1000 - 350t$, $y = -25t$, $z = 210t$. $x = 0$ ⟹ $1000 - 350t = 0$ ⟹ $t = \frac{20}{7}$, corresponding to the point

$\left(0, -\frac{500}{7}, 600\right)$. Thus the projection of the clipped line is the line segment between the points $\left(0, -400, \frac{600}{7}\right)$ and

$\left(0, -\frac{500}{7}, 600\right)$.

3. From Equation 13.5.4 [ET 12.5.4], equations for the four sides of

the screen are $\mathbf{r}_1(t) = (1 - t)\langle 0, -400, 0 \rangle + t \langle 0, -400, 600 \rangle$,

$\mathbf{r}_2(t) = (1 - t)\langle 0, -400, 600 \rangle + t \langle 0, 400, 600 \rangle$,

$\mathbf{r}_3(t) = (1 - t)\langle 0, 400, 0 \rangle + t \langle 0, 400, 600 \rangle$, and

$\mathbf{r}_4(t) = (1 - t)\langle 0, -400, 0 \rangle + t \langle 0, 400, 0 \rangle$. The clipped line

segment connects the points $(125, -350, 75)$ and

$(650, -25, 210)$, so an equation for the segment is

$\mathbf{r}_5(t) = (1 - t)\langle 125, -350, 75 \rangle + t \langle 650, -25, 210 \rangle$.

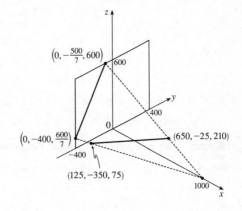

The projection of the clipped segment connects the points

$\left(0, -400, \frac{600}{7}\right)$ and $\left(0, -\frac{500}{7}, 600\right)$, so an equation is $\mathbf{r}_6(t) = (1 - t)\left\langle 0, -400, \frac{600}{7}\right\rangle + t \left\langle 0, -\frac{500}{7}, 600\right\rangle$.

The sight line on the left connects the points $(1000, 0, 0)$ and $\left(0, -400, \frac{600}{7}\right)$, so an equation is

$\mathbf{r}_7(t) = (1 - t)\langle 1000, 0, 0 \rangle + t \left\langle 0, -400, \frac{600}{7}\right\rangle$. The other sight line connects $(1000, 0, 0)$ to $\left(0, -\frac{500}{7}, 600\right)$, so an equation

is $\mathbf{r}_8(t) = (1 - t)\langle 1000, 0, 0 \rangle + t \left\langle 0, -\frac{500}{7}, 600\right\rangle$.

4. The vector from $(621, -147, 206)$ to $(563, 31, 242)$, $\mathbf{v}_1 = \langle -58, 178, 36 \rangle$, lies in the plane of the rectangle, as does the

vector from $(621, -147, 206)$ to $(657, -111, 86)$, $\mathbf{v}_2 = \langle 36, 36, -120 \rangle$. A normal vector for the plane is

$\mathbf{v}_1 \times \mathbf{v}_2 = \langle -1888, -142, -708 \rangle$ or $\langle 8, 2, 3 \rangle$, and an equation of the plane is $8x + 2y + 3z = 5292$. The line L intersects

this plane when $8(230 + 630t) + 2(-285 + 390t) + 3(102 + 162t) = 5292$ ⟹ $t = \frac{1858}{3153} \approx 0.589$. The corresponding point is approximately $(601.25, -55.18, 197.46)$. Starting at this point, a portion of the line is hidden behind the rectangle. The line becomes visible again at the left edge of the rectangle, specifically the edge between the points $(621, -147, 206)$ and $(657, -111, 86)$. (This is most easily determined by graphing the rectangle and the line.) A plane through these two points and the camera's location, $(1000, 0, 0)$, will clip the line at the point it becomes visible. Two vectors in this plane are $\mathbf{v}_1 = \langle -379, -147, 206 \rangle$ and $\mathbf{v}_2 = \langle -343, -111, 86 \rangle$. A normal vector for the plane is $\mathbf{v}_1 \times \mathbf{v}_2 = \langle 10224, -38064, -8352 \rangle$ and an equation of the plane is $213x - 793y - 174z = 213{,}000$. L intersects this plane when $213(230 + 630t) - 793(-285 + 390t) - 174(102 + 162t) = 213{,}000 \Rightarrow t = \frac{44{,}247}{203{,}268} \approx 0.2177$. The corresponding point is approximately $(367.14, -200.11, 137.26)$. Thus the portion of L that should be removed is the segment between the points $(601.25, -55.18, 197.46)$ and $(367.14, -200.11, 137.26)$.

13.6 Cylinders and Quadric Surfaces

1. (a) In $\mathbb{R}^2$, the equation $y = x^2$ represents a parabola.

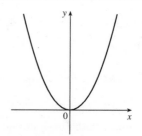

(b) In $\mathbb{R}^3$, the equation $y = x^2$ doesn't involve z, so any horizontal plane with equation $z = k$ intersects the graph in a curve with equation $y = x^2$. Thus, the surface is a parabolic cylinder, made up of infinitely many shifted copies of the same parabola. The rulings are parallel to the z-axis.

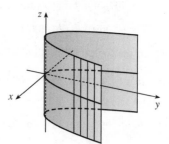

(c) In $\mathbb{R}^3$, the equation $z = y^2$ also represents a parabolic cylinder. Since x doesn't appear, the graph is formed by moving the parabola $z = y^2$ in the direction of the x-axis. Thus, the rulings of the cylinder are parallel to the x-axis.

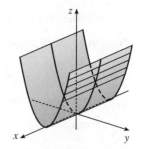

2. (a)

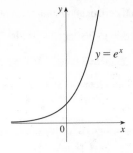

(b) Since the equation $y = e^x$ doesn't involve z, horizontal traces are copies of the curve $y = e^x$. The rulings are parallel to the z-axis.

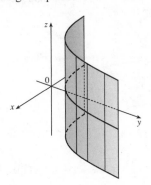

(c) The equation $z = e^y$ doesn't involve x, so vertical traces in $x = k$ (parallel to the yz-plane) are copies of the curve $z = e^y$. The rulings are parallel to the x-axis.

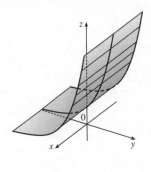

3. Since x is missing from the equation, the vertical traces $y^2 + 4z^2 = 4$, $x = k$, are copies of the same ellipse in the plane $x = k$. Thus, the surface $y^2 + 4z^2 = 4$ is an elliptic cylinder with rulings parallel to the x-axis.

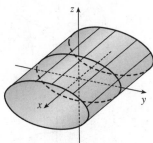

4. Since y is missing from the equation, each vertical trace $z = 4 - x^2$, $y = k$, is a copy of the same parabola in the plane $y = k$. Thus, the surface $z = 4 - x^2$ is a parabolic cylinder with rulings parallel to the y-axis.

5. Since z is missing, each horizontal trace $x = y^2$, $z = k$, is a copy of the same parabola in the plane $z = k$. Thus, the surface $x - y^2 = 0$ is a parabolic cylinder with rulings parallel to the z-axis.

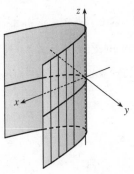

6. Since x is missing, each vertical trace $yz = 4$, $x = k$ is a copy of the same hyperbola in the plane $x = k$. Thus, the surface $yz = 4$ is a hyperbolic cylinder with rulings parallel to the x-axis.

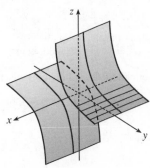

7. Since y is missing, each vertical trace $z = \cos x$, $y = k$ is a copy of a cosine curve in the plane $y = k$. Thus, the surface $z = \cos x$ is a cylindrical surface with rulings parallel to the y-axis.

8. Since z is missing, each horizontal trace $x^2 - y^2 = 1$, $z = k$ is a copy of the same hyperbola in the plane $z = k$. Thus, the surface $x^2 - y^2 = 1$ is a hyperbolic cylinder with rulings parallel to the z-axis.

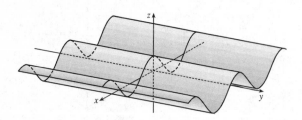

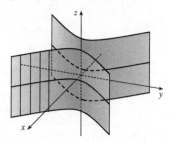

9. (a) The traces of $x^2 + y^2 - z^2 = 1$ in $x = k$ are $y^2 - z^2 = 1 - k^2$, a family of hyperbolas. (Note that the hyperbolas are oriented differently for $-1 < k < 1$ than for $k < -1$ or $k > 1$.) The traces in $y = k$ are $x^2 - z^2 = 1 - k^2$, a similar family of hyperbolas. The traces in $z = k$ are $x^2 + y^2 = 1 + k^2$, a family of circles. For $k = 0$, the trace in the xy-plane, the circle is of radius 1. As $|k|$ increases, so does the radius of the circle. This behavior, combined with the hyperbolic vertical traces, gives the graph of the hyperboloid of one sheet in Table 1.

(b) The shape of the surface is unchanged, but the hyperboloid is rotated so that its axis is the y-axis. Traces in $y = k$ are circles, while traces in $x = k$ and $z = k$ are hyperbolas.

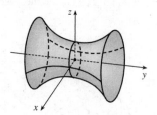

(c) Completing the square in y gives $x^2 + (y + 1)^2 - z^2 = 1$. The surface is a hyperboloid identical to the one in part (a) but shifted one unit in the negative y-direction.

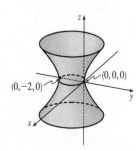

10. (a) The traces of $-x^2 - y^2 + z^2 = 1$ in $x = k$ are $-y^2 + z^2 = 1 + k^2$, a family of hyperbolas, as are the traces in $y = k$, $-x^2 + z^2 = 1 + k^2$. The traces in $z = k$ are $x^2 + y^2 = k^2 - 1$, a family of circles for $|k| > 1$. As $|k|$ increases, the radii of the circles increase; the traces are empty for $|k| < 1$. This behavior, combined with the vertical traces, gives the graph of the hyperboloid of two sheets in Table 1.

(b) The graph has the same shape as the hyperboloid in part (a) but is rotated so that its axis is the x-axis. Traces in $x = k$, $|k| > 1$, are circles, while traces in $y = k$ and $z = k$ are hyperbolas.

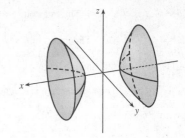

11. For $x = y^2 + 4z^2$, the traces in $x = k$ are $y^2 + 4z^2 = k$. When $k > 0$ we have a family of ellipses. When $k = 0$ we have just a point at the origin, and the trace is empty for $k < 0$. The traces in $y = k$ are $x = 4z^2 + k^2$, a family of parabolas opening in the positive x-direction. Similarly, the traces in $z = k$ are $x = y^2 + 4k^2$, a family of parabolas opening in the positive x-direction. We recognize the graph as an elliptic paraboloid with axis the x-axis and vertex the origin.

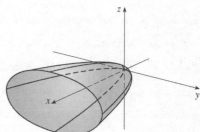

12. $9x^2 - y^2 + z^2 = 0$. The traces in $x = k$ are $y^2 - z^2 = 9k^2$, a family of hyperbolas if $k \neq 0$ and two intersecting lines if $k = 0$. The traces in $y = k$ are $9x^2 + z^2 = k^2$, $k \geq 0$, a family of ellipses; the traces in $z = k$ are $y^2 - 9x^2 = k^2$, a family of hyperbolas for $k \neq 0$ and two intersecting lines for $k = 0$. We recognize the graph as an elliptic cone with axis the y-axis and vertex the origin.

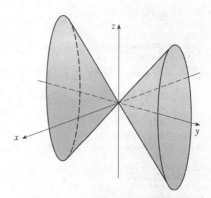

13. $x^2 = y^2 + 4z^2$. The traces in $x = k$ are the ellipses $y^2 + 4z^2 = k^2$. The traces in $y = k$ are $x^2 - 4z^2 = k^2$, hyperbolas for $k \neq 0$ and two intersecting lines if $k = 0$. Similarly, the traces in $z = k$ are $x^2 - y^2 = 4k^2$, hyperbolas for $k \neq 0$ and two intersecting lines if $k = 0$. We recognize the graph as an elliptic cone with axis the x-axis and vertex the origin.

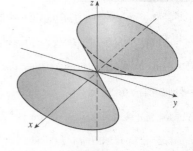

14. $25x^2 + 4y^2 + z^2 = 100$. The traces in $x = k$ are $4y^2 + z^2 = 100 - 25k^2$, a family of ellipses for $|k| < 2$. (The traces are a single point for $|k| = 2$ and are empty for $|k| > 2$.) Similarly, the traces in $y = k$ are the ellipses $25x^2 + z^2 = 100 - 4k^2$, $|k| < 5$, and the traces in $z = k$ are the ellipses $25x^2 + 4y^2 = 100 - k^2$, $|k| < 10$. The graph is an ellipsoid centered at the origin with intercepts $x = \pm 2$, $y = \pm 5$, $z = \pm 10$.

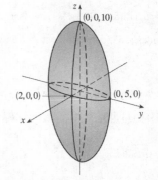

15. $-x^2 + 4y^2 - z^2 = 4$. The traces in $x = k$ are the hyperbolas $4y^2 - z^2 = 4 + k^2$. The traces in $y = k$ are $x^2 + z^2 = 4k^2 - 4$, a family of circles for $|k| > 1$, and the traces in $z = k$ are $4y^2 - x^2 = 4 + k^2$, a family of hyperbolas. Thus the surface is a hyperboloid of two sheets with axis the y-axis.

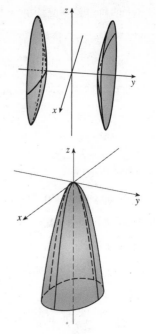

16. $4x^2 + 9y^2 + z = 0$. The traces in $x = k$ are the parabolas $z = -9y^2 - 4k^2$ which open downward. Similarly, the traces in $y = k$ are the parabolas $z = -4x^2 - 9k^2$, also opening downward, and the traces in $z = k$ are $4x^2 + 9y^2 = -k$, $k \leq 0$, a family of ellipses. The graph is an elliptic paraboloid with axis the z-axis, opening downward, and vertex the origin.

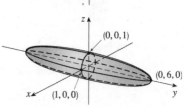

17. $36x^2 + y^2 + 36z^2 = 36$. The traces in $x = k$ are $y^2 + 36z^2 = 36(1 - k^2)$, a family of ellipses for $|k| < 1$. (The traces are a single point for $|k| = 1$ and are empty for $|k| > 1$.) The traces in $y = k$ are the circles $36x^2 + 36z^2 = 36 - k^2 \iff x^2 + z^2 = 1 - \frac{1}{36}k^2$, $|k| < 6$, and the traces in $z = k$ are the ellipses $36x^2 + y^2 = 36(1 - k^2)$, $|k| < 1$. The graph is an ellipsoid centered at the origin with intercepts $x = \pm 1$, $y = \pm 6$, $z = \pm 1$.

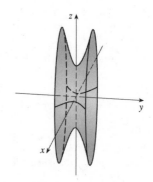

18. $4x^2 - 16y^2 + z^2 = 16$. The traces in $x = k$ are $z^2 - 16y^2 = 16 - 4k^2$, a family of hyperbolas for $|k| \neq 2$ and two intersecting lines when $|k| = 2$. (Note that the hyperbolas are oriented differently for $|k| < 2$ than for $|k| > 2$.) The traces in $y = k$ are $4x^2 + z^2 = 16(1 + k^2)$, a family of ellipses, and the traces in $z = k$ are $4x^2 - 16y^2 = 16 - k^2$, two intersecting lines when $|k| = 4$ and a family of hyperbolas when $|k| \neq 4$ (oriented differently for $|k| < 4$ than for $|k| > 4$). We recognize the graph as a hyperboloid of one sheet with axis the y-axis.

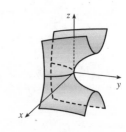

19. $y = z^2 - x^2$. The traces in $x = k$ are the parabolas $y = z^2 - k^2$; the traces in $y = k$ are $k = z^2 - x^2$, which are hyperbolas (note the hyperbolas are oriented differently for $k > 0$ than for $k < 0$); and the traces in $z = k$ are the parabolas $y = k^2 - x^2$. Thus, $\dfrac{y}{1} = \dfrac{z^2}{1^2} - \dfrac{x^2}{1^2}$ is a hyperbolic paraboloid.

20. $x = y^2 - z^2$. The traces in $x = k$ are $y^2 - z^2 = k$, two intersecting lines

when $k = 0$ and a family of hyperbolas for $k \neq 0$ (oriented differently for

$k > 0$ than for $k < 0$). The traces in $y = k$ are the parabolas

$x = -z^2 + k^2$, opening in the negative x-direction, and the traces in $z = k$

are the parabolas $x = y^2 - k^2$ which open in the positive x-direction. The

graph is a hyperbolic paraboloid with saddle point $(0,0,0)$.

21. This is the equation of an ellipsoid: $x^2 + 4y^2 + 9z^2 = x^2 + \dfrac{y^2}{(1/2)^2} + \dfrac{z^2}{(1/3)^2} = 1$, with x-intercepts ± 1, y-intercepts $\pm\frac{1}{2}$

and z-intercepts $\pm\frac{1}{3}$. So the major axis is the x-axis and the only possible graph is VII.

22. This is the equation of an ellipsoid: $9x^2 + 4y^2 + z^2 = \dfrac{x^2}{(1/3)^2} + \dfrac{y^2}{(1/2)^2} + z^2 = 1$, with x-intercepts $\pm\frac{1}{3}$, y-intercepts $\pm\frac{1}{2}$

and z-intercepts ± 1. So the major axis is the z-axis and the only possible graph is IV.

23. This is the equation of a hyperboloid of one sheet, with $a = b = c = 1$. Since the coefficient of y^2 is negative, the axis of the

hyperboloid is the y-axis, hence the correct graph is II.

24. This is a hyperboloid of two sheets, with $a = b = c = 1$. This surface does not intersect the xz-plane at all, so the axis of the

hyperboloid is the y-axis and the graph is III.

25. There are no real values of x and z that satisfy this equation for $y < 0$, so this surface does not extend to the left of the

xz-plane. The surface intersects the plane $y = k > 0$ in an ellipse. Notice that y occurs to the first power whereas x and z

occur to the second power. So the surface is an elliptic paraboloid with axis the y-axis. Its graph is VI.

26. This is the equation of a cone with axis the y-axis, so the graph is I.

27. This surface is a cylinder because the variable y is missing from the equation. The intersection of the surface and the xz-plane

is an ellipse. So the graph is VIII.

28. This is the equation of a hyperbolic paraboloid. The trace in the xy-plane is the parabola $y = x^2$. So the correct graph is V.

29. $z^2 = 4x^2 + 9y^2 + 36$ or $-4x^2 - 9y^2 + z^2 = 36$ or

$-\dfrac{x^2}{9} - \dfrac{y^2}{4} + \dfrac{z^2}{36} = 1$ represents a hyperboloid of two

sheets with axis the z-axis.

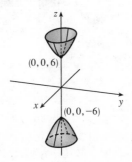

30. $x^2 = 2y^2 + 3z^2$ or $x^2 = \dfrac{y^2}{1/2} + \dfrac{z^2}{1/3}$ or $\dfrac{x^2}{6} = \dfrac{y^2}{3} + \dfrac{z^2}{2}$

represents an elliptic cone with vertex $(0,0,0)$ and axis

the x-axis.

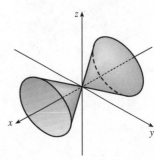

31. $x = 2y^2 + 3z^2$ or $x = \dfrac{y^2}{1/2} + \dfrac{z^2}{1/3}$ or $\dfrac{x}{6} = \dfrac{y^2}{3} + \dfrac{z^2}{2}$

represents an elliptic paraboloid with vertex $(0,0,0)$ and

axis the x-axis.

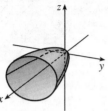

32. $4x - y^2 + 4z^2 = 0$ or $4x = y^2 - 4z^2$ or $x = \dfrac{y^2}{4} - z^2$

represents a hyperbolic paraboloid with center $(0,0,0)$.

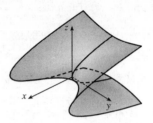

33. Completing squares in y and z gives

$4x^2 + (y-2)^2 + 4(z-3)^2 = 4$ or

$x^2 + \dfrac{(y-2)^2}{4} + (z-3)^2 = 1$, an ellipsoid with

center $(0,2,3)$.

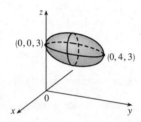

34. Completing squares in y and z gives

$4(y-2)^2 + (z-2)^2 - x = 0$ or

$\dfrac{x}{4} = (y-2)^2 + \dfrac{(z-2)^2}{4}$, an elliptic paraboloid with

vertex $(0,2,2)$ and axis the horizontal line $y = 2$, $z = 2$.

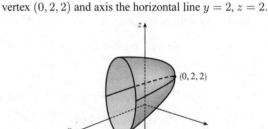

35. Completing squares in all three variables gives

$(x-2)^2 - (y+1)^2 + (z-1)^2 = 0$ or

$(y+1)^2 = (x-2)^2 + (z-1)^2$, a circular cone with

center $(2,-1,1)$ and axis the horizontal line $x = 2$,

$z = 1$.

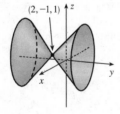

36. Completing squares in all three variables gives

$(x-1)^2 - (y-1)^2 + (z+2)^2 = 2$ or

$\dfrac{(x-1)^2}{2} - \dfrac{(y-1)^2}{2} + \dfrac{(z+2)^2}{2} = 1$, a hyperboloid of

one sheet with center $(1,1,-2)$ and axis the horizontal

line $x = 1$, $z = -2$.

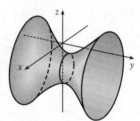

37. Solving the equation for z we get $z = \pm\sqrt{1 + 4x^2 + y^2}$, so we plot separately $z = \sqrt{1 + 4x^2 + y^2}$ and

$z = -\sqrt{1 + 4x^2 + y^2}$.

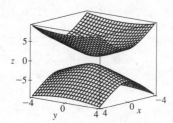 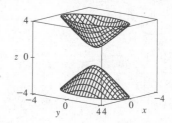

To restrict the z-range as in the second graph, we can use the option `view = -4..4` in Maple's `plot3d` command, or

`PlotRange -> {-4,4}` in Mathematica's `Plot3D` command.

38. We plot the surface $z = x^2 - y^2$.

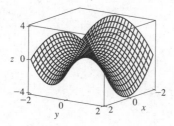

39. Solving the equation for z we get $z = \pm\sqrt{4x^2 + y^2}$, so we plot separately $z = \sqrt{4x^2 + y^2}$ and $z = -\sqrt{4x^2 + y^2}$.

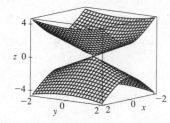 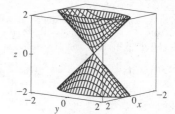

40. We plot the surface $z = x^2 - 6x + 4y^2$.

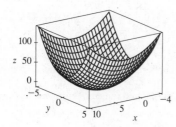 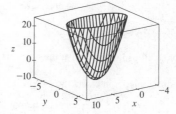

41.

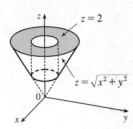

42.

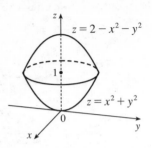

43. The surface is a paraboloid of revolution (circular paraboloid) with vertex at the origin, axis the y-axis and opens to the right. Thus the trace in the yz-plane is also a parabola: $y = z^2$, $x = 0$. The equation is $y = x^2 + z^2$.

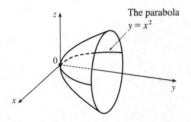

44. The surface is a right circular cone with vertex at $(0, 0, 0)$ and axis the x-axis. For $x = k \neq 0$, the trace is a circle with center $(k, 0, 0)$ and radius $r = y = \dfrac{x}{3} = \dfrac{k}{3}$. Thus the equation is $(x/3)^2 = y^2 + z^2$ or $x^2 = 9y^2 + 9z^2$.

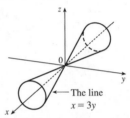

45. Let $P = (x, y, z)$ be an arbitrary point equidistant from $(-1, 0, 0)$ and the plane $x = 1$. Then the distance from P to $(-1, 0, 0)$ is $\sqrt{(x+1)^2 + y^2 + z^2}$ and the distance from P to the plane $x = 1$ is $|x - 1| / \sqrt{1^2} = |x - 1|$ (by Equation 13.5.9 [ET 12.5.9]). So $|x - 1| = \sqrt{(x+1)^2 + y^2 + z^2}$ $\Leftrightarrow$ $(x - 1)^2 = (x + 1)^2 + y^2 + z^2$ $\Leftrightarrow$ $x^2 - 2x + 1 = x^2 + 2x + 1 + y^2 + z^2$ $\Leftrightarrow$ $-4x = y^2 + z^2$. Thus the collection of all such points P is a circular paraboloid with vertex at the origin, axis the x-axis, which opens in the negative direction.

46. Let $P = (x, y, z)$ be an arbitrary point whose distance from the x-axis is twice its distance from the yz-plane. The distance from P to the x-axis is $\sqrt{(x - x)^2 + y^2 + z^2} = \sqrt{y^2 + z^2}$ and the distance from P to the yz-plane ($x = 0$) is $|x| / 1 = |x|$. Thus $\sqrt{y^2 + z^2} = 2|x|$ $\Leftrightarrow$ $y^2 + z^2 = 4x^2$ $\Leftrightarrow$ $x^2 = (y^2/2^2) + (z^2/2^2)$. So the surface is a right circular cone with vertex the origin and axis the x-axis.

47. (a) An equation for an ellipsoid centered at the origin with intercepts $x = \pm a$, $y = \pm b$, and $z = \pm c$ is $\dfrac{x^2}{a^2} + \dfrac{y^2}{b^2} + \dfrac{z^2}{c^2} = 1$.

Here the poles of the model intersect the z-axis at $z = \pm 6356.523$ and the equator intersects the x- and y-axes at $x = \pm 6378.137$, $y = \pm 6378.137$, so an equation is

$$\frac{x^2}{(6378.137)^2} + \frac{y^2}{(6378.137)^2} + \frac{z^2}{(6356.523)^2} = 1$$

(b) Traces in $z = k$ are the circles $\dfrac{x^2}{(6378.137)^2} + \dfrac{y^2}{(6378.137)^2} = 1 - \dfrac{k^2}{(6356.523)^2}$ $\Leftrightarrow$

$$x^2 + y^2 = (6378.137)^2 - \left(\frac{6378.137}{6356.523}\right)^2 k^2.$$

(c) To identify the traces in $y = mx$ we substitute $y = mx$ into the equation of the ellipsoid:

$$\frac{x^2}{(6378.137)^2} + \frac{(mx)^2}{(6378.137)^2} + \frac{z^2}{(6356.523)^2} = 1$$

$$\frac{(1+m^2)x^2}{(6378.137)^2} + \frac{z^2}{(6356.523)^2} = 1$$

$$\frac{x^2}{(6378.137)^2/(1+m^2)} + \frac{z^2}{(6356.523)^2} = 1$$

As expected, this is a family of ellipses.

48. If we position the hyperboloid on coordinate axes so that it is centered at the origin with axis the z-axis then its equation is

given by $\dfrac{x^2}{a^2} + \dfrac{y^2}{b^2} - \dfrac{z^2}{c^2} = 1$. Horizontal traces in $z = k$ are $\dfrac{x^2}{a^2} + \dfrac{y^2}{b^2} = 1 + \dfrac{k^2}{c^2}$, a family of ellipses, but we know that the

traces are circles so we must have $a = b$. The trace in $z = 0$ is $\dfrac{x^2}{a^2} + \dfrac{y^2}{a^2} = 1 \iff x^2 + y^2 = a^2$ and since the minimum

radius of 100 m occurs there, we must have $a = 100$. The base of the tower is the trace in $z = -500$ given by

$\dfrac{x^2}{a^2} + \dfrac{y^2}{a^2} = 1 + \dfrac{(-500)^2}{c^2}$ but $a = 100$ so the trace is $x^2 + y^2 = 100^2 + 50{,}000^2 \dfrac{1}{c^2}$. We know the base is a circle of

radius 140, so we must have $100^2 + 50{,}000^2 \dfrac{1}{c^2} = 140^2 \implies c^2 = \dfrac{50{,}000^2}{140^2 - 100^2} = \dfrac{781{,}250}{3}$ and an equation for the

tower is $\dfrac{x^2}{100^2} + \dfrac{y^2}{100^2} - \dfrac{z^2}{(781{,}250)/3} = 1$ or $\dfrac{x^2}{10{,}000} + \dfrac{y^2}{10{,}000} - \dfrac{3z^2}{781{,}250} = 1$, $-500 \le z \le 500$.

49. If (a, b, c) satisfies $z = y^2 - x^2$, then $c = b^2 - a^2$. $L_1: x = a + t, \; y = b + t, \; z = c + 2(b - a)t$,

$L_2: x = a + t, \; y = b - t, \; z = c - 2(b + a)t$. Substitute the parametric equations of L_1 into the equation

of the hyperbolic paraboloid in order to find the points of intersection: $z = y^2 - x^2 \implies$

$c + 2(b - a)t = (b + t)^2 - (a + t)^2 = b^2 - a^2 + 2(b - a)t \implies c = b^2 - a^2$. As this is true for all values of t,

L_1 lies on $z = y^2 - x^2$. Performing similar operations with L_2 gives: $z = y^2 - x^2 \implies$

$c - 2(b + a)t = (b - t)^2 - (a + t)^2 = b^2 - a^2 - 2(b + a)t \implies c = b^2 - a^2$. This tells us that all of L_2 also lies on

$z = y^2 - x^2$.

50. Any point on the curve of intersection must satisfy both $2x^2 + 4y^2 - 2z^2 + 6x = 2$ and $2x^2 + 4y^2 - 2z^2 - 5y = 0$.

Subtracting, we get $6x + 5y = 2$, which is linear and therefore the equation of a plane. Thus the curve of intersection lies in

this plane.

51.

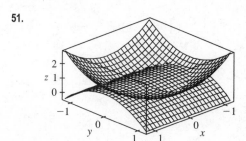

The curve of intersection looks like a bent ellipse. The projection

of this curve onto the xy-plane is the set of points $(x, y, 0)$ which

satisfy $x^2 + y^2 = 1 - y^2 \iff x^2 + 2y^2 = 1 \iff$

$x^2 + \dfrac{y^2}{\left(1/\sqrt{2}\right)^2} = 1$. This is an equation of an ellipse.

13 Review

CONCEPT CHECK

1. A scalar is a real number, while a vector is a quantity that has both a real-valued magnitude and a direction.

2. To add two vectors geometrically, we can use either the Triangle Law or the Parallelogram Law, as illustrated in Figures 3 and 4 in Section 13.2 [ET 12.2]. Algebraically, we add the corresponding components of the vectors.

3. For $c > 0$, $c\,\mathbf{a}$ is a vector with the same direction as $\mathbf{a}$ and length c times the length of $\mathbf{a}$. If $c < 0$, $c\mathbf{a}$ points in the opposite direction as $\mathbf{a}$ and has length $|c|$ times the length of $\mathbf{a}$. (See Figures 7 and 15 in Section 13.2 [ET 12.2].) Algebraically, to find $c\,\mathbf{a}$ we multiply each component of $\mathbf{a}$ by c.

4. See (1) in Section 13.2 [ET 12.2].

5. See Theorem 13.3.3 [ET 12.3.3] and Definition 13.3.1 [ET 12.3.1].

6. The dot product can be used to find the angle between two vectors and the scalar projection of one vector onto another. In particular, the dot product can determine if two vectors are orthogonal. Also, the dot product can be used to determine the work done moving an object given the force and displacement vectors.

7. See the boxed equations on page 819 [ET 783] as well as Figures 4 and 5 and the accompanying discussion on pages 818–19 [ET 782–83].

8. See Theorem 13.4.6 [ET 12.4.6] and the preceding discussion; use either (1) or (4) in Section 13.4 [ET 12.4].

9. The cross product can be used to create a vector orthogonal to two given vectors as well as to determine if two vectors are parallel. The cross product can also be used to find the area of a parallelogram determined by two vectors. In addition, the cross product can be used to determine torque if the force and position vectors are known.

10. (a) The area of the parallelogram determined by $\mathbf{a}$ and $\mathbf{b}$ is the length of the cross product: $|\mathbf{a} \times \mathbf{b}|$.

    (b) The volume of the parallelepiped determined by $\mathbf{a}$, $\mathbf{b}$, and $\mathbf{c}$ is the magnitude of their scalar triple product: $|\mathbf{a} \cdot (\mathbf{b} \times \mathbf{c})|$.

11. If an equation of the plane is known, it can be written as $ax + by + cz + d = 0$. A normal vector, which is perpendicular to the plane, is $\langle a, b, c \rangle$ (or any scalar multiple of $\langle a, b, c \rangle$). If an equation is not known, we can use points on the plane to find two non-parallel vectors which lie in the plane. The cross product of these vectors is a vector perpendicular to the plane.

12. The angle between two intersecting planes is defined as the acute angle between their normal vectors. We can find this angle using Corollary 13.3.6 [ET 12.3.6].

13. See (1), (2), and (3) in Section 13.5 [ET 12.5].

14. See (5), (6), and (7) in Section 13.5 [ET 12.5].

15. (a) Two (nonzero) vectors are parallel if and only if one is a scalar multiple of the other. In addition, two nonzero vectors are parallel if and only if their cross product is $\mathbf{0}$.

    (b) Two vectors are perpendicular if and only if their dot product is 0.

    (c) Two planes are parallel if and only if their normal vectors are parallel.

16. (a) Determine the vectors $\overrightarrow{PQ} = \langle a_1, a_2, a_3 \rangle$ and $\overrightarrow{PR} = \langle b_1, b_2, b_3 \rangle$. If there is a scalar t such that $\langle a_1, a_2, a_3 \rangle = t \langle b_1, b_2, b_3 \rangle$, then the vectors are parallel and the points must all lie on the same line.

Alternatively, if $\overrightarrow{PQ} \times \overrightarrow{PR} = \mathbf{0}$, then $\overrightarrow{PQ}$ and $\overrightarrow{PR}$ are parallel, so P, Q, and R are collinear.

Thirdly, an algebraic method is to determine an equation of the line joining two of the points, and then check whether or not the third point satisfies this equation.

(b) Find the vectors $\overrightarrow{PQ} = \mathbf{a}$, $\overrightarrow{PR} = \mathbf{b}$, $\overrightarrow{PS} = \mathbf{c}$. $\mathbf{a} \times \mathbf{b}$ is normal to the plane formed by P, Q and R, and so S lies on this plane if $\mathbf{a} \times \mathbf{b}$ and $\mathbf{c}$ are orthogonal, that is, if $(\mathbf{a} \times \mathbf{b}) \cdot \mathbf{c} = 0$. (Or use the reasoning in Example 5 in Section 13.4 [ET 12.4].)

Alternatively, find an equation for the plane determined by three of the points and check whether or not the fourth point satisfies this equation.

17. (a) See Exercise 13.4.43 [ET 12.4.43].

(b) See Example 8 in Section 13.5 [ET 12.5].

(c) See Example 10 in Section 13.5 [ET 12.5].

18. The traces of a surface are the curves of intersection of the surface with planes parallel to the coordinate planes. We can find the trace in the plane $x = k$ (parallel to the yz-plane) by setting $x = k$ and determining the curve represented by the resulting equation. Traces in the planes $y = k$ (parallel to the xz-plane) and $z = k$ (parallel to the xy-plane) are found similarly.

19. See Table 1 in Section 13.6 [ET 12.6].

TRUE-FALSE QUIZ

1. True, by Theorem 13.3.2 [ET 12.3.2], property 2.

2. False. Property 1 of Theorem 13.4.8 [ET 12.4.8] says that $\mathbf{u} \times \mathbf{v} = -\mathbf{v} \times \mathbf{u}$.

3. True. If θ is the angle between $\mathbf{u}$ and $\mathbf{v}$, then by Theorem 13.4.6 [ET 12.4.6],
$|\mathbf{u} \times \mathbf{v}| = |\mathbf{u}|\,|\mathbf{v}| \sin \theta = |\mathbf{v}|\,|\mathbf{u}| \sin \theta = |\mathbf{v} \times \mathbf{u}|$.
(Or, by Theorem 13.4.8 [ET 12.4.8], $|\mathbf{u} \times \mathbf{v}| = |-\mathbf{v} \times \mathbf{u}| = |-1|\,|\mathbf{v} \times \mathbf{u}| = |\mathbf{v} \times \mathbf{u}|$.)

4. This is true by Theorem 13.3.2 [ET 12.3.2], property 4.

5. Theorem 13.4.8 [ET 12.4.8], property 2 tells us that this is true.

6. This is true by Theorem 13.4.8 [ET 12.4.8], property 4.

7. This is true by Theorem 13.4.8 [ET 12.4.8], property 5.

8. In general, this assertion is false; a counterexample is $\mathbf{i} \times (\mathbf{i} \times \mathbf{j}) \neq (\mathbf{i} \times \mathbf{i}) \times \mathbf{j}$. (See the paragraph preceding Theorem 13.4.8 [ET 12.4.8].)

9. This is true because $\mathbf{u} \times \mathbf{v}$ is orthogonal to $\mathbf{u}$ (see Theorem 13.4.5 [ET 12.4.5]), and the dot product of two orthogonal vectors is 0.

10. $(\mathbf{u} + \mathbf{v}) \times \mathbf{v} = \mathbf{u} \times \mathbf{v} + \mathbf{v} \times \mathbf{v}$ [by Theorem 13.4.8 [ET 12.4.8], property 4]

              $= \mathbf{u} \times \mathbf{v} + \mathbf{0}$ [by Example 13.4.2 [ET 12.4.2]]

              $= \mathbf{u} \times \mathbf{v}$, so this is true.

11. If $|\mathbf{u}| = 1$, $|\mathbf{v}| = 1$ and θ is the angle between these two vectors (so $0 \leq \theta \leq \pi$), then by Theorem 13.4.6 [ET 12.4.6],

$|\mathbf{u} \times \mathbf{v}| = |\mathbf{u}| \, |\mathbf{v}| \sin \theta = \sin \theta$, which is equal to 1 if and only if $\theta = \frac{\pi}{2}$ (that is, if and only if the two vectors are orthogonal).

Therefore, the assertion that the cross product of two unit vectors is a unit vector is false.

12. This is false, because according to Equation 13.5.8 [ET 12.5.8], $ax + by + cz + d = 0$ is the general equation of a plane.

13. This is false. In $\mathbb{R}^2$, $x^2 + y^2 = 1$ represents a circle, but $\{(x, y, z) \mid x^2 + y^2 = 1\}$ represents a *three-dimensional surface*,

namely, a circular cylinder with axis the z-axis.

14. This is false, as the dot product of two vectors is a scalar, not a vector.

15. False. For example, $\mathbf{i} \cdot \mathbf{j} = 0$ but $\mathbf{i} \neq \mathbf{0}$ and $\mathbf{j} \neq \mathbf{0}$.

16. This is false. By Corollary 13.4.7 [ET 12.4.7], $\mathbf{u} \times \mathbf{v} = \mathbf{0}$ for any nonzero parallel vectors $\mathbf{u}$, $\mathbf{v}$. For instance, $\mathbf{i} \times \mathbf{i} = \mathbf{0}$.

17. This is true. If $\mathbf{u}$ and $\mathbf{v}$ are both nonzero, then by (7) in Section 13.3 [ET 12.3], $\mathbf{u} \cdot \mathbf{v} = 0$ implies that $\mathbf{u}$ and $\mathbf{v}$ are orthogonal.

But $\mathbf{u} \times \mathbf{v} = \mathbf{0}$ implies that $\mathbf{u}$ and $\mathbf{v}$ are parallel (see Corollary 13.4.7 [ET 12.4.7]). Two nonzero vectors can't be both parallel

and orthogonal, so at least one of $\mathbf{u}$, $\mathbf{v}$ must be $\mathbf{0}$.

18. This is true. We know $\mathbf{u} \cdot \mathbf{v} = |\mathbf{u}| \, |\mathbf{v}| \cos \theta$ where $|\mathbf{u}| \geq 0$, $|\mathbf{v}| \geq 0$, and $|\cos \theta| \leq 1$, so $|\mathbf{u} \cdot \mathbf{v}| = |\mathbf{u}| \, |\mathbf{v}| \, |\cos \theta| \leq |\mathbf{u}| \, |\mathbf{v}|$.

EXERCISES

1. (a) The radius of the sphere is the distance between the points $(-1, 2, 1)$ and $(6, -2, 3)$, namely,

$\sqrt{[6 - (-1)]^2 + (-2 - 2)^2 + (3 - 1)^2} = \sqrt{69}$. By the formula for an equation of a sphere (see page 804 [ET 768]),

an equation of the sphere with center $(-1, 2, 1)$ and radius $\sqrt{69}$ is $(x + 1)^2 + (y - 2)^2 + (z - 1)^2 = 69$.

(b) The intersection of this sphere with the yz-plane is the set of points on the sphere whose x-coordinate is 0. Putting $x = 0$

into the equation, we have $(y - 2)^2 + (z - 1)^2 = 68$, $x = 0$ which represents a circle in the yz-plane with center $(0, 2, 1)$

and radius $\sqrt{68}$.

(c) Completing squares gives $(x - 4)^2 + (y + 1)^2 + (z + 3)^2 = -1 + 16 + 1 + 9 = 25$. Thus the sphere is centered at

$(4, -1, -3)$ and has radius 5.

2. (a)

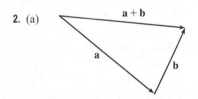

(b)

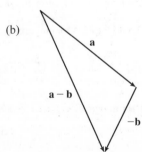

(c)
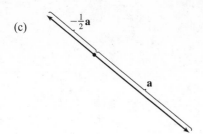

(d)
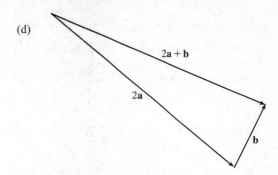

3. $\mathbf{u} \cdot \mathbf{v} = |\mathbf{u}|\,|\mathbf{v}| \cos 45° = (2)(3)\frac{\sqrt{2}}{2} = 3\sqrt{2}.$ $|\mathbf{u} \times \mathbf{v}| = |\mathbf{u}|\,|\mathbf{v}| \sin 45° = (2)(3)\frac{\sqrt{2}}{2} = 3\sqrt{2}.$
By the right-hand rule, $\mathbf{u} \times \mathbf{v}$ is directed out of the page.

4. (a) $2\mathbf{a} + 3\mathbf{b} = 2\mathbf{i} + 2\mathbf{j} - 4\mathbf{k} + 9\mathbf{i} - 6\mathbf{j} + 3\mathbf{k} = 11\mathbf{i} - 4\mathbf{j} - \mathbf{k}$

(b) $|\mathbf{b}| = \sqrt{9 + 4 + 1} = \sqrt{14}$

(c) $\mathbf{a} \cdot \mathbf{b} = (1)(3) + (1)(-2) + (-2)(1) = -1$

(d) $\mathbf{a} \times \mathbf{b} = \begin{vmatrix} \mathbf{i} & \mathbf{j} & \mathbf{k} \\ 1 & 1 & -2 \\ 3 & -2 & 1 \end{vmatrix} = (1-4)\mathbf{i} - (1+6)\mathbf{j} + (-2-3)\mathbf{k} = -3\mathbf{i} - 7\mathbf{j} - 5\mathbf{k}$

(e) $\mathbf{b} \times \mathbf{c} = \begin{vmatrix} \mathbf{i} & \mathbf{j} & \mathbf{k} \\ 3 & -2 & 1 \\ 0 & 1 & -5 \end{vmatrix} = 9\mathbf{i} + 15\mathbf{j} + 3\mathbf{k},$ $|\mathbf{b} \times \mathbf{c}| = 3\sqrt{9 + 25 + 1} = 3\sqrt{35}$

(f) $\mathbf{a} \cdot (\mathbf{b} \times \mathbf{c}) = \begin{vmatrix} 1 & 1 & -2 \\ 3 & -2 & 1 \\ 0 & 1 & -5 \end{vmatrix} = \begin{vmatrix} -2 & 1 \\ 1 & -5 \end{vmatrix} - \begin{vmatrix} 3 & 1 \\ 0 & -5 \end{vmatrix} - 2\begin{vmatrix} 3 & -2 \\ 0 & 1 \end{vmatrix} = 9 + 15 - 6 = 18$

(g) $\mathbf{c} \times \mathbf{c} = \mathbf{0}$ for any $\mathbf{c}$.

(h) From part (e),

$$\mathbf{a} \times (\mathbf{b} \times \mathbf{c}) = \mathbf{a} \times (9\mathbf{i} + 15\mathbf{j} + 3\mathbf{k}) = \begin{vmatrix} \mathbf{i} & \mathbf{j} & \mathbf{k} \\ 1 & 1 & -2 \\ 9 & 15 & 3 \end{vmatrix}$$

$$= (3 + 30)\mathbf{i} - (3 + 18)\mathbf{j} + (15 - 9)\mathbf{k} = 33\mathbf{i} - 21\mathbf{j} + 6\mathbf{k}$$

(i) The scalar projection is $\text{comp}_\mathbf{a}\,\mathbf{b} = |\mathbf{b}| \cos\theta = \mathbf{a} \cdot \mathbf{b}/|\mathbf{a}| = -\frac{1}{\sqrt{6}}.$

(j) The vector projection is $\text{proj}_\mathbf{a}\,\mathbf{b} = -\frac{1}{\sqrt{6}}\left(\frac{\mathbf{a}}{|\mathbf{a}|}\right) = -\frac{1}{6}(\mathbf{i} + \mathbf{j} - 2\mathbf{k}).$

(k) $\cos\theta = \frac{\mathbf{a} \cdot \mathbf{b}}{|\mathbf{a}|\,|\mathbf{b}|} = \frac{-1}{\sqrt{6}\sqrt{14}} = \frac{-1}{2\sqrt{21}}$ and $\theta = \cos^{-1}\left(\frac{-1}{2\sqrt{21}}\right) \approx 96°.$

5. For the two vectors to be orthogonal, we need $\langle 3, 2, x \rangle \cdot \langle 2x, 4, x \rangle = 0$ $\Leftrightarrow$ $(3)(2x) + (2)(4) + (x)(x) = 0$ $\Leftrightarrow$
$x^2 + 6x + 8 = 0$ $\Leftrightarrow$ $(x+2)(x+4) = 0$ $\Leftrightarrow$ $x = -2$ or $x = -4.$

6. We know that the cross product of two vectors is orthogonal to both. So we calculate

   $(\mathbf{j} + 2\mathbf{k}) \times (\mathbf{i} - 2\mathbf{j} + 3\mathbf{k}) = [3 - (-4)]\mathbf{i} - (0 - 2)\mathbf{j} + (0 - 1)\mathbf{k} = 7\mathbf{i} + 2\mathbf{j} - \mathbf{k}.$

   Then two unit vectors orthogonal to both given vectors are $\pm\dfrac{7\mathbf{i} + 2\mathbf{j} - \mathbf{k}}{\sqrt{7^2 + 2^2 + (-1)^2}} = \pm\dfrac{1}{3\sqrt{6}}(7\mathbf{i} + 2\mathbf{j} - \mathbf{k}),$

   that is, $\dfrac{7}{3\sqrt{6}}\mathbf{i} + \dfrac{2}{3\sqrt{6}}\mathbf{j} - \dfrac{1}{3\sqrt{6}}\mathbf{k}$ and $-\dfrac{7}{3\sqrt{6}}\mathbf{i} - \dfrac{2}{3\sqrt{6}}\mathbf{j} + \dfrac{1}{3\sqrt{6}}\mathbf{k}.$

7. (a) $(\mathbf{u} \times \mathbf{v}) \cdot \mathbf{w} = \mathbf{u} \cdot (\mathbf{v} \times \mathbf{w}) = 2$

   (b) $\mathbf{u} \cdot (\mathbf{w} \times \mathbf{v}) = \mathbf{u} \cdot [-(\mathbf{v} \times \mathbf{w})] = -\mathbf{u} \cdot (\mathbf{v} \times \mathbf{w}) = -2$

   (c) $\mathbf{v} \cdot (\mathbf{u} \times \mathbf{w}) = (\mathbf{v} \times \mathbf{u}) \cdot \mathbf{w} = -(\mathbf{u} \times \mathbf{v}) \cdot \mathbf{w} = -2$

   (d) $(\mathbf{u} \times \mathbf{v}) \cdot \mathbf{v} = \mathbf{u} \cdot (\mathbf{v} \times \mathbf{v}) = \mathbf{u} \cdot \mathbf{0} = 0$

8. $(\mathbf{a} \times \mathbf{b}) \cdot [(\mathbf{b} \times \mathbf{c}) \times (\mathbf{c} \times \mathbf{a})] = (\mathbf{a} \times \mathbf{b}) \cdot ([(\mathbf{b} \times \mathbf{c}) \cdot \mathbf{a}]\mathbf{c} - [(\mathbf{b} \times \mathbf{c}) \cdot \mathbf{c}]\mathbf{a})$

   $\text{(by Property 6 of Theorem 13.4.8 [ET 12.4.8])}$

   $= (\mathbf{a} \times \mathbf{b}) \cdot [(\mathbf{b} \times \mathbf{c}) \cdot \mathbf{a}]\mathbf{c} = [\mathbf{a} \cdot (\mathbf{b} \times \mathbf{c})](\mathbf{a} \times \mathbf{b}) \cdot \mathbf{c}$

   $= [\mathbf{a} \cdot (\mathbf{b} \times \mathbf{c})][\mathbf{a} \cdot (\mathbf{b} \times \mathbf{c})] = [\mathbf{a} \cdot (\mathbf{b} \times \mathbf{c})]^2$

9. For simplicity, consider a unit cube positioned with its back left corner at the origin. Vector representations of the diagonals joining the points $(0, 0, 0)$ to $(1, 1, 1)$ and $(1, 0, 0)$ to $(0, 1, 1)$ are $\langle 1, 1, 1 \rangle$ and $\langle -1, 1, 1 \rangle$. Let θ be the angle between these two vectors. $\langle 1, 1, 1 \rangle \cdot \langle -1, 1, 1 \rangle = -1 + 1 + 1 = 1 = |\langle 1, 1, 1 \rangle||\langle -1, 1, 1 \rangle| \cos \theta = 3 \cos \theta \Rightarrow \cos \theta = \frac{1}{3} \Rightarrow$ $\theta = \cos^{-1}\left(\frac{1}{3}\right) \approx 71°.$

10. $\overrightarrow{AB} = \langle 1, 3, -1 \rangle$, $\overrightarrow{AC} = \langle -2, 1, 3 \rangle$ and $\overrightarrow{AD} = \langle -1, 3, 1 \rangle$. By Equation 13.4.10 [ET 12.4.10],

    $\overrightarrow{AB} \cdot \left(\overrightarrow{AC} \times \overrightarrow{AD}\right) = \begin{vmatrix} 1 & 3 & -1 \\ -2 & 1 & 3 \\ -1 & 3 & 1 \end{vmatrix} = \begin{vmatrix} 1 & 3 \\ 3 & 1 \end{vmatrix} - 3\begin{vmatrix} -2 & 3 \\ -1 & 1 \end{vmatrix} - \begin{vmatrix} -2 & 1 \\ -1 & 3 \end{vmatrix} = -8 - 3 + 5 = -6.$

    The volume is $\left|\overrightarrow{AB} \cdot \left(\overrightarrow{AC} \times \overrightarrow{AD}\right)\right| = 6$ cubic units.

11. $\overrightarrow{AB} = \langle 1, 0, -1 \rangle$, $\overrightarrow{AC} = \langle 0, 4, 3 \rangle$, so

    (a) a vector perpendicular to the plane is $\overrightarrow{AB} \times \overrightarrow{AC} = \langle 0 + 4, -(3 + 0), 4 - 0 \rangle = \langle 4, -3, 4 \rangle.$

    (b) $\frac{1}{2}\left|\overrightarrow{AB} \times \overrightarrow{AC}\right| = \frac{1}{2}\sqrt{16 + 9 + 16} = \frac{\sqrt{41}}{2}.$

12. $\mathbf{D} = 4\mathbf{i} + 3\mathbf{j} + 6\mathbf{k}$, $W = \mathbf{F} \cdot \mathbf{D} = 12 + 15 + 60 = 87$ J

13. Let F_1 be the magnitude of the force directed $20°$ away from the direction of shore, and let F_2 be the magnitude of the other force. Separating these forces into components parallel to the direction of the resultant force and perpendicular to it gives

    $F_1 \cos 20° + F_2 \cos 30° = 255$ **(1)**, and $F_1 \sin 20° - F_2 \sin 30° = 0 \Rightarrow F_1 = F_2 \dfrac{\sin 30°}{\sin 20°}$ **(2)**. Substituting **(2)**

    into **(1)** gives $F_2(\sin 30° \cot 20° + \cos 30°) = 255 \Rightarrow F_2 \approx 114$ N. Substituting this into **(2)** gives $F_1 \approx 166$ N.

14. $|\boldsymbol{\tau}| = |\mathbf{r}|\,|\mathbf{F}|\sin\theta = (0.40)(50)\sin(90° - 30°) \approx 17.3\,\text{N}\cdot\text{m}.$

15. The line has direction $\mathbf{v} = \langle -3, 2, 3 \rangle$. Letting $P_0 = (4, -1, 2)$, parametric equations are

$x = 4 - 3t,\ y = -1 + 2t,\ z = 2 + 3t.$

16. A direction vector for the line is $\mathbf{v} = \langle 3, 2, 1 \rangle$, so parametric equations for the line are $x = 1 + 3t,\ y = 2t,\ z = -1 + t.$

17. A direction vector for the line is a normal vector for the plane, $\mathbf{n} = \langle 2, -1, 5 \rangle$, and parametric equations for the line are

$x = -2 + 2t,\ y = 2 - t,\ z = 4 + 5t.$

18. Since the two planes are parallel, they will have the same normal vectors. Then we can take $\mathbf{n} = \langle 1, 4, -3 \rangle$ and an equation of

the plane is $1(x - 2) + 4(y - 1) - 3(z - 0) = 0$ or $x + 4y - 3z = 6.$

19. Here the vectors $\mathbf{a} = \langle 4 - 3, 0 - (-1), 2 - 1 \rangle = \langle 1, 1, 1 \rangle$ and $\mathbf{b} = \langle 6 - 3, 3 - (-1), 1 - 1 \rangle = \langle 3, 4, 0 \rangle$ lie in the plane,

so $\mathbf{n} = \mathbf{a} \times \mathbf{b} = \langle -4, 3, 1 \rangle$ is a normal vector to the plane and an equation of the plane is

$-4(x - 3) + 3(y - (-1)) + 1(z - 1) = 0$ or $-4x + 3y + z = -14.$

20. If we first find two nonparallel vectors in the plane, their cross product will be a normal vector to the plane. Since the given

line lies in the plane, its direction vector $\mathbf{a} = \langle 2, -1, 3 \rangle$ is one vector in the plane. We can verify that the given point $(1, 2, -2)$

does not lie on this line. The point $(0, 3, 1)$ is on the line (obtained by putting $t = 0$) and hence in the plane, so the vector

$\mathbf{b} = \langle 0 - 1, 3 - 2, 1 - (-2) \rangle = \langle -1, 1, 3 \rangle$ lies in the plane, and a normal vector is $\mathbf{n} = \mathbf{a} \times \mathbf{b} = \langle -6, -9, 1 \rangle$. Thus an

equation of the plane is $-6(x - 1) - 9(y - 2) + (z + 2) = 0$ or $6x + 9y - z = 26.$

21. Substitution of the parametric equations into the equation of the plane gives $2x - y + z = 2(2 - t) - (1 + 3t) + 4t = 2 \quad \Rightarrow$

$-t + 3 = 2 \quad \Rightarrow \quad t = 1.$ When $t = 1$, the parametric equations give $x = 2 - 1 = 1,\ y = 1 + 3 = 4$ and $z = 4$. Therefore,

the point of intersection is $(1, 4, 4).$

22. Use the formula proven in Exercise 13.4.43(a) [ET 12.4.43(a)]. In the notation used in that exercise, $\mathbf{a}$ is just the direction of

the line; that is, $\mathbf{a} = \langle 1, -1, 2 \rangle$. A point on the line is $(1, 2, -1)$ (setting $t = 0$), and therefore

$\mathbf{b} = \langle 1 - 0, 2 - 0, -1 - 0 \rangle = \langle 1, 2, -1 \rangle.$ Hence $d = \dfrac{|\mathbf{a} \times \mathbf{b}|}{|\mathbf{a}|} = \dfrac{|\langle 1, -1, 2 \rangle \times \langle 1, 2, -1 \rangle|}{\sqrt{1 + 1 + 4}} = \dfrac{|\langle -3, 3, 3 \rangle|}{\sqrt{6}} = \sqrt{\dfrac{27}{6}} = \dfrac{3}{\sqrt{2}}.$

23. Since the direction vectors $\langle 2, 3, 4 \rangle$ and $\langle 6, -1, 2 \rangle$ aren't parallel, neither are the lines. For the lines to intersect, the three

equations $1 + 2t = -1 + 6s,\ 2 + 3t = 3 - s,\ 3 + 4t = -5 + 2s$ must be satisfied simultaneously. Solving the first two

equations gives $t = \frac{1}{5},\ s = \frac{2}{5}$ and checking we see these values don't satisfy the third equation. Thus the lines aren't parallel

and they don't intersect, so they must be skew.

24. (a) The normal vectors are $\langle 1, 1, -1 \rangle$ and $\langle 2, -3, 4 \rangle$. Since these vectors aren't parallel, neither are the planes parallel.

Also $\langle 1, 1, -1 \rangle \cdot \langle 2, -3, 4 \rangle = 2 - 3 - 4 = -5 \neq 0$ so the normal vectors, and thus the planes, are not perpendicular.

(b) $\cos\theta = \dfrac{\langle 1, 1, -1 \rangle \cdot \langle 2, -3, 4 \rangle}{\sqrt{3}\sqrt{29}} = -\dfrac{5}{\sqrt{87}}$ and $\theta = \cos^{-1}\left(-\frac{5}{\sqrt{87}}\right) \approx 122°$ [or we can say $\approx 58°$].

25. $\mathbf{n}_1 = \langle 1, 0, -1 \rangle$ and $\mathbf{n}_2 = \langle 0, 1, 2 \rangle$. Setting $z = 0$, it is easy to see that $(1, 3, 0)$ is a point on the line of intersection of

$x - z = 1$ and $y + 2z = 3$. The direction of this line is $\mathbf{v}_1 = \mathbf{n}_1 \times \mathbf{n}_2 = \langle 1, -2, 1 \rangle$. A second vector parallel to the desired

plane is $\mathbf{v}_2 = \langle 1, 1, -2 \rangle$, since it is perpendicular to $x + y - 2z = 1$. Therefore, the normal of the plane in question is

$\mathbf{n} = \mathbf{v}_1 \times \mathbf{v}_2 = \langle 4 - 1, 1 + 2, 1 + 2 \rangle = 3\langle 1, 1, 1 \rangle$. Taking $(x_0, y_0, z_0) = (1, 3, 0)$, the equation we are looking for is

$(x - 1) + (y - 3) + z = 0 \iff x + y + z = 4$.

26. (a) The vectors $\overrightarrow{AB} = \langle -1 - 2, -1 - 1, 10 - 1 \rangle = \langle -3, -2, 9 \rangle$ and $\overrightarrow{AC} = \langle 1 - 2, 3 - 1, -4 - 1 \rangle = \langle -1, 2, -5 \rangle$ lie in the

plane, so $\mathbf{n} = \overrightarrow{AB} \times \overrightarrow{AC} = \langle -3, -2, 9 \rangle \times \langle -1, 2, -5 \rangle = \langle -8, -24, -8 \rangle$ or equivalently $\langle 1, 3, 1 \rangle$ is a normal vector to

the plane. The point $A(2, 1, 1)$ lies on the plane so an equation of the plane is $1(x - 2) + 3(y - 1) + 1(z - 1) = 0$ or

$x + 3y + z = 6$.

(b) The line is perpendicular to the plane so it is parallel to a normal vector for the plane, namely $\langle 1, 3, 1 \rangle$. If the line passes

through $B(-1, -1, 10)$ then symmetric equations are $\dfrac{x - (-1)}{1} = \dfrac{y - (-1)}{3} = \dfrac{z - 10}{1}$ or $x + 1 = \dfrac{y + 1}{3} = z - 10$.

(c) Normal vectors for the two planes are $\mathbf{n}_1 = \langle 1, 3, 1 \rangle$ and $\mathbf{n}_2 = \langle 2, -4, -3 \rangle$. The angle θ between the planes is given by

$$\cos\theta = \frac{\mathbf{n}_1 \cdot \mathbf{n}_2}{|\mathbf{n}_1|\ |\mathbf{n}_2|} = \frac{\langle 1, 3, 1 \rangle \cdot \langle 2, -4, -3 \rangle}{\sqrt{1^2 + 3^2 + 1^2}\ \sqrt{2^2 + (-4)^2 + (-3)^2}} = \frac{2 - 12 - 3}{\sqrt{11}\ \sqrt{29}} = -\frac{13}{\sqrt{319}}$$

Thus $\theta = \cos^{-1}\left(-\dfrac{13}{\sqrt{319}}\right) \approx 137°$ or $180° - 137° = 43°$.

(d) From part (c), the point $(2, 0, 4)$ lies on the second plane, but notice that the point also satisfies the equation of the first

plane, so the point lies on the line of intersection of the planes. A vector $\mathbf{v}$ in the direction of this intersecting line is

perpendicular to the normal vectors of both planes, so take $\mathbf{v} = \mathbf{n}_1 \times \mathbf{n}_2 = \langle 1, 3, 1 \rangle \times \langle 2, -4, -3 \rangle = \langle -5, 5, -10 \rangle$ or

equivalently we can take $\mathbf{v} = \langle 1, -1, 2 \rangle$. Parametric equations for the line are $x = 2 + t, y = -t, z = 4 + 2t$.

27. By Exercise 13.5.73 [ET 12.5.73], $D = \dfrac{|2 - 24|}{\sqrt{26}} = \dfrac{22}{\sqrt{26}}$.

28. The equation $x = 3$ represents a plane parallel to the yz-plane and 3 units in front of it.

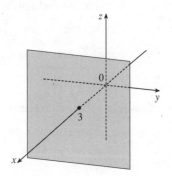

29. The equation $x = z$ represents a plane perpendicular to the xz-plane and intersecting the xz-plane in the line $x = z, y = 0$.

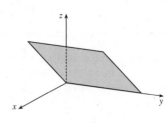

30. The equation $y = z^2$ represents a parabolic cylinder whose trace in the xz-plane is the x-axis and which opens to the right.

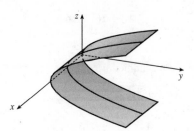

31. The equation $x^2 = y^2 + 4z^2$ represents a (right elliptical) cone with vertex at the origin and axis the x-axis.

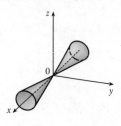

32. $4x - y + 2z = 4$ is a plane with intercepts $(1, 0, 0)$, $(0, -4, 0)$, and $(0, 0, 2)$.

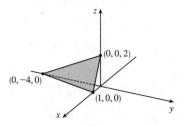

33. An equivalent equation is $-x^2 + \dfrac{y^2}{4} - z^2 = 1$, a hyperboloid of two sheets with axis the y-axis. For $|y| > 2$, traces parallel to the xz-plane are circles.

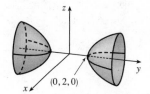

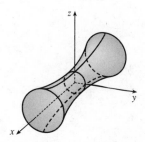

34. An equivalent equation is $-x^2 + y^2 + z^2 = 1$, a hyperboloid of one sheet with axis the x-axis.

35. Completing the square in y gives

$$4x^2 + 4(y - 1)^2 + z^2 = 4 \text{ or } x^2 + (y - 1)^2 + \frac{z^2}{4} = 1,$$

an ellipsoid centered at $(0, 1, 0)$.

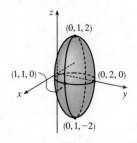

36. Completing the square in y and z gives

$x = (y - 1)^2 + (z - 2)^2$, a circular paraboloid with vertex $(0, 1, 2)$ and axis the horizontal line $y = 1$, $z = 2$.

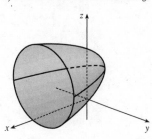

37. $4x^2 + y^2 = 16$ $\Leftrightarrow$ $\dfrac{x^2}{4} + \dfrac{y^2}{16} = 1$. The equation of the ellipsoid is $\dfrac{x^2}{4} + \dfrac{y^2}{16} + \dfrac{z^2}{c^2} = 1$, since the horizontal trace in the

plane $z = 0$ must be the original ellipse. The traces of the ellipsoid in the yz-plane must be circles since the surface is obtained

by rotation about the x-axis. Therefore, $c^2 = 16$ and the equation of the ellipsoid is $\dfrac{x^2}{4} + \dfrac{y^2}{16} + \dfrac{z^2}{16} = 1$ $\Leftrightarrow$

$4x^2 + y^2 + z^2 = 16$.

38. The distance from a point $P\,(x, y, z)$ to the plane $y = 1$ is $|y - 1|$, so the given condition becomes

$|y - 1| = 2\sqrt{(x - 0)^2 + (y + 1)^2 + (z - 0)^2}$ $\Rightarrow$ $|y - 1| = 2\sqrt{x^2 + (y + 1)^2 + z^2}$ $\Rightarrow$

$(y - 1)^2 = 4x^2 + 4(y + 1)^2 + 4z^2$ $\Leftrightarrow$ $-3 = 4x^2 + (3y^2 + 10y) + 4z^2$ $\Leftrightarrow$

$\frac{16}{3} = 4x^2 + 3\left(y + \frac{5}{3}\right)^2 + 4z^2$ $\Rightarrow$ $\frac{3}{4}x^2 + \frac{9}{16}\left(y + \frac{5}{3}\right)^2 + \frac{3}{4}z^2 = 1$.

This is the equation of an ellipsoid whose center is $\left(0, -\frac{5}{3}, 0\right)$.

☐ PROBLEMS PLUS

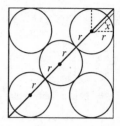

1. Since three-dimensional situations are often difficult to visualize and work with, let us first try to find an analogous problem in two dimensions. The analogue of a cube is a square and the analogue of a sphere is a circle. Thus a similar problem in two dimensions is the following: if five circles with the same radius r are contained in a square of side 1 m so that the circles touch each other and four of the circles touch two sides of the square, find r.

The diagonal of the square is $\sqrt{2}$. The diagonal is also $4r + 2x$. But x is the diagonal of a smaller square of side r. Therefore

$$x = \sqrt{2}\,r \quad \Rightarrow \quad \sqrt{2} = 4r + 2x = 4r + 2\sqrt{2}\,r = \left(4 + 2\sqrt{2}\right)r \quad \Rightarrow \quad r = \frac{\sqrt{2}}{4 + 2\sqrt{2}}.$$

Let's use these ideas to solve the original three-dimensional problem. The diagonal of the cube is $\sqrt{1^2 + 1^2 + 1^2} = \sqrt{3}$. The diagonal of the cube is also $4r + 2x$ where x is the diagonal of a smaller cube with edge r. Therefore

$$x = \sqrt{r^2 + r^2 + r^2} = \sqrt{3}\,r \quad \Rightarrow \quad \sqrt{3} = 4r + 2x = 4r + 2\sqrt{3}\,r = \left(4 + 2\sqrt{3}\right)r. \text{ Thus } r = \frac{\sqrt{3}}{4 + 2\sqrt{3}} = \frac{2\sqrt{3} - 3}{2}.$$

The radius of each ball is $\left(\sqrt{3} - \frac{3}{2}\right)$ m.

2. Try an analogous problem in two dimensions. Consider a rectangle with length L and width W and find the area of S in terms of L and W. Since S contains B, it has area

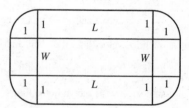

$$A(S) = LW + \text{ the area of two } L \times 1 \text{ rectangles}$$

$$+ \text{ the area of two } 1 \times W \text{ rectangles}$$

$$+ \text{ the area of four quarter-circles of radius 1}$$

as seen in the diagram. So $A(S) = LW + 2L + 2W + \pi \cdot 1^2$.

Now in three dimensions, the volume of S is

$$LWH + 2(L \times W \times 1) + 2(1 \times W \times H) + 2(L \times 1 \times H)$$

$$+ \text{ the volume of 4 quarter-cylinders with radius 1 and height } W$$

$$+ \text{ the volume of 4 quarter-cylinders with radius 1 and height } L$$

$$+ \text{ the volume of 4 quarter-cylinders with radius 1 and height } H$$

$$+ \text{ the volume of 8 eighths of a sphere of radius 1}$$

So

$$V(S) = LWH + 2LW + 2WH + 2LH + \pi \cdot 1^2 \cdot W + \pi \cdot 1^2 \cdot L + \pi \cdot 1^2 \cdot H + \tfrac{4}{3}\pi \cdot 1^3$$

$$= LWH + 2(LW + WH + LH) + \pi(L + W + H) + \tfrac{4}{3}\pi.$$

3. (a) We find the line of intersection L as in Example 13.5.7(b) [ET 12.5.7(b)]. Observe that the point $(-1, c, c)$ lies on both planes. Now since L lies in both planes, it is perpendicular to both of the normal vectors $\mathbf{n}_1$ and $\mathbf{n}_2$, and thus parallel to

their cross product $\mathbf{n}_1 \times \mathbf{n}_2 = \begin{vmatrix} \mathbf{i} & \mathbf{j} & \mathbf{k} \\ c & 1 & 1 \\ 1 & -c & c \end{vmatrix} = \langle 2c, -c^2 + 1, -c^2 - 1 \rangle$. So symmetric equations of L can be written as

$\dfrac{x+1}{-2c} = \dfrac{y-c}{c^2 - 1} = \dfrac{z-c}{c^2 + 1}$, provided that $c \neq 0, \pm 1$.

If $c = 0$, then the two planes are given by $y + z = 0$ and $x = -1$, so symmetric equations of L are $x = -1$, $y = -z$. If $c = -1$, then the two planes are given by $-x + y + z = -1$ and $x + y + z = -1$, and they intersect in the line $x = 0$, $y = -z - 1$. If $c = 1$, then the two planes are given by $x + y + z = 1$ and $x - y + z = 1$, and they intersect in the line $y = 0$, $x = 1 - z$.

(b) If we set $z = t$ in the symmetric equations and solve for x and y separately, we get $x + 1 = \dfrac{(t - c)(-2c)}{c^2 + 1}$,

$y - c = \dfrac{(t - c)(c^2 - 1)}{c^2 + 1} \quad \Rightarrow \quad x = \dfrac{-2ct + (c^2 - 1)}{c^2 + 1}$, $y = \dfrac{(c^2 - 1)t + 2c}{c^2 + 1}$. Eliminating c from these equations, we

have $x^2 + y^2 = t^2 + 1$. So the curve traced out by L in the plane $z = t$ is a circle with center at $(0, 0, t)$ and radius $\sqrt{t^2 + 1}$.

(c) The area of a horizontal cross-section of the solid is $A(z) = \pi(z^2 + 1)$, so $V = \int_0^1 A(z)\,dz = \pi \left[\frac{1}{3} z^3 + z \right]_0^1 = \frac{4\pi}{3}$.

4. (a) We consider velocity vectors for the plane and the wind. Let $\mathbf{v}_i$ be the initial, intended velocity for the plane and $\mathbf{v}_g$ the actual velocity relative to the ground. If $\mathbf{w}$ is the velocity of the wind, $\mathbf{v}_g$ is the resultant, that is, the vector sum $\mathbf{v}_i + \mathbf{w}$ as shown in the figure. We know $\mathbf{v}_i = 180\,\mathbf{j}$, and since the plane actually flew 80 km in $\frac{1}{2}$ hour, $|\mathbf{v}_g| = 160$. Thus

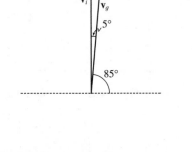

$\mathbf{v}_g = (160 \cos 85°)\,\mathbf{i} + (160 \sin 85°)\,\mathbf{j} \approx 13.9\,\mathbf{i} + 159.4\,\mathbf{j}$. Finally,

$\mathbf{v}_i + \mathbf{w} = \mathbf{v}_g$, so $\mathbf{w} = \mathbf{v}_g - \mathbf{v}_i \approx 13.9\,\mathbf{i} - 20.6\,\mathbf{j}$. Thus, the wind velocity is about $13.9\,\mathbf{i} - 20.6\,\mathbf{j}$, and the wind speed is

$|\mathbf{w}| \approx \sqrt{(13.9)^2 + (-20.6)^2} \approx 24.9$ km/h.

(b) Let $\mathbf{v}$ be the velocity the pilot should take. With the effect of wind, the actual velocity (with respect to the ground) will be $\mathbf{v} + \mathbf{w}$, which we want to be $\mathbf{v}_i$. Thus $\mathbf{v} = \mathbf{v}_i - \mathbf{w} \approx 180\,\mathbf{j} - (13.9\,\mathbf{i} - 20.6\,\mathbf{j}) \approx -13.9\,\mathbf{i} + 200.6\,\mathbf{j}$. The angle for this vector can be found by $\tan \theta \approx \frac{200.6}{-13.9} \quad \Rightarrow \quad \theta \approx 94.0°$, or $4.0°$ west of north.

5. (a) When $\theta = \theta_s$, the block is not moving, so the sum of the forces on the block must be $\mathbf{0}$, thus $\mathbf{N} + \mathbf{F} + \mathbf{W} = \mathbf{0}$. This relationship is illustrated geometrically in the figure. Since the vectors form a right triangle, we have

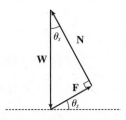

$\tan(\theta_s) = \dfrac{|\mathbf{F}|}{|\mathbf{N}|} = \dfrac{\mu_s n}{n} = \mu_s$.

(b) We place the block at the origin and sketch the force vectors acting on the block, including the additional horizontal force **H**, with initial points at the origin. We then rotate this system so that **F** lies along the positive x-axis and the inclined plane is parallel to the x-axis.

$|\mathbf{F}|$ is maximal, so $|\mathbf{F}| = \mu_s n$ for $\theta > \theta_s$. Then the vectors, in terms of components parallel and perpendicular to the inclined plane, are

$$\mathbf{N} = n\,\mathbf{j} \qquad \mathbf{F} = (\mu_s n)\,\mathbf{i}$$

$$\mathbf{W} = (-mg\sin\theta)\,\mathbf{i} + (-mg\cos\theta)\,\mathbf{j} \qquad \mathbf{H} = (h_{\min}\cos\theta)\,\mathbf{i} + (-h_{\min}\sin\theta)\,\mathbf{j}$$

Equating components, we have

$$\mu_s n - mg\sin\theta + h_{\min}\cos\theta = 0 \quad \Rightarrow \quad h_{\min}\cos\theta + \mu_s n = mg\sin\theta \tag{1}$$

$$n - mg\cos\theta - h_{\min}\sin\theta = 0 \quad \Rightarrow \quad h_{\min}\sin\theta + mg\cos\theta = n \tag{2}$$

(c) Since **(2)** is solved for n, we substitute into **(1)**:

$$h_{\min}\cos\theta + \mu_s(h_{\min}\sin\theta + mg\cos\theta) = mg\sin\theta \quad \Rightarrow$$

$$h_{\min}\cos\theta + h_{\min}\mu_s\sin\theta = mg\sin\theta - mg\mu_s\cos\theta \quad \Rightarrow$$

$$h_{\min} = mg\left(\frac{\sin\theta - \mu_s\cos\theta}{\cos\theta + \mu_s\sin\theta}\right) = mg\left(\frac{\tan\theta - \mu_s}{1 + \mu_s\tan\theta}\right)$$

From part (a) we know $\mu_s = \tan\theta_s$, so this becomes $h_{\min} = mg\left(\dfrac{\tan\theta - \tan\theta_s}{1 + \tan\theta_s\tan\theta}\right)$ and using a trigonometric identity, this is $mg\tan(\theta - \theta_s)$ as desired.

Note for $\theta = \theta_s$, $h_{\min} = mg\tan 0 = 0$, which makes sense since the block is at rest for θ_s, thus no additional force **H** is necessary to prevent it from moving. As θ increases, the factor $\tan(\theta - \theta_s)$, and hence the value of $h_{\min}$, increases slowly for small values of $\theta - \theta_s$ but much more rapidly as $\theta - \theta_s$ becomes significant. This seems reasonable, as the steeper the inclined plane, the less the horizontal components of the various forces affect the movement of the block, so we would need a much larger magnitude of horizontal force to keep the block motionless. If we allow $\theta \to 90°$, corresponding to the inclined plane being placed vertically, the value of $h_{\min}$ is quite large; this is to be expected, as it takes a great amount of horizontal force to keep an object from moving vertically. In fact, without friction (so $\theta_s = 0$), we would have $\theta \to 90° \quad \Rightarrow \quad h_{\min} \to \infty$, and it would be impossible to keep the block from slipping.

(d) Since h_{max} is the largest value of h that keeps the block from slipping, the force of friction is keeping the block from moving *up* the inclined plane; thus, **F** is directed *down* the plane. Our system of forces is similar to that in part (b), then, except that we have $\mathbf{F} = -(\mu_s n)\,\mathbf{i}$. (Note that $|\mathbf{F}|$ is again maximal.) Following our procedure in parts (b) and (c), we equate components:

$$-\mu_s n - mg\sin\theta + h_{max}\cos\theta = 0 \quad \Rightarrow \quad h_{max}\cos\theta - \mu_s n = mg\sin\theta$$

$$n - mg\cos\theta - h_{max}\sin\theta = 0 \quad \Rightarrow \quad h_{max}\sin\theta + mg\cos\theta = n$$

Then substituting,

$$h_{max}\cos\theta - \mu_s(h_{max}\sin\theta + mg\cos\theta) = mg\sin\theta \quad \Rightarrow$$

$$h_{max}\cos\theta - h_{max}\mu_s\sin\theta = mg\sin\theta + mg\mu_s\cos\theta \quad \Rightarrow$$

$$h_{max} = mg\left(\frac{\sin\theta + \mu_s\cos\theta}{\cos\theta - \mu_s\sin\theta}\right) = mg\left(\frac{\tan\theta + \mu_s}{1 - \mu_s\tan\theta}\right)$$

$$= mg\left(\frac{\tan\theta + \tan\theta_s}{1 - \tan\theta_s\tan\theta}\right) = mg\tan(\theta + \theta_s)$$

We would expect h_{max} to increase as θ increases, with similar behavior as we established for h_{min}, but with h_{max} values always larger than h_{min}. We can see that this is the case if we graph h_{max} as a function of θ, as the curve is the graph of h_{min} translated $2\theta_s$ to the left, so the equation does seem reasonable. Notice that the equation predicts $h_{max} \to \infty$ as $\theta \to (90° - \theta_s)$. In fact, as h_{max} increases, the normal force increases as well. When $(90° - \theta_s) \le \theta \le 90°$, the horizontal force is completely counteracted by the sum of the normal and frictional forces, so no part of the horizontal force contributes to moving the block up the plane no matter how large its magnitude.

14 □ VECTOR FUNCTIONS

14.1 Vector Functions and Space Curves

1. The component functions $\sqrt{4-t^2}$, e^{-3t}, and $\ln(t+1)$ are all defined when $4-t^2 \geq 0 \Rightarrow -2 \leq t \leq 2$ and

$t+1 > 0 \Rightarrow t > -1$, so the domain of $\mathbf{r}$ is $(-1, 2]$.

2. The component functions $\dfrac{t-2}{t+2}$, $\sin t$, and $\ln(9-t^2)$ are all defined when $t \neq -2$ and $9-t^2 > 0 \Rightarrow -3 < t < 3$,

so the domain of $\mathbf{r}$ is $(-3, -2) \cup (-2, 3)$.

3. $\lim\limits_{t\to 0^+} \cos t = \cos 0 = 1$, $\lim\limits_{t\to 0^+} \sin t = \sin 0 = 0$, $\lim\limits_{t\to 0^+} t \ln t = \lim\limits_{t\to 0^+} \dfrac{\ln t}{1/t} = \lim\limits_{t\to 0^+} \dfrac{1/t}{-1/t^2} = \lim\limits_{t\to 0^+} -t = 0$

[by l'Hospital's Rule]. Thus $\lim\limits_{t\to 0^+} \langle \cos t, \sin t, t \ln t \rangle = \left\langle \lim\limits_{t\to 0^+} \cos t, \lim\limits_{t\to 0^+} \sin t, \lim\limits_{t\to 0^+} t \ln t \right\rangle = \langle 1, 0, 0 \rangle$.

4. $\lim\limits_{t\to 0} \dfrac{e^t - 1}{t} = \lim\limits_{t\to 0} \dfrac{e^t}{1} = 1$ [using l'Hospital's Rule],

$\lim\limits_{t\to 0} \dfrac{\sqrt{1+t}-1}{t} = \lim\limits_{t\to 0} \dfrac{\sqrt{1+t}-1}{t} \cdot \dfrac{\sqrt{1+t}+1}{\sqrt{1+t}+1} = \lim\limits_{t\to 0} \dfrac{1}{\sqrt{1+t}+1} = \dfrac{1}{2}, \lim\limits_{t\to 0} \dfrac{3}{1+t} = 3$.

Thus the given limit equals $\langle 1, \frac{1}{2}, 3 \rangle$.

5. $\lim\limits_{t\to 0} e^{-3t} = e^0 = 1$, $\lim\limits_{t\to 0} \dfrac{t^2}{\sin^2 t} = \lim\limits_{t\to 0} \dfrac{1}{\dfrac{\sin^2 t}{t^2}} = \dfrac{1}{\lim\limits_{t\to 0} \dfrac{\sin^2 t}{t^2}} = \dfrac{1}{\left(\lim\limits_{t\to 0} \dfrac{\sin t}{t}\right)^2} = \dfrac{1}{1^2} = 1$

and $\lim\limits_{t\to 0} \cos 2t = \cos 0 = 1$. Thus the given limit equals $\mathbf{i} + \mathbf{j} + \mathbf{k}$.

6. $\lim\limits_{t\to\infty} \arctan t = \frac{\pi}{2}$, $\lim\limits_{t\to\infty} e^{-2t} = 0$, $\lim\limits_{t\to\infty} \dfrac{\ln t}{t} = \lim\limits_{t\to\infty} \dfrac{1/t}{1} = 0$ [by l'Hospital's Rule].

Thus $\lim\limits_{t\to\infty} \left\langle \arctan t, e^{-2t}, \dfrac{\ln t}{t} \right\rangle = \langle \frac{\pi}{2}, 0, 0 \rangle$.

7. The corresponding parametric equations for this curve are $x = \sin t$, $y = t$.

We can make a table of values, or we can eliminate the parameter: $t = y \Rightarrow$

$x = \sin y$, with $y \in \mathbb{R}$. By comparing different values of t, we find the direction in

which t increases as indicated in the graph.

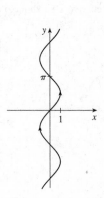

8. The corresponding parametric equations for this curve are $x = t^3$, $y = t^2$.

We can make a table of values, or we can eliminate the parameter:

$x = t^3 \quad \Rightarrow \quad t = \sqrt[3]{x} \quad \Rightarrow \quad y = t^2 = (\sqrt[3]{x})^2 = x^{2/3}$,

with $t \in \mathbb{R} \quad \Rightarrow \quad x \in \mathbb{R}$. By comparing different values of t, we find the

direction in which t increases as indicated in the graph.

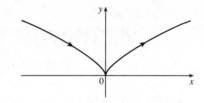

9. The corresponding parametric equations are $x = t$, $y = \cos 2t$, $z = \sin 2t$.

Note that $y^2 + z^2 = \cos^2 2t + \sin^2 2t = 1$, so the curve lies on the circular

cylinder $y^2 + z^2 = 1$. Since $x = t$, the curve is a helix.

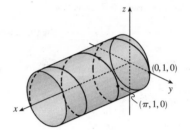

10. The corresponding parametric equations are $x = 1 + t$, $y = 3t$, $z = -t$,

which are parametric equations of a line through the point $(1, 0, 0)$ and with

direction vector $\langle 1, 3, -1 \rangle$.

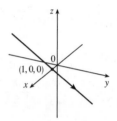

11. The corresponding parametric equations are $x = 1$, $y = \cos t$, $z = 2 \sin t$.

Eliminating the parameter in y and z gives $y^2 + (z/2)^2 = \cos^2 t + \sin^2 t = 1$

or $y^2 + z^2/4 = 1$. Since $x = 1$, the curve is an ellipse centered at $(1, 0, 0)$ in

the plane $x = 1$.

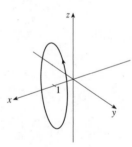

12. The parametric equations are $x = t^2$, $y = t$, $z = 2$, so we have $x = y^2$ with $z = 2$.

Thus the curve is a parabola in the plane $z = 2$ with vertex $(0, 0, 2)$.

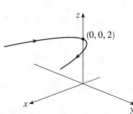

13. The parametric equations are $x = t^2$, $y = t^4$, $z = t^6$. These are positive

for $t \neq 0$ and 0 when $t = 0$. So the curve lies entirely in the first quadrant.

The projection of the graph onto the xy-plane is $y = x^2$, $y > 0$, a half parabola.

On the xz-plane $z = x^3$, $z > 0$, a half cubic, and the yz-plane, $y^3 = z^2$.

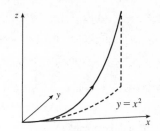

14. If $x = \cos t$, $y = -\cos t$, $z = \sin t$, then $x^2 + z^2 = 1$ and $y^2 + z^2 = 1$,

so the curve is contained in the intersection of circular cylinders along the

x- and y-axes. Furthermore, $y = -x$, so the curve is an ellipse in the

plane $y = -x$, centered at the origin.

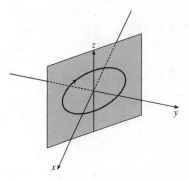

15. Taking $\mathbf{r}_0 = \langle 0, 0, 0 \rangle$ and $\mathbf{r}_1 = \langle 1, 2, 3 \rangle$, we have from Equation 13.5.4 [ET 12.5.4]

$\mathbf{r}(t) = (1 - t)\,\mathbf{r}_0 + t\,\mathbf{r}_1 = (1 - t)\,\langle 0, 0, 0 \rangle + t\,\langle 1, 2, 3 \rangle, 0 \leq t \leq 1$ or $\mathbf{r}(t) = \langle t, 2t, 3t \rangle, 0 \leq t \leq 1$.

Parametric equations are $x = t$, $y = 2t$, $z = 3t$, $0 \leq t \leq 1$.

16. Taking $\mathbf{r}_0 = \langle 1, 0, 1 \rangle$ and $\mathbf{r}_1 = \langle 2, 3, 1 \rangle$, we have from Equation 13.5.4 [ET 12.5.4]

$\mathbf{r}(t) = (1 - t)\,\mathbf{r}_0 + t\,\mathbf{r}_1 = (1 - t)\,\langle 1, 0, 1 \rangle + t\,\langle 2, 3, 1 \rangle, 0 \leq t \leq 1$ or $\mathbf{r}(t) = \langle 1 + t, 3t, 1 \rangle, 0 \leq t \leq 1$.

Parametric equations are $x = 1 + t$, $y = 3t$, $z = 1$, $0 \leq t \leq 1$.

17. Taking $\mathbf{r}_0 = \langle 1, -1, 2 \rangle$ and $\mathbf{r}_1 = \langle 4, 1, 7 \rangle$, we have

$\mathbf{r}(t) = (1 - t)\,\mathbf{r}_0 + t\,\mathbf{r}_1 = (1 - t)\,\langle 1, -1, 2 \rangle + t\,\langle 4, 1, 7 \rangle, 0 \leq t \leq 1$ or $\mathbf{r}(t) = \langle 1 + 3t, -1 + 2t, 2 + 5t \rangle, 0 \leq t \leq 1$.

Parametric equations are $x = 1 + 3t$, $y = -1 + 2t$, $z = 2 + 5t$, $0 \leq t \leq 1$.

18. Taking $\mathbf{r}_0 = \langle -2, 4, 0 \rangle$ and $\mathbf{r}_1 = \langle 6, -1, 2 \rangle$, we have

$\mathbf{r}(t) = (1 - t)\,\mathbf{r}_0 + t\,\mathbf{r}_1 = (1 - t)\,\langle -2, 4, 0 \rangle + t\,\langle 6, -1, 2 \rangle, 0 \leq t \leq 1$ or $\mathbf{r}(t) = \langle -2 + 8t, 4 - 5t, 2t \rangle, 0 \leq t \leq 1$.

Parametric equations are $x = -2 + 8t$, $y = 4 - 5t$, $z = 2t$, $0 \leq t \leq 1$.

19. $x = \cos 4t$, $y = t$, $z = \sin 4t$.   At any point (x, y, z) on the curve, $x^2 + z^2 = \cos^2 4t + \sin^2 4t = 1$. So the curve lies on a

circular cylinder with axis the y-axis. Since $y = t$, this is a helix. So the graph is VI.

20. $x = t$, $y = t^2$, $z = e^{-t}$.   At any point on the curve, $y = x^2$. So the curve lies on the parabolic cylinder $y = x^2$. Note that y

and z are positive for all t, and the point $(0, 0, 1)$ is on the curve (when $t = 0$). As $t \to \infty$, $(x, y, z) \to (\infty, \infty, 0)$, while

as $t \to -\infty$, $(x, y, z) \to (-\infty, \infty, \infty)$, so the graph must be II.

21. $x = t$, $y = 1/(1 + t^2)$, $z = t^2$.   Note that y and z are positive for all t. The curve passes through $(0, 1, 0)$ when $t = 0$.

As $t \to \infty$, $(x, y, z) \to (\infty, 0, \infty)$, and as $t \to -\infty$, $(x, y, z) \to (-\infty, 0, \infty)$. So the graph is IV.

22. $x = e^{-t}\cos 10t$, $y = e^{-t}\sin 10t$, $z = e^{-t}$.

$x^2 + y^2 = e^{-2t}\cos^2 10t + e^{-2t}\sin^2 10t = e^{-2t}(\cos^2 10t + \sin^2 10t) = e^{-2t} = z^2$, so the curve lies on the cone

$x^2 + y^2 = z^2$. Also, z is always positive; the graph must be I.

23. $x = \cos t$, $y = \sin t$, $z = \sin 5t$. $x^2 + y^2 = \cos^2 t + \sin^2 t = 1$, so the curve lies on a circular cylinder with axis the

z-axis. Each of x, y and z is periodic, and at $t = 0$ and $t = 2\pi$ the curve passes through the same point, so the curve repeats

itself and the graph is V.

24. $x = \cos t$, $y = \sin t$, $z = \ln t$. $x^2 + y^2 = \cos^2 t + \sin^2 t = 1$, so the curve lies on a circular cylinder with axis the z-axis.

As $t \to 0$, $z \to -\infty$, so the graph is III.

25. If $x = t\cos t$, $y = t\sin t$, $z = t$, then $x^2 + y^2 = t^2\cos^2 t + t^2\sin^2 t = t^2 = z^2$,

so the curve lies on the cone $z^2 = x^2 + y^2$. Since $z = t$, the curve is a spiral on

this cone.

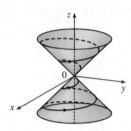

26. Here $x^2 = \sin^2 t = z$ and $x^2 + y^2 = \sin^2 t + \cos^2 t = 1$, so the

curve is contained in the intersection of the parabolic cylinder

$z = x^2$ with the circular cylinder $x^2 + y^2 = 1$. We get the complete

intersection for $0 \le t \le 2\pi$.

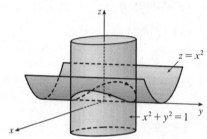

27. Parametric equations for the curve are $x = t$, $y = 0$, $z = 2t - t^2$. Substituting into the equation of the paraboloid gives

$2t - t^2 = t^2 \quad \Rightarrow \quad 2t = 2t^2 \quad \Rightarrow \quad t = 0, 1$. Since $\mathbf{r}(0) = \mathbf{0}$ and $\mathbf{r}(1) = \mathbf{i} + \mathbf{k}$, the points of intersection

are $(0, 0, 0)$ and $(1, 0, 1)$.

28. Parametric equations for the helix are $x = \sin t$, $y = \cos t$, $z = t$. Substituting into the equation of the sphere gives

$\sin^2 t + \cos^2 t + t^2 = 5 \quad \Rightarrow \quad 1 + t^2 = 5 \quad \Rightarrow \quad t = \pm 2$. Since $\mathbf{r}(2) = \langle \sin 2, \cos 2, 2 \rangle$ and

$\mathbf{r}(-2) = \langle \sin(-2), \cos(-2), -2 \rangle$, the points of intersection are $(\sin 2, \cos 2, 2) \approx (0.909, -0.416, 2)$ and

$(\sin(-2), \cos(-2), -2) \approx (-0.909, -0.416, -2)$.

29. $\mathbf{r}(t) = \langle \cos t \sin 2t, \sin t \sin 2t, \cos 2t \rangle$.

We include both a regular plot and a plot

showing a tube of radius 0.08 around the

curve.

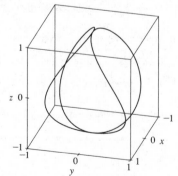

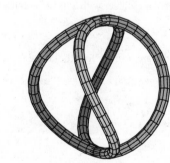

30. $\mathbf{r}(t) = \langle t^2, \ln t, t \rangle$

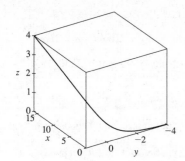

31. $\mathbf{r}(t) = \langle t, t \sin t, t \cos t \rangle$

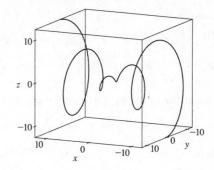

32. $\mathbf{r}(t) = \langle t, e^t, \cos t \rangle$

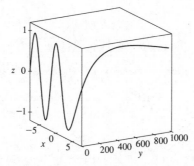

33.

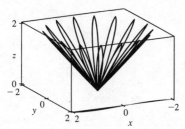

$x = (1 + \cos 16t) \cos t$, $y = (1 + \cos 16t) \sin t$, $z = 1 + \cos 16t$. At any point on the graph,

$$x^2 + y^2 = (1 + \cos 16t)^2 \cos^2 t + (1 + \cos 16t)^2 \sin^2 t$$
$$= (1 + \cos 16t)^2 = z^2,$$

so the graph lies on the cone $x^2 + y^2 = z^2$. From the graph at left, we see that this curve looks like the projection of a leaved two-dimensional curve onto a cone.

34.

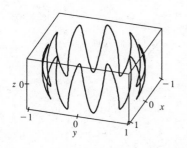

$x = \sqrt{1 - 0.25 \cos^2 10t} \cos t$, $y = \sqrt{1 - 0.25 \cos^2 10t} \sin t$, $z = 0.5 \cos 10t$. At any point on the graph,

$$x^2 + y^2 + z^2 = (1 - 0.25 \cos^2 10t) \cos^2 t$$
$$+ (1 - 0.25 \cos^2 10t) \sin^2 t + 0.25 \cos^2 t$$
$$= 1 - 0.25 \cos^2 10t + 0.25 \cos^2 10t = 1,$$

so the graph lies on the sphere $x^2 + y^2 + z^2 = 1$, and since $z = 0.5 \cos 10t$ the graph resembles a trigonometric curve with ten peaks projected onto the sphere. The graph is generated by $t \in [0, 2\pi]$.

35. If $t = -1$, then $x = 1$, $y = 4$, $z = 0$, so the curve passes through the point $(1, 4, 0)$. If $t = 3$, then $x = 9$, $y = -8$, $z = 28$, so the curve passes through the point $(9, -8, 28)$. For the point $(4, 7, -6)$ to be on the curve, we require $y = 1 - 3t = 7 \Rightarrow t = -2$. But then $z = 1 + (-2)^3 = -7 \neq -6$, so $(4, 7, -6)$ is not on the curve.

36. The projection of the curve C of intersection onto the xy-plane is the circle $x^2 + y^2 = 4$, $z = 0$.

Then we can write $x = 2\cos t$, $y = 2\sin t$, $0 \le t \le 2\pi$. Since C also lies on the surface $z = xy$, we have

$z = xy = (2\cos t)(2\sin t) = 4\cos t \sin t$, or $2\sin(2t)$. Then parametric equations for C are $x = 2\cos t$, $y = 2\sin t$,

$z = 2\sin(2t)$, $0 \le t \le 2\pi$, and the corresponding vector function is $\mathbf{r}(t) = 2\cos t\,\mathbf{i} + 2\sin t\,\mathbf{j} + 2\sin(2t)\,\mathbf{k}$, $0 \le t \le 2\pi$.

37. Both equations are solved for z, so we can substitute to eliminate z: $\sqrt{x^2 + y^2} = 1 + y \;\Rightarrow\; x^2 + y^2 = 1 + 2y + y^2 \;\Rightarrow\;$

$x^2 = 1 + 2y \;\Rightarrow\; y = \frac{1}{2}(x^2 - 1)$. We can form parametric equations for the curve C of intersection by choosing a

parameter $x = t$, then $y = \frac{1}{2}(t^2 - 1)$ and $z = 1 + y = 1 + \frac{1}{2}(t^2 - 1) = \frac{1}{2}(t^2 + 1)$. Thus a vector function representing C

is $\mathbf{r}(t) = t\,\mathbf{i} + \frac{1}{2}(t^2 - 1)\,\mathbf{j} + \frac{1}{2}(t^2 + 1)\,\mathbf{k}$.

38. The projection of the curve C of intersection onto the xy-plane is the parabola $y = x^2$, $z = 0$. Then we can choose the

parameter $x = t \;\Rightarrow\; y = t^2$. Since C also lies on the surface $z = 4x^2 + y^2$, we have $z = 4x^2 + y^2 = 4t^2 + (t^2)^2$.

Then parametric equations for C are $x = t$, $y = t^2$, $z = 4t^2 + t^4$, and the corresponding vector function

is $\mathbf{r}(t) = t\,\mathbf{i} + t^2\,\mathbf{j} + (4t^2 + t^4)\,\mathbf{k}$.

39.

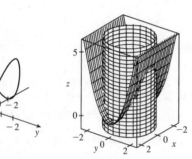
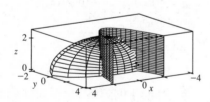

The projection of the curve C of intersection onto the

xy-plane is the circle $x^2 + y^2 = 4$, $z = 0$. Then we can write

$x = 2\cos t$, $y = 2\sin t$, $0 \le t \le 2\pi$. Since C also lies on

the surface $z = x^2$, we have $z = x^2 = (2\cos t)^2 = 4\cos^2 t$.

Then parametric equations for C are $x = 2\cos t$, $y = 2\sin t$,

$z = 4\cos^2 t$, $0 \le t \le 2\pi$.

40.

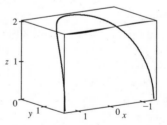

$x = t \;\Rightarrow\; y = t^2 \;\Rightarrow\; 4z^2 = 16 - x^2 - 4y^2 = 16 - t^2 - 4t^4 \;\Rightarrow\; z = \sqrt{4 - \left(\frac{1}{2}t\right)^2 - t^4}$.

Note that z is positive because the intersection is with the top half of the ellipsoid. Hence the curve is given

by $x = t$, $y = t^2$, $z = \sqrt{4 - \frac{1}{4}t^2 - t^4}$.

41. For the particles to collide, we require $\mathbf{r}_1(t) = \mathbf{r}_2(t)$ ⟺ $\langle t^2, 7t - 12, t^2 \rangle = \langle 4t - 3, t^2, 5t - 6 \rangle$. Equating components

gives $t^2 = 4t - 3$, $7t - 12 = t^2$, and $t^2 = 5t - 6$. From the first equation, $t^2 - 4t + 3 = 0$ ⟺ $(t-3)(t-1) = 0$ so $t = 1$

or $t = 3$. $t = 1$ does not satisfy the other two equations, but $t = 3$ does. The particles collide when $t = 3$, at the

point $(9, 9, 9)$.

42. The particles collide provided $\mathbf{r}_1(t) = \mathbf{r}_2(t)$ ⟺ $\langle t, t^2, t^3 \rangle = \langle 1 + 2t, 1 + 6t, 1 + 14t \rangle$. Equating components gives

$t = 1 + 2t$, $t^2 = 1 + 6t$, and $t^3 = 1 + 14t$. The first equation gives $t = -1$, but this does not satisfy the other equations, so

the particles do not collide. For the paths to intersect, we need to find a value for t and a value for s where $\mathbf{r}_1(t) = \mathbf{r}_2(s)$ ⟺

$\langle t, t^2, t^3 \rangle = \langle 1 + 2s, 1 + 6s, 1 + 14s \rangle$. Equating components, $t = 1 + 2s$, $t^2 = 1 + 6s$, and $t^3 = 1 + 14s$. Substituting the

first equation into the second gives $(1 + 2s)^2 = 1 + 6s$ ⟹ $4s^2 - 2s = 0$ ⟹ $2s(2s - 1) = 0$ ⟹ $s = 0$ or $s = \frac{1}{2}$.

From the first equation, $s = 0$ ⟹ $t = 1$ and $s = \frac{1}{2}$ ⟹ $t = 2$. Checking, we see that both pairs of values satisfy the

third equation. Thus the paths intersect twice, at the point $(1, 1, 1)$ when $s = 0$ and $t = 1$, and at $(2, 4, 8)$ when $s = \frac{1}{2}$

and $t = 2$.

43. (a) $\lim\limits_{t \to a} \mathbf{u}(t) + \lim\limits_{t \to a} \mathbf{v}(t) = \left\langle \lim\limits_{t \to a} u_1(t), \lim\limits_{t \to a} u_2(t), \lim\limits_{t \to a} u_3(t) \right\rangle + \left\langle \lim\limits_{t \to a} v_1(t), \lim\limits_{t \to a} v_2(t), \lim\limits_{t \to a} v_3(t) \right\rangle$ and the limits of these

component functions must each exist since the vector functions both possess limits as $t \to a$. Then adding the two vectors

and using the addition property of limits for real-valued functions, we have that

$$\lim_{t \to a} \mathbf{u}(t) + \lim_{t \to a} \mathbf{v}(t) = \left\langle \lim_{t \to a} u_1(t) + \lim_{t \to a} v_1(t), \lim_{t \to a} u_2(t) + \lim_{t \to a} v_2(t), \lim_{t \to a} u_3(t) + \lim_{t \to a} v_3(t) \right\rangle$$

$$= \left\langle \lim_{t \to a} [u_1(t) + v_1(t)], \lim_{t \to a} [u_2(t) + v_2(t)], \lim_{t \to a} [u_3(t) + v_3(t)] \right\rangle$$

$$= \lim_{t \to a} \langle u_1(t) + v_1(t), u_2(t) + v_2(t), u_3(t) + v_3(t) \rangle \qquad \text{[using (1) backward]}$$

$$= \lim_{t \to a} [\mathbf{u}(t) + \mathbf{v}(t)]$$

(b) $\lim\limits_{t \to a} c\mathbf{u}(t) = \lim\limits_{t \to a} \langle cu_1(t), cu_2(t), cu_3(t) \rangle = \left\langle \lim\limits_{t \to a} cu_1(t), \lim\limits_{t \to a} cu_2(t), \lim\limits_{t \to a} cu_3(t) \right\rangle$

$$= \left\langle c \lim_{t \to a} u_1(t), c \lim_{t \to a} u_2(t), c \lim_{t \to a} u_3(t) \right\rangle = c \left\langle \lim_{t \to a} u_1(t), \lim_{t \to a} u_2(t), \lim_{t \to a} u_3(t) \right\rangle$$

$$= c \lim_{t \to a} \langle u_1(t), u_2(t), u_3(t) \rangle = c \lim_{t \to a} \mathbf{u}(t)$$

(c) $\lim\limits_{t \to a} \mathbf{u}(t) \cdot \lim\limits_{t \to a} \mathbf{v}(t) = \left\langle \lim\limits_{t \to a} u_1(t), \lim\limits_{t \to a} u_2(t), \lim\limits_{t \to a} u_3(t) \right\rangle \cdot \left\langle \lim\limits_{t \to a} v_1(t), \lim\limits_{t \to a} v_2(t), \lim\limits_{t \to a} v_3(t) \right\rangle$

$$= \left[\lim_{t \to a} u_1(t) \right] \left[\lim_{t \to a} v_1(t) \right] + \left[\lim_{t \to a} u_2(t) \right] \left[\lim_{t \to a} v_2(t) \right] + \left[\lim_{t \to a} u_3(t) \right] \left[\lim_{t \to a} v_3(t) \right]$$

$$= \lim_{t \to a} u_1(t)v_1(t) + \lim_{t \to a} u_2(t)v_2(t) + \lim_{t \to a} u_3(t)v_3(t)$$

$$= \lim_{t \to a} [u_1(t)v_1(t) + u_2(t)v_2(t) + u_3(t)v_3(t)] = \lim_{t \to a} [\mathbf{u}(t) \cdot \mathbf{v}(t)]$$

(d) $\displaystyle\lim_{t\to a}\mathbf{u}(t)\times\lim_{t\to a}\mathbf{v}(t)=\left\langle\lim_{t\to a}u_1(t),\lim_{t\to a}u_2(t),\lim_{t\to a}u_3(t)\right\rangle\times\left\langle\lim_{t\to a}v_1(t),\lim_{t\to a}v_2(t),\lim_{t\to a}v_3(t)\right\rangle$

$\qquad = \left\langle\left[\lim_{t\to a}u_2(t)\right]\left[\lim_{t\to a}v_3(t)\right]-\left[\lim_{t\to a}u_3(t)\right]\left[\lim_{t\to a}v_2(t)\right],\right.$

$\qquad\qquad \left[\lim_{t\to a}u_3(t)\right]\left[\lim_{t\to a}v_1(t)\right]-\left[\lim_{t\to a}u_1(t)\right]\left[\lim_{t\to a}v_3(t)\right],$

$\qquad\qquad \left.\left[\lim_{t\to a}u_1(t)\right]\left[\lim_{t\to a}v_2(t)\right]-\left[\lim_{t\to a}u_2(t)\right]\left[\lim_{t\to a}v_1(t)\right]\right\rangle$

$\qquad = \left\langle\lim_{t\to a}[u_2(t)v_3(t)-u_3(t)v_2(t)],\lim_{t\to a}[u_3(t)v_1(t)-u_1(t)v_3(t)],\right.$

$\qquad\qquad \left.\lim_{t\to a}[u_1(t)v_2(t)-u_2(t)v_1(t)]\right\rangle$

$\qquad = \lim_{t\to a}\left\langle u_2(t)v_3(t)-u_3(t)v_2(t),u_3(t)v_1(t)-u_1(t)v_3(t),u_1(t)v_2(t)-u_2(t)v_1(t)\right\rangle$

$\qquad = \lim_{t\to a}[\mathbf{u}(t)\times\mathbf{v}(t)]$

44. The projection of the curve onto the xy-plane is given by the parametric equations $x=(2+\cos 1.5t)\cos t$,

$y=(2+\cos 1.5t)\sin t$. If we convert to polar coordinates, we have

$r^2=x^2+y^2=[(2+\cos 1.5t)\cos t]^2+[(2+\cos 1.5t)\sin t]^2=(2+\cos 1.5t)^2(\cos^2 t+\sin^2 t)=(2+\cos 1.5t)^2\quad\Rightarrow$

$r=2+\cos 1.5t$. Also, $\tan\theta=\dfrac{y}{x}=\dfrac{(2+\cos 1.5t)\sin t}{(2+\cos 1.5t)\cos t}=\tan t\quad\Rightarrow\quad\theta=t$.

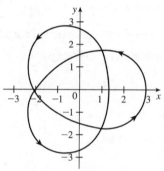

Thus the polar equation of the curve is $r=2+\cos 1.5\theta$. At $\theta=0$, we have

$r=3$, and r decreases to 1 as θ increases to $\frac{2\pi}{3}$. For $\frac{2\pi}{3}\le\theta\le\frac{4\pi}{3}$, r

increases to 3; r decreases to 1 again at $\theta=2\pi$, increases to 3 at $\theta=\frac{8\pi}{3}$,

decreases to 1 at $\theta=\frac{10\pi}{3}$, and completes the closed curve by increasing

to 3 at $\theta=4\pi$. We sketch an approximate graph as shown in the figure.

We can determine how the curve passes over itself by investigating the maximum and minimum values of z for

$t=\theta\in[0,4\pi]$. Since $z=\sin 1.5t$, z is maximized where $\sin 1.5t=1\quad\Rightarrow\quad 1.5t=\frac{\pi}{2},\frac{5\pi}{2},$ or $\frac{9\pi}{2}\quad\Rightarrow$

$t=\frac{\pi}{3},\frac{5\pi}{3},$ or 3π. z is minimized where $\sin 1.5t=-1\quad\Rightarrow$

$1.5t=\frac{3\pi}{2},\frac{7\pi}{2},$ or $\frac{11\pi}{2}\quad\Rightarrow\quad t=\pi,\frac{7\pi}{3},$ or $\frac{11\pi}{3}$. Note that these are

precisely the values for which $\cos 1.5t=0\quad\Rightarrow\quad r=2$, and on the graph

of the projection, these six points appear to be at the three self-intersections

we see. Comparing the maximum and minimum values of z at these

intersections, we can determine where the curve passes over itself, as

indicated in the figure.

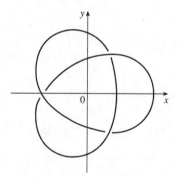

We show a computer-drawn graph of the curve from above, as well as views from the front and from the right side.

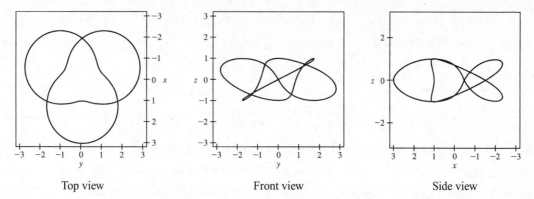

Top view　　　　　　　　Front view　　　　　　　　Side view

The top view graph shows a more accurate representation of the projection of the trefoil knot on the xy-plane (the axes are rotated $90°$). Notice the indentations the graph exhibits at the points corresponding to $r = 1$. Finally, we graph several additional viewpoints of the trefoil knot, along with two plots showing a tube of radius 0.2 around the curve.

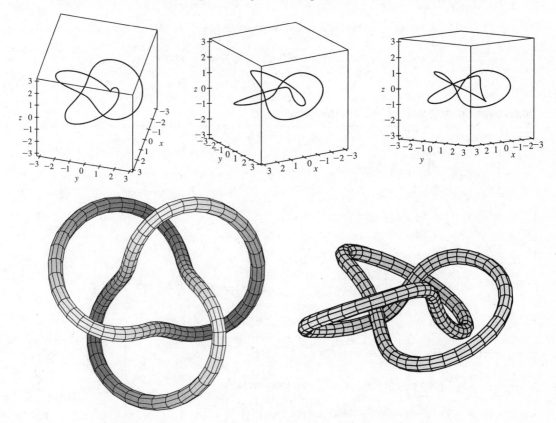

45. Let $\mathbf{r}(t) = \langle f(t), g(t), h(t) \rangle$ and $\mathbf{b} = \langle b_1, b_2, b_3 \rangle$. If $\lim\limits_{t \to a} \mathbf{r}(t) = \mathbf{b}$, then $\lim\limits_{t \to a} \mathbf{r}(t)$ exists, so by (1),

$\mathbf{b} = \lim\limits_{t \to a} \mathbf{r}(t) = \left\langle \lim\limits_{t \to a} f(t), \lim\limits_{t \to a} g(t), \lim\limits_{t \to a} h(t) \right\rangle$. By the definition of equal vectors we have $\lim\limits_{t \to a} f(t) = b_1, \lim\limits_{t \to a} g(t) = b_2$

and $\lim\limits_{t \to a} h(t) = b_3$. But these are limits of real-valued functions, so by the definition of limits, for every $\varepsilon > 0$ there exists

$\delta_1 > 0$, $\delta_2 > 0$, $\delta_3 > 0$ so that if $0 < |t - a| < \delta_1$ then $|f(t) - b_1| < \varepsilon/3$, if $0 < |t - a| < \delta_2$ then $|g(t) - b_2| < \varepsilon/3$, and

if $0 < |t - a| < \delta_3$ then $|h(t) - b_3| < \varepsilon/3$. Letting $\delta = $ minimum of $\{\delta_1, \delta_2, \delta_3\}$, then if $0 < |t - a| < \delta$ we have

$|f(t) - b_1| + |g(t) - b_2| + |h(t) - b_3| < \varepsilon/3 + \varepsilon/3 + \varepsilon/3 = \varepsilon$. But

$$|\mathbf{r}(t) - \mathbf{b}| = |\langle f(t) - b_1, g(t) - b_2, h(t) - b_3\rangle| = \sqrt{(f(t) - b_1)^2 + (g(t) - b_2)^2 + (h(t) - b_3)^2}$$

$$\le \sqrt{[f(t) - b_1]^2} + \sqrt{[g(t) - b_2]^2} + \sqrt{[h(t) - b_3]^2} = |f(t) - b_1| + |g(t) - b_2| + |h(t) - b_3|$$

Thus for every $\varepsilon > 0$ there exists $\delta > 0$ such that if $0 < |t - a| < \delta$ then

$|\mathbf{r}(t) - \mathbf{b}| \le |f(t) - b_1| + |g(t) - b_2| + |h(t) - b_3| < \varepsilon$. Conversely, suppose for every $\varepsilon > 0$, there exists $\delta > 0$ such

that if $0 < |t - a| < \delta$ then $|\mathbf{r}(t) - \mathbf{b}| < \varepsilon$ $\Leftrightarrow$ $|\langle f(t) - b_1, g(t) - b_2, h(t) - b_3\rangle| < \varepsilon$ $\Leftrightarrow$

$\sqrt{[f(t) - b_1]^2 + [g(t) - b_2]^2 + [h(t) - b_3]^2} < \varepsilon$ $\Leftrightarrow$ $[f(t) - b_1]^2 + [g(t) - b_2]^2 + [h(t) - b_3]^2 < \varepsilon^2$. But each term

on the left side of the last inequality is positive, so if $0 < |t - a| < \delta$, then $[f(t) - b_1]^2 < \varepsilon^2$, $[g(t) - b_2]^2 < \varepsilon^2$ and

$[h(t) - b_3]^2 < \varepsilon^2$ or, taking the square root of both sides in each of the above, $|f(t) - b_1| < \varepsilon$, $|g(t) - b_2| < \varepsilon$ and

$|h(t) - b_3| < \varepsilon$. And by definition of limits of real-valued functions we have $\lim\limits_{t \to a} f(t) = b_1$, $\lim\limits_{t \to a} g(t) = b_2$ and

$\lim\limits_{t \to a} h(t) = b_3$. But by (1), $\lim\limits_{t \to a} \mathbf{r}(t) = \left\langle \lim\limits_{t \to a} f(t), \lim\limits_{t \to a} g(t), \lim\limits_{t \to a} h(t) \right\rangle$, so $\lim\limits_{t \to a} \mathbf{r}(t) = \langle b_1, b_2, b_3 \rangle = \mathbf{b}$.

14.2 Derivatives and Integrals of Vector Functions

ET 13.2

1. (a)

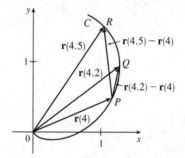

(b) $\dfrac{\mathbf{r}(4.5) - \mathbf{r}(4)}{0.5} = 2[\mathbf{r}(4.5) - \mathbf{r}(4)]$, so we draw a vector in the same

direction but with twice the length of the vector $\mathbf{r}(4.5) - \mathbf{r}(4)$.

$\dfrac{\mathbf{r}(4.2) - \mathbf{r}(4)}{0.2} = 5[\mathbf{r}(4.2) - \mathbf{r}(4)]$, so we draw a vector in the same

direction but with 5 times the length of the vector $\mathbf{r}(4.2) - \mathbf{r}(4)$.

(c) By Definition 1, $\mathbf{r}'(4) = \lim\limits_{h \to 0} \dfrac{\mathbf{r}(4 + h) - \mathbf{r}(4)}{h}$. $\quad \mathbf{T}(4) = \dfrac{\mathbf{r}'(4)}{|\mathbf{r}'(4)|}$.

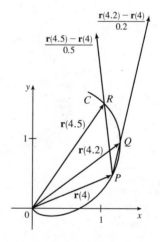

(d) $\mathbf{T}(4)$ is a unit vector in the same direction as $\mathbf{r}'(4)$, that is, parallel to the tangent line to the curve at $\mathbf{r}(4)$ with length 1.

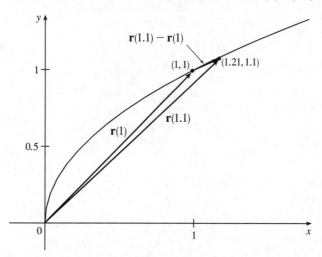

2. (a) The curve can be represented by the parametric equations $x = t^2$, $y = t$, $0 \le t \le 2$. Eliminating the parameter, we have $x = y^2$, $0 \le y \le 2$, a portion of which we graph here, along with the vectors $\mathbf{r}(1)$, $\mathbf{r}(1.1)$, and $\mathbf{r}(1.1) - \mathbf{r}(1)$.

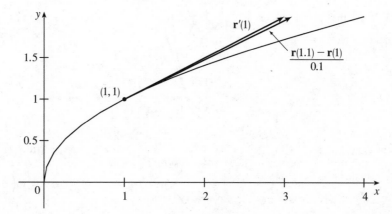

(b) Since $\mathbf{r}(t) = \langle t^2, t \rangle$, we differentiate components, giving $\mathbf{r}'(t) = \langle 2t, 1 \rangle$, so $\mathbf{r}'(1) = \langle 2, 1 \rangle$.

$$\frac{\mathbf{r}(1.1) - \mathbf{r}(1)}{0.1} = \frac{\langle 1.21, 1.1 \rangle - \langle 1, 1 \rangle}{0.1} = 10 \langle 0.21, 0.1 \rangle = \langle 2.1, 1 \rangle.$$

As we can see from the graph, these vectors are very close in length and direction. $\mathbf{r}'(1)$ is defined to be $\displaystyle\lim_{h \to 0} \frac{\mathbf{r}(1+h) - \mathbf{r}(1)}{h}$, and we recognize $\dfrac{\mathbf{r}(1.1) - \mathbf{r}(1)}{0.1}$ as the expression after the limit sign with $h = 0.1$. Since h is close to 0, we would expect $\dfrac{\mathbf{r}(1.1) - \mathbf{r}(1)}{0.1}$ to be a vector close to $\mathbf{r}'(1)$.

3. Since $(x+2)^2 = t^2 = y - 1 \Rightarrow$
$y = (x+2)^2 - 1$, the curve is a
parabola.

(a), (c)

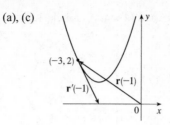

(b) $\mathbf{r}'(t) = \langle 1, 2t \rangle$,
$\mathbf{r}'(-1) = \langle 1, -2 \rangle$

4. Since $x = 1 + t = 1 + y^2$, the
curve is part of a parabola. Here we
have $y \geq 0$.

(a), (c)

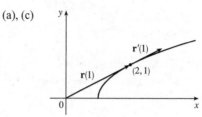

(b) $\mathbf{r}'(t) = \left\langle 1, \dfrac{1}{2\sqrt{t}} \right\rangle$,
$\mathbf{r}'(1) = \langle 1, \tfrac{1}{2} \rangle$

5. $x = \sin t$, $y = 2\cos t$ so
$x^2 + (y/2)^2 = 1$ and the curve is
an ellipse.

(a), (c)

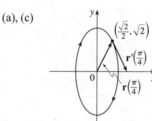

(b) $\mathbf{r}'(t) = \cos t\,\mathbf{i} - 2\sin t\,\mathbf{j}$,
$\mathbf{r}'\left(\dfrac{\pi}{4}\right) = \dfrac{\sqrt{2}}{2}\,\mathbf{i} - \sqrt{2}\,\mathbf{j}$

6. Since $y = e^{-t} = \dfrac{1}{e^t} = \dfrac{1}{x}$ the
curve is part of the hyperbola
$y = \dfrac{1}{x}$. Note that $x > 0$, $y > 0$.

(a), (c)

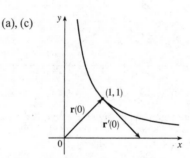

(b) $\mathbf{r}'(t) = e^t\,\mathbf{i} - e^{-t}\,\mathbf{j}$,
$\mathbf{r}'(0) = \mathbf{i} - \mathbf{j}$

7. Since $y = e^{3t} = (e^t)^3 = x^3$, the
curve is part of a cubic cuve. Note
that here, $x > 0$.

(a), (c)

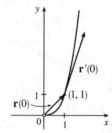

(b) $\mathbf{r}'(t) = e^t\,\mathbf{i} + 3e^{3t}\,\mathbf{j}$,
$\mathbf{r}'(0) = \mathbf{i} + 3\,\mathbf{j}$

8. $x = 1 + \cos t$, $y = 2 + \sin t$ so
$(x-1)^2 + (y-2)^2 = 1$ and the
curve is a circle.

(a), (c)

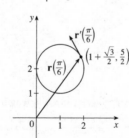

(b) $\mathbf{r}'(t) = -\sin t\,\mathbf{i} + \cos t\,\mathbf{j}$,
$\mathbf{r}'\left(\dfrac{\pi}{6}\right) = -\dfrac{1}{2}\,\mathbf{i} + \dfrac{\sqrt{3}}{2}\,\mathbf{j}$

9. $\mathbf{r}'(t) = \left\langle \dfrac{d}{dt}\left[t\sin t\right], \dfrac{d}{dt}\left[t^2\right], \dfrac{d}{dt}\left[t\cos 2t\right] \right\rangle = \langle t\cos t + \sin t, 2t, t(-\sin 2t)\cdot 2 + \cos 2t\rangle$

$\qquad = \langle t\cos t + \sin t, 2t, \cos 2t - 2t\sin 2t\rangle$

10. $\mathbf{r}(t) = \langle \tan t, \sec t, 1/t^2\rangle \;\; \Rightarrow \;\; \mathbf{r}'(t) = \langle \sec^2 t, \sec t\tan t, -2/t^3\rangle$

11. $\mathbf{r}(t) = \mathbf{i} - \mathbf{j} + e^{4t}\,\mathbf{k} \;\; \Rightarrow \;\; \mathbf{r}'(t) = 0\,\mathbf{i} + 0\,\mathbf{j} + 4e^{4t}\,\mathbf{k} = 4e^{4t}\,\mathbf{k}$

12. $\mathbf{r}(t) = \sin^{-1} t\,\mathbf{i} + \sqrt{1-t^2}\,\mathbf{j} + \mathbf{k} \;\; \Rightarrow \;\; \mathbf{r}'(t) = \dfrac{1}{\sqrt{1-t^2}}\,\mathbf{i} - \dfrac{t}{\sqrt{1-t^2}}\,\mathbf{j}$

13. $\mathbf{r}(t) = e^{t^2}\,\mathbf{i} - \mathbf{j} + \ln(1+3t)\,\mathbf{k} \;\; \Rightarrow \;\; \mathbf{r}'(t) = 2te^{t^2}\,\mathbf{i} + \dfrac{3}{1+3t}\,\mathbf{k}$

14. $\mathbf{r}'(t) = [at(-3\sin 3t) + a\cos 3t]\,\mathbf{i} + b\cdot 3\sin^2 t\cos t\,\mathbf{j} + c\cdot 3\cos^2 t(-\sin t)\,\mathbf{k}$

$\qquad = (a\cos 3t - 3at\sin 3t)\,\mathbf{i} + 3b\sin^2 t\,\cos t\,\mathbf{j} - 3c\cos^2 t\,\sin t\,\mathbf{k}$

15. $\mathbf{r}'(t) = \mathbf{0} + \mathbf{b} + 2t\,\mathbf{c} = \mathbf{b} + 2t\,\mathbf{c}$ by Formulas 1 and 3 of Theorem 3.

16. To find $\mathbf{r}'(t)$, we first expand $\mathbf{r}(t) = t\,\mathbf{a}\times(\mathbf{b}+t\,\mathbf{c}) = t(\mathbf{a}\times\mathbf{b}) + t^2(\mathbf{a}\times\mathbf{c})$, so $\mathbf{r}'(t) = \mathbf{a}\times\mathbf{b} + 2t(\mathbf{a}\times\mathbf{c})$.

17. $\mathbf{r}'(t) = \langle -te^{-t} + e^{-t}, 2/(1+t^2), 2e^t\rangle \;\; \Rightarrow \;\; \mathbf{r}'(0) = \langle 1, 2, 2\rangle$. So $|\mathbf{r}'(0)| = \sqrt{1^2 + 2^2 + 2^2} = \sqrt{9} = 3$ and

$\qquad \mathbf{T}(0) = \dfrac{\mathbf{r}'(0)}{|\mathbf{r}'(0)|} = \frac{1}{3}\langle 1, 2, 2\rangle = \left\langle \frac{1}{3}, \frac{2}{3}, \frac{2}{3}\right\rangle$.

18. $\mathbf{r}'(t) = \dfrac{2}{\sqrt{t}}\,\mathbf{i} + 2t\,\mathbf{j} + \mathbf{k} \;\; \Rightarrow \;\; \mathbf{r}'(1) = 2\,\mathbf{i} + 2\,\mathbf{j} + \mathbf{k}$. Thus

$\qquad \mathbf{T}(1) = \dfrac{\mathbf{r}'(1)}{|\mathbf{r}'(1)|} = \dfrac{1}{\sqrt{2^2 + 2^2 + 1^2}}\,(2\,\mathbf{i} + 2\,\mathbf{j} + \mathbf{k}) = \frac{1}{3}(2\,\mathbf{i} + 2\,\mathbf{j} + \mathbf{k}) = \frac{2}{3}\,\mathbf{i} + \frac{2}{3}\,\mathbf{j} + \frac{1}{3}\,\mathbf{k}$.

19. $\mathbf{r}'(t) = -\sin t\,\mathbf{i} + 3\,\mathbf{j} + 4\cos 2t\,\mathbf{k} \;\; \Rightarrow \;\; \mathbf{r}'(0) = 3\,\mathbf{j} + 4\,\mathbf{k}$. Thus

$\qquad \mathbf{T}(0) = \dfrac{\mathbf{r}'(0)}{|\mathbf{r}'(0)|} = \dfrac{1}{\sqrt{0^2 + 3^2 + 4^2}}\,(3\,\mathbf{j} + 4\,\mathbf{k}) = \frac{1}{5}(3\,\mathbf{j} + 4\,\mathbf{k}) = \frac{3}{5}\,\mathbf{j} + \frac{4}{5}\,\mathbf{k}$.

20. $\mathbf{r}'(t) = 2\cos t\,\mathbf{i} - 2\sin t\,\mathbf{j} + \sec^2 t\,\mathbf{k} \;\; \Rightarrow \;\; \mathbf{r}'\left(\frac{\pi}{4}\right) = \sqrt{2}\,\mathbf{i} - \sqrt{2}\,\mathbf{j} + 2\,\mathbf{k}$ and $\left|\mathbf{r}'\left(\frac{\pi}{4}\right)\right| = \sqrt{2 + 2 + 4} = 2\sqrt{2}$. Thus

$\qquad \mathbf{T}\left(\frac{\pi}{4}\right) = \dfrac{\mathbf{r}'\left(\frac{\pi}{4}\right)}{\left|\mathbf{r}'\left(\frac{\pi}{4}\right)\right|} = \dfrac{1}{2\sqrt{2}}\left(\sqrt{2}\,\mathbf{i} - \sqrt{2}\,\mathbf{j} + 2\,\mathbf{k}\right) = \frac{1}{2}\,\mathbf{i} - \frac{1}{2}\,\mathbf{j} + \frac{1}{\sqrt{2}}\,\mathbf{k}$.

21. $\mathbf{r}(t) = \langle t, t^2, t^3\rangle \;\; \Rightarrow \;\; \mathbf{r}'(t) = \langle 1, 2t, 3t^2\rangle$. Then $\mathbf{r}'(1) = \langle 1, 2, 3\rangle$ and $|\mathbf{r}'(1)| = \sqrt{1^2 + 2^2 + 3^2} = \sqrt{14}$, so

$\qquad \mathbf{T}(1) = \dfrac{\mathbf{r}'(1)}{|\mathbf{r}'(1)|} = \frac{1}{\sqrt{14}}\langle 1, 2, 3\rangle = \left\langle \frac{1}{\sqrt{14}}, \frac{2}{\sqrt{14}}, \frac{3}{\sqrt{14}}\right\rangle$. $\mathbf{r}''(t) = \langle 0, 2, 6t\rangle$, so

$$\mathbf{r}'(t)\times\mathbf{r}''(t) = \begin{vmatrix} \mathbf{i} & \mathbf{j} & \mathbf{k} \\ 1 & 2t & 3t^2 \\ 0 & 2 & 6t \end{vmatrix} = \begin{vmatrix} 2t & 3t^2 \\ 2 & 6t \end{vmatrix}\mathbf{i} - \begin{vmatrix} 1 & 3t^2 \\ 0 & 6t \end{vmatrix}\mathbf{j} + \begin{vmatrix} 1 & 2t \\ 0 & 2 \end{vmatrix}\mathbf{k}$$

$$= (12t^2 - 6t^2)\,\mathbf{i} - (6t - 0)\,\mathbf{j} + (2 - 0)\,\mathbf{k} = \langle 6t^2, -6t, 2\rangle$$

22. $\mathbf{r}(t) = \langle e^{2t}, e^{-2t}, te^{2t} \rangle$ $\Rightarrow$ $\mathbf{r}'(t) = \langle 2e^{2t}, -2e^{-2t}, (2t+1)e^{2t} \rangle$ $\Rightarrow$ $\mathbf{r}'(0) = \langle 2e^0, -2e^0, (0+1)e^0 \rangle = \langle 2, -2, 1 \rangle$

and $|\mathbf{r}'(0)| = \sqrt{2^2 + (-2)^2 + 1^2} = 3$. Then $\mathbf{T}(0) = \dfrac{\mathbf{r}'(0)}{|\mathbf{r}'(0)|} = \frac{1}{3}\langle 2, -2, 1 \rangle = \langle \frac{2}{3}, -\frac{2}{3}, \frac{1}{3} \rangle$.

$\mathbf{r}''(t) = \langle 4e^{2t}, 4e^{-2t}, (4t+4)e^{2t} \rangle$ $\Rightarrow$ $\mathbf{r}''(0) = \langle 4e^0, 4e^0, (0+4)e^0 \rangle = \langle 4, 4, 4 \rangle$.

$\mathbf{r}'(t) \cdot \mathbf{r}''(t) = \langle 2e^{2t}, -2e^{-2t}, (2t+1)e^{2t} \rangle \cdot \langle 4e^{2t}, 4e^{-2t}, (4t+4)e^{2t} \rangle$

$\qquad = (2e^{2t})(4e^{2t}) + (-2e^{-2t})(4e^{-2t}) + ((2t+1)e^{2t})((4t+4)e^{2t})$

$\qquad = 8e^{4t} - 8e^{-4t} + (8t^2 + 12t + 4)e^{4t} = (8t^2 + 12t + 12)e^{4t} - 8e^{-4t}$

23. The vector equation for the curve is $\mathbf{r}(t) = \langle 1 + 2\sqrt{t}, t^3 - t, t^3 + t \rangle$, so $\mathbf{r}'(t) = \langle 1/\sqrt{t}, 3t^2 - 1, 3t^2 + 1 \rangle$. The point

$(3, 0, 2)$ corresponds to $t = 1$, so the tangent vector there is $\mathbf{r}'(1) = \langle 1, 2, 4 \rangle$. Thus, the tangent line goes through the point

$(3, 0, 2)$ and is parallel to the vector $\langle 1, 2, 4 \rangle$. Parametric equations are $x = 3 + t$, $y = 2t$, $z = 2 + 4t$.

24. The vector equation for the curve is $\mathbf{r}(t) = \langle e^t, te^t, te^{t^2} \rangle$, so $\mathbf{r}'(t) = \langle e^t, te^t + e^t, 2t^2 e^{t^2} + e^{t^2} \rangle$. The point $(1, 0, 0)$

corresponds to $t = 0$, so the tangent vector there is $\mathbf{r}'(0) = \langle 1, 1, 1 \rangle$. Thus, the tangent line is parallel to the vector $\langle 1, 1, 1 \rangle$

and includes the point $(1, 0, 0)$. Parametric equations are $x = 1 + 1 \cdot t = 1 + t$, $y = 0 + 1 \cdot t = t$, $z = 0 + 1 \cdot t = t$.

25. The vector equation for the curve is $\mathbf{r}(t) = \langle e^{-t} \cos t, e^{-t} \sin t, e^{-t} \rangle$, so

$$\mathbf{r}'(t) = \langle e^{-t}(-\sin t) + (\cos t)(-e^{-t}), e^{-t} \cos t + (\sin t)(-e^{-t}), (-e^{-t}) \rangle$$

$$= \langle -e^{-t}(\cos t + \sin t), e^{-t}(\cos t - \sin t), -e^{-t} \rangle$$

The point $(1, 0, 1)$ corresponds to $t = 0$, so the tangent vector there is

$\mathbf{r}'(0) = \langle -e^0(\cos 0 + \sin 0), e^0(\cos 0 - \sin 0), -e^0 \rangle = \langle -1, 1, -1 \rangle$. Thus, the tangent line is parallel to the vector

$\langle -1, 1, -1 \rangle$ and parametric equations are $x = 1 + (-1)t = 1 - t$, $y = 0 + 1 \cdot t = t$, $z = 1 + (-1)t = 1 - t$.

26. $\mathbf{r}(t) = \langle \ln t, 2\sqrt{t}, t^2 \rangle$, $\mathbf{r}'(t) = \langle 1/t, 1/\sqrt{t}, 2t \rangle$. At $(0, 2, 1)$, $t = 1$ and $\mathbf{r}'(1) = \langle 1, 1, 2 \rangle$. Thus, parametric equations of the

tangent line are $x = t$, $y = 2 + t$, $z = 1 + 2t$.

27. $\mathbf{r}(t) = \langle t, e^{-t}, 2t - t^2 \rangle$ $\Rightarrow$ $\mathbf{r}'(t) = \langle 1, -e^{-t}, 2 - 2t \rangle$. At $(0, 1, 0)$,

$t = 0$ and $\mathbf{r}'(0) = \langle 1, -1, 2 \rangle$. Thus, parametric equations of the tangent

line are $x = t$, $y = 1 - t$, $z = 2t$.

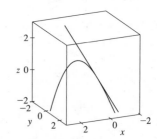

28. $\mathbf{r}(t) = \langle 2 \cos t, 2 \sin t, 4 \cos 2t \rangle$,

$\mathbf{r}'(t) = \langle -2 \sin t, 2 \cos t, -8 \sin 2t \rangle$. At $(\sqrt{3}, 1, 2)$, $t = \frac{\pi}{6}$ and

$\mathbf{r}'(\frac{\pi}{6}) = \langle -1, \sqrt{3}, -4\sqrt{3} \rangle$. Thus, parametric equations of the

tangent line are $x = \sqrt{3} - t$, $y = 1 + \sqrt{3}t$, $z = 2 - 4\sqrt{3}t$.

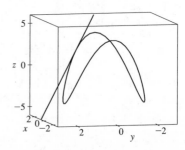

29. $\mathbf{r}(t) = \langle t \cos t, t, t \sin t \rangle \quad \Rightarrow \quad \mathbf{r}'(t) = \langle \cos t - t \sin t, 1, t \cos t + \sin t \rangle$.

At $(-\pi, \pi, 0)$, $t = \pi$ and $\mathbf{r}'(\pi) = \langle -1, 1, -\pi \rangle$. Thus, parametric equations

of the tangent line are $x = -\pi - t$, $y = \pi + t$, $z = -\pi t$.

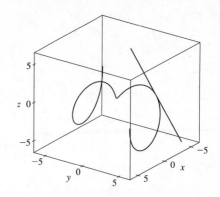

30. (a) The tangent line at $t = 0$ is the line through the point with position vector $\mathbf{r}(0) = \langle \sin 0, 2 \sin 0, \cos 0 \rangle = \langle 0, 0, 1 \rangle$, and in

the direction of the tangent vector, $\mathbf{r}'(0) = \langle \pi \cos 0, 2\pi \cos 0, -\pi \sin 0 \rangle = \langle \pi, 2\pi, 0 \rangle$. So an equation of the line is

$\langle x, y, z \rangle = \mathbf{r}(0) + u\,\mathbf{r}'(0) = \langle 0 + \pi u, 0 + 2\pi u, 1 \rangle = \langle \pi u, 2\pi u, 1 \rangle$.

$\mathbf{r}\left(\frac{1}{2}\right) = \langle \sin \frac{\pi}{2}, 2 \sin \frac{\pi}{2}, \cos \frac{\pi}{2} \rangle = \langle 1, 2, 0 \rangle$,

(b)

$\mathbf{r}'\left(\frac{1}{2}\right) = \langle \pi \cos \frac{\pi}{2}, 2\pi \cos \frac{\pi}{2}, -\pi \sin \frac{\pi}{2} \rangle = \langle 0, 0, -\pi \rangle$.

So the equation of the second line is

$\langle x, y, z \rangle = \langle 1, 2, 0 \rangle + v \langle 0, 0, -\pi \rangle = \langle 1, 2, -\pi v \rangle$.

The lines intersect where $\langle \pi u, 2\pi u, 1 \rangle = \langle 1, 2, -\pi v \rangle$,

so the point of intersection is $(1, 2, 1)$.

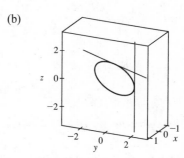

31. The angle of intersection of the two curves is the angle between the two tangent vectors to the curves at the point of

intersection. Since $\mathbf{r}_1'(t) = \langle 1, 2t, 3t^2 \rangle$ and $t = 0$ at $(0, 0, 0)$, $\mathbf{r}_1'(0) = \langle 1, 0, 0 \rangle$ is a tangent vector to $\mathbf{r}_1$ at $(0, 0, 0)$. Similarly,

$\mathbf{r}_2'(t) = \langle \cos t, 2 \cos 2t, 1 \rangle$ and since $\mathbf{r}_2(0) = \langle 0, 0, 0 \rangle$, $\mathbf{r}_2'(0) = \langle 1, 2, 1 \rangle$ is a tangent vector to $\mathbf{r}_2$ at $(0, 0, 0)$. If θ is the angle

between these two tangent vectors, then $\cos \theta = \frac{1}{\sqrt{1}\sqrt{6}} \langle 1, 0, 0 \rangle \cdot \langle 1, 2, 1 \rangle = \frac{1}{\sqrt{6}}$ and $\theta = \cos^{-1}\left(\frac{1}{\sqrt{6}}\right) \approx 66°$.

32. To find the point of intersection, we must find the values of t and s which satisfy the following three equations simultaneously:

$t = 3 - s$, $1 - t = s - 2$, $3 + t^2 = s^2$. Solving the last two equations gives $t = 1$, $s = 2$ (check these in the first equation).

Thus the point of intersection is $(1, 0, 4)$. To find the angle θ of intersection, we proceed as in Exercise 31. The tangent

vectors to the respective curves at $(1, 0, 4)$ are $\mathbf{r}_1'(1) = \langle 1, -1, 2 \rangle$ and $\mathbf{r}_2'(2) = \langle -1, 1, 4 \rangle$. So

$\cos \theta = \frac{1}{\sqrt{6}\sqrt{18}} (-1 - 1 + 8) = \frac{6}{6\sqrt{3}} = \frac{1}{\sqrt{3}}$ and $\theta = \cos^{-1}\left(\frac{1}{\sqrt{3}}\right) \approx 55°$.

Note: In Exercise 31, the curves intersect when the value of both parameters is zero. However, as seen in this exercise, it is not

necessary for the parameters to be of equal value at the point of intersection.

33. $\int_0^1 (16t^3\,\mathbf{i} - 9t^2\,\mathbf{j} + 25t^4\,\mathbf{k})\,dt = \left(\int_0^1 16t^3\,dt\right)\mathbf{i} - \left(\int_0^1 9t^2\,dt\right)\mathbf{j} + \left(\int_0^1 25t^4\,dt\right)\mathbf{k}$

$= \left[4t^4\right]_0^1\,\mathbf{i} - \left[3t^3\right]_0^1\,\mathbf{j} + \left[5t^5\right]_0^1\,\mathbf{k} = 4\,\mathbf{i} - 3\,\mathbf{j} + 5\,\mathbf{k}$

34. $\int_0^1 \left(\dfrac{4}{1+t^2}\,\mathbf{j} + \dfrac{2t}{1+t^2}\,\mathbf{k} \right) dt = \left[4\tan^{-1}t\,\mathbf{j} + \ln(1+t^2)\,\mathbf{k} \right]_0^1 = \left[4\tan^{-1}1\,\mathbf{j} + \ln 2\,\mathbf{k} \right] - \left[4\tan^{-1}0\,\mathbf{j} + \ln 1\,\mathbf{k} \right]$

$$= 4\left(\tfrac{\pi}{4}\right)\mathbf{j} + \ln 2\,\mathbf{k} - 0\,\mathbf{j} - 0\,\mathbf{k} = \pi\,\mathbf{j} + \ln 2\,\mathbf{k}$$

35. $\int_0^{\pi/2} (3\sin^2 t\,\cos t\,\mathbf{i} + 3\sin t\,\cos^2 t\,\mathbf{j} + 2\sin t\,\cos t\,\mathbf{k})\,dt$

$$= \left(\int_0^{\pi/2} 3\sin^2 t\,\cos t\,dt \right)\mathbf{i} + \left(\int_0^{\pi/2} 3\sin t\,\cos^2 t\,dt \right)\mathbf{j} + \left(\int_0^{\pi/2} 2\sin t\,\cos t\,dt \right)\mathbf{k}$$

$$= \left[\sin^3 t\right]_0^{\pi/2}\mathbf{i} + \left[-\cos^3 t\right]_0^{\pi/2}\mathbf{j} + \left[\sin^2 t\right]_0^{\pi/2}\mathbf{k} = (1-0)\,\mathbf{i} + (0+1)\,\mathbf{j} + (1-0)\,\mathbf{k} = \mathbf{i} + \mathbf{j} + \mathbf{k}$$

36. $\int_1^2 \left(t^2\,\mathbf{i} + t\sqrt{t-1}\,\mathbf{j} + t\sin\pi t\,\mathbf{k} \right) dt = \left[\tfrac{1}{3}t^3\,\mathbf{i} + \left(\tfrac{2}{5}(t-1)^{5/2} + \tfrac{2}{3}(t-1)^{3/2} \right)\mathbf{j} \right]_1^2 + \left(\left[-\tfrac{1}{\pi}t\cos\pi t \right]_1^2 + \int_1^2 \tfrac{1}{\pi}\cos\pi t\,dt \right)\mathbf{k}$

$$= \tfrac{7}{3}\,\mathbf{i} + \tfrac{16}{15}\,\mathbf{j} + \left(-\tfrac{3}{\pi} + \left[\tfrac{1}{\pi^2}\sin\pi t \right]_1^2 \right)\mathbf{k} = \tfrac{7}{3}\,\mathbf{i} + \tfrac{16}{15}\,\mathbf{j} - \tfrac{3}{\pi}\,\mathbf{k}$$

37. $\int \left(e^t\,\mathbf{i} + 2t\,\mathbf{j} + \ln t\,\mathbf{k} \right) dt = \left(\int e^t\,dt \right)\mathbf{i} + \left(\int 2t\,dt \right)\mathbf{j} + \left(\int \ln t\,dt \right)\mathbf{k}$

$$= e^t\,\mathbf{i} + t^2\,\mathbf{j} + (t\ln t - t)\,\mathbf{k} + \mathbf{C}, \ \text{ where } \mathbf{C} \text{ is a vector constant of integration.}$$

38. $\int (\cos\pi t\,\mathbf{i} + \sin\pi t\,\mathbf{j} + t\,\mathbf{k})\,dt = \left(\int \cos\pi t\,dt \right)\mathbf{i} + \left(\int \sin\pi t\,dt \right)\mathbf{j} + \left(\int t\,dt \right)\mathbf{k}$

$$= \tfrac{1}{\pi}\sin\pi t\,\mathbf{i} - \tfrac{1}{\pi}\cos\pi t\,\mathbf{j} + \tfrac{1}{2}t^2\,\mathbf{k} + \mathbf{C}$$

39. $\mathbf{r}'(t) = 2t\,\mathbf{i} + 3t^2\,\mathbf{j} + \sqrt{t}\,\mathbf{k} \ \Rightarrow \ \mathbf{r}(t) = t^2\,\mathbf{i} + t^3\,\mathbf{j} + \tfrac{2}{3}t^{3/2}\,\mathbf{k} + \mathbf{C}, \ \text{ where } \mathbf{C} \text{ is a constant vector.}$

But $\mathbf{i} + \mathbf{j} = \mathbf{r}\,(1) = \mathbf{i} + \mathbf{j} + \tfrac{2}{3}\mathbf{k} + \mathbf{C}$. Thus $\mathbf{C} = -\tfrac{2}{3}\mathbf{k}$ and $\mathbf{r}(t) = t^2\,\mathbf{i} + t^3\,\mathbf{j} + \left(\tfrac{2}{3}t^{3/2} - \tfrac{2}{3} \right)\mathbf{k}$.

40. $\mathbf{r}'(t) = t\,\mathbf{i} + e^t\,\mathbf{j} + te^t\,\mathbf{k} \ \Rightarrow \ \mathbf{r}(t) = \tfrac{1}{2}t^2\,\mathbf{i} + e^t\,\mathbf{j} + (te^t - e^t)\,\mathbf{k} + \mathbf{C}$. But $\mathbf{i} + \mathbf{j} + \mathbf{k} = \mathbf{r}\,(0) = \mathbf{j} - \mathbf{k} + \mathbf{C}$.

Thus $\mathbf{C} = \mathbf{i} + 2\,\mathbf{k}$ and $\mathbf{r}(t) = \left(\tfrac{1}{2}t^2 + 1 \right)\mathbf{i} + e^t\,\mathbf{j} + (te^t - e^t + 2)\,\mathbf{k}$.

For Exercises 41–44, let $\mathbf{u}(t) = \langle u_1(t), u_2(t), u_3(t) \rangle$ and $\mathbf{v}(t) = \langle v_1(t), v_2(t), v_3(t) \rangle$. In each of these exercises, the procedure is to apply Theorem 2 so that the corresponding properties of derivatives of real-valued functions can be used.

41. $\dfrac{d}{dt}\left[\mathbf{u}(t) + \mathbf{v}(t) \right] = \dfrac{d}{dt}\,\langle u_1(t) + v_1(t), u_2(t) + v_2(t), u_3(t) + v_3(t) \rangle$

$$= \left\langle \dfrac{d}{dt}\left[u_1(t) + v_1\,(t) \right], \dfrac{d}{dt}\left[u_2(t) + v_2(t) \right], \dfrac{d}{dt}\left[u_3(t) + v_3(t) \right] \right\rangle$$

$$= \langle u_1'(t) + v_1'(t), u_2'(t) + v_2'(t), u_3'(t) + v_3'(t) \rangle$$

$$= \langle u_1'(t), u_2'\,(t)\,, u_3'(t) \rangle + \langle v_1'(t), v_2'(t), v_3'(t) \rangle = \mathbf{u}'(t) + \mathbf{v}'(t)$$

42. $\dfrac{d}{dt}\left[f(t)\,\mathbf{u}(t) \right] = \dfrac{d}{dt}\,\langle f(t)u_1(t), f(t)u_2(t), f(t)u_3(t) \rangle$

$$= \left\langle \dfrac{d}{dt}\left[f(t)u_1(t) \right], \dfrac{d}{dt}\left[f(t)u_2(t) \right], \dfrac{d}{dt}\left[f(t)u_3(t) \right] \right\rangle$$

$$= \langle f'(t)u_1(t) + f(t)u_1'(t), f'(t)u_2(t) + f(t)u_2'(t), f'(t)u_3(t) + f(t)u_3'(t) \rangle$$

$$= f'(t)\,\langle u_1(t), u_2(t), u_3(t) \rangle + f(t)\,\langle u_1'(t), u_2'(t), u_3'(t) \rangle = f'(t)\,\mathbf{u}(t) + f(t)\,\mathbf{u}'(t)$$

43. $\dfrac{d}{dt}\left[\mathbf{u}(t) \times \mathbf{v}(t)\right] = \dfrac{d}{dt}\langle u_2(t)v_3(t) - u_3(t)v_2(t), u_3(t)v_1(t) - u_1(t)v_3(t), u_1(t)v_2(t) - u_2(t)v_1(t)\rangle$

$$= \langle u_2'v_3(t) + u_2(t)v_3'(t) - u_3'(t)v_2(t) - u_3(t)v_2'(t),$$

$$u_3'(t)v_1(t) + u_3(t)v_1'(t) - u_1'(t)v_3(t) - u_1(t)v_3'(t),$$

$$u_1'(t)v_2(t) + u_1(t)v_2'(t) - u_2'(t)v_1(t) - u_2(t)v_1'(t)\rangle$$

$$= \langle u_2'(t)v_3(t) - u_3'(t)v_2(t), u_3'(t)v_1(t) - u_1'(t)v_3(t), u_1'(t)v_2(t) - u_2'(t)v_1(t)\rangle$$

$$+ \langle u_2(t)v_3'(t) - u_3(t)v_2'(t), u_3(t)v_1'(t) - u_1(t)v_3'(t), u_1(t)v_2'(t) - u_2(t)v_1'(t)\rangle$$

$$= \mathbf{u}'(t) \times \mathbf{v}(t) + \mathbf{u}(t) \times \mathbf{v}'(t)$$

Alternate solution: Let $\mathbf{r}(t) = \mathbf{u}(t) \times \mathbf{v}(t)$. Then

$$\mathbf{r}(t+h) - \mathbf{r}(t) = [\mathbf{u}(t+h) \times \mathbf{v}(t+h)] - [\mathbf{u}(t) \times \mathbf{v}(t)]$$

$$= [\mathbf{u}(t+h) \times \mathbf{v}(t+h)] - [\mathbf{u}(t) \times \mathbf{v}(t)] + [\mathbf{u}(t+h) \times \mathbf{v}(t)] - [\mathbf{u}(t+h) \times \mathbf{v}(t)]$$

$$= \mathbf{u}(t+h) \times [\mathbf{v}(t+h) - \mathbf{v}(t)] + [\mathbf{u}(t+h) - \mathbf{u}(t)] \times \mathbf{v}(t)$$

(Be careful of the order of the cross product.) Dividing through by h and taking the limit as $h \to 0$ we have

$$\mathbf{r}'(t) = \lim_{h \to 0}\frac{\mathbf{u}(t+h) \times [\mathbf{v}(t+h) - \mathbf{v}(t)]}{h} + \lim_{h \to 0}\frac{[\mathbf{u}(t+h) - \mathbf{u}(t)] \times \mathbf{v}(t)}{h} = \mathbf{u}(t) \times \mathbf{v}'(t) + \mathbf{u}'(t) \times \mathbf{v}(t)$$

by Exercise 14.1.43(a) [ET 13.1.43(a)] and Definition 1.

44. $\dfrac{d}{dt}\left[\mathbf{u}(f(t))\right] = \dfrac{d}{dt}\langle u_1(f(t)), u_2(f(t)), u_3(f(t))\rangle = \left\langle \dfrac{d}{dt}\left[u_1(f(t))\right], \dfrac{d}{dt}\left[u_2(f(t))\right], \dfrac{d}{dt}\left[u_3(f(t))\right]\right\rangle$

$$= \langle f'(t)u_1'(f(t)), f'(t)u_2'(f(t)), f'(t)u_3'(f(t))\rangle = f'(t)\,\mathbf{u}'(t)$$

45. $\dfrac{d}{dt}\left[\mathbf{u}(t) \cdot \mathbf{v}(t)\right] = \mathbf{u}'(t) \cdot \mathbf{v}(t) + \mathbf{u}(t) \cdot \mathbf{v}'(t)$ [by Formula 4 of Theorem 3]

$$= \langle \cos t, -\sin t, 1\rangle \cdot \langle t, \cos t, \sin t\rangle + \langle \sin t, \cos t, t\rangle \cdot \langle 1, -\sin t, \cos t\rangle$$

$$= t\cos t - \cos t \,\sin t + \sin t + \sin t - \cos t \,\sin t + t\cos t$$

$$= 2t\cos t + 2\sin t - 2\cos t \,\sin t$$

46. $\dfrac{d}{dt}\left[\mathbf{u}(t) \times \mathbf{v}(t)\right] = \mathbf{u}'(t) \times \mathbf{v}(t) + \mathbf{u}(t) \times \mathbf{v}'(t)$ [by Formula 5 of Theorem 3]

$$= \langle \cos t, -\sin t, 1\rangle \times \langle t, \cos t, \sin t\rangle + \langle \sin t, \cos t, t\rangle \times \langle 1, -\sin t, \cos t\rangle$$

$$= \langle -\sin^2 t - \cos t, t - \cos t \,\sin t, \cos^2 t + t\sin t\rangle$$

$$+ \langle \cos^2 t + t\sin t, t - \cos t \,\sin t, -\sin^2 t - \cos t\rangle$$

$$= \langle \cos^2 t - \sin^2 t - \cos t + t\sin t, 2t - 2\cos t \,\sin t, \cos^2 t - \sin^2 t - \cos t + t\sin t\rangle$$

47. $\dfrac{d}{dt}\left[\mathbf{r}(t) \times \mathbf{r}'(t)\right] = \mathbf{r}'(t) \times \mathbf{r}'(t) + \mathbf{r}(t) \times \mathbf{r}''(t)$ by Formula 5 of Theorem 3. But $\mathbf{r}'(t) \times \mathbf{r}'(t) = \mathbf{0}$ (by Example 2 in

Section 13.4 [ET 12.4]). Thus, $\dfrac{d}{dt}\left[\mathbf{r}(t) \times \mathbf{r}'(t)\right] = \mathbf{r}(t) \times \mathbf{r}''(t)$.

48. $\dfrac{d}{dt}(\mathbf{u}(t) \cdot [\mathbf{v}(t) \times \mathbf{w}(t)]) = \mathbf{u}'(t) \cdot [\mathbf{v}(t) \times \mathbf{w}(t)] + \mathbf{u}(t) \cdot \dfrac{d}{dt}[\mathbf{v}(t) \times \mathbf{w}(t)]$

$$= \mathbf{u}'(t) \cdot [\mathbf{v}(t) \times \mathbf{w}(t)] + \mathbf{u}(t) \cdot [\mathbf{v}'(t) \times \mathbf{w}(t) + \mathbf{v}(t) \times \mathbf{w}'(t)]$$

$$= \mathbf{u}'(t) \cdot [\mathbf{v}(t) \times \mathbf{w}(t)] + \mathbf{u}(t) \cdot [\mathbf{v}'(t) \times \mathbf{w}(t)] + \mathbf{u}(t) \cdot [\mathbf{v}(t) \times \mathbf{w}'(t)]$$

$$= \mathbf{u}'(t) \cdot [\mathbf{v}(t) \times \mathbf{w}(t)] - \mathbf{v}'(t) \cdot [\mathbf{u}(t) \times \mathbf{w}(t)] + \mathbf{w}'(t) \cdot [\mathbf{u}(t) \times \mathbf{v}(t)]$$

49. $\dfrac{d}{dt}|\mathbf{r}(t)| = \dfrac{d}{dt}[\mathbf{r}(t) \cdot \mathbf{r}(t)]^{1/2} = \tfrac{1}{2}[\mathbf{r}(t) \cdot \mathbf{r}(t)]^{-1/2}[2\mathbf{r}(t) \cdot \mathbf{r}'(t)] = \dfrac{1}{|\mathbf{r}(t)|}\mathbf{r}(t) \cdot \mathbf{r}'(t)$

50. Since $\mathbf{r}(t) \cdot \mathbf{r}'(t) = 0$, we have $0 = 2\mathbf{r}(t) \cdot \mathbf{r}'(t) = \dfrac{d}{dt}[\mathbf{r}(t) \cdot \mathbf{r}(t)] = \dfrac{d}{dt}|\mathbf{r}(t)|^2$. Thus $|\mathbf{r}(t)|^2$, and so $|\mathbf{r}(t)|$, is a constant,

and hence the curve lies on a sphere with center the origin.

51. Since $\mathbf{u}(t) = \mathbf{r}(t) \cdot [\mathbf{r}'(t) \times \mathbf{r}''(t)]$,

$$\mathbf{u}'(t) = \mathbf{r}'(t) \cdot [\mathbf{r}'(t) \times \mathbf{r}''(t)] + \mathbf{r}(t) \cdot \dfrac{d}{dt}[\mathbf{r}'(t) \times \mathbf{r}''(t)]$$

$$= 0 + \mathbf{r}(t) \cdot [\mathbf{r}''(t) \times \mathbf{r}''(t) + \mathbf{r}'(t) \times \mathbf{r}'''(t)] \qquad \text{[since } \mathbf{r}'(t) \perp \mathbf{r}'(t) \times \mathbf{r}''(t)]$$

$$= \mathbf{r}(t) \cdot [\mathbf{r}'(t) \times \mathbf{r}'''(t)] \qquad \text{[since } \mathbf{r}''(t) \times \mathbf{r}''(t) = \mathbf{0}]$$

14.3 Arc Length and Curvature

1. $\mathbf{r}(t) = \langle 2\sin t, 5t, 2\cos t \rangle \;\Rightarrow\; \mathbf{r}'(t) = \langle 2\cos t, 5, -2\sin t \rangle \;\Rightarrow\; |\mathbf{r}'(t)| = \sqrt{(2\cos t)^2 + 5^2 + (-2\sin t)^2} = \sqrt{29}.$

Then using Formula 3, we have $L = \int_{-10}^{10} |\mathbf{r}'(t)|\,dt = \int_{-10}^{10} \sqrt{29}\,dt = \sqrt{29}\,t\Big]_{-10}^{10} = 20\sqrt{29}.$

2. $\mathbf{r}(t) = \langle 2t, t^2, \tfrac{1}{3}t^3 \rangle \;\Rightarrow\; \mathbf{r}'(t) = \langle 2, 2t, t^2 \rangle \;\Rightarrow$

$|\mathbf{r}'(t)| = \sqrt{2^2 + (2t)^2 + (t^2)^2} = \sqrt{4 + 4t^2 + t^4} = \sqrt{(2+t^2)^2} = 2 + t^2$ for $0 \le t \le 1$. Then using Formula 3, we have

$L = \int_0^1 |\mathbf{r}'(t)|\,dt = \int_0^1 (2 + t^2)\,dt = 2t + \tfrac{1}{3}t^3\Big]_0^1 = \tfrac{7}{3}.$

3. $\mathbf{r}(t) = \sqrt{2}\,t\,\mathbf{i} + e^t\mathbf{j} + e^{-t}\mathbf{k} \;\Rightarrow\; \mathbf{r}'(t) = \sqrt{2}\,\mathbf{i} + e^t\mathbf{j} - e^{-t}\mathbf{k} \;\Rightarrow$

$|\mathbf{r}'(t)| = \sqrt{\left(\sqrt{2}\right)^2 + (e^t)^2 + (-e^{-t})^2} = \sqrt{2 + e^{2t} + e^{-2t}} = \sqrt{(e^t + e^{-t})^2} = e^t + e^{-t}$ [since $e^t + e^{-t} > 0$].

Then $L = \int_0^1 |\mathbf{r}'(t)|\,dt = \int_0^1 (e^t + e^{-t})\,dt = [e^t - e^{-t}]_0^1 = e - e^{-1}.$

4. $\mathbf{r}(t) = \cos t\,\mathbf{i} + \sin t\,\mathbf{j} + \ln\cos t\,\mathbf{k} \;\Rightarrow\; \mathbf{r}'(t) = -\sin t\,\mathbf{i} + \cos t\,\mathbf{j} + \dfrac{-\sin t}{\cos t}\,\mathbf{k} = -\sin t\,\mathbf{i} + \cos t\,\mathbf{j} - \tan t\,\mathbf{k},$

$|\mathbf{r}'(t)| = \sqrt{(-\sin t)^2 + \cos^2 t + (-\tan t)^2} = \sqrt{1 + \tan^2 t} = \sqrt{\sec^2 t} = |\sec t|.$ Since $\sec t > 0$ for $0 \le t \le \pi/4$, here we

can say $|\mathbf{r}'(t)| = \sec t$. Then

$$L = \int_0^{\pi/4} \sec t\,dt = \Big[\ln|\sec t + \tan t|\Big]_0^{\pi/4} = \ln\left|\sec\frac{\pi}{4} + \tan\frac{\pi}{4}\right| - \ln|\sec 0 + \tan 0|$$

$$= \ln\left|\sqrt{2} + 1\right| - \ln|1 + 0| = \ln(\sqrt{2} + 1).$$

5. $\mathbf{r}(t) = \mathbf{i} + t^2\mathbf{j} + t^3\mathbf{k}$ $\Rightarrow$ $\mathbf{r}'(t) = 2t\mathbf{j} + 3t^2\mathbf{k}$ $\Rightarrow$ $|\mathbf{r}'(t)| = \sqrt{4t^2 + 9t^4} = t\sqrt{4 + 9t^2}$ [since $t \geq 0$].

Then $L = \int_0^1 |\mathbf{r}'(t)|\,dt = \int_0^1 t\sqrt{4 + 9t^2}\,dt = \frac{1}{18} \cdot \frac{2}{3}(4 + 9t^2)^{3/2}\Big]_0^1 = \frac{1}{27}(13^{3/2} - 4^{3/2}) = \frac{1}{27}(13^{3/2} - 8)$.

6. $\mathbf{r}(t) = 12t\,\mathbf{i} + 8t^{3/2}\,\mathbf{j} + 3t^2\,\mathbf{k}$ $\Rightarrow$ $\mathbf{r}'(t) = 12\,\mathbf{i} + 12\sqrt{t}\,\mathbf{j} + 6t\,\mathbf{k}$ $\Rightarrow$

$|\mathbf{r}'(t)| = \sqrt{144 + 144t + 36t^2} = \sqrt{36(t + 2)^2} = 6\,|t + 2| = 6(t + 2)$ for $0 \leq t \leq 1$. Then

$L = \int_0^1 |\mathbf{r}'(t)|\,dt = \int_0^1 6(t + 2)\,dt = \Big[3t^2 + 12t\Big]_0^1 = 15$.

7. $\mathbf{r}(t) = \langle \sqrt{t}, t, t^2 \rangle$ $\Rightarrow$ $\mathbf{r}'(t) = \left\langle \frac{1}{2\sqrt{t}}, 1, 2t \right\rangle$ $\Rightarrow$ $|\mathbf{r}'(t)| = \sqrt{\left(\frac{1}{2\sqrt{t}}\right)^2 + 1^2 + (2t)^2} = \sqrt{\frac{1}{4t} + 1 + 4t^2}$, so

$L = \int_1^4 |\mathbf{r}'(t)|\,dt = \int_1^4 \sqrt{\frac{1}{4t} + 1 + 4t^2}\,dt \approx 15.3841$.

8. $\mathbf{r}(t) = \langle t, \ln t, t\ln t \rangle$ $\Rightarrow$ $\mathbf{r}'(t) = \langle 1, 1/t, 1 + \ln t \rangle$ $\Rightarrow$

$|\mathbf{r}'(t)| = \sqrt{1^2 + (1/t)^2 + (1 + \ln t)^2} = \sqrt{2 + 1/t^2 + 2\ln t + (\ln t)^2}$ and

$L = \int_1^2 |\mathbf{r}'(t)|\,dt = \int_1^2 \sqrt{2 + 1/t^2 + 2\ln t + (\ln t)^2}\,dt \approx 1.8581$.

9. $\mathbf{r}(t) = \langle \sin t, \cos t, \tan t \rangle$ $\Rightarrow$ $\mathbf{r}'(t) = \langle \cos t, -\sin t, \sec^2 t \rangle$ $\Rightarrow$

$|\mathbf{r}'(t)| = \sqrt{\cos^2 t + (-\sin t)^2 + (\sec^2 t)^2} = \sqrt{1 + \sec^4 t}$ and $L = \int_0^{\pi/4} |\mathbf{r}'(t)|\,dt = \int_0^{\pi/4} \sqrt{1 + \sec^4 t}\,dt \approx 1.2780$.

10. We plot two different views of the curve with parametric equations $x = \sin t$, $y = \sin 2t$, $z = \sin 3t$. To help visualize the curve, we also include a plot showing a tube of radius 0.07 around the curve.

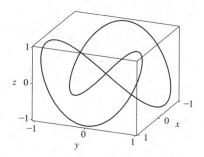
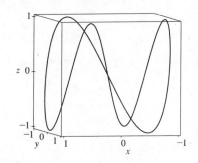

The complete curve is given by the parameter interval $[0, 2\pi]$ and we have $\mathbf{r}'(t) = \langle \cos t, 2\cos 2t, 3\cos 3t \rangle$ $\Rightarrow$

$|\mathbf{r}'(t)| = \sqrt{\cos^2 t + 4\cos^2 2t + 9\cos^2 3t}$, so $L = \int_0^{2\pi} |\mathbf{r}'(t)|\,dt = \int_0^{2\pi} \sqrt{\cos^2 t + 4\cos^2 2t + 9\cos^2 3t}\,dt \approx 16.0264$.

11. The projection of the curve C onto the xy-plane is the curve $x^2 = 2y$ or $y = \frac{1}{2}x^2$, $z = 0$. Then we can choose the parameter $x = t$ $\Rightarrow$ $y = \frac{1}{2}t^2$. Since C also lies on the surface $3z = xy$, we have $z = \frac{1}{3}xy = \frac{1}{3}(t)(\frac{1}{2}t^2) = \frac{1}{6}t^3$. Then parametric equations for C are $x = t$, $y = \frac{1}{2}t^2$, $z = \frac{1}{6}t^3$ and the corresponding vector equation is $\mathbf{r}(t) = \langle t, \frac{1}{2}t^2, \frac{1}{6}t^3 \rangle$. The origin corresponds to $t = 0$ and the point $(6, 18, 36)$ corresponds to $t = 6$, so

$$L = \int_0^6 |\mathbf{r}'(t)|\,dt = \int_0^6 \left|\langle 1, t, \frac{1}{2}t^2 \rangle\right|\,dt = \int_0^6 \sqrt{1^2 + t^2 + \left(\frac{1}{2}t^2\right)^2}\,dt = \int_0^6 \sqrt{1 + t^2 + \frac{1}{4}t^4}\,dt$$

$$= \int_0^6 \sqrt{(1 + \frac{1}{2}t^2)^2}\,dt = \int_0^6 (1 + \frac{1}{2}t^2)\,dt = \Big[t + \frac{1}{6}t^3\Big]_0^6 = 6 + 36 = 42$$

12. Let C be the curve of intersection. The projection of C onto the xy-plane is the ellipse $4x^2 + y^2 = 4$ or $x^2 + y^2/4 = 1$,

$z = 0$. Then we can write $x = \cos t$, $y = 2\sin t$, $0 \le t \le 2\pi$. Since C also lies on the plane $x + y + z = 2$, we have

$z = 2 - x - y = 2 - \cos t - 2\sin t$. Then parametric equations for C are $x = \cos t$, $y = 2\sin t$, $z = 2 - \cos t - 2\sin t$,

$0 \le t \le 2\pi$, and the corresponding vector equation is $\mathbf{r}(t) = \langle \cos t, 2\sin t, 2 - \cos t - 2\sin t \rangle$. Differentiating gives

$\mathbf{r}'(t) = \langle -\sin t, 2\cos t, \sin t - 2\cos t \rangle \quad \Rightarrow$

$|\mathbf{r}'(t)| = \sqrt{(-\sin t)^2 + (2\cos t)^2 + (\sin t - 2\cos t)^2} = \sqrt{2\sin^2 t + 8\cos^2 t - 4\sin t \cos t}$. The length of C is

$L = \int_0^{2\pi} |\mathbf{r}'(t)|\, dt = \int_0^{2\pi} \sqrt{2\sin^2 t + 8\cos^2 t - 4\sin t \cos t}\, dt \approx 13.5191$.

13. $\mathbf{r}(t) = 2t\,\mathbf{i} + (1 - 3t)\,\mathbf{j} + (5 + 4t)\,\mathbf{k} \quad \Rightarrow \quad \mathbf{r}'(t) = 2\,\mathbf{i} - 3\,\mathbf{j} + 4\,\mathbf{k}$ and $\frac{ds}{dt} = |\mathbf{r}'(t)| = \sqrt{4 + 9 + 16} = \sqrt{29}$. Then

$s = s(t) = \int_0^t |\mathbf{r}'(u)|\, du = \int_0^t \sqrt{29}\, du = \sqrt{29}\, t$. Therefore, $t = \frac{1}{\sqrt{29}}s$, and substituting for t in the original equation, we

have $\mathbf{r}(t(s)) = \frac{2}{\sqrt{29}}s\,\mathbf{i} + \left(1 - \frac{3}{\sqrt{29}}s\right)\mathbf{j} + \left(5 + \frac{4}{\sqrt{29}}s\right)\mathbf{k}$.

14. $\mathbf{r}(t) = e^{2t}\cos 2t\,\mathbf{i} + 2\,\mathbf{j} + e^{2t}\sin 2t\,\mathbf{k} \quad \Rightarrow \quad \mathbf{r}'(t) = 2e^{2t}(\cos 2t - \sin 2t)\,\mathbf{i} + 2e^{2t}(\cos 2t + \sin 2t)\,\mathbf{k}$,

$\frac{ds}{dt} = |\mathbf{r}'(t)| = 2e^{2t}\sqrt{(\cos 2t - \sin 2t)^2 + (\cos 2t + \sin 2t)^2} = 2e^{2t}\sqrt{2\cos^2 2t + 2\sin^2 2t} = 2\sqrt{2}\,e^{2t}$.

$s = s(t) = \int_0^t |\mathbf{r}'(u)|\, du = \int_0^t 2\sqrt{2}\,e^{2u}\, du = \sqrt{2}\,e^{2u}\Big]_0^t = \sqrt{2}\,(e^{2t} - 1) \quad \Rightarrow \quad \frac{s}{\sqrt{2}} + 1 = e^{2t} \quad \Rightarrow \quad t = \frac{1}{2}\ln\left(\frac{s}{\sqrt{2}} + 1\right)$.

Substituting, we have

$$\mathbf{r}(t(s)) = e^{2\left(\frac{1}{2}\ln\left(\frac{s}{\sqrt{2}}+1\right)\right)}\cos 2\left(\tfrac{1}{2}\ln\left(\tfrac{s}{\sqrt{2}}+1\right)\right)\mathbf{i} + 2\,\mathbf{j} + e^{2\left(\frac{1}{2}\ln\left(\frac{s}{\sqrt{2}}+1\right)\right)}\sin 2\left(\tfrac{1}{2}\ln\left(\tfrac{s}{\sqrt{2}}+1\right)\right)\mathbf{k}$$

$$= \left(\tfrac{s}{\sqrt{2}}+1\right)\cos\left(\ln\left(\tfrac{s}{\sqrt{2}}+1\right)\right)\mathbf{i} + 2\,\mathbf{j} + \left(\tfrac{s}{\sqrt{2}}+1\right)\sin\left(\ln\left(\tfrac{s}{\sqrt{2}}+1\right)\right)\mathbf{k}$$

15. Here $\mathbf{r}(t) = \langle 3\sin t, 4t, 3\cos t \rangle$, so $\mathbf{r}'(t) = \langle 3\cos t, 4, -3\sin t \rangle$ and $|\mathbf{r}'(t)| = \sqrt{9\cos^2 t + 16 + 9\sin^2 t} = \sqrt{25} = 5$.

The point $(0, 0, 3)$ corresponds to $t = 0$, so the arc length function beginning at $(0, 0, 3)$ and measuring in the positive

direction is given by $s(t) = \int_0^t |\mathbf{r}'(u)|\, du = \int_0^t 5\, du = 5t$. $\quad s(t) = 5 \quad \Rightarrow \quad 5t = 5 \quad \Rightarrow \quad t = 1$, thus your location after

moving 5 units along the curve is $(3\sin 1, 4, 3\cos 1)$.

16. $\mathbf{r}(t) = \left(\dfrac{2}{t^2 + 1} - 1\right)\mathbf{i} + \dfrac{2t}{t^2 + 1}\,\mathbf{j} \quad \Rightarrow \quad \mathbf{r}'(t) = \dfrac{-4t}{(t^2 + 1)^2}\,\mathbf{i} + \dfrac{-2t^2 + 2}{(t^2 + 1)^2}\,\mathbf{j}$,

$\dfrac{ds}{dt} = |\mathbf{r}'(t)| = \sqrt{\left[\dfrac{-4t}{(t^2+1)^2}\right]^2 + \left[\dfrac{-2t^2 + 2}{(t^2+1)^2}\right]^2} = \sqrt{\dfrac{4t^4 + 8t^2 + 4}{(t^2+1)^4}} = \sqrt{\dfrac{4(t^2+1)^2}{(t^2+1)^4}} = \sqrt{\dfrac{4}{(t^2+1)^2}} = \dfrac{2}{t^2+1}$.

Since the initial point $(1, 0)$ corresponds to $t = 0$, the arc length function

$s(t) = \int_0^t |\mathbf{r}'(u)|\, du = \int_0^t \dfrac{2}{u^2 + 1}\, du = 2\arctan t$. Then $\arctan t = \tfrac{1}{2}s \quad \Rightarrow \quad t = \tan\tfrac{1}{2}s$. Substituting, we have

$$\mathbf{r}(t(s)) = \left[\dfrac{2}{\tan^2\left(\frac{1}{2}s\right) + 1} - 1\right]\mathbf{i} + \dfrac{2\tan\left(\frac{1}{2}s\right)}{\tan^2\left(\frac{1}{2}s\right) + 1}\,\mathbf{j} = \dfrac{1 - \tan^2\left(\frac{1}{2}s\right)}{1 + \tan^2\left(\frac{1}{2}s\right)}\,\mathbf{i} + \dfrac{2\tan\left(\frac{1}{2}s\right)}{\sec^2\left(\frac{1}{2}s\right)}\,\mathbf{j}$$

$$= \dfrac{1 - \tan^2\left(\frac{1}{2}s\right)}{\sec^2\left(\frac{1}{2}s\right)}\,\mathbf{i} + 2\tan\left(\tfrac{1}{2}s\right)\cos^2\left(\tfrac{1}{2}s\right)\mathbf{j} = \left[\cos^2\left(\tfrac{1}{2}s\right) - \sin^2\left(\tfrac{1}{2}s\right)\right]\mathbf{i} + 2\sin\left(\tfrac{1}{2}s\right)\cos\left(\tfrac{1}{2}s\right)\mathbf{j} = \cos s\,\mathbf{i} + \sin s\,\mathbf{j}$$

With this parametrization, we recognize the function as representing the unit circle. Note here that the curve approaches, but

does not include, the point $(-1, 0)$, since $\cos s = -1$ for $s = \pi + 2k\pi$ (k an integer) but then $t = \tan\left(\frac{1}{2}s\right)$ is undefined.

17. (a) $\mathbf{r}(t) = \langle 2\sin t, 5t, 2\cos t\rangle \;\Rightarrow\; \mathbf{r}'(t) = \langle 2\cos t, 5, -2\sin t\rangle \;\Rightarrow\; |\mathbf{r}'(t)| = \sqrt{4\cos^2 t + 25 + 4\sin^2 t} = \sqrt{29}.$

Then $\mathbf{T}(t) = \dfrac{\mathbf{r}'(t)}{|\mathbf{r}'(t)|} = \dfrac{1}{\sqrt{29}}\langle 2\cos t, 5, -2\sin t\rangle$ or $\left\langle \frac{2}{\sqrt{29}}\cos t, \frac{5}{\sqrt{29}}, -\frac{2}{\sqrt{29}}\sin t\right\rangle.$

$\mathbf{T}'(t) = \frac{1}{\sqrt{29}}\langle -2\sin t, 0, -2\cos t\rangle \;\Rightarrow\; |\mathbf{T}'(t)| = \frac{1}{\sqrt{29}}\sqrt{4\sin^2 t + 0 + 4\cos^2 t} = \frac{2}{\sqrt{29}}.$ Thus

$\mathbf{N}(t) = \dfrac{\mathbf{T}'(t)}{|\mathbf{T}'(t)|} = \dfrac{1/\sqrt{29}}{2/\sqrt{29}}\langle -2\sin t, 0, -2\cos t\rangle = \langle -\sin t, 0, -\cos t\rangle.$

(b) $\kappa(t) = \dfrac{|\mathbf{T}'(t)|}{|\mathbf{r}'(t)|} = \dfrac{2/\sqrt{29}}{\sqrt{29}} = \dfrac{2}{29}$

18. (a) $\mathbf{r}(t) = \langle t^2, \sin t - t\cos t, \cos t + t\sin t\rangle \;\Rightarrow$

$\mathbf{r}'(t) = \langle 2t, \cos t + t\sin t - \cos t, -\sin t + t\cos t + \sin t\rangle = \langle 2t, t\sin t, t\cos t\rangle \;\Rightarrow$

$|\mathbf{r}'(t)| = \sqrt{4t^2 + t^2\sin^2 t + t^2\cos^2 t} = \sqrt{4t^2 + t^2(\cos^2 t + \sin^2 t)} = \sqrt{5t^2} = \sqrt{5}\,t$ [since $t > 0$]. Then

$\mathbf{T}(t) = \dfrac{\mathbf{r}'(t)}{|\mathbf{r}'(t)|} = \dfrac{1}{\sqrt{5}\,t}\langle 2t, t\sin t, t\cos t\rangle = \frac{1}{\sqrt{5}}\langle 2, \sin t, \cos t\rangle.$ $\mathbf{T}'(t) = \frac{1}{\sqrt{5}}\langle 0, \cos t, -\sin t\rangle \;\Rightarrow$

$|\mathbf{T}'(t)| = \frac{1}{\sqrt{5}}\sqrt{0 + \cos^2 t + \sin^2 t} = \frac{1}{\sqrt{5}}.$ Thus $\mathbf{N}(t) = \dfrac{\mathbf{T}'(t)}{|\mathbf{T}'(t)|} = \dfrac{1/\sqrt{5}}{1/\sqrt{5}}\langle 0, \cos t, -\sin t\rangle = \langle 0, \cos t, -\sin t\rangle.$

(b) $\kappa(t) = \dfrac{|\mathbf{T}'(t)|}{|\mathbf{r}'(t)|} = \dfrac{1/\sqrt{5}}{\sqrt{5}\,t} = \dfrac{1}{5t}$

19. (a) $\mathbf{r}(t) = \langle \sqrt{2}\,t, e^t, e^{-t}\rangle \;\Rightarrow\; \mathbf{r}'(t) = \langle \sqrt{2}, e^t, -e^{-t}\rangle \;\Rightarrow\; |\mathbf{r}'(t)| = \sqrt{2 + e^{2t} + e^{-2t}} = \sqrt{(e^t + e^{-t})^2} = e^t + e^{-t}.$

Then

$\mathbf{T}(t) = \dfrac{\mathbf{r}'(t)}{|\mathbf{r}'(t)|} = \dfrac{1}{e^t + e^{-t}}\langle \sqrt{2}, e^t, -e^{-t}\rangle = \dfrac{1}{e^{2t} + 1}\langle \sqrt{2}e^t, e^{2t}, -1\rangle \quad \left[\text{after multiplying by }\dfrac{e^t}{e^t}\right]$ and

$\mathbf{T}'(t) = \dfrac{1}{e^{2t}+1}\langle \sqrt{2}e^t, 2e^{2t}, 0\rangle - \dfrac{2e^{2t}}{(e^{2t}+1)^2}\langle \sqrt{2}e^t, e^{2t}, -1\rangle$

$\qquad = \dfrac{1}{(e^{2t}+1)^2}\big[(e^{2t}+1)\langle \sqrt{2}e^t, 2e^{2t}, 0\rangle - 2e^{2t}\langle \sqrt{2}e^t, e^{2t}, -1\rangle\big] = \dfrac{1}{(e^{2t}+1)^2}\langle \sqrt{2}e^t(1 - e^{2t}), 2e^{2t}, 2e^{2t}\rangle$

Then

$|\mathbf{T}'(t)| = \dfrac{1}{(e^{2t}+1)^2}\sqrt{2e^{2t}(1 - 2e^{2t} + e^{4t}) + 4e^{4t} + 4e^{4t}} = \dfrac{1}{(e^{2t}+1)^2}\sqrt{2e^{2t}(1 + 2e^{2t} + e^{4t})}$

$\qquad = \dfrac{1}{(e^{2t}+1)^2}\sqrt{2e^{2t}(1 + e^{2t})^2} = \dfrac{\sqrt{2}e^t(1 + e^{2t})}{(e^{2t}+1)^2} = \dfrac{\sqrt{2}e^t}{e^{2t}+1}$

Therefore

$\mathbf{N}(t) = \dfrac{\mathbf{T}'(t)}{|\mathbf{T}'(t)|} = \dfrac{e^{2t}+1}{\sqrt{2}e^t}\dfrac{1}{(e^{2t}+1)^2}\langle \sqrt{2}e^t(1 - e^{2t}), 2e^{2t}, 2e^{2t}\rangle$

$\qquad = \dfrac{1}{\sqrt{2}e^t(e^{2t}+1)}\langle \sqrt{2}e^t(1 - e^{2t}), 2e^{2t}, 2e^{2t}\rangle = \dfrac{1}{e^{2t}+1}\langle 1 - e^{2t}, \sqrt{2}e^t, \sqrt{2}e^t\rangle$

(b) $\kappa(t) = \dfrac{|\mathbf{T}'(t)|}{|\mathbf{r}'(t)|} = \dfrac{\sqrt{2}e^t}{e^{2t}+1}\cdot\dfrac{1}{e^t + e^{-t}} = \dfrac{\sqrt{2}e^t}{e^{3t} + 2e^t + e^{-t}} = \dfrac{\sqrt{2}e^{2t}}{e^{4t} + 2e^{2t} + 1} = \dfrac{\sqrt{2}e^{2t}}{(e^{2t}+1)^2}$

20. (a) $\mathbf{r}(t) = \langle t, \frac{1}{2}t^2, t^2 \rangle$ $\Rightarrow$ $\mathbf{r}'(t) = \langle 1, t, 2t \rangle$ $\Rightarrow$ $|\mathbf{r}'(t)| = \sqrt{1 + t^2 + 4t^2} = \sqrt{1 + 5t^2}$. Then

$$\mathbf{T}(t) = \frac{\mathbf{r}'(t)}{|\mathbf{r}'(t)|} = \frac{1}{\sqrt{1 + 5t^2}} \langle 1, t, 2t \rangle.$$

$$\mathbf{T}'(t) = \frac{-5t}{(1 + 5t^2)^{3/2}} \langle 1, t, 2t \rangle + \frac{1}{\sqrt{1 + 5t^2}} \langle 0, 1, 2 \rangle \quad \text{(by Formula 3 of Theorem 14.2.3 [ET 13.2.3])}$$

$$= \frac{1}{(1 + 5t^2)^{3/2}} (\langle -5t, -5t^2, -10t^2 \rangle + \langle 0, 1 + 5t^2, 2 + 10t^2 \rangle) = \frac{1}{(1 + 5t^2)^{3/2}} \langle -5t, 1, 2 \rangle$$

$$|\mathbf{T}'(t)| = \frac{1}{(1 + 5t^2)^{3/2}} \sqrt{25t^2 + 1 + 4} = \frac{1}{(1 + 5t^2)^{3/2}} \sqrt{25t^2 + 5} = \frac{\sqrt{5}\sqrt{5t^2 + 1}}{(1 + 5t^2)^{3/2}} = \frac{\sqrt{5}}{1 + 5t^2}$$

Thus $\mathbf{N}(t) = \dfrac{\mathbf{T}'(t)}{|\mathbf{T}'(t)|} = \dfrac{1 + 5t^2}{\sqrt{5}} \cdot \dfrac{1}{(1 + 5t^2)^{3/2}} \langle -5t, 1, 2 \rangle = \dfrac{1}{\sqrt{5 + 25t^2}} \langle -5t, 1, 2 \rangle$.

(b) $\kappa(t) = \dfrac{|\mathbf{T}'(t)|}{|\mathbf{r}'(t)|} = \dfrac{\sqrt{5}/(1 + 5t^2)}{\sqrt{1 + 5t^2}} = \dfrac{\sqrt{5}}{(1 + 5t^2)^{3/2}}$

21. $\mathbf{r}(t) = t^2\,\mathbf{i} + t\,\mathbf{k}$ $\Rightarrow$ $\mathbf{r}'(t) = 2t\,\mathbf{i} + \mathbf{k}$, $\mathbf{r}''(t) = 2\,\mathbf{i}$, $|\mathbf{r}'(t)| = \sqrt{(2t)^2 + 0^2 + 1^2} = \sqrt{4t^2 + 1}$, $\mathbf{r}'(t) \times \mathbf{r}''(t) = 2\,\mathbf{j}$,

$|\mathbf{r}'(t) \times \mathbf{r}''(t)| = 2$. Then $\kappa(t) = \dfrac{|\mathbf{r}'(t) \times \mathbf{r}''(t)|}{|\mathbf{r}'(t)|^3} = \dfrac{2}{\left(\sqrt{4t^2 + 1}\right)^3} = \dfrac{2}{(4t^2 + 1)^{3/2}}$.

22. $\mathbf{r}(t) = t\,\mathbf{i} + t\,\mathbf{j} + (1 + t^2)\,\mathbf{k}$ $\Rightarrow$ $\mathbf{r}'(t) = \mathbf{i} + \mathbf{j} + 2t\,\mathbf{k}$, $\mathbf{r}''(t) = 2\,\mathbf{k}$, $|\mathbf{r}'(t)| = \sqrt{1^2 + 1^2 + (2t)^2} = \sqrt{4t^2 + 2}$,

$\mathbf{r}'(t) \times \mathbf{r}''(t) = 2\,\mathbf{i} - 2\,\mathbf{j}$, $|\mathbf{r}'(t) \times \mathbf{r}''(t)| = \sqrt{2^2 + 2^2 + 0^2} = \sqrt{8} = 2\sqrt{2}$.

Then $\kappa(t) = \dfrac{|\mathbf{r}'(t) \times \mathbf{r}''(t)|}{|\mathbf{r}'(t)|^3} = \dfrac{2\sqrt{2}}{\left(\sqrt{4t^2 + 2}\right)^3} = \dfrac{2\sqrt{2}}{\left(\sqrt{2}\sqrt{2t^2 + 1}\right)^3} = \dfrac{1}{(2t^2 + 1)^{3/2}}$.

23. $\mathbf{r}(t) = 3t\,\mathbf{i} + 4\sin t\,\mathbf{j} + 4\cos t\,\mathbf{k}$ $\Rightarrow$ $\mathbf{r}'(t) = 3\,\mathbf{i} + 4\cos t\,\mathbf{j} - 4\sin t\,\mathbf{k}$, $\mathbf{r}''(t) = -4\sin t\,\mathbf{j} - 4\cos t\,\mathbf{k}$,

$|\mathbf{r}'(t)| = \sqrt{9 + 16\cos^2 t + 16\sin^2 t} = \sqrt{9 + 16} = 5$, $\mathbf{r}'(t) \times \mathbf{r}''(t) = -16\,\mathbf{i} + 12\cos t\,\mathbf{j} - 12\sin t\,\mathbf{k}$,

$|\mathbf{r}'(t) \times \mathbf{r}''(t)| = \sqrt{256 + 144\cos^2 t + 144\sin^2 t} = \sqrt{400} = 20$. Then $\kappa(t) = \dfrac{|\mathbf{r}'(t) \times \mathbf{r}''(t)|}{|\mathbf{r}'(t)|^3} = \dfrac{20}{5^3} = \dfrac{4}{25}$.

24. $\mathbf{r}(t) = \langle e^t \cos t, e^t \sin t, t \rangle$ $\Rightarrow$ $\mathbf{r}'(t) = \langle e^t \cos t - e^t \sin t, e^t \cos t + e^t \sin t, 1 \rangle$. The point $(1, 0, 0)$ corresponds to

$t = 0$, and $\mathbf{r}'(0) = \langle 1, 1, 1 \rangle$ $\Rightarrow$ $|\mathbf{r}'(0)| = \sqrt{1^2 + 1^2 + 1^2} = \sqrt{3}$.

$\mathbf{r}''(t) = \langle e^t \cos t - e^t \sin t - e^t \cos t - e^t \sin t, e^t \cos t - e^t \sin t + e^t \cos t + e^t \sin t, 0 \rangle = \langle -2e^t \sin t, 2e^t \cos t, 0 \rangle$ $\Rightarrow$

$\mathbf{r}''(0) = \langle 0, 2, 0 \rangle$. $\mathbf{r}'(0) \times \mathbf{r}''(0) = \langle -2, 0, 2 \rangle$. $|\mathbf{r}'(0) \times \mathbf{r}''(0)| = \sqrt{(-2)^2 + 0^2 + 2^2} = \sqrt{8} = 2\sqrt{2}$.

Then $\kappa(0) = \dfrac{|\mathbf{r}'(0) \times \mathbf{r}''(0)|}{|\mathbf{r}'(0)|^3} = \dfrac{2\sqrt{2}}{\left(\sqrt{3}\right)^3} = \dfrac{2\sqrt{2}}{3\sqrt{3}}$ or $\dfrac{2\sqrt{6}}{9}$.

25. $\mathbf{r}(t) = \langle t, t^2, t^3 \rangle$ $\Rightarrow$ $\mathbf{r}'(t) = \langle 1, 2t, 3t^2 \rangle$. The point $(1, 1, 1)$ corresponds to $t = 1$, and $\mathbf{r}'(1) = \langle 1, 2, 3 \rangle$ $\Rightarrow$

$|\mathbf{r}'(1)| = \sqrt{1 + 4 + 9} = \sqrt{14}$. $\mathbf{r}''(t) = \langle 0, 2, 6t \rangle$ $\Rightarrow$ $\mathbf{r}''(1) = \langle 0, 2, 6 \rangle$. $\mathbf{r}'(1) \times \mathbf{r}''(1) = \langle 6, -6, 2 \rangle$, so

$|\mathbf{r}'(1) \times \mathbf{r}''(1)| = \sqrt{36 + 36 + 4} = \sqrt{76}$. Then $\kappa(1) = \dfrac{|\mathbf{r}'(1) \times \mathbf{r}''(1)|}{|\mathbf{r}'(1)|^3} = \dfrac{\sqrt{76}}{\sqrt{14}^3} = \dfrac{1}{7}\sqrt{\dfrac{19}{14}}$.

26.

$$\mathbf{r}(t) = \left\langle t, 4t^{3/2}, -t^2 \right\rangle \quad \Rightarrow \quad \mathbf{r}'(t) = \left\langle 1, 6t^{1/2}, -2t \right\rangle,$$

$$\mathbf{r}''(t) = \left\langle 0, 3t^{-1/2}, -2 \right\rangle, \quad |\mathbf{r}'(t)|^3 = \left(1 + 36t + 4t^2\right)^{3/2},$$

$$\mathbf{r}'(t) \times \mathbf{r}''(t) = \left\langle -12t^{1/2} + 6t^{1/2}, 2, 3t^{-1/2} \right\rangle \quad \Rightarrow$$

$$|\mathbf{r}'(t) \times \mathbf{r}''(t)| = \sqrt{36t + 4 + 9t^{-1}} = \left[\frac{36t^2 + 4t + 9}{t}\right]^{1/2}$$

$$\kappa(t) = \frac{|\mathbf{r}'(t) \times \mathbf{r}''(t)|}{|\mathbf{r}'(t)|^3} = \left(\frac{36t^2 + 4t + 9}{t}\right)^{1/2} \frac{1}{(1 + 36t + 4t^2)^{3/2}} = \frac{\sqrt{36t^2 + 4t + 9}}{t^{1/2}(1 + 36t + 4t^2)^{3/2}}.$$

The point $(1, 4, -1)$ corresponds to $t = 1$, so the curvature at this point is $\kappa(1) = \dfrac{\sqrt{36 + 4 + 9}}{(1 + 36 + 4)^{3/2}} = \dfrac{7}{41\sqrt{41}}$.

27. $f(x) = 2x - x^2$, $\quad f'(x) = 2 - 2x$, $\quad f''(x) = -2$,

$$\kappa(x) = \frac{|f''(x)|}{[1 + (f'(x))^2]^{3/2}} = \frac{|-2|}{[1 + (2 - 2x)^2]^{3/2}} = \frac{2}{(4x^2 - 8x + 5)^{3/2}}$$

28. $f(x) = \cos x$, $\quad f'(x) = -\sin x$, $\quad f''(x) = -\cos x$,

$$\kappa(x) = \frac{|f''(x)|}{[1 + (f'(x))^2]^{3/2}} = \frac{|-\cos x|}{[1 + (-\sin x)^2]^{3/2}} = \frac{|\cos x|}{(1 + \sin^2 x)^{3/2}}$$

29. $f(x) = 4x^{5/2}$, $\quad f'(x) = 10x^{3/2}$, $\quad f''(x) = 15x^{1/2}$,

$$\kappa(x) = \frac{|f''(x)|}{[1 + (f'(x))^2]^{3/2}} = \frac{\left|15x^{1/2}\right|}{[1 + (10x^{3/2})^2]^{3/2}} = \frac{15\sqrt{x}}{(1 + 100x^3)^{3/2}}$$

30. $y = \ln x \quad \Rightarrow \quad y' = \dfrac{1}{x}, \quad y'' = -\dfrac{1}{x^2}$,

$$\kappa(x) = \frac{|y''(x)|}{\left[1 + (y'(x))^2\right]^{3/2}} = \left|\frac{-1}{x^2}\right| \frac{1}{(1 + 1/x^2)^{3/2}} = \frac{1}{x^2} \frac{(x^2)^{3/2}}{(x^2 + 1)^{3/2}} = \frac{|x|}{(x^2 + 1)^{3/2}} = \frac{x}{(x^2 + 1)^{3/2}} \quad \text{[since } x > 0\text{]}.$$

To find the maximum curvature, we first find the critical numbers of $\kappa(x)$:

$$\kappa'(x) = \frac{(x^2 + 1)^{3/2} - x\left(\frac{3}{2}\right)(x^2 + 1)^{1/2}(2x)}{[(x^2 + 1)^{3/2}]^2} = \frac{(x^2 + 1)^{1/2}[(x^2 + 1) - 3x^2]}{(x^2 + 1)^3} = \frac{1 - 2x^2}{(x^2 + 1)^{5/2}};$$

$\kappa'(x) = 0 \quad \Rightarrow \quad 1 - 2x^2 = 0$, so the only critical number in the domain is $x = \frac{1}{\sqrt{2}}$. Since $\kappa'(x) > 0$ for $0 < x < \frac{1}{\sqrt{2}}$

and $\kappa'(x) < 0$ for $x > \frac{1}{\sqrt{2}}$, $\kappa(x)$ attains its maximum at $x = \frac{1}{\sqrt{2}}$. Thus, the maximum curvature occurs at $\left(\frac{1}{\sqrt{2}}, \ln \frac{1}{\sqrt{2}}\right)$.

Since $\lim\limits_{x \to \infty} \dfrac{x}{(x^2 + 1)^{3/2}} = 0$, $\kappa(x)$ approaches 0 as $x \to \infty$.

31. Since $y' = y'' = e^x$, the curvature is $\kappa(x) = \dfrac{|y''(x)|}{[1 + (y'(x))^2]^{3/2}} = \dfrac{e^x}{(1 + e^{2x})^{3/2}} = e^x(1 + e^{2x})^{-3/2}$.

To find the maximum curvature, we first find the critical numbers of $\kappa(x)$:

$$\kappa'(x) = e^x(1 + e^{2x})^{-3/2} + e^x\left(-\frac{3}{2}\right)(1 + e^{2x})^{-5/2}(2e^{2x}) = e^x \frac{1 + e^{2x} - 3e^{2x}}{(1 + e^{2x})^{5/2}} = e^x \frac{1 - 2e^{2x}}{(1 + e^{2x})^{5/2}}.$$

$\kappa'(x) = 0$ when $1 - 2e^{2x} = 0$, so $e^{2x} = \frac{1}{2}$ or $x = -\frac{1}{2}\ln 2$. And since $1 - 2e^{2x} > 0$ for $x < -\frac{1}{2}\ln 2$ and

$1 - 2e^{2x} < 0$ for $x > -\frac{1}{2}\ln 2$, the maximum curvature is attained at the point $\left(-\frac{1}{2}\ln 2, e^{(-\ln 2)/2}\right) = \left(-\frac{1}{2}\ln 2, \frac{1}{\sqrt{2}}\right)$.

Since $\lim\limits_{x \to \infty} e^x(1 + e^{2x})^{-3/2} = 0$, $\kappa(x)$ approaches 0 as $x \to \infty$.

32. We can take the parabola as having its vertex at the origin and opening upward, so the equation is $f(x) = ax^2$, $a > 0$. Then by

Equation 11, $\kappa(x) = \dfrac{|f''(x)|}{[1 + (f'(x))^2]^{3/2}} = \dfrac{|2a|}{[1 + (2ax)^2]^{3/2}} = \dfrac{2a}{(1 + 4a^2x^2)^{3/2}}$, thus $\kappa(0) = 2a$. We want $\kappa(0) = 4$, so

$a = 2$ and the equation is $y = 2x^2$.

33. (a) C appears to be changing direction more quickly at P than Q, so we would expect the curvature to be greater at P.

(b) First we sketch approximate osculating circles at P and Q. Using the

axes scale as a guide, we measure the radius of the osculating circle

at P to be approximately 0.8 units, thus $\rho = \dfrac{1}{\kappa}$ $\Rightarrow$

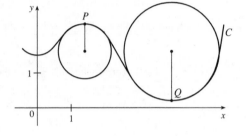

$\kappa = \dfrac{1}{\rho} \approx \dfrac{1}{0.8} \approx 1.3$. Similarly, we estimate the radius of the

osculating circle at Q to be 1.4 units, so $\kappa = \dfrac{1}{\rho} \approx \dfrac{1}{1.4} \approx 0.7$.

34. $y = x^4 - 2x^2$ $\Rightarrow$ $y' = 4x^3 - 4x$, $y'' = 12x^2 - 4$, and

$\kappa(x) = \dfrac{|y''|}{\left[1 + (y')^2\right]^{3/2}} = \dfrac{|12x^2 - 4|}{\left[1 + (4x^3 - 4x)^2\right]^{3/2}}$. The graph of the

curvature here is what we would expect. The graph of $y = x^4 - 2x^2$

appears to be bending most sharply at the origin and near $x = \pm 1$.

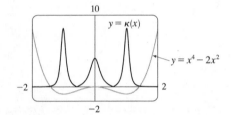

35. $y = x^{-2}$ $\Rightarrow$ $y' = -2x^{-3}$, $y'' = 6x^{-4}$, and

$\kappa(x) = \dfrac{|y''|}{[1 + (y')^2]^{3/2}} = \dfrac{|6x^{-4}|}{[1 + (-2x^{-3})^2]^{3/2}} = \dfrac{6}{x^4(1 + 4x^{-6})^{3/2}}$.

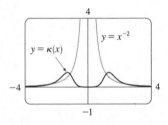

The appearance of the two humps in this graph is perhaps a little surprising, but it is

explained by the fact that $y = x^{-2}$ increases asymptotically at the origin from both

directions, and so its graph has very little bend there. (Note that $\kappa(0)$ is undefined.)

36. Notice that the curve a is highest for the same x-values at which curve b is turning more sharply, and a is 0 or near 0 where b is nearly straight. So, a must be the graph of $y = \kappa(x)$, and b is the graph of $y = f(x)$.

37. Notice that the curve b has two inflection points at which the graph appears almost straight. We would expect the curvature to be 0 or nearly 0 at these values, but the curve a isn't near 0 there. Thus, a must be the graph of $y = f(x)$ rather than the graph of curvature, and b is the graph of $y = \kappa(x)$.

38. (a) The complete curve is given by $0 \le t \le 2\pi$. Curvature

appears to have a local (or absolute) maximum at 6

points. (Look at points where the curve appears to turn

more sharply.)

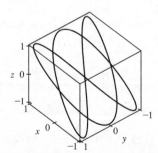

(b) Using a CAS, we find (after simplifying)

$$\kappa(t) = \frac{3\sqrt{2}\,\sqrt{(5\sin t + \sin 5t)^2}}{(9\cos 6t + 2\cos 4t + 11)^{3/2}}.$$ (To compute cross

products in Maple, use the `VectorCalculus` package and

the `CrossProduct(a,b)` command; in Mathematica, use

`Cross[a,b]`.) The graph shows 6 local (or absolute)

maximum points for $0 \le t \le 2\pi$, as observed in part (a).

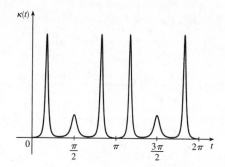

39. Using a CAS, we find (after simplifying)

$$\kappa(t) = \frac{6\sqrt{4\cos^2 t - 12\cos t + 13}}{(17 - 12\cos t)^{3/2}}.$$ (To compute cross

products in Maple, use the `VectorCalculus` package and

the `CrossProduct(a,b)` command; in Mathematica, use

`Cross[a,b]`.) Curvature is largest at integer multiples of 2π.

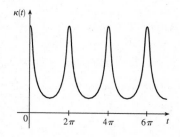

40. Here $\mathbf{r}(t) = \langle f(t), g(t)\rangle$, $\mathbf{r}'(t) = \langle f'(t), g'(t)\rangle$, $\mathbf{r}''(t) = \langle f''(t), g''(t)\rangle$,

$$|\mathbf{r}'(t)|^3 = \left[\sqrt{(f'(t))^2 + (g'(t))^2}\right]^3 = [(f'(t))^2 + (g'(t))^2]^{3/2} = (\dot{x}^2 + \dot{y}^2)^{3/2}, \text{ and}$$

$$|\mathbf{r}'(t) \times \mathbf{r}''(t)| = |\langle 0, 0, f'(t)\,g''(t) - f''(t)\,g'(t)\rangle| = [(\dot{x}\ddot{y} - \ddot{x}\dot{y})^2]^{1/2} = |\dot{x}\ddot{y} - \dot{y}\ddot{x}|. \text{ Thus } \kappa(t) = \frac{|\dot{x}\ddot{y} - \dot{y}\ddot{x}|}{[\dot{x}^2 + \dot{y}^2]^{3/2}}.$$

41. $x = e^t \cos t \Rightarrow \dot{x} = e^t(\cos t - \sin t) \Rightarrow \ddot{x} = e^t(-\sin t - \cos t) + e^t(\cos t - \sin t) = -2e^t \sin t,$

$y = e^t \sin t \Rightarrow \dot{y} = e^t(\cos t + \sin t) \Rightarrow \ddot{y} = e^t(-\sin t + \cos t) + e^t(\cos t + \sin t) = 2e^t \cos t.$ Then

$$\kappa(t) = \frac{|\dot{x}\ddot{y} - \dot{y}\ddot{x}|}{[\dot{x}^2 + \dot{y}^2]^{3/2}} = \frac{\left|e^t(\cos t - \sin t)(2e^t \cos t) - e^t(\cos t + \sin t)(-2e^t \sin t)\right|}{([e^t(\cos t - \sin t)]^2 + [e^t(\cos t + \sin t)]^2)^{3/2}}$$

$$= \frac{\left|2e^{2t}(\cos^2 t - \sin t \cos t + \sin t \cos t + \sin^2 t)\right|}{\left[e^{2t}(\cos^2 t - 2\cos t \sin t + \sin^2 t + \cos^2 t + 2\cos t \sin t + \sin^2 t)\right]^{3/2}} = \frac{|2e^{2t}(1)|}{[e^{2t}(1+1)]^{3/2}} = \frac{2e^{2t}}{e^{3t}(2)^{3/2}} = \frac{1}{\sqrt{2}\,e^t}$$

42. $x = 1 + t^3 \Rightarrow \dot{x} = 3t^2 \Rightarrow \ddot{x} = 6t,$ $y = t + t^2 \Rightarrow \dot{y} = 1 + 2t \Rightarrow \ddot{y} = 2.$

Then $\kappa(t) = \dfrac{|\dot{x}\ddot{y} - \dot{y}\ddot{x}|}{[\dot{x}^2 + \dot{y}^2]^{3/2}} = \dfrac{|(3t^2)(2) - (1+2t)(6t)|}{[(3t^2)^2 + (1+2t)^2]^{3/2}} = \dfrac{|-6t^2 - 6t|}{(9t^4 + 4t^2 + 4t + 1)^{3/2}} = \dfrac{6\,|t^2 + t|}{(9t^4 + 4t^2 + 4t + 1)^{3/2}}.$

43. $\left(1, \frac{2}{3}, 1\right)$ corresponds to $t = 1$. $\mathbf{T}(t) = \dfrac{\mathbf{r}'(t)}{|\mathbf{r}'(t)|} = \dfrac{\langle 2t, 2t^2, 1\rangle}{\sqrt{4t^2 + 4t^4 + 1}} = \dfrac{\langle 2t, 2t^2, 1\rangle}{2t^2 + 1},$ so $\mathbf{T}(1) = \left\langle \frac{2}{3}, \frac{2}{3}, \frac{1}{3}\right\rangle.$

$\mathbf{T}'(t) = -4t(2t^2 + 1)^{-2}\langle 2t, 2t^2, 1\rangle + (2t^2 + 1)^{-1}\langle 2, 4t, 0\rangle$ (by Formula 3 of Theorem 14.2 [ET 13.2])

$\qquad = (2t^2 + 1)^{-2}\langle -8t^2 + 4t^2 + 2, -8t^3 + 8t^3 + 4t, -4t\rangle = 2(2t^2 + 1)^{-2}\langle 1 - 2t^2, 2t, -2t\rangle$

$$\mathbf{N}(t) = \frac{\mathbf{T}'(t)}{|\mathbf{T}'(t)|} = \frac{2(2t^2 + 1)^{-2}\langle 1 - 2t^2, 2t, -2t\rangle}{2(2t^2 + 1)^{-2}\sqrt{(1 - 2t^2)^2 + (2t)^2 + (-2t)^2}} = \frac{\langle 1 - 2t^2, 2t, -2t\rangle}{\sqrt{1 - 4t^2 + 4t^4 + 8t^2}} = \frac{\langle 1 - 2t^2, 2t, -2t\rangle}{1 + 2t^2}$$

$\mathbf{N}(1) = \left\langle -\frac{1}{3}, \frac{2}{3}, -\frac{2}{3}\right\rangle$ and $\mathbf{B}(1) = \mathbf{T}(1) \times \mathbf{N}(1) = \left\langle -\frac{4}{9} - \frac{2}{9}, -\left(-\frac{4}{9} + \frac{1}{9}\right), \frac{4}{9} + \frac{2}{9}\right\rangle = \left\langle -\frac{2}{3}, \frac{1}{3}, \frac{2}{3}\right\rangle.$

44. $(1, 0, 0)$ corresponds to $t = 0$. $\mathbf{r}(t) = \langle \cos t, \sin t, \ln \cos t \rangle$, and in Exercise 4 we found that $\mathbf{r}'(t) = \langle -\sin t, \cos t, -\tan t \rangle$

and $|\mathbf{r}'(t)| = |\sec t|$. Here we can assume $-\frac{\pi}{2} < t < \frac{\pi}{2}$ and then $\sec t > 0$ $\Rightarrow$ $|\mathbf{r}'(t)| = \sec t$.

$$\mathbf{T}(t) = \frac{\mathbf{r}'(t)}{|\mathbf{r}'(t)|} = \frac{\langle -\sin t, \cos t, -\tan t \rangle}{\sec t} = \langle -\sin t \cos t, \cos^2 t, -\sin t \rangle \quad \text{and} \quad \mathbf{T}(0) = \langle 0, 1, 0 \rangle.$$

$\mathbf{T}'(t) = \langle -[(\sin t)(-\sin t) + (\cos t)(\cos t)], 2(\cos t)(-\sin t), -\cos t \rangle = \langle \sin^2 t - \cos^2 t, -2\sin t \cos t, -\cos t \rangle$, so

$$\mathbf{N}(0) = \frac{\mathbf{T}'(0)}{|\mathbf{T}'(0)|} = \frac{\langle -1, 0, -1 \rangle}{\sqrt{1 + 0 + 1}} = \frac{1}{\sqrt{2}} \langle -1, 0, -1 \rangle = \left\langle -\frac{1}{\sqrt{2}}, 0, -\frac{1}{\sqrt{2}} \right\rangle.$$

Finally, $\mathbf{B}(0) = \mathbf{T}(0) \times \mathbf{N}(0) = \langle 0, 1, 0 \rangle \times \left\langle -\frac{1}{\sqrt{2}}, 0, -\frac{1}{\sqrt{2}} \right\rangle = \left\langle -\frac{1}{\sqrt{2}}, 0, \frac{1}{\sqrt{2}} \right\rangle.$

45. $(0, \pi, -2)$ corresponds to $t = \pi$. $\mathbf{r}(t) = \langle 2\sin 3t, t, 2\cos 3t \rangle$ $\Rightarrow$

$$\mathbf{T}(t) = \frac{\mathbf{r}'(t)}{|\mathbf{r}'(t)|} = \frac{\langle 6\cos 3t, 1, -6\sin 3t \rangle}{\sqrt{36\cos^2 3t + 1 + 36\sin^2 3t}} = \frac{1}{\sqrt{37}} \langle 6\cos 3t, 1, -6\sin 3t \rangle.$$

$\mathbf{T}(\pi) = \frac{1}{\sqrt{37}} \langle -6, 1, 0 \rangle$ is a normal vector for the normal plane, and so $\langle -6, 1, 0 \rangle$ is also normal. Thus an equation for the

plane is $-6(x - 0) + 1(y - \pi) + 0(z + 2) = 0$ or $y - 6x = \pi$.

$$\mathbf{T}'(t) = \frac{1}{\sqrt{37}} \langle -18\sin 3t, 0, -18\cos 3t \rangle \quad \Rightarrow \quad |\mathbf{T}'(t)| = \frac{\sqrt{18^2 \sin^2 3t + 18^2 \cos^2 3t}}{\sqrt{37}} = \frac{18}{\sqrt{37}} \quad \Rightarrow$$

$$\mathbf{N}(t) = \frac{\mathbf{T}'(t)}{|\mathbf{T}'(t)|} = \langle -\sin 3t, 0, -\cos 3t \rangle. \text{ So } \mathbf{N}(\pi) = \langle 0, 0, 1 \rangle \text{ and } \mathbf{B}(\pi) = \frac{1}{\sqrt{37}} \langle -6, 1, 0 \rangle \times \langle 0, 0, 1 \rangle = \frac{1}{\sqrt{37}} \langle 1, 6, 0 \rangle.$$

Since $\mathbf{B}(\pi)$ is a normal to the osculating plane, so is $\langle 1, 6, 0 \rangle$.

An equation for the plane is $1(x - 0) + 6(y - \pi) + 0(z + 2) = 0$ or $x + 6y = 6\pi$.

46. $t = 1$ at $(1, 1, 1)$. $\mathbf{r}'(t) = \langle 1, 2t, 3t^2 \rangle$. $\mathbf{r}'(1) = \langle 1, 2, 3 \rangle$ is normal to the normal plane, so an equation for this plane

is $1(x - 1) + 2(y - 1) + 3(z - 1) = 0$, or $x + 2y + 3z = 6$.

$$\mathbf{T}(t) = \frac{\mathbf{r}'(t)}{|\mathbf{r}'(t)|} = \frac{1}{\sqrt{1 + 4t^2 + 9t^4}} \langle 1, 2t, 3t^2 \rangle. \text{ Using the product rule on each term of } \mathbf{T}(t) \text{ gives}$$

$$\mathbf{T}'(t) = \frac{1}{(1 + 4t^2 + 9t^4)^{3/2}} \langle -\tfrac{1}{2}(8t + 36t^3), 2(1 + 4t^2 + 9t^4) - \tfrac{1}{2}(8t + 36t^3)2t,$$

$$6t(1 + 4t^2 + 9t^4) - \tfrac{1}{2}(8t + 36t^3)3t^2 \rangle$$

$$= \frac{1}{(1 + 4t^2 + 9t^4)^{3/2}} \langle -4t - 18t^3, 2 - 18t^4, 6t + 12t^3 \rangle = \frac{-2}{(14)^{3/2}} \langle 11, 8, -9 \rangle \text{ when } t = 1.$$

$\mathbf{N}(1) \parallel \mathbf{T}'(1) \parallel \langle 11, 8, -9 \rangle$ and $\mathbf{T}(1) \parallel \mathbf{r}'(1) = \langle 1, 2, 3 \rangle$ $\Rightarrow$ a normal vector to the osculating plane is

$\langle 11, 8, -9 \rangle \times \langle 1, 2, 3 \rangle = \langle 42, -42, 14 \rangle$ or equivalently $\langle 3, -3, 1 \rangle$.

An equation for the plane is $3(x - 1) - 3(y - 1) + (z - 1) = 0$ or $3x - 3y + z = 1$.

47. The ellipse $9x^2 + 4y^2 = 36$ is given by the parametric equations $x = 2\cos t$, $y = 3\sin t$, so using the result from

Exercise 40,

$$\kappa(t) = \frac{|\dot{x}\ddot{y} - \ddot{x}\dot{y}|}{[\dot{x}^2 + \dot{y}^2]^{3/2}} = \frac{|(-2\sin t)(-3\sin t) - (3\cos t)(-2\cos t)|}{(4\sin^2 t + 9\cos^2 t)^{3/2}} = \frac{6}{(4\sin^2 t + 9\cos^2 t)^{3/2}}.$$

At $(2, 0)$, $t = 0$. Now $\kappa(0) = \frac{6}{27} = \frac{2}{9}$, so the radius of the osculating circle is

$1/\kappa(0) = \frac{9}{2}$ and its center is $\left(-\frac{5}{2}, 0\right)$. Its equation is therefore $\left(x + \frac{5}{2}\right)^2 + y^2 = \frac{81}{4}$.

At $(0, 3)$, $t = \frac{\pi}{2}$, and $\kappa\left(\frac{\pi}{2}\right) = \frac{6}{8} = \frac{3}{4}$. So the radius of the osculating circle is $\frac{4}{3}$ and

its center is $\left(0, \frac{5}{3}\right)$. Hence its equation is $x^2 + \left(y - \frac{5}{3}\right)^2 = \frac{16}{9}$.

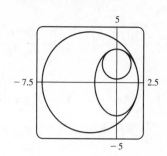

48. $y = \frac{1}{2}x^2$ $\Rightarrow$ $y' = x$ and $y'' = 1$, so Formula 11 gives $\kappa(x) = \dfrac{1}{(1 + x^2)^{3/2}}$. So the curvature at $(0, 0)$ is $\kappa(0) = 1$ and

the osculating circle has radius 1 and center $(0, 1)$, and hence equation $x^2 + (y - 1)^2 = 1$. The curvature at $\left(1, \frac{1}{2}\right)$

is $\kappa(1) = \dfrac{1}{(1 + 1^2)^{3/2}} = \dfrac{1}{2\sqrt{2}}$. The tangent line to the parabola at $\left(1, \frac{1}{2}\right)$

has slope 1, so the normal line has slope -1. Thus the center of the

osculating circle lies in the direction of the unit vector $\left\langle -\frac{1}{\sqrt{2}}, \frac{1}{\sqrt{2}} \right\rangle$.

The circle has radius $2\sqrt{2}$, so its center has position vector

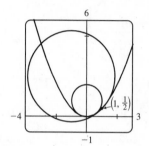

$\left\langle 1, \frac{1}{2} \right\rangle + 2\sqrt{2} \left\langle -\frac{1}{\sqrt{2}}, \frac{1}{\sqrt{2}} \right\rangle = \left\langle -1, \frac{5}{2} \right\rangle$. So the equation of the circle

is $(x + 1)^2 + \left(y - \frac{5}{2}\right)^2 = 8$.

49. The tangent vector is normal to the normal plane, and the vector $\langle 6, 6, -8 \rangle$ is normal to the given plane.

But $\mathbf{T}(t) \parallel \mathbf{r}'(t)$ and $\langle 6, 6, -8 \rangle \parallel \langle 3, 3, -4 \rangle$, so we need to find t such that $\mathbf{r}'(t) \parallel \langle 3, 3, -4 \rangle$.

$\mathbf{r}(t) = \left\langle t^3, 3t, t^4 \right\rangle$ $\Rightarrow$ $\mathbf{r}'(t) = \left\langle 3t^2, 3, 4t^3 \right\rangle \parallel \langle 3, 3, -4 \rangle$ when $t = -1$. So the planes are parallel at the point $(-1, -3, 1)$.

50. To find the osculating plane, we first calculate the unit tangent and normal vectors.

In Maple, we use the `VectorCalculus` package and set `r:=<t^3,3*t,t^4>;`. After differentiating, the

`Normalize` command converts the tangent vector to the unit tangent vector: `T:=Normalize(diff(r,t));`. After

simplifying, we find that $\mathbf{T}(t) = \dfrac{\left\langle 3t^2, 3, 4t^3 \right\rangle}{\sqrt{16t^6 + 9t^4 + 9}}$. We use a similar procedure to compute the unit normal vector,

`N:=Normalize(diff(T,t));`. After simplifying, we have $\mathbf{N}(t) = \dfrac{\left\langle -t(8t^6 - 9), -3t^3(3 + 8t^2), 6t^2(t^4 + 3) \right\rangle}{\sqrt{t^2(4t^6 + 36t^2 + 9)(16t^6 + 9t^4 + 9)}}$. Then

we use the command `B:=CrossProduct(T,N);`. After simplification, we find that $\mathbf{B}(t) = \dfrac{\left\langle 6t^2, -2t^4, -3t \right\rangle}{\sqrt{t^2(4t^6 + 36t^2 + 9)}}$.

In Mathematica, we define the vector function `r={t^3,3*t,t^4}` and use the command `Dt` to differentiate. We find

$\mathbf{T}(t)$ by dividing the result by its magnitude, computed using the `Norm` command. (You may wish to include the option

`Element[t,Reals]` to obtain simpler expressions.) $\mathbf{N}(t)$ is found similarly, and we use `Cross[T,N]` to find $\mathbf{B}(t)$.

Now $\mathbf{B}(t)$ is parallel to $\left\langle 6t^2, -2t^4, -3t \right\rangle$, so if $\mathbf{B}(t)$ is parallel to $\langle 1, 1, 1 \rangle$ for some $t \neq 0$ [since $\mathbf{B}(0) = \mathbf{0}$], then

$\left\langle 6t^2, -2t^4, -3t \right\rangle = k \left\langle 1, 1, 1 \right\rangle$ for some value of k. But then $6t^2 = -2t^4 = -3t$ which has no solution for $t \neq 0$. So there is

no such osculating plane.

51. $\kappa = \left|\dfrac{d\mathbf{T}}{ds}\right| = \left|\dfrac{d\mathbf{T}/dt}{ds/dt}\right| = \dfrac{|d\mathbf{T}/dt|}{ds/dt}$ and $\mathbf{N} = \dfrac{d\mathbf{T}/dt}{|d\mathbf{T}/dt|}$, so $\kappa\mathbf{N} = \dfrac{\left|\dfrac{d\mathbf{T}}{dt}\right|\dfrac{d\mathbf{T}}{dt}}{\left|\dfrac{d\mathbf{T}}{dt}\right|\dfrac{ds}{dt}} = \dfrac{d\mathbf{T}/dt}{ds/dt} = \dfrac{d\mathbf{T}}{ds}$ by the Chain Rule.

52. For a plane curve, $\mathbf{T} = |\mathbf{T}|\cos\phi\,\mathbf{i} + |\mathbf{T}|\sin\phi\,\mathbf{j} = \cos\phi\,\mathbf{i} + \sin\phi\,\mathbf{j}$. Then

$$\dfrac{d\mathbf{T}}{ds} = \left(\dfrac{d\mathbf{T}}{d\phi}\right)\left(\dfrac{d\phi}{ds}\right) = (-\sin\phi\,\mathbf{i} + \cos\phi\,\mathbf{j})\left(\dfrac{d\phi}{ds}\right) \text{ and } \left|\dfrac{d\mathbf{T}}{ds}\right| = |-\sin\phi\,\mathbf{i} + \cos\phi\,\mathbf{j}|\left|\dfrac{d\phi}{ds}\right| = \left|\dfrac{d\phi}{ds}\right|. \text{ Hence for a plane}$$

curve, the curvature is $\kappa = |d\phi/ds|$.

53. (a) $|\mathbf{B}| = 1 \;\Rightarrow\; \mathbf{B}\cdot\mathbf{B} = 1 \;\Rightarrow\; \dfrac{d}{ds}(\mathbf{B}\cdot\mathbf{B}) = 0 \;\Rightarrow\; 2\dfrac{d\mathbf{B}}{ds}\cdot\mathbf{B} = 0 \;\Rightarrow\; \dfrac{d\mathbf{B}}{ds}\perp\mathbf{B}$

(b) $\mathbf{B} = \mathbf{T}\times\mathbf{N} \;\Rightarrow$

$$\dfrac{d\mathbf{B}}{ds} = \dfrac{d}{ds}(\mathbf{T}\times\mathbf{N}) = \dfrac{d}{dt}(\mathbf{T}\times\mathbf{N})\dfrac{1}{ds/dt} = \dfrac{d}{dt}(\mathbf{T}\times\mathbf{N})\dfrac{1}{|\mathbf{r}'(t)|} = [(\mathbf{T}'\times\mathbf{N}) + (\mathbf{T}\times\mathbf{N}')]\dfrac{1}{|\mathbf{r}'(t)|}$$

$$= \left[\left(\mathbf{T}'\times\dfrac{\mathbf{T}'}{|\mathbf{T}'|}\right) + (\mathbf{T}\times\mathbf{N}')\right]\dfrac{1}{|\mathbf{r}'(t)|} = \dfrac{\mathbf{T}\times\mathbf{N}'}{|\mathbf{r}'(t)|} \;\Rightarrow\; \dfrac{d\mathbf{B}}{ds}\perp\mathbf{T}$$

(c) $\mathbf{B} = \mathbf{T}\times\mathbf{N} \;\Rightarrow\; \mathbf{T}\perp\mathbf{N}, \mathbf{B}\perp\mathbf{T}$ and $\mathbf{B}\perp\mathbf{N}$. So $\mathbf{B}, \mathbf{T}$ and $\mathbf{N}$ form an orthogonal set of vectors in the three-dimensional space $\mathbb{R}^3$. From parts (a) and (b), $d\mathbf{B}/ds$ is perpendicular to both $\mathbf{B}$ and $\mathbf{T}$, so $d\mathbf{B}/ds$ is parallel to $\mathbf{N}$. Therefore, $d\mathbf{B}/ds = -\tau(s)\mathbf{N}$, where $\tau(s)$ is a scalar.

(d) Since $\mathbf{B} = \mathbf{T}\times\mathbf{N}, \mathbf{T}\perp\mathbf{N}$ and both $\mathbf{T}$ and $\mathbf{N}$ are unit vectors, $\mathbf{B}$ is a unit vector mutually perpendicular to both $\mathbf{T}$ and $\mathbf{N}$. For a plane curve, $\mathbf{T}$ and $\mathbf{N}$ always lie in the plane of the curve, so that $\mathbf{B}$ is a constant unit vector always perpendicular to the plane. Thus $d\mathbf{B}/ds = \mathbf{0}$, but $d\mathbf{B}/ds = -\tau(s)\mathbf{N}$ and $\mathbf{N}\neq\mathbf{0}$, so $\tau(s) = 0$.

54. $\mathbf{N} = \mathbf{B}\times\mathbf{T} \;\Rightarrow$

$$\dfrac{d\mathbf{N}}{ds} = \dfrac{d}{ds}(\mathbf{B}\times\mathbf{T}) = \dfrac{d\mathbf{B}}{ds}\times\mathbf{T} + \mathbf{B}\times\dfrac{d\mathbf{T}}{ds} \qquad \text{(by Formula 5 of Theorem 14.2.3 [ET 13.2.3])}$$

$$= -\tau\mathbf{N}\times\mathbf{T} + \mathbf{B}\times\kappa\mathbf{N} \qquad \text{(by Formulas 3 and 1)}$$

$$= -\tau(\mathbf{N}\times\mathbf{T}) + \kappa(\mathbf{B}\times\mathbf{N}) \qquad \text{(by Property 2 of Theorem 13.4.8 [ET 12.4.8])}$$

But $\mathbf{B}\times\mathbf{N} = \mathbf{B}\times(\mathbf{B}\times\mathbf{T}) = (\mathbf{B}\cdot\mathbf{T})\mathbf{B} - (\mathbf{B}\cdot\mathbf{B})\mathbf{T}$ (by Property 6 of Theorem 13.4.8 [ET 12.4.8]) $= -\mathbf{T} \;\Rightarrow$ $d\mathbf{N}/ds = \tau(\mathbf{T}\times\mathbf{N}) - \kappa\mathbf{T} = -\kappa\mathbf{T} + \tau\mathbf{B}$.

55. (a) $\mathbf{r}' = s'\,\mathbf{T} \;\Rightarrow\; \mathbf{r}'' = s''\,\mathbf{T} + s'\,\mathbf{T}' = s''\,\mathbf{T} + s'\dfrac{d\mathbf{T}}{ds}s' = s''\,\mathbf{T} + \kappa(s')^2\,\mathbf{N}$ by the first Serret-Frenet formula.

(b) Using part (a), we have

$$\mathbf{r}'\times\mathbf{r}'' = (s'\,\mathbf{T})\times[s''\,\mathbf{T} + \kappa(s')^2\,\mathbf{N}]$$

$$= [(s'\,\mathbf{T})\times(s''\,\mathbf{T})] + [(s'\mathbf{T})\times(\kappa(s')^2\,\mathbf{N})] \qquad \text{(by Property 3 of Theorem 13.4.8 [ET 12.4.8])}$$

$$= (s's'')(\mathbf{T}\times\mathbf{T}) + \kappa(s')^3(\mathbf{T}\times\mathbf{N}) = \mathbf{0} + \kappa(s')^3\,\mathbf{B} = \kappa(s')^3\,\mathbf{B}$$

(c) Using part (a), we have

$$\mathbf{r}''' = [s''\,\mathbf{T} + \kappa(s')^2\,\mathbf{N}]' = s'''\,\mathbf{T} + s''\,\mathbf{T}' + \kappa'(s')^2\,\mathbf{N} + 2\kappa s's''\,\mathbf{N} + \kappa(s')^2\,\mathbf{N}'$$

$$= s'''\,\mathbf{T} + s''\frac{d\mathbf{T}}{ds}\,s' + \kappa'(s')^2\,\mathbf{N} + 2\kappa s's''\,\mathbf{N} + \kappa(s')^2\frac{d\mathbf{N}}{ds}\,s'$$

$$= s'''\,\mathbf{T} + s''s'\kappa\,\mathbf{N} + \kappa'(s')^2\,\mathbf{N} + 2\kappa s's''\,\mathbf{N} + \kappa(s')^3(-\kappa\,\mathbf{T} + \tau\,\mathbf{B}) \qquad \text{[by the second formula]}$$

$$= [s''' - \kappa^2(s')^3]\,\mathbf{T} + [3\kappa s's'' + \kappa'(s')^2]\,\mathbf{N} + \kappa\tau(s')^3\,\mathbf{B}$$

(d) Using parts (b) and (c) and the facts that $\mathbf{B}\cdot\mathbf{T} = 0$, $\mathbf{B}\cdot\mathbf{N} = 0$, and $\mathbf{B}\cdot\mathbf{B} = 1$, we get

$$\frac{(\mathbf{r}'\times\mathbf{r}'')\cdot\mathbf{r}'''}{|\mathbf{r}'\times\mathbf{r}''|^2} = \frac{\kappa(s')^3\,\mathbf{B}\cdot\{[s''' - \kappa^2(s')^3]\,\mathbf{T} + [3\kappa s's'' + \kappa'(s')^2]\,\mathbf{N} + \kappa\tau(s')^3\,\mathbf{B}\}}{|\kappa(s')^3\,\mathbf{B}|^2} = \frac{\kappa(s')^3\kappa\tau(s')^3}{[\kappa(s')^3]^2} = \tau.$$

56. First we find the quantities required to compute κ:

$$\mathbf{r}'(t) = \langle -a\sin t, a\cos t, b\rangle \;\Rightarrow\; \mathbf{r}''(t) = \langle -a\cos t, -a\sin t, 0\rangle \;\Rightarrow\; \mathbf{r}'''(t) = \langle a\sin t, -a\cos t, 0\rangle$$

$$|\mathbf{r}'(t)| = \sqrt{(-a\sin t)^2 + (a\cos t)^2 + b^2} = \sqrt{a^2 + b^2}$$

$$\mathbf{r}'(t)\times\mathbf{r}''(t) = \begin{vmatrix} \mathbf{i} & \mathbf{j} & \mathbf{k} \\ -a\sin t & a\cos t & b \\ -a\cos t & -a\sin t & 0 \end{vmatrix} = ab\sin t\,\mathbf{i} - ab\cos t\,\mathbf{j} + a^2\,\mathbf{k}$$

$$|\mathbf{r}'(t)\times\mathbf{r}''(t)| = \sqrt{(ab\sin t)^2 + (-ab\cos t)^2 + (a^2)^2} = \sqrt{a^2b^2 + a^4}$$

$$(\mathbf{r}'(t)\times\mathbf{r}''(t))\cdot\mathbf{r}'''(t) = (ab\sin t)(a\sin t) + (-ab\cos t)(-a\cos t) + (a^2)(0) = a^2 b$$

Then by Theorem 10, $\kappa(t) = \dfrac{|\mathbf{r}'(t)\times\mathbf{r}''(t)|}{|\mathbf{r}'(t)|^3} = \dfrac{\sqrt{a^2b^2 + a^4}}{\left(\sqrt{a^2 + b^2}\right)^3} = \dfrac{a\sqrt{a^2 + b^2}}{\left(\sqrt{a^2 + b^2}\right)^3} = \dfrac{a}{a^2 + b^2}$ which is a constant.

From Exercise 55(d), the torsion τ is given by $\tau = \dfrac{(\mathbf{r}'\times\mathbf{r}'')\cdot\mathbf{r}'''}{|\mathbf{r}'\times\mathbf{r}''|^2} = \dfrac{a^2 b}{\left(\sqrt{a^2b^2 + a^4}\right)^2} = \dfrac{b}{a^2 + b^2}$ which is also a constant.

57. $\mathbf{r} = \left\langle t, \frac{1}{2}t^2, \frac{1}{3}t^3\right\rangle \;\Rightarrow\; \mathbf{r}' = \langle 1, t, t^2\rangle$, $\mathbf{r}'' = \langle 0, 1, 2t\rangle$, $\mathbf{r}''' = \langle 0, 0, 2\rangle \;\Rightarrow\; \mathbf{r}'\times\mathbf{r}'' = \langle t^2, -2t, 1\rangle \;\Rightarrow$

$$\tau = \frac{(\mathbf{r}'\times\mathbf{r}'')\cdot\mathbf{r}'''}{|\mathbf{r}'\times\mathbf{r}''|^2} = \frac{\langle t^2, -2t, 1\rangle\cdot\langle 0, 0, 2\rangle}{t^4 + 4t^2 + 1} = \frac{2}{t^4 + 4t^2 + 1}$$

58. $\mathbf{r} = \langle\sinh t, \cosh t, t\rangle \;\Rightarrow\; \mathbf{r}' = \langle\cosh t, \sinh t, 1\rangle$, $\mathbf{r}'' = \langle\sinh t, \cosh t, 0\rangle$, $\mathbf{r}''' = \langle\cosh t, \sinh t, 0\rangle \;\Rightarrow$

$$\mathbf{r}'\times\mathbf{r}'' = \left\langle -\cosh t, \sinh t, \cosh^2 t - \sinh^2 t\right\rangle = \langle -\cosh t, \sinh t, 1\rangle \;\Rightarrow$$

$$\kappa = \frac{|\mathbf{r}'\times\mathbf{r}''|}{|\mathbf{r}'|^3} = \frac{|\langle -\cosh t, \sinh t, 1\rangle|}{|\langle\cosh t, \sinh t, 1\rangle|^3} = \frac{\sqrt{\cosh^2 t + \sinh^2 t + 1}}{\left(\cosh^2 t + \sinh^2 t + 1\right)^{3/2}} = \frac{1}{\cosh^2 t + \sinh^2 t + 1} = \frac{1}{2\cosh^2 t},$$

$$\tau = \frac{(\mathbf{r}'\times\mathbf{r}'')\cdot\mathbf{r}'''}{|\mathbf{r}'\times\mathbf{r}''|^2} = \frac{\langle -\cosh t, \sinh t, 1\rangle\cdot\langle\cosh t, \sinh t, 0\rangle}{\cosh^2 t + \sinh^2 t + 1} = \frac{-\cosh^2 t + \sinh^2 t}{2\cosh^2 t} = \frac{-1}{2\cosh^2 t}$$

So at the point $(0, 1, 0)$, $t = 0$, and $\kappa = \frac{1}{2}$ and $\tau = -\frac{1}{2}$.

59. For one helix, the vector equation is $\mathbf{r}(t) = \langle 10\cos t, 10\sin t, 34t/(2\pi)\rangle$ (measuring in angstroms), because the radius of each helix is 10 angstroms, and z increases by 34 angstroms for each increase of 2π in t. Using the arc length formula, letting t go from 0 to $2.9 \times 10^8 \times 2\pi$, we find the approximate length of each helix to be

$$L = \int_0^{2.9 \times 10^8 \times 2\pi} |\mathbf{r}'(t)| \, dt = \int_0^{2.9 \times 10^8 \times 2\pi} \sqrt{(-10\sin t)^2 + (10\cos t)^2 + \left(\tfrac{34}{2\pi}\right)^2} \, dt = \sqrt{100 + \left(\tfrac{34}{2\pi}\right)^2} \Big]_0^{2.9 \times 10^8 \times 2\pi}$$

$$= 2.9 \times 10^8 \times 2\pi \sqrt{100 + \left(\tfrac{34}{2\pi}\right)^2} \approx 2.07 \times 10^{10} \text{ Å} \text{ — more than two meters!}$$

60. (a) For the function $F(x) = \begin{cases} 0 & \text{if } x < 0 \\ P(x) & \text{if } 0 < x < 1 \\ 1 & \text{if } x \geq 1 \end{cases}$ to be continuous, we must have $P(0) = 0$ and $P(1) = 1$.

For F' to be continuous, we must have $P'(0) = P'(1) = 0$. The curvature of the curve $y = F(x)$ at the point $(x, F(x))$

is $\kappa(x) = \dfrac{|F''(x)|}{\left(1 + [F'(x)]^2\right)^{3/2}}$. For $\kappa(x)$ to be continuous, we must have $P''(0) = P''(1) = 0$.

Write $P(x) = ax^5 + bx^4 + cx^3 + dx^2 + ex + f$. Then $P'(x) = 5ax^4 + 4bx^3 + 3cx^2 + 2dx + e$ and

$P''(x) = 20ax^3 + 12bx^2 + 6cx + 2d$. Our six conditions are:

$$P(0) = 0 \quad \Rightarrow \quad f = 0 \quad \textbf{(1)} \qquad\qquad P(1) = 1 \quad \Rightarrow \quad a + b + c + d + e + f = 1 \quad \textbf{(2)}$$

$$P'(0) = 0 \quad \Rightarrow \quad e = 0 \quad \textbf{(3)} \qquad\qquad P'(1) = 0 \quad \Rightarrow \quad 5a + 4b + 3c + 2d + e = 0 \quad \textbf{(4)}$$

$$P''(0) = 0 \quad \Rightarrow \quad d = 0 \quad \textbf{(5)} \qquad\qquad P''(1) = 0 \quad \Rightarrow \quad 20a + 12b + 6c + 2d = 0 \quad \textbf{(6)}$$

From **(1)**, **(3)**, and **(5)**, we have $d = e = f = 0$. Thus **(2)**, **(4)** and **(6)** become **(7)** $a + b + c = 1$, **(8)** $5a + 4b + 3c = 0$,

and **(9)** $10a + 6b + 3c = 0$. Subtracting **(8)** from **(9)** gives **(10)** $5a + 2b = 0$. Multiplying **(7)** by 3 and subtracting from

(8) gives **(11)** $2a + b = -3$. Multiplying **(11)** by 2 and subtracting from **(10)** gives $a = 6$. By **(10)**, $b = -15$.

By **(7)**, $c = 10$. Thus, $P(x) = 6x^5 - 15x^4 + 10x^3$.

(b)

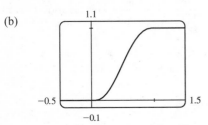

14.4 Motion in Space: Velocity and Acceleration

1. (a) If $\mathbf{r}(t) = x(t)\,\mathbf{i} + y(t)\,\mathbf{j} + z(t)\,\mathbf{k}$ is the position vector of the particle at time t, then the average velocity over the time

interval $[0, 1]$ is

$$\mathbf{v}_{\text{ave}} = \frac{\mathbf{r}(1) - \mathbf{r}(0)}{1 - 0} = \frac{(4.5\,\mathbf{i} + 6.0\,\mathbf{j} + 3.0\,\mathbf{k}) - (2.7\,\mathbf{i} + 9.8\,\mathbf{j} + 3.7\,\mathbf{k})}{1} = 1.8\,\mathbf{i} - 3.8\,\mathbf{j} - 0.7\,\mathbf{k}. \text{ Similarly, over the other}$$

intervals we have

$$[0.5, 1]: \quad \mathbf{v}_{\text{ave}} = \frac{\mathbf{r}(1) - \mathbf{r}(0.5)}{1 - 0.5} = \frac{(4.5\,\mathbf{i} + 6.0\,\mathbf{j} + 3.0\,\mathbf{k}) - (3.5\,\mathbf{i} + 7.2\,\mathbf{j} + 3.3\,\mathbf{k})}{0.5} = 2.0\,\mathbf{i} - 2.4\,\mathbf{j} - 0.6\,\mathbf{k}$$

$$[1, 2]: \quad \mathbf{v}_{\text{ave}} = \frac{\mathbf{r}(2) - \mathbf{r}(1)}{2 - 1} = \frac{(7.3\,\mathbf{i} + 7.8\,\mathbf{j} + 2.7\,\mathbf{k}) - (4.5\,\mathbf{i} + 6.0\,\mathbf{j} + 3.0\,\mathbf{k})}{1} = 2.8\,\mathbf{i} + 1.8\,\mathbf{j} - 0.3\,\mathbf{k}$$

$$[1, 1.5]: \quad \mathbf{v}_{\text{ave}} = \frac{\mathbf{r}(1.5) - \mathbf{r}(1)}{1.5 - 1} = \frac{(5.9\,\mathbf{i} + 6.4\,\mathbf{j} + 2.8\,\mathbf{k}) - (4.5\,\mathbf{i} + 6.0\,\mathbf{j} + 3.0\,\mathbf{k})}{0.5} = 2.8\,\mathbf{i} + 0.8\,\mathbf{j} - 0.4\,\mathbf{k}$$

(b) We can estimate the velocity at $t = 1$ by averaging the average velocities over the time intervals $[0.5, 1]$ and $[1, 1.5]$:

$\mathbf{v}(1) \approx \frac{1}{2}[(2\mathbf{i} - 2.4\mathbf{j} - 0.6\mathbf{k}) + (2.8\mathbf{i} + 0.8\mathbf{j} - 0.4\mathbf{k})] = 2.4\mathbf{i} - 0.8\mathbf{j} - 0.5\mathbf{k}$. Then the speed is

$|\mathbf{v}(1)| \approx \sqrt{(2.4)^2 + (-0.8)^2 + (-0.5)^2} \approx 2.58$.

2. (a) The average velocity over $2 \le t \le 2.4$ is

$\dfrac{\mathbf{r}(2.4) - \mathbf{r}(2)}{2.4 - 2} = 2.5\,[\mathbf{r}(2.4) - \mathbf{r}(2)]$, so we sketch a vector in the same

direction but 2.5 times the length of $[\mathbf{r}(2.4) - \mathbf{r}(2)]$.

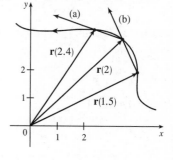

(b) The average velocity over $1.5 \le t \le 2$ is

$\dfrac{\mathbf{r}(2) - \mathbf{r}(1.5)}{2 - 1.5} = 2[\mathbf{r}(2) - \mathbf{r}(1.5)]$, so we sketch a vector in the

same direction but twice the length of $[\mathbf{r}(2) - \mathbf{r}(1.5)]$.

(c) Using Equation 2 we have $\mathbf{v}(2) = \lim\limits_{h \to 0} \dfrac{\mathbf{r}(2 + h) - \mathbf{r}(2)}{h}$.

(d) $\mathbf{v}(2)$ is tangent to the curve at $\mathbf{r}(2)$ and points in the direction of

increasing t. Its length is the speed of the particle at $t = 2$. We can

estimate the speed by averaging the lengths of the vectors found in

parts (a) and (b) which represent the average speed over $2 \le t \le 2.4$ and

$1.5 \le t \le 2$ respectively. Using the axes scale as a guide, we estimate the

vectors to have lengths 2.8 and 2.7. Thus, we estimate the speed at $t = 2$

to be $|\mathbf{v}(2)| \approx \frac{1}{2}(2.8 + 2.7) = 2.75$ and we draw the velocity vector $\mathbf{v}(2)$

with this length.

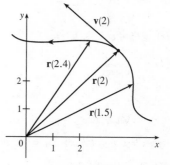

3. $\mathbf{r}(t) = \left\langle -\frac{1}{2}t^2, t \right\rangle \quad \Rightarrow$

$\mathbf{v}(t) = \mathbf{r}'(t) = \langle -t, 1 \rangle$

$\mathbf{a}(t) = \mathbf{r}''(t) = \langle -1, 0 \rangle$

$|\mathbf{v}(t)| = \sqrt{t^2 + 1}$

At $t = 2$:

$\mathbf{v}(2) = \langle -2, 1 \rangle$

$\mathbf{a}(2) = \langle -1, 0 \rangle$

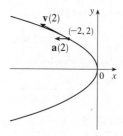

4. $\mathbf{r}(t) = \left\langle 2 - t, 4\sqrt{t} \right\rangle \quad \Rightarrow$

$\mathbf{v}(t) = \mathbf{r}'(t) = \left\langle -1, 2/\sqrt{t} \right\rangle$

$\mathbf{a}(t) = \mathbf{r}''(t) = \left\langle 0, -1/t^{3/2} \right\rangle$

$|\mathbf{v}(t)| = \sqrt{1 + 4/t}$

At $t = 1$:

$\mathbf{v}(1) = \langle -1, 2 \rangle$

$\mathbf{a}(1) = \langle 0, -1 \rangle$

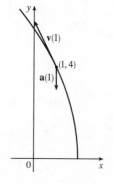

5. $(t) = 3\cos t\,\mathbf{i} + 2\sin t\,\mathbf{j}$ ⇒

$\mathbf{v}(t) = -3\sin t\,\mathbf{i} + 2\cos t\,\mathbf{j}$

$\mathbf{a}(t) = -3\cos t\,\mathbf{i} - 2\sin t\,\mathbf{j}$

At $t = \pi/3$:

$\mathbf{v}\left(\frac{\pi}{3}\right) = -\frac{3\sqrt{3}}{2}\mathbf{i} + \mathbf{j}$

$\mathbf{a}\left(\frac{\pi}{3}\right) = -\frac{3}{2}\mathbf{i} - \sqrt{3}\mathbf{j}$

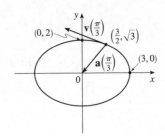

$|\mathbf{v}(t)| = \sqrt{9\sin^2 t + 4\cos^2 t} = \sqrt{4 + 5\sin^2 t}$

Notice that $x^2/9 + y^2/4 = \sin^2 t + \cos^2 t = 1$, so the path is an ellipse.

6. $\mathbf{r}(t) = e^t\,\mathbf{i} + e^{2t}\,\mathbf{j}$ ⇒

$\mathbf{v}(t) = e^t\,\mathbf{i} + 2e^{2t}\,\mathbf{j}$

$\mathbf{a}(t) = e^t\,\mathbf{i} + 4e^{2t}\,\mathbf{j}$

At $t = 0$:

$\mathbf{v}(0) = \mathbf{i} + 2\mathbf{j}$

$\mathbf{a}(0) = \mathbf{i} + 4\mathbf{j}$

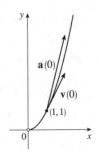

$|\mathbf{v}(t)| = \sqrt{e^{2t} + 4e^{4t}} = e^t\sqrt{1 + 4e^{2t}}$

Notice that $y = e^{2t} = \left(e^t\right)^2 = x^2$, so the particle travels along a parabola,

but $x = e^t$, so $x > 0$.

7. $\mathbf{r}(t) = t\,\mathbf{i} + t^2\,\mathbf{j} + 2\,\mathbf{k}$ ⇒

$\mathbf{v}(t) = \mathbf{i} + 2t\,\mathbf{j}$

$\mathbf{a}(t) = 2\,\mathbf{j}$

At $t = 1$:

$\mathbf{v}(1) = \mathbf{i} + 2\mathbf{j}$

$\mathbf{a}(1) = 2\mathbf{j}$

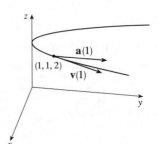

$|\mathbf{v}(t)| = \sqrt{1 + 4t^2}$

Here $x = t,\, y = t^2$ ⇒ $y = x^2$ and $z = 2$, so the path of the particle is a

parabola in the plane $z = 2$.

8. $\mathbf{r}(t) = t\,\mathbf{i} + 2\cos t\,\mathbf{j} + \sin t\,\mathbf{k}$ ⇒

$\mathbf{v}(t) = \mathbf{i} - 2\sin t\,\mathbf{j} + \cos t\,\mathbf{k}$

$\mathbf{a}(t) = -2\cos t\,\mathbf{j} - \sin t\,\mathbf{k}$

At $t = 0$:

$\mathbf{v}(0) = \mathbf{i} + \mathbf{k}$

$\mathbf{a}(0) = -2\mathbf{j}$

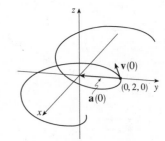

$|\mathbf{v}(t)| = \sqrt{1 + 4\sin^2 t + \cos^2 t} = \sqrt{2 + 3\sin^2 t}$

Since $y^2/4 + z^2 = 1,\, x = t$, the path of the particle is an elliptical helix

about the x-axis.

9. $\mathbf{r}(t) = \langle t^2 + 1, t^3, t^2 - 1\rangle$ ⇒ $\mathbf{v}(t) = \mathbf{r}'(t) = \langle 2t, 3t^2, 2t\rangle$, $\mathbf{a}(t) = \mathbf{v}'(t) = \langle 2, 6t, 2\rangle$,

$|\mathbf{v}(t)| = \sqrt{(2t)^2 + (3t^2)^2 + (2t)^2} = \sqrt{9t^4 + 8t^2} = |t|\sqrt{9t^2 + 8}$.

10. $\mathbf{r}(t) = \langle 2\cos t, 3t, 2\sin t\rangle$ ⇒ $\mathbf{v}(t) = \mathbf{r}'(t) = \langle -2\sin t, 3, 2\cos t\rangle$, $\mathbf{a}(t) = \mathbf{v}'(t) = \langle -2\cos t, 0, -2\sin t\rangle$,

$|\mathbf{v}(t)| = \sqrt{4\sin^2 t + 9 + 4\cos^2 t} = \sqrt{13}$.

11. $\mathbf{r}(t) = \sqrt{2}\,t\,\mathbf{i} + e^t\,\mathbf{j} + e^{-t}\,\mathbf{k} \;\Rightarrow\; \mathbf{v}(t) = \mathbf{r}'(t) = \sqrt{2}\,\mathbf{i} + e^t\,\mathbf{j} - e^{-t}\,\mathbf{k}, \quad \mathbf{a}(t) = \mathbf{v}'(t) = e^t\,\mathbf{j} + e^{-t}\,\mathbf{k},$

$|\mathbf{v}(t)| = \sqrt{2 + e^{2t} + e^{-2t}} = \sqrt{(e^t + e^{-t})^2} = e^t + e^{-t}.$

12. $\mathbf{r}(t) = t^2\,\mathbf{i} + \ln t\,\mathbf{j} + t\,\mathbf{k} \;\Rightarrow\; \mathbf{v}(t) = \mathbf{r}'(t) = 2t\,\mathbf{i} + t^{-1}\,\mathbf{j} + \mathbf{k}, \mathbf{a}(t) = \mathbf{v}'(t) = 2\,\mathbf{i} - t^{-2}\,\mathbf{j}, |\mathbf{v}(t)| = \sqrt{4t^2 + t^{-2} + 1}.$

13. $\mathbf{r}(t) = e^t\langle \cos t, \sin t, t\rangle \;\Rightarrow\;$

$\mathbf{v}(t) = \mathbf{r}'(t) = e^t\langle \cos t, \sin t, t\rangle + e^t\langle -\sin t, \cos t, 1\rangle = e^t\langle \cos t - \sin t, \sin t + \cos t, t + 1\rangle$

$\mathbf{a}(t) = \mathbf{v}'(t) = e^t\langle \cos t - \sin t - \sin t - \cos t, \sin t + \cos t + \cos t - \sin t, t + 1 + 1\rangle$

$\qquad = e^t\langle -2\sin t, 2\cos t, t + 2\rangle$

$|\mathbf{v}(t)| = e^t\sqrt{\cos^2 t + \sin^2 t - 2\cos t\sin t + \sin^2 t + \cos^2 t + 2\sin t\cos t + t^2 + 2t + 1}$

$\qquad = e^t\sqrt{t^2 + 2t + 3}$

14. $\mathbf{r}(t) = t\sin t\,\mathbf{i} + t\cos t\,\mathbf{j} + t^2\,\mathbf{k} \;\Rightarrow\; \mathbf{v}(t) = \mathbf{r}'(t) = (\sin t + t\cos t)\,\mathbf{i} + (\cos t - t\sin t)\,\mathbf{j} + 2t\,\mathbf{k},$

$\mathbf{a}(t) = \mathbf{v}'(t) = (2\cos t - t\sin t)\,\mathbf{i} + (-2\sin t - t\cos t)\,\mathbf{j} + 2\,\mathbf{k},$

$|\mathbf{v}(t)| = \sqrt{(\sin^2 t + 2t\sin t\cos t + t^2\cos^2 t) + (\cos^2 t - 2t\sin t\cos t + t^2\sin^2 t) + 4t^2} = \sqrt{5t^2 + 1}.$

15. $\mathbf{a}(t) = \mathbf{i} + 2\,\mathbf{j} \;\Rightarrow\; \mathbf{v}(t) = \int \mathbf{a}(t)\,dt = \int (\mathbf{i} + 2\,\mathbf{j})\,dt = t\,\mathbf{i} + 2t\,\mathbf{j} + \mathbf{C} \text{ and } \mathbf{k} = \mathbf{v}\,(0) = \mathbf{C},$

so $\mathbf{C} = \mathbf{k}$ and $\mathbf{v}(t) = t\,\mathbf{i} + 2t\,\mathbf{j} + \mathbf{k}.\quad \mathbf{r}(t) = \int \mathbf{v}(t)\,dt = \int (t\,\mathbf{i} + 2t\,\mathbf{j} + \mathbf{k})\,dt = \tfrac{1}{2}t^2\,\mathbf{i} + t^2\,\mathbf{j} + t\,\mathbf{k} + \mathbf{D}.$

But $\mathbf{i} = \mathbf{r}\,(0) = \mathbf{D}$, so $\mathbf{D} = \mathbf{i}$ and $\mathbf{r}(t) = \left(\tfrac{1}{2}t^2 + 1\right)\mathbf{i} + t^2\,\mathbf{j} + t\,\mathbf{k}.$

16. $\mathbf{a}(t) = 2\,\mathbf{i} + 6t\,\mathbf{j} + 12t^2\,\mathbf{k} \;\Rightarrow\; \mathbf{v}(t) = \int (2\,\mathbf{i} + 6t\,\mathbf{j} + 12t^2\,\mathbf{k})\,dt = 2t\,\mathbf{i} + 3t^2\,\mathbf{j} + 4t^3\,\mathbf{k} + \mathbf{C}, \text{ and } \mathbf{i} = \mathbf{v}(0) = \mathbf{C},$

so $\mathbf{C} = \mathbf{i}$ and $\mathbf{v}(t) = (2t + 1)\mathbf{i} + 3t^2\,\mathbf{j} + 4t^3\,\mathbf{k}.\quad \mathbf{r}(t) = \int \left[(2t + 1)\mathbf{i} + 3t^2\,\mathbf{j} + 4t^3\,\mathbf{k}\right]dt = (t^2 + t)\,\mathbf{i} + t^3\,\mathbf{j} + t^4\,\mathbf{k} + \mathbf{D}.$

But $\mathbf{j} - \mathbf{k} = \mathbf{r}(0) = \mathbf{D}$, so $\mathbf{D} = \mathbf{j} - \mathbf{k}$ and $\mathbf{r}(t) = (t^2 + t)\,\mathbf{i} + (t^3 + 1)\,\mathbf{j} + (t^4 - 1)\,\mathbf{k}.$

17. (a) $\mathbf{a}(t) = 2t\,\mathbf{i} + \sin t\,\mathbf{j} + \cos 2t\,\mathbf{k} \;\Rightarrow\;$

$\mathbf{v}(t) = \int (2t\,\mathbf{i} + \sin t\,\mathbf{j} + \cos 2t\,\mathbf{k})\,dt = t^2\,\mathbf{i} - \cos t\,\mathbf{j} + \tfrac{1}{2}\sin 2t\,\mathbf{k} + \mathbf{C}$

and $\mathbf{i} = \mathbf{v}\,(0) = -\mathbf{j} + \mathbf{C}$, so $\mathbf{C} = \mathbf{i} + \mathbf{j}$

and $\mathbf{v}(t) = (t^2 + 1)\,\mathbf{i} + (1 - \cos t)\,\mathbf{j} + \tfrac{1}{2}\sin 2t\,\mathbf{k}.$

$\mathbf{r}(t) = \int [(t^2 + 1)\,\mathbf{i} + (1 - \cos t)\,\mathbf{j} + \tfrac{1}{2}\sin 2t\,\mathbf{k}]\,dt$

$\qquad = \left(\tfrac{1}{3}t^3 + t\right)\mathbf{i} + (t - \sin t)\,\mathbf{j} - \tfrac{1}{4}\cos 2t\,\mathbf{k} + \mathbf{D}$

(b)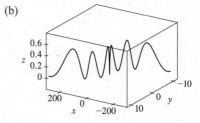

But $\mathbf{j} = \mathbf{r}\,(0) = -\tfrac{1}{4}\mathbf{k} + \mathbf{D}$, so $\mathbf{D} = \mathbf{j} + \tfrac{1}{4}\mathbf{k}$ and $\mathbf{r}(t) = \left(\tfrac{1}{3}t^3 + t\right)\mathbf{i} + (t - \sin t + 1)\,\mathbf{j} + \left(\tfrac{1}{4} - \tfrac{1}{4}\cos 2t\right)\mathbf{k}.$

18. (a) $\mathbf{a}(t) = t\,\mathbf{i} + e^t\,\mathbf{j} + e^{-t}\,\mathbf{k} \;\Rightarrow$

$\mathbf{v}(t) = \int (t\,\mathbf{i} + e^t\,\mathbf{j} + e^{-t}\,\mathbf{k})\,dt = \frac{1}{2}t^2\,\mathbf{i} + e^t\,\mathbf{j} - e^{-t}\,\mathbf{k} + \mathbf{C}$

and $\mathbf{k} = \mathbf{v}(0) = \mathbf{j} - \mathbf{k} + \mathbf{C}$, so $\mathbf{C} = -\mathbf{j} + 2\,\mathbf{k}$

and $\mathbf{v}(t) = \frac{1}{2}t^2\,\mathbf{i} + (e^t - 1)\,\mathbf{j} + (2 - e^{-t})\,\mathbf{k}$.

$\mathbf{r}(t) = \int \left[\frac{1}{2}t^2\,\mathbf{i} + (e^t - 1)\,\mathbf{j} + (2 - e^{-t})\,\mathbf{k}\right] dt$

$\qquad = \frac{1}{6}t^3\,\mathbf{i} + (e^t - t)\,\mathbf{j} + (e^{-t} + 2t)\,\mathbf{k} + \mathbf{D}$

(b)

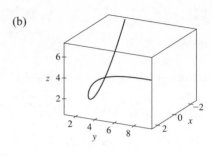

But $\mathbf{j} + \mathbf{k} = \mathbf{r}(0) = \mathbf{j} + \mathbf{k} + \mathbf{D}$, so $\mathbf{D} = \mathbf{0}$ and $\mathbf{r}(t) = \frac{1}{6}t^3\,\mathbf{i} + (e^t - t)\,\mathbf{j} + (e^{-t} + 2t)\,\mathbf{k}$.

19. $\mathbf{r}(t) = \langle t^2, 5t, t^2 - 16t \rangle \;\Rightarrow\; \mathbf{v}(t) = \langle 2t, 5, 2t - 16 \rangle, |\mathbf{v}(t)| = \sqrt{4t^2 + 25 + 4t^2 - 64t + 256} = \sqrt{8t^2 - 64t + 281}$

and $\dfrac{d}{dt}\,|\mathbf{v}(t)| = \frac{1}{2}(8t^2 - 64t + 281)^{-1/2}(16t - 64)$. This is zero if and only if the numerator is zero, that is,

$16t - 64 = 0$ or $t = 4$. Since $\dfrac{d}{dt}\,|\mathbf{v}(t)| < 0$ for $t < 4$ and $\dfrac{d}{dt}\,|\mathbf{v}(t)| > 0$ for $t > 4$, the minimum speed of $\sqrt{153}$ is attained

at $t = 4$ units of time.

20. Since $\mathbf{r}(t) = t^3\,\mathbf{i} + t^2\,\mathbf{j} + t^3\,\mathbf{k}$, $\mathbf{a}(t) = \mathbf{r}''(t) = 6t\,\mathbf{i} + 2\,\mathbf{j} + 6t\,\mathbf{k}$. By Newton's Second Law,

$\mathbf{F}(t) = m\,\mathbf{a}(t) = 6mt\,\mathbf{i} + 2m\,\mathbf{j} + 6mt\,\mathbf{k}$ is the required force.

21. $|\mathbf{F}(t)| = 20$ N in the direction of the positive z-axis, so $\mathbf{F}(t) = 20\,\mathbf{k}$. Also $m = 4$ kg, $\mathbf{r}(0) = \mathbf{0}$ and $\mathbf{v}(0) = \mathbf{i} - \mathbf{j}$.

Since $20\,\mathbf{k} = \mathbf{F}(t) = 4\,\mathbf{a}(t)$, $\mathbf{a}(t) = 5\,\mathbf{k}$. Then $\mathbf{v}(t) = 5t\,\mathbf{k} + \mathbf{c}_1$ where $\mathbf{c}_1 = \mathbf{i} - \mathbf{j}$ so $\mathbf{v}(t) = \mathbf{i} - \mathbf{j} + 5t\,\mathbf{k}$ and the

speed is $|\mathbf{v}(t)| = \sqrt{1 + 1 + 25t^2} = \sqrt{25t^2 + 2}$. Also $\mathbf{r}(t) = t\,\mathbf{i} - t\,\mathbf{j} + \frac{5}{2}t^2\,\mathbf{k} + \mathbf{c}_2$ and $\mathbf{0} = \mathbf{r}(0)$, so $\mathbf{c}_2 = \mathbf{0}$

and $\mathbf{r}(t) = t\,\mathbf{i} - t\,\mathbf{j} + \frac{5}{2}t^2\,\mathbf{k}$.

22. The argument here is the same as that in Example 14.2.4 [ET 13.2.4] with $\mathbf{r}(t)$ replaced by $\mathbf{v}(t)$ and $\mathbf{r}'(t)$ replaced by $\mathbf{a}(t)$.

23. $|\mathbf{v}(0)| = 500$ m/s and since the angle of elevation is $30°$, the direction of the velocity is $\frac{1}{2}(\sqrt{3}\,\mathbf{i} + \mathbf{j})$. Thus

$\mathbf{v}(0) = 250(\sqrt{3}\,\mathbf{i} + \mathbf{j})$ and if we set up the axes so the projectile starts at the origin, then $\mathbf{r}(0) = \mathbf{0}$. Ignoring air resistance, the

only force is that due to gravity, so $\mathbf{F}(t) = -mg\,\mathbf{j}$ where $g \approx 9.8$ m/s^2. Thus $\mathbf{a}(t) = -g\,\mathbf{j}$ and $\mathbf{v}(t) = -gt\,\mathbf{j} + \mathbf{c}_1$. But

$250(\sqrt{3}\,\mathbf{i} + \mathbf{j}) = \mathbf{v}(0) = \mathbf{c}_1$, so $\mathbf{v}(t) = 250\sqrt{3}\,\mathbf{i} + (250 - gt)\,\mathbf{j}$ and $\mathbf{r}(t) = 250\sqrt{3}\,t\,\mathbf{i} + (250t - \frac{1}{2}gt^2)\,\mathbf{j} + \mathbf{c}_2$ where

$\mathbf{0} = \mathbf{r}(0) = \mathbf{c}_2$. Thus $\mathbf{r}(t) = 250\sqrt{3}\,t\,\mathbf{i} + (250t - \frac{1}{2}gt^2)\,\mathbf{j}$.

(a) Setting $250t - \frac{1}{2}gt^2 = 0$ gives $t = 0$ or $t = \frac{500}{g} \approx 51.0$ s. So the range is $250\sqrt{3} \cdot \frac{500}{g} \approx 22$ km.

(b) $0 = \dfrac{d}{dt}\left(250t - \frac{1}{2}gt^2\right) = 250 - gt$ implies that the maximum height is attained when $t = 250/g \approx 25.5$ s.

Thus, the maximum height is $(250)(250/g) - g(250/g)^2\frac{1}{2} = (250)^2/(2g) \approx 3.2$ km.

(c) From part (a), impact occurs at $t = 500/g \approx 51.0$. Thus, the velocity at impact is

$\mathbf{v}(500/g) = 250\sqrt{3}\,\mathbf{i} + [250 - g(500/g)]\,\mathbf{j} = 250\sqrt{3}\,\mathbf{i} - 250\,\mathbf{j}$ and the speed is $|\mathbf{v}(500/g)| = 250\sqrt{3 + 1} = 500$ m/s.

24. As in Exercise 23, $\mathbf{v}(t) = 250\sqrt{3}\,\mathbf{i} + (250 - gt)\,\mathbf{j}$ and $\mathbf{r}(t) = 250\sqrt{3}\,t\,\mathbf{i} + \left(250t - \frac{1}{2}gt^2\right)\mathbf{j} + \mathbf{c}_2$.

But $\mathbf{r}(0) = 200\,\mathbf{j}$, so $\mathbf{c}_2 = 200\,\mathbf{j}$ and $\mathbf{r}(t) = 250\sqrt{3}\,t\,\mathbf{i} + \left(200 + 250t - \frac{1}{2}gt^2\right)\mathbf{j}$.

(a) $200 + 250t - \frac{1}{2}gt^2 = 0$ implies that $gt^2 - 500t - 400 = 0$ or $t = \dfrac{500 \pm \sqrt{500^2 + 1600g}}{2g}$. Taking the positive t-value

gives $t = \dfrac{500 + \sqrt{250{,}000 + 1600g}}{2g} \approx 51.8$ s. Thus the range is $(250\sqrt{3})\,\dfrac{500 + \sqrt{250{,}000 + 1600g}}{2g} \approx 22.4$ km.

(b) $0 = \dfrac{d}{dt}\left(200 + 250t - \frac{1}{2}gt^2\right) = 250 - gt$ implies that the maximum height is attained when $t = 250/g \approx 25.5$ s and

thus the maximum height is $\left[200 + (250)\left(\dfrac{250}{g}\right) - \dfrac{g}{2}\left(\dfrac{250}{g}\right)^2\right] = 200 + \dfrac{(250)^2}{2g} \approx 3.4$ km.

Alternate solution: Because the projectile is fired in the same direction and with the same velocity as in Exercise 23, but from a point 200 m higher, the maximum height reached is 200 m higher than that found in Exercise 23, that is, $3.2\text{ km} + 200\text{ m} = 3.4$ km.

(c) From part (a), impact occurs at $t = \dfrac{500 + \sqrt{250{,}000 + 1600g}}{2g}$. Thus the velocity at impact is

$250\sqrt{3}\,\mathbf{i} + \left[250 - g\,\dfrac{500 + \sqrt{250{,}000 + 1600g}}{2g}\right]\mathbf{j}$, so $|\mathbf{v}| \approx \sqrt{(250)^2(3) + (250 - 51.8g)^2} \approx 504$ m/s.

25. As in Example 5, $\mathbf{r}(t) = (v_0 \cos 45°)t\,\mathbf{i} + \left[(v_0 \sin 45°)t - \frac{1}{2}gt^2\right]\mathbf{j} = \frac{1}{2}\left[v_0\sqrt{2}\,t\,\mathbf{i} + \left(v_0\sqrt{2}\,t - gt^2\right)\mathbf{j}\right]$. Then the ball

lands at $t = \dfrac{v_0\sqrt{2}}{g}$ s. Now since it lands 90 m away, $90 = \frac{1}{2}v_0\sqrt{2}\,\dfrac{v_0\sqrt{2}}{g}$ or $v_0^2 = 90g$ and the initial velocity

is $v_0 = \sqrt{90g} \approx 30$ m/s.

26. As in Example 5, $\mathbf{r}(t) = (v_0 \cos 30°)t\,\mathbf{i} + \left[(v_0 \sin 30°)t - \frac{1}{2}gt^2\right]\mathbf{j} = \frac{1}{2}\left[v_0\sqrt{3}\,t\,\mathbf{i} + \left(v_0\,t - gt^2\right)\mathbf{j}\right]$ and then

$\mathbf{v}(t) = \mathbf{r}'(t) = \frac{1}{2}\left[v_0\sqrt{3}\,\mathbf{i} + (v_0 - 2gt)\,\mathbf{j}\right]$. The shell reaches its maximum height when the vertical component of velocity

is zero, so $\frac{1}{2}(v_0 - 2gt) = 0 \;\Rightarrow\; t = \dfrac{v_0}{2g}$. The vertical height of the shell at that time is 500 m,

so $\dfrac{1}{2}\left[v_0\left(\dfrac{v_0}{2g}\right) - g\left(\dfrac{v_0}{2g}\right)^2\right] = 500 \;\Rightarrow\; \dfrac{v_0^2}{8g} = 500 \;\Rightarrow\; v_0 = \sqrt{4000g} = \sqrt{4000(9.8)} \approx 198$ m/s.

27. Let α be the angle of elevation. Then $v_0 = 150$ m/s and from Example 5, the horizontal distance traveled by the projectile is

$d = \dfrac{v_0^2 \sin 2\alpha}{g}$. Thus $\dfrac{150^2 \sin 2\alpha}{g} = 800 \;\Rightarrow\; \sin 2\alpha = \dfrac{800g}{150^2} \approx 0.3484 \;\Rightarrow\; 2\alpha \approx 20.4°$ or $180 - 20.4 = 159.6°$.

Two angles of elevation then are $\alpha \approx 10.2°$ and $\alpha \approx 79.8°$.

28. Here $v_0 = 115$ ft/s, the angle of elevation is $\alpha = 50°$, and if we place the origin at home plate, then $\mathbf{r}(0) = 3\,\mathbf{j}$.

As in Example 5, we have $\mathbf{r}(t) = -\frac{1}{2}gt^2\,\mathbf{j} + t\,\mathbf{v}_0 + \mathbf{D}$ where $\mathbf{D} = \mathbf{r}(0) = 3\,\mathbf{j}$ and $\mathbf{v}_0 = v_0 \cos \alpha\,\mathbf{i} + v_0 \sin \alpha\,\mathbf{j}$,

so $\mathbf{r}(t) = (v_0 \cos \alpha)t\,\mathbf{i} + \left[(v_0 \sin \alpha)t - \frac{1}{2}gt^2 + 3\right]\mathbf{j}$. Thus, parametric equations for the trajectory of the ball are

$x = (v_0 \cos \alpha)t$, $y = (v_0 \sin \alpha)t - \frac{1}{2}gt^2 + 3$. The ball reaches the fence when $x = 400 \;\Rightarrow$

$(v_0 \cos \alpha)t = 400 \quad \Rightarrow \quad t = \dfrac{400}{v_0 \cos \alpha} = \dfrac{400}{115 \cos 50°} \approx 5.41$ s. At this time, the height of the ball is

$y = (v_0 \sin \alpha)t - \frac{1}{2}gt^2 + 3 \approx (115 \sin 50°)(5.41) - \frac{1}{2}(32)(5.41)^2 + 3 \approx 11.2$ ft. Since the fence is 10 ft high, the ball

clears the fence.

29. Place the catapult at the origin and assume the catapult is 100 meters from the city, so the city lies between $(100, 0)$

    and $(600, 0)$. The initial speed is $v_0 = 80$ m/s and let θ be the angle the catapult is set at. As in Example 5, the trajectory of

    the catapulted rock is given by $\mathbf{r}(t) = (80 \cos \theta)t \, \mathbf{i} + \left[(80 \sin \theta)t - 4.9t^2 \right] \mathbf{j}$. The top of the near city wall is at $(100, 15)$,

    which the rock will hit when $(80 \cos \theta) t = 100 \quad \Rightarrow \quad t = \dfrac{5}{4 \cos \theta}$ and $(80 \sin \theta)t - 4.9t^2 = 15 \quad \Rightarrow$

    $80 \sin \theta \cdot \dfrac{5}{4 \cos \theta} - 4.9 \left(\dfrac{5}{4 \cos \theta} \right)^2 = 15 \quad \Rightarrow \quad 100 \tan \theta - 7.65625 \sec^2 \theta = 15$. Replacing $\sec^2 \theta$ with $\tan^2 \theta + 1$ gives

    $7.65625 \tan^2 \theta - 100 \tan \theta + 22.62625 = 0$. Using the quadratic formula, we have $\tan \theta \approx 0.230324, 12.8309 \quad \Rightarrow$

    $\theta \approx 13.0°, 85.5°$. So for $13.0° < \theta < 85.5°$, the rock will land beyond the near city wall. The base of the far wall is

    located at $(600, 0)$ which the rock hits if $(80 \cos \theta)t = 600 \quad \Rightarrow \quad t = \dfrac{15}{2 \cos \theta}$ and $(80 \sin \theta)t - 4.9t^2 = 0 \quad \Rightarrow$

    $80 \sin \theta \cdot \dfrac{15}{2 \cos \theta} - 4.9 \left(\dfrac{15}{2 \cos \theta} \right)^2 = 0 \quad \Rightarrow \quad 600 \tan \theta - 275.625 \sec^2 \theta = 0 \quad \Rightarrow$

    $275.625 \tan^2 \theta - 600 \tan \theta + 275.625 = 0$. Solutions are $\tan \theta \approx 0.658678, 1.51819 \quad \Rightarrow \quad \theta \approx 33.4°, 56.6°$. Thus the

    rock lands beyond the enclosed city ground for $33.4° < \theta < 56.6°$, and the angles that allow the rock to land on city ground

    are $13.0° < \theta < 33.4°, 56.6° < \theta < 85.5°$. If you consider that the rock can hit the far wall and bounce back into the city, we

    calculate the angles that cause the rock to hit the top of the wall at $(600, 15)$: $(80 \cos \theta)t = 600 \quad \Rightarrow \quad t = \dfrac{15}{2 \cos \theta}$ and

    $(80 \sin \theta)t - 4.9t^2 = 15 \quad \Rightarrow \quad 600 \tan \theta - 275.625 \sec^2 \theta = 15 \quad \Rightarrow \quad 275.625 \tan^2 \theta - 600 \tan \theta + 290.625 = 0$.

    Solutions are $\tan \theta \approx 0.727506, 1.44936 \quad \Rightarrow \quad \theta \approx 36.0°, 55.4°$, so the catapult should be set with angle θ where

    $13.0° < \theta < 36.0°, 55.4° < \theta < 85.5°$.

30. Place the ball at the origin and consider $\mathbf{j}$ to be pointing in the northward direction with $\mathbf{i}$ pointing east and $\mathbf{k}$ pointing

    upward. Force $=$ mass $\times$ acceleration $\quad \Rightarrow \quad$ acceleration $=$ force/mass, so the wind applies a constant acceleration of

    $4 \text{ N}/0.8 \text{ kg} = 5 \text{ m/s}^2$ in the easterly direction. Combined with the acceleration due to gravity, the acceleration acting

    on the ball is $\mathbf{a}(t) = 5 \, \mathbf{i} - 9.8 \, \mathbf{k}$. Then $\mathbf{v}(t) = \int \mathbf{a}(t) \, dt = 5t \, \mathbf{i} - 9.8t \, \mathbf{k} + \mathbf{C}$ where $\mathbf{C}$ is a constant vector.

    We know $\mathbf{v}(0) = \mathbf{C} = -30 \cos 30° \, \mathbf{j} + 30 \sin 30° \, \mathbf{k} = -15 \sqrt{3} \, \mathbf{j} + 15 \, \mathbf{k} \quad \Rightarrow \quad \mathbf{C} = -15 \sqrt{3} \, \mathbf{j} + 15 \, \mathbf{k}$ and

    $\mathbf{v}(t) = 5t \, \mathbf{i} - 15 \sqrt{3} \, \mathbf{j} + (15 - 9.8t) \, \mathbf{k}$. $\quad \mathbf{r}(t) = \int \mathbf{v}(t) \, dt = 2.5t^2 \, \mathbf{i} - 15 \sqrt{3} t \, \mathbf{j} + \left(15t - 4.9t^2 \right) \mathbf{k} + \mathbf{D}$ but $\mathbf{r}(0) = \mathbf{D} = \mathbf{0}$

    so $\mathbf{r}(t) = 2.5t^2 \, \mathbf{i} - 15 \sqrt{3} t \, \mathbf{j} + \left(15t - 4.9t^2 \right) \mathbf{k}$. The ball lands when $15t - 4.9t^2 = 0 \quad \Rightarrow \quad t = 0, t = 15/4.9 \approx 3.0612$ s,

    so the ball lands at approximately $\mathbf{r}(3.0612) \approx 23.43 \, \mathbf{i} - 79.53 \, \mathbf{j}$ which is 82.9 m away in the direction S $16.4°$E. Its speed is

    approximately $|\mathbf{v}(3.0612)| \approx |15.306 \, \mathbf{i} - 15 \sqrt{3} \, \mathbf{j} - 15 \, \mathbf{k}| \approx 33.68$ m/s.

31. (a) After t seconds, the boat will be $5t$ meters west of point A. The velocity of the water at that location is $\frac{3}{400}(5t)(40-5t)\mathbf{j}$. The velocity of the boat in still water is $5\,\mathbf{i}$, so the resultant velocity of the boat is

$\mathbf{v}(t) = 5\,\mathbf{i} + \frac{3}{400}(5t)(40-5t)\,\mathbf{j} = 5\,\mathbf{i} + \left(\frac{3}{2}t - \frac{3}{16}t^2\right)\mathbf{j}$. Integrating, we obtain

$\mathbf{r}(t) = 5t\,\mathbf{i} + \left(\frac{3}{4}t^2 - \frac{1}{16}t^3\right)\mathbf{j} + \mathbf{C}$. If we place the origin at A (and consider $\mathbf{j}$

to coincide with the northern direction) then $\mathbf{r}(0) = \mathbf{0} \;\Rightarrow\; \mathbf{C} = \mathbf{0}$ and we have $\mathbf{r}(t) = 5t\,\mathbf{i} + \left(\frac{3}{4}t^2 - \frac{1}{16}t^3\right)\mathbf{j}$. The boat

reaches the east bank after 8 s, and it is located at $\mathbf{r}(8) = 5(8)\mathbf{i} + \left(\frac{3}{4}(8)^2 - \frac{1}{16}(8)^3\right)\mathbf{j} = 40\,\mathbf{i} + 16\,\mathbf{j}$. Thus the boat is 16 m

downstream.

(b) Let α be the angle north of east that the boat heads. Then the velocity of the boat in still water is given by

$5(\cos\alpha)\,\mathbf{i} + 5(\sin\alpha)\,\mathbf{j}$. At t seconds, the boat is $5(\cos\alpha)t$ meters from the west bank, at which point the velocity

of the water is $\frac{3}{400}[5(\cos\alpha)t][40 - 5(\cos\alpha)t]\,\mathbf{j}$. The resultant velocity of the boat is given by

$\mathbf{v}(t) = 5(\cos\alpha)\,\mathbf{i} + \left[5\sin\alpha + \frac{3}{400}(5t\cos\alpha)(40 - 5t\cos\alpha)\right]\mathbf{j} = (5\cos\alpha)\,\mathbf{i} + \left(5\sin\alpha + \frac{3}{2}t\cos\alpha - \frac{3}{16}t^2\cos^2\alpha\right)\mathbf{j}$.

Integrating, $\mathbf{r}(t) = (5t\cos\alpha)\,\mathbf{i} + \left(5t\sin\alpha + \frac{3}{4}t^2\cos\alpha - \frac{1}{16}t^3\cos^2\alpha\right)\mathbf{j}$ (where we have again placed

the origin at A). The boat will reach the east bank when $5t\cos\alpha = 40 \;\Rightarrow\; t = \dfrac{40}{5\cos\alpha} = \dfrac{8}{\cos\alpha}$.

In order to land at point $B(40,0)$ we need $5t\sin\alpha + \frac{3}{4}t^2\cos\alpha - \frac{1}{16}t^3\cos^2\alpha = 0 \;\Rightarrow$

$5\left(\dfrac{8}{\cos\alpha}\right)\sin\alpha + \frac{3}{4}\left(\dfrac{8}{\cos\alpha}\right)^2\cos\alpha - \frac{1}{16}\left(\dfrac{8}{\cos\alpha}\right)^3\cos^2\alpha = 0 \;\Rightarrow\; \dfrac{1}{\cos\alpha}(40\sin\alpha + 48 - 32) = 0 \;\Rightarrow$

$40\sin\alpha + 16 = 0 \;\Rightarrow\; \sin\alpha = -\frac{2}{5}$. Thus $\alpha = \sin^{-1}\left(-\frac{2}{5}\right) \approx -23.6°$, so the boat should head 23.6° south of

east (upstream). The path does seem realistic. The boat initially heads
upstream to counteract the effect of the current. Near the center of the river,
the current is stronger and the boat is pushed downstream. When the boat
nears the eastern bank, the current is slower and the boat is able to progress
upstream to arrive at point B.

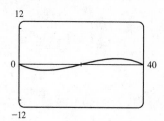

32. As in Exercise 31(b), let α be the angle north of east that the boat heads, so the velocity of the boat in still water is given by

$5(\cos\alpha)\,\mathbf{i} + 5(\sin\alpha)\,\mathbf{j}$. At t seconds, the boat is $5(\cos\alpha)t$ meters from the west bank, at which point the velocity of the water

is $3\sin(\pi x/40)\,\mathbf{j} = 3\sin[\pi \cdot 5(\cos\alpha)t/40]\,\mathbf{j} = 3\sin\left(\frac{\pi}{8}t\cos\alpha\right)\mathbf{j}$. The resultant velocity of the boat then is given by

$\mathbf{v}(t) = 5(\cos\alpha)\,\mathbf{i} + \left[5\sin\alpha + 3\sin\left(\frac{\pi}{8}t\cos\alpha\right)\right]\mathbf{j}$. Integrating,

$\mathbf{r}(t) = (5t\cos\alpha)\,\mathbf{i} + \left[5t\sin\alpha - \dfrac{24}{\pi\cos\alpha}\cos\left(\frac{\pi}{8}t\cos\alpha\right)\right]\mathbf{j} + \mathbf{C}$.

If we place the origin at A then $\mathbf{r}(0) = \mathbf{0} \;\Rightarrow\; -\dfrac{24}{\pi\cos\alpha}\,\mathbf{j} + \mathbf{C} = \mathbf{0} \;\Rightarrow\; \mathbf{C} = \dfrac{24}{\pi\cos\alpha}\,\mathbf{j}$ and

$$\mathbf{r}(t) = (5t \cos \alpha)\,\mathbf{i} + \left[5t \sin \alpha - \frac{24}{\pi \cos \alpha}\cos\left(\tfrac{\pi}{8}t \cos \alpha\right) + \frac{24}{\pi \cos \alpha}\right]\mathbf{j}.$$

The boat will reach the east bank when $5t \cos \alpha = 40 \;\Rightarrow\; t = \dfrac{8}{\cos \alpha}$.

In order to land at point $B(40,0)$ we need $5t \sin \alpha - \dfrac{24}{\pi \cos \alpha}\cos\left(\tfrac{\pi}{8}t \cos \alpha\right) + \dfrac{24}{\pi \cos \alpha} = 0 \;\Rightarrow\;$

$$5\left(\frac{8}{\cos \alpha}\right)\sin \alpha - \frac{24}{\pi \cos \alpha}\cos\left[\frac{\pi}{8}\left(\frac{8}{\cos \alpha}\right)\cos \alpha\right] + \frac{24}{\pi \cos \alpha} = 0 \;\Rightarrow\; \frac{1}{\cos \alpha}\left(40 \sin \alpha - \frac{24}{\pi}\cos \pi + \frac{24}{\pi}\right) = 0 \;\Rightarrow\;$$

$40 \sin \alpha + \dfrac{48}{\pi} = 0 \;\Rightarrow\; \sin \alpha = -\dfrac{6}{5\pi}$. Thus $\alpha = \sin^{-1}\left(-\dfrac{6}{5\pi}\right) \approx -22.5°$, so the boat should head $22.5°$ south of east.

33. $\mathbf{r}(t) = (3t - t^3)\,\mathbf{i} + 3t^2\,\mathbf{j} \;\Rightarrow\; \mathbf{r}'(t) = (3 - 3t^2)\,\mathbf{i} + 6t\,\mathbf{j},$

$|\mathbf{r}'(t)| = \sqrt{(3-3t^2)^2 + (6t)^2} = \sqrt{9 + 18t^2 + 9t^4} = \sqrt{(3-3t^2)^2} = 3 + 3t^2,$

$\mathbf{r}''(t) = -6t\,\mathbf{i} + 6\,\mathbf{j}$, $\mathbf{r}'(t) \times \mathbf{r}''(t) = (18 + 18t^2)\,\mathbf{k}$. Then Equation 9 gives

$$a_T = \frac{\mathbf{r}'(t) \cdot \mathbf{r}''(t)}{|\mathbf{r}'(t)|} = \frac{(3-3t^2)(-6t) + (6t)(6)}{3 + 3t^2} = \frac{18t + 18t^3}{3 + 3t^2} = \frac{18t(1+t^2)}{3(1+t^2)} = 6t \quad \Big[\text{or by Equation 8,}$$

$$a_T = v' = \frac{d}{dt}\left[3 + 3t^2\right] = 6t\Big] \quad \text{and Equation 10 gives } a_N = \frac{|\mathbf{r}'(t) \times \mathbf{r}''(t)|}{|\mathbf{r}'(t)|} = \frac{18 + 18t^2}{3 + 3t^2} = \frac{18(1+t^2)}{3(1+t^2)} = 6.$$

34. $\mathbf{r}(t) = (1+t)\,\mathbf{i} + (t^2 - 2t)\,\mathbf{j} \;\Rightarrow\; \mathbf{r}'(t) = \mathbf{i} + (2t-2)\,\mathbf{j}, \quad |\mathbf{r}'(t)| = \sqrt{1^2 + (2t-2)^2} = \sqrt{4t^2 - 8t + 5},$

$\mathbf{r}''(t) = 2\,\mathbf{j}, \quad \mathbf{r}'(t) \times \mathbf{r}''(t) = 2\,\mathbf{k}$.   Then Equation 9 gives $a_T = \dfrac{\mathbf{r}'(t) \cdot \mathbf{r}''(t)}{|\mathbf{r}'(t)|} = \dfrac{2(2t-2)}{\sqrt{4t^2 - 8t + 5}}$ and Equation 10

gives $a_N = \dfrac{|\mathbf{r}'(t) \times \mathbf{r}''(t)|}{|\mathbf{r}'(t)|} = \dfrac{2}{\sqrt{4t^2 - 8t + 5}}$.

35. $\mathbf{r}(t) = \cos t\,\mathbf{i} + \sin t\,\mathbf{j} + t\,\mathbf{k} \;\Rightarrow\; \mathbf{r}'(t) = -\sin t\,\mathbf{i} + \cos t\,\mathbf{j} + \mathbf{k}, \quad |\mathbf{r}'(t)| = \sqrt{\sin^2 t + \cos^2 t + 1} = \sqrt{2},$

$\mathbf{r}''(t) = -\cos t\,\mathbf{i} - \sin t\,\mathbf{j}, \quad \mathbf{r}'(t) \times \mathbf{r}''(t) = \sin t\,\mathbf{i} - \cos t\,\mathbf{j} + \mathbf{k}.$

Then $a_T = \dfrac{\mathbf{r}'(t) \cdot \mathbf{r}''(t)}{|\mathbf{r}'(t)|} = \dfrac{\sin t \cos t - \sin t \cos t}{\sqrt{2}} = 0$ and $a_N = \dfrac{|\mathbf{r}'(t) \times \mathbf{r}''(t)|}{|\mathbf{r}'(t)|} = \dfrac{\sqrt{\sin^2 t + \cos^2 t + 1}}{\sqrt{2}} = \dfrac{\sqrt{2}}{\sqrt{2}} = 1.$

36. $\mathbf{r}(t) = t\,\mathbf{i} + t^2\,\mathbf{j} + 3t\,\mathbf{k} \;\Rightarrow\; \mathbf{r}'(t) = \mathbf{i} + 2t\,\mathbf{j} + 3\,\mathbf{k}, \quad |\mathbf{r}'(t)| = \sqrt{1^2 + (2t)^2 + 3^2} = \sqrt{4t^2 + 10},$

$\mathbf{r}''(t) = 2\,\mathbf{j}, \quad \mathbf{r}'(t) \times \mathbf{r}''(t) = -6\,\mathbf{i} + 2\,\mathbf{k}.$

Then $a_T = \dfrac{\mathbf{r}'(t) \cdot \mathbf{r}''(t)}{|\mathbf{r}'(t)|} = \dfrac{4t}{\sqrt{4t^2 + 10}}$ and $a_N = \dfrac{|\mathbf{r}'(t) \times \mathbf{r}''(t)|}{|\mathbf{r}'(t)|} = \dfrac{2\sqrt{10}}{\sqrt{4t^2 + 10}}.$

37. $\mathbf{r}(t) = e^t\,\mathbf{i} + \sqrt{2}\,t\,\mathbf{j} + e^{-t}\,\mathbf{k} \;\Rightarrow\; \mathbf{r}'(t) = e^t\,\mathbf{i} + \sqrt{2}\,\mathbf{j} - e^{-t}\,\mathbf{k}, \quad |\mathbf{r}(t)| = \sqrt{e^{2t} + 2 + e^{-2t}} = \sqrt{(e^t + e^{-t})^2} = e^t + e^{-t},$

$\mathbf{r}''(t) = e^t\,\mathbf{i} + e^{-t}\,\mathbf{k}$. Then $a_T = \dfrac{e^{2t} - e^{-2t}}{e^t + e^{-t}} = \dfrac{(e^t + e^{-t})(e^t - e^{-t})}{e^t + e^{-t}} = e^t - e^{-t} = 2 \sinh t$

and $a_N = \dfrac{|\sqrt{2}\,e^{-t}\,\mathbf{i} - 2\,\mathbf{j} - \sqrt{2}\,e^t\,\mathbf{k}|}{e^t + e^{-t}} = \dfrac{\sqrt{2(e^{-2t} + 2 + e^{2t})}}{e^t + e^{-t}} = \sqrt{2}\,\dfrac{e^t + e^{-t}}{e^t + e^{-t}} = \sqrt{2}.$

38. $\mathbf{r}(t) = t\,\mathbf{i} + \cos^2 t\,\mathbf{j} + \sin^2 t\,\mathbf{k} \;\Rightarrow\; \mathbf{r}'(t) = \mathbf{i} - 2\cos t \sin t\,\mathbf{j} + 2\sin t \cos t\,\mathbf{k} = \mathbf{i} - \sin 2t\,\mathbf{j} + \sin 2t\,\mathbf{k}$,

$|\mathbf{r}'(t)| = \sqrt{1 + 2\sin^2 2t}$, $\mathbf{r}''(t) = 2(\sin^2 t - \cos^2 t)\,\mathbf{j} + 2(\cos^2 t - \sin^2 t)\,\mathbf{k} = -2\cos 2t\,\mathbf{j} + 2\cos 2t\,\mathbf{k}$. So

$$a_T = \frac{2\sin 2t \cos 2t + 2\sin 2t \cos 2t}{\sqrt{1 + 2\sin^2 2t}} = \frac{4\sin 2t \cos 2t}{\sqrt{1 + 2\sin^2 2t}} \text{ and } a_N = \frac{|-2\cos 2t\,\mathbf{j} - 2\cos 2t\,\mathbf{k}|}{\sqrt{1 + 2\sin^2 2t}} = \frac{2\sqrt{2}\,|\cos 2t|}{\sqrt{1 + 2\sin^2 t}}.$$

39. The tangential component of $\mathbf{a}$ is the length of the projection of $\mathbf{a}$ onto $\mathbf{T}$, so we sketch
the scalar projection of $\mathbf{a}$ in the tangential direction to the curve and estimate its length to
be 4.5 (using the fact that $\mathbf{a}$ has length 10 as a guide). Similarly, the normal component of
$\mathbf{a}$ is the length of the projection of $\mathbf{a}$ onto $\mathbf{N}$, so we sketch the scalar projection of $\mathbf{a}$ in the
normal direction to the curve and estimate its length to be 9.0. Thus $a_T \approx 4.5 \text{ cm/s}^2$ and
$a_N \approx 9.0 \text{ cm/s}^2$.

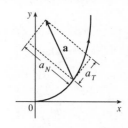

40. $\mathbf{L}(t) = m\,\mathbf{r}(t) \times \mathbf{v}(t) \;\Rightarrow\;$

$$\mathbf{L}'(t) = m[\mathbf{r}'(t) \times \mathbf{v}(t) + \mathbf{r}(t) \times \mathbf{v}'(t)] \qquad \text{(by Formula 5 of Theorem 14.2.3 [ET 13.2.3])}$$

$$= m[\mathbf{v}(t) \times \mathbf{v}(t) + \mathbf{r}(t) \times \mathbf{v}'(t)] = m[\mathbf{0} + \mathbf{r}(t) \times \mathbf{a}(t)] = \boldsymbol{\tau}(t)$$

So if the torque is always $\mathbf{0}$, then $\mathbf{L}'(t) = \mathbf{0}$ for all t, and so $\mathbf{L}(t)$ is constant.

41. If the engines are turned off at time t, then the spacecraft will continue to travel in the direction of $\mathbf{v}(t)$, so we need a t such
that for some scalar $s > 0$, $\mathbf{r}(t) + s\,\mathbf{v}(t) = \langle 6, 4, 9 \rangle$. $\mathbf{v}(t) = \mathbf{r}'(t) = \mathbf{i} + \dfrac{1}{t}\,\mathbf{j} + \dfrac{8t}{(t^2+1)^2}\,\mathbf{k} \;\Rightarrow\;$

$$\mathbf{r}(t) + s\,\mathbf{v}(t) = \left\langle 3 + t + s,\ 2 + \ln t + \frac{s}{t},\ 7 - \frac{4}{t^2+1} + \frac{8st}{(t^2+1)^2} \right\rangle \;\Rightarrow\; 3 + t + s = 6 \;\Rightarrow\; s = 3 - t,$$

so $7 - \dfrac{4}{t^2+1} + \dfrac{8(3-t)t}{(t^2+1)^2} = 9 \;\Leftrightarrow\; \dfrac{24t - 12t^2 - 4}{(t^2+1)^2} = 2 \;\Leftrightarrow\; t^4 + 8t^2 - 12t + 3 = 0.$

It is easily seen that $t = 1$ is a root of this polynomial. Also $2 + \ln 1 + \dfrac{3-1}{1} = 4$, so $t = 1$ is the desired solution.

42. (a) $m\dfrac{d\mathbf{v}}{dt} = \dfrac{dm}{dt}\,\mathbf{v}_e \;\Leftrightarrow\; \dfrac{d\mathbf{v}}{dt} = \dfrac{1}{m}\dfrac{dm}{dt}\,\mathbf{v}_e$. Integrating both sides of this equation with respect to t gives

$$\int_0^t \frac{d\mathbf{v}}{du}\,du = \mathbf{v}_e \int_0^t \frac{1}{m}\frac{dm}{du}\,du \;\Rightarrow\; \int_{\mathbf{v}(0)}^{\mathbf{v}(t)} d\mathbf{v} = \mathbf{v}_e \int_{m(0)}^{m(t)} \frac{dm}{m} \quad \text{[Substitution Rule]} \;\Rightarrow$$

$$\mathbf{v}(t) - \mathbf{v}(0) = \ln\!\left(\frac{m(t)}{m(0)}\right)\mathbf{v}_e \;\Rightarrow\; \mathbf{v}(t) = \mathbf{v}(0) - \ln\!\left(\frac{m(0)}{m(t)}\right)\mathbf{v}_e.$$

(b) $|\mathbf{v}(t)| = 2\,|\mathbf{v}_e|$, and $|\mathbf{v}(0)| = 0$. Therefore, by part (a), $2\,|\mathbf{v}_e| = \left|-\ln\!\left(\dfrac{m(0)}{m(t)}\right)\mathbf{v}_e\right| \;\Rightarrow$

$$2\,|\mathbf{v}_e| = \ln\!\left(\frac{m(0)}{m(t)}\right)|\mathbf{v}_e|. \quad \left[\textit{Note: }\, m(0) > m(t) \text{ so that } \ln\!\left(\frac{m(0)}{m(t)}\right) > 0\right] \;\Rightarrow\; m(t) = e^{-2}m(0).$$

Thus $\dfrac{m(0) - e^{-2}m(0)}{m(0)} = 1 - e^{-2}$ is the fraction of the initial mass that is burned as fuel.

APPLIED PROJECT Kepler's Laws

1. With $\mathbf{r} = (r\cos\theta)\,\mathbf{i} + (r\sin\theta)\,\mathbf{j}$ and $\mathbf{h} = \alpha\,\mathbf{k}$ where $\alpha > 0$,

   (a) $\mathbf{h} = \mathbf{r} \times \mathbf{r}' = [(r\cos\theta)\,\mathbf{i} + (r\sin\theta)\,\mathbf{j}] \times \left[\left(r'\cos\theta - r\sin\theta\,\dfrac{d\theta}{dt}\right)\mathbf{i} + \left(r'\sin\theta + r\cos\theta\,\dfrac{d\theta}{dt}\right)\mathbf{j}\right]$

   $= \left[rr'\cos\theta\sin\theta + r^2\cos^2\theta\,\dfrac{d\theta}{dt} - rr'\cos\theta\sin\theta + r^2\sin^2\theta\,\dfrac{d\theta}{dt}\right]\mathbf{k} = r^2\dfrac{d\theta}{dt}\,\mathbf{k}$

   (b) Since $\mathbf{h} = \alpha\,\mathbf{k}$, $\alpha > 0$, $\alpha = |\mathbf{h}|$. But by part (a), $\alpha = |\mathbf{h}| = r^2\,(d\theta/dt)$.

   (c) $A(t) = \frac{1}{2}\int_{\theta_0}^{\theta} |\mathbf{r}|^2\,d\theta = \frac{1}{2}\int_{t_0}^{t} r^2\,(d\theta/dt)\,dt$ in polar coordinates. Thus, by the Fundamental Theorem of Calculus,

   $\dfrac{dA}{dt} = \dfrac{r^2}{2}\dfrac{d\theta}{dt}$.

   (d) $\dfrac{dA}{dt} = \dfrac{r^2}{2}\dfrac{d\theta}{dt} = \dfrac{h}{2} = $ constant since $\mathbf{h}$ is a constant vector and $h = |\mathbf{h}|$.

2. (a) Since $dA/dt = \frac{1}{2}h$, a constant, $A(t) = \frac{1}{2}ht + c_1$. But $A(0) = 0$, so $A(t) = \frac{1}{2}ht$. But $A(T) = $ area of the ellipse $= \pi ab$
   and $A(T) = \frac{1}{2}hT$, so $T = 2\pi ab/h$.

   (b) $h^2/(GM) = ed$ where e is the eccentricity of the ellipse. But $a = ed/(1 - e^2)$ or $ed = a(1 - e^2)$ and $1 - e^2 = b^2/a^2$.
   Hence $h^2/(GM) = ed = b^2/a$.

   (c) $T^2 = \dfrac{4\pi^2 a^2 b^2}{h^2} = 4\pi^2 a^2 b^2\,\dfrac{a}{GMb^2} = \dfrac{4\pi^2}{GM}a^3$.

3. From Problem 2, $T^2 = \dfrac{4\pi^2}{GM}a^3$. $T \approx 365.25$ days $\times\, 24 \cdot 60^2\,\dfrac{\text{seconds}}{\text{day}} \approx 3.1558 \times 10^7$ seconds. Therefore

   $a^3 = \dfrac{GMT^2}{4\pi^2} \approx \dfrac{(6.67 \times 10^{-11})(1.99 \times 10^{30})(3.1558 \times 10^7)^2}{4\pi^2} \approx 3.348 \times 10^{33}\ \text{m}^3 \quad\Rightarrow\quad a \approx 1.496 \times 10^{11}$ m. Thus, the

   length of the major axis of the earth's orbit (that is, $2a$) is approximately 2.99×10^{11} m $= 2.99 \times 10^8$ km.

4. We can adapt the equation $T^2 = \dfrac{4\pi^2}{GM}a^3$ from Problem 2(c) with the earth at the center of the system, so T is the period of the

   satellite's orbit about the earth, M is the mass of the earth, and a is the length of the semimajor axis of the satellite's orbit

   (measured from the earth's center). Since we want the satellite to remain fixed above a particular point on the earth's equator,

   T must coincide with the period of the earth's own rotation, so $T = 24$ h $= 86{,}400$ s. The mass of the earth is

   $M = 5.98 \times 10^{24}$ kg, so $a = \left(\dfrac{T^2 GM}{4\pi^2}\right)^{1/3} \approx \left[\dfrac{(86{,}400)^2(6.67 \times 10^{-11})(5.98 \times 10^{24})}{4\pi^2}\right]^{1/3} \approx 4.23 \times 10^7$ m. If we

   assume a circular orbit, the radius of the orbit is a, and since the radius of the earth is 6.37×10^6 m, the required altitude

   above the earth's surface for the satellite is $4.23 \times 10^7 - 6.37 \times 10^6 \approx 3.59 \times 10^7$ m, or $35{,}900$ km.

14 Review

CONCEPT CHECK

1. A vector function is a function whose domain is a set of real numbers and whose range is a set of vectors. To find the derivative or integral, we can differentiate or integrate each component of the vector function.

2. The tip of the moving vector $\mathbf{r}(t)$ of a continuous vector function traces out a space curve.

3. The tangent vector to a smooth curve at a point P with position vector $\mathbf{r}(t)$ is the vector $\mathbf{r}'(t)$. The tangent line at P is the line through P parallel to the tangent vector $\mathbf{r}'(t)$. The unit tangent vector is $\mathbf{T}(t) = \dfrac{\mathbf{r}'(t)}{|\mathbf{r}'(t)|}$.

4. (a)–(f) See Theorem 14.2.3 [ET 13.2.3].

5. Use Formula 14.3.2 [ET 13.3.2], or equivalently, 14.3.3 [ET 13.3.3].

6. (a) The curvature of a curve is $\kappa = \left| \dfrac{d\mathbf{T}}{ds} \right|$ where $\mathbf{T}$ is the unit tangent vector.

   (b) $\kappa(t) = \left| \dfrac{\mathbf{T}'(t)}{\mathbf{r}'(t)} \right|$
   (c) $\kappa(t) = \dfrac{|\mathbf{r}'(t) \times \mathbf{r}''(t)|}{|\mathbf{r}'(t)|^3}$
   (d) $\kappa(x) = \dfrac{|f''(x)|}{[1 + (f'(x))^2]^{3/2}}$

7. (a) The unit normal vector: $\mathbf{N}(t) = \dfrac{\mathbf{T}'(t)}{|\mathbf{T}'(t)|}$.   The binormal vector: $\mathbf{B}(t) = \mathbf{T}(t) \times \mathbf{N}(t)$.

   (b) See the discussion preceding Example 7 in Section 14.3 [ET 13.3].

8. (a) If $\mathbf{r}(t)$ is the position vector of the particle on the space curve, the velocity $\mathbf{v}(t) = \mathbf{r}'(t)$, the speed is given by $|\mathbf{v}(t)|$, and the acceleration $\mathbf{a}(t) = \mathbf{v}'(t) = \mathbf{r}''(t)$.

   (b) $\mathbf{a} = a_T \mathbf{T} + a_N \mathbf{N}$ where $a_T = v'$ and $a_N = \kappa v^2$.

9. See the statement of Kepler's Laws on page 880 [ET page 844].

TRUE-FALSE QUIZ

1. True. If we reparametrize the curve by replacing $u = t^3$, we have $\mathbf{r}(u) = u\,\mathbf{i} + 2u\,\mathbf{j} + 3u\,\mathbf{k}$, which is a line through the origin with direction vector $\mathbf{i} + 2\,\mathbf{j} + 3\,\mathbf{k}$.

2. True. See Theorem 14.2.2 [ET 13.2.2].

3. False. By Formula 5 of Theorem 14.2.3[ET 13.2.3], $\dfrac{d}{dt}\,[\mathbf{u}(t) \times \mathbf{v}(t)] = \mathbf{u}'(t) \times \mathbf{v}(t) + \mathbf{u}(t) \times \mathbf{v}'(t)$.

4. False. For example, let $\mathbf{r}(t) = \langle \cos t, \sin t \rangle$. Then $|\mathbf{r}(t)| = \sqrt{\cos^2 t + \sin^2 t} = 1 \;\Rightarrow\; \dfrac{d}{dt}\,|\mathbf{r}(t)| = 0$, but $|\mathbf{r}'(t)| = |\langle -\sin t, \cos t \rangle| = \sqrt{(-\sin t)^2 + \cos^2 t} = 1$.

5. False. κ is the magnitude of the rate of change of the unit tangent vector $\mathbf{T}$ with respect to arc length s, not with respect to t.

6. False. The binormal vector, by the definition given in Section 14.3 [ET 13.3], is $\mathbf{B}(t) = \mathbf{T}(t) \times \mathbf{N}(t) = -[\mathbf{N}(t) \times \mathbf{T}(t)]$.

7. True. At an inflection point where f is twice continuously differentiable we must have $f''(x) = 0$, and by Equation 14.3.11 [ET 13.3.11], the curvature is 0 there.

8. True. From Equation 14.3.9 [ET 13.3.9], $\kappa(t) = 0$ ⟺ $|\mathbf{T}'(t)| = 0$ ⟺ $\mathbf{T}'(t) = \mathbf{0}$ for all t. But then $\mathbf{T}(t) = \mathbf{C}$, a constant vector, which is true only for a straight line.

9. False. If $\mathbf{r}(t)$ is the position of a moving particle at time t and $|\mathbf{r}(t)| = 1$ then the particle lies on the unit circle or the unit sphere, but this does not mean that the speed $|\mathbf{r}'(t)|$ must be constant. As a counterexample, let $\mathbf{r}(t) = \langle t, \sqrt{1 - t^2} \rangle$, then $\mathbf{r}'(t) = \langle 1, -t/\sqrt{1 - t^2} \rangle$ and $|\mathbf{r}(t)| = \sqrt{t^2 + 1 - t^2} = 1$ but $|\mathbf{r}'(t)| = \sqrt{1 + t^2/(1 - t^2)} = 1/\sqrt{1 - t^2}$ which is not constant.

10. True. See Example 4 in Section 14.2 [ET 13.2].

11. True. See the discussion preceding Example 7 in Section 14.3 [ET 13.3].

12. False. For example, $\mathbf{r}_1(t) = \langle t, t \rangle$ and $\mathbf{r}_2(t) = \langle 2t, 2t \rangle$ both represent the same plane curve (the line $y = x$), but the tangent vector $\mathbf{r}_1'(t) = \langle 1, 1 \rangle$ for all t, while $\mathbf{r}_2'(t) = \langle 2, 2 \rangle$. In fact, different parametrizations give parallel tangent vectors at a point, but their magnitudes may differ.

EXERCISES

1. (a) The corresponding parametric equations for the curve are $x = t$, $y = \cos \pi t$, $z = \sin \pi t$. Since $y^2 + z^2 = 1$, the curve is contained in a circular cylinder with axis the x-axis. Since $x = t$, the curve is a helix.

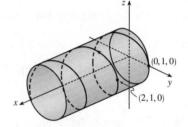

(b) $\mathbf{r}(t) = t\,\mathbf{i} + \cos \pi t\,\mathbf{j} + \sin \pi t\,\mathbf{k}$ ⟹

$\mathbf{r}'(t) = \mathbf{i} - \pi \sin \pi t\,\mathbf{j} + \pi \cos \pi t\,\mathbf{k}$ ⟹

$\mathbf{r}''(t) = -\pi^2 \cos \pi t\,\mathbf{j} - \pi^2 \sin \pi t\,\mathbf{k}$

2. (a) The expressions $\sqrt{2 - t}$, $(e^t - 1)/t$, and $\ln(t + 1)$ are all defined when $2 - t \geq 0$ ⟹ $t \leq 2, t \neq 0$, and $t + 1 > 0$ ⟹ $t > -1$. Thus the domain of $\mathbf{r}$ is $(-1, 0) \cup (0, 2]$.

(b) $\displaystyle\lim_{t \to 0} \mathbf{r}(t) = \left\langle \lim_{t \to 0} \sqrt{2 - t}, \lim_{t \to 0} \frac{e^t - 1}{t}, \lim_{t \to 0} \ln(t + 1) \right\rangle = \left\langle \sqrt{2 - 0}, \lim_{t \to 0} \frac{e^t}{1}, \ln(0 + 1) \right\rangle$

$= \langle \sqrt{2}, 1, 0 \rangle$ [using l'Hospital's Rule in the y-component]

(c) $\mathbf{r}'(t) = \left\langle \dfrac{d}{dt} \sqrt{2 - t}, \dfrac{d}{dt} \dfrac{e^t - 1}{t}, \dfrac{d}{dt} \ln(t + 1) \right\rangle = \left\langle -\dfrac{1}{2\sqrt{2 - t}}, \dfrac{te^t - e^t + 1}{t^2}, \dfrac{1}{t + 1} \right\rangle$

3. The projection of the curve C of intersection onto the xy-plane is the circle $x^2 + y^2 = 16, z = 0$. So we can write $x = 4\cos t$, $y = 4\sin t$, $0 \leq t \leq 2\pi$. From the equation of the plane, we have $z = 5 - x = 5 - 4\cos t$, so parametric equations for C are $x = 4\cos t$, $y = 4\sin t$, $z = 5 - 4\cos t, 0 \leq t \leq 2\pi$, and the corresponding vector function is $\mathbf{r}(t) = 4\cos t\,\mathbf{i} + 4\sin t\,\mathbf{j} + (5 - 4\cos t)\,\mathbf{k}, 0 \leq t \leq 2\pi$.

4. The curve is given by $\mathbf{r}(t) = \langle 2\sin t, 2\sin 2t, 2\sin 3t \rangle$, so

$\mathbf{r}'(t) = \langle 2\cos t, 4\cos 2t, 6\cos 3t \rangle$. The point $\left(1, \sqrt{3}, 2\right)$ corresponds to $t = \frac{\pi}{6}$

(or $\frac{\pi}{6} + 2k\pi$, k an integer), so the tangent vector there is $\mathbf{r}'\left(\frac{\pi}{6}\right) = \langle \sqrt{3}, 2, 0 \rangle$.

Then the tangent line has direction vector $\langle \sqrt{3}, 2, 0 \rangle$ and includes the point

$\left(1, \sqrt{3}, 2\right)$, so parametric equations are $x = 1 + \sqrt{3}\,t$, $y = \sqrt{3} + 2t$, $z = 2$.

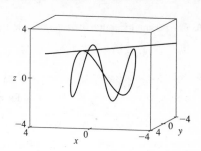

5. $\int_0^1 (t^2\,\mathbf{i} + t\cos \pi t\,\mathbf{j} + \sin \pi t\,\mathbf{k})\,dt = \left(\int_0^1 t^2\,dt\right)\mathbf{i} + \left(\int_0^1 t\cos \pi t\,dt\right)\mathbf{j} + \left(\int_0^1 \sin \pi t\,dt\right)\mathbf{k}$

$$= \left[\tfrac{1}{3}t^3\right]_0^1 \mathbf{i} + \left(\left[\tfrac{t}{\pi}\sin \pi t\right]_0^1 - \int_0^1 \tfrac{1}{\pi}\sin \pi t\,dt\right)\mathbf{j} + \left[-\tfrac{1}{\pi}\cos \pi t\right]_0^1 \mathbf{k}$$

$$= \tfrac{1}{3}\mathbf{i} + \left[\tfrac{1}{\pi^2}\cos \pi t\right]_0^1 \mathbf{j} + \tfrac{2}{\pi}\mathbf{k} = \tfrac{1}{3}\mathbf{i} - \tfrac{2}{\pi^2}\mathbf{j} + \tfrac{2}{\pi}\mathbf{k}$$

where we integrated by parts in the y-component.

6. (a) C intersects the xz-plane where $y = 0 \;\Rightarrow\; 2t - 1 = 0 \;\Rightarrow\; t = \frac{1}{2}$, so the point

is $\left(2 - \left(\frac{1}{2}\right)^3, 0, \ln \frac{1}{2}\right) = \left(\frac{15}{8}, 0, -\ln 2\right)$.

(b) The curve is given by $\mathbf{r}(t) = \langle 2 - t^3, 2t - 1, \ln t \rangle$, so $\mathbf{r}'(t) = \langle -3t^2, 2, 1/t \rangle$. The point $(1, 1, 0)$ corresponds to $t = 1$, so

the tangent vector there is $\mathbf{r}'(1) = \langle -3, 2, 1 \rangle$. Then the tangent line has direction vector $\langle -3, 2, 1 \rangle$ and includes the point

$(1, 1, 0)$, so parametric equations are $x = 1 - 3t$, $y = 1 + 2t$, $z = t$.

(c) The normal plane has normal vector $\mathbf{r}'(1) = \langle -3, 2, 1 \rangle$ and equation $-3(x - 1) + 2(y - 1) + z = 0$ or $3x - 2y - z = 1$.

7. $\mathbf{r}(t) = \langle t^2, t^3, t^4 \rangle \;\Rightarrow\; \mathbf{r}'(t) = \langle 2t, 3t^2, 4t^3 \rangle \;\Rightarrow\; |\mathbf{r}'(t)| = \sqrt{4t^2 + 9t^4 + 16t^6}$ and

$L = \int_0^3 |\mathbf{r}'(t)|\,dt = \int_0^3 \sqrt{4t^2 + 9t^4 + 16t^6}\,dt$. Using Simpson's Rule with $f(t) = \sqrt{4t^2 + 9t^4 + 16t^6}$ and $n = 6$ we

have $\Delta t = \frac{3-0}{6} = \frac{1}{2}$ and

$$L \approx \tfrac{\Delta t}{3}\left[f(0) + 4f\left(\tfrac{1}{2}\right) + 2f(1) + 4f\left(\tfrac{3}{2}\right) + 2f(2) + 4f\left(\tfrac{5}{2}\right) + f(3)\right]$$

$$= \tfrac{1}{6}\left[\sqrt{0 + 0 + 0} + 4 \cdot \sqrt{4\left(\tfrac{1}{2}\right)^2 + 9\left(\tfrac{1}{2}\right)^4 + 16\left(\tfrac{1}{2}\right)^6} + 2 \cdot \sqrt{4(1)^2 + 9(1)^4 + 16(1)^6}\right.$$

$$+ 4 \cdot \sqrt{4\left(\tfrac{3}{2}\right)^2 + 9\left(\tfrac{3}{2}\right)^4 + 16\left(\tfrac{3}{2}\right)^6} + 2 \cdot \sqrt{4(2)^2 + 9(2)^4 + 16(2)^6}$$

$$\left. + 4 \cdot \sqrt{4\left(\tfrac{5}{2}\right)^2 + 9\left(\tfrac{5}{2}\right)^4 + 16\left(\tfrac{5}{2}\right)^6} + \sqrt{4(3)^2 + 9(3)^4 + 16(3)^6}\right]$$

$$\approx 86.631$$

8. $\mathbf{r}'(t) = \left\langle 3t^{1/2}, -2\sin 2t, 2\cos 2t \right\rangle$, $\;|\mathbf{r}'(t)| = \sqrt{9t + 4(\sin^2 2t + \cos^2 2t)} = \sqrt{9t + 4}$.

Thus $L = \int_0^1 \sqrt{9t + 4}\,dt = \int_4^{13} \tfrac{1}{9}u^{1/2}\,du = \tfrac{1}{9} \cdot \tfrac{2}{3}u^{3/2}\Big]_4^{13} = \tfrac{2}{27}(13^{3/2} - 8)$.

9. The angle of intersection of the two curves, θ, is the angle between their respective tangents at the point of intersection.

For both curves the point $(1, 0, 0)$ occurs when $t = 0$.

$\mathbf{r}_1'(t) = -\sin t\,\mathbf{i} + \cos t\,\mathbf{j} + \mathbf{k} \;\Rightarrow\; \mathbf{r}_1'(0) = \mathbf{j} + \mathbf{k}$ and $\mathbf{r}_2'(t) = \mathbf{i} + 2t\,\mathbf{j} + 3t^2\,\mathbf{k} \;\Rightarrow\; \mathbf{r}_2'(0) = \mathbf{i}$.

$\mathbf{r}_1'(0) \cdot \mathbf{r}_2'(0) = (\mathbf{j} + \mathbf{k}) \cdot \mathbf{i} = 0$. Therefore, the curves intersect in a right angle, that is, $\theta = \frac{\pi}{2}$.

10. The parametric value corresponding to the point $(1, 0, 1)$ is $t = 0$.

$\mathbf{r}'(t) = e^t\,\mathbf{i} + e^t(\cos t + \sin t)\,\mathbf{j} + e^t(\cos t - \sin t)\,\mathbf{k} \quad \Rightarrow \quad |\mathbf{r}'(t)| = e^t\sqrt{1 + (\cos t + \sin t)^2 + (\cos t - \sin t)^2} = \sqrt{3}\,e^t$

and $s(t) = \int_0^t e^u\sqrt{3}\,du = \sqrt{3}(e^t - 1) \quad \Rightarrow \quad t = \ln\!\left(1 + \tfrac{1}{\sqrt{3}}s\right)$.

Therefore, $\mathbf{r}(t(s)) = \left(1 + \tfrac{1}{\sqrt{3}}s\right)\mathbf{i} + \left(1 + \tfrac{1}{\sqrt{3}}s\right)\sin\ln\!\left(1 + \tfrac{1}{\sqrt{3}}s\right)\mathbf{j} + \left(1 + \tfrac{1}{\sqrt{3}}s\right)\cos\ln\!\left(1 + \tfrac{1}{\sqrt{3}}s\right)\mathbf{k}$.

11. (a) $\mathbf{T}(t) = \dfrac{\mathbf{r}'(t)}{|\mathbf{r}'(t)|} = \dfrac{\langle t^2, t, 1\rangle}{|\langle t^2, t, 1\rangle|} = \dfrac{\langle t^2, t, 1\rangle}{\sqrt{t^4 + t^2 + 1}}$

(b) $\mathbf{T}'(t) = -\tfrac{1}{2}(t^4 + t^2 + 1)^{-3/2}(4t^3 + 2t)\,\langle t^2, t, 1\rangle + (t^4 + t^2 + 1)^{-1/2}\langle 2t, 1, 0\rangle$

$= \dfrac{-2t^3 - t}{(t^4 + t^2 + 1)^{3/2}}\,\langle t^2, t, 1\rangle + \dfrac{1}{(t^4 + t^2 + 1)^{1/2}}\,\langle 2t, 1, 0\rangle$

$= \dfrac{\langle -2t^5 - t^3,\, -2t^4 - t^2,\, -2t^3 - t\rangle + \langle 2t^5 + 2t^3 + 2t,\, t^4 + t^2 + 1,\, 0\rangle}{(t^4 + t^2 + 1)^{3/2}} = \dfrac{\langle 2t,\, -t^4 + 1,\, -2t^3 - t\rangle}{(t^4 + t^2 + 1)^{3/2}}$

$|\mathbf{T}'(t)| = \dfrac{\sqrt{4t^2 + t^8 - 2t^4 + 1 + 4t^6 + 4t^4 + t^2}}{(t^4 + t^2 + 1)^{3/2}} = \dfrac{\sqrt{t^8 + 4t^6 + 2t^4 + 5t^2}}{(t^4 + t^2 + 1)^{3/2}}$ and $\mathbf{N}(t) = \dfrac{\langle 2t,\, 1 - t^4,\, -2t^3 - t\rangle}{\sqrt{t^8 + 4t^6 + 2t^4 + 5t^2}}$.

(c) $\kappa(t) = \dfrac{|\mathbf{T}'(t)|}{|\mathbf{r}'(t)|} = \dfrac{\sqrt{t^8 + 4t^6 + 2t^4 + 5t^2}}{(t^4 + t^2 + 1)^2}$

12. Using Exercise 14.3.40 [ET 13.3.40], we have $\mathbf{r}'(t) = \langle -3\sin t, 4\cos t\rangle$, $\mathbf{r}''(t) = \langle -3\cos t, -4\sin t\rangle$,

$|\mathbf{r}'(t)|^3 = \left(\sqrt{9\sin^2 t + 4\cos^2 t}\right)^3$ and then

$\kappa(t) = \dfrac{|(-3\sin t)(-4\sin t) - (4\cos t)(-3\cos t)|}{(9\sin^2 t + 16\cos^2 t)^{3/2}} = \dfrac{12}{(9\sin^2 t + 16\cos^2 t)^{3/2}}$.

At $(3, 0)$, $t = 0$ and $\kappa(0) = 12/(16)^{3/2} = \tfrac{12}{64} = \tfrac{3}{16}$. At $(0, 4)$, $t = \tfrac{\pi}{2}$ and $\kappa\!\left(\tfrac{\pi}{2}\right) = 12/9^{3/2} = \tfrac{12}{27} = \tfrac{4}{9}$.

13. $y' = 4x^3$, $y'' = 12x^2$ and $\kappa(x) = \dfrac{|y''|}{[1 + (y')^2]^{3/2}} = \dfrac{|12x^2|}{(1 + 16x^6)^{3/2}}$, so $\kappa(1) = \dfrac{12}{17^{3/2}}$.

14. $\kappa(x) = \dfrac{|12x^2 - 2|}{[1 + (4x^3 - 2x)^2]^{3/2}} \quad \Rightarrow \quad \kappa(0) = 2$.

So the osculating circle has radius $\tfrac{1}{2}$ and center $\left(0, -\tfrac{1}{2}\right)$.

Thus its equation is $x^2 + \left(y + \tfrac{1}{2}\right)^2 = \tfrac{1}{4}$.

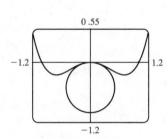

15. $\mathbf{r}(t) = \langle \sin 2t, t, \cos 2t\rangle \quad \Rightarrow \quad \mathbf{r}'(t) = \langle 2\cos 2t, 1, -2\sin 2t\rangle \quad \Rightarrow \quad \mathbf{T}(t) = \tfrac{1}{\sqrt{5}}\langle 2\cos 2t, 1, -2\sin 2t\rangle \quad \Rightarrow$

$\mathbf{T}'(t) = \tfrac{1}{\sqrt{5}}\langle -4\sin 2t, 0, -4\cos 2t\rangle \quad \Rightarrow \quad \mathbf{N}(t) = \langle -\sin 2t, 0, -\cos 2t\rangle$. So $\mathbf{N} = \mathbf{N}(\pi) = \langle 0, 0, -1\rangle$ and

$\mathbf{B} = \mathbf{T} \times \mathbf{N} = \tfrac{1}{\sqrt{5}}\langle -1, 2, 0\rangle$. So a normal to the osculating plane is $\langle -1, 2, 0\rangle$ and an equation is

$-1(x - 0) + 2(y - \pi) + 0(z - 1) = 0$ or $x - 2y + 2\pi = 0$.

16. (a) The average velocity over $[3, 3.2]$ is given by

$$\frac{\mathbf{r}(3.2) - \mathbf{r}(3)}{3.2 - 3} = 5[\mathbf{r}(3.2) - \mathbf{r}(3)], \text{ so we draw a}$$

vector with the same direction but 5 times the length

of the vector $[\mathbf{r}(3.2) - \mathbf{r}(3)]$.

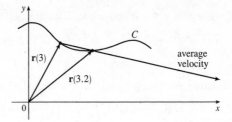

(b) $\mathbf{v}(3) = \mathbf{r}'(3) = \lim\limits_{h \to 0} \dfrac{\mathbf{r}(3 + h) - \mathbf{r}(3)}{h}$

(c) $\mathbf{T}(3) = \dfrac{\mathbf{r}'(3)}{|\mathbf{r}'(3)|}$, a unit vector in the same direction as

$\mathbf{r}'(3)$, that is, parallel to the tangent line to the curve at

$\mathbf{r}(3)$, pointing in the direction corresponding to

increasing t, and with length 1.

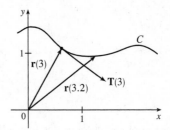

17. $\mathbf{r}(t) = t \ln t \, \mathbf{i} + t \mathbf{j} + e^{-t} \mathbf{k}$, $\mathbf{v}(t) = \mathbf{r}'(t) = (1 + \ln t) \mathbf{i} + \mathbf{j} - e^{-t} \mathbf{k}$,

$|\mathbf{v}(t)| = \sqrt{(1 + \ln t)^2 + 1^2 + (-e^{-t})^2} = \sqrt{2 + 2 \ln t + (\ln t)^2 + e^{-2t}}$, $\mathbf{a}(t) = \mathbf{v}'(t) = \frac{1}{t} \mathbf{i} + e^{-t} \mathbf{k}$

18. $\mathbf{v}(t) = \int \mathbf{a}(t)\, dt = \int (6t \, \mathbf{i} + 12t^2 \, \mathbf{j} - 6t \, \mathbf{k})\, dt = 3t^2 \, \mathbf{i} + 4t^3 \, \mathbf{j} - 3t^2 \, \mathbf{k} + \mathbf{C}$, but $\mathbf{i} - \mathbf{j} + 3 \mathbf{k} = \mathbf{v}(0) = \mathbf{0} + \mathbf{C}$,

so $\mathbf{C} = \mathbf{i} - \mathbf{j} + 3 \mathbf{k}$ and $\mathbf{v}(t) = (3t^2 + 1) \mathbf{i} + (4t^3 - 1) \mathbf{j} + (3 - 3t^2) \mathbf{k}$.

$\mathbf{r}(t) = \int \mathbf{v}(t)\, dt = (t^3 + t) \mathbf{i} + (t^4 - t) \mathbf{j} + (3t - t^3) \mathbf{k} + \mathbf{D}$.

But $\mathbf{r}(0) = \mathbf{0}$, so $\mathbf{D} = \mathbf{0}$ and $\mathbf{r}(t) = (t^3 + t) \mathbf{i} + (t^4 - t) \mathbf{j} + (3t - t^3) \mathbf{k}$.

19. We set up the axes so that the shot leaves the athlete's hand 7 ft above the origin. Then we are given $\mathbf{r}(0) = 7\mathbf{j}$,

$|\mathbf{v}(0)| = 43$ ft/s, and $\mathbf{v}(0)$ has direction given by a $45°$ angle of elevation. Then a unit vector in the direction of $\mathbf{v}(0)$ is

$\frac{1}{\sqrt{2}}(\mathbf{i} + \mathbf{j}) \ \Rightarrow \ \mathbf{v}(0) = \frac{43}{\sqrt{2}}(\mathbf{i} + \mathbf{j})$. Assuming air resistance is negligible, the only external force is due to gravity, so as

in Example 14.4.5 [ET 13.4.5] we have $\mathbf{a} = -g\mathbf{j}$ where here $g \approx 32$ ft/s^2. Since $\mathbf{v}'(t) = \mathbf{a}(t)$, we integrate, giving

$\mathbf{v}(t) = -gt\mathbf{j} + \mathbf{C}$ where $\mathbf{C} = \mathbf{v}(0) = \frac{43}{\sqrt{2}}(\mathbf{i} + \mathbf{j}) \ \Rightarrow \ \mathbf{v}(t) = \frac{43}{\sqrt{2}}\mathbf{i} + \left(\frac{43}{\sqrt{2}} - gt\right)\mathbf{j}$. Since $\mathbf{r}'(t) = \mathbf{v}(t)$ we integrate

again, so $\mathbf{r}(t) = \frac{43}{\sqrt{2}}t\mathbf{i} + \left(\frac{43}{\sqrt{2}}t - \frac{1}{2}gt^2\right)\mathbf{j} + \mathbf{D}$. But $\mathbf{D} = \mathbf{r}(0) = 7\mathbf{j} \ \Rightarrow \ \mathbf{r}(t) = \frac{43}{\sqrt{2}}t\mathbf{i} + \left(\frac{43}{\sqrt{2}}t - \frac{1}{2}gt^2 + 7\right)\mathbf{j}$.

(a) At 2 seconds, the shot is at $\mathbf{r}(2) = \frac{43}{\sqrt{2}}(2)\mathbf{i} + \left(\frac{43}{\sqrt{2}}(2) - \frac{1}{2}g(2)^2 + 7\right)\mathbf{j} \approx 60.8\,\mathbf{i} + 3.8\,\mathbf{j}$, so the shot is about 3.8 ft above

the ground, at a horizontal distance of 60.8 ft from the athlete.

(b) The shot reaches its maximum height when the vertical component of velocity is 0: $\frac{43}{\sqrt{2}} - gt = 0 \ \Rightarrow$

$t = \dfrac{43}{\sqrt{2}\,g} \approx 0.95$ s. Then $\mathbf{r}(0.95) \approx 28.9\,\mathbf{i} + 21.4\,\mathbf{j}$, so the maximum height is approximately 21.4 ft.

(c) The shot hits the ground when the vertical component of $\mathbf{r}(t)$ is 0, so $\frac{43}{\sqrt{2}}t - \frac{1}{2}gt^2 + 7 = 0 \ \Rightarrow$

$-16t^2 + \frac{43}{\sqrt{2}}t + 7 = 0 \ \Rightarrow \ t \approx 2.11$ s. $\mathbf{r}(2.11) \approx 64.2\,\mathbf{i} - 0.08\,\mathbf{j}$, thus the shot lands approximately 64.2 ft from the

athlete.

20. $\mathbf{r}'(t) = \mathbf{i} + 2\mathbf{j} + 2t\,\mathbf{k}$, $\quad \mathbf{r}''(t) = 2\,\mathbf{k}$, $\quad |\mathbf{r}'(t)| = \sqrt{1 + 4 + 4t^2} = \sqrt{4t^2 + 5}$.

Then $a_T = \dfrac{\mathbf{r}'(t) \cdot \mathbf{r}''(t)}{|\mathbf{r}'(t)|} = \dfrac{4t}{\sqrt{4t^2 + 5}}$ and $a_N = \dfrac{|\mathbf{r}'(t) \times \mathbf{r}''(t)|}{|\mathbf{r}'(t)|} = \dfrac{|4\,\mathbf{i} - 2\,\mathbf{j}|}{\sqrt{4t^2 + 5}} = \dfrac{2\sqrt{5}}{\sqrt{4t^2 + 5}}$.

21. (a) Instead of proceeding directly, we use Formula 3 of Theorem 14.2.3 [ET 13.2.3]: $\mathbf{r}(t) = t\,\mathbf{R}(t)$ $\Rightarrow$

$\mathbf{v} = \mathbf{r}'(t) = \mathbf{R}(t) + t\,\mathbf{R}'(t) = \cos \omega t\,\mathbf{i} + \sin \omega t\,\mathbf{j} + t\,\mathbf{v}_d$.

(b) Using the same method as in part (a) and starting with $\mathbf{v} = \mathbf{R}(t) + t\,\mathbf{R}'(t)$, we have

$\mathbf{a} = \mathbf{v}' = \mathbf{R}'(t) + \mathbf{R}'(t) + t\,\mathbf{R}''(t) = 2\,\mathbf{R}'(t) + t\,\mathbf{R}''(t) = 2\,\mathbf{v}_d + t\,\mathbf{a}_d$.

(c) Here we have $\mathbf{r}(t) = e^{-t} \cos \omega t\,\mathbf{i} + e^{-t} \sin \omega t\,\mathbf{j} = e^{-t}\,\mathbf{R}(t)$. So, as in parts (a) and (b),

$\mathbf{v} = \mathbf{r}'(t) = e^{-t}\,\mathbf{R}'(t) - e^{-t}\,\mathbf{R}(t) = e^{-t}[\mathbf{R}'(t) - \mathbf{R}(t)]$ $\Rightarrow$

$\mathbf{a} = \mathbf{v}' = e^{-t}[\mathbf{R}''(t) - \mathbf{R}'(t)] - e^{-t}[\mathbf{R}'(t) - \mathbf{R}(t)] = e^{-t}[\mathbf{R}''(t) - 2\,\mathbf{R}'(t) + \mathbf{R}(t)]$

$\qquad = e^{-t}\,\mathbf{a}_d - 2e^{-t}\,\mathbf{v}_d + e^{-t}\,\mathbf{R}$

Thus, the Coriolis acceleration (the sum of the "extra" terms not involving $\mathbf{a}_d$) is $-2e^{-t}\,\mathbf{v}_d + e^{-t}\,\mathbf{R}$.

22. (a) $F(x) = \begin{cases} 1 & \text{if } x \le 0 \\ \sqrt{1 - x^2} & \text{if } 0 < x < \frac{1}{\sqrt{2}} \\ \sqrt{2} - x & \text{if } x \ge \frac{1}{\sqrt{2}} \end{cases}$ $\Rightarrow$ $F'(x) = \begin{cases} 0 & \text{if } x < 0 \\ -x/\sqrt{1 - x^2} & \text{if } 0 < x < \frac{1}{\sqrt{2}} \\ -1 & \text{if } x > \frac{1}{\sqrt{2}} \end{cases}$ $\Rightarrow$

$F''(x) = \begin{cases} 0 & \text{if } x < 0 \\ -1/(1 - x^2)^{3/2} & \text{if } 0 < x < \frac{1}{\sqrt{2}} \\ 0 & \text{if } x > \frac{1}{\sqrt{2}} \end{cases}$

since $\dfrac{d}{dx}[-x(1 - x^2)^{-1/2}] = -(1 - x^2)^{-1/2} - x^2(1 - x^2)^{-3/2} = -(1 - x^2)^{-3/2}$.

Now $\lim\limits_{x \to 0^+} \sqrt{1 - x^2} = 1 = F(0)$ and $\lim\limits_{x \to (1/\sqrt{2})^-} \sqrt{1 - x^2} = \frac{1}{\sqrt{2}} = F\left(\frac{1}{\sqrt{2}}\right)$, so F is continuous. Also, since

$\lim\limits_{x \to 0^+} F'(x) = 0 = \lim\limits_{x \to 0^-} F'(x)$ and $\lim\limits_{x \to (1/\sqrt{2})^-} F'(x) = -1 = \lim\limits_{x \to (1/\sqrt{2})^+} F'(x)$, F' is continuous. But

$\lim\limits_{x \to 0^+} F''(x) = -1 \ne 0 = \lim\limits_{x \to 0^-} F''(x)$, so F'' is not continuous at $x = 0$. (The same is true at $x = \frac{1}{\sqrt{2}}$.)

So F does not have continuous curvature.

(b) Set $P(x) = ax^5 + bx^4 + cx^3 + dx^2 + ex + f$. The continuity conditions on P are $P(0) = 0$, $P(1) = 1$, $P'(0) = 0$ and

$P'(1) = 1$. Also the curvature must be continuous. For $x \le 0$ and $x \ge 1$, $\kappa(x) = 0$; elsewhere

$\kappa(x) = \dfrac{|P''(x)|}{(1 + [P'(x)]^2)^{3/2}}$, so we need $P''(0) = 0$ and $P''(1) = 0$.

The conditions $P(0) = P'(0) = P''(0) = 0$ imply that $d = e = f = 0$.

The other conditions imply that $a + b + c = 1$, $5a + 4b + 3c = 1$, and

$10a + 6b + 3c = 0$. From these, we find that $a = 3$, $b = -8$, and $c = 6$.

Therefore $P(x) = 3x^5 - 8x^4 + 6x^3$. Since there was no solution with

$a = 0$, this could not have been done with a polynomial of degree 4.

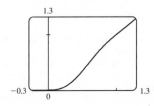

☐ PROBLEMS PLUS

1. (a) $\mathbf{r}(t) = R\cos\omega t\,\mathbf{i} + R\sin\omega t\,\mathbf{j} \;\Rightarrow\; \mathbf{v} = \mathbf{r}'(t) = -\omega R\sin\omega t\,\mathbf{i} + \omega R\cos\omega t\,\mathbf{j}$, so $\mathbf{r} = R(\cos\omega t\,\mathbf{i} + \sin\omega t\,\mathbf{j})$ and

$\mathbf{v} = \omega R(-\sin\omega t\,\mathbf{i} + \cos\omega t\,\mathbf{j})$. $\mathbf{v}\cdot\mathbf{r} = \omega R^2(-\cos\omega t\sin\omega t + \sin\omega t\cos\omega t) = 0$, so $\mathbf{v}\perp\mathbf{r}$. Since $\mathbf{r}$ points along a

radius of the circle, and $\mathbf{v}\perp\mathbf{r}$, $\mathbf{v}$ is tangent to the circle. Because it is a velocity vector, $\mathbf{v}$ points in the direction of motion.

(b) In (a), we wrote $\mathbf{v}$ in the form $\omega R\,\mathbf{u}$, where $\mathbf{u}$ is the unit vector $-\sin\omega t\,\mathbf{i} + \cos\omega t\,\mathbf{j}$. Clearly $|\mathbf{v}| = \omega R\,|\mathbf{u}| = \omega R$. At

speed ωR, the particle completes one revolution, a distance $2\pi R$, in time $T = \dfrac{2\pi R}{\omega R} = \dfrac{2\pi}{\omega}$.

(c) $\mathbf{a} = \dfrac{d\mathbf{v}}{dt} = -\omega^2 R\cos\omega t\,\mathbf{i} - \omega^2 R\sin\omega t\,\mathbf{j} = -\omega^2 R(\cos\omega t\,\mathbf{i} + \sin\omega t\,\mathbf{j})$, so $\mathbf{a} = -\omega^2\mathbf{r}$. This shows that $\mathbf{a}$ is proportional

to $\mathbf{r}$ and points in the opposite direction (toward the origin). Also, $|\mathbf{a}| = \omega^2\,|\mathbf{r}| = \omega^2 R$.

(d) By Newton's Second Law (see Section 14.4 [ET 13.4]), $\mathbf{F} = m\mathbf{a}$, so $|\mathbf{F}| = m\,|\mathbf{a}| = mR\omega^2 = \dfrac{m\,(\omega R)^2}{R} = \dfrac{m\,|\mathbf{v}|^2}{R}$.

2. (a) Dividing the equation $|\mathbf{F}|\sin\theta = \dfrac{mv_R^2}{R}$ by the equation $|\mathbf{F}|\cos\theta = mg$, we obtain $\tan\theta = \dfrac{v_R^2}{Rg}$, so $v_R^2 = Rg\tan\theta$.

(b) $R = 400$ ft and $\theta = 12°$, so $v_R = \sqrt{Rg\tan\theta} \approx \sqrt{400\cdot 32\cdot\tan 12°} \approx 52.16$ ft/s ≈ 36 mi/h.

(c) We want to choose a new radius R_1 for which the new rated speed is $\frac{3}{2}$ of the old one: $\sqrt{R_1 g\tan 12°} = \frac{3}{2}\sqrt{Rg\tan 12°}$.

Squaring, we get $R_1 g\tan 12° = \frac{9}{4}Rg\tan 12°$, so $R_1 = \frac{9}{4}R = \frac{9}{4}(400) = 900$ ft.

3. (a) The projectile reaches maximum height when $0 = \dfrac{dy}{dt} = \dfrac{d}{dt}\left[(v_0\sin\alpha)t - \frac{1}{2}gt^2\right] = v_0\sin\alpha - gt$; that is, when

$t = \dfrac{v_0\sin\alpha}{g}$ and $y = (v_0\sin\alpha)\left(\dfrac{v_0\sin\alpha}{g}\right) - \dfrac{1}{2}g\left(\dfrac{v_0\sin\alpha}{g}\right)^2 = \dfrac{v_0^2\sin^2\alpha}{2g}$. This is the maximum height attained when

the projectile is fired with an angle of elevation α. This maximum height is largest when $\alpha = \frac{\pi}{2}$. In that case, $\sin\alpha = 1$

and the maximum height is $\dfrac{v_0^2}{2g}$.

(b) Let $R = v_0^2/g$. We are asked to consider the parabola $x^2 + 2Ry - R^2 = 0$ which can be rewritten as $y = -\dfrac{1}{2R}x^2 + \dfrac{R}{2}$.

The points on or inside this parabola are those for which $-R\leq x\leq R$ and $0\leq y\leq \dfrac{-1}{2R}x^2 + \dfrac{R}{2}$. When the projectile is

fired at angle of elevation α, the points (x, y) along its path satisfy the relations $x = (v_0\cos\alpha)\,t$ and

$y = (v_0\sin\alpha)t - \frac{1}{2}gt^2$, where $0\leq t\leq (2v_0\sin\alpha)/g$ (as in Example 14.4.5 [ET 13.4.5]). Thus

$|x| \leq \left|v_0\cos\alpha\left(\dfrac{2v_0\sin\alpha}{g}\right)\right| = \left|\dfrac{v_0^2}{g}\sin 2\alpha\right| \leq \left|\dfrac{v_0^2}{g}\right| = |R|$. This shows that $-R\leq x\leq R$.

For t in the specified range, we also have $y = t\left(v_0\sin\alpha - \frac{1}{2}gt\right) = \frac{1}{2}gt\left(\dfrac{2v_0\sin\alpha}{g} - t\right) \geq 0$ and

$y = (v_0\sin\alpha)\dfrac{x}{v_0\cos\alpha} - \dfrac{g}{2}\left(\dfrac{x}{v_0\cos\alpha}\right)^2 = (\tan\alpha)\,x - \dfrac{g}{2v_0^2\cos^2\alpha}x^2 = -\dfrac{1}{2R\cos^2\alpha}x^2 + (\tan\alpha)\,x$. Thus

$$y - \left(\frac{-1}{2R}x^2 + \frac{R}{2}\right) = \frac{-1}{2R\cos^2\alpha}x^2 + \frac{1}{2R}x^2 + (\tan\alpha)\,x - \frac{R}{2}$$

$$= \frac{x^2}{2R}\left(1 - \frac{1}{\cos^2\alpha}\right) + (\tan\alpha)\,x - \frac{R}{2} = \frac{x^2(1 - \sec^2\alpha) + 2R\,(\tan\alpha)\,x - R^2}{2R}$$

$$= \frac{-(\tan^2\alpha)\,x^2 + 2R\,(\tan\alpha)\,x - R^2}{2R} = \frac{-[(\tan\alpha)\,x - R]^2}{2R} \le 0$$

We have shown that every target that can be hit by the projectile lies on or inside the parabola $y = -\dfrac{1}{2R}x^2 + \dfrac{R}{2}$.

Now let (a, b) be any point on or inside the parabola $y = -\dfrac{1}{2R}x^2 + \dfrac{R}{2}$. Then $-R \le a \le R$ and $0 \le b \le -\dfrac{1}{2R}a^2 + \dfrac{R}{2}$.

We seek an angle α such that (a, b) lies in the path of the projectile; that is, we wish to find an angle α such that

$b = -\dfrac{1}{2R\cos^2\alpha}a^2 + (\tan\alpha)\,a$ or equivalently $b = \dfrac{-1}{2R}(\tan^2\alpha + 1)a^2 + (\tan\alpha)\,a$. Rearranging this equation we get

$\dfrac{a^2}{2R}\tan^2\alpha - a\tan\alpha + \left(\dfrac{a^2}{2R} + b\right) = 0$ or $a^2(\tan\alpha)^2 - 2aR(\tan\alpha) + (a^2 + 2bR) = 0$ (∗). This quadratic equation

for $\tan\alpha$ has real solutions exactly when the discriminant is nonnegative. Now $B^2 - 4AC \ge 0 \iff$

$(-2aR)^2 - 4a^2(a^2 + 2bR) \ge 0 \iff 4a^2(R^2 - a^2 - 2bR) \ge 0 \iff -a^2 - 2bR + R^2 \ge 0 \iff$

$b \le \dfrac{1}{2R}(R^2 - a^2) \iff b \le \dfrac{-1}{2R}a^2 + \dfrac{R}{2}$. This condition is satisfied since (a, b) is on or inside the parabola

$y = -\dfrac{1}{2R}x^2 + \dfrac{R}{2}$. It follows that (a, b) lies in the path of the projectile when $\tan\alpha$ satisfies (∗), that is, when

$$\tan\alpha = \frac{2aR \pm \sqrt{4a^2(R^2 - a^2 - 2bR)}}{2a^2} = \frac{R \pm \sqrt{R^2 - 2bR - a^2}}{a}.$$

(c)

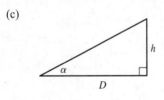

If the gun is pointed at a target with height h at a distance D downrange, then

$\tan\alpha = h/D$. When the projectile reaches a distance D downrange (remember

we are assuming that it doesn't hit the ground first), we have $D = x = (v_0\cos\alpha)t$,

$$\text{so } t = \frac{D}{v_0\cos\alpha} \text{ and } y = (v_0\sin\alpha)t - \tfrac{1}{2}gt^2 = D\tan\alpha - \frac{gD^2}{2v_0^2\cos^2\alpha}.$$

Meanwhile, the target, whose x-coordinate is also D, has fallen from height h to height

$h - \tfrac{1}{2}gt^2 = D\tan\alpha - \dfrac{gD^2}{2v_0^2\cos^2\alpha}$. Thus the projectile hits the target.

4. (a) As in Problem 3, $\mathbf{r}(t) = (v_0\cos\alpha)t\,\mathbf{i} + \left[(v_0\sin\alpha)t - \tfrac{1}{2}gt^2\right]\mathbf{j}$, so $x = (v_0\cos\alpha)t$ and $y = (v_0\sin\alpha)t - \tfrac{1}{2}gt^2$. The

difference here is that the projectile travels until it reaches a point where $x > 0$ and $y = -(\tan\theta)x$. (Here $0 \le \theta \le \tfrac{\pi}{2}$.)

From the parametric equations, we obtain $t = \dfrac{x}{v_0\cos\alpha}$ and $y = \dfrac{(v_0\sin\alpha)x}{v_0\cos\alpha} - \dfrac{gx^2}{2v_0^2\cos^2\alpha} = (\tan\alpha)x - \dfrac{gx^2}{2v_0^2\cos^2\alpha}$.

Thus the projectile hits the inclined plane at the point where $(\tan\alpha)x - \dfrac{gx^2}{2v_0^2\cos^2\alpha} = -(\tan\theta)x$. Since

$\dfrac{gx^2}{2v_0^2\cos^2\alpha} = (\tan\alpha + \tan\theta)x$ and $x > 0$, we must have $\dfrac{gx}{2v_0^2\cos^2\alpha} = \tan\alpha + \tan\theta$. It follows that

$x = \dfrac{2v_0^2 \cos^2 \alpha}{g}(\tan \alpha + \tan \theta)$ and $t = \dfrac{x}{v_0 \cos \alpha} = \dfrac{2v_0 \cos \alpha}{g}(\tan \alpha + \tan \theta)$. This means that the parametric

equations are defined for t in the interval $\left[0, \dfrac{2v_0 \cos \alpha}{g}(\tan \alpha + \tan \theta)\right]$.

(b) The downhill range (that is, the distance to the projectile's landing point as
measured along the inclined plane) is $R(\alpha) = x \sec \theta$, where x is the
coordinate of the landing point calculated in part (a). Thus

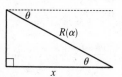

$$R(\alpha) = \frac{2v_0^2 \cos^2 \alpha}{g}(\tan \alpha + \tan \theta)\sec \theta = \frac{2v_0^2}{g}\left(\frac{\sin \alpha \cos \alpha}{\cos \theta} + \frac{\cos^2 \alpha \sin \theta}{\cos^2 \theta}\right)$$

$$= \frac{2v_0^2 \cos \alpha}{g \cos^2 \theta}(\sin \alpha \cos \theta + \cos \alpha \sin \theta) = \frac{2v_0^2 \cos \alpha \sin(\alpha + \theta)}{g \cos^2 \theta}$$

$R(\alpha)$ is maximized when

$$0 = R'(\alpha) = \frac{2v_0^2}{g \cos^2 \theta}[-\sin \alpha \sin(\alpha + \theta) + \cos \alpha \cos(\alpha + \theta)]$$

$$= \frac{2v_0^2}{g \cos^2 \theta}\cos[(\alpha + \theta) + \alpha] = \frac{2v_0^2 \cos(2\alpha + \theta)}{g \cos^2 \theta}$$

This condition implies that $\cos(2\alpha + \theta) = 0 \;\Rightarrow\; 2\alpha + \theta = \frac{\pi}{2} \;\Rightarrow\; \alpha = \frac{1}{2}\left(\frac{\pi}{2} - \theta\right)$.

(c) The solution is similar to the solutions to parts (a) and (b). This time the projectile travels until it reaches a point where
$x > 0$ and $y = (\tan \theta)x$. Since $\tan \theta = -\tan(-\theta)$, we obtain the solution from the previous one by replacing θ with $-\theta$.
The desired angle is $\alpha = \frac{1}{2}\left(\frac{\pi}{2} + \theta\right)$.

(d) As observed in part (c), firing the projectile up an inclined plane with angle of inclination θ involves the same equations as
in parts (a) and (b) but with θ replaced by $-\theta$. So if R is the distance up an inclined plane, we know from part (b) that

$R = \dfrac{2v_0^2 \cos \alpha \sin(\alpha - \theta)}{g \cos^2(-\theta)} \;\Rightarrow\; v_0^2 = \dfrac{Rg \cos^2 \theta}{2 \cos \alpha \sin(\alpha - \theta)}$. v_0^2 is minimized

(and hence v_0 is minimized) with respect to α when

$$0 = \frac{d}{d\alpha}\left(v_0^2\right) = \frac{Rg \cos^2 \theta}{2} \cdot \frac{-(\cos \alpha \cos(\alpha - \theta) - \sin \alpha \sin(\alpha - \theta))}{[\cos \alpha \sin(\alpha - \theta)]^2}$$

$$= \frac{-Rg \cos^2 \theta}{2} \cdot \frac{\cos[\alpha + (\alpha - \theta)]}{[\cos \alpha \sin(\alpha - \theta)]^2} = \frac{-Rg \cos^2 \theta}{2} \cdot \frac{\cos(2\alpha - \theta)}{[\cos \alpha \sin(\alpha - \theta)]^2}$$

Since $\theta < \alpha < \frac{\pi}{2}$, this implies $\cos(2\alpha - \theta) = 0 \;\Leftrightarrow\; 2\alpha - \theta = \frac{\pi}{2} \;\Rightarrow\; \alpha = \frac{1}{2}\left(\frac{\pi}{2} + \theta\right)$. Thus the initial speed, and

hence the energy required, is minimized for $\alpha = \frac{1}{2}\left(\frac{\pi}{2} + \theta\right)$.

5. (a) $\mathbf{a} = -g\mathbf{j} \;\Rightarrow\; \mathbf{v} = \mathbf{v}_0 - gt\mathbf{j} = 2\mathbf{i} - gt\mathbf{j} \;\Rightarrow\; \mathbf{s} = \mathbf{s}_0 + 2t\mathbf{i} - \frac{1}{2}gt^2\mathbf{j} = 3.5\mathbf{j} + 2t\mathbf{i} - \frac{1}{2}gt^2\mathbf{j} \;\Rightarrow\;$

$\mathbf{s} = 2t\mathbf{i} + \left(3.5 - \frac{1}{2}gt^2\right)\mathbf{j}$. Therefore $y = 0$ when $t = \sqrt{7/g}$ seconds. At that instant, the ball is $2\sqrt{7/g} \approx 0.94$ ft to the
right of the table top. Its coordinates (relative to an origin on the floor directly under the table's edge) are $(0.94, 0)$. At
impact, the velocity is $\mathbf{v} = 2\mathbf{i} - \sqrt{7g}\,\mathbf{j}$, so the speed is $|\mathbf{v}| = \sqrt{4 + 7g} \approx 15$ ft/s.

(b) The slope of the curve when $t = \sqrt{\dfrac{7}{g}}$ is $\dfrac{dy}{dx} = \dfrac{dy/dt}{dx/dt} = \dfrac{-gt}{2} = \dfrac{-g\sqrt{7/g}}{2} = \dfrac{-\sqrt{7g}}{2}$. Thus $\cot\theta = \dfrac{\sqrt{7g}}{2}$

and $\theta \approx 7.6°$.

(c) From (a), $|\mathbf{v}| = \sqrt{4+7g}$. So the ball rebounds with speed $0.8\sqrt{4+7g} \approx 12.08$ ft/s at angle of inclination

$90° - \theta \approx 82.3886°$. By Example 14.4.5 [ET 13.4.5], the horizontal distance traveled between bounces is

$d = \dfrac{v_0^2 \sin 2\alpha}{g}$, where $v_0 \approx 12.08$ ft/s and $\alpha \approx 82.3886°$. Therefore, $d \approx 1.197$ ft. So the ball strikes the floor at

about $2\sqrt{7/g} + 1.197 \approx 2.13$ ft to the right of the table's edge.

6. By the Fundamental Theorem of Calculus, $\mathbf{r}'(t) = \langle \sin\left(\tfrac{1}{2}\pi t^2\right), \cos\left(\tfrac{1}{2}\pi t^2\right)\rangle$, $|\mathbf{r}'(t)| = 1$ and so $\mathbf{T}(t) = \mathbf{r}'(t)$.

Thus $\mathbf{T}'(t) = \pi t \langle \cos\left(\tfrac{1}{2}\pi t^2\right), -\sin\left(\tfrac{1}{2}\pi t^2\right)\rangle$ and the curvature is $\kappa = |\mathbf{T}'(t)| = \sqrt{(\pi t)^2(1)} = \pi|t|$.

7. The trajectory of the projectile is given by $\mathbf{r}(t) = (v\cos\alpha)t\,\mathbf{i} + \left[(v\sin\alpha)t - \tfrac{1}{2}gt^2\right]\mathbf{j}$, so

$\mathbf{v}(t) = \mathbf{r}'(t) = v\cos\alpha\,\mathbf{i} + (v\sin\alpha - gt)\,\mathbf{j}$ and

$$|\mathbf{v}(t)| = \sqrt{(v\cos\alpha)^2 + (v\sin\alpha - gt)^2} = \sqrt{v^2 - (2vg\sin\alpha)t + g^2t^2} = \sqrt{g^2\left(t^2 - \dfrac{2v}{g}(\sin\alpha)t + \dfrac{v^2}{g^2}\right)}$$

$$= g\sqrt{\left(t - \dfrac{v}{g}\sin\alpha\right)^2 + \dfrac{v^2}{g^2} - \dfrac{v^2}{g^2}\sin^2\alpha} = g\sqrt{\left(t - \dfrac{v}{g}\sin\alpha\right)^2 + \dfrac{v^2}{g^2}\cos^2\alpha}$$

The projectile hits the ground when $(v\sin\alpha)t - \tfrac{1}{2}gt^2 = 0 \Rightarrow t = \dfrac{2v}{g}\sin\alpha$, so the distance traveled by the projectile is

$$L(\alpha) = \int_0^{(2v/g)\sin\alpha} |\mathbf{v}(t)|\,dt = \int_0^{(2v/g)\sin\alpha} g\sqrt{\left(t - \dfrac{v}{g}\sin\alpha\right)^2 + \dfrac{v^2}{g^2}\cos^2\alpha}\,dt$$

$$= g\left[\dfrac{t - (v/g)\sin\alpha}{2}\sqrt{\left(t - \dfrac{v}{g}\sin\alpha\right)^2 + \left(\dfrac{v}{g}\cos\alpha\right)^2}\right.$$

$$\left.+ \dfrac{[(v/g)\cos\alpha]^2}{2}\ln\left(t - \dfrac{v}{g}\sin\alpha + \sqrt{\left(t - \dfrac{v}{g}\sin\alpha\right)^2 + \left(\dfrac{v}{g}\cos\alpha\right)^2}\right)\right]_0^{(2v/g)\sin\alpha}$$

[using Formula 21 in the Table of Integrals]

$$= \dfrac{g}{2}\left[\dfrac{v}{g}\sin\alpha\sqrt{\left(\dfrac{v}{g}\sin\alpha\right)^2 + \left(\dfrac{v}{g}\cos\alpha\right)^2} + \left(\dfrac{v}{g}\cos\alpha\right)^2\ln\left(\dfrac{v}{g}\sin\alpha + \sqrt{\left(\dfrac{v}{g}\sin\alpha\right)^2 + \left(\dfrac{v}{g}\cos\alpha\right)^2}\right)\right.$$

$$\left.+ \dfrac{v}{g}\sin\alpha\sqrt{\left(\dfrac{v}{g}\sin\alpha\right)^2 + \left(\dfrac{v}{g}\cos\alpha\right)^2} - \left(\dfrac{v}{g}\cos\alpha\right)^2\ln\left(-\dfrac{v}{g}\sin\alpha + \sqrt{\left(\dfrac{v}{g}\sin\alpha\right)^2 + \left(\dfrac{v}{g}\cos\alpha\right)^2}\right)\right]$$

$$= \dfrac{g}{2}\left[\dfrac{v}{g}\sin\alpha \cdot \dfrac{v}{g} + \dfrac{v^2}{g^2}\cos^2\alpha\ln\left(\dfrac{v}{g}\sin\alpha + \dfrac{v}{g}\right) + \dfrac{v}{g}\sin\alpha \cdot \dfrac{v}{g} - \dfrac{v^2}{g^2}\cos^2\alpha\ln\left(-\dfrac{v}{g}\sin\alpha + \dfrac{v}{g}\right)\right]$$

$$= \dfrac{v^2}{g}\sin\alpha + \dfrac{v^2}{2g}\cos^2\alpha\ln\left(\dfrac{(v/g)\sin\alpha + v/g}{-(v/g)\sin\alpha + v/g}\right) = \dfrac{v^2}{g}\sin\alpha + \dfrac{v^2}{2g}\cos^2\alpha\ln\left(\dfrac{1 + \sin\alpha}{1 - \sin\alpha}\right)$$

We want to maximize $L(\alpha)$ for $0 \le \alpha \le \pi/2$.

$$L'(\alpha) = \frac{v^2}{g} \cos \alpha + \frac{v^2}{2g} \left[\cos^2 \alpha \cdot \frac{1 - \sin \alpha}{1 + \sin \alpha} \cdot \frac{2 \cos \alpha}{(1 - \sin \alpha)^2} - 2 \cos \alpha \sin \alpha \ln \left(\frac{1 + \sin \alpha}{1 - \sin \alpha} \right) \right]$$

$$= \frac{v^2}{g} \cos \alpha + \frac{v^2}{2g} \left[\cos^2 \alpha \cdot \frac{2}{\cos \alpha} - 2 \cos \alpha \sin \alpha \ln \left(\frac{1 + \sin \alpha}{1 - \sin \alpha} \right) \right]$$

$$= \frac{v^2}{g} \cos \alpha + \frac{v^2}{g} \cos \alpha \left[1 - \sin \alpha \ln \left(\frac{1 + \sin \alpha}{1 - \sin \alpha} \right) \right] = \frac{v^2}{g} \cos \alpha \left[2 - \sin \alpha \ln \left(\frac{1 + \sin \alpha}{1 - \sin \alpha} \right) \right]$$

$L(\alpha)$ has critical points for $0 < \alpha < \pi/2$ when $L'(\alpha) = 0 \quad \Rightarrow \quad 2 - \sin \alpha \ln \left(\frac{1 + \sin \alpha}{1 - \sin \alpha} \right) = 0$ (since $\cos \alpha \ne 0$).

Solving by graphing (or using a CAS) gives $\alpha \approx 0.9855$. Compare values at the critical point and the endpoints:

$L(0) = 0$, $L(\pi/2) = v^2/g$, and $L(0.9855) \approx 1.20 v^2/g$. Thus the distance traveled by the projectile is maximized

for $\alpha \approx 0.9855$ or $\approx 56°$.

8. As the cable is wrapped around the spool, think of the top or bottom of the

cable forming a helix of radius $R + r$. Let h be the vertical distance

between coils. Then, from similar triangles, $\dfrac{2r}{\sqrt{h^2 - 4r^2}} = \dfrac{2\pi(r + R)}{h} \Rightarrow$

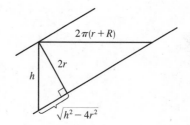

$h^2 r^2 = \pi^2 (r + R)^2 (h^2 - 4r^2) \quad \Rightarrow \quad h = \dfrac{2\pi r(r + R)}{\sqrt{\pi^2(r + R)^2 - r^2}}.$

If we parametrize the helix by $x(t) = (R + r) \cos t$, $y(t) = (R + r) \sin t$, then we must have $z(t) = [h/(2\pi)]t$. The length of

one complete cycle is

$$\ell = \int_0^{2\pi} \sqrt{[x'(t)]^2 + [y'(t)]^2 + [z'(t)]^2}\, dt = \int_0^{2\pi} \sqrt{(R + r)^2 + \left(\frac{h}{2\pi} \right)^2}\, dt = 2\pi \sqrt{(R + r)^2 + \left(\frac{h}{2\pi} \right)^2}$$

$$= 2\pi \sqrt{(R + r)^2 + \frac{r^2(R + r)^2}{\pi^2(R + r)^2 - r^2}} = 2\pi(R + r) \sqrt{1 + \frac{r^2}{\pi^2(R + r)^2 - r^2}} = \frac{2\pi^2(R + r)^2}{\sqrt{\pi^2(R + r)^2 - r^2}}$$

The number of complete cycles is $[\![L/\ell]\!]$, and so the shortest length along the spool is

$$h \left[\!\!\left[\frac{L}{\ell} \right]\!\!\right] = \frac{2\pi r(R + r)}{\sqrt{\pi^2(R + r)^2 - r^2}} \left[\!\!\left[\frac{L\sqrt{\pi^2(R + r)^2 - r^2}}{2\pi^2(R + r)^2} \right]\!\!\right]$$

15 ☐ PARTIAL DERIVATIVES

☐ ET 14

15.1 Functions of Several Variables

1. (a) From Table 1, $f(-15, 40) = -27$, which means that if the temperature is $-15°C$ and the wind speed is 40 km/h, then the air would feel equivalent to approximately $-27°C$ without wind.

(b) The question is asking: when the temperature is $-20°C$, what wind speed gives a wind-chill index of $-30°C$? From Table 1, the speed is 20 km/h.

(c) The question is asking: when the wind speed is 20 km/h, what temperature gives a wind-chill index of $-49°C$? From Table 1, the temperature is $-35°C$.

(d) The function $W = f(-5, v)$ means that we fix T at -5 and allow v to vary, resulting in a function of one variable. In other words, the function gives wind-chill index values for different wind speeds when the temperature is $-5°C$. From Table 1 (look at the row corresponding to $T = -5$), the function decreases and appears to approach a constant value as v increases.

(e) The function $W = f(T, 50)$ means that we fix v at 50 and allow T to vary, again giving a function of one variable. In other words, the function gives wind-chill index values for different temperatures when the wind speed is 50 km/h . From Table 1 (look at the column corresponding to $v = 50$), the function increases almost linearly as T increases.

2. (a) From Table 3, $f(95, 70) = 124$, which means that when the actual temperature is $95°F$ and the relative humidity is 70%, the perceived air temperature is approximately $124°F$.

(b) Looking at the row corresponding to $T = 90$, we see that $f(90, h) = 100$ when $h = 60$.

(c) Looking at the column corresponding to $h = 50$, we see that $f(T, 50) = 88$ when $T = 85$.

(d) $I = f(80, h)$ means that T is fixed at 80 and h is allowed to vary, resulting in a function of h that gives the humidex values for different relative humidities when the actual temperature is $80°F$. Similarly, $I = f(100, h)$ is a function of one variable that gives the humidex values for different relative humidities when the actual temperature is $100°F$. Looking at the rows of the table corresponding to $T = 80$ and $T = 100$, we see that $f(80, h)$ increases at a relatively constant rate of approximately $1°F$ per 10% relative humidity, while $f(100, h)$ increases more quickly (at first with an average rate of change of $5°F$ per 10% relative humidity) and at an increasing rate (approximately $12°F$ per 10% relative humidity for larger values of h).

3. If the amounts of labor and capital are both doubled, we replace L, K in the function with $2L, 2K$, giving

$$P(2L, 2K) = 1.01(2L)^{0.75}(2K)^{0.25} = 1.01(2^{0.75})(2^{0.25})L^{0.75}K^{0.25} = (2^1)1.01L^{0.75}K^{0.25} = 2P(L, K)$$

Thus, the production is doubled. It is also true for the general case $P(L, K) = bL^\alpha K^{1-\alpha}$:

$$P(2L, 2K) = b(2L)^\alpha(2K)^{1-\alpha} = b(2^\alpha)(2^{1-\alpha})L^\alpha K^{1-\alpha} = (2^{\alpha+1-\alpha})bL^\alpha K^{1-\alpha} = 2P(L, K).$$

4. We compare the values for the wind-chill index given by Table 1 with those given by the model function:

Modeled Wind-Chill Index Values $W(T, v)$

Wind Speed (km/h)

T \ V	5	10	15	20	25	30	40	50	60	70	80
5	4.08	2.66	1.74	1.07	0.52	0.05	−0.71	−1.33	−1.85	−2.30	−2.70
0	−1.59	−3.31	−4.42	−5.24	−5.91	−6.47	−7.40	−8.14	−8.77	−9.32	−9.80
−5	−7.26	−9.29	−10.58	−11.55	−12.34	−13.00	−14.08	−14.96	−15.70	−16.34	−16.91
−10	−12.93	−15.26	−16.75	−17.86	−18.76	−19.52	−20.77	−21.77	−22.62	−23.36	−24.01
−15	−18.61	−21.23	−22.91	−24.17	−25.19	−26.04	−27.45	−28.59	−29.54	−30.38	−31.11
−20	−24.28	−27.21	−29.08	−30.48	−31.61	−32.57	−34.13	−35.40	−36.47	−37.40	−38.22
−25	−29.95	−33.18	−35.24	−36.79	−38.04	−39.09	−40.82	−42.22	−43.39	−44.42	−45.32
−30	−35.62	−39.15	−41.41	−43.10	−44.46	−45.62	−47.50	−49.03	−50.32	−51.44	−52.43
−35	−41.30	−45.13	−47.57	−49.41	−50.89	−52.14	−54.19	−55.84	−57.24	−58.46	−59.53
−40	−46.97	−51.10	−53.74	−55.72	−57.31	−58.66	−60.87	−62.66	−64.17	−65.48	−66.64

Actual temperature (°C)

The values given by the function appear to be fairly close (within 0.5) to the values in Table 1.

5. (a) According to Table 4, $f(40, 15) = 25$, which means that if a 40-knot wind has been blowing in the open sea for 15 hours, it will create waves with estimated heights of 25 feet.

(b) $h = f(30, t)$ means we fix v at 30 and allow t to vary, resulting in a function of one variable. Thus here, $h = f(30, t)$ gives the wave heights produced by 30-knot winds blowing for t hours. From the table (look at the row corresponding to $v = 30$), the function increases but at a declining rate as t increases. In fact, the function values appear to be approaching a limiting value of approximately 19, which suggests that 30-knot winds cannot produce waves higher than about 19 feet.

(c) $h = f(v, 30)$ means we fix t at 30, again giving a function of one variable. So, $h = f(v, 30)$ gives the wave heights produced by winds of speed v blowing for 30 hours. From the table (look at the column corresponding to $t = 30$), the function appears to increase at an increasing rate, with no apparent limiting value. This suggests that faster winds (lasting 30 hours) always create higher waves.

6. (a) $f(1, 1) = \ln(1 + 1 - 1) = \ln 1 = 0$

(b) $f(e, 1) = \ln(e + 1 - 1) = \ln e = 1$

(c) $\ln(x + y - 1)$ is defined only when $x + y - 1 > 0$, that is, $y > 1 - x$. So the domain of f is $\{(x, y) \mid y > 1 - x\}$.

(d) Since $\ln(x + y - 1)$ can be any real number, the range is $\mathbb{R}$.

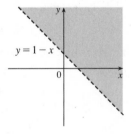

7. (a) $f(2,0) = 2^2 e^{3(2)(0)} = 4(1) = 4$

(b) Since both x^2 and the exponential function are defined everywhere, $x^2 e^{3xy}$ is defined for all choices of values for x and y. Thus the domain of f is $\mathbb{R}^2$.

(c) Because the range of $g(x,y) = 3xy$ is $\mathbb{R}$, and the range of e^x is $(0, \infty)$, the range of $e^{g(x,y)} = e^{3xy}$ is $(0, \infty)$. The range of x^2 is $[0, \infty)$, so the range of the product $x^2 e^{3xy}$ is $[0, \infty)$.

8. $\sqrt{1 + x - y^2}$ is defined only when $1 + x - y^2 \geq 0 \Rightarrow x \geq y^2 - 1$, so the domain of f is $\{(x, y) \mid x \geq y^2 - 1\}$, all those points on or to the right of the parabola $x = y^2 - 1$. The range of f is $[0, \infty)$.

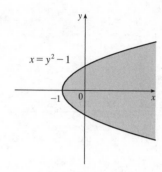

9. (a) $f(2, -1, 6) = e^{\sqrt{6 - 2^2 - (-1)^2}} = e^{\sqrt{1}} = e$.

(b) $e^{\sqrt{z - x^2 - y^2}}$ is defined when $z - x^2 - y^2 \geq 0 \Rightarrow z \geq x^2 + y^2$. Thus the domain of f is $\{(x, y, z) \mid z \geq x^2 + y^2\}$.

(c) Since $\sqrt{z - x^2 - y^2} \geq 0$, we have $e^{\sqrt{z - x^2 - y^2}} \geq 1$. Thus the range of f is $[1, \infty)$.

10. (a) $g(2, -2, 4) = \ln(25 - 2^2 - (-2)^2 - 4^2) = \ln 1 = 0$.

(b) For the logarithmic function to be defined, we need $25 - x^2 - y^2 - z^2 > 0$. Thus the domain of g is $\{(x, y, z) \mid x^2 + y^2 + z^2 < 25\}$, the interior of the sphere $x^2 + y^2 + z^2 = 25$.

(c) Since $0 < 25 - x^2 - y^2 - z^2 \leq 25$ for (x, y, z) in the domain of g, $\ln(25 - x^2 - y^2 - z^2) \leq \ln 25$. Thus the range of g is $(-\infty, \ln 25]$.

11. $\sqrt{x + y}$ is defined only when $x + y \geq 0$, or $y \geq -x$. So the domain of f is $\{(x, y) \mid y \geq -x\}$.

12. We need $xy \geq 0$, so $D = \{(x, y) \mid xy \geq 0\}$, the first and third quadrants.

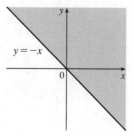

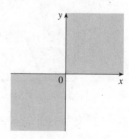

13. $\ln(9 - x^2 - 9y^2)$ is defined only when

$9 - x^2 - 9y^2 > 0$, or $\frac{1}{9}x^2 + y^2 < 1$. So the domain of f

is $\left\{ (x, y) \mid \frac{1}{9}x^2 + y^2 < 1 \right\}$, the interior of an ellipse.

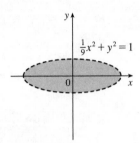

14. We need $y - x \geq 0$ or $y \geq x$ and $y + x > 0$ or $x > -y$.

Thus $D = \{ (x, y) \mid -y < x \leq y, y > 0 \}$.

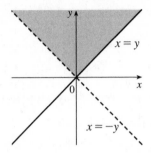

15. $\sqrt{1 - x^2}$ is defined only when $1 - x^2 \geq 0$, or $x^2 \leq 1$

$\Leftrightarrow\ -1 \leq x \leq 1$, and $\sqrt{1 - y^2}$ is defined only when

$1 - y^2 \geq 0$, or $y^2 \leq 1\ \Leftrightarrow\ -1 \leq y \leq 1$. Thus the

domain of f is $\{ (x, y) \mid -1 \leq x \leq 1,\ -1 \leq y \leq 1 \}$.

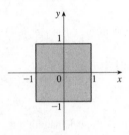

16. $\sqrt{y} + \sqrt{25 - x^2 - y^2}$ is defined only when $y \geq 0$ and

$25 - x^2 - y^2 \geq 0\ \Leftrightarrow\ x^2 + y^2 \leq 25$. So the domain

of f is $\{ (x, y) \mid x^2 + y^2 \leq 25,\ y \geq 0 \}$, a half disk of

radius 5.

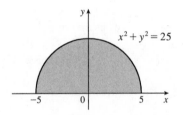

17. $\sqrt{y - x^2}$ is defined only when $y - x^2 \geq 0$, or $y \geq x^2$.

In addition, f is not defined if $1 - x^2 = 0\ \Rightarrow$

$x = \pm 1$. Thus the domain of f is

$\{ (x, y) \mid y \geq x^2,\ x \neq \pm 1 \}$.

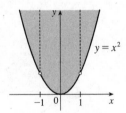

18. $\arcsin(x^2 + y^2 - 2)$ is defined only when

$-1 \leq x^2 + y^2 - 2 \leq 1\ \Leftrightarrow\ 1 \leq x^2 + y^2 \leq 3$. Thus

the domain of f is $\{ (x, y) \mid 1 \leq x^2 + y^2 \leq 3 \}$.

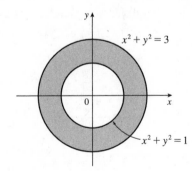

19. We need $1 - x^2 - y^2 - z^2 \geq 0$ or $x^2 + y^2 + z^2 \leq 1$,

so $D = \left\{ (x, y, z) \mid x^2 + y^2 + z^2 \leq 1 \right\}$ (the points inside

or on the sphere of radius 1, center the origin).

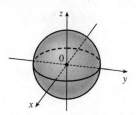

20. f is defined only when $16 - 4x^2 - 4y^2 - z^2 > 0$ ⇒

$$\frac{x^2}{4} + \frac{y^2}{4} + \frac{z^2}{16} < 1. \text{ Thus,}$$

$$D = \left\{ (x, y, z) \,\middle|\, \frac{x^2}{4} + \frac{y^2}{4} + \frac{z^2}{16} < 1 \right\}, \text{ that is, the points}$$

inside the ellipsoid $\dfrac{x^2}{4} + \dfrac{y^2}{4} + \dfrac{z^2}{16} = 1$.

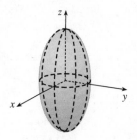

21. $z = 3$, a horizontal plane through the point $(0, 0, 3)$.

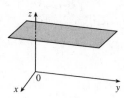

22. $z = y$, a plane which intersects the yz-plane in the line

$z = y$, $x = 0$. The portion of this plane that lies in the

first octant is shown.

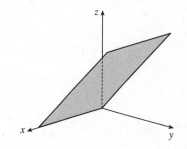

23. $z = 10 - 4x - 5y$ or $4x + 5y + z = 10$, a plane with

intercepts 2.5, 2, and 10.

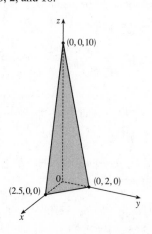

24. $z = \cos x$, a "wave."

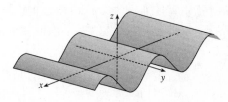

25. $z = y^2 + 1$, a parabolic cylinder

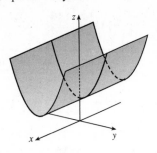

26. $z = 3 - x^2 - y^2$, a circular paraboloid with vertex at $(0, 0, 3)$.

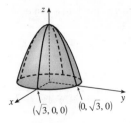

27. $z = 4x^2 + y^2 + 1$, an elliptic paraboloid with vertex at $(0, 0, 1)$.

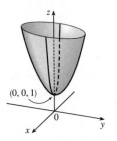

28. $z = \sqrt{16 - x^2 - 16y^2}$ so $z \geq 0$ and $z^2 + x^2 + 16y^2 = 16$, the top half of an ellipsoid.

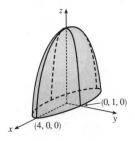

29. $z = \sqrt{x^2 + y^2}$ so $x^2 + y^2 = z^2$ and $z \geq 0$, the top half of a right circular cone.

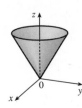

30. All six graphs have different traces in the planes $x = 0$ and $y = 0$, so we investigate these for each function.

(a) $f(x, y) = |x| + |y|$. The trace in $x = 0$ is $z = |y|$, and in $y = 0$ is $z = |x|$, so it must be graph VI.

(b) $f(x, y) = |xy|$. The trace in $x = 0$ is $z = 0$, and in $y = 0$ is $z = 0$, so it must be graph V.

(c) $f(x, y) = \dfrac{1}{1 + x^2 + y^2}$. The trace in $x = 0$ is $z = \dfrac{1}{1 + y^2}$, and in $y = 0$ is $z = \dfrac{1}{1 + x^2}$. In addition, we can see that f is close to 0 for large values of x and y, so this is graph I.

(d) $f(x, y) = (x^2 - y^2)^2$. The trace in $x = 0$ is $z = y^4$, and in $y = 0$ is $z = x^4$. Both graph II and graph IV seem plausible; notice the trace in $z = 0$ is $0 = (x^2 - y^2)^2 \ \Rightarrow \ y = \pm x$, so it must be graph IV.

(e) $f(x, y) = (x - y)^2$. The trace in $x = 0$ is $z = y^2$, and in $y = 0$ is $z = x^2$. Both graph II and graph IV seem plausible; notice the trace in $z = 0$ is $0 = (x - y)^2 \ \Rightarrow \ y = x$, so it must be graph II.

(f) $f(x, y) = \sin(|x| + |y|)$. The trace in $x = 0$ is $z = \sin|y|$, and in $y = 0$ is $z = \sin|x|$. In addition, notice that the oscillating nature of the graph is characteristic of trigonometric functions. So this is graph III.

31. The point $(-3, 3)$ lies between the level curves with z-values 50 and 60. Since the point is a little closer to the level curve with $z = 60$, we estimate that $f(-3, 3) \approx 56$. The point $(3, -2)$ appears to be just about halfway between the level curves with z-values 30 and 40, so we estimate $f(3, -2) \approx 35$. The graph rises as we approach the origin, gradually from above, steeply from below.

32. If we start at the origin and move along the x-axis, for example, the z-values of a cone centered at the origin increase at a constant rate, so we would expect its level curves to be equally spaced. A paraboloid with vertex the origin, on the other hand, has z-values which change slowly near the origin and more quickly as we move farther away. Thus, we would expect its level curves near the origin to be spaced more widely apart than those farther from the origin. Therefore contour map I must correspond to the paraboloid, and contour map II the cone.

33. Near A, the level curves are very close together, indicating that the terrain is quite steep. At B, the level curves are much farther apart, so we would expect the terrain to be much less steep than near A, perhaps almost flat.

34.

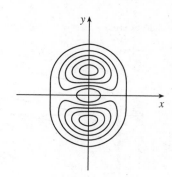

35.

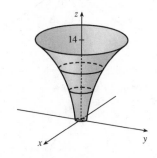

36.

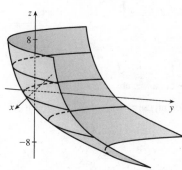

37.

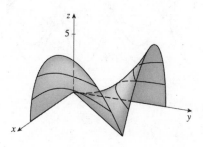

38.

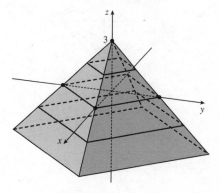

39. The level curves are $(y - 2x)^2 = k$ or $y = 2x \pm \sqrt{k}$, $k \geq 0$, a family of pairs of parallel lines.

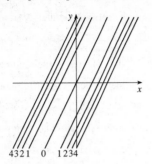

40. The level curves are $x^3 - y = k$ or $y = x^3 - k$, a family of cubic curves.

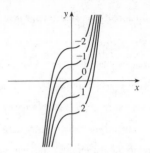

41. The level curves are $y - \ln x = k$ or $y = \ln x + k$.

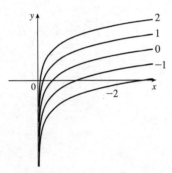

42. The level curves are $e^{y/x} = k$ or equivalently $y = x \ln k$ ($x \neq 0$), a family of lines with slope $\ln k$ ($k > 0$) without the origin.

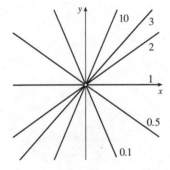

43. The level curves are $ye^x = k$ or $y = ke^{-x}$, a family of exponential curves.

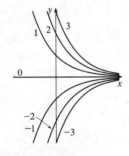

44. $k = y \sec x$ or $y = k \cos x$, $x \neq \frac{\pi}{2} + n\pi$ [n an integer].

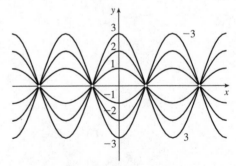

45. The level curves are $\sqrt{y^2 - x^2} = k$ or $y^2 - x^2 = k^2$, $k \geq 0$. When $k = 0$ the level curve is the pair of lines $y = \pm x$. For $k > 0$, the level curves are hyperbolas with axis the y-axis.

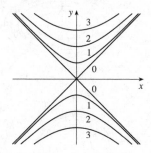

46. For $k \neq 0$ and $(x, y) \neq (0, 0)$, $k = \dfrac{y}{x^2 + y^2}$ ⇔

$$x^2 + y^2 - \frac{y}{k} = 0 \quad\Leftrightarrow\quad x^2 + \left(y - \frac{1}{2k}\right)^2 = \frac{1}{4k^2},$$ a family of circles with center $\left(0, \frac{1}{2k}\right)$ and radius $\frac{1}{2k}$ (without the origin). If $k = 0$, the level curve is the x-axis.

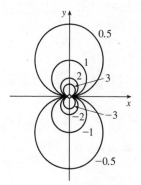

47. The contour map consists of the level curves $k = x^2 + 9y^2$, a family of ellipses with major axis the x-axis. (Or, if $k = 0$, the origin.)

The graph of $f(x, y)$ is the surface $z = x^2 + 9y^2$, an elliptic paraboloid.

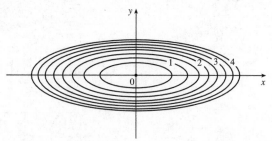

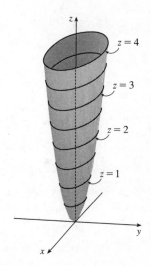

If we visualize lifting each ellipse $k = x^2 + 9y^2$ of the contour map to the plane $z = k$, we have horizontal traces that indicate the shape of the graph of f.

48.

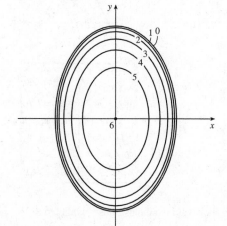

The contour map consists of the level curves $k = \sqrt{36 - 9x^2 - 4y^2}$ ⇒ $9x^2 + 4y^2 = 36 - k^2$, $k \geq 0$, a family of ellipses with major axis the y-axis. (Or, if $k = 6$, the origin.)

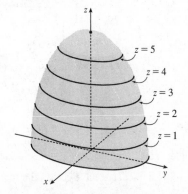

The graph of $f(x, y)$ is the surface $z = \sqrt{36 - 9x^2 - 4y^2}$, or equivalently the upper half of the ellipsoid

$9x^2 + 4y^2 + z^2 = 36$. If we visualize lifting each ellipse $k = \sqrt{36 - 9x^2 - 4y^2}$ of the contour map to the plane $z = k$, we have horizontal traces that indicate the shape of the graph of f.

49. The isothermals are given by $k = 100/(1 + x^2 + 2y^2)$ or

$x^2 + 2y^2 = (100 - k)/k$ [$0 < k \le 100$], a family of ellipses.

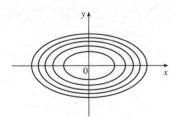

50. The equipotential curves are $k = \dfrac{c}{\sqrt{r^2 - x^2 - y^2}}$ or

$x^2 + y^2 = r^2 - \left(\dfrac{c}{k}\right)^2$, a family of circles ($k \ge c/r$).

Note: As $k \to \infty$, the radius of the circle approaches r.

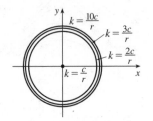

51. $f(x, y) = e^{-x^2} + e^{-2y^2}$

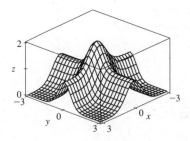
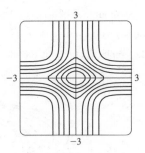

52. $f(x, y) = (1 - 3x^2 + y^2)e^{1 - x^2 - y^2}$

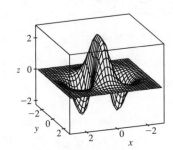
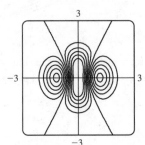

53. $f(x, y) = xy^2 - x^3$

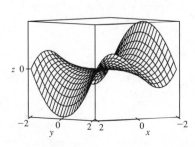
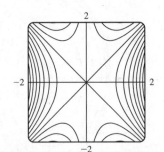

The traces parallel to the yz-plane (such as the left-front trace in the graph above) are parabolas; those parallel to the xz-plane (such as the right-front trace) are cubic curves. The surface is called a monkey saddle because a monkey sitting on the surface near the origin has places for both legs and tail to rest.

54. $f(x, y) = xy^3 - yx^3$

The traces parallel to either the yz-plane or the xz-plane are cubic curves.

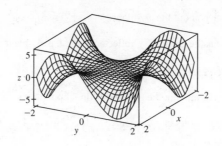 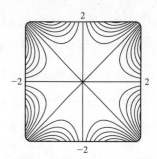

55. (a) C   (b) II

Reasons: This function is periodic in both x and y, and the function is the same when x is interchanged with y, so its graph is symmetric about the plane $y = x$. In addition, the function is 0 along the x- and y-axes. These conditions are satisfied only by C and II.

56. (a) A   (b) IV

Reasons: This function is periodic in y but not x, a condition satisfied only by A and IV. Also, note that traces in $x = k$ are cosine curves with amplitude that increases as x increases.

57. (a) F   (b) I

Reasons: This function is periodic in both x and y but is constant along the lines $y = x + k$, a condition satisfied only by F and I.

58. (a) E   (b) III

Reasons: This function is periodic in both x and y, but unlike the function in Exercise 57, it is not constant along lines such as $y = x + \pi$, so the contour map is III. Also notice that traces in $y = k$ are vertically shifted copies of the sine wave $z = \sin x$, so the graph must be E.

59. (a) B   (b) VI

Reasons: This function is 0 along the lines $x = \pm 1$ and $y = \pm 1$. The only contour map in which this could occur is VI. Also note that the trace in the xz-plane is the parabola $z = 1 - x^2$ and the trace in the yz-plane is the parabola $z = 1 - y^2$, so the graph is B.

60. (a) D   (b) V

Reasons: This function is not periodic, ruling out the graphs in A, C, E, and F. Also, the values of z approach 0 as we use points farther from the origin. The only graph that shows this behavior is D, which corresponds to V.

61. $k = x + 3y + 5z$ is a family of parallel planes with normal vector $\langle 1, 3, 5 \rangle$.

62. $k = x^2 + 3y^2 + 5z^2$ is a family of ellipsoids for $k > 0$ and the origin for $k = 0$.

63. $k = x^2 - y^2 + z^2$ are the equations of the level surfaces. For $k = 0$, the surface is a right circular cone with vertex the origin and axis the y-axis. For $k > 0$, we have a family of hyperboloids of one sheet with axis the y-axis. For $k < 0$, we have a family of hyperboloids of two sheets with axis the y-axis.

64. $k = x^2 - y^2$ is a family of hyperbolic cylinders oriented vertically. The cross section of each level surface in the xy-plane is a hyperbola with axis the x-axis when $k > 0$ and y-axis when $k < 0$. (When $k = 0$, the level surface is two intersecting vertical planes.)

65. (a) The graph of g is the graph of f shifted upward 2 units.

(b) The graph of g is the graph of f stretched vertically by a factor of 2.

(c) The graph of g is the graph of f reflected about the xy-plane.

(d) The graph of $g(x, y) = -f(x, y) + 2$ is the graph of f reflected about the xy-plane and then shifted upward 2 units.

66. (a) The graph of g is the graph of f shifted 2 units in the positive x-direction.

(b) The graph of g is the graph of f shifted 2 units in the negative y-direction.

(c) The graph of g is the graph of f shifted 3 units in the negative x-direction and 4 units in the positive y-direction.

67. $f(x, y) = 3x - x^4 - 4y^2 - 10xy$

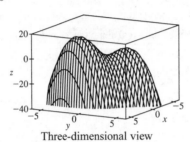

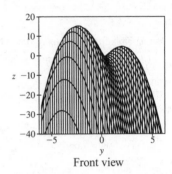

Three-dimensional view   Front view

It does appear that the function has a maximum value, at the higher of the two "hilltops." From the front view graph, the maximum value appears to be approximately 15. Both hilltops could be considered local maximum points, as the values of f there are larger than at the neighboring points. There does not appear to be any local minimum point; although the valley shape between the two peaks looks like a minimum of some kind, some neighboring points have lower function values.

68. $f(x, y) = xye^{-x^2-y^2}$

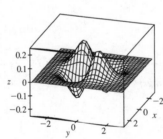

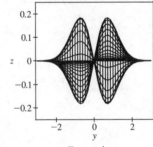

Three-dimensional view   Front view

The function does have a maximum value, which it appears to achieve at two different points (the two "hilltops"). From the front view graph, we can estimate the maximum value to be approximately 0.18. These same two points can also be considered local maximum points. The two "valley bottoms" visible in the graph can be considered local minimum points, as all the neighboring points give greater values of f.

69.

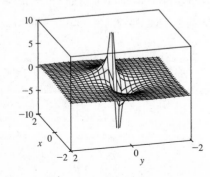

$f(x, y) = \dfrac{x + y}{x^2 + y^2}$. As both x and y become large, the function values appear to approach 0, regardless of which direction is considered. As (x, y) approaches the origin, the graph exhibits asymptotic behavior. From some directions, $f(x, y) \to \infty$, while in others $f(x, y) \to -\infty$. (These are the vertical spikes visible in the graph.) If the graph is examined carefully, however, one can see that $f(x, y)$ approaches 0 along the line $y = -x$.

70.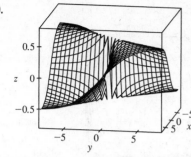

$f(x, y) = \dfrac{xy}{x^2 + y^2}$. The graph exhibits different limiting values as x and y become large or as (x, y) approaches the origin, depending on the direction being examined. For example, although f is undefined at the origin, the function values appear to be $\frac{1}{2}$ along the line $y = x$, regardless of the distance from the origin. Along the line $y = -x$, the value is always $-\frac{1}{2}$. Along the axes, $f(x, y) = 0$ for all values of (x, y) except the origin. Other directions, heading toward the origin or away from the origin, give various limiting values between $-\frac{1}{2}$ and $\frac{1}{2}$.

71. $f(x, y) = e^{cx^2 + y^2}$. First, if $c = 0$, the graph is the cylindrical surface $z = e^{y^2}$ (whose level curves are parallel lines). When $c > 0$, the vertical trace above the y-axis remains fixed while the sides of the surface in the x-direction "curl" upward, giving the graph a shape resembling an elliptic paraboloid. The level curves of the surface are ellipses centered at the origin.

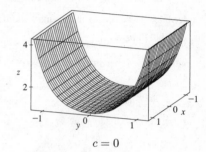

$c = 0$

For $0 < c < 1$, the ellipses have major axis the x-axis and the eccentricity increases as $c \to 0$.

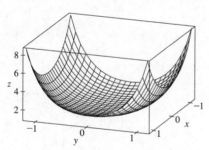

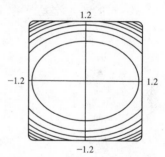

$c = 0.5$ (level curves in increments of 1)

For $c = 1$ the level curves are circles centered at the origin.

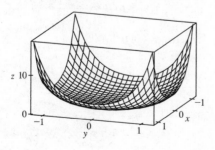

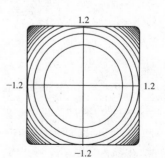

$c = 1$ (level curves in increments of 1)

When $c > 1$, the level curves are ellipses with major axis the y-axis, and the eccentricity increases as c increases.

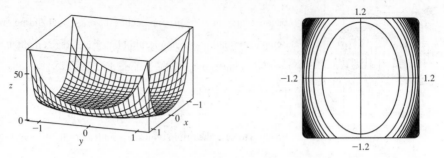

$c = 2$ (level curves in increments of 4)

For values of $c < 0$, the sides of the surface in the x-direction curl downward and approach the xy-plane (while the vertical trace $x = 0$ remains fixed), giving a saddle-shaped appearance to the graph near the point $(0, 0, 1)$. The level curves consist of a family of hyperbolas. As c decreases, the surface becomes flatter in the x-direction and the surface's approach to the curve in the trace $x = 0$ becomes steeper, as the graphs demonstrate.

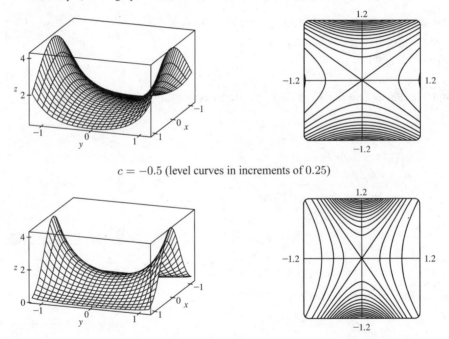

$c = -0.5$ (level curves in increments of 0.25)

$c = -2$ (level curves in increments of 0.25)

72. $z = (ax^2 + by^2)e^{-x^2 - y^2}$. There are only three basic shapes which can be obtained (the fourth and fifth graphs are the reflections of the first and second ones in the xy-plane). Interchanging a and b rotates the graph by $90°$ about the z-axis.

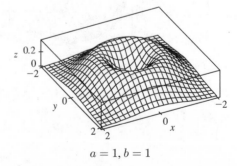

$a = 1, b = 1$

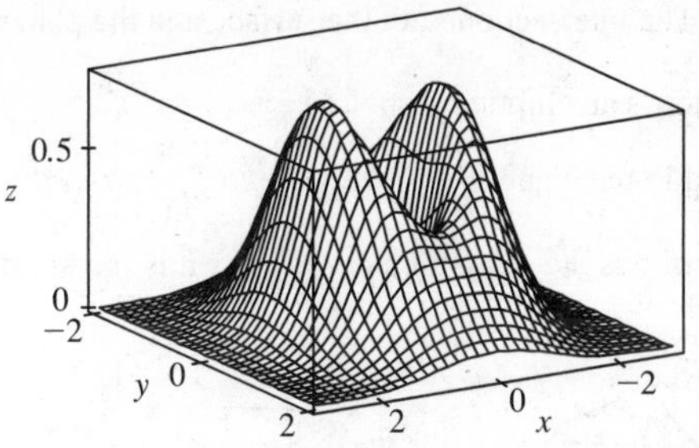

$a = 2, b = 1$

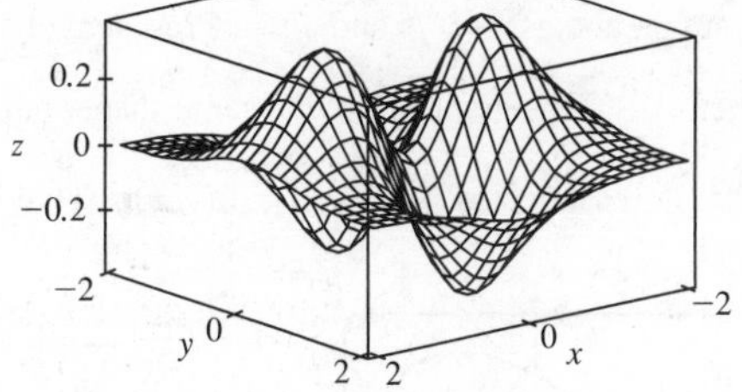

$a = 1, b = -1$

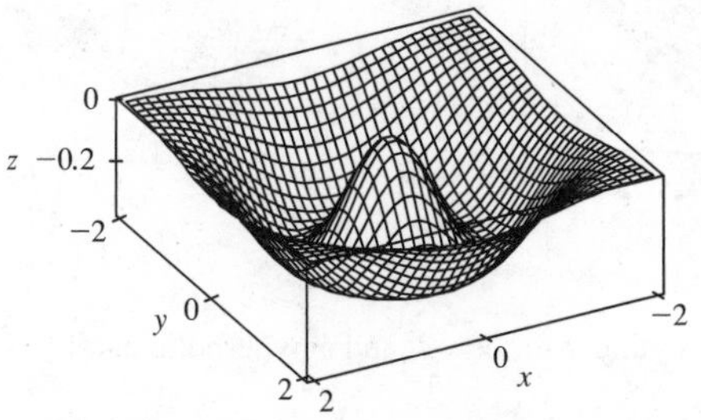

$a = -1, b = -1$

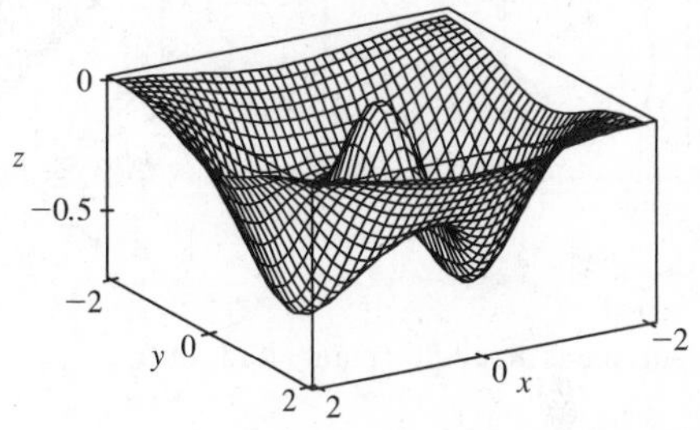

$a = -2, b = -1$

If a and b are both positive $(a \neq b)$, we see that the graph has two maximum points whose height increases as a and b increase. If a and b have opposite signs, the graph has two maximum points and two minimum points, and if a and b are both negative, the graph has one maximum point and two minimum points.

73. $z = x^2 + y^2 + cxy$. When $c < -2$, the surface intersects the plane $z = k \neq 0$ in a hyperbola. (See graph below.) It intersects the plane $x = y$ in the parabola $z = (2 + c)x^2$, and the plane $x = -y$ in the parabola $z = (2 - c)x^2$. These parabolas open in opposite directions, so the surface is a hyperbolic paraboloid.

When $c = -2$ the surface is $z = x^2 + y^2 - 2xy = (x - y)^2$. So the surface is constant along each line $x - y = k$. That is, the surface is a cylinder with axis $x - y = 0$, $z = 0$. The shape of the cylinder is determined by its intersection with the plane $x + y = 0$, where $z = 4x^2$, and hence the cylinder is parabolic with minima of 0 on the line $y = x$.

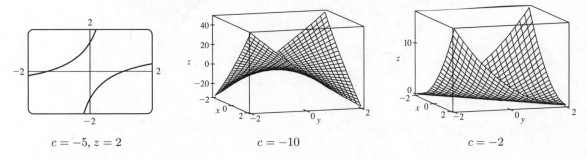

$c = -5, z = 2$

$c = -10$

$c = -2$

When $-2 < c \leq 0$, $z \geq 0$ for all x and y. If x and y have the same sign, then $x^2 + y^2 + cxy \geq x^2 + y^2 - 2xy = (x - y)^2 \geq 0$. If they have opposite signs, then $cxy \geq 0$. The intersection with the surface and the plane $z = k > 0$ is an ellipse (see graph below). The intersection with the surface and the planes $x = 0$ and

$y = 0$ are parabolas $z = y^2$ and $z = x^2$ respectively, so the surface is an elliptic paraboloid.

When $c > 0$ the graphs have the same shape, but are reflected in the plane $x = 0$, because

$x^2 + y^2 + cxy = (-x)^2 + y^2 + (-c)(-x)y$. That is, the value of z is the same for c at (x, y) as it is for $-c$ at $(-x, y)$.

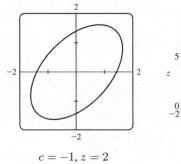

$c = -1, z = 2$

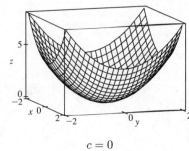

$c = 0$

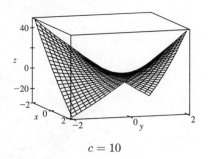

$c = 10$

So the surface is an elliptic paraboloid for $0 < c < 2$, a parabolic cylinder for $c = 2$, and a hyperbolic paraboloid for $c > 2$.

74. First, we graph $f(x, y) = \sqrt{x^2 + y^2}$.

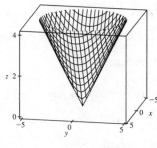

$$f(x, y) = \sqrt{x^2 + y^2}$$

Graphs of the other four functions follow.

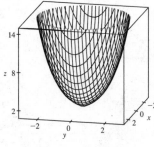

$$f(x, y) = e^{\sqrt{x^2 + y^2}}$$

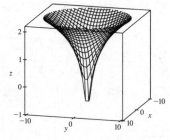

$$f(x, y) = \ln \sqrt{x^2 + y^2}$$

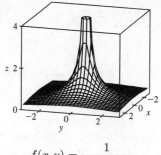

$$f(x, y) = \sin\left(\sqrt{x^2 + y^2}\right) \qquad\qquad f(x, y) = \frac{1}{\sqrt{x^2 + y^2}}$$

Notice that each graph $f(x, y) = g\left(\sqrt{x^2 + y^2}\right)$ exhibits radial symmetry about the z-axis and the trace in the xz-plane for

$x \geq 0$ is the graph of $z = g(x)$, $x \geq 0$. This suggests that the graph of $f(x, y) = g\left(\sqrt{x^2 + y^2}\right)$ is obtained from the graph

of g by graphing $z = g(x)$ in the xz-plane and rotating the curve about the z-axis.

75. (a) $P = bL^\alpha K^{1-\alpha} \;\Rightarrow\; \dfrac{P}{K} = bL^\alpha K^{-\alpha} \;\Rightarrow\; \dfrac{P}{K} = b\left(\dfrac{L}{K}\right)^\alpha \;\Rightarrow\; \ln\dfrac{P}{K} = \ln\left(b\left(\dfrac{L}{K}\right)^\alpha\right) \;\Rightarrow\;$

$\ln\dfrac{P}{K} = \ln b + \alpha \ln\left(\dfrac{L}{K}\right)$

(b) We list the values for $\ln(L/K)$ and $\ln(P/K)$ for the years 1899–1922. (Historically, these values were rounded to
2 decimal places.)

Year	$x = \ln(L/K)$	$y = \ln(P/K)$
1899	0	0
1900	−0.02	−0.06
1901	−0.04	−0.02
1902	−0.04	0
1903	−0.07	−0.05
1904	−0.13	−0.12
1905	−0.18	−0.04
1906	−0.20	−0.07
1907	−0.23	−0.15
1908	−0.41	−0.38
1909	−0.33	−0.24
1910	−0.35	−0.27

Year	$x = \ln(L/K)$	$y = \ln(P/K)$
1911	−0.38	−0.34
1912	−0.38	−0.24
1913	−0.41	−0.25
1914	−0.47	−0.37
1915	−0.53	−0.34
1916	−0.49	−0.28
1917	−0.53	−0.39
1918	−0.60	−0.50
1919	−0.68	−0.57
1920	−0.74	−0.57
1921	−1.05	−0.85
1922	−0.98	−0.59

After entering the (x, y) pairs into a calculator or CAS, the resulting least squares regression line through the points is
approximately $y = 0.75136x + 0.01053$, which we round to $y = 0.75x + 0.01$.

(c) Comparing the regression line from part (b) to the equation $y = \ln b + \alpha x$ with $x = \ln(L/K)$ and $y = \ln(P/K)$, we have

$\alpha = 0.75$ and $\ln b = 0.01 \;\Rightarrow\; b = e^{0.01} \approx 1.01$. Thus, the Cobb-Douglas production function is

$P = bL^\alpha K^{1-\alpha} = 1.01 L^{0.75} K^{0.25}$.

15.2   Limits and Continuity

1. In general, we can't say anything about $f(3, 1)$! $\lim\limits_{(x,y)\to(3,1)} f(x,y) = 6$ means that the values of $f(x,y)$ approach 6 as

(x, y) approaches, but is not equal to, $(3, 1)$. If f is continuous, we know that $\lim\limits_{(x,y)\to(a,b)} f(x,y) = f(a,b)$, so

$\lim\limits_{(x,y)\to(3,1)} f(x,y) = f(3,1) = 6$.

2. (a) The outdoor temperature as a function of longitude, latitude, and time is continuous. Small changes in longitude, latitude, or time can produce only small changes in temperature, as the temperature doesn't jump abruptly from one value to another.

(b) Elevation is not necessarily continuous. If we think of a cliff with a sudden drop-off, a very small change in longitude or latitude can produce a comparatively large change in elevation, without all the intermediate values being attained. Elevation *can* jump from one value to another.

(c) The cost of a taxi ride is usually discontinuous. The cost normally increases in jumps, so small changes in distance traveled or time can produce a jump in cost. A graph of the function would show breaks in the surface.

3. We make a table of values of

$$f(x, y) = \frac{x^2 y^3 + x^3 y^2 - 5}{2 - xy} \text{ for a set}$$

of (x, y) points near the origin.

x \ y	−0.2	−0.1	−0.05	0	0.05	0.1	0.2
−0.2	−2.551	−2.525	−2.513	−2.500	−2.488	−2.475	−2.451
−0.1	−2.525	−2.513	−2.506	−2.500	−2.494	−2.488	−2.475
−0.05	−2.513	−2.506	−2.503	−2.500	−2.497	−2.494	−2.488
0	−2.500	−2.500	−2.500		−2.500	−2.500	−2.500
0.05	−2.488	−2.494	−2.497	−2.500	−2.503	−2.506	−2.513
0.1	−2.475	−2.488	−2.494	−2.500	−2.506	−2.513	−2.525
0.2	−2.451	−2.475	−2.488	−2.500	−2.513	−2.525	−2.551

As the table shows, the values of $f(x, y)$ seem to approach -2.5 as (x, y) approaches the origin from a variety of different directions. This suggests that $\lim\limits_{(x,y)\to(0,0)} f(x,y) = -2.5$. Since f is a rational function, it is continuous on its domain. f is

defined at $(0, 0)$, so we can use direct substitution to establish that $\lim\limits_{(x,y)\to(0,0)} f(x,y) = \dfrac{0^2 0^3 + 0^3 0^2 - 5}{2 - 0 \cdot 0} = -\dfrac{5}{2}$, verifying

our guess.

4. We make a table of values of

$$f(x, y) = \frac{2xy}{x^2 + 2y^2} \text{ for a set of } (x, y)$$

points near the origin.

x \ y	−0.3	−0.2	−0.1	0	0.1	0.2	0.3
−0.3	0.667	0.706	0.545	0.000	−0.545	−0.706	−0.667
−0.2	0.545	0.667	0.667	0.000	−0.667	−0.667	−0.545
−0.1	0.316	0.444	0.667	0.000	−0.667	−0.444	−0.316
0	0.000	0.000	0.000		0.000	0.000	0.000
0.1	−0.316	−0.444	−0.667	0.000	0.667	0.444	0.316
0.2	−0.545	−0.667	−0.667	0.000	0.667	0.667	0.545
0.3	−0.667	−0.706	−0.545	0.000	0.545	0.706	0.667

It appears from the table that the values of $f(x, y)$ are not approaching a single value as (x, y) approaches the origin. For verification, if we first approach $(0, 0)$ along the x-axis, we have $f(x, 0) = 0$, so $f(x, y) \to 0$. But if we approach $(0, 0)$ along the line $y = x$, $f(x, x) = \dfrac{2x^2}{x^2 + 2x^2} = \dfrac{2}{3}$ $(x \neq 0)$, so $f(x, y) \to \dfrac{2}{3}$. Since f approaches different values along different paths to the origin, this limit does not exist.

5. $f(x, y) = 5x^3 - x^2 y^2$ is a polynomial, and hence continuous, so $\displaystyle\lim_{(x,y) \to (1,2)} f(x, y) = f(1, 2) = 5(1)^3 - (1)^2 (2)^2 = 1.$

6. $-xy$ is a polynomial and therefore continuous. Since e^t is a continuous function, the composition e^{-xy} is also continuous. Similarly, $x + y$ is a polynomial and $\cos t$ is a continuous function, so the composition $\cos(x + y)$ is continuous. The product of continuous functions is continuous, so $f(x, y) = e^{-xy} \cos(x + y)$ is a continuous function and

$\displaystyle\lim_{(x,y) \to (1,-1)} f(x, y) = f(1, -1) = e^{-(1)(-1)} \cos(1 + (-1)) = e^1 \cos 0 = e.$

7. $f(x, y) = \dfrac{4 - xy}{x^2 + 3y^2}$ is a rational function and hence continuous on its domain.

$(2, 1)$ is in the domain of f, so f is continuous there and $\displaystyle\lim_{(x,y) \to (2,1)} f(x, y) = f(2, 1) = \dfrac{4 - (2)(1)}{(2)^2 + 3(1)^2} = \dfrac{2}{7}.$

8. $\dfrac{1 + y^2}{x^2 + xy}$ is a rational function and hence continuous on its domain, which includes $(1, 0)$. $\ln t$ is a continuous function for $t > 0$, so the composition $f(x, y) = \ln\left(\dfrac{1 + y^2}{x^2 + xy}\right)$ is continuous wherever $\dfrac{1 + y^2}{x^2 + xy} > 0$. In particular, f is continuous at $(1, 0)$ and so $\displaystyle\lim_{(x,y) \to (1,0)} f(x, y) = f(1, 0) = \ln\left(\dfrac{1 + 0^2}{1^2 + 1 \cdot 0}\right) = \ln\dfrac{1}{1} = 0.$

9. $f(x, y) = y^4 / (x^4 + 3y^4)$. First approach $(0, 0)$ along the x-axis. Then $f(x, 0) = 0/x^4 = 0$ for $x \neq 0$, so $f(x, y) \to 0$. Now approach $(0, 0)$ along the y-axis. Then for $y \neq 0$, $f(0, y) = y^4 / 3y^4 = 1/3$, so $f(x, y) \to 1/3$. Since f has two different limits along two different lines, the limit does not exist.

10. $f(x, y) = (x^2 + \sin^2 y) / (2x^2 + y^2)$. First approach $(0, 0)$ along the x-axis. Then $f(x, 0) = x^2 / 2x^2 = \dfrac{1}{2}$ for $x \neq 0$, so $f(x, y) \to \dfrac{1}{2}$. Next approach $(0, 0)$ along the y-axis. For $y \neq 0$, $f(0, y) = \dfrac{\sin^2 y}{y^2} = \left(\dfrac{\sin y}{y}\right)^2$ and $\displaystyle\lim_{y \to 0} \dfrac{\sin y}{y} = 1$, so $f(x, y) \to 1$. Since f has two different limits along two different lines, the limit does not exist.

11. $f(x, y) = (xy \cos y) / (3x^2 + y^2)$. On the x-axis, $f(x, 0) = 0$ for $x \neq 0$, so $f(x, y) \to 0$ as $(x, y) \to (0, 0)$ along the x-axis. Approaching $(0, 0)$ along the line $y = x$, $f(x, x) = (x^2 \cos x) / 4x^2 = \dfrac{1}{4} \cos x$ for $x \neq 0$, so $f(x, y) \to \dfrac{1}{4}$ along this line. Thus the limit does not exist.

12. $f(x, y) = 6x^3 y / (2x^4 + y^4)$. On the x-axis, $f(x, 0) = 0$ for $x \neq 0$, so $f(x, y) \to 0$ as $(x, y) \to (0, 0)$ along the x-axis. Approaching $(0, 0)$ along the line $y = x$ gives $f(x, x) = 6x^4 / (3x^4) = 2$ for $x \neq 0$, so along this line $f(x, y) \to 2$ as $(x, y) \to (0, 0)$. Thus the limit does not exist.

13. $f(x, y) = \dfrac{xy}{\sqrt{x^2 + y^2}}$. We can see that the limit along any line through $(0, 0)$ is 0, as well as along other paths through

$(0, 0)$ such as $x = y^2$ and $y = x^2$. So we suspect that the limit exists and equals 0; we use the Squeeze Theorem to prove our

assertion. $0 \leq \left| \dfrac{xy}{\sqrt{x^2 + y^2}} \right| \leq |x|$ since $|y| \leq \sqrt{x^2 + y^2}$, and $|x| \to 0$ as $(x, y) \to (0, 0)$. So $\displaystyle\lim_{(x,y) \to (0,0)} f(x, y) = 0$.

14. $f(x, y) = \dfrac{x^4 - y^4}{x^2 + y^2} = \dfrac{(x^2 + y^2)(x^2 - y^2)}{x^2 + y^2} = x^2 - y^2$ for $(x, y) \neq (0, 0)$. Thus the limit as $(x, y) \to (0, 0)$ is 0.

15. Let $f(x, y) = \dfrac{x^2 y e^y}{x^4 + 4y^2}$. Then $f(x, 0) = 0$ for $x \neq 0$, so $f(x, y) \to 0$ as $(x, y) \to (0, 0)$ along the x-axis. Approaching

$(0, 0)$ along the y-axis or the line $y = x$ also gives a limit of 0. But $f(x, x^2) = \dfrac{x^2 x^2 e^{x^2}}{x^4 + 4(x^2)^2} = \dfrac{x^4 e^{x^2}}{5x^4} = \dfrac{e^{x^2}}{5}$ for $x \neq 0$, so

$f(x, y) \to e^0 / 5 = \frac{1}{5}$ as $(x, y) \to (0, 0)$ along the parabola $y = x^2$. Thus the limit doesn't exist.

16. We can use the Squeeze Theorem to show that $\displaystyle\lim_{(x,y) \to (0,0)} \dfrac{x^2 \sin^2 y}{x^2 + 2y^2} = 0$:

$0 \leq \dfrac{x^2 \sin^2 y}{x^2 + 2y^2} \leq \sin^2 y$ since $\dfrac{x^2}{x^2 + 2y^2} \leq 1$, and $\sin^2 y \to 0$ as $(x, y) \to (0, 0)$, so $\displaystyle\lim_{(x,y) \to (0,0)} \dfrac{x^2 \sin^2 y}{x^2 + 2y^2} = 0$.

17. $\displaystyle\lim_{(x,y) \to (0,0)} \dfrac{x^2 + y^2}{\sqrt{x^2 + y^2 + 1} - 1} = \lim_{(x,y) \to (0,0)} \dfrac{x^2 + y^2}{\sqrt{x^2 + y^2 + 1} - 1} \cdot \dfrac{\sqrt{x^2 + y^2 + 1} + 1}{\sqrt{x^2 + y^2 + 1} + 1}$

$= \displaystyle\lim_{(x,y) \to (0,0)} \dfrac{(x^2 + y^2)\left(\sqrt{x^2 + y^2 + 1} + 1\right)}{x^2 + y^2} = \lim_{(x,y) \to (0,0)} \left(\sqrt{x^2 + y^2 + 1} + 1\right) = 2$

18. $f(x, y) = xy^4 / (x^2 + y^8)$. On the x-axis, $f(x, 0) = 0$ for $x \neq 0$, so $f(x, y) \to 0$ as $(x, y) \to (0, 0)$ along the x-axis.

Approaching $(0, 0)$ along the curve $x = y^4$ gives $f(y^4, y) = y^8 / 2y^8 = \frac{1}{2}$ for $y \neq 0$, so along this path $f(x, y) \to \frac{1}{2}$ as

$(x, y) \to (0, 0)$. Thus the limit does not exist.

19. e^{-xy} and $\sin(\pi z/2)$ are each compositions of continuous functions, and hence continuous, so their product

$f(x, y, z) = e^{-xy} \sin(\pi z/2)$ is a continuous function. Then

$\displaystyle\lim_{(x,y,z) \to (3,0,1)} f(x, y, z) = f(3, 0, 1) = e^{-(3)(0)} \sin(\pi \cdot 1/2) = 1$.

20. $f(x, y, z) = \dfrac{x^2 + 2y^2 + 3z^2}{x^2 + y^2 + z^2}$. Then $f(x, 0, 0) = \dfrac{x^2 + 0 + 0}{x^2 + 0 + 0} = 1$ for $x \neq 0$, so $f(x, y, z) \to 1$ as $(x, y, z) \to (0, 0, 0)$

along the x-axis. But $f(0, y, 0) = \dfrac{0 + 2y^2 + 0}{0 + y^2 + 0} = 2$ for $y \neq 0$, so $f(x, y, z) \to 2$ as $(x, y, z) \to (0, 0, 0)$ along the y-axis.

Thus, the limit doesn't exist.

21. $f(x, y, z) = \dfrac{xy + yz^2 + xz^2}{x^2 + y^2 + z^4}$. Then $f(x, 0, 0) = 0/x^2 = 0$ for $x \neq 0$, so as $(x, y, z) \to (0, 0, 0)$ along the x-axis,

$f(x, y, z) \to 0$. But $f(x, x, 0) = x^2 / (2x^2) = \frac{1}{2}$ for $x \neq 0$, so as $(x, y, z) \to (0, 0, 0)$ along the line $y = x$, $z = 0$,

$f(x, y, z) \to \frac{1}{2}$. Thus the limit doesn't exist.

22. $f(x, y, z) = \dfrac{yz}{x^2 + 4y^2 + 9z^2}$. Then $f(x, 0, 0) = 0$ for $x \neq 0$, so as $(x, y, z) \to (0, 0, 0)$ along the x-axis, $f(x, y, z) \to 0$.

But $f(0, y, y) = y^2/(13y^2) = \frac{1}{13}$ for $y \neq 0$, so as $(x, y, z) \to (0, 0, 0)$ along the line $z = y$, $x = 0$, $f(x, y, z) \to \frac{1}{13}$.

Thus the limit doesn't exist.

23.

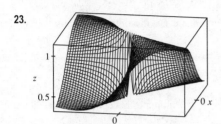

From the ridges on the graph, we see that as $(x, y) \to (0, 0)$ along the lines under the two ridges, $f(x, y)$ approaches different values. So the limit does not exist.

24.

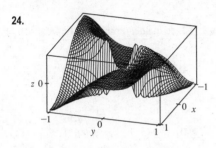

From the graph, it appears that as we approach the origin along the lines $x = 0$ or $y = 0$, the function is everywhere 0, whereas if we approach the origin along a certain curve it has a constant value of about $\frac{1}{2}$. [In fact,

$$f(y^3, y) = y^6/(2y^6) = \frac{1}{2} \text{ for } y \neq 0, \text{ so } f(x, y) \to \frac{1}{2} \text{ as } (x, y) \to (0, 0)$$

along the curve $x = y^3$.] Since the function approaches different values depending on the path of approach, the limit does not exist.

25. $h(x, y) = g(f(x, y)) = (2x + 3y - 6)^2 + \sqrt{2x + 3y - 6}$. Since f is a polynomial, it is continuous on $\mathbb{R}^2$ and g is continuous on its domain $\{t \mid t \geq 0\}$. Thus h is continuous on its domain.

$D = \{(x, y) \mid 2x + 3y - 6 \geq 0\} = \{(x, y) \mid y \geq -\frac{2}{3}x + 2\}$, which consists of all points on or above the line $y = -\frac{2}{3}x + 2$.

26. $h(x, y) = g(f(x, y)) = \dfrac{1 - xy}{1 + x^2 y^2} + \ln\left(\dfrac{1 - xy}{1 + x^2 y^2}\right)$. f is a rational function, so it is continuous on its domain. Because

$1 + x^2 y^2 > 0$, the domain of f is $\mathbb{R}^2$, so f is continuous everywhere. g is continuous on its domain $\{t \mid t > 0\}$. Thus h is

continuous on its domain $\left\{(x, y) \,\middle|\, \dfrac{1 - xy}{1 + x^2 y^2} > 0\right\} = \{(x, y) \mid xy < 1\}$ which consists of all points between (but not on)

the two branches of the hyperbola $y = 1/x$.

27.

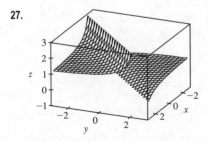

From the graph, it appears that f is discontinuous along the line $y = x$.

If we consider $f(x, y) = e^{1/(x-y)}$ as a composition of functions,

$g(x, y) = 1/(x - y)$ is a rational function and therefore continuous except where $x - y = 0 \;\Rightarrow\; y = x$. Since the function $h(t) = e^t$ is continuous everywhere, the composition $h(g(x, y)) = e^{1/(x-y)} = f(x, y)$ is continuous except along the line $y = x$, as we suspected.

28.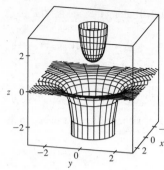

We can see a circular break in the graph, corresponding approximately to the unit circle, where f is discontinuous. Since $f(x, y) = \dfrac{1}{1 - x^2 - y^2}$ is a rational function, it is continuous except where $1 - x^2 - y^2 = 0 \Rightarrow x^2 + y^2 = 1$, confirming our observation that f is discontinuous on the circle $x^2 + y^2 = 1$.

29. The functions $\sin(xy)$ and $e^x - y^2$ are continuous everywhere, so $F(x, y) = \dfrac{\sin(xy)}{e^x - y^2}$ is continuous except where $e^x - y^2 = 0 \Rightarrow y^2 = e^x \Rightarrow y = \pm\sqrt{e^x} = \pm e^{\frac{1}{2}x}$. Thus F is continuous on its domain $\{(x, y) \mid y \neq \pm e^{x/2}\}$.

30. $F(x, y) = \dfrac{x - y}{1 + x^2 + y^2}$ is a rational function and thus is continuous on its domain $\mathbb{R}^2$ (since the denominator is never zero).

31. $F(x, y) = \arctan\left(x + \sqrt{y}\right) = g(f(x, y))$ where $f(x, y) = x + \sqrt{y}$, continuous on its domain $\{(x, y) \mid y \geq 0\}$, and $g(t) = \arctan t$ is continuous everywhere. Thus F is continuous on its domain $\{(x, y) \mid y \geq 0\}$.

32. $e^{x^2 y}$ is continuous on $\mathbb{R}^2$ and $\sqrt{x + y^2}$ is continuous on its domain $\{(x, y) \mid x + y^2 \geq 0\} = \{(x, y) \mid x \geq -y^2\}$, so $F(x, y) = e^{x^2 y} + \sqrt{x + y^2}$ is continuous on the set $\{(x, y) \mid x \geq -y^2\}$.

33. $G(x, y) = \ln\left(x^2 + y^2 - 4\right) = g(f(x, y))$ where $f(x, y) = x^2 + y^2 - 4$, continuous on $\mathbb{R}^2$, and $g(t) = \ln t$, continuous on its domain $\{t \mid t > 0\}$. Thus G is continuous on its domain $\{(x, y) \mid x^2 + y^2 - 4 > 0\} = \{(x, y) \mid x^2 + y^2 > 4\}$, the exterior of the circle $x^2 + y^2 = 4$.

34. $G(x, y) = g(f(x, y))$ where $f(x, y) = (x + y)^{-2}$, a rational function that is continuous on $\mathbb{R}^2$ except where $x + y = 0$, and $g(t) = \tan^{-1} t$, continuous everywhere. Thus G is continuous on its domain $\{(x, y) \mid x + y \neq 0\} = \{(x, y) \mid y \neq -x\}$.

35. $\sqrt{y}$ is continuous on its domain $\{y \mid y \geq 0\}$ and $x^2 - y^2 + z^2$ is continuous everywhere, so $f(x, y, z) = \dfrac{\sqrt{y}}{x^2 - y^2 + z^2}$ is continuous for $y \geq 0$ and $x^2 - y^2 + z^2 \neq 0 \Rightarrow y^2 \neq x^2 + z^2$, that is, $\left\{(x, y, z) \mid y \geq 0, y \neq \sqrt{x^2 + z^2}\right\}$.

36. $f(x, y, z) = \sqrt{x + y + z} = h(g(x, y, z))$ where $g(x, y, z) = x + y + z$, continuous everywhere, and $h(t) = \sqrt{t}$ is continuous on its domain $\{t \mid t \geq 0\}$. Thus f is continuous on its domain $\{(x, y, z) \mid x + y + z \geq 0\}$, so f is continuous on and above the plane $z = -x - y$.

37. $f(x, y) = \begin{cases} \dfrac{x^2 y^3}{2x^2 + y^2} & \text{if } (x, y) \neq (0, 0) \\ 1 & \text{if } (x, y) = (0, 0) \end{cases}$ The first piece of f is a rational function defined everywhere except at the origin, so f is continuous on $\mathbb{R}^2$ except possibly at the origin. Since $x^2 \leq 2x^2 + y^2$, we have $\left|x^2 y^3 / (2x^2 + y^2)\right| \leq \left|y^3\right|$. We know that $\left|y^3\right| \to 0$ as $(x, y) \to (0, 0)$. So, by the Squeeze Theorem, $\displaystyle\lim_{(x,y)\to(0,0)} f(x, y) = \lim_{(x,y)\to(0,0)} \dfrac{x^2 y^3}{2x^2 + y^2} = 0$. But $f(0, 0) = 1$, so f is discontinuous at $(0, 0)$. Therefore, f is continuous on the set $\{(x, y) \mid (x, y) \neq (0, 0)\}$.

38. $f(x, y) = \begin{cases} \dfrac{xy}{x^2 + xy + y^2} & \text{if } (x, y) \neq (0, 0) \\ 0 & \text{if } (x, y) = (0, 0) \end{cases}$ The first piece of f is a rational function defined everywhere except

at the origin, so f is continuous on $\mathbb{R}^2$ except possibly at the origin. $f(x, 0) = 0/x^2 = 0$ for $x \neq 0$, so $f(x, y) \to 0$ as

$(x, y) \to (0, 0)$ along the x-axis. But $f(x, x) = x^2/(3x^2) = \frac{1}{3}$ for $x \neq 0$, so $f(x, y) \to \frac{1}{3}$ as $(x, y) \to (0, 0)$ along the

line $y = x$. Thus $\lim\limits_{(x,y) \to (0,0)} f(x, y)$ doesn't exist, so f is not continuous at $(0, 0)$ and the largest set on which f is continuous

is $\{(x, y) \mid (x, y) \neq (0, 0)\}$.

39. $\lim\limits_{(x,y) \to (0,0)} \dfrac{x^3 + y^3}{x^2 + y^2} = \lim\limits_{r \to 0^+} \dfrac{(r\cos\theta)^3 + (r\sin\theta)^3}{r^2} = \lim\limits_{r \to 0^+} (r\cos^3\theta + r\sin^3\theta) = 0$

40. $\lim\limits_{(x,y) \to (0,0)} (x^2 + y^2)\ln(x^2 + y^2) = \lim\limits_{r \to 0^+} r^2 \ln r^2 = \lim\limits_{r \to 0^+} \dfrac{\ln r^2}{1/r^2} = \lim\limits_{r \to 0^+} \dfrac{(1/r^2)(2r)}{-2/r^3}$ [using l'Hospital's Rule]

$$= \lim\limits_{r \to 0^+} (-r^2) = 0$$

41. $\lim\limits_{(x,y) \to (0,0)} \dfrac{e^{-x^2-y^2} - 1}{x^2 + y^2} = \lim\limits_{r \to 0^+} \dfrac{e^{-r^2} - 1}{r^2} = \lim\limits_{r \to 0^+} \dfrac{e^{-r^2}(-2r)}{2r}$ [using l'Hospital's Rule]

$$= \lim\limits_{r \to 0^+} -e^{-r^2} = -e^0 = -1$$

42. $\lim\limits_{(x,y) \to (0,0)} \dfrac{\sin(x^2 + y^2)}{x^2 + y^2} = \lim\limits_{r \to 0^+} \dfrac{\sin(r^2)}{r^2}$, which is an

indeterminate form of type $0/0$. Using l'Hospital's Rule, we get

$$\lim\limits_{r \to 0^+} \dfrac{\sin(r^2)}{r^2} \overset{\text{H}}{=} \lim\limits_{r \to 0^+} \dfrac{2r\cos(r^2)}{2r} = \lim\limits_{r \to 0^+} \cos(r^2) = 1.$$

Or: Use the fact that $\lim\limits_{\theta \to 0} \dfrac{\sin\theta}{\theta} = 1$.

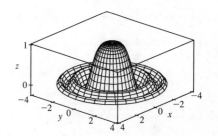

43. $f(x, y) = \begin{cases} \dfrac{\sin(xy)}{xy} & \text{if } (x, y) \neq (0, 0) \\ 1 & \text{if } (x, y) = (0, 0) \end{cases}$

From the graph, it appears that f is continuous everywhere. We know

xy is continuous on $\mathbb{R}^2$ and $\sin t$ is continuous everywhere, so

$\sin(xy)$ is continuous on $\mathbb{R}^2$ and $\dfrac{\sin(xy)}{xy}$ is continuous on $\mathbb{R}^2$

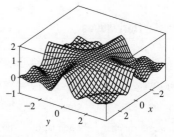

except possibly where $xy = 0$. To show that f is continuous at those points, consider any point (a, b) in $\mathbb{R}^2$ where $ab = 0$.

Because xy is continuous, $xy \to ab = 0$ as $(x, y) \to (a, b)$. If we let $t = xy$, then $t \to 0$ as $(x, y) \to (a, b)$ and

$\lim\limits_{(x,y) \to (a,b)} \dfrac{\sin(xy)}{xy} = \lim\limits_{t \to 0} \dfrac{\sin(t)}{t} = 1$ by Equation 3.4.2 [ET 3.3.2]. Thus $\lim\limits_{(x,y) \to (a,b)} f(x, y) = f(a, b)$ and f is continuous

on $\mathbb{R}^2$.

44. (a) $f(x, y) = \begin{cases} 0 & \text{if } y \leq 0 \text{ or } y \geq x^4 \\ 1 & \text{if } 0 < y < x^4 \end{cases}$ Consider the path $y = mx^a$, $0 < a < 4$. [The path does not pass through

$(0, 0)$ if $a \leq 0$ except for the trivial case where $m = 0$.] If $mx^a \leq 0$ then $f(x, mx^a) = 0$. If $mx^a > 0$ then

$mx^a = |mx^a| = |m|\,|x^a|$ and $mx^a \geq x^4 \iff |m|\,|x^a| \geq x^4 \iff \dfrac{x^4}{|x^a|} \leq |m| \iff |x|^{4-a} \leq |m|$ whenever x^a

is defined. Then $mx^a \geq x^4 \iff |x| \leq |m|^{1/(4-a)}$ so $f(x, mx^a) = 0$ for $|x| \leq |m|^{1/(4-a)}$ and $f(x, y) \to 0$ as

$(x, y) \to (0, 0)$ along this path.

(b) If we approach $(0, 0)$ along the path $y = x^5$, $x > 0$ then we have $f(x, x^5) = 1$ for $0 < x < 1$ because $0 < x^5 < x^4$ there.

Thus $f(x, y) \to 1$ as $(x, y) \to (0, 0)$ along this path, but in part (a) we found a limit of 0 along other paths, so

$\displaystyle\lim_{(x,y)\to(0,0)} f(x, y)$ doesn't exist and f is discontinuous at $(0, 0)$.

(c) First we show that f is discontinuous at any point $(a, 0)$ on the x-axis. If we approach $(a, 0)$ along the path $x = a, y > 0$

then $f(a, y) = 1$ for $0 < y < a^4$, so $f(x, y) \to 1$ as $(x, y) \to (a, 0)$ along this path. If we approach $(a, 0)$ along the path

$x = a, y < 0$ then $f(a, y) = 0$ since $y < 0$ and $f(x, y) \to 0$ as $(x, y) \to (a, 0)$. Thus the limit does not exist and f is

discontinuous on the line $y = 0$. f is also discontinuous on the curve $y = x^4$: For any point (a, a^4) on this curve,

approaching the point along the path $x = a, y > a^4$ gives $f(a, y) = 0$ since $y > a^4$, so $f(x, y) \to 0$ as $(x, y) \to (a, a^4)$.

But approaching the point along the path $x = a, y < a^4$ gives $f(a, y) = 1$ for $y > 0$, so $f(x, y) \to 1$ as $(x, y) \to (a, a^4)$

and the limit does not exist there.

45. Since $|\mathbf{x} - \mathbf{a}|^2 = |\mathbf{x}|^2 + |\mathbf{a}|^2 - 2\,|\mathbf{x}|\,|\mathbf{a}|\cos\theta \geq |\mathbf{x}|^2 + |\mathbf{a}|^2 - 2\,|\mathbf{x}|\,|\mathbf{a}| = (|\mathbf{x}| - |\mathbf{a}|)^2$, we have $\big||\mathbf{x}| - |\mathbf{a}|\big| \leq |\mathbf{x} - \mathbf{a}|$. Let

$\epsilon > 0$ be given and set $\delta = \epsilon$. Then if $0 < |\mathbf{x} - \mathbf{a}| < \delta$, $\big||\mathbf{x}| - |\mathbf{a}|\big| \leq |\mathbf{x} - \mathbf{a}| < \delta = \epsilon$. Hence $\lim_{\mathbf{x}\to\mathbf{a}} |\mathbf{x}| = |\mathbf{a}|$ and

$f(\mathbf{x}) = |\mathbf{x}|$ is continuous on $\mathbb{R}^n$.

46. Let $\epsilon > 0$ be given. We need to find $\delta > 0$ such that if $0 < |\mathbf{x} - \mathbf{a}| < \delta$ then $|f(\mathbf{x}) - f(\mathbf{a})| = |\mathbf{c}\cdot\mathbf{x} - \mathbf{c}\cdot\mathbf{a}| < \epsilon$.

But $|\mathbf{c}\cdot\mathbf{x} - \mathbf{c}\cdot\mathbf{a}| = |\mathbf{c}\cdot(\mathbf{x} - \mathbf{a})|$ and $|\mathbf{c}\cdot(\mathbf{x} - \mathbf{a})| \leq |\mathbf{c}|\,|\mathbf{x} - \mathbf{a}|$ by Exercise 13.3.57 [ET 12.3.57]

(the Cauchy-Schwartz Inequality). Set $\delta = \epsilon/|\mathbf{c}|$. Then if $0 < |\mathbf{x} - \mathbf{a}| < \delta$,

$|f(\mathbf{x}) - f(\mathbf{a})| = |\mathbf{c}\cdot\mathbf{x} - \mathbf{c}\cdot\mathbf{a}| \leq |\mathbf{c}|\,|\mathbf{x} - \mathbf{a}| < |\mathbf{c}|\,\delta = |\mathbf{c}|\,(\epsilon/|\mathbf{c}|) = \epsilon$. So f is continuous on $\mathbb{R}^n$.

15.3 Partial Derivatives ET 14.3

1. (a) $\partial T/\partial x$ represents the rate of change of T when we fix y and t and consider T as a function of the single variable x, which

describes how quickly the temperature changes when longitude changes but latitude and time are constant. $\partial T/\partial y$

represents the rate of change of T when we fix x and t and consider T as a function of y, which describes how quickly the

temperature changes when latitude changes but longitude and time are constant. $\partial T/\partial t$ represents the rate of change of T

when we fix x and y and consider T as a function of t, which describes how quickly the temperature changes over time for

a constant longitude and latitude.

(b) $f_x(158, 21, 9)$ represents the rate of change of temperature at longitude $158°$W, latitude $21°$N at 9:00 AM when only

longitude varies. Since the air is warmer to the west than to the east, increasing longitude results in an increased air

temperature, so we would expect $f_x(158, 21, 9)$ to be positive. $f_y(158, 21, 9)$ represents the rate of change of temperature

at the same time and location when only latitude varies. Since the air is warmer to the south and cooler to the north,

increasing latitude results in a decreased air temperature, so we would expect $f_y(158, 21, 9)$ to be negative. $f_t(158, 21, 9)$

represents the rate of change of temperature at the same time and location when only time varies. Since typically air temperature increases from the morning to the afternoon as the sun warms it, we would expect $f_t(158, 21, 9)$ to be positive.

2. By Definition 4, $f_T(92, 60) = \lim\limits_{h \to 0} \dfrac{f(92 + h, 60) - f(92, 60)}{h}$, which we can approximate by considering $h = 2$ and

$h = -2$ and using the values given in Table 1: $f_T(92, 60) \approx \dfrac{f(94, 60) - f(92, 60)}{2} = \dfrac{111 - 105}{2} = 3$,

$f_T(92, 60) \approx \dfrac{f(90, 60) - f(92, 60)}{-2} = \dfrac{100 - 105}{-2} = 2.5$. Averaging these values, we estimate $f_T(92, 60)$ to be

approximately 2.75. Thus, when the actual temperature is $92°$F and the relative humidity is 60%, the apparent temperature rises by about $2.75°$F for every degree that the actual temperature rises.

Similarly, $f_H(92, 60) = \lim\limits_{h \to 0} \dfrac{f(92, 60 + h) - f(92, 60)}{h}$ which we can approximate by considering $h = 5$ and $h = -5$:

$f_H(92, 60) \approx \dfrac{f(92, 65) - f(92, 60)}{5} = \dfrac{108 - 105}{5} = 0.6$, $f_H(92, 60) \approx \dfrac{f(92, 55) - f(92, 60)}{-5} = \dfrac{103 - 105}{-5} = 0.4$.

Averaging these values, we estimate $f_H(92, 60)$ to be approximately 0.5. Thus, when the actual temperature is $92°$F and the relative humidity is 60%, the apparent temperature rises by about $0.5°$F for every percent that the relative humidity increases.

3. (a) By Definition 4, $f_T(-15, 30) = \lim\limits_{h \to 0} \dfrac{f(-15 + h, 30) - f(-15, 30)}{h}$, which we can approximate by considering $h = 5$

and $h = -5$ and using the values given in the table:

$f_T(-15, 30) \approx \dfrac{f(-10, 30) - f(-15, 30)}{5} = \dfrac{-20 - (-26)}{5} = \dfrac{6}{5} = 1.2$,

$f_T(-15, 30) \approx \dfrac{f(-20, 30) - f(-15, 30)}{-5} = \dfrac{-33 - (-26)}{-5} = \dfrac{-7}{-5} = 1.4$. Averaging these values, we estimate

$f_T(-15, 30)$ to be approximately 1.3. Thus, when the actual temperature is $-15°$C and the wind speed is 30 km/h, the apparent temperature rises by about $1.3°$C for every degree that the actual temperature rises.

Similarly, $f_v(-15, 30) = \lim\limits_{h \to 0} \dfrac{f(-15, 30 + h) - f(-15, 30)}{h}$ which we can approximate by considering $h = 10$ and

$h = -10$: $f_v(-15, 30) \approx \dfrac{f(-15, 40) - f(-15, 30)}{10} = \dfrac{-27 - (-26)}{10} = \dfrac{-1}{10} = -0.1$,

$f_v(-15, 30) \approx \dfrac{f(-15, 20) - f(-15, 30)}{-10} = \dfrac{-24 - (-26)}{-10} = \dfrac{2}{-10} = -0.2$. Averaging these values, we estimate

$f_v(-15, 30)$ to be approximately -0.15. Thus, when the actual temperature is $-15°$C and the wind speed is 30 km/h, the apparent temperature decreases by about $0.15°$C for every km/h that the wind speed increases.

(b) For a fixed wind speed v, the values of the wind-chill index W increase as temperature T increases (look at a column of

the table), so $\dfrac{\partial W}{\partial T}$ is positive. For a fixed temperature T, the values of W decrease (or remain constant) as v increases

(look at a row of the table), so $\dfrac{\partial W}{\partial v}$ is negative (or perhaps 0).

(c) For fixed values of T, the function values $f(T, v)$ appear to become constant (or nearly constant) as v increases, so the

corresponding rate of change is 0 or near 0 as v increases. This suggests that $\lim\limits_{v \to \infty} (\partial W / \partial v) = 0$.

4. (a) $\partial h/\partial v$ represents the rate of change of h when we fix t and consider h as a function of v, which describes how quickly the wave heights change when the wind speed changes for a fixed time duration. $\partial h/\partial t$ represents the rate of change of h when we fix v and consider h as a function of t, which describes how quickly the wave heights change when the duration of time changes, but the wind speed is constant.

(b) By Definition 4, $f_v(40, 15) = \lim\limits_{h \to 0} \dfrac{f(40 + h, 15) - f(40, 15)}{h}$ which we can approximate by considering

$h = 10$ and $h = -10$ and using the values given in the table: $f_v(40, 15) \approx \dfrac{f(50, 15) - f(40, 15)}{10} = \dfrac{36 - 25}{10} = 1.1$,

$f_v(40, 15) \approx \dfrac{f(30, 15) - f(40, 15)}{-10} = \dfrac{16 - 25}{-10} = 0.9$. Averaging these values, we have $f_v(40, 15) \approx 1.0$. Thus, when

a 40-knot wind has been blowing for 15 hours, the wave heights should increase by about 1 foot for every knot that the

wind speed increases (with the same time duration). Similarly, $f_t(40, 15) = \lim\limits_{h \to 0} \dfrac{f(40, 15 + h) - f(40, 15)}{h}$ which we

can approximate by considering $h = 5$ and $h = -5$: $f_t(40, 15) \approx \dfrac{f(40, 20) - f(40, 15)}{5} = \dfrac{28 - 25}{5} = 0.6$,

$f_t(40, 15) \approx \dfrac{f(40, 10) - f(40, 15)}{-5} = \dfrac{21 - 25}{-5} = 0.8$. Averaging these values, we have $f_t(40, 15) \approx 0.7$. Thus, when a

40-knot wind has been blowing for 15 hours, the wave heights increase by about 0.7 feet for every additional hour that the

wind blows.

(c) For fixed values of v, the function values $f(v, t)$ appear to increase in smaller and smaller increments, becoming nearly constant as t increases. Thus, the corresponding rate of change is nearly 0 as t increases, suggesting that

$\lim\limits_{t \to \infty} (\partial h/\partial t) = 0$.

5. (a) If we start at $(1, 2)$ and move in the positive x-direction, the graph of f increases. Thus $f_x(1, 2)$ is positive.

(b) If we start at $(1, 2)$ and move in the positive y-direction, the graph of f decreases. Thus $f_y(1, 2)$ is negative.

6. (a) The graph of f decreases if we start at $(-1, 2)$ and move in the positive x-direction, so $f_x(-1, 2)$ is negative.

(b) The graph of f decreases if we start at $(-1, 2)$ and move in the positive y-direction, so $f_y(-1, 2)$ is negative.

7. (a) $f_{xx} = \frac{\partial}{\partial x}(f_x)$, so f_{xx} is the rate of change of f_x in the x-direction. f_x is negative at $(-1, 2)$ and if we move in the positive x-direction, the surface becomes less steep. Thus the values of f_x are increasing and $f_{xx}(-1, 2)$ is positive.

(b) f_{yy} is the rate of change of f_y in the y-direction. f_y is negative at $(-1, 2)$ and if we move in the positive y-direction, the surface becomes steeper. Thus the values of f_y are decreasing, and $f_{yy}(-1, 2)$ is negative.

8. (a) $f_{xy} = \frac{\partial}{\partial y}(f_x)$, so f_{xy} is the rate of change of f_x in the y-direction. f_x is positive at $(1, 2)$ and if we move in the positive y-direction, the surface becomes steeper, looking in the positive x-direction. Thus the values of f_x are increasing and $f_{xy}(1, 2)$ is positive.

(b) f_x is negative at $(-1, 2)$ and if we move in the positive y-direction, the surface gets steeper (with negative slope), looking in the positive x-direction. This means that the values of f_x are decreasing as y increases, so $f_{xy}(-1, 2)$ is negative.

9. First of all, if we start at the point $(3, -3)$ and move in the positive y-direction, we see that both b and c decrease, while a

increases. Both b and c have a low point at about $(3, -1.5)$, while a is 0 at this point. So a is definitely the graph of f_y, and

one of b and c is the graph of f. To see which is which, we start at the point $(-3, -1.5)$ and move in the positive x-direction.

b traces out a line with negative slope, while c traces out a parabola opening downward. This tells us that b is the x-derivative

of c. So c is the graph of f, b is the graph of f_x, and a is the graph of f_y.

10. $f_x(2, 1)$ is the rate of change of f at $(2, 1)$ in the x-direction. If we start at $(2, 1)$, where $f(2, 1) = 10$, and move in the

positive x-direction, we reach the next contour line (where $f(x, y) = 12$) after approximately 0.6 units. This represents an

average rate of change of about $\frac{2}{0.6}$. If we approach the point $(2, 1)$ from the left (moving in the positive x-direction) the

output values increase from 8 to 10 with an increase in x of approximately 0.9 units, corresponding to an average rate of

change of $\frac{2}{0.9}$. A good estimate for $f_x(2, 1)$ would be the average of these two, so $f_x(2, 1) \approx 2.8$. Similarly, $f_y(2, 1)$ is the

rate of change of f at $(2, 1)$ in the y-direction. If we approach $(2, 1)$ from below, the output values decrease from 12 to 10 with

a change in y of approximately 1 unit, corresponding to an average rate of change of -2. If we start at $(2, 1)$ and move in the

positive y-direction, the output values decrease from 10 to 8 after approximately 0.9 units, a rate of change of $\frac{-2}{0.9}$. Averaging

these two results, we estimate $f_y(2, 1) \approx -2.1$.

11. $f(x, y) = 16 - 4x^2 - y^2 \implies f_x(x, y) = -8x$ and $f_y(x, y) = -2y \implies f_x(1, 2) = -8$ and $f_y(1, 2) = -4$. The graph

of f is the paraboloid $z = 16 - 4x^2 - y^2$ and the vertical plane $y = 2$ intersects it in the parabola $z = 12 - 4x^2$, $y = 2$

(the curve C_1 in the first figure). The slope of the tangent line

to this parabola at $(1, 2, 8)$ is $f_x(1, 2) = -8$. Similarly the

plane $x = 1$ intersects the paraboloid in the parabola

$z = 12 - y^2$, $x = 1$ (the curve C_2 in the second figure) and

the slope of the tangent line at $(1, 2, 8)$ is $f_y(1, 2) = -4$.

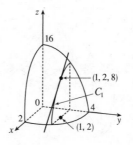
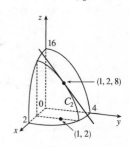

12. $f(x, y) = (4 - x^2 - 4y^2)^{1/2} \implies f_x(x, y) = -x(4 - x^2 - 4y^2)^{-1/2}$ and $f_y(x, y) = -4y(4 - x^2 - 4y^2)^{-1/2} \implies$

$f_x(1, 0) = -\frac{1}{\sqrt{3}}$, $f_y(1, 0) = 0$. The graph of f is the upper half of the ellipsoid $z^2 + x^2 + 4y^2 = 4$ and the plane $y = 0$

intersects the graph in the semicircle $x^2 + z^2 = 4$, $z \geq 0$ and the slope of the tangent line T_1 to this semicircle

at $\left(1, 0, \sqrt{3}\right)$ is $f_x(1, 0) = -\frac{1}{\sqrt{3}}$. Similarly the plane $x = 1$

intersects the graph in the semi-ellipse $z^2 + 4y^2 = 3$, $z \geq 0$

and the slope of the tangent line T_2 to this semi-ellipse at

$\left(1, 0, \sqrt{3}\right)$ is $f_y(1, 0) = 0$.

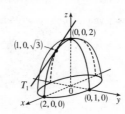
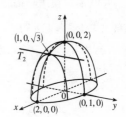

13. $f(x, y) = x^2 + y^2 + x^2 y \Rightarrow f_x = 2x + 2xy, \quad f_y = 2y + x^2$

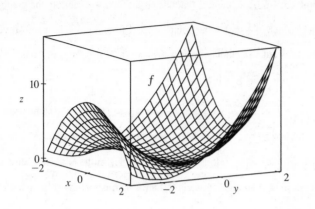

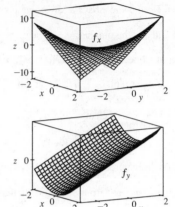

Note that the traces of f in planes parallel to the xz-plane are parabolas which open downward for $y < -1$ and upward for $y > -1$, and the traces of f_x in these planes are straight lines, which have negative slopes for $y < -1$ and positive slopes for $y > -1$. The traces of f in planes parallel to the yz-plane are parabolas which always open upward, and the traces of f_y in these planes are straight lines with positive slopes.

14. $f(x, y) = xe^{-x^2-y^2} \Rightarrow f_x = x\left(-2xe^{-x^2-y^2}\right) + e^{-x^2-y^2} = e^{-x^2-y^2}(1 - 2x^2), \quad f_y = -2xye^{-x^2-y^2}$

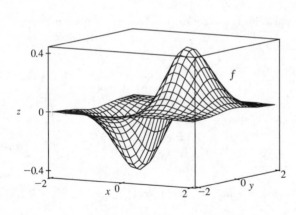

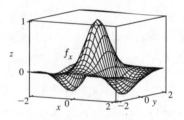

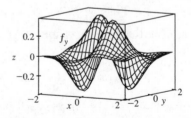

Note that traces of f in planes parallel to the xz-plane have two extreme values, while traces of f_x in these planes have two zeros. Traces of f in planes parallel to the yz-plane have only one extreme value (a minimum if $x < 0$, a maximum if $x > 0$), and traces of f_y in these planes have only one zero (going from negative to positive if $x < 0$ and from positive to negative if $x > 0$).

15. $f(x, y) = y^5 - 3xy \Rightarrow f_x(x, y) = 0 - 3y = -3y, f_y(x, y) = 5y^4 - 3x$

16. $f(x, y) = x^4 y^3 + 8x^2 y \Rightarrow$
$f_x(x, y) = 4x^3 \cdot y^3 + 8 \cdot 2x \cdot y = 4x^3 y^3 + 16xy, \quad f_y(x, y) = x^4 \cdot 3y^2 + 8x^2 \cdot 1 = 3x^4 y^2 + 8x^2$

17. $f(x, t) = e^{-t} \cos \pi x \Rightarrow f_x(x, t) = e^{-t}(-\sin \pi x)(\pi) = -\pi e^{-t} \sin \pi x, f_t(x, t) = e^{-t}(-1) \cos \pi x = -e^{-t} \cos \pi x$

18. $f(x,t) = \sqrt{x}\,\ln t \Rightarrow f_x(x,t) = \frac{1}{2}x^{-1/2}\ln t = (\ln t)/(2\sqrt{x}),\ f_t(x,t) = \sqrt{x}\cdot\frac{1}{t} = \sqrt{x}/t$

19. $z = (2x+3y)^{10} \Rightarrow \dfrac{\partial z}{\partial x} = 10(2x+3y)^9\cdot 2 = 20(2x+3y)^9,\ \dfrac{\partial z}{\partial y} = 10(2x+3y)^9\cdot 3 = 30(2x+3y)^9$

20. $z = \tan xy \Rightarrow \dfrac{\partial z}{\partial x} = (\sec^2 xy)(y) = y\sec^2 xy,\ \dfrac{\partial z}{\partial y} = (\sec^2 xy)(x) = x\sec^2 xy$

21. $f(x,y) = \dfrac{x-y}{x+y} \Rightarrow f_x(x,y) = \dfrac{(1)(x+y)-(x-y)(1)}{(x+y)^2} = \dfrac{2y}{(x+y)^2},$

$f_y(x,y) = \dfrac{(-1)(x+y)-(x-y)(1)}{(x+y)^2} = -\dfrac{2x}{(x+y)^2}$

22. $f(x,y) = x^y \Rightarrow f_x(x,y) = yx^{y-1},\ f_y(x,y) = x^y\ln x$

23. $w = \sin\alpha\cos\beta \Rightarrow \dfrac{\partial w}{\partial\alpha} = \cos\alpha\cos\beta,\ \dfrac{\partial w}{\partial\beta} = -\sin\alpha\sin\beta$

24. $w = \dfrac{e^v}{u+v^2} \Rightarrow \dfrac{\partial w}{\partial u} = \dfrac{0(u+v^2)-e^v(1)}{(u+v^2)^2} = -\dfrac{e^v}{(u+v^2)^2},\ \dfrac{\partial w}{\partial v} = \dfrac{e^v(u+v^2)-e^v(2v)}{(u+v^2)^2} = \dfrac{e^v(u+v^2-2v)}{(u+v^2)^2}$

25. $f(r,s) = r\ln(r^2+s^2) \Rightarrow f_r(r,s) = r\cdot\dfrac{2r}{r^2+s^2} + \ln(r^2+s^2)\cdot 1 = \dfrac{2r^2}{r^2+s^2} + \ln(r^2+s^2),$

$f_s(r,s) = r\cdot\dfrac{2s}{r^2+s^2} + 0 = \dfrac{2rs}{r^2+s^2}$

26. $f(x,t) = \arctan\left(x\sqrt{t}\right) \Rightarrow f_x(x,t) = \dfrac{1}{1+\left(x\sqrt{t}\right)^2}\cdot\sqrt{t} = \dfrac{\sqrt{t}}{1+x^2t},$

$f_t(x,t) = \dfrac{1}{1+\left(x\sqrt{t}\right)^2}\cdot x\left(\dfrac{1}{2}t^{-1/2}\right) = \dfrac{x}{2\sqrt{t}\,(1+x^2t)}$

27. $u = te^{w/t} \Rightarrow \dfrac{\partial u}{\partial t} = t\cdot e^{w/t}(-wt^{-2}) + e^{w/t}\cdot 1 = e^{w/t} - \dfrac{w}{t}e^{w/t} = e^{w/t}\left(1-\dfrac{w}{t}\right),\ \dfrac{\partial u}{\partial w} = te^{w/t}\cdot\dfrac{1}{t} = e^{w/t}$

28. $f(x,y) = \displaystyle\int_y^x \cos(t^2)\,dt \Rightarrow f_x(x,y) = \dfrac{\partial}{\partial x}\displaystyle\int_y^x \cos(t^2)\,dt = \cos(x^2)$ by the Fundamental Theorem of Calculus, Part 1;

$f_y(x,y) = \dfrac{\partial}{\partial y}\displaystyle\int_y^x \cos(t^2)\,dt = -\dfrac{\partial}{\partial y}\cos(t^2)\,dt = -\cos(y^2).$

29. $f(x,y,z) = xz - 5x^2y^3z^4 \Rightarrow f_x(x,y,z) = z - 10xy^3z^4,\ f_y(x,y,z) = -15x^2y^2z^4,\ f_z(x,y,z) = x - 20x^2y^3z^3$

30. $f(x,y,z) = x\sin(y-z) \Rightarrow f_x(x,y,z) = \sin(y-z),\ f_y(x,y,z) = x\cos(y-z),$

$f_z(x,y,z) = x\cos(y-z)(-1) = -x\cos(y-z)$

31. $w = \ln(x+2y+3z) \Rightarrow \dfrac{\partial w}{\partial x} = \dfrac{1}{x+2y+3z},\ \dfrac{\partial w}{\partial y} = \dfrac{2}{x+2y+3z},\ \dfrac{\partial w}{\partial z} = \dfrac{3}{x+2y+3z}$

32. $w = ze^{xyz} \Rightarrow$

$\dfrac{\partial w}{\partial x} = ze^{xyz}\cdot yz = yz^2e^{xyz},\ \dfrac{\partial w}{\partial y} = ze^{xyz}\cdot xz = xz^2e^{xyz},\ \dfrac{\partial w}{\partial z} = ze^{xyz}\cdot xy + e^{xyz}\cdot 1 = (xyz+1)e^{xyz}$

33. $u = xy\sin^{-1}(yz)$ $\Rightarrow$ $\dfrac{\partial u}{\partial x} = y\sin^{-1}(yz)$, $\dfrac{\partial u}{\partial y} = xy \cdot \dfrac{1}{\sqrt{1-(yz)^2}}(z) + \sin^{-1}(yz) \cdot x = \dfrac{xyz}{\sqrt{1-y^2z^2}} + x\sin^{-1}(yz)$,

$\dfrac{\partial u}{\partial z} = xy \cdot \dfrac{1}{\sqrt{1-(yz)^2}}(y) = \dfrac{xy^2}{\sqrt{1-y^2z^2}}$

34. $u = x^{y/z}$ $\Rightarrow$ $u_x = \dfrac{y}{z}x^{(y/z)-1}$, $u_y = x^{y/z}\ln x \cdot \dfrac{1}{z} = \dfrac{x^{y/z}}{z}\ln x$, $u_z = x^{y/z}\ln x \cdot \dfrac{-y}{z^2} = -\dfrac{yx^{y/z}}{z^2}\ln x$

35. $f(x,y,z,t) = xyz^2\tan(yt)$ $\Rightarrow$ $f_x(x,y,z,t) = yz^2\tan(yt)$,

$f_y(x,y,z,t) = xyz^2 \cdot \sec^2(yt) \cdot t + xz^2\tan(yt) = xyz^2t\sec^2(yt) + xz^2\tan(yt)$,

$f_z(x,y,z,t) = 2xyz\tan(yt)$, $f_t(x,y,z,t) = xyz^2\sec^2(yt) \cdot y = xy^2z^2\sec^2(yt)$

36. $f(x,y,z,t) = \dfrac{xy^2}{t+2z}$ $\Rightarrow$ $f_x(x,y,z,t) = \dfrac{y^2}{t+2z}$, $f_y(x,y,z,t) = \dfrac{2xy}{t+2z}$,

$f_z(x,y,z,t) = xy^2(-1)(t+2z)^{-2}(2) = -\dfrac{2xy^2}{(t+2z)^2}$, $f_t(x,y,z,t) = xy^2(-1)(t+2z)^{-2}(1) = -\dfrac{xy^2}{(t+2z)^2}$

37. $u = \sqrt{x_1^2 + x_2^2 + \cdots + x_n^2}$. For each $i = 1, \ldots, n$, $u_{x_i} = \frac{1}{2}\left(x_1^2 + x_2^2 + \cdots + x_n^2\right)^{-1/2}(2x_i) = \dfrac{x_i}{\sqrt{x_1^2 + x_2^2 + \cdots + x_n^2}}$.

38. $u = \sin(x_1 + 2x_2 + \cdots + nx_n)$. For each $i = 1, \ldots, n$, $u_{x_i} = i\cos(x_1 + 2x_2 + \cdots + nx_n)$.

39. $f(x,y) = \ln\left(x + \sqrt{x^2 + y^2}\right)$ $\Rightarrow$

$f_x(x,y) = \dfrac{1}{x + \sqrt{x^2+y^2}}\left[1 + \frac{1}{2}(x^2+y^2)^{-1/2}(2x)\right] = \dfrac{1}{x + \sqrt{x^2+y^2}}\left(1 + \dfrac{x}{\sqrt{x^2+y^2}}\right)$,

so $f_x(3,4) = \dfrac{1}{3 + \sqrt{3^2+4^2}}\left(1 + \dfrac{3}{\sqrt{3^2+4^2}}\right) = \frac{1}{8}\left(1 + \frac{3}{5}\right) = \frac{1}{5}$.

40. $f(x,y) = \arctan(y/x)$ $\Rightarrow$ $f_x(x,y) = \dfrac{1}{1+(y/x)^2}(-yx^{-2}) = \dfrac{-y}{x^2(1+y^2/x^2)} = -\dfrac{y}{x^2+y^2}$,

so $f_x(2,3) = -\dfrac{3}{2^2+3^2} = -\dfrac{3}{13}$.

41. $f(x,y,z) = \dfrac{y}{x+y+z}$ $\Rightarrow$ $f_y(x,y,z) = \dfrac{1(x+y+z) - y(1)}{(x+y+z)^2} = \dfrac{x+z}{(x+y+z)^2}$,

so $f_y(2,1,-1) = \dfrac{2+(-1)}{(2+1+(-1))^2} = \dfrac{1}{4}$.

42. $f(x,y,z) = \sqrt{\sin^2 x + \sin^2 y + \sin^2 z}$ $\Rightarrow$

$f_z(x,y,z) = \frac{1}{2}\left(\sin^2 x + \sin^2 y + \sin^2 z\right)^{-1/2}(0 + 0 + 2\sin z \cdot \cos z) = \dfrac{\sin z\cos z}{\sqrt{\sin^2 x + \sin^2 y + \sin^2 z}}$,

so $f_z\left(0,0,\frac{\pi}{4}\right) = \dfrac{\sin\frac{\pi}{4}\cos\frac{\pi}{4}}{\sqrt{\sin^2 0 + \sin^2 0 + \sin^2\frac{\pi}{4}}} = \dfrac{\frac{\sqrt{2}}{2} \cdot \frac{\sqrt{2}}{2}}{\sqrt{0 + 0 + \left(\frac{\sqrt{2}}{2}\right)^2}} = \dfrac{\frac{1}{2}}{\frac{\sqrt{2}}{2}} = \dfrac{1}{\sqrt{2}}$ or $\dfrac{\sqrt{2}}{2}$.

43. $f(x,y) = xy^2 - x^3y \Rightarrow$

$$f_x(x,y) = \lim_{h \to 0} \frac{f(x+h,y) - f(x,y)}{h} = \lim_{h \to 0} \frac{(x+h)y^2 - (x+h)^3y - (xy^2 - x^3y)}{h}$$

$$= \lim_{h \to 0} \frac{h(y^2 - 3x^2y - 3xyh - yh^2)}{h} = \lim_{h \to 0}(y^2 - 3x^2y - 3xyh - yh^2) = y^2 - 3x^2y$$

$$f_y(x,y) = \lim_{h \to 0} \frac{f(x,y+h) - f(x,y)}{h} = \lim_{h \to 0} \frac{x(y+h)^2 - x^3(y+h) - (xy^2 - x^3y)}{h} = \lim_{h \to 0} \frac{h(2xy + xh - x^3)}{h}$$

$$= \lim_{h \to 0}(2xy + xh - x^3) = 2xy - x^3$$

44. $f(x,y) = \dfrac{x}{x+y^2} \Rightarrow$

$$f_x(x,y) = \lim_{h \to 0} \frac{f(x+h,y) - f(x,y)}{h} = \lim_{h \to 0} \frac{\frac{x+h}{x+h+y^2} - \frac{x}{x+y^2}}{h} \cdot \frac{(x+h+y^2)(x+y^2)}{(x+h+y^2)(x+y^2)}$$

$$= \lim_{h \to 0} \frac{(x+h)(x+y^2) - x(x+h+y^2)}{h(x+h+y^2)(x+y^2)} = \lim_{h \to 0} \frac{y^2h}{h(x+h+y^2)(x+y^2)}$$

$$= \lim_{h \to 0} \frac{y^2}{(x+h+y^2)(x+y^2)} = \frac{y^2}{(x+y^2)^2}$$

$$f_y(x,y) = \lim_{h \to 0} \frac{f(x,y+h) - f(x,y)}{h} = \lim_{h \to 0} \frac{\frac{x}{x+(y+h)^2} - \frac{x}{x+y^2}}{h} \cdot \frac{[x+(y+h)^2](x+y^2)}{[x+(y+h)^2](x+y^2)}$$

$$= \lim_{h \to 0} \frac{x(x+y^2) - x[x+(y+h)^2]}{h[x+(y+h)^2](x+y^2)} = \lim_{h \to 0} \frac{h(-2xy - xh)}{h[x+(y+h)^2](x+y^2)}$$

$$= \lim_{h \to 0} \frac{-2xy - xh}{[x+(y+h)^2](x+y^2)} = \frac{-2xy}{(x+y^2)^2}$$

45. $x^2 + y^2 + z^2 = 3xyz \Rightarrow \dfrac{\partial}{\partial x}(x^2 + y^2 + z^2) = \dfrac{\partial}{\partial x}(3xyz) \Rightarrow 2x + 0 + 2z\dfrac{\partial z}{\partial x} = 3y\left(x\dfrac{\partial z}{\partial x} + z \cdot 1\right) \Leftrightarrow$

$2z\dfrac{\partial z}{\partial x} - 3xy\dfrac{\partial z}{\partial x} = 3yz - 2x \Leftrightarrow (2z - 3xy)\dfrac{\partial z}{\partial x} = 3yz - 2x$, so $\dfrac{\partial z}{\partial x} = \dfrac{3yz - 2x}{2z - 3xy}$.

$\dfrac{\partial}{\partial y}(x^2 + y^2 + z^2) = \dfrac{\partial}{\partial y}(3xyz) \Rightarrow 0 + 2y + 2z\dfrac{\partial z}{\partial y} = 3x\left(y\dfrac{\partial z}{\partial y} + z \cdot 1\right) \Leftrightarrow 2z\dfrac{\partial z}{\partial y} - 3xy\dfrac{\partial z}{\partial y} = 3xz - 2y \Leftrightarrow$

$(2z - 3xy)\dfrac{\partial z}{\partial y} = 3xz - 2y$, so $\dfrac{\partial z}{\partial y} = \dfrac{3xz - 2y}{2z - 3xy}$.

46. $yz = \ln(x+z) \Rightarrow \dfrac{\partial}{\partial x}(yz) = \dfrac{\partial}{\partial x}(\ln(x+z)) \Rightarrow y\dfrac{\partial z}{\partial x} = \dfrac{1}{x+z}\left(1 + \dfrac{\partial z}{\partial x}\right) \Leftrightarrow \left(y - \dfrac{1}{x+z}\right)\dfrac{\partial z}{\partial x} = \dfrac{1}{x+z}$,

so $\dfrac{\partial z}{\partial x} = \dfrac{1/(x+z)}{y - 1/(x+z)} = \dfrac{1}{y(x+z) - 1}$.

$\dfrac{\partial}{\partial y}(yz) = \dfrac{\partial}{\partial y}(\ln(x+z)) \Rightarrow y\dfrac{\partial z}{\partial y} + z \cdot 1 = \dfrac{1}{x+z}\left(0 + \dfrac{\partial z}{\partial y}\right) \Leftrightarrow \left(y - \dfrac{1}{x+z}\right)\dfrac{\partial z}{\partial y} = -z$,

so $\dfrac{\partial z}{\partial y} = \dfrac{-z}{y - 1/(x+z)} = \dfrac{z(x+z)}{1 - y(x+z)}$.

47. $x - z = \arctan(yz)$ $\Rightarrow$ $\dfrac{\partial}{\partial x}(x - z) = \dfrac{\partial}{\partial x}(\arctan(yz))$ $\Rightarrow$ $1 - \dfrac{\partial z}{\partial x} = \dfrac{1}{1 + (yz)^2} \cdot y\dfrac{\partial z}{\partial x}$ $\Leftrightarrow$

$1 = \left(\dfrac{y}{1 + y^2 z^2} + 1\right)\dfrac{\partial z}{\partial x}$ $\Leftrightarrow$ $1 = \left(\dfrac{y + 1 + y^2 z^2}{1 + y^2 z^2}\right)\dfrac{\partial z}{\partial x}$, so $\dfrac{\partial z}{\partial x} = \dfrac{1 + y^2 z^2}{1 + y + y^2 z^2}$.

$\dfrac{\partial}{\partial y}(x - z) = \dfrac{\partial}{\partial y}(\arctan(yz))$ $\Rightarrow$ $0 - \dfrac{\partial z}{\partial y} = \dfrac{1}{1 + (yz)^2} \cdot \left(y\dfrac{\partial z}{\partial y} + z \cdot 1\right)$ $\Leftrightarrow$

$-\dfrac{z}{1 + y^2 z^2} = \left(\dfrac{y}{1 + y^2 z^2} + 1\right)\dfrac{\partial z}{\partial y}$ $\Leftrightarrow$ $-\dfrac{z}{1 + y^2 z^2} = \left(\dfrac{y + 1 + y^2 z^2}{1 + y^2 z^2}\right)\dfrac{\partial z}{\partial y}$ $\Leftrightarrow$ $\dfrac{\partial z}{\partial y} = -\dfrac{z}{1 + y + y^2 z^2}$.

48. $\sin(xyz) = x + 2y + 3z$ $\Rightarrow$ $\dfrac{\partial}{\partial x}(\sin(xyz)) = \dfrac{\partial}{\partial x}(x + 2y + 3z)$ $\Rightarrow$ $\cos(xyz) \cdot y\left(x\dfrac{\partial z}{\partial x} + z\right) = 1 + 3\dfrac{\partial z}{\partial x}$ $\Leftrightarrow$

$(xy\cos(xyz) - 3)\dfrac{\partial z}{\partial x} = 1 - yz\cos(xyz)$, so $\dfrac{\partial z}{\partial x} = \dfrac{1 - yz\cos(xyz)}{xy\cos(xyz) - 3}$.

$\dfrac{\partial}{\partial y}(\sin(xyz)) = \dfrac{\partial}{\partial y}(x + 2y + 3z)$ $\Rightarrow$ $\cos(xyz) \cdot x\left(y\dfrac{\partial z}{\partial y} + z\right) = 2 + 3\dfrac{\partial z}{\partial y}$ $\Leftrightarrow$

$(xy\cos(xyz) - 3)\dfrac{\partial z}{\partial y} = 2 - xz\cos(xyz)$, so $\dfrac{\partial z}{\partial y} = \dfrac{2 - xz\cos(xyz)}{xy\cos(xyz) - 3}$.

49. (a) $z = f(x) + g(y)$ $\Rightarrow$ $\dfrac{\partial z}{\partial x} = f'(x)$, $\dfrac{\partial z}{\partial y} = g'(y)$

(b) $z = f(x + y)$. Let $u = x + y$. Then $\dfrac{\partial z}{\partial x} = \dfrac{df}{du}\dfrac{\partial u}{\partial x} = \dfrac{df}{du}(1) = f'(u) = f'(x + y)$,

$\dfrac{\partial z}{\partial y} = \dfrac{df}{du}\dfrac{\partial u}{\partial y} = \dfrac{df}{du}(1) = f'(u) = f'(x + y)$.

50. (a) $z = f(x)g(y)$ $\Rightarrow$ $\dfrac{\partial z}{\partial x} = f'(x)g(y)$, $\dfrac{\partial z}{\partial y} = f(x)g'(y)$

(b) $z = f(xy)$. Let $u = xy$. Then $\dfrac{\partial u}{\partial x} = y$ and $\dfrac{\partial u}{\partial y} = x$. Hence $\dfrac{\partial z}{\partial x} = \dfrac{df}{du}\dfrac{\partial u}{\partial x} = \dfrac{df}{du} \cdot y = yf'(u) = yf'(xy)$

and $\dfrac{\partial z}{\partial y} = \dfrac{df}{du}\dfrac{\partial u}{\partial y} = \dfrac{df}{du} \cdot x = xf'(u) = xf'(xy)$.

(c) $z = f\left(\dfrac{x}{y}\right)$. Let $u = \dfrac{x}{y}$. Then $\dfrac{\partial u}{\partial x} = \dfrac{1}{y}$ and $\dfrac{\partial u}{\partial y} = -\dfrac{x}{y^2}$. Hence $\dfrac{\partial z}{\partial x} = \dfrac{df}{du}\dfrac{\partial u}{\partial x} = f'(u)\dfrac{1}{y} = \dfrac{f'(x/y)}{y}$

and $\dfrac{\partial z}{\partial y} = \dfrac{df}{du}\dfrac{\partial u}{\partial y} = f'(u)\left(-\dfrac{x}{y^2}\right) = -\dfrac{xf'(x/y)}{y^2}$.

51. $f(x, y) = x^3 y^5 + 2x^4 y$ $\Rightarrow$ $f_x(x, y) = 3x^2 y^5 + 8x^3 y$, $f_y(x, y) = 5x^3 y^4 + 2x^4$. Then $f_{xx}(x, y) = 6xy^5 + 24x^2 y$,

$f_{xy}(x, y) = 15x^2 y^4 + 8x^3$, $f_{yx}(x, y) = 15x^2 y^4 + 8x^3$, and $f_{yy}(x, y) = 20x^3 y^3$.

52. $f(x, y) = \sin^2(mx + ny)$ $\Rightarrow$ $f_x(x, y) = 2\sin(mx + ny)\cos(mx + ny) \cdot m = m\sin(2mx + 2ny)$ [using the

identity $\sin 2\theta = 2\sin\theta\cos\theta$], $f_y(x, y) = 2\sin(mx + ny)\cos(mx + ny) \cdot n = n\sin(2mx + 2ny)$.

Then $f_{xx}(x, y) = m\cos(2mx + 2ny) \cdot 2m = 2m^2\cos(2mx + 2ny)$,

$f_{xy}(x, y) = m\cos(2mx + 2ny) \cdot 2n = 2mn\cos(2mx + 2ny)$,

$f_{yx}(x, y) = n \cos(2mx + 2ny) \cdot 2m = 2mn \cos(2mx + 2ny)$, and

$f_{yy}(x, y) = n \cos(2mx + 2ny) \cdot 2n = 2n^2 \cos(2mx + 2ny)$.

53. $w = \sqrt{u^2 + v^2} \quad \Rightarrow \quad w_u = \frac{1}{2}(u^2 + v^2)^{-1/2} \cdot 2u = \frac{u}{\sqrt{u^2 + v^2}}, w_v = \frac{1}{2}(u^2 + v^2)^{-1/2} \cdot 2v = \frac{v}{\sqrt{u^2 + v^2}}$. Then

$$w_{uu} = \frac{1 \cdot \sqrt{u^2 + v^2} - u \cdot \frac{1}{2}(u^2 + v^2)^{-1/2}(2u)}{\left(\sqrt{u^2 + v^2}\right)^2} = \frac{\sqrt{u^2 + v^2} - u^2/\sqrt{u^2 + v^2}}{u^2 + v^2} = \frac{u^2 + v^2 - u^2}{(u^2 + v^2)^{3/2}} = \frac{v^2}{(u^2 + v^2)^{3/2}},$$

$$w_{uv} = u\left(-\frac{1}{2}\right)(u^2 + v^2)^{-3/2}(2v) = -\frac{uv}{(u^2 + v^2)^{3/2}}, w_{vu} = v\left(-\frac{1}{2}\right)(u^2 + v^2)^{-3/2}(2u) = -\frac{uv}{(u^2 + v^2)^{3/2}},$$

$$w_{vv} = \frac{1 \cdot \sqrt{u^2 + v^2} - v \cdot \frac{1}{2}(u^2 + v^2)^{-1/2}(2v)}{\left(\sqrt{u^2 + v^2}\right)^2} = \frac{\sqrt{u^2 + v^2} - v^2/\sqrt{u^2 + v^2}}{u^2 + v^2} = \frac{u^2 + v^2 - v^2}{(u^2 + v^2)^{3/2}} = \frac{u^2}{(u^2 + v^2)^{3/2}}.$$

54. $v = \frac{xy}{x - y} \quad \Rightarrow \quad v_x = \frac{y(x - y) - xy(1)}{(x - y)^2} = -\frac{y^2}{(x - y)^2}$,

$v_y = \frac{x(x - y) - xy(-1)}{(x - y)^2} = \frac{x^2}{(x - y)^2}$. Then $v_{xx} = -y^2(-2)(x - y)^{-3}(1) = \frac{2y^2}{(x - y)^3}$,

$$v_{xy} = -\frac{2y(x - y)^2 - y^2 \cdot 2(x - y)(-1)}{[(x - y)^2]^2} = -\frac{2y(x - y) + 2y^2}{(x - y)^3} = -\frac{2xy}{(x - y)^3},$$

$$v_{yx} = \frac{2x(x - y)^2 - x^2 \cdot 2(x - y)(1)}{[(x - y)^2]^2} = \frac{2x(x - y) - 2x^2}{(x - y)^3} = -\frac{2xy}{(x - y)^3}, \quad v_{yy} = x^2(-2)(x - y)^{-3}(-1) = \frac{2x^2}{(x - y)^3}.$$

55. $z = \arctan\frac{x + y}{1 - xy} \quad \Rightarrow$

$$z_x = \frac{1}{1 + \left(\frac{x+y}{1-xy}\right)^2} \cdot \frac{(1)(1 - xy) - (x + y)(-y)}{(1 - xy)^2} = \frac{1 + y^2}{(1 - xy)^2 + (x + y)^2} = \frac{1 + y^2}{1 + x^2 + y^2 + x^2y^2}$$

$$= \frac{1 + y^2}{(1 + x^2)(1 + y^2)} = \frac{1}{1 + x^2},$$

$$z_y = \frac{1}{1 + \left(\frac{x+y}{1-xy}\right)^2} \cdot \frac{(1)(1 - xy) - (x + y)(-x)}{(1 - xy)^2} = \frac{1 + x^2}{(1 - xy)^2 + (x + y)^2} = \frac{1 + x^2}{(1 + x^2)(1 + y^2)} = \frac{1}{1 + y^2}.$$

Then $z_{xx} = -(1 + x^2)^{-2} \cdot 2x = -\frac{2x}{(1 + x^2)^2}$, $z_{xy} = 0$, $z_{yx} = 0$, $z_{yy} = -(1 + y^2)^{-2} \cdot 2y = -\frac{2y}{(1 + y^2)^2}$.

56. $v = e^{xe^y} \quad \Rightarrow \quad v_x = e^{xe^y} \cdot e^y = e^{y+xe^y}, v_y = e^{xe^y} \cdot xe^y = xe^{y+xe^y}$. Then $v_{xx} = e^{y+xe^y} \cdot e^y = e^{2y+xe^y}$,

$v_{xy} = e^{y+xe^y}(1 + xe^y), \quad v_{yx} = xe^{y+xe^y}(e^y) + e^{y+xe^y}(1) = e^{y+xe^y}(1 + xe^y)$,

$v_{yy} = xe^{y+xe^y}(1 + xe^y) = e^{y+xe^y}(x + x^2e^y)$.

57. $u = x\sin(x + 2y) \quad \Rightarrow \quad u_x = x \cdot \cos(x + 2y)(1) + \sin(x + 2y) \cdot 1 = x\cos(x + 2y) + \sin(x + 2y)$,

$u_{xy} = x(-\sin(x + 2y)(2)) + \cos(x + 2y)(2) = 2\cos(x + 2y) - 2x\sin(x + 2y)$,

$u_y = x\cos(x + 2y)(2) = 2x\cos(x + 2y)$,

$u_{yx} = 2x \cdot (-\sin(x + 2y)(1)) + \cos(x + 2y) \cdot 2 = 2\cos(x + 2y) - 2x\sin(x + 2y)$. Thus $u_{xy} = u_{yx}$.

58. $u = x^4y^2 - 2xy^5 \quad\Rightarrow\quad u_x = 4x^3y^2 - 2y^5, \ u_{xy} = 8x^3y - 10y^4$ and $u_y = 2x^4y - 10xy^4, \ u_{yx} = 8x^3y - 10y^4$.

Thus $u_{xy} = u_{yx}$.

59. $u = \ln\sqrt{x^2 + y^2} = \ln(x^2 + y^2)^{1/2} = \frac{1}{2}\ln(x^2 + y^2) \quad\Rightarrow\quad u_x = \frac{1}{2}\frac{1}{x^2 + y^2} \cdot 2x = \frac{x}{x^2 + y^2}$,

$u_{xy} = x(-1)(x^2 + y^2)^{-2}(2y) = -\dfrac{2xy}{(x^2 + y^2)^2}$ and $u_y = \dfrac{1}{2}\dfrac{1}{x^2 + y^2} \cdot 2y = \dfrac{y}{x^2 + y^2}$,

$u_{yx} = y(-1)(x^2 + y^2)^{-2}(2x) = -\dfrac{2xy}{(x^2 + y^2)^2}$. Thus $u_{xy} = u_{yx}$.

60. $u = xye^y \quad\Rightarrow\quad u_x = ye^y, \ u_{xy} = ye^y + e^y = (y+1)e^y$ and $u_y = x(ye^y + e^y) = x(y+1)e^y, \ u_{yx} = (y+1)e^y$.

Thus $u_{xy} = u_{yx}$.

61. $f(x,y) = 3xy^4 + x^3y^2 \quad\Rightarrow\quad f_x = 3y^4 + 3x^2y^2, \ f_{xx} = 6xy^2, \ f_{xxy} = 12xy$ and

$f_y = 12xy^3 + 2x^3y, \ f_{yy} = 36xy^2 + 2x^3, \ f_{yyy} = 72xy$.

62. $f(x,t) = x^2e^{-ct} \quad\Rightarrow\quad f_t = x^2(-ce^{-ct}), \ f_{tt} = x^2(c^2e^{-ct}), \ f_{ttt} = x^2(-c^3e^{-ct}) = -c^3x^2e^{-ct}$ and

$f_{tx} = 2x(-ce^{-ct}), \ f_{txx} = 2(-ce^{-ct}) = -2ce^{-ct}$.

63. $f(x,y,z) = \cos(4x + 3y + 2z) \quad\Rightarrow$

$f_x = -\sin(4x + 3y + 2z)(4) = -4\sin(4x + 3y + 2z), \ f_{xy} = -4\cos(4x + 3y + 2z)(3) = -12\cos(4x + 3y + 2z)$,

$f_{xyz} = -12(-\sin(4x + 3y + 2z))(2) = 24\sin(4x + 3y + 2z)$ and

$f_y = -\sin(4x + 3y + 2z)(3) = -3\sin(4x + 3y + 2z)$,

$f_{yz} = -3\cos(4x + 3y + 2z)(2) = -6\cos(4x + 3y + 2z), \ f_{yzz} = -6(-\sin(4x + 3y + 2z))(2) = 12\sin(4x + 3y + 2z)$.

64. $f(r,s,t) = r\ln(rs^2t^3) \quad\Rightarrow\quad f_r = r \cdot \dfrac{1}{rs^2t^3}(s^2t^3) + \ln(rs^2t^3) \cdot 1 = \dfrac{rs^2t^3}{rs^2t^3} + \ln(rs^2t^3) = 1 + \ln(rs^2t^3)$,

$f_{rs} = \dfrac{1}{rs^2t^3}(2rst^3) = \dfrac{2}{s} = 2s^{-1}, \ f_{rss} = -2s^{-2} = -\dfrac{2}{s^2}$ and $f_{rst} = 0$.

65. $u = e^{r\theta}\sin\theta \quad\Rightarrow\quad \dfrac{\partial u}{\partial \theta} = e^{r\theta}\cos\theta + \sin\theta \cdot e^{r\theta}(r) = e^{r\theta}(\cos\theta + r\sin\theta)$,

$\dfrac{\partial^2 u}{\partial r\,\partial \theta} = e^{r\theta}(\sin\theta) + (\cos\theta + r\sin\theta)e^{r\theta}(\theta) = e^{r\theta}(\sin\theta + \theta\cos\theta + r\theta\sin\theta)$,

$\dfrac{\partial^3 u}{\partial r^2\,\partial \theta} = e^{r\theta}(\theta\sin\theta) + (\sin\theta + \theta\cos\theta + r\theta\sin\theta) \cdot e^{r\theta}(\theta) = \theta e^{r\theta}(2\sin\theta + \theta\cos\theta + r\theta\sin\theta)$.

66. $z = u\sqrt{v-w} = u(v-w)^{1/2} \quad\Rightarrow\quad \dfrac{\partial z}{\partial w} = u\left[\frac{1}{2}(v-w)^{-1/2}(-1)\right] = -\frac{1}{2}u(v-w)^{-1/2}$,

$\dfrac{\partial^2 z}{\partial v\,\partial w} = -\frac{1}{2}u\left(-\frac{1}{2}(v-w)^{-3/2}(1)\right) = \frac{1}{4}u(v-w)^{-3/2}, \quad \dfrac{\partial^3 z}{\partial u\,\partial v\,\partial w} = \frac{1}{4}(v-w)^{-3/2}$.

67. $w = \dfrac{x}{y + 2z} = x(y + 2z)^{-1} \quad\Rightarrow\quad \dfrac{\partial w}{\partial x} = (y + 2z)^{-1}, \quad \dfrac{\partial^2 w}{\partial y\,\partial x} = -(y + 2z)^{-2}(1) = -(y + 2z)^{-2}$,

$\dfrac{\partial^3 w}{\partial z\,\partial y\,\partial x} = -(-2)(y + 2z)^{-3}(2) = 4(y + 2z)^{-3} = \dfrac{4}{(y + 2z)^3}$ and $\dfrac{\partial w}{\partial y} = x(-1)(y + 2z)^{-2}(1) = -x(y + 2z)^{-2}$,

$\dfrac{\partial^2 w}{\partial x\,\partial y} = -(y + 2z)^{-2}, \quad \dfrac{\partial^3 w}{\partial x^2\,\partial y} = 0$.

68. $u = x^a y^b z^c$. If $a = 0$, or if $b = 0$ or 1, or if $c = 0, 1,$ or 2, then $\dfrac{\partial^6 u}{\partial x\, \partial y^2\, \partial z^3} = 0$. Otherwise $\dfrac{\partial u}{\partial z} = cx^a y^b z^{c-1}$,

$\dfrac{\partial^2 u}{\partial z^2} = c(c-1)x^a y^b z^{c-2}$, $\dfrac{\partial^3 u}{\partial z^3} = c(c-1)(c-2)x^a y^b z^{c-3}$, $\dfrac{\partial^4 u}{\partial y\, \partial z^3} = bc(c-1)(c-2)x^a y^{b-1} z^{c-3}$,

$\dfrac{\partial^5 u}{\partial y^2\, \partial z^3} = b(b-1)c(c-1)(c-2)x^a y^{b-2} z^{c-3}$, and $\dfrac{\partial^6 u}{\partial x\, \partial y^2\, \partial z^3} = ab(b-1)c(c-1)(c-2)x^{a-1} y^{b-2} z^{c-3}$.

69. By Definition 4, $f_x(3,2) = \displaystyle\lim_{h \to 0} \dfrac{f(3+h, 2) - f(3,2)}{h}$ which we can approximate by considering $h = 0.5$ and $h = -0.5$:

$f_x(3,2) \approx \dfrac{f(3.5,2) - f(3,2)}{0.5} = \dfrac{22.4 - 17.5}{0.5} = 9.8$, $f_x(3,2) \approx \dfrac{f(2.5,2) - f(3,2)}{-0.5} = \dfrac{10.2 - 17.5}{-0.5} = 14.6$. Averaging

these values, we estimate $f_x(3,2)$ to be approximately 12.2. Similarly, $f_x(3,2.2) = \displaystyle\lim_{h \to 0} \dfrac{f(3+h, 2.2) - f(3, 2.2)}{h}$ which

we can approximate by considering $h = 0.5$ and $h = -0.5$: $f_x(3, 2.2) \approx \dfrac{f(3.5, 2.2) - f(3, 2.2)}{0.5} = \dfrac{26.1 - 15.9}{0.5} = 20.4$,

$f_x(3, 2.2) \approx \dfrac{f(2.5, 2.2) - f(3, 2.2)}{-0.5} = \dfrac{9.3 - 15.9}{-0.5} = 13.2$. Averaging these values, we have $f_x(3, 2.2) \approx 16.8$.

To estimate $f_{xy}(3, 2)$, we first need an estimate for $f_x(3, 1.8)$:

$f_x(3, 1.8) \approx \dfrac{f(3.5, 1.8) - f(3, 1.8)}{0.5} = \dfrac{20.0 - 18.1}{0.5} = 3.8$, $f_x(3, 1.8) \approx \dfrac{f(2.5, 1.8) - f(3, 1.8)}{-0.5} = \dfrac{12.5 - 18.1}{-0.5} = 11.2$.

Averaging these values, we get $f_x(3, 1.8) \approx 7.5$. Now $f_{xy}(x, y) = \dfrac{\partial}{\partial y}[f_x(x,y)]$ and $f_x(x,y)$ is itself a function of two

variables, so Definition 4 says that $f_{xy}(x, y) = \dfrac{\partial}{\partial y}[f_x(x,y)] = \displaystyle\lim_{h \to 0} \dfrac{f_x(x, y+h) - f_x(x, y)}{h} \Rightarrow$

$f_{xy}(3, 2) = \displaystyle\lim_{h \to 0} \dfrac{f_x(3, 2+h) - f_x(3, 2)}{h}$.

We can estimate this value using our previous work with $h = 0.2$ and $h = -0.2$:

$f_{xy}(3, 2) \approx \dfrac{f_x(3, 2.2) - f_x(3, 2)}{0.2} = \dfrac{16.8 - 12.2}{0.2} = 23$, $f_{xy}(3, 2) \approx \dfrac{f_x(3, 1.8) - f_x(3, 2)}{-0.2} = \dfrac{7.5 - 12.2}{-0.2} = 23.5$.

Averaging these values, we estimate $f_{xy}(3, 2)$ to be approximately 23.25.

70. (a) If we fix y and allow x to vary, the level curves indicate that the value of f decreases as we move through P in the positive
x-direction, so f_x is negative at P.

(b) If we fix x and allow y to vary, the level curves indicate that the value of f increases as we move through P in the positive
y-direction, so f_y is positive at P.

(c) $f_{xx} = \dfrac{\partial}{\partial x}(f_x)$, so if we fix y and allow x to vary, f_{xx} is the rate of change of f_x as x increases. Note that at points to the
right of P the level curves are spaced farther apart (in the x-direction) than at points to the left of P, demonstrating that f
decreases less quickly with respect to x to the right of P. So as we move through P in the positive x-direction the
(negative) value of f_x increases, hence $\dfrac{\partial}{\partial x}(f_x) = f_{xx}$ is positive at P.

(d) $f_{xy} = \dfrac{\partial}{\partial y}(f_x)$, so if we fix x and allow y to vary, f_{xy} is the rate of change of f_x as y increases. The level curves are
closer together (in the x-direction) at points above P than at those below P, demonstrating that f decreases more quickly

with respect to x for y-values above P. So as we move through P in the positive y-direction, the (negative) value of f_x decreases, hence f_{xy} is negative.

(e) $f_{yy} = \dfrac{\partial}{\partial y}(f_y)$, so if we fix x and allow y to vary, f_{yy} is the rate of change of f_y as y increases. The level curves are closer together (in the y-direction) at points above P than at those below P, demonstrating that f increases more quickly with respect to y above P. So as we move through P in the positive y-direction the (positive) value of f_y increases, hence $\dfrac{\partial}{\partial y}(f_y) = f_{yy}$ is positive at P.

71. $u = e^{-\alpha^2 k^2 t} \sin kx \quad \Rightarrow \quad u_x = ke^{-\alpha^2 k^2 t} \cos kx,\ u_{xx} = -k^2 e^{-\alpha^2 k^2 t} \sin kx,$ and $u_t = -\alpha^2 k^2 e^{-\alpha^2 k^2 t} \sin kx$. Thus $\alpha^2 u_{xx} = u_t$.

72. (a) $u = x^2 + y^2 \quad \Rightarrow \quad u_x = 2x,\ u_{xx} = 2;\quad u_y = 2y,\ u_{yy} = 2$. Thus $u_{xx} + u_{yy} \neq 0$ and $u = x^2 + y^2$ does not satisfy Laplace's Equation.

(b) $u = x^2 - y^2$ is a solution: $u_{xx} = 2,\ u_{yy} = -2$ so $u_{xx} + u_{yy} = 0$.

(c) $u = x^3 + 3xy^2$ is not a solution: $u_x = 3x^2 + 3y^2,\ u_{xx} = 6x;\quad u_y = 6xy,\ u_{yy} = 6x$.

(d) $u = \ln\sqrt{x^2 + y^2}$ is a solution: $u_x = \dfrac{1}{\sqrt{x^2 + y^2}}\left(\dfrac{1}{2}\right)(x^2 + y^2)^{-1/2}(2x) = \dfrac{x}{x^2 + y^2}$,

$u_{xx} = \dfrac{(x^2 + y^2) - x(2x)}{(x^2 + y^2)^2} = \dfrac{y^2 - x^2}{(x^2 + y^2)^2}$. By symmetry, $u_{yy} = \dfrac{x^2 - y^2}{(x^2 + y^2)^2}$, so $u_{xx} + u_{yy} = 0$.

(e) $u = \sin x \cosh y + \cos x \sinh y$ is a solution: $u_x = \cos x \cosh y - \sin x \sinh y,\ u_{xx} = -\sin x \cosh y - \cos x \sinh y$, and $u_y = \sin x \sinh y + \cos x \cosh y,\ u_{yy} = \sin x \cosh y + \cos x \sinh y$.

(f) $u = e^{-x}\cos y - e^{-y}\cos x$ is a solution: $u_x = -e^{-x}\cos y + e^{-y}\sin x,\quad u_{xx} = e^{-x}\cos y + e^{-y}\cos x$, and $u_y = -e^{-x}\sin y + e^{-y}\cos x,\quad u_{yy} = -e^{-x}\cos y - e^{-y}\cos x$.

73. $u = \dfrac{1}{\sqrt{x^2 + y^2 + z^2}} \quad \Rightarrow \quad u_x = \left(-\tfrac{1}{2}\right)(x^2 + y^2 + z^2)^{-3/2}(2x) = -x(x^2 + y^2 + z^2)^{-3/2}$ and

$u_{xx} = -(x^2 + y^2 + z^2)^{-3/2} - x\left(-\tfrac{3}{2}\right)(x^2 + y^2 + z^2)^{-5/2}(2x) = \dfrac{2x^2 - y^2 - z^2}{(x^2 + y^2 + z^2)^{5/2}}$.

By symmetry, $u_{yy} = \dfrac{2y^2 - x^2 - z^2}{(x^2 + y^2 + z^2)^{5/2}}$ and $u_{zz} = \dfrac{2z^2 - x^2 - y^2}{(x^2 + y^2 + z^2)^{5/2}}$.

Thus $u_{xx} + u_{yy} + u_{zz} = \dfrac{2x^2 - y^2 - z^2 + 2y^2 - x^2 - z^2 + 2z^2 - x^2 - y^2}{(x^2 + y^2 + z^2)^{5/2}} = 0$.

74. (a) $u = \sin(kx)\sin(akt) \quad \Rightarrow \quad u_t = ak\sin(kx)\cos(akt),\ u_{tt} = -a^2 k^2 \sin(kx)\sin(akt),\ u_x = k\cos(kx)\sin(akt)$, $u_{xx} = -k^2 \sin(kx)\sin(akt)$. Thus $u_{tt} = a^2 u_{xx}$.

(b) $u = \dfrac{t}{a^2 t^2 - x^2} \quad \Rightarrow \quad u_t = \dfrac{(a^2 t^2 - x^2) - t(2a^2 t)}{(a^2 t^2 - x^2)^2} = -\dfrac{a^2 t^2 + x^2}{(a^2 t^2 - x^2)^2}$,

$u_{tt} = \dfrac{-2a^2 t(a^2 t^2 - x^2)^2 + (a^2 t^2 - x^2)(2)(a^2 t^2 - x^2)(2a^2 t)}{(a^2 t^2 - x^2)^4} = \dfrac{2a^4 t^3 + 6a^2 tx^2}{(a^2 t^2 - x^2)^3}$,

$$u_x = t(-1)(a^2t^2 - x^2)^{-2}(2x) = \frac{2tx}{(a^2t^2 - x^2)^2},$$

$$u_{xx} = \frac{2t(a^2t^2 - x^2)^2 - 2tx\,(2)(a^2t^2 - x^2)(-2x)}{(a^2t^2 - x^2)^4} = \frac{2a^2t^3 - 2tx^2 + 8tx^2}{(a^2t^2 - x^2)^3} = \frac{2a^2t^3 + 6tx^2}{(a^2t^2 - x^2)^3}.$$

Thus $u_{tt} = a^2 u_{xx}$.

(c) $u = (x - at)^6 + (x + at)^6 \Rightarrow u_t = -6a(x - at)^5 + 6a(x + at)^5,\ u_{tt} = 30a^2(x - at)^4 + 30a^2(x + at)^4,$

$u_x = 6(x - at)^5 + 6(x + at)^5,\ u_{xx} = 30(x - at)^4 + 30(x + at)^4$. Thus $u_{tt} = a^2 u_{xx}$.

(d) $u = \sin(x - at) + \ln(x + at) \Rightarrow u_t = -a\cos(x - at) + \dfrac{a}{x + at},\quad u_{tt} = -a^2\sin(x - at) - \dfrac{a^2}{(x + at)^2},$

$u_x = \cos(x - at) + \dfrac{1}{x + at},\quad u_{xx} = -\sin(x - at) - \dfrac{1}{(x + at)^2}$. Thus $u_{tt} = a^2 u_{xx}$.

75. Let $v = x + at,\ w = x - at$. Then $u_t = \dfrac{\partial[f(v) + g(w)]}{\partial t} = \dfrac{df(v)}{dv}\dfrac{\partial v}{\partial t} + \dfrac{dg(w)}{dw}\dfrac{\partial w}{\partial t} = af'(v) - ag'(w)$ and

$$u_{tt} = \frac{\partial[af'(v) - ag'(w)]}{\partial t} = a[af''(v) + ag''(w)] = a^2[f''(v) + g''(w)].$$ Similarly, by using the Chain Rule we have

$u_x = f'(v) + g'(w)$ and $u_{xx} = f''(v) + g''(w)$. Thus $u_{tt} = a^2 u_{xx}$.

76. For each i, $i = 1, \ldots, n$, $\partial u/\partial x_i = a_i e^{a_1 x_1 + a_2 x_2 + \cdots + a_n x_n}$ and $\partial^2 u/\partial x_i^2 = a_i^2 e^{a_1 x_1 + a_2 x_2 + \cdots + a_n x_n}$.

Then $\dfrac{\partial^2 u}{\partial x_1^2} + \dfrac{\partial^2 u}{\partial x_2^2} + \cdots + \dfrac{\partial^2 u}{\partial x_n^2} = \left(a_1^2 + a_2^2 + \cdots + a_n^2\right)e^{a_1 x_1 + a_2 x_2 + \cdots + a_n x_n} = e^{a_1 x_1 + a_2 x_2 + \cdots + a_n x_n} = u$

since $a_1^2 + a_2^2 + \cdots + a_n^2 = 1$.

77. $z = \ln(e^x + e^y) \Rightarrow \dfrac{\partial z}{\partial x} = \dfrac{e^x}{e^x + e^y}$ and $\dfrac{\partial z}{\partial y} = \dfrac{e^y}{e^x + e^y}$, so $\dfrac{\partial z}{\partial x} + \dfrac{\partial z}{\partial y} = \dfrac{e^x}{e^x + e^y} + \dfrac{e^y}{e^x + e^y} = \dfrac{e^x + e^y}{e^x + e^y} = 1.$

$$\frac{\partial^2 z}{\partial x^2} = \frac{e^x(e^x + e^y) - e^x(e^x)}{(e^x + e^y)^2} = \frac{e^{x+y}}{(e^x + e^y)^2},\quad \frac{\partial^2 z}{\partial x\,\partial y} = \frac{0 - e^y(e^x)}{(e^x + e^y)^2} = -\frac{e^{x+y}}{(e^x + e^y)^2},\quad \text{and}$$

$$\frac{\partial^2 z}{\partial y^2} = \frac{e^y(e^x + e^y) - e^y(e^y)}{(e^x + e^y)^2} = \frac{e^{x+y}}{(e^x + e^y)^2}.\ \text{Thus}$$

$$\frac{\partial^2 z}{\partial x^2}\frac{\partial^2 z}{\partial y^2} - \left(\frac{\partial^2 z}{\partial x\,\partial y}\right)^2 = \frac{e^{x+y}}{(e^x + e^y)^2}\cdot\frac{e^{x+y}}{(e^x + e^y)^2} - \left(-\frac{e^{x+y}}{(e^x + e^y)^2}\right)^2 = \frac{(e^{x+y})^2}{(e^x + e^y)^4} - \frac{(e^{x+y})^2}{(e^x + e^y)^4} = 0$$

78. $P = bL^\alpha K^\beta$, so $\dfrac{\partial P}{\partial L} = \alpha bL^{\alpha-1}K^\beta$ and $\dfrac{\partial P}{\partial K} = \beta bL^\alpha K^{\beta-1}$. Then

$$L\frac{\partial P}{\partial L} + K\frac{\partial P}{\partial K} = L(\alpha bL^{\alpha-1}K^\beta) + K(\beta bL^\alpha K^{\beta-1}) = \alpha bL^{1+\alpha-1}K^\beta + \beta bL^\alpha K^{1+\beta-1} = (\alpha + \beta)bL^\alpha K^\beta = (\alpha + \beta)P$$

79. If we fix $K = K_0$, $P(L, K_0)$ is a function of a single variable L, and $\dfrac{dP}{dL} = \alpha\dfrac{P}{L}$ is a separable differential equation. Then

$$\frac{dP}{P} = \alpha\frac{dL}{L} \Rightarrow \int\frac{dP}{P} = \int\alpha\frac{dL}{L} \Rightarrow \ln|P| = \alpha\ln|L| + C(K_0),\ \text{where } C(K_0) \text{ can depend on } K_0. \text{ Then}$$

$|P| = e^{\alpha\ln|L| + C(K_0)}$, and since $P > 0$ and $L > 0$, we have $P = e^{\alpha\ln L}e^{C(K_0)} = e^{C(K_0)}e^{\ln L^\alpha} = C_1(K_0)L^\alpha$ where

$C_1(K_0) = e^{C(K_0)}$.

80. (a) $\partial T/\partial x = -60(2x)/(1+x^2+y^2)^2$, so at $(2, 1)$, $T_x = -240/(1+4+1)^2 = -\frac{20}{3}$.

(b) $\partial T/\partial y = -60(2y)/(1+x^2+y^2)^2$, so at $(2, 1)$, $T_y = -120/36 = -\frac{10}{3}$. Thus from the point $(2, 1)$ the temperature is

decreasing at a rate of $\frac{20}{3}°$C/m in the x-direction and is decreasing at a rate of $\frac{10}{3}°$C/m in the y-direction.

81. By the Chain Rule, taking the partial derivative of both sides with respect to R_1 gives

$$\frac{\partial R^{-1}}{\partial R}\frac{\partial R}{\partial R_1} = \frac{\partial\left[(1/R_1)+(1/R_2)+(1/R_3)\right]}{\partial R_1} \quad \text{or} \quad -R^{-2}\frac{\partial R}{\partial R_1} = -R_1^{-2}. \text{ Thus } \frac{\partial R}{\partial R_1} = \frac{R^2}{R_1^2}.$$

82. $P = \dfrac{mRT}{V}$ so $\dfrac{\partial P}{\partial V} = \dfrac{-mRT}{V^2}$; $V = \dfrac{mRT}{P}$, so $\dfrac{\partial V}{\partial T} = \dfrac{mR}{P}$; $T = \dfrac{PV}{mR}$, so $\dfrac{\partial T}{\partial P} = \dfrac{V}{mR}$.

Thus $\dfrac{\partial P}{\partial V}\dfrac{\partial V}{\partial T}\dfrac{\partial T}{\partial P} = \dfrac{-mRT}{V^2}\dfrac{mR}{P}\dfrac{V}{mR} = \dfrac{-mRT}{PV} = -1$, since $PV = mRT$.

83. By Exercise 82, $PV = mRT \Rightarrow P = \dfrac{mRT}{V}$, so $\dfrac{\partial P}{\partial T} = \dfrac{mR}{V}$. Also, $PV = mRT \Rightarrow V = \dfrac{mRT}{P}$ and $\dfrac{\partial V}{\partial T} = \dfrac{mR}{P}$.

Since $T = \dfrac{PV}{mR}$, we have $T\dfrac{\partial P}{\partial T}\dfrac{\partial V}{\partial T} = \dfrac{PV}{mR}\cdot\dfrac{mR}{V}\cdot\dfrac{mR}{P} = mR$.

84. $\dfrac{\partial W}{\partial T} = 0.6215 + 0.3965v^{0.16}$. When $T = -15°$C and $v = 30$ km/h, $\dfrac{\partial W}{\partial T} = 0.6215 + 0.3965(30)^{0.16} \approx 1.3048$, so we

would expect the apparent temperature to drop by approximately $1.3°$C if the actual temperature decreases by $1°$C.

$\dfrac{\partial W}{\partial v} = -11.37(0.16)v^{-0.84} + 0.3965T(0.16)v^{-0.84}$ and when $T = -15°$C and $v = 30$ km/h,

$\dfrac{\partial W}{\partial v} = -11.37(0.16)(30)^{-0.84} + 0.3965(-15)(0.16)(30)^{-0.84} \approx -0.1592$, so we would expect the apparent temperature

to drop by approximately $0.16°$C if the wind speed increases by 1 km/h.

85. $\dfrac{\partial K}{\partial m} = \frac{1}{2}v^2$, $\dfrac{\partial K}{\partial v} = mv$, $\dfrac{\partial^2 K}{\partial v^2} = m$. Thus $\dfrac{\partial K}{\partial m}\cdot\dfrac{\partial^2 K}{\partial v^2} = \frac{1}{2}v^2 m = K$.

86. The Law of Cosines says that $a^2 = b^2 + c^2 - 2bc\cos A$. Thus $\dfrac{\partial(a^2)}{\partial a} = \dfrac{\partial(b^2 + c^2 - 2ab\cos A)}{\partial a}$ or

$2a = -2bc(-\sin A)\dfrac{\partial A}{\partial a}$, implying that $\dfrac{\partial A}{\partial a} = \dfrac{a}{bc\sin A}$. Taking the partial derivative of both sides with respect to b gives

$0 = 2b - 2c(\cos A) - 2bc(-\sin A)\dfrac{\partial A}{\partial b}$. Thus $\dfrac{\partial A}{\partial b} = \dfrac{c\cos A - b}{bc\sin A}$. By symmetry, $\dfrac{\partial A}{\partial c} = \dfrac{b\cos A - c}{bc\sin A}$.

87. $f_x(x, y) = x + 4y \Rightarrow f_{xy}(x, y) = 4$ and $f_y(x, y) = 3x - y \Rightarrow f_{yx}(x, y) = 3$. Since f_{xy} and f_{yx} are continuous

everywhere but $f_{xy}(x, y) \neq f_{yx}(x, y)$, Clairaut's Theorem implies that such a function $f(x, y)$ does not exist.

88. Setting $x = 1$, the equation of the parabola of intersection is

$z = 6 - 1 - 1 - 2y^2 = 4 - 2y^2$. The slope of the tangent is

$\partial z/\partial y = -4y$, so at $(1, 2, -4)$ the slope is -8. Parametric

equations for the line are therefore $x = 1$, $y = 2 + t$,

$z = -4 - 8t$.

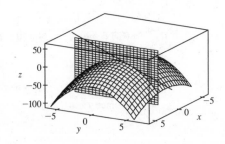

89. By the geometry of partial derivatives, the slope of the tangent line is $f_x(1, 2)$. By implicit differentiation of

$4x^2 + 2y^2 + z^2 = 16$, we get $8x + 2z\,(\partial z/\partial x) = 0 \;\Rightarrow\; \partial z/\partial x = -4x/z$, so when $x = 1$ and $z = 2$ we have

$\partial z/\partial x = -2$. So the slope is $f_x(1, 2) = -2$. Thus the tangent line is given by $z - 2 = -2(x - 1)$, $y = 2$. Taking the

parameter to be $t = x - 1$, we can write parametric equations for this line: $x = 1 + t$, $y = 2$, $z = 2 - 2t$.

90. $T(x, t) = T_0 + T_1 e^{-\lambda x} \sin(\omega t - \lambda x)$

(a) $\partial T/\partial x = T_1 e^{-\lambda x} [\cos(\omega t - \lambda x)(-\lambda)] + T_1(-\lambda e^{-\lambda x}) \sin(\omega t - \lambda x) = -\lambda T_1 e^{-\lambda x} [\sin(\omega t - \lambda x) + \cos(\omega t - \lambda x)]$.

This quantity represents the rate of change of temperature with respect to depth below the surface, at a given time t.

(b) $\partial T/\partial t = T_1 e^{-\lambda x} [\cos(\omega t - \lambda x)(\omega)] = \omega T_1 e^{-\lambda x} \cos(\omega t - \lambda x)$. This quantity represents the rate of change of

temperature with respect to time at a fixed depth x.

(c) $T_{xx} = \dfrac{\partial}{\partial x}\left(\dfrac{\partial T}{\partial x}\right)$

$= -\lambda T_1 \left(e^{-\lambda x} [\cos(\omega t - \lambda x)(-\lambda) - \sin(\omega t - \lambda x)(-\lambda)] + e^{-\lambda x}(-\lambda) [\sin(\omega t - \lambda x) + \cos(\omega t - \lambda x)]\right)$

$= 2\lambda^2 T_1 e^{-\lambda x} \cos(\omega t - \lambda x)$

But from part (b), $T_t = \omega T_1 e^{-\lambda x} \cos(\omega t - \lambda x) = \dfrac{\omega}{2\lambda^2} T_{xx}$. So with $k = \dfrac{\omega}{2\lambda^2}$, the function T satisfies the heat equation.

(d)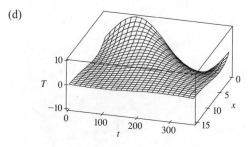

Note that near the surface (that is, for small x) the

temperature varies greatly as t changes, but deeper

(for large x) the temperature is more stable.

(e) The term $-\lambda x$ is a phase shift: it represents the fact that since heat diffuses slowly through soil, it takes time for changes

in the surface temperature to affect the temperature at deeper points. As x increases, the phase shift also increases. For

example, at the surface the highest temperature is reached at $t \approx 100$, whereas at a depth of 5 feet the peak temperature is

attained at $t \approx 150$, and at a depth of 10 feet, at $t \approx 220$.

91. By Clairaut's Theorem, $f_{xyy} = (f_{xy})_y = (f_{yx})_y = f_{yxy} = (f_y)_{xy} = (f_y)_{yx} = f_{yyx}$.

92. (a) Since we are differentiating n times, with two choices of variable at each differentiation, there are 2^n nth-order partial

derivatives.

(b) If these partial derivatives are all continuous, then the order in which the partials are taken doesn't affect the value of the

result, that is, all nth-order partial derivatives with p partials with respect to x and $n - p$ partials with respect to y are

equal. Since the number of partials taken with respect to x for an nth-order partial derivative can range from 0 to n, a

function of two variables has $n + 1$ distinct partial derivatives of order n if these partial derivatives are all continuous.

(c) Since n differentiations are to be performed with three choices of variable at each differentiation, there are 3^n nth-order

partial derivatives of a function of three variables.

93. Let $g(x) = f(x, 0) = x(x^2)^{-3/2}e^0 = x|x|^{-3}$. But we are using the point $(1, 0)$, so near $(1, 0)$, $g(x) = x^{-2}$. Then $g'(x) = -2x^{-3}$ and $g'(1) = -2$, so using (1) we have $f_x(1, 0) = g'(1) = -2$.

94. $f_x(0, 0) = \lim\limits_{h \to 0} \dfrac{f(0 + h, 0) - f(0, 0)}{h} = \lim\limits_{h \to 0} \dfrac{(h^3 + 0)^{1/3} - 0}{h} = \lim\limits_{h \to 0} \dfrac{h}{h} = 1.$

Or: Let $g(x) = f(x, 0) = \sqrt[3]{x^3 + 0} = x$. Then $g'(x) = 1$ and $g'(0) = 1$ so, by (1), $f_x(0, 0) = g'(0) = 1$.

95. (a)

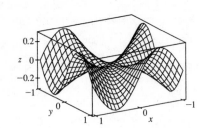

(b) For $(x, y) \neq (0, 0)$,

$$f_x(x, y) = \frac{(3x^2y - y^3)(x^2 + y^2) - (x^3y - xy^3)(2x)}{(x^2 + y^2)^2}$$

$$= \frac{x^4y + 4x^2y^3 - y^5}{(x^2 + y^2)^2}$$

and by symmetry $f_y(x, y) = \dfrac{x^5 - 4x^3y^2 - xy^4}{(x^2 + y^2)^2}$.

(c) $f_x(0, 0) = \lim\limits_{h \to 0} \dfrac{f(h, 0) - f(0, 0)}{h} = \lim\limits_{h \to 0} \dfrac{(0/h^2) - 0}{h} = 0$ and $f_y(0, 0) = \lim\limits_{h \to 0} \dfrac{f(0, h) - f(0, 0)}{h} = 0.$

(d) By (3), $f_{xy}(0, 0) = \dfrac{\partial f_x}{\partial y} = \lim\limits_{h \to 0} \dfrac{f_x(0, h) - f_x(0, 0)}{h} = \lim\limits_{h \to 0} \dfrac{(-h^5 - 0)/h^4}{h} = -1$ while by (2),

$f_{yx}(0, 0) = \dfrac{\partial f_y}{\partial x} = \lim\limits_{h \to 0} \dfrac{f_y(h, 0) - f_y(0, 0)}{h} = \lim\limits_{h \to 0} \dfrac{h^5/h^4}{h} = 1.$

(e) For $(x, y) \neq (0, 0)$, we use a CAS to compute

$$f_{xy}(x, y) = \frac{x^6 + 9x^4y^2 - 9x^2y^4 - y^6}{(x^2 + y^2)^3}$$

Now as $(x, y) \to (0, 0)$ along the x-axis, $f_{xy}(x, y) \to 1$ while as $(x, y) \to (0, 0)$ along the y-axis, $f_{xy}(x, y) \to -1$. Thus f_{xy} isn't continuous at $(0, 0)$ and Clairaut's Theorem doesn't apply, so there is no contradiction. The graphs of f_{xy} and f_{yx} are identical except at the origin, where we observe the discontinuity.

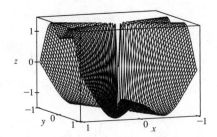

15.4 Tangent Planes and Linear Approximations   ET 14.4

1. $z = f(x, y) = 4x^2 - y^2 + 2y \;\Rightarrow\; f_x(x, y) = 8x,\; f_y(x, y) = -2y + 2$, so $f_x(-1, 2) = -8,\; f_y(-1, 2) = -2$. By Equation 2, an equation of the tangent plane is $z - 4 = f_x(-1, 2)[x - (-1)] + f_y(-1, 2)(y - 2) \;\Rightarrow$ $z - 4 = -8(x + 1) - 2(y - 2)$ or $z = -8x - 2y$.

2. $z = f(x, y) = 3(x - 1)^2 + 2(y + 3)^2 + 7 \;\Rightarrow\; f_x(x, y) = 6(x - 1),\; f_y(x, y) = 4(y + 3)$, so $f_x(2, -2) = 6$ and $f_y(2, -2) = 4$. By Equation 2, an equation of the tangent plane is $z - 12 = f_x(2, -2)(x - 2) + f_y(2, -2)\,[y - (-2)] \;\Rightarrow$ $z - 12 = 6(x - 2) + 4(y + 2)$ or $z = 6x + 4y + 8$.

3. $z = f(x, y) = \sqrt{xy} \;\Rightarrow\; f_x(x, y) = \frac{1}{2}(xy)^{-1/2} \cdot y = \frac{1}{2}\sqrt{y/x},\; f_y(x, y) = \frac{1}{2}(xy)^{-1/2} \cdot x = \frac{1}{2}\sqrt{x/y}$, so $f_x(1, 1) = \frac{1}{2}$ and $f_y(1, 1) = \frac{1}{2}$. Thus an equation of the tangent plane is $z - 1 = f_x(1, 1)(x - 1) + f_y(1, 1)(y - 1) \;\Rightarrow$ $z - 1 = \frac{1}{2}(x - 1) + \frac{1}{2}(y - 1)$ or $x + y - 2z = 0$.

4. $z = f(x, y) = y \ln x$ $\Rightarrow$ $f_x(x, y) = y/x$, $f_y(x, y) = \ln x$, so $f_x(1, 4) = 4$, $f_y(1, 4) = 0$, and an equation of the tangent

plane is $z - 0 = f_x(1, 4)(x - 1) + f_y(1, 4)(y - 4)$ $\Rightarrow$ $z = 4(x - 1) + 0(y - 4)$ or $z = 4x - 4$.

5. $z = f(x, y) = y \cos(x - y)$ $\Rightarrow$ $f_x = y(-\sin(x - y)(1)) = -y\sin(x - y)$,

$f_y = y(-\sin(x - y)(-1)) + \cos(x - y) = y\sin(x - y) + \cos(x - y)$, so $f_x(2, 2) = -2\sin(0) = 0$,

$f_y(2, 2) = 2\sin(0) + \cos(0) = 1$ and an equation of the tangent plane is $z - 2 = 0(x - 2) + 1(y - 2)$ or $z = y$.

6. $z = f(x, y) = e^{x^2 - y^2}$ $\Rightarrow$ $f_x(x, y) = 2xe^{x^2 - y^2}$, $f_y(x, y) = -2ye^{x^2 - y^2}$, so $f_x(1, -1) = 2$, $f_y(1, -1) = 2$.

By Equation 2, an equation of the tangent plane is $z - 1 = f_x(1, -1)(x - 1) + f_y(1, -1)[y - (-1)]$ $\Rightarrow$

$z - 1 = 2(x - 1) + 2(y + 1)$ or $z = 2x + 2y + 1$.

7. $z = f(x, y) = x^2 + xy + 3y^2$, so $f_x(x, y) = 2x + y$ $\Rightarrow$ $f_x(1, 1) = 3$, $f_y(x, y) = x + 6y$ $\Rightarrow$ $f_y(1, 1) = 7$ and an

equation of the tangent plane is $z - 5 = 3(x - 1) + 7(y - 1)$ or $z = 3x + 7y - 5$. After zooming in, the surface and the

tangent plane become almost indistinguishable. (Here, the tangent plane is below the surface.) If we zoom in farther, the

surface and the tangent plane will appear to coincide.

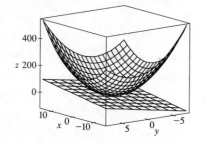 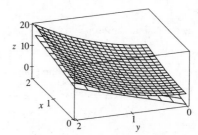

8. $z = f(x, y) = \arctan(xy^2)$ $\Rightarrow$ $f_x = \dfrac{1}{1 + (xy^2)^2}(y^2) = \dfrac{y^2}{1 + x^2 y^4}$, $f_y = \dfrac{1}{1 + (xy^2)^2}(2xy) = \dfrac{2xy}{1 + x^2 y^4}$,

$f_x(1, 1) = \frac{1}{1+1} = \frac{1}{2}$, $f_y(1, 1) = \frac{2}{1+1} = 1$, so an equation of the tangent plane is $z - \frac{\pi}{4} = \frac{1}{2}(x - 1) + 1(y - 1)$ or

$z = \frac{1}{2}x + y - \frac{3}{2} + \frac{\pi}{4}$. After zooming in, the surface and the tangent plane become almost indistinguishable. (Here the

tangent plane is above the surface.) If we zoom in farther, the surface and the tangent plane will appear to coincide.

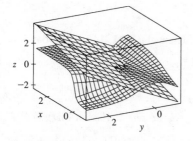 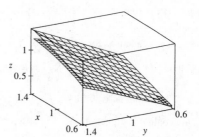

9. $f(x, y) = \dfrac{xy \sin(x - y)}{1 + x^2 + y^2}$.   A CAS gives $f_x(x, y) = \dfrac{y \sin(x - y) + xy \cos(x - y)}{1 + x^2 + y^2} - \dfrac{2x^2 y \sin(x - y)}{(1 + x^2 + y^2)^2}$ and

$f_y(x, y) = \dfrac{x \sin(x - y) - xy \cos(x - y)}{1 + x^2 + y^2} - \dfrac{2xy^2 \sin(x - y)}{(1 + x^2 + y^2)^2}$. We use the CAS to evaluate these at $(1, 1)$, and then

substitute the results into Equation 2 to compute an equation of the tangent plane: $z = \frac{1}{3}x - \frac{1}{3}y$. The surface and tangent plane are shown in the first graph below. After zooming in, the surface and the tangent plane become almost indistinguishable, as shown in the second graph. (Here, the tangent plane is shown with fewer traces than the surface.) If we zoom in farther, the surface and the tangent plane will appear to coincide.

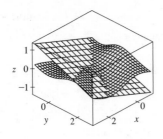
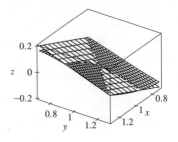

10. $f(x, y) = e^{-xy/10} \left(\sqrt{x} + \sqrt{y} + \sqrt{xy} \right)$. A CAS gives

$$f_x(x, y) = -\tfrac{1}{10} y e^{-xy/10} \left(\sqrt{x} + \sqrt{y} + \sqrt{xy} \right) + e^{-xy/10} \left(\frac{1}{2\sqrt{x}} + \frac{y}{2\sqrt{xy}} \right) \text{ and}$$

$$f_y = -\tfrac{1}{10} x e^{-xy/10} \left(\sqrt{x} + \sqrt{y} + \sqrt{xy} \right) + e^{-xy/10} \left(\frac{1}{2\sqrt{y}} + \frac{x}{2\sqrt{xy}} \right).$$ We use the CAS to evaluate these at $(1, 1)$, and

then substitute the results into Equation 2 to get an equation of the tangent plane: $z = 0.7e^{-0.1}x + 0.7e^{-0.1}y + 1.6e^{-0.1}$. The surface and tangent plane are shown in the first graph below. After zooming in, the surface and the tangent plane become almost indistinguishable, as shown in the second graph. (Here, the tangent plane is above the surface.) If we zoom in farther, the surface and the tangent plane will appear to coincide.

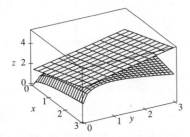
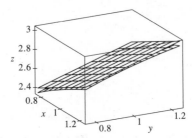

11. $f(x, y) = x\sqrt{y}$. The partial derivatives are $f_x(x, y) = \sqrt{y}$ and $f_y(x, y) = \dfrac{x}{2\sqrt{y}}$, so $f_x(1, 4) = 2$ and $f_y(1, 4) = \tfrac{1}{4}$. Both

f_x and f_y are continuous functions for $y > 0$, so by Theorem 8, f is differentiable at $(1, 4)$. By Equation 3, the linearization of f at $(1, 4)$ is given by $L(x, y) = f(1, 4) + f_x(1, 4)(x - 1) + f_y(1, 4)(y - 4) = 2 + 2(x - 1) + \tfrac{1}{4}(y - 4) = 2x + \tfrac{1}{4}y - 1$.

12. $f(x, y) = x^3 y^4$. The partial derivatives are $f_x(x, y) = 3x^2 y^4$ and $f_y(x, y) = 4x^3 y^3$, so $f_x(1, 1) = 3$ and $f_y(1, 1) = 4$. Both f_x and f_y are continuous functions, so f is differentiable at $(1, 1)$ by Theorem 8. The linearization of f at $(1, 1)$ is given by $L(x, y) = f(1, 1) + f_x(1, 1)(x - 1) + f_y(1, 1)(y - 1) = 1 + 3(x - 1) + 4(y - 1) = 3x + 4y - 6$.

13. $f(x, y) = \dfrac{x}{x + y}$. The partial derivatives are $f_x(x, y) = \dfrac{1(x + y) - x(1)}{(x + y)^2} = y/(x + y)^2$ and

$f_y(x, y) = x(-1)(x + y)^{-2} \cdot 1 = -x/(x + y)^2$, so $f_x(2, 1) = \tfrac{1}{9}$ and $f_y(2, 1) = -\tfrac{2}{9}$. Both f_x and f_y are continuous

functions for $y \neq -x$, so f is differentiable at $(2, 1)$ by Theorem 8. The linearization of f at $(2, 1)$ is given by

$$L(x, y) = f(2, 1) + f_x(2, 1)(x - 2) + f_y(2, 1)(y - 1) = \tfrac{2}{3} + \tfrac{1}{9}(x - 2) - \tfrac{2}{9}(y - 1) = \tfrac{1}{9}x - \tfrac{2}{9}y + \tfrac{2}{3}.$$

14. $f(x, y) = \sqrt{x + e^{4y}} = (x + e^{4y})^{1/2}$. The partial derivatives are $f_x(x, y) = \frac{1}{2}(x + e^{4y})^{-1/2}$ and

$f_y(x, y) = \frac{1}{2}(x + e^{4y})^{-1/2}(4e^{4y}) = 2e^{4y}(x + e^{4y})^{-1/2}$, so $f_x(3, 0) = \frac{1}{2}(3 + e^0)^{-1/2} = \frac{1}{4}$ and

$f_y(3, 0) = 2e^0(3 + e^0)^{-1/2} = 1$. Both f_x and f_y are continuous functions near $(3, 0)$, so f is

differentiable at $(3, 0)$ by Theorem 8. The linearization of f at $(3, 0)$ is

$L(x, y) = f(3, 0) + f_x(3, 0)(x - 3) + f_y(3, 0)(y - 0) = 2 + \frac{1}{4}(x - 3) + 1(y - 0) = \frac{1}{4}x + y + \frac{5}{4}$.

15. $f(x, y) = e^{-xy} \cos y$. The partial derivatives are $f_x(x, y) = e^{-xy}(-y) \cos y = -ye^{-xy} \cos y$ and

$f_y(x, y) = e^{-xy}(-\sin y) + (\cos y)e^{-xy}(-x) = -e^{-xy}(\sin y + x \cos y)$, so $f_x(\pi, 0) = 0$ and $f_y(\pi, 0) = -\pi$.

Both f_x and f_y are continuous functions, so f is differentiable at $(\pi, 0)$, and the linearization of f at $(\pi, 0)$ is

$L(x, y) = f(\pi, 0) + f_x(\pi, 0)(x - \pi) + f_y(\pi, 0)(y - 0) = 1 + 0(x - \pi) - \pi(y - 0) = 1 - \pi y$.

16. $f(x, y) = \sin(2x + 3y)$. The partial derivatives are $f_x(x, y) = 2\cos(2x + 3y)$ and $f_y(x, y) = 3\cos(2x + 3y)$, so

$f_x(-3, 2) = 2$ and $f_y(-3, 2) = 3$. Both f_x and f_y are continuous functions, so f is differentiable at $(-3, 2)$, and the

linearization of f at $(-3, 2)$ is

$L(x, y) = f(-3, 2) + f_x(-3, 2)(x + 3) + f_y(-3, 2)(y - 2) = 0 + 2(x + 3) + 3(y - 2) = 2x + 3y$.

17. Let $f(x, y) = \dfrac{2x + 3}{4y + 1}$. Then $f_x(x, y) = \dfrac{2}{4y + 1}$ and $f_y(x, y) = (2x + 3)(-1)(4y + 1)^{-2}(4) = \dfrac{-8x - 12}{(4y + 1)^2}$. Both f_x and f_y

are continuous functions for $y \neq -\frac{1}{4}$, so by Theorem 8, f is differentiable at $(0, 0)$. We have $f_x(0, 0) = 2$, $f_y(0, 0) = -12$

and the linear approximation of f at $(0, 0)$ is $f(x, y) \approx f(0, 0) + f_x(0, 0)(x - 0) + f_y(0, 0)(y - 0) = 3 + 2x - 12y$.

18. Let $f(x, y) = \sqrt{y + \cos^2 x}$. Then $f_x(x, y) = \frac{1}{2}(y + \cos^2 x)^{-1/2}(2\cos x)(-\sin x) = -\cos x \sin x / \sqrt{y + \cos^2 x}$ and

$f_y(x, y) = \frac{1}{2}(y + \cos^2 x)^{-1/2}(1) = 1/\left(2\sqrt{y + \cos^2 x}\right)$. Both f_x and f_y are continuous functions for $y > -\cos^2 x$, so f

is differentiable at $(0, 0)$ by Theorem 8. We have $f_x(0, 0) = 0$ and $f_y(0, 0) = \frac{1}{2}$, so the linear approximation of f at $(0, 0)$ is

$f(x, y) \approx f(0, 0) + f_x(0, 0)(x - 0) + f_y(0, 0)(y - 0) = 1 + 0x + \frac{1}{2}y = 1 + \frac{1}{2}y$.

19. $f(x, y) = \sqrt{20 - x^2 - 7y^2} \quad \Rightarrow \quad f_x(x, y) = -\dfrac{x}{\sqrt{20 - x^2 - 7y^2}}$ and $f_y(x, y) = -\dfrac{7y}{\sqrt{20 - x^2 - 7y^2}}$,

so $f_x(2, 1) = -\frac{2}{3}$ and $f_y(2, 1) = -\frac{7}{3}$. Then the linear approximation of f at $(2, 1)$ is given by

$f(x, y) \approx f(2, 1) + f_x(2, 1)(x - 2) + f_y(2, 1)(y - 1) = 3 - \frac{2}{3}(x - 2) - \frac{7}{3}(y - 1) = -\frac{2}{3}x - \frac{7}{3}y + \frac{20}{3}$.

Thus $f(1.95, 1.08) \approx -\frac{2}{3}(1.95) - \frac{7}{3}(1.08) + \frac{20}{3} = 2.84\overline{6}$.

20. $f(x, y) = \ln(x - 3y) \quad \Rightarrow \quad f_x(x, y) = \dfrac{1}{x - 3y}$ and $f_y(x, y) = -\dfrac{3}{x - 3y}$, so $f_x(7, 2) = 1$ and $f_y(7, 2) = -3$.

Then the linear approximation of f at $(7, 2)$ is given by

$\qquad f(x, y) \approx f(7, 2) + f_x(7, 2)(x - 7) + f_y(7, 2)(y - 2)$

$\qquad\qquad = 0 + 1(x - 7) - 3(y - 2) = x - 3y - 1$

Thus $f(6.9, 2.06) \approx 6.9 - 3(2.06) - 1 = -0.28$. The graph shows

that our approximated value is slightly greater than the actual value.

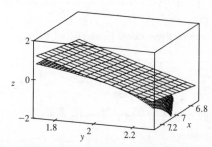

21. $f(x, y, z) = \sqrt{x^2 + y^2 + z^2} \Rightarrow f_x(x, y, z) = \dfrac{x}{\sqrt{x^2 + y^2 + z^2}}$, $f_y(x, y, z) = \dfrac{y}{\sqrt{x^2 + y^2 + z^2}}$, and

$f_z(x, y, z) = \dfrac{z}{\sqrt{x^2 + y^2 + z^2}}$, so $f_x(3, 2, 6) = \frac{3}{7}$, $f_y(3, 2, 6) = \frac{2}{7}$, $f_z(3, 2, 6) = \frac{6}{7}$. Then the linear approximation of f at

$(3, 2, 6)$ is given by

$$f(x, y, z) \approx f(3, 2, 6) + f_x(3, 2, 6)(x - 3) + f_y(3, 2, 6)(y - 2) + f_z(3, 2, 6)(z - 6)$$

$$= 7 + \tfrac{3}{7}(x - 3) + \tfrac{2}{7}(y - 2) + \tfrac{6}{7}(z - 6) = \tfrac{3}{7}x + \tfrac{2}{7}y + \tfrac{6}{7}z$$

Thus $\sqrt{(3.02)^2 + (1.97)^2 + (5.99)^2} = f(3.02, 1.97, 5.99) \approx \frac{3}{7}(3.02) + \frac{2}{7}(1.97) + \frac{6}{7}(5.99) \approx 6.9914$.

22. From the table, $f(40, 20) = 28$. To estimate $f_v(40, 20)$ and $f_t(40, 20)$ we follow the procedure used in Exercise 15.3.4

[ET 14.3.4]. Since $f_v(40, 20) = \lim\limits_{h \to 0} \dfrac{f(40 + h, 20) - f(40, 20)}{h}$, we approximate this quantity with $h = \pm 10$ and use the

values given in the table:

$$f_v(40, 20) \approx \dfrac{f(50, 20) - f(40, 20)}{10} = \dfrac{40 - 28}{10} = 1.2, \quad f_v(40, 20) \approx \dfrac{f(30, 20) - f(40, 20)}{-10} = \dfrac{17 - 28}{-10} = 1.1$$

Averaging these values gives $f_v(40, 20) \approx 1.15$. Similarly, $f_t(40, 20) = \lim\limits_{h \to 0} \dfrac{f(40, 20 + h) - f(40, 20)}{h}$, so we use $h = 10$

and $h = -5$:

$$f_t(40, 20) \approx \dfrac{f(40, 30) - f(40, 20)}{10} = \dfrac{31 - 28}{10} = 0.3, \quad f_t(40, 20) \approx \dfrac{f(40, 15) - f(40, 20)}{-5} = \dfrac{25 - 28}{-5} = 0.6$$

Averaging these values gives $f_t(40, 15) \approx 0.45$. The linear approximation, then, is

$$f(v, t) \approx f(40, 20) + f_v(40, 20)(v - 40) + f_t(40, 20)(t - 20) \approx 28 + 1.15(v - 40) + 0.45(t - 20)$$

When $v = 43$ and $t = 24$, we estimate $f(43, 24) \approx 28 + 1.15(43 - 40) + 0.45(24 - 20) = 33.25$, so we would expect the

wave heights to be approximately 33.25 ft.

23. From the table, $f(94, 80) = 127$. To estimate $f_T(94, 80)$ and $f_H(94, 80)$ we follow the procedure used in Section 15.3

[ET 14.3]. Since $f_T(94, 80) = \lim\limits_{h \to 0} \dfrac{f(94 + h, 80) - f(94, 80)}{h}$, we approximate this quantity with $h = \pm 2$ and use the

values given in the table:

$$f_T(94, 80) \approx \dfrac{f(96, 80) - f(94, 80)}{2} = \dfrac{135 - 127}{2} = 4, \quad f_T(94, 80) \approx \dfrac{f(92, 80) - f(94, 80)}{-2} = \dfrac{119 - 127}{-2} = 4$$

Averaging these values gives $f_T(94, 80) \approx 4$. Similarly, $f_H(94, 80) = \lim\limits_{h \to 0} \dfrac{f(94, 80 + h) - f(94, 80)}{h}$, so we use $h = \pm 5$:

$$f_H(94, 80) \approx \dfrac{f(94, 85) - f(94, 80)}{5} = \dfrac{132 - 127}{5} = 1, \quad f_H(94, 80) \approx \dfrac{f(94, 75) - f(94, 80)}{-5} = \dfrac{122 - 127}{-5} = 1$$

Averaging these values gives $f_H(94, 80) \approx 1$. The linear approximation, then, is

$$f(T, H) \approx f(94, 80) + f_T(94, 80)(T - 94) + f_H(94, 80)(H - 80)$$

$$\approx 127 + 4(T - 94) + 1(H - 80) \qquad [\text{or } 4T + H - 329]$$

Thus when $T = 95$ and $H = 78$, $f(95, 78) \approx 127 + 4(95 - 94) + 1(78 - 80) = 129$, so we estimate the heat index to be

approximately $129°$F.

24. From the table, $f(-15, 50) = -29$. To estimate $f_T(-15, 50)$ and $f_v(-15, 50)$ we follow the procedure used in Section 15.3

[ET 14.3]. Since $f_T(-15, 50) = \lim\limits_{h \to 0} \dfrac{f(-15 + h, 50) - f(-15, 50)}{h}$, we approximate this quantity with $h = \pm 5$ and use the

values given in the table:

$$f_T(-15, 50) \approx \frac{f(-10, 50) - f(-15, 50)}{5} = \frac{-22 - (-29)}{5} = 1.4$$

$$f_T(-15, 50) \approx \frac{f(-20, 50) - f(-15, 50)}{-5} = \frac{-35 - (-29)}{-5} = 1.2$$

Averaging these values gives $f_T(-15, 50) \approx 1.3$. Similarly $f_v(-15, 50) = \lim\limits_{h \to 0} \dfrac{f(-15, 50 + h) - f(-15, 50)}{h}$,

so we use $h = \pm 10$:

$$f_v(-15, 50) \approx \frac{f(-15, 60) - f(-15, 50)}{10} = \frac{-30 - (-29)}{10} = -0.1$$

$$f_v(-15, 50) \approx \frac{f(-15, 40) - f(-15, 50)}{-10} = \frac{-27 - (-29)}{-10} = -0.2$$

Averaging these values gives $f_v(-15, 50) \approx -0.15$. The linear approximation to the wind-chill index function, then, is

$f(T, v) \approx f(-15, 50) + f_T(-15, 50)(T - (-15)) + f_v(-15, 50)(v - 50) \approx -29 + (1.3)(T + 15) - (0.15)(v - 50)$.

Thus when $T = -17°\text{C}$ and $v = 55$ km/h, $f(-17, 55) \approx -29 + (1.3)(-17 + 15) - (0.15)(55 - 50) = -32.35$, so we

estimate the wind-chill index to be approximately $-32.35°\text{C}$.

25. $z = x^3 \ln(y^2) \quad \Rightarrow \quad dz = \dfrac{\partial z}{\partial x} dx + \dfrac{\partial z}{\partial y} dy = 3x^2 \ln(y^2)\, dx + x^3 \cdot \dfrac{1}{y^2}(2y)\, dy = 3x^2 \ln(y^2)\, dx + \dfrac{2x^3}{y} dy$

26. $v = y \cos xy \quad \Rightarrow$

$dv = \dfrac{\partial v}{\partial x} dx + \dfrac{\partial v}{\partial y} dy = y(-\sin xy)y\, dx + [y(-\sin xy)x + \cos xy]\, dy = -y^2 \sin xy\, dx + (\cos xy - xy \sin xy)\, dy$

27. $m = p^5 q^3 \quad \Rightarrow \quad dm = \dfrac{\partial m}{\partial p} dp + \dfrac{\partial m}{\partial q} dq = 5p^4 q^3\, dp + 3p^5 q^2\, dq$

28. $T = \dfrac{v}{1 + uvw} \quad \Rightarrow$

$$dT = \frac{\partial T}{\partial u}\, du + \frac{\partial T}{\partial v}\, dv + \frac{\partial T}{\partial w}\, dw$$

$$= v(-1)(1 + uvw)^{-2}(vw)\, du + \frac{1(1 + uvw) - v(uw)}{(1 + uvw)^2}\, dv + v(-1)(1 + uvw)^{-2}(uv)\, dw$$

$$= -\frac{v^2 w}{(1 + uvw)^2}\, du + \frac{1}{(1 + uvw)^2}\, dv - \frac{uv^2}{(1 + uvw)^2}\, dw$$

29. $R = \alpha \beta^2 \cos \gamma \quad \Rightarrow \quad dR = \dfrac{\partial R}{\partial \alpha}\, d\alpha + \dfrac{\partial R}{\partial \beta}\, d\beta + \dfrac{\partial R}{\partial \gamma}\, d\gamma = \beta^2 \cos \gamma\, d\alpha + 2\alpha\beta \cos \gamma\, d\beta - \alpha\beta^2 \sin \gamma\, d\gamma$

30. $w = xye^{xz} \quad \Rightarrow$

$dw = \dfrac{\partial w}{\partial x} dx + \dfrac{\partial w}{\partial y} dy + \dfrac{\partial w}{\partial z} dz = (xyze^{xz} + ye^{xz})\, dx + xe^{xz}\, dy + x^2 ye^{xz}\, dz = (xz + 1)ye^{xz}\, dx + xe^{xz}\, dy + x^2 ye^{xz}\, dz$

31. $dx = \Delta x = 0.05$, $dy = \Delta y = 0.1$, $z = 5x^2 + y^2$, $z_x = 10x$, $z_y = 2y$. Thus when $x = 1$ and $y = 2$,

$dz = z_x(1, 2)\, dx + z_y(1, 2)\, dy = (10)(0.05) + (4)(0.1) = 0.9$ while

$\Delta z = f(1.05, 2.1) - f(1, 2) = 5(1.05)^2 + (2.1)^2 - 5 - 4 = 0.9225$.

32. $dx = \Delta x = -0.04$, $dy = \Delta y = 0.05$, $z = x^2 - xy + 3y^2$, $z_x = 2x - y$, $z_y = 6y - x$. Thus when $x = 3$ and $y = -1$,

$dz = (7)(-0.04) + (-9)(0.05) = -0.73$ while $\Delta z = (2.96)^2 - (2.96)(-0.95) + 3(-0.95)^2 - (9 + 3 + 3) = -0.7189$.

33. $dA = \dfrac{\partial A}{\partial x}\, dx + \dfrac{\partial A}{\partial y}\, dy = y\, dx + x\, dy$ and $|\Delta x| \le 0.1$, $|\Delta y| \le 0.1$. We use $dx = 0.1$, $dy = 0.1$ with $x = 30$, $y = 24$;

then the maximum error in the area is about $dA = 24(0.1) + 30(0.1) = 5.4 \text{ cm}^2$.

34. Let S be surface area. Then $S = 2(xy + xz + yz)$ and $dS = 2(y + z)\, dx + 2(x + z)\, dy + 2(x + y)\, dz$. The maximum error

occurs with $\Delta x = \Delta y = \Delta z = 0.2$. Using $dx = \Delta x$, $dy = \Delta y$, $dz = \Delta z$ we find the maximum error in calculated surface

area to be about $dS = (220)(0.2) + (260)(0.2) + (280)(0.2) = 152 \text{ cm}^2$.

35. The volume of a can is $V = \pi r^2 h$ and $\Delta V \approx dV$ is an estimate of the amount of tin. Here $dV = 2\pi rh\, dr + \pi r^2\, dh$, so put

$dr = 0.04$, $dh = 0.08$ (0.04 on top, 0.04 on bottom) and then $\Delta V \approx dV = 2\pi(48)(0.04) + \pi(16)(0.08) \approx 16.08 \text{ cm}^3$.

Thus the amount of tin is about 16 cm³.

36. Let V be the volume. Then $V = \pi r^2 h$ and $\Delta V \approx dV = 2\pi rh\, dr + \pi r^2\, dh$ is an estimate of the amount of metal. With

$dr = 0.05$ and $dh = 0.2$ we get $dV = 2\pi(2)(10)(0.05) + \pi(2)^2(0.2) = 2.80\pi \approx 8.8 \text{ cm}^3$.

37. The area of the rectangle is $A = xy$, and $\Delta A \approx dA$ is an estimate of the area of paint in the stripe. Here $dA = y\, dx + x\, dy$,

so with $dx = dy = \frac{3+3}{12} = \frac{1}{2}$, $\Delta A \approx dA = (100)\left(\frac{1}{2}\right) + (200)\left(\frac{1}{2}\right) = 150 \text{ ft}^2$. Thus there are approximately 150 ft² of paint

in the stripe.

38. Here $dV = \Delta V = 0.3$, $dT = \Delta T = -5$, $P = 8.31\dfrac{T}{V}$, so

$$dP = \left(\frac{8.31}{V}\right)dT - \frac{8.31 \cdot T}{V^2}\,dV = 8.31\left[-\frac{5}{12} - \frac{310}{144}\cdot\frac{3}{10}\right] \approx -8.83. \text{ Thus the pressure will drop by about } 8.83 \text{ kPa.}$$

39. First we find $\dfrac{\partial R}{\partial R_1}$ implicitly by taking partial derivatives of both sides with respect to R_1:

$$\frac{\partial}{\partial R_1}\left(\frac{1}{R}\right) = \frac{\partial\left[(1/R_1) + (1/R_2) + (1/R_3)\right]}{\partial R_1} \;\Rightarrow\; -R^{-2}\frac{\partial R}{\partial R_1} = -R_1^{-2} \;\Rightarrow\; \frac{\partial R}{\partial R_1} = \frac{R^2}{R_1^2}. \text{ Then by symmetry,}$$

$\dfrac{\partial R}{\partial R_2} = \dfrac{R^2}{R_2^2}$, $\dfrac{\partial R}{\partial R_3} = \dfrac{R^2}{R_3^2}$. When $R_1 = 25$, $R_2 = 40$ and $R_3 = 50$, $\dfrac{1}{R} = \dfrac{17}{200}$ $\Leftrightarrow$ $R = \frac{200}{17}\ \Omega$.

Since the possible error for each R_i is 0.5%, the maximum error of R is attained by setting $\Delta R_i = 0.005R_i$. So

$$\Delta R \approx dR = \frac{\partial R}{\partial R_1}\Delta R_1 + \frac{\partial R}{\partial R_2}\Delta R_2 + \frac{\partial R}{\partial R_3}\Delta R_3 = (0.005)R^2\left(\frac{1}{R_1} + \frac{1}{R_2} + \frac{1}{R_3}\right) = (0.005)R = \tfrac{1}{17} \approx 0.059\ \Omega.$$

40. Let x, y, z and w be the four numbers with $p(x, y, z, w) = xyzw$. Since the largest error due to rounding

for each number is 0.05, the maximum error in the calculated product is approximated by

$dp = (yzw)(0.05) + (xzw)(0.05) + (xyw)(0.05) + (xyz)(0.05)$. Furthermore, each of the numbers is positive but less than

50, so the product of any three is between 0 and $(50)^3$. Thus $dp \le 4(50)^3(0.05) = 25{,}000$.

41. The errors in measurement are at most 2%, so $\left|\dfrac{\Delta w}{w}\right| \le 0.02$ and $\left|\dfrac{\Delta h}{h}\right| \le 0.02$. The relative error in the calculated surface

area is

$$\frac{\Delta S}{S} \approx \frac{dS}{S} = \frac{0.1091(0.425 w^{0.425-1})h^{0.725}\,dw + 0.1091 w^{0.425}(0.725 h^{0.725-1})\,dh}{0.1091 w^{0.425} h^{0.725}} = 0.425\frac{dw}{w} + 0.725\frac{dh}{h}$$

To estimate the maximum relative error, we use $\dfrac{dw}{w} = \left|\dfrac{\Delta w}{w}\right| = 0.02$ and $\dfrac{dh}{h} = \left|\dfrac{\Delta h}{h}\right| = 0.02 \quad \Rightarrow$

$\dfrac{dS}{S} = 0.425\,(0.02) + 0.725\,(0.02) = 0.023$. Thus the maximum percentage error is approximately 2.3%.

42. $\mathbf{r}_1(t) = \langle 2 + 3t, 1 - t^2, 3 - 4t + t^2\rangle \quad \Rightarrow \quad \mathbf{r}_1'(t) = \langle 3, -2t, -4 + 2t\rangle,\ \mathbf{r}_2(u) = \langle 1 + u^2, 2u^3 - 1, 2u + 1\rangle \quad \Rightarrow$

$\mathbf{r}_2'(u) = \langle 2u, 6u^2, 2\rangle$. Both curves pass through P since $\mathbf{r}_1(0) = \mathbf{r}_2(1) = \langle 2, 1, 3\rangle$, so the tangent vectors

$\mathbf{r}_1'(0) = \langle 3, 0, -4\rangle$ and $\mathbf{r}_2'(1) = \langle 2, 6, 2\rangle$ are both parallel to the tangent plane to S at P. A normal vector for the tangent

plane is $\mathbf{r}_1'(0) \times \mathbf{r}_2'(1) = \langle 3, 0, -4\rangle \times \langle 2, 6, 2\rangle = \langle 24, -14, 18\rangle$, so an equation of the tangent plane is

$24(x - 2) - 14(y - 1) + 18(z - 3) = 0$ or $12x - 7y + 9z = 44$.

43. $\Delta z = f(a + \Delta x, b + \Delta y) - f(a, b) = (a + \Delta x)^2 + (b + \Delta y)^2 - (a^2 + b^2)$

$\qquad = a^2 + 2a\,\Delta x + (\Delta x)^2 + b^2 + 2b\,\Delta y + (\Delta y)^2 - a^2 - b^2 = 2a\,\Delta x + (\Delta x)^2 + 2b\,\Delta y + (\Delta y)^2$

But $f_x(a, b) = 2a$ and $f_y(a, b) = 2b$ and so $\Delta z = f_x(a, b)\,\Delta x + f_y(a, b)\,\Delta y + \Delta x\,\Delta x + \Delta y\,\Delta y$, which is Definition 7

with $\varepsilon_1 = \Delta x$ and $\varepsilon_2 = \Delta y$. Hence f is differentiable.

44. $\Delta z = f(a + \Delta x, b + \Delta y) - f(a, b) = (a + \Delta x)(b + \Delta y) - 5(b + \Delta y)^2 - (ab - 5b^2)$

$\qquad = ab + a\,\Delta y + b\,\Delta x + \Delta x\,\Delta y - 5b^2 - 10b\,\Delta y - 5(\Delta y)^2 - ab + 5b^2$

$\qquad = (a - 10b)\,\Delta y + b\,\Delta x + \Delta x\,\Delta y - 5\,\Delta y\,\Delta y,$

but $f_x(a, b) = b$ and $f_y(a, b) = a - 10b$ and so $\Delta z = f_x(a, b)\,\Delta x + f_y(a, b)\,\Delta y + \Delta x\,\Delta y - 5\Delta y\,\Delta y$, which is Definition 7

with $\varepsilon_1 = \Delta y$ and $\varepsilon_2 = -5\,\Delta y$. Hence f is differentiable.

45. To show that f is continuous at (a, b) we need to show that $\displaystyle\lim_{(x,y)\to(a,b)} f(x, y) = f(a, b)$ or

equivalently $\displaystyle\lim_{(\Delta x, \Delta y)\to(0,0)} f(a + \Delta x, b + \Delta y) = f(a, b)$. Since f is differentiable at (a, b),

$f(a + \Delta x, b + \Delta y) - f(a, b) = \Delta z = f_x(a, b)\,\Delta x + f_y(a, b)\,\Delta y + \varepsilon_1\,\Delta x + \varepsilon_2\,\Delta y$, where ε_1 and $\varepsilon_2 \to 0$ as

$(\Delta x, \Delta y) \to (0, 0)$. Thus $f(a + \Delta x, b + \Delta y) = f(a, b) + f_x(a, b)\,\Delta x + f_y(a, b)\,\Delta y + \varepsilon_1\,\Delta x + \varepsilon_2\,\Delta y$. Taking the limit of

both sides as $(\Delta x, \Delta y) \to (0, 0)$ gives $\displaystyle\lim_{(\Delta x, \Delta y)\to(0,0)} f(a + \Delta x, b + \Delta y) = f(a, b)$. Thus f is continuous at (a, b).

46. (a) $\displaystyle\lim_{h\to 0} \frac{f(h, 0) - f(0, 0)}{h} = \lim_{h\to 0} \frac{0 - 0}{h} = 0$ and $\displaystyle\lim_{h\to 0} \frac{f(0, h) - f(0, 0)}{h} = \lim_{h\to 0} \frac{0 - 0}{h} = 0$. Thus $f_x(0, 0) = f_y(0, 0) = 0$.

To show that f isn't differentiable at $(0, 0)$ we need only show that f is not continuous at $(0, 0)$ and apply Exercise 45. As

$(x, y) \to (0, 0)$ along the x-axis $f(x, y) = 0/x^2 = 0$ for $x \ne 0$ so $f(x, y) \to 0$ as $(x, y) \to (0, 0)$ along the x-axis. But

as $(x, y) \to (0, 0)$ along the line $y = x$, $f(x, x) = x^2/(2x^2) = \frac{1}{2}$ for $x \ne 0$ so $f(x, y) \to \frac{1}{2}$ as $(x, y) \to (0, 0)$ along this

line. Thus $\displaystyle\lim_{(x,y)\to(0,0)} f(x, y)$ doesn't exist, so f is discontinuous at $(0, 0)$ and thus not differentiable there.

(b) For $(x, y) \neq (0, 0)$, $f_x(x, y) = \dfrac{(x^2 + y^2)y - xy(2x)}{(x^2 + y^2)^2} = \dfrac{y(y^2 - x^2)}{(x^2 + y^2)^2}$. If we approach $(0, 0)$ along the y-axis, then

$$f_x(x, y) = f_x(0, y) = \frac{y^3}{y^4} = \frac{1}{y}, \text{ so } f_x(x, y) \to \pm\infty \text{ as } (x, y) \to (0, 0). \text{ Thus } \lim_{(x,y)\to(0,0)} f_x(x, y) \text{ does not exist and}$$

$f_x(x, y)$ is not continuous at $(0, 0)$. Similarly, $f_y(x, y) = \dfrac{(x^2 + y^2)x - xy(2y)}{(x^2 + y^2)^2} = \dfrac{x(x^2 - y^2)}{(x^2 + y^2)^2}$ for $(x, y) \neq (0, 0)$, and

if we approach $(0, 0)$ along the x-axis, then $f_y(x, y) = f_x(x, 0) = \dfrac{x^3}{x^4} = \dfrac{1}{x}$. Thus $\lim\limits_{(x,y)\to(0,0)} f_y(x, y)$ does not exist and

$f_y(x, y)$ is not continuous at $(0, 0)$.

15.5   The Chain Rule   ET 14.5

1. $z = x^2 + y^2 + xy$, $x = \sin t$, $y = e^t$ $\Rightarrow$ $\dfrac{dz}{dt} = \dfrac{\partial z}{\partial x}\dfrac{dx}{dt} + \dfrac{\partial z}{\partial y}\dfrac{dy}{dt} = (2x + y)\cos t + (2y + x)e^t$

2. $z = \cos(x + 4y)$, $x = 5t^4$, $y = 1/t$ $\Rightarrow$

$$\frac{dz}{dt} = \frac{\partial z}{\partial x}\frac{dx}{dt} + \frac{\partial z}{\partial y}\frac{dy}{dt} = -\sin(x + 4y)(1)(20t^3) + [-\sin(x + 4y)(4)](-t^{-2})$$

$$= -20t^3 \sin(x + 4y) + \frac{4}{t^2}\sin(x + 4y) = \left(\frac{4}{t^2} - 20t^3\right)\sin(x + 4y)$$

3. $z = \sqrt{1 + x^2 + y^2}$, $x = \ln t$, $y = \cos t$ $\Rightarrow$

$$\frac{dz}{dt} = \frac{\partial z}{\partial x}\frac{dx}{dt} + \frac{\partial z}{\partial y}\frac{dy}{dt} = \tfrac{1}{2}(1 + x^2 + y^2)^{-1/2}(2x) \cdot \frac{1}{t} + \tfrac{1}{2}(1 + x^2 + y^2)^{-1/2}(2y)(-\sin t) = \frac{1}{\sqrt{1 + x^2 + y^2}}\left(\frac{x}{t} - y\sin t\right)$$

4. $z = \tan^{-1}(y/x)$, $x = e^t$, $y = 1 - e^{-t}$ $\Rightarrow$

$$\frac{dz}{dt} = \frac{\partial z}{\partial x}\frac{dx}{dt} + \frac{\partial z}{\partial y}\frac{dy}{dt} = \frac{1}{1 + (y/x)^2}(-yx^{-2}) \cdot e^t + \frac{1}{1 + (y/x)^2}(1/x) \cdot (-e^{-t})(-1)$$

$$= -\frac{y}{x^2 + y^2} \cdot e^t + \frac{1}{x + y^2/x} \cdot e^{-t} = \frac{xe^{-t} - ye^t}{x^2 + y^2}$$

5. $w = xe^{y/z}$, $x = t^2$, $y = 1 - t$, $z = 1 + 2t$ $\Rightarrow$

$$\frac{dw}{dt} = \frac{\partial w}{\partial x}\frac{dx}{dt} + \frac{\partial w}{\partial y}\frac{dy}{dt} + \frac{\partial w}{\partial z}\frac{dz}{dt} = e^{y/z} \cdot 2t + xe^{y/z}\left(\frac{1}{z}\right) \cdot (-1) + xe^{y/z}\left(-\frac{y}{z^2}\right) \cdot 2 = e^{y/z}\left(2t - \frac{x}{z} - \frac{2xy}{z^2}\right)$$

6. $w = \ln\sqrt{x^2 + y^2 + z^2} = \tfrac{1}{2}\ln(x^2 + y^2 + z^2)$, $x = \sin t$, $y = \cos t$, $z = \tan t$ $\Rightarrow$

$$\frac{dw}{dt} = \frac{\partial w}{\partial x}\frac{dx}{dt} + \frac{\partial w}{\partial y}\frac{dy}{dt} + \frac{\partial w}{\partial z}\frac{dz}{dt} = \frac{1}{2} \cdot \frac{2x}{x^2 + y^2 + z^2} \cdot \cos t + \frac{1}{2} \cdot \frac{2y}{x^2 + y^2 + z^2} \cdot (-\sin t) + \frac{1}{2} \cdot \frac{2z}{x^2 + y^2 + z^2} \cdot \sec^2 t$$

$$= \frac{x\cos t - y\sin t + z\sec^2 t}{x^2 + y^2 + z^2}$$

7. $z = x^2 y^3$, $x = s\cos t$, $y = s\sin t$ $\Rightarrow$

$$\frac{\partial z}{\partial s} = \frac{\partial z}{\partial x}\frac{\partial x}{\partial s} + \frac{\partial z}{\partial y}\frac{\partial y}{\partial s} = 2xy^3\cos t + 3x^2 y^2 \sin t$$

$$\frac{\partial z}{\partial t} = \frac{\partial z}{\partial x}\frac{\partial x}{\partial t} + \frac{\partial z}{\partial y}\frac{\partial y}{\partial t} = (2xy^3)(-s\sin t) + (3x^2 y^2)(s\cos t) = -2sxy^3 \sin t + 3sx^2 y^2 \cos t$$

8. $z = \arcsin(x - y)$, $x = s^2 + t^2$, $y = 1 - 2st$ $\Rightarrow$

$$\frac{\partial z}{\partial s} = \frac{\partial z}{\partial x}\frac{\partial x}{\partial s} + \frac{\partial z}{\partial y}\frac{\partial y}{\partial s} = \frac{1}{\sqrt{1 - (x - y)^2}}(1) \cdot 2s + \frac{1}{\sqrt{1 - (x - y)^2}}(-1) \cdot (-2t) = \frac{2s + 2t}{\sqrt{1 - (x - y)^2}}$$

$$\frac{\partial z}{\partial t} = \frac{\partial z}{\partial x}\frac{\partial x}{\partial t} + \frac{\partial z}{\partial y}\frac{\partial y}{\partial t} = \frac{1}{\sqrt{1 - (x - y)^2}}(1) \cdot 2t + \frac{1}{\sqrt{1 - (x - y)^2}}(-1) \cdot (-2s) = \frac{2s + 2t}{\sqrt{1 - (x - y)^2}}$$

9. $z = \sin\theta\cos\phi$, $\theta = st^2$, $\phi = s^2 t$ $\Rightarrow$

$$\frac{\partial z}{\partial s} = \frac{\partial z}{\partial \theta}\frac{\partial \theta}{\partial s} + \frac{\partial z}{\partial \phi}\frac{\partial \phi}{\partial s} = (\cos\theta\,\cos\phi)(t^2) + (-\sin\theta\,\sin\phi)(2st) = t^2\cos\theta\,\cos\phi - 2st\sin\theta\,\sin\phi$$

$$\frac{\partial z}{\partial t} = \frac{\partial z}{\partial \theta}\frac{\partial \theta}{\partial t} + \frac{\partial z}{\partial \phi}\frac{\partial \phi}{\partial t} = (\cos\theta\,\cos\phi)(2st) + (-\sin\theta\,\sin\phi)(s^2) = 2st\cos\theta\,\cos\phi - s^2\sin\theta\,\sin\phi$$

10. $z = e^{x+2y}$, $x = s/t$, $y = t/s$ $\Rightarrow$

$$\frac{\partial z}{\partial s} = \frac{\partial z}{\partial x}\frac{\partial x}{\partial s} + \frac{\partial z}{\partial y}\frac{\partial y}{\partial s} = (e^{x+2y})(1/t) + (2e^{x+2y})(-ts^{-2}) = e^{x+2y}\left(\frac{1}{t} - \frac{2t}{s^2}\right)$$

$$\frac{\partial z}{\partial t} = \frac{\partial z}{\partial x}\frac{\partial x}{\partial t} + \frac{\partial z}{\partial y}\frac{\partial y}{\partial t} = (e^{x+2y})(-st^{-2}) + (2e^{x+2y})(1/s) = e^{x+2y}\left(\frac{2}{s} - \frac{s}{t^2}\right)$$

11. $z = e^r\cos\theta$, $r = st$, $\theta = \sqrt{s^2 + t^2}$ $\Rightarrow$

$$\frac{\partial z}{\partial s} = \frac{\partial z}{\partial r}\frac{\partial r}{\partial s} + \frac{\partial z}{\partial \theta}\frac{\partial \theta}{\partial s} = e^r\cos\theta \cdot t + e^r(-\sin\theta) \cdot \tfrac{1}{2}(s^2 + t^2)^{-1/2}(2s) = te^r\cos\theta - e^r\sin\theta \cdot \frac{s}{\sqrt{s^2 + t^2}}$$

$$= e^r\left(t\cos\theta - \frac{s}{\sqrt{s^2 + t^2}}\sin\theta\right)$$

$$\frac{\partial z}{\partial t} = \frac{\partial z}{\partial r}\frac{\partial r}{\partial t} + \frac{\partial z}{\partial \theta}\frac{\partial \theta}{\partial t} = e^r\cos\theta \cdot s + e^r(-\sin\theta) \cdot \tfrac{1}{2}(s^2 + t^2)^{-1/2}(2t) = se^r\cos\theta - e^r\sin\theta \cdot \frac{t}{\sqrt{s^2 + t^2}}$$

$$= e^r\left(s\cos\theta - \frac{t}{\sqrt{s^2 + t^2}}\sin\theta\right)$$

12. $z = \tan(u/v)$, $u = 2s + 3t$, $v = 3s - 2t$ $\Rightarrow$

$$\frac{\partial z}{\partial s} = \frac{\partial z}{\partial u}\frac{\partial u}{\partial s} + \frac{\partial z}{\partial v}\frac{\partial v}{\partial s} = \sec^2(u/v)(1/v) \cdot 2 + \sec^2(u/v)(-uv^{-2}) \cdot 3$$

$$= \frac{2}{v}\sec^2\left(\frac{u}{v}\right) - \frac{3u}{v^2}\sec^2\left(\frac{u}{v}\right) = \frac{2v - 3u}{v^2}\sec^2\left(\frac{u}{v}\right)$$

$$\frac{\partial z}{\partial t} = \frac{\partial z}{\partial u}\frac{\partial u}{\partial t} + \frac{\partial z}{\partial v}\frac{\partial v}{\partial t} = \sec^2(u/v)(1/v) \cdot 3 + \sec^2(u/v)(-uv^{-2}) \cdot (-2)$$

$$= \frac{3}{v}\sec^2\left(\frac{u}{v}\right) + \frac{2u}{v^2}\sec^2\left(\frac{u}{v}\right) = \frac{2u + 3v}{v^2}\sec^2\left(\frac{u}{v}\right)$$

13. When $t = 3$, $x = g(3) = 2$ and $y = h(3) = 7$. By the Chain Rule (2),

$$\frac{dz}{dt} = \frac{\partial f}{\partial x}\frac{dx}{dt} + \frac{\partial f}{\partial y}\frac{dy}{dt} = f_x(2, 7)g'(3) + f_y(2, 7)\,h'(3) = (6)(5) + (-8)(-4) = 62.$$

14. By the Chain Rule (3), $\dfrac{\partial W}{\partial s} = \dfrac{\partial W}{\partial u}\dfrac{\partial u}{\partial s} + \dfrac{\partial W}{\partial v}\dfrac{\partial v}{\partial s}$ $\Rightarrow$

$$W_s(1, 0) = F_u(u(1, 0), v(1, 0))\,u_s(1, 0) + F_v(u(1, 0), v(1, 0))\,v_s(1, 0) = F_u(2, 3)u_s(1, 0) + F_v(2, 3)v_s(1, 0)$$

$$= (-1)(-2) + (10)(5) = 52$$

Similarly, $\dfrac{\partial W}{\partial t} = \dfrac{\partial W}{\partial u}\dfrac{\partial u}{\partial t} + \dfrac{\partial W}{\partial v}\dfrac{\partial v}{\partial t} \;\Rightarrow$

$$W_t(1,0) = F_u(u(1,0), v(1,0))\, u_t(1,0) + F_v(u(1,0), v(1,0))\, v_t(1,0) = F_u(2,3)u_t(1,0) + F_v(2,3)v_t(1,0)$$

$$= (-1)(6) + (10)(4) = 34$$

15. $g(u,v) = f(x(u,v), y(u,v))$ where $x = e^u + \sin v,\; y = e^u + \cos v \;\Rightarrow$

$\dfrac{\partial x}{\partial u} = e^u,\; \dfrac{\partial x}{\partial v} = \cos v,\; \dfrac{\partial y}{\partial u} = e^u,\; \dfrac{\partial y}{\partial v} = -\sin v$. By the Chain Rule (3), $\dfrac{\partial g}{\partial u} = \dfrac{\partial f}{\partial x}\dfrac{\partial x}{\partial u} + \dfrac{\partial f}{\partial y}\dfrac{\partial y}{\partial u}$. Then

$$g_u(0,0) = f_x(x(0,0), y(0,0))\, x_u(0,0) + f_y(x(0,0), y(0,0))\, y_u(0,0) = f_x(1,2)(e^0) + f_y(1,2)(e^0) = 2(1) + 5(1) = 7.$$

Similarly, $\dfrac{\partial g}{\partial v} = \dfrac{\partial f}{\partial x}\dfrac{\partial x}{\partial v} + \dfrac{\partial f}{\partial y}\dfrac{\partial y}{\partial v}$. Then

$$g_v(0,0) = f_x(x(0,0), y(0,0))\, x_v(0,0) + f_y(x(0,0), y(0,0))\, y_v(0,0) = f_x(1,2)(\cos 0) + f_y(1,2)(-\sin 0)$$

$$= 2(1) + 5(0) = 2$$

16. $g(r,s) = f(x(r,s), y(r,s))$ where $x = 2r - s,\; y = s^2 - 4r \;\Rightarrow\; \dfrac{\partial x}{\partial r} = 2,\; \dfrac{\partial x}{\partial s} = -1,\; \dfrac{\partial y}{\partial r} = -4,\; \dfrac{\partial y}{\partial s} = 2s.$

By the Chain Rule (3), $\dfrac{\partial g}{\partial r} = \dfrac{\partial f}{\partial x}\dfrac{\partial x}{\partial r} + \dfrac{\partial f}{\partial y}\dfrac{\partial y}{\partial r}$. Then

$$g_r(1,2) = f_x(x(1,2), y(1,2))\, x_r(1,2) + f_y(x(1,2), y(1,2))\, y_r(1,2) = f_x(0,0)(2) + f_y(0,0)(-4)$$

$$= 4(2) + 8(-4) = -24$$

Similarly, $\dfrac{\partial g}{\partial s} = \dfrac{\partial f}{\partial x}\dfrac{\partial x}{\partial s} + \dfrac{\partial f}{\partial y}\dfrac{\partial y}{\partial s}$. Then

$$g_s(1,2) = f_x(x(1,2), y(1,2))\, x_s(1,2) + f_y(x(1,2), y(1,2))\, y_s(1,2) = f_x(0,0)(-1) + f_y(0,0)(4)$$

$$= 4(-1) + 8(4) = 28$$

17.

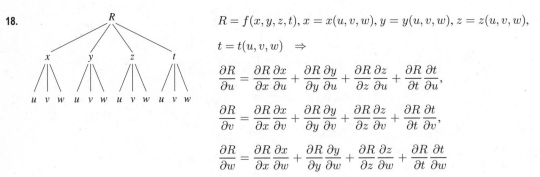

$u = f(x,y),\; x = x(r,s,t),\; y = y(r,s,t) \;\Rightarrow$

$$\dfrac{\partial u}{\partial r} = \dfrac{\partial u}{\partial x}\dfrac{\partial x}{\partial r} + \dfrac{\partial u}{\partial y}\dfrac{\partial y}{\partial r},\; \dfrac{\partial u}{\partial s} = \dfrac{\partial u}{\partial x}\dfrac{\partial x}{\partial s} + \dfrac{\partial u}{\partial y}\dfrac{\partial y}{\partial s},\; \dfrac{\partial u}{\partial t} = \dfrac{\partial u}{\partial x}\dfrac{\partial x}{\partial t} + \dfrac{\partial u}{\partial y}\dfrac{\partial y}{\partial t}$$

18.

$R = f(x,y,z,t),\; x = x(u,v,w),\; y = y(u,v,w),\; z = z(u,v,w),$

$t = t(u,v,w) \;\Rightarrow$

$$\dfrac{\partial R}{\partial u} = \dfrac{\partial R}{\partial x}\dfrac{\partial x}{\partial u} + \dfrac{\partial R}{\partial y}\dfrac{\partial y}{\partial u} + \dfrac{\partial R}{\partial z}\dfrac{\partial z}{\partial u} + \dfrac{\partial R}{\partial t}\dfrac{\partial t}{\partial u},$$

$$\dfrac{\partial R}{\partial v} = \dfrac{\partial R}{\partial x}\dfrac{\partial x}{\partial v} + \dfrac{\partial R}{\partial y}\dfrac{\partial y}{\partial v} + \dfrac{\partial R}{\partial z}\dfrac{\partial z}{\partial v} + \dfrac{\partial R}{\partial t}\dfrac{\partial t}{\partial v},$$

$$\dfrac{\partial R}{\partial w} = \dfrac{\partial R}{\partial x}\dfrac{\partial x}{\partial w} + \dfrac{\partial R}{\partial y}\dfrac{\partial y}{\partial w} + \dfrac{\partial R}{\partial z}\dfrac{\partial z}{\partial w} + \dfrac{\partial R}{\partial t}\dfrac{\partial t}{\partial w}$$

19.

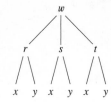

$w = f(r, s, t), \quad r = r(x, y), \quad s = s(x, y), \quad t = t(x, y) \quad \Rightarrow$

$$\frac{\partial w}{\partial x} = \frac{\partial w}{\partial r}\frac{\partial r}{\partial x} + \frac{\partial w}{\partial s}\frac{\partial s}{\partial x} + \frac{\partial w}{\partial t}\frac{\partial t}{\partial x}, \quad \frac{\partial w}{\partial y} = \frac{\partial w}{\partial r}\frac{\partial r}{\partial y} + \frac{\partial w}{\partial s}\frac{\partial s}{\partial y} + \frac{\partial w}{\partial t}\frac{\partial t}{\partial y}$$

20.

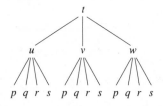

$t = f(u, v, w), \quad u = u(p, q, r, s), \quad v = v(p, q, r, s), \quad w = w(p, q, r, s) \quad \Rightarrow$

$$\frac{\partial t}{\partial p} = \frac{\partial t}{\partial u}\frac{\partial u}{\partial p} + \frac{\partial t}{\partial v}\frac{\partial v}{\partial p} + \frac{\partial t}{\partial w}\frac{\partial w}{\partial p}, \quad \frac{\partial t}{\partial q} = \frac{\partial t}{\partial u}\frac{\partial u}{\partial q} + \frac{\partial t}{\partial v}\frac{\partial v}{\partial q} + \frac{\partial t}{\partial w}\frac{\partial w}{\partial q},$$

$$\frac{\partial t}{\partial r} = \frac{\partial t}{\partial u}\frac{\partial u}{\partial r} + \frac{\partial t}{\partial v}\frac{\partial v}{\partial r} + \frac{\partial t}{\partial w}\frac{\partial w}{\partial r}, \quad \frac{\partial t}{\partial s} = \frac{\partial t}{\partial u}\frac{\partial u}{\partial s} + \frac{\partial t}{\partial v}\frac{\partial v}{\partial s} + \frac{\partial t}{\partial w}\frac{\partial w}{\partial s}$$

21. $z = x^2 + xy^3, \quad x = uv^2 + w^3, \quad y = u + ve^w \quad \Rightarrow$

$$\frac{\partial z}{\partial u} = \frac{\partial z}{\partial x}\frac{\partial x}{\partial u} + \frac{\partial z}{\partial y}\frac{\partial y}{\partial u} = (2x + y^3)(v^2) + (3xy^2)(1),$$

$$\frac{\partial z}{\partial v} = \frac{\partial z}{\partial x}\frac{\partial x}{\partial v} + \frac{\partial z}{\partial y}\frac{\partial y}{\partial v} = (2x + y^3)(2uv) + (3xy^2)(e^w),$$

$$\frac{\partial z}{\partial w} = \frac{\partial z}{\partial x}\frac{\partial x}{\partial w} + \frac{\partial z}{\partial y}\frac{\partial y}{\partial w} = (2x + y^3)(3w^2) + (3xy^2)(ve^w).$$

When $u = 2$, $v = 1$, and $w = 0$, we have $x = 2$, $y = 3$,

so $\dfrac{\partial z}{\partial u} = (31)(1) + (54)(1) = 85, \quad \dfrac{\partial z}{\partial v} = (31)(4) + (54)(1) = 178, \quad \dfrac{\partial z}{\partial w} = (31)(0) + (54)(1) = 54.$

22. $u = (r^2 + s^2)^{1/2}, \quad r = y + x\cos t, \quad s = x + y\sin t \quad \Rightarrow$

$$\frac{\partial u}{\partial x} = \frac{\partial u}{\partial r}\frac{\partial r}{\partial x} + \frac{\partial u}{\partial s}\frac{\partial s}{\partial x} = \tfrac{1}{2}(r^2 + s^2)^{-1/2}(2r)(\cos t) + \tfrac{1}{2}(r^2 + s^2)^{-1/2}(2s)(1) = (r\cos t + s)/\sqrt{r^2 + s^2},$$

$$\frac{\partial u}{\partial y} = \frac{\partial u}{\partial r}\frac{\partial r}{\partial y} + \frac{\partial u}{\partial s}\frac{\partial s}{\partial y} = \tfrac{1}{2}(r^2 + s^2)^{-1/2}(2r)(1) + \tfrac{1}{2}(r^2 + s^2)^{-1/2}(2s)(\sin t) = (r + s\sin t)/\sqrt{r^2 + s^2},$$

$$\frac{\partial u}{\partial t} = \frac{\partial u}{\partial r}\frac{\partial r}{\partial t} + \frac{\partial u}{\partial s}\frac{\partial s}{\partial t} = \tfrac{1}{2}(r^2 + s^2)^{-1/2}(2r)(-x\sin t) + \tfrac{1}{2}(r^2 + s^2)^{-1/2}(2s)(y\cos t) = \frac{-rx\sin t + sy\cos t}{\sqrt{r^2 + s^2}}.$$

When $x = 1$, $y = 2$, and $t = 0$ we have $r = 3$ and $s = 1$, so $\dfrac{\partial u}{\partial x} = \dfrac{4}{\sqrt{10}}, \quad \dfrac{\partial u}{\partial y} = \dfrac{3}{\sqrt{10}}, \quad$ and $\dfrac{\partial u}{\partial t} = \dfrac{2}{\sqrt{10}}.$

23. $R = \ln(u^2 + v^2 + w^2), \quad u = x + 2y, \quad v = 2x - y, \quad w = 2xy \quad \Rightarrow$

$$\frac{\partial R}{\partial x} = \frac{\partial R}{\partial u}\frac{\partial u}{\partial x} + \frac{\partial R}{\partial v}\frac{\partial v}{\partial x} + \frac{\partial R}{\partial w}\frac{\partial w}{\partial x} = \frac{2u}{u^2 + v^2 + w^2}(1) + \frac{2v}{u^2 + v^2 + w^2}(2) + \frac{2w}{u^2 + v^2 + w^2}(2y)$$

$$= \frac{2u + 4v + 4wy}{u^2 + v^2 + w^2},$$

$$\frac{\partial R}{\partial y} = \frac{\partial R}{\partial u}\frac{\partial u}{\partial y} + \frac{\partial R}{\partial v}\frac{\partial v}{\partial y} + \frac{\partial R}{\partial w}\frac{\partial w}{\partial y} = \frac{2u}{u^2 + v^2 + w^2}(2) + \frac{2v}{u^2 + v^2 + w^2}(-1) + \frac{2w}{u^2 + v^2 + w^2}(2x)$$

$$= \frac{4u - 2v + 4wx}{u^2 + v^2 + w^2}.$$

When $x = y = 1$ we have $u = 3$, $v = 1$, and $w = 2$, so $\dfrac{\partial R}{\partial x} = \dfrac{9}{7}$ and $\dfrac{\partial R}{\partial y} = \dfrac{9}{7}.$

24. $M = xe^{y-z^2}$, $x = 2uv$, $y = u - v$, $z = u + v$ $\Rightarrow$

$$\frac{\partial M}{\partial u} = \frac{\partial M}{\partial x}\frac{\partial x}{\partial u} + \frac{\partial M}{\partial y}\frac{\partial y}{\partial u} + \frac{\partial M}{\partial z}\frac{\partial z}{\partial u} = e^{y-z^2}(2v) + xe^{y-z^2}(1) + x(-2z)e^{y-z^2}(1) = e^{y-z^2}(2v + x - 2xz),$$

$$\frac{\partial M}{\partial v} = \frac{\partial M}{\partial x}\frac{\partial x}{\partial v} + \frac{\partial M}{\partial y}\frac{\partial y}{\partial v} + \frac{\partial M}{\partial z}\frac{\partial z}{\partial v} = e^{y-z^2}(2u) + xe^{y-z^2}(-1) + x(-2z)e^{y-z^2}(1) = e^{y-z^2}(2u - x - 2xz).$$

When $u = 3$, $v = -1$ we have $x = -6$, $y = 4$, and $z = 2$, so $\dfrac{\partial M}{\partial u} = 16$ and $\dfrac{\partial M}{\partial v} = 36$.

25. $u = x^2 + yz$, $x = pr\cos\theta$, $y = pr\sin\theta$, $z = p + r$ $\Rightarrow$

$$\frac{\partial u}{\partial p} = \frac{\partial u}{\partial x}\frac{\partial x}{\partial p} + \frac{\partial u}{\partial y}\frac{\partial y}{\partial p} + \frac{\partial u}{\partial z}\frac{\partial z}{\partial p} = (2x)(r\cos\theta) + (z)(r\sin\theta) + (y)(1) = 2xr\cos\theta + zr\sin\theta + y,$$

$$\frac{\partial u}{\partial r} = \frac{\partial u}{\partial x}\frac{\partial x}{\partial r} + \frac{\partial u}{\partial y}\frac{\partial y}{\partial r} + \frac{\partial u}{\partial z}\frac{\partial z}{\partial r} = (2x)(p\cos\theta) + (z)(p\sin\theta) + (y)(1) = 2xp\cos\theta + zp\sin\theta + y,$$

$$\frac{\partial u}{\partial \theta} = \frac{\partial u}{\partial x}\frac{\partial x}{\partial \theta} + \frac{\partial u}{\partial y}\frac{\partial y}{\partial \theta} + \frac{\partial u}{\partial z}\frac{\partial z}{\partial \theta} = (2x)(-pr\sin\theta) + (z)(pr\cos\theta) + (y)(0) = -2xpr\sin\theta + zpr\cos\theta.$$

When $p = 2$, $r = 3$, and $\theta = 0$ we have $x = 6$, $y = 0$, and $z = 5$, so $\dfrac{\partial u}{\partial p} = 36$, $\dfrac{\partial u}{\partial r} = 24$, and $\dfrac{\partial u}{\partial \theta} = 30$.

26. $Y = w\tan^{-1}(uv)$, $u = r + s$, $v = s + t$, $w = t + r$ $\Rightarrow$

$$\frac{\partial Y}{\partial r} = \frac{\partial Y}{\partial u}\frac{\partial u}{\partial r} + \frac{\partial Y}{\partial v}\frac{\partial v}{\partial r} + \frac{\partial Y}{\partial w}\frac{\partial w}{\partial r} = \frac{w}{1+(uv)^2}(v)(1) + \frac{w}{1+(uv)^2}(u)(0) + \tan^{-1}(uv)(1) = \frac{vw}{1+u^2v^2} + \tan^{-1}(uv),$$

$$\frac{\partial Y}{\partial s} = \frac{\partial Y}{\partial u}\frac{\partial u}{\partial s} + \frac{\partial Y}{\partial v}\frac{\partial v}{\partial s} + \frac{\partial Y}{\partial w}\frac{\partial w}{\partial s} = \frac{wv}{1+u^2v^2}(1) + \frac{wu}{1+u^2v^2}(1) + \tan^{-1}(uv)(0) = \frac{w(v+u)}{1+u^2v^2},$$

$$\frac{\partial Y}{\partial t} = \frac{\partial Y}{\partial u}\frac{\partial u}{\partial t} + \frac{\partial Y}{\partial v}\frac{\partial v}{\partial t} + \frac{\partial Y}{\partial w}\frac{\partial w}{\partial t} = \frac{wv}{1+u^2v^2}(0) + \frac{wu}{1+u^2v^2}(1) + \tan^{-1}(uv)(1) = \frac{wu}{1+u^2v^2} + \tan^{-1}(uv).$$

When $r = 1$, $s = 0$, and $t = 1$, we have $u = 1$, $v = 1$, and $w = 2$, so $\dfrac{\partial Y}{\partial r} = 1 + \dfrac{\pi}{4}$, $\dfrac{\partial Y}{\partial s} = 2$, and $\dfrac{\partial Y}{\partial t} = 1 + \dfrac{\pi}{4}$.

27. $\sqrt{xy} = 1 + x^2 y$, so let $F(x, y) = (xy)^{1/2} - 1 - x^2 y = 0$. Then by Equation 6

$$\frac{dy}{dx} = -\frac{F_x}{F_y} = -\frac{\frac{1}{2}(xy)^{-1/2}(y) - 2xy}{\frac{1}{2}(xy)^{-1/2}(x) - x^2} = -\frac{y - 4xy\sqrt{xy}}{x - 2x^2\sqrt{xy}} = \frac{4(xy)^{3/2} - y}{x - 2x^2\sqrt{xy}}.$$

28. $y^5 + x^2 y^3 = 1 + ye^{x^2}$, so let $F(x, y) = y^5 + x^2 y^3 - 1 - ye^{x^2} = 0$. Then

$$\frac{dy}{dx} = -\frac{F_x}{F_y} = -\frac{2xy^3 - 2xye^{x^2}}{5y^4 + 3x^2y^2 - e^{x^2}} = \frac{2xye^{x^2} - 2xy^3}{5y^4 + 3x^2y^2 - e^{x^2}}.$$

29. $\cos(x - y) = xe^y$, so let $F(x, y) = \cos(x - y) - xe^y = 0$.

Then $\dfrac{dy}{dx} = -\dfrac{F_x}{F_y} = -\dfrac{-\sin(x - y) - e^y}{-\sin(x - y)(-1) - xe^y} = \dfrac{\sin(x - y) + e^y}{\sin(x - y) - xe^y}.$

30. $\sin x + \cos y = \sin x \cos y$, so let $F(x, y) = \sin x + \cos y - \sin x \cos y = 0$. Then

$$\frac{dy}{dx} = -\frac{F_x}{F_y} = -\frac{\cos x - \cos x \cos y}{-\sin y + \sin x \sin y} = \frac{\cos x(\cos y - 1)}{\sin y(\sin x - 1)}.$$

31. $x^2 + y^2 + z^2 = 3xyz$, so let $F(x, y, z) = x^2 + y^2 + z^2 - 3xyz = 0$. Then by Equations 7

$$\frac{\partial z}{\partial x} = -\frac{F_x}{F_z} = -\frac{2x - 3yz}{2z - 3xy} = \frac{3yz - 2x}{2z - 3xy} \quad \text{and} \quad \frac{\partial z}{\partial y} = -\frac{F_y}{F_z} = -\frac{2y - 3xz}{2z - 3xy} = \frac{3xz - 2y}{2z - 3xy}.$$

32. $xyz = \cos(x + y + z)$. Let $F(x, y, z) = xyz - \cos(x + y + z) = 0$, so

$$\frac{\partial z}{\partial x} = -\frac{F_x}{F_z} = -\frac{yz + \sin(x + y + z)}{xy + \sin(x + y + z)} \quad \text{and} \quad \frac{\partial z}{\partial y} = -\frac{F_y}{F_z} = -\frac{xz + \sin(x + y + z)}{xy + \sin(x + y + z)}.$$

33. $x - z = \arctan(yz)$, so let $F(x, y, z) = x - z - \arctan(yz) = 0$. Then

$$\frac{\partial z}{\partial x} = -\frac{F_x}{F_z} = -\frac{1}{-1 - \dfrac{1}{1 + (yz)^2}(y)} = \frac{1 + y^2 z^2}{1 + y + y^2 z^2}$$

$$\frac{\partial z}{\partial y} = -\frac{F_y}{F_z} = -\frac{-\dfrac{1}{1 + (yz)^2}(z)}{-1 - \dfrac{1}{1 + (yz)^2}(y)} = -\frac{\dfrac{z}{1 + y^2 z^2}}{\dfrac{1 + y^2 z^2 + y}{1 + y^2 z^2}} = -\frac{z}{1 + y + y^2 z^2}$$

34. $yz = \ln(x + z)$, so let $F(x, y, z) = yz - \ln(x + z) = 0$. Then

$$\frac{\partial z}{\partial x} = -\frac{F_x}{F_z} = -\frac{-\dfrac{1}{x + z}(1)}{y - \dfrac{1}{x + z}(1)} = \frac{1}{y(x + z) - 1} \quad \text{and} \quad \frac{\partial z}{\partial y} = -\frac{F_y}{F_z} = -\frac{z}{y - \dfrac{1}{x + z}} = -\frac{z(x + z)}{y(x + z) - 1}$$

35. Since x and y are each functions of t, $T(x, y)$ is a function of t, so by the Chain Rule, $\dfrac{dT}{dt} = \dfrac{\partial T}{\partial x}\dfrac{dx}{dt} + \dfrac{\partial T}{\partial y}\dfrac{dy}{dt}$. After

3 seconds, $x = \sqrt{1 + t} = \sqrt{1 + 3} = 2$, $y = 2 + \frac{1}{3}t = 2 + \frac{1}{3}(3) = 3$, $\dfrac{dx}{dt} = \dfrac{1}{2\sqrt{1 + t}} = \dfrac{1}{2\sqrt{1 + 3}} = \dfrac{1}{4}$, and $\dfrac{dy}{dt} = \dfrac{1}{3}$.

Then $\dfrac{dT}{dt} = T_x(2, 3)\dfrac{dx}{dt} + T_y(2, 3)\dfrac{dy}{dt} = 4\left(\frac{1}{4}\right) + 3\left(\frac{1}{3}\right) = 2$. Thus the temperature is rising at a rate of $2°\text{C/s}$.

36. (a) Since $\partial W/\partial T$ is negative, a rise in average temperature (while annual rainfall remains constant) causes a decrease in

wheat production at the current production levels. Since $\partial W/\partial R$ is positive, an increase in annual rainfall (while the

average temperature remains constant) causes an increase in wheat production.

(b) Since the average temperature is rising at a rate of $0.15°\text{C/year}$, we know that $dT/dt = 0.15$. Since rainfall is

decreasing at a rate of 0.1 cm/year, we know $dR/dt = -0.1$. Then, by the Chain Rule,

$\dfrac{dW}{dt} = \dfrac{\partial W}{\partial T}\dfrac{dT}{dt} + \dfrac{\partial W}{\partial R}\dfrac{dR}{dt} = (-2)(0.15) + (8)(-0.1) = -1.1$. Thus we estimate that wheat production will decrease

at a rate of 1.1 units/year.

37. $C = 1449.2 + 4.6T - 0.055T^2 + 0.00029T^3 + 0.016D$, so $\dfrac{\partial C}{\partial T} = 4.6 - 0.11T + 0.00087T^2$ and $\dfrac{\partial C}{\partial D} = 0.016$.

According to the graph, the diver is experiencing a temperature of approximately $12.5°\text{C}$ at $t = 20$ minutes, so

$\dfrac{\partial C}{\partial T} = 4.6 - 0.11(12.5) + 0.00087(12.5)^2 \approx 3.36$. By sketching tangent lines at $t = 20$ to the graphs given, we estimate

$\dfrac{dD}{dt} \approx \dfrac{1}{2}$ and $\dfrac{dT}{dt} \approx -\dfrac{1}{10}$. Then, by the Chain Rule, $\dfrac{dC}{dt} = \dfrac{\partial C}{\partial T}\dfrac{dT}{dt} + \dfrac{\partial C}{\partial D}\dfrac{dD}{dt} \approx (3.36)\left(-\frac{1}{10}\right) + (0.016)\left(\frac{1}{2}\right) \approx -0.33$.

Thus the speed of sound experienced by the diver is decreasing at a rate of approximately 0.33 m/s per minute.

38. $V = \pi r^2 h / 3$, so $\dfrac{dV}{dt} = \dfrac{\partial V}{\partial r}\dfrac{dr}{dt} + \dfrac{\partial V}{\partial h}\dfrac{dh}{dt} = \dfrac{2\pi r h}{3} 1.8 + \dfrac{\pi r^2}{3}(-2.5) = 20{,}160\pi - 12{,}000\pi = 8160\pi$ in^3/s.

39. (a) $V = \ell w h$, so by the Chain Rule,

$$\frac{dV}{dt} = \frac{\partial V}{\partial \ell}\frac{d\ell}{dt} + \frac{\partial V}{\partial w}\frac{dw}{dt} + \frac{\partial V}{\partial h}\frac{dh}{dt} = wh\frac{d\ell}{dt} + \ell h\frac{dw}{dt} + \ell w\frac{dh}{dt} = 2\cdot2\cdot2 + 1\cdot2\cdot2 + 1\cdot2\cdot(-3) = 6 \text{ m}^3/\text{s}.$$

(b) $S = 2(\ell w + \ell h + wh)$, so by the Chain Rule,

$$\frac{dS}{dt} = \frac{\partial S}{\partial \ell}\frac{d\ell}{dt} + \frac{\partial S}{\partial w}\frac{dw}{dt} + \frac{\partial S}{\partial h}\frac{dh}{dt} = 2(w + h)\frac{d\ell}{dt} + 2(\ell + h)\frac{dw}{dt} + 2(\ell + w)\frac{dh}{dt}$$

$$= 2(2 + 2)2 + 2(1 + 2)2 + 2(1 + 2)(-3) = 10 \text{ m}^2/\text{s}$$

(c) $L^2 = \ell^2 + w^2 + h^2 \;\Rightarrow\; 2L\dfrac{dL}{dt} = 2\ell\dfrac{d\ell}{dt} + 2w\dfrac{dw}{dt} + 2h\dfrac{dh}{dt} = 2(1)(2) + 2(2)(2) + 2(2)(-3) = 0 \;\Rightarrow$

$dL/dt = 0$ m/s.

40. $I = \dfrac{V}{R} \;\Rightarrow$

$$\frac{dI}{dt} = \frac{\partial I}{\partial V}\frac{dV}{dt} + \frac{\partial I}{\partial R}\frac{dR}{dt} = \frac{1}{R}\frac{dV}{dt} - \frac{V}{R^2}\frac{dR}{dt} = \frac{1}{R}\frac{dV}{dt} - \frac{I}{R}\frac{dR}{dt} = \frac{1}{400}(-0.01) - \frac{0.08}{400}(0.03) = -0.000031 \text{ A/s}$$

41. $\dfrac{dP}{dt} = 0.05$, $\dfrac{dT}{dt} = 0.15$, $V = 8.31\dfrac{T}{P}$ and $\dfrac{dV}{dt} = \dfrac{8.31}{P}\dfrac{dT}{dt} - 8.31\dfrac{T}{P^2}\dfrac{dP}{dt}$. Thus when $P = 20$ and $T = 320$,

$$\frac{dV}{dt} = 8.31\left[\frac{0.15}{20} - \frac{(0.05)(320)}{400}\right] \approx -0.27 \text{ L/s}.$$

42. Let x and y be the respective distances of car A and car B from the intersection and let z be the distance between the two cars.

Then $dx/dt = -90$, $dy/dt = -80$ and $z^2 = x^2 + y^2$. When $x = 0.3$ and $y = 0.4$, $z = \sqrt{0.25} = 0.5$ and

$2z\,(dz/dt) = 2x\,(dx/dt) + 2y\,(dy/dt)$ or $dz/dt = 0.6(-90) + 0.8(-80) = -118$ km/h.

43. Let x be the length of the first side of the triangle and y the length of the second side. The area A of the triangle is given by

$A = \frac{1}{2}xy\sin\theta$ where θ is the angle between the two sides. Thus A is a function of x, y, and θ, and x, y, and θ are each in

turn functions of time t. We are given that $\dfrac{dx}{dt} = 3$, $\dfrac{dy}{dt} = -2$, and because A is constant, $\dfrac{dA}{dt} = 0$. By the Chain Rule,

$$\frac{dA}{dt} = \frac{\partial A}{\partial x}\frac{dx}{dt} + \frac{\partial A}{\partial y}\frac{dy}{dt} + \frac{\partial A}{\partial \theta}\frac{d\theta}{dt} \;\Rightarrow\; \frac{dA}{dt} = \frac{1}{2}y\sin\theta \cdot \frac{dx}{dt} + \frac{1}{2}x\sin\theta \cdot \frac{dy}{dt} + \frac{1}{2}xy\cos\theta \cdot \frac{d\theta}{dt}.$$ When $x = 20$, $y = 30$,

and $\theta = \pi/6$ we have

$$0 = \tfrac{1}{2}(30)\left(\sin\tfrac{\pi}{6}\right)(3) + \tfrac{1}{2}(20)\left(\sin\tfrac{\pi}{6}\right)(-2) + \tfrac{1}{2}(20)(30)\left(\cos\tfrac{\pi}{6}\right)\frac{d\theta}{dt}$$

$$= 45\cdot\tfrac{1}{2} - 20\cdot\tfrac{1}{2} + 300\cdot\frac{\sqrt{3}}{2}\cdot\frac{d\theta}{dt} = \tfrac{25}{2} + 150\sqrt{3}\frac{d\theta}{dt}$$

Solving for $\dfrac{d\theta}{dt}$ gives $\dfrac{d\theta}{dt} = \dfrac{-25/2}{150\sqrt{3}} = -\dfrac{1}{12\sqrt{3}}$, so the angle between the sides is decreasing at a rate of

$1/(12\sqrt{3}) \approx 0.048$ rad/s.

44. $f_o = \left(\dfrac{c + v_o}{c - v_s}\right) f_s = \left(\dfrac{332 + 34}{332 - 40}\right) 460 \approx 576.6$ Hz. v_o and v_s are functions of time t, so

$$\frac{df_o}{dt} = \frac{\partial f_o}{\partial v_o} \frac{dv_o}{dt} + \frac{\partial f_o}{\partial v_s} \frac{dv_s}{dt} = \left(\frac{1}{c - v_s}\right) f_s \cdot \frac{dv_o}{dt} + \frac{c + v_o}{(c - v_s)^2} f_s \cdot \frac{dv_s}{dt}$$

$$= \left(\tfrac{1}{332 - 40}\right)(460)(1.2) + \tfrac{332 + 34}{(332 - 40)^2}(460)(1.4) \approx 4.65 \text{ Hz/s}$$

45. (a) By the Chain Rule, $\dfrac{\partial z}{\partial r} = \dfrac{\partial z}{\partial x} \cos\theta + \dfrac{\partial z}{\partial y} \sin\theta$, $\dfrac{\partial z}{\partial \theta} = \dfrac{\partial z}{\partial x}(-r\sin\theta) + \dfrac{\partial z}{\partial y} r\cos\theta$.

(b) $\left(\dfrac{\partial z}{\partial r}\right)^2 = \left(\dfrac{\partial z}{\partial x}\right)^2 \cos^2\theta + 2\dfrac{\partial z}{\partial x}\dfrac{\partial z}{\partial y}\cos\theta\,\sin\theta + \left(\dfrac{\partial z}{\partial y}\right)^2 \sin^2\theta$,

$\left(\dfrac{\partial z}{\partial \theta}\right)^2 = \left(\dfrac{\partial z}{\partial x}\right)^2 r^2 \sin^2\theta - 2\dfrac{\partial z}{\partial x}\dfrac{\partial z}{\partial y} r^2 \cos\theta\,\sin\theta + \left(\dfrac{\partial z}{\partial y}\right)^2 r^2 \cos^2\theta$. Thus

$\left(\dfrac{\partial z}{\partial r}\right)^2 + \dfrac{1}{r^2}\left(\dfrac{\partial z}{\partial \theta}\right)^2 = \left[\left(\dfrac{\partial z}{\partial x}\right)^2 + \left(\dfrac{\partial z}{\partial y}\right)^2\right](\cos^2\theta + \sin^2\theta) = \left(\dfrac{\partial z}{\partial x}\right)^2 + \left(\dfrac{\partial z}{\partial y}\right)^2$.

46. By the Chain Rule, $\dfrac{\partial u}{\partial s} = \dfrac{\partial u}{\partial x} e^s \cos t + \dfrac{\partial u}{\partial y} e^s \sin t$, $\dfrac{\partial u}{\partial t} = \dfrac{\partial u}{\partial x}(-e^s \sin t) + \dfrac{\partial u}{\partial y} e^s \cos t$. Then

$\left(\dfrac{\partial u}{\partial s}\right)^2 = \left(\dfrac{\partial u}{\partial x}\right)^2 e^{2s} \cos^2 t + 2\dfrac{\partial u}{\partial x}\dfrac{\partial u}{\partial y} e^{2s} \cos t\,\sin t + \left(\dfrac{\partial u}{\partial y}\right)^2 e^{2s} \sin^2 t$ and

$\left(\dfrac{\partial u}{\partial t}\right)^2 = \left(\dfrac{\partial u}{\partial x}\right)^2 e^{2s} \sin^2 t - 2\dfrac{\partial u}{\partial x}\dfrac{\partial u}{\partial y} e^{2s} \cos t\,\sin t + \left(\dfrac{\partial u}{\partial y}\right)^2 e^{2s} \sin^2 t$. Thus

$\left[\left(\dfrac{\partial u}{\partial s}\right)^2 + \left(\dfrac{\partial u}{\partial t}\right)^2\right] e^{-2s} = \left(\dfrac{\partial u}{\partial x}\right)^2 + \left(\dfrac{\partial u}{\partial y}\right)^2$.

47. Let $u = x - y$. Then $\dfrac{\partial z}{\partial x} = \dfrac{dz}{du}\dfrac{\partial u}{\partial x} = \dfrac{dz}{du}$ and $\dfrac{\partial z}{\partial y} = \dfrac{dz}{du}(-1)$. Thus $\dfrac{\partial z}{\partial x} + \dfrac{\partial z}{\partial y} = 0$.

48. $\dfrac{\partial z}{\partial s} = \dfrac{\partial z}{\partial x} + \dfrac{\partial z}{\partial y}$ and $\dfrac{\partial z}{\partial t} = \dfrac{\partial z}{\partial x} - \dfrac{\partial z}{\partial y}$. Thus $\dfrac{\partial z}{\partial s}\dfrac{\partial z}{\partial t} = \left(\dfrac{\partial z}{\partial x}\right)^2 - \left(\dfrac{\partial z}{\partial y}\right)^2$.

49. Let $u = x + at$, $v = x - at$. Then $z = f(u) + g(v)$, so $\partial z/\partial u = f'(u)$ and $\partial z/\partial v = g'(v)$.

Thus $\dfrac{\partial z}{\partial t} = \dfrac{\partial z}{\partial u}\dfrac{\partial u}{\partial t} + \dfrac{\partial z}{\partial v}\dfrac{\partial v}{\partial t} = af'(u) - ag'(v)$ and

$\dfrac{\partial^2 z}{\partial t^2} = a\dfrac{\partial}{\partial t}[f'(u) - g'(v)] = a\left(\dfrac{df'(u)}{du}\dfrac{\partial u}{\partial t} - \dfrac{dg'(v)}{dv}\dfrac{\partial v}{\partial t}\right) = a^2 f''(u) + a^2 g''(v)$.

Similarly $\dfrac{\partial z}{\partial x} = f'(u) + g'(v)$ and $\dfrac{\partial^2 z}{\partial x^2} = f''(u) + g''(v)$. Thus $\dfrac{\partial^2 z}{\partial t^2} = a^2 \dfrac{\partial^2 z}{\partial x^2}$.

50. By the Chain Rule, $\dfrac{\partial u}{\partial s} = e^s \cos t \dfrac{\partial u}{\partial x} + e^s \sin t \dfrac{\partial u}{\partial y}$ and $\dfrac{\partial u}{\partial t} = -e^s \sin t \dfrac{\partial u}{\partial x} + e^s \cos t \dfrac{\partial u}{\partial y}$.

Then $\dfrac{\partial^2 u}{\partial s^2} = e^s \cos t \dfrac{\partial u}{\partial x} + e^s \cos t \dfrac{\partial}{\partial s}\left(\dfrac{\partial u}{\partial x}\right) + e^s \sin t \dfrac{\partial u}{\partial y} + e^s \sin t \dfrac{\partial}{\partial s}\left(\dfrac{\partial u}{\partial y}\right)$. But

$\dfrac{\partial}{\partial s}\left(\dfrac{\partial u}{\partial x}\right) = \dfrac{\partial^2 u}{\partial x^2}\dfrac{\partial x}{\partial s} + \dfrac{\partial^2 u}{\partial y\,\partial x}\dfrac{\partial y}{\partial s} = e^s \cos t \dfrac{\partial^2 u}{\partial x^2} + e^s \sin t \dfrac{\partial^2 u}{\partial y\,\partial x}$ and

$\dfrac{\partial}{\partial s}\left(\dfrac{\partial u}{\partial y}\right) = \dfrac{\partial^2 u}{\partial y^2}\dfrac{\partial y}{\partial s} + \dfrac{\partial^2 u}{\partial x\,\partial y}\dfrac{\partial x}{\partial s} = e^s \sin t \dfrac{\partial^2 u}{\partial y^2} + e^s \cos t \dfrac{\partial^2 u}{\partial x\,\partial y}$.

Also, by continuity of the partials, $\dfrac{\partial^2 u}{\partial x\,\partial y} = \dfrac{\partial^2 u}{\partial y\,\partial x}$. Thus

$$\frac{\partial^2 u}{\partial s^2} = e^s\cos t\,\frac{\partial u}{\partial x} + e^s\cos t\left(e^s\cos t\,\frac{\partial^2 u}{\partial x^2} + e^s\sin t\,\frac{\partial^2 u}{\partial x\,\partial y}\right) + e^s\sin t\,\frac{\partial u}{\partial y} + e^s\sin t\left(e^s\sin t\,\frac{\partial^2 u}{\partial y^2} + e^s\cos t\,\frac{\partial^2 u}{\partial x\,\partial y}\right)$$

$$= e^s\cos t\,\frac{\partial u}{\partial x} + e^s\sin t\,\frac{\partial u}{\partial y} + e^{2s}\cos^2 t\,\frac{\partial^2 u}{\partial x^2} + 2e^{2s}\cos t\sin t\,\frac{\partial^2 u}{\partial x\,\partial y} + e^{2s}\sin^2 t\,\frac{\partial^2 u}{\partial y^2}$$

Similarly

$$\frac{\partial^2 u}{\partial t^2} = -e^s\cos t\,\frac{\partial u}{\partial x} - e^s\sin t\,\frac{\partial}{\partial t}\left(\frac{\partial u}{\partial x}\right) - e^s\sin t\,\frac{\partial u}{\partial y} + e^s\cos t\,\frac{\partial}{\partial t}\left(\frac{\partial u}{\partial y}\right)$$

$$= -e^s\cos t\,\frac{\partial u}{\partial x} - e^s\sin t\left(-e^s\sin t\,\frac{\partial^2 u}{\partial x^2} + e^s\cos t\,\frac{\partial^2 u}{\partial x\,\partial y}\right)$$

$$\qquad -e^s\sin t\,\frac{\partial u}{\partial y} + e^s\cos t\left(e^s\cos t\,\frac{\partial^2 u}{\partial y^2} - e^s\sin t\,\frac{\partial^2 u}{\partial x\,\partial y}\right)$$

$$= -e^s\cos t\,\frac{\partial u}{\partial x} - e^s\sin t\,\frac{\partial u}{\partial y} + e^{2s}\sin^2 t\,\frac{\partial^2 u}{\partial x^2} - 2e^{2s}\cos t\sin t\,\frac{\partial^2 u}{\partial x\,\partial y} + e^{2s}\cos^2 t\,\frac{\partial^2 u}{\partial y^2}$$

Thus $e^{-2s}\left(\dfrac{\partial^2 u}{\partial s^2} + \dfrac{\partial^2 u}{\partial t^2}\right) = (\cos^2 t + \sin^2 t)\left(\dfrac{\partial^2 u}{\partial x^2} + \dfrac{\partial^2 u}{\partial y^2}\right) = \dfrac{\partial^2 u}{\partial x^2} + \dfrac{\partial^2 u}{\partial y^2}$, as desired.

51. $\dfrac{\partial z}{\partial s} = \dfrac{\partial z}{\partial x}\,2s + \dfrac{\partial z}{\partial y}\,2r$. Then

$$\frac{\partial^2 z}{\partial r\,\partial s} = \frac{\partial}{\partial r}\left(\frac{\partial z}{\partial x}\,2s\right) + \frac{\partial}{\partial r}\left(\frac{\partial z}{\partial y}\,2r\right)$$

$$= \frac{\partial^2 z}{\partial x^2}\frac{\partial x}{\partial r}\,2s + \frac{\partial}{\partial y}\left(\frac{\partial z}{\partial x}\right)\frac{\partial y}{\partial r}\,2s + \frac{\partial z}{\partial x}\frac{\partial}{\partial r}\,2s + \frac{\partial^2 z}{\partial y^2}\frac{\partial y}{\partial r}\,2r + \frac{\partial}{\partial x}\left(\frac{\partial z}{\partial y}\right)\frac{\partial x}{\partial r}\,2r + \frac{\partial z}{\partial y}\,2$$

$$= 4rs\,\frac{\partial^2 z}{\partial x^2} + \frac{\partial^2 z}{\partial y\,\partial x}\,4s^2 + 0 + 4rs\,\frac{\partial^2 z}{\partial y^2} + \frac{\partial^2 z}{\partial x\,\partial y}\,4r^2 + 2\,\frac{\partial z}{\partial y}$$

By the continuity of the partials, $\dfrac{\partial^2 z}{\partial r\partial s} = 4rs\,\dfrac{\partial^2 z}{\partial x^2} + 4rs\,\dfrac{\partial^2 z}{\partial y^2} + (4r^2 + 4s^2)\,\dfrac{\partial^2 z}{\partial x\,\partial y} + 2\,\dfrac{\partial z}{\partial y}$.

52. By the Chain Rule,

(a) $\dfrac{\partial z}{\partial r} = \dfrac{\partial z}{\partial x}\cos\theta + \dfrac{\partial z}{\partial y}\sin\theta$ 　　　　　　(b) $\dfrac{\partial z}{\partial\theta} = -\dfrac{\partial z}{\partial x}\,r\sin\theta + \dfrac{\partial z}{\partial y}\,r\cos\theta$

(c) $\dfrac{\partial^2 z}{\partial r\,\partial\theta} = \dfrac{\partial^2 z}{\partial\theta\,\partial r} = \dfrac{\partial}{\partial\theta}\left(\dfrac{\partial z}{\partial x}\cos\theta + \dfrac{\partial z}{\partial y}\sin\theta\right) = -\sin\theta\,\dfrac{\partial z}{\partial x} + \cos\theta\,\dfrac{\partial}{\partial\theta}\left(\dfrac{\partial z}{\partial x}\right) + \cos\theta\,\dfrac{\partial z}{\partial y} + \sin\theta\,\dfrac{\partial}{\partial\theta}\left(\dfrac{\partial z}{\partial y}\right)$

$$= -\sin\theta\,\frac{\partial z}{\partial x} + \cos\theta\left(\frac{\partial^2 z}{\partial x^2}\frac{\partial x}{\partial\theta} + \frac{\partial^2 z}{\partial y\,\partial x}\frac{\partial y}{\partial\theta}\right) + \cos\theta\,\frac{\partial z}{\partial y} + \sin\theta\left(\frac{\partial^2 z}{\partial y^2}\frac{\partial y}{\partial\theta} + \frac{\partial^2 z}{\partial x\,\partial y}\frac{\partial x}{\partial\theta}\right)$$

$$= -\sin\theta\,\frac{\partial z}{\partial x} + \cos\theta\left(-r\sin\theta\,\frac{\partial^2 z}{\partial x^2} + r\cos\theta\,\frac{\partial^2 z}{\partial y\,\partial x}\right) + \cos\theta\,\frac{\partial z}{\partial y} + \sin\theta\left(r\cos\theta\,\frac{\partial^2 z}{\partial y^2} - r\sin\theta\,\frac{\partial^2 z}{\partial x\,\partial y}\right)$$

$$= -\sin\theta\,\frac{\partial z}{\partial x} - r\cos\theta\sin\theta\,\frac{\partial^2 z}{\partial x^2} + r\cos^2\theta\,\frac{\partial^2 z}{\partial y\,\partial x} + \cos\theta\,\frac{\partial z}{\partial y} + r\cos\theta\sin\theta\,\frac{\partial^2 z}{\partial y^2} - r\sin^2\theta\,\frac{\partial^2 z}{\partial y\,\partial x}$$

$$= \cos\theta\,\frac{\partial z}{\partial y} - \sin\theta\,\frac{\partial z}{\partial x} + r\cos\theta\sin\theta\left(\frac{\partial^2 z}{\partial y^2} - \frac{\partial^2 z}{\partial x^2}\right) + r(\cos^2\theta - \sin^2\theta)\,\frac{\partial^2 z}{\partial y\,\partial x}$$

53. $\dfrac{\partial z}{\partial r} = \dfrac{\partial z}{\partial x}\cos\theta + \dfrac{\partial z}{\partial y}\sin\theta$ and $\dfrac{\partial z}{\partial\theta} = -\dfrac{\partial z}{\partial x}r\sin\theta + \dfrac{\partial z}{\partial y}r\cos\theta$. Then

$$\frac{\partial^2 z}{\partial r^2} = \cos\theta\left(\frac{\partial^2 z}{\partial x^2}\cos\theta + \frac{\partial^2 z}{\partial y\,\partial x}\sin\theta\right) + \sin\theta\left(\frac{\partial^2 z}{\partial y^2}\sin\theta + \frac{\partial^2 z}{\partial x\,\partial y}\cos\theta\right)$$

$$= \cos^2\theta\,\frac{\partial^2 z}{\partial x^2} + 2\cos\theta\,\sin\theta\,\frac{\partial^2 z}{\partial x\,\partial y} + \sin^2\theta\,\frac{\partial^2 z}{\partial y^2}$$

and

$$\frac{\partial^2 z}{\partial\theta^2} = -r\cos\theta\,\frac{\partial z}{\partial x} + (-r\sin\theta)\left(\frac{\partial^2 z}{\partial x^2}(-r\sin\theta) + \frac{\partial^2 z}{\partial y\,\partial x}r\cos\theta\right)$$

$$-r\sin\theta\,\frac{\partial z}{\partial y} + r\cos\theta\left(\frac{\partial^2 z}{\partial y^2}r\cos\theta + \frac{\partial^2 z}{\partial x\,\partial y}(-r\sin\theta)\right)$$

$$= -r\cos\theta\,\frac{\partial z}{\partial x} - r\sin\theta\,\frac{\partial z}{\partial y} + r^2\sin^2\theta\,\frac{\partial^2 z}{\partial x^2} - 2r^2\cos\theta\,\sin\theta\,\frac{\partial^2 z}{\partial x\,\partial y} + r^2\cos^2\theta\,\frac{\partial^2 z}{\partial y^2}$$

Thus

$$\frac{\partial^2 z}{\partial r^2} + \frac{1}{r^2}\frac{\partial^2 z}{\partial\theta^2} + \frac{1}{r}\frac{\partial z}{\partial r} = (\cos^2\theta + \sin^2\theta)\frac{\partial^2 z}{\partial x^2} + (\sin^2\theta + \cos^2\theta)\frac{\partial^2 z}{\partial y^2}$$

$$-\frac{1}{r}\cos\theta\,\frac{\partial z}{\partial x} - \frac{1}{r}\sin\theta\,\frac{\partial z}{\partial y} + \frac{1}{r}\left(\cos\theta\,\frac{\partial z}{\partial x} + \sin\theta\,\frac{\partial z}{\partial y}\right)$$

$$= \frac{\partial^2 z}{\partial x^2} + \frac{\partial^2 z}{\partial y^2} \text{ as desired.}$$

54. (a) $\dfrac{\partial z}{\partial t} = \dfrac{\partial z}{\partial x}\dfrac{\partial x}{\partial t} + \dfrac{\partial z}{\partial y}\dfrac{\partial y}{\partial t}$. Then

$$\frac{\partial^2 z}{\partial t^2} = \frac{\partial}{\partial t}\left(\frac{\partial z}{\partial x}\frac{\partial x}{\partial t}\right) + \frac{\partial}{\partial t}\left(\frac{\partial z}{\partial y}\frac{\partial y}{\partial t}\right) = \frac{\partial}{\partial t}\left(\frac{\partial z}{\partial x}\right)\frac{\partial x}{\partial t} + \frac{\partial^2 x}{\partial t^2}\frac{\partial z}{\partial x} + \frac{\partial}{\partial t}\left(\frac{\partial z}{\partial y}\right)\frac{\partial y}{\partial t} + \frac{\partial^2 y}{\partial t^2}\frac{\partial z}{\partial y}$$

$$= \frac{\partial^2 z}{\partial x^2}\left(\frac{\partial x}{\partial t}\right)^2 + \frac{\partial^2 z}{\partial y\,\partial x}\frac{\partial x}{\partial t}\frac{\partial y}{\partial t} + \frac{\partial^2 x}{\partial t^2}\frac{\partial z}{\partial x} + \frac{\partial^2 z}{\partial y^2}\left(\frac{\partial y}{\partial t}\right)^2 + \frac{\partial^2 z}{\partial x\,\partial y}\frac{\partial y}{\partial t}\frac{\partial x}{\partial t} + \frac{\partial^2 y}{\partial t^2}\frac{\partial z}{\partial y}$$

$$= \frac{\partial^2 z}{\partial x^2}\left(\frac{\partial x}{\partial t}\right)^2 + 2\frac{\partial^2 z}{\partial x\,\partial y}\frac{\partial x}{\partial t}\frac{\partial y}{\partial t} + \frac{\partial^2 z}{\partial y^2}\left(\frac{\partial y}{\partial t}\right)^2 + \frac{\partial^2 x}{\partial t^2}\frac{\partial z}{\partial x} + \frac{\partial^2 y}{\partial t^2}\frac{\partial z}{\partial y}$$

(b) $\dfrac{\partial^2 z}{\partial s\,\partial t} = \dfrac{\partial}{\partial s}\left(\dfrac{\partial z}{\partial x}\dfrac{\partial x}{\partial t} + \dfrac{\partial z}{\partial y}\dfrac{\partial y}{\partial t}\right)$

$$= \left(\frac{\partial^2 z}{\partial x^2}\frac{\partial x}{\partial s} + \frac{\partial^2 z}{\partial y\,\partial x}\frac{\partial y}{\partial s}\right)\frac{\partial x}{\partial t} + \frac{\partial z}{\partial x}\frac{\partial^2 x}{\partial s\,\partial t} + \left(\frac{\partial^2 z}{\partial y^2}\frac{\partial y}{\partial s} + \frac{\partial^2 z}{\partial x\,\partial y}\frac{\partial x}{\partial s}\right)\frac{\partial y}{\partial t} + \frac{\partial z}{\partial y}\frac{\partial^2 y}{\partial s\,\partial t}$$

$$= \frac{\partial^2 z}{\partial x^2}\frac{\partial x}{\partial s}\frac{\partial x}{\partial t} + \frac{\partial^2 z}{\partial x\,\partial y}\left(\frac{\partial y}{\partial s}\frac{\partial x}{\partial t} + \frac{\partial y}{\partial t}\frac{\partial x}{\partial s}\right) + \frac{\partial z}{\partial x}\frac{\partial^2 x}{\partial s\,\partial t} + \frac{\partial z}{\partial y}\frac{\partial^2 y}{\partial s\,\partial t} + \frac{\partial^2 z}{\partial y^2}\frac{\partial y}{\partial s}\frac{\partial y}{\partial t}$$

55. (a) Since f is a polynomial, it has continuous second-order partial derivatives, and

$$f(tx, ty) = (tx)^2(ty) + 2(tx)(ty)^2 + 5(ty)^3 = t^3 x^2 y + 2t^3 xy^2 + 5t^3 y^3 = t^3(x^2 y + 2xy^2 + 5y^3) = t^3 f(x, y).$$

Thus, f is homogeneous of degree 3.

(b) Differentiating both sides of $f(tx, ty) = t^n f(x, y)$ with respect to t using the Chain Rule, we get

$$\frac{\partial}{\partial t}f(tx, ty) = \frac{\partial}{\partial t}\left[t^n f(x, y)\right] \quad\Leftrightarrow$$

$$\frac{\partial}{\partial(tx)}f(tx, ty)\cdot\frac{\partial(tx)}{\partial t} + \frac{\partial}{\partial(ty)}f(tx, ty)\cdot\frac{\partial(ty)}{\partial t} = x\frac{\partial}{\partial(tx)}f(tx, ty) + y\frac{\partial}{\partial(ty)}f(tx, ty) = nt^{n-1}f(x, y).$$

Setting $t = 1$: $x\dfrac{\partial}{\partial x}f(x, y) + y\dfrac{\partial}{\partial y}f(x, y) = nf(x, y)$.

56. Differentiating both sides of $f(tx, ty) = t^n f(x, y)$ with respect to t using the Chain Rule, we get

$$\frac{\partial}{\partial(tx)} f(tx, ty) \cdot \frac{\partial(tx)}{\partial t} + \frac{\partial}{\partial(ty)} f(tx, ty) \cdot \frac{\partial(ty)}{\partial t} = x \frac{\partial}{\partial(tx)} f(tx, ty) + y \frac{\partial}{\partial(ty)} f(tx, ty) = nt^{n-1} f(x, y) \text{ and}$$

differentiating again with respect to t gives

$$x \left[\frac{\partial^2}{\partial(tx)^2} f(tx, ty) \cdot \frac{\partial(tx)}{\partial t} + \frac{\partial^2}{\partial(ty)\partial(tx)} f(tx, ty) \cdot \frac{\partial(ty)}{\partial t} \right]$$

$$+ y \left[\frac{\partial^2}{\partial(tx)\partial(ty)} f(tx, ty) \cdot \frac{\partial(tx)}{\partial t} + \frac{\partial^2}{\partial(ty)^2} f(tx, ty) \cdot \frac{\partial(ty)}{\partial t} \right] = n(n-1)t^{n-1} f(x, y).$$

Setting $t = 1$ and using the fact that $f_{yx} = f_{xy}$, we have $x^2 f_{xx} + 2xy f_{xy} + y^2 f_{yy} = n(n-1) f(x, y)$.

57. Differentiating both sides of $f(tx, ty) = t^n f(x, y)$ with respect to x using the Chain Rule, we get

$$\frac{\partial}{\partial x} f(tx, ty) = \frac{\partial}{\partial x} [t^n f(x, y)] \quad \Leftrightarrow$$

$$\frac{\partial}{\partial(tx)} f(tx, ty) \cdot \frac{\partial(tx)}{\partial x} + \frac{\partial}{\partial(ty)} f(tx, ty) \cdot \frac{\partial(ty)}{\partial x} = t^n \frac{\partial}{\partial x} f(x, y) \quad \Leftrightarrow \quad t f_x(tx, ty) = t^n f_x(x, y).$$

Thus $f_x(tx, ty) = t^{n-1} f_x(x, y)$.

58. $F(x, y, z) = 0$ is assumed to define z as a function of x and y, that is, $z = f(x, y)$. So by (7), $\dfrac{\partial z}{\partial x} = -\dfrac{F_x}{F_z}$ since $F_z \neq 0$.

Similarly, it is assumed that $F(x, y, z) = 0$ defines x as a function of y and z, that is $x = h(x, z)$. Then $F(h(y, z), y, z) = 0$

and by the Chain Rule, $F_x \dfrac{\partial x}{\partial y} + F_y \dfrac{\partial y}{\partial y} + F_z \dfrac{\partial z}{\partial y} = 0$. But $\dfrac{\partial z}{\partial y} = 0$ and $\dfrac{\partial y}{\partial y} = 1$, so $F_x \dfrac{\partial x}{\partial y} + F_y = 0 \quad \Rightarrow \quad \dfrac{\partial x}{\partial y} = -\dfrac{F_y}{F_x}$.

A similar calculation shows that $\dfrac{\partial y}{\partial z} = -\dfrac{F_z}{F_y}$. Thus $\dfrac{\partial z}{\partial x} \dfrac{\partial x}{\partial y} \dfrac{\partial y}{\partial z} = \left(-\dfrac{F_x}{F_z} \right) \left(-\dfrac{F_y}{F_x} \right) \left(-\dfrac{F_z}{F_y} \right) = -1$.

15.6 Directional Derivatives and the Gradient Vector

1. We can approximate the directional derivative of the pressure function at K in the direction of S by the average rate of change of pressure between the points where the red line intersects the contour lines closest to K (extend the red line slightly at the left). In the direction of S, the pressure changes from 1000 millibars to 996 millibars and we estimate the distance between these two points to be approximately 50 km (using the fact that the distance from K to S is 300 km). Then the rate of change of pressure in the direction given is approximately $\frac{996 - 1000}{50} = -0.08$ millibar/km.

2. First we draw a line passing through Dubbo and Sydney. We approximate the directional derivative at Dubbo in the direction of Sydney by the average rate of change of temperature between the points where the line intersects the contour lines closest to Dubbo. In the direction of Sydney, the temperature changes from $30°C$ to $27°C$. We estimate the distance between these two points to be approximately 120 km, so the rate of change of maximum temperature in the direction given is approximately $\frac{27 - 30}{120} = -0.025°C/km$.

3. $D_{\mathbf{u}} f(-20, 30) = \nabla f(-20, 30) \cdot \mathbf{u} = f_T(-20, 30) \left(\frac{1}{\sqrt{2}} \right) + f_v(-20, 30) \left(\frac{1}{\sqrt{2}} \right)$.

$f_T(-20, 30) = \lim\limits_{h \to 0} \dfrac{f(-20 + h, 30) - f(-20, 30)}{h}$, so we can approximate $f_T(-20, 30)$ by considering $h = \pm 5$ and using the values given in the table: $f_T(-20, 30) \approx \dfrac{f(-15, 30) - f(-20, 30)}{5} = \dfrac{-26 - (-33)}{5} = 1.4,$

$f_T(-20, 30) \approx \dfrac{f(-25, 30) - f(-20, 30)}{-5} = \dfrac{-39 - (-33)}{-5} = 1.2.$ Averaging these values gives $f_T(-20, 30) \approx 1.3.$

Similarly, $f_v(-20, 30) = \lim\limits_{h \to 0} \dfrac{f(-20, 30 + h) - f(-20, 30)}{h}$, so we can approximate $f_v(-20, 30)$ with $h = \pm 10$:

$f_v(-20, 30) \approx \dfrac{f(-20, 40) - f(-20, 30)}{10} = \dfrac{-34 - (-33)}{10} = -0.1,$

$f_v(-20, 30) \approx \dfrac{f(-20, 20) - f(-20, 30)}{-10} = \dfrac{-30 - (-33)}{-10} = -0.3.$ Averaging these values gives $f_v(-20, 30) \approx -0.2.$

Then $D_{\mathbf{u}} f(-20, 30) \approx 1.3\left(\dfrac{1}{\sqrt{2}}\right) + (-0.2)\left(\dfrac{1}{\sqrt{2}}\right) \approx 0.778.$

4. $f(x, y) = x^2 y^3 - y^4 \;\Rightarrow\; f_x(x, y) = 2xy^3$ and $f_y(x, y) = 3x^2 y^2 - 4y^3$. If $\mathbf{u}$ is a unit vector in the direction of $\theta = \frac{\pi}{4}$, then from Equation 6, $D_{\mathbf{u}} f(2, 1) = f_x(2, 1) \cos\left(\frac{\pi}{4}\right) + f_y(2, 1) \sin\left(\frac{\pi}{4}\right) = 4 \cdot \frac{\sqrt{2}}{2} + 8 \cdot \frac{\sqrt{2}}{2} = 6\sqrt{2}.$

5. $f(x, y) = ye^{-x} \;\Rightarrow\; f_x(x, y) = -ye^{-x}$ and $f_y(x, y) = e^{-x}$. If $\mathbf{u}$ is a unit vector in the direction of $\theta = 2\pi/3$, then from Equation 6, $D_{\mathbf{u}} f(0, 4) = f_x(0, 4) \cos\left(\frac{2\pi}{3}\right) + f_y(0, 4) \sin\left(\frac{2\pi}{3}\right) = -4 \cdot \left(-\frac{1}{2}\right) + 1 \cdot \frac{\sqrt{3}}{2} = 2 + \frac{\sqrt{3}}{2}.$

6. $f(x, y) = x \sin(xy) \;\Rightarrow\; f_x(x, y) = x \cos(xy) \cdot y + \sin(xy) = xy \cos(xy) + \sin(xy)$ and $f_y(x, y) = x \cos(xy) \cdot x = x^2 \cos(xy)$. If $\mathbf{u}$ is a unit vector in the direction of $\theta = \frac{\pi}{3}$, then from Equation 6, $D_{\mathbf{u}} f(2, 0) = f_x(2, 0) \cos \frac{\pi}{3} + f_y(2, 0) \sin \frac{\pi}{3} = 0 + 4\left(\frac{\sqrt{3}}{2}\right) = 2\sqrt{3}.$

7. $f(x, y) = \sin(2x + 3y)$

(a) $\nabla f(x, y) = \dfrac{\partial f}{\partial x}\mathbf{i} + \dfrac{\partial f}{\partial y}\mathbf{j} = [\cos(2x + 3y) \cdot 2]\mathbf{i} + [\cos(2x + 3y) \cdot 3]\mathbf{j} = 2\cos(2x + 3y)\mathbf{i} + 3\cos(2x + 3y)\mathbf{j}$

(b) $\nabla f(-6, 4) = (2\cos 0)\mathbf{i} + (3\cos 0)\mathbf{j} = 2\mathbf{i} + 3\mathbf{j}$

(c) By Equation 9, $D_{\mathbf{u}} f(-6, 4) = \nabla f(-6, 4) \cdot \mathbf{u} = (2\mathbf{i} + 3\mathbf{j}) \cdot \frac{1}{2}(\sqrt{3}\mathbf{i} - \mathbf{j}) = \frac{1}{2}(2\sqrt{3} - 3) = \sqrt{3} - \frac{3}{2}.$

8. $f(x, y) = y^2/x$

(a) $\nabla f(x, y) = \dfrac{\partial f}{\partial x}\mathbf{i} + \dfrac{\partial f}{\partial y}\mathbf{j} = y^2(-x^{-2})\mathbf{i} + (2y/x)\mathbf{j} = -\dfrac{y^2}{x^2}\mathbf{i} + \dfrac{2y}{x}\mathbf{j}$

(b) $\nabla f(1, 2) = -4\mathbf{i} + 4\mathbf{j}$

(c) By Equation 9, $D_{\mathbf{u}} f(1, 2) = \nabla f(1, 2) \cdot \mathbf{u} = (-4\mathbf{i} + 4\mathbf{j}) \cdot \frac{1}{3}(2\mathbf{i} + \sqrt{5}\mathbf{j}) = \frac{1}{3}(-8 + 4\sqrt{5}) = \frac{4}{3}(\sqrt{5} - 2).$

9. $f(x, y, z) = xe^{2yz}$

(a) $\nabla f(x, y, z) = \langle f_x(x, y, z), f_y(x, y, z), f_z(x, y, z) \rangle = \langle e^{2yz}, 2xze^{2yz}, 2xye^{2yz} \rangle$

(b) $\nabla f(3, 0, 2) = \langle 1, 12, 0 \rangle$

(c) By Equation 14, $D_{\mathbf{u}} f(3, 0, 2) = \nabla f(3, 0, 2) \cdot \mathbf{u} = \langle 1, 12, 0 \rangle \cdot \langle \frac{2}{3}, -\frac{2}{3}, \frac{1}{3} \rangle = \frac{2}{3} - \frac{24}{3} + 0 = -\frac{22}{3}.$

10. $f(x, y, z) = \sqrt{x + yz} = (x + yz)^{1/2}$

(a) $\nabla f(x, y, z) = \left\langle \frac{1}{2}(x + yz)^{-1/2}(1), \frac{1}{2}(x + yz)^{-1/2}(z), \frac{1}{2}(x + yz)^{-1/2}(y) \right\rangle$

$\qquad = \langle 1/(2\sqrt{x + yz}), z/(2\sqrt{x + yz}), y/(2\sqrt{x + yz}) \rangle$

(b) $\nabla f(1, 3, 1) = \langle \frac{1}{4}, \frac{1}{4}, \frac{3}{4} \rangle$

(c) $D_{\mathbf{u}} f(1, 3, 1) = \nabla f(1, 3, 1) \cdot \mathbf{u} = \langle \frac{1}{4}, \frac{1}{4}, \frac{3}{4} \rangle \cdot \langle \frac{2}{7}, \frac{3}{7}, \frac{6}{7} \rangle = \frac{2}{28} + \frac{3}{28} + \frac{18}{28} = \frac{23}{28}$

11. $f(x, y) = 1 + 2x\sqrt{y} \implies \nabla f(x, y) = \left\langle 2\sqrt{y}, 2x \cdot \frac{1}{2}y^{-1/2} \right\rangle = \left\langle 2\sqrt{y}, x/\sqrt{y} \right\rangle$, $\nabla f(3, 4) = \left\langle 4, \frac{3}{2} \right\rangle$, and a unit vector in

the direction of **v** is $\mathbf{u} = \frac{1}{\sqrt{4^2 + (-3)^2}} \langle 4, -3 \rangle = \left\langle \frac{4}{5}, -\frac{3}{5} \right\rangle$, so $D_{\mathbf{u}} f(3, 4) = \nabla f(3, 4) \cdot \mathbf{u} = \left\langle 4, \frac{3}{2} \right\rangle \cdot \left\langle \frac{4}{5}, -\frac{3}{5} \right\rangle = \frac{23}{10}$.

12. $f(x, y) = \ln(x^2 + y^2) \implies \nabla f(x, y) = \left\langle \frac{2x}{x^2 + y^2}, \frac{2y}{x^2 + y^2} \right\rangle$, $\nabla f(2, 1) = \left\langle \frac{4}{5}, \frac{2}{5} \right\rangle$, and

a unit vector in the direction of $\mathbf{v} = \langle -1, 2 \rangle$ is $\mathbf{u} = \frac{1}{\sqrt{1+4}} \langle -1, 2 \rangle = \left\langle -\frac{1}{\sqrt{5}}, \frac{2}{\sqrt{5}} \right\rangle$, so

$D_{\mathbf{u}} f(2, 1) = \nabla f(2, 1) \cdot \mathbf{u} = \left\langle \frac{4}{5}, \frac{2}{5} \right\rangle \cdot \left\langle -\frac{1}{\sqrt{5}}, \frac{2}{\sqrt{5}} \right\rangle = -\frac{4}{5\sqrt{5}} + \frac{4}{5\sqrt{5}} = 0$.

13. $g(p, q) = p^4 - p^2 q^3 \implies \nabla g(p, q) = \left(4p^3 - 2pq^3 \right) \mathbf{i} + \left(-3p^2 q^2 \right) \mathbf{j}$, $\nabla g(2, 1) = 28\mathbf{i} - 12\mathbf{j}$, and a unit

vector in the direction of **v** is $\mathbf{u} = \frac{1}{\sqrt{1^2 + 3^2}} (\mathbf{i} + 3\mathbf{j}) = \frac{1}{\sqrt{10}} (\mathbf{i} + 3\mathbf{j})$, so

$D_{\mathbf{u}} g(2, 1) = \nabla g(2, 1) \cdot \mathbf{u} = (28\mathbf{i} - 12\mathbf{j}) \cdot \frac{1}{\sqrt{10}} (\mathbf{i} + 3\mathbf{j}) = \frac{1}{\sqrt{10}} (28 - 36) = -\frac{8}{\sqrt{10}}$ or $-\frac{4\sqrt{10}}{5}$.

14. $g(r, s) = \tan^{-1}(rs) \implies \nabla g(r, s) = \left(\frac{1}{1 + (rs)^2} \cdot s \right) \mathbf{i} + \left(\frac{1}{1 + (rs)^2} \cdot r \right) \mathbf{j} = \frac{s}{1 + r^2 s^2} \mathbf{i} + \frac{r}{1 + r^2 s^2} \mathbf{j}$,

$\nabla g(1, 2) = \frac{2}{5}\mathbf{i} + \frac{1}{5}\mathbf{j}$, and a unit vector in the direction of **v** is $\mathbf{u} = \frac{1}{\sqrt{5^2 + 10^2}} (5\mathbf{i} + 10\mathbf{j}) = \frac{1}{5\sqrt{5}} (5\mathbf{i} + 10\mathbf{j}) = \frac{1}{\sqrt{5}}\mathbf{i} + \frac{2}{\sqrt{5}}\mathbf{j}$,

so $D_{\mathbf{u}} g(1, 2) = \nabla g(1, 2) \cdot \mathbf{u} = (\frac{2}{5}\mathbf{i} + \frac{1}{5}\mathbf{j}) \cdot (\frac{1}{\sqrt{5}}\mathbf{i} + \frac{2}{\sqrt{5}}\mathbf{j}) = \frac{2}{5\sqrt{5}} + \frac{2}{5\sqrt{5}} = \frac{4}{5\sqrt{5}}$ or $\frac{4\sqrt{5}}{25}$.

15. $f(x, y, z) = xe^y + ye^z + ze^x \implies \nabla f(x, y, z) = \langle e^y + ze^x, xe^y + e^z, ye^z + e^x \rangle$, $\nabla f(0, 0, 0) = \langle 1, 1, 1 \rangle$, and a unit

vector in the direction of **v** is $\mathbf{u} = \frac{1}{\sqrt{25 + 1 + 4}} \langle 5, 1, -2 \rangle = \frac{1}{\sqrt{30}} \langle 5, 1, -2 \rangle$, so

$D_{\mathbf{u}} f(0, 0, 0) = \nabla f(0, 0, 0) \cdot \mathbf{u} = \langle 1, 1, 1 \rangle \cdot \frac{1}{\sqrt{30}} \langle 5, 1, -2 \rangle = \frac{4}{\sqrt{30}}$.

16. $f(x, y, z) = \sqrt{xyz} \implies$

$\nabla f(x, y, z) = \left\langle \frac{1}{2}(xyz)^{-1/2} \cdot yz, \frac{1}{2}(xyz)^{-1/2} \cdot xz, \frac{1}{2}(xyz)^{-1/2} \cdot xy \right\rangle = \left\langle \frac{yz}{2\sqrt{xyz}}, \frac{xz}{2\sqrt{xyz}}, \frac{xy}{2\sqrt{xyz}} \right\rangle$,

$\nabla f(3, 2, 6) = \left\langle \frac{12}{2\sqrt{36}}, \frac{18}{2\sqrt{36}}, \frac{6}{2\sqrt{36}} \right\rangle = \left\langle 1, \frac{3}{2}, \frac{1}{2} \right\rangle$, and a unit vector in the

direction of **v** is $\mathbf{u} = \frac{1}{\sqrt{1 + 4 + 4}} \langle -1, -2, 2 \rangle = \left\langle -\frac{1}{3}, -\frac{2}{3}, \frac{2}{3} \right\rangle$, so

$D_{\mathbf{u}} f(3, 2, 6) = \nabla f(3, 2, 6) \cdot \mathbf{u} = \left\langle 1, \frac{3}{2}, \frac{1}{2} \right\rangle \cdot \left\langle -\frac{1}{3}, -\frac{2}{3}, \frac{2}{3} \right\rangle = -\frac{1}{3} - 1 + \frac{1}{3} = -1$.

17. $g(x, y, z) = (x + 2y + 3z)^{3/2} \implies$

$\nabla g(x, y, z) = \left\langle \frac{3}{2}(x + 2y + 3z)^{1/2}(1), \frac{3}{2}(x + 2y + 3z)^{1/2}(2), \frac{3}{2}(x + 2y + 3z)^{1/2}(3) \right\rangle$

$= \left\langle \frac{3}{2}\sqrt{x + 2y + 3z}, 3\sqrt{x + 2y + 3z}, \frac{9}{2}\sqrt{x + 2y + 3z} \right\rangle$, $\nabla g(1, 1, 2) = \left\langle \frac{9}{2}, 9, \frac{27}{2} \right\rangle$,

and a unit vector in the direction of $\mathbf{v} = 2\mathbf{j} - \mathbf{k}$ is $\mathbf{u} = \frac{2}{\sqrt{5}}\mathbf{j} - \frac{1}{\sqrt{5}}\mathbf{k}$, so

$D_{\mathbf{u}} g(1, 1, 2) = \left\langle \frac{9}{2}, 9, \frac{27}{2} \right\rangle \cdot \left\langle 0, \frac{2}{\sqrt{5}}, -\frac{1}{\sqrt{5}} \right\rangle = \frac{18}{\sqrt{5}} - \frac{27}{2\sqrt{5}} = \frac{9}{2\sqrt{5}}$.

18. $D_{\mathbf{u}} f(2,2) = \nabla f(2,2) \cdot \mathbf{u}$, the scalar projection of $\nabla f(2,2)$ onto $\mathbf{u}$, so we draw a
perpendicular from the tip of $\nabla f(2,2)$ to the line containing $\mathbf{u}$. We can use the
point $(2,2)$ to determine the scale of the axes, and we estimate the length of the
projection to be approximately 3.0 units. Since the angle between $\nabla f(2,2)$ and $\mathbf{u}$
is greater than $90°$, the scalar projection is negative. Thus $D_{\mathbf{u}} f(2,2) \approx -3$.

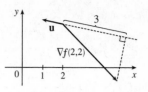

19. $f(x,y) = \sqrt{xy} \quad \Rightarrow \quad \nabla f(x,y) = \left\langle \frac{1}{2}(xy)^{-1/2}(y), \frac{1}{2}(xy)^{-1/2}(x) \right\rangle = \left\langle \dfrac{y}{2\sqrt{xy}}, \dfrac{x}{2\sqrt{xy}} \right\rangle$, so $\nabla f(2,8) = \left\langle 1, \frac{1}{4} \right\rangle$.

The unit vector in the direction of $\overrightarrow{PQ} = \langle 5-2, 4-8 \rangle = \langle 3, -4 \rangle$ is $\mathbf{u} = \left\langle \frac{3}{5}, -\frac{4}{5} \right\rangle$, so

$D_{\mathbf{u}} f(2,8) = \nabla f(2,8) \cdot \mathbf{u} = \left\langle 1, \frac{1}{4} \right\rangle \cdot \left\langle \frac{3}{5}, -\frac{4}{5} \right\rangle = \frac{2}{5}$.

20. $f(x,y,z) = xy + yz + zx \quad \Rightarrow \quad \nabla f(x,y,z) = \langle y+z, x+z, y+x \rangle$, so $\nabla f(1,-1,3) = \langle 2,4,0 \rangle$. The unit vector in the

direction of $\overrightarrow{PQ} = \langle 1,5,2 \rangle$ is $\mathbf{u} = \frac{1}{\sqrt{30}}\langle 1,5,2 \rangle$, so $D_{\mathbf{u}} f(1,-1,3) = \nabla f(1,-1,3) \cdot \mathbf{u} = \langle 2,4,0 \rangle \cdot \frac{1}{\sqrt{30}}\langle 1,5,2 \rangle = \frac{22}{\sqrt{30}}$.

21. $f(x,y) = y^2/x = y^2 x^{-1} \quad \Rightarrow \quad \nabla f(x,y) = \langle -y^2 x^{-2}, 2yx^{-1} \rangle = \langle -y^2/x^2, 2y/x \rangle$.

$\nabla f(2,4) = \langle -4,4 \rangle$, or equivalently $\langle -1,1 \rangle$, is the direction of maximum rate of change, and the maximum rate
is $|\nabla f(2,4)| = \sqrt{16+16} = 4\sqrt{2}$.

22. $f(p,q) = qe^{-p} + pe^{-q} \quad \Rightarrow \quad \nabla f(p,q) = \langle -qe^{-p} + e^{-q}, e^{-p} - pe^{-q} \rangle$.

$\nabla f(0,0) = \langle 1,1 \rangle$ is the direction of maximum rate of change and the maximum rate is $|\nabla f(0,0)| = \sqrt{2}$.

23. $f(x,y) = \sin(xy) \quad \Rightarrow \quad \nabla f(x,y) = \langle y\cos(xy), x\cos(xy) \rangle$, $\nabla f(1,0) = \langle 0,1 \rangle$. Thus the maximum rate of change is
$|\nabla f(1,0)| = 1$ in the direction $\langle 0,1 \rangle$.

24. $f(x,y,z) = \dfrac{x+y}{z} \quad \Rightarrow \quad \nabla f(x,y,z) = \left\langle \dfrac{1}{z}, \dfrac{1}{z}, -\dfrac{x+y}{z^2} \right\rangle$, $\nabla f(1,1,-1) = \langle -1,-1,-2 \rangle$. Thus the maximum rate of

change is $|\nabla f(1,1,-1)| = \sqrt{1+1+4} = \sqrt{6}$ in the direction $\langle -1,-1,-2 \rangle$.

25. $f(x,y,z) = \sqrt{x^2 + y^2 + z^2} \quad \Rightarrow$

$\nabla f(x,y,z) = \left\langle \frac{1}{2}(x^2+y^2+z^2)^{-1/2} \cdot 2x, \frac{1}{2}(x^2+y^2+z^2)^{-1/2} \cdot 2y, \frac{1}{2}(x^2+y^2+z^2)^{-1/2} \cdot 2z \right\rangle$

$= \left\langle \dfrac{x}{\sqrt{x^2+y^2+z^2}}, \dfrac{y}{\sqrt{x^2+y^2+z^2}}, \dfrac{z}{\sqrt{x^2+y^2+z^2}} \right\rangle$,

$\nabla f(3,6,-2) = \left\langle \frac{3}{\sqrt{49}}, \frac{6}{\sqrt{49}}, \frac{-2}{\sqrt{49}} \right\rangle = \left\langle \frac{3}{7}, \frac{6}{7}, -\frac{2}{7} \right\rangle$. Thus the maximum rate of change is

$|\nabla f(3,6,-2)| = \sqrt{\left(\frac{3}{7}\right)^2 + \left(\frac{6}{7}\right)^2 + \left(-\frac{2}{7}\right)^2} = \sqrt{\frac{9+36+4}{49}} = 1$ in the direction $\left\langle \frac{3}{7}, \frac{6}{7}, -\frac{2}{7} \right\rangle$ or equivalently $\langle 3,6,-2 \rangle$.

26. $f(x,y,z) = \tan(x+2y+3z) \quad \Rightarrow$

$\nabla f(x,y,z) = \langle \sec^2(x+2y+3z)(1), \sec^2(x+2y+3z)(2), \sec^2(x+2y+3z)(3) \rangle$.

$\nabla f(-5,1,1) = \langle \sec^2(0), 2\sec^2(0), 3\sec^2(0) \rangle = \langle 1,2,3 \rangle$ is the direction of maximum rate of change and the maximum rate
is $|\nabla f(-5,1,1)| = \sqrt{14}$.

27. (a) As in the proof of Theorem 15, $D_{\mathbf{u}}\, f = |\nabla f| \cos\theta$. Since the minimum value of $\cos\theta$ is -1 occurring when $\theta = \pi$, the minimum value of $D_{\mathbf{u}}\, f$ is $-|\nabla f|$ occurring when $\theta = \pi$, that is when $\mathbf{u}$ is in the opposite direction of ∇f (assuming $\nabla f \neq \mathbf{0}$).

(b) $f(x,y) = x^4 y - x^2 y^3 \;\Rightarrow\; \nabla f(x,y) = \langle 4x^3 y - 2xy^3, x^4 - 3x^2 y^2\rangle$, so f decreases fastest at the point $(2,-3)$ in the direction $-\nabla f(2,-3) = -\langle 12, -92\rangle = \langle -12, 92\rangle$.

28. $f(x,y) = ye^{-xy} \;\Rightarrow\; f_x(x,y) = ye^{-xy}(-y) = -y^2 e^{-xy}, \;\; f_y(x,y) = ye^{-xy}(-x) + e^{-xy} = (1-xy)e^{-xy}$ and $f_x(0,2) = -4e^0 = -4, \;\; f_y(0,2) = (1-0)e^0 = 1$. If $\mathbf{u}$ is a unit vector which makes an angle θ with the positive x-axis, then $D_{\mathbf{u}} f(0,2) = f_x(0,2)\cos\theta + f_y(0,2)\sin\theta = -4\cos\theta + \sin\theta$. We want $D_{\mathbf{u}} f(0,2) = 1$, so $-4\cos\theta + \sin\theta = 1 \;\Rightarrow\;$ $\sin\theta = 1 + 4\cos\theta \;\Rightarrow\; \sin^2\theta = (1 + 4\cos\theta)^2 \;\Rightarrow\; 1 - \cos^2\theta = 1 + 8\cos\theta + 16\cos^2\theta \;\Rightarrow\;$ $17\cos^2\theta + 8\cos\theta = 0 \;\Rightarrow\; \cos\theta(17\cos\theta + 8) = 0 \;\Rightarrow\; \cos\theta = 0$ or $\cos\theta = -\frac{8}{17}$. If $\cos\theta = 0$ then $\theta = \frac{\pi}{2}$ or $\theta = \frac{3\pi}{2}$ but $\frac{3\pi}{2}$ does not satisfy the original equation. If $\cos\theta = -\frac{8}{17}$ then $\theta = \cos^{-1}\left(-\frac{8}{17}\right)$ or $\theta = 2\pi - \cos^{-1}\left(-\frac{8}{17}\right)$ but $\theta = \cos^{-1}\left(-\frac{8}{17}\right)$ is not a solution of the original equation. Thus the directions are $\theta = \frac{\pi}{2}$ or $\theta = 2\pi - \cos^{-1}\left(-\frac{8}{17}\right) \approx 4.22$ rad.

29. The direction of fastest change is $\nabla f(x,y) = (2x-2)\,\mathbf{i} + (2y-4)\,\mathbf{j}$, so we need to find all points (x,y) where $\nabla f(x,y)$ is parallel to $\mathbf{i} + \mathbf{j} \;\Leftrightarrow\; (2x-2)\,\mathbf{i} + (2y-4)\,\mathbf{j} = k\,(\mathbf{i} + \mathbf{j}) \;\Leftrightarrow\; k = 2x-2$ and $k = 2y-4$. Then $2x - 2 = 2y - 4 \;\Rightarrow\;$ $y = x+1$, so the direction of fastest change is $\mathbf{i} + \mathbf{j}$ at all points on the line $y = x+1$.

30. The fisherman is traveling in the direction $\langle -80, -60\rangle$. A unit vector in this direction is $\mathbf{u} = \frac{1}{100}\langle -80, -60\rangle = \langle -\frac{4}{5}, -\frac{3}{5}\rangle$, and if the depth of the lake is given by $f(x,y) = 200 + 0.02x^2 - 0.001y^3$, then $\nabla f(x,y) = \langle 0.04x, -0.003y^2\rangle$. $D_{\mathbf{u}}\, f(80,60) = \nabla f(80,60) \cdot \mathbf{u} = \langle 3.2, -10.8\rangle \cdot \langle -\frac{4}{5}, -\frac{3}{5}\rangle = 3.92$. Since $D_{\mathbf{u}}\, f(80,60)$ is positive, the depth of the lake is increasing near $(80,60)$ in the direction toward the buoy.

31. $T = \dfrac{k}{\sqrt{x^2 + y^2 + z^2}}$ and $120 = T(1,2,2) = \dfrac{k}{3}$ so $k = 360$.

(a) $\mathbf{u} = \dfrac{\langle 1, -1, 1\rangle}{\sqrt{3}}$,

$D_{\mathbf{u}}T(1,2,2) = \nabla T(1,2,2) \cdot \mathbf{u} = \left[-360(x^2 + y^2 + z^2)^{-3/2}\langle x,y,z\rangle\right]_{(1,2,2)} \cdot \mathbf{u} = -\frac{40}{3}\langle 1,2,2\rangle \cdot \frac{1}{\sqrt{3}}\langle 1,-1,1\rangle = -\frac{40}{3\sqrt{3}}$

(b) From (a), $\nabla T = -360(x^2 + y^2 + z^2)^{-3/2}\langle x,y,z\rangle$, and since $\langle x,y,z\rangle$ is the position vector of the point (x,y,z), the vector $-\langle x,y,z\rangle$, and thus ∇T, always points toward the origin.

32. $\nabla T = -400e^{-x^2 - 3y^2 - 9z^2}\langle x, 3y, 9z\rangle$

(a) $\mathbf{u} = \frac{1}{\sqrt{6}}\langle 1, -2, 1\rangle$, $\nabla T(2,-1,2) = -400e^{-43}\langle 2, -3, 18\rangle$ and

$D_{\mathbf{u}}\, T(2,-1,2) = \left(-\dfrac{400e^{-43}}{\sqrt{6}}\right)(26) = -\dfrac{5200\sqrt{6}}{3e^{43}}\ {}^\circ\mathrm{C/m}.$

(b) $\nabla T(2,-1,2) = 400e^{-43}\langle -2, 3, -18\rangle$ or equivalently $\langle -2, 3, -18\rangle$.

(c) $|\nabla T| = 400e^{-x^2-3y^2-9z^2}\sqrt{x^2+9y^2+81z^2}$ °C/m is the maximum rate of increase. At $(2,-1,2)$ the maximum rate

of increase is $400e^{-43}\sqrt{337}$ °C/m.

33. $\nabla V(x,y,z) = \langle 10x - 3y + yz, xz - 3x, xy \rangle$, $\nabla V(3,4,5) = \langle 38, 6, 12 \rangle$

(a) $D_{\mathbf{u}} V(3,4,5) = \langle 38, 6, 12 \rangle \cdot \frac{1}{\sqrt{3}}\langle 1, 1, -1 \rangle = \frac{32}{\sqrt{3}}$

(b) $\nabla V(3,4,5) = \langle 38, 6, 12 \rangle$, or equivalently, $\langle 19, 3, 6 \rangle$.

(c) $|\nabla V(3,4,5)| = \sqrt{38^2 + 6^2 + 12^2} = \sqrt{1624} = 2\sqrt{406}$

34. $z = f(x,y) = 1000 - 0.005x^2 - 0.01y^2 \quad \Rightarrow \quad \nabla f(x,y) = \langle -0.01x, -0.02y \rangle$ and $\nabla f(60,40) = \langle -0.6, -0.8 \rangle$.

(a) Due south is in the direction of the unit vector $\mathbf{u} = -\mathbf{j}$ and

$D_{\mathbf{u}} f(60,40) = \nabla f(60,40) \cdot \langle 0, -1 \rangle = \langle -0.6, -0.8 \rangle \cdot \langle 0, -1 \rangle = 0.8$. Thus, if you walk due south from $(60, 40, 966)$

you will ascend at a rate of 0.8 vertical meters per horizontal meter.

(b) Northwest is in the direction of the unit vector $\mathbf{u} = \frac{1}{\sqrt{2}}\langle -1, 1 \rangle$ and

$D_{\mathbf{u}} f(60,40) = \nabla f(60,40) \cdot \frac{1}{\sqrt{2}}\langle -1, 1 \rangle = \langle -0.6, -0.8 \rangle \cdot \frac{1}{\sqrt{2}}\langle -1, 1 \rangle = -\frac{0.2}{\sqrt{2}} \approx -0.14$. Thus, if you walk northwest

from $(60, 40, 966)$ you will descend at a rate of approximately 0.14 vertical meters per horizontal meter.

(c) $\nabla f(60,40) = \langle -0.6, -0.8 \rangle$ is the direction of largest slope with a rate of ascent given by

$|\nabla f(60,40)| = \sqrt{(-0.6)^2 + (-0.8)^2} = 1$. The angle above the horizontal in which the path begins is given by

$\tan\theta = 1 \quad \Rightarrow \quad \theta = 45°$.

35. A unit vector in the direction of $\overrightarrow{AB}$ is $\mathbf{i}$ and a unit vector in the direction of $\overrightarrow{AC}$ is $\mathbf{j}$. Thus $D_{\overrightarrow{AB}} f(1,3) = f_x(1,3) = 3$ and

$D_{\overrightarrow{AC}} f(1,3) = f_y(1,3) = 26$. Therefore $\nabla f(1,3) = \langle f_x(1,3), f_y(1,3) \rangle = \langle 3, 26 \rangle$, and by definition,

$D_{\overrightarrow{AD}} f(1,3) = \nabla f \cdot \mathbf{u}$ where $\mathbf{u}$ is a unit vector in the direction of $\overrightarrow{AD}$, which is $\langle \frac{5}{13}, \frac{12}{13} \rangle$. Therefore,

$D_{\overrightarrow{AD}} f(1,3) = \langle 3, 26 \rangle \cdot \langle \frac{5}{13}, \frac{12}{13} \rangle = 3 \cdot \frac{5}{13} + 26 \cdot \frac{12}{13} = \frac{327}{13}$.

36. The curve of steepest ascent is perpendicular to all

of the contour lines.

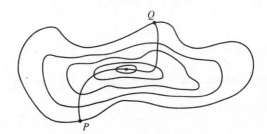

37. (a) $\nabla(au + bv) = \left\langle \frac{\partial(au+bv)}{\partial x}, \frac{\partial(au+bv)}{\partial y} \right\rangle = \left\langle a\frac{\partial u}{\partial x} + b\frac{\partial v}{\partial x}, a\frac{\partial u}{\partial y} + b\frac{\partial v}{\partial y} \right\rangle = a\left\langle \frac{\partial u}{\partial x}, \frac{\partial u}{\partial y} \right\rangle + b\left\langle \frac{\partial v}{\partial x}, \frac{\partial v}{\partial y} \right\rangle$

$= a\nabla u + b\nabla v$

(b) $\nabla(uv) = \left\langle v\frac{\partial u}{\partial x} + u\frac{\partial v}{\partial x}, v\frac{\partial u}{\partial y} + u\frac{\partial v}{\partial y} \right\rangle = v\left\langle \frac{\partial u}{\partial x}, \frac{\partial u}{\partial y} \right\rangle + u\left\langle \frac{\partial v}{\partial x}, \frac{\partial v}{\partial y} \right\rangle = v\nabla u + u\nabla v$

(c) $\nabla\left(\dfrac{u}{v}\right) = \left\langle \dfrac{v\dfrac{\partial u}{\partial x} - u\dfrac{\partial v}{\partial x}}{v^2},\ \dfrac{v\dfrac{\partial u}{\partial y} - u\dfrac{\partial v}{\partial y}}{v^2} \right\rangle = \dfrac{v\left\langle \dfrac{\partial u}{\partial x}, \dfrac{\partial u}{\partial y}\right\rangle - u\left\langle \dfrac{\partial v}{\partial x}, \dfrac{\partial v}{\partial y}\right\rangle}{v^2} = \dfrac{v\,\nabla u - u\,\nabla v}{v^2}$

(d) $\nabla u^n = \left\langle \dfrac{\partial(u^n)}{\partial x}, \dfrac{\partial(u^n)}{\partial y}\right\rangle = \left\langle nu^{n-1}\dfrac{\partial u}{\partial x}, nu^{n-1}\dfrac{\partial u}{\partial y}\right\rangle = nu^{n-1}\,\nabla u$

38. If we place the initial point of the gradient vector $\nabla f(4,6)$ at $(4,6)$, the vector is perpendicular to the level curve of f that

includes $(4,6)$, so we sketch a portion of the level curve through $(4,6)$ (using the nearby level curves as a guideline)

and draw a line perpendicular to the curve at $(4,6)$. The gradient vector is

parallel to this line, pointing in the direction of increasing function values, and

with length equal to the maximum value of the directional derivative of f at

$(4,6)$. We can estimate this length by finding the average rate of change in the

direction of the gradient. The line intersects the contour lines corresponding to

-2 and -3 with an estimated distance of 0.5 units. Thus the rate of change is

approximately $\dfrac{-2-(-3)}{0.5} = 2$, and we sketch the gradient vector with

length 2.

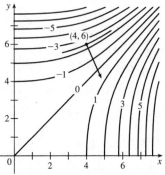

39. Let $F(x,y,z) = 2(x-2)^2 + (y-1)^2 + (z-3)^2$. Then $2(x-2)^2 + (y-1)^2 + (z-3)^2 = 10$ is a level surface of F.

$F_x(x,y,z) = 4(x-2) \ \Rightarrow \ F_x(3,3,5) = 4$, $F_y(x,y,z) = 2(y-1) \ \Rightarrow \ F_y(3,3,5) = 4$, and

$F_z(x,y,z) = 2(z-3) \ \Rightarrow \ F_z(3,3,5) = 4$.

(a) Equation 19 gives an equation of the tangent plane at $(3,3,5)$ as $4(x-3) + 4(y-3) + 4(z-5) = 0 \ \Leftrightarrow$

$4x + 4y + 4z = 44$ or equivalently $x + y + z = 11$.

(b) By Equation 20, the normal line has symmetric equations $\dfrac{x-3}{4} = \dfrac{y-3}{4} = \dfrac{z-5}{4}$ or equivalently

$x - 3 = y - 3 = z - 5$. Corresponding parametric equations are $x = 3 + t$, $y = 3 + t$, $z = 5 + t$.

40. Let $F(x,y,z) = x^2 - z^2 - y$. Then $y = x^2 - z^2 \ \Leftrightarrow \ x^2 - z^2 - y = 0$ is a level surface of F. $F_x(x,y,z) = 2x \ \Rightarrow$

$F_x(4,7,3) = 8$, $F_y(x,y,z) = -1 \ \Rightarrow \ F_y(4,7,3) = -1$, and $F_z(x,y,z) = -2z \ \Rightarrow \ F_z(4,7,3) = -6$.

(a) An equation of the tangent plane at $(4,7,3)$ is $8(x-4) - 1(y-7) - 6(z-3) = 0$ or $8x - y - 6z = 7$.

(b) The normal line has symmetric equations $\dfrac{x-4}{8} = \dfrac{y-7}{-1} = \dfrac{z-3}{-6}$ and parametric equations $x = 4 + 8t$, $y = 7 - t$,

$z = 3 - 6t$.

41. Let $F(x,y,z) = x^2 - 2y^2 + z^2 + yz$. Then $x^2 - 2y^2 + z^2 + yz = 2$ is a level surface of F

and $\nabla F(x,y,z) = \langle 2x, -4y + z, 2z + y\rangle$.

(a) $\nabla F(2,1,-1) = \langle 4, -5, -1\rangle$ is a normal vector for the tangent plane at $(2,1,-1)$, so an equation of the tangent plane

is $4(x-2) - 5(y-1) - 1(z+1) = 0$ or $4x - 5y - z = 4$.

(b) The normal line has direction $\langle 4, -5, -1\rangle$, so parametric equations are $x = 2 + 4t$, $y = 1 - 5t$, $z = -1 - t$, and

symmetric equations are $\dfrac{x-2}{4} = \dfrac{y-1}{-5} = \dfrac{z+1}{-1}$.

42. Let $F(x, y, z) = x - z - 4\arctan(yz)$. Then $x - z = 4\arctan(yz)$ is the level surface $F(x, y, z) = 0$,

and $\nabla F(x, y, z) = \left\langle 1, -\dfrac{4z}{1 + y^2 z^2}, -1 - \dfrac{4y}{1 + y^2 z^2} \right\rangle$.

(a) $\nabla F(1 + \pi, 1, 1) = \langle 1, -2, -3 \rangle$ and an equation of the tangent plane is $1(x - (1 + \pi)) - 2(y - 1) - 3(z - 1) = 0$
or $x - 2y - 3z = -4 + \pi$.

(b) The normal line has direction $\langle 1, -2, -3 \rangle$, so parametric equations are $x = 1 + \pi + t$, $y = 1 - 2t$, $z = 1 - 3t$, and

symmetric equations are $x - 1 - \pi = \dfrac{y - 1}{-2} = \dfrac{z - 1}{-3}$.

43. $F(x, y, z) = -z + xe^y \cos z \quad \Rightarrow \quad \nabla F(x, y, z) = \langle e^y \cos z, xe^y \cos z, -1 - xe^y \sin z \rangle$ and $\nabla F(1, 0, 0) = \langle 1, 1, -1 \rangle$.

(a) $1(x - 1) + 1(y - 0) - 1(z - 0) = 0$ or $x + y - z = 1$

(b) $x - 1 = y = -z$

44. $F(x, y, z) = yz - \ln(x + z) \quad \Rightarrow \quad \nabla F(x, y, z) = \left\langle -\dfrac{1}{x + z}, z, y - \dfrac{1}{x + z} \right\rangle$ and $\nabla F(0, 0, 1) = \langle -1, 1, -1 \rangle$.

(a) $(-1)(x - 0) + (1)(y - 0) - 1(z - 1) = 0$ or $x - y + z = 1$

(b) Parametric equations are $x = -t$, $y = t$, $z = 1 - t$ and symmetric equations are $\dfrac{x}{-1} = \dfrac{y}{1} = \dfrac{z - 1}{-1}$ or $-x = y = 1 - z$.

45. $F(x, y, z) = xy + yz + zx$,

$\nabla F(x, y, z) = \langle y + z, x + z, y + x \rangle$,

$\nabla F(1, 1, 1) = \langle 2, 2, 2 \rangle$, so an equation of the tangent

plane is $2x + 2y + 2z = 6$ or $x + y + z = 3$, and the

normal line is given by $x - 1 = y - 1 = z - 1$ or

$x = y = z$. To graph the surface we solve for z:

$z = \dfrac{3 - xy}{x + y}$.

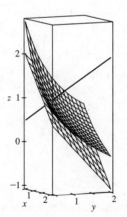

46. $F(x, y, z) = xyz$,

$\nabla F(x, y, z) = \langle yz, xz, yx \rangle, \nabla F(1, 2, 3) = \langle 6, 3, 2 \rangle$, so

an equation of the tangent plane is $6x + 3y + 2z = 18$,

and the normal line is given by $\dfrac{x - 1}{6} = \dfrac{y - 2}{3} = \dfrac{z - 3}{2}$

or $x = 1 + 6t$, $y = 2 + 3t$, $z = 3 + 2t$. To graph the

surface we solve for z: $z = \dfrac{6}{xy}$.

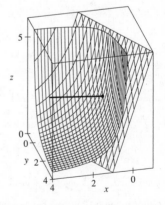

47. $f(x, y) = xy \quad \Rightarrow \quad \nabla f(x, y) = \langle y, x \rangle, \nabla f(3, 2) = \langle 2, 3 \rangle$. $\nabla f(3, 2)$

is perpendicular to the tangent line, so the tangent line has equation

$\nabla f(3, 2) \cdot \langle x - 3, y - 2 \rangle = 0 \quad \Rightarrow \quad \langle 2, 3 \rangle \cdot \langle x - 3, x - 2 \rangle = 0 \quad \Rightarrow$

$2(x - 3) + 3(y - 2) = 0$ or $2x + 3y = 12$.

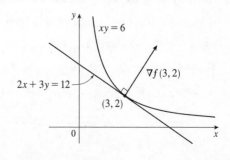

48. $g(x, y) = x^2 + y^2 - 4x \Rightarrow \nabla g(x, y) = \langle 2x - 4, 2y \rangle$,

$\nabla g(1, 2) = \langle -2, 4 \rangle$. $\nabla g(1, 2)$ is perpendicular to the tangent line, so

the tangent line has equation $\nabla g(1, 2) \cdot \langle x - 1, y - 2 \rangle = 0 \Rightarrow$

$\langle -2, 4 \rangle \cdot \langle x - 1, y - 2 \rangle = 0 \Rightarrow -2(x - 1) + 4(y - 2) = 0 \Leftrightarrow$

$-2x + 4y = 6$ or equivalently $-x + 2y = 3$.

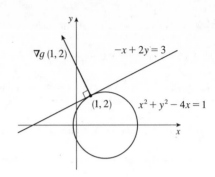

49. $\nabla F(x_0, y_0, z_0) = \left\langle \dfrac{2x_0}{a^2}, \dfrac{2y_0}{b^2}, \dfrac{2z_0}{c^2} \right\rangle$. Thus an equation of the tangent plane at (x_0, y_0, z_0) is

$$\frac{2x_0}{a^2} x + \frac{2y_0}{b^2} y + \frac{2z_0}{c^2} z = 2\left(\frac{x_0^2}{a^2} + \frac{y_0^2}{b^2} + \frac{z_0^2}{c^2} \right) = 2(1) = 2 \text{ since } (x_0, y_0, z_0) \text{ is a point on the ellipsoid. Hence}$$

$\dfrac{x_0}{a^2} x + \dfrac{y_0}{b^2} y + \dfrac{z_0}{c^2} z = 1$ is an equation of the tangent plane.

50. $\nabla F(x_0, y_0, z_0) = \left\langle \dfrac{2x_0}{a^2}, \dfrac{2y_0}{b^2}, \dfrac{-2z_0}{c^2} \right\rangle$, so an equation of the tangent plane at (x_0, y_0, z_0) is

$$\frac{2x_0}{a^2} x + \frac{2y_0}{b^2} y - \frac{2z_0}{c^2} z = 2\left(\frac{x_0^2}{a^2} + \frac{y_0^2}{b^2} - \frac{z_0^2}{c^2} \right) = 2 \text{ or } \frac{x_0}{a^2} x + \frac{y_0}{b^2} y - \frac{z_0}{c^2} z = 1.$$

51. $\nabla F(x_0, y_0, z_0) = \left\langle \dfrac{2x_0}{a^2}, \dfrac{2y_0}{b^2}, \dfrac{-1}{c} \right\rangle$, so an equation of the tangent plane is $\dfrac{2x_0}{a^2} x + \dfrac{2y_0}{b^2} y - \dfrac{1}{c} z = \dfrac{2x_0^2}{a^2} + \dfrac{2y_0^2}{b^2} - \dfrac{z_0}{c}$

or $\dfrac{2x_0}{a^2} x + \dfrac{2y_0}{b^2} y = \dfrac{z}{c} + 2\left(\dfrac{x_0^2}{a^2} + \dfrac{y_0^2}{b^2} \right) - \dfrac{z_0}{c}$. But $\dfrac{z_0}{c} = \dfrac{x_0^2}{a^2} + \dfrac{y_0^2}{b^2}$, so the equation can be written as

$\dfrac{2x_0}{a^2} x + \dfrac{2y_0}{b^2} y = \dfrac{z + z_0}{c}$.

52. Let $F(x, y, z) = x^2 + z^2 - y$; then the paraboloid $y = x^2 + z^2$ is a level surface of F. $\nabla F(x, y, z) = \langle 2x, -1, 2z \rangle$ is a

normal vector to the surface at (x, y, z) and so it is a normal vector for the tangent plane there. The tangent plane is parallel

to the plane $x + 2y + 3z = 1$ when the normal vectors of the planes are parallel, so we need a point (x_0, y_0, z_0) on the

paraboloid where $\langle 2x_0, -1, 2z_0 \rangle = k \langle 1, 2, 3 \rangle$. Comparing y-components we have $k = -\frac{1}{2}$, so

$\langle 2x_0, -1, 2z_0 \rangle = \langle -\frac{1}{2}, -1, -\frac{3}{2} \rangle$ and $2x_0 = -\frac{1}{2} \Rightarrow x_0 = -\frac{1}{4}, 2z_0 = -\frac{3}{2} \Rightarrow z_0 = -\frac{3}{4}$. Then

$y_0 = x_0^2 + z_0^2 = \left(-\frac{1}{4} \right)^2 + \left(-\frac{3}{4} \right)^2 = \frac{5}{8}$ and the point is $\left(-\frac{1}{4}, \frac{5}{8}, -\frac{3}{4} \right)$.

53. The hyperboloid $x^2 - y^2 - z^2 = 1$ is a level surface of $F(x, y, z) = x^2 - y^2 - z^2$ and $\nabla F(x, y, z) = \langle 2x, -2y, -2z \rangle$ is a

normal vector to the surface and hence a normal vector for the tangent plane at (x, y, z). The tangent plane is parallel to the

plane $z = x + y$ or $x + y - z = 0$ if and only if the corresponding normal vectors are parallel, so we need a point (x_0, y_0, z_0)

on the hyperboloid where $\langle 2x_0, -2y_0, -2z_0 \rangle = c \langle 1, 1, -1 \rangle$ or equivalently $\langle x_0, -y_0, -z_0 \rangle = k \langle 1, 1, -1 \rangle$ for some $k \neq 0$.

Then we must have $x_0 = k$, $y_0 = -k$, $z_0 = k$ and substituting into the equation of the hyperboloid gives

$k^2 - (-k)^2 - k^2 = 1 \Leftrightarrow -k^2 = 1$, an impossibility. Thus there is no such point on the hyperboloid.

54. First note that the point $(1, 1, 2)$ is on both surfaces. For the ellipsoid, an equation of the tangent plane at $(1, 1, 2)$ is

$6x + 4y + 4z = 18$ or $3x + 2y + 2z = 9$, and for the sphere, an equation of the tangent plane at $(1, 1, 2)$ is

$(2 - 8)x + (2 - 6)y + (4 - 8)z = -18$ or $-6x - 4y - 4z = -18$ or $3x + 2y + 2z = 9$. Since these tangent planes

are the same, the surfaces are tangent to each other at the point $(1, 1, 2)$.

55. Let (x_0, y_0, z_0) be a point on the cone [other than $(0, 0, 0)$]. Then an equation of the tangent plane to the cone at this point is

$2x_0 x + 2y_0 y - 2z_0 z = 2(x_0^2 + y_0^2 - z_0^2)$. But $x_0^2 + y_0^2 = z_0^2$ so the tangent plane is given by $x_0 x + y_0 y - z_0 z = 0$, a plane

which always contains the origin.

56. Let (x_0, y_0, z_0) be a point on the sphere. Then the normal line is given by $\dfrac{x - x_0}{2x_0} = \dfrac{y - y_0}{2y_0} = \dfrac{z - z_0}{2z_0}$. For the center

$(0, 0, 0)$ to be on the line, we need $-\dfrac{x_0}{2x_0} = -\dfrac{y_0}{2y_0} = -\dfrac{z_0}{2z_0}$ or equivalently $1 = 1 = 1$, which is true.

57. Let (x_0, y_0, z_0) be a point on the surface. Then an equation of the tangent plane at the point is

$$\frac{x}{2\sqrt{x_0}} + \frac{y}{2\sqrt{y_0}} + \frac{z}{2\sqrt{z_0}} = \frac{\sqrt{x_0} + \sqrt{y_0} + \sqrt{z_0}}{2}.$$ But $\sqrt{x_0} + \sqrt{y_0} + \sqrt{z_0} = \sqrt{c}$, so the equation is

$$\frac{x}{\sqrt{x_0}} + \frac{y}{\sqrt{y_0}} + \frac{z}{\sqrt{z_0}} = \sqrt{c}.$$ The x-, y-, and z-intercepts are $\sqrt{cx_0}$, $\sqrt{cy_0}$ and $\sqrt{cz_0}$ respectively. (The x-intercept is found

by setting $y = z = 0$ and solving the resulting equation for x, and the y- and z-intercepts are found similarly.) So the sum of

the intercepts is $\sqrt{c}\left(\sqrt{x_0} + \sqrt{y_0} + \sqrt{z_0}\right) = c$, a constant.

58. The surface $xyz = 1$ is a level surface of $F(x, y, z) = xyz$ and $\nabla F(x, y, z) = \langle yz, xz, xy \rangle$ is normal to the surface, so a

normal vector for the tangent plane to the surface at (x_0, y_0, z_0) is $\langle y_0 z_0, x_0 z_0, x_0 y_0 \rangle$. An equation for the tangent plane is

$y_0 z_0 (x - x_0) + x_0 z_0 (y - y_0) + x_0 y_0 (z - z_0) = 0 \quad \Rightarrow \quad y_0 z_0 x + x_0 z_0 y + x_0 y_0 z = 3x_0 y_0 z_0$ or $\dfrac{x}{x_0} + \dfrac{y}{y_0} + \dfrac{z}{z_0} = 3$.

If (x_0, y_0, z_0) is in the first octant, then the tangent plane cuts off a pyramid in the first octant with vertices $(0, 0, 0)$,

$(3x_0, 0, 0)$, $(0, 3y_0, 0)$, $(0, 0, 3z_0)$. The base in the xy-plane is a triangle with area $\frac{1}{2}(3x_0)(3y_0)$ and the height (along the

z-axis) of the pyramid is $3z_0$. The volume of the pyramid for any (x_0, y_0, z_0) on the surface $xyz = 1$ in the first octant is

$\frac{1}{3}$ (base) (height) $= \frac{1}{3} \cdot \frac{1}{2}(3x_0)(3y_0) \cdot 3z_0 = \frac{9}{2} x_0 y_0 z_0 = \frac{9}{2}$ since $x_0 y_0 z_0 = 1$.

59. If $f(x, y, z) = z - x^2 - y^2$ and $g(x, y, z) = 4x^2 + y^2 + z^2$, then the tangent line is perpendicular to both ∇f and ∇g

at $(-1, 1, 2)$. The vector $\mathbf{v} = \nabla f \times \nabla g$ will therefore be parallel to the tangent line.

We have $\nabla f(x, y, z) = \langle -2x, -2y, 1 \rangle \quad \Rightarrow \quad \nabla f(-1, 1, 2) = \langle 2, -2, 1 \rangle$, and $\nabla g(x, y, z) = \langle 8x, 2y, 2z \rangle \quad \Rightarrow$

$\nabla g(-1, 1, 2) = \langle -8, 2, 4 \rangle$. Hence $\mathbf{v} = \nabla f \times \nabla g = \begin{vmatrix} \mathbf{i} & \mathbf{j} & \mathbf{k} \\ 2 & -2 & 1 \\ -8 & 2 & 4 \end{vmatrix} = -10\,\mathbf{i} - 16\,\mathbf{j} - 12\,\mathbf{k}.$

Parametric equations are: $x = -1 - 10t$, $y = 1 - 16t$, $z = 2 - 12t$.

60. (a) Let $f(x, y, z) = y + z$ and $g(x, y, z) = x^2 + y^2$. Then the required

tangent line is perpendicular to both ∇f and ∇g at $(1, 2, 1)$ and the

vector $\mathbf{v} = \nabla f \times \nabla g$ is parallel to the tangent line. We have

$\nabla f(x, y, z) = \langle 0, 1, 1 \rangle \quad \Rightarrow \quad \nabla f(1, 2, 1) = \langle 0, 1, 1 \rangle$, and

$\nabla g(x, y, z) = \langle 2x, 2y, 0 \rangle \quad \Rightarrow \quad \nabla g(1, 2, 1) = \langle 2, 4, 0 \rangle$. Hence

$$\mathbf{v} = \nabla f \times \nabla g = \begin{vmatrix} \mathbf{i} & \mathbf{j} & \mathbf{k} \\ 0 & 1 & 1 \\ 2 & 4 & 0 \end{vmatrix} = -4\,\mathbf{i} + 2\,\mathbf{j} - 2\,\mathbf{k}. \text{ So parametric equations}$$

(b)

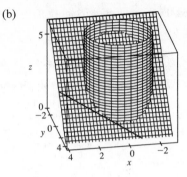

of the desired tangent line are $x = 1 - 4t, \; y = 2 + 2t, \; z = 1 - 2t$.

61. (a) The direction of the normal line of F is given by ∇F, and that of G by ∇G. Assuming that

$\nabla F \neq 0 \neq \nabla G$, the two normal lines are perpendicular at P if $\nabla F \cdot \nabla G = 0$ at $P \quad \Leftrightarrow$

$\langle \partial F / \partial x, \partial F / \partial y, \partial F / \partial z \rangle \cdot \langle \partial G / \partial x, \partial G / \partial y, \partial G / \partial z \rangle = 0$ at $P \quad \Leftrightarrow \quad F_x G_x + F_y G_y + F_z G_z = 0$ at P.

(b) Here $F = x^2 + y^2 - z^2$ and $G = x^2 + y^2 + z^2 - r^2$, so

$\nabla F \cdot \nabla G = \langle 2x, 2y, -2z \rangle \cdot \langle 2x, 2y, 2z \rangle = 4x^2 + 4y^2 - 4z^2 = 4F = 0$, since the point (x, y, z) lies on the graph of

$F = 0$. To see that this is true without using calculus, note that $G = 0$ is the equation of a sphere centered at the origin and

$F = 0$ is the equation of a right circular cone with vertex at the origin (which is generated by lines through the origin).

At any point of intersection, the sphere's normal line (which passes through the origin) lies on the cone, and thus is

perpendicular to the cone's normal line. So the surfaces with equations $F = 0$ and $G = 0$ are everywhere orthogonal.

62. (a) The function $f(x, y) = (xy)^{1/3}$ is continuous on $\mathbb{R}^2$ since it is a composition of a polynomial and the cube root function,

both of which are continuous. (See the text just after Example 8 in Section 15.2 [ET 14.2].)

$$f_x(0, 0) = \lim_{h \to 0} \frac{f(0 + h, 0) - f(0, 0)}{h} = \lim_{h \to 0} \frac{(h \cdot 0)^{1/3} - 0}{h} = 0,$$

$$f_y(0, 0) = \lim_{h \to 0} \frac{f(0, 0 + h) - f(0, 0)}{h} = \lim_{h \to 0} \frac{(0 \cdot h)^{1/3} - 0}{h} = 0.$$

Therefore, $f_x(0, 0)$ and $f_y(0, 0)$ do exist and are equal to 0. Now let $\mathbf{u}$ be any unit vector other than $\mathbf{i}$ and $\mathbf{j}$

(these correspond to f_x and f_y respectively.) Then $\mathbf{u} = a\,\mathbf{i} + b\,\mathbf{j}$ where $a \neq 0$ and $b \neq 0$. Thus

$$D_{\mathbf{u}} f(0, 0) = \lim_{h \to 0} \frac{f(0 + ha, 0 + hb) - f(0, 0)}{h} = \lim_{h \to 0} \frac{\sqrt[3]{(ha)(hb)}}{h} = \lim_{h \to 0} \frac{\sqrt[3]{ab}}{h^{1/3}} \text{ and this limit does not exist,}$$

so $D_{\mathbf{u}} f(0, 0)$ does not exist.

(b)

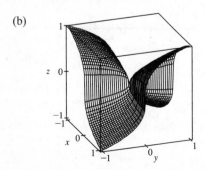

Notice that if we start at the origin and proceed in the direction of

the x- or y-axis, then the graph is flat. But if we proceed in any

other direction, then the graph is extremely steep.

63. Let $\mathbf{u} = \langle a, b \rangle$ and $\mathbf{v} = \langle c, d \rangle$. Then we know that at the given point, $D_{\mathbf{u}} f = \nabla f \cdot \mathbf{u} = af_x + bf_y$ and

$D_{\mathbf{v}} f = \nabla f \cdot \mathbf{v} = cf_x + df_y$. But these are just two linear equations in the two unknowns f_x and f_y, and since $\mathbf{u}$ and $\mathbf{v}$

are not parallel, we can solve the equations to find $\nabla f = \langle f_x, f_y \rangle$ at the given point. In fact,

$$\nabla f = \left\langle \frac{d\, D_{\mathbf{u}}\, f - b\, D_{\mathbf{v}}\, f}{ad - bc}, \frac{a\, D_{\mathbf{v}}\, f - c\, D_{\mathbf{u}}\, f}{ad - bc} \right\rangle.$$

64. Since $z = f(x, y)$ is differentiable at $\mathbf{x}_0 = (x_0, y_0)$, by Definition 15.4.7 [ET 14.4.7] we have

$\Delta z = f_x(x_0, y_0)\, \Delta x + f_y(x_0, y_0)\, \Delta y + \varepsilon_1\, \Delta x + \varepsilon_2\, \Delta y$ where $\varepsilon_1, \varepsilon_2 \to 0$ as $(\Delta x, \Delta y) \to (0, 0)$. Now

$\Delta z = f(\mathbf{x}) - f(\mathbf{x}_0)$, $\langle \Delta x, \Delta y \rangle = \mathbf{x} - \mathbf{x}_0$ so $(\Delta x, \Delta y) \to (0, 0)$ is equivalent to $\mathbf{x} \to \mathbf{x}_0$ and

$\langle f_x(x_0, y_0), f_y(x_0, y_0) \rangle = \nabla f(\mathbf{x}_0)$. Substituting into (15.4.7) [ET (14.4.7)] gives

$f(\mathbf{x}) - f(\mathbf{x}_0) = \nabla f(\mathbf{x}_0) \cdot (\mathbf{x} - \mathbf{x}_0) + \langle \varepsilon_1, \varepsilon_2 \rangle \cdot \langle \Delta x, \Delta y \rangle$ or $\langle \varepsilon_1, \varepsilon_2 \rangle \cdot (\mathbf{x} - \mathbf{x}_0) = f(\mathbf{x}) - f(\mathbf{x}_0) - \nabla f(\mathbf{x}_0) \cdot (\mathbf{x} - \mathbf{x}_0)$,

and so $\dfrac{f(\mathbf{x}) - f(\mathbf{x}_0) - \nabla f(\mathbf{x}_0) \cdot (\mathbf{x} - \mathbf{x}_0)}{|\mathbf{x} - \mathbf{x}_0|} = \dfrac{\langle \varepsilon_1, \varepsilon_2 \rangle \cdot (\mathbf{x} - \mathbf{x}_0)}{|\mathbf{x} - \mathbf{x}_0|}$. But $\dfrac{\mathbf{x} - \mathbf{x}_0}{|\mathbf{x} - \mathbf{x}_0|}$ is a unit vector so

$\displaystyle\lim_{\mathbf{x} \to \mathbf{x}_0} \frac{\langle \varepsilon_1, \varepsilon_2 \rangle \cdot (\mathbf{x} - \mathbf{x}_0)}{|\mathbf{x} - \mathbf{x}_0|} = 0$ since $\varepsilon_1, \varepsilon_2 \to 0$ as $\mathbf{x} \to \mathbf{x}_0$. Hence $\displaystyle\lim_{\mathbf{x} \to \mathbf{x}_0} \frac{f(\mathbf{x}) - f(\mathbf{x}_0) - \nabla f(\mathbf{x}_0) \cdot (\mathbf{x} - \mathbf{x}_0)}{|\mathbf{x} - \mathbf{x}_0|} = 0$.

15.7   Maximum and Minimum Values   ET 14.7

1. (a) First we compute $D(1, 1) = f_{xx}(1, 1)\, f_{yy}(1, 1) - [f_{xy}(1, 1)]^2 = (4)(2) - (1)^2 = 7$. Since $D(1, 1) > 0$ and

$f_{xx}(1, 1) > 0$, f has a local minimum at $(1, 1)$ by the Second Derivatives Test.

(b) $D(1, 1) = f_{xx}(1, 1)\, f_{yy}(1, 1) - [f_{xy}(1, 1)]^2 = (4)(2) - (3)^2 = -1$. Since $D(1, 1) < 0$, f has a saddle point at $(1, 1)$

by the Second Derivatives Test.

2. (a) $D = g_{xx}(0, 2)\, g_{yy}(0, 2) - [g_{xy}(0, 2)]^2 = (-1)(1) - (6)^2 = -37$. Since $D < 0$, g has a saddle point at $(0, 2)$ by the

Second Derivatives Test.

(b) $D = g_{xx}(0, 2)\, g_{yy}(0, 2) - [g_{xy}(0, 2)]^2 = (-1)(-8) - (2)^2 = 4$. Since $D > 0$ and $g_{xx}(0, 2) < 0$, g has a local

maximum at $(0, 2)$ by the Second Derivatives Test.

(c) $D = g_{xx}(0, 2)\, g_{yy}(0, 2) - [g_{xy}(0, 2)]^2 = (4)(9) - (6)^2 = 0$. In this case the Second Derivatives Test gives no

information about g at the point $(0, 2)$.

3. In the figure, a point at approximately $(1, 1)$ is enclosed by level curves which are oval in shape and indicate that as we move

away from the point in any direction the values of f are increasing. Hence we would expect a local minimum at or near $(1, 1)$.

The level curves near $(0, 0)$ resemble hyperbolas, and as we move away from the origin, the values of f increase in some

directions and decrease in others, so we would expect to find a saddle point there.

To verify our predictions, we have $f(x, y) = 4 + x^3 + y^3 - 3xy \;\Rightarrow\; f_x(x, y) = 3x^2 - 3y$, $f_y(x, y) = 3y^2 - 3x$. We

have critical points where these partial derivatives are equal to 0: $3x^2 - 3y = 0$, $3y^2 - 3x = 0$. Substituting $y = x^2$ from the first equation into the second equation gives $3(x^2)^2 - 3x = 0 \Rightarrow 3x(x^3 - 1) = 0 \Rightarrow x = 0$ or $x = 1$. Then we have

two critical points, $(0,0)$ and $(1,1)$. The second partial derivatives are $f_{xx}(x,y) = 6x$, $f_{xy}(x,y) = -3$, and $f_{yy}(x,y) = 6y$,

so $D(x,y) = f_{xx}(x,y)\,f_{yy}(x,y) - [f_{xy}(x,y)]^2 = (6x)(6y) - (-3)^2 = 36xy - 9$. Then $D(0,0) = 36(0)(0) - 9 = -9$,

and $D(1,1) = 36(1)(1) - 9 = 27$. Since $D(0,0) < 0$, f has a saddle point at $(0,0)$ by the Second Derivatives Test. Since

$D(1,1) > 0$ and $f_{xx}(1,1) > 0$, f has a local minimum at $(1,1)$.

4. In the figure, points at approximately $(-1,1)$ and $(-1,-1)$ are enclosed by oval-shaped level curves which indicate that as we

move away from either point in any direction, the values of f are increasing. Hence we would expect local minima at or near

$(-1, \pm 1)$. Similarly, the point $(1,0)$ appears to be enclosed by oval-shaped level curves which indicate that as we move away

from the point in any direction the values of f are decreasing, so we should have a local maximum there. We also show

hyperbola-shaped level curves near the points $(-1,0)$, $(1,1)$, and $(1,-1)$. The values of f increase along some paths leaving

these points and decrease in others, so we should have a saddle point at each of these points.

To confirm our predictions, we have $f(x,y) = 3x - x^3 - 2y^2 + y^4 \Rightarrow f_x(x,y) = 3 - 3x^2$, $f_y(x,y) = -4y + 4y^3$.

Setting these partial derivatives equal to 0, we have $3 - 3x^2 = 0 \Rightarrow x = \pm 1$ and $-4y + 4y^3 = 0 \Rightarrow$

$y(y^2 - 1) = 0 \Rightarrow y = 0, \pm 1$. So our critical points are $(\pm 1, 0)$, $(\pm 1, \pm 1)$. The second partial

derivatives are $f_{xx}(x,y) = -6x$, $f_{xy}(x,y) = 0$, and $f_{yy}(x,y) = 12y^2 - 4$, so

$$D(x,y) = f_{xx}(x,y)\,f_{yy}(x,y) - [f_{xy}(x,y)]^2 = (-6x)(12y^2 - 4) - (0)^2 = -72xy^2 + 24x.$$

We use the Second Derivatives Test to classify the 6 critical points:

Critical Point	D	f_{xx}	Conclusion
$(1,0)$	24	-6	$D > 0$, $f_{xx} < 0 \Rightarrow f$ has a local maximum at $(1,0)$
$(1,1)$	-48		$D < 0 \Rightarrow f$ has a saddle point at $(1,1)$
$(1,-1)$	-48		$D < 0 \Rightarrow f$ has a saddle point at $(1,-1)$
$(-1,0)$	-24		$D < 0 \Rightarrow f$ has a saddle point at $(-1,0)$
$(-1,1)$	48	6	$D > 0$, $f_{xx} > 0 \Rightarrow f$ has a local minimum at $(-1,1)$
$(-1,-1)$	48	6	$D > 0$, $f_{xx} > 0 \Rightarrow f$ has a local minimum at $(-1,-1)$

5. $f(x,y) = 9 - 2x + 4y - x^2 - 4y^2 \Rightarrow f_x = -2 - 2x$, $f_y = 4 - 8y$,

$f_{xx} = -2$, $f_{xy} = 0$, $f_{yy} = -8$. Then $f_x = 0$ and $f_y = 0$ imply

$x = -1$ and $y = \frac{1}{2}$, and the only critical point is $\left(-1, \frac{1}{2}\right)$.

$D(x,y) = f_{xx}f_{yy} - (f_{xy})^2 = (-2)(-8) - 0^2 = 16$, and since

$D\left(-1, \frac{1}{2}\right) = 16 > 0$ and $f_{xx}\left(-1, \frac{1}{2}\right) = -2 < 0$, $f\left(-1, \frac{1}{2}\right) = 11$ is a

local maximum by the Second Derivatives Test.

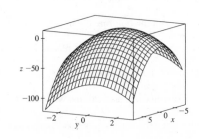

6. $f(x, y) = x^3 y + 12x^2 - 8y \Rightarrow f_x = 3x^2 y + 24x$,

$f_y = x^3 - 8$, $f_{xx} = 6xy + 24$, $f_{xy} = 3x^2$, $f_{yy} = 0$.

Then $f_y = 0$ implies $x = 2$, and substitution into $f_x = 0$ gives

$12y + 48 = 0 \Rightarrow y = -4$. Thus, the only critical point is $(2, -4)$.

$D(2, -4) = (-24)(0) - 12^2 = -144 < 0$, so $(2, -4)$ is a saddle point.

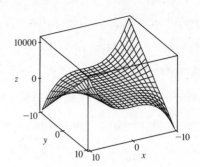

7. $f(x, y) = x^4 + y^4 - 4xy + 2 \Rightarrow f_x = 4x^3 - 4y$, $f_y = 4y^3 - 4x$,

$f_{xx} = 12x^2$, $f_{xy} = -4$, $f_{yy} = 12y^2$. Then $f_x = 0$ implies $y = x^3$,

and substitution into $f_y = 0 \Rightarrow x = y^3$ gives $x^9 - x = 0 \Rightarrow$

$x(x^8 - 1) = 0 \Rightarrow x = 0$ or $x = \pm 1$. Thus the critical points are $(0, 0)$,

$(1, 1)$, and $(-1, -1)$. Now $D(0, 0) = 0 \cdot 0 - (-4)^2 = -16 < 0$,

so $(0, 0)$ is a saddle point. $D(1, 1) = (12)(12) - (-4)^2 > 0$ and

$f_{xx}(1, 1) = 12 > 0$, so $f(1, 1) = 0$ is a local minimum. $D(-1, -1) = (12)(12) - (-4)^2 > 0$ and

$f_{xx} = (-1, -1) = 12 > 0$, so $f(-1, -1) = 0$ is also a local minimum.

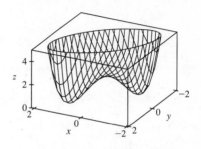

8. $f(x, y) = e^{4y - x^2 - y^2} \Rightarrow f_x = -2xe^{4y - x^2 - y^2}$,

$f_y = (4 - 2y)e^{4y - x^2 - y^2}$, $f_{xx} = (4x^2 - 2)e^{4y - x^2 - y^2}$,

$f_{xy} = -2x(4 - 2y)e^{4y - x^2 - y^2}$, $f_{yy} = (4y^2 - 16y + 14)e^{4y - x^2 - y^2}$.

Then $f_x = 0$ and $f_y = 0$ implies $x = 0$ and $y = 2$, so the only critical

point is $(0, 2)$. Now $D(0, 2) = (-2e^4)(-2e^4) - 0^2 = 4e^8 > 0$ and

$f_{xx}(0, 2) = -2e^4 < 0$, so $f(0, 2) = e^4$ is a local maximum.

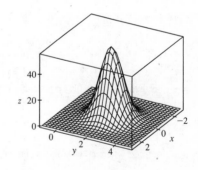

9. $f(x, y) = (1 + xy)(x + y) = x + y + x^2 y + xy^2 \Rightarrow$

$f_x = 1 + 2xy + y^2$, $f_y = 1 + x^2 + 2xy$, $f_{xx} = 2y$, $f_{xy} = 2x + 2y$,

$f_{yy} = 2x$. Then $f_x = 0$ implies $1 + 2xy + y^2 = 0$ and $f_y = 0$ implies

$1 + x^2 + 2xy = 0$. Subtracting the second equation from the first gives

$y^2 - x^2 = 0 \Rightarrow y = \pm x$, but if $y = x$ then $1 + 2xy + y^2 = 0 \Rightarrow$

$1 + 3x^2 = 0$ which has no real solution. If $y = -x$ then

$1 + 2xy + y^2 = 0 \Rightarrow 1 - x^2 = 0 \Rightarrow x = \pm 1$, so critical points are $(1, -1)$ and $(-1, 1)$.

$D(1, -1) = (-2)(2) - 0 < 0$ and $D(-1, 1) = (2)(-2) - 0 < 0$, so $(-1, 1)$ and $(1, -1)$ are saddle points.

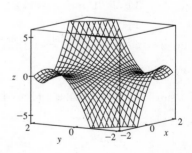

10. $f(x,y) = 2x^3 + xy^2 + 5x^2 + y^2 \Rightarrow f_x = 6x^2 + y^2 + 10x$,

$f_y = 2xy + 2y$, $f_{xx} = 12x + 10$, $f_{yy} = 2x + 2$, $f_{xy} = 2y$. Then

$f_y = 0$ implies $y = 0$ or $x = -1$. Substituting into $f_x = 0$ gives the

critical points $(0,0)$, $(-\frac{5}{3}, 0)$, $(-1, \pm 2)$. Now $D(0,0) = 20 > 0$

and $f_{xx}(0,0) = 10 > 0$, so $f(0,0) = 0$ is a local minimum.

Also $f_{xx}(-\frac{5}{3}, 0) < 0$, $D(-\frac{5}{3}, 0) > 0$, and $D(-1, \pm 2) < 0$.

Hence $f(-\frac{5}{3}, 0) = \frac{125}{27}$ is a local maximum while $(-1, \pm 2)$ are saddle points.

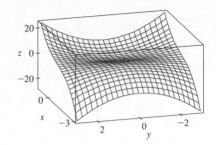

11. $f(x,y) = x^3 - 12xy + 8y^3 \Rightarrow f_x = 3x^2 - 12y$, $f_y = -12x + 24y^2$,

$f_{xx} = 6x$, $f_{xy} = -12$, $f_{yy} = 48y$. Then $f_x = 0$ implies $x^2 = 4y$ and

$f_y = 0$ implies $x = 2y^2$. Substituting the second equation into the first

gives $(2y^2)^2 = 4y \Rightarrow 4y^4 = 4y \Rightarrow 4y(y^3 - 1) = 0 \Rightarrow y = 0$ or

$y = 1$. If $y = 0$ then $x = 0$ and if $y = 1$ then $x = 2$, so the critical points

are $(0,0)$ and $(2,1)$. $D(0,0) = (0)(0) - (-12)^2 = -144 < 0$, so $(0,0)$ is a saddle point.

$D(2,1) = (12)(48) - (-12)^2 = 432 > 0$ and $f_{xx}(2,1) = 12 > 0$ so $f(2,1) = -8$ is a local minimum.

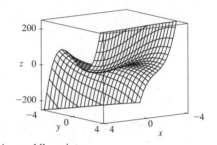

12. $f(x,y) = xy + \dfrac{1}{x} + \dfrac{1}{y} \Rightarrow f_x = y - \dfrac{1}{x^2}$, $f_y = x - \dfrac{1}{y^2}$, $f_{xx} = \dfrac{2}{x^3}$,

$f_{xy} = 1$, $f_{yy} = \dfrac{2}{y^3}$. Then $f_x = 0$ implies $y = \dfrac{1}{x^2}$ and $f_y = 0$ implies

$x = \dfrac{1}{y^2}$. Substituting the first equation into the second gives

$x = \dfrac{1}{(1/x^2)^2} \Rightarrow x = x^4 \Rightarrow x(x^3 - 1) = 0 \Rightarrow x = 0$ or $x = 1$.

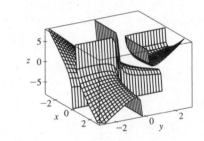

f is not defined when $x = 0$, and when $x = 1$ we have $y = 1$, so the only critical point is $(1,1)$.

$D(1,1) = (2)(2) - 1^2 = 3 > 0$ and $f_{xx}(1,1) = 2 > 0$, so $f(1,1) = 3$ is a local minimum.

13. $f(x,y) = e^x \cos y \Rightarrow f_x = e^x \cos y$, $f_y = -e^x \sin y$.

Now $f_x = 0$ implies $\cos y = 0$ or $y = \frac{\pi}{2} + n\pi$ for n an integer.

But $\sin\left(\frac{\pi}{2} + n\pi\right) \neq 0$, so there are no critical points.

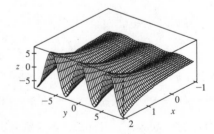

14. $f(x,y) = y\cos x \Rightarrow f_x = -y\sin x,\ f_y = \cos x,\ f_{xx} = -y\cos x,$

$f_{xy} = -\sin x,\ f_{yy} = 0.$ Then $f_y = 0$ if and only if $x = \frac{\pi}{2} + n\pi$ for n an

integer. But $\sin\left(\frac{\pi}{2} + n\pi\right) \neq 0$, so $f_x = 0 \Rightarrow y = 0$ and the critical

points are $\left(\frac{\pi}{2} + n\pi, 0\right)$, n an integer.

$D\left(\frac{\pi}{2} + n\pi, 0\right) = (0)(0) - (\pm 1)^2 = -1 < 0$, so each critical point is

a saddle point.

15. $f(x,y) = (x^2 + y^2)e^{y^2 - x^2} \Rightarrow$

$f_x = (x^2 + y^2)e^{y^2 - x^2}(-2x) + 2xe^{y^2 - x^2} = 2xe^{y^2 - x^2}(1 - x^2 - y^2),$

$f_y = (x^2 + y^2)e^{y^2 - x^2}(2y) + 2ye^{y^2 - x^2} = 2ye^{y^2 - x^2}(1 + x^2 + y^2),$

$f_{xx} = 2xe^{y^2 - x^2}(-2x) + (1 - x^2 - y^2)\left(2x\left(-2xe^{y^2 - x^2}\right) + 2e^{y^2 - x^2}\right) = 2e^{y^2 - x^2}\left((1 - x^2 - y^2)(1 - 2x^2) - 2x^2\right),$

$f_{xy} = 2xe^{y^2 - x^2}(-2y) + 2x(2y)e^{y^2 - x^2}(1 - x^2 - y^2) = -4xye^{y^2 - x^2}(x^2 + y^2),$

$f_{yy} = 2ye^{y^2 - x^2}(2y) + (1 + x^2 + y^2)\left(2y\left(2ye^{y^2 - x^2}\right) + 2e^{y^2 - x^2}\right) = 2e^{y^2 - x^2}\left((1 + x^2 + y^2)(1 + 2y^2) + 2y^2\right).$

$f_y = 0$ implies $y = 0$, and substituting into $f_x = 0$ gives

$2xe^{-x^2}(1 - x^2) = 0 \Rightarrow x = 0$ or $x = \pm 1$. Thus the critical points are

$(0,0)$ and $(\pm 1, 0)$. Now $D(0,0) = (2)(2) - 0 > 0$ and $f_{xx}(0,0) = 2 > 0$,

so $f(0,0) = 0$ is a local minimum. $D(\pm 1, 0) = (-4e^{-1})(4e^{-1}) - 0 < 0$

so $(\pm 1, 0)$ are saddle points.

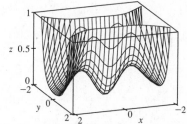

16. $f(x,y) = e^y(y^2 - x^2) \Rightarrow f_x = -2xe^y,\ f_y = (2y + y^2 - x^2)e^y,$

$f_{xx} = -2e^y,\ f_{xy} = -2xe^y,\ f_{yy} = (2 + 4y + y^2 - x^2)e^y.$ Then $f_x = 0$

implies $x = 0$ and substituting into $f_y = 0$ gives $(2y + y^2)e^y = 0 \Rightarrow$

$y(2 + y) = 0 \Rightarrow y = 0$ or $y = -2$, so the critical points are $(0,0)$ and

$(0, -2)$. $D(0,0) = (-2)(2) - (0)^2 = -4 < 0$ so $(0,0)$ is a saddle point.

$D(0, -2) = (-2e^{-2})(-2e^{-2}) - (0)^2 = 4e^{-4} > 0$ and $f_{xx}(0, -2) = -2e^{-2} < 0$, so $f(0, -2) = 4e^{-2}$ is a local

maximum.

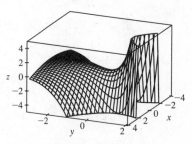

17. $f(x,y) = y^2 - 2y\cos x \Rightarrow f_x = 2y\sin x,\ f_y = 2y - 2\cos x,$

$f_{xx} = 2y\cos x,\ f_{xy} = 2\sin x,\ f_{yy} = 2.$ Then $f_x = 0$ implies $y = 0$ or

$\sin x = 0 \Rightarrow x = 0,\ \pi,$ or 2π for $-1 \leq x \leq 7$. Substituting $y = 0$ into

$f_y = 0$ gives $\cos x = 0 \Rightarrow x = \frac{\pi}{2}$ or $\frac{3\pi}{2}$, substituting $x = 0$ or $x = 2\pi$

into $f_y = 0$ gives $y = 1$, and substituting $x = \pi$ into $f_y = 0$ gives $y = -1$.

Thus the critical points are $(0, 1),\ \left(\frac{\pi}{2}, 0\right),\ (\pi, -1),\ \left(\frac{3\pi}{2}, 0\right),$ and $(2\pi, 1)$.

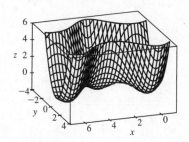

$D\left(\frac{\pi}{2}, 0\right) = D\left(\frac{3\pi}{2}, 0\right) = -4 < 0$ so $\left(\frac{\pi}{2}, 0\right)$ and $\left(\frac{3\pi}{2}, 0\right)$ are saddle points. $D(0, 1) = D(\pi, -1) = D(2\pi, 1) = 4 > 0$ and

$f_{xx}(0, 1) = f_{xx}(\pi, -1) = f_{xx}(2\pi, 1) = 2 > 0$, so $f(0, 1) = f(\pi, -1) = f(2\pi, 1) = -1$ are local minima.

18. $f(x, y) = \sin x \sin y \Rightarrow f_x = \cos x \sin y, f_y = \sin x \cos y, f_{xx} = -\sin x \sin y, f_{xy} = \cos x \cos y,$

$f_{yy} = -\sin x \sin y$. Here we have $-\pi < x < \pi$ and $-\pi < y < \pi$, so $f_x = 0$ implies $\cos x = 0$ or $\sin y = 0$. If $\cos x = 0$

then $x = -\frac{\pi}{2}$ or $\frac{\pi}{2}$, and if $\sin y = 0$ then $y = 0$. Substituting $x = \pm\frac{\pi}{2}$ into $f_y = 0$ gives $\cos y = 0 \Rightarrow y = -\frac{\pi}{2}$ or $\frac{\pi}{2}$, and

substituting $y = 0$ into $f_y = 0$ gives $\sin x = 0 \Rightarrow x = 0$. Thus the critical points are $\left(-\frac{\pi}{2}, \pm\frac{\pi}{2}\right), \left(\frac{\pi}{2}, \pm\frac{\pi}{2}\right)$, and $(0, 0)$.

$D(0, 0) = -1 < 0$ so $(0, 0)$ is a saddle point.

$D\left(-\frac{\pi}{2}, \pm\frac{\pi}{2}\right) = D\left(\frac{\pi}{2}, \pm\frac{\pi}{2}\right) = 1 > 0$ and

$f_{xx}\left(-\frac{\pi}{2}, -\frac{\pi}{2}\right) = f_{xx}\left(\frac{\pi}{2}, \frac{\pi}{2}\right) = -1 < 0$ while

$f_{xx}\left(-\frac{\pi}{2}, \frac{\pi}{2}\right) = f_{xx}\left(\frac{\pi}{2}, -\frac{\pi}{2}\right) = 1 > 0$, so $f\left(-\frac{\pi}{2}, -\frac{\pi}{2}\right) = f\left(\frac{\pi}{2}, \frac{\pi}{2}\right) = 1$

are local maxima and $f\left(-\frac{\pi}{2}, \frac{\pi}{2}\right) = f\left(\frac{\pi}{2}, -\frac{\pi}{2}\right) = 1$ are local minima.

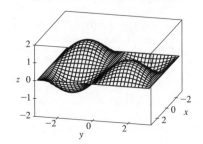

19. $f(x, y) = x^2 + 4y^2 - 4xy + 2 \Rightarrow f_x = 2x - 4y, f_y = 8y - 4x, f_{xx} = 2, f_{xy} = -4, f_{yy} = 8$. Then $f_x = 0$

and $f_y = 0$ each implies $y = \frac{1}{2}x$, so all points of the form $\left(x_0, \frac{1}{2}x_0\right)$ are critical points and for each of these we have

$D\left(x_0, \frac{1}{2}x_0\right) = (2)(8) - (-4)^2 = 0$. The Second Derivatives Test gives no information, but

$f(x, y) = x^2 + 4y^2 - 4xy + 2 = (x - 2y)^2 + 2 \geq 2$ with equality if and only if $y = \frac{1}{2}x$. Thus $f\left(x_0, \frac{1}{2}x_0\right) = 2$ are all local

(and absolute) minima.

20. $f(x, y) = x^2 y e^{-x^2 - y^2} \Rightarrow$

$f_x = x^2 y e^{-x^2 - y^2}(-2x) + 2xy e^{-x^2 - y^2} = 2xy(1 - x^2)e^{-x^2 - y^2},$

$f_y = x^2 y e^{-x^2 - y^2}(-2y) + x^2 e^{-x^2 - y^2} = x^2(1 - 2y^2)e^{-x^2 - y^2},$

$f_{xx} = 2y(2x^4 - 5x^2 + 1)e^{-x^2 - y^2},$

$f_{xy} = 2x(1 - x^2)(1 - 2y^2)e^{-x^2 - y^2}, \quad f_{yy} = 2x^2 y(2y^2 - 3)e^{-x^2 - y^2}.$

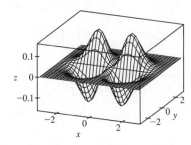

$f_x = 0$ implies $x = 0$, $y = 0$, or $x = \pm 1$. If $x = 0$ then $f_y = 0$ for any y-value, so all points of the form $(0, y)$ are critical

points. If $y = 0$ then $f_y = 0 \Rightarrow x^2 e^{-x^2} = 0 \Rightarrow x = 0$, so $(0, 0)$ (already included above) is a critical point. If $x = \pm 1$

then $(1 - 2y^2)e^{-1 - y^2} = 0 \Rightarrow y = \pm\frac{1}{\sqrt{2}}$, so $\left(\pm 1, \frac{1}{\sqrt{2}}\right)$ and $\left(\pm 1, -\frac{1}{\sqrt{2}}\right)$ are critical points. Now

$D\left(\pm 1, \frac{1}{\sqrt{2}}\right) = 8e^{-3} > 0, f_{xx}\left(\pm 1, \frac{1}{\sqrt{2}}\right) = -2\sqrt{2}\,e^{-3/2} < 0$ and $D\left(\pm 1, -\frac{1}{\sqrt{2}}\right) = 8e^{-3} > 0,$

$f_{xx}\left(\pm 1, -\frac{1}{\sqrt{2}}\right) = 2\sqrt{2}\,e^{-3/2} > 0$, so $f\left(\pm 1, \frac{1}{\sqrt{2}}\right) = \frac{1}{\sqrt{2}}e^{-3/2}$ are local maximum points while

$f\left(\pm 1, -\frac{1}{\sqrt{2}}\right) = -\frac{1}{\sqrt{2}}e^{-3/2}$ are local minimum points. At all critical points $(0, y)$ we have $D(0, y) = 0$, so the Second

Derivatives Test gives no information. However, if $y > 0$ then $x^2 y e^{-x^2 - y^2} \geq 0$ with equality only when $x = 0$, so we have

local minimum values $f(0, y) = 0, y > 0$. Similarly, if $y < 0$ then $x^2 y e^{-x^2 - y^2} \leq 0$ with equality when $x = 0$ so

$f(0, y) = 0, y < 0$ are local maximum values, and $(0, 0)$ is a saddle point.

21. $f(x,y) = x^2 + y^2 + x^{-2}y^{-2}$

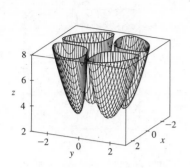
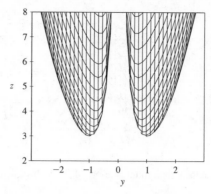
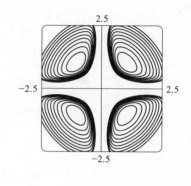

From the graphs, there appear to be local minima of about $f(1, \pm 1) = f(-1, \pm 1) \approx 3$ (and no local maxima or saddle points). $f_x = 2x - 2x^{-3}y^{-2}$, $f_y = 2y - 2x^{-2}y^{-3}$, $f_{xx} = 2 + 6x^{-4}y^{-2}$, $f_{xy} = 4x^{-3}y^{-3}$, $f_{yy} = 2 + 6x^{-2}y^{-4}$. Then $f_x = 0$ implies $2x^4y^2 - 2 = 0$ or $x^4y^2 = 1$ or $y^2 = x^{-4}$. Note that neither x nor y can be zero. Now $f_y = 0$ implies $2x^2y^4 - 2 = 0$, and with $y^2 = x^{-4}$ this implies $2x^{-6} - 2 = 0$ or $x^6 = 1$. Thus $x = \pm 1$ and if $x = 1$, $y = \pm 1$; if $x = -1$, $y = \pm 1$. So the critical points are $(1,1)$, $(1,-1)$, $(-1,1)$ and $(-1,-1)$. Now $D(1, \pm 1) = D(-1, \pm 1) = 64 - 16 > 0$ and $f_{xx} > 0$ always, so $f(1, \pm 1) = f(-1, \pm 1) = 3$ are local minima.

22. $f(x,y) = xye^{-x^2-y^2}$

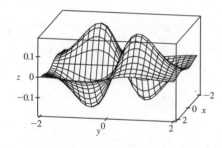
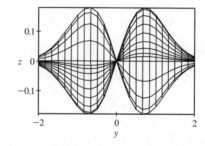
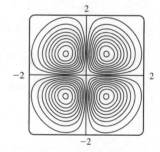

There appear to be local maxima of about $f(\pm 0.7, \pm 0.7) \approx 0.18$ and local minima of about $f(\pm 0.7, \mp 0.7) \approx -0.18$. Also, there seems to be a saddle point at the origin.

$f_x = ye^{-x^2-y^2}(1 - 2x^2)$, $f_y = xe^{-x^2-y^2}(1 - 2y^2)$, $f_{xx} = 2xye^{-x^2-y^2}(2x^2 - 3)$, $f_{yy} = 2xye^{-x^2-y^2}(2y^2 - 3)$, $f_{xy} = (1 - 2x^2)e^{-x^2-y^2}(1 - 2y^2)$. Then $f_x = 0$ implies $y = 0$ or $x = \pm\frac{1}{\sqrt{2}}$.

Substituting these values into $f_y = 0$ gives the critical points $(0,0)$, $\left(\frac{1}{\sqrt{2}}, \pm\frac{1}{\sqrt{2}}\right)$, $\left(-\frac{1}{\sqrt{2}}, \pm\frac{1}{\sqrt{2}}\right)$. Then

$D(x,y) = e^{2(-x^2-y^2)}\left[4x^2y^2(2x^2 - 3)(2y^2 - 3) - (1 - 2x^2)^2(1 - 2y^2)^2\right]$, so $D(0,0) = -1$, while $D\left(\frac{1}{\sqrt{2}}, \pm\frac{1}{\sqrt{2}}\right) > 0$

and $D\left(-\frac{1}{\sqrt{2}}, \pm\frac{1}{\sqrt{2}}\right) > 0$. But $f_{xx}\left(\frac{1}{\sqrt{2}}, \frac{1}{\sqrt{2}}\right) < 0$, $f_{xx}\left(\frac{1}{\sqrt{2}}, -\frac{1}{\sqrt{2}}\right) > 0$, $f_{xx}\left(-\frac{1}{\sqrt{2}}, \frac{1}{\sqrt{2}}\right) > 0$, $f_{xx}\left(-\frac{1}{\sqrt{2}}, -\frac{1}{\sqrt{2}}\right) < 0$.

Hence $(0,0)$ is a saddle point; $f\left(\frac{1}{\sqrt{2}}, -\frac{1}{\sqrt{2}}\right) = f\left(-\frac{1}{\sqrt{2}}, \frac{1}{\sqrt{2}}\right) = -\frac{1}{2e}$ are local minima and

$f\left(\frac{1}{\sqrt{2}}, \frac{1}{\sqrt{2}}\right) = f\left(-\frac{1}{\sqrt{2}}, -\frac{1}{\sqrt{2}}\right) = \frac{1}{2e}$ are local maxima.

23. $f(x, y) = \sin x + \sin y + \sin(x + y)$, $0 \le x \le 2\pi$, $0 \le y \le 2\pi$

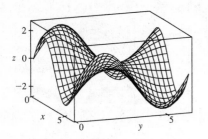 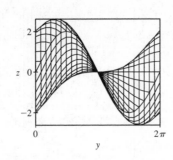 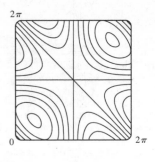

From the graphs it appears that f has a local maximum at about $(1, 1)$ with value approximately 2.6, a local minimum

at about $(5, 5)$ with value approximately -2.6, and a saddle point at about $(3, 3)$.

$f_x = \cos x + \cos(x + y)$, $f_y = \cos y + \cos(x + y)$, $f_{xx} = -\sin x - \sin(x + y)$, $f_{yy} = -\sin y - \sin(x + y)$,

$f_{xy} = -\sin(x + y)$. Setting $f_x = 0$ and $f_y = 0$ and subtracting gives $\cos x - \cos y = 0$ or $\cos x = \cos y$. Thus $x = y$

or $x = 2\pi - y$. If $x = y$, $f_x = 0$ becomes $\cos x + \cos 2x = 0$ or $2\cos^2 x + \cos x - 1 = 0$, a quadratic in $\cos x$. Thus

$\cos x = -1$ or $\frac{1}{2}$ and $x = \pi$, $\frac{\pi}{3}$, or $\frac{5\pi}{3}$, yielding the critical points (π, π), $\left(\frac{\pi}{3}, \frac{\pi}{3}\right)$ and $\left(\frac{5\pi}{3}, \frac{5\pi}{3}\right)$. Similarly if

$x = 2\pi - y$, $f_x = 0$ becomes $(\cos x) + 1 = 0$ and the resulting critical point is (π, π). Now

$D(x, y) = \sin x \sin y + \sin x \sin(x + y) + \sin y \sin(x + y)$. So $D(\pi, \pi) = 0$ and the Second Derivatives Test doesn't apply.

However, along the line $y = x$ we have $f(x, x) = 2\sin x + \sin 2x = 2\sin x + 2\sin x \cos x = 2\sin x(1 + \cos x)$, and

$f(x, x) > 0$ for $0 < x < \pi$ while $f(x, x) < 0$ for $\pi < x < 2\pi$. Thus every disk with center (π, π) contains points where f is

positive as well as points where f is negative, so the graph crosses its tangent plane $(z = 0)$ there and (π, π) is a saddle point.

$D\left(\frac{\pi}{3}, \frac{\pi}{3}\right) = \frac{9}{4} > 0$ and $f_{xx}\left(\frac{\pi}{3}, \frac{\pi}{3}\right) < 0$ so $f\left(\frac{\pi}{3}, \frac{\pi}{3}\right) = \frac{3\sqrt{3}}{2}$ is a local maximum while $D\left(\frac{5\pi}{3}, \frac{5\pi}{3}\right) = \frac{9}{4} > 0$ and

$f_{xx}\left(\frac{5\pi}{3}, \frac{5\pi}{3}\right) > 0$, so $f\left(\frac{5\pi}{3}, \frac{5\pi}{3}\right) = -\frac{3\sqrt{3}}{2}$ is a local minimum.

24. $f(x, y) = \sin x + \sin y + \cos(x + y)$, $0 \le x \le \frac{\pi}{4}$, $0 \le y \le \frac{\pi}{4}$

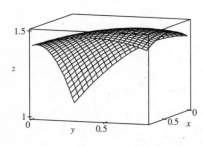 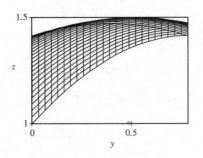 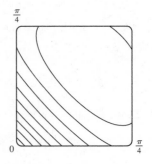

From the graphs, it seems that f has a local maximum at about $(0.5, 0.5)$. $f_x = \cos x - \sin(x + y)$,

$f_y = \cos y - \sin(x + y)$, $f_{xx} = -\sin x - \cos(x + y)$, $f_{yy} = -\sin y - \cos(x + y)$, $f_{xy} = -\cos(x + y)$. Setting $f_x = 0$

and $f_y = 0$ and subtracting gives $\cos x = \cos y$. Thus $x = y$. Substituting $x = y$ into $f_x = 0$ gives $\cos x - \sin 2x = 0$ or

$\cos x(1 - 2\sin x) = 0$. But $\cos x \neq 0$ for $0 \leq x \leq \frac{\pi}{4}$ and $1 - 2\sin x = 0$ implies $x = \frac{\pi}{6}$, so the only critical point is $\left(\frac{\pi}{6}, \frac{\pi}{6}\right)$.

Here $f_{xx}\left(\frac{\pi}{6}, \frac{\pi}{6}\right) = -1 < 0$ and $D\left(\frac{\pi}{6}, \frac{\pi}{6}\right) = (-1)^2 - \frac{1}{4} > 0$. Thus $f\left(\frac{\pi}{6}, \frac{\pi}{6}\right) = \frac{3}{2}$ is a local maximum.

25. $f(x, y) = x^4 - 5x^2 + y^2 + 3x + 2 \implies f_x(x, y) = 4x^3 - 10x + 3$ and $f_y(x, y) = 2y$. $f_y = 0 \implies y = 0$, and the graph

of f_x shows that the roots of $f_x = 0$ are approximately $x = -1.714, 0.312$ and 1.402. (Alternatively, we could have used a

calculator or a CAS to find these roots.) So to three decimal places, the critical points are $(-1.714, 0)$, $(1.402, 0)$, and

$(0.312, 0)$. Now since $f_{xx} = 12x^2 - 10$, $f_{xy} = 0$, $f_{yy} = 2$, and $D = 24x^2 - 20$, we have $D(-1.714, 0) > 0$,

$f_{xx}(-1.714, 0) > 0$, $D(1.402, 0) > 0$, $f_{xx}(1.402, 0) > 0$, and $D(0.312, 0) < 0$. Therefore $f(-1.714, 0) \approx -9.200$ and

$f(1.402, 0) \approx 0.242$ are local minima, and $(0.312, 0)$ is a saddle point. The lowest point on the graph is approximately

$(-1.714, 0, -9.200)$.

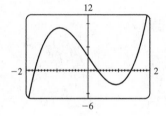 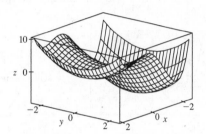

26. $f(x, y) = 5 - 10xy - 4x^2 + 3y - y^4 \implies f_x(x, y) = -10y - 8x$, $f_y(x, y) = -10x + 3 - 4y^3$.

Now $f_x = 0 \implies x = -\frac{5}{4}y$, so using a graph, we find solutions to

$0 = f_y\left(-\frac{5}{4}y, y\right) = -10\left(-\frac{5}{4}y\right) + 3 - 4y^3 = -4y^3 + \frac{25}{2}y + 3$. (Alternatively, we could have found the roots of $f_x = f_y = 0$

directly, using a calculator or a CAS.) To three decimal places, the solutions are $y \approx 1.877, -0.245$, and -1.633, so f has

critical points at approximately $(-2.347, 1.877)$, $(0.306, -0.245)$, and $(2.041, -1.633)$. Now since $f_{xx} = -8$, $f_{xy} = -10$,

$f_{yy} = -12y^2$, and $D = 96y^2 - 100$, we have $D(-2.347, 1.877) > 0$, $D(0.306, -0.245) < 0$, and $D(2.041, -1.633) > 0$.

Therefore, since $f_{xx} < 0$ everywhere, $f(-2.347, 1.877) \approx 20.238$ and $f(2.041, -1.633) \approx 9.657$

are local maxima, and $(0.306, -0.245)$ is a saddle point. The highest point on the graph is approximately

$(-2.347, 1.877, 20.238)$.

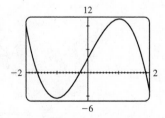 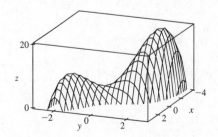

27. $f(x, y) = 2x + 4x^2 - y^2 + 2xy^2 - x^4 - y^4 \Rightarrow f_x(x, y) = 2 + 8x + 2y^2 - 4x^3, f_y(x, y) = -2y + 4xy - 4y^3$.

Now $f_y = 0 \Leftrightarrow 2y(2y^2 - 2x + 1) = 0 \Leftrightarrow y = 0$ or $y^2 = x - \frac{1}{2}$. The first of these implies that $f_x = -4x^3 + 8x + 2$,

and the second implies that $f_x = 2 + 8x + 2\left(x - \frac{1}{2}\right) - 4x^3 = -4x^3 + 10x + 1$. From the graphs, we see that the first

possibility for f_x has roots at approximately $-1.267, -0.259$, and 1.526, and the second has a root at approximately 1.629

(the negative roots do not give critical points, since $y^2 = x - \frac{1}{2}$ must be positive). So to three decimal places, f has critical

points at $(-1.267, 0)$, $(-0.259, 0)$, $(1.526, 0)$, and $(1.629, \pm 1.063)$. Now since $f_{xx} = 8 - 12x^2$, $f_{xy} = 4y$,

$f_{yy} = 4x - 12y^2$, and $D = (8 - 12x^2)(4x - 12y^2) - 16y^2$, we have $D(-1.267, 0) > 0$, $f_{xx}(-1.267, 0) > 0$,

$D(-0.259, 0) < 0$, $D(1.526, 0) < 0$, $D(1.629, \pm 1.063) > 0$, and $f_{xx}(1.629, \pm 1.063) < 0$. Therefore, to three decimal

places, $f(-1.267, 0) \approx 1.310$ and $f(1.629, \pm 1.063) \approx 8.105$ are local maxima, and $(-0.259, 0)$ and $(1.526, 0)$ are saddle

points. The highest points on the graph are approximately $(1.629, \pm 1.063, 8.105)$.

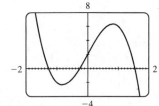 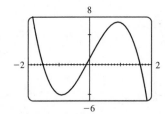 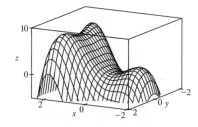

28. $f(x, y) = e^x + y^4 - x^3 + 4\cos y \Rightarrow f_x(x, y) = e^x - 3x^2$ and $f_y(x, y) = 4y^3 - 4\sin y$. From the graphs, we see that to

three decimal places, $f_x = 0$ when $x \approx -0.459, 0.910$, or 3.733, and $f_y = 0$ when $y \approx 0$ or ± 0.929. (Alternatively, we

could have used a calculator or a CAS to find the roots of $f_x = 0$ and $f_y = 0$.) So, to three decimal places, f has critical points

at $(-0.459, 0)$, $(-0.459, \pm 0.929)$, $(0.910, 0)$, $(0.910, \pm 0.929)$, $(3.733, 0)$, and $(3.733, \pm 0.929)$. Now $f_{xx} = e^x - 6x$,

$f_{xy} = 0$, $f_{yy} = 12y^2 - 4\cos y$, and $D = (e^x - 6x)(12y^2 - 4\cos y)$. Therefore $D(-0.459, 0) < 0$,

$D(-0.459, \pm 0.929) > 0$, $f_{xx}(-0.459, \pm 0.929) > 0$, $D(0.910, 0) > 0$, $f_{xx}(0.910, 0) < 0$, $D(0.910, \pm 0.929) < 0$,

$D(3.733, 0) < 0$, $D(3.733, \pm 0.929) > 0$, and $f_{xx}(3.733, \pm 0.929) > 0$. So $f(-0.459, \pm 0.929) \approx 3.868$ and

$f(3.733, \pm 0.929) \approx -7.077$ are local minima, $f(0.910, 0) \approx 5.731$ is a local maximum, and $(-0.459, 0)$, $(0.910, \pm 0.929)$,

and $(3.733, 0)$ are saddle points. The lowest points on the graph are approximately $(3.733, \pm 0.929, -7.077)$.

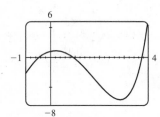 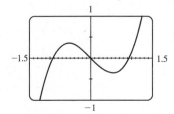 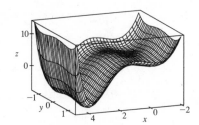

29. Since f is a polynomial it is continuous on D, so an absolute maximum and minimum exist. Here $f_x = 4$, $f_y = -5$ so there are no critical points inside D. Thus the absolute extrema must both occur on the boundary. Along L_1: $x = 0$ and $f(0, y) = 1 - 5y$ for $0 \le y \le 3$, a decreasing function in y, so the maximum value is $f(0, 0) = 1$ and the minimum value is $f(0, 3) = -14$. Along L_2: $y = 0$ and $f(x, 0) = 1 + 4x$ for $0 \le x \le 2$, an increasing function in x, so the minimum value is $f(0, 0) = 1$ and the maximum value is $f(2, 0) = 9$. Along L_3: $y = -\frac{3}{2}x + 3$ and $f\left(x, -\frac{3}{2}x + 3\right) = \frac{23}{2}x - 14$ for $0 \le x \le 2$, an increasing function in x, so the minimum value is $f(0, 3) = -14$ and the maximum value is $f(2, 0) = 9$. Thus the absolute maximum of f on D is $f(2, 0) = 9$ and the absolute minimum is $f(0, 3) = -14$.

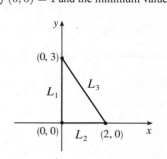

30. Since f is a polynomial it is continuous on D, so an absolute maximum and minimum exist. $f_x = y - 1$, $f_y = x - 2$, and setting $f_x = f_y = 0$ gives $(2, 1)$ as the only critical point, where $f(2, 1) = 1$. Along L_1: $x = 1$ and $f(1, y) = 2 - y$ for $0 \le y \le 4$, a decreasing function in y, so the maximum value is $f(1, 0) = 2$ and the minimum value is $f(1, 4) = -2$. Along L_2: $y = 0$ and $f(x, 0) = 3 - x$ for $1 \le x \le 5$, a decreasing function in x, so the maximum value is $f(1, 0) = 2$ and the minimum value is $f(5, 0) = -2$. Along L_3: $y = 5 - x$ and

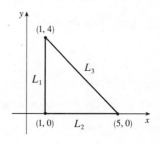

$f(x, 5 - x) = -x^2 + 6x - 7 = -(x - 3)^2 + 2$ for $1 \le x \le 5$, which has a maximum at $x = 3$ where $f(3, 2) = 2$ and a minimum at both $x = 1$ and $x = 5$, where $f(1, 4) = f(5, 0) = -2$. Thus the absolute maximum of f on D is $f(1, 0) = f(3, 2) = 2$ and the absolute minimum is $f(1, 4) = f(5, 0) = -2$.

31. $f_x(x, y) = 2x + 2xy$, $f_y(x, y) = 2y + x^2$, and setting $f_x = f_y = 0$ gives $(0, 0)$ as the only critical point in D, with $f(0, 0) = 4$.

On L_1: $y = -1$, $f(x, -1) = 5$, a constant.

On L_2: $x = 1$, $f(1, y) = y^2 + y + 5$, a quadratic in y which attains its maximum at $(1, 1)$, $f(1, 1) = 7$ and its minimum at $\left(1, -\frac{1}{2}\right)$, $f\left(1, -\frac{1}{2}\right) = \frac{19}{4}$.

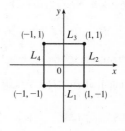

On L_3: $f(x, 1) = 2x^2 + 5$ which attains its maximum at $(-1, 1)$ and $(1, 1)$ with $f(\pm 1, 1) = 7$ and its minimum at $(0, 1)$, $f(0, 1) = 5$.

On L_4: $f(-1, y) = y^2 + y + 5$ with maximum at $(-1, 1)$, $f(-1, 1) = 7$ and minimum at $\left(-1, -\frac{1}{2}\right)$, $f\left(-1, -\frac{1}{2}\right) = \frac{19}{4}$.

Thus the absolute maximum is attained at both $(\pm 1, 1)$ with $f(\pm 1, 1) = 7$ and the absolute minimum on D is attained at $(0, 0)$ with $f(0, 0) = 4$.

32. $f_x(x, y) = 4 - 2x$ and $f_y(x, y) = 6 - 2y$, so the only critical point is $(2, 3)$ (which is in D) where $f(2, 3) = 13$.

Along L_1: $y = 0$, so $f(x, 0) = 4x - x^2 = -(x - 2)^2 + 4$, $0 \leq x \leq 4$, which has a maximum value when $x = 2$ where

$f(2, 0) = 4$ and a minimum value both when $x = 0$ and $x = 4$, where $f(0, 0) = f(4, 0) = 0$. Along L_2: $x = 4$, so

$f(4, y) = 6y - y^2 = -(y - 3)^2 + 9$, $0 \leq y \leq 5$, which has a maximum value when $y = 3$ where $f(4, 3) = 9$ and a

minimum value when $y = 0$ where $f(4, 0) = 0$. Along L_3: $y = 5$, so $f(x, 5) = -x^2 + 4x + 5 = -(x - 2)^2 + 9$,

$0 \leq x \leq 4$, which has a maximum value when $x = 2$ where $f(2, 5) = 9$ and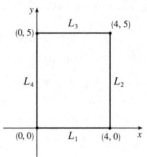

a minimum value both when $x = 0$ and $x = 4$, where $f(0, 5) = f(4, 5) = 5$.

Along L_4: $x = 0$, so $f(0, y) = 6y - y^2 = -(y - 3)^2 + 9$, $0 \leq y \leq 5$,

which has a maximum value when $y = 3$ where $f(0, 3) = 9$ and a minimum

value when $y = 0$ where $f(0, 0) = 0$. Thus the absolute maximum is

$f(2, 3) = 13$ and the absolute minimum is attained at both $(0, 0)$ and $(4, 0)$,

where $f(0, 0) = f(4, 0) = 0$.

33. $f(x, y) = x^4 + y^4 - 4xy + 2$ is a polynomial and hence continuous on D, so

it has an absolute maximum and minimum on D. In Exercise 7, we found the

critical points of f; only $(1, 1)$ with $f(1, 1) = 0$ is inside D. On L_1: $y = 0$,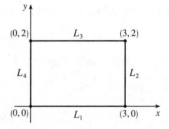

$f(x, 0) = x^4 + 2$, $0 \leq x \leq 3$, a polynomial in x which attains its maximum

at $x = 3$, $f(3, 0) = 83$, and its minimum at $x = 0$, $f(0, 0) = 2$.

On L_2: $x = 3$, $f(3, y) = y^4 - 12y + 83$, $0 \leq y \leq 2$, a polynomial in y

which attains its minimum at $y = \sqrt[3]{3}$, $f(3, \sqrt[3]{3}) = 83 - 9\sqrt[3]{3} \approx 70.0$, and its maximum at $y = 0$, $f(3, 0) = 83$.

On L_3: $y = 2$, $f(x, 2) = x^4 - 8x + 18$, $0 \leq x \leq 3$, a polynomial in x which attains its minimum at $x = \sqrt[3]{2}$,

$f(\sqrt[3]{2}, 2) = 18 - 6\sqrt[3]{2} \approx 10.4$, and its maximum at $x = 3$, $f(3, 2) = 75$. On L_4: $x = 0$, $f(0, y) = y^4 + 2$, $0 \leq y \leq 2$, a

polynomial in y which attains its maximum at $y = 2$, $f(0, 2) = 18$, and its minimum at $y = 0$, $f(0, 0) = 2$. Thus the absolute

maximum of f on D is $f(3, 0) = 83$ and the absolute minimum is $f(1, 1) = 0$.

34. $f_x = y^2$ and $f_y = 2xy$, and since $f_x = 0 \iff y = 0$, there are no critical

points in the interior of D. Along L_1: $y = 0$ and $f(x, 0) = 0$.

Along L_2: $x = 0$ and $f(0, y) = 0$. Along L_3: $y = \sqrt{3 - x^2}$, so let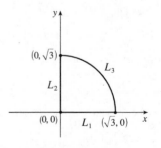

$g(x) = f(x, \sqrt{3 - x^2}) = 3x - x^3$ for $0 \leq x \leq \sqrt{3}$. Then

$g'(x) = 3 - 3x^2 = 0 \iff x = 1$. The maximum value is $f(1, \sqrt{2}) = 2$

and the minimum occurs both at $x = 0$ and $x = \sqrt{3}$ where

$f(0, \sqrt{3}) = f(\sqrt{3}, 0) = 0$. Thus the absolute maximum of f on D is $f(1, \sqrt{2}) = 2$, and the absolute minimum is 0 which

occurs at all points along L_1 and L_2.

35. $f_x(x, y) = 6x^2$ and $f_y(x, y) = 4y^3$. And so $f_x = 0$ and $f_y = 0$ only occur when $x = y = 0$. Hence, the only critical point

inside the disk is at $x = y = 0$ where $f(0, 0) = 0$. Now on the circle $x^2 + y^2 = 1$, $y^2 = 1 - x^2$ so let

$g(x) = f(x, y) = 2x^3 + (1 - x^2)^2 = x^4 + 2x^3 - 2x^2 + 1, -1 \leq x \leq 1$. Then $g'(x) = 4x^3 + 6x^2 - 4x = 0 \Rightarrow x = 0$,

-2, or $\frac{1}{2}$. $f(0, \pm 1) = g(0) = 1$, $f\left(\frac{1}{2}, \pm\frac{\sqrt{3}}{2}\right) = g\left(\frac{1}{2}\right) = \frac{13}{16}$, and $(-2, -3)$ is not in D. Checking the endpoints, we get

$f(-1, 0) = g(-1) = -2$ and $f(1, 0) = g(1) = 2$. Thus the absolute maximum and minimum of f on D are $f(1, 0) = 2$ and

$f(-1, 0) = -2$.

Another method: On the boundary $x^2 + y^2 = 1$ we can write $x = \cos\theta$, $y = \sin\theta$, so $f(\cos\theta, \sin\theta) = 2\cos^3\theta + \sin^4\theta$,

$0 \leq \theta \leq 2\pi$.

36. $f_x(x, y) = 3x^2 - 3$ and $f_y(x, y) = -3y^2 + 12$ and the critical

points are $(1, 2)$, $(1, -2)$, $(-1, 2)$, and $(-1, -2)$. But only $(1, 2)$

and $(-1, 2)$ are in D and $f(1, 2) = 14$, $f(-1, 2) = 18$. Along L_1:

$x = -2$ and $f(-2, y) = -2 - y^3 + 12y$, $-2 \leq y \leq 3$, which has

a maximum at $y = 2$ where $f(-2, 2) = 14$ and a minimum at

$y = -2$ where $f(-2, -2) = -18$. Along L_2: $x = 2$ and

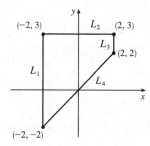

$f(2, y) = 2 - y^3 + 12y$, $2 \leq y \leq 3$, which has a maximum at $y = 2$ where $f(2, 2) = 18$ and a minimum at $y = 3$ where

$f(2, 3) = 11$. Along L_3: $y = 3$ and $f(x, 3) = x^3 - 3x + 9$, $-2 \leq x \leq 2$, which has a maximum at $x = -1$ and $x = 2$ where

$f(-1, 3) = f(2, 3) = 11$ and a minimum at $x = 1$ and $x = -2$ where $f(1, 3) = f(-2, 3) = 7$.

Along L_4: $y = x$ and $f(x, x) = 9x$, $-2 \leq x \leq 2$, which has a maximum at $x = 2$ where $f(2, 2) = 18$ and a minimum at

$x = -2$ where $f(-2, -2) = -18$. So the absolute maximum value of f on D is $f(2, 2) = 18$ and the minimum is

$f(-2, -2) = -18$.

37. $f(x, y) = -(x^2 - 1)^2 - (x^2 y - x - 1)^2 \Rightarrow f_x(x, y) = -2(x^2 - 1)(2x) - 2(x^2 y - x - 1)(2xy - 1)$ and

$f_y(x, y) = -2(x^2 y - x - 1)x^2$. Setting $f_y(x, y) = 0$ gives either $x = 0$ or $x^2 y - x - 1 = 0$.

There are no critical points for $x = 0$, since $f_x(0, y) = -2$, so we set $x^2 y - x - 1 = 0 \Leftrightarrow y = \dfrac{x + 1}{x^2} \quad [x \neq 0]$,

so $f_x\left(x, \dfrac{x + 1}{x^2}\right) = -2(x^2 - 1)(2x) - 2\left(x^2 \dfrac{x + 1}{x^2} - x - 1\right)\left(2x \dfrac{x + 1}{x^2} - 1\right) = -4x(x^2 - 1)$. Therefore

$f_x(x, y) = f_y(x, y) = 0$ at the points $(1, 2)$ and $(-1, 0)$. To classify these critical points, we calculate

$f_{xx}(x, y) = -12x^2 - 12x^2 y^2 + 12xy + 4y + 2$, $f_{yy}(x, y) = -2x^4$,

and $f_{xy}(x, y) = -8x^3 y + 6x^2 + 4x$. In order to use the Second

Derivatives Test we calculate

$D(-1, 0) = f_{xx}(-1, 0) f_{yy}(-1, 0) - [f_{xy}(-1, 0)]^2 = 16 > 0$,

$f_{xx}(-1, 0) = -10 < 0$, $D(1, 2) = 16 > 0$, and $f_{xx}(1, 2) = -26 < 0$, so

both $(-1, 0)$ and $(1, 2)$ give local maxima.

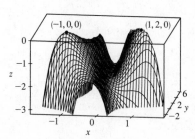

38. $f(x, y) = 3xe^y - x^3 - e^{3y}$ is differentiable everywhere, so the requirement

for critical points is that $f_x = 3e^y - 3x^2 = 0$ **(1)** and

$f_y = 3xe^y - 3e^{3y} = 0$ **(2)**. From **(1)** we obtain $e^y = x^2$, and then **(2)** gives

$3x^3 - 3x^6 = 0 \Rightarrow x = 1$ or 0, but only $x = 1$ is valid, since $x = 0$

makes **(1)** impossible. So substituting $x = 1$ into **(1)** gives $y = 0$, and the

only critical point is $(1, 0)$.

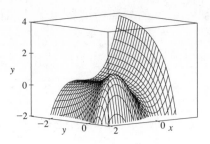

The Second Derivatives Test shows that this gives a local maximum, since

$D(1, 0) = \left[-6x(3xe^y - 9e^{3y}) - (3e^y)^2\right]_{(1,0)} = 27 > 0$ and $f_{xx}(1, 0) = [-6x]_{(1,0)} = -6 < 0$. But $f(1, 0) = 1$ is not an

absolute maximum because, for instance, $f(-3, 0) = 17$. This can also be seen from the graph.

39. Let d be the distance from $(2, 1, -1)$ to any point (x, y, z) on the plane $x + y - z = 1$, so

$d = \sqrt{(x - 2)^2 + (y - 1)^2 + (z + 1)^2}$ where $z = x + y - 1$, and we minimize

$d^2 = f(x, y) = (x - 2)^2 + (y - 1)^2 + (x + y)^2$. Then $f_x(x, y) = 2(x - 2) + 2(x + y) = 4x + 2y - 4$,

$f_y(x, y) = 2(y - 1) + 2(x + y) = 2x + 4y - 2$. Solving $4x + 2y - 4 = 0$ and $2x + 4y - 2 = 0$ simultaneously gives $x = 1$,

$y = 0$. An absolute minimum exists (since there is a minimum distance from the point to the plane) and it must occur at a

critical point, so the shortest distance occurs for $x = 1$, $y = 0$ for which $d = \sqrt{(1 - 2)^2 + (0 - 1)^2 + (0 + 1)^2} = \sqrt{3}$.

40. Here the distance d from a point on the plane to the point $(1, 2, 3)$ is $d = \sqrt{(x - 1)^2 + (y - 2)^2 + (z - 3)^2}$,

where $z = 4 - x + y$. We can minimize $d^2 = f(x, y) = (x - 1)^2 + (y - 2)^2 + (1 - x + y)^2$, so

$f_x(x, y) = 2(x - 1) + 2(1 - x + y)(-1) = 4x - 2y - 4$ and $f_y(x, y) = 2(y - 2) + 2(1 - x + y) = 4y - 2x - 2$.

Solving $4x - 2y - 4 = 0$ and $4y - 2x - 2 = 0$ simultaneously gives $x = \frac{5}{3}$ and $y = \frac{4}{3}$, so the only critical point is $\left(\frac{5}{3}, \frac{4}{3}\right)$.

This point must correspond to the minimum distance, so the point on the plane closest to $(1, 2, 3)$ is $\left(\frac{5}{3}, \frac{4}{3}, \frac{11}{3}\right)$.

41. Let d be the distance from the point $(4, 2, 0)$ to any point (x, y, z) on the cone, so $d = \sqrt{(x - 4)^2 + (y - 2)^2 + z^2}$ where

$z^2 = x^2 + y^2$, and we minimize $d^2 = (x - 4)^2 + (y - 2)^2 + x^2 + y^2 = f(x, y)$. Then

$f_x(x, y) = 2(x - 4) + 2x = 4x - 8$, $f_y(x, y) = 2(y - 2) + 2y = 4y - 4$, and the critical points occur when $f_x = 0 \Rightarrow$

$x = 2$, $f_y = 0 \Rightarrow y = 1$. Thus the only critical point is $(2, 1)$. An absolute minimum exists (since there is a minimum

distance from the cone to the point) which must occur at a critical point, so the points on the cone closest

to $(4, 2, 0)$ are $\left(2, 1, \pm\sqrt{5}\right)$.

42. The distance from the origin to a point (x, y, z) on the surface is $d = \sqrt{x^2 + y^2 + z^2}$ where $y^2 = 9 + xz$, so we minimize

$d^2 = x^2 + 9 + xz + z^2 = f(x, z)$. Then $f_x = 2x + z$, $f_z = x + 2z$, and $f_x = 0$, $f_z = 0 \Rightarrow x = 0$, $z = 0$, so the only

critical point is $(0, 0)$. $D(0, 0) = (2)(2) - 1 = 3 > 0$ with $f_{xx}(0, 0) = 2 > 0$, so this is a minimum. Thus $y^2 = 9 + 0 \Rightarrow$

$y = \pm 3$ and the points on the surface closest to the origin are $(0, \pm 3, 0)$.

43. $x + y + z = 100$, so maximize $f(x, y) = xy(100 - x - y)$. $f_x = 100y - 2xy - y^2$, $f_y = 100x - x^2 - 2xy$,

$f_{xx} = -2y$, $f_{yy} = -2x$, $f_{xy} = 100 - 2x - 2y$. Then $f_x = 0$ implies $y = 0$ or $y = 100 - 2x$. Substituting $y = 0$ into

$f_y = 0$ gives $x = 0$ or $x = 100$ and substituting $y = 100 - 2x$ into $f_y = 0$ gives $3x^2 - 100x = 0$ so $x = 0$ or $\frac{100}{3}$.

Thus the critical points are $(0, 0)$, $(100, 0)$, $(0, 100)$ and $\left(\frac{100}{3}, \frac{100}{3}\right)$.

$D(0, 0) = D(100, 0) = D(0, 100) = -10{,}000$ while $D\left(\frac{100}{3}, \frac{100}{3}\right) = \frac{10{,}000}{3}$ and $f_{xx}\left(\frac{100}{3}, \frac{100}{3}\right) = -\frac{200}{3} < 0$. Thus $(0, 0)$,

$(100, 0)$ and $(0, 100)$ are saddle points whereas $f\left(\frac{100}{3}, \frac{100}{3}\right)$ is a local maximum. Thus the numbers are $x = y = z = \frac{100}{3}$.

44. Let x, y, z, be the positive numbers. Then $x + y + z = 12$ and we want to minimize

$x^2 + y^2 + z^2 = x^2 + y^2 + (12 - x - y)^2 = f(x, y)$ for $0 < x, y < 12$. $f_x = 2x + 2(12 - x - y)(-1) = 4x + 2y - 24$,

$f_y = 2y + 2(12 - x - y)(-1) = 2x + 4y - 24$, $f_{xx} = 4$, $f_{xy} = 2$, $f_{yy} = 4$. Then $f_x = 0$ implies $4x + 2y = 24$ or

$y = 12 - 2x$ and substituting into $f_y = 0$ gives $2x + 4(12 - 2x) = 24$ $\Rightarrow$ $6x = 24$ $\Rightarrow$ $x = 4$ and then $y = 4$, so

the only critical point is $(4, 4)$. $D(4, 4) = 16 - 4 > 0$ and $f_{xx}(4, 4) = 4 > 0$, so $f(4, 4)$ is a local minimum. $f(4, 4)$ is also

the absolute minimum [compare to the values of f as $x, y \to 0$ or 12] so the numbers are $x = y = z = 4$.

45. Center the sphere at the origin so that its equation is $x^2 + y^2 + z^2 = r^2$, and orient the inscribed rectangular box so that its

edges are parallel to the coordinate axes. Any vertex of the box satisfies $x^2 + y^2 + z^2 = r^2$, so take (x, y, z) to be the vertex

in the first octant. Then the box has length $2x$, width $2y$, and height $2z = 2\sqrt{r^2 - x^2 - y^2}$ with volume given by

$V(x, y) = (2x)(2y)\left(2\sqrt{r^2 - x^2 - y^2}\right) = 8xy\sqrt{r^2 - x^2 - y^2}$ for $0 < x < r$, $0 < y < r$. Then

$V_x = (8xy) \cdot \frac{1}{2}(r^2 - x^2 - y^2)^{-1/2}(-2x) + \sqrt{r^2 - x^2 - y^2} \cdot 8y = \dfrac{8y(r^2 - 2x^2 - y^2)}{\sqrt{r^2 - x^2 - y^2}}$ and $V_y = \dfrac{8x(r^2 - x^2 - 2y^2)}{\sqrt{r^2 - x^2 - y^2}}$.

Setting $V_x = 0$ gives $y = 0$ or $2x^2 + y^2 = r^2$, but $y > 0$ so only the latter solution applies. Similarly, $V_y = 0$ with $x > 0$

implies $x^2 + 2y^2 = r^2$. Substituting, we have $2x^2 + y^2 = x^2 + 2y^2$ $\Rightarrow$ $x^2 = y^2$ $\Rightarrow$ $y = x$. Then $x^2 + 2y^2 = r^2$ $\Rightarrow$

$3x^2 = r^2$ $\Rightarrow$ $x = \sqrt{r^2/3} = r/\sqrt{3} = y$. Thus the only critical point is $\left(r/\sqrt{3}, r/\sqrt{3}\right)$. There must be a maximum

volume and here it must occur at a critical point, so the maximum volume occurs when $x = y = r/\sqrt{3}$ and the maximum

volume is $V\left(\frac{r}{\sqrt{3}}, \frac{r}{\sqrt{3}}\right) = 8\left(\frac{r}{\sqrt{3}}\right)\left(\frac{r}{\sqrt{3}}\right)\sqrt{r^2 - \left(\frac{r}{\sqrt{3}}\right)^2 - \left(\frac{r}{\sqrt{3}}\right)^2} = \dfrac{8}{3\sqrt{3}}r^3$.

46. Let x, y, and z be the dimensions of the box. We wish to minimize surface area $= 2xy + 2xz + 2yz$, but we have

volume $= xyz = 1000$ $\Rightarrow$ $z = \dfrac{1000}{xy}$ so we minimize

$f(x, y) = 2xy + 2x\left(\dfrac{1000}{xy}\right) + 2y\left(\dfrac{1000}{xy}\right) = 2xy + \dfrac{2000}{y} + \dfrac{2000}{x}$. Then $f_x = 2y - \dfrac{2000}{x^2}$ and $f_y = 2x - \dfrac{2000}{y^2}$. Setting

$f_x = 0$ implies $y = \dfrac{1000}{x^2}$ and substituting into $f_y = 0$ gives $x - \dfrac{x^4}{1000} = 0$ $\Rightarrow$ $x^3 = 1000$ [since $x \neq 0$] $\Rightarrow$ $x = 10$.

The surface area has a minimum but no maximum and it must occur at a critical point, so the minimal surface area occurs for a

box with dimensions $x = 10$ cm, $y = 1000/10^2 = 10$ cm, $z = 1000/10^2 = 10$ cm.

47. Maximize $f(x, y) = \dfrac{xy}{3}(6 - x - 2y)$, then the maximum volume is $V = xyz$.

$f_x = \frac{1}{3}(6y - 2xy - y^2) = \frac{1}{3}y(6 - 2x - 2y)$ and $f_y = \frac{1}{3}x(6 - x - 4y)$. Setting $f_x = 0$ and $f_y = 0$ gives the critical point

$(2, 1)$ which geometrically must yield a maximum. Thus the volume of the largest such box is $V = (2)(1)\left(\frac{2}{3}\right) = \frac{4}{3}$.

48. Surface area $= 2(xy + xz + yz) = 64 \text{ cm}^2$, so $xy + xz + yz = 32$ or $z = \dfrac{32 - xy}{x + y}$. Maximize the volume

$f(x, y) = xy\,\dfrac{32 - xy}{x + y}$. Then $f_x = \dfrac{32y^2 - 2xy^3 - x^2y^2}{(x + y)^2} = y^2\,\dfrac{32 - 2xy - x^2}{(x + y)^2}$ and $f_y = x^2\,\dfrac{32 - 2xy - y^2}{(x + y)^2}$. Setting

$f_x = 0$ implies $y = \dfrac{32 - x^2}{2x}$ and substituting into $f_y = 0$ gives $32(4x^2) - (32 - x^2)(4x^2) - (32 - x^2)^2 = 0$ or

$3x^4 + 64x^2 - (32)^2 = 0$. Thus $x^2 = \frac{64}{6}$ or $x = \frac{8}{\sqrt{6}}$, $y = \frac{64/3}{16/\sqrt{6}} = \frac{8}{\sqrt{6}}$ and $z = \frac{8}{\sqrt{6}}$. Thus the box is a cube with edge

length $\frac{8}{\sqrt{6}}$ cm.

49. Let the dimensions be x, y, and z; then $4x + 4y + 4z = c$ and the volume is

$V = xyz = xy\left(\frac{1}{4}c - x - y\right) = \frac{1}{4}cxy - x^2y - xy^2$, $x > 0$, $y > 0$. Then $V_x = \frac{1}{4}cy - 2xy - y^2$ and $V_y = \frac{1}{4}cx - x^2 - 2xy$,

so $V_x = 0 = V_y$ when $2x + y = \frac{1}{4}c$ and $x + 2y = \frac{1}{4}c$. Solving, we get $x = \frac{1}{12}c$, $y = \frac{1}{12}c$ and $z = \frac{1}{4}c - x - y = \frac{1}{12}c$. From

the geometrical nature of the problem, this critical point must give an absolute maximum. Thus the box is a cube with edge

length $\frac{1}{12}c$.

50. The cost equals $5xy + 2(xz + yz)$ and $xyz = V$, so $C(x, y) = 5xy + 2V(x + y)/(xy) = 5xy + 2V(x^{-1} + y^{-1})$. Then

$C_x = 5y - 2Vx^{-2}$, $C_y = 5x - 2Vy^{-2}$, $f_x = 0$ implies $y = 2V/(5x^2)$, $f_y = 0$ implies $x = \sqrt[3]{\frac{2}{5}V} = y$. Thus the

dimensions of the aquarium which minimize the cost are $x = y = \sqrt[3]{\frac{2}{5}V}$ units, $z = V^{1/3}\left(\frac{5}{2}\right)^{2/3}$.

51. Let the dimensions be x, y and z, then minimize $xy + 2(xz + yz)$ if $xyz = 32{,}000 \text{ cm}^3$. Then

$f(x, y) = xy + [64{,}000(x + y)/xy] = xy + 64{,}000(x^{-1} + y^{-1})$, $f_x = y - 64{,}000x^{-2}$, $f_y = x - 64{,}000y^{-2}$.

And $f_x = 0$ implies $y = 64{,}000/x^2$; substituting into $f_y = 0$ implies $x^3 = 64{,}000$ or $x = 40$ and then $y = 40$. Now

$D(x, y) = [(2)(64{,}000)]^2 x^{-3}y^{-3} - 1 > 0$ for $(40, 40)$ and $f_{xx}(40, 40) > 0$ so this is indeed a minimum. Thus the

dimensions of the box are $x = y = 40$ cm, $z = 20$ cm.

52. Let x be the length of the north and south walls, y the length of the east and west walls, and z the height of the building.

The heat loss is given by $h = 10(2yz) + 8(2xz) + 1(xy) + 5(xy) = 6xy + 16xz + 20yz$.

The volume is 4000 m^3, so $xyz = 4000$, and we substitute $z = \frac{4000}{xy}$ to

obtain the heat loss function $h(x, y) = 6xy + 80{,}000/x + 64{,}000/y$.

(a) Since $z = \frac{4000}{xy} \geq 4$, $xy \leq 1000 \;\Rightarrow\; y \leq 1000/x$. Also $x \geq 30$ and

$y \geq 30$, so the domain of h is $D = \{(x, y) \mid x \geq 30, 30 \leq y \leq 1000/x\}$.

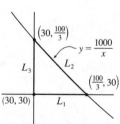

(b) $h(x,y) = 6xy + 80{,}000x^{-1} + 64{,}000y^{-1}$ $\Rightarrow$ $h_x = 6y - 80{,}000x^{-2}$, $h_y = 6x - 64{,}000y^{-2}$. $h_x = 0$ implies

$6x^2y = 80{,}000$ $\Rightarrow$ $y = \dfrac{80{,}000}{6x^2}$ and substituting into $h_y = 0$ gives $6x = 64{,}000\left(\dfrac{6x^2}{80{,}000}\right)^2$ $\Rightarrow$

$x^3 = \dfrac{80{,}000^2}{6 \cdot 64{,}000} = \dfrac{50{,}000}{3}$, so $x = \sqrt[3]{\dfrac{50{,}000}{3}} = 10\sqrt[3]{\dfrac{50}{3}}$ $\Rightarrow$ $y = \dfrac{80}{\sqrt[3]{60}}$, and the only critical point of h is

$\left(10\sqrt[3]{\dfrac{50}{3}}, \dfrac{80}{\sqrt[3]{60}}\right) \approx (25.54, 20.43)$ which is not in D. Next we check the boundary of D.

On L_1: $y = 30$, $h(x, 30) = 180x + 80{,}000/x + 6400/3$, $30 \le x \le \frac{100}{3}$. Since $h'(x, 30) = 180 - 80{,}000/x^2 > 0$ for

$30 \le x \le \frac{100}{3}$, $h(x, 30)$ is an increasing function with minimum $h(30, 30) = 10{,}200$ and maximum

$h\left(\frac{100}{3}, 30\right) \approx 10{,}533$.

On L_2: $y = 1000/x$, $h(x, 1000/x) = 6000 + 64x + 80{,}000/x$, $30 \le x \le \frac{100}{3}$.

Since $h'(x, 1000/x) = 64 - 80{,}000/x^2 < 0$ for $30 \le x \le \frac{100}{3}$, $h(x, 1000/x)$ is a decreasing function with minimum

$h\left(\frac{100}{3}, 30\right) \approx 10{,}533$ and maximum $h\left(30, \frac{100}{3}\right) \approx 10{,}587$.

On L_3: $x = 30$, $h(30, y) = 180y + 64{,}000/y + 8000/3$, $30 \le y \le \frac{100}{3}$. $h'(30, y) = 180 - 64{,}000/y^2 > 0$ for

$30 \le y \le \frac{100}{3}$, so $h(30, y)$ is an increasing function of y with minimum $h(30, 30) = 10{,}200$ and maximum

$h\left(30, \frac{100}{3}\right) \approx 10{,}587$.

Thus the absolute minimum of h is $h(30, 30) = 10{,}200$, and the dimensions of the building that minimize heat loss are

walls 30 m in length and height $\frac{4000}{30^2} = \frac{40}{9} \approx 4.44$ m.

(c) From part (b), the only critical point of h, which gives a local (and absolute) minimum, is approximately

$h(25.54, 20.43) \approx 9396$. So a building of volume 4000 m^2 with dimensions $x \approx 25.54$ m, $y \approx 20.43$ m,

$z \approx \frac{4000}{(25.54)(20.43)} \approx 7.67$ m has the least amount of heat loss.

53. Let x, y, z be the dimensions of the rectangular box. Then the volume of the box is xyz and

$L = \sqrt{x^2 + y^2 + z^2}$ $\Rightarrow$ $L^2 = x^2 + y^2 + z^2$ $\Rightarrow$ $z = \sqrt{L^2 - x^2 - y^2}$.

Substituting, we have volume $V(x, y) = xy\sqrt{L^2 - x^2 - y^2}$, $(x, y > 0)$.

$V_x = xy \cdot \frac{1}{2}(L^2 - x^2 - y^2)^{-1/2}(-2x) + y\sqrt{L^2 - x^2 - y^2} = y\sqrt{L^2 - x^2 - y^2} - \dfrac{x^2y}{\sqrt{L^2 - x^2 - y^2}}$,

$V_y = x\sqrt{L^2 - x^2 - y^2} - \dfrac{xy^2}{\sqrt{L^2 - x^2 - y^2}}$. $V_x = 0$ implies $y(L^2 - x^2 - y^2) = x^2y$ $\Rightarrow$ $y(L^2 - 2x^2 - y^2) = 0$ $\Rightarrow$

$2x^2 + y^2 = L^2$ (since $y > 0$), and $V_y = 0$ implies $x(L^2 - x^2 - y^2) = xy^2$ $\Rightarrow$ $x(L^2 - x^2 - 2y^2) = 0$ $\Rightarrow$

$x^2 + 2y^2 = L^2$ (since $x > 0$). Substituting $y^2 = L^2 - 2x^2$ into $x^2 + 2y^2 = L^2$ gives $x^2 + 2L^2 - 4x^2 = L^2$ $\Rightarrow$

$3x^2 = L^2$ $\Rightarrow$ $x = L/\sqrt{3}$ (since $x > 0$) and then $y = \sqrt{L^2 - 2(L/\sqrt{3})^2} = L/\sqrt{3}$. So the only critical point is

$\left(L/\sqrt{3}, L/\sqrt{3}\right)$ which, from the geometrical nature of the problem, must give an absolute maximum. Thus the maximum

volume is $V\left(L/\sqrt{3}, L/\sqrt{3}\right) = \left(L/\sqrt{3}\right)^2\sqrt{L^2 - \left(L/\sqrt{3}\right)^2 - \left(L/\sqrt{3}\right)^2} = L^3/(3\sqrt{3})$ cubic units.

54. Since $p + q + r = 1$ we can substitute $p = 1 - r - q$ into P giving

$P = P(q, r) = 2(1 - r - q)q + 2(1 - r - q)r + 2rq = 2q - 2q^2 + 2r - 2r^2 - 2rq$. Since p, q and r represent proportions

and $p + q + r = 1$, we know $q \geq 0$, $r \geq 0$, and $q + r \leq 1$. Thus, we want to find the absolute maximum of the continuous

function $P(q, r)$ on the closed set D enclosed by the lines $q = 0$, $r = 0$, and $q + r = 1$. To find any critical points, we set the

partial derivatives equal to zero: $P_q(q, r) = 2 - 4q - 2r = 0$ and $P_r(q, r) = 2 - 4r - 2q = 0$. The first equation gives

$r = 1 - 2q$, and substituting into the second equation we have $2 - 4(1 - 2q) - 2q = 0 \Rightarrow q = \frac{1}{3}$. Then we have one

critical point, $\left(\frac{1}{3}, \frac{1}{3}\right)$, where $P\left(\frac{1}{3}, \frac{1}{3}\right) = \frac{2}{3}$. Next we find the maximum values of P on the boundary of D which consists of

three line segments. For the segment given by $r = 0$, $0 \leq q \leq 1$, $P(q, r) = P(q, 0) = 2q - 2q^2$, $0 \leq q \leq 1$. This represents

a parabola with maximum value $P\left(\frac{1}{2}, 0\right) = \frac{1}{2}$. On the segment $q = 0$, $0 \leq r \leq 1$ we have $P(0, r) = 2r - 2r^2$, $0 \leq r \leq 1$.

This represents a parabola with maximum value $P\left(0, \frac{1}{2}\right) = \frac{1}{2}$. Finally, on the segment $q + r = 1$, $0 \leq q \leq 1$,

$P(q, r) = P(q, 1 - q) = 2q - 2q^2$, $0 \leq q \leq 1$ which has a maximum value of $P\left(\frac{1}{2}, \frac{1}{2}\right) = \frac{1}{2}$. Comparing these values with

the value of P at the critical point, we see that the absolute maximum value of $P(q, r)$ on D is $\frac{2}{3}$.

55. Note that here the variables are m and b, and $f(m, b) = \sum\limits_{i=1}^{n} [y_i - (mx_i + b)]^2$. Then $f_m = \sum\limits_{i=1}^{n} -2x_i[y_i - (mx_i + b)] = 0$

implies $\sum\limits_{i=1}^{n} \left(x_i y_i - mx_i^2 - bx_i\right) = 0$ or $\sum\limits_{i=1}^{n} x_i y_i = m \sum\limits_{i=1}^{n} x_i^2 + b \sum\limits_{i=1}^{n} x_i$ and $f_b = \sum\limits_{i=1}^{n} -2[y_i - (mx_i + b)] = 0$ implies

$\sum\limits_{i=1}^{n} y_i = m \sum\limits_{i=1}^{n} x_i + \sum\limits_{i=1}^{n} b = m \left(\sum\limits_{i=1}^{n} x_i\right) + nb$. Thus we have the two desired equations.

Now $f_{mm} = \sum\limits_{i=1}^{n} 2x_i^2$, $f_{bb} = \sum\limits_{i=1}^{n} 2 = 2n$ and $f_{mb} = \sum\limits_{i=1}^{n} 2x_i$. And $f_{mm}(m, b) > 0$ always and

$D(m, b) = 4n\left(\sum\limits_{i=1}^{n} x_i^2\right) - 4\left(\sum\limits_{i=1}^{n} x_i\right)^2 = 4\left[n\left(\sum\limits_{i=1}^{n} x_i^2\right) - \left(\sum\limits_{i=1}^{n} x_i\right)^2\right] > 0$ always so the solutions of these two

equations do indeed minimize $\sum\limits_{i=1}^{n} d_i^2$.

56. Any such plane must cut out a tetrahedron in the first octant. We need to minimize the volume of the tetrahedron that passes

through the point $(1, 2, 3)$. Writing the equation of the plane as $\dfrac{x}{a} + \dfrac{y}{b} + \dfrac{z}{c} = 1$, the volume of the tetrahedron is given by

$V = \dfrac{abc}{6}$. But $(1, 2, 3)$ must lie on the plane, so we need $\dfrac{1}{a} + \dfrac{2}{b} + \dfrac{3}{c} = 1$ $(\star)$ and thus can think of c as a function of a and b.

Then $V_a = \dfrac{b}{6}\left(c + a\dfrac{\partial c}{\partial a}\right)$ and $V_b = \dfrac{a}{6}\left(c + b\dfrac{\partial c}{\partial b}\right)$. Differentiating $(\star)$ with respect to a we get $-a^{-2} - 3c^{-2}\dfrac{\partial c}{\partial a} = 0 \Rightarrow$

$\dfrac{\partial c}{\partial a} = \dfrac{-c^2}{3a^2}$, and differentiating $(\star)$ with respect to b gives $-2b^{-2} - 3c^{-2}\dfrac{\partial c}{\partial b} = 0 \Rightarrow \dfrac{\partial c}{\partial b} = \dfrac{-2c^2}{3b^2}$. Then

$V_a = \dfrac{b}{6}\left(c + a\dfrac{-c^2}{3a^2}\right) = 0 \Rightarrow c = 3a$, and $V_b = \dfrac{a}{6}\left(c + b\dfrac{-2c^2}{3b^2}\right) = 0 \Rightarrow c = \frac{3}{2}b$. Thus $3a = \frac{3}{2}b$ or $b = 2a$. Putting

these into $(\star)$ gives $\dfrac{3}{a} = 1$ or $a = 3$ and then $b = 6$, $c = 9$. Thus the equation of the required plane is $\dfrac{x}{3} + \dfrac{y}{6} + \dfrac{z}{9} = 1$

or $6x + 3y + 2z = 18$.

APPLIED PROJECT   Designing a Dumpster

Note: The difficulty and results of this project vary widely with the type of container studied. In addition to the variation of basic shapes of containers, dumpsters may include additional constructed parts such as supports, lift pockets, wheels, etc. Also, a CAS or graphing utility may be needed to solve the resulting equations.

Here we present a typical solution for one particular trash dumpster.

1. The basic shape and dimensions (in inches) of an actual trash Dumpster are as shown in the figure.

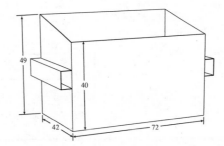

The front and back, as well as both sides, have an extra one-inch-wide flap that is folded under and welded to the base. In addition, the side panels each fold over one inch onto the front and back pieces where they are welded. Each side has a rectangular lift pocket, with cross-section 5 by 8 inches, made of the same material. These are attached with an extra one-inch width of steel on both top and bottom where each pocket is welded to the side sheet. All four sides have a "lip" at the top; the front and back panels have an extra 5 inches of steel at the top which is folded outward in three creases to form a rectangular tube. The edge is then welded back to the main sheet. The two sides form a top lip with separate sheets of steel 5 inches wide, similarly bent into three sides and welded to the main sheets (requiring two welds each). These extend beyond the main side sheets by 1.5 inches at each end in order to join with the lips on the front and back panels. The container has a hinged lid, extra steel supports on the base at each corner, metal "fins" serving as extra support for the side lift pockets, and wheels underneath. The volume of the container is $V = \frac{1}{2}(40 + 49) \times 42 \times 72 = 134{,}568$ in^3 or 77.875 ft^3.

2. First, we assume that some aspects of the construction do not change with different dimensions, so they may be considered fixed costs. This includes the lid (with hinges), wheels, and extra steel supports. Also, the upper "lip" we previously described extends beyond the side width to connect to the other pieces. We can safely assume that this extra portion, including any associated welds, costs the same regardless of the container's dimensions, so we will consider just the portion matching the measurement of the side panels in our calculations. We will further assume that the angle of the top of the container should be preserved. Then to compute the variable costs, let x be the width, y the length, and z the height of the front of the container. The back of the container is 9 inches, or $\frac{3}{4}$ ft, taller than the front, so using similar triangles we can say the back panel has height $z + \frac{3}{14}x$. Measuring in feet, we want the volume to remain constant, so

$V = \frac{1}{2}\left(z + z + \frac{3}{14}x\right)(x)(y) = xyz + \frac{3}{28}x^2y = 77.875$. To determine a function for the variable cost, we first find the area of each sheet of metal needed. The base has area xy ft^2. The front panel has visible area yz plus $\frac{1}{12}y$ for the portion folded onto the base and $\frac{5}{12}y$ for the steel at the top used to form the lip, so $\left(yz + \frac{1}{2}y\right)$ ft^2 in total. Similarly, the back sheet has area $y\left(z + \frac{3}{14}x\right) + \frac{1}{12}y + \frac{5}{12}y = yz + \frac{3}{14}xy + \frac{1}{2}y$. Each side has visible area $\frac{1}{2}\left[z + \left(z + \frac{3}{14}x\right)\right](x)$, and the sheet includes one-inch flaps folding onto the front and back panels, so with area $\frac{1}{12}z$ and $\frac{1}{12}\left(z + \frac{3}{14}x\right)$, and a one-inch flap to fold onto the base with area $\frac{1}{12}x$. The lift pocket is constructed of a piece of steel 20 inches by x ft (including the 2 extra inches used by the welds). The additional metal used to make the lip at the top of the panel has width 5 inches and length that we can determine using the Pythagorean Theorem: $x^2 + \left(\frac{3}{14}x\right)^2 = \text{length}^2$, so length $= \frac{\sqrt{205}}{14}x \approx 1.0227x$. Thus the area of steel needed for each side panel is approximately

$$\tfrac{1}{2}\left[z + \left(z + \tfrac{3}{14}x\right)\right](x) + \tfrac{1}{12}z + \tfrac{1}{12}\left(z + \tfrac{3}{14}x\right) + \tfrac{1}{12}x + \tfrac{5}{3}x + \tfrac{5}{12}(1.0227x) \approx xz + \tfrac{3}{28}x^2 + \tfrac{1}{6}z + 2.194x$$

We also have the following welds:

Weld	Length
Front, back welded to base	$2y$
Sides welded to base	$2x$
Sides welded to front	$2z$
Sides welded to back	$2\left(z + \frac{3}{14}x\right)$
Weld on front and back lip	$2y$
Two welds on each side lip	$4(1.0227x)$
Two welds for each lift pocket	$4x$

Thus the total length of welds needed is

$$2y + 2x + 2z + 2\left(z + \tfrac{3}{14}x\right) + 2y + 4(1.0227x) + 4x \approx 10.519x + 4y + 4z$$

Finally, the total variable cost is approximately

$$0.90(xy) + 0.70\left[\left(yz + \tfrac{1}{2}y\right) + \left(yz + \tfrac{3}{14}xy + \tfrac{1}{2}y\right) + 2\left(xz + \tfrac{3}{28}x^2 + \tfrac{1}{6}z + 2.194x\right)\right] + 0.18(10.519x + 4y + 4z)$$
$$\approx 1.05xy + 1.4yz + 1.42y + 1.4xz + 0.15x^2 + 0.953z + 4.965x$$

We would like to minimize this function while keeping volume constant, so since $xyz + \frac{3}{28}x^2y = 77.875$

we can substitute $z = \dfrac{77.875}{xy} - \dfrac{3}{28}x$ giving variable cost as a function of x and y:

$$C(x, y) \approx 0.9xy + \frac{109.0}{x} + 1.42y + \frac{109.0}{y} + \frac{74.2}{xy} + 4.86x$$

Using a CAS, we solve the system of equations $C_x(x, y) = 0$ and $C_y(x, y) = 0$; the only critical point within an appropriate domain is approximately $(3.58, 5.29)$. From the nature of the function C (or from a graph) we can determine that C has an absolute minimum at $(3.58, 5.29)$, and so the minimum cost is attained for $x \approx 3.58$ ft (or 43.0 in), $y \approx 5.29$ ft (or 63.5 in), and $z \approx \frac{77.875}{3.58(5.29)} - \frac{3}{28}(3.58) \approx 3.73$ ft (or 44.8 in).

3. The fixed cost aspects of the container which we did not include in our calculations, such as the wheels and lid, don't affect the validity of our results. Some of our other assumptions, however, may influence the accuracy of our findings. We simplified the price of the steel sheets to include cuts and bends, and we simplified the price of welding to include the labor and materials. This may not be accurate for areas of the container, such as the lip and lift pockets, that require several cuts, bends, and welds in a relatively small surface area. Consequently, increasing some dimensions of the container may not increase the cost in the

same manner as our computations predict. If we do not assume that the angle of the sloped top of the container must be preserved, it is likely that we could further improve our cost. Finally, our results show that the length of the container should be changed to minimize cost; this may not be possible if the two lift pockets must remain a fixed distance apart for handling by machinery.

4. The minimum variable cost using our values found in Problem 2 is $C(3.58, 5.29) \approx \$96.95$, while the current dimensions give an estimated variable cost of $C(3.5, 6.0) \approx \$97.30$. If we determine that our assumptions and simplifications are acceptable, our work shows that a slight savings can be gained by adjusting the dimensions of the container. However, the difference in cost is modest, and may not justify changes in the manufacturing process.

DISCOVERY PROJECT Quadratic Approximations and Critical Points

1. $Q(x, y) = f(a, b) + f_x(a, b)(x - a) + f_y(a, b)(y - b) + \frac{1}{2}f_{xx}(a, b)(x - a)^2$
$$+ f_{xy}(a, b)(x - a)(y - b) + \frac{1}{2}f_{yy}(a, b)(y - b)^2,$$

so

$$Q_x(x, y) = f_x(a, b) + \frac{1}{2}f_{xx}(a, b)(2)(x - a) + f_{xy}(a, b)(y - b) = f_x(a, b) + f_{xx}(a, b)(x - a) + f_{xy}(a, b)(y - b)$$

At (a, b) we have $Q_x(a, b) = f_x(a, b) + f_{xx}(a, b)(a - a) + f_{xy}(a, b)(b - b) = f_x(a, b)$.

Similarly, $Q_y(x, y) = f_y(a, b) + f_{xy}(a, b)(x - a) + f_{yy}(a, b)(y - b) \Rightarrow$

$Q_y(a, b) = f_y(a, b) + f_{xy}(a, b)(a - a) + f_{yy}(a, b)(b - b) = f_y(a, b)$.

For the second-order partial derivatives we have

$$Q_{xx}(x, y) = \frac{\partial}{\partial x}[f_x(a, b) + f_{xx}(a, b)(x - a) + f_{xy}(a, b)(y - b)] = f_{xx}(a, b) \quad \Rightarrow \quad Q_{xx}(a, b) = f_{xx}(a, b)$$

$$Q_{xy}(x, y) = \frac{\partial}{\partial y}[f_x(a, b) + f_{xx}(a, b)(x - a) + f_{xy}(a, b)(y - b)] = f_{xy}(a, b) \quad \Rightarrow \quad Q_{xy}(a, b) = f_{xy}(a, b)$$

$$Q_{yy}(x, y) = \frac{\partial}{\partial y}[f_y(a, b) + f_{xy}(a, b)(x - a) + f_{yy}(a, b)(y - b)] = f_{yy}(a, b) \quad \Rightarrow \quad Q_{yy}(a, b) = f_{yy}(a, b)$$

2. (a) First we find the partial derivatives and values that will be needed:

$$f(x, y) = e^{-x^2 - y^2} \qquad\qquad f(0, 0) = 1$$
$$f_x(x, y) = -2xe^{-x^2 - y^2} \qquad\qquad f_x(0, 0) = 0$$
$$f_y(x, y) = -2ye^{-x^2 - y^2} \qquad\qquad f_y(0, 0) = 0$$
$$f_{xx}(x, y) = (4x^2 - 2)e^{-x^2 - y^2} \qquad\qquad f_{xx}(0, 0) = -2$$
$$f_{xy}(x, y) = 4xye^{-x^2 - y^2} \qquad\qquad f_{xy}(0, 0) = 0$$
$$f_{yy}(x, y) = (4y^2 - 2)e^{-x^2 - y^2} \qquad\qquad f_{yy}(0, 0) = -2$$

Then the first-degree Taylor polynomial of f at $(0, 0)$ is

$$L(x, y) = f(0, 0) + f_x(0, 0)(x - 0) + f_y(0, 0)(y - 0) = 1 + (0)(x - 0) + (0)(y - 0) = 1$$

The second-degree Taylor polynomial is given by

$$Q(x,y) = f(0,0) + f_x(0,0)(x-0) + f_y(0,0)(y-0) + \tfrac{1}{2}f_{xx}(0,0)(x-0)^2$$
$$+ f_{xy}(0,0)(x-0)(y-0) + \tfrac{1}{2}f_{yy}(0,0)(y-0)^2$$
$$= 1 - x^2 - y^2$$

(b)

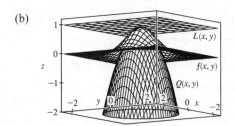

As we see from the graph, L approximates f well only for points (x,y) extremely close to the origin. Q is a much better approximation; the shape of its graph looks similar to that of the graph of f near the origin, and the values of Q appear to be good estimates for the values of f within a significant radius of the origin.

3. (a) First we find the partial derivatives and values that will be needed:

$$f(x,y) = xe^y \qquad f(1,0) = 1 \qquad\qquad f_{xx}(x,y) = 0 \qquad f_{xx}(1,0) = 0$$
$$f_x(x,y) = e^y \qquad f_x(1,0) = 1 \qquad\qquad f_{xy}(x,y) = e^y \qquad f_{xy}(1,0) = 1$$
$$f_y(x,y) = xe^y \qquad f_y(1,0) = 1 \qquad\qquad f_{yy}(x,y) = xe^y \qquad f_{yy}(1,0) = 1$$

Then the first-degree Taylor polynomial of f at $(1,0)$ is

$$L(x,y) = f(1,0) + f_x(1,0)(x-1) + f_y(1,0)(y-0) = 1 + (1)(x-1) + (1)(y-0) = x + y$$

The second-degree Taylor polynomial is given by

$$Q(x,y) = f(1,0) + f_x(1,0)(x-1) + f_y(1,0)(y-0) + \tfrac{1}{2}f_{xx}(1,0)(x-1)^2$$
$$+ f_{xy}(1,0)(x-1)(y-0) + \tfrac{1}{2}f_{yy}(1,0)(y-0)^2$$
$$= \tfrac{1}{2}y^2 + x + xy$$

(b) $L(0.9, 0.1) = 0.9 + 0.1 = 1.0$

$Q(0.9, 0.1) = \tfrac{1}{2}(0.1)^2 + 0.9 + (0.9)(0.1) = 0.995$

$f(0.9, 0.1) = 0.9e^{0.1} \approx 0.9947$

(c)

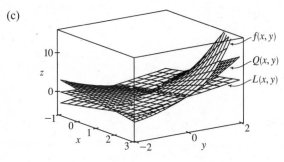

As we see from the graph, L and Q both approximate f reasonably well near the point $(1, 0)$. As we venture farther from the point, the graph of Q follows the shape of the graph of f more closely than L.

4. (a) $f(x,y) = ax^2 + bxy + cy^2 = a\left[x^2 + \dfrac{b}{a}xy + \dfrac{c}{a}y^2\right] = a\left[x^2 + \dfrac{b}{a}xy + \left(\dfrac{b}{2a}y\right)^2 - \left(\dfrac{b}{2a}y\right)^2 + \dfrac{c}{a}y^2\right]$

$= a\left[\left(x + \dfrac{b}{2a}y\right)^2 - \dfrac{b^2}{4a^2}y^2 + \dfrac{c}{a}y^2\right] = a\left[\left(x + \dfrac{b}{2a}y\right)^2 + \left(\dfrac{4ac - b^2}{4a^2}\right)y^2\right]$

(b) For $D = 4ac - b^2$, from part (a) we have $f(x, y) = a\left[\left(x + \dfrac{b}{2a}y\right)^2 + \left(\dfrac{D}{4a^2}\right)y^2\right]$. If $D > 0$,

$\left(\dfrac{D}{4a^2}\right)y^2 \geq 0$ and $\left(x + \dfrac{b}{2a}y\right)^2 \geq 0$, so $\left[\left(x + \dfrac{b}{2a}y\right)^2 + \left(\dfrac{D}{4a^2}\right)y^2\right] \geq 0$. Here $a > 0$, thus

$f(x, y) = a\left[\left(x + \dfrac{b}{2a}y\right)^2 + \left(\dfrac{D}{4a^2}\right)y^2\right] \geq 0$. We know $f(0, 0) = 0$, so $f(0, 0) \leq f(x, y)$ for all (x, y), and by

definition f has a local minimum at $(0, 0)$.

(c) As in part (b), $\left[\left(x + \dfrac{b}{2a}y\right)^2 + \left(\dfrac{D}{4a^2}\right)y^2\right] \geq 0$, and since $a < 0$ we have

$f(x, y) = a\left[\left(x + \dfrac{b}{2a}y\right)^2 + \left(\dfrac{D}{4a^2}\right)y^2\right] \leq 0$. Since $f(0, 0) = 0$, we must have $f(0, 0) \geq f(x, y)$ for all (x, y), so by

definition f has a local maximum at $(0, 0)$.

(d) $f(x, y) = ax^2 + bxy + cy^2$, so $f_x(x, y) = 2ax + by \;\Rightarrow\; f_x(0, 0) = 0$ and $f_y(x, y) = bx + 2cy \;\Rightarrow\; f_y(0, 0) = 0$.

Since $f(0, 0) = 0$ and f and its partial derivatives are continuous, we know from Equation 15.4.2 [ET 14.4.2] that the

tangent plane to the graph of f at $(0, 0)$ is the plane $z = 0$. Then f has a saddle point at $(0, 0)$ if the graph of f crosses the

tangent plane at $(0, 0)$, or equivalently, if some paths to the origin have positive function values while other paths have

negative function values. Suppose we approach the origin along the x-axis; then we have $y = 0 \;\Rightarrow\; f(x, 0) = ax^2$

which has the same sign as a. We must now find at least one path to the origin where $f(x, y)$ gives values with sign

opposite that of a. Since $f(x, y) = a\left[\left(x + \dfrac{b}{2a}y\right)^2 + \left(\dfrac{D}{4a^2}\right)y^2\right]$, if we approach the origin along the line $x = -\dfrac{b}{2a}y$,

we have $f\left(-\dfrac{b}{2a}y, y\right) = a\left[\left(-\dfrac{b}{2a}y + \dfrac{b}{2a}y\right)^2 + \left(\dfrac{D}{4a^2}\right)y^2\right] = \dfrac{D}{4a}y^2$. Since $D < 0$, these values have signs opposite

that of a. Thus, f has a saddle point at $(0, 0)$.

5. (a) Since the partial derivatives of f exist at $(0, 0)$ and $(0, 0)$ is a critical point, we know $f_x(0, 0) = 0$ and $f_y(0, 0) = 0$. Then

the second-degree Taylor polynomial of f at $(0, 0)$ can be expressed as

$$Q(x, y) = f(0, 0) + f_x(0, 0)(x - 0) + f_y(0, 0)(y - 0) + \tfrac{1}{2}f_{xx}(0, 0)(x - 0)^2$$
$$+ f_{xy}(0, 0)(x - 0)(y - 0) + \tfrac{1}{2}f_{yy}(0, 0)(y - 0)^2$$
$$= \tfrac{1}{2}f_{xx}(0, 0)x^2 + f_{xy}(0, 0)xy + \tfrac{1}{2}f_{yy}(0, 0)y^2$$

(b) $Q(x, y) = \tfrac{1}{2}f_{xx}(0, 0)x^2 + f_{xy}(0, 0)xy + \tfrac{1}{2}f_{yy}(0, 0)y^2$ fits the form of the polynomial function in

Problem 4 with $a = \tfrac{1}{2}f_{xx}(0, 0)$, $b = f_{xy}(0, 0)$, and $c = \tfrac{1}{2}f_{yy}(0, 0)$. Then we know Q is a paraboloid, and

that Q has a local maximum, local minimum, or saddle point at $(0, 0)$. Here,

$D = 4ac - b^2 = 4\left(\tfrac{1}{2}\right)f_{xx}(0, 0)\left(\tfrac{1}{2}\right)f_{yy}(0, 0) - [f_{xy}(0, 0)]^2 = f_{xx}(0, 0)f_{yy}(0, 0) - [f_{xy}(0, 0)]^2$, and if $D > 0$ with

$a = \tfrac{1}{2}f_{xx}(0, 0) > 0 \;\Rightarrow\; f_{xx}(0, 0) > 0$, we know from Problem 4 that Q has a local minimum at $(0, 0)$. Similarly, if

$D > 0$ and $a < 0 \;\Rightarrow\; f_{xx}(0, 0) < 0$, Q has a local maximum at $(0, 0)$, and if $D < 0$, Q has a saddle point at $(0, 0)$.

(c) Since $f(x, y) \approx Q(x, y)$ near $(0, 0)$, part (b) suggests that for $D = f_{xx}(0, 0)f_{yy}(0, 0) - [f_{xy}(0, 0)]^2$, if $D > 0$ and

$f_{xx}(0, 0) > 0$, f has a local minimum at $(0, 0)$. If $D > 0$ and $f_{xx}(0, 0) < 0$, f has a local maximum at $(0, 0)$, and if

$D < 0$, f has a saddle point at $(0, 0)$. Together with the conditions given in part (a), this is precisely the Second

Derivatives Test from Section 15.7 [ET 14.7].

15.8 Lagrange Multipliers   ET 14.8

1. At the extreme values of f, the level curves of f just touch the curve $g(x, y) = 8$ with a common tangent line. (See Figure 1

and the accompanying discussion.) We can observe several such occurrences on the contour map, but the level curve

$f(x, y) = c$ with the largest value of c which still intersects the curve $g(x, y) = 8$ is approximately $c = 59$, and the smallest

value of c corresponding to a level curve which intersects $g(x, y) = 8$ appears to be $c = 30$. Thus we estimate the maximum

value of f subject to the constraint $g(x, y) = 8$ to be about 59 and the minimum to be 30.

2. (a) The values $c = \pm 1$ and $c = 1.25$ seem to give curves which are

tangent to the circle. These values represent possible extreme values

of the function $x^2 + y$ subject to the constraint $x^2 + y^2 = 1$.

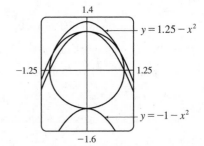

(b) $\nabla f = \langle 2x, 1 \rangle$, $\lambda \nabla g = \langle 2\lambda x, 2\lambda y \rangle$. So $2x = 2\lambda x$ ⇒ either

$\lambda = 1$ or $x = 0$. If $\lambda = 1$, then $y = \frac{1}{2}$ and so $x = \pm\frac{\sqrt{3}}{2}$ (from the

constraint). If $x = 0$, then $y = \pm 1$. Therefore f has possible extreme

values at the points $(0, \pm 1)$ and $\left(\pm\frac{\sqrt{3}}{2}, \frac{1}{2}\right)$. We calculate $f\left(\pm\frac{\sqrt{3}}{2}, \frac{1}{2}\right) = \frac{5}{4}$ (the maximum value), $f(0, 1) = 1$, and

$f(0, -1) = -1$ (the minimum value). These are our answers from part (a).

3. $f(x, y) = x^2 + y^2$, $g(x, y) = xy = 1$, and $\nabla f = \lambda \nabla g$ ⇒ $\langle 2x, 2y \rangle = \langle \lambda y, \lambda x \rangle$, so $2x = \lambda y$, $2y = \lambda x$, and $xy = 1$.

From the last equation, $x \neq 0$ and $y \neq 0$, so $2x = \lambda y$ ⇒ $\lambda = 2x/y$. Substituting, we have $2y = (2x/y) x$ ⇒

$y^2 = x^2$ ⇒ $y = \pm x$. But $xy = 1$, so $x = y = \pm 1$ and the possible points for the extreme values of f are $(1, 1)$ and

$(-1, -1)$. Here there is no maximum value, since the constraint $xy = 1$ allows x or y to become arbitrarily large, and hence

$f(x, y) = x^2 + y^2$ can be made arbitrarily large. The minimum value is $f(1, 1) = f(-1, -1) = 2$.

4. $f(x, y) = 4x + 6y$, $g(x, y) = x^2 + y^2 = 13$ ⇒ $\nabla f = \langle 4, 6 \rangle$, $\lambda \nabla g = \langle 2\lambda x, 2\lambda y \rangle$. Then $2\lambda x = 4$ and $2\lambda y = 6$ imply

$x = \frac{2}{\lambda}$ and $y = \frac{3}{\lambda}$. But $13 = x^2 + y^2 = \left(\frac{2}{\lambda}\right)^2 + \left(\frac{3}{\lambda}\right)^2$ ⇒ $13 = \frac{13}{\lambda^2}$ ⇒ $\lambda = \pm 1$, so f has possible extreme values

at the points $(2, 3)$, $(-2, -3)$. We compute $f(2, 3) = 26$ and $f(-2, -3) = -26$, so the maximum value of f on

$x^2 + y^2 = 13$ is $f(2, 3) = 26$ and the minimum value is $f(-2, -3) = -26$.

5. $f(x, y) = x^2 y$, $g(x, y) = x^2 + 2y^2 = 6$ ⇒ $\nabla f = \langle 2xy, x^2 \rangle$, $\lambda \nabla g = \langle 2\lambda x, 4\lambda y \rangle$. Then $2xy = 2\lambda x$ implies $x = 0$ or

$\lambda = y$. If $x = 0$, then $x^2 = 4\lambda y$ implies $\lambda = 0$ or $y = 0$. However, if $y = 0$ then $g(x, y) = 0$, a contradiction. So $\lambda = 0$ and

then $g(x, y) = 6$ ⇒ $y = \pm\sqrt{3}$. If $\lambda = y$, then $x^2 = 4\lambda y$ implies $x^2 = 4y^2$, and so $g(x, y) = 6$ ⇒

$4y^2 + 2y^2 = 6 \implies y^2 = 1 \implies y = \pm 1$. Thus f has possible extreme values at the points $(0, \pm\sqrt{3})$, $(\pm 2, 1)$, and $(\pm 2, -1)$. After evaluating f at these points, we find the maximum value to be $f(\pm 2, 1) = 4$ and the minimum to be $f(\pm 2, -1) = -4$.

6. $f(x, y) = e^{xy}$, $g(x, y) = x^3 + y^3 = 16$, and $\nabla f = \lambda \nabla g \implies \langle ye^{xy}, xe^{xy} \rangle = \langle 3\lambda x^2, 3\lambda y^2 \rangle$, so $ye^{xy} = 3\lambda x^2$ and $xe^{xy} = 3\lambda y^2$. Note that $x = 0 \iff y = 0$ which contradicts $x^3 + y^3 = 16$, so we may assume $x \neq 0$, $y \neq 0$, and then $\lambda = ye^{xy}/(3x^2) = xe^{xy}/(3y^2) \implies x^3 = y^3 \implies x = y$. But $x^3 + y^3 = 16$, so $2x^3 = 16 \implies x = 2 = y$. Here there is no minimum value, since we can choose points satisfying the constraint $x^3 + y^3 = 16$ that make $f(x, y) = e^{xy}$ arbitrarily close to 0 (but never equal to 0). The maximum value is $f(2, 2) = e^4$.

7. $f(x, y, z) = 2x + 6y + 10z$, $g(x, y, z) = x^2 + y^2 + z^2 = 35 \implies \nabla f = \langle 2, 6, 10 \rangle$, $\lambda \nabla g = \langle 2\lambda x, 2\lambda y, 2\lambda z \rangle$. Then $2\lambda x = 2$, $2\lambda y = 6$, $2\lambda z = 10$ imply $x = \dfrac{1}{\lambda}$, $y = \dfrac{3}{\lambda}$, and $z = \dfrac{5}{\lambda}$. But $35 = x^2 + y^2 + z^2 = \left(\dfrac{1}{\lambda}\right)^2 + \left(\dfrac{3}{\lambda}\right)^2 + \left(\dfrac{5}{\lambda}\right)^2 \implies 35 = \dfrac{35}{\lambda^2} \implies \lambda = \pm 1$, so f has possible extreme values at the points $(1, 3, 5)$, $(-1, -3, -5)$. The maximum value of f on $x^2 + y^2 + z^2 = 35$ is $f(1, 3, 5) = 70$, and the minimum is $f(-1, -3, -5) = -70$.

8. $f(x, y, z) = 8x - 4z$, $g(x, y, z) = x^2 + 10y^2 + z^2 = 5 \implies \nabla f = \langle 8, 0, -4 \rangle$, $\lambda \nabla g = \langle 2\lambda x, 20\lambda y, 2\lambda z \rangle$. Then $2\lambda x = 8$, $20\lambda y = 0$, $2\lambda z = -4$ imply $x = \dfrac{4}{\lambda}$, $y = 0$, and $z = -\dfrac{2}{\lambda}$. But $5 = x^2 + 10y^2 + z^2 = \left(\dfrac{4}{\lambda}\right)^2 + 10(0)^2 + \left(-\dfrac{2}{\lambda}\right)^2 \implies 5 = \dfrac{20}{\lambda^2} \implies \lambda = \pm 2$, so f has possible extreme values at the points $(2, 0, -1)$, $(-2, 0, 1)$. The maximum of f on $x^2 + 10y^2 + z^2 = 5$ is $f(2, 0, -1) = 20$, and the minimum is $f(-2, 0, 1) = -20$.

9. $f(x, y, z) = xyz$, $g(x, y, z) = x^2 + 2y^2 + 3z^2 = 6 \implies \nabla f = \langle yz, xz, xy \rangle$, $\lambda \nabla g = \langle 2\lambda x, 4\lambda y, 6\lambda z \rangle$. If $\lambda = 0$ then at least one of the coordinates is 0, in which case $f(x, y, z) = 0$. (None of these ends up giving a maximum or minimum.) If $\lambda \neq 0$, then $\nabla f = \lambda \nabla g$ implies $\lambda = (yz)/(2x) = (xz)/(4y) = (xy)/(6z)$ or $x^2 = 2y^2$ and $z^2 = \frac{2}{3}y^2$. Thus $x^2 + 2y^2 + 3z^2 = 6$ implies $6y^2 = 6$ or $y = \pm 1$. Thus the possible remaining points are $\left(\sqrt{2}, \pm 1, \sqrt{\frac{2}{3}}\right)$, $\left(\sqrt{2}, \pm 1, -\sqrt{\frac{2}{3}}\right)$, $\left(-\sqrt{2}, \pm 1, \sqrt{\frac{2}{3}}\right)$, $\left(-\sqrt{2}, \pm 1, -\sqrt{\frac{2}{3}}\right)$. The maximum value of f on the ellipsoid is $\frac{2}{\sqrt{3}}$, occurring when all coordinates are positive or exactly two are negative and the minimum is $-\frac{2}{\sqrt{3}}$ occurring when 1 or 3 of the coordinates are negative.

10. $f(x, y, z) = x^2 y^2 z^2$, $g(x, y, z) = x^2 + y^2 + z^2 = 1 \implies \nabla f = \langle 2xy^2 z^2, 2yx^2 z^2, 2zx^2 y^2 \rangle$, $\lambda \nabla g = \langle 2\lambda x, 2\lambda y, 2\lambda z \rangle$. Then $\nabla f = \lambda \nabla g$ implies **(1)** $\lambda = y^2 z^2 = x^2 z^2 = x^2 y^2$ and $\lambda \neq 0$, or **(2)** $\lambda = 0$ and one or two (but not three) of the coordinates are 0. If **(1)** then $x^2 = y^2 = z^2 = \frac{1}{3}$. The minimum value of f on the sphere occurs in case **(2)** with a value of 0 and the maximum value is $\frac{1}{27}$ which arises from all the points from **(1)**, that is, the points $\left(\pm\frac{1}{\sqrt{3}}, \frac{1}{\sqrt{3}}, \frac{1}{\sqrt{3}}\right)$, $\left(\pm\frac{1}{\sqrt{3}}, -\frac{1}{\sqrt{3}}, \frac{1}{\sqrt{3}}\right)$, $\left(\pm\frac{1}{\sqrt{3}}, -\frac{1}{\sqrt{3}}, -\frac{1}{\sqrt{3}}\right)$.

11. $f(x,y,z) = x^2 + y^2 + z^2$, $g(x,y,z) = x^4 + y^4 + z^4 = 1$ $\Rightarrow$ $\nabla f = \langle 2x, 2y, 2z \rangle$, $\lambda \nabla g = \langle 4\lambda x^3, 4\lambda y^3, 4\lambda z^3 \rangle$.

Case 1: If $x \neq 0$, $y \neq 0$ and $z \neq 0$, then $\nabla f = \lambda \nabla g$ implies $\lambda = 1/(2x^2) = 1/(2y^2) = 1/(2z^2)$ or $x^2 = y^2 = z^2$ and

$3x^4 = 1$ or $x = \pm \frac{1}{\sqrt[4]{3}}$ giving the points $\left(\pm \frac{1}{\sqrt[4]{3}}, \frac{1}{\sqrt[4]{3}}, \frac{1}{\sqrt[4]{3}} \right)$, $\left(\pm \frac{1}{\sqrt[4]{3}}, -\frac{1}{\sqrt[4]{3}}, \frac{1}{\sqrt[4]{3}} \right)$, $\left(\pm \frac{1}{\sqrt[4]{3}}, \frac{1}{\sqrt[4]{3}}, -\frac{1}{\sqrt[4]{3}} \right)$, $\left(\pm \frac{1}{\sqrt[4]{3}}, -\frac{1}{\sqrt[4]{3}}, -\frac{1}{\sqrt[4]{3}} \right)$

all with an f-value of $\sqrt{3}$.

Case 2: If one of the variables equals zero and the other two are not zero, then the squares of the two nonzero coordinates are equal with common value $\frac{1}{\sqrt{2}}$ and corresponding f value of $\sqrt{2}$.

Case 3: If exactly two of the variables are zero, then the third variable has value ± 1 with the corresponding f value of 1. Thus on $x^4 + y^4 + z^4 = 1$, the maximum value of f is $\sqrt{3}$ and the minimum value is 1.

12. $f(x,y,z) = x^4 + y^4 + z^4$, $g(x,y,z) = x^2 + y^2 + z^2 = 1$ $\Rightarrow$ $\nabla f = \langle 4x^3, 4y^3, 4z^3 \rangle$, $\lambda \nabla g = \langle 2\lambda x, 2\lambda y, 2\lambda z \rangle$.

Case 1: If $x \neq 0$, $y \neq 0$ and $z \neq 0$ then $\nabla f = \lambda \nabla g$ implies $\lambda = 2x^2 = 2y^2 = 2z^2$ or $x^2 = y^2 = z^2 = \frac{1}{3}$ yielding 8 points each with an f-value of $\frac{1}{3}$.

Case 2: If one of the variables is 0 and the other two are not, then the squares of the two nonzero coordinates are equal with common value $\frac{1}{2}$ and the corresponding f-value is $\frac{1}{2}$.

Case 3: If exactly two of the variables are 0, then the third variable has value ± 1 with corresponding f-value of 1. Thus on $x^2 + y^2 + z^2 = 1$, the maximum value of f is 1 and the minimum value is $\frac{1}{3}$.

13. $f(x,y,z,t) = x + y + z + t$, $g(x,y,z,t) = x^2 + y^2 + z^2 + t^2 = 1$ $\Rightarrow$ $\langle 1,1,1,1 \rangle = \langle 2\lambda x, 2\lambda y, 2\lambda z, 2\lambda t \rangle$, so $\lambda = 1/(2x) = 1/(2y) = 1/(2z) = 1/(2t)$ and $x = y = z = t$. But $x^2 + y^2 + z^2 + t^2 = 1$, so the possible points are $\left(\pm \frac{1}{2}, \pm \frac{1}{2}, \pm \frac{1}{2}, \pm \frac{1}{2} \right)$. Thus the maximum value of f is $f\left(\frac{1}{2}, \frac{1}{2}, \frac{1}{2}, \frac{1}{2} \right) = 2$ and the minimum value is $f\left(-\frac{1}{2}, -\frac{1}{2}, -\frac{1}{2}, -\frac{1}{2} \right) = -2$.

14. $f(x_1, x_2, \ldots, x_n) = x_1 + x_2 + \cdots + x_n$, $g(x_1, x_2, \ldots, x_n) = x_1^2 + x_2^2 + \cdots + x_n^2 = 1$ $\Rightarrow$ $\langle 1, 1, \ldots, 1 \rangle = \langle 2\lambda x_1, 2\lambda x_2, \ldots, 2\lambda x_n \rangle$, so $\lambda = 1/(2x_1) = 1/(2x_2) = \cdots = 1/(2x_n)$ and $x_1 = x_2 = \cdots = x_n$. But $x_1^2 + x_2^2 + \cdots + x_n^2 = 1$, so $x_i = \pm 1/\sqrt{n}$ for $i = 1, \ldots, n$. Thus the maximum value of f is $f(1/\sqrt{n}, 1/\sqrt{n}, \ldots, 1/\sqrt{n}) = \sqrt{n}$ and the minimum value is $f(-1/\sqrt{n}, -1/\sqrt{n}, \ldots, -1/\sqrt{n}) = -\sqrt{n}$.

15. $f(x,y,z) = x + 2y$, $g(x,y,z) = x + y + z = 1$, $h(x,y,z) = y^2 + z^2 = 4$ $\Rightarrow$ $\nabla f = \langle 1, 2, 0 \rangle$, $\lambda \nabla g = \langle \lambda, \lambda, \lambda \rangle$ and $\mu \nabla h = \langle 0, 2\mu y, 2\mu z \rangle$. Then $1 = \lambda$, $2 = \lambda + 2\mu y$ and $0 = \lambda + 2\mu z$ so $\mu y = \frac{1}{2} = -\mu z$ or $y = 1/(2\mu)$, $z = -1/(2\mu)$. Thus $x + y + z = 1$ implies $x = 1$ and $y^2 + z^2 = 4$ implies $\mu = \pm \frac{1}{2\sqrt{2}}$. Then the possible points are $\left(1, \pm\sqrt{2}, \mp\sqrt{2} \right)$ and the maximum value is $f\left(1, \sqrt{2}, -\sqrt{2} \right) = 1 + 2\sqrt{2}$ and the minimum value is $f\left(1, -\sqrt{2}, \sqrt{2} \right) = 1 - 2\sqrt{2}$.

16. $f(x,y,z) = 3x - y - 3z$, $g(x,y,z) = x + y - z = 0$, $h(x,y,z) = x^2 + 2z^2 = 1$ $\Rightarrow$ $\nabla f = \langle 3, -1, -3 \rangle$, $\lambda \nabla g = \langle \lambda, \lambda, -\lambda \rangle$, $\mu \nabla h = \langle 2\mu x, 0, 4\mu z \rangle$. Then $3 = \lambda + 2\mu x$, $-1 = \lambda$ and $-3 = -\lambda + 4\mu z$, so $\lambda = -1$, $\mu z = -1$, $\mu x = 2$. Thus $h(x,y,z) = 1$ implies $\frac{4}{\mu^2} + 2\left(\frac{1}{\mu^2} \right) = 1$ or $\mu = \pm\sqrt{6}$, so $z = \mp\frac{1}{\sqrt{6}}$; $x = \pm\frac{2}{\sqrt{6}}$; and $g(x,y,z) = 0$

implies $y = \mp\frac{3}{\sqrt{6}}$. Hence the maximum of f subject to the constraints is $f\left(\frac{\sqrt{6}}{3}, -\frac{\sqrt{6}}{2}, -\frac{\sqrt{6}}{6}\right) = 2\sqrt{6}$ and the minimum

is $f\left(-\frac{\sqrt{6}}{3}, \frac{\sqrt{6}}{2}, \frac{\sqrt{6}}{6}\right) = -2\sqrt{6}$.

17. $f(x, y, z) = yz + xy$, $g(x, y, z) = xy = 1$, $h(x, y, z) = y^2 + z^2 = 1$ $\Rightarrow$ $\nabla f = \langle y, x + z, y \rangle$, $\lambda\nabla g = \langle \lambda y, \lambda x, 0 \rangle$,

$\mu\nabla h = \langle 0, 2\mu y, 2\mu z \rangle$. Then $y = \lambda y$ implies $\lambda = 1$ [$y \neq 0$ since $g(x, y, z) = 1$], $x + z = \lambda x + 2\mu y$ and $y = 2\mu z$. Thus

$\mu = z/(2y) = y/(2y)$ or $y^2 = z^2$, and so $y^2 + z^2 = 1$ implies $y = \pm\frac{1}{\sqrt{2}}$, $z = \pm\frac{1}{\sqrt{2}}$. Then $xy = 1$ implies $x = \pm\sqrt{2}$ and

the possible points are $\left(\pm\sqrt{2}, \pm\frac{1}{\sqrt{2}}, \frac{1}{\sqrt{2}}\right)$, $\left(\pm\sqrt{2}, \pm\frac{1}{\sqrt{2}}, -\frac{1}{\sqrt{2}}\right)$. Hence the maximum of f subject to the constraints is

$f\left(\pm\sqrt{2}, \pm\frac{1}{\sqrt{2}}, \pm\frac{1}{\sqrt{2}}\right) = \frac{3}{2}$ and the minimum is $f\left(\pm\sqrt{2}, \pm\frac{1}{\sqrt{2}}, \mp\frac{1}{\sqrt{2}}\right) = \frac{1}{2}$.

Note: Since $xy = 1$ is one of the constraints we could have solved the problem by solving $f(y, z) = yz + 1$ subject to

$y^2 + z^2 = 1$.

18. $f(x, y) = 2x^2 + 3y^2 - 4x - 5$ $\Rightarrow$ $\nabla f = \langle 4x - 4, 6y \rangle = \langle 0, 0 \rangle$ $\Rightarrow$ $x = 1$, $y = 0$. Thus $(1, 0)$ is the only critical point

of f, and it lies in the region $x^2 + y^2 < 16$. On the boundary, $g(x, y) = x^2 + y^2 = 16$ $\Rightarrow$ $\lambda\nabla g = \langle 2\lambda x, 2\lambda y \rangle$, so

$6y = 2\lambda y$ $\Rightarrow$ either $y = 0$ or $\lambda = 3$. If $y = 0$, then $x = \pm4$; if $\lambda = 3$, then $4x - 4 = 2\lambda x$ $\Rightarrow$ $x = -2$ and

$y = \pm2\sqrt{3}$. Now $f(1, 0) = -7$, $f(4, 0) = 11$, $f(-4, 0) = 43$, and $f\left(-2, \pm2\sqrt{3}\right) = 47$. Thus the maximum value of

$f(x, y)$ on the disk $x^2 + y^2 \leq 16$ is $f\left(-2, \pm2\sqrt{3}\right) = 47$, and the minimum value is $f(1, 0) = -7$.

19. $f(x, y) = e^{-xy}$. For the interior of the region, we find the critical points: $f_x = -ye^{-xy}$, $f_y = -xe^{-xy}$, so the only

critical point is $(0, 0)$, and $f(0, 0) = 1$. For the boundary, we use Lagrange multipliers. $g(x, y) = x^2 + 4y^2 = 1$ $\Rightarrow$

$\lambda\nabla g = \langle 2\lambda x, 8\lambda y \rangle$, so setting $\nabla f = \lambda\nabla g$ we get $-ye^{-xy} = 2\lambda x$ and $-xe^{-xy} = 8\lambda y$. The first of these gives

$e^{-xy} = -2\lambda x/y$, and then the second gives $-x(-2\lambda x/y) = 8\lambda y$ $\Rightarrow$ $x^2 = 4y^2$. Solving this last equation with the

constraint $x^2 + 4y^2 = 1$ gives $x = \pm\frac{1}{\sqrt{2}}$ and $y = \pm\frac{1}{2\sqrt{2}}$. Now $f\left(\pm\frac{1}{\sqrt{2}}, \mp\frac{1}{2\sqrt{2}}\right) = e^{1/4} \approx 1.284$ and

$f\left(\pm\frac{1}{\sqrt{2}}, \pm\frac{1}{2\sqrt{2}}\right) = e^{-1/4} \approx 0.779$. The former are the maxima on the region and the latter are the minima.

20. (a) $f(x, y) = 2x + 3y$, $g(x, y) = \sqrt{x} + \sqrt{y} = 5$ $\Rightarrow$ $\nabla f = \langle 2, 3 \rangle = \lambda\nabla g = \lambda\left\langle \frac{1}{2\sqrt{x}}, \frac{1}{2\sqrt{y}} \right\rangle$. Then

$2 = \dfrac{\lambda}{2\sqrt{x}}$ and $3 = \dfrac{\lambda}{2\sqrt{y}}$ so $4\sqrt{x} = \lambda = 6\sqrt{y}$ $\Rightarrow$ $\sqrt{y} = \frac{2}{3}\sqrt{x}$. With $\sqrt{x} + \sqrt{y} = 5$ we have $\sqrt{x} + \frac{2}{3}\sqrt{x} = 5$ $\Rightarrow$

$\sqrt{x} = 3$ $\Rightarrow$ $x = 9$. Substituting into $\sqrt{y} = \frac{2}{3}\sqrt{x}$ gives $\sqrt{y} = 2$ or $y = 4$. Thus the only possible extreme value

subject to the constraint is $f(9, 4) = 30$. (The question remains whether this is indeed the maximum of f.)

(b) $f(25, 0) = 50$ which is larger than the result of part (a).

(c)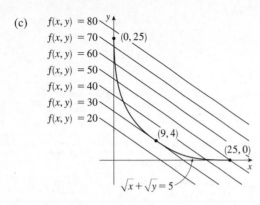

We can see from the level curves of f that the maximum occurs at the left endpoint $(0, 25)$ of the constraint curve g.

The maximum value is $f(0, 25) = 75$.

(d) Here ∇g does not exist if $x = 0$ or $y = 0$, so the method will not locate any associated points. Also, the method of Lagrange multipliers identifies points where the level curves of f share a common tangent line with the constraint curve g. This normally does not occur at an endpoint, although an absolute maximum or minimum may occur there.

(e) Here $f(9, 4)$ is the absolute *minimum* of f subject to g.

21. (a) $f(x, y) = x$, $g(x, y) = y^2 + x^4 - x^3 = 0$ $\Rightarrow$ $\nabla f = \langle 1, 0 \rangle = \lambda \nabla g = \lambda \langle 4x^3 - 3x^2, 2y \rangle$. Then

$1 = \lambda(4x^3 - 3x^2)$ **(1)** and $0 = 2\lambda y$ **(2)**. We have $\lambda \neq 0$ from **(1)**, so **(2)** gives $y = 0$. Then, from the constraint equation,

$x^4 - x^3 = 0$ $\Rightarrow$ $x^3(x - 1) = 0$ $\Rightarrow$ $x = 0$ or $x = 1$. But $x = 0$ contradicts **(1)**, so the only possible extreme value subject to the constraint is $f(1, 0) = 1$. (The question remains whether this is indeed the minimum of f.)

(b) The constraint is $y^2 + x^4 - x^3 = 0$ $\Leftrightarrow$ $y^2 = x^3 - x^4$. The left side is non-negative, so we must have $x^3 - x^4 \geq 0$ which is true only for $0 \leq x \leq 1$. Therefore the minimum possible value for $f(x, y) = x$ is 0 which occurs for $x = y = 0$. However, $\lambda \nabla g(0, 0) = \lambda \langle 0 - 0, 0 \rangle = \langle 0, 0 \rangle$ and $\nabla f(0, 0) = \langle 1, 0 \rangle$, so $\nabla f(0, 0) \neq \lambda \nabla g(0, 0)$ for all values of λ.

(c) Here $\nabla g(0, 0) = \mathbf{0}$ but the method of Lagrange multipliers requires that $\nabla g \neq \mathbf{0}$ everywhere on the constraint curve.

22. (a) The graphs of $f(x, y) = 3.7$ and $f(x, y) = 350$ seem to be tangent to the circle, and so 3.7 and 350 are the approximate minimum and maximum values of the function $f(x, y)$ subject to the constraint $(x - 3)^2 + (y - 3)^2 = 9$.

(b) Let $g(x, y) = (x - 3)^2 + (y - 3)^2$. We calculate $f_x(x, y) = 3x^2 + 3y$, $f_y(x, y) = 3y^2 + 3x$, $g_x(x, y) = 2x - 6$, and $g_y(x, y) = 2y - 6$, and use a CAS to search for solutions to the equations $g(x, y) = (x - 3)^2 + (y - 3)^2 = 9$,

$f_x = \lambda g_x$, and $f_y = \lambda g_y$. The solutions are $(x, y) = \left(3 - \frac{3}{2}\sqrt{2}, 3 - \frac{3}{2}\sqrt{2}\right) \approx (0.879, 0.879)$ and

$(x, y) = \left(3 + \frac{3}{2}\sqrt{2}, 3 + \frac{3}{2}\sqrt{2}\right) \approx (5.121, 5.121)$. These give $f\left(3 - \frac{3}{2}\sqrt{2}, 3 - \frac{3}{2}\sqrt{2}\right) = \frac{351}{2} - \frac{243}{2}\sqrt{2} \approx 3.673$ and

$f\left(3 + \frac{3}{2}\sqrt{2}, 3 + \frac{3}{2}\sqrt{2}\right) = \frac{351}{2} + \frac{243}{2}\sqrt{2} \approx 347.33$, in accordance with part (a).

23. $P(L, K) = bL^\alpha K^{1-\alpha}$, $g(L, K) = mL + nK = p$ $\Rightarrow$ $\nabla P = \left\langle \alpha b L^{\alpha-1} K^{1-\alpha}, (1 - \alpha)bL^\alpha K^{-\alpha} \right\rangle$, $\lambda \nabla g = \langle \lambda m, \lambda n \rangle$.

Then $\alpha b(K/L)^{1-\alpha} = \lambda m$ and $(1 - \alpha)b(L/K)^\alpha = \lambda n$ and $mL + nK = p$, so $\alpha b(K/L)^{1-\alpha}/m = (1 - \alpha)b(L/K)^\alpha/n$ or

$n\alpha/[m(1 - \alpha)] = (L/K)^\alpha (L/K)^{1-\alpha}$ or $L = Kn\alpha/[m(1 - \alpha)]$. Substituting into $mL + nK = p$ gives $K = (1 - \alpha)p/n$

and $L = \alpha p/m$ for the maximum production.

24. $C(L, K) = mL + nK$, $g(L, K) = bL^\alpha K^{1-\alpha} = Q$ $\Rightarrow$ $\nabla C = \langle m, n \rangle$, $\lambda \nabla g = \langle \lambda \alpha b L^{\alpha-1} K^{1-\alpha}, \lambda(1-\alpha) b L^\alpha K^{-\alpha} \rangle$.

Then $\dfrac{m}{\alpha b} \left(\dfrac{L}{K} \right)^{1-\alpha} = \dfrac{n}{(1-\alpha)b} \left(\dfrac{K}{L} \right)^{\alpha}$ and $bL^\alpha K^{1-\alpha} = Q$ $\Rightarrow$ $\dfrac{n\alpha}{m(1-\alpha)} = \left(\dfrac{L}{K} \right)^{1-\alpha} \left(\dfrac{L}{K} \right)^{\alpha}$ $\Rightarrow$

$L = \dfrac{Kn\alpha}{m(1-\alpha)}$ and so $b \left[\dfrac{Kn\alpha}{m(1-\alpha)} \right]^{\alpha} K^{1-\alpha} = Q$. Hence $K = \dfrac{Q}{b(n\alpha/[m(1-\alpha)])^\alpha} = \dfrac{Qm^\alpha(1-\alpha)^\alpha}{bn^\alpha \alpha^\alpha}$

and $L = \dfrac{Qm^{\alpha-1}(1-\alpha)^{\alpha-1}}{bn^{\alpha-1}\alpha^{\alpha-1}} = \dfrac{Qn^{1-\alpha}\alpha^{1-\alpha}}{bm^{1-\alpha}(1-\alpha)^{1-\alpha}}$ minimizes cost.

25. Let the sides of the rectangle be x and y. Then $f(x, y) = xy$, $g(x, y) = 2x + 2y = p$ $\Rightarrow$ $\nabla f(x, y) = \langle y, x \rangle$,

$\lambda \nabla g = \langle 2\lambda, 2\lambda \rangle$. Then $\lambda = \frac{1}{2}y = \frac{1}{2}x$ implies $x = y$ and the rectangle with maximum area is a square with side length $\frac{1}{4}p$.

26. Let $f(x, y, z) = s(s-x)(s-y)(s-z)$, $g(x, y, z) = x + y + z$. Then

$\nabla f = \langle -s(s-y)(s-z), -s(s-x)(s-z), -s(s-x)(s-y) \rangle$, $\lambda \nabla g = \langle \lambda, \lambda, \lambda \rangle$. Thus

$(s-y)(s-z) = (s-x)(s-z)$ **(1)**, and $(s-x)(s-z) = (s-x)(s-y)$ **(2)**.

(1) implies $x = y$ while **(2)** implies $y = z$, so $x = y = z = p/3$ and the triangle with maximum area is equilateral.

27. Let $f(x, y, z) = d^2 = (x-2)^2 + (y-1)^2 + (z+1)^2$, then we want to minimize f subject to the constraint

$g(x, y, z) = x + y - z = 1$. $\nabla f = \lambda \nabla g$ $\Rightarrow$ $\langle 2(x-2), 2(y-1), 2(z+1) \rangle = \lambda \langle 1, 1, -1 \rangle$, so $x = (\lambda + 4)/2$,

$y = (\lambda+2)/2$, $z = -(\lambda+2)/2$. Substituting into the constraint equation gives $\dfrac{\lambda+4}{2} + \dfrac{\lambda+2}{2} + \dfrac{\lambda+2}{2} = 1$ $\Rightarrow$

$3\lambda + 8 = 2$ $\Rightarrow$ $\lambda = -2$, so $x = 1$, $y = 0$, and $z = 0$. This must correspond to a minimum, so the shortest distance is

$d = \sqrt{(1-2)^2 + (0-1)^2 + (0+1)^2} = \sqrt{3}$.

28. Let $f(x, y, z) = d^2 = (x-1)^2 + (y-2)^2 + (z-3)^2$, then we want to minimize f subject to the constraint

$g(x, y, z) = x - y + z = 4$. $\nabla f = \lambda \nabla g$ $\Rightarrow$ $\langle 2(x-1), 2(y-2), 2(z-3) \rangle = \lambda \langle 1, -1, 1 \rangle$, so $x = (\lambda+2)/2$,

$y = (4-\lambda)/2$, $z = (\lambda+6)/2$. Substituting into the constraint equation gives $\dfrac{\lambda+2}{2} - \dfrac{4-\lambda}{2} + \dfrac{\lambda+6}{2} = 4$ $\Rightarrow$ $\lambda = \frac{4}{3}$,

so $x = \frac{5}{3}$, $y = \frac{4}{3}$, and $z = \frac{11}{3}$. This must correspond to a minimum, so the point on the plane closest to the point $(1, 2, 3)$

is $\left(\frac{5}{3}, \frac{4}{3}, \frac{11}{3} \right)$.

29. Let $f(x, y, z) = d^2 = (x-4)^2 + (y-2)^2 + z^2$. Then we want to minimize f subject to the constraint

$g(x, y, z) = x^2 + y^2 - z^2 = 0$. $\nabla f = \lambda \nabla g$ $\Rightarrow$ $\langle 2(x-4), 2(y-2), 2z \rangle = \langle 2\lambda x, 2\lambda y, -2\lambda z \rangle$, so $x - 4 = \lambda x$,

$y - 2 = \lambda y$, and $z = -\lambda z$. From the last equation we have $z + \lambda z = 0$ $\Rightarrow$ $z(1+\lambda) = 0$, so either $z = 0$ or $\lambda = -1$.

But from the constraint equation we have $z = 0$ $\Rightarrow$ $x^2 + y^2 = 0$ $\Rightarrow$ $x = y = 0$ which is not possible from the first two

equations. So $\lambda = -1$ and $x - 4 = \lambda x$ $\Rightarrow$ $x = 2$, $y - 2 = \lambda y$ $\Rightarrow$ $y = 1$, and $x^2 + y^2 - z^2 = 0$ $\Rightarrow$

$4 + 1 - z^2 = 0$ $\Rightarrow$ $z = \pm\sqrt{5}$. This must correspond to a minimum, so the points on the cone closest to $(4, 2, 0)$

are $(2, 1, \pm\sqrt{5})$.

30. Let $f(x, y, z) = d^2 = x^2 + y^2 + z^2$. Then we want to minimize f subject to the constraint $g(x, y, z) = y^2 - xz = 9$.

$\nabla f = \lambda \nabla g$ $\Rightarrow$ $\langle 2x, 2y, 2z \rangle = \langle -\lambda z, 2\lambda y, -\lambda x \rangle$, so $2x = -\lambda z$, $y = \lambda y$, and $2z = -\lambda x$. If $x = 0$ then the last equation

implies $z = 0$, and from the constraint $y^2 - xz = 9$ we have $y = \pm 3$. If $x \neq 0$, then the first and third equations give

$\lambda = -2x/z = -2z/x$ $\Rightarrow$ $x^2 = z^2$. From the second equation we have $y = 0$ or $\lambda = 1$. If $y = 0$ then $y^2 - xz = 9$ $\Rightarrow$

$z = -9/x$ and $x^2 = z^2$ $\Rightarrow$ $x^2 = 81/x^2$ $\Rightarrow$ $x = \pm 3$. Since $z = -9/x$, $x = 3$ $\Rightarrow$ $z = -3$ and $x = -3$ $\Rightarrow$

$z = 3$. If $\lambda = 1$, then $2x = -z$ and $2z = -x$ which implies $x = z = 0$, contradicting the assumption that $x \neq 0$. Thus the possible points are $(0, \pm 3, 0)$, $(3, 0, -3)$, $(-3, 0, 3)$. We have $f(0, \pm 3, 0) = 9$ and $f(3, 0, -3) = f(-3, 0, 3) = 18$, so the points on the surface that are closest to the origin are $(0, \pm 3, 0)$.

31. $f(x, y, z) = xyz$, $g(x, y, z) = x + y + z = 100$ $\Rightarrow$ $\nabla f = \langle yz, xz, xy \rangle = \lambda \nabla g = \langle \lambda, \lambda, \lambda \rangle$. Then $\lambda = yz = xz = xy$ implies $x = y = z = \frac{100}{3}$.

32. Minimize $f(x, y, z) = x^2 + y^2 + z^2$ subject to $g(x, y, z) = x + y + z = 12$ with $x > 0$, $y > 0$, $z > 0$. Then
$\nabla f = \lambda \nabla g$ $\Rightarrow$ $\langle 2x, 2y, 2z \rangle = \lambda \langle 1, 1, 1 \rangle$ $\Rightarrow$ $2x = \lambda$, $2y = \lambda$, $2z = \lambda$ $\Rightarrow$ $x = y = z$, so
$x + y + z = 12$ $\Rightarrow$ $3x = 12$ $\Rightarrow$ $x = 4 = y = z$. By comparing nearby values we can confirm that this gives a minimum and not a maximum. Thus the three numbers are 4, 4, and 4.

33. If the dimensions are $2x$, $2y$, and $2z$, then maximize $f(x, y, z) = (2x)(2y)(2z) = 8xyz$ subject to
$g(x, y, z) = x^2 + y^2 + z^2 = r^2$ $(x > 0, y > 0, z > 0)$. Then $\nabla f = \lambda \nabla g$ $\Rightarrow$ $\langle 8yz, 8xz, 8xy \rangle = \lambda \langle 2x, 2y, 2z \rangle$ $\Rightarrow$
$8yz = 2\lambda x$, $8xz = 2\lambda y$, and $8xy = 2\lambda z$, so $\lambda = \dfrac{4yz}{x} = \dfrac{4xz}{y} = \dfrac{4xy}{z}$. This gives $x^2 z = y^2 z$ $\Rightarrow$ $x^2 = y^2$ (since $z \neq 0$)
and $xy^2 = xz^2$ $\Rightarrow$ $z^2 = y^2$, so $x^2 = y^2 = z^2$ $\Rightarrow$ $x = y = z$, and substituting into the constraint
equation gives $3x^2 = r^2$ $\Rightarrow$ $x = r/\sqrt{3} = y = z$. Thus the largest volume of such a box is
$$f\left(\frac{r}{\sqrt{3}}, \frac{r}{\sqrt{3}}, \frac{r}{\sqrt{3}}\right) = 8\left(\frac{r}{\sqrt{3}}\right)\left(\frac{r}{\sqrt{3}}\right)\left(\frac{r}{\sqrt{3}}\right) = \frac{8}{3\sqrt{3}} r^3.$$

34. If the dimensions of the box are x, y, and z then minimize $f(x, y, z) = 2xy + 2xz + 2yz$ subject to $g(x, y, z) = xyz = 1000$
$(x > 0, y > 0, z > 0)$. Then $\nabla f = \lambda \nabla g$ $\Rightarrow$ $\langle 2y + 2z, 2x + 2z, 2x + 2y \rangle = \lambda \langle yz, xz, xy \rangle$ $\Rightarrow$ $2y + 2z = \lambda yz$,
$2x + 2z = \lambda xz$, $2x + 2y = \lambda xy$. Solving for λ in each equation gives $\lambda = \dfrac{2}{z} + \dfrac{2}{y} = \dfrac{2}{z} + \dfrac{2}{x} = \dfrac{2}{y} + \dfrac{2}{x}$ $\Rightarrow$ $x = y = z$.
From $xyz = 1000$ we have $x^3 = 1000$ $\Rightarrow$ $x = 10$ and the dimensions of the box are $x = y = z = 10$ cm.

35. $f(x, y, z) = xyz$, $g(x, y, z) = x + 2y + 3z = 6$ $\Rightarrow$ $\nabla f = \langle yz, xz, xy \rangle = \lambda \nabla g = \langle \lambda, 2\lambda, 3\lambda \rangle$.
Then $\lambda = yz = \frac{1}{2} xz = \frac{1}{3} xy$ implies $x = 2y$, $z = \frac{2}{3} y$. But $2y + 2y + 2y = 6$ so $y = 1$, $x = 2$, $z = \frac{2}{3}$ and the volume is $V = \frac{4}{3}$.

36. $f(x, y, z) = xyz$, $g(x, y, z) = xy + yz + xz = 32$ $\Rightarrow$ $\nabla f = \langle yz, xz, xy \rangle = \lambda \nabla g = \langle \lambda(y + z), \lambda(x + z), \lambda(x + y) \rangle$.
Then $\lambda(y + z) = yz$ **(1)**, $\lambda(x + z) = xz$ **(2)**, and $\lambda(x + y) = xy$ **(3)**. And **(1)** minus **(2)** implies $\lambda(y - x) = z(y - x)$
so $x = y$ or $\lambda = z$. If $\lambda = z$, then **(1)** implies $z(y + z) = yz$ or $z = 0$ which is false. Thus $x = y$. Similarly **(2)** minus **(3)**
implies $\lambda(z - y) = x(z - y)$ so $y = z$ or $\lambda = x$. As above, $\lambda \neq x$, so $x = y = z$ and $3x^2 = 32$ or $x = y = z = \frac{8}{\sqrt{6}}$ cm.

37. $f(x, y, z) = xyz$, $g(x, y, z) = 4(x + y + z) = c$ $\Rightarrow$ $\nabla f = \langle yz, xz, xy \rangle$, $\lambda \nabla g = \langle 4\lambda, 4\lambda, 4\lambda \rangle$. Thus
$4\lambda = yz = xz = xy$ or $x = y = z = \frac{1}{12} c$ are the dimensions giving the maximum volume.

38. $C(x, y, z) = 5xy + 2xz + 2yz$, $g(x, y, z) = xyz = V$ $\Rightarrow$
$\nabla C = \langle 5y + 2z, 5x + 2z, 2x + 2y \rangle = \lambda \nabla g = \langle \lambda yz, \lambda xz, \lambda xy \rangle$. Then $\lambda yz = 5y + 2z$ **(1)**, $\lambda xz = 5x + 2z$ **(2)**,
$\lambda xy = 2(x + y)$ **(3)**, and $xyz = V$ **(4)**. Now **(1)** $-$ **(2)** implies $\lambda z(y - x) = 5(y - x)$, so $x = y$ or $\lambda = 5/z$, but z can't
be 0, so $x = y$. Then twice **(2)** minus five times **(3)** together with $x = y$ implies $\lambda y(2x - 5y) = 2(2z - 5y)$ which gives
$z = \frac{5}{2} y$ [again $\lambda \neq 2/y$ or else **(3)** implies $y = 0$]. Hence $\frac{5}{2} y^3 = V$ and the dimensions which minimize cost are
$x = y = \sqrt[3]{\frac{2}{5} V}$ units, $z = V^{1/3} \left(\frac{5}{2}\right)^{2/3}$ units.

39. If the dimensions of the box are given by x, y, and z, then we need to find the maximum value of $f(x, y, z) = xyz$
$[x, y, z > 0]$ subject to the constraint $L = \sqrt{x^2 + y^2 + z^2}$ or $g(x, y, z) = x^2 + y^2 + z^2 = L^2$. $\nabla f = \lambda \nabla g$ $\Rightarrow$

$\langle yz, xz, xy \rangle = \lambda \langle 2x, 2y, 2z \rangle$, so $yz = 2\lambda x \Rightarrow \lambda = \dfrac{yz}{2x}$, $xz = 2\lambda y \Rightarrow \lambda = \dfrac{xz}{2y}$, and $xy = 2\lambda z \Rightarrow \lambda = \dfrac{xy}{2z}$. Thus

$\lambda = \dfrac{yz}{2x} = \dfrac{xz}{2y} \Rightarrow x^2 = y^2$ [since $z \neq 0$] $\Rightarrow x = y$ and $\lambda = \dfrac{yz}{2x} = \dfrac{xy}{2z} \Rightarrow x = z$ [since $y \neq 0$].

Substituting into the constraint equation gives $x^2 + x^2 + x^2 = L^2 \Rightarrow x^2 = L^2/3 \Rightarrow x = L/\sqrt{3} = y = z$ and the

maximum volume is $\left(L/\sqrt{3}\right)^3 = L^3/\left(3\sqrt{3}\right)$.

40. Let the dimensions of the box be x, y, and z, so its volume is $f(x, y, z) = xyz$, its surface area is $2xy + 2yz + 2xz = 1500$

and its total edge length is $4x + 4y + 4z = 200$. We find the extreme values of $f(x, y, z)$ subject to the constraints

$g(x, y, z) = xy + yz + xz = 750$ and $h(x, y, z) = x + y + z = 50$. Then

$\nabla f = \langle yz, xz, xy \rangle = \lambda \nabla g + \mu \nabla h = \langle \lambda(y + z), \lambda(x + z), \lambda(x + y) \rangle + \langle \mu, \mu, \mu \rangle$. So $yz = \lambda(y + z) + \mu$ **(1)**,

$xz = \lambda(x + z) + \mu$ **(2)**, and $xy = \lambda(x + y) + \mu$ **(3)**. Notice that the box can't be a cube or else $x = y = z = \frac{50}{3}$

but then $xy + yz + xz = \frac{2500}{3} \neq 750$. Assume x is the distinct side, that is, $x \neq y$, $x \neq z$. Then **(1)** minus **(2)** implies

$z(y - x) = \lambda(y - x)$ or $\lambda = z$, and **(1)** minus **(3)** implies $y(z - x) = \lambda(z - x)$ or $\lambda = y$. So $y = z = \lambda$ and $x + y + z = 50$

implies $x = 50 - 2\lambda$; also $xy + yz + xz = 750$ implies $x(2\lambda) + \lambda^2 = 750$. Hence $50 - 2\lambda = \dfrac{750 - \lambda^2}{2\lambda}$ or

$3\lambda^2 - 100\lambda + 750 = 0$ and $\lambda = \dfrac{50 \pm 5\sqrt{10}}{3}$, giving the points $\left(\frac{1}{3}\left(50 \mp 10\sqrt{10}\right), \frac{1}{3}\left(50 \pm 5\sqrt{10}\right), \frac{1}{3}\left(50 \pm 5\sqrt{10}\right)\right)$.

Thus the minimum of f is $f\left(\frac{1}{3}\left(50 - 10\sqrt{3}\right), \frac{1}{3}\left(50 + 5\sqrt{10}\right), \frac{1}{3}\left(50 + 5\sqrt{10}\right)\right) = \frac{1}{27}\left(87{,}500 - 2500\sqrt{10}\right)$, and its

maximum is $f\left(\frac{1}{3}\left(50 + 10\sqrt{10}\right), \frac{1}{3}\left(50 - 5\sqrt{10}\right), \frac{1}{3}\left(50 - 5\sqrt{10}\right)\right) = \frac{1}{27}\left(87{,}500 + 2500\sqrt{10}\right)$.

Note: If either y or z is the distinct side, then symmetry gives the same result.

41. We need to find the extreme values of $f(x, y, z) = x^2 + y^2 + z^2$ subject to the two constraints $g(x, y, z) = x + y + 2z = 2$

and $h(x, y, z) = x^2 + y^2 - z = 0$. $\nabla f = \langle 2x, 2y, 2z \rangle$, $\lambda \nabla g = \langle \lambda, \lambda, 2\lambda \rangle$ and $\mu \nabla h = \langle 2\mu x, 2\mu y, -\mu \rangle$. Thus we need

$2x = \lambda + 2\mu x$ **(1)**, $2y = \lambda + 2\mu y$ **(2)**, $2z = 2\lambda - \mu$ **(3)**, $x + y + 2z = 2$ **(4)**, and $x^2 + y^2 - z = 0$ **(5)**.

From **(1)** and **(2)**, $2(x - y) = 2\mu(x - y)$, so if $x \neq y$, $\mu = 1$. Putting this in **(3)** gives $2z = 2\lambda - 1$ or $\lambda = z + \frac{1}{2}$, but putting

$\mu = 1$ into **(1)** says $\lambda = 0$. Hence $z + \frac{1}{2} = 0$ or $z = -\frac{1}{2}$. Then **(4)** and **(5)** become $x + y - 3 = 0$ and $x^2 + y^2 + \frac{1}{2} = 0$. The

last equation cannot be true, so this case gives no solution. So we must have $x = y$. Then **(4)** and **(5)** become $2x + 2z = 2$ and

$2x^2 - z = 0$ which imply $z = 1 - x$ and $z = 2x^2$. Thus $2x^2 = 1 - x$ or $2x^2 + x - 1 = (2x - 1)(x + 1) = 0$ so $x = \frac{1}{2}$ or

$x = -1$. The two points to check are $\left(\frac{1}{2}, \frac{1}{2}, \frac{1}{2}\right)$ and $(-1, -1, 2)$: $f\left(\frac{1}{2}, \frac{1}{2}, \frac{1}{2}\right) = \frac{3}{4}$ and $f(-1, -1, 2) = 6$. Thus $\left(\frac{1}{2}, \frac{1}{2}, \frac{1}{2}\right)$ is

the point on the ellipse nearest the origin and $(-1, -1, 2)$ is the one farthest from the origin.

42. (a) After plotting $z = \sqrt{x^2 + y^2}$, the top half of the cone, and the plane

$z = (5 - 4x + 3y)/8$ we see the ellipse formed by the intersection of the

surfaces. The ellipse can be plotted explicitly using cylindrical coordinates

(see Section 16.7 [ET 15.7]): The cone is given by $z = r$, and the plane is

$4r \cos\theta - 3r \sin\theta + 8z = 5$. Substituting $z = r$ into the plane equation

gives $4r \cos\theta - 3r \sin\theta + 8r = 5 \Rightarrow r = \dfrac{5}{4\cos\theta - 3\sin\theta + 8}$.

Since $z = r$ on the ellipse, parametric equations (in cylindrical coordinates)

are $\theta = t$, $r = z = \dfrac{5}{4\cos t - 3\sin t + 8}$, $0 \leq t \leq 2\pi$.

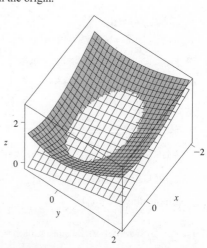

(b) We need to find the extreme values of $f(x, y, z) = z$ subject to the two constraints $g(x, y, z) = 4x - 3y + 8z = 5$ and

$h(x, y, z) = x^2 + y^2 - z^2 = 0$. $\nabla f = \lambda \nabla g + \mu \nabla h$ ⇒ $\langle 0, 0, 1 \rangle = \lambda \langle 4, -3, 8 \rangle + \mu \langle 2x, 2y, -2z \rangle$, so we need

$4\lambda + 2\mu x = 0$ ⇒ $x = -\frac{2\lambda}{\mu}$ **(1)**, $-3\lambda + 2\mu y = 0$ ⇒ $y = \frac{3\lambda}{2\mu}$ **(2)**, $8\lambda - 2\mu z = 1$ ⇒ $z = \frac{8\lambda - 1}{2\mu}$ **(3)**,

$4x - 3y + 8z = 5$ **(4)**, and $x^2 + y^2 = z^2$ **(5)**. [Note that $\mu \neq 0$, else $\lambda = 0$ from **(1)**, but substitution into **(3)** gives a

contradiction.] Substituting **(1)**, **(2)**, and **(3)** into **(4)** gives $4\left(-\frac{2\lambda}{\mu}\right) - 3\left(\frac{3\lambda}{2\mu}\right) + 8\left(\frac{8\lambda - 1}{2\mu}\right) = 5$ ⇒ $\mu = \frac{39\lambda - 8}{10}$ and into

(5) gives $\left(-\frac{2\lambda}{\mu}\right)^2 + \left(\frac{3\lambda}{2\mu}\right)^2 = \left(\frac{8\lambda - 1}{2\mu}\right)^2$ ⇒ $16\lambda^2 + 9\lambda^2 = (8\lambda - 1)^2$ ⇒ $39\lambda^2 - 16\lambda + 1 = 0$ ⇒ $\lambda = \frac{1}{13}$

or $\lambda = \frac{1}{3}$. If $\lambda = \frac{1}{13}$ then $\mu = -\frac{1}{2}$ and $x = \frac{4}{13}$, $y = -\frac{3}{13}$, $z = \frac{5}{13}$. If $\lambda = \frac{1}{3}$ then $\mu = \frac{1}{2}$ and $x = -\frac{4}{3}$, $y = 1$, $z = \frac{5}{3}$.

Thus the highest point on the ellipse is $\left(-\frac{4}{3}, 1, \frac{5}{3}\right)$ and the lowest point is $\left(\frac{4}{13}, -\frac{3}{13}, \frac{5}{13}\right)$.

43. $f(x, y, z) = ye^{x-z}$, $g(x, y, z) = 9x^2 + 4y^2 + 36z^2 = 36$, $h(x, y, z) = xy + yz = 1$. $\nabla f = \lambda \nabla g + \mu \nabla h$ ⇒

$\langle ye^{x-z}, e^{x-z}, -ye^{x-z} \rangle = \lambda \langle 18x, 8y, 72z \rangle + \mu \langle y, x + z, y \rangle$, so $ye^{x-z} = 18\lambda x + \mu y$, $e^{x-z} = 8\lambda y + \mu(x + z)$,

$-ye^{x-z} = 72\lambda z + \mu y$, $9x^2 + 4y^2 + 36z^2 = 36$, $xy + yz = 1$. Using a CAS to solve these 5 equations simultaneously for x,

y, z, λ, and μ (in Maple, use the `allvalues` command), we get 4 real-valued solutions:

$$x \approx 0.222444, \qquad y \approx -2.157012, \qquad z \approx -0.686049, \qquad \lambda \approx -0.200401, \qquad \mu \approx 2.108584$$
$$x \approx -1.951921, \qquad y \approx -0.545867, \qquad z \approx 0.119973, \qquad \lambda \approx 0.003141, \qquad \mu \approx -0.076238$$
$$x \approx 0.155142, \qquad y \approx 0.904622, \qquad z \approx 0.950293, \qquad \lambda \approx -0.012447, \qquad \mu \approx 0.489938$$
$$x \approx 1.138731, \qquad y \approx 1.768057, \qquad z \approx -0.573138, \qquad \lambda \approx 0.317141, \qquad \mu \approx 1.862675$$

Substituting these values into f gives $f(0.222444, -2.157012, -0.686049) \approx -5.3506$,

$f(-1.951921, -0.545867, 0.119973) \approx -0.0688$, $f(0.155142, 0.904622, 0.950293) \approx 0.4084$,

$f(1.138731, 1.768057, -0.573138) \approx 9.7938$. Thus the maximum is approximately 9.7938, and the mininum is

approximately -5.3506.

44. $f(x, y, z) = x + y + z$, $g(x, y, z) = x^2 - y^2 - z = 0$, $h(x, y, z) = x^2 + z^2 = 4$.

$\nabla f = \lambda \nabla g + \mu \nabla h$ ⇒ $\langle 1, 1, 1 \rangle = \lambda \langle 2x, -2y, -1 \rangle + \mu \langle 2x, 0, 2z \rangle$, so $1 = 2\lambda x + 2\mu x$, $1 = -2\lambda y$, $1 = -\lambda + 2\mu z$,

$x^2 - y^2 = z$, $x^2 + z^2 = 4$. Using a CAS to solve these 5 equations simultaneously for x, y, z, λ, and μ, we get 4 real-valued

solutions:

$$x \approx -1.652878, \qquad y \approx -1.964194, \qquad z \approx -1.126052, \qquad \lambda \approx 0.254557, \qquad \mu \approx -0.557060$$
$$x \approx -1.502800, \qquad y \approx 0.968872, \qquad z \approx 1.319694, \qquad \lambda \approx -0.516064, \qquad \mu \approx 0.183352$$
$$x \approx -0.992513, \qquad y \approx 1.649677, \qquad z \approx -1.736352, \qquad \lambda \approx -0.303090, \qquad \mu \approx -0.200682$$
$$x \approx 1.895178, \qquad y \approx 1.718347, \qquad z \approx 0.638984, \qquad \lambda \approx -0.290977, \qquad \mu \approx 0.554805$$

Substituting these values into f gives $f(-1.652878, -1.964194, -1.126052) \approx -4.7431$,

$f(-1.502800, 0.968872, 1.319694) \approx 0.7858$, $f(-0.992513, 1.649677, -1.736352) \approx -1.0792$,

$f(1.895178, 1.718347, 0.638984) \approx 4.2525$. Thus the maximum is approximately 4.2525, and the mininum is

approximately -4.7431.

45. (a) We wish to maximize $f(x_1, x_2, \ldots, x_n) = \sqrt[n]{x_1 x_2 \cdots x_n}$ subject to $g(x_1, x_2, \ldots, x_n) = x_1 + x_2 + \cdots + x_n = c$ and $x_i > 0$.

$$\nabla f = \left\langle \tfrac{1}{n}(x_1 x_2 \cdots x_n)^{\frac{1}{n}-1}(x_2 \cdots x_n), \ \tfrac{1}{n}(x_1 x_2 \cdots x_n)^{\frac{1}{n}-1}(x_1 x_3 \cdots x_n), \ \ldots, \ \tfrac{1}{n}(x_1 x_2 \cdots x_n)^{\frac{1}{n}-1}(x_1 \cdots x_{n-1}) \right\rangle$$

and $\lambda \nabla g = \langle \lambda, \lambda, \ldots, \lambda \rangle$, so we need to solve the system of equations

$$\tfrac{1}{n}(x_1 x_2 \cdots x_n)^{\frac{1}{n}-1}(x_2 \cdots x_n) = \lambda \quad \Rightarrow \quad x_1^{1/n} x_2^{1/n} \cdots x_n^{1/n} = n\lambda x_1$$

$$\tfrac{1}{n}(x_1 x_2 \cdots x_n)^{\frac{1}{n}-1}(x_1 x_3 \cdots x_n) = \lambda \quad \Rightarrow \quad x_1^{1/n} x_2^{1/n} \cdots x_n^{1/n} = n\lambda x_2$$

$$\vdots$$

$$\tfrac{1}{n}(x_1 x_2 \cdots x_n)^{\frac{1}{n}-1}(x_1 \cdots x_{n-1}) = \lambda \quad \Rightarrow \quad x_1^{1/n} x_2^{1/n} \cdots x_n^{1/n} = n\lambda x_n$$

This implies $n\lambda x_1 = n\lambda x_2 = \cdots = n\lambda x_n$. Note $\lambda \neq 0$, otherwise we can't have all $x_i > 0$. Thus $x_1 = x_2 = \cdots = x_n$. But $x_1 + x_2 + \cdots + x_n = c \ \Rightarrow \ nx_1 = c \ \Rightarrow \ x_1 = \dfrac{c}{n} = x_2 = x_3 = \cdots = x_n$. Then the only point where f can have an extreme value is $\left(\dfrac{c}{n}, \dfrac{c}{n}, \ldots, \dfrac{c}{n}\right)$. Since we can choose values for $(x_1, x_2, \ldots, x_n)$ that make f as close to zero (but not equal) as we like, f has no minimum value. Thus the maximum value is

$$f\left(\frac{c}{n}, \frac{c}{n}, \ldots, \frac{c}{n}\right) = \sqrt[n]{\frac{c}{n} \cdot \frac{c}{n} \cdots \frac{c}{n}} = \frac{c}{n}.$$

(b) From part (a), $\dfrac{c}{n}$ is the maximum value of f. Thus $f(x_1, x_2, \ldots, x_n) = \sqrt[n]{x_1 x_2 \cdots x_n} \le \dfrac{c}{n}$. But $x_1 + x_2 + \cdots + x_n = c$, so $\sqrt[n]{x_1 x_2 \cdots x_n} \le \dfrac{x_1 + x_2 + \cdots + x_n}{n}$. These two means are equal when f attains its maximum value $\dfrac{c}{n}$, but this can occur only at the point $\left(\dfrac{c}{n}, \dfrac{c}{n}, \ldots, \dfrac{c}{n}\right)$ we found in part (a). So the means are equal only when $x_1 = x_2 = x_3 = \cdots = x_n = \dfrac{c}{n}$.

46. (a) Let $f(x_1, \ldots, x_n, y_1, \ldots, y_n) = \sum_{i=1}^{n} x_i y_i$, $g(x_1, \ldots, x_n) = \sum_{i=1}^{n} x_i^2$, and $h(x_1, \ldots, x_n) = \sum_{i=1}^{n} y_i^2$. Then

$$\nabla f = \nabla \sum_{i=1}^{n} x_i y_i = \langle y_1, y_2, \ldots, y_n, x_1, x_2, \ldots, x_n \rangle, \ \nabla g = \nabla \sum_{i=1}^{n} x_i^2 = \langle 2x_1, 2x_2, \ldots, 2x_n, 0, 0, \ldots, 0 \rangle \text{ and}$$

$$\nabla h = \nabla \sum_{i=1}^{n} y_i^2 = \langle 0, 0, \ldots, 0, 2y_1, 2y_2, \ldots, 2y_n \rangle. \text{ So } \nabla f = \lambda \nabla g + \mu \nabla h \ \Leftrightarrow \ y_i = 2\lambda x_i \text{ and } x_i = 2\mu y_i,$$

$1 \le i \le n$. Then $1 = \sum_{i=1}^{n} y_i^2 = \sum_{i=1}^{n} 4\lambda^2 x_i^2 = 4\lambda^2 \sum_{i=1}^{n} x_i^2 = 4\lambda^2 \ \Rightarrow \ \lambda = \pm\tfrac{1}{2}$. If $\lambda = \tfrac{1}{2}$ then $y_i = 2\left(\tfrac{1}{2}\right)x_i = x_i$,

$1 \le i \le n$. Thus $\sum_{i=1}^{n} x_i y_i = \sum_{i=1}^{n} x_i^2 = 1$. Similarly if $\lambda = -\tfrac{1}{2}$ we get $y_i = -x_i$ and $\sum_{i=1}^{n} x_i y_i = -1$. Similarly we get

$\mu = \pm\tfrac{1}{2}$ giving $y_i = \pm x_i$, $1 \le i \le n$, and $\sum_{i=1}^{n} x_i y_i = \pm 1$. Thus the maximum value of $\sum_{i=1}^{n} x_i y_i$ is 1.

(b) Here we assume $\sum_{i=1}^{n} a_i^2 \neq 0$ and $\sum_{i=1}^{n} b_i^2 \neq 0$. (If $\sum_{i=1}^{n} a_i^2 = 0$, then each $a_i = 0$ and so the inequality is trivially true.)

$$x_i = \frac{a_i}{\sqrt{\sum a_j^2}} \ \Rightarrow \ \sum x_i^2 = \frac{\sum a_i^2}{\sum a_j^2} = 1, \text{ and } y_i = \frac{b_i}{\sqrt{\sum b_j^2}} \ \Rightarrow \ \sum y_i^2 = \frac{\sum b_i^2}{\sum b_j^2} = 1. \text{ Therefore, from part (a),}$$

$$\sum x_i y_i = \sum \frac{a_i b_i}{\sqrt{\sum a_j^2} \sqrt{\sum b_j^2}} \le 1 \ \Leftrightarrow \ \sum a_i b_i \le \sqrt{\sum a_j^2} \sqrt{\sum b_j^2}.$$

APPLIED PROJECT Rocket Science

1. Initially the rocket engine has mass $M_r = M_1$ and payload mass $P = M_2 + M_3 + A$. Then the change in velocity resulting from the first stage is $\Delta V_1 = -c \ln\left(1 - \dfrac{(1-S)M_1}{M_2 + M_3 + A + M_1}\right)$. After the first stage is jettisoned we can consider the rocket engine to have mass $M_r = M_2$ and the payload to have mass $P = M_3 + A$. The resulting change in velocity from the second stage is $\Delta V_2 = -c \ln\left(1 - \dfrac{(1-S)M_2}{M_3 + A + M_2}\right)$. When only the third stage remains, we have $M_r = M_3$ and $P = A$, so the resulting change in velocity is $\Delta V_3 = -c \ln\left(1 - \dfrac{(1-S)M_3}{A + M_3}\right)$. Since the rocket started from rest, the final velocity attained is

$$
\begin{aligned}
v_f &= \Delta V_1 + \Delta V_2 + \Delta V_3 \\
&= -c\ln\left(1 - \frac{(1-S)M_1}{M_2+M_3+A+M_1}\right) + (-c)\ln\left(1 - \frac{(1-S)M_2}{M_3+A+M_2}\right) + (-c)\ln\left(1 - \frac{(1-S)M_3}{A+M_3}\right) \\
&= -c\left[\ln\left(\frac{M_1+M_2+M_3+A-(1-S)M_1}{M_1+M_2+M_3+A}\right) + \ln\left(\frac{M_2+M_3+A-(1-S)M_2}{M_2+M_3+A}\right) \right.\\
&\qquad\qquad \left. + \ln\left(\frac{M_3+A-(1-S)M_3}{M_3+A}\right)\right] \\
&= c\left[\ln\left(\frac{M_1+M_2+M_3+A}{SM_1+M_2+M_3+A}\right) + \ln\left(\frac{M_2+M_3+A}{SM_2+M_3+A}\right) + \ln\left(\frac{M_3+A}{SM_3+A}\right)\right]
\end{aligned}
$$

2. Define $N_1 = \dfrac{M_1+M_2+M_3+A}{SM_1+M_2+M_3+A}$, $N_2 = \dfrac{M_2+M_3+A}{SM_2+M_3+A}$, and $N_3 = \dfrac{M_3+A}{SM_3+A}$. Then

$$
\begin{aligned}
\frac{(1-S)N_1}{1-SN_1} &= \frac{(1-S)\dfrac{M_1+M_2+M_3+A}{SM_1+M_2+M_3+A}}{1-S\dfrac{M_1+M_2+M_3+A}{SM_1+M_2+M_3+A}} = \frac{(1-S)(M_1+M_2+M_3+A)}{SM_1+M_2+M_3+A-S(M_1+M_2+M_3+A)} \\
&= \frac{(1-S)(M_1+M_2+M_3+A)}{(1-S)(M_2+M_3+A)} = \frac{M_1+M_2+M_3+A}{M_2+M_3+A}
\end{aligned}
$$

as desired.

Similarly,

$$
\frac{(1-S)N_2}{1-SN_2} = \frac{(1-S)(M_2+M_3+A)}{SM_2+M_3+A-S(M_2+M_3+A)} = \frac{(1-S)(M_2+M_3+A)}{(1-S)(M_3+A)} = \frac{M_2+M_3+A}{M_3+A}
$$

and

$$
\frac{(1-S)N_3}{1-SN_3} = \frac{(1-S)(M_3+A)}{SM_3+A-S(M_3+A)} = \frac{(1-S)(M_3+A)}{(1-S)(A)} = \frac{M_3+A}{A}
$$

Then

$$
\begin{aligned}
\frac{M+A}{A} &= \frac{M_1+M_2+M_3+A}{A} = \frac{M_1+M_2+M_3+A}{M_2+M_3+A} \cdot \frac{M_2+M_3+A}{M_3+A} \cdot \frac{M_3+A}{A} \\
&= \frac{(1-S)N_1}{1-SN_1} \cdot \frac{(1-S)N_2}{1-SN_2} \cdot \frac{(1-S)N_3}{1-SN_3} = \frac{(1-S)^3 N_1 N_2 N_3}{(1-SN_1)(1-SN_2)(1-SN_3)}
\end{aligned}
$$

3. Since $A > 0$, $M + A$ and consequently $\dfrac{M + A}{A}$ is minimized for the same values as M. $\ln x$ is a strictly increasing function,

so $\ln\left(\dfrac{M + A}{A}\right)$ must give a minimum for the same values as $\dfrac{M + A}{A}$ and hence M. We then wish to minimize

$\ln\left(\dfrac{M + A}{A}\right)$ subject to the constraint $c\left(\ln N_1 + \ln N_2 + \ln N_3\right) = v_f$. From Problem 2,

$$
\begin{aligned}
\ln\left(\frac{M + A}{A}\right) &= \ln\left(\frac{(1 - S)^3\, N_1 N_2 N_3}{(1 - SN_1)\,(1 - SN_2)\,(1 - SN_3)}\right) \\
&= 3\ln(1 - S) + \ln N_1 + \ln N_2 + \ln N_3 - \ln(1 - SN_1) - \ln(1 - SN_2) - \ln(1 - SN_3)
\end{aligned}
$$

Using the method of Lagrange multipliers, we need to solve $\nabla\left[\ln\left(\dfrac{M + A}{A}\right)\right] = \lambda \nabla[c(\ln N_1 + \ln N_2 + \ln N_3)]$ with

$c(\ln N_1 + \ln N_2 + \ln N_3) = v_f$ in terms of N_1, N_2, and N_3. The resulting system is

$$
\frac{1}{N_1} + \frac{S}{1 - SN_1} = \lambda\,\frac{c}{N_1} \qquad \frac{1}{N_2} + \frac{S}{1 - SN_2} = \lambda\,\frac{c}{N_2} \qquad \frac{1}{N_3} + \frac{S}{1 - SN_3} = \lambda\,\frac{c}{N_3}
$$

$$
c\left(\ln N_1 + \ln N_2 + \ln N_3\right) = v_f
$$

One approach to solving the system is isolating $c\lambda$ in the first three equations which gives

$$
1 + \frac{SN_1}{1 - SN_1} = c\lambda = 1 + \frac{SN_2}{1 - SN_2} = 1 + \frac{SN_3}{1 - SN_3} \quad \Rightarrow \quad \frac{N_1}{1 - SN_1} = \frac{N_2}{1 - SN_2} = \frac{N_3}{1 - SN_3} \quad \Rightarrow
$$

$N_1 = N_2 = N_3$ (Verify!). This says the fourth equation can be expressed as $c(\ln N_1 + \ln N_1 + \ln N_1) = v_f \;\Rightarrow$

$3c \ln N_1 = v_f \;\Rightarrow\; \ln N_1 = \dfrac{v_f}{3c}$. Thus the minimum mass M of the rocket engine is attained for

$N_1 = N_2 = N_3 = e^{v_f/(3c)}$.

4. Using the previous results, $\dfrac{M + A}{A} = \dfrac{(1 - S)^3\, N_1 N_2 N_3}{(1 - SN_1)(1 - SN_2)(1 - SN_3)} = \dfrac{(1 - S)^3 \left[e^{v_f/(3c)}\right]^3}{\left[1 - Se^{v_f/(3c)}\right]^3} = \dfrac{(1 - S)^3 e^{v_f/c}}{\left[1 - Se^{v_f/(3c)}\right]^3}.$

Then $M = \dfrac{A(1 - S)^3 e^{v_f/c}}{\left[1 - Se^{v_f/(3c)}\right]^3} - A.$

5. (a) From Problem 4, $M = \dfrac{A(1 - 0.2)^3 e^{(17{,}500/6000)}}{\left(1 - 0.2 e^{[17{,}500/(3 \cdot 6000)]}\right)^3} - A \approx 90.4A - A = 89.4A.$

(b) First, $N_3 = \dfrac{M_3 + A}{SM_3 + A} \;\Rightarrow\; e^{[17{,}500/(3 \cdot 6000)]} = \dfrac{M_3 + A}{0.2 M_3 + A} \;\Rightarrow\; M_3 = \dfrac{A(1 - e^{35/36})}{0.2 e^{35/36} - 1} \approx 3.49A.$

Then $N_2 = \dfrac{M_2 + M_3 + A}{SM_2 + M_3 + A} = \dfrac{M_2 + 3.49A + A}{0.2 M_2 + 3.49A + A} \;\Rightarrow\; M_2 = \dfrac{4.49A(1 - e^{35/36})}{0.2 e^{35/36} - 1} \approx 15.67A$ and

$N_3 = \dfrac{M_1 + M_2 + M_3 + A}{SM_1 + M_2 + M_3 + A} = \dfrac{M_1 + 15.67A + 3.49A + A}{0.2 M_1 + 15.67A + 3.49A + A} \;\Rightarrow\; M_1 = \dfrac{20.16A(1 - e^{35/36})}{0.2 e^{35/36} - 1} \approx 70.36A.$

6. As in Problem 5, $N_3 = \dfrac{M_3 + A}{SM_3 + A} \;\Rightarrow\; e^{24{,}700/(3 \cdot 6000)} = \dfrac{M_3 + A}{0.2 M_3 + A} \;\Rightarrow\; M_3 = \dfrac{A(1 - e^{247/180})}{0.2 e^{247/180} - 1} \approx 13.9A,$

$N_2 = \dfrac{M_2 + M_3 + A}{SM_2 + M_3 + A} = \dfrac{M_2 + 13.9A + A}{0.2 M_2 + 13.9A + A} \;\Rightarrow\; M_2 = \dfrac{14.9A(1 - e^{247/180})}{0.2 e^{247/180} - 1} \approx 208A,$ and

$$N_3 = \frac{M_1 + M_2 + M_3 + A}{SM_1 + M_2 + M_3 + A} = \frac{M_1 + 208A + 13.9A + A}{0.2M_1 + 208A + 13.9A + A} \quad \Rightarrow \quad M_1 = \frac{222.9A(1 - e^{247/180})}{0.2e^{247/180} - 1} \approx 3110A.$$

Here $A = 500$, so the mass of each stage of the rocket engine is approximately $M_1 = 3110(500) = 1{,}550{,}000$ lb,

$M_2 = 208(500) = 104{,}000$ lb, and $M_3 = 13.9(500) = 6950$ lb.

APPLIED PROJECT   Hydro-Turbine Optimization

1. We wish to maximize the total energy production for a given total flow, so we can say Q_T is fixed and we want to maximize $KW_1 + KW_2 + KW_3$. Notice each KW_i has a constant factor $\left(170 - 1.6 \cdot 10^{-6}Q_T^2\right)$, so to simplify the computations we can equivalently maximize

$$
\begin{aligned}
f(Q_1, Q_2, Q_3) &= \frac{KW_1 + KW_2 + KW_3}{170 - 1.6 \cdot 10^{-6}Q_T^2} \\
&= \left(-18.89 + 0.1277Q_1 - 4.08 \cdot 10^{-5}Q_1^2\right) + \left(-24.51 + 0.1358Q_2 - 4.69 \cdot 10^{-5}Q_2^2\right) \\
&\qquad\qquad + \left(-27.02 + 0.1380Q_3 - 3.84 \cdot 10^{-5}Q_3^2\right)
\end{aligned}
$$

subject to the constraint $g(Q_1, Q_2, Q_3) = Q_1 + Q_2 + Q_3 = Q_T$. So first we find the values of Q_1, Q_2, Q_3 where $\nabla f(Q_1, Q_2, Q_3) = \lambda \nabla g(Q_1, Q_2, Q_3)$ and $Q_1 + Q_2 + Q_3 = Q_T$ which is equivalent to solving the system

$$0.1277 - 2(4.08 \cdot 10^{-5})Q_1 = \lambda$$

$$0.1358 - 2(4.69 \cdot 10^{-5})Q_2 = \lambda$$

$$0.1380 - 2(3.84 \cdot 10^{-5})Q_3 = \lambda$$

$$Q_1 + Q_2 + Q_3 = Q_T$$

Comparing the first and third equations, we have $0.1277 - 2(4.08 \cdot 10^{-5})Q_1 = 0.1380 - 2(3.84 \cdot 10^{-5})Q_3 \Rightarrow Q_1 = -126.2255 + 0.9412Q_3$. From the second and third equations,

$0.1358 - 2(4.69 \cdot 10^{-5})Q_2 = 0.1380 - 2(3.84 \cdot 10^{-5})Q_3 \Rightarrow Q_2 = -23.4542 + 0.8188Q_3$. Substituting

into $Q_1 + Q_2 + Q_3 = Q_T$ gives $(-126.2255 + 0.9412Q_3) + (-23.4542 + 0.8188Q_3) + Q_3 = Q_T \Rightarrow$

$2.76Q_3 = Q_T + 149.6797 \Rightarrow Q_3 = 0.3623Q_T + 54.23$. Then

$Q_1 = -126.2255 + 0.9412Q_3 = -126.2255 + 0.9412(0.3623Q_T + 54.23) = 0.3410Q_T - 75.18$ and

$Q_2 = -23.4542 + 0.8188(0.3623Q_T + 54.23) = 0.2967Q_T + 20.95$. As long as we maintain $250 \le Q_1 \le 1110$,

$250 \le Q_2 \le 1110$, and $250 \le Q_3 \le 1225$, we can reason from the nature of the functions KW_i that these values give a maximum of f, and hence a maximum energy production, and not a minimum.

2. From Problem 1, the value of Q_1 that maximizes energy production is $0.3410Q_T - 75.18$, but since $250 \le Q_1 \le 1110$, we must have $250 \le 0.3410Q_T - 75.18 \le 1110 \Rightarrow 325.18 \le 0.3410Q_T \le 1185.18 \Rightarrow 953.6 \le Q_T \le 3475.6$.

Similarly, $250 \le Q_2 \le 1110 \Rightarrow 250 \le 0.2967Q_T + 20.95 \le 1110 \Rightarrow 772.0 \le Q_T \le 3670.5$, and

$250 \le Q_3 \le 1225 \Rightarrow 250 \le 0.3623Q_T + 54.23 \le 1225 \Rightarrow 540.4 \le Q_T \le 3231.5$. Consolidating these results, we see that the values from Problem 1 are applicable only for $953.6 \le Q_T \le 3231.5$.

3. If $Q_T = 2500$, the results from Problem 1 show that the maximum energy production occurs for

$$Q_1 = 0.3410Q_T - 75.18 = 0.3410(2500) - 75.18 = 777.3$$

$$Q_2 = 0.2967Q_T + 20.95 = 0.2967(2500) + 20.95 = 762.7$$

$$Q_3 = 0.3623Q_T + 54.23 = 0.3623(2500) + 54.23 = 960.0$$

The energy produced for these values is $KW_1 + KW_2 + KW_3 \approx 8915.2 + 8285.1 + 11{,}211.3 \approx 28{,}411.6$.

We compute the energy production for a nearby distribution, $Q_1 = 770$, $Q_2 = 760$, and $Q_3 = 970$:

$KW_1 + KW_2 + KW_3 \approx 8839.8 + 8257.4 + 11{,}313.5 = 28{,}410.7$. For another example, we take $Q_1 = 780$, $Q_2 = 765$,

and $Q_3 = 955$: $KW_1 + KW_2 + KW_3 \approx 8942.9 + 8308.8 + 11{,}159.7 = 28{,}411.4$. These distributions are both close to the

distribution from Problem 1 and both give slightly lower energy productions, suggesting that $Q_1 = 777.3$, $Q_2 = 762.7$, and

$Q_3 = 960.0$ is indeed the optimal distribution.

4. First we graph each power function in its domain if all of the
flow is directed to that turbine (so $Q_i = Q_T$). If we use only one
turbine, the graph indicates that for a water flow of 1000 ft^3/s,
Turbine 3 produces the most power, approximately 12,200 kW.
In comparison, if we use all three turbines, the results of
Problem 1 with $Q_T = 1000$ give $Q_1 = 265.8$, $Q_2 = 317.7$,
and $Q_3 = 416.5$, resulting in a total energy production of

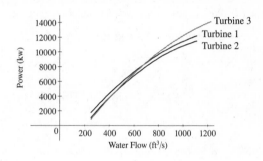

$KW_1 + KW_2 + KW_3 \approx 8397.4$ kW. Here, using only one turbine produces significantly more energy! If the flow is only

600 ft^3/s, we do not have the option of using all three turbines, as the domain restrictions require a minimum of 250 ft^3/s

in each turbine. We can use just one turbine, then, and from the graph Turbine 1 produces the most energy for a water flow

of 600 ft^3.

5. If we examine the graph from Problem 4, we see that for water flows above approximately 450 ft^3/s, Turbine 2 produces the

least amount of power. Therefore it seems reasonable to assume that we should distribute the incoming flow of 1500 ft^3/s

between Turbines 1 and 3. (This can be verified by computing the power produced with the other pairs of turbines for

comparison.) So now we wish to maximize $KW_1 + KW_3$ subject to the constraint $Q_1 + Q_3 = Q_T$ where $Q_T = 1500$.

As in Problem 1, we can equivalently maximize

$$f(Q_1, Q_3) = \frac{KW_1 + KW_3}{170 - 1.6 \cdot 10^{-6}Q_T^2}$$

$$= \left(-18.89 + 0.1277Q_1 - 4.08 \cdot 10^{-5}Q_1^2\right) + \left(-27.02 + 0.1380Q_3 - 3.84 \cdot 10^{-5}Q_3^2\right)$$

subject to the constraint $g(Q_1, Q_3) = Q_1 + Q_3 = Q_T$.

Then we solve $\nabla f(Q_1, Q_3) = \lambda \nabla g(Q_1, Q_3) \Rightarrow 0.1277 - 2\left(4.08 \cdot 10^{-5}\right)Q_1 = \lambda$ and

$0.1380 - 2(3.84 \cdot 10^{-5})Q_3 = \lambda$, thus $0.1277 - 2(4.08 \cdot 10^{-5})Q_1 = 0.1380 - 2(3.84 \cdot 10^{-5})Q_3 \Rightarrow$

$Q_1 = -126.2255 + 0.9412Q_3$. Substituting into $Q_1 + Q_3 = Q_T$ gives $-126.2255 + 0.9412Q_3 + Q_3 = 1500 \Rightarrow$

$Q_3 \approx 837.7$, and then $Q_1 = Q_T - Q_3 \approx 1500 - 837.7 = 662.3$. So we should apportion approximately 662.3 ft^3/s to

Turbine 1 and the remaining 837.7 ft³/s to Turbine 3. The resulting energy production is

$KW_1 + KW_3 \approx 7952.1 + 10,256.2 = 18,208.3$ kW. (We can verify that this is indeed a maximum energy production by checking nearby distributions.) In comparison, if we use all three turbines with $Q_T = 1500$ we get $Q_1 = 436.3$, $Q_2 = 466.0$, and $Q_3 = 597.7$, resulting in a total energy production of $KW_1 + KW_2 + KW_3 \approx 16,538.7$ kW. Clearly, for this flow level it is beneficial to use only two turbines.

6. Note that an incoming flow of 3400 ft³/s is not within the domain we established in Problem 2, so we cannot simply use our previous work to give the optimal distribution. We will need to use all three turbines, due to the capacity limitations of each individual turbine, but 3400 is less than the maximum combined capacity of 3445 ft³/s, so we still must decide how to distribute the flows. From the graph in Problem 4, Turbine 3 produces the most power for the higher flows, so it seems reasonable to use Turbine 3 at its maximum capacity of 1225 and distribute the remaining 2175 ft³/s flow between Turbines 1 and 2. We can again use the technique of Lagrange multipliers to determine the optimal distribution. Following the procedure we used in Problem 5, we wish to maximize $KW_1 + KW_2$ subject to the constraint $Q_1 + Q_2 = Q_T$ where $Q_T = 2175$. We can equivalently maximize

$$f(Q_1, Q_2) = \frac{KW_1 + KW_2}{170 - 1.6 \cdot 10^{-6} Q_T^2}$$

$$= \left(-18.89 + 0.1277 Q_1 - 4.08 \cdot 10^{-5} Q_1^2\right) + \left(-24.51 + 0.1358 Q_2 - 4.69 \cdot 10^{-5} Q_2^2\right)$$

subject to the constraint $g(Q_1, Q_2) = Q_1 + Q_2 = Q_T$. Then we solve $\nabla f(Q_1, Q_2) = \lambda \nabla g(Q_1, Q_2)$ ⇒

$0.1277 - 2(4.08 \cdot 10^{-5})Q_1 = \lambda$ and $0.1358 - 2(4.69 \cdot 10^{-5})Q_2 = \lambda$, thus

$0.1277 - 2(4.08 \cdot 10^{-5})Q_1 = 0.1358 - 2(4.69 \cdot 10^{-5})Q_2$ ⇒ $Q_1 = -99.2647 + 1.1495 Q_2$. Substituting

into $Q_1 + Q_2 = Q_T$ gives $-99.2647 + 1.1495 Q_2 + Q_2 = 2175$ ⇒ $Q_2 \approx 1058.0$, and then $Q_1 \approx 1117.0$. This value for Q_1 is larger than the allowable maximum flow to Turbine 1, but the result indicates that the flow to Turbine 1 should be maximized. Thus we should recommend that the company apportion the maximum allowable flows to Turbines 1 and 3, 1110 and 1225 ft³/s, and the remaining 1065 ft³/s to Turbine 2. Checking nearby distributions within the domain verifies that we have indeed found the optimal distribution.

15 Review

ET 14

CONCEPT CHECK

1. (a) A function f of two variables is a rule that assigns to each ordered pair (x, y) of real numbers in its domain a unique real number denoted by $f(x, y)$.

   (b) One way to visualize a function of two variables is by graphing it, resulting in the surface $z = f(x, y)$. Another method for visualizing a function of two variables is a contour map. The contour map consists of level curves of the function which are horizontal traces of the graph of the function projected onto the xy-plane. Also, we can use an arrow diagram such as Figure 1 in Section 15.1 [ET 14.1].

2. A function f of three variables is a rule that assigns to each ordered triple (x, y, z) in its domain a unique real number $f(x, y, z)$. We can visualize a function of three variables by examining its level surfaces $f(x, y, z) = k$, where k is a constant.

3. $\lim\limits_{(x,y)\to(a,b)} f(x, y) = L$ means the values of $f(x, y)$ approach the number L as the point (x, y) approaches the point (a, b) along any path that is within the domain of f. We can show that a limit at a point does not exist by finding two different paths approaching the point along which $f(x, y)$ has different limits.

4. (a) See Definition 15.2.4 [ET 14.2.4].

(b) If f is continuous on $\mathbb{R}^2$, its graph will appear as a surface without holes or breaks.

5. (a) See (2) and (3) in Section 15.3 [ET 14.3].

(b) See "Interpretations of Partial Derivatives" on page 917 [ET 881].

(c) To find f_x, regard y as a constant and differentiate $f(x, y)$ with respect to x. To find f_y, regard x as a constant and differentiate $f(x, y)$ with respect to y.

6. See the statement of Clairaut's Theorem on page 921 [ET 885].

7. (a) See (2) in Section 15.4 [ET 14.4]

(b) See (19) and the preceding discussion in Section 15.6 [ET 14.6].

8. See (3) and (4) and the accompanying discussion in Section 15.4 [ET 14.4]. We can interpret the linearization of f at (a, b) geometrically as the linear function whose graph is the tangent plane to the graph of f at (a, b). Thus it is the linear function which best approximates f near (a, b).

9. (a) See Definition 15.4.7 [ET 14.4.7].

(b) Use Theorem 15.4.8 [ET 14.4.8].

10. See (10) and the associated discussion in Section 15.4 [ET 14.4].

11. See (2) and (3) in Section 15.5 [ET 14.5].

12. See (7) and the preceding discussion in Section 15.5 [ET 14.5].

13. (a) See Definition 15.6.2 [ET 14.6.2]. We can interpret it as the rate of change of f at (x_0, y_0) in the direction of **u**. Geometrically, if P is the point $(x_0, y_0, f(x_0, y_0))$ on the graph of f and C is the curve of intersection of the graph of f with the vertical plane that passes through P in the direction **u**, the directional derivative of f at (x_0, y_0) in the direction of **u** is the slope of the tangent line to C at P. (See Figure 5 in Section 15.6 [ET 14.6].)

(b) See Theorem 15.6.3 [ET 14.6.3].

14. (a) See (8) and (13) in Section 15.6 [ET 14.6].

(b) $D_{\mathbf{u}} f(x, y) = \nabla f(x, y) \cdot \mathbf{u}$ or $D_{\mathbf{u}} f(x, y, z) = \nabla f(x, y, z) \cdot \mathbf{u}$

(c) The gradient vector of a function points in the direction of maximum rate of increase of the function. On a graph of the function, the gradient points in the direction of steepest ascent.

15. (a) f has a local maximum at (a, b) if $f(x, y) \le f(a, b)$ when (x, y) is near (a, b).

(b) f has an absolute maximum at (a, b) if $f(x, y) \le f(a, b)$ for all points (x, y) in the domain of f.

(c) f has a local minimum at (a, b) if $f(x, y) \geq f(a, b)$ when (x, y) is near (a, b).

(d) f has an absolute minimum at (a, b) if $f(x, y) \geq f(a, b)$ for all points (x, y) in the domain of f.

(e) f has a saddle point at (a, b) if $f(a, b)$ is a local maximum in one direction but a local minimum in another.

16. (a) By Theorem 15.7.2 [ET 14.7.2], if f has a local maximum at (a, b) and the first-order partial derivatives of f exist there, then $f_x(a, b) = 0$ and $f_y(a, b) = 0$.

(b) A critical point of f is a point (a, b) such that $f_x(a, b) = 0$ and $f_y(a, b) = 0$ or one of these partial derivatives does not exist.

17. See (3) in Section 15.7 [ET 14.7]

18. (a) See Figure 11 and the accompanying discussion in Section 15.7 [ET 14.7].

(b) See Theorem 15.7.8 [ET 14.7.8].

(c) See the procedure outlined in (9) in Section 15.7 [ET 14.7].

19. See the discussion beginning on page 970 [ET 934]; see "Two Constraints" on page 974 [ET 938].

TRUE-FALSE QUIZ

1. True. $f_y(a, b) = \lim\limits_{h \to 0} \dfrac{f(a, b+h) - f(a, b)}{h}$ from Equation 15.3.3 [ET 14.3.3]. Let $h = y - b$. As $h \to 0$, $y \to b$. Then by

substituting, we get $f_y(a, b) = \lim\limits_{y \to b} \dfrac{f(a, y) - f(a, b)}{y - b}$.

2. False. If there were such a function, then $f_{xy} = 2y$ and $f_{yx} = 1$. So $f_{xy} \neq f_{yx}$, which contradicts Clairaut's Theorem.

3. False. $f_{xy} = \dfrac{\partial^2 f}{\partial y\, \partial x}$.

4. True. From Equation 15.6.14 [ET 14.6.14] we get $D_{\mathbf{k}} f(x, y, z) = \nabla f(x, y, z) \cdot \langle 0, 0, 1 \rangle = f_z(x, y, z)$.

5. False. See Example 15.2.3 [ET 14.2.3].

6. False. See Exercise 15.4.46(a) [ET 14.4.46(a)].

7. True. If f has a local minimum and f is differentiable at (a, b) then by Theorem 15.7.2 [ET 14.7.2], $f_x(a, b) = 0$ and $f_y(a, b) = 0$, so $\nabla f(a, b) = \langle f_x(a, b), f_y(a, b) \rangle = \langle 0, 0 \rangle = \mathbf{0}$.

8. False. If f is not continuous at $(2, 5)$, then we can have $\lim\limits_{(x,y) \to (2,5)} f(x, y) \neq f(2, 5)$. (See Example 15.2.7 [ET 14.2.7].)

9. False. $\nabla f(x, y) = \langle 0, 1/y \rangle$.

10. True. This is part (c) of the Second Derivatives Test (15.7.3) [ET (14.7.3)].

11. True. $\nabla f = \langle \cos x, \cos y \rangle$, so $|\nabla f| = \sqrt{\cos^2 x + \cos^2 y}$. But $|\cos \theta| \leq 1$, so $|\nabla f| \leq \sqrt{2}$. Now

$D_{\mathbf{u}} f(x, y) = \nabla f \cdot \mathbf{u} = |\nabla f|\, |\mathbf{u}| \cos \theta$, but $\mathbf{u}$ is a unit vector, so $|D_{\mathbf{u}} f(x, y)| \leq \sqrt{2} \cdot 1 \cdot 1 = \sqrt{2}$.

12. False. See Exercise 15.7.37 [ET 14.7.37].

EXERCISES

1. $\ln(x + y + 1)$ is defined only when $x + y + 1 > 0$ $\Rightarrow$ $y > -x - 1$,
so the domain of f is $\{(x, y) \mid y > -x - 1\}$, all those points above the
line $y = -x - 1$.

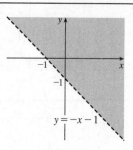

2. $\sqrt{4 - x^2 - y^2}$ is defined only when $4 - x^2 - y^2 \geq 0$ $\Leftrightarrow$ $x^2 + y^2 \leq 4$, and
$\sqrt{1 - x^2}$ is defined only when $1 - x^2 \geq 0$ $\Leftrightarrow$ $-1 \leq x \leq 1$, so the domain of
f is $\{(x, y) \mid -1 \leq x \leq 1, -\sqrt{4 - x^2} \leq y \leq \sqrt{4 - x^2}\,\}$, which consists of those
points on or inside the circle $x^2 + y^2 = 4$ for $-1 \leq x \leq 1$.

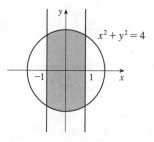

3. $z = f(x, y) = 1 - y^2$, a parabolic cylinder

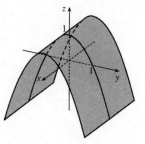

4. $z = f(x, y) = x^2 + (y - 2)^2$, a circular paraboloid with
vertex $(0, 2, 0)$ and axis parallel to the z-axis

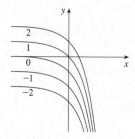

5. The level curves are $\sqrt{4x^2 + y^2} = k$ or $4x^2 + y^2 = k^2$,
$k \geq 0$, a family of ellipses.

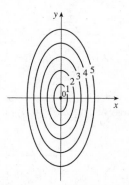

6. The level curves are $e^x + y = k$ or $y = -e^x + k$, a
family of exponential curves.

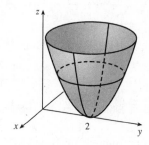

7.

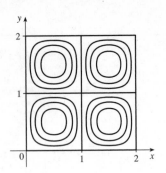

8.

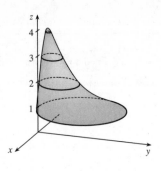

9. f is a rational function, so it is continuous on its domain. Since f is defined at $(1, 1)$, we use direct substitution to evaluate

the limit: $\displaystyle\lim_{(x,y)\to(1,1)} \frac{2xy}{x^2 + 2y^2} = \frac{2(1)(1)}{1^2 + 2(1)^2} = \frac{2}{3}$.

10. As $(x, y) \to (0, 0)$ along the x-axis, $f(x, 0) = 0/x^2 = 0$ for $x \neq 0$, so $f(x, y) \to 0$ along this line. But

$f(x, x) = 2x^2/(3x^2) = \frac{2}{3}$, so as $(x, y) \to (0, 0)$ along the line $x = y$, $f(x, y) \to \frac{2}{3}$. Thus the limit doesn't exist.

11. (a) $T_x(6, 4) = \displaystyle\lim_{h \to 0} \frac{T(6+h, 4) - T(6, 4)}{h}$, so we can approximate $T_x(6, 4)$ by considering $h = \pm 2$ and

using the values given in the table: $T_x(6, 4) \approx \dfrac{T(8, 4) - T(6, 4)}{2} = \dfrac{86 - 80}{2} = 3$,

$T_x(6, 4) \approx \dfrac{T(4, 4) - T(6, 4)}{-2} = \dfrac{72 - 80}{-2} = 4$. Averaging these values, we estimate $T_x(6, 4)$ to be approximately

$3.5°\text{C}/\text{m}$. Similarly, $T_y(6, 4) = \displaystyle\lim_{h \to 0} \frac{T(6, 4+h) - T(6, 4)}{h}$, which we can approximate with $h = \pm 2$:

$T_y(6, 4) \approx \dfrac{T(6, 6) - T(6, 4)}{2} = \dfrac{75 - 80}{2} = -2.5$, $T_y(6, 4) \approx \dfrac{T(6, 2) - T(6, 4)}{-2} = \dfrac{87 - 80}{-2} = -3.5$. Averaging these

values, we estimate $T_y(6, 4)$ to be approximately $-3.0°\text{C}/\text{m}$.

(b) Here $\mathbf{u} = \left\langle \frac{1}{\sqrt{2}}, \frac{1}{\sqrt{2}} \right\rangle$, so by Equation 15.6.9 [ET 14.6.9], $D_{\mathbf{u}} T(6, 4) = \nabla T(6, 4) \cdot \mathbf{u} = T_x(6, 4) \frac{1}{\sqrt{2}} + T_y(6, 4) \frac{1}{\sqrt{2}}$.

Using our estimates from part (a), we have $D_{\mathbf{u}} T(6, 4) \approx (3.5) \frac{1}{\sqrt{2}} + (-3.0) \frac{1}{\sqrt{2}} = \frac{1}{2\sqrt{2}} \approx 0.35$. This means that as we

move through the point $(6, 4)$ in the direction of $\mathbf{u}$, the temperature increases at a rate of approximately $0.35°\text{C}/\text{m}$.

Alternatively, we can use Definition 15.6.2 [ET 14.6.2]: $D_{\mathbf{u}} T(6, 4) = \displaystyle\lim_{h \to 0} \frac{T\left(6 + h \frac{1}{\sqrt{2}}, 4 + h \frac{1}{\sqrt{2}}\right) - T(6, 4)}{h}$,

which we can estimate with $h = \pm 2\sqrt{2}$. Then $D_{\mathbf{u}} T(6, 4) \approx \dfrac{T(8, 6) - T(6, 4)}{2\sqrt{2}} = \dfrac{80 - 80}{2\sqrt{2}} = 0$,

$D_{\mathbf{u}} T(6, 4) \approx \dfrac{T(4, 2) - T(6, 4)}{-2\sqrt{2}} = \dfrac{74 - 80}{-2\sqrt{2}} = \dfrac{3}{\sqrt{2}}$. Averaging these values, we have $D_{\mathbf{u}} T(6, 4) \approx \frac{3}{2\sqrt{2}} \approx 1.1°\text{C}/\text{m}$.

(c) $T_{xy}(x, y) = \dfrac{\partial}{\partial y} [T_x(x, y)] = \displaystyle\lim_{h \to 0} \frac{T_x(x, y+h) - T_x(x, y)}{h}$, so $T_{xy}(6, 4) = \displaystyle\lim_{h \to 0} \frac{T_x(6, 4+h) - T_x(6, 4)}{h}$ which we can

estimate with $h = \pm 2$. We have $T_x(6, 4) \approx 3.5$ from part (a), but we will also need values for $T_x(6, 6)$ and $T_x(6, 2)$. If we

use $h = \pm 2$ and the values given in the table, we have

$T_x(6, 6) \approx \dfrac{T(8, 6) - T(6, 6)}{2} = \dfrac{80 - 75}{2} = 2.5$, $T_x(6, 6) \approx \dfrac{T(4, 6) - T(6, 6)}{-2} = \dfrac{68 - 75}{-2} = 3.5$.

Averaging these values, we estimate $T_x(6,6) \approx 3.0$. Similarly,

$$T_x(6,2) \approx \frac{T(8,2) - T_x(6,2)}{2} = \frac{90 - 87}{2} = 1.5, \; T_x(6,2) \approx \frac{T(4,2) - T(6,2)}{-2} = \frac{74 - 87}{-2} = 6.5.$$

Averaging these values, we estimate $T_x(6,2) \approx 4.0$. Finally, we estimate $T_{xy}(6,4)$:

$$T_{xy}(6,4) \approx \frac{T_x(6,6) - T_x(6,4)}{2} = \frac{3.0 - 3.5}{2} = -0.25, \; T_{xy}(6,4) \approx \frac{T_x(6,2) - T_x(6,4)}{-2} = \frac{4.0 - 3.5}{-2} = -0.25.$$

Averaging these values, we have $T_{xy}(6,4) \approx -0.25$.

12. From the table, $T(6,4) = 80$, and from Exercise 11 we estimated $T_x(6,4) \approx 3.5$ and $T_y(6,4) \approx -3.0$. The linear approximation then is

$$T(x,y) \approx T(6,4) + T_x(6,4)(x-6) + T_y(6,4)(y-4) \approx 80 + 3.5(x-6) - 3(y-4) = 3.5x - 3y + 71$$

Thus at the point $(5, 3.8)$, we can use the linear approximation to estimate $T(5, 3.8) \approx 3.5(5) - 3(3.8) + 71 \approx 77.1°\text{C}$.

13. $f(x,y) = \sqrt{2x + y^2} \; \Rightarrow \; f_x = \frac{1}{2}(2x + y^2)^{-1/2}(2) = \frac{1}{\sqrt{2x + y^2}}, \; f_y = \frac{1}{2}(2x + y^2)^{-1/2}(2y) = \frac{y}{\sqrt{2x + y^2}}$

14. $u = e^{-r}\sin 2\theta \; \Rightarrow \; u_r = -e^{-r}\sin 2\theta, \; u_\theta = 2e^{-r}\cos 2\theta$

15. $g(u,v) = u\tan^{-1} v \; \Rightarrow \; g_u = \tan^{-1} v, \; g_v = \frac{u}{1 + v^2}$

16. $w = \frac{x}{y - z} \; \Rightarrow \; w_x = \frac{1}{y - z}, \; w_y = x(-1)(y-z)^{-2} = -\frac{x}{(y-z)^2}, \; w_z = x(-1)(y-z)^{-2}(-1) = \frac{x}{(y-z)^2}$

17. $T(p,q,r) = p\ln(q + e^r) \; \Rightarrow \; T_p = \ln(q + e^r), \; T_q = \frac{p}{q + e^r}, \; T_r = \frac{pe^r}{q + e^r}$

18. $C = 1449.2 + 4.6T - 0.055T^2 + 0.00029T^3 + (1.34 - 0.01T)(S - 35) + 0.016D \; \Rightarrow$

$\partial C/\partial T = 4.6 - 0.11T + 0.00087T^2 - 0.01(S - 35), \; \partial C/\partial S = 1.34 - 0.01T$, and $\partial C/\partial D = 0.016$. When $T = 10$, $S = 35$, and $D = 100$ we have $\partial C/\partial T = 4.6 - 0.11(10) + 0.00087(10)^2 - 0.01(35 - 35) \approx 3.587$, thus in $10°\text{C}$ water with salinity 35 parts per thousand and a depth of 100 m, the speed of sound increases by about 3.59 m/s for every degree Celsius that the water temperature rises. Similarly, $\partial C/\partial S = 1.34 - 0.01(10) = 1.24$, so the speed of sound increases by about 1.24 m/s for every part per thousand the salinity of the water increases. $\partial C/\partial D = 0.016$, so the speed of sound increases by about 0.016 m/s for every meter that the depth is increased.

19. $f(x,y) = 4x^3 - xy^2 \; \Rightarrow \; f_x = 12x^2 - y^2, \; f_y = -2xy, \; f_{xx} = 24x, \; f_{yy} = -2x, \; f_{xy} = f_{yx} = -2y$

20. $z = xe^{-2y} \; \Rightarrow \; z_x = e^{-2y}, \; z_y = -2xe^{-2y}, \; z_{xx} = 0, \; z_{yy} = 4xe^{-2y}, \; z_{xy} = z_{yx} = -2e^{-2y}$

21. $f(x,y,z) = x^k y^l z^m \; \Rightarrow \; f_x = kx^{k-1}y^l z^m, \; f_y = lx^k y^{l-1}z^m, \; f_z = mx^k y^l z^{m-1}, \; f_{xx} = k(k-1)x^{k-2}y^l z^m,$

$f_{yy} = l(l-1)x^k y^{l-2}z^m, \; f_{zz} = m(m-1)x^k y^l z^{m-2}, \; f_{xy} = f_{yx} = klx^{k-1}y^{l-1}z^m, \; f_{xz} = f_{zx} = kmx^{k-1}y^l z^{m-1},$

$f_{yz} = f_{zy} = lmx^k y^{l-1}z^{m-1}$

22. $v = r\cos(s + 2t) \; \Rightarrow \; v_r = \cos(s + 2t), \; v_s = -r\sin(s + 2t), \; v_t = -2r\sin(s + 2t), \; v_{rr} = 0, \; v_{ss} = -r\cos(s + 2t),$

$v_{tt} = -4r\cos(s + 2t), \; v_{rs} = v_{sr} = -\sin(s + 2t), \; v_{rt} = v_{tr} = -2\sin(s + 2t), \; v_{st} = v_{ts} = -2r\cos(s + 2t)$

23. $z = xy + xe^{y/x}$ $\Rightarrow$ $\dfrac{\partial z}{\partial x} = y - \dfrac{y}{x}e^{y/x} + e^{y/x}$, $\dfrac{\partial z}{\partial y} = x + e^{y/x}$ and

$x\dfrac{\partial z}{\partial x} + y\dfrac{\partial z}{\partial y} = x\left(y - \dfrac{y}{x}e^{y/x} + e^{y/x}\right) + y\left(x + e^{y/x}\right) = xy - ye^{y/x} + xe^{y/x} + xy + ye^{y/x} = xy + xy + xe^{y/x} = xy + z.$

24. $z = \sin(x + \sin t)$ $\Rightarrow$ $\dfrac{\partial z}{\partial x} = \cos(x + \sin t)$, $\dfrac{\partial z}{\partial t} = \cos(x + \sin t)\cos t$,

$\dfrac{\partial^2 z}{\partial x \partial t} = -\sin(x + \sin t)\cos t$, $\dfrac{\partial^2 z}{\partial x^2} = -\sin(x + \sin t)$ and

$\dfrac{\partial z}{\partial x}\dfrac{\partial^2 z}{\partial x \partial t} = \cos(x + \sin t)\left[-\sin(x + \sin t)\cos t\right] = \cos(x + \sin t)(\cos t)\left[-\sin(x + \sin t)\right] = \dfrac{\partial z}{\partial t}\dfrac{\partial^2 z}{\partial x^2}.$

25. (a) $z_x = 6x + 2$ $\Rightarrow$ $z_x(1, -2) = 8$ and $z_y = -2y$ $\Rightarrow$ $z_y(1, -2) = 4$, so an equation of the tangent plane is

$z - 1 = 8(x - 1) + 4(y + 2)$ or $z = 8x + 4y + 1$.

(b) A normal vector to the tangent plane (and the surface) at $(1, -2, 1)$ is $\langle 8, 4, -1\rangle$. Then parametric equations for the normal

line there are $x = 1 + 8t$, $y = -2 + 4t$, $z = 1 - t$, and symmetric equations are $\dfrac{x - 1}{8} = \dfrac{y + 2}{4} = \dfrac{z - 1}{-1}$.

26. (a) $z_x = e^x \cos y$ $\Rightarrow$ $z_x(0, 0) = 1$ and $z_y = -e^x \sin y$ $\Rightarrow$ $z_y(0, 0) = 0$, so an equation of the tangent plane is

$z - 1 = 1(x - 0) + 0(y - 0)$ or $z = x + 1$.

(b) A normal vector to the tangent plane (and the surface) at $(0, 0, 1)$ is $\langle 1, 0, -1\rangle$. Then parametric equations for the normal

line there are $x = t$, $y = 0$, $z = 1 - t$, and symmetric equations are $x = 1 - z$, $y = 0$.

27. (a) Let $F(x, y, z) = x^2 + 2y^2 - 3z^2$. Then $F_x = 2x$, $F_y = 4y$, $F_z = -6z$, so $F_x(2, -1, 1) = 4$, $F_y(2, -1, 1) = -4$,

$F_z(2, -1, 1) = -6$. From Equation 15.6.19 [ET 14.6.19], an equation of the tangent plane is

$4(x - 2) - 4(y + 1) - 6(z - 1) = 0$ or, equivalently, $2x - 2y - 3z = 3$.

(b) From Equations 15.6.20 [ET 14.6.20], symmetric equations for the normal line are $\dfrac{x - 2}{4} = \dfrac{y + 1}{-4} = \dfrac{z - 1}{-6}$.

28. (a) Let $F(x, y, z) = xy + yz + zx$. Then $F_x = y + z$, $F_y = x + z$, $F_z = x + y$, so

$F_x(1, 1, 1) = F_y(1, 1, 1) = F_z(1, 1, 1) = 2$. From Equation 15.6.19 [ET 14.6.19], an equation of the tangent plane is

$2(x - 1) + 2(y - 1) + 2(z - 1) = 0$ or, equivalently, $x + y + z = 3$.

(b) From Equations 15.6.20 [ET 14.6.20], symmetric equations for the normal line are $\dfrac{x - 1}{2} = \dfrac{y - 1}{2} = \dfrac{z - 1}{2}$ or,

equivalently, $x = y = z$.

29. (a) $\mathbf{r}(u, v) = (u + v)\mathbf{i} + u^2\mathbf{j} + v^2\mathbf{k}$ and the point $(3, 4, 1)$ corresponds to $u = 2$, $v = 1$. Then $\mathbf{r}_u = \mathbf{i} + 2u\mathbf{j}$ $\Rightarrow$

$\mathbf{r}_u(2, 1) = \mathbf{i} + 4\mathbf{j}$ and $\mathbf{r}_v = \mathbf{i} + 2v\mathbf{k}$ $\Rightarrow$ $\mathbf{r}_v(2, 1) = \mathbf{i} + 2\mathbf{j}$. A normal vector to the surface at $(3, 4, 1)$ is

$\mathbf{r}_u \times \mathbf{r}_v = 8\mathbf{i} - 2\mathbf{j} - 4\mathbf{k}$, so an equation of the tangent plane there is $8(x - 3) - 2(y - 4) - 4(z - 1) = 0$ or equivalently

$4x - y - 2z = 6$.

(b) A direction vector for the normal line through $(3, 4, 1)$ is $8\mathbf{i} - 2\mathbf{j} - 4\mathbf{k}$, so a vector equation is

$\mathbf{r}(t) = (3\mathbf{i} + 4\mathbf{j} + \mathbf{k}) + t(8\mathbf{i} - 2\mathbf{j} - 4\mathbf{k})$, and the corresponding parametric equations are $x = 3 + 8t$, $y = 4 - 2t$,

$z = 1 - 4t$.

30. Let $f(x, y) = x^2 + y^4$. Then $f_x(x, y) = 2x$ and $f_y(x, y) = 4y^3$, so $f_x(1, 1) = 2$,

$f_y(1, 1) = 4$ and an equation of the tangent plane is $z - 2 = 2(x - 1) + 4(y - 1)$

or $2x + 4y - z = 4$. A normal vector to the tangent plane is $\langle 2, 4, -1 \rangle$ so the

normal line is given by $\dfrac{x - 1}{2} = \dfrac{y - 1}{4} = \dfrac{z - 2}{-1}$ or $x = 1 + 2t$, $y = 1 + 4t$,

$z = 2 - t$.

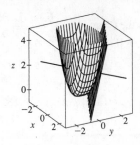

31. The hyperboloid is a level surface of the function $F(x, y, z) = x^2 + 4y^2 - z^2$, so a normal vector to the surface at (x_0, y_0, z_0)

is $\nabla F(x_0, y_0, z_0) = \langle 2x_0, 8y_0, -2z_0 \rangle$. A normal vector for the plane $2x + 2y + z = 5$ is $\langle 2, 2, 1 \rangle$. For the planes to be

parallel, we need the normal vectors to be parallel, so $\langle 2x_0, 8y_0, -2z_0 \rangle = k \langle 2, 2, 1 \rangle$, or $x_0 = k$, $y_0 = \frac{1}{4}k$, and $z_0 = -\frac{1}{2}k$.

But $x_0^2 + 4y_0^2 - z_0^2 = 4 \quad \Rightarrow \quad k^2 + \frac{1}{4}k^2 - \frac{1}{4}k^2 = 4 \quad \Rightarrow \quad k^2 = 4 \quad \Rightarrow \quad k = \pm 2$. So there are two such points:

$\left(2, \frac{1}{2}, -1\right)$ and $\left(-2, -\frac{1}{2}, 1\right)$.

32. $u = \ln(1 + se^{2t}) \quad \Rightarrow \quad du = \dfrac{\partial u}{\partial s}\, ds + \dfrac{\partial u}{\partial t}\, dt = \dfrac{e^{2t}}{1 + se^{2t}}\, ds + \dfrac{2se^{2t}}{1 + se^{2t}}\, dt$

33. $f(x, y, z) = x^3 \sqrt{y^2 + z^2} \quad \Rightarrow \quad f_x(x, y, z) = 3x^2 \sqrt{y^2 + z^2},\ f_y(x, y, z) = \dfrac{yx^3}{\sqrt{y^2 + z^2}},\ f_z(x, y, z) = \dfrac{zx^3}{\sqrt{y^2 + z^2}}$,

so $f(2, 3, 4) = 8(5) = 40$, $f_x(2, 3, 4) = 3(4)\sqrt{25} = 60$, $f_y(2, 3, 4) = \frac{3(8)}{\sqrt{25}} = \frac{24}{5}$, and $f_z(2, 3, 4) = \frac{4(8)}{\sqrt{25}} = \frac{32}{5}$. Then the

linear approximation of f at $(2, 3, 4)$ is

$$f(x, y, z) \approx f(2, 3, 4) + f_x(2, 3, 4)(x - 2) + f_y(2, 3, 4)(y - 3) + f_z(2, 3, 4)(z - 4)$$

$$= 40 + 60(x - 2) + \tfrac{24}{5}(y - 3) + \tfrac{32}{5}(z - 4) = 60x + \tfrac{24}{5}y + \tfrac{32}{5}z - 120$$

Then $(1.98)^3 \sqrt{(3.01)^2 + (3.97)^2} = f(1.98, 3.01, 3.97) \approx 60(1.98) + \frac{24}{5}(3.01) + \frac{32}{5}(3.97) - 120 = 38.656$.

34. (a) $dA = \dfrac{\partial A}{\partial x}\, dx + \dfrac{\partial A}{\partial y}\, dy = \frac{1}{2}y\, dx + \frac{1}{2}x\, dy$ and $|\Delta x| \leq 0.002$, $|\Delta y| \leq 0.002$. Thus the maximum error in the calculated

area is about $dA = 6(0.002) + \frac{5}{2}(0.002) = 0.017$ m^2 or 170 cm^2.

(b) $z = \sqrt{x^2 + y^2}$, $dz = \dfrac{x}{\sqrt{x^2 + y^2}}\, dx + \dfrac{y}{\sqrt{x^2 + y^2}}\, dy$ and $|\Delta x| \leq 0.002$, $|\Delta y| \leq 0.002$. Thus the maximum error in the

calculated hypotenuse length is about $dz = \frac{5}{13}(0.002) + \frac{12}{13}(0.002) = \frac{0.17}{65} \approx 0.0026$ m or 0.26 cm.

35. $\dfrac{du}{dp} = \dfrac{\partial u}{\partial x}\dfrac{dx}{dp} + \dfrac{\partial u}{\partial y}\dfrac{dy}{dp} + \dfrac{\partial u}{\partial z}\dfrac{dz}{dp} = 2xy^3(1 + 6p) + 3x^2y^2(pe^p + e^p) + 4z^3(p\cos p + \sin p)$

36. $\dfrac{\partial v}{\partial s} = \dfrac{\partial v}{\partial x}\dfrac{\partial x}{\partial s} + \dfrac{\partial v}{\partial y}\dfrac{\partial y}{\partial s} = \left(2x \sin y + y^2 e^{xy}\right)(1) + \left(x^2 \cos y + xye^{xy} + e^{xy}\right)(t)$.

$s = 0, t = 1 \quad \Rightarrow \quad x = 2, y = 0$, so $\dfrac{\partial v}{\partial s} = 0 + (4 + 1)(1) = 5$.

$\dfrac{\partial v}{\partial t} = \dfrac{\partial v}{\partial x}\dfrac{\partial x}{\partial t} + \dfrac{\partial v}{\partial y}\dfrac{\partial y}{\partial t} = \left(2x \sin y + y^2 e^{xy}\right)(2) + \left(x^2 \cos y + xye^{xy} + e^{xy}\right)(s) = 0 + 0 = 0$.

37. By the Chain Rule, $\dfrac{\partial z}{\partial s} = \dfrac{\partial z}{\partial x}\dfrac{\partial x}{\partial s} + \dfrac{\partial z}{\partial y}\dfrac{\partial y}{\partial s}$. When $s = 1$ and $t = 2$, $x = g(1,2) = 3$ and $y = h(1,2) = 6$, so

$\dfrac{\partial z}{\partial s} = f_x(3,6)g_s(1,2) + f_y(3,6)h_s(1,2) = (7)(-1) + (8)(-5) = -47$. Similarly, $\dfrac{\partial z}{\partial t} = \dfrac{\partial z}{\partial x}\dfrac{\partial x}{\partial t} + \dfrac{\partial z}{\partial y}\dfrac{\partial y}{\partial t}$, so

$\dfrac{\partial z}{\partial t} = f_x(3,6)g_t(1,2) + f_y(3,6)h_t(1,2) = (7)(4) + (8)(10) = 108.$

38.

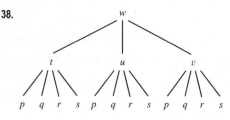

Using the tree diagram as a guide, we have

$\dfrac{\partial w}{\partial p} = \dfrac{\partial w}{\partial t}\dfrac{\partial t}{\partial p} + \dfrac{\partial w}{\partial u}\dfrac{\partial u}{\partial p} + \dfrac{\partial w}{\partial v}\dfrac{\partial v}{\partial p} \qquad \dfrac{\partial w}{\partial q} = \dfrac{\partial w}{\partial t}\dfrac{\partial t}{\partial q} + \dfrac{\partial w}{\partial u}\dfrac{\partial u}{\partial q} + \dfrac{\partial w}{\partial v}\dfrac{\partial v}{\partial q}$

$\dfrac{\partial w}{\partial r} = \dfrac{\partial w}{\partial t}\dfrac{\partial t}{\partial r} + \dfrac{\partial w}{\partial u}\dfrac{\partial u}{\partial r} + \dfrac{\partial w}{\partial v}\dfrac{\partial v}{\partial r} \qquad \dfrac{\partial w}{\partial s} = \dfrac{\partial w}{\partial t}\dfrac{\partial t}{\partial s} + \dfrac{\partial w}{\partial u}\dfrac{\partial u}{\partial s} + \dfrac{\partial w}{\partial v}\dfrac{\partial v}{\partial s}$

39. $\dfrac{\partial z}{\partial x} = 2xf'(x^2 - y^2)$, $\quad \dfrac{\partial z}{\partial y} = 1 - 2yf'(x^2 - y^2) \quad \left[\text{where } f' = \dfrac{df}{d(x^2 - y^2)} \right]$. Then

$y\dfrac{\partial z}{\partial x} + x\dfrac{\partial z}{\partial y} = 2xyf'(x^2 - y^2) + x - 2xyf'(x^2 - y^2) = x.$

40. $A = \frac{1}{2}xy\sin\theta$, $dx/dt = 3$, $dy/dt = -2$, $d\theta/dt = 0.05$, and $\dfrac{dA}{dt} = \dfrac{1}{2}\left[(y\sin\theta)\dfrac{dx}{dt} + (x\sin\theta)\dfrac{dy}{dt} + (xy\cos\theta)\dfrac{d\theta}{dt} \right]$.

So when $x = 40$, $y = 50$ and $\theta = \frac{\pi}{6}$, $\dfrac{dA}{dt} = \dfrac{1}{2}\left[(25)(3) + (20)(-2) + (1000\sqrt{3})(0.05) \right] = \dfrac{35 + 50\sqrt{3}}{2} \approx 60.8 \text{ in}^2/\text{s}.$

41. $\dfrac{\partial z}{\partial x} = \dfrac{\partial z}{\partial u}y + \dfrac{\partial z}{\partial v}\dfrac{-y}{x^2}$ and

$\dfrac{\partial^2 z}{\partial x^2} = y\dfrac{\partial}{\partial x}\left(\dfrac{\partial z}{\partial u}\right) + \dfrac{2y}{x^3}\dfrac{\partial z}{\partial v} + \dfrac{-y}{x^2}\dfrac{\partial}{\partial x}\left(\dfrac{\partial z}{\partial v}\right) = \dfrac{2y}{x^3}\dfrac{\partial z}{\partial v} + y\left(\dfrac{\partial^2 z}{\partial u^2}y + \dfrac{\partial^2 z}{\partial v\,\partial u}\dfrac{-y}{x^2}\right) + \dfrac{-y}{x^2}\left(\dfrac{\partial^2 z}{\partial v^2}\dfrac{-y}{x^2} + \dfrac{\partial^2 z}{\partial u\,\partial v}y\right)$

$\qquad = \dfrac{2y}{x^3}\dfrac{\partial z}{\partial v} + y^2\dfrac{\partial^2 z}{\partial u^2} - \dfrac{2y^2}{x^2}\dfrac{\partial^2 z}{\partial u\,\partial v} + \dfrac{y^2}{x^4}\dfrac{\partial^2 z}{\partial v^2}$

Also $\dfrac{\partial z}{\partial y} = x\dfrac{\partial z}{\partial u} + \dfrac{1}{x}\dfrac{\partial z}{\partial v}$ and

$\dfrac{\partial^2 z}{\partial y^2} = x\dfrac{\partial}{\partial y}\left(\dfrac{\partial z}{\partial u}\right) + \dfrac{1}{x}\dfrac{\partial}{\partial y}\left(\dfrac{\partial z}{\partial v}\right) = x\left(\dfrac{\partial^2 z}{\partial u^2}x + \dfrac{\partial^2 z}{\partial v\,\partial u}\dfrac{1}{x}\right) + \dfrac{1}{x}\left(\dfrac{\partial^2 z}{\partial v^2}\dfrac{1}{x} + \dfrac{\partial^2 z}{\partial u\,\partial v}x\right) = x^2\dfrac{\partial^2 z}{\partial u^2} + 2\dfrac{\partial^2 z}{\partial u\,\partial v} + \dfrac{1}{x^2}\dfrac{\partial^2 z}{\partial v^2}$

Thus

$x^2\dfrac{\partial^2 z}{\partial x^2} - y^2\dfrac{\partial^2 z}{\partial y^2} = \dfrac{2y}{x}\dfrac{\partial z}{\partial v} + x^2y^2\dfrac{\partial^2 z}{\partial u^2} - 2y^2\dfrac{\partial^2 z}{\partial u\,\partial v} + \dfrac{y^2}{x^2}\dfrac{\partial^2 z}{\partial v^2} - x^2y^2\dfrac{\partial^2 z}{\partial u^2} - 2y^2\dfrac{\partial^2 z}{\partial u\,\partial v} - \dfrac{y^2}{x^2}\dfrac{\partial^2 z}{\partial v^2}$

$\qquad = \dfrac{2y}{x}\dfrac{\partial z}{\partial v} - 4y^2\dfrac{\partial^2 z}{\partial u\,\partial v} = 2v\dfrac{\partial z}{\partial v} - 4uv\dfrac{\partial^2 z}{\partial u\,\partial v}$

since $y = xv = \dfrac{uv}{y}$ or $y^2 = uv$.

42. $F(x,y,z) = e^{xyz} - yz^4 - x^2z^3 = 0$, so $\dfrac{\partial z}{\partial x} = -\dfrac{F_x}{F_z} = -\dfrac{yze^{xyz} - 2xz^3}{xye^{xyz} - 4yz^3 - 3x^2z^2} = \dfrac{2xz^3 - yze^{xyz}}{xye^{xyz} - 4yz^3 - 3x^2z^2}$ and

$\dfrac{\partial z}{\partial y} = -\dfrac{F_y}{F_z} = -\dfrac{xze^{xyz} - z^4}{xye^{xyz} - 4yz^3 - 3x^2z^2} = \dfrac{z^4 - xze^{xyz}}{xye^{xyz} - 4yz^3 - 3x^2z^2}.$

43. $\nabla f = \left\langle z^2 \sqrt{y}\, e^{x\sqrt{y}}, \dfrac{xz^2 e^{x\sqrt{y}}}{2\sqrt{y}}, 2ze^{x\sqrt{y}} \right\rangle = ze^{x\sqrt{y}} \left\langle z\sqrt{y}, \dfrac{xz}{2\sqrt{y}}, 2 \right\rangle$

44. (a) By Theorem 15.6.15 [ET 14.6.15], the maximum value of the directional derivative occurs when **u** has the same direction as the gradient vector.

(b) It is a minimum when **u** is in the direction opposite to that of the gradient vector (that is, **u** is in the direction of $-\nabla f$), since $D_{\mathbf{u}} f = |\nabla f| \cos\theta$ (see the proof of Theorem 15.6.15 [ET 14.6.15]) has a minimum when $\theta = \pi$.

(c) The directional derivative is 0 when **u** is perpendicular to the gradient vector, since then $D_{\mathbf{u}} f = \nabla f \cdot \mathbf{u} = 0$.

(d) The directional derivative is half of its maximum value when $D_{\mathbf{u}} f = |\nabla f| \cos\theta = \frac{1}{2}|\nabla f| \;\Leftrightarrow\; \cos\theta = \frac{1}{2} \;\Leftrightarrow$
$\theta = \frac{\pi}{3}$.

45. $\nabla f = \langle 1/\sqrt{x}, -2y \rangle$, $\nabla f(1,5) = \langle 1, -10 \rangle$, $\mathbf{u} = \frac{1}{5}\langle 3, -4 \rangle$. Then $D_{\mathbf{u}} f(1,5) = \frac{43}{5}$.

46. $\nabla f = \left\langle 2xy + \sqrt{1+z}, x^2, x/(2\sqrt{1+z}) \right\rangle$, $\nabla f(1,2,3) = \langle 6, 1, \frac{1}{4} \rangle$, $\mathbf{u} = \langle \frac{2}{3}, \frac{1}{3}, -\frac{2}{3} \rangle$. Then $D_{\mathbf{u}} f(1,2,3) = \frac{25}{6}$.

47. $\nabla f = \langle 2xy, x^2 + 1/(2\sqrt{y}) \rangle$, $|\nabla f(2,1)| = |\langle 4, \frac{9}{2} \rangle|$. Thus the maximum rate of change of f at $(2,1)$ is $\frac{\sqrt{145}}{2}$ in the direction $\langle 4, \frac{9}{2} \rangle$.

48. $\nabla f = \langle zye^{xy}, zxe^{xy}, e^{xy} \rangle$, $\nabla f(0,1,2) = \langle 2,0,1 \rangle$ is the direction of most rapid increase while the rate is $|\langle 2,0,1 \rangle| = \sqrt{5}$.

49. First we draw a line passing through Homestead and the eye of the hurricane. We can approximate the directional derivative at Homestead in the direction of the eye of the hurricane by the average rate of change of wind speed between the points where this line intersects the contour lines closest to Homestead. In the direction of the eye of the hurricane, the wind speed changes from 45 to 50 knots. We estimate the distance between these two points to be approximately 8 miles, so the rate of change of wind speed in the direction given is approximately $\frac{50-45}{8} = \frac{5}{8} = 0.625$ knot/mi.

50. The surfaces are $f(x,y,z) = z - 2x^2 + y^2 = 0$ and $g(x,y,z) = z - 4 = 0$. The tangent line is perpendicular to both ∇f and ∇g at $(-2,2,4)$. The vector $\mathbf{v} = \nabla f \times \nabla g$ is therefore parallel to the line. $\nabla f(x,y,z) = \langle -4x, 2y, 1 \rangle \;\Rightarrow$
$\nabla f(-2,2,4) = \langle 8,4,1 \rangle$, $\nabla g(x,y,z) = \langle 0,0,1 \rangle \;\Rightarrow\; \nabla g\langle -2,2,4 \rangle = \langle 0,0,1 \rangle$. Hence

$$\mathbf{v} = \nabla f \times \nabla g = \begin{vmatrix} \mathbf{i} & \mathbf{j} & \mathbf{k} \\ 8 & 4 & 1 \\ 0 & 0 & 1 \end{vmatrix} = 4\mathbf{i} - 8\mathbf{j}.$$ Thus, parametric equations are: $x = -2 + 4t$, $y = 2 - 8t$, $z = 4$.

51. $f(x,y) = x^2 - xy + y^2 + 9x - 6y + 10 \;\Rightarrow\; f_x = 2x - y + 9$,
$f_y = -x + 2y - 6$, $f_{xx} = 2 = f_{yy}$, $f_{xy} = -1$. Then $f_x = 0$ and $f_y = 0$
imply $y = 1$, $x = -4$. Thus the only critical point is $(-4, 1)$ and
$f_{xx}(-4,1) > 0$, $D(-4,1) = 3 > 0$, so $f(-4,1) = -11$ is a local minimum.

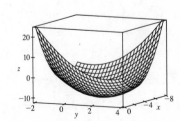

52. $f(x, y) = x^3 - 6xy + 8y^3 \Rightarrow f_x = 3x^2 - 6y$, $f_y = -6x + 24y^2$,

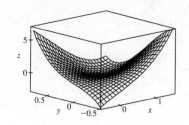

$f_{xx} = 6x$, $f_{yy} = 48y$, $f_{xy} = -6$. Then $f_x = 0$ implies $y = x^2/2$, substituting

into $f_y = 0$ implies $6x(x^3 - 1) = 0$, so the critical points are $(0, 0)$, $\left(1, \frac{1}{2}\right)$.

$D(0, 0) = -36 < 0$ so $(0, 0)$ is a saddle point while $f_{xx}\left(1, \frac{1}{2}\right) = 6 > 0$ and

$D\left(1, \frac{1}{2}\right) = 108 > 0$ so $f\left(1, \frac{1}{2}\right) = -1$ is a local minimum.

53. $f(x, y) = 3xy - x^2y - xy^2 \Rightarrow f_x = 3y - 2xy - y^2$,

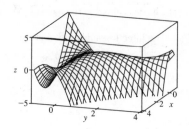

$f_y = 3x - x^2 - 2xy$, $f_{xx} = -2y$, $f_{yy} = -2x$, $f_{xy} = 3 - 2x - 2y$. Then

$f_x = 0$ implies $y(3 - 2x - y) = 0$ so $y = 0$ or $y = 3 - 2x$. Substituting into

$f_y = 0$ implies $x(3 - x) = 0$ or $3x(-1 + x) = 0$. Hence the critical points are

$(0, 0)$, $(3, 0)$, $(0, 3)$ and $(1, 1)$. $D(0, 0) = D(3, 0) = D(0, 3) = -9 < 0$ so

$(0, 0)$, $(3, 0)$, and $(0, 3)$ are saddle points. $D(1, 1) = 3 > 0$ and

$f_{xx}(1, 1) = -2 < 0$, so $f(1, 1) = 1$ is a local maximum.

54. $f(x, y) = (x^2 + y)e^{y/2} \Rightarrow f_x = 2xe^{y/2}$, $f_y = e^{y/2}(2 + x^2 + y)/2$,

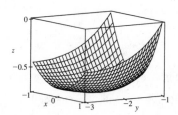

$f_{xx} = 2e^{y/2}$, $f_{yy} = e^{y/2}(4 + x^2 + y)/4$, $f_{xy} = xe^{y/2}$. Then $f_x = 0$ implies

$x = 0$, so $f_y = 0$ implies $y = -2$. But $f_{xx}(0, -2) > 0$,

$D(0, -2) = e^{-2} - 0 > 0$ so $f(0, -2) = -2/e$ is a local minimum.

55. First solve inside D. Here $f_x = 4y^2 - 2xy^2 - y^3$, $f_y = 8xy - 2x^2y - 3xy^2$.

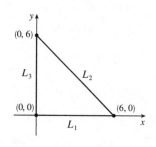

Then $f_x = 0$ implies $y = 0$ or $y = 4 - 2x$, but $y = 0$ isn't inside D. Substituting

$y = 4 - 2x$ into $f_y = 0$ implies $x = 0$, $x = 2$ or $x = 1$, but $x = 0$ isn't inside D,

and when $x = 2$, $y = 0$ but $(2, 0)$ isn't inside D. Thus the only critical point inside

D is $(1, 2)$ and $f(1, 2) = 4$. Secondly we consider the boundary of D.

On L_1: $f(x, 0) = 0$ and so $f = 0$ on L_1. On L_2: $x = -y + 6$ and

$f(-y + 6, y) = y^2(6 - y)(-2) = -2(6y^2 - y^3)$ which has critical pointsat $y = 0$ and $y = 4$. Then $f(6, 0) = 0$ while

$f(2, 4) = -64$. On L_3: $f(0, y) = 0$, so $f = 0$ on L_3. Thus on D the absolute maximum of f is $f(1, 2) = 4$ while the

absolute minimum is $f(2, 4) = -64$.

56. Inside D: $f_x = 2xe^{-x^2-y^2}(1 - x^2 - 2y^2) = 0$ implies $x = 0$ or $x^2 + 2y^2 = 1$. Then if $x = 0$,

$f_y = 2ye^{-x^2-y^2}(2 - x^2 - 2y^2) = 0$ implies $y = 0$ or $2 - 2y^2 = 0$ giving the critical points $(0, 0)$, $(0, \pm 1)$. If

$x^2 + 2y^2 = 1$, then $f_y = 0$ implies $y = 0$ giving the critical points $(\pm 1, 0)$. Now $f(0, 0) = 0$, $f(\pm 1, 0) = e^{-1}$ and

$f(0, \pm 1) = 2e^{-1}$. On the boundary of D: $x^2 + y^2 = 4$, so $f(x, y) = e^{-4}(4 + y^2)$ and f is smallest when $y = 0$ and largest

when $y^2 = 4$. But $f(\pm 2, 0) = 4e^{-4}$, $f(0, \pm 2) = 8e^{-4}$. Thus on D the absolute maximum of f is $f(0, \pm 1) = 2e^{-1}$ and the

absolute minimum is $f(0, 0) = 0$.

57. $f(x, y) = x^3 - 3x + y^4 - 2y^2$

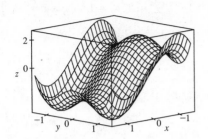
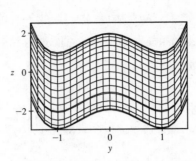
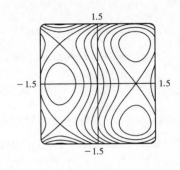

From the graphs, it appears that f has a local maximum $f(-1, 0) \approx 2$, local minima $f(1, \pm1) \approx -3$, and saddle points at $(-1, \pm1)$ and $(1, 0)$.

To find the exact quantities, we calculate $f_x = 3x^2 - 3 = 0 \Leftrightarrow x = \pm1$ and $f_y = 4y^3 - 4y = 0 \Leftrightarrow y = 0, \pm1$, giving the critical points estimated above. Also $f_{xx} = 6x$, $f_{xy} = 0$, $f_{yy} = 12y^2 - 4$, so using the Second Derivatives Test, $D(-1, 0) = 24 > 0$ and $f_{xx}(-1, 0) = -6 < 0$ indicating a local maximum $f(-1, 0) = 2$; $D(1, \pm1) = 48 > 0$ and $f_{xx}(1, \pm1) = 6 > 0$ indicating local minima $f(1, \pm1) = -3$; and $D(-1, \pm1) = -48$ and $D(1, 0) = -24$, indicating saddle points.

58. $f(x, y) = 12 + 10y - 2x^2 - 8xy - y^4 \Rightarrow f_x(x, y) = -4x - 8y$, $f_y(x, y) = 10 - 8x - 4y^3$. Now $f_x(x, y) = 0 \Rightarrow x = -2x$, and substituting this into $f_y(x, y) = 0$ gives $10 + 16y - 4y^3 = 0 \Leftrightarrow 5 + 8y - 2y^3 = 0$.

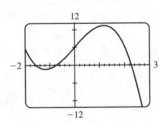
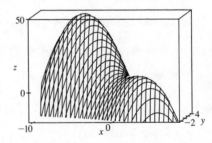

From the first graph, we see that this is true when $y \approx -1.542$, -0.717, or 2.260. (Alternatively, we could have found the solutions to $f_x = f_y = 0$ using a CAS.) So to three decimal places, the critical points are $(3.085, -1.542)$, $(1.434, -0.717)$, and $(-4.519, 2.260)$. Now in order to use the Second Derivatives Test, we calculate $f_{xx} = -4$, $f_{xy} = -8$, $f_{yy} = -12y^2$, and $D = 48y^2 - 64$. So since $D(3.085, -1.542) > 0$, $D(1.434, -0.717) < 0$, and $D(-4.519, 2.260) > 0$, and f_{xx} is always negative, $f(x, y)$ has local maxima $f(-4.519, 2.260) \approx 49.373$ and $f(3.085, -1.542) \approx 9.948$, and a saddle point at approximately $(1.434, -0.717)$. The highest point on the graph is approximately $(-4.519, 2.260, 49.373)$.

59. $f(x, y) = x^2 y$, $g(x, y) = x^2 + y^2 = 1 \Rightarrow \nabla f = \langle 2xy, x^2 \rangle = \lambda \nabla g = \langle 2\lambda x, 2\lambda y \rangle$. Then $2xy = 2\lambda x$ and $x^2 = 2\lambda y$ imply $\lambda = x^2/(2y)$ and $\lambda = y$ if $x \neq 0$ and $y \neq 0$. Hence $x^2 = 2y^2$. Then $x^2 + y^2 = 1$ implies $3y^2 = 1$ so $y = \pm\frac{1}{\sqrt{3}}$ and $x = \pm\sqrt{\frac{2}{3}}$. [Note if $x = 0$ then $x^2 = 2\lambda y$ implies $y = 0$ and $f(0, 0) = 0$.] Thus the possible points are $\left(\pm\sqrt{\frac{2}{3}}, \pm\frac{1}{\sqrt{3}}\right)$ and the absolute maxima are $f\left(\pm\sqrt{\frac{2}{3}}, \frac{1}{\sqrt{3}}\right) = \frac{2}{3\sqrt{3}}$ while the absolute minima are $f\left(\pm\sqrt{\frac{2}{3}}, -\frac{1}{\sqrt{3}}\right) = -\frac{2}{3\sqrt{3}}$.

60. $f(x,y) = 1/x + 1/y$, $g(x,y) = 1/x^2 + 1/y^2 = 1$ $\Rightarrow$ $\nabla f = \langle -x^{-2}, -y^{-2} \rangle = \lambda \nabla g = \langle -2\lambda x^{-3}, -2\lambda y^{-3} \rangle$. Then $-x^{-2} = -2\lambda x^3$ or $x = 2\lambda$ and $-y^{-2} = -2\lambda y^{-3}$ or $y = 2\lambda$. Thus $x = y$, so $1/x^2 + 1/y^2 = 2/x^2 = 1$ implies $x = \pm\sqrt{2}$ and the possible points are $(\pm\sqrt{2}, \pm\sqrt{2})$. The absolute maximum of f subject to $x^{-2} + y^{-2} = 1$ is then $f(\sqrt{2}, \sqrt{2}) = \sqrt{2}$ and the absolute minimum is $f(-\sqrt{2}, -\sqrt{2}) = -\sqrt{2}$.

61. $f(x,y,z) = xyz$, $g(x,y,z) = x^2 + y^2 + z^2 = 3$. $\nabla f = \lambda \nabla g$ $\Rightarrow$ $\langle yz, xz, xy \rangle = \lambda \langle 2x, 2y, 2z \rangle$. If any of x, y, or z is zero, then $x = y = z = 0$ which contradicts $x^2 + y^2 + z^2 = 3$. Then $\lambda = \frac{yz}{2x} = \frac{xz}{2y} = \frac{xy}{2z}$ $\Rightarrow$ $2y^2 z = 2x^2 z$ $\Rightarrow$ $y^2 = x^2$, and similarly $2yz^2 = 2x^2 y$ $\Rightarrow$ $z^2 = x^2$. Substituting into the constraint equation gives $x^2 + x^2 + x^2 = 3$ $\Rightarrow$ $x^2 = 1 = y^2 = z^2$. Thus the possible points are $(1, 1, \pm 1)$, $(1, -1, \pm 1)$, $(-1, 1, \pm 1)$, $(-1, -1, \pm 1)$. The absolute maximum is $f(1,1,1) = f(1,-1,-1) = f(-1,1,-1) = f(-1,-1,1) = 1$ and the absolute minimum is $f(1,1,-1) = f(1,-1,1) = f(-1,1,1) = f(-1,-1,-1) = -1$.

62. $f(x,y,z) = x^2 + 2y^2 + 3z^2$, $g(x,y,z) = x + y + z = 1$, $h(x,y,z) = x - y + 2z = 2$ $\Rightarrow$ $\nabla f = \langle 2x, 4y, 6z \rangle = \lambda \nabla g + \mu \nabla h = \langle \lambda + \mu, \lambda - \mu, \lambda + 2\mu \rangle$ and $2x = \lambda + \mu$ **(1)**, $4y = \lambda - \mu$ **(2)**, $6z = \lambda + 2\mu$ **(3)**, $x + y + z = 1$ **(4)**, $x - y + 2z = 2$ **(5)**. Then six times **(1)** plus three times **(2)** plus two times **(3)** implies $12(x + y + z) = 11\lambda + 7\mu$, so **(4)** gives $11\lambda + 7\mu = 12$. Also six times **(1)** minus three times **(2)** plus four times **(3)** implies $12(x - y + 2z) = 7\lambda + 17\mu$, so **(5)** gives $7\lambda + 17\mu = 24$. Solving $11\lambda + 7\mu = 12$, $7\lambda + 17\mu = 24$ simultaneously gives $\lambda = \frac{6}{23}$, $\mu = \frac{30}{23}$. Substituting into **(1)**, **(2)**, and **(3)** implies $x = \frac{18}{23}$, $y = -\frac{6}{23}$, $z = \frac{11}{23}$ giving only one point. Then $f\left(\frac{18}{23}, -\frac{6}{23}, \frac{11}{23}\right) = \frac{33}{23}$. Now since $(0, 0, 1)$ satisfies both constraints and $f(0, 0, 1) = 3 > \frac{33}{23}$, $f\left(\frac{18}{23}, -\frac{6}{23}, \frac{11}{23}\right) = \frac{33}{23}$ is an absolute minimum, and there is no absolute maximum.

63. $f(x,y,z) = x^2 + y^2 + z^2$, $g(x,y,z) = xy^2 z^3 = 2$ $\Rightarrow$ $\nabla f = \langle 2x, 2y, 2z \rangle = \lambda \nabla g = \langle \lambda y^2 z^3, 2\lambda xyz^3, 3\lambda xy^2 z^2 \rangle$. Since $xy^2 z^3 = 2$, $x \neq 0$, $y \neq 0$ and $z \neq 0$, so $2x = \lambda y^2 z^3$ **(1)**, $1 = \lambda xz^3$ **(2)**, $2 = 3\lambda xy^2 z$ **(3)**. Then **(2)** and **(3)** imply $\frac{1}{xz^3} = \frac{2}{3xy^2 z}$ or $y^2 = \frac{2}{3} z^2$ so $y = \pm z\sqrt{\frac{2}{3}}$. Similarly **(1)** and **(3)** imply $\frac{2x}{y^2 z^3} = \frac{2}{3xy^2 z}$ or $3x^2 = z^2$ so $x = \pm\frac{1}{\sqrt{3}} z$. But $xy^2 z^3 = 2$ so x and z must have the same sign, that is, $x = \frac{1}{\sqrt{3}} z$. Thus $g(x,y,z) = 2$ implies $\frac{1}{\sqrt{3}} z \left(\frac{2}{3} z^2\right) z^3 = 2$ or $z = \pm 3^{1/4}$ and the possible points are $(\pm 3^{-1/4}, 3^{-1/4}\sqrt{2}, \pm 3^{1/4})$, $(\pm 3^{-1/4}, -3^{-1/4}\sqrt{2}, \pm 3^{1/4})$. However at each of these points f takes on the same value, $2\sqrt{3}$. But $(2, 1, 1)$ also satisfies $g(x,y,z) = 2$ and $f(2,1,1) = 6 > 2\sqrt{3}$. Thus f has an absolute minimum value of $2\sqrt{3}$ and no absolute maximum subject to the constraint $xy^2 z^3 = 2$.

Alternate solution: $g(x,y,z) = xy^2 z^3 = 2$ implies $y^2 = \dfrac{2}{xz^3}$, so minimize $f(x, z) = x^2 + \dfrac{2}{xz^3} + z^2$. Then $f_x = 2x - \dfrac{2}{x^2 z^3}$, $f_z = -\dfrac{6}{xz^4} + 2z$, $f_{xx} = 2 + \dfrac{4}{x^3 z^3}$, $f_{zz} = \dfrac{24}{xz^5} + 2$ and $f_{xz} = \dfrac{6}{x^2 z^4}$. Now $f_x = 0$ implies $2x^3 z^3 - 2 = 0$ or $z = 1/x$. Substituting into $f_y = 0$ implies $-6x^3 + 2x^{-1} = 0$ or $x = \frac{1}{\sqrt[4]{3}}$, so the two critical points are

$\left(\pm\frac{1}{\sqrt[4]{3}}, \pm\sqrt[4]{3}\right)$. Then $D\left(\pm\frac{1}{\sqrt[4]{3}}, \pm\sqrt[4]{3}\right) = (2+4)\left(2+\frac{24}{3}\right) - \left(\frac{6}{\sqrt{3}}\right)^2 > 0$ and $f_{xx}\left(\pm\frac{1}{\sqrt[4]{3}}, \pm\sqrt[4]{3}\right) = 6 > 0$, so each point

is a minimum. Finally, $y^2 = \dfrac{2}{xz^3}$, so the four points closest to the origin are $\left(\pm\frac{1}{\sqrt[4]{3}}, \frac{\sqrt{2}}{\sqrt[4]{3}}, \pm\sqrt[4]{3}\right)$, $\left(\pm\frac{1}{\sqrt[4]{3}}, -\frac{\sqrt{2}}{\sqrt[4]{3}}, \pm\sqrt[4]{3}\right)$.

64. $V = xyz$, say x is the length and $x + 2y + 2z \le 108$, $x > 0$, $y > 0$, $z > 0$. First maximize V subject to $x + 2y + 2z = 108$

with x, y, z all positive. Then $\langle yz, xz, xy \rangle = \langle \lambda, 2\lambda, 2\lambda \rangle$ implies $2yz = xz$ or $x = 2y$ and $xz = xy$ or $z = y$. Thus

$g(x, y, z) = 108$ implies $6y = 108$ or $y = 18 = z$, $x = 36$, so the volume is $V = 11{,}664$ cubic units. Since $(104, 1, 1)$ also

satisfies $g(x, y, z) = 108$ and $V(104, 1, 1) = 104$ cubic units, $(36, 18, 18)$ gives an absolute maximum of V subject to

$g(x, y, z) = 108$. But if $x + 2y + 2z < 108$, there exists $\alpha > 0$ such that $x + 2y + 2z = 108 - \alpha$ and as above

$6y = 108 - \alpha$ implies $y = (108 - \alpha)/6 = z$, $x = (108 - \alpha)/3$ with $V = (108 - \alpha)^3/(6^2 \cdot 3) < (108)^3/(6^2 \cdot 3) = 11{,}664$.

Hence we have shown that the maximum of V subject to $g(x, y, z) \le 108$ is the maximum of V subject to $g(x, y, z) = 108$

(an intuitively obvious fact).

65.

The area of the triangle is $\frac{1}{2}ca\sin\theta$ and the area of the rectangle is bc. Thus, the

area of the whole object is $f(a, b, c) = \frac{1}{2}ca\sin\theta + bc$. The perimeter of the object

is $g(a, b, c) = 2a + 2b + c = P$. To simplify $\sin\theta$ in terms of a, b, and c notice

that $a^2\sin^2\theta + \left(\frac{1}{2}c\right)^2 = a^2 \;\Rightarrow\; \sin\theta = \dfrac{1}{2a}\sqrt{4a^2 - c^2}$. Thus

$$f(a, b, c) = \frac{c}{4}\sqrt{4a^2 - c^2} + bc.$$ (Instead of using θ, we could just have used the

Pythagorean Theorem.) As a result, by Lagrange's method, we must find a, b, c, and λ by solving $\nabla f = \lambda \nabla g$ which gives the

following equations: $ca(4a^2 - c^2)^{-1/2} = 2\lambda$ **(1)**, $c = 2\lambda$ **(2)**, $\frac{1}{4}(4a^2 - c^2)^{1/2} - \frac{1}{4}c^2(4a^2 - c^2)^{-1/2} + b = \lambda$ **(3)**, and

$2a + 2b + c = P$ **(4)**. From **(2)**, $\lambda = \frac{1}{2}c$ and so **(1)** produces $ca(4a^2 - c^2)^{-1/2} = c \;\Rightarrow\; (4a^2 - c^2)^{1/2} = a \;\Rightarrow$

$4a^2 - c^2 = a^2 \;\Rightarrow\; c = \sqrt{3}\,a$ **(5)**. Similarly, since $(4a^2 - c^2)^{1/2} = a$ and $\lambda = \frac{1}{2}c$, **(3)** gives $\dfrac{a}{4} - \dfrac{c^2}{4a} + b = \dfrac{c}{2}$, so from

(5), $\dfrac{a}{4} - \dfrac{3a}{4} + b = \dfrac{\sqrt{3}\,a}{2} \;\Rightarrow\; -\dfrac{a}{2} - \dfrac{\sqrt{3}\,a}{2} = -b \;\Rightarrow\; b = \dfrac{a}{2}\left(1 + \sqrt{3}\right)$ **(6)**. Substituting **(5)** and **(6)** into **(4)** we get:

$2a + a\left(1 + \sqrt{3}\right) + \sqrt{3}\,a = P \;\Rightarrow\; 3a + 2\sqrt{3}\,a = P \;\Rightarrow\; a = \dfrac{P}{3 + 2\sqrt{3}} = \dfrac{2\sqrt{3} - 3}{3}P$ and thus

$b = \dfrac{\left(2\sqrt{3} - 3\right)\left(1 + \sqrt{3}\right)}{6}P = \dfrac{3 - \sqrt{3}}{6}P$ and $c = \left(2 - \sqrt{3}\right)P$.

66. (a) $\mathbf{r}(t) = x(t)\,\mathbf{i} + y(t)\,\mathbf{j} + f(x(t), y(t))\,\mathbf{k} \;\Rightarrow\; \mathbf{v} = \dfrac{d\mathbf{r}}{dt} = \dfrac{dx}{dt}\,\mathbf{i} + \dfrac{dy}{dt}\,\mathbf{j} + \left(f_x\dfrac{dx}{dt} + f_y\dfrac{dy}{dt}\right)\mathbf{k}$ (by the Chain Rule).

Therefore $\qquad K = \frac{1}{2}m\,|\mathbf{v}|^2 = \dfrac{m}{2}\left[\left(\dfrac{dx}{dt}\right)^2 + \left(\dfrac{dy}{dt}\right)^2 + \left(f_x\dfrac{dx}{dt} + f_y\dfrac{dy}{dt}\right)^2\right]$

$$= \dfrac{m}{2}\left[\left(1 + f_x^2\right)\left(\dfrac{dx}{dt}\right)^2 + 2f_xf_y\left(\dfrac{dx}{dt}\right)\left(\dfrac{dy}{dt}\right) + \left(1 + f_y^2\right)\left(\dfrac{dy}{dt}\right)^2\right]$$

(b) $\mathbf{a} = \dfrac{d\mathbf{v}}{dt} = \dfrac{d^2x}{dt^2}\mathbf{i} + \dfrac{d^2y}{dt^2}\mathbf{j} + \left[f_{xx}\left(\dfrac{dx}{dt}\right)^2 + 2f_{xy}\dfrac{dx}{dt}\dfrac{dy}{dt} + f_{yy}\left(\dfrac{dy}{dt}\right)^2 + f_x\dfrac{d^2x}{dt^2} + f_y\dfrac{d^2y}{dt^2} \right]\mathbf{k}$

(c) If $z = x^2 + y^2$, where $x = t\cos t$ and $y = t\sin t$, then $z = f(x, y) = t^2$.

$\mathbf{r} = t\cos t\,\mathbf{i} + t\sin t\,\mathbf{j} + t^2\,\mathbf{k} \quad \Rightarrow \quad \mathbf{v} = (\cos t - t\sin t)\,\mathbf{i} + (\sin t + t\cos t)\,\mathbf{j} + 2t\,\mathbf{k}$,

$K = \dfrac{m}{2}[(\cos t - t\sin t)^2 + (\sin t + t\cos t)^2 + (2t)^2] = \dfrac{m}{2}(1 + t^2 + 4t^2) = \dfrac{m}{2}(1 + 5t^2)$, and

$\mathbf{a} = (-2\sin t - t\cos t)\,\mathbf{i} + (2\cos t - t\sin t)\,\mathbf{j} + 2\,\mathbf{k}$. Notice that it is easier not to use the formulas in (a) and (b).

☐ PROBLEMS PLUS

1. The areas of the smaller rectangles are $A_1 = xy$, $A_2 = (L-x)y$,

$A_3 = (L-x)(W-y)$, $A_4 = x(W-y)$. For $0 \leq x \leq L$, $0 \leq y \leq W$, let

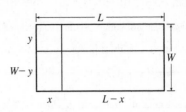

$$f(x,y) = A_1^2 + A_2^2 + A_3^2 + A_4^2$$

$$= x^2 y^2 + (L-x)^2 y^2 + (L-x)^2 (W-y)^2 + x^2 (W-y)^2$$

$$= [x^2 + (L-x)^2][y^2 + (W-y)^2]$$

Then we need to find the maximum and minimum values of $f(x,y)$. Here

$f_x(x,y) = [2x - 2(L-x)][y^2 + (W-y)^2] = 0 \quad \Rightarrow \quad 4x - 2L = 0$ or $x = \frac{1}{2}L$, and

$f_y(x,y) = [x^2 + (L-x)^2][2y - 2(W-y)] = 0 \quad \Rightarrow \quad 4y - 2W = 0$ or $y = W/2$. Also

$f_{xx} = 4[y^2 + (W-y)^2]$, $f_{yy} = 4[x^2 + (L-x)^2]$, and $f_{xy} = (4x - 2L)(4y - 2W)$. Then

$D = 16[y^2 + (W-y)^2][x^2 + (L-x)^2] - (4x - 2L)^2(4y - 2W)^2$. Thus when $x = \frac{1}{2}L$ and $y = \frac{1}{2}W$, $D > 0$ and

$f_{xx} = 2W^2 > 0$. Thus a minimum of f occurs at $\left(\frac{1}{2}L, \frac{1}{2}W\right)$ and this minimum value is $f\left(\frac{1}{2}L, \frac{1}{2}W\right) = \frac{1}{4}L^2 W^2$.

There are no other critical points, so the maximum must occur on the boundary. Now along the width of the rectangle let

$g(y) = f(0,y) = f(L,y) = L^2[y^2 + (W-y)^2]$, $0 \leq y \leq W$. Then $g'(y) = L^2[2y - 2(W-y)] = 0 \quad \Leftrightarrow \quad y = \frac{1}{2}W$.

And $g\left(\frac{1}{2}\right) = \frac{1}{2}L^2 W^2$. Checking the endpoints, we get $g(0) = g(W) = L^2 W^2$. Along the length of the rectangle let

$h(x) = f(x,0) = f(x,W) = W^2[x^2 + (L-x)^2]$, $0 \leq x \leq L$. By symmetry $h'(x) = 0 \quad \Leftrightarrow \quad x = \frac{1}{2}L$ and

$h\left(\frac{1}{2}L\right) = \frac{1}{2}L^2 W^2$. At the endpoints we have $h(0) = h(L) = L^2 W^2$. Therefore $L^2 W^2$ is the maximum value of f.

This maximum value of f occurs when the "cutting" lines correspond to sides of the rectangle.

2. (a) The level curves of the function $C(x,y) = e^{-(x^2 + 2y^2)/10^4}$ are the

curves $e^{-(x^2 + 2y^2)/10^4} = k$ (k is a positive constant). This equation is

equivalent to $x^2 + 2y^2 = K \quad \Rightarrow \quad \dfrac{x^2}{\left(\sqrt{K}\right)^2} + \dfrac{y^2}{\left(\sqrt{K/2}\right)^2} = 1$, where

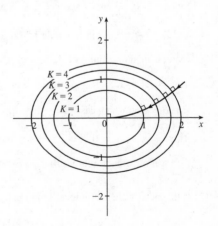

$K = -10^4 \ln k$, a family of ellipses. We sketch level curves for $K = 1$,

2, 3, and 4. If the shark always swims in the direction of maximum

increase of blood concentration, its direction at any point would coincide

with the gradient vector. Then we know the shark's path is perpendicular

to the level curves it intersects. We sketch one example of such a path.

(b) $\nabla C = -\dfrac{2}{10^4} e^{-(x^2 + 2y^2)/10^4} (x\,\mathbf{i} + 2y\,\mathbf{j})$. And ∇C points in the direction of most rapid increase in concentration; that is,

∇C is tangent to the most rapid increase curve. If $r(t) = x(t)\,\mathbf{i} + y(t)\,\mathbf{j}$ is a parametrization of the most rapid increase curve, then $\dfrac{d\mathbf{r}}{dt} = \dfrac{dx}{dt}\,\mathbf{i} + \dfrac{dy}{dt}\,\mathbf{j}$ is tangent to the curve, so $\dfrac{d\mathbf{r}}{dt} = \lambda\nabla C \;\Rightarrow\; \dfrac{dx}{dt} = \lambda\left[-\dfrac{2}{10^4}\,e^{-(x^2+2y^2)/10^4}\right]x$ and

$\dfrac{dy}{dt} = \lambda\left[-\dfrac{2}{10^4}e^{-(x^2+2y^2)/10^4}\right](2y)$. Therefore $\dfrac{dy}{dx} = \dfrac{dy/dt}{dx/dt} = 2\,\dfrac{y}{x} \;\Rightarrow\; \dfrac{dy}{y} = 2\,\dfrac{dx}{x} \;\Rightarrow\; \ln|y| = 2\ln|x|$ so that

$y = kx^2$ for some constant k. But $y(x_0) = y_0 \;\Rightarrow\; y_0 = kx_0^2 \;\Rightarrow\; k = y_0/x_0^2 \quad (x_0 = 0 \;\Rightarrow\; y_0 = 0 \;\Rightarrow\;$ the

shark is already at the origin, so we can assume $x_0 \neq 0$.) Therefore the path the shark will follow is along the parabola

$y = y_0(x/x_0)^2$.

3. (a) The area of a trapezoid is $\frac{1}{2}h(b_1 + b_2)$, where h is the height (the distance between the two parallel sides) and b_1, b_2 are

the lengths of the bases (the parallel sides). From the figure in the text, we see that $h = x\sin\theta$, $b_1 = w - 2x$, and

$b_2 = w - 2x + 2x\cos\theta$. Therefore the cross-sectional area of the rain gutter is

$$A(x,\theta) = \tfrac{1}{2}x\sin\theta\,[(w-2x) + (w-2x+2x\cos\theta)] = (x\sin\theta)(w-2x+x\cos\theta)$$
$$= wx\sin\theta - 2x^2\sin\theta + x^2\sin\theta\cos\theta,\; 0 < x \le \tfrac{1}{2}w,\, 0 < \theta \le \tfrac{\pi}{2}$$

We look for the critical points of A: $\partial A/\partial x = w\sin\theta - 4x\sin\theta + 2x\sin\theta\cos\theta$ and

$\partial A/\partial\theta = wx\cos\theta - 2x^2\cos\theta + x^2(\cos^2\theta - \sin^2\theta)$, so $\partial A/\partial x = 0 \;\Leftrightarrow\; \sin\theta\,(w - 4x + 2x\cos\theta) = 0 \;\Leftrightarrow\;$

$\cos\theta = \dfrac{4x - w}{2x} = 2 - \dfrac{w}{2x} \quad (0 < \theta \le \tfrac{\pi}{2} \;\Rightarrow\; \sin\theta > 0)$. If, in addition, $\partial A/\partial\theta = 0$, then

$$0 = wx\cos\theta - 2x^2\cos\theta + x^2(2\cos^2\theta - 1)$$
$$= wx\left(2 - \dfrac{w}{2x}\right) - 2x^2\left(2 - \dfrac{w}{2x}\right) + x^2\left[2\left(2 - \dfrac{w}{2x}\right)^2 - 1\right]$$
$$= 2wx - \tfrac{1}{2}w^2 - 4x^2 + wx + x^2\left[8 - \dfrac{4w}{x} + \dfrac{w^2}{2x^2} - 1\right] = -wx + 3x^2 = x(3x - w)$$

Since $x > 0$, we must have $x = \frac{1}{3}w$, in which case $\cos\theta = \frac{1}{2}$, so $\theta = \frac{\pi}{3}$, $\sin\theta = \frac{\sqrt{3}}{2}$, $k = \frac{\sqrt{3}}{6}w$, $b_1 = \frac{1}{3}w$, $b_2 = \frac{2}{3}w$,

and $A = \frac{\sqrt{3}}{12}w^2$. As in Example 15.7.6 [ET 14.7.6], we can argue from the physical nature of this problem that we have

found a local maximum of A. Now checking the boundary of A, let

$g(\theta) = A(w/2, \theta) = \frac{1}{2}w^2\sin\theta - \frac{1}{2}w^2\sin\theta + \frac{1}{4}w^2\sin\theta\cos\theta = \frac{1}{8}w^2\sin 2\theta$, $0 < \theta \le \frac{\pi}{2}$. Clearly g is maximized when

$\sin 2\theta = 1$ in which case $A = \frac{1}{8}w^2$. Also along the line $\theta = \frac{\pi}{2}$, let $h(x) = A\left(x, \frac{\pi}{2}\right) = wx - 2x^2$, $0 < x < \frac{1}{2}w \;\Rightarrow\;$

$h'(x) = w - 4x = 0 \;\Leftrightarrow\; x = \frac{1}{4}w$, and $h\left(\frac{1}{4}w\right) = w\left(\frac{1}{4}w\right) - 2\left(\frac{1}{4}w\right)^2 = \frac{1}{8}w^2$. Since $\frac{1}{8}w^2 < \frac{\sqrt{3}}{12}w^2$, we conclude that

the local maximum found earlier was an absolute maximum.

(b) If the metal were bent into a semi-circular gutter of radius r, we would have $w = \pi r$ and $A = \frac{1}{2}\pi r^2 = \frac{1}{2}\pi\left(\dfrac{w}{\pi}\right)^2 = \dfrac{w^2}{2\pi}$.

Since $\dfrac{w^2}{2\pi} > \dfrac{\sqrt{3}\,w^2}{12}$, it *would* be better to bend the metal into a gutter with a semicircular cross-section.

4. Since $(x + y + z)^r/(x^2 + y^2 + z^2)$ is a rational function with domain $\{(x, y, z) \mid (x, y, z) \neq (0, 0, 0)\}$, f is continuous on

$\mathbb{R}^3$ if and only if $\displaystyle\lim_{(x,y,z)\to(0,0,0)} f(x, y, z) = f(0, 0, 0) = 0$. Recall that $(a + b)^2 \le 2a^2 + 2b^2$ and a double application of this inequality to $(x + y + z)^2$ gives $(x + y + z)^2 \leq 4x^2 + 4y^2 + 2z^2 \leq 4(x^2 + y^2 + z^2)$. Now for each r,

$$|(x + y + z)^r| = \left(|x + y + z|^2\right)^{r/2} = \left[(x + y + z)^2\right]^{r/2} \leq \left[4(x^2 + y^2 + z^2)\right]^{r/2} = 2^r(x^2 + y^2 + z^2)^{r/2}$$

for $(x, y, z) \neq (0, 0, 0)$. Thus

$$|f(x, y, z) - 0| = \left|\frac{(x + y + z)^r}{x^2 + y^2 + z^2}\right| = \frac{|(x + y + z)^r|}{x^2 + y^2 + z^2} \leq 2^r \frac{(x^2 + y^2 + z^2)^{r/2}}{x^2 + y^2 + z^2} = 2^r(x^2 + y^2 + z^2)^{(r/2)-1}$$

for $(x, y, z) \neq (0, 0, 0)$. Thus if $(r/2) - 1 > 0$, that is $r > 2$, then $2^r(x^2 + y^2 + z^2)^{(r/2)-1} \to 0$ as $(x, y, z) \to (0, 0, 0)$

and so $\lim\limits_{(x,y,z) \to (0,0,0)} (x + y + z)^r / (x^2 + y^2 + z^2) = 0$. Hence for $r > 2$, f is continuous on $\mathbb{R}^3$. Now if $r \leq 2$, then as

$(x, y, z) \to (0, 0, 0)$ along the x-axis, $f(x, 0, 0) = x^r/x^2 = x^{r-2}$ for $x \neq 0$. So when $r = 2$, $f(x, y, z) \to 1 \neq 0$ as

$(x, y, z) \to (0, 0, 0)$ along the x-axis and when $r < 2$ the limit of $f(x, y, z)$ as $(x, y, z) \to (0, 0, 0)$ along the x-axis doesn't

exist and thus can't be zero. Hence for $r \leq 2$ f isn't continuous at $(0, 0, 0)$ and thus is not continuous on $\mathbb{R}^3$.

5. Let $g(x, y) = xf\left(\dfrac{y}{x}\right)$. Then $g_x(x, y) = f\left(\dfrac{y}{x}\right) + xf'\left(\dfrac{y}{x}\right)\left(-\dfrac{y}{x^2}\right) = f\left(\dfrac{y}{x}\right) - \dfrac{y}{x}f'\left(\dfrac{y}{x}\right)$ and

$g_y(x, y) = xf'\left(\dfrac{y}{x}\right)\left(\dfrac{1}{x}\right) = f'\left(\dfrac{y}{x}\right)$. Thus the tangent plane at (x_0, y_0, z_0) on the surface has equation

$$z - x_0 f\left(\frac{y_0}{x_0}\right) = \left[f\left(\frac{y_0}{x_0}\right) - y_0 x_0^{-1} f'\left(\frac{y_0}{x_0}\right)\right](x - x_0) + f'\left(\frac{y_0}{x_0}\right)(y - y_0) \quad \Rightarrow$$

$$\left[f\left(\frac{y_0}{x_0}\right) - y_0 x_0^{-1} f'\left(\frac{y_0}{x_0}\right)\right]x + \left[f'\left(\frac{y_0}{x_0}\right)\right]y - z = 0.$$ But any plane whose equation is of the form $ax + by + cz = 0$

passes through the origin. Thus the origin is the common point of intersection.

6. (a) At $(x_1, y_1, 0)$ the equations of the tangent planes to $z = f(x, y)$ and $z = g(x, y)$ are

$$P_1: \ z - f(x_1, y_1) = f_x(x_1, y_1)(x - x_1) + f_y(x_1, y_1)(y - y_1)$$

and

$$P_2: \ z - g(x_1, y_1) = g_x(x_1, y_1)(x - x_1) + g_y(x_1, y_1)(y - y_1)$$

respectively. P_1 intersects the xy-plane in the line given by $f_x(x_1, y_1)(x - x_1) + f_y(x_1, y_1)(y - y_1) = -f(x_1, y_1)$,

$z = 0$; and P_2 intersects the xy-plane in the line given by $g_x(x_1, y_1)(x - x_1) + g_y(x_1, y_1)(y - y_1) = -g(x_1, y_1)$,

$z = 0$. The point $(x_2, y_2, 0)$ is the point of intersection of these two lines, since $(x_2, y_2, 0)$ is the point where the line of

intersection of the two tangent planes intersects the xy-plane. Thus (x_2, y_2) is the solution of the simultaneous equations

$$f_x(x_1, y_1)(x_2 - x_1) + f_y(x_1, y_1)(y_2 - y_1) = -f(x_1, y_1)$$

and

$$g_x(x_1, y_1)(x_2 - x_1) + g_y(x_1, y_1)(y_2 - y_1) = -g(x_1, y_1)$$

For simplicity, rewrite $f_x(x_1, y_1)$ as f_x and similarly for f_y, g_x, g_y, f and g and solve the equations

$(f_x)(x_2 - x_1) + (f_y)(y_2 - y_1) = -f$ and $(g_x)(x_2 - x_1) + (g_y)(y_2 - y_1) = -g$ simultaneously for $(x_2 - x_1)$ and

$(y_2 - y_1)$. Then $y_2 - y_1 = \dfrac{gf_x - fg_x}{g_x f_y - f_x g_y}$ or $y_2 = y_1 - \dfrac{gf_x - fg_x}{f_x g_y - g_x f_y}$ and $(f_x)(x_2 - x_1) + \dfrac{(f_y)(gf_x - fg_x)}{g_x f_y - f_x g_y} = -f$ so

$$x_2 - x_1 = \frac{-f - [(f_y)(gf_x - fg_x)/(g_x f_y - f_x g_y)]}{f_x} = \frac{fg_y - f_y g}{g_x f_y - f_x g_y}. \text{ Hence } x_2 = x_1 - \frac{fg_y - f_y g}{f_x g_y - g_x f_y}.$$

(b) Let $f(x, y) = x^x + y^y - 1000$ and $g(x, y) = x^y + y^x - 100$. Then we wish to solve the system of equations $f(x, y) = 0$,

$g(x, y) = 0$. Recall $\dfrac{d}{dx}[x^x] = x^x(1 + \ln x)$ (differentiate logarithmically), so $f_x(x, y) = x^x(1 + \ln x)$,

$f_y(x, y) = y^y(1 + \ln y)$, $g_x(x, y) = yx^{y-1} + y^x \ln y$, and $g_y(x, y) = x^y \ln x + xy^{x-1}$. Looking at the graph, we

estimate the first point of intersection of the curves, and thus the solution to the system, to be approximately $(2.5, 4.5)$.

Then following the method of part (a), $x_1 = 2.5$, $y_1 = 4.5$ and

$$x_2 = 2.5 - \frac{f(2.5, 4.5)\, g_y(2.5, 4.5) - f_y(2.5, 4.5)\, g(2.5, 4.5)}{f_x(2.5, 4.5)\, g_y(2.5, 4.5) - f_y(2.5, 4.5)\, g_x(2.5, 4.5)} \approx 2.447674117$$

$$y_2 = 4.5 - \frac{f_x(2.5, 4.5)\, g(2.5, 4.5) - f(2.5, 4.5)\, g_x(2.5, 4.5)}{f_x(2.5, 4.5)\, g_y(2.5, 4.5) - f_y(2.5, 4.5)\, g_x(2.5, 4.5)} \approx 4.555657467$$

Continuing this procedure, we arrive at the following values. (If you use a CAS, you may need to increase its

computational precision.)

$x_1 = 2.5$	$y_1 = 4.5$
$x_2 = 2.447674117$	$y_2 = 4.555657467$
$x_3 = 2.449614877$	$y_3 = 4.551969333$
$x_4 = 2.449624628$	$y_4 = 4.551951420$
$x_5 = 2.449624628$	$y_5 = 4.551951420$

Thus, to six decimal places, the point of intersection is $(2.449625, 4.551951)$. The second point of intersection can be

found similarly, or, by symmetry it is approximately $(4.551951, 2.449625)$.

7. Since we are minimizing the area of the ellipse, and the circle lies above the x-axis,

the ellipse will intersect the circle for only one value of y. This y-value must

satisfy both the equation of the circle and the equation of the ellipse. Now

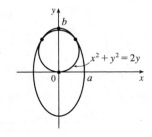

$\dfrac{x^2}{a^2} + \dfrac{y^2}{b^2} = 1 \;\Rightarrow\; x^2 = \dfrac{a^2}{b^2}(b^2 - y^2)$. Substituting into the equation of the

circle gives $\dfrac{a^2}{b^2}(b^2 - y^2) + y^2 - 2y = 0 \;\Rightarrow\; \left(\dfrac{b^2 - a^2}{b^2}\right)y^2 - 2y + a^2 = 0$.

In order for there to be only one solution to this quadratic equation, the discriminant must be 0, so $4 - 4a^2\dfrac{b^2 - a^2}{b^2} = 0 \;\Rightarrow$

$b^2 - a^2 b^2 + a^4 = 0$. The area of the ellipse is $A(a, b) = \pi ab$, and we minimize this function subject to the constraint

$g(a, b) = b^2 - a^2 b^2 + a^4 = 0$.

Now $\nabla A = \lambda \nabla g \;\Leftrightarrow\; \pi b = \lambda(4a^3 - 2ab^2)$, $\pi a = \lambda(2b - 2ba^2) \;\Rightarrow\; \lambda = \dfrac{\pi b}{2a(2a^2 - b^2)}$ **(1)**,

$\lambda = \dfrac{\pi a}{2b(1 - a^2)}$ **(2)**, $b^2 - a^2 b^2 + a^4 = 0$ **(3)**. Comparing **(1)** and **(2)** gives $\dfrac{\pi b}{2a(2a^2 - b^2)} = \dfrac{\pi a}{2b(1 - a^2)} \;\Rightarrow$

$2\pi b^2 = 4\pi a^4 \;\Leftrightarrow\; a^2 = \frac{1}{\sqrt{2}}b$. Substitute this into **(3)** to get $b = \frac{3}{\sqrt{2}} \;\Rightarrow\; a = \sqrt{\frac{3}{2}}$.

8. The tangent plane to the surface $xy^2z^2 = 1$, at the point (x_0, y_0, z_0) is

$$y_0^2 z_0^2 (x - x_0) + 2x_0 y_0 z_0^2 (y - y_0) + 2x_0 y_0^2 z_0 (z - z_0) = 0 \quad \Rightarrow \quad (y_0^2 z_0^2)x + (2x_0 y_0 z_0^2)y + (2x_0 y_0^2 z_0)z = 5x_0 y_0^2 z_0^2 = 5.$$

Using the formula derived in Example 13.5.8 [ET 12.5.8], we find that the distance from $(0,0,0)$ to this tangent plane is

$$D(x_0, y_0, z_0) = \frac{|5x_0 y_0^2 z_0^2|}{\sqrt{(y_0^2 z_0^2)^2 + (2x_0 y_0 z_0^2)^2 + (2x_0 y_0^2 z_0)^2}}.$$

When D is a maximum, D^2 is a maximum and $\nabla D^2 = \mathbf{0}$. Dropping the subscripts, let

$$f(x, y, z) = D^2 = \frac{25(xyz)^2}{y^2 z^2 + 4x^2 z^2 + 4x^2 y^2}. \text{ Now use the fact that for points on the surface } xy^2 z^2 = 1 \text{ we have } z^2 = \frac{1}{xy^2},$$

to get $f(x, y) = D^2 = \dfrac{25x}{\dfrac{1}{x} + \dfrac{4x}{y^2} + 4x^2 y^2} = \dfrac{25x^2 y^2}{y^2 + 4x^2 + 4x^3 y^4}$. Now $\nabla D^2 = \mathbf{0} \quad \Rightarrow \quad f_x = 0$ and $f_y = 0$.

$$f_x = 0 \quad \Rightarrow \quad \frac{50xy^2(y^2 + 4x^2 + 4x^3 y^4) - (8x + 12x^2 y^4)(25x^2 y^2)}{(y^2 + 4x^2 + 4x^3 y^4)^2} = 0 \quad \Rightarrow$$

$xy^2(y^2 + 4x^2 + 4x^3 y^4) - (4x + 6x^2 y^4)x^2 y^2 = 0 \quad \Rightarrow \quad xy^4 - 2x^4 y^6 = 0 \quad \Rightarrow \quad xy^4(1 - 2x^3 y^2) = 0 \quad \Rightarrow$

$1 = 2y^2 x^3$ (since $x = 0$, $y = 0$ both give a minimum distance of 0). Also $f_y = 0 \quad \Rightarrow$

$$\frac{50x^2 y(y^2 + 4x^2 + 4x^3 y^4) - (2y + 16x^3 y^3)25x^2 y^2}{(y^2 + 4x^2 + 4x^3 y^4)^2} = 0 \quad \Rightarrow \quad 4x^4 y - 4x^5 y^5 = 0 \quad \Rightarrow \quad x^4 y(1 - xy^4) = 0 \quad \Rightarrow$$

$1 = xy^4$. Now substituting $x = 1/y^4$ into $1 = 2y^2 x^3$, we get $1 = 2y^{-10} \quad \Rightarrow \quad y = \pm 2^{1/10} \quad \Rightarrow \quad x = 2^{-2/5} \quad \Rightarrow$

$z^2 = \dfrac{1}{xy^2} = \dfrac{1}{(2^{-2/5})(2^{1/5})} = 2^{1/5} \quad \Rightarrow \quad z = \pm 2^{1/10}$.

Therefore the tangent planes that are farthest from the origin are at the four points $(2^{-2/5}, \pm 2^{1/10}, \pm 2^{1/10})$. These points all give a maximum since the minimum distance occurs when $x_0 = 0$ or $y_0 = 0$ in which case $D = 0$. The equations are

$$(2^{1/5}2^{1/5})x \pm [(2)(2^{-2/5})(2^{1/10})(2^{1/5})]y \pm [(2)(2^{-2/5})(2^{1/5})(2^{1/10})]z = 5 \quad \Rightarrow \quad (2^{2/5})x \pm (2^{9/10})y \pm (2^{9/10})z = 5.$$

16 ☐ MULTIPLE INTEGRALS

16.1 Double Integrals over Rectangles

1. (a) The subrectangles are shown in the figure.

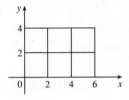

The surface is the graph of $f(x, y) = xy$ and $\Delta A = 4$, so we estimate

$$V \approx \sum_{i=1}^{3} \sum_{j=1}^{2} f(x_i, y_j) \, \Delta A$$

$$= f(2, 2) \, \Delta A + f(2, 4) \, \Delta A + f(4, 2) \, \Delta A + f(4, 4) \, \Delta A + f(6, 2) \, \Delta A + f(6, 4) \, \Delta A$$

$$= 4(4) + 8(4) + 8(4) + 16(4) + 12(4) + 24(4) = 288$$

(b) $V \approx \sum_{i=1}^{3} \sum_{j=1}^{2} f(\overline{x}_i, \overline{y}_j) \, \Delta A = f(1, 1) \, \Delta A + f(1, 3) \, \Delta A + f(3, 1) \, \Delta A + f(3, 3) \, \Delta A + f(5, 1) \, \Delta A + f(5, 3) \, \Delta A$

$$= 1(4) + 3(4) + 3(4) + 9(4) + 5(4) + 15(4) = 144$$

2. The subrectangles are shown in the figure.

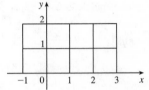

Since $\Delta A = 1$, we estimate

$$\iint_R \left(y^2 - 2x^2\right) dA \approx \sum_{i=1}^{4} \sum_{j=1}^{2} f\left(x_{ij}^*, y_{ij}^*\right) \Delta A$$

$$= f(-1, 1) \, \Delta A + f(-1, 2) \, \Delta A + f(0, 1) \, \Delta A + f(0, 2) \, \Delta A$$

$$+ f(1, 1) \, \Delta A + f(1, 2) \, \Delta A + f(2, 1) \, \Delta A + f(2, 2) \, \Delta A$$

$$= -1(1) + 2(1) + 1(1) + 4(1) - 1(1) + 2(1) - 7(1) - 4(1) = -4$$

3. (a) The subrectangles are shown in the figure. Since $\Delta A = \pi^2/4$, we estimate

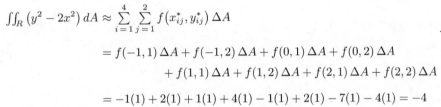

$$\iint_R \sin(x + y) \, dA \approx \sum_{i=1}^{2} \sum_{j=1}^{2} f\left(x_{ij}^*, y_{ij}^*\right) \Delta A$$

$$= f(0, 0) \, \Delta A + f\left(0, \tfrac{\pi}{2}\right) \Delta A + f\left(\tfrac{\pi}{2}, 0\right) \Delta A + f\left(\tfrac{\pi}{2}, \tfrac{\pi}{2}\right) \Delta A$$

$$= 0\left(\tfrac{\pi^2}{4}\right) + 1\left(\tfrac{\pi^2}{4}\right) + 1\left(\tfrac{\pi^2}{4}\right) + 0\left(\tfrac{\pi^2}{4}\right) = \tfrac{\pi^2}{2} \approx 4.935$$

(b) $\iint_R \sin(x + y) \, dA \approx \sum_{i=1}^{2} \sum_{j=1}^{2} f(\overline{x}_i, \overline{y}_j) \, \Delta A$

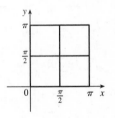

$$= f\left(\tfrac{\pi}{4}, \tfrac{\pi}{4}\right) \Delta A + f\left(\tfrac{\pi}{4}, \tfrac{3\pi}{4}\right) \Delta A + f\left(\tfrac{3\pi}{4}, \tfrac{\pi}{4}\right) \Delta A + f\left(\tfrac{3\pi}{4}, \tfrac{3\pi}{4}\right) \Delta A$$

$$= 1\left(\tfrac{\pi^2}{4}\right) + 0\left(\tfrac{\pi^2}{4}\right) + 0\left(\tfrac{\pi^2}{4}\right) + (-1)\left(\tfrac{\pi^2}{4}\right) = 0$$

4. (a) The subrectangles are shown in the figure.

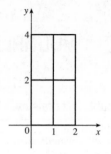

The surface is the graph of $f(x,y) = x + 2y^2$ and $\Delta A = 2$, so we estimate

$$V = \iint_R (x + 2y^2)\, dA \approx \sum_{i=1}^{2} \sum_{j=1}^{2} f(x_{ij}^*, y_{ij}^*)\, \Delta A$$

$$= f(1,0)\, \Delta A + f(1,2)\, \Delta A + f(2,0)\, \Delta A + f(2,2)\, \Delta A$$

$$= 1(2) + 9(2) + 2(2) + 10(2) = 44$$

(b) $V = \iint_R (x + 2y^2)\, dA \approx \sum_{i=1}^{2} \sum_{j=1}^{2} f(\overline{x}_i, \overline{y}_j)\, \Delta A$

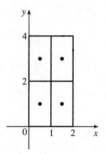

$$= f\left(\tfrac{1}{2}, 1\right) \Delta A + f\left(\tfrac{1}{2}, 3\right) \Delta A + f\left(\tfrac{3}{2}, 1\right) \Delta A + f\left(\tfrac{3}{2}, 3\right) \Delta A$$

$$= \tfrac{5}{2}(2) + \tfrac{37}{2}(2) + \tfrac{7}{2}(2) + \tfrac{39}{2}(2) = 88$$

5. (a) Each subrectangle and its midpoint are shown in the figure. The area of each subrectangle is $\Delta A = 2$, so we evaluate f at each midpoint and estimate

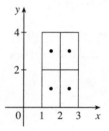

$$\iint_R f(x,y)\, dA \approx \sum_{i=1}^{2} \sum_{j=1}^{2} f(\overline{x}_i, \overline{y}_j)\, \Delta A$$

$$= f(1.5, 1)\, \Delta A + f(1.5, 3)\, \Delta A$$

$$\qquad + f(2.5, 1)\, \Delta A + f(2.5, 3)\, \Delta A$$

$$= 1(2) + (-8)(2) + 5(2) + (-1)(2) = -6$$

(b) The subrectangles are shown in the figure. In each subrectangle, the sample point farthest from the origin is the upper right corner, and the area of each subrectangle is $\Delta A = \tfrac{1}{2}$. Thus we estimate

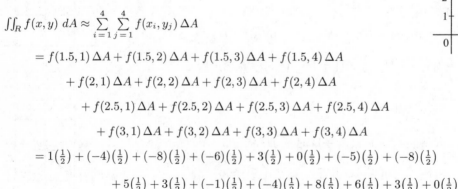

$$\iint_R f(x,y)\, dA \approx \sum_{i=1}^{4} \sum_{j=1}^{4} f(x_i, y_j)\, \Delta A$$

$$= f(1.5, 1)\, \Delta A + f(1.5, 2)\, \Delta A + f(1.5, 3)\, \Delta A + f(1.5, 4)\, \Delta A$$

$$\quad + f(2, 1)\, \Delta A + f(2, 2)\, \Delta A + f(2, 3)\, \Delta A + f(2, 4)\, \Delta A$$

$$\quad + f(2.5, 1)\, \Delta A + f(2.5, 2)\, \Delta A + f(2.5, 3)\, \Delta A + f(2.5, 4)\, \Delta A$$

$$\quad + f(3, 1)\, \Delta A + f(3, 2)\, \Delta A + f(3, 3)\, \Delta A + f(3, 4)\, \Delta A$$

$$= 1\left(\tfrac{1}{2}\right) + (-4)\left(\tfrac{1}{2}\right) + (-8)\left(\tfrac{1}{2}\right) + (-6)\left(\tfrac{1}{2}\right) + 3\left(\tfrac{1}{2}\right) + 0\left(\tfrac{1}{2}\right) + (-5)\left(\tfrac{1}{2}\right) + (-8)\left(\tfrac{1}{2}\right)$$

$$\qquad + 5\left(\tfrac{1}{2}\right) + 3\left(\tfrac{1}{2}\right) + (-1)\left(\tfrac{1}{2}\right) + (-4)\left(\tfrac{1}{2}\right) + 8\left(\tfrac{1}{2}\right) + 6\left(\tfrac{1}{2}\right) + 3\left(\tfrac{1}{2}\right) + 0\left(\tfrac{1}{2}\right)$$

$$= -3.5$$

6. To approximate the volume, let R be the planar region corresponding to the surface of the water in the pool, and place R on coordinate axes so that x and y correspond to the dimensions given. Then we define $f(x, y)$ to be the depth of the water at (x, y), so the volume of water in the pool is the volume of the solid that lies above the rectangle $R = [0, 20] \times [0, 30]$ and below the graph of $f(x, y)$. We can estimate this volume using the Midpoint Rule with $m = 2$ and $n = 3$, so $\Delta A = 100$. Each subrectangle with its midpoint is shown in the figure. Then

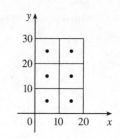

$$V \approx \sum_{i=1}^{2} \sum_{j=1}^{3} f(\overline{x}_i, \overline{y}_j)\,\Delta A = \Delta A[f(5, 5) + f(5, 15) + f(5, 25) + f(15, 5) + f(15, 15) + f(15, 25)]$$

$$= 100(3 + 7 + 10 + 3 + 5 + 8) = 3600$$

Thus, we estimate that the pool contains 3600 cubic feet of water.

Alternatively, we can approximate the volume with a Riemann sum where $m = 4$, $n = 6$ and the sample points are taken to be, for example, the upper right corner of each subrectangle. Then $\Delta A = 25$ and

$$V \approx \sum_{i=1}^{4} \sum_{j=1}^{6} f(x_i, y_j)\,\Delta A$$

$$= 25[3 + 4 + 7 + 8 + 10 + 8 + 4 + 6 + 8 + 10 + 12 + 10 + 3 + 4 + 5 + 6 + 8 + 7 + 2 + 2 + 2 + 3 + 4 + 4]$$

$$= 25(140) = 3500$$

So we estimate that the pool contains 3500 ft^3 of water.

7. The values of $f(x, y) = \sqrt{52 - x^2 - y^2}$ get smaller as we move farther from the origin, so on any of the subrectangles in the problem, the function will have its largest value at the lower left corner of the subrectangle and its smallest value at the upper right corner, and any other value will lie between these two. So using these subrectangles we have $U < V < L$. (Note that this is true no matter how R is divided into subrectangles.)

8. Divide R into 4 equal rectangles (squares) and identify the midpoint of each subrectangle as shown in the figure.

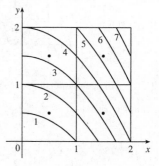

The area of each subrectangle is $\Delta A = 1$, so using the contour map to estimate the function values at each midpoint, we have

$$\iint_R f(x, y)\,dA \approx \sum_{i=1}^{2} \sum_{j=1}^{2} f(\overline{x}_i, \overline{y}_j)\,\Delta A = f\left(\tfrac{1}{2}, \tfrac{1}{2}\right)\Delta A + f\left(\tfrac{1}{2}, \tfrac{3}{2}\right)\Delta A + f\left(\tfrac{3}{2}, \tfrac{1}{2}\right)\Delta A + f\left(\tfrac{3}{2}, \tfrac{3}{2}\right)\Delta A$$

$$\approx (1.3)(1) + (3.3)(1) + (3.2)(1) + (5.2)(1) = 13.0$$

You could improve the estimate by increasing m and n to use a larger number of smaller subrectangles.

9. (a) With $m = n = 2$, we have $\Delta A = 4$. Using the contour map to estimate the value of f at the center of each subrectangle, we have

$$\iint_R f(x,y)\, dA \approx \sum_{i=1}^{2} \sum_{j=1}^{2} f(\overline{x}_i, \overline{y}_j)\, \Delta A = \Delta A[f(1,1) + f(1,3) + f(3,1) + f(3,3)] \approx 4(27 + 4 + 14 + 17) = 248$$

(b) $f_{\text{ave}} = \frac{1}{A(R)} \iint_R f(x,y)\, dA \approx \frac{1}{16}(248) = 15.5$

10. As in Example 4, we place the origin at the southwest corner of the state. Then $R = [0, 388] \times [0, 276]$ (in miles) is the rectangle corresponding to Colorado and we define $f(x,y)$ to be the temperature at the location (x,y). The average temperature is given by

$$f_{\text{ave}} = \frac{1}{A(R)} \iint_R f(x,y)\, dA = \frac{1}{388 \cdot 276} \iint_R f(x,y)\, dA$$

To use the Midpoint Rule with $m = n = 4$, we divide R into 16 regions of equal size, as shown in the figure, with the center of each subrectangle indicated.

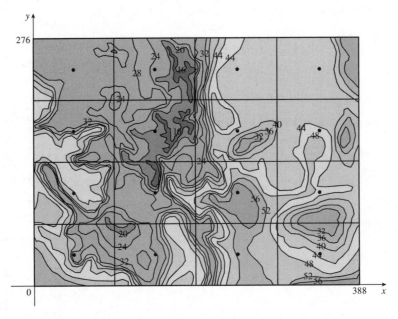

The area of each subrectangle is $\Delta A = \frac{388}{4} \cdot \frac{276}{4} = 6693$, so using the contour map to estimate the function values at each midpoint, we have

$$\iint_R f(x,y)\, dA \sum_{i=1}^{4} \sum_{j=1}^{4} f(\overline{x}_i, \overline{y}_j)\, \Delta A$$

$$\approx \Delta A\, [31 + 28 + 52 + 43 + 43 + 25 + 57 + 46 + 36 + 20 + 42 + 45 + 30 + 23 + 43 + 41]$$

$$= 6693(605)$$

Therefore, $f_{\text{ave}} \approx \dfrac{6693 \cdot 605}{388 \cdot 276} \approx 37.8$, so the average temperature in Colorado at 4:00 PM on February 26, 2007, was approximately $37.8°\text{F}$.

11. $z = 3 > 0$, so we can interpret the integral as the volume of the solid S that lies below the plane $z = 3$ and above the rectangle $[-2, 2] \times [1, 6]$. S is a rectangular solid, thus $\iint_R 3 \, dA = 4 \cdot 5 \cdot 3 = 60$.

12. $z = 5 - x \geq 0$ for $0 \leq x \leq 5$, so we can interpret the integral as the volume of the solid S that lies below the plane $z = 5 - x$ and above the rectangle $[0, 5] \times [0, 3]$. S is a triangular cylinder whose volume is $3(\text{area of triangle}) = 3\left(\frac{1}{2} \cdot 5 \cdot 5\right) = 37.5$. Thus

$$\iint_R (5 - x) \, dA = 37.5$$

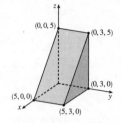

13. $z = f(x, y) = 4 - 2y \geq 0$ for $0 \leq y \leq 1$. Thus the integral represents the volume of that part of the rectangular solid $[0, 1] \times [0, 1] \times [0, 4]$ which lies below the plane $z = 4 - 2y$. So

$$\iint_R (4 - 2y) \, dA = (1)(1)(2) + \tfrac{1}{2}(1)(1)(2) = 3$$

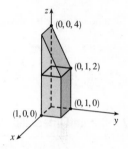

14. Here $z = \sqrt{9 - y^2}$, so $z^2 + y^2 = 9$, $z \geq 0$. Thus the integral represents the volume of the top half of the part of the circular cylinder $z^2 + y^2 = 9$ that lies above the rectangle $[0, 4] \times [0, 2]$.

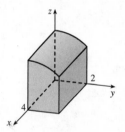

15. To calculate the estimates using a programmable calculator, we can use an algorithm similar to that of Exercise 5.1.7 [ET 5.1.7]. In Maple, we can define the function $f(x, y) = \sqrt{1 + xe^{-y}}$ (calling it f), load the student package, and then use the command

```
middlesum(middlesum(f,x=0..1,m),
          y=0..1,m);
```

to get the estimate with $n = m^2$ squares of equal size. Mathematica has no special Riemann sum command, but we can define f and then use nested Sum commands to calculate the estimates.

n	estimate
1	1.141606
4	1.143191
16	1.143535
64	1.143617
256	1.143637
1024	1.143642

16.

n	estimate
1	0.934591
4	0.881991
16	0.865750

n	estimate
64	0.860490
256	0.858745
1024	0.858157

17. If we divide R into mn subrectangles, $\iint_R k \, dA \approx \sum_{i=1}^{m} \sum_{j=1}^{n} f\left(x_{ij}^*, y_{ij}^*\right) \Delta A$ for any choice of sample points $\left(x_{ij}^*, y_{ij}^*\right)$.

But $f\left(x_{ij}^*, y_{ij}^*\right) = k$ always and $\sum_{i=1}^{m} \sum_{j=1}^{n} \Delta A = $ area of $R = (b-a)(d-c)$. Thus, no matter how we choose the sample

points, $\sum_{i=1}^{m} \sum_{j=1}^{n} f\left(x_{ij}^*, y_{ij}^*\right) \Delta A = k \sum_{i=1}^{m} \sum_{j=1}^{n} \Delta A = k(b-a)(d-c)$ and so

$$\iint_R k \, dA = \lim_{m,n \to \infty} \sum_{i=1}^{m} \sum_{j=1}^{n} f\left(x_{ij}^*, y_{ij}^*\right) \Delta A = \lim_{m,n \to \infty} k \sum_{i=1}^{m} \sum_{j=1}^{n} \Delta A = \lim_{m,n \to \infty} k(b-a)(d-c) = k(b-a)(d-c).$$

18. Because $\sin \pi x$ is an increasing function for $0 \le x \le \frac{1}{4}$, we have $\sin 0 \le \sin \pi x \le \sin \frac{\pi}{4} \;\; \Rightarrow \;\; 0 \le \sin \pi x \le \frac{\sqrt{2}}{2}$.

Similarly, $\cos \pi y$ is a decreasing function for $\frac{1}{4} \le y \le \frac{1}{2}$, so $0 = \cos \frac{\pi}{2} \le \cos \pi y \le \cos \frac{\pi}{4} = \frac{\sqrt{2}}{2}$. Thus on R,

$0 \le \sin \pi x \, \cos \pi y \le \frac{\sqrt{2}}{2} \cdot \frac{\sqrt{2}}{2} = \frac{1}{2}$. Property (9) gives $\iint_R 0 \, dA \le \iint_R \sin \pi x \cos \pi y \, dA \le \iint_R \frac{1}{2} \, dA$, so by Exercise 17 we

have $0 \le \iint_R \sin \pi x \, \cos \pi y \, dA \le \frac{1}{2} \left(\frac{1}{4} - 0\right) \left(\frac{1}{2} - \frac{1}{4}\right) = \frac{1}{32}$.

16.2 Iterated Integrals

1. $\int_0^5 12x^2 y^3 \, dx = \left[12 \dfrac{x^3}{3} y^3\right]_{x=0}^{x=5} = 4x^3 y^3 \big]_{x=0}^{x=5} = 4(5)^3 \, y^3 - 4(0)^3 \, y^3 = 500y^3$,

$\int_0^1 12x^2 y^3 \, dy = \left[12x^2 \dfrac{y^4}{4}\right]_{y=0}^{y=1} = 3x^2 y^4 \big]_{y=0}^{y=1} = 3x^2 (1)^4 - 3x^2 (0)^4 = 3x^2$

2. $\int_0^5 (y + xe^y) \, dx = \left[xy + \dfrac{x^2}{2} e^y\right]_{x=0}^{x=5} = \left(5y + \frac{25}{2} e^y\right) - (0 + 0) = 5y + \frac{25}{2} e^y$,

$\int_0^1 (y + xe^y) \, dy = \left[\dfrac{y^2}{2} + xe^y\right]_{y=0}^{y=1} = \left(\frac{1}{2} + xe^1\right) - (0 + xe^0) = \frac{1}{2} + ex - x$

3. $\int_1^3 \int_0^1 (1 + 4xy) \, dx \, dy = \int_1^3 \left[x + 2x^2 y\right]_{x=0}^{x=1} dy = \int_1^3 (1 + 2y) \, dy = \left[y + y^2\right]_1^3 = (3 + 9) - (1 + 1) = 10$

4. $\int_0^1 \int_1^2 (4x^3 - 9x^2 y^2) \, dy \, dx = \int_0^1 \left[4x^3 y - 3x^2 y^3\right]_{y=1}^{y=2} dx = \int_0^1 \left[(8x^3 - 24x^2) - (4x^3 - 3x^2)\right] dx$

$\qquad = \int_0^1 (4x^3 - 21x^2) \, dx = \left[x^4 - 7x^3\right]_0^1 = (1 - 7) - (0 - 0) = -6$

5. $\int_0^2 \int_0^{\pi/2} x \sin y \, dy \, dx = \int_0^2 x \, dx \int_0^{\pi/2} \sin y \, dy$ [as in Example 5] $= \left[\dfrac{x^2}{2}\right]_0^2 \left[-\cos y\right]_0^{\pi/2} = (2 - 0)(0 + 1) = 2$

6. $\int_{\pi/6}^{\pi/2} \int_{-1}^{5} \cos y \, dx \, dy = \int_{-1}^{5} dx \int_{\pi/6}^{\pi/2} \cos y \, dy$ [by Equation 5]

$\qquad = \left[x\right]_{-1}^{5} \left[\sin y\right]_{\pi/6}^{\pi/2} = [5 - (-1)](\sin \frac{\pi}{2} - \sin \frac{\pi}{6}) = 6(1 - \frac{1}{2}) = 3$

7. $\int_0^2 \int_0^1 (2x + y)^8 \, dx \, dy = \int_0^2 \left[\dfrac{1}{2} \dfrac{(2x + y)^9}{9}\right]_{x=0}^{x=1} dy$ [substitute $u = 2x + y \;\; \Rightarrow \;\; dx = \frac{1}{2} du$]

$\qquad = \dfrac{1}{18} \int_0^2 \left[(2 + y)^9 - (0 + y)^9\right] dy = \dfrac{1}{18} \left[\dfrac{(2 + y)^{10}}{10} - \dfrac{y^{10}}{10}\right]_0^2$

$\qquad = \frac{1}{180} \left[(4^{10} - 2^{10}) - (2^{10} - 0^{10})\right] = \frac{1,046,528}{180} = \frac{261,632}{45}$

8. $\int_0^1 \int_1^2 \dfrac{xe^x}{y}\, dy\, dx = \int_0^1 xe^x\, dx \int_1^2 \dfrac{1}{y}\, dy$ [as in Example 5] $= \big[xe^x - e^x\big]_0^1 \big[\ln|y|\big]_1^2$ [by integrating by parts]

$$= [(e - e) - (0 - 1)](\ln 2 - 0) = \ln 2$$

9. $\int_1^4 \int_1^2 \left(\dfrac{x}{y} + \dfrac{y}{x}\right) dy\, dx = \int_1^4 \left[x \ln|y| + \dfrac{1}{x} \cdot \dfrac{1}{2}\, y^2\right]_{y=1}^{y=2} dx = \int_1^4 \left(x \ln 2 + \dfrac{3}{2x}\right) dx = \left[\tfrac{1}{2} x^2 \ln 2 + \tfrac{3}{2} \ln|x|\right]_1^4$

$$= 8\ln 2 + \tfrac{3}{2} \ln 4 - \tfrac{1}{2} \ln 2 = \tfrac{15}{2} \ln 2 + 3 \ln 4^{1/2} = \tfrac{21}{2} \ln 2$$

10. $\int_0^1 \int_0^3 e^{x+3y}\, dx\, dy = \int_0^1 \int_0^3 e^x e^{3y}\, dx\, dy = \int_0^3 e^x\, dx \int_0^1 e^{3y}\, dy = \big[e^x\big]_0^3 \left[\tfrac{1}{3} e^{3y}\right]_0^1$

$$= (e^3 - e^0) \cdot \tfrac{1}{3}(e^3 - e^0) = \tfrac{1}{3}(e^3 - 1)^2 \text{ or } \tfrac{1}{3}(e^6 - 2e^3 + 1)$$

11. $\int_0^1 \int_0^1 (u - v)^5\, du\, dv = \int_0^1 \left[\tfrac{1}{6}(u - v)^6\right]_{u=0}^{u=1} dv = \tfrac{1}{6} \int_0^1 \left[(1 - v)^6 - (0 - v)^6\right] dv$

$$= \tfrac{1}{6} \int_0^1 \left[(1 - v)^6 - v^6\right] dv = \tfrac{1}{6}\left[-\tfrac{1}{7}(1 - v)^7 - \tfrac{1}{7} v^7\right]_0^1$$

$$= -\tfrac{1}{42}\big[(0 + 1) - (1 + 0)\big] = 0$$

12. $\int_0^1 \int_0^1 xy\sqrt{x^2 + y^2}\, dy\, dx = \int_0^1 x\left[\tfrac{1}{3}(x^2 + y^2)^{3/2}\right]_{y=0}^{y=1} dx = \tfrac{1}{3} \int_0^1 x\left[(x^2 + 1)^{3/2} - x^3\right] dx = \tfrac{1}{3} \int_0^1 [x(x^2 + 1)^{3/2} - x^4]\, dx$

$$= \tfrac{1}{3}\left[\tfrac{1}{5}(x^2 + 1)^{5/2} - \tfrac{1}{5}x^5\right]_0^1 = \tfrac{1}{15}\left[2^{5/2} - 1 - 1 + 0\right] = \tfrac{2}{15}(2\sqrt{2} - 1)$$

13. $\int_0^2 \int_0^\pi r \sin^2 \theta\, d\theta\, dr = \int_0^2 r\, dr \int_0^\pi \sin^2 \theta\, d\theta$ [as in Example 5] $= \int_0^2 r\, dr \int_0^\pi \tfrac{1}{2}(1 - \cos 2\theta)\, d\theta$

$$= \left[\tfrac{1}{2}r^2\right]_0^2 \cdot \tfrac{1}{2}\left[\theta - \tfrac{1}{2}\sin 2\theta\right]_0^\pi = (2 - 0) \cdot \tfrac{1}{2}\left[\left(\pi - \tfrac{1}{2}\sin 2\pi\right) - \left(0 - \tfrac{1}{2}\sin 0\right)\right]$$

$$= 2 \cdot \tfrac{1}{2}[(\pi - 0) - (0 - 0)] = \pi$$

14. $\int_0^1 \int_0^1 \sqrt{s + t}\, ds\, dt = \int_0^1 \left[\tfrac{2}{3}(s + t)^{3/2}\right]_{s=0}^{s=1} dt = \tfrac{2}{3} \int_0^1 [(1 + t)^{3/2} - t^{3/2}]\, dt = \tfrac{2}{3}\left[\tfrac{2}{5}(1 + t)^{5/2} - \tfrac{2}{5}t^{5/2}\right]_0^1$

$$= \tfrac{4}{15}[(2^{5/2} - 1) - (1 - 0)] = \tfrac{4}{15}\left(2^{5/2} - 2\right) \text{ or } \tfrac{8}{15}(2\sqrt{2} - 1)$$

15. $\iint_R (6x^2 y^3 - 5y^4)\, dA = \int_0^3 \int_0^1 (6x^2 y^3 - 5y^4)\, dy\, dx = \int_0^3 \left[\tfrac{3}{2} x^2 y^4 - y^5\right]_{y=0}^{y=1} dx = \int_0^3 \left(\tfrac{3}{2} x^2 - 1\right) dx$

$$= \left[\tfrac{1}{2} x^3 - x\right]_0^3 = \tfrac{27}{2} - 3 = \tfrac{21}{2}$$

16. $\iint_R \cos(x + 2y)\, dA = \int_0^\pi \int_0^{\pi/2} \cos(x + 2y)\, dy\, dx = \int_0^\pi \left[\tfrac{1}{2}\sin(x + 2y)\right]_{y=0}^{y=\pi/2} dx = \tfrac{1}{2} \int_0^\pi (\sin(x + \pi) - \sin x)\, dx$

$$= \tfrac{1}{2}\left[-\cos(x + \pi) + \cos x\right]_0^\pi = \tfrac{1}{2}\left[-\cos 2\pi + \cos \pi - (-\cos \pi + \cos 0)\right]$$

$$= \tfrac{1}{2}(-1 - 1 - (1 + 1)) = -2$$

17. $\iint_R \dfrac{xy^2}{x^2 + 1}\, dA = \int_0^1 \int_{-3}^3 \dfrac{xy^2}{x^2 + 1}\, dy\, dx = \int_0^1 \dfrac{x}{x^2 + 1}\, dx \int_{-3}^3 y^2\, dy = \left[\dfrac{1}{2} \ln(x^2 + 1)\right]_0^1 \left[\dfrac{1}{3} y^3\right]_{-3}^3$

$$= \tfrac{1}{2}(\ln 2 - \ln 1) \cdot \tfrac{1}{3}(27 + 27) = 9 \ln 2$$

18. $\iint_R \dfrac{1 + x^2}{1 + y^2}\, dA = \int_0^1 \int_0^1 \dfrac{1 + x^2}{1 + y^2}\, dy\, dx = \int_0^1 (1 + x^2)\, dx \int_0^1 \dfrac{1}{1 + y^2}\, dy = \left[x + \dfrac{1}{3} x^3\right]_0^1 \left[\tan^{-1} y\right]_0^1$

$$= \left(1 + \tfrac{1}{3} - 0\right)\left(\tfrac{\pi}{4} - 0\right) = \tfrac{\pi}{3}$$

19. $\int_0^{\pi/6} \int_0^{\pi/3} x \sin(x+y) \, dy \, dx$

$= \int_0^{\pi/6} \left[-x \cos(x+y) \right]_{y=0}^{y=\pi/3} dx = \int_0^{\pi/6} \left[x \cos x - x \cos\left(x + \frac{\pi}{3}\right) \right] dx$

$= x \left[\sin x - \sin\left(x + \frac{\pi}{3}\right) \right]_0^{\pi/6} - \int_0^{\pi/6} \left[\sin x - \sin\left(x + \frac{\pi}{3}\right) \right] dx$ [by integrating by parts separately for each term]

$= \frac{\pi}{6} \left[\frac{1}{2} - 1 \right] - \left[-\cos x + \cos\left(x + \frac{\pi}{3}\right) \right]_0^{\pi/6} = -\frac{\pi}{12} - \left[-\frac{\sqrt{3}}{2} + 0 - \left(-1 + \frac{1}{2}\right) \right] = \frac{\sqrt{3}-1}{2} - \frac{\pi}{12}$

20. $\iint_R \frac{x}{1+xy} \, dA = \int_0^1 \int_0^1 \frac{x}{1+xy} \, dy \, dx = \int_0^1 \left[\ln(1+xy) \right]_{y=0}^{y=1} dx = \int_0^1 \left[\ln(1+x) - \ln 1 \right] dx$

$= \int_0^1 \ln(1+x) \, dx = \left[(1+x)\ln(1+x) - x \right]_0^1$ [by integrating by parts]

$= (2\ln 2 - 1) - (\ln 1 - 0) = 2\ln 2 - 1$

21. $\iint_R xye^{x^2 y} \, dA = \int_0^2 \int_0^1 xye^{x^2 y} \, dx \, dy = \int_0^2 \left[\frac{1}{2} e^{x^2 y} \right]_{x=0}^{x=1} dy = \frac{1}{2} \int_0^2 (e^y - 1) \, dy = \frac{1}{2} \left[e^y - y \right]_0^2$

$= \frac{1}{2} [(e^2 - 2) - (1 - 0)] = \frac{1}{2}(e^2 - 3)$

22. $\int_0^1 \int_1^2 \frac{x}{x^2 + y^2} \, dx \, dy = \int_0^1 \left[\frac{1}{2} \ln(x^2 + y^2) \right]_{x=1}^{x=2} dy = \frac{1}{2} \int_0^1 \left[\ln(4 + y^2) - \ln(1 + y^2) \right] dy$

To evaluate the first term, we integrate by parts with $u = \ln(4 + y^2) \implies du = \dfrac{2y}{4 + y^2} \, dy$ and

$dv = dy \implies v = y$. Then

$$\int \ln(4 + y^2) \, dy = y\ln(4 + y^2) - \int \frac{2y^2}{4 + y^2} \, dy = y\ln(4 + y^2) - \int \left(2 - \frac{8}{4 + y^2} \right) dy$$

$$= y\ln(4 + y^2) - 2y + 8 \cdot \frac{1}{2} \tan^{-1}\left(\frac{y}{2}\right) = y\ln(4 + y^2) - 2y + 4\tan^{-1}\left(\frac{y}{2}\right)$$

Similarly, $\int \ln(1 + y^2) \, dy = y\ln(1 + y^2) - 2y + 2\tan^{-1} y$. Thus,

$$\int_0^1 \int_1^2 \frac{x}{x^2 + y^2} \, dx \, dy = \frac{1}{2} \int_0^1 \left[\ln(4 + y^2) - \ln(1 + y^2) \right] dy$$

$$= \frac{1}{2} \left[y\ln(4 + y^2) - 2y + 4\tan^{-1}\left(\frac{y}{2}\right) - y\ln(1 + y^2) + 2y - 2\tan^{-1} y \right]_0^1$$

$$= \frac{1}{2} \left[\left(\ln 5 + 4\tan^{-1}\left(\frac{1}{2}\right) - \ln 2 - 2\tan^{-1} 1 \right) - 0 \right]$$

$$= \frac{1}{2} \left[\ln 5 - \ln 2 + 4\tan^{-1}\left(\frac{1}{2}\right) - 2\left(\frac{\pi}{4}\right) \right] = \frac{1}{2} \ln \frac{5}{2} + 2\tan^{-1}\left(\frac{1}{2}\right) - \frac{\pi}{4}$$

23. $z = f(x,y) = 4 - x - 2y \geq 0$ for $0 \leq x \leq 1$ and $0 \leq y \leq 1$. So the solid

is the region in the first octant which lies below the plane $z = 4 - x - 2y$

and above $[0, 1] \times [0, 1]$.

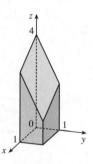

24. $z = 2 - x^2 - y^2 \geq 0$ for $0 \leq x \leq 1$ and $0 \leq y \leq 1$. So the solid is the

region in the first octant which lies below the circular paraboloid

$z = 2 - x^2 - y^2$ and above $[0, 1] \times [0, 1]$.

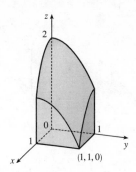

25. $V = \iint_R (12 - 3x - 2y)\, dA = \int_{-2}^{3} \int_{0}^{1} (12 - 3x - 2y)\, dx\, dy = \int_{-2}^{3} \left[12x - \frac{3}{2}x^2 - 2xy \right]_{x=0}^{x=1} dy$

$= \int_{-2}^{3} \left(\frac{21}{2} - 2y \right) dy = \left[\frac{21}{2}y - y^2 \right]_{-2}^{3} = \frac{95}{2}$

26. $V = \iint_R (4 + x^2 - y^2)\, dA = \int_{-1}^{1} \int_{0}^{2} (4 + x^2 - y^2)\, dy\, dx = \int_{-1}^{1} \left[4y + x^2 y - \frac{1}{3}y^3 \right]_{y=0}^{y=2} dx$

$= \int_{-1}^{1} \left(2x^2 + \frac{16}{3} \right) dx = \left[\frac{2}{3}x^3 + \frac{16}{3}x \right]_{-1}^{1} = \frac{2}{3} + \frac{16}{3} + \frac{2}{3} + \frac{16}{3} = 12$

27. $V = \int_{-2}^{2} \int_{-1}^{1} \left(1 - \frac{1}{4}x^2 - \frac{1}{9}y^2 \right) dx\, dy = 4 \int_{0}^{2} \int_{0}^{1} \left(1 - \frac{1}{4}x^2 - \frac{1}{9}y^2 \right) dx\, dy$

$= 4 \int_{0}^{2} \left[x - \frac{1}{12}x^3 - \frac{1}{9}y^2 x \right]_{x=0}^{x=1} dy = 4 \int_{0}^{2} \left(\frac{11}{12} - \frac{1}{9}y^2 \right) dy = 4 \left[\frac{11}{12}y - \frac{1}{27}y^3 \right]_{0}^{2} = 4 \cdot \frac{83}{54} = \frac{166}{27}$

28. $V = \int_{-1}^{1} \int_{0}^{\pi} (1 + e^x \sin y)\, dy\, dx = \int_{-1}^{1} \left[y - e^x \cos y \right]_{y=0}^{y=\pi} dx = \int_{-1}^{1} (\pi + e^x - 0 + e^x)\, dx$

$= \int_{-1}^{1} (\pi + 2e^x)\, dx = \left[\pi x + 2e^x \right]_{-1}^{1} = 2\pi + 2e - \frac{2}{e}$

29. Here we need the volume of the solid lying under the surface $z = x \sec^2 y$ and above the rectangle $R = [0, 2] \times [0, \pi/4]$ in the

xy-plane.

$$V = \int_{0}^{2} \int_{0}^{\pi/4} x \sec^2 y\, dy\, dx = \int_{0}^{2} x\, dx \int_{0}^{\pi/4} \sec^2 y\, dy = \left[\frac{1}{2}x^2 \right]_{0}^{2} \left[\tan y \right]_{0}^{\pi/4}$$

$$= (2 - 0)(\tan \tfrac{\pi}{4} - \tan 0) = 2(1 - 0) = 2$$

30. The cylinder intersects the xy-plane along the line $x = 4$, so in the first octant, the solid lies below the surface $z = 16 - x^2$

and above the rectangle $R = [0, 4] \times [0, 5]$ in the xy-plane.

$$V = \int_{0}^{5} \int_{0}^{4} (16 - x^2)\, dx\, dy = \int_{0}^{4} (16 - x^2)\, dx \int_{0}^{5} dy = \left[16x - \frac{1}{3}x^3 \right]_{0}^{4} \left[y \right]_{0}^{5} = \left(64 - \frac{64}{3} - 0 \right)(5 - 0) = \frac{640}{3}$$

31. The solid lies below the surface $z = 2 + x^2 + (y - 2)^2$ and above the plane $z = 1$ for $-1 \leq x \leq 1, 0 \leq y \leq 4$. The volume

of the solid is the difference in volumes between the solid that lies under $z = 2 + x^2 + (y - 2)^2$ over the rectangle

$R = [-1, 1] \times [0, 4]$ and the solid that lies under $z = 1$ over R.

$$V = \int_{0}^{4} \int_{-1}^{1} [2 + x^2 + (y - 2)^2]\, dx\, dy - \int_{0}^{4} \int_{-1}^{1} (1)\, dx\, dy = \int_{0}^{4} \left[2x + \frac{1}{3}x^3 + x(y - 2)^2 \right]_{x=-1}^{x=1} dy - \int_{-1}^{1} dx \int_{0}^{4} dy$$

$$= \int_{0}^{4} \left[(2 + \tfrac{1}{3} + (y - 2)^2) - (-2 - \tfrac{1}{3} - (y - 2)^2) \right] dy - \left[x \right]_{-1}^{1} \left[y \right]_{0}^{4}$$

$$= \int_{0}^{4} \left[\tfrac{14}{3} + 2(y - 2)^2 \right] dy - [1 - (-1)][4 - 0] = \left[\tfrac{14}{3}y + \tfrac{2}{3}(y - 2)^3 \right]_{0}^{4} - (2)(4)$$

$$= \left[\left(\tfrac{56}{3} + \tfrac{16}{3} \right) - \left(0 - \tfrac{16}{3} \right) \right] - 8 = \frac{88}{3} - 8 = \frac{64}{3}$$

32.

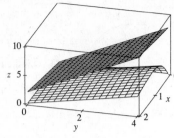

The solid lies below the plane $z = x + 2y$ and above the surface

$z = \dfrac{2xy}{x^2 + 1}$ for $0 \leq x \leq 2$, $0 \leq y \leq 4$. The volume of the solid is

the difference in volumes between the solid that lies under

$z = x + 2y$ over the rectangle $R = [0, 2] \times [0, 4]$ and the solid that

lies under $z = \dfrac{2xy}{x^2 + 1}$ over R.

$$V = \int_0^2 \int_0^4 (x + 2y)\,dy\,dx - \int_0^2 \int_0^4 \frac{2xy}{x^2 + 1}\,dy\,dx = \int_0^2 \left[xy + y^2 \right]_{y=0}^{y=4} dx - \int_0^2 \frac{2x}{x^2 + 1}\,dx \int_0^4 y\,dy$$

$$= \int_0^2 \left[(4x + 16) - (0 + 0) \right] dx - \left[\ln \left| x^2 + 1 \right| \right]_0^2 \left[\tfrac{1}{2} y^2 \right]_0^4$$

$$= \left[2x^2 + 16x \right]_0^2 - (\ln 5 - \ln 1)(8 - 0) = (8 + 32 - 0) - 8\ln 5 = 40 - 8\ln 5$$

33. In Maple, we can calculate the integral by defining the integrand as `f`

and then using the command `int(int(f,x=0..1),y=0..1);`.

In Mathematica, we can use the command

$$\texttt{Integrate[f,\{x,0,1\},\{y,0,1\}]}$$

We find that $\iint_R x^5 y^3 e^{xy}\,dA = 21e - 57 \approx 0.0839$. We can use `plot3d`

(in Maple) or `Plot3D` (in Mathematica) to graph the function.

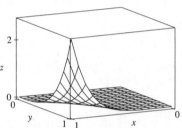

34. In Maple, we can calculate the integral by defining

`f:=exp(-x^2)*cos(x^2+y^2);` and `g:=2-x^2-y^2;`

and then [since $2 - x^2 - y^2 > e^{-x^2} \cos(x^2 + y^2)$ for

$-1 \leq x \leq 1, -1 \leq y \leq 1$] using the command

`evalf(int(int(g-f,x=-1..1),y=-1..1),5);`. The 5 indicates

that we want only five significant digits; this speeds up the calculation

considerably.

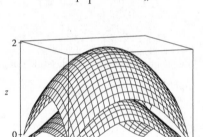

In Mathematica, we can use the command `NIntegrate[g-f,{x,-1,1},{y,-1,1}]`. We find that

$\iint_R \left[(2 - x^2 - y^2) - \left(e^{-x^2} \cos(x^2 + y^2) \right) \right] dA \approx 3.0271$. We can use the `plot3d` command (in Maple) or `Plot3D`

(in Mathematica) to graph both functions on the same screen.

35. R is the rectangle $[-1, 1] \times [0, 5]$. Thus, $A(R) = 2 \cdot 5 = 10$ and

$$f_{ave} = \frac{1}{A(R)} \iint_R f(x, y)\,dA = \tfrac{1}{10} \int_0^5 \int_{-1}^1 x^2 y\,dx\,dy = \tfrac{1}{10} \int_0^5 \left[\tfrac{1}{3} x^3 y \right]_{x=-1}^{x=1} dy = \tfrac{1}{10} \int_0^5 \tfrac{2}{3} y\,dy = \tfrac{1}{10} \left[\tfrac{1}{3} y^2 \right]_0^5 = \tfrac{5}{6}.$$

36. $A(R) = 4 \cdot 1 = 4$, so

$$f_{ave} = \frac{1}{A(R)} \iint_R f(x, y)\,dA = \tfrac{1}{4} \int_0^4 \int_0^1 e^y \sqrt{x + e^y}\,dy\,dx = \tfrac{1}{4} \int_0^4 \left[\tfrac{2}{3} (x + e^y)^{3/2} \right]_{y=0}^{y=1} dx$$

$$= \tfrac{1}{4} \cdot \tfrac{2}{3} \int_0^4 \left[(x + e)^{3/2} - (x + 1)^{3/2} \right] dx = \tfrac{1}{6} \left[\tfrac{2}{5} (x + e)^{5/2} - \tfrac{2}{5} (x + 1)^{5/2} \right]_0^4$$

$$= \tfrac{1}{6} \cdot \tfrac{2}{5} \left[(4 + e)^{5/2} - 5^{5/2} - e^{5/2} + 1 \right] = \tfrac{1}{15} \left[(4 + e)^{5/2} - e^{5/2} - 5^{5/2} + 1 \right] \approx 3.327$$

37. Let $f(x, y) = \dfrac{x - y}{(x + y)^3}$. Then a CAS gives $\int_0^1 \int_0^1 f(x, y)\, dy\, dx = \frac{1}{2}$ and $\int_0^1 \int_0^1 f(x, y)\, dx\, dy = -\frac{1}{2}$.

To explain the seeming violation of Fubini's Theorem, note that f has an infinite discontinuity at $(0, 0)$ and thus does not satisfy the conditions of Fubini's Theorem. In fact, both iterated integrals involve improper integrals which diverge at their lower limits of integration.

38. (a) Loosely speaking, Fubini's Theorem says that the order of integration of a function of two variables does not affect the value of the double integral, while Clairaut's Theorem says that the order of differentiation of such a function does not affect the value of the second-order derivative. Also, both theorems require continuity (though Fubini's allows a finite number of smooth curves to contain discontinuities).

(b) To find g_{xy}, we first hold y constant and use the single-variable Fundamental Theorem of Calculus, Part 1:

$$g_x = \frac{d}{dx}\, g(x, y) = \frac{d}{dx} \int_a^x \left(\int_c^y f(s, t)\, dt \right) ds = \int_c^y f(x, t)\, dt.$$ Now we use the Fundamental Theorem again:

$$g_{xy} = \frac{d}{dy} \int_c^y f(x, t)\, dt = f(x, y).$$

To find g_{yx}, we first use Fubini's Theorem to find that $\int_a^x \int_c^y f(s, t)\, dt\, ds = \int_c^y \int_a^x f(s, t)\, dt\, ds$, and then use the Fundamental Theorem twice, as above, to get $g_{yx} = f(x, y)$. So $g_{xy} = g_{yx} = f(x, y)$.

16.3 Double Integrals over General Regions

ET 15.3

1. $\int_0^4 \int_0^{\sqrt{y}} xy^2\, dx\, dy = \int_0^4 \left[\frac{1}{2} x^2 y^2 \right]_{x=0}^{x=\sqrt{y}} dy = \int_0^4 \frac{1}{2} y^2 [(\sqrt{y})^2 - 0^2] dy = \frac{1}{2} \int_0^4 y^3\, dy = \frac{1}{2} \left[\frac{1}{4} y^4 \right]_0^4 = \frac{1}{2}(64 - 0) = 32$

2. $\int_0^1 \int_{2x}^2 (x - y)\, dy\, dx = \int_0^1 \left[xy - \frac{1}{2} y^2 \right]_{y=2x}^{y=2} dx = \int_0^1 \left[x(2) - \frac{1}{2}(2)^2 - x(2x) + \frac{1}{2}(2x)^2 \right] dx$

$$= \int_0^1 (2x - 2)\, dx = \left[x^2 - 2x \right]_0^1 = 1 - 2 - 0 + 0 = -1$$

3. $\int_0^1 \int_{x^2}^x (1 + 2y)\, dy\, dx = \int_0^1 \left[y + y^2 \right]_{y=x^2}^{y=x} dx = \int_0^1 \left[x + x^2 - x^2 - (x^2)^2 \right] dx$

$$= \int_0^1 (x - x^4)\, dx = \left[\frac{1}{2} x^2 - \frac{1}{5} x^5 \right]_0^1 = \frac{1}{2} - \frac{1}{5} - 0 + 0 = \frac{3}{10}$$

4. $\int_0^2 \int_y^{2y} xy\, dx\, dy = \int_0^2 \left[\frac{1}{2} x^2 y \right]_{x=y}^{x=2y} dy = \int_0^2 \frac{1}{2} y(4y^2 - y^2)\, dy = \frac{1}{2} \int_0^2 3y^3\, dy = \frac{3}{2} \left[\frac{1}{4} y^4 \right]_0^2 = \frac{3}{2}(4 - 0) = 6$

5. $\int_0^{\pi/2} \int_0^{\cos\theta} e^{\sin\theta}\, dr\, d\theta = \int_0^{\pi/2} \left[r e^{\sin\theta} \right]_{r=0}^{r=\cos\theta} d\theta = \int_0^{\pi/2} (\cos\theta)\, e^{\sin\theta}\, d\theta = e^{\sin\theta} \Big]_0^{\pi/2} = e^{\sin(\pi/2)} - e^0 = e - 1$

6. $\int_0^1 \int_0^v \sqrt{1 - v^2}\, du\, dv = \int_0^1 \left[u \sqrt{1 - v^2} \right]_{u=0}^{u=v} dv = \int_0^1 v \sqrt{1 - v^2}\, dv = -\frac{1}{3}(1 - v^2)^{3/2} \Big]_0^1 = -\frac{1}{3}(0 - 1) = \frac{1}{3}$

7. $\iint_D y^2\, dA = \int_{-1}^1 \int_{-y-2}^y y^2\, dx\, dy = \int_{-1}^1 \left[xy^2 \right]_{x=-y-2}^{x=y} dy = \int_{-1}^1 y^2 \left[y - (-y - 2) \right] dy$

$$= \int_{-1}^1 (2y^3 + 2y^2)\, dy = \left[\frac{1}{2} y^4 + \frac{2}{3} y^3 \right]_{-1}^1 = \frac{1}{2} + \frac{2}{3} - \frac{1}{2} + \frac{2}{3} = \frac{4}{3}$$

8. $\iint_D \frac{y}{x^5 + 1}\, dA = \int_0^1 \int_0^{x^2} \frac{y}{x^5 + 1}\, dy\, dx = \int_0^1 \frac{1}{x^5 + 1} \left[\frac{y^2}{2} \right]_{y=0}^{y=x^2} dx = \frac{1}{2} \int_0^1 \frac{x^4}{x^5 + 1}\, dx = \frac{1}{2} \left[\frac{1}{5} \ln \left| x^5 + 1 \right| \right]_0^1$

$$= \frac{1}{10}(\ln 2 - \ln 1) = \frac{1}{10} \ln 2$$

9. $\iint_D x \, dA = \int_0^\pi \int_0^{\sin x} x \, dy \, dx = \int_0^\pi [xy]_{y=0}^{y=\sin x} \, dx = \int_0^\pi x \sin x \, dx \quad \left[\begin{array}{c} \text{integrate by parts} \\ \text{with } u = x, \, dv = \sin x \, dx \end{array} \right]$

$\qquad = [-x \cos x + \sin x]_0^\pi = -\pi \cos \pi + \sin \pi + 0 - \sin 0 = \pi$

10. $\iint_D x^3 \, dA = \int_1^e \int_0^{\ln x} x^3 \, dy \, dx = \int_1^e [x^3 y]_{y=0}^{y=\ln x} \, dx = \int_1^e x^3 \ln x \, dx \quad \left[\begin{array}{c} \text{integrate by parts} \\ \text{with } u = \ln x, \, dv = x^3 \, dx \end{array} \right]$

$\qquad = \left[\frac{1}{4} x^4 \ln x - \frac{1}{16} x^4 \right]_1^e = \frac{1}{4} e^4 - \frac{1}{16} e^4 - 0 + \frac{1}{16} = \frac{3}{16} e^4 + \frac{1}{16}$

11. $\iint_D y^2 e^{xy} \, dA = \int_0^4 \int_0^y y^2 e^{xy} \, dx \, dy = \int_0^4 [y e^{xy}]_{x=0}^{x=y} \, dy = \int_0^4 \left(y e^{y^2} - y \right) dy$

$\qquad = \left[\frac{1}{2} e^{y^2} - \frac{1}{2} y^2 \right]_0^4 = \frac{1}{2} e^{16} - 8 - \frac{1}{2} + 0 = \frac{1}{2} e^{16} - \frac{17}{2}$

12. $\int_0^1 \int_0^y x \sqrt{y^2 - x^2} \, dx \, dy = \int_0^1 \left[-\frac{1}{3} (y^2 - x^2)^{3/2} \right]_{x=0}^{x=y} \, dy = \frac{1}{3} \int_0^1 y^3 \, dy = \frac{1}{3} \cdot \frac{1}{4} y^4 \big]_0^1 = \frac{1}{12}$

13. $\int_0^1 \int_0^{x^2} x \cos y \, dy \, dx = \int_0^1 [x \sin y]_{y=0}^{y=x^2} \, dx = \int_0^1 x \sin x^2 \, dx = -\frac{1}{2} \cos x^2 \big]_0^1 = \frac{1}{2} (1 - \cos 1)$

14.

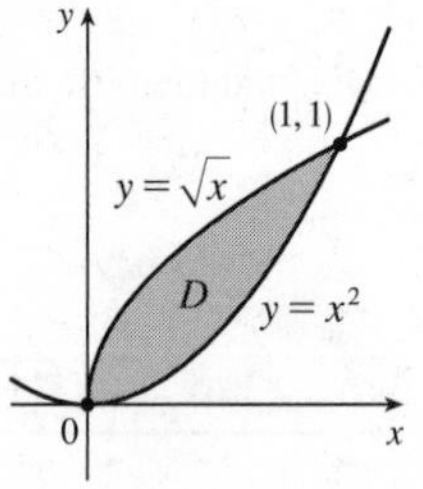

$\int_0^1 \int_{x^2}^{\sqrt{x}} (x + y) \, dy \, dx = \int_0^1 \left[xy + \frac{1}{2} y^2 \right]_{y=x^2}^{y=\sqrt{x}} \, dx$

$\qquad = \int_0^1 \left(x^{3/2} + \frac{1}{2} x - x^3 - \frac{1}{2} x^4 \right) dx$

$\qquad = \left[\frac{2}{5} x^{5/2} + \frac{1}{4} x^2 - \frac{1}{4} x^4 - \frac{1}{10} x^5 \right]_0^1 = \frac{3}{10}$

15.

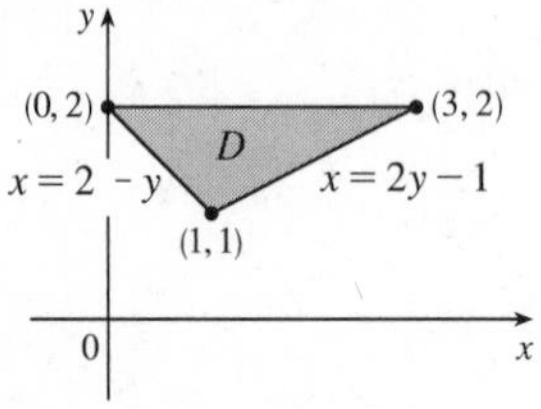

$\int_1^2 \int_{2-y}^{2y-1} y^3 \, dx \, dy = \int_1^2 [xy^3]_{x=2-y}^{x=2y-1} \, dy = \int_1^2 [(2y - 1) - (2 - y)] y^3 \, dy$

$\qquad = \int_1^2 (3y^4 - 3y^3) \, dy = \left[\frac{3}{5} y^5 - \frac{3}{4} y^4 \right]_1^2$

$\qquad = \frac{96}{5} - 12 - \frac{3}{5} + \frac{3}{4} = \frac{147}{20}$

16.

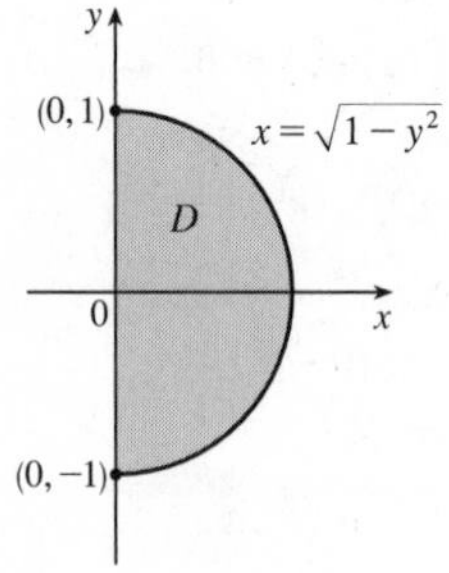

$\iint_D xy^2 \, dA = \int_{-1}^1 \int_0^{\sqrt{1-y^2}} xy^2 \, dx \, dy$

$\qquad = \int_{-1}^1 y^2 \left[\frac{1}{2} x^2 \right]_{x=0}^{x=\sqrt{1-y^2}} \, dy = \frac{1}{2} \int_{-1}^1 y^2 (1 - y^2) \, dy$

$\qquad = \frac{1}{2} \int_{-1}^1 (y^2 - y^4) \, dy = \frac{1}{2} \left[\frac{1}{3} y^3 - \frac{1}{5} y^5 \right]_{-1}^1$

$\qquad = \frac{1}{2} \left(\frac{1}{3} - \frac{1}{5} + \frac{1}{3} - \frac{1}{5} \right) = \frac{2}{15}$

17.

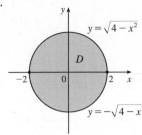

$$\int_{-2}^{2} \int_{-\sqrt{4-x^2}}^{\sqrt{4-x^2}} (2x - y) \, dy \, dx$$

$$= \int_{-2}^{2} \left[2xy - \tfrac{1}{2}y^2 \right]_{y=-\sqrt{4-x^2}}^{y=\sqrt{4-x^2}} dx$$

$$= \int_{-2}^{2} \left[2x\sqrt{4-x^2} - \tfrac{1}{2}(4-x^2) + 2x\sqrt{4-x^2} + \tfrac{1}{2}(4-x^2) \right] dx$$

$$= \int_{-2}^{2} 4x\sqrt{4-x^2} \, dx = -\tfrac{4}{3}(4-x^2)^{3/2} \Big]_{-2}^{2} = 0$$

[Or, note that $4x\sqrt{4-x^2}$ is an odd function, so $\int_{-2}^{2} 4x\sqrt{4-x^2} \, dx = 0$.]

18.

$$\iint_D 2xy \, dA = \int_0^1 \int_{2x}^{3-x} 2xy \, dy \, dx = \int_0^1 \left[xy^2 \right]_{y=2x}^{y=3-x} dx$$

$$= \int_0^1 x[(3-x)^2 - (2x)^2] \, dx = \int_0^1 (-3x^3 - 6x^2 + 9x) \, dx$$

$$= \left[-\tfrac{3}{4}x^4 - 2x^3 + \tfrac{9}{2}x^2 \right]_0^1 = -\tfrac{3}{4} - 2 + \tfrac{9}{2} = \tfrac{7}{4}$$

19.

$$V = \int_0^1 \int_{x^4}^{x} (x + 2y) \, dy \, dx$$

$$= \int_0^1 \left[xy + y^2 \right]_{y=x^4}^{y=x} dx = \int_0^1 (2x^2 - x^5 - x^8) \, dx$$

$$= \left[\tfrac{2}{3}x^3 - \tfrac{1}{6}x^6 - \tfrac{1}{9}x^9 \right]_0^1 = \tfrac{2}{3} - \tfrac{1}{6} - \tfrac{1}{9} = \tfrac{7}{18}$$

20.

$$V = \int_0^1 \int_{y^3}^{y^2} (2x + y^2) \, dx \, dy$$

$$= \int_0^1 \left[x^2 + xy^2 \right]_{x=y^3}^{x=y^2} dy = \int_0^1 (2y^4 - y^6 - y^5) \, dy$$

$$= \left[\tfrac{2}{5}y^5 - \tfrac{1}{7}y^7 - \tfrac{1}{6}y^6 \right]_0^1 = \tfrac{19}{210}$$

21.

$$V = \int_1^2 \int_1^{7-3y} xy \, dx \, dy = \int_1^2 \left[\tfrac{1}{2}x^2 y \right]_{x=1}^{x=7-3y} dy$$

$$= \tfrac{1}{2} \int_1^2 (48y - 42y^2 + 9y^3) \, dy$$

$$= \tfrac{1}{2} \left[24y^2 - 14y^3 + \tfrac{9}{4}y^4 \right]_1^2 = \tfrac{31}{8}$$

22.

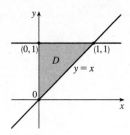

$V = \int_0^1 \int_x^1 (x^2 + 3y^2)\, dy\, dx$

$= \int_0^1 \left[x^2 y + y^3 \right]_{y=x}^{y=1} dx = \int_0^1 (x^2 + 1 - 2x^3)\, dx$

$= \left[\frac{1}{3} x^3 + x - \frac{1}{2} x^4 \right]_0^1 = \frac{5}{6}$

23.

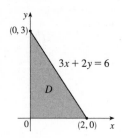

$V = \int_0^2 \int_0^{3 - \frac{3}{2} x} (6 - 3x - 2y)\, dy\, dx$

$= \int_0^2 \left[6y - 3xy - y^2 \right]_{y=0}^{y = 3 - \frac{3}{2} x} dx$

$= \int_0^2 \left[6(3 - \frac{3}{2} x) - 3x(3 - \frac{3}{2} x) - (3 - \frac{3}{2} x)^2 \right] dx$

$= \int_0^2 \left(\frac{9}{4} x^2 - 9x + 9 \right) dx = \left[\frac{3}{4} x^3 - \frac{9}{2} x^2 + 9x \right]_0^2 = 6 - 0 = 6$

24.

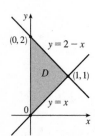

$V = \int_0^1 \int_x^{2-x} x\, dy\, dx$

$= \int_0^1 x \left[y \right]_{y=x}^{y=2-x} dx = \int_0^1 (2x - 2x^2)\, dx$

$= \left[x^2 - \frac{2}{3} x^3 \right]_0^1 = \frac{1}{3}$

25.

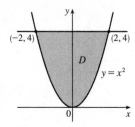

$V = \int_{-2}^2 \int_{x^2}^4 x^2\, dy\, dx$

$= \int_{-2}^2 x^2 \left[y \right]_{y=x^2}^{y=4} dx = \int_{-2}^2 (4x^2 - x^4)\, dx$

$= \left[\frac{4}{3} x^3 - \frac{1}{5} x^5 \right]_{-2}^2 = \frac{32}{3} - \frac{32}{5} + \frac{32}{3} - \frac{32}{5} = \frac{128}{15}$

26.

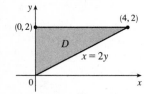

$V = \int_0^2 \int_0^{2y} \sqrt{4 - y^2}\, dx\, dy = \int_0^2 \left[x \sqrt{4 - y^2} \right]_{x=0}^{x=2y} dy$

$= \int_0^2 2y \sqrt{4 - y^2}\, dy = \left[-\frac{2}{3} \left(4 - y^2 \right)^{3/2} \right]_0^2 = 0 + \frac{16}{3} = \frac{16}{3}$

27.

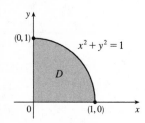

$V = \int_0^1 \int_0^{\sqrt{1 - x^2}} y\, dy\, dx = \int_0^1 \left[\frac{y^2}{2} \right]_{y=0}^{y = \sqrt{1 - x^2}} dx$

$= \int_0^1 \frac{1 - x^2}{2}\, dx = \frac{1}{2} \left[x - \frac{1}{3} x^3 \right]_0^1 = \frac{1}{3}$

28.

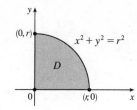

By symmetry, the desired volume V is 8 times the volume V_1 in the first octant. Now

$$V_1 = \int_0^r \int_0^{\sqrt{r^2-y^2}} \sqrt{r^2-y^2}\, dx\, dy = \int_0^r \left[x\sqrt{r^2-y^2} \right]_{x=0}^{x=\sqrt{r^2-y^2}} dy$$

$$= \int_0^r (r^2-y^2)\, dy = \left[r^2 y - \tfrac{1}{3}y^3 \right]_0^r = \tfrac{2}{3}r^3$$

Thus $V = \tfrac{16}{3}r^3$.

29.

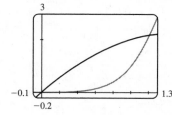

From the graph, it appears that the two curves intersect at $x = 0$ and at $x \approx 1.213$. Thus the desired integral is

$$\iint_D x\, dA \approx \int_0^{1.213} \int_{x^4}^{3x-x^2} x\, dy\, dx = \int_0^{1.213} \left[xy \right]_{y=x^4}^{y=3x-x^2} dx$$

$$= \int_0^{1.213} (3x^2 - x^3 - x^5)\, dx = \left[x^3 - \tfrac{1}{4}x^4 - \tfrac{1}{6}x^6 \right]_0^{1.213}$$

$$\approx 0.713$$

30.

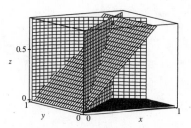 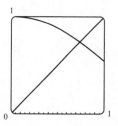

The desired solid is shown in the first graph. From the second graph, we estimate that $y = \cos x$ intersects $y = x$ at $x \approx 0.7391$. Therefore the volume of the solid is

$$V \approx \int_0^{0.7391} \int_x^{\cos x} z\, dy\, dx = \int_0^{0.7391} \int_x^{\cos x} x\, dy\, dx = \int_0^{0.7391} \left[xy \right]_{y=x}^{y=\cos x} dx$$

$$= \int_0^{0.7391} (x\cos x - x^2)\, dx = \left[\cos x + x\sin x - \tfrac{1}{3}x^3 \right]_0^{0.7391} \approx 0.1024$$

Note: There is a different solid which can also be construed to satisfy the conditions stated in the exercise. This is the solid bounded by all of the given surfaces, as well as the plane $y = 0$. In case you calculated the volume of this solid and want to check your work, its volume is $V \approx \int_0^{0.7391} \int_0^x x\, dy\, dx + \int_{0.7391}^{\pi/2} \int_0^{\cos x} x\, dy\, dx \approx 0.4684$.

31. The two bounding curves $y = 1 - x^2$ and $y = x^2 - 1$ intersect at $(\pm 1, 0)$ with $1 - x^2 \geq x^2 - 1$ on $[-1, 1]$. Within this region, the plane $z = 2x + 2y + 10$ is above the plane $z = 2 - x - y$, so

$$V = \int_{-1}^1 \int_{x^2-1}^{1-x^2} (2x + 2y + 10)\, dy\, dx - \int_{-1}^1 \int_{x^2-1}^{1-x^2} (2 - x - y)\, dy\, dx$$

$$= \int_{-1}^1 \int_{x^2-1}^{1-x^2} (2x + 2y + 10 - (2 - x - y))\, dy\, dx$$

$$= \int_{-1}^1 \int_{x^2-1}^{1-x^2} (3x + 3y + 8)\, dy\, dx = \int_{-1}^1 \left[3xy + \tfrac{3}{2}y^2 + 8y \right]_{y=x^2-1}^{y=1-x^2} dx$$

$$= \int_{-1}^1 \left[3x(1-x^2) + \tfrac{3}{2}(1-x^2)^2 + 8(1-x^2) - 3x(x^2-1) - \tfrac{3}{2}(x^2-1)^2 - 8(x^2-1) \right] dx$$

$$= \int_{-1}^1 (-6x^3 - 16x^2 + 6x + 16)\, dx = \left[-\tfrac{3}{2}x^4 - \tfrac{16}{3}x^3 + 3x^2 + 16x \right]_{-1}^1$$

$$= -\tfrac{3}{2} - \tfrac{16}{3} + 3 + 16 + \tfrac{3}{2} - \tfrac{16}{3} - 3 + 16 = \tfrac{64}{3}$$

32. The two planes intersect in the line $y = 1$, $z = 3$, so the region of

integration is the plane region enclosed by the parabola $y = x^2$ and the

line $y = 1$. We have $2 + y \geq 3y$ for $0 \leq y \leq 1$, so the solid region is

bounded above by $z = 2 + y$ and bounded below by $z = 3y$.

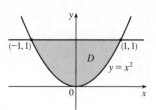

$$V = \int_{-1}^{1} \int_{x^2}^{1} (2 + y)\, dy\, dx - \int_{-1}^{1} \int_{x^2}^{1} (3y)\, dy\, dx = \int_{-1}^{1} \int_{x^2}^{1} (2 + y - 3y)\, dy\, dx = \int_{-1}^{1} \int_{x^2}^{1} (2 - 2y)\, dy\, dx$$

$$= \int_{-1}^{1} \left[2y - y^2 \right]_{y=x^2}^{y=1} dx = \int_{-1}^{1} (1 - 2x^2 + x^4)\, dx = x - \tfrac{2}{3}x^3 + \tfrac{1}{5}x^5 \Big]_{-1}^{1} = \tfrac{16}{15}$$

33. The solid lies below the plane $z = 1 - x - y$

or $x + y + z = 1$ and above the region

$D = \{(x, y) \mid 0 \leq x \leq 1, 0 \leq y \leq 1 - x\}$

in the xy-plane. The solid is a tetrahedron.

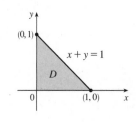

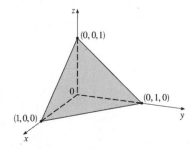

34. The solid lies below the plane $z = 1 - x$

and above the region

$D = \left\{ (x, y) \mid 0 \leq x \leq 1, 0 \leq y \leq 1 - x^2 \right\}$

in the xy-plane.

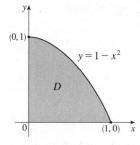

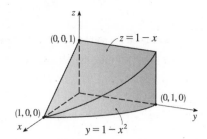

35. The two bounding curves $y = x^3 - x$ and $y = x^2 + x$ intersect at the origin and at $x = 2$, with $x^2 + x > x^3 - x$ on $(0, 2)$.
Using a CAS, we find that the volume is

$$V = \int_{0}^{2} \int_{x^3 - x}^{x^2 + x} z\, dy\, dx = \int_{0}^{2} \int_{x^3 - x}^{x^2 + x} (x^3 y^4 + xy^2)\, dy\, dx = \frac{13{,}984{,}735{,}616}{14{,}549{,}535}$$

36. For $|x| \leq 1$ and $|y| \leq 1$, $2x^2 + y^2 < 8 - x^2 - 2y^2$. Also, the cylinder is described by the inequalities $-1 \leq x \leq 1$,
$-\sqrt{1 - x^2} \leq y \leq \sqrt{1 - x^2}$. So the volume is given by

$$V = \int_{-1}^{1} \int_{-\sqrt{1 - x^2}}^{\sqrt{1 - x^2}} \left[(8 - x^2 - 2y^2) - (2x^2 + y^2) \right] dy\, dx = \frac{13\pi}{2} \qquad \text{[using a CAS]}$$

37. The two surfaces intersect in the circle $x^2 + y^2 = 1$, $z = 0$ and the region of integration is the disk $D: x^2 + y^2 \leq 1$.

Using a CAS, the volume is $\displaystyle\iint_{D} (1 - x^2 - y^2)\, dA = \int_{-1}^{1} \int_{-\sqrt{1 - x^2}}^{\sqrt{1 - x^2}} (1 - x^2 - y^2)\, dy\, dx = \frac{\pi}{2}.$

38. The projection onto the xy-plane of the intersection of the two surfaces is the circle $x^2 + y^2 = 2y$ $\Rightarrow$

$x^2 + y^2 - 2y = 0$ $\Rightarrow$ $x^2 + (y-1)^2 = 1$, so the region of integration is given by $-1 \le x \le 1$,

$1 - \sqrt{1-x^2} \le y \le 1 + \sqrt{1-x^2}$. In this region, $2y \ge x^2 + y^2$ so, using a CAS, the volume is

$$V = \int_{-1}^{1} \int_{1-\sqrt{1-x^2}}^{1+\sqrt{1-x^2}} [2y - (x^2 + y^2)] \, dy \, dx = \frac{\pi}{2}$$

39.

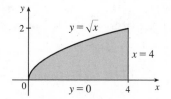

Because the region of integration is

$D = \{(x,y) \mid 0 \le y \le \sqrt{x}, 0 \le x \le 4\} = \{(x,y) \mid y^2 \le x \le 4, 0 \le y \le 2\}$

we have $\int_0^4 \int_0^{\sqrt{x}} f(x,y) \, dy \, dx = \iint_D f(x,y) \, dA = \int_0^2 \int_{y^2}^4 f(x,y) \, dx \, dy$.

40.

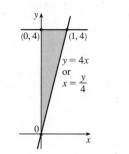

Because the region of integration is

$D = \{(x,y) \mid 4x \le y \le 4, 0 \le x \le 1\} = \{(x,y) \mid 0 \le x \le \frac{y}{4}, 0 \le y \le 4\}$

we have $\int_0^1 \int_{4x}^4 f(x,y) \, dy \, dx = \iint_D f(x,y) \, dA = \int_0^4 \int_0^{y/4} f(x,y) \, dx \, dy$.

41.

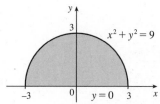

Because the region of integration is

$$D = \left\{(x,y) \mid -\sqrt{9-y^2} \le x \le \sqrt{9-y^2}, 0 \le y \le 3\right\}$$
$$= \{(x,y) \mid 0 \le y \le \sqrt{9-x^2}, -3 \le x \le 3\}$$

we have

$$\int_0^3 \int_{-\sqrt{9-y^2}}^{\sqrt{9-y^2}} f(x,y) \, dx \, dy = \iint_D f(x,y) \, dA$$
$$= \int_{-3}^3 \int_0^{\sqrt{9-x^2}} f(x,y) \, dy \, dx$$

42.

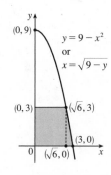

To reverse the order, we must break the region into two separate type I regions.

Because the region of integration is

$D = \{(x,y) \mid 0 \le x \le \sqrt{9-y}, 0 \le y \le 3\}$

$= \{(x,y) \mid 0 \le y \le 3, 0 \le x \le \sqrt{6}\} \cup \{(x,y) \mid 0 \le y \le 9-x^2, \sqrt{6} \le x \le 3\}$

we have

$$\int_0^3 \int_0^{\sqrt{9-y}} f(x,y) \, dx \, dy = \iint_D f(x,y) \, dA$$
$$= \int_0^{\sqrt{6}} \int_0^3 f(x,y) \, dy \, dx + \int_{\sqrt{6}}^3 \int_0^{9-x^2} f(x,y) \, dy \, dx$$

43.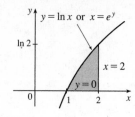

Because the region of integration is

$$D = \{(x,y) \mid 0 \le y \le \ln x, 1 \le x \le 2\} = \{(x,y) \mid e^y \le x \le 2, 0 \le y \le \ln 2\}$$

we have

$$\int_1^2 \int_0^{\ln x} f(x,y)\, dy\, dx = \iint_D f(x,y)\, dA = \int_0^{\ln 2} \int_{e^y}^2 f(x,y)\, dx\, dy$$

44.

Because the region of integration is

$$D = \left\{(x,y) \mid \arctan x \le y \le \tfrac{\pi}{4}, 0 \le x \le 1\right\}$$
$$= \left\{(x,y) \mid 0 \le x \le \tan y, 0 \le y \le \tfrac{\pi}{4}\right\}$$

we have

$$\int_0^1 \int_{\arctan x}^{\pi/4} f(x,y)\, dy\, dx = \iint_D f(x,y)\, dA = \int_0^{\pi/4} \int_0^{\tan y} f(x,y)\, dx\, dy$$

45.

$$\int_0^1 \int_{3y}^3 e^{x^2}\, dx\, dy = \int_0^3 \int_0^{x/3} e^{x^2}\, dy\, dx = \int_0^3 \left[e^{x^2} y\right]_{y=0}^{y=x/3} dx$$

$$= \int_0^3 \left(\frac{x}{3}\right) e^{x^2}\, dx = \tfrac{1}{6}\, e^{x^2}\Big]_0^3 = \frac{e^9 - 1}{6}$$

46.

$$\int_0^{\sqrt{\pi}} \int_y^{\sqrt{\pi}} \cos(x^2)\, dx\, dy = \int_0^{\sqrt{\pi}} \int_0^x \cos(x^2)\, dy\, dx$$

$$= \int_0^{\sqrt{\pi}} \cos(x^2) \left[y\right]_{y=0}^{y=x} dx = \int_0^{\sqrt{\pi}} x \cos(x^2)\, dx$$

$$= \tfrac{1}{2} \sin(x^2)\Big]_0^{\sqrt{\pi}} = \tfrac{1}{2}(\sin \pi - \sin 0) = 0$$

47.

$$\int_0^4 \int_{\sqrt{x}}^2 \frac{1}{y^3 + 1}\, dy\, dx = \int_0^2 \int_0^{y^2} \frac{1}{y^3 + 1}\, dx\, dy$$

$$= \int_0^2 \frac{1}{y^3 + 1} \left[x\right]_{x=0}^{x=y^2} dy = \int_0^2 \frac{y^2}{y^3 + 1}\, dy$$

$$= \tfrac{1}{3} \ln\left|y^3 + 1\right|\,\Big]_0^2 = \tfrac{1}{3}(\ln 9 - \ln 1) = \tfrac{1}{3} \ln 9$$

48.

$$\int_0^1 \int_x^1 e^{x/y}\, dy\, dx = \int_0^1 \int_0^y e^{x/y}\, dx\, dy = \int_0^1 \left[y e^{x/y}\right]_{x=0}^{x=y} dy$$

$$= \int_0^1 (e - 1)y\, dy = \tfrac{1}{2}(e - 1)y^2\Big]_0^1$$

$$= \tfrac{1}{2}(e - 1)$$

49.

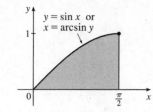

$y = \sin x$ or $x = \arcsin y$

$$\int_0^1 \int_{\arcsin y}^{\pi/2} \cos x \sqrt{1 + \cos^2 x} \, dx \, dy$$

$$= \int_0^{\pi/2} \int_0^{\sin x} \cos x \sqrt{1 + \cos^2 x} \, dy \, dx$$

$$= \int_0^{\pi/2} \cos x \sqrt{1 + \cos^2 x} \left[y \right]_{y=0}^{y=\sin x} dx$$

$$= \int_0^{\pi/2} \cos x \sqrt{1 + \cos^2 x} \sin x \, dx \qquad \begin{bmatrix} \text{Let } u = \cos x, \, du = -\sin x \, dx, \\ dx = du/(-\sin x) \end{bmatrix}$$

$$= \int_1^0 -u \sqrt{1 + u^2} \, du = -\tfrac{1}{3}\left(1 + u^2\right)^{3/2} \Big]_1^0$$

$$= \tfrac{1}{3}\left(\sqrt{8} - 1\right) = \tfrac{1}{3}\left(2\sqrt{2} - 1\right)$$

50.

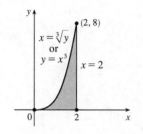

$(2, 8)$

$x = \sqrt[3]{y}$ or $y = x^3$

$x = 2$

$$\int_0^8 \int_{\sqrt[3]{y}}^2 e^{x^4} \, dx \, dy = \int_0^2 \int_0^{x^3} e^{x^4} \, dy \, dx$$

$$= \int_0^2 e^{x^4} \left[y \right]_{y=0}^{y=x^3} dx = \int_0^2 x^3 e^{x^4} \, dx$$

$$= \tfrac{1}{4} e^{x^4} \Big]_0^2 = \tfrac{1}{4}(e^{16} - 1)$$

51. $D = \{(x,y) \mid 0 \le x \le 1, \, -x + 1 \le y \le 1\} \cup \{(x,y) \mid -1 \le x \le 0, \, x + 1 \le y \le 1\}$

$\qquad \cup \{(x,y) \mid 0 \le x \le 1, \, -1 \le y \le x - 1\} \cup \{(x,y) \mid -1 \le x \le 0, \, -1 \le y \le -x - 1\}$, all type I.

$$\iint_D x^2 \, dA = \int_0^1 \int_{1-x}^1 x^2 \, dy \, dx + \int_{-1}^0 \int_{x+1}^1 x^2 \, dy \, dx + \int_0^1 \int_{-1}^{x-1} x^2 \, dy \, dx + \int_{-1}^0 \int_{-1}^{-x-1} x^2 \, dy \, dx$$

$$= 4 \int_0^1 \int_{1-x}^1 x^2 \, dy \, dx \qquad \text{[by symmetry of the regions and because } f(x,y) = x^2 \ge 0]$$

$$= 4 \int_0^1 x^3 \, dx = 4 \left[\tfrac{1}{4} x^4 \right]_0^1 = 1$$

52. $D = \{(x,y) \mid -1 \le y \le 0, \, -1 \le x \le y - y^3\} \cup \{(x,y) \mid 0 \le y \le 1, \, \sqrt{y} - 1 \le x \le y - y^3\}$, both type II.

$$\iint_D y \, dA = \int_{-1}^0 \int_{-1}^{y-y^3} y \, dx \, dy + \int_0^1 \int_{\sqrt{y}-1}^{y-y^3} y \, dx \, dy = \int_{-1}^0 [xy]_{x=-1}^{x=y-y^3} dy + \int_0^1 [xy]_{x=\sqrt{y}-1}^{x=y-y^3} dy$$

$$= \int_{-1}^0 (y^2 - y^4 + y) \, dy + \int_0^1 (y^2 - y^4 - y^{3/2} + y) \, dy$$

$$= \left[\tfrac{1}{3} y^3 - \tfrac{1}{5} y^5 + \tfrac{1}{2} y^2 \right]_{-1}^0 + \left[\tfrac{1}{3} y^3 - \tfrac{1}{5} y^5 - \tfrac{2}{5} y^{5/2} + \tfrac{1}{2} y^2 \right]_0^1$$

$$= \left(0 - \tfrac{11}{30}\right) + \left(\tfrac{7}{30} - 0\right) = -\tfrac{2}{15}$$

53. Here $Q = \{(x,y) \mid x^2 + y^2 \le \tfrac{1}{4}, x \ge 0, y \ge 0\}$, and $0 \le (x^2 + y^2)^2 \le \left(\tfrac{1}{4}\right)^2 \quad \Rightarrow \quad -\tfrac{1}{16} \le -(x^2 + y^2)^2 \le 0$ so

$e^{-1/16} \le e^{-(x^2+y^2)^2} \le e^0 = 1$ since e^t is an increasing function. We have $A(Q) = \tfrac{1}{4} \pi \left(\tfrac{1}{2}\right)^2 = \tfrac{\pi}{16}$, so by Property 11,

$e^{-1/16} A(Q) \le \iint_Q e^{-(x^2+y^2)^2} dA \le 1 \cdot A(Q) \quad \Rightarrow \quad \tfrac{\pi}{16} e^{-1/16} \le \iint_Q e^{-(x^2+y^2)^2} dA \le \tfrac{\pi}{16}$ or we can say

$0.1844 < \iint_Q e^{-(x^2+y^2)^2} dA < 0.1964$. (We have rounded the lower bound down and the upper bound up to preserve the

inequalities.)

54. T is the triangle with vertices $(0,0)$, $(1,0)$, and $(1,2)$ so $A(T) = \tfrac{1}{2}(1)(2) = 1$. We have $0 \le \sin^4(x+y) \le 1$ for all x, y,

and Property 11 gives $0 \cdot A(T) \le \iint_T \sin^4(x+y) dA \le 1 \cdot A(T) \quad \Rightarrow \quad 0 \le \iint_T \sin^4(x+y) dA \le 1$.

55. The average value of a function f of two variables defined on a rectangle R was

defined in Section 16.1 [ET 15.1] as $f_{ave} = \frac{1}{A(R)} \iint_R f(x,y)\,dA$. Extending

this definition to general regions D, we have $f_{ave} = \frac{1}{A(D)} \iint_D f(x,y)\,dA$.

Here $D = \{(x,y) \mid 0 \le x \le 1, 0 \le y \le 3x\}$, so $A(D) = \frac{1}{2}(1)(3) = \frac{3}{2}$ and

$$f_{ave} = \frac{1}{A(D)} \iint_D f(x,y)\,dA = \frac{1}{3/2} \int_0^1 \int_0^{3x} xy\,dy\,dx$$

$$= \frac{2}{3} \int_0^1 \left[\frac{1}{2}xy^2\right]_{y=0}^{y=3x} dx = \frac{1}{3} \int_0^1 9x^3\,dx = \frac{3}{4}x^4\Big]_0^1 = \frac{3}{4}$$

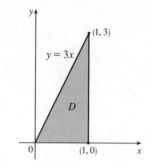

56. Here $D = \{(x,y) \mid 0 \le x \le 1, 0 \le y \le x^2\}$, so

$A(D) = \int_0^1 x^2\,dx = \frac{1}{3}x^3\Big]_0^1 = \frac{1}{3}$ and

$$f_{ave} = \frac{1}{A(D)} \iint_D f(x,y)\,dA = \frac{1}{1/3} \int_0^1 \int_0^{x^2} x\sin y\,dy\,dx$$

$$= 3 \int_0^1 \left[-x\cos y\right]_{y=0}^{y=x^2} dx$$

$$= 3 \int_0^1 \left[x - x\cos(x^2)\right] dx = 3\left[\frac{1}{2}x^2 - \frac{1}{2}\sin(x^2)\right]_0^1$$

$$= 3\left(\frac{1}{2} - \frac{1}{2}\sin 1 - 0\right) = \frac{3}{2}(1 - \sin 1)$$

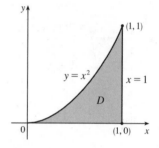

57. Since $m \le f(x,y) \le M$, $\iint_D m\,dA \le \iint_D f(x,y)\,dA \le \iint_D M\,dA$ by (8) $\Rightarrow$

$m\iint_D 1\,dA \le \iint_D f(x,y)\,dA \le M\iint_D 1\,dA$ by (7) $\Rightarrow$ $mA(D) \le \iint_D f(x,y)\,dA \le MA(D)$ by (10).

58.

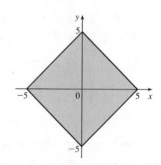

$$\iint_D f(x,y)\,dA = \int_0^1 \int_0^{2y} f(x,y)\,dx\,dy + \int_1^3 \int_0^{3-y} f(x,y)\,dx\,dy$$

$$= \int_0^2 \int_{x/2}^{3-x} f(x,y)\,dy\,dx$$

59. $\iint_D (x^2\tan x + y^3 + 4)\,dA = \iint_D x^2\tan x\,dA + \iint_D y^3\,dA + \iint_D 4\,dA$. But $x^2\tan x$ is an odd function of x and D is

symmetric with respect to the y-axis, so $\iint_D x^2\tan x\,dA = 0$. Similarly, y^3 is an odd function of y and D is symmetric with

respect to the x-axis, so $\iint_D y^3\,dA = 0$. Thus

$$\iint_D (x^2\tan x + y^3 + 4)\,dA = 4\iint_D dA = 4(\text{area of } D) = 4 \cdot \pi(\sqrt{2})^2 = 8\pi$$

60. First, $\iint_D (2 - 3x + 4y)\,dA = \iint_D 2\,dA - \iint_D 3x\,dA + \iint_D 4y\,dA$. The region

D, shown in the figure, is symmetric with respect to the y-axis and $3x$ is an odd

function of x, so $\iint_D 3x\,dA = 0$. Similarly, $4y$ is an odd function of y and D is

symmetric with respect to the x-axis, so $\iint_D 4y\,dA = 0$. Then

$$\iint_D (2 - 3x + 4y)\,dA = \iint_D 2\,dA = 2\iint_D dA$$

$$= 2(\text{area of } D) = 2(50) = 100$$

61. Since $\sqrt{1 - x^2 - y^2} \geq 0$, we can interpret $\iint_D \sqrt{1 - x^2 - y^2}\, dA$ as the volume of the solid that lies below the graph of $z = \sqrt{1 - x^2 - y^2}$ and above the region D in the xy-plane. $z = \sqrt{1 - x^2 - y^2}$ is equivalent to $x^2 + y^2 + z^2 = 1$, $z \geq 0$ which meets the xy-plane in the circle $x^2 + y^2 = 1$, the boundary of D. Thus, the solid is an upper hemisphere of radius 1 which has volume $\frac{1}{2}\left[\frac{4}{3}\pi (1)^3\right] = \frac{2}{3}\pi$.

62. To find the equations of the boundary curves, we require that the z-values of the two surfaces be the same. In Maple, we use the command
`solve(4-x^2-y^2=1-x-y,y);` and in Mathematica, we use
`Solve[4-x^2-y^2==1-x-y,y]`. We find that the curves have equations $y = \dfrac{1 \pm \sqrt{13 + 4x - 4x^2}}{2}$. To find the two points of intersection

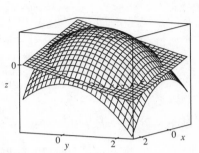

of these curves, we use the CAS to solve $13 + 4x - 4x^2 = 0$, finding that $x = \frac{1 \pm \sqrt{14}}{2}$. So, using the CAS to evaluate the integral, the volume of intersection is

$$V = \int_{(1-\sqrt{14})/2}^{(1+\sqrt{14})/2} \int_{(1-\sqrt{13+4x-4x^2})/2}^{(1+\sqrt{13+4x-4x^2})/2} [(4 - x^2 - y^2) - (1 - x - y)]\, dy\, dx = \frac{49\pi}{8}$$

16.4 Double Integrals in Polar Coordinates

ET 15.4

1. The region R is more easily described by polar coordinates: $R = \left\{(r, \theta) \mid 0 \leq r \leq 4, 0 \leq \theta \leq \frac{3\pi}{2}\right\}$.

Thus $\iint_R f(x, y)\, dA = \int_0^{3\pi/2} \int_0^4 f(r\cos\theta, r\sin\theta)\, r\, dr\, d\theta$.

2. The region R is more easily described by rectangular coordinates: $R = \left\{(x, y) \mid -1 \leq x \leq 1, 0 \leq y \leq 1 - x^2\right\}$.

Thus $\iint_R f(x, y)\, dA = \int_{-1}^1 \int_0^{1-x^2} f(x, y)\, dy\, dx$.

3. The region R is more easily described by rectangular coordinates: $R = \left\{(x, y) \mid -1 \leq x \leq 1, 0 \leq y \leq \frac{1}{2}x + \frac{1}{2}\right\}$.

Thus $\iint_R f(x, y)\, dA = \int_{-1}^1 \int_0^{(x+1)/2} f(x, y)\, dy\, dx$.

4. The region R is more easily described by polar coordinates: $R = \left\{(r, \theta) \mid 3 \leq r \leq 6,\ -\frac{\pi}{2} \leq \theta \leq \frac{\pi}{2}\right\}$.

Thus $\iint_R f(x, y)\, dA = \int_{-\pi/2}^{\pi/2} \int_3^6 f(r\cos\theta, r\sin\theta)\, r\, dr\, d\theta$.

5. The integral $\int_\pi^{2\pi} \int_4^7 r\, dr\, d\theta$ represents the area of the region

$R = \{(r, \theta) \mid 4 \leq r \leq 7, \pi \leq \theta \leq 2\pi\}$, the lower half of a ring.

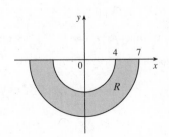

$$\int_\pi^{2\pi} \int_4^7 r\, dr\, d\theta = \left(\int_\pi^{2\pi} d\theta\right)\left(\int_4^7 r\, dr\right)$$

$$= \left[\theta\right]_\pi^{2\pi} \left[\tfrac{1}{2}r^2\right]_4^7 = \pi \cdot \tfrac{1}{2}(49 - 16) = \tfrac{33\pi}{2}$$

6. The integral $\int_0^{\pi/2} \int_0^{4\cos\theta} r\,dr\,d\theta$ represents the area of the region

$R = \{(r, \theta) \mid 0 \le r \le 4\cos\theta, 0 \le \theta \le \pi/2\}$. Since $r = 4\cos\theta \iff$

$r^2 = 4r\cos\theta \iff x^2 + y^2 = 4x \iff (x-2)^2 + y^2 = 4$, R is the

portion in the first quadrant of a circle of radius 2 with center $(2, 0)$.

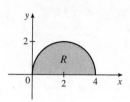

$\int_0^{\pi/2} \int_0^{4\cos\theta} r\,dr\,d\theta = \int_0^{\pi/2} \left[\frac{1}{2}r^2\right]_{r=0}^{r=4\cos\theta} d\theta = \int_0^{\pi/2} 8\cos^2\theta\,d\theta = \int_0^{\pi/2} 4(1+\cos 2\theta)\,d\theta = 4\left[\theta + \frac{1}{2}\sin 2\theta\right]_0^{\pi/2} = 2\pi$

7. The disk D can be described in polar coordinates as $D = \{(r, \theta) \mid 0 \le r \le 3, 0 \le \theta \le 2\pi\}$. Then

$\iint_D xy\,dA = \int_0^{2\pi} \int_0^3 (r\cos\theta)(r\sin\theta)\,r\,dr\,d\theta = \left(\int_0^{2\pi} \sin\theta\cos\theta\,d\theta\right)\left(\int_0^3 r^3\,dr\right) = \left[\frac{1}{2}\sin^2\theta\right]_0^{2\pi} \left[\frac{1}{4}r^4\right]_0^3 = 0.$

8. $\iint_R (x+y)\,dA = \int_{\pi/2}^{3\pi/2} \int_1^2 (r\cos\theta + r\sin\theta)\,r\,dr\,d\theta = \int_{\pi/2}^{3\pi/2} \int_1^2 r^2(\cos\theta + \sin\theta)\,dr\,d\theta$

$\qquad = \left(\int_{\pi/2}^{3\pi/2} (\cos\theta + \sin\theta)\,d\theta\right)\left(\int_1^2 r^2\,dr\right) = \left[\sin\theta - \cos\theta\right]_{\pi/2}^{3\pi/2} \left[\frac{1}{3}r^3\right]_1^2$

$\qquad = (-1 - 0 - 1 + 0)\left(\frac{8}{3} - \frac{1}{3}\right) = -\frac{14}{3}$

9. $\iint_R \cos(x^2 + y^2)\,dA = \int_0^\pi \int_0^3 \cos(r^2)\,r\,dr\,d\theta = \left(\int_0^\pi d\theta\right)\left(\int_0^3 r\cos(r^2)\,dr\right)$

$\qquad = \left[\theta\right]_0^\pi \left[\frac{1}{2}\sin(r^2)\right]_0^3 = \pi \cdot \frac{1}{2}(\sin 9 - \sin 0) = \frac{\pi}{2}\sin 9$

10. $\iint_R \sqrt{4 - x^2 - y^2}\,dA = \int_{-\pi/2}^{\pi/2} \int_0^2 \sqrt{4 - r^2}\,r\,dr\,d\theta = \left(\int_{-\pi/2}^{\pi/2} d\theta\right)\left(\int_0^2 r\sqrt{4 - r^2}\,dr\right)$

$\qquad = \left[\theta\right]_{-\pi/2}^{\pi/2} \left[-\frac{1}{2} \cdot \frac{2}{3}(4 - r^2)^{3/2}\right]_0^2 = \left(\frac{\pi}{2} + \frac{\pi}{2}\right)\left(-\frac{1}{3}(0 - 4^{3/2})\right) = \frac{8}{3}\pi$

11. $\iint_D e^{-x^2 - y^2}\,dA = \int_{-\pi/2}^{\pi/2} \int_0^2 e^{-r^2}\,r\,dr\,d\theta = \left(\int_{-\pi/2}^{\pi/2} d\theta\right)\left(\int_0^2 re^{-r^2}\,dr\right)$

$\qquad = \left[\theta\right]_{-\pi/2}^{\pi/2} \left[-\frac{1}{2}e^{-r^2}\right]_0^2 = \pi\left(-\frac{1}{2}\right)(e^{-4} - e^0) = \frac{\pi}{2}(1 - e^{-4})$

12. $\iint_R ye^x\,dA = \int_0^{\pi/2} \int_0^5 (r\sin\theta)\,e^{r\cos\theta}\,r\,dr\,d\theta = \int_0^5 \int_0^{\pi/2} r^2\sin\theta\,e^{r\cos\theta}\,d\theta\,dr$. First we integrate $\int_0^{\pi/2} r^2\sin\theta\,e^{r\cos\theta}\,d\theta$:

Let $u = r\cos\theta \implies du = -r\sin\theta\,d\theta$, and $\int_0^{\pi/2} r^2\sin\theta\,e^{r\cos\theta}\,d\theta = \int_{u=r}^{u=0} -re^u\,du = -r[e^0 - e^r] = re^r - r$.

Then $\int_0^5 \int_0^{\pi/2} r^2\sin\theta\,e^{r\cos\theta}\,d\theta\,dr = \int_0^5 (re^r - r)\,dr = \left[re^r - e^r - \frac{1}{2}r^2\right]_0^5 = 4e^5 - \frac{23}{2}$, where we integrated by parts in the first term.

13. R is the region shown in the figure, and can be described

by $R = \{(r, \theta) \mid 0 \le \theta \le \pi/4, 1 \le r \le 2\}$. Thus

$\iint_R \arctan(y/x)\,dA = \int_0^{\pi/4} \int_1^2 \arctan(\tan\theta)\,r\,dr\,d\theta$ since $y/x = \tan\theta$.

Also, $\arctan(\tan\theta) = \theta$ for $0 \le \theta \le \pi/4$, so the integral becomes

$\int_0^{\pi/4} \int_1^2 \theta\,r\,dr\,d\theta = \int_0^{\pi/4} \theta\,d\theta \int_1^2 r\,dr = \left[\frac{1}{2}\theta^2\right]_0^{\pi/4} \left[\frac{1}{2}r^2\right]_1^2 = \frac{\pi^2}{32} \cdot \frac{3}{2} = \frac{3}{64}\pi^2.$

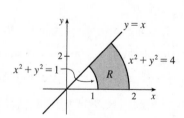

14.

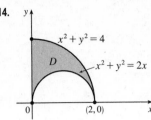

$$\iint_D x\,dA = \iint_{\substack{x^2+y^2\le 4\\ x\ge 0, y\ge 0}} x\,dA - \iint_{\substack{(x-1)^2+y^2\le 1\\ y\ge 0}} x\,dA$$

$$= \int_0^{\pi/2}\int_0^2 r^2\cos\theta\,dr\,d\theta - \int_0^{\pi/2}\int_0^{2\cos\theta} r^2\cos\theta\,dr\,d\theta$$

$$= \int_0^{\pi/2} \tfrac{1}{3}(8\cos\theta)\,d\theta - \int_0^{\pi/2}\tfrac{1}{3}(8\cos^4\theta)\,d\theta$$

$$= \tfrac{8}{3} - \tfrac{8}{12}\left[\cos^3\theta\sin\theta + \tfrac{3}{2}(\theta+\sin\theta\cos\theta)\right]_0^{\pi/2}$$

$$= \tfrac{8}{3} - \tfrac{2}{3}\left[0 + \tfrac{3}{2}\left(\tfrac{\pi}{2}\right)\right] = \tfrac{16-3\pi}{6}$$

15. One loop is given by the region

$D = \{(r,\theta)\,|-\pi/6\le\theta\le\pi/6,\, 0\le r\le\cos 3\theta\}$, so the area is

$$\iint_D dA = \int_{-\pi/6}^{\pi/6}\int_0^{\cos 3\theta} r\,dr\,d\theta = \int_{-\pi/6}^{\pi/6}\left[\tfrac{1}{2}r^2\right]_{r=0}^{r=\cos 3\theta} d\theta$$

$$= \int_{-\pi/6}^{\pi/6}\tfrac{1}{2}\cos^2 3\theta\,d\theta = 2\int_0^{\pi/6}\tfrac{1}{2}\left(\frac{1+\cos 6\theta}{2}\right) d\theta$$

$$= \tfrac{1}{2}\left[\theta + \tfrac{1}{6}\sin 6\theta\right]_0^{\pi/6} = \frac{\pi}{12}$$

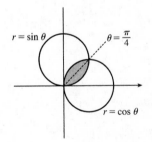

16. $D = \{(r,\theta)\mid 0\le\theta\le 2\pi,\, 0\le r\le 4+3\cos\theta\}$, so

$$A(D) = \iint_D dA = \int_0^{2\pi}\int_0^{4+3\cos\theta} r\,dr\,d\theta = \int_0^{2\pi}\left[\tfrac{1}{2}r^2\right]_{r=0}^{r=4+3\cos\theta} d\theta = \tfrac{1}{2}\int_0^{2\pi}(4+3\cos\theta)^2 d\theta$$

$$= \tfrac{1}{2}\int_0^{2\pi}(16+24\cos\theta + 9\cos^2\theta)\,d\theta = \tfrac{1}{2}\int_0^{2\pi}\left(16+24\cos\theta + 9\cdot\frac{1+\cos 2\theta}{2}\right) d\theta$$

$$= \tfrac{1}{2}\left[16\theta + 24\sin\theta + \tfrac{9}{2}\theta + \tfrac{9}{4}\sin 2\theta\right]_0^{2\pi} = \tfrac{41}{2}\pi$$

17. By symmetry,

$$A = 2\int_0^{\pi/4}\int_0^{\sin\theta} r\,dr\,d\theta = 2\int_0^{\pi/4}\left[\tfrac{1}{2}r^2\right]_{r=0}^{r=\sin\theta} d\theta$$

$$= \int_0^{\pi/4}\sin^2\theta\,d\theta = \int_0^{\pi/4}\tfrac{1}{2}(1-\cos 2\theta)\,d\theta$$

$$= \tfrac{1}{2}\left[\theta - \tfrac{1}{2}\sin 2\theta\right]_0^{\pi/4}$$

$$= \tfrac{1}{2}\left[\tfrac{\pi}{4} - \tfrac{1}{2}\sin\tfrac{\pi}{2} - 0 + \tfrac{1}{2}\sin 0\right] = \tfrac{1}{8}(\pi - 2)$$

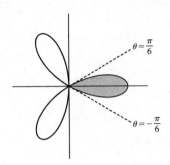

18. The region lies between the two polar curves in quadrants I and IV, but in quadrants II and III the region is enclosed by the cardiod. In the first quadrant, $1+\cos\theta = 3\cos\theta$ when $\cos\theta = \tfrac{1}{2}$ $\Rightarrow$ $\theta = \tfrac{\pi}{3}$, so the area of the region inside the cardiod and outside the circle is

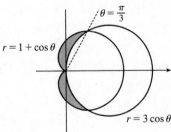

$$A_1 = \int_{\pi/3}^{\pi/2}\int_{3\cos\theta}^{1+\cos\theta} r\,dr\,d\theta = \int_{\pi/3}^{\pi/2}\left[\tfrac{1}{2}r^2\right]_{r=3\cos\theta}^{r=1+\cos\theta} d\theta$$

$$= \tfrac{1}{2}\int_{\pi/3}^{\pi/2}(1+2\cos\theta - 8\cos^2\theta)\,d\theta = \tfrac{1}{2}\left[\theta + 2\sin\theta - 8\left(\tfrac{1}{2}\theta + \tfrac{1}{4}\sin 2\theta\right)\right]_{\pi/3}^{\pi/2}$$

$$= \left[-\tfrac{3}{2}\theta + \sin\theta - \sin 2\theta\right]_{\pi/3}^{\pi/2} = \left(-\tfrac{3\pi}{4} + 1 - 0\right) - \left(-\tfrac{\pi}{2} + \tfrac{\sqrt{3}}{2} - \tfrac{\sqrt{3}}{2}\right) = 1 - \tfrac{\pi}{4}.$$

The area of the region in the second quadrant is

$$A_2 = \int_{\pi/2}^{\pi} \int_0^{1+\cos\theta} r\, dr\, d\theta = \int_{\pi/2}^{\pi} \left[\tfrac{1}{2}r^2\right]_{r=0}^{r=1+\cos\theta} d\theta = \tfrac{1}{2}\int_{\pi/2}^{\pi}(1+2\cos\theta+\cos^2\theta)d\theta$$

$$= \tfrac{1}{2}\left[\theta + 2\sin\theta + \tfrac{1}{2}\theta + \tfrac{1}{4}\sin 2\theta\right]_{\pi/2}^{\pi} = \tfrac{1}{2}\left(\tfrac{3\pi}{4} - 2\right) = \tfrac{3\pi}{8} - 1.$$

By symmetry, the total area is $A = 2(A_1 + A_2) = 2\left(1 - \tfrac{\pi}{4} + \tfrac{3\pi}{8} - 1\right) = \tfrac{\pi}{4}$.

19. $V = \iint_{x^2+y^2 \le 4} \sqrt{x^2+y^2}\, dA = \int_0^{2\pi} \int_0^2 \sqrt{r^2}\, r\, dr\, d\theta = \int_0^{2\pi} d\theta \int_0^2 r^2\, dr = \left[\theta\right]_0^{2\pi} \left[\tfrac{1}{3}r^3\right]_0^2 = 2\pi\left(\tfrac{8}{3}\right) = \tfrac{16}{3}\pi$

20. The paraboloid $z = 18 - 2x^2 - 2y^2$ intersects the xy-plane in the circle $x^2 + y^2 = 9$, so

$$V = \iint_{x^2+y^2 \le 9} \left(18 - 2x^2 - 2y^2\right) dA = \iint_{x^2+y^2 \le 9} \left[18 - 2\left(x^2+y^2\right)\right] dA = \int_0^{2\pi} \int_0^3 \left(18 - 2r^2\right) r\, dr\, d\theta$$

$$= \int_0^{2\pi} d\theta \int_0^3 \left(18r - 2r^3\right) dr = \left[\theta\right]_0^{2\pi} \left[9r^2 - \tfrac{1}{2}r^4\right]_0^3 = (2\pi)\left(81 - \tfrac{81}{2}\right) = 81\pi$$

21. The hyperboloid of two sheets $-x^2 - y^2 + z^2 = 1$ intersects the plane $z = 2$ when $-x^2 - y^2 + 4 = 1$ or $x^2 + y^2 = 3$. So the solid region lies above the surface $z = \sqrt{1 + x^2 + y^2}$ and below the plane $z = 2$ for $x^2 + y^2 \le 3$, and its volume is

$$V = \iint_{x^2+y^2 \le 3} \left(2 - \sqrt{1+x^2+y^2}\right) dA = \int_0^{2\pi} \int_0^{\sqrt{3}} (2 - \sqrt{1+r^2})\, r\, dr\, d\theta$$

$$= \int_0^{2\pi} d\theta \int_0^{\sqrt{3}} (2r - r\sqrt{1+r^2})dr = \left[\theta\right]_0^{2\pi} \left[r^2 - \tfrac{1}{3}(1+r^2)^{3/2}\right]_0^{\sqrt{3}}$$

$$= 2\pi\left(3 - \tfrac{8}{3} - 0 + \tfrac{1}{3}\right) = \tfrac{4}{3}\pi$$

22. The sphere $x^2 + y^2 + z^2 = 16$ intersects the xy-plane in the circle $x^2 + y^2 = 16$, so

$$V = 2 \iint_{4 \le x^2+y^2 \le 16} \sqrt{16 - x^2 - y^2}\, dA \quad \text{[by symmetry]} \quad = 2\int_0^{2\pi} \int_2^4 \sqrt{16 - r^2}\, r\, dr\, d\theta = 2\int_0^{2\pi} d\theta \int_2^4 r(16 - r^2)^{1/2}dr$$

$$= 2\left[\theta\right]_0^{2\pi} \left[-\tfrac{1}{3}(16 - r^2)^{3/2}\right]_2^4 = -\tfrac{2}{3}(2\pi)(0 - 12^{3/2}) = \tfrac{4\pi}{3}\left(12\sqrt{12}\right) = 32\sqrt{3}\,\pi$$

23. By symmetry,

$$V = 2 \iint_{x^2+y^2 \le a^2} \sqrt{a^2 - x^2 - y^2}\, dA = 2\int_0^{2\pi} \int_0^a \sqrt{a^2 - r^2}\, r\, dr\, d\theta = 2\int_0^{2\pi} d\theta \int_0^a r\sqrt{a^2 - r^2}\, dr$$

$$= 2\left[\theta\right]_0^{2\pi} \left[-\tfrac{1}{3}(a^2 - r^2)^{3/2}\right]_0^a = 2(2\pi)\left(0 + \tfrac{1}{3}a^3\right) = \tfrac{4\pi}{3}a^3$$

24. The paraboloid $z = 1 + 2x^2 + 2y^2$ intersects the plane $z = 7$ when $7 = 1 + 2x^2 + 2y^2$ or $x^2 + y^2 = 3$ and we are restricted to the first octant, so

$$V = \iint_{\substack{x^2+y^2 \le 3, \\ x \ge 0, y \ge 0}} \left[7 - \left(1 + 2x^2 + 2y^2\right)\right] dA = \int_0^{\pi/2} \int_0^{\sqrt{3}} \left[7 - (1 + 2r^2)\right] r\, dr\, d\theta$$

$$= \int_0^{\pi/2} d\theta \int_0^{\sqrt{3}} \left(6r - 2r^3\right) dr = \left[\theta\right]_0^{\pi/2} \left[3r^2 - \tfrac{1}{2}r^4\right]_0^{\sqrt{3}} = \tfrac{\pi}{2} \cdot \tfrac{9}{2} = \tfrac{9}{4}\pi$$

25. The cone $z = \sqrt{x^2 + y^2}$ intersects the sphere $x^2 + y^2 + z^2 = 1$ when $x^2 + y^2 + \left(\sqrt{x^2 + y^2}\right)^2 = 1$ or $x^2 + y^2 = \frac{1}{2}$. So

$$V = \iint\limits_{x^2 + y^2 \le 1/2} \left(\sqrt{1 - x^2 - y^2} - \sqrt{x^2 + y^2}\right) dA = \int_0^{2\pi} \int_0^{1/\sqrt{2}} \left(\sqrt{1 - r^2} - r\right) r \, dr \, d\theta$$

$$= \int_0^{2\pi} d\theta \int_0^{1/\sqrt{2}} \left(r\sqrt{1 - r^2} - r^2\right) dr = \left[\theta\right]_0^{2\pi} \left[-\frac{1}{3}(1 - r^2)^{3/2} - \frac{1}{3}r^3\right]_0^{1/\sqrt{2}} = 2\pi\left(-\frac{1}{3}\right)\left(\frac{1}{\sqrt{2}} - 1\right) = \frac{\pi}{3}\left(2 - \sqrt{2}\right)$$

26. The two paraboloids intersect when $3x^2 + 3y^2 = 4 - x^2 - y^2$ or $x^2 + y^2 = 1$. So

$$V = \iint\limits_{x^2 + y^2 \le 1} [(4 - x^2 - y^2) - 3(x^2 + y^2)] \, dA = \int_0^{2\pi} \int_0^1 4(1 - r^2) \, r \, dr \, d\theta$$

$$= \int_0^{2\pi} d\theta \int_0^1 (4r - 4r^3) \, dr = \left[\theta\right]_0^{2\pi} \left[2r^2 - r^4\right]_0^1 = 2\pi$$

27. The given solid is the region inside the cylinder $x^2 + y^2 = 4$ between the surfaces $z = \sqrt{64 - 4x^2 - 4y^2}$

and $z = -\sqrt{64 - 4x^2 - 4y^2}$. So

$$V = \iint\limits_{x^2 + y^2 \le 4} \left[\sqrt{64 - 4x^2 - 4y^2} - \left(-\sqrt{64 - 4x^2 - 4y^2}\right)\right] dA = \iint\limits_{x^2 + y^2 \le 4} 2\sqrt{64 - 4x^2 - 4y^2} \, dA$$

$$= 4 \int_0^{2\pi} \int_0^2 \sqrt{16 - r^2} \, r \, dr \, d\theta = 4 \int_0^{2\pi} d\theta \int_0^2 r\sqrt{16 - r^2} \, dr = 4 \left[\theta\right]_0^{2\pi} \left[-\frac{1}{3}(16 - r^2)^{3/2}\right]_0^2$$

$$= 8\pi\left(-\frac{1}{3}\right)(12^{3/2} - 16^{2/3}) = \frac{8\pi}{3}\left(64 - 24\sqrt{3}\right)$$

28. (a) Here the region in the xy-plane is the annular region $r_1^2 \le x^2 + y^2 \le r_2^2$ and the desired volume is twice that above the xy-plane. Hence

$$V = 2 \iint\limits_{r_1^2 \le x^2 + y^2 \le r_2^2} \sqrt{r_2^2 - x^2 - y^2} \, dA = 2 \int_0^{2\pi} \int_{r_1}^{r_2} \sqrt{r_2^2 - r^2} \, r \, dr \, d\theta = 2 \int_0^{2\pi} d\theta \int_{r_1}^{r_2} \sqrt{r_2^2 - r^2} \, r \, dr$$

$$= \frac{4\pi}{3} \left[-(r_2^2 - r^2)^{3/2}\right]_{r_1}^{r_2} = \frac{4\pi}{3}(r_2^2 - r_1^2)^{3/2}$$

(b) A cross-sectional cut is shown in the figure.

So $r_2^2 = \left(\frac{1}{2}h\right)^2 + r_1^2$ or $\frac{1}{4}h^2 = r_2^2 - r_1^2$.

Thus the volume in terms of h is $V = \frac{4\pi}{3}\left(\frac{1}{4}h^2\right)^{3/2} = \frac{\pi}{6}h^3$.

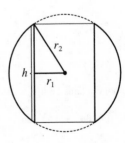

29.

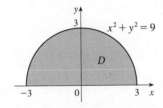

$$\int_{-3}^3 \int_0^{\sqrt{9 - x^2}} \sin(x^2 + y^2) \, dy \, dx = \int_0^{\pi} \int_0^3 \sin\left(r^2\right) r \, dr \, d\theta$$

$$= \int_0^{\pi} d\theta \int_0^3 r \sin\left(r^2\right) dr = \left[\theta\right]_0^{\pi} \left[-\frac{1}{2}\cos\left(r^2\right)\right]_0^3$$

$$= \pi\left(-\frac{1}{2}\right)(\cos 9 - 1) = \frac{\pi}{2}(1 - \cos 9)$$

30.

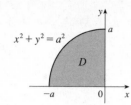

$$\int_{\pi/2}^{\pi} \int_0^a (r\cos\theta)^2 (r\sin\theta)\, r\, dr\, d\theta = \int_{\pi/2}^{\pi} \int_0^a r^4 \cos^2\theta \sin\theta\, dr\, d\theta$$
$$= \int_{\pi/2}^{\pi} \cos^2\theta \sin\theta\, d\theta \int_0^a r^4\, dr$$
$$= \left[-\tfrac{1}{3}\cos^3\theta\right]_{\pi/2}^{\pi} \left[\tfrac{1}{5}r^5\right]_0^a$$
$$= -\tfrac{1}{3}\left(\cos^3\pi - \cos^3\tfrac{\pi}{2}\right)\left(\tfrac{1}{5}a^5\right) = \tfrac{1}{15}a^5$$

31.

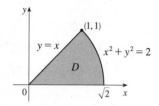

$$\int_0^{\pi/4} \int_0^{\sqrt{2}} (r\cos\theta + r\sin\theta)\, r\, dr\, d\theta = \int_0^{\pi/4} (\cos\theta + \sin\theta)\, d\theta \int_0^{\sqrt{2}} r^2\, dr$$
$$= \left[\sin\theta - \cos\theta\right]_0^{\pi/4} \left[\tfrac{1}{3}r^3\right]_0^{\sqrt{2}}$$
$$= \left[\tfrac{\sqrt{2}}{2} - \tfrac{\sqrt{2}}{2} - 0 + 1\right] \cdot \tfrac{1}{3}\left(2\sqrt{2} - 0\right) = \tfrac{2\sqrt{2}}{3}$$

32.

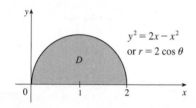

$$\int_0^{\pi/2} \int_0^{2\cos\theta} r^2\, dr\, d\theta = \int_0^{\pi/2} \left[\tfrac{1}{3}r^3\right]_{r=0}^{r=2\cos\theta} d\theta = \int_0^{\pi/2} \left(\tfrac{8}{3}\cos^3\theta\right) d\theta$$
$$= \tfrac{8}{3}\int_0^{\pi/2} (1 - \sin^2\theta)\cos\theta\, d\theta$$
$$= \tfrac{8}{3}\left[\sin\theta - \tfrac{1}{3}\sin^3\theta\right]_0^{\pi/2} = \tfrac{16}{9}$$

33. The surface of the water in the pool is a circular disk D with radius 20 ft. If we place D on coordinate axes with the origin at the center of D and define $f(x, y)$ to be the depth of the water at (x, y), then the volume of water in the pool is the volume of the solid that lies above $D = \{(x, y) \mid x^2 + y^2 \le 400\}$ and below the graph of $f(x, y)$. We can associate north with the positive y-direction, so we are given that the depth is constant in the x-direction and the depth increases linearly in the y-direction from $f(0, -20) = 2$ to $f(0, 20) = 7$. The trace in the yz-plane is a line segment from $(0, -20, 2)$ to $(0, 20, 7)$. The slope of this line is $\frac{7-2}{20-(-20)} = \frac{1}{8}$, so an equation of the line is $z - 7 = \frac{1}{8}(y - 20) \Rightarrow z = \frac{1}{8}y + \frac{9}{2}$. Since $f(x, y)$ is independent of x, $f(x, y) = \frac{1}{8}y + \frac{9}{2}$. Thus the volume is given by $\iint_D f(x, y)\, dA$, which is most conveniently evaluated using polar coordinates. Then $D = \{(r, \theta) \mid 0 \le r \le 20, 0 \le \theta \le 2\pi\}$ and substituting $x = r\cos\theta$, $y = r\sin\theta$ the integral becomes

$$\int_0^{2\pi} \int_0^{20} \left(\tfrac{1}{8}r\sin\theta + \tfrac{9}{2}\right) r\, dr\, d\theta = \int_0^{2\pi} \left[\tfrac{1}{24}r^3\sin\theta + \tfrac{9}{4}r^2\right]_{r=0}^{r=20} d\theta = \int_0^{2\pi} \left(\tfrac{1000}{3}\sin\theta + 900\right) d\theta$$
$$= \left[-\tfrac{1000}{3}\cos\theta + 900\theta\right]_0^{2\pi} = 1800\pi$$

Thus the pool contains $1800\pi \approx 5655 \text{ ft}^3$ of water.

34. (a) If $R \le 100$, the total amount of water supplied each hour to the region within R feet of the sprinkler is

$$V = \int_0^{2\pi} \int_0^R e^{-r} r\, dr\, d\theta = \int_0^{2\pi} d\theta \int_0^R r e^{-r}\, dr = \left[\theta\right]_0^{2\pi} \left[-re^{-r} - e^{-r}\right]_0^R$$
$$= 2\pi[-Re^{-R} - e^{-R} + 0 + 1] = 2\pi(1 - Re^{-R} - e^{-R}) \text{ ft}^3$$

(b) The average amount of water per hour per square foot supplied to the region within R feet of the sprinkler is

$$\frac{V}{\text{area of region}} = \frac{V}{\pi R^2} = \frac{2(1 - Re^{-R} - e^{-R})}{R^2} \text{ ft}^3 \text{ (per hour per square foot). See the definition of the average value of a}$$

function on page 992 [ET 956].

35. $\int_{1/\sqrt{2}}^{1}\int_{\sqrt{1-x^2}}^{x} xy\, dy\, dx + \int_{1}^{\sqrt{2}}\int_{0}^{x} xy\, dy\, dx + \int_{\sqrt{2}}^{2}\int_{0}^{\sqrt{4-x^2}} xy\, dy\, dx$

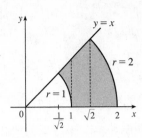

$$= \int_{0}^{\pi/4}\int_{1}^{2} r^3 \cos\theta \sin\theta \, dr\, d\theta = \int_{0}^{\pi/4}\left[\frac{r^4}{4}\cos\theta\sin\theta\right]_{r=1}^{r=2} d\theta$$

$$= \frac{15}{4}\int_{0}^{\pi/4}\sin\theta\cos\theta \, d\theta = \frac{15}{4}\left[\frac{\sin^2\theta}{2}\right]_{0}^{\pi/4} = \frac{15}{16}$$

36. (a) $\iint_{D_a} e^{-(x^2+y^2)}\, dA = \int_{0}^{2\pi}\int_{0}^{a} re^{-r^2}\, dr\, d\theta = 2\pi\left[-\frac{1}{2}e^{-r^2}\right]_{0}^{a} = \pi\left(1 - e^{-a^2}\right)$ for each a. Then $\lim_{a\to\infty}\pi\left(1 - e^{-a^2}\right) = \pi$

since $e^{-a^2} \to 0$ as $a \to \infty$. Hence $\int_{-\infty}^{\infty}\int_{-\infty}^{\infty} e^{-(x^2+y^2)}\, dA = \pi$.

(b) $\iint_{S_a} e^{-(x^2+y^2)}\, dA = \int_{-a}^{a}\int_{-a}^{a} e^{-x^2}e^{-y^2}\, dx\, dy = \left(\int_{-a}^{a} e^{-x^2}\, dx\right)\left(\int_{-a}^{a} e^{-y^2}\, dy\right)$ for each a.

Then, from (a), $\pi = \iint_{\mathbb{R}^2} -(x^2+y^2)\, dA$, so

$$\pi = \lim_{a\to\infty}\iint_{S_a} e^{-(x^2+y^2)}\, dA = \lim_{a\to\infty}\left(\int_{-a}^{a} e^{-x^2}\, dx\right)\left(\int_{-a}^{a} e^{-y^2}\, dy\right) = \left(\int_{-\infty}^{\infty} e^{-x^2}\, dx\right)\left(\int_{-\infty}^{\infty} e^{-y^2}\, dy\right).$$

To evaluate $\lim_{a\to\infty}\left(\int_{-a}^{a} e^{-x^2}\, dx\right)\left(\int_{-a}^{a} e^{-y^2}\, dy\right)$, we are using the fact that these integrals are bounded. This is true since

on $[-1,1]$, $0 < e^{-x^2} \le 1$ while on $(-\infty, -1)$, $0 < e^{-x^2} \le e^x$ and on $(1, \infty)$, $0 < e^{-x^2} < e^{-x}$. Hence

$$0 \le \int_{-\infty}^{\infty} e^{-x^2}\, dx \le \int_{-\infty}^{-1} e^x\, dx + \int_{-1}^{1} dx + \int_{1}^{\infty} e^{-x}\, dx = 2(e^{-1}+1).$$

(c) Since $\left(\int_{-\infty}^{\infty} e^{-x^2}\, dx\right)\left(\int_{-\infty}^{\infty} e^{-y^2}\, dy\right) = \pi$ and y can be replaced by x, $\left(\int_{-\infty}^{\infty} e^{-x^2}\, dx\right)^2 = \pi$ implies that

$\int_{-\infty}^{\infty} e^{-x^2}\, dx = \pm\sqrt{\pi}$. But $e^{-x^2} \ge 0$ for all x, so $\int_{-\infty}^{\infty} e^{-x^2}\, dx = \sqrt{\pi}$.

(d) Letting $t = \sqrt{2}\, x$, $\int_{-\infty}^{\infty} e^{-x^2}\, dx = \int_{-\infty}^{\infty} \frac{1}{\sqrt{2}}\left(e^{-t^2/2}\right) dt$, so that $\sqrt{\pi} = \frac{1}{\sqrt{2}}\int_{-\infty}^{\infty} e^{-t^2/2}\, dt$ or $\int_{-\infty}^{\infty} e^{-t^2/2}\, dt = \sqrt{2\pi}$.

37. (a) We integrate by parts with $u = x$ and $dv = xe^{-x^2}\, dx$. Then $du = dx$ and $v = -\frac{1}{2}e^{-x^2}$, so

$$\int_{0}^{\infty} x^2 e^{-x^2}\, dx = \lim_{t\to\infty}\int_{0}^{t} x^2 e^{-x^2}\, dx = \lim_{t\to\infty}\left(-\frac{1}{2}xe^{-x^2}\Big]_{0}^{t} + \int_{0}^{t}\frac{1}{2}e^{-x^2}\, dx\right)$$

$$= \lim_{t\to\infty}\left(-\frac{1}{2}te^{-t^2}\right) + \frac{1}{2}\int_{0}^{\infty} e^{-x^2}\, dx = 0 + \frac{1}{2}\int_{0}^{\infty} e^{-x^2}\, dx \qquad \text{[by l'Hospital's Rule]}$$

$$= \frac{1}{4}\int_{-\infty}^{\infty} e^{-x^2}\, dx \qquad \text{[since } e^{-x^2} \text{ is an even function]}$$

$$= \frac{1}{4}\sqrt{\pi} \qquad \text{[by Exercise 36(c)]}$$

(b) Let $u = \sqrt{x}$. Then $u^2 = x \Rightarrow dx = 2u\, du \Rightarrow$

$$\int_{0}^{\infty}\sqrt{x}\, e^{-x}\, dx = \lim_{t\to\infty}\int_{0}^{t}\sqrt{x}\, e^{-x}\, dx = \lim_{t\to\infty}\int_{0}^{\sqrt{t}} u e^{-u^2} 2u\, du = 2\int_{0}^{\infty} u^2 e^{-u^2}\, du = 2\left(\frac{1}{4}\sqrt{\pi}\right) \ \text{[by part(a)]} \ = \frac{1}{2}\sqrt{\pi}.$$

16.5 Applications of Double Integrals

ET 15.5

1. $Q = \iint_{D} \sigma(x,y)\, dA = \int_{1}^{3}\int_{0}^{2}(2xy + y^2)\, dy\, dx = \int_{1}^{3}\left[xy^2 + \frac{1}{3}y^3\right]_{y=0}^{y=2} dx$

$$= \int_{1}^{3}\left(4x + \frac{8}{3}\right) dx = \left[2x^2 + \frac{8}{3}x\right]_{1}^{3} = 16 + \frac{16}{3} = \frac{64}{3}\ \text{C}$$

2. $Q = \iint_D \sigma(x,y)\,dA = \iint_D (x + y + x^2 + y^2)\,dA = \int_0^{2\pi} \int_0^2 (r\cos\theta + r\sin\theta + r^2)\,r\,dr\,d\theta$

$= \int_0^{2\pi} \int_0^2 [r^2(\cos\theta + \sin\theta) + r^3]\,dr\,d\theta = \int_0^{2\pi} \left[\frac{1}{3}r^3(\cos\theta + \sin\theta) + \frac{1}{4}r^4\right]_{r=0}^{r=2}\,d\theta$

$= \int_0^{2\pi} \left[\frac{8}{3}(\cos\theta + \sin\theta) + 4\right]d\theta = \left[\frac{8}{3}(\sin\theta - \cos\theta) + 4\theta\right]_0^{2\pi} = 8\pi$ C

3. $m = \iint_D \rho(x,y)\,dA = \int_0^2 \int_{-1}^1 xy^2\,dy\,dx = \int_0^2 x\,dx \int_{-1}^1 y^2\,dy = \left[\frac{1}{2}x^2\right]_0^2 \left[\frac{1}{3}y^3\right]_{-1}^1 = 2 \cdot \frac{2}{3} = \frac{4}{3}$,

$\overline{x} = \frac{1}{m}\iint_D x\rho(x,y)\,dA = \frac{3}{4}\int_0^2 \int_{-1}^1 x^2 y^2\,dy\,dx = \frac{3}{4}\int_0^2 x^2\,dx \int_{-1}^1 y^2\,dy = \frac{3}{4}\left[\frac{1}{3}x^3\right]_0^2 \left[\frac{1}{3}y^3\right]_{-1}^1 = \frac{3}{4} \cdot \frac{8}{3} \cdot \frac{2}{3} = \frac{4}{3}$,

$\overline{y} = \frac{1}{m}\iint_D y\rho(x,y)\,dA = \frac{3}{4}\int_0^2 \int_{-1}^1 xy^3\,dy\,dx = \frac{3}{4}\int_0^2 x\,dx \int_{-1}^1 y^3\,dy = \frac{3}{4}\left[\frac{1}{2}x^2\right]_0^2 \left[\frac{1}{4}y^4\right]_{-1}^1 = \frac{3}{4} \cdot 2 \cdot 0 = 0$.

Hence, $(\overline{x}, \overline{y}) = \left(\frac{4}{3}, 0\right)$.

4. $m = \iint_D \rho(x,y)\,dA = \int_0^a \int_0^b cxy\,dy\,dx = c\int_0^a x\,dx \int_0^b y\,dy = c\left[\frac{1}{2}x^2\right]_0^a \left[\frac{1}{2}y^2\right]_0^b = \frac{1}{4}a^2 b^2 c$,

$M_y = \iint_D x\rho(x,y)\,dA = \int_0^a \int_0^b cx^2 y\,dy\,dx = c\int_0^a x^2\,dx \int_0^b y\,dy = c\left[\frac{1}{3}x^3\right]_0^a \left[\frac{1}{2}y^2\right]_0^b = \frac{1}{6}a^3 b^2 c$, and

$M_x = \iint_D y\rho(x,y)\,dA = \int_0^a \int_0^b cxy^2\,dy\,dx = c\int_0^a x\,dx \int_0^b y^2\,dy = c\left[\frac{1}{2}x^2\right]_0^a \left[\frac{1}{3}y^3\right]_0^b = \frac{1}{6}a^2 b^3 c$.

Hence, $(\overline{x}, \overline{y}) = \left(\dfrac{M_y}{m}, \dfrac{M_x}{m}\right) = \left(\frac{2}{3}a, \frac{2}{3}b\right)$.

5. $m = \int_0^2 \int_{x/2}^{3-x} (x + y)\,dy\,dx = \int_0^2 \left[xy + \frac{1}{2}y^2\right]_{y=x/2}^{y=3-x}\,dx = \int_0^2 \left[x\left(3 - \frac{3}{2}x\right) + \frac{1}{2}(3 - x)^2 - \frac{1}{8}x^2\right]dx$

$= \int_0^2 \left(-\frac{9}{8}x^2 + \frac{9}{2}\right)dx = \left[-\frac{9}{8}\left(\frac{1}{3}x^3\right) + \frac{9}{2}x\right]_0^2 = 6$,

$M_y = \int_0^2 \int_{x/2}^{3-x} (x^2 + xy)\,dy\,dx = \int_0^2 \left[x^2 y + \frac{1}{2}xy^2\right]_{y=x/2}^{y=3-x}\,dx = \int_0^2 \left(\frac{9}{2}x - \frac{9}{8}x^3\right)dx = \frac{9}{2}$,

$M_x = \int_0^2 \int_{x/2}^{3-y} (xy + y^2)\,dy\,dx = \int_0^2 \left[\frac{1}{2}xy^2 + \frac{1}{3}y^3\right]_{y=x/2}^{y=3-x}\,dx = \int_0^2 \left(9 - \frac{9}{2}x\right)dx = 9$.

Hence $m = 6$, $(\overline{x}, \overline{y}) = \left(\dfrac{M_y}{m}, \dfrac{M_x}{m}\right) = \left(\dfrac{3}{4}, \dfrac{3}{2}\right)$.

6. Here $D = \{(x,y) \mid 0 \le x \le 2, x \le y \le 6 - 2x\}$.

$m = \int_0^2 \int_x^{6-2x} x^2\,dy\,dx = \int_0^2 x^2 (6 - 2x - x)\,dx = \int_0^2 (6x^2 - 3x^3)\,dx = \left[2x^3 - \frac{3}{4}x^4\right]_0^2 = 4$,

$M_y = \int_0^2 \int_x^{6-2x} x \cdot x^2\,dy\,dx = \int_0^2 x^3(6 - 2x - x)\,dx = \int_0^2 (6x^3 - 3x^4)\,dx = \left[\frac{3}{2}x^4 - \frac{3}{5}x^5\right]_0^2 = \frac{24}{5}$,

$M_x = \int_0^2 \int_x^{6-2x} y \cdot x^2\,dy\,dx = \int_0^2 x^2 \left[\frac{1}{2}(6 - 2x)^2 - \frac{1}{2}x^2\right]dx = \frac{1}{2}\int_0^2 (3x^4 - 24x^3 + 36x^2)\,dx$

$= \frac{1}{2}\left[\frac{3}{5}x^5 - 6x^4 + 12x^3\right]_0^2 = \frac{48}{5}$.

Hence $m = 4$, $(\overline{x}, \overline{y}) = \left(\dfrac{24/5}{4}, \dfrac{48/5}{4}\right) = \left(\frac{6}{5}, \frac{12}{5}\right)$.

7. $m = \int_0^1 \int_0^{e^x} y\,dy\,dx = \int_0^1 \left[\frac{1}{2}y^2\right]_{y=0}^{y=e^x}\,dx = \frac{1}{2}\int_0^1 e^{2x}\,dx = \frac{1}{4}e^{2x}\Big]_0^1 = \frac{1}{4}(e^2 - 1)$,

$M_y = \int_0^1 \int_0^{e^x} xy\,dy\,dx = \frac{1}{2}\int_0^1 xe^{2x}\,dx = \frac{1}{2}\left[\frac{1}{2}xe^{2x} - \frac{1}{4}e^{2x}\right]_0^1 = \frac{1}{8}(e^2 + 1)$,

$M_x = \int_0^1 \int_0^{e^x} y^2\,dy\,dx = \int_0^1 \left[\frac{1}{3}y^3\right]_{y=0}^{y=e^x}\,dx = \frac{1}{3}\int_0^1 e^{3x}\,dx = \frac{1}{3}\left[\frac{1}{3}e^{3x}\right]_0^1 = \frac{1}{9}(e^3 - 1)$.

Hence $m = \frac{1}{4}(e^2 - 1)$, $(\overline{x}, \overline{y}) = \left(\dfrac{\frac{1}{8}(e^2 + 1)}{\frac{1}{4}(e^2 - 1)}, \dfrac{\frac{1}{9}(e^3 - 1)}{\frac{1}{4}(e^2 - 1)}\right) = \left(\dfrac{e^2 + 1}{2(e^2 - 1)}, \dfrac{4(e^3 - 1)}{9(e^2 - 1)}\right)$.

8. $m = \int_0^1 \int_0^{\sqrt{x}} x \, dy \, dx = \int_0^1 x\big[y\big]_{y=0}^{y=\sqrt{x}} dx = \int_0^1 x^{3/2} \, dx = \frac{2}{5}x^{5/2}\Big]_0^1 = \frac{2}{5}$,

$M_y = \int_0^1 \int_0^{\sqrt{x}} x^2 \, dy \, dx = \int_0^1 x^2\big[y\big]_{y=0}^{y=\sqrt{x}} dx = \int_0^1 x^{5/2} \, dx = \frac{2}{7}x^{7/2}\Big]_0^1 = \frac{2}{7}$,

$M_x = \int_0^1 \int_0^{\sqrt{x}} yx \, dy \, dx = \int_0^1 x\big[\frac{1}{2}y^2\big]_{y=0}^{y=\sqrt{x}} dx = \frac{1}{2}\int_0^1 x^2 \, dx = \frac{1}{2}\big[\frac{1}{3}x^3\big]_0^1 = \frac{1}{6}$.

Hence $m = \frac{2}{5}$, $(\overline{x}, \overline{y}) = \left(\frac{2/7}{2/5}, \frac{1/6}{2/5}\right) = \left(\frac{5}{7}, \frac{5}{12}\right)$.

9. Note that $\sin(\pi x/L) \geq 0$ for $0 \leq x \leq L$.

$m = \int_0^L \int_0^{\sin(\pi x/L)} y \, dy \, dx = \int_0^L \frac{1}{2}\sin^2(\pi x/L) \, dx = \frac{1}{2}\big[\frac{1}{2}x - \frac{L}{4\pi}\sin(2\pi x/L)\big]_0^L = \frac{1}{4}L$,

$M_y = \int_0^L \int_0^{\sin(\pi x/L)} x \cdot y \, dy \, dx = \frac{1}{2}\int_0^L x \sin^2(\pi x/L) \, dx \quad \begin{bmatrix} \text{integrate by parts with} \\ u = x, dv = \sin^2(\pi x/L) \, dx \end{bmatrix}$

$\qquad = \frac{1}{2} \cdot x\big(\frac{1}{2}x - \frac{L}{4\pi}\sin(2\pi x/L)\big)\big]_0^L - \frac{1}{2}\int_0^L \big[\frac{1}{2}x - \frac{L}{4\pi}\sin(2\pi x/L)\big] \, dx$

$\qquad = \frac{1}{4}L^2 - \frac{1}{2}\big[\frac{1}{4}x^2 + \frac{L^2}{4\pi^2}\cos(2\pi x/L)\big]_0^L = \frac{1}{4}L^2 - \frac{1}{2}\left(\frac{1}{4}L^2 + \frac{L^2}{4\pi^2} - \frac{L^2}{4\pi^2}\right) = \frac{1}{8}L^2$,

$M_x = \int_0^L \int_0^{\sin(\pi x/L)} y \cdot y \, dy \, dx = \int_0^L \frac{1}{3}\sin^3(\pi x/L) \, dx = \frac{1}{3}\int_0^L \big[1 - \cos^2(\pi x/L)\big]\sin(\pi x/L) \, dx$

$\qquad \qquad \qquad [\text{substitute } u = \cos(\pi x/L) \quad \Rightarrow \quad du = -\frac{\pi}{L}\sin(\pi x/L)]$

$\qquad = \frac{1}{3}\left(-\frac{L}{\pi}\right)\big[\cos(\pi x/L) - \frac{1}{3}\cos^3(\pi x/L)\big]_0^L = -\frac{L}{3\pi}\left(-1 + \frac{1}{3} - 1 + \frac{1}{3}\right) = \frac{4}{9\pi}L$.

Hence $m = \dfrac{L}{4}$, $(\overline{x}, \overline{y}) = \left(\dfrac{L^2/8}{L/4}, \dfrac{4L/(9\pi)}{L/4}\right) = \left(\dfrac{L}{2}, \dfrac{16}{9\pi}\right)$.

10.

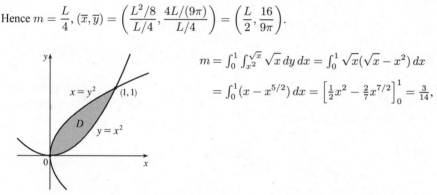

$m = \int_0^1 \int_{x^2}^{\sqrt{x}} \sqrt{x} \, dy \, dx = \int_0^1 \sqrt{x}(\sqrt{x} - x^2) \, dx$

$\qquad = \int_0^1 (x - x^{5/2}) \, dx = \big[\frac{1}{2}x^2 - \frac{2}{7}x^{7/2}\big]_0^1 = \frac{3}{14}$,

$M_y = \int_0^1 \int_{x^2}^{\sqrt{x}} x\sqrt{x} \, dy \, dx = \int_0^1 x\sqrt{x}(\sqrt{x} - x^2)dx = \int_0^1 (x^2 - x^{7/2})dx = \big[\frac{1}{3}x^3 - \frac{2}{9}x^{9/2}\big]_0^1 = \frac{1}{9}$

$M_x = \int_0^1 \int_{x^2}^{\sqrt{x}} y\sqrt{x} \, dy \, dx = \int_0^1 \sqrt{x} \cdot \frac{1}{2}(x - x^4)dx = \frac{1}{2}\int_0^1 (x^{3/2} - x^{9/2})dx$

$\qquad = \frac{1}{2}\big[\frac{2}{5}x^{5/2} - \frac{2}{11}x^{11/2}\big]_0^1 = \frac{1}{2} \cdot \frac{12}{55} = \frac{6}{55}$.

Hence $m = \frac{3}{14}$, $(\overline{x}, \overline{y}) = \left(\frac{1/9}{3/14}, \frac{6/55}{3/14}\right) = \left(\frac{14}{27}, \frac{28}{55}\right)$.

11. $\rho(x, y) = ky = kr\sin\theta$, $m = \int_0^{\pi/2} \int_0^1 kr^2 \sin\theta \, dr \, d\theta = \frac{1}{3}k\int_0^{\pi/2} \sin\theta \, d\theta = \frac{1}{3}k\big[-\cos\theta\big]_0^{\pi/2} = \frac{1}{3}k$,

$M_y = \int_0^{\pi/2} \int_0^1 kr^3 \sin\theta\cos\theta \, dr \, d\theta = \frac{1}{4}k\int_0^{\pi/2} \sin\theta\cos\theta \, d\theta = \frac{1}{8}k\big[-\cos 2\theta\big]_0^{\pi/2} = \frac{1}{8}k$,

$M_x = \int_0^{\pi/2} \int_0^1 kr^3 \sin^2 \theta \, dr \, d\theta = \frac{1}{4}k \int_0^{\pi/2} \sin^2 \theta \, d\theta = \frac{1}{8}k\left[\theta + \sin 2\theta\right]_0^{\pi/2} = \frac{\pi}{16}k.$

Hence $(\overline{x}, \overline{y}) = \left(\frac{3}{8}, \frac{3\pi}{16}\right)$.

12. $\rho(x, y) = k(x^2 + y^2) = kr^2$, $m = \int_0^{\pi/2} \int_0^1 kr^3 \, dr \, d\theta = \frac{\pi}{8}k$,

$M_y = \int_0^{\pi/2} \int_0^1 kr^4 \cos \theta \, dr \, d\theta = \frac{1}{5}k \int_0^{\pi/2} \cos \theta \, d\theta = \frac{1}{5}k\left[\sin \theta\right]_0^{\pi/2} = \frac{1}{5}k,$

$M_x = \int_0^{\pi/2} \int_0^1 kr^4 \sin \theta \, dr \, d\theta = \frac{1}{5}k \int_0^{\pi/2} \sin \theta \, d\theta = \frac{1}{5}k\left[-\cos \theta\right]_0^{\pi/2} = \frac{1}{5}k.$

Hence $(\overline{x}, \overline{y}) = \left(\frac{8}{5\pi}, \frac{8}{5\pi}\right)$.

13.

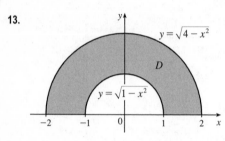

$\rho(x, y) = k\sqrt{x^2 + y^2} = kr,$

$m = \iint_D \rho(x, y) \, dA = \int_0^\pi \int_1^2 kr \cdot r \, dr \, d\theta$

$= k \int_0^\pi d\theta \int_1^2 r^2 \, dr = k(\pi)\left[\frac{1}{3}r^3\right]_1^2 = \frac{7}{3}\pi k,$

$M_y = \iint_D x\rho(x, y) \, dA = \int_0^\pi \int_1^2 (r\cos \theta)(kr) \, r \, dr \, d\theta = k\int_0^\pi \cos \theta \, d\theta \int_1^2 r^3 \, dr$

$= k\left[\sin \theta\right]_0^\pi \left[\frac{1}{4}r^4\right]_1^2 = k(0)\left(\frac{15}{4}\right) = 0$ [this is to be expected as the region and density function are symmetric about the y-axis]

$M_x = \iint_D y\rho(x, y) \, dA = \int_0^\pi \int_1^2 (r\sin \theta)(kr) \, r \, dr \, d\theta = k\int_0^\pi \sin \theta \, d\theta \int_1^2 r^3 \, dr$

$= k\left[-\cos \theta\right]_0^\pi \left[\frac{1}{4}r^4\right]_1^2 = k(1+1)\left(\frac{15}{4}\right) = \frac{15}{2}k.$

Hence $(\overline{x}, \overline{y}) = \left(0, \frac{15k/2}{7\pi k/3}\right) = \left(0, \frac{45}{14\pi}\right).$

14. Now $\rho(x, y) = k/\sqrt{x^2 + y^2} = k/r$, so

$m = \iint_D \rho(x, y) \, dA = \int_0^\pi \int_1^2 (k/r) \, r \, dr \, d\theta = k\int_0^\pi d\theta \int_1^2 dr = k(\pi)(1) = \pi k,$

$M_y = \iint_D x\rho(x, y) \, dA = \int_0^\pi \int_1^2 (r\cos \theta)(k/r) \, r \, dr \, d\theta = k\int_0^\pi \cos \theta \, d\theta \int_1^2 r \, dr$

$= k\left[\sin \theta\right]_0^\pi \left[\frac{1}{2}r^2\right]_1^2 = k(0)\left(\frac{3}{2}\right) = 0,$

$M_x = \iint_D y\rho(x, y) \, dA = \int_0^\pi \int_1^2 (r\sin \theta)(k/r) \, r \, dr \, d\theta = k\int_0^\pi \sin \theta \, d\theta \int_1^2 r \, dr$

$= k\left[-\cos \theta\right]_0^\pi \left[\frac{1}{2}r^2\right]_1^2 = k(1+1)\left(\frac{3}{2}\right) = 3k.$

Hence $(\overline{x}, \overline{y}) = \left(0, \frac{3k}{\pi k}\right) = \left(0, \frac{3}{\pi}\right).$

15. Placing the vertex opposite the hypotenuse at $(0, 0)$, $\rho(x, y) = k(x^2 + y^2)$. Then

$m = \int_0^a \int_0^{a-x} k(x^2 + y^2) \, dy \, dx = k\int_0^a \left[ax^2 - x^3 + \frac{1}{3}(a - x)^3\right] dx = k\left[\frac{1}{3}ax^3 - \frac{1}{4}x^4 - \frac{1}{12}(a - x)^4\right]_0^a = \frac{1}{6}ka^4.$

By symmetry,

$M_y = M_x = \int_0^a \int_0^{a-x} ky(x^2 + y^2) \, dy \, dx = k\int_0^a \left[\frac{1}{2}(a - x)^2 x^2 + \frac{1}{4}(a - x)^4\right] dx$

$= k\left[\frac{1}{6}a^2x^3 - \frac{1}{4}ax^4 + \frac{1}{10}x^5 - \frac{1}{20}(a - x)^5\right]_0^a = \frac{1}{15}ka^5$

Hence $(\overline{x}, \overline{y}) = \left(\frac{2}{5}a, \frac{2}{5}a\right).$

16. $\rho(x, y) = k/\sqrt{x^2 + y^2} = k/r.$

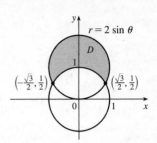

$$m = \int_{\pi/6}^{5\pi/6} \int_1^{2\sin\theta} \frac{k}{r}\, r\, dr\, d\theta = k\int_{\pi/6}^{5\pi/6} [(2\sin\theta) - 1]\, d\theta$$

$$= k\big[-2\cos\theta - \theta\big]_{\pi/6}^{5\pi/6} = 2k\big(\sqrt{3} - \tfrac{\pi}{3}\big)$$

By symmetry of D and $f(x) = x$, $M_y = 0$, and

$$M_x = \int_{\pi/6}^{5\pi/6}\int_1^{2\sin\theta} kr\sin\theta\, dr\, d\theta = \tfrac{1}{2}k\int_{\pi/6}^{5\pi/6}(4\sin^3\theta - \sin\theta)\, d\theta$$

$$= \tfrac{1}{2}k\big[-3\cos\theta + \tfrac{4}{3}\cos^3\theta\big]_{\pi/6}^{5\pi/6} = \sqrt{3}\,k$$

Hence $(\overline{x}, \overline{y}) = \left(0, \dfrac{3\sqrt{3}}{2(3\sqrt{3} - \pi)}\right).$

17. $I_x = \iint_D y^2\rho(x,y)\, dA = \int_0^1\int_0^{e^x} y^2 \cdot y\, dy\, dx = \int_0^1 \big[\tfrac{1}{4}y^4\big]_{y=0}^{y=e^x}\, dx = \tfrac{1}{4}\int_0^1 e^{4x}\, dx = \tfrac{1}{4}\big[\tfrac{1}{4}e^{4x}\big]_0^1 = \tfrac{1}{16}(e^4 - 1),$

$I_y = \iint_D x^2\rho(x,y)\, dA = \int_0^1\int_0^{e^x} x^2 y\, dy\, dx = \int_0^1 x^2\big[\tfrac{1}{2}y^2\big]_{y=0}^{y=e^x}\, dx = \tfrac{1}{2}\int_0^1 x^2 e^{2x}\, dx$

$$= \tfrac{1}{2}\big[\big(\tfrac{1}{2}x^2 - \tfrac{1}{2}x + \tfrac{1}{4}\big)e^{2x}\big]_0^1 \quad \text{[integrate by parts twice]} \quad = \tfrac{1}{8}(e^2 - 1),$$

and $I_0 = I_x + I_y = \tfrac{1}{16}(e^4 - 1) + \tfrac{1}{8}(e^2 - 1) = \tfrac{1}{16}(e^4 + 2e^2 - 3).$

18. $I_x = \int_0^{\pi/2}\int_0^1 (r^2\sin^2\theta)(kr^2)\, r\, dr\, d\theta = \tfrac{1}{6}k\int_0^{\pi/2}\sin^2\theta\, d\theta = \tfrac{1}{6}k\big[\tfrac{1}{4}(2\theta - \sin 2\theta)\big]_0^{\pi/2} = \tfrac{\pi}{24}k,$

$I_y = \int_0^{\pi/2}\int_0^1 (r^2\cos^2\theta)(kr^2)\, r\, dr\, d\theta = \tfrac{1}{6}k\int_0^{\pi/6}\cos^2\theta\, d\theta = \tfrac{1}{6}k\big[\tfrac{1}{4}(2\theta + \sin 2\theta)\big]_0^{\pi/2} = \tfrac{\pi}{24}k,$

and $I_0 = I_x + I_y = \tfrac{\pi}{12}k.$

19. As in Exercise 15, we place the vertex opposite the hypotenuse at $(0, 0)$ and the equal sides along the positive axes.

$I_x = \int_0^a\int_0^{a-x} y^2 k(x^2 + y^2)\, dy\, dx = k\int_0^a\int_0^{a-x}(x^2 y^2 + y^4)\, dy\, dx = k\int_0^a \big[\tfrac{1}{3}x^2 y^3 + \tfrac{1}{5}y^5\big]_{y=0}^{y=a-x}\, dx$

$$= k\int_0^a\big[\tfrac{1}{3}x^2(a-x)^3 + \tfrac{1}{5}(a-x)^5\big]\, dx = k\big[\tfrac{1}{3}\big(\tfrac{1}{3}a^3 x^3 - \tfrac{3}{4}a^2 x^4 + \tfrac{3}{5}a x^5 - \tfrac{1}{6}x^6\big) - \tfrac{1}{30}(a-x)^6\big]_0^a = \tfrac{7}{180}ka^6,$$

$I_y = \int_0^a\int_0^{a-x} x^2 k(x^2 + y^2)\, dy\, dx = k\int_0^a\int_0^{a-x}(x^4 + x^2 y^2)\, dy\, dx = k\int_0^a\big[x^4 y + \tfrac{1}{3}x^2 y^3\big]_{y=0}^{y=a-x}\, dx$

$$= k\int_0^a\big[x^4(a-x) + \tfrac{1}{3}x^2(a-x)^3\big]\, dx = k\big[\tfrac{1}{5}ax^5 - \tfrac{1}{6}x^6 + \tfrac{1}{3}\big(\tfrac{1}{3}a^3 x^3 - \tfrac{3}{4}a^2 x^4 + \tfrac{3}{5}ax^5 - \tfrac{1}{6}x^6\big)\big]_0^a = \tfrac{7}{180}ka^6,$$

and $I_0 = I_x + I_y = \tfrac{7}{90}ka^6.$

20. If we find the moments of inertia about the x- and y-axes, we can determine in which direction rotation will be more difficult. (See the explanation following Example 4.) The moment of inertia about the x-axis is given by

$$I_x = \iint_D y^2\rho(x,y)\, dA = \int_0^2\int_0^2 y^2(1 + 0.1x)\, dy\, dx = \int_0^2(1 + 0.1x)\big[\tfrac{1}{3}y^3\big]_{y=0}^{y=2}\, dx$$

$$= \tfrac{8}{3}\int_0^2(1 + 0.1x)\, dx = \tfrac{8}{3}\big[x + 0.1\cdot\tfrac{1}{2}x^2\big]_0^2 = \tfrac{8}{3}(2.2) \approx 5.87$$

Similarly, the moment of inertia about the y-axis is given by

$$I_y = \iint_D x^2\rho(x,y)\, dA = \int_0^2\int_0^2 x^2(1 + 0.1x)\, dy\, dx = \int_0^2 x^2(1 + 0.1x)\big[y\big]_{y=0}^{y=2}\, dx$$

$$= 2\int_0^2(x^2 + 0.1x^3)\, dx = 2\big[\tfrac{1}{3}x^3 + 0.1\cdot\tfrac{1}{4}x^4\big]_0^2 = 2\big(\tfrac{8}{3} + 0.4\big) \approx 6.13$$

Since $I_y > I_x$, more force is required to rotate the fan blade about the y-axis.

21. Using a CAS, we find $m = \iint_D \rho(x, y)\, dA = \int_0^\pi \int_0^{\sin x} xy\, dy\, dx = \dfrac{\pi^2}{8}$. Then

$$\bar{x} = \frac{1}{m} \iint_D x\rho(x, y)\, dA = \frac{8}{\pi^2} \int_0^\pi \int_0^{\sin x} x^2 y\, dy\, dx = \frac{2\pi}{3} - \frac{1}{\pi}\ \text{and}$$

$$\bar{y} = \frac{1}{m} \iint_D y\rho(x, y)\, dA = \frac{8}{\pi^2} \int_0^\pi \int_0^{\sin x} xy^2\, dy\, dx = \frac{16}{9\pi},\ \text{so}\ (\bar{x}, \bar{y}) = \left(\frac{2\pi}{3} - \frac{1}{\pi}, \frac{16}{9\pi} \right).$$

The moments of inertia are $I_x = \iint_D y^2 \rho(x, y)\, dA = \int_0^\pi \int_0^{\sin x} xy^3\, dy\, dx = \dfrac{3\pi^2}{64}$,

$I_y = \iint_D x^2 \rho(x, y)\, dA = \int_0^\pi \int_0^{\sin x} x^3 y\, dy\, dx = \dfrac{\pi^2}{16}(\pi^2 - 3)$, and $I_0 = I_x + I_y = \dfrac{\pi^2}{64}(4\pi^2 - 9)$.

22. Using a CAS, we find $m = \iint_D \sqrt{x^2 + y^2}\, dA = \int_0^{2\pi} \int_0^{1+\cos\theta} r^2\, dr\, d\theta = \frac{5}{3}\pi$,

$\bar{x} = \frac{1}{m} \iint_D x\sqrt{x^2 + y^2}\, dA = \frac{3}{5\pi} \int_0^{2\pi} \int_0^{1+\cos\theta} r^3 \cos\theta\, dr\, d\theta = \frac{21}{20}$ and

$\bar{y} = \frac{1}{m} \iint_D y\sqrt{x^2 + y^2}\, dA = \frac{3}{5\pi} \int_0^{2\pi} \int_0^{1+\cos\theta} r^3 \sin\theta\, dr\, d\theta = 0$, so $(\bar{x}, \bar{y}) = \left(\frac{21}{20}, 0 \right)$.

The moments of inertia are $I_x = \iint_D y^2 \sqrt{x^2 + y^2}\, dA = \int_0^{2\pi} \int_0^{1+\cos\theta} r^4 \sin^2\theta\, dr\, d\theta = \frac{33}{40}\pi$,

$I_y = \iint_D x^2 \sqrt{x^2 + y^2}\, dA = \int_0^{2\pi} \int_0^{1+\cos\theta} r^4 \cos^2\theta\, dr\, d\theta = \frac{93}{40}\pi$, and $I_0 = I_x + I_y = \frac{63}{20}\pi$.

23. $I_x = \iint_D y^2 \rho(x, y)\, dA = \int_0^h \int_0^b \rho y^2\, dx\, dy = \rho \int_0^b dx \int_0^h y^2\, dy = \rho\left[x \right]_0^b \left[\frac{1}{3}y^3 \right]_0^h = \rho b\left(\frac{1}{3}h^3 \right) = \frac{1}{3}\rho b h^3$,

$I_y = \iint_D x^2 \rho(x, y)\, dA = \int_0^h \int_0^b \rho x^2\, dx\, dy = \rho \int_0^b x^2\, dx \int_0^h dy = \rho\left[\frac{1}{3}x^3 \right]_0^b [y]_0^h = \frac{1}{3}\rho b^3 h$,

and $m = \rho$ (area of rectangle) $= \rho b h$ since the lamina is homogeneous. Hence $\bar{\bar{x}}^2 = \dfrac{I_y}{m} = \dfrac{\frac{1}{3}\rho b^3 h}{\rho b h} = \dfrac{b^2}{3} \Rightarrow \bar{\bar{x}} = \dfrac{b}{\sqrt{3}}$

and $\bar{\bar{y}}^2 = \dfrac{I_x}{m} = \dfrac{\frac{1}{3}\rho b h^3}{\rho b h} = \dfrac{h^2}{3} \Rightarrow \bar{\bar{y}} = \dfrac{h}{\sqrt{3}}$.

24. Here we assume $b > 0$, $h > 0$ but note that we arrive at the same results if $b < 0$ or $h < 0$. We have

$D = \left\{ (x, y) \mid 0 \le x \le b, 0 \le y \le h - \frac{h}{b}x \right\}$, so

$I_x = \int_0^b \int_0^{h-hx/b} y^2 \rho\, dy\, dx = \rho \int_0^b \left[\frac{1}{3}y^3 \right]_{y=0}^{y=h-hx/b} dx = \frac{1}{3}\rho \int_0^b \left(h - \frac{h}{b}x \right)^3 dx$

$= \frac{1}{3}\rho \left[-\frac{b}{h}\left(\frac{1}{4} \right)\left(h - \frac{h}{b}x \right)^4 \right]_0^b = -\frac{b}{12h}\rho(0 - h^4) = \frac{1}{12}\rho b h^3$,

$I_y = \int_0^b \int_0^{h-hx/b} x^2 \rho\, dy\, dx = \rho \int_0^b x^2 \left(h - \frac{h}{b}x \right) dx = \rho \int_0^b \left(hx^2 - \frac{h}{b}x^3 \right) dx$

$= \rho \left[\frac{h}{3}x^3 - \frac{h}{4b}x^4 \right]_0^b = \rho\left(\frac{hb^3}{3} - \frac{hb^3}{4} \right) = \frac{1}{12}\rho b^3 h$,

and $m = \int_0^b \int_0^{h-hx/b} \rho\, dy\, dx = \rho \int_0^b \left(h - \frac{h}{b}x \right) dx = \rho \left[hx - \frac{h}{2b}x^2 \right]_0^b = \frac{1}{2}\rho b h$. Hence $\bar{\bar{x}}^2 = \dfrac{I_y}{m} = \dfrac{\frac{1}{12}\rho b^3 h}{\frac{1}{2}\rho b h} = \dfrac{b^2}{6} \Rightarrow$

$\bar{\bar{x}} = \dfrac{b}{\sqrt{6}}$ and $\bar{\bar{y}}^2 = \dfrac{I_x}{m} = \dfrac{\frac{1}{12}\rho b h^3}{\frac{1}{2}\rho b h} = \dfrac{h^2}{6} \Rightarrow \bar{\bar{y}} = \dfrac{h}{\sqrt{6}}$.

25. In polar coordinates, the region is $D = \left\{ (r, \theta) \mid 0 \le r \le a, 0 \le \theta \le \frac{\pi}{2} \right\}$, so

$I_x = \iint_D y^2 \rho\, dA = \int_0^{\pi/2} \int_0^a \rho(r\sin\theta)^2 r\, dr\, d\theta = \rho \int_0^{\pi/2} \sin^2\theta\, d\theta \int_0^a r^3\, dr$

$= \rho\left[\frac{1}{2}\theta - \frac{1}{4}\sin 2\theta \right]_0^{\pi/2} \left[\frac{1}{4}r^4 \right]_0^a = \rho\left(\frac{\pi}{4} \right)\left(\frac{1}{4}a^4 \right) = \frac{1}{16}\rho a^4 \pi$,

$I_y = \iint_D x^2 \rho \, dA = \int_0^{\pi/2} \int_0^a \rho (r \cos \theta)^2 \, r \, dr \, d\theta = \rho \int_0^{\pi/2} \cos^2 \theta \, d\theta \int_0^a r^3 \, dr$

$= \rho \left[\frac{1}{2}\theta + \frac{1}{4} \sin 2\theta \right]_0^{\pi/2} \left[\frac{1}{4} r^4 \right]_0^a = \rho \left(\frac{\pi}{4} \right) \left(\frac{1}{4} a^4 \right) = \frac{1}{16} \rho a^4 \pi$,

and $m = \rho \cdot A(D) = \rho \cdot \frac{1}{4} \pi a^2$ since the lamina is homogeneous. Hence $\overline{\overline{x}}^2 = \overline{\overline{y}}^2 = \dfrac{\frac{1}{16} \rho a^4 \pi}{\frac{1}{4} \rho a^2 \pi} = \dfrac{a^2}{4} \quad \Rightarrow \quad \overline{\overline{x}} = \overline{\overline{y}} = \dfrac{a}{2}$.

26. $m = \int_0^\pi \int_0^{\sin x} \rho \, dy \, dx = \rho \int_0^\pi \sin x \, dx = \rho \left[-\cos x \right]_0^\pi = 2\rho$,

$I_x = \int_0^\pi \int_0^{\sin x} \rho y^2 \, dy \, dx = \frac{1}{3} \rho \int_0^\pi \sin^3 x \, dx = \frac{1}{3} \rho \int_0^\pi (1 - \cos^2 x) \sin x \, dx = \frac{1}{3} \rho \left[-\cos \theta + \frac{1}{3} \cos^3 \theta \right]_0^\pi = \frac{4}{9} \rho$,

$I_y = \int_0^\pi \int_0^{\sin x} \rho x^2 \, dy \, dx = \rho \int_0^\pi x^2 \sin x \, dx = \rho \left[-x^2 \cos x + 2x \sin x + 2 \cos x \right]_0^\pi \quad$ [by integrating by parts twice]

$= \rho (\pi^2 - 4)$.

Then $\overline{\overline{y}}^2 = \dfrac{I_x}{m} = \dfrac{2}{9}$, so $\overline{\overline{y}} = \dfrac{\sqrt{2}}{3}$ and $\overline{\overline{x}}^2 = \dfrac{I_y}{m} = \dfrac{\pi^2 - 4}{2}$, so $\overline{\overline{x}} = \sqrt{\dfrac{\pi^2 - 4}{2}}$.

27. (a) $f(x, y)$ is a joint density function, so we know $\iint_{\mathbb{R}^2} f(x, y) \, dA = 1$. Since $f(x, y) = 0$ outside the rectangle $[0, 1] \times [0, 2]$, we can say

$$\iint_{\mathbb{R}^2} f(x, y) \, dA = \int_{-\infty}^\infty \int_{-\infty}^\infty f(x, y) \, dy \, dx = \int_0^1 \int_0^2 Cx(1 + y) \, dy \, dx$$
$$= C \int_0^1 x \left[y + \frac{1}{2} y^2 \right]_{y=0}^{y=2} dx = C \int_0^1 4x \, dx = C \left[2x^2 \right]_0^1 = 2C$$

Then $2C = 1 \quad \Rightarrow \quad C = \frac{1}{2}$.

(b) $P(X \leq 1, Y \leq 1) = \int_{-\infty}^1 \int_{-\infty}^1 f(x, y) \, dy \, dx = \int_0^1 \int_0^1 \frac{1}{2} x(1 + y) \, dy \, dx$

$= \int_0^1 \frac{1}{2} x \left[y + \frac{1}{2} y^2 \right]_{y=0}^{y=1} dx = \int_0^1 \frac{1}{2} x \left(\frac{3}{2} \right) dx = \frac{3}{4} \left[\frac{1}{2} x^2 \right]_0^1 = \frac{3}{8}$ or 0.375

(c) $P(X + Y \leq 1) = P((X, Y) \in D)$ where D is the triangular region shown in the figure. Thus

$P(X + Y \leq 1) = \iint_D f(x, y) \, dA = \int_0^1 \int_0^{1-x} \frac{1}{2} x(1 + y) \, dy \, dx$

$= \int_0^1 \frac{1}{2} x \left[y + \frac{1}{2} y^2 \right]_{y=0}^{y=1-x} dx = \int_0^1 \frac{1}{2} x \left(\frac{1}{2} x^2 - 2x + \frac{3}{2} \right) dx$

$= \frac{1}{4} \int_0^1 (x^3 - 4x^2 + 3x) \, dx = \frac{1}{4} \left[\frac{x^4}{4} - 4\frac{x^3}{3} + 3\frac{x^2}{2} \right]_0^1$

$= \frac{5}{48} \approx 0.1042$

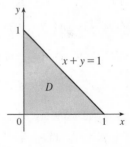

28. (a) $f(x, y) \geq 0$, so f is a joint density function if $\iint_{\mathbb{R}^2} f(x, y) \, dA = 1$. Here, $f(x, y) = 0$ outside the square $[0, 1] \times [0, 1]$,

so $\iint_{\mathbb{R}^2} f(x, y) \, dA = \int_0^1 \int_0^1 4xy \, dy \, dx = \int_0^1 \left[2xy^2 \right]_{y=0}^{y=1} dx = \int_0^1 2x \, dx = x^2 \Big]_0^1 = 1$.

Thus, $f(x, y)$ is a joint density function.

(b) (i) No restriction is placed on Y, so

$P\left(X \geq \frac{1}{2} \right) = \int_{1/2}^\infty \int_{-\infty}^\infty f(x, y) \, dy \, dx = \int_{1/2}^1 \int_0^1 4xy \, dy \, dx = \int_{1/2}^1 \left[2xy^2 \right]_{y=0}^{y=1} dx = \int_{1/2}^1 2x \, dx = x^2 \Big]_{1/2}^1 = \frac{3}{4}$.

(ii) $P\left(X \geq \frac{1}{2}, Y \leq \frac{1}{2} \right) = \int_{1/2}^\infty \int_{-\infty}^{1/2} f(x, y) \, dy \, dx = \int_{1/2}^1 \int_0^{1/2} 4xy \, dy \, dx$

$= \int_{1/2}^1 \left[2xy^2 \right]_{y=0}^{y=1/2} dx = \int_{1/2}^1 \frac{1}{2} x \, dx = \frac{1}{2} \cdot \frac{1}{2} x^2 \Big]_{1/2}^1 = \frac{3}{16}$

(c) The expected value of X is given by

$$\mu_1 = \iint_{\mathbb{R}^2} x\, f(x,y)\, dA = \int_0^1 \int_0^1 x(4xy)\, dy\, dx = \int_0^1 2x^2 \big[y^2\big]_{y=0}^{y=1}\, dx = 2\int_0^1 x^2\, dx = 2\big[\tfrac{1}{3}x^3\big]_0^1 = \tfrac{2}{3}$$

The expected value of Y is

$$\mu_2 = \iint_{\mathbb{R}^2} y\, f(x,y)\, dA = \int_0^1 \int_0^1 y(4xy)\, dy\, dx = \int_0^1 4x\big[\tfrac{1}{3}y^3\big]_{y=0}^{y=1}\, dx = \tfrac{4}{3}\int_0^1 x\, dx = \tfrac{4}{3}\big[\tfrac{1}{2}x^2\big]_0^1 = \tfrac{2}{3}$$

29. (a) $f(x,y) \geq 0$, so f is a joint density function if $\iint_{\mathbb{R}^2} f(x,y)\, dA = 1$. Here, $f(x,y) = 0$ outside the first quadrant, so

$$\iint_{\mathbb{R}^2} f(x,y)\, dA = \int_0^\infty \int_0^\infty 0.1e^{-(0.5x+0.2y)}\, dy\, dx = 0.1\int_0^\infty \int_0^\infty e^{-0.5x}e^{-0.2y}\, dy\, dx = 0.1\int_0^\infty e^{-0.5x}\, dx \int_0^\infty e^{-0.2y}\, dy$$

$$= 0.1 \lim_{t\to\infty} \int_0^t e^{-0.5x}\, dx \lim_{t\to\infty} \int_0^t e^{-0.2y}\, dy = 0.1 \lim_{t\to\infty} \big[-2e^{-0.5x}\big]_0^t \lim_{t\to\infty} \big[-5e^{-0.2y}\big]_0^t$$

$$= 0.1 \lim_{t\to\infty} \big[-2(e^{-0.5t} - 1)\big] \lim_{t\to\infty} \big[-5(e^{-0.2t} - 1)\big] = (0.1)\cdot(-2)(0-1)\cdot(-5)(0-1) = 1$$

Thus $f(x,y)$ is a joint density function.

(b) (i) No restriction is placed on X, so

$$P(Y \geq 1) = \int_{-\infty}^\infty \int_1^\infty f(x,y)\, dy\, dx = \int_0^\infty \int_1^\infty 0.1e^{-(0.5x+0.2y)}\, dy\, dx$$

$$= 0.1\int_0^\infty e^{-0.5x}\, dx \int_1^\infty e^{-0.2y}\, dy = 0.1 \lim_{t\to\infty} \int_0^t e^{-0.5x}\, dx \lim_{t\to\infty} \int_1^t e^{-0.2y}\, dy$$

$$= 0.1 \lim_{t\to\infty} \big[-2e^{-0.5x}\big]_0^t \lim_{t\to\infty} \big[-5e^{-0.2y}\big]_1^t = 0.1 \lim_{t\to\infty} \big[-2(e^{-0.5t} - 1)\big] \lim_{t\to\infty} \big[-5(e^{-0.2t} - e^{-0.2})\big]$$

$$(0.1)\cdot(-2)(0-1)\cdot(-5)(0-e^{-0.2}) = e^{-0.2} \approx 0.8187$$

(ii) $P(X \leq 2, Y \leq 4) = \int_{-\infty}^2 \int_{-\infty}^4 f(x,y)\, dy\, dx = \int_0^2 \int_0^4 0.1e^{-(0.5x+0.2y)}\, dy\, dx$

$$= 0.1\int_0^2 e^{-0.5x}\, dx \int_0^4 e^{-0.2y}\, dy = 0.1\big[-2e^{-0.5x}\big]_0^2 \big[-5e^{-0.2y}\big]_0^4$$

$$= (0.1)\cdot(-2)(e^{-1} - 1)\cdot(-5)(e^{-0.8} - 1)$$

$$= (e^{-1} - 1)(e^{-0.8} - 1) = 1 + e^{-1.8} - e^{-0.8} - e^{-1} \approx 0.3481$$

(c) The expected value of X is given by

$$\mu_1 = \iint_{\mathbb{R}^2} x\, f(x,y)\, dA = \int_0^\infty \int_0^\infty x\Big[0.1e^{-(0.5x+0.2y)}\Big]\, dy\, dx$$

$$= 0.1\int_0^\infty xe^{-0.5x}\, dx \int_0^\infty e^{-0.2y}\, dy = 0.1 \lim_{t\to\infty} \int_0^t xe^{-0.5x}\, dx \lim_{t\to\infty} \int_0^t e^{-0.2y}\, dy$$

To evaluate the first integral, we integrate by parts with $u = x$ and $dv = e^{-0.5x}\, dx$ (or we can use Formula 96 in the Table of Integrals): $\int xe^{-0.5x}\, dx = -2xe^{-0.5x} - \int -2e^{-0.5x}\, dx = -2xe^{-0.5x} - 4e^{-0.5x} = -2(x+2)e^{-0.5x}$. Thus

$$\mu_1 = 0.1 \lim_{t\to\infty} \big[-2(x+2)e^{-0.5x}\big]_0^t \lim_{t\to\infty} \big[-5e^{-0.2y}\big]_0^t$$

$$= 0.1 \lim_{t\to\infty} (-2)\big[(t+2)e^{-0.5t} - 2\big] \lim_{t\to\infty} (-5)\big[e^{-0.2t} - 1\big]$$

$$= 0.1(-2)\Big(\lim_{t\to\infty} \frac{t+2}{e^{0.5t}} - 2\Big)(-5)(-1) = 2 \qquad \text{[by l'Hospital's Rule]}$$

The expected value of Y is given by

$$\mu_2 = \iint_{\mathbb{R}^2} y\, f(x,y)\, dA = \int_0^\infty \int_0^\infty y\Big[0.1e^{-(0.5+0.2y)}\Big]\, dy\, dx$$

$$= 0.1\int_0^\infty e^{-0.5x}\, dx \int_0^\infty ye^{-0.2y}\, dy = 0.1 \lim_{t\to\infty} \int_0^t e^{-0.5x}\, dx \lim_{t\to\infty} \int_0^t ye^{-0.2y}\, dy$$

To evaluate the second integral, we integrate by parts with $u = y$ and $dv = e^{-0.2y}\, dy$ (or again we can use Formula 96 in the Table of Integrals) which gives $\int ye^{-0.2y}\,dy = -5ye^{-0.2y} + \int 5e^{-0.2y}\,dy = -5(y+5)e^{-0.2y}$. Then

$$\mu_2 = 0.1 \lim_{t\to\infty} \left[-2e^{-0.5x}\right]_0^t \lim_{t\to\infty} \left[-5(y+5)e^{-0.2y}\right]_0^t$$

$$= 0.1 \lim_{t\to\infty} \left[-2(e^{-0.5t}-1)\right] \lim_{t\to\infty} \left(-5\left[(t+5)e^{-0.2t}-5\right]\right)$$

$$= 0.1(-2)(-1)\cdot(-5)\left(\lim_{t\to\infty}\frac{t+5}{e^{0.2t}}-5\right) = 5 \qquad \text{[by l'Hospital's Rule]}$$

30. (a) The lifetime of each bulb has exponential density function

$$f(t) = \begin{cases} 0 & \text{if } t < 0 \\ \frac{1}{1000}e^{-t/1000} & \text{if } t \geq 0 \end{cases}$$

If X and Y are the lifetimes of the individual bulbs, then X and Y are independent, so the joint density function is the product of the individual density functions:

$$f(x,y) = \begin{cases} 10^{-6}e^{-(x+y)/1000} & \text{if } x \geq 0,\, y \geq 0 \\ 0 & \text{otherwise} \end{cases}$$

The probability that both of the bulbs fail within 1000 hours is

$$P(X \leq 1000, Y \leq 1000) = \int_{-\infty}^{1000} \int_{-\infty}^{1000} f(x,y)\,dy\,dx = \int_0^{1000}\int_0^{1000} 10^{-6}e^{-(x+y)/1000}\,dy\,dx$$

$$= 10^{-6}\int_0^{1000} e^{-x/1000}\,dx \int_0^{1000} e^{-y/1000}\,dy$$

$$= 10^{-6}\left[-1000e^{-x/1000}\right]_0^{1000} \left[-1000e^{-y/1000}\right]_0^{1000}$$

$$= \left(e^{-1}-1\right)^2 \approx 0.3996$$

(b) Now we are asked for the probability that the combined lifetimes of both bulbs is 1000 hours or less. Thus we want to find $P(X+Y \leq 1000)$, or equivalently $P((X,Y)\in D)$ where D is the triangular region shown in the figure. Then

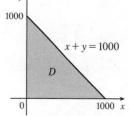

$$P(X+Y \leq 1000) = \iint_D f(x,y)\,dA$$

$$= \int_0^{1000}\int_0^{1000-x} 10^{-6}e^{-(x+y)/1000}\,dy\,dx$$

$$= 10^{-6}\int_0^{1000}\left[-1000e^{-(x+y)/1000}\right]_{y=0}^{y=1000-x}\,dx = -10^{-3}\int_0^{1000}\left(e^{-1}-e^{-x/1000}\right)\,dx$$

$$= -10^{-3}\left[e^{-1}x + 1000e^{-x/1000}\right]_0^{1000} = 1 - 2e^{-1} \approx 0.2642$$

31. (a) The random variables X and Y are normally distributed with $\mu_1 = 45$, $\mu_2 = 20$, $\sigma_1 = 0.5$, and $\sigma_2 = 0.1$.

The individual density functions for X and Y, then, are $f_1(x) = \dfrac{1}{0.5\sqrt{2\pi}}\,e^{-(x-45)^2/0.5}$ and

$f_2(y) = \dfrac{1}{0.1\sqrt{2\pi}}\,e^{-(y-20)^2/0.02}$. Since X and Y are independent, the joint density function is the product

$$f(x,y) = f_1(x)f_2(y) = \frac{1}{0.5\sqrt{2\pi}}\,e^{-(x-45)^2/0.5}\,\frac{1}{0.1\sqrt{2\pi}}\,e^{-(y-20)^2/0.02} = \frac{10}{\pi}e^{-2(x-45)^2-50(y-20)^2}.$$

Then $P(40 \leq X \leq 50, 20 \leq Y \leq 25) = \int_{40}^{50}\int_{20}^{25} f(x,y)\,dy\,dx = \dfrac{10}{\pi}\int_{40}^{50}\int_{20}^{25} e^{-2(x-45)^2-50(y-20)^2}\,dy\,dx$.

Using a CAS or calculator to evaluate the integral, we get $P(40 \leq X \leq 50, 20 \leq Y \leq 25) \approx 0.500$.

(b) $P(4(X-45)^2 + 100(Y-20)^2 \le 2) = \iint_D \frac{10}{\pi} e^{-2(x-45)^2 - 50(y-20)^2} \, dA$, where D is the region enclosed by the ellipse

$4(x-45)^2 + 100(y-20)^2 = 2$. Solving for y gives $y = 20 \pm \frac{1}{10}\sqrt{2-4(x-45)^2}$, the upper and lower halves of the

ellipse, and these two halves meet where $y = 20$ [since the ellipse is centered at $(45, 20)$] $\Rightarrow$ $4(x-45)^2 = 2$ $\Rightarrow$

$x = 45 \pm \frac{1}{\sqrt{2}}$. Thus

$$\iint_D \frac{10}{\pi} e^{-2(x-45)^2-50(y-20)^2} \, dA = \frac{10}{\pi} \int_{45-1/\sqrt{2}}^{45+1/\sqrt{2}} \int_{20-\frac{1}{10}\sqrt{2-4(x-45)^2}}^{20+\frac{1}{10}\sqrt{2-4(x-45)^2}} e^{-2(x-45)^2-50(y-20)^2} \, dy \, dx.$$

Using a CAS or calculator to evaluate the integral, we get $P(4(X-45)^2 + 100(Y-20)^2 \le 2) \approx 0.632$.

32. Because X and Y are independent, the joint density function for Xavier's and Yolanda's arrival times is the product of the

individual density functions:

$$f(x,y) = f_1(x)f_2(y) = \begin{cases} \frac{1}{50}e^{-x}y & \text{if } x \ge 0, 0 \le y \le 10 \\ 0 & \text{otherwise} \end{cases}$$

Since Xavier won't wait for Yolanda, they won't meet unless $X \ge Y$.

Additionally, Yolanda will wait up to half an hour but no longer, so they

won't meet unless $X - Y \le 30$. Thus the probability that they meet is

$P((X,Y) \in D)$ where D is the parallelogram shown in the figure. The

integral is simpler to evaluate if we consider D as a type II region, so

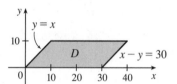

$$P((X,Y) \in D) = \iint_D f(x,y) \, dx \, dy = \int_0^{10} \int_y^{y+30} \frac{1}{50} e^{-x} y \, dx \, dy = \frac{1}{50} \int_0^{10} y \left[-e^{-x}\right]_{x=y}^{x=y+30} dy$$

$$= \frac{1}{50} \int_0^{10} y(-e^{-(y+30)} + e^{-y}) \, dy = \frac{1}{50}(1 - e^{-30}) \int_0^{10} y e^{-y} \, dy$$

By integration by parts (or Formula 96 in the Table of Integrals), this is

$\frac{1}{50}(1-e^{-30})\left[-(y+1)e^{-y}\right]_0^{10} = \frac{1}{50}(1-e^{-30})(1-11e^{-10}) \approx 0.020$. Thus there is only about a 2% chance they will meet.

Such is student life!

33. (a) If $f(P,A)$ is the probability that an individual at A will be infected by an individual at P, and $k\,dA$ is the number of

infected individuals in an element of area dA, then $f(P,A)k\,dA$ is the number of infections that should result from

exposure of the individual at A to infected people in the element of area dA. Integration over D gives the number of

infections of the person at A due to all the infected people in D. In rectangular coordinates (with the origin at the city's

center), the exposure of a person at A is

$$E = \iint_D kf(P,A) \, dA = k \iint_D \frac{20 - d(P,A)}{20} \, dA = k \iint_D \left[1 - \frac{\sqrt{(x-x_0)^2 + (y-y_0)^2}}{20}\right] dx \, dy$$

(b) If $A = (0,0)$, then

$$E = k \iint_D \left[1 - \frac{1}{20}\sqrt{x^2 + y^2}\right] dx \, dy$$

$$= k \int_0^{2\pi} \int_0^{10} \left(1 - \frac{r}{20}\right) r \, dr \, d\theta = 2\pi k \left[\frac{r^2}{2} - \frac{r^3}{60}\right]_0^{10}$$

$$= 2\pi k \left(50 - \frac{50}{3}\right) = \frac{200}{3}\pi k \approx 209k$$

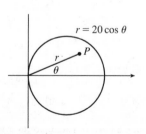

For A at the edge of the city, it is convenient to use a polar coordinate system centered at A. Then the polar equation for the circular boundary of the city becomes $r = 20 \cos \theta$ instead of $r = 10$, and the distance from A to a point P in the city is again r (see the figure). So

$$E = k \int_{-\pi/2}^{\pi/2} \int_0^{20 \cos \theta} \left(1 - \frac{r}{20}\right) r \, dr \, d\theta = k \int_{-\pi/2}^{\pi/2} \left[\frac{r^2}{2} - \frac{r^3}{60}\right]_{r=0}^{r=20 \cos \theta} d\theta$$

$$= k \int_{-\pi/2}^{\pi/2} \left(200 \cos^2 \theta - \frac{400}{3} \cos^3 \theta\right) d\theta = 200k \int_{-\pi/2}^{\pi/2} \left[\frac{1}{2} + \frac{1}{2} \cos 2\theta - \frac{2}{3}\left(1 - \sin^2 \theta\right) \cos \theta\right] d\theta$$

$$= 200k \left[\frac{1}{2}\theta + \frac{1}{4} \sin 2\theta - \frac{2}{3} \sin \theta + \frac{2}{3} \cdot \frac{1}{3} \sin^3 \theta\right]_{-\pi/2}^{\pi/2} = 200k \left[\frac{\pi}{4} + 0 - \frac{2}{3} + \frac{2}{9} + \frac{\pi}{4} + 0 - \frac{2}{3} + \frac{2}{9}\right]$$

$$= 200k \left(\frac{\pi}{2} - \frac{8}{9}\right) \approx 136k$$

Therefore the risk of infection is much lower at the edge of the city than in the middle, so it is better to live at the edge.

16.6   Triple Integrals

ET 15.6

1. $\iiint_B xyz^2 \, dV = \int_0^1 \int_0^3 \int_{-1}^2 xyz^2 \, dy \, dz \, dx = \int_0^1 \int_0^3 \left[\frac{1}{2}xy^2z^2\right]_{y=-1}^{y=2} dz \, dx = \int_0^1 \int_0^3 \frac{3}{2}xz^2 \, dz \, dx$

$= \int_0^1 \left[\frac{1}{2}xz^3\right]_{z=0}^{z=3} dx = \int_0^1 \frac{27}{2} x \, dx = \frac{27}{4}x^2\big]_0^1 = \frac{27}{4}$

2. There are six different possible orders of integration.

$\iiint_E (xz - y^3) \, dV = \int_{-1}^1 \int_0^2 \int_0^1 (xz - y^3) \, dz \, dy \, dx = \int_{-1}^1 \int_0^2 \left[\frac{1}{2}xz^2 - y^3z\right]_{z=0}^{z=1} dy \, dx = \int_{-1}^1 \int_0^2 \left(\frac{1}{2}x - y^3\right) dy \, dx$

$\qquad = \int_{-1}^1 \left[\frac{1}{2}xy - \frac{1}{4}y^4\right]_{y=0}^{y=2} dx = \int_{-1}^1 (x - 4) \, dx = \left[\frac{1}{2}x^2 - 4x\right]_{-1}^1 = -8$

$\iiint_E (xz - y^3) \, dV = \int_0^2 \int_{-1}^1 \int_0^1 (xz - y^3) \, dz \, dx \, dy = \int_0^2 \int_{-1}^1 \left[\frac{1}{2}xz^2 - y^3z\right]_{z=0}^{z=1} dx \, dy$

$\qquad = \int_0^2 \int_{-1}^1 \left(\frac{1}{2}x - y^3\right) dx \, dy = \int_0^2 \left[\frac{1}{4}x^2 - xy^3\right]_{x=-1}^{x=1} dy = \int_0^2 -2y^3 \, dy = -\frac{1}{2}y^4\big]_0^2 = -8$

$\iiint_E (xz - y^3) \, dV = \int_{-1}^1 \int_0^1 \int_0^2 (xz - y^3) \, dy \, dz \, dx = \int_{-1}^1 \int_0^1 \left[xyz - \frac{1}{4}y^4\right]_{y=0}^{y=2} dz \, dx$

$\qquad = \int_{-1}^1 \int_0^1 (2xz - 4) \, dz \, dx = \int_{-1}^1 \left[xz^2 - 4z\right]_{z=0}^{z=1} dx = \int_{-1}^1 (x - 4) \, dx = \left[\frac{1}{2}x^2 - 4x\right]_{-1}^1 = -8$

$\iiint_E (xz - y^3) \, dV = \int_0^1 \int_{-1}^1 \int_0^2 (xz - y^3) \, dy \, dx \, dz = \int_0^1 \int_{-1}^1 \left[xyz - \frac{1}{4}y^4\right]_{y=0}^{y=2} dx \, dz$

$\qquad = \int_0^1 \int_{-1}^1 (2xz - 4) \, dx \, dz = \int_0^1 \left[x^2z - 4x\right]_{x=-1}^{x=1} dz = \int_0^1 -8 \, dz = -8z\big]_0^1 = -8$

$\iiint_E (xz - y^3) \, dV = \int_0^2 \int_0^1 \int_{-1}^1 (xz - y^3) \, dx \, dz \, dy = \int_0^2 \int_0^1 \left[\frac{1}{2}x^2z - xy^3\right]_{x=-1}^{x=1} dz \, dy$

$\qquad = \int_0^2 \int_0^1 -2y^3 \, dz \, dy = \int_0^2 \left[-2y^3z\right]_{z=0}^{z=1} dy = \int_0^2 -2y^3 \, dy = -\frac{1}{2}y^4\big]_0^2 = -8$

$\iiint_E (xz - y^3) \, dV = \int_0^1 \int_0^2 \int_{-1}^1 (xz - y^3) \, dx \, dy \, dz = \int_0^1 \int_0^2 \left[\frac{1}{2}x^2z - xy^3\right]_{x=-1}^{x=1} dy \, dz$

$\qquad = \int_0^1 \int_0^2 -2y^3 \, dy \, dz = \int_0^1 \left[-\frac{1}{2}y^4\right]_{y=0}^{y=2} dz = \int_0^1 -8 \, dz = -8z\big]_0^1 = -8$

3. $\int_0^1 \int_0^z \int_0^{x+z} 6xz \, dy \, dx \, dz = \int_0^1 \int_0^z \left[6xyz\right]_{y=0}^{y=x+z} dx \, dz = \int_0^1 \int_0^z 6xz(x + z) \, dx \, dz$

$\qquad = \int_0^1 \left[2x^3z + 3x^2z^2\right]_{x=0}^{x=z} dz = \int_0^1 (2z^4 + 3z^4) \, dz = \int_0^1 5z^4 \, dz = z^5\big]_0^1 = 1$

4. $\int_0^1 \int_x^{2x} \int_0^y 2xyz \, dz \, dy \, dx = \int_0^1 \int_x^{2x} \left[xyz^2\right]_{z=0}^{z=y} dy \, dx = \int_0^1 \int_x^{2x} xy^3 \, dy \, dx$

$\qquad = \int_0^1 \left[\frac{1}{4}xy^4\right]_{y=x}^{y=2x} dx = \int_0^1 \frac{15}{4}x^5 \, dx = \frac{5}{8}x^6\big]_0^1 = \frac{5}{8}$

5. $\int_0^3 \int_0^1 \int_0^{\sqrt{1-z^2}} ze^y \, dx \, dz \, dy = \int_0^3 \int_0^1 \left[xze^y \right]_{x=0}^{x=\sqrt{1-z^2}} dz \, dy = \int_0^3 \int_0^1 ze^y \sqrt{1-z^2} \, dz \, dy$

$$= \int_0^3 \left[-\tfrac{1}{3}(1-z^2)^{3/2} e^y \right]_{z=0}^{z=1} dy = \int_0^3 \tfrac{1}{3} e^y \, dy = \tfrac{1}{3} e^y \big]_0^3 = \tfrac{1}{3}(e^3 - 1)$$

6. $\int_0^1 \int_0^z \int_0^y ze^{-y^2} \, dx \, dy \, dz = \int_0^1 \int_0^z \left[xze^{-y^2} \right]_{x=0}^{x=y} dy \, dz = \int_0^1 \int_0^z yze^{-y^2} \, dy \, dz = \int_0^1 \left[-\tfrac{1}{2} ze^{-y^2} \right]_{y=0}^{y=z} dz$

$$= \int_0^1 -\tfrac{1}{2} z\left(e^{-z^2} - 1 \right) dz = \tfrac{1}{2} \int_0^1 \left(z - ze^{-z^2} \right) dz$$

$$= \tfrac{1}{2} \left[\tfrac{1}{2} z^2 + \tfrac{1}{2} e^{-z^2} \right]_0^1 = \tfrac{1}{4}(1 + e^{-1} - 0 - 1) = \tfrac{1}{4e}$$

7. $\int_0^{\pi/2} \int_0^y \int_0^x \cos(x+y+z) dz \, dx \, dy = \int_0^{\pi/2} \int_0^y \left[\sin(x+y+z) \right]_{z=0}^{z=x} dx \, dy$

$$= \int_0^{\pi/2} \int_0^y \left[\sin(2x+y) - \sin(x+y) \right] dx \, dy$$

$$= \int_0^{\pi/2} \left[-\tfrac{1}{2}\cos(2x+y) + \cos(x+y) \right]_{x=0}^{x=y} dy$$

$$= \int_0^{\pi/2} \left[-\tfrac{1}{2}\cos 3y + \cos 2y + \tfrac{1}{2}\cos y - \cos y \right] dy$$

$$= \left[-\tfrac{1}{6}\sin 3y + \tfrac{1}{2}\sin 2y - \tfrac{1}{2}\sin y \right]_0^{\pi/2} = \tfrac{1}{6} - \tfrac{1}{2} = -\tfrac{1}{3}$$

8. $\int_0^{\sqrt{\pi}} \int_0^x \int_0^{xz} x^2 \sin y \, dy \, dz \, dx = \int_0^{\sqrt{\pi}} \int_0^x \left[-x^2 \cos y \right]_{y=0}^{y=xz} dz \, dx = \int_0^{\sqrt{\pi}} \int_0^x (x^2 - x^2 \cos xz) \, dz \, dx$

$$= \int_0^{\sqrt{\pi}} \left[x^2 z - x \sin xz \right]_{z=0}^{z=x} dx = \int_0^{\sqrt{\pi}} (x^3 - x \sin x^2) \, dx$$

$$= \left[\tfrac{1}{4} x^4 + \tfrac{1}{2} \cos x^2 \right]_0^{\sqrt{\pi}} = \tfrac{1}{4}\pi^2 - \tfrac{1}{2} - \tfrac{1}{2} = \tfrac{1}{4}\pi^2 - 1$$

9. $\iiint_E 2x \, dV = \int_0^2 \int_0^{\sqrt{4-y^2}} \int_0^y 2x \, dz \, dx \, dy = \int_0^2 \int_0^{\sqrt{4-y^2}} \left[2xz \right]_{z=0}^{z=y} dx \, dy = \int_0^2 \int_0^{\sqrt{4-y^2}} 2xy \, dx \, dy$

$$= \int_0^2 \left[x^2 y \right]_{x=0}^{x=\sqrt{4-y^2}} dy = \int_0^2 (4 - y^2)y \, dy = \left[2y^2 - \tfrac{1}{4} y^4 \right]_0^2 = 4$$

10. $\iiint_E yz \cos(x^5) \, dV = \int_0^1 \int_0^x \int_x^{2x} yz \cos(x^5) \, dz \, dy \, dx = \int_0^1 \int_0^x \left[\tfrac{1}{2} yz^2 \cos(x^5) \right]_{z=x}^{z=2x} dy \, dx$

$$= \tfrac{1}{2} \int_0^1 \int_0^x 3x^2 y \cos(x^5) \, dy \, dx = \tfrac{1}{2} \int_0^1 \left[\tfrac{3}{2} x^2 y^2 \cos(x^5) \right]_{y=0}^{y=x} dx$$

$$= \tfrac{3}{4} \int_0^1 x^4 \cos(x^5) \, dx = \tfrac{3}{4} \left[\tfrac{1}{5} \sin(x^5) \right]_0^1 = \tfrac{3}{20}(\sin 1 - \sin 0) = \tfrac{3}{20} \sin 1$$

11. Here $E = \{(x, y, z) \mid 0 \le x \le 1, 0 \le y \le \sqrt{x}, 0 \le z \le 1 + x + y\}$, so

$$\iiint_E 6xy \, dV = \int_0^1 \int_0^{\sqrt{x}} \int_0^{1+x+y} 6xy \, dz \, dy \, dx = \int_0^1 \int_0^{\sqrt{x}} \left[6xyz \right]_{z=0}^{z=1+x+y} dy \, dx = \int_0^1 \int_0^{\sqrt{x}} 6xy(1+x+y) \, dy \, dx$$

$$= \int_0^1 \left[3xy^2 + 3x^2 y^2 + 2xy^3 \right]_{y=0}^{y=\sqrt{x}} dx = \int_0^1 (3x^2 + 3x^3 + 2x^{5/2}) \, dx = \left[x^3 + \tfrac{3}{4} x^4 + \tfrac{4}{7} x^{7/2} \right]_0^1 = \tfrac{65}{28}$$

12. Here E is the region in the first octant that lies below the plane $2x + 2y + z = 4$ (and above the region in the xy-plane bounded by the lines $x = 0$, $y = 0$, $x + y = 2$). So

$$\iiint_E y \, dV = \int_0^2 \int_0^{2-x} \int_0^{4-2x-2y} y \, dz \, dy \, dx = \int_0^2 \int_0^{2-x} y(4 - 2x - 2y) \, dy \, dx = \int_0^2 \int_0^{2-x} (4y - 2xy - 2y^2) \, dy \, dx$$

$$= \int_0^2 \left[2y^2 - xy^2 - \tfrac{2}{3} y^3 \right]_{y=0}^{y=2-x} dx = \int_0^2 \left[2(2-x)^2 - x(2-x)^2 - \tfrac{2}{3}(2-x)^3 \right] dx$$

$$= \int_0^2 \left[(2-x)(2-x)^2 - \tfrac{2}{3}(2-x)^3 \right] dx = \tfrac{1}{3} \int_0^2 (2-x)^3 \, dx$$

$$= \tfrac{1}{3} \left[-\tfrac{1}{4}(2-x)^4 \right]_0^2 = -\tfrac{1}{12}(0 - 16) = \tfrac{4}{3}$$

13.

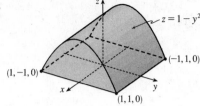

E is the region below the parabolic cylinder $z = 1 - y^2$ and above the square $[-1, 1] \times [-1, 1]$ in the xy-plane.

$$\iiint_E x^2 e^y \, dV = \int_{-1}^1 \int_{-1}^1 \int_0^{1-y^2} x^2 e^y \, dz \, dy \, dx$$

$$= \int_{-1}^1 \int_{-1}^1 x^2 e^y (1 - y^2) \, dy \, dx$$

$$= \int_{-1}^1 x^2 \, dx \int_{-1}^1 (e^y - y^2 e^y) \, dy$$

$$= \left[\tfrac{1}{3} x^3 \right]_{-1}^1 \left[e^y - (y^2 - 2y + 2) e^y \right]_{-1}^1 \qquad \begin{bmatrix} \text{integrate by} \\ \text{parts twice} \end{bmatrix}$$

$$= \tfrac{1}{3}(2)[e - e - e^{-1} + 5e^{-1}] = \tfrac{8}{3e}$$

14.

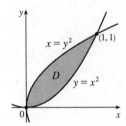

E is the solid above the region shown in the xy-plane and below the plane $z = x + y$. Thus,

$$\iiint_E xy \, dV = \int_0^1 \int_{x^2}^{\sqrt{x}} \int_0^{x+y} xy \, dz \, dy \, dx = \int_0^1 \int_{x^2}^{\sqrt{x}} xy(x + y) \, dy \, dx$$

$$= \int_0^1 \int_{x^2}^{\sqrt{x}} (x^2 y + xy^2) \, dy \, dx = \int_0^1 \left[\tfrac{1}{2} x^2 y^2 + \tfrac{1}{3} xy^3 \right]_{y=x^2}^{y=\sqrt{x}} \, dx$$

$$= \int_0^1 \left(\tfrac{1}{2} x^3 + \tfrac{1}{3} x^{5/2} - \tfrac{1}{2} x^6 - \tfrac{1}{3} x^7 \right) \, dx$$

$$= \left[\tfrac{1}{8} x^4 + \tfrac{2}{21} x^{7/2} - \tfrac{1}{14} x^7 - \tfrac{1}{24} x^8 \right]_0^1 = \tfrac{1}{8} + \tfrac{2}{21} - \tfrac{1}{14} - \tfrac{1}{24} = \tfrac{3}{28}$$

15.

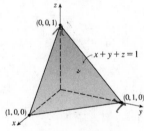

Here $T = \{(x, y, z) \mid 0 \le x \le 1, 0 \le y \le 1 - x, 0 \le z \le 1 - x - y\}$, so

$$\iiint_T x^2 \, dV = \int_0^1 \int_0^{1-x} \int_0^{1-x-y} x^2 \, dz \, dy \, dx = \int_0^1 \int_0^{1-x} x^2 (1 - x - y) \, dy \, dx$$

$$= \int_0^1 \int_0^{1-x} (x^2 - x^3 - x^2 y) \, dy \, dx = \int_0^1 \left[x^2 y - x^3 y - \tfrac{1}{2} x^2 y^2 \right]_{y=0}^{y=1-x} \, dx$$

$$= \int_0^1 \left[x^2 (1 - x) - x^3 (1 - x) - \tfrac{1}{2} x^2 (1 - x)^2 \right] \, dx$$

$$= \int_0^1 \left(\tfrac{1}{2} x^4 - x^3 + \tfrac{1}{2} x^2 \right) \, dx = \left[\tfrac{1}{10} x^5 - \tfrac{1}{4} x^4 + \tfrac{1}{6} x^3 \right]_0^1$$

$$= \tfrac{1}{10} - \tfrac{1}{4} + \tfrac{1}{6} = \tfrac{1}{60}$$

16.

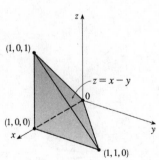

Here $T = \{(x, y, z) \mid 0 \le x \le 1, 0 \le y \le x, 0 \le z \le x - y\}$, so

$$\iiint_T xyz \, dV = \int_0^1 \int_0^x \int_0^{x-y} xyz \, dz \, dy \, dx = \int_0^1 \int_0^x \left[\tfrac{1}{2} xyz^2 \right]_{z=0}^{z=x-y} \, dy \, dx$$

$$= \int_0^1 \int_0^x \tfrac{1}{2} xy(x - y)^2 \, dy \, dx = \tfrac{1}{2} \int_0^1 \int_0^x (x^3 y - 2x^2 y^2 + xy^3) \, dy \, dx$$

$$= \tfrac{1}{2} \int_0^1 \left[\tfrac{1}{2} x^3 y^2 - \tfrac{2}{3} x^2 y^3 + \tfrac{1}{4} xy^4 \right]_{y=0}^{y=x} \, dx$$

$$= \tfrac{1}{2} \int_0^1 \left(\tfrac{1}{2} x^5 - \tfrac{2}{3} x^5 + \tfrac{1}{4} x^5 \right) \, dx$$

$$= \tfrac{1}{2} \int_0^1 \tfrac{1}{12} x^5 \, dx = \tfrac{1}{144} x^6 \Big]_0^1 = \tfrac{1}{144}$$

17.

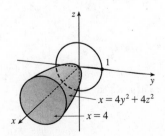

The projection E on the yz-plane is the disk $y^2 + z^2 \le 1$. Using polar coordinates $y = r \cos \theta$ and $z = r \sin \theta$, we get

$$\iiint_E x \, dV = \iint_D \left[\int_{4y^2 + 4z^2}^4 x \, dx \right] dA = \tfrac{1}{2} \iint_D \left[4^2 - (4y^2 + 4z^2)^2 \right] dA$$

$$= 8 \int_0^{2\pi} \int_0^1 (1 - r^4) \, r \, dr \, d\theta = 8 \int_0^{2\pi} d\theta \int_0^1 (r - r^5) \, dr$$

$$= 8(2\pi) \left[\tfrac{1}{2} r^2 - \tfrac{1}{6} r^6 \right]_0^1 = \tfrac{16\pi}{3}$$

18.

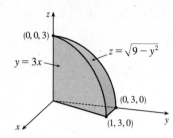

$$\int_0^1 \int_{3x}^3 \int_0^{\sqrt{9-y^2}} z \, dz \, dy \, dx = \int_0^1 \int_{3x}^3 \tfrac{1}{2}(9-y^2) \, dy \, dx$$

$$= \int_0^1 \left[\tfrac{9}{2}y - \tfrac{1}{6}y^3\right]_{y=3x}^{y=3} dx$$

$$= \int_0^1 \left[9 - \tfrac{27}{2}x + \tfrac{9}{2}x^3\right] dx$$

$$= \left[9x - \tfrac{27}{4}x^2 + \tfrac{9}{8}x^4\right]_0^1 = \tfrac{27}{8}$$

19. The plane $2x + y + z = 4$ intersects the xy-plane when

$2x + y + 0 = 4 \;\Rightarrow\; y = 4 - 2x$, so

$E = \{(x, y, z) \mid 0 \le x \le 2, 0 \le y \le 4 - 2x, 0 \le z \le 4 - 2x - y\}$ and

$$V = \int_0^2 \int_0^{4-2x} \int_0^{4-2x-y} dz \, dy \, dx = \int_0^2 \int_0^{4-2x} (4 - 2x - y) \, dy \, dx$$

$$= \int_0^2 \left[4y - 2xy - \tfrac{1}{2}y^2\right]_{y=0}^{y=4-2x} dx$$

$$= \int_0^2 \left[4(4 - 2x) - 2x(4 - 2x) - \tfrac{1}{2}(4 - 2x)^2\right] dx$$

$$= \int_0^2 (2x^2 - 8x + 8) \, dx = \left[\tfrac{2}{3}x^3 - 4x^2 + 8x\right]_0^2 = \tfrac{16}{3}$$

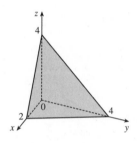

20.

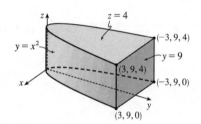

$$V = \iiint_E dV = \int_{-3}^3 \int_{x^2}^9 \int_0^4 dz \, dy \, dx$$

$$= 4\int_{-3}^3 \int_{x^2}^9 dy \, dx = 4\int_{-3}^3 (9 - x^2) \, dx$$

$$= 4\left[9x - \tfrac{1}{3}x^3\right]_{-3}^3 = 4(27 - 9 + 27 - 9) = 144$$

21. $V = \displaystyle\int_{-3}^3 \int_{-\sqrt{9-x^2}}^{\sqrt{9-x^2}} \int_1^{5-y} dz \, dy \, dx = \int_{-3}^3 \int_{-\sqrt{9-x^2}}^{\sqrt{9-x^2}} (5 - y - 1) \, dy \, dx = \int_{-3}^3 \left[4y - \tfrac{1}{2}y^2\right]_{y=-\sqrt{9-x^2}}^{y=\sqrt{9-x^2}} dx$

$$= \int_{-3}^3 8\sqrt{9-x^2} \, dx = 8\left[\tfrac{x}{2}\sqrt{9-x^2} + \tfrac{9}{2}\sin^{-1}\left(\tfrac{x}{3}\right)\right]_{-3}^3 \quad \begin{bmatrix} \text{using trigonometric substitution or} \\ \text{Formula 30 in the Table of Integrals} \end{bmatrix}$$

$$= 8\left[\tfrac{9}{2}\sin^{-1}(1) - \tfrac{9}{2}\sin^{-1}(-1)\right] = 36\left(\tfrac{\pi}{2} - \left(-\tfrac{\pi}{2}\right)\right) = 36\pi$$

Alternatively, use polar coordinates to evaluate the double integral:

$$\int_{-3}^3 \int_{-\sqrt{9-x^2}}^{\sqrt{9-x^2}} (4 - y) \, dy \, dx = \int_0^{2\pi} \int_0^3 (4 - r\sin\theta) \, r \, dr \, d\theta$$

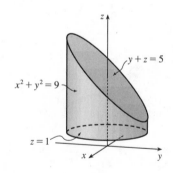

$$= \int_0^{2\pi} \left[2r^2 - \tfrac{1}{3}r^3\sin\theta\right]_{r=0}^{r=3} d\theta$$

$$= \int_0^{2\pi} (18 - 9\sin\theta) \, d\theta$$

$$= 18\theta + 9\cos\theta \Big]_0^{2\pi} = 36\pi$$

22. The paraboloid $x = y^2 + z^2$ intersects the plane $x = 16$ in the circle $y^2 + z^2 = 16$, $x = 16$. Thus,

$E = \{(x, y, z) \mid y^2 + z^2 \le x \le 16, y^2 + z^2 \le 16\}$.

Let $D = \{(y, z) \mid y^2 + z^2 \le 16\}$. Then using polar coordinates $y = r\cos\theta$ and $z = r\sin\theta$, we have

$$V = \iint_D \left(\int_{y^2+z^2}^{16} dx \right) dA = \iint_D (16 - (y^2 + z^2)) \, dA$$

$$= \int_0^{2\pi} \int_0^4 (16 - r^2) \, r \, dr \, d\theta = \int_0^{2\pi} d\theta \int_0^4 (16r - r^3) \, dr$$

$$= \left[\theta \right]_0^{2\pi} \left[8r^2 - \tfrac{1}{4}r^4 \right]_0^4 = 2\pi(128 - 64) = 128\pi$$

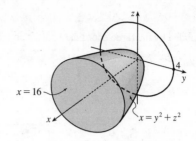

23. (a) The wedge can be described as the region

$$D = \left\{ (x, y, z) \mid y^2 + z^2 \le 1, 0 \le x \le 1, 0 \le y \le x \right\}$$
$$= \left\{ (x, y, z) \mid 0 \le x \le 1, 0 \le y \le x, 0 \le z \le \sqrt{1 - y^2} \right\}$$

So the integral expressing the volume of the wedge is

$$\iiint_D dV = \int_0^1 \int_0^x \int_0^{\sqrt{1-y^2}} dz \, dy \, dx.$$

(b) A CAS gives $\int_0^1 \int_0^x \int_0^{\sqrt{1-y^2}} dz \, dy \, dx = \tfrac{\pi}{4} - \tfrac{1}{3}$.

(Or use Formulas 30 and 87 from the Table of Integrals.)

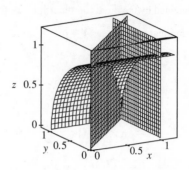

24. (a) Divide B into 8 cubes of size $\Delta V = 8$. With $f(x, y, z) = \sqrt{x^2 + y^2 + z^2}$, the Midpoint Rule gives

$$\iiint_B \sqrt{x^2 + y^2 + z^2} \, dV \approx \sum_{i=1}^2 \sum_{j=1}^2 \sum_{k=1}^2 f(\overline{x}_i, \overline{y}_j, \overline{z}_k) \, \Delta V$$

$$= 8[f(1,1,1) + f(1,1,3) + f(1,3,1) + f(1,3,3) + f(3,1,1)$$
$$+ f(3,1,3) + f(3,3,1) + f(3,3,3)]$$

$$\approx 239.64$$

(b) Using a CAS we have $\iiint_B \sqrt{x^2 + y^2 + z^2} \, dV = \int_0^4 \int_0^4 \int_0^4 \sqrt{x^2 + y^2 + z^2} \, dz \, dy \, dx \approx 245.91$. This differs from the estimate in part (a) by about 2.5%.

25. Here $f(x, y, z) = \dfrac{1}{\ln(1 + x + y + z)}$ and $\Delta V = 2 \cdot 4 \cdot 2 = 16$, so the Midpoint Rule gives

$$\iiint_B f(x, y, z) \, dV \approx \sum_{i=1}^l \sum_{j=1}^m \sum_{k=1}^n f(\overline{x}_i, \overline{y}_j, \overline{z}_k) \, \Delta V$$

$$= 16[f(1,2,1) + f(1,2,3) + f(1,6,1) + f(1,6,3)$$
$$+ f(3,2,1) + f(3,2,3) + f(3,6,1) + f(3,6,3)]$$

$$= 16\left[\tfrac{1}{\ln 5} + \tfrac{1}{\ln 7} + \tfrac{1}{\ln 9} + \tfrac{1}{\ln 11} + \tfrac{1}{\ln 7} + \tfrac{1}{\ln 9} + \tfrac{1}{\ln 11} + \tfrac{1}{\ln 13} \right] \approx 60.533$$

26. Here $f(x, y, z) = \sin(xy^2z^3)$ and $\Delta V = 2 \cdot 1 \cdot \tfrac{1}{2} = 1$, so the Midpoint Rule gives

$$\iiint_B f(x, y, z) \, dV \approx \sum_{i=1}^l \sum_{j=1}^m \sum_{k=1}^n f(\overline{x}_i, \overline{y}_j, \overline{z}_k) \, \Delta V$$

$$= 1\left[f\left(1, \tfrac{1}{2}, \tfrac{1}{4}\right) + f\left(1, \tfrac{1}{2}, \tfrac{3}{4}\right) + f\left(1, \tfrac{3}{2}, \tfrac{1}{4}\right) + f\left(1, \tfrac{3}{2}, \tfrac{3}{4}\right) \right.$$
$$\left. + f\left(3, \tfrac{1}{2}, \tfrac{1}{4}\right) + f\left(3, \tfrac{1}{2}, \tfrac{3}{4}\right) + f\left(3, \tfrac{3}{2}, \tfrac{1}{4}\right) + f\left(3, \tfrac{3}{2}, \tfrac{3}{4}\right) \right]$$

$$= \sin \tfrac{1}{256} + \sin \tfrac{27}{256} + \sin \tfrac{9}{256} + \sin \tfrac{243}{256} + \sin \tfrac{3}{256} + \sin \tfrac{81}{256} + \sin \tfrac{27}{256} + \sin \tfrac{729}{256} \approx 1.675$$

27. $E = \{(x, y, z) \mid 0 \le x \le 1, 0 \le z \le 1 - x, 0 \le y \le 2 - 2z\}$,

the solid bounded by the three coordinate planes and the planes

$z = 1 - x$, $y = 2 - 2z$.

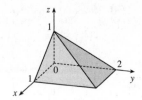

28. $E = \left\{(x, y, z) \mid 0 \le y \le 2, 0 \le z \le 2 - y, 0 \le x \le 4 - y^2\right\}$,

the solid bounded by the three coordinate planes, the plane $z = 2 - y$,

and the cylindrical surface $x = 4 - y^2$.

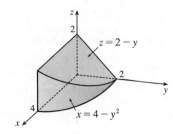

29.

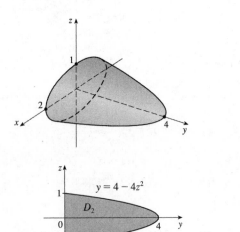

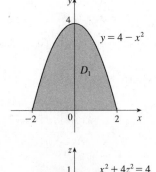

If D_1, D_2, D_3 are the projections of E on the xy-, yz-, and xz-planes, then

$$D_1 = \left\{(x, y) \mid -2 \le x \le 2, 0 \le y \le 4 - x^2\right\} = \left\{(x, y) \mid 0 \le y \le 4, \, -\sqrt{4-y} \le x \le \sqrt{4-y}\right\}$$

$$D_2 = \left\{(y, z) \mid 0 \le y \le 4, \, -\tfrac{1}{2}\sqrt{4-y} \le z \le \tfrac{1}{2}\sqrt{4-y}\right\} = \left\{(y, z) \mid -1 \le z \le 1, 0 \le y \le 4 - 4z^2\right\}$$

$$D_3 = \left\{(x, z) \mid x^2 + 4z^2 \le 4\right\}$$

Therefore

$$E = \left\{(x, y, z) \mid -2 \le x \le 2, 0 \le y \le 4 - x^2, \, -\tfrac{1}{2}\sqrt{4 - x^2 - y} \le z \le \tfrac{1}{2}\sqrt{4 - x^2 - y}\right\}$$

$$= \left\{(x, y, z) \mid 0 \le y \le 4, \, -\sqrt{4-y} \le x \le \sqrt{4-y}, \, -\tfrac{1}{2}\sqrt{4 - x^2 - y} \le z \le \tfrac{1}{2}\sqrt{4 - x^2 - y}\right\}$$

$$= \left\{(x, y, z) \mid -1 \le z \le 1, 0 \le y \le 4 - 4z^2, \, -\sqrt{4 - y - 4z^2} \le x \le \sqrt{4 - y - 4z^2}\right\}$$

$$= \left\{(x, y, z) \mid 0 \le y \le 4, \, -\tfrac{1}{2}\sqrt{4-y} \le z \le \tfrac{1}{2}\sqrt{4-y}, \, -\sqrt{4 - y - 4z^2} \le x \le \sqrt{4 - y - 4z^2}\right\}$$

$$= \left\{(x, y, z) \mid -2 \le x \le 2, \, -\tfrac{1}{2}\sqrt{4 - x^2} \le z \le \tfrac{1}{2}\sqrt{4 - x^2}, 0 \le y \le 4 - x^2 - 4z^2\right\}$$

$$= \left\{(x, y, z) \mid -1 \le z \le 1, \, -\sqrt{4 - 4z^2} \le x \le \sqrt{4 - 4z^2}, 0 \le y \le 4 - x^2 - 4z^2\right\}$$

Then

$$\iiint_E f(x, y, z)\, dV = \int_{-2}^{2} \int_{0}^{4-x^2} \int_{-\sqrt{4-x^2-y}/2}^{\sqrt{4-x^2-y}/2} f(x, y, z)\, dz\, dy\, dx = \int_{0}^{4} \int_{-\sqrt{4-y}}^{\sqrt{4-y}} \int_{-\sqrt{4-x^2-y}/2}^{\sqrt{4-x^2-y}/2} f(x, y, z)\, dz\, dx\, dy$$

$$= \int_{-1}^{1} \int_{0}^{4-4z^2} \int_{-\sqrt{4-y-4z^2}}^{\sqrt{4-y-4z^2}} f(x, y, z)\, dx\, dy\, dz = \int_{0}^{4} \int_{-\sqrt{4-y}/2}^{\sqrt{4-y}/2} \int_{-\sqrt{4-y-4z^2}}^{\sqrt{4-y-4z^2}} f(x, y, z)\, dx\, dz\, dy$$

$$= \int_{-2}^{2} \int_{-\sqrt{4-x^2}/2}^{\sqrt{4-x^2}/2} \int_{0}^{4-x^2-4z^2} f(x, y, z)\, dy\, dz\, dx = \int_{-1}^{1} \int_{-\sqrt{4-4z^2}}^{\sqrt{4-4z^2}} \int_{0}^{4-x^2-4z^2} f(x, y, z)\, dy\, dx\, dz$$

30.

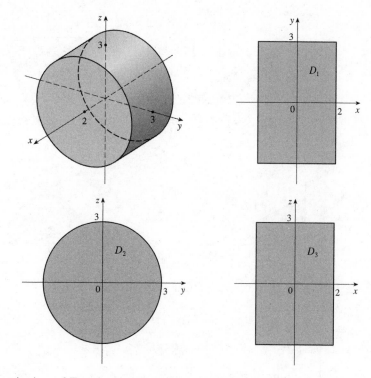

If D_1, D_2, D_3 are the projections of E on the xy-, yz-, and xz-planes, then

$$D_1 = \{(x, y) \mid -2 \le x \le 2, \; -3 \le y \le 3\}$$
$$D_2 = \{(y, z) \mid y^2 + z^2 \le 9\}$$
$$D_3 = \{(x, z) \mid -2 \le x \le 2, \; -3 \le z \le 3\}$$

Therefore

$$E = \left\{(x, y, z) \mid -2 \le x \le 2, \; -3 \le y \le 3, \; -\sqrt{9 - y^2} \le z \le \sqrt{9 - y^2}\right\}$$
$$= \left\{(x, y, z) \mid -3 \le y \le 3, \; -\sqrt{9 - y^2} \le z \le \sqrt{9 - y^2}, \; -2 \le x \le 2\right\}$$
$$= \left\{(x, y, z) \mid -3 \le z \le 3, \; -\sqrt{9 - z^2} \le y \le \sqrt{9 - z^2}, \; -2 \le x \le 2\right\}$$
$$= \left\{(x, y, z) \mid -2 \le x \le 2, \; -3 \le z \le 3, \; -\sqrt{9 - z^2} \le y \le \sqrt{9 - z^2}\right\}$$

and

$$\iiint_E f(x,y,z)\,dV = \int_{-2}^{2}\int_{-3}^{3}\int_{-\sqrt{9-y^2}}^{\sqrt{9-y^2}} f(x,y,z)\,dz\,dy\,dx = \int_{-3}^{3}\int_{-2}^{2}\int_{-\sqrt{9-y^2}}^{\sqrt{9-y^2}} f(x,y,z)\,dz\,dx\,dy$$

$$= \int_{-3}^{3}\int_{-\sqrt{9-y^2}}^{\sqrt{9-y^2}}\int_{-2}^{2} f(x,y,z)\,dx\,dz\,dy = \int_{-3}^{3}\int_{-\sqrt{9-z^2}}^{\sqrt{9-z^2}}\int_{-2}^{2} f(x,y,z)\,dx\,dy\,dz$$

$$= \int_{-2}^{2}\int_{-3}^{3}\int_{-\sqrt{9-z^2}}^{\sqrt{9-z^2}} f(x,y,z)\,dy\,dz\,dx = \int_{-3}^{3}\int_{-2}^{2}\int_{-\sqrt{9-z^2}}^{\sqrt{9-z^2}} f(x,y,z)\,dy\,dx\,dz$$

31.

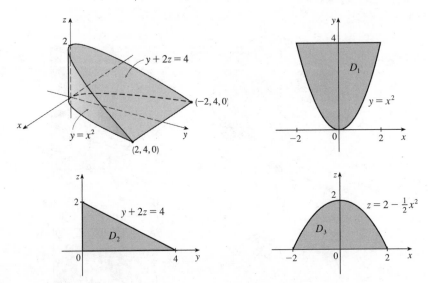

If D_1, D_2, and D_3 are the projections of E on the xy-, yz-, and xz-planes, then

$$D_1 = \left\{(x,y)\mid -2 \le x \le 2, x^2 \le y \le 4\right\} = \left\{(x,y)\mid 0 \le y \le 4, -\sqrt{y} \le x \le \sqrt{y}\right\},$$

$$D_2 = \left\{(y,z)\mid 0 \le y \le 4, 0 \le z \le 2 - \tfrac{1}{2}y\right\} = \left\{(y,z)\mid 0 \le z \le 2, 0 \le y \le 4 - 2z\right\}, \text{ and}$$

$$D_3 = \left\{(x,z)\mid -2 \le x \le 2, 0 \le z \le 2 - \tfrac{1}{2}x^2\right\} = \left\{(x,z)\mid 0 \le z \le 2, -\sqrt{4-2z} \le x \le \sqrt{4-2z}\right\}$$

Therefore
$$E = \left\{(x,y,z)\mid -2 \le x \le 2, x^2 \le y \le 4, 0 \le z \le 2 - \tfrac{1}{2}y\right\}$$

$$= \left\{(x,y,z)\mid 0 \le y \le 4, -\sqrt{y} \le x \le \sqrt{y}, 0 \le z \le 2 - \tfrac{1}{2}y\right\}$$

$$= \left\{(x,y,z)\mid 0 \le y \le 4, 0 \le z \le 2 - \tfrac{1}{2}y, -\sqrt{y} \le x \le \sqrt{y}\right\}$$

$$= \left\{(x,y,z)\mid 0 \le z \le 2, 0 \le y \le 4 - 2z, -\sqrt{y} \le x \le \sqrt{y}\right\}$$

$$= \left\{(x,y,z)\mid -2 \le x \le 2, 0 \le z \le 2 - \tfrac{1}{2}x^2, x^2 \le y \le 4 - 2z\right\}$$

$$= \left\{(x,y,z)\mid 0 \le z \le 2, -\sqrt{4-2z} \le x \le \sqrt{4-2z}, x^2 \le y \le 4 - 2z\right\}$$

Then
$$\iiint_E f(x,y,z)\,dV = \int_{-2}^{2}\int_{x^2}^{4}\int_{0}^{2-y/2} f(x,y,z)\,dz\,dy\,dx = \int_{0}^{4}\int_{-\sqrt{y}}^{\sqrt{y}}\int_{0}^{2-y/2} f(x,y,z)\,dz\,dx\,dy$$

$$= \int_{0}^{4}\int_{0}^{2-y/2}\int_{-\sqrt{y}}^{\sqrt{y}} f(x,y,z)\,dx\,dz\,dy = \int_{0}^{2}\int_{0}^{4-2z}\int_{-\sqrt{y}}^{\sqrt{y}} f(x,y,z)\,dx\,dy\,dz$$

$$= \int_{-2}^{2}\int_{0}^{2-x^2/2}\int_{x^2}^{4-2z} f(x,y,z)\,dy\,dz\,dx = \int_{0}^{2}\int_{-\sqrt{4-2z}}^{\sqrt{4-2z}}\int_{x^2}^{4-2z} f(x,y,z)\,dy\,dx\,dz$$

32.

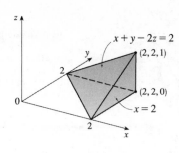

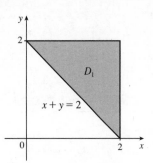

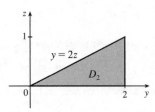

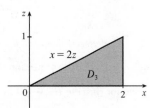

If D_1, D_2, and D_3 are the projections of E on the xy-, yz-, and xz-planes, then

$$D_1 = \left\{(x, y) \mid 0 \le x \le 2, 2 - x \le y \le 2\right\} = \left\{(x, y) \mid 0 \le y \le 2, 2 - y \le x \le 2\right\},$$

$$D_2 = \left\{(y, z) \mid 0 \le y \le 2, 0 \le z \le \tfrac{1}{2}y\right\} = \left\{(y, z) \mid 0 \le z \le 1, 2z \le y \le 2\right\}, \text{ and}$$

$$D_3 = \left\{(x, z) \mid 0 \le x \le 2, 0 \le z \le \tfrac{1}{2}x\right\} = \left\{(x, z) \mid 0 \le z \le 1, 2z \le x \le 2\right\}$$

Therefore

$$E = \left\{(x, y, z) \mid 0 \le x \le 2, 2 - x \le y \le 2, 0 \le z \le \tfrac{1}{2}(x + y - 2)\right\}$$

$$= \left\{(x, y, z) \mid 0 \le y \le 2, 2 - y \le x \le 2, 0 \le z \le \tfrac{1}{2}(x + y - 2)\right\}$$

$$= \left\{(x, y, z) \mid 0 \le y \le 2, 0 \le z \le \tfrac{1}{2}y, 2 - y + 2z \le x \le 2\right\}$$

$$= \left\{(x, y, z) \mid 0 \le z \le 1, 2z \le y \le 2, 2 - y + 2z \le x \le 2\right\}$$

$$= \left\{(x, y, z) \mid 0 \le x \le 2, 0 \le z \le \tfrac{1}{2}x, 2 - x + 2z \le y \le 2\right\}$$

$$= \left\{(x, y, z) \mid 0 \le z \le 1, 2z \le x \le 2, 2 - x + 2z \le y \le 2\right\}$$

Then $\quad \iiint_E f(x, y, z)\, dV = \int_0^2 \int_{2-x}^2 \int_0^{(x+y-2)/2} f(x, y, z)\, dz\, dy\, dx = \int_0^2 \int_{2-y}^2 \int_0^{(x+y-2)/2} f(x, y, z)\, dz\, dx\, dy$

$$= \int_0^2 \int_0^{y/2} \int_{2-y+2z}^2 f(x, y, z)\, dx\, dz\, dy = \int_0^1 \int_{2z}^2 \int_{2-y+2z}^2 f(x, y, z)\, dx\, dy\, dz$$

$$= \int_0^2 \int_0^{x/2} \int_{2-x+2z}^2 f(x, y, z)\, dy\, dz\, dx = \int_0^1 \int_{2z}^2 \int_{2-x+2z}^2 f(x, y, z)\, dy\, dx\, dz$$

33.

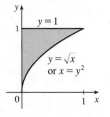

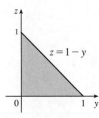

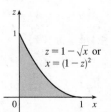

The diagrams show the projections of E on the xy-, yz-, and xz-planes. Therefore

$$\int_0^1 \int_{\sqrt{x}}^1 \int_0^{1-y} f(x, y, z)\, dz\, dy\, dx = \int_0^1 \int_0^{y^2} \int_0^{1-y} f(x, y, z)\, dz\, dx\, dy = \int_0^1 \int_0^{1-z} \int_0^{y^2} f(x, y, z)\, dx\, dy\, dz$$

$$= \int_0^1 \int_0^{1-y} \int_0^{y^2} f(x, y, z)\, dx\, dz\, dy = \int_0^1 \int_0^{1-\sqrt{x}} \int_{\sqrt{x}}^{1-z} f(x, y, z)\, dy\, dz\, dx$$

$$= \int_0^1 \int_0^{(1-z)^2} \int_{\sqrt{x}}^{1-z} f(x, y, z)\, dy\, dx\, dz$$

34.

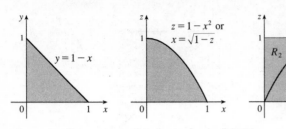

The projections of E onto the xy- and xz-planes are as in the first two diagrams and so

$$\int_0^1 \int_0^{1-x^2} \int_0^{1-x} f(x,y,z)\, dy\, dz\, dx = \int_0^1 \int_0^{\sqrt{1-z}} \int_0^{1-x} f(x,y,z)\, dy\, dx\, dz$$

$$= \int_0^1 \int_0^{1-y} \int_0^{1-x^2} f(x,y,z)\, dz\, dx\, dy = \int_0^1 \int_0^{1-x} \int_0^{1-x^2} f(x,y,z)\, dz\, dy\, dx$$

Now the surface $z = 1 - x^2$ intersects the plane $y = 1 - x$ in a curve whose projection in the yz-plane is $z = 1 - (1-y)^2$ or $z = 2y - y^2$. So we must split up the projection of E on the yz-plane into two regions as in the third diagram. For (y, z) in R_1, $0 \le x \le 1 - y$ and for (y, z) in R_2, $0 \le x \le \sqrt{1-z}$, and so the given integral is also equal to

$$\int_0^1 \int_0^{1-\sqrt{1-z}} \int_0^{\sqrt{1-z}} f(x,y,z)\, dx\, dy\, dz + \int_0^1 \int_{1-\sqrt{1-z}}^1 \int_0^{1-y} f(x,y,z)\, dx\, dy\, dz$$

$$= \int_0^1 \int_0^{2y-y^2} \int_0^{1-y} f(x,y,z)\, dx\, dz\, dy + \int_0^1 \int_{2y-y^2}^1 \int_0^{\sqrt{1-z}} f(x,y,z)\, dx\, dz\, dy.$$

35.

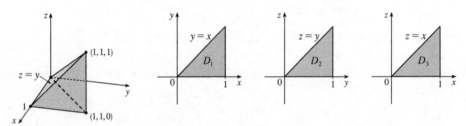

$$\int_0^1 \int_y^1 \int_0^y f(x,y,z)\, dz\, dx\, dy = \iiint_E f(x,y,z)\, dV \text{ where } E = \{(x,y,z) \mid 0 \le z \le y, y \le x \le 1, 0 \le y \le 1\}.$$

If D_1, D_2, and D_3 are the projections of E on the xy-, yz- and xz-planes then

$$D_1 = \{(x,y) \mid 0 \le y \le 1, y \le x \le 1\} = \{(x,y) \mid 0 \le x \le 1, 0 \le y \le x\},$$

$$D_2 = \{(y,z) \mid 0 \le y \le 1, 0 \le z \le y\} = \{(y,z) \mid 0 \le z \le 1, z \le y \le 1\}, \text{ and}$$

$$D_3 = \{(x,z) \mid 0 \le x \le 1, 0 \le z \le x\} = \{(x,z) \mid 0 \le z \le 1, z \le x \le 1\}.$$

Thus we also have

$$E = \{(x,y,z) \mid 0 \le x \le 1, 0 \le y \le x, 0 \le z \le y\} = \{(x,y,z) \mid 0 \le y \le 1, 0 \le z \le y, y \le x \le 1\}$$

$$= \{(x,y,z) \mid 0 \le z \le 1, z \le y \le 1, y \le x \le 1\} = \{(x,y,z) \mid 0 \le x \le 1, 0 \le z \le x, z \le y \le x\}$$

$$= \{(x,y,z) \mid 0 \le z \le 1, z \le x \le 1, z \le y \le x\}.$$

Then

$$\int_0^1 \int_y^1 \int_0^y f(x,y,z)\, dz\, dx\, dy = \int_0^1 \int_0^x \int_0^y f(x,y,z)\, dz\, dy\, dx = \int_0^1 \int_0^y \int_y^1 f(x,y,z)\, dx\, dz\, dy$$

$$= \int_0^1 \int_z^1 \int_y^1 f(x,y,z)\, dx\, dy\, dz = \int_0^1 \int_0^x \int_z^x f(x,y,z)\, dy\, dz\, dx$$

$$= \int_0^1 \int_z^1 \int_z^x f(x,y,z)\, dy\, dx\, dz$$

36.

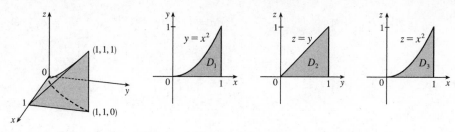

$\int_0^1 \int_0^{x^2} \int_0^y f(x, y, z)\, dz\, dy\, dx = \iiint_E f(x, y, z)\, dV$ where $E = \{(x, y, z) \mid 0 \le x \le 1, 0 \le y \le x^2, 0 \le z \le y\}$.

If D_1, D_2, D_3 are the projections of E on the xy-, yz-, and xz-planes, then

$$D_1 = \{(x, y) \mid 0 \le x \le 1, 0 \le y \le x^2\} = \{(x, y) \mid 0 \le y \le 1, \sqrt{y} \le x \le 1\},$$
$$D_2 = \{(y, z) \mid 0 \le y \le 1, 0 \le z \le y\} = \{(y, z) \mid 0 \le z \le 1, z \le y \le 1\},$$
$$D_3 = \{(x, z) \mid 0 \le x \le 1, 0 \le z \le x^2\} = \{(x, z) \mid 0 \le z \le 1, \sqrt{z} \le x \le 1\}.$$

Thus we also have

$$E = \{(x, y, z) \mid 0 \le y \le 1, \sqrt{y} \le x \le 1, 0 \le z \le y\} = \{(x, y, z) \mid 0 \le y \le 1, 0 \le z \le y, \sqrt{y} \le x \le 1\}$$
$$= \{(x, y, z) \mid 0 \le z \le 1, z \le y \le 1, \sqrt{y} \le x \le 1\} = \{(x, y, z) \mid 0 \le x \le 1, 0 \le z \le x^2, z \le y \le x^2\}$$
$$= \{(x, y, z) \mid 0 \le z \le 1, \sqrt{z} \le x \le 1, z \le y \le x^2\}$$

Then

$$\int_0^1 \int_0^{x^2} \int_0^y f(x, y, z)\, dz\, dy\, dx = \int_0^1 \int_{\sqrt{y}}^1 \int_0^y f(x, y, z)\, dz\, dx\, dy = \int_0^1 \int_0^y \int_{\sqrt{y}}^1 f(x, y, z)\, dx\, dz\, dy$$
$$= \int_0^1 \int_z^1 \int_{\sqrt{y}}^1 f(x, y, z)\, dx\, dy\, dz = \int_0^1 \int_0^{x^2} \int_z^{x^2} f(x, y, z)\, dy\, dz\, dx$$
$$= \int_0^1 \int_{\sqrt{z}}^1 \int_z^{x^2} f(x, y, z)\, dy\, dx\, dz$$

37. $m = \iiint_E \rho(x, y, z)\, dV = \int_0^1 \int_0^{\sqrt{x}} \int_0^{1+x+y} 2\, dz\, dy\, dx = \int_0^1 \int_0^{\sqrt{x}} 2(1 + x + y)\, dy\, dx$

$= \int_0^1 \left[2y + 2xy + y^2\right]_{y=0}^{y=\sqrt{x}} dx = \int_0^1 \left(2\sqrt{x} + 2x^{3/2} + x\right) dx = \left[\frac{4}{3}x^{3/2} + \frac{4}{5}x^{5/2} + \frac{1}{2}x^2\right]_0^1 = \frac{79}{30}$

$M_{yz} = \iiint_E x\rho(x, y, z)\, dV = \int_0^1 \int_0^{\sqrt{x}} \int_0^{1+x+y} 2x\, dz\, dy\, dx = \int_0^1 \int_0^{\sqrt{x}} 2x(1 + x + y)\, dy\, dx$

$= \int_0^1 \left[2xy + 2x^2y + xy^2\right]_{y=0}^{y=\sqrt{x}} dx = \int_0^1 (2x^{3/2} + 2x^{5/2} + x^2)\, dx = \left[\frac{4}{5}x^{5/2} + \frac{4}{7}x^{7/2} + \frac{1}{3}x^3\right]_0^1 = \frac{179}{105}$

$M_{xz} = \iiint_E y\rho(x, y, z)\, dV = \int_0^1 \int_0^{\sqrt{x}} \int_0^{1+x+y} 2y\, dz\, dy\, dx = \int_0^1 \int_0^{\sqrt{x}} 2y(1 + x + y)\, dy\, dx$

$= \int_0^1 \left[y^2 + xy^2 + \frac{2}{3}y^3\right]_{y=0}^{y=\sqrt{x}} dx = \int_0^1 \left(x + x^2 + \frac{2}{3}x^{3/2}\right) dx = \left[\frac{1}{2}x^2 + \frac{1}{3}x^3 + \frac{4}{15}x^{5/2}\right]_0^1 = \frac{11}{10}$

$M_{xy} = \iiint_E z\rho(x, y, z)\, dV = \int_0^1 \int_0^{\sqrt{x}} \int_0^{1+x+y} 2z\, dz\, dy\, dx = \int_0^1 \int_0^{\sqrt{x}} \left[z^2\right]_{z=0}^{z=1+x+y} dy\, dx = \int_0^1 \int_0^{\sqrt{x}} (1 + x + y)^2\, dy\, dx$

$= \int_0^1 \int_0^{\sqrt{x}} (1 + 2x + 2y + 2xy + x^2 + y^2)\, dy\, dx = \int_0^1 \left[y + 2xy + y^2 + xy^2 + x^2y + \frac{1}{3}y^3\right]_{y=0}^{y=\sqrt{x}} dx$

$= \int_0^1 \left(\sqrt{x} + \frac{7}{3}x^{3/2} + x + x^2 + x^{5/2}\right) dx = \left[\frac{2}{3}x^{3/2} + \frac{14}{15}x^{5/2} + \frac{1}{2}x^2 + \frac{1}{3}x^3 + \frac{2}{7}x^{7/2}\right]_0^1 = \frac{571}{210}$

Thus the mass is $\frac{79}{30}$ and the center of mass is $(\overline{x}, \overline{y}, \overline{z}) = \left(\dfrac{M_{yz}}{m}, \dfrac{M_{xz}}{m}, \dfrac{M_{xy}}{m}\right) = \left(\dfrac{358}{553}, \dfrac{33}{79}, \dfrac{571}{553}\right)$.

38. $m = \int_{-1}^{1} \int_{0}^{1-y^2} \int_{0}^{1-z} 4 \, dx \, dz \, dy = 4 \int_{-1}^{1} \int_{0}^{1-y^2} (1-z) \, dz \, dy = 4 \int_{-1}^{1} \left[z - \frac{1}{2}z^2 \right]_{z=0}^{z=1-y^2} dy = 2 \int_{-1}^{1} (1-y^4) \, dy = \frac{16}{5}$,

$M_{yz} = \int_{-1}^{1} \int_{0}^{1-y^2} \int_{0}^{1-z} 4x \, dx \, dz \, dy = 2 \int_{-1}^{1} \int_{0}^{1-y^2} (1-z)^2 \, dz \, dy = 2 \int_{-1}^{1} \left[-\frac{1}{3}(1-z)^3 \right]_{z=0}^{z=1-y^2} dy$

$\quad = \frac{2}{3} \int_{-1}^{1} (1-y^6) \, dy = \left(\frac{4}{3} \right)\left(\frac{6}{7} \right) = \frac{24}{21}$

$M_{xz} = \int_{-1}^{1} \int_{0}^{1-y^2} \int_{0}^{1-z} 4y \, dx \, dz \, dy = \int_{-1}^{1} \int_{0}^{1-y^2} 4y(1-z) \, dz \, dy$

$\quad = \int_{-1}^{1} \left[4y(1-y^2) - 2y(1-y^2)^2 \right] dy = \int_{-1}^{1} (2y - 2y^5) \, dy = 0$ [the integrand is odd]

$M_{xy} = \int_{-1}^{1} \int_{0}^{1-y^2} \int_{0}^{1-z} 4z \, dx \, dz \, dy = \int_{-1}^{1} \int_{0}^{1-y^2} (4z - 4z^2) \, dz \, dy = 2 \int_{-1}^{1} \left[(1-y^2)^2 - \frac{2}{3}(1-y^2)^3 \right] dy$

$\quad = 2 \int_{-1}^{1} \left[\frac{1}{3} - y^4 + \frac{2}{3}y^6 \right] dy = \left[\frac{4}{3}y - \frac{4}{5}y^5 + \frac{8}{21}y^7 \right]_{0}^{1} = \frac{96}{105} = \frac{32}{35}$

Thus, $(\overline{x}, \overline{y}, \overline{z}) = \left(\frac{5}{14}, 0, \frac{2}{7} \right)$

39. $m = \int_{0}^{a} \int_{0}^{a} \int_{0}^{a} (x^2 + y^2 + z^2) \, dx \, dy \, dz = \int_{0}^{a} \int_{0}^{a} \left[\frac{1}{3}x^3 + xy^2 + xz^2 \right]_{x=0}^{x=a} dy \, dz = \int_{0}^{a} \int_{0}^{a} \left(\frac{1}{3}a^3 + ay^2 + az^2 \right) dy \, dz$

$\quad = \int_{0}^{a} \left[\frac{1}{3}a^3 y + \frac{1}{3}ay^3 + ayz^2 \right]_{y=0}^{y=a} dz = \int_{0}^{a} \left(\frac{2}{3}a^4 + a^2 z^2 \right) dz = \left[\frac{2}{3}a^4 z + \frac{1}{3}a^2 z^3 \right]_{0}^{a} = \frac{2}{3}a^5 + \frac{1}{3}a^5 = a^5$

$M_{yz} = \int_{0}^{a} \int_{0}^{a} \int_{0}^{a} \left[x^3 + x(y^2 + z^2) \right] dx \, dy \, dz = \int_{0}^{a} \int_{0}^{a} \left[\frac{1}{4}a^4 + \frac{1}{2}a^2(y^2 + z^2) \right] dy \, dz$

$\quad = \int_{0}^{a} \left(\frac{1}{4}a^5 + \frac{1}{6}a^5 + \frac{1}{2}a^3 z^2 \right) dz = \frac{1}{4}a^6 + \frac{1}{3}a^6 = \frac{7}{12}a^6 = M_{xz} = M_{xy}$ by symmetry of E and $\rho(x, y, z)$

Hence $(\overline{x}, \overline{y}, \overline{z}) = \left(\frac{7}{12}a, \frac{7}{12}a, \frac{7}{12}a \right)$.

40. $m = \int_{0}^{1} \int_{0}^{1-x} \int_{0}^{1-x-y} y \, dz \, dy \, dx = \int_{0}^{1} \int_{0}^{1-x} \left[(1-x)y - y^2 \right] dy \, dx$

$\quad = \int_{0}^{1} \left[\frac{1}{2}(1-x)^3 - \frac{1}{3}(1-x)^3 \right] dx = \frac{1}{6} \int_{0}^{1} (1-x)^3 \, dx = \frac{1}{24}$

$M_{yz} = \int_{0}^{1} \int_{0}^{1-x} \int_{0}^{1-x-y} xy \, dz \, dy \, dx = \int_{0}^{1} \int_{0}^{1-x} \left[(x - x^2)y - xy^2 \right] dy \, dx$

$\quad = \int_{0}^{1} \left[\frac{1}{2}x(1-x)^3 - \frac{1}{3}x(1-x)^3 \right] dx = \frac{1}{6} \int_{0}^{1} \left(x - 3x^2 + 3x^3 - x^4 \right) dx = \frac{1}{6} \left(\frac{1}{2} - 1 + \frac{3}{4} - \frac{1}{5} \right) = \frac{1}{120}$

$M_{xz} = \int_{0}^{1} \int_{0}^{1-x} \int_{0}^{1-x-y} y^2 \, dz \, dy \, dx = \int_{0}^{1} \int_{0}^{1-x} \left[(1-x)y^2 - y^3 \right] dy \, dx$

$\quad = \int_{0}^{1} \left[\frac{1}{3}(1-x)^4 - \frac{1}{4}(1-x)^4 \right] dx = \frac{1}{12} \left[-\frac{1}{5}(1-x)^5 \right]_{0}^{1} = \frac{1}{60}$

$M_{xy} = \int_{0}^{1} \int_{0}^{1-x} \int_{0}^{1-x-y} yz \, dz \, dy \, dx = \int_{0}^{1} \int_{0}^{1-x} \left[\frac{1}{2}y(1 - x - y)^2 \right] dy \, dx$

$\quad = \frac{1}{2} \int_{0}^{1} \int_{0}^{1-x} \left[(1-x)^2 y - 2(1-x)y^2 + y^3 \right] dy \, dx = \frac{1}{2} \int_{0}^{1} \left[\frac{1}{2}(1-x)^4 - \frac{2}{3}(1-x)^4 + \frac{1}{4}(1-x)^4 \right] dx$

$\quad = \frac{1}{24} \int_{0}^{1} (1-x)^4 \, dx = -\frac{1}{24} \left[\frac{1}{5}(1-x)^5 \right]_{0}^{1} = \frac{1}{120}$

Hence $(\overline{x}, \overline{y}, \overline{z}) = \left(\frac{1}{5}, \frac{2}{5}, \frac{1}{5} \right)$.

41. $I_x = \int_{0}^{L} \int_{0}^{L} \int_{0}^{L} k(y^2 + z^2) \, dz \, dy \, dx = k \int_{0}^{L} \int_{0}^{L} \left(Ly^2 + \frac{1}{3}L^3 \right) dy \, dx = k \int_{0}^{L} \frac{2}{3}L^4 \, dx = \frac{2}{3}kL^5$.

By symmetry, $I_x = I_y = I_z = \frac{2}{3}kL^5$.

42. Let k be the density. Then

$$I_x = \int_{-c/2}^{c/2} \int_{-b/2}^{b/2} \int_{-a/2}^{a/2} k(y^2 + z^2)\, dx\, dy\, dz = ka \int_{-c/2}^{c/2} \int_{-b/2}^{b/2} (y^2 + z^2)\, dy\, dz$$

$$= ak \int_{-c/2}^{c/2} \left[\tfrac{1}{3}y^3 + z^2 y\right]_{y=-b/2}^{y=b/2} dz = ak \int_{-c/2}^{c/2} \left(\tfrac{1}{12}b^3 + bz^2\right) dz = ak\left[\tfrac{1}{12}b^3 z + \tfrac{1}{3}bz^3\right]_{-c/2}^{c/2}$$

$$= ak\left(\tfrac{1}{12}b^3 c + \tfrac{1}{12}bc^3\right) = \tfrac{1}{12}kabc(b^2 + c^2)$$

By symmetry, $I_y = \tfrac{1}{12}kabc(a^2 + c^2)$ and $I_z = \tfrac{1}{12}kabc(a^2 + b^2)$.

43. $I_z = \iiint_E (x^2 + y^2)\, \rho(x, y, z)\, dV = \iint_{x^2 + y^2 \le a^2} \left[\int_0^h k(x^2 + y^2)\, dz\right] dA = \iint_{x^2 + y^2 \le a^2} k(x^2 + y^2)h\, dA$

$$= kh \int_0^{2\pi} \int_0^a (r^2)\, r\, dr\, d\theta = kh \int_0^{2\pi} d\theta \int_0^a r^3\, dr = kh(2\pi)\left[\tfrac{1}{4}r^4\right]_0^a = 2\pi kh \cdot \tfrac{1}{4}a^4 = \tfrac{1}{2}\pi kh a^4$$

44. $I_z = \iiint_E (x^2 + y^2)\rho(x, y, z)\, dV = \iint_{x^2 + y^2 \le h^2} \left[\int_{\sqrt{x^2 + y^2}}^h k(x^2 + y^2)\, dz\right] dA$

$$= \iint_{x^2 + y^2 \le h^2} k(x^2 + y^2)\left(h - \sqrt{x^2 + y^2}\right) dA = k \int_0^{2\pi} \int_0^h r^2(h - r)\, r\, dr\, d\theta$$

$$= k \int_0^{2\pi} d\theta \int_0^h \left(r^3 h - r^4\right) dr = k(2\pi)\left[\tfrac{1}{4}r^4 h - \tfrac{1}{5}r^5\right]_0^h = 2\pi k \left(\tfrac{1}{4}h^5 - \tfrac{1}{5}h^5\right) = \tfrac{1}{10}\pi kh^5$$

45. (a) $m = \int_{-3}^3 \int_{-\sqrt{9-x^2}}^{\sqrt{9-x^2}} \int_1^{5-y} \sqrt{x^2 + y^2}\, dz\, dy\, dx$

(b) $(\overline{x}, \overline{y}, \overline{z}) = \left(\dfrac{M_{yz}}{m}, \dfrac{M_{xz}}{m}, \dfrac{M_{xy}}{m}\right)$ where

$$M_{yz} = \int_{-3}^3 \int_{-\sqrt{9-x^2}}^{\sqrt{9-x^2}} \int_1^{5-y} x\sqrt{x^2 + y^2}\, dz\, dy\, dx, \quad M_{xz} = \int_{-3}^3 \int_{-\sqrt{9-x^2}}^{\sqrt{9-x^2}} \int_1^{5-y} y\sqrt{x^2 + y^2}\, dz\, dy\, dx, \text{ and}$$

$$M_{xy} = \int_{-3}^3 \int_{-\sqrt{9-x^2}}^{\sqrt{9-x^2}} \int_1^{5-y} z\sqrt{x^2 + y^2}\, dz\, dy\, dx.$$

(c) $I_z = \int_{-3}^3 \int_{-\sqrt{9-x^2}}^{\sqrt{9-x^2}} \int_1^{5-y} (x^2 + y^2)\sqrt{x^2 + y^2}\, dz\, dy\, dx = \int_{-3}^3 \int_{-\sqrt{9-x^2}}^{\sqrt{9-x^2}} \int_1^{5-y} (x^2 + y^2)^{3/2}\, dz\, dy\, dx$

46. (a) $m = \int_{-1}^1 \int_{-\sqrt{1-y^2}}^{\sqrt{1-y^2}} \int_0^{\sqrt{1-x^2-y^2}} \sqrt{x^2 + y^2 + z^2}\, dz\, dx\, dy$

(b) $(\overline{x}, \overline{y}, \overline{z})$ where $\overline{x} = m^{-1} \int_{-1}^1 \int_{-\sqrt{1-y^2}}^{\sqrt{1-y^2}} \int_0^{\sqrt{1-x^2-y^2}} x\sqrt{x^2 + y^2 + z^2}\, dz\, dx\, dy$,

$$\overline{y} = m^{-1} \int_{-1}^1 \int_{-\sqrt{1-y^2}}^{\sqrt{1-y^2}} \int_0^{\sqrt{1-x^2-y^2}} y\sqrt{x^2 + y^2 + z^2}\, dz\, dx\, dy,$$

$$\overline{z} = m^{-1} \int_{-1}^1 \int_{-\sqrt{1-y^2}}^{\sqrt{1-y^2}} \int_0^{\sqrt{1-x^2-y^2}} z\sqrt{x^2 + y^2 + z^2}\, dz\, dx\, dy$$

(c) $I_z = \int_{-1}^1 \int_{-\sqrt{1-y^2}}^{\sqrt{1-y^2}} \int_0^{\sqrt{1-x^2-y^2}} (x^2 + y^2)(1 + x + y + z)\, dz\, dx\, dy$

47. (a) $m = \int_0^1 \int_0^{\sqrt{1-x^2}} \int_0^y (1 + x + y + z)\, dz\, dy\, dx = \frac{3\pi}{32} + \frac{11}{24}$

(b) $(\overline{x}, \overline{y}, \overline{z}) = \left(m^{-1} \int_0^1 \int_0^{\sqrt{1-x^2}} \int_0^y x(1 + x + y + z)\, dz\, dy\, dx, \right.$

$$m^{-1} \int_0^1 \int_0^{\sqrt{1-x^2}} \int_0^y y(1 + x + y + z)\, dz\, dy\, dx,$$

$$\left. m^{-1} \int_0^1 \int_0^{\sqrt{1-x^2}} \int_0^y z(1 + x + y + z)\, dz\, dy\, dx \right)$$

$$= \left(\frac{28}{9\pi + 44}, \frac{30\pi + 128}{45\pi + 220}, \frac{45\pi + 208}{135\pi + 660} \right)$$

(c) $I_z = \int_0^1 \int_0^{\sqrt{1-x^2}} \int_0^y (x^2 + y^2)(1 + x + y + z)\, dz\, dy\, dx = \dfrac{68 + 15\pi}{240}$

48. (a) $m = \int_0^1 \int_{3x}^3 \int_0^{\sqrt{9-y^2}} (x^2 + y^2)\, dz\, dy\, dx = \frac{56}{5} = 11.2$

(b) $(\overline{x}, \overline{y}, \overline{z})$ where $\overline{x} = m^{-1} \int_0^1 \int_{3x}^3 \int_0^{\sqrt{9-y^2}} x(x^2 + y^2)\, dz\, dy\, dx \approx 0.375$,

$\overline{y} = m^{-1} \int_0^1 \int_{3x}^3 \int_0^{\sqrt{9-y^2}} y(x^2 + y^2)\, dz\, dy\, dx = \frac{45\pi}{64} \approx 2.209$,

$\overline{z} = m^{-1} \int_0^1 \int_{3x}^3 \int_0^{\sqrt{9-y^2}} z(x^2 + y^2)\, dz\, dy\, dx = \frac{15}{16} = 0.9375$.

(c) $I_z = \int_0^1 \int_{3x}^3 \int_0^{\sqrt{9-y^2}} (x^2 + y^2)^2\, dz\, dy\, dx = \frac{10{,}464}{175} \approx 59.79$

49. (a) $f(x, y, z)$ is a joint density function, so we know $\iiint_{\mathbb{R}^3} f(x, y, z)\, dV = 1$. Here we have

$$\iiint_{\mathbb{R}^3} f(x, y, z)\, dV = \int_{-\infty}^\infty \int_{-\infty}^\infty \int_{-\infty}^\infty f(x, y, z)\, dz\, dy\, dx = \int_0^2 \int_0^2 \int_0^2 Cxyz\, dz\, dy\, dx$$

$$= C \int_0^2 x\, dx \int_0^2 y\, dy \int_0^2 z\, dz = C \left[\tfrac{1}{2}x^2\right]_0^2 \left[\tfrac{1}{2}y^2\right]_0^2 \left[\tfrac{1}{2}z^2\right]_0^2 = 8C$$

Then we must have $8C = 1 \;\Rightarrow\; C = \frac{1}{8}$.

(b) $P(X \le 1, Y \le 1, Z \le 1) = \int_{-\infty}^1 \int_{-\infty}^1 \int_{-\infty}^1 f(x, y, z)\, dz\, dy\, dx = \int_0^1 \int_0^1 \int_0^1 \tfrac{1}{8}xyz\, dz\, dy\, dx$

$$= \tfrac{1}{8} \int_0^1 x\, dx \int_0^1 y\, dy \int_0^1 z\, dz = \tfrac{1}{8} \left[\tfrac{1}{2}x^2\right]_0^1 \left[\tfrac{1}{2}y^2\right]_0^1 \left[\tfrac{1}{2}z^2\right]_0^1 = \tfrac{1}{8}\left(\tfrac{1}{2}\right)^3 = \tfrac{1}{64}$$

(c) $P(X + Y + Z \le 1) = P((X, Y, Z) \in E)$ where E is the solid region in the first octant bounded by the coordinate planes and the plane $x + y + z = 1$. The plane $x + y + z = 1$ meets the xy-plane in the line $x + y = 1$, so we have

$$P(X + Y + Z \le 1) = \iiint_E f(x, y, z)\, dV = \int_0^1 \int_0^{1-x} \int_0^{1-x-y} \tfrac{1}{8}xyz\, dz\, dy\, dx$$

$$= \tfrac{1}{8} \int_0^1 \int_0^{1-x} xy\left[\tfrac{1}{2}z^2\right]_{z=0}^{z=1-x-y} dy\, dx = \tfrac{1}{16} \int_0^1 \int_0^{1-x} xy(1 - x - y)^2\, dy\, dx$$

$$= \tfrac{1}{16} \int_0^1 \int_0^{1-x} [(x^3 - 2x^2 + x)y + (2x^2 - 2x)y^2 + xy^3]\, dy\, dx$$

$$= \tfrac{1}{16} \int_0^1 \left[(x^3 - 2x^2 + x)\tfrac{1}{2}y^2 + (2x^2 - 2x)\tfrac{1}{3}y^3 + x\left(\tfrac{1}{4}y^4\right)\right]_{y=0}^{y=1-x} dx$$

$$= \tfrac{1}{192} \int_0^1 (x - 4x^2 + 6x^3 - 4x^4 + x^5)\, dx = \tfrac{1}{192}\left(\tfrac{1}{30}\right) = \tfrac{1}{5760}$$

50. (a) $f(x, y, z)$ is a joint density function, so we know $\iiint_{\mathbb{R}^3} f(x, y, z)\, dV = 1$. Here we have

$$\iiint_{\mathbb{R}^3} f(x, y, z)\, dV = \int_{-\infty}^{\infty}\int_{-\infty}^{\infty}\int_{-\infty}^{\infty} f(x, y, z)\, dz\, dy\, dx = \int_0^{\infty}\int_0^{\infty}\int_0^{\infty} Ce^{-(0.5x+0.2y+0.1z)}\, dz\, dy\, dx$$

$$= C\int_0^{\infty} e^{-0.5x}\, dx \int_0^{\infty} e^{-0.2y}\, dy \int_0^{\infty} e^{-0.1z}\, dz$$

$$= C \lim_{t\to\infty}\int_0^t e^{-0.5x}\, dx \lim_{t\to\infty}\int_0^t e^{-0.2y}\, dy \lim_{t\to\infty}\int_0^t e^{-0.1z}\, dz$$

$$= C \lim_{t\to\infty}\left[-2e^{-0.5x}\right]_0^t \lim_{t\to\infty}\left[-5e^{-0.2y}\right]_0^t \lim_{t\to\infty}\left[-10e^{-0.1z}\right]_0^t$$

$$= C \lim_{t\to\infty}\left[-2(e^{-0.5t}-1)\right] \lim_{t\to\infty}\left[-5(e^{-0.2t}-1)\right] \lim_{t\to\infty}\left[-10(e^{-0.1t}-1)\right]$$

$$= C \cdot (-2)(0-1) \cdot (-5)(0-1) \cdot (-10)(0-1) = 100C$$

So we must have $100C = 1 \;\Rightarrow\; C = \frac{1}{100}$.

(b) We have no restriction on Z, so

$$P(X \le 1, Y \le 1) = \int_{-\infty}^1\int_{-\infty}^1\int_{-\infty}^{\infty} f(x, y, z)\, dz\, dy\, dx = \int_0^1\int_0^1\int_0^{\infty} \tfrac{1}{100}e^{-(0.5x+0.2y+0.1z)}\, dz\, dy\, dx$$

$$= \tfrac{1}{100}\int_0^1 e^{-0.5x}\, dx \int_0^1 e^{-0.2y}\, dy \int_0^{\infty} e^{-0.1z}\, dz$$

$$= \tfrac{1}{100}\left[-2e^{-0.5x}\right]_0^1 \left[-5e^{-0.2y}\right]_0^1 \lim_{t\to\infty}\left[-10e^{-0.1z}\right]_0^t \qquad \text{[by part (a)]}$$

$$= \tfrac{1}{100}(2 - 2e^{-0.5})(5 - 5e^{-0.2})(10) = (1 - e^{-0.5})(1 - e^{-0.2}) \approx 0.07132$$

(c) $P(X \le 1, Y \le 1, Z \le 1) = \int_{-\infty}^1\int_{-\infty}^1\int_{-\infty}^1 f(x, y, z)\, dz\, dy\, dx = \int_0^1\int_0^1\int_0^1 \tfrac{1}{100}e^{-(0.5x+0.2y+0.1z)}\, dz\, dy\, dx$

$$= \tfrac{1}{100}\int_0^1 e^{-0.5x}\, dx \int_0^1 e^{-0.2y}\, dy \int_0^1 e^{-0.1z}\, dz$$

$$= \tfrac{1}{100}\left[-2e^{-0.5x}\right]_0^1 \left[-5e^{-0.2y}\right]_0^1 \left[-10e^{-0.1z}\right]_0^1$$

$$= (1 - e^{-0.5})(1 - e^{-0.2})(1 - e^{-0.1}) \approx 0.006787$$

51. $V(E) = L^3 \;\Rightarrow\; f_{\text{ave}} = \dfrac{1}{L^3}\displaystyle\int_0^L\int_0^L\int_0^L xyz\, dx\, dy\, dz = \dfrac{1}{L^3}\int_0^L x\, dx \int_0^L y\, dy \int_0^L z\, dz$

$$= \dfrac{1}{L^3}\left[\dfrac{x^2}{2}\right]_0^L \left[\dfrac{y^2}{2}\right]_0^L \left[\dfrac{z^2}{2}\right]_0^L = \dfrac{1}{L^3}\dfrac{L^2}{2}\dfrac{L^2}{2}\dfrac{L^2}{2} = \dfrac{L^3}{8}$$

52. $V(E) = \displaystyle\int_{-1}^1\int_{-\sqrt{1-x^2}}^{\sqrt{1-x^2}}\int_0^{1-x^2-y^2} dz\, dy\, dx = \int_{-1}^1\int_{-\sqrt{1-x^2}}^{\sqrt{1-x^2}}(1 - x^2 - y^2)\, dy\, dx$

$$= \int_0^{2\pi}\int_0^1 (1 - r^2)\, r\, dr\, d\theta = \int_0^{2\pi} d\theta \int_0^1 (r - r^3)\, dr = 2\pi\left(\dfrac{r^2}{2} - \dfrac{r^4}{4}\right)\Big|_0^1 = \dfrac{\pi}{2}.$$

Then $\quad f_{\text{ave}} = \dfrac{1}{\pi/2}\displaystyle\iiint_E (x^2 z + y^2 z)\, dV = \dfrac{2}{\pi}\int_{-1}^1\int_{-\sqrt{1-x^2}}^{\sqrt{1-x^2}}\int_0^{1-x^2-y^2}(x^2 + y^2)\, z\, dz\, dy\, dx$

$$= \dfrac{2}{\pi}\int_{-1}^1\int_{-\sqrt{1-x^2}}^{\sqrt{1-x^2}}(x^2 + y^2)\cdot \tfrac{1}{2}(1 - x^2 - y^2)^2\, dy\, dx = \dfrac{1}{\pi}\int_0^{2\pi}\int_0^1 r^2(1 - r^2)^2\, r\, dr\, d\theta$$

$$= \dfrac{1}{\pi}\int_0^{2\pi} d\theta \int_0^1 (r^3 - 2r^5 + r^7)\, dr = \dfrac{1}{\pi}(2\pi)\left[\tfrac{1}{4}r^4 - \tfrac{1}{3}r^6 + \tfrac{1}{8}r^8\right]_0^1 = 2\left(\tfrac{1}{24}\right) = \tfrac{1}{12}$$

53. The triple integral will attain its maximum when the integrand $1 - x^2 - 2y^2 - 3z^2$ is positive in the region E and negative everywhere else. For if E contains some region F where the integrand is negative, the integral could be increased by excluding F from E, and if E fails to contain some part G of the region where the integrand is positive, the integral could be increased by including G in E. So we require that $x^2 + 2y^2 + 3z^2 \le 1$. This describes the region bounded by the ellipsoid $x^2 + 2y^2 + 3z^2 = 1$.

DISCOVERY PROJECT   Volumes of Hyperspheres

In this project we use V_n to denote the n-dimensional volume of an n-dimensional hypersphere.

1. The interior of the circle is the set of points $\left\{ (x, y) \mid -r \leq y \leq r, \ -\sqrt{r^2 - y^2} \leq x \leq \sqrt{r^2 - y^2} \right\}$. So, substituting $y = r \sin \theta$ and then using Formula 64 to evaluate the integral, we get

$$V_2 = \int_{-r}^{r} \int_{-\sqrt{r^2 - y^2}}^{\sqrt{r^2 - y^2}} dx \, dy = \int_{-r}^{r} 2 \sqrt{r^2 - y^2} \, dy = \int_{-\pi/2}^{\pi/2} 2r \sqrt{1 - \sin^2 \theta} \, (r \cos \theta \, d\theta)$$

$$= 2r^2 \int_{-\pi/2}^{\pi/2} \cos^2 \theta \, d\theta = 2r^2 \left[\tfrac{1}{2}\theta + \tfrac{1}{4} \sin 2\theta \right]_{-\pi/2}^{\pi/2} = 2r^2 \left(\tfrac{\pi}{2} \right) = \pi r^2$$

2. The region of integration is

$\left\{ (x, y, z) \mid -r \leq z \leq r, -\sqrt{r^2 - z^2} \leq y \leq \sqrt{r^2 - z^2}, -\sqrt{r^2 - z^2 - y^2} \leq x \leq \sqrt{r^2 - z^2 - y^2} \right\}$. Substituting $y = \sqrt{r^2 - z^2} \sin \theta$ and using Formula 64 to integrate $\cos^2 \theta$, we get

$$V_3 = \int_{-r}^{r} \int_{-\sqrt{r^2 - z^2}}^{\sqrt{r^2 - z^2}} \int_{-\sqrt{r^2 - z^2 - y^2}}^{\sqrt{r^2 - z^2 - y^2}} dx \, dy \, dz = \int_{-r}^{r} \int_{-\sqrt{r^2 - z^2}}^{\sqrt{r^2 - z^2}} 2 \sqrt{r^2 - z^2 - y^2} \, dy \, dz$$

$$= \int_{-r}^{r} \int_{-\pi/2}^{\pi/2} 2 \sqrt{r^2 - z^2} \sqrt{1 - \sin^2 \theta} \left(\sqrt{r^2 - z^2} \cos \theta \, d\theta \right) dz$$

$$= 2 \left[\int_{-r}^{r} (r^2 - z^2) \, dz \right] \left[\int_{-\pi/2}^{\pi/2} \cos^2 \theta \, d\theta \right] = 2 \left(\frac{4r^3}{3} \right) \left(\frac{\pi}{2} \right) = \frac{4\pi r^3}{3}$$

3. Here we substitute $y = \sqrt{r^2 - w^2 - z^2} \sin \theta$ and, later, $w = r \sin \phi$. Because $\int_{-\pi/2}^{\pi/2} \cos^p \theta \, d\theta$ seems to occur frequently in these calculations, it is useful to find a general formula for that integral. From Exercises 45 and 46 in Section 8.1 [ET 7.1], we have

$$\int_{0}^{\pi/2} \sin^{2k} x \, dx = \frac{1 \cdot 3 \cdot 5 \cdot \cdots \cdot (2k - 1)}{2 \cdot 4 \cdot 6 \cdot \cdots \cdot 2k} \frac{\pi}{2} \qquad \text{and} \qquad \int_{0}^{\pi/2} \sin^{2k+1} x \, dx = \frac{2 \cdot 4 \cdot 6 \cdot \cdots \cdot 2k}{1 \cdot 3 \cdot 5 \cdot \cdots \cdot (2k + 1)}$$

and from the symmetry of the sine and cosine functions, we can conclude that

$$\int_{-\pi/2}^{\pi/2} \cos^{2k} x \, dx = 2 \int_{0}^{\pi/2} \sin^{2k} x \, dx = \frac{1 \cdot 3 \cdot 5 \cdot \cdots \cdot (2k - 1)\pi}{2 \cdot 4 \cdot 6 \cdot \cdots \cdot 2k} \tag{1}$$

$$\int_{-\pi/2}^{\pi/2} \cos^{2k+1} x \, dx = 2 \int_{0}^{\pi/2} \sin^{2k+1} x \, dx = \frac{2 \cdot 2 \cdot 4 \cdot 6 \cdot \cdots \cdot 2k}{1 \cdot 3 \cdot 5 \cdot \cdots \cdot (2k + 1)} \tag{2}$$

Thus $\quad V_4 = \int_{-r}^{r} \int_{-\sqrt{r^2 - w^2}}^{\sqrt{r^2 - w^2}} \int_{-\sqrt{r^2 - w^2 - z^2}}^{\sqrt{r^2 - w^2 - z^2}} \int_{-\sqrt{r^2 - w^2 - z^2 - y^2}}^{\sqrt{r^2 - w^2 - z^2 - y^2}} dx \, dy \, dz \, dw$

$$= 2 \int_{-r}^{r} \int_{-\sqrt{r^2 - w^2}}^{\sqrt{r^2 - w^2}} \int_{-\sqrt{r^2 - w^2 - z^2}}^{\sqrt{r^2 - w^2 - z^2}} \sqrt{r^2 - w^2 - z^2 - y^2} \, dy \, dz \, dw$$

$$= 2 \int_{-r}^{r} \int_{-\sqrt{r^2 - w^2}}^{\sqrt{r^2 - w^2}} \int_{-\pi/2}^{\pi/2} (r^2 - w^2 - z^2) \cos^2 \theta \, d\theta \, dz \, dw$$

$$= 2 \left[\int_{-r}^{r} \int_{-\sqrt{r^2 - w^2}}^{\sqrt{r^2 - w^2}} (r^2 - w^2 - z^2) \, dz \, dw \right] \left[\int_{-\pi/2}^{\pi/2} \cos^2 \theta \, d\theta \right]$$

$$= 2 \left(\tfrac{\pi}{2} \right) \left[\int_{-r}^{r} \tfrac{4}{3} (r^2 - w^2)^{3/2} \, dw \right] = \pi \left(\tfrac{4}{3} \right) \int_{-\pi/2}^{\pi/2} r^4 \cos^4 \phi \, d\phi = \frac{4\pi}{3} r^4 \cdot \frac{1 \cdot 3 \cdot \pi}{2 \cdot 4} = \frac{\pi^2 r^4}{2}$$

4. By using the substitutions $x_i = \sqrt{r^2 - x_n^2 - x_{n-1}^2 - \cdots - x_{i+1}^2}\,\cos\theta_i$ and then applying Formulas 1 and 2 from Problem 3, we can write

$$V_n = \int_{-r}^{r}\int_{-\sqrt{r^2-x_n^2}}^{\sqrt{r^2-x_n^2}}\cdots\int_{-\sqrt{r^2-x_n^2-x_{n-1}^2-\cdots-x_3^2}}^{\sqrt{r^2-x_n^2-x_{n-1}^2-\cdots-x_3^2}}\int_{-\sqrt{r^2-x_n^2-x_{n-1}^2-\cdots-x_3^2-x_2^2}}^{\sqrt{r^2-x_n^2-x_{n-1}^2-\cdots-x_3^2-x_2^2}} dx_1\,dx_2\cdots dx_{n-1}\,dx_n$$

$$= 2\left[\int_{-\pi/2}^{\pi/2}\cos^2\theta_2\,d\theta_2\right]\left[\int_{-\pi/2}^{\pi/2}\cos^3\theta_3\,d\theta_3\right]\cdots\left[\int_{-\pi/2}^{\pi/2}\cos^{n-1}\theta_{n-1}\,d\theta_{n-1}\right]\left[\int_{-\pi/2}^{\pi/2}\cos^n\theta_n\,d\theta_n\right]r^n$$

$$= \begin{cases} \left[2\cdot\dfrac{\pi}{2}\right]\left[\dfrac{2\cdot2}{1\cdot3}\cdot\dfrac{1\cdot3\pi}{2\cdot4}\right]\left[\dfrac{2\cdot2\cdot4}{1\cdot3\cdot5}\cdot\dfrac{1\cdot3\cdot5\pi}{2\cdot4\cdot6}\right]\cdots\left[\dfrac{2\cdots(n-2)}{1\cdots(n-1)}\cdot\dfrac{1\cdots(n-1)\pi}{2\cdots n}\right]r^n & n \text{ even} \\[2ex] 2\left[\dfrac{\pi}{2}\cdot\dfrac{2\cdot2}{1\cdot3}\right]\left[\dfrac{1\cdot3\pi}{2\cdot4}\cdot\dfrac{2\cdot2\cdot4}{1\cdot3\cdot5}\right]\cdots\left[\dfrac{1\cdots(n-2)\pi}{2\cdots(n-1)}\cdot\dfrac{2\cdots(n-1)}{1\cdots n}\right]r^n & n \text{ odd} \end{cases}$$

By canceling within each set of brackets, we find that

$$V_n = \begin{cases} \dfrac{2\pi}{2}\cdot\dfrac{2\pi}{4}\cdot\dfrac{2\pi}{6}\cdots\cdot\dfrac{2\pi}{n}r^n = \dfrac{(2\pi)^{n/2}}{2\cdot4\cdot6\cdots\cdot n}r^n = \dfrac{\pi^{n/2}}{(\frac{1}{2}n)!}r^n & n \text{ even} \\[2ex] 2\cdot\dfrac{2\pi}{3}\cdot\dfrac{2\pi}{5}\cdot\dfrac{2\pi}{7}\cdots\cdot\dfrac{2\pi}{n}r^n = \dfrac{2(2\pi)^{(n-1)/2}}{3\cdot5\cdot7\cdots\cdot n}r^n = \dfrac{2^n\left[\frac{1}{2}(n-1)\right]!\,\pi^{(n-1)/2}}{n!}r^n & n \text{ odd} \end{cases}$$

16.7   Triple Integrals in Cylindrical Coordinates

ET 15.7

1. (a)

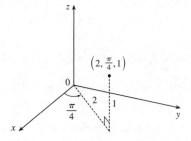

$x = 2\cos\dfrac{\pi}{4} = \sqrt{2}$, $y = 2\sin\dfrac{\pi}{4} = \sqrt{2}$, $z = 1$,

so the point is $\left(\sqrt{2}, \sqrt{2}, 1\right)$ in rectangular coordinates.

(b)

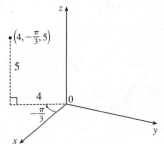

$x = 4\cos\left(-\frac{\pi}{3}\right) = 2$, $y = 4\sin\left(-\frac{\pi}{3}\right) = -2\sqrt{3}$,

and $z = 5$, so the point is $\left(2, -2\sqrt{3}, 5\right)$ in rectangular coordinates.

2. (a)

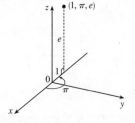

$x = 1\cos\pi = -1$, $y = 1\sin\pi = 0$, and $z = e$,

so the point is $(-1, 0, e)$ in rectangular coordinates.

(b)

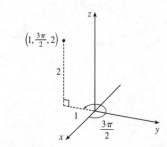

$x = 1\cos\dfrac{3\pi}{2} = 0$, $y = 1\sin\dfrac{3\pi}{2} = -1$, $z = 2$,

so the point is $(0, -1, 2)$ in rectangular coordinates.

3. (a) $r^2 = x^2 + y^2 = 1^2 + (-1)^2 = 2$ so $r = \sqrt{2}$; $\tan \theta = \dfrac{y}{x} = \dfrac{-1}{1} = -1$ and the point $(1, -1)$ is in the fourth quadrant of

the xy-plane, so $\theta = \frac{7\pi}{4} + 2n\pi$; $z = 4$. Thus, one set of cylindrical coordinates is $\left(\sqrt{2}, \frac{7\pi}{4}, 4\right)$.

(b) $r^2 = (-1)^2 + \left(-\sqrt{3}\right)^2 = 4$ so $r = 2$; $\tan \theta = \frac{-\sqrt{3}}{-1} = \sqrt{3}$ and the point $\left(-1, -\sqrt{3}\right)$ is in the third quadrant of the

xy-plane, so $\theta = \frac{4\pi}{3} + 2n\pi$; $z = 2$. Thus, one set of cylindrical coordinates is $\left(2, \frac{4\pi}{3}, 2\right)$.

4. (a) $r^2 = \left(2\sqrt{3}\right)^2 + 2^2 = 16$ so $r = 4$; $\tan \theta = \frac{2}{2\sqrt{3}} = \frac{1}{\sqrt{3}}$ and the point $\left(2\sqrt{3}, 2\right)$ is in the first quadrant of the xy-plane, so

$\theta = \frac{\pi}{6} + 2n\pi$; $z = -1$. Thus, one set of cylindrical coordinates is $\left(4, \frac{\pi}{6}, -1\right)$.

(b) $r^2 = 4^2 + (-3)^2 = 25$ so $r = 5$; $\tan \theta = \frac{-3}{4}$ and the point $(4, -3)$ is in the fourth quadrant of the xy-plane, so

$\theta = \tan^{-1}\left(-\frac{3}{4}\right) + 2n\pi \approx -0.64 + 2n\pi$; $z = 2$. Thus, one set of cylindrical coordinates is

$\left(5, \tan^{-1}\left(-\frac{3}{4}\right) + 2\pi, 2\right) \approx (5, 5.64, 2)$.

5. Since $\theta = \frac{\pi}{4}$ but r and z may vary, the surface is a vertical half-plane including the z-axis and intersecting the xy-plane in the

half-line $y = x$, $x \geq 0$.

6. Since $r = 5$, $x^2 + y^2 = 25$ and the surface is a circular cylinder with radius 5 and axis the z-axis.

7. $z = 4 - r^2 = 4 - (x^2 + y^2)$ or $4 - x^2 - y^2$, so the surface is a circular paraboloid with vertex $(0, 0, 4)$, axis the z-axis, and

opening downward.

8. Since $2r^2 + z^2 = 1$ and $r^2 = x^2 + y^2$, we have $2(x^2 + y^2) + z^2 = 1$ or $2x^2 + 2y^2 + z^2 = 1$, an ellipsoid centered at the

origin with intercepts $x = \pm\frac{1}{\sqrt{2}}$, $y = \pm\frac{1}{\sqrt{2}}$, $z = \pm1$.

9. (a) $x^2 + y^2 = r^2$, so the equation becomes $z = r^2$.

(b) Substituting $x^2 + y^2 = r^2$ and $y = r \sin \theta$, the equation $x^2 + y^2 = 2y$ becomes $r^2 = 2r \sin \theta$ or $r = 2 \sin \theta$.

10. (a) Substituting $x = r \cos \theta$ and $y = r \sin \theta$, the equation $3x + 2y + z = 6$ becomes $3r \cos \theta + 2r \sin \theta + z = 6$ or

$z = 6 - r(3 \cos \theta + 2 \sin \theta)$.

(b) The equation $-x^2 - y^2 + z^2 = 1$ can be written as $-(x^2 + y^2) + z^2 = 1$ which becomes $-r^2 + z^2 = 1$ or $z^2 = 1 + r^2$

in cylindrical coordinates.

11.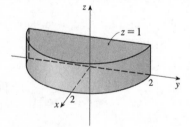

$0 \leq r \leq 2$ and $0 \leq z \leq 1$ describe a solid circular cylinder with

radius 2, axis the z-axis, and height 1, but $-\pi/2 \leq \theta \leq \pi/2$ restricts

the solid to the first and fourth quadrants of the xy-plane, so we have

a half-cylinder.

12.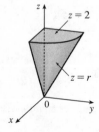

$z = r = \sqrt{x^2 + y^2}$ is a cone that opens upward. Thus $r \leq z \leq 2$ is the region above this

cone and beneath the horizontal plane $z = 2$. $0 \leq \theta \leq \frac{\pi}{2}$ restricts the solid to that part of

this region in the first octant.

13. We can position the cylindrical shell vertically so that its axis coincides with the z-axis and its base lies in the xy-plane. If we use centimeters as the unit of measurement, then cylindrical coordinates conveniently describe the shell as $6 \le r \le 7$, $0 \le \theta \le 2\pi, 0 \le z \le 20$.

14. In cylindrical coordinates, the equations are $z = r^2$ and $z = 5 - r^2$. The curve of intersection is $r^2 = 5 - r^2$ or $r = \sqrt{5/2}$. So we graph the surfaces in cylindrical coordinates, with $0 \le r \le \sqrt{5/2}$. In Maple, we can use the `coords=cylindrical` option in a regular `plot3d` command. In Mathematica, we can use `ParametricPlot3D`.

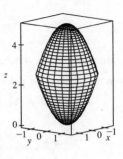

15.

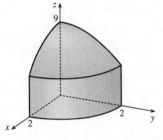

The region of integration is given in cylindrical coordinates by $E = \{(r, \theta, z) \mid 0 \le \theta \le 2\pi, 0 \le r \le 4, r \le z \le 4\}$. This represents the solid region bounded below by the cone $z = r$ and above by the horizontal plane $z = 4$.

$$\int_0^4 \int_0^{2\pi} \int_r^4 r \, dz \, d\theta \, dr = \int_0^4 \int_0^{2\pi} [rz]_{z=r}^{z=4} \, d\theta \, dr = \int_0^4 \int_0^{2\pi} r(4 - r) \, d\theta \, dr$$

$$= \int_0^4 (4r - r^2) \, dr \int_0^{2\pi} d\theta = \left[2r^2 - \tfrac{1}{3}r^3\right]_0^4 \left[\theta\right]_0^{2\pi}$$

$$= \left(32 - \tfrac{64}{3}\right)(2\pi) = \tfrac{64\pi}{3}$$

16.

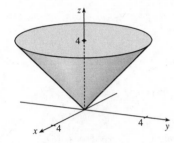

The region of integration is given in cylindrical coordinates by $E = \{(r, \theta, z) \mid 0 \le \theta \le \pi/2, 0 \le r \le 2, 0 \le z \le 9 - r^2\}$. This represents the solid region in the first octant enclosed by the circular cylinder $r = 2$, bounded above by $z = 9 - r^2$, a circular paraboloid, and bounded below by the xy-plane.

$$\int_0^{\pi/2} \int_0^2 \int_0^{9-r^2} r \, dz \, dr \, d\theta = \int_0^{\pi/2} \int_0^2 [rz]_{z=0}^{z=9-r^2} \, dr \, d\theta$$

$$= \int_0^{\pi/2} \int_0^2 r(9 - r^2) \, dr \, d\theta = \int_0^{\pi/2} d\theta \int_0^2 (9r - r^3) \, dr$$

$$= \left[\theta\right]_0^{\pi/2} \left[\tfrac{9}{2}r^2 - \tfrac{1}{4}r^4\right]_0^2 = \tfrac{\pi}{2}(18 - 4) = 7\pi$$

17. In cylindrical coordinates, E is given by $\{(r, \theta, z) \mid 0 \le \theta \le 2\pi, 0 \le r \le 4, -5 \le z \le 4\}$. So

$$\iiint_E \sqrt{x^2 + y^2} \, dV = \int_0^{2\pi} \int_0^4 \int_{-5}^4 \sqrt{r^2} \, r \, dz \, dr \, d\theta = \int_0^{2\pi} d\theta \int_0^4 r^2 \, dr \int_{-5}^4 dz$$

$$= \left[\theta\right]_0^{2\pi} \left[\tfrac{1}{3}r^3\right]_0^4 \left[z\right]_{-5}^4 = (2\pi)\left(\tfrac{64}{3}\right)(9) = 384\pi$$

18. The paraboloid $z = 1 - x^2 - y^2$ intersects the xy-plane in the circle $x^2 + y^2 = r^2 = 1$ or $r = 1$, so in cylindrical coordinates, E is given by $\{(r, \theta, z) \mid 0 \le \theta \le \tfrac{\pi}{2}, 0 \le r \le 1, 0 \le z \le 1 - r^2\}$. Thus

$$\iiint_E (x^3 + xy^2) \, dV = \int_0^{\pi/2} \int_0^1 \int_0^{1-r^2} (r^3 \cos^3 \theta + r^3 \cos \theta \sin^2 \theta) \, r \, dz \, dr \, d\theta = \int_0^{\pi/2} \int_0^1 \int_0^{1-r^2} r^4 \cos \theta \, dz \, dr \, d\theta$$

$$= \int_0^{\pi/2} \int_0^1 r^4 \cos \theta \left[z\right]_{z=0}^{z=1-r^2} \, dr \, d\theta = \int_0^{\pi/2} \int_0^1 r^4 (1 - r^2) \cos \theta \, dr \, d\theta$$

$$= \int_0^{\pi/2} \cos \theta \left[\tfrac{1}{5}r^5 - \tfrac{1}{7}r^7\right]_{r=0}^{r=1} \, d\theta = \int_0^{\pi/2} \tfrac{2}{35} \cos \theta \, d\theta = \tfrac{2}{35} \left[\sin \theta\right]_0^{\pi/2} = \tfrac{2}{35}$$

19. In cylindrical coordinates E is bounded by the paraboloid $z = 1 + r^2$, the cylinder $r^2 = 5$ or $r = \sqrt{5}$, and the xy-plane,

so E is given by $\{(r, \theta, z) \mid 0 \le \theta \le 2\pi, 0 \le r \le \sqrt{5}, 0 \le z \le 1 + r^2\}$. Thus

$$\iiint_E e^z \, dV = \int_0^{2\pi} \int_0^{\sqrt{5}} \int_0^{1+r^2} e^z \, r \, dz \, dr \, d\theta = \int_0^{2\pi} \int_0^{\sqrt{5}} r \left[e^z \right]_{z=0}^{z=1+r^2} dr \, d\theta = \int_0^{2\pi} \int_0^{\sqrt{5}} r(e^{1+r^2} - 1) \, dr \, d\theta$$

$$= \int_0^{2\pi} d\theta \int_0^{\sqrt{5}} \left(re^{1+r^2} - r \right) dr = 2\pi \left[\tfrac{1}{2} e^{1+r^2} - \tfrac{1}{2} r^2 \right]_0^{\sqrt{5}} = \pi(e^6 - e - 5)$$

20. In cylindrical coordinates E is bounded by the planes $z = 0$, $z = r\cos\theta + r\sin\theta + 5$ and the cylinders $r = 2$ and $r = 3$, so

E is given by $\{(r, \theta, z) \mid 0 \le \theta \le 2\pi, 2 \le r \le 3, 0 \le z \le r\cos\theta + r\sin\theta + 5\}$. Thus

$$\iiint_E x \, dV = \int_0^{2\pi} \int_2^3 \int_0^{r\cos\theta + r\sin\theta + 5} (r\cos\theta) \, r \, dz \, dr \, d\theta = \int_0^{2\pi} \int_2^3 (r^2\cos\theta)[z]_{z=0}^{z=r\cos\theta + r\sin\theta + 5} \, dr \, d\theta$$

$$= \int_0^{2\pi} \int_2^3 (r^2\cos\theta)(r\cos\theta + r\sin\theta + 5) \, dr \, d\theta = \int_0^{2\pi} \int_2^3 (r^3(\cos^2\theta + \cos\theta\sin\theta) + 5r^2\cos\theta) \, dr \, d\theta$$

$$= \int_0^{2\pi} \left[\tfrac{1}{4} r^4(\cos^2\theta + \cos\theta\sin\theta) + \tfrac{5}{3} r^3 \cos\theta \right]_{r=2}^{r=3} d\theta$$

$$= \int_0^{2\pi} \left[\left(\tfrac{81}{4} - \tfrac{16}{4} \right)(\cos^2\theta + \cos\theta\sin\theta) + \tfrac{5}{3}(27 - 8)\cos\theta \right] d\theta$$

$$= \int_0^{2\pi} \left(\tfrac{65}{4} \left(\tfrac{1}{2}(1 + \cos 2\theta) + \cos\theta\sin\theta \right) + \tfrac{95}{3}\cos\theta \right) d\theta = \left[\tfrac{65}{8}\theta + \tfrac{65}{16}\sin 2\theta + \tfrac{65}{8}\sin^2\theta + \tfrac{95}{3}\sin\theta \right]_0^{2\pi} = \tfrac{65}{4}\pi$$

21. In cylindrical coordinates, E is bounded by the cylinder $r = 1$, the plane $z = 0$, and the cone $z = 2r$. So

$E = \{(r, \theta, z) \mid 0 \le \theta \le 2\pi, 0 \le r \le 1, 0 \le z \le 2r\}$ and

$$\iiint_E x^2 \, dV = \int_0^{2\pi} \int_0^1 \int_0^{2r} r^2 \cos^2\theta \, r \, dz \, dr \, d\theta = \int_0^{2\pi} \int_0^1 \left[r^3 \cos^2\theta \, z \right]_{z=0}^{z=2r} dr \, d\theta = \int_0^{2\pi} \int_0^1 2r^4 \cos^2\theta \, dr \, d\theta$$

$$= \int_0^{2\pi} \left[\tfrac{2}{5} r^5 \cos^2\theta \right]_{r=0}^{r=1} d\theta = \tfrac{2}{5} \int_0^{2\pi} \cos^2\theta \, d\theta = \frac{2}{5} \int_0^{2\pi} \frac{1 + \cos 2\theta}{2} \, d\theta = \frac{1}{5} \left[\theta + \frac{1}{2}\sin 2\theta \right]_0^{2\pi} = \frac{2\pi}{5}$$

22. In cylindrical coordinates E is the solid region within the cylinder $r = 1$ bounded above and below by the sphere $r^2 + z^2 = 4$,

so $E = \{(r, \theta, z) \mid 0 \le \theta \le 2\pi, 0 \le r \le 1, -\sqrt{4 - r^2} \le z \le \sqrt{4 - r^2}\}$. Thus the volume is

$$\iiint_E dV = \int_0^{2\pi} \int_0^1 \int_{-\sqrt{4-r^2}}^{\sqrt{4-r^2}} r \, dz \, dr \, d\theta = \int_0^{2\pi} \int_0^1 2r\sqrt{4 - r^2} \, dr \, d\theta$$

$$= \int_0^{2\pi} d\theta \int_0^1 2r\sqrt{4 - r^2} \, dr = 2\pi \left[-\tfrac{2}{3}(4 - r^2)^{3/2} \right]_0^1 = \tfrac{4}{3}\pi(8 - 3^{3/2})$$

23. (a) The paraboloids intersect when $x^2 + y^2 = 36 - 3x^2 - 3y^2 \;\Rightarrow\; x^2 + y^2 = 9$, so the region of integration

is $D = \{(x, y) \mid x^2 + y^2 \le 9\}$. Then, in cylindrical coordinates,

$E = \{(r, \theta, z) \mid r^2 \le z \le 36 - 3r^2, 0 \le r \le 3, 0 \le \theta \le 2\pi\}$ and

$$V = \int_0^{2\pi} \int_0^3 \int_{r^2}^{36-3r^2} r \, dz \, dr \, d\theta = \int_0^{2\pi} \int_0^3 (36r - 4r^3) \, dr \, d\theta = \int_0^{2\pi} \left[18r^2 - r^4 \right]_{r=0}^{r=3} d\theta = \int_0^{2\pi} 81 \, d\theta = 162\pi.$$

(b) For constant density K, $m = KV = 162\pi K$ from part (a). Since the region is homogeneous and symmetric,

$M_{yz} = M_{xz} = 0$ and

$$M_{xy} = \int_0^{2\pi} \int_0^3 \int_{r^2}^{36-3r^2} (zK) \, r \, dz \, dr \, d\theta = K \int_0^{2\pi} \int_0^3 r \left[\tfrac{1}{2} z^2 \right]_{z=r^2}^{z=36-3r^2} dr \, d\theta$$

$$= \tfrac{K}{2} \int_0^{2\pi} \int_0^3 r((36 - 3r^2)^2 - r^4) \, dr \, d\theta = \tfrac{K}{2} \int_0^{2\pi} d\theta \int_0^3 (8r^5 - 216r^3 + 1296r) \, dr$$

$$= \tfrac{K}{2}(2\pi) \left[\tfrac{8}{6} r^6 - \tfrac{216}{4} r^4 + \tfrac{1296}{2} r^2 \right]_0^3 = \pi K(2430) = 2430\pi K$$

Thus $(\overline{x}, \overline{y}, \overline{z}) = \left(\dfrac{M_{yz}}{m}, \dfrac{M_{xz}}{m}, \dfrac{M_{xy}}{m} \right) = \left(0, 0, \dfrac{2430\pi K}{162\pi K} \right) = (0, 0, 15).$

24. (a) $V = \int_{-\pi/2}^{\pi/2} \int_0^{a\cos\theta} \int_{-\sqrt{a^2-r^2}}^{\sqrt{a^2-r^2}} r\,dz\,dr\,d\theta$

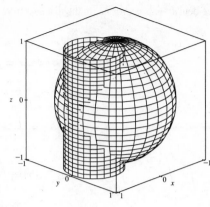

(b)

$$= 4 \int_0^{\pi/2} \int_0^{a\cos\theta} \int_0^{\sqrt{a^2-r^2}} r\,dz\,dr\,d\theta$$

$$= 4 \int_0^{\pi/2} \int_0^{a\cos\theta} r\sqrt{a^2-r^2}\,dr\,d\theta$$

$$= -\tfrac{4}{3} \int_0^{\pi/2} \left[\left(a^2-r^2\right)^{3/2} \right]_{r=0}^{r=a\cos\theta} d\theta$$

$$= -\tfrac{4}{3} \int_0^{\pi/2} \left[\left(a^2 - a^2\cos^2\theta\right)^{3/2} - a^3 \right] d\theta$$

$$= -\tfrac{4}{3} \int_0^{\pi/2} \left[\left(a^2\sin^2\theta\right)^{3/2} - a^3 \right] d\theta$$

$$= -\tfrac{4}{3} \int_0^{\pi/2} \left(a^3\sin^3\theta - a^3 \right) d\theta$$

$$= -\frac{4a^3}{3} \int_0^{\pi/2} \left[\sin\theta\left(1 - \cos^2\theta\right) - 1 \right] d\theta$$

$$= -\frac{4a^3}{3} \left[-\cos\theta + \tfrac{1}{3}\cos^3\theta - \theta \right]_0^{\pi/2} = -\frac{4a^3}{3}\left(-\tfrac{\pi}{2} + \tfrac{2}{3}\right) = \tfrac{2}{9}a^3(3\pi - 4)$$

To plot the cylinder and the sphere on the same screen in Maple, we can use the sequence of commands

```
sphere:=plot3d(sqrt(1-z^2),theta=0..2*Pi,z=-1..1,coords=cylindrical):
cylinder:=plot3d(cos(theta),theta=-Pi/2..Pi/2,z=-1..1,coords=cylindrical):
with(plots):  display3d({sphere,cylinder});
```

In Mathematica, we can use

```
sphere=ParametricPlot3D[{Sqrt[1-z^2]*Cos[theta],Sqrt[1-z^2]*Sin[theta],z},
                {theta,0,2Pi},{z-1,1}]
cylinder=ParametricPlot3D[{(Cos[theta])^2,Cos[theta]*Sin[theta],z},
                {theta,-Pi/2,Pi/2},{z,-1,1}]
Show[{sphere,cylinder}]
```

25. The paraboloid $z = 4x^2 + 4y^2$ intersects the plane $z = a$ when $a = 4x^2 + 4y^2$ or $x^2 + y^2 = \tfrac{1}{4}a$. So, in cylindrical coordinates, $E = \left\{(r,\theta,z) \mid 0 \le r \le \tfrac{1}{2}\sqrt{a}, 0 \le \theta \le 2\pi, 4r^2 \le z \le a\right\}$. Thus

$$m = \int_0^{2\pi} \int_0^{\sqrt{a}/2} \int_{4r^2}^a Kr\,dz\,dr\,d\theta = K \int_0^{2\pi} \int_0^{\sqrt{a}/2} \left(ar - 4r^3\right)\,dr\,d\theta$$

$$= K \int_0^{2\pi} \left[\tfrac{1}{2}ar^2 - r^4 \right]_{r=0}^{r=\sqrt{a}/2} d\theta = K \int_0^{2\pi} \tfrac{1}{16}a^2\,d\theta = \tfrac{1}{8}a^2\pi K$$

Since the region is homogeneous and symmetric, $M_{yz} = M_{xz} = 0$ and

$$M_{xy} = \int_0^{2\pi} \int_0^{\sqrt{a}/2} \int_{4r^2}^a Krz\,dz\,dr\,d\theta = K \int_0^{2\pi} \int_0^{\sqrt{a}/2} \left(\tfrac{1}{2}a^2 r - 8r^5\right)\,dr\,d\theta$$

$$= K \int_0^{2\pi} \left[\tfrac{1}{4}a^2 r^2 - \tfrac{4}{3}r^6 \right]_{r=0}^{r=\sqrt{a}/2} d\theta = K \int_0^{2\pi} \tfrac{1}{24}a^3\,d\theta = \tfrac{1}{12}a^3\pi K$$

Hence $(\overline{x}, \overline{y}, \overline{z}) = \left(0, 0, \tfrac{2}{3}a\right)$.

26. Since density is proportional to the distance from the z-axis, we can say $\rho(x, y, z) = K\sqrt{x^2 + y^2}$. Then

$$m = 2\int_0^{2\pi}\int_0^a\int_0^{\sqrt{a^2-r^2}} Kr^2\,dz\,dr\,d\theta = 2K\int_0^{2\pi}\int_0^a r^2\sqrt{a^2-r^2}\,dr\,d\theta$$

$$= 2K\int_0^{2\pi}\left[\tfrac{1}{8}r(2r^2-a^2)\sqrt{a^2-r^2} + \tfrac{1}{8}a^4\sin^{-1}(r/a)\right]_{r=0}^{r=a}\,d\theta = 2K\int_0^{2\pi}\left[\left(\tfrac{1}{8}a^4\right)\left(\tfrac{\pi}{2}\right)\right]\,d\theta = \tfrac{1}{4}a^4\pi^2 K$$

27. The region of integration is the region above the cone $z = \sqrt{x^2 + y^2}$, or $z = r$, and below the plane $z = 2$. Also, we have

$-2 \le y \le 2$ with $-\sqrt{4 - y^2} \le x \le \sqrt{4 - y^2}$ which describes a circle of radius 2 in the xy-plane centered at $(0, 0)$. Thus,

$$\int_{-2}^2\int_{-\sqrt{4-y^2}}^{\sqrt{4-y^2}}\int_{\sqrt{x^2+y^2}}^2 xz\,dz\,dx\,dy = \int_0^{2\pi}\int_0^2\int_r^2 (r\cos\theta)\,z\,r\,dz\,dr\,d\theta = \int_0^{2\pi}\int_0^2\int_r^2 r^2(\cos\theta)\,z\,dz\,dr\,d\theta$$

$$= \int_0^{2\pi}\int_0^2 r^2(\cos\theta)\left[\tfrac{1}{2}z^2\right]_{z=r}^{z=2}\,dr\,d\theta = \tfrac{1}{2}\int_0^{2\pi}\int_0^2 r^2(\cos\theta)\left(4 - r^2\right)\,dr\,d\theta$$

$$= \tfrac{1}{2}\int_0^{2\pi}\cos\theta\,d\theta\int_0^2 \left(4r^2 - r^4\right)\,dr = \tfrac{1}{2}\left[\sin\theta\right]_0^{2\pi}\left[\tfrac{4}{3}r^3 - \tfrac{1}{5}r^5\right]_0^2 = 0$$

28. The region of integration is the region above the plane $z = 0$ and below the paraboloid $z = 9 - x^2 - y^2$. Also, we have

$-3 \le x \le 3$ with $0 \le y \le \sqrt{9 - x^2}$ which describes the upper half of a circle of radius 3 in the xy-plane centered at $(0, 0)$.

Thus,

$$\int_{-3}^3\int_0^{\sqrt{9-x^2}}\int_0^{9-x^2-y^2}\sqrt{x^2+y^2}\,dz\,dy\,dx = \int_0^\pi\int_0^3\int_0^{9-r^2}\sqrt{r^2}\,r\,dz\,dr\,d\theta = \int_0^\pi\int_0^3\int_0^{9-r^2} r^2\,dz\,dr\,d\theta$$

$$= \int_0^\pi\int_0^3 r^2\left(9 - r^2\right)\,dr\,d\theta = \int_0^\pi d\theta\int_0^3 \left(9r^2 - r^4\right)\,dr$$

$$= \left[\theta\right]_0^\pi\left[3r^3 - \tfrac{1}{5}r^5\right]_0^3 = \pi\left(81 - \tfrac{243}{5}\right) = \tfrac{162}{5}\pi$$

29. (a) The mountain comprises a solid conical region C. The work done in lifting a small volume of material ΔV with density

$g(P)$ to a height $h(P)$ above sea level is $h(P)g(P)\,\Delta V$. Summing over the whole mountain we get

$W = \iiint_C h(P)g(P)\,dV$.

(b) Here C is a solid right circular cone with radius $R = 62{,}000$ ft, height $H = 12{,}400$ ft,

and density $g(P) = 200$ lb/ft^3 at all points P in C. We use cylindrical coordinates:

$$W = \int_0^{2\pi}\int_0^H\int_0^{R(1-z/H)} z\cdot 200r\,dr\,dz\,d\theta = 2\pi\int_0^H 200z\left[\tfrac{1}{2}r^2\right]_{r=0}^{r=R(1-z/H)}\,dz$$

$$= 400\pi\int_0^H z\,\frac{R^2}{2}\left(1 - \frac{z}{H}\right)^2\,dz = 200\pi R^2\int_0^H\left(z - \frac{2z^2}{H} + \frac{z^3}{H^2}\right)\,dz$$

$$= 200\pi R^2\left[\frac{z^2}{2} - \frac{2z^3}{3H} + \frac{z^4}{4H^2}\right]_0^H = 200\pi R^2\left(\frac{H^2}{2} - \frac{2H^2}{3} + \frac{H^2}{4}\right)$$

$$= \tfrac{50}{3}\pi R^2 H^2 = \tfrac{50}{3}\pi (62{,}000)^2(12{,}400)^2 \approx 3.1\times 10^{19}\text{ ft-lb}$$

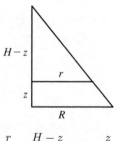

$H - z$

r

z

R

$$\frac{r}{R} = \frac{H-z}{H} = 1 - \frac{z}{H}$$

DISCOVERY PROJECT   The Intersection of Three Cylinders

1. The three cylinders in the illustration in the text can be
visualized as representing the surfaces $x^2 + y^2 = 1$,
$x^2 + z^2 = 1$, and $y^2 + z^2 = 1$. Then we sketch the solid
of intersection with the coordinate axes and equations
indicated. To be more precise, we start by finding the
bounding curves of the solid (shown in the first graph
below) enclosed by the two cylinders $x^2 + z^2 = 1$ and
$y^2 + z^2 = 1$: $x = \pm y = \pm\sqrt{1 - z^2}$ are the symmetric

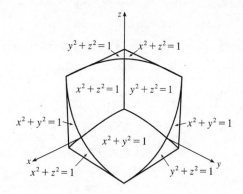

equations, and these can be expressed parametrically as $x = s$, $y = \pm s$, $z = \pm\sqrt{1 - s^2}$, $-1 \le s \le 1$. Now the cylinder
$x^2 + y^2 = 1$ intersects these curves at the eight points $\left(\pm\frac{1}{\sqrt{2}}, \pm\frac{1}{\sqrt{2}}, \pm\frac{1}{\sqrt{2}}\right)$. The resulting solid has twelve curved faces
bounded by "edges" which are arcs of circles, as shown in the third diagram. Each cylinder defines four of the twelve faces.

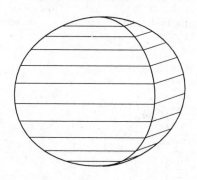
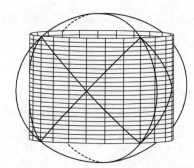
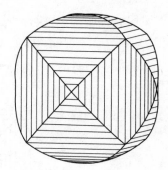

2. To find the volume, we split the solid into sixteen congruent
pieces, one of which lies in the part of the first octant with
$0 \le \theta \le \frac{\pi}{4}$. (Naturally, we use cylindrical coordinates!)
This piece is described by
$$\left\{ (r, \theta, z) \mid 0 \le r \le 1, 0 \le \theta \le \tfrac{\pi}{4}, 0 \le z \le \sqrt{1 - x^2} \right\},$$
and so, substituting $x = r\cos\theta$, the volume of the entire
solid is

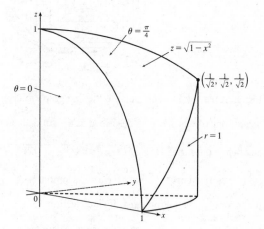

$$V = 16 \int_0^{\pi/4} \int_0^1 \int_0^{\sqrt{1-x^2}} r \, dz \, dr \, d\theta$$

$$= 16 \int_0^{\pi/4} \int_0^1 r \sqrt{1 - r^2 \cos^2\theta} \, dr \, d\theta$$

$$= 16 - 8\sqrt{2} \approx 4.6863$$

3. To graph the edges of the solid, we use parametrized curves similar to those found in Problem 1 for the intersection of two cylinders. We must restrict the parameter intervals so that each arc extends exactly to the desired vertex. One possible set of parametric equations (with all sign choices allowed) is

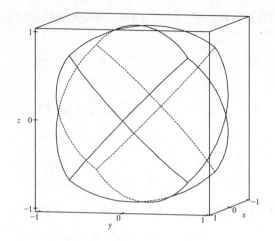

$x = r, y = \pm r, z = \pm\sqrt{1 - r^2}, -\frac{1}{\sqrt{2}} \le r \le \frac{1}{\sqrt{2}}$;

$x = \pm s, y = \pm\sqrt{1 - s^2}, z = s, -\frac{1}{\sqrt{2}} \le s \le \frac{1}{\sqrt{2}}$;

$x = \pm\sqrt{1 - t^2}, y = t, z = \pm t, -\frac{1}{\sqrt{2}} \le t \le \frac{1}{\sqrt{2}}$.

4. Let the three cylinders be $x^2 + y^2 = a^2$, $x^2 + z^2 = 1$, and $y^2 + z^2 = 1$.

If $a < 1$, then the four faces defined by the cylinder $x^2 + y^2 = 1$ in Problem 1 collapse into a single face, as in the first graph. If $1 < a < \sqrt{2}$, then each pair of vertically opposed faces, defined by one of the other two cylinders, collapse into a single face, as in the second graph. If $a \ge \sqrt{2}$, then the vertical cylinder encloses the solid of intersection of the other two cylinders completely, so the solid of intersection coincides with the solid of intersection of the two cylinders $x^2 + z^2 = 1$ and $y^2 + z^2 = 1$, as illustrated in Problem 1.

If we were to vary b or c instead of a, we would get solids with the same shape, but differently oriented.

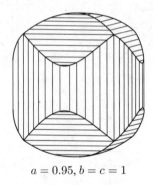

$a = 0.95, b = c = 1$

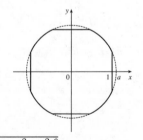

$a = 1.1, b = c = 1$

5. If $a < 1$, the solid looks similar to the first graph in Problem 4. As in Problem 2, we split the solid into sixteen congruent pieces, one of which can be described as the solid above the polar region $\left\{(r, \theta) \mid 0 \le r \le a, 0 \le \theta \le \frac{\pi}{4}\right\}$ in the xy-plane and below the surface $z = \sqrt{1 - x^2} = \sqrt{1 - r^2 \cos^2 \theta}$. Thus, the total volume is $V = 16 \int_0^{\pi/4} \int_0^a \sqrt{1 - r^2 \cos^2 \theta}\, r\, dr\, d\theta$.

If $a > 1$ and $a < \sqrt{2}$, we have a solid similar to the second graph in Problem 4. Its intersection with the xy-plane is graphed at the right. Again we split the solid into sixteen congruent pieces, one of which is the solid above the region shown in the second figure and below the surface $z = \sqrt{1 - x^2} = \sqrt{1 - r^2 \cos^2 \theta}$.

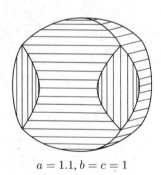

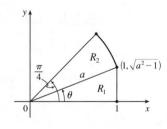

We split the region of integration where the outside boundary changes from the vertical line $x = 1$ to the circle $x^2 + y^2 = a^2$ or $r = 1$. R_1 is a right triangle, so $\cos\theta = \frac{1}{a}$. Thus, the boundary between R_1 and R_2 is $\theta = \cos^{-1}\left(\frac{1}{a}\right)$ in polar coordinates, or $y = \sqrt{a^2 - 1}\, x$ in rectangular coordinates. Using rectangular coordinates for the region R_1 and polar coordinates for R_2, we find the total volume of the solid to be

$$V = 16\left[\int_0^1 \int_0^{\sqrt{a^2-1}\,x} \sqrt{1 - x^2}\,dy\,dx + \int_{\cos^{-1}(1/a)}^{\pi/4} \int_0^a \sqrt{1 - r^2 \cos^2\theta}\, r\,dr\,d\theta\right]$$

If $a \geq \sqrt{2}$, the cylinder $x^2 + y^2 = 1$ completely encloses the intersection of the other two cylinders, so the solid of intersection of the three cylinders coincides with the intersection of $x^2 + z^2 = 1$ and $y^2 + z^2 = 1$ as illustrated in Exercise 17.6.58 [ET 16.6.58]. Its volume is $V = 16\int_0^1 \int_0^x \sqrt{1 - x^2}\,dy\,dx$.

16.8   Triple Integrals in Spherical Coordinates   ET 15.8

1. (a)

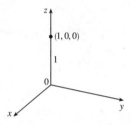

$x = \rho\sin\phi\cos\theta = (1)\sin 0\cos 0 = 0$,
$y = \rho\sin\phi\sin\theta = (1)\sin 0\sin 0 = 0$, and
$z = \rho\cos\phi = (1)\cos 0 = 1$ so the point is
$(0, 0, 1)$ in rectangular coordinates.

(b)

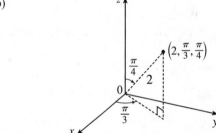

$x = 2\sin\frac{\pi}{4}\cos\frac{\pi}{3} = \frac{\sqrt{2}}{2}$, $y = 2\sin\frac{\pi}{4}\sin\frac{\pi}{3} = \frac{\sqrt{6}}{2}$,
$z = 2\cos\frac{\pi}{4} = \sqrt{2}$ so the point is $\left(\frac{\sqrt{2}}{2}, \frac{\sqrt{6}}{2}, \sqrt{2}\right)$ in rectangular coordinates.

2. (a)

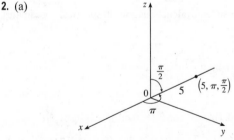

$x = 5\sin\frac{\pi}{2}\cos\pi = -5$, $y = 5\sin\frac{\pi}{2}\sin\pi = 0$,
$z = 5\cos\frac{\pi}{2} = 0$ so the point is $(-5, 0, 0)$ in
rectangular coordinates.

(b)

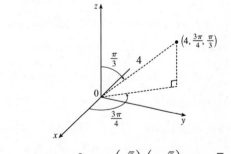

$x = 4\sin\frac{\pi}{3}\cos\frac{3\pi}{4} = 4\left(\frac{\sqrt{3}}{2}\right)\left(-\frac{\sqrt{2}}{2}\right) = -\sqrt{6}$,
$y = 4\sin\frac{\pi}{3}\sin\frac{3\pi}{4} = 4\left(\frac{\sqrt{3}}{2}\right)\left(\frac{\sqrt{2}}{2}\right) = \sqrt{6}$,
$z = 4\cos\frac{\pi}{3} = 4\left(\frac{1}{2}\right) = 2$ so the point is $\left(-\sqrt{6}, \sqrt{6}, 2\right)$
in rectangular coordinates.

3. (a) $\rho = \sqrt{x^2 + y^2 + z^2} = \sqrt{1 + 3 + 12} = 4$, $\cos\phi = \dfrac{z}{\rho} = \dfrac{2\sqrt{3}}{4} = \dfrac{\sqrt{3}}{2}$ $\Rightarrow$ $\phi = \dfrac{\pi}{6}$, and

$\cos\theta = \dfrac{x}{\rho\sin\phi} = \dfrac{1}{4\sin(\pi/6)} = \dfrac{1}{2}$ $\Rightarrow$ $\theta = \dfrac{\pi}{3}$ [since $y > 0$]. Thus spherical coordinates are $\left(4, \dfrac{\pi}{3}, \dfrac{\pi}{6}\right)$.

(b) $\rho = \sqrt{0 + 1 + 1} = \sqrt{2}$, $\cos\phi = \dfrac{-1}{\sqrt{2}}$ $\Rightarrow$ $\phi = \dfrac{3\pi}{4}$, and $\cos\theta = \dfrac{0}{\sqrt{2}\sin(3\pi/4)} = 0$ $\Rightarrow$ $\theta = \dfrac{3\pi}{2}$ [since $y < 0$].

Thus spherical coordinates are $\left(\sqrt{2}, \dfrac{3\pi}{2}, \dfrac{3\pi}{4}\right)$.

4. (a) $\rho = \sqrt{x^2 + y^2 + z^2} = \sqrt{0 + 3 + 1} = 2$, $\cos\phi = \dfrac{z}{\rho} = \dfrac{1}{2}$ $\Rightarrow$ $\phi = \dfrac{\pi}{3}$, and $\cos\theta = \dfrac{x}{\rho\sin\phi} = \dfrac{0}{2\sin(\pi/3)} = 0$ $\Rightarrow$

$\theta = \dfrac{\pi}{2}$ [since $y > 0$]. Thus spherical coordinates are $\left(2, \dfrac{\pi}{2}, \dfrac{\pi}{3}\right)$.

(b) $\rho = \sqrt{1 + 1 + 6} = 2\sqrt{2}$, $\cos\phi = \dfrac{\sqrt{6}}{2\sqrt{2}} = \dfrac{\sqrt{3}}{2}$ $\Rightarrow$ $\phi = \dfrac{\pi}{6}$, and $\cos\theta = \dfrac{-1}{2\sqrt{2}\sin(\pi/6)} = -\dfrac{1}{\sqrt{2}}$ $\Rightarrow$

$\theta = \dfrac{3\pi}{4}$ [since $y > 0$]. Thus spherical coordinates are $\left(2\sqrt{2}, \dfrac{3\pi}{4}, \dfrac{\pi}{6}\right)$.

5. Since $\phi = \dfrac{\pi}{3}$, the surface is the top half of the right circular cone with vertex at the origin and axis the positive z-axis.

6. Since $\rho = 3$, $x^2 + y^2 + z^2 = 9$ and the surface is a sphere with center the origin and radius 3.

7. $\rho = \sin\theta\sin\phi$ $\Rightarrow$ $\rho^2 = \rho\sin\theta\sin\phi$ $\Leftrightarrow$ $x^2 + y^2 + z^2 = y$ $\Leftrightarrow$ $x^2 + y^2 - y + \frac{1}{4} + z^2 = \frac{1}{4}$ $\Leftrightarrow$

$x^2 + (y - \frac{1}{2})^2 + z^2 = \frac{1}{4}$. Therefore, the surface is a sphere of radius $\frac{1}{2}$ centered at $\left(0, \frac{1}{2}, 0\right)$.

8. $\rho^2\left(\sin^2\phi\sin^2\theta + \cos^2\phi\right) = 9$ $\Leftrightarrow$ $(\rho\sin\phi\sin\theta)^2 + (\rho\cos\phi)^2 = 9$ $\Leftrightarrow$ $y^2 + z^2 = 9$. Thus the surface is a circular

cylinder of radius 3 with axis the x-axis.

9. (a) $x = \rho\sin\phi\cos\theta$, $y = \rho\sin\phi\sin\theta$, and $z = \rho\cos\phi$, so the equation $z^2 = x^2 + y^2$ becomes

$(\rho\cos\phi)^2 = (\rho\sin\phi\cos\theta)^2 + (\rho\sin\phi\sin\theta)^2$ or $\rho^2\cos^2\phi = \rho^2\sin^2\phi$. If $\rho \neq 0$, this becomes $\cos^2\phi = \sin^2\phi$. ($\rho = 0$

corresponds to the origin which is included in the surface.) There are many equivalent equations in spherical coordinates,

such as $\tan^2\phi = 1$, $2\cos^2\phi = 1$, $\cos 2\phi = 0$, or even $\phi = \dfrac{\pi}{4}$, $\phi = \dfrac{3\pi}{4}$.

(b) $x^2 + z^2 = 9$ $\Leftrightarrow$ $(\rho\sin\phi\cos\theta)^2 + (\rho\cos\phi)^2 = 9$ $\Leftrightarrow$ $\rho^2\sin^2\phi\cos^2\theta + \rho^2\cos^2\phi = 9$ or

$\rho^2\left(\sin^2\phi\cos^2\theta + \cos^2\phi\right) = 9$.

10. (a) $x^2 - 2x + y^2 + z^2 = 0$ $\Leftrightarrow$ $(x^2 + y^2 + z^2) - 2x = 0$ $\Leftrightarrow$ $\rho^2 - 2(\rho\sin\phi\cos\theta) = 0$ or $\rho = 2\sin\phi\cos\theta$.

(b) $x + 2y + 3z = 1$ $\Leftrightarrow$ $\rho\sin\phi\cos\theta + 2\rho\sin\phi\sin\theta + 3\rho\cos\phi = 1$ or $\rho = 1/(\sin\phi\cos\theta + 2\sin\phi\sin\theta + 3\cos\phi)$.

11. $\rho = 2$ represents a sphere of radius 2, centered at the origin, so $\rho \leq 2$ is this

sphere and its interior. $0 \leq \phi \leq \dfrac{\pi}{2}$ restricts the solid to that portion of the

region that lies on or above the xy-plane, and $0 \leq \theta \leq \dfrac{\pi}{2}$ further restricts the

solid to the first octant. Thus the solid is the portion in the first octant of the

solid ball centered at the origin with radius 2.

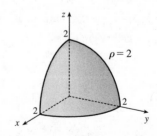

12. $2 \leq \rho \leq 3$ represents the solid region between and including the spheres of
radii 2 and 3, centered at the origin. $\frac{\pi}{2} \leq \phi \leq \pi$ restricts the solid to that
portion on or below the xy-plane.

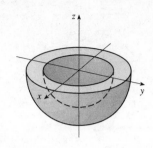

13. $\rho \leq 1$ represents the solid sphere of radius 1 centered at the origin.
$\frac{3\pi}{4} \leq \phi \leq \pi$ restricts the solid to that portion on or below the cone $\phi = \frac{3\pi}{4}$.

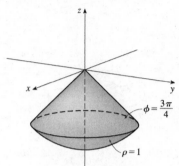

14. $\rho \leq 2$ represents the solid sphere of radius 2 centered at the origin. Notice
that $x^2 + y^2 = (\rho \sin \phi \cos \theta)^2 + (\rho \sin \phi \sin \theta)^2 = \rho^2 \sin^2 \phi$. Then
$\rho = \csc \phi \;\Rightarrow\; \rho \sin \phi = 1 \;\Rightarrow\; \rho^2 \sin^2 \phi = x^2 + y^2 = 1$, so $\rho \leq \csc \phi$
restricts the solid to that portion on or inside the circular cylinder
$x^2 + y^2 = 1$.

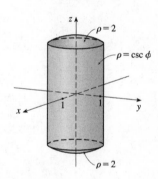

15. $z \geq \sqrt{x^2 + y^2}$ because the solid lies above the cone. Squaring both sides of this inequality gives $z^2 \geq x^2 + y^2 \;\Rightarrow\;$
$2z^2 \geq x^2 + y^2 + z^2 = \rho^2 \;\Rightarrow\; z^2 = \rho^2 \cos^2 \phi \geq \frac{1}{2}\rho^2 \;\Rightarrow\; \cos^2 \phi \geq \frac{1}{2}$. The cone opens upward so that the inequality is
$\cos \phi \geq \frac{1}{\sqrt{2}}$, or equivalently $0 \leq \phi \leq \frac{\pi}{4}$. In spherical coordinates the sphere $z = x^2 + y^2 + z^2$ is $\rho \cos \phi = \rho^2 \;\Rightarrow\;$
$\rho = \cos \phi$. $0 \leq \rho \leq \cos \phi$ because the solid lies below the sphere. The solid can therefore be described as the region in
spherical coordinates satisfying $0 \leq \rho \leq \cos \phi, 0 \leq \phi \leq \frac{\pi}{4}$.

16. (a) The hollow ball is a spherical shell with outer radius 15 cm and inner radius 14.5 cm. If we center the ball at the origin of
the coordinate system and use centimeters as the unit of measurement, then spherical coordinates conveniently describe the
hollow ball as $14.5 \leq \rho \leq 15, 0 \leq \theta \leq 2\pi, 0 \leq \phi \leq \pi$.

(b) If we position the ball as in part (a), one possibility is to take the half of the ball that is above the xy-plane which is
described by $14.5 \leq \rho \leq 15, 0 \leq \theta \leq 2\pi, 0 \leq \phi \leq \pi/2$.

17.

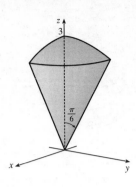

The region of integration is given in spherical coordinates by
$E = \{(\rho, \theta, \phi) \mid 0 \le \rho \le 3, 0 \le \theta \le \pi/2, 0 \le \phi \le \pi/6\}$. This represents the solid region in the first octant bounded above by the sphere $\rho = 3$ and below by the cone $\phi = \pi/6$.

$$\int_0^{\pi/6} \int_0^{\pi/2} \int_0^3 \rho^2 \sin\phi \, d\rho \, d\theta \, d\phi = \int_0^{\pi/6} \sin\phi \, d\phi \int_0^{\pi/2} d\theta \int_0^3 \rho^2 \, d\rho$$

$$= \left[-\cos\phi\right]_0^{\pi/6} \left[\theta\right]_0^{\pi/2} \left[\tfrac{1}{3}\rho^3\right]_0^3$$

$$= \left(1 - \frac{\sqrt{3}}{2}\right)\left(\frac{\pi}{2}\right)(9) = \frac{9\pi}{4}\left(2 - \sqrt{3}\right)$$

18.

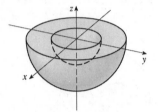

The region of integration is given in spherical coordinates by
$E = \{(\rho, \theta, \phi) \mid 1 \le \rho \le 2, 0 \le \theta \le 2\pi, \pi/2 \le \phi \le \pi\}$. This represents the solid region between the spheres $\rho = 1$ and $\rho = 2$ and below the xy-plane.

$$\int_0^{2\pi} \int_{\pi/2}^{\pi} \int_1^2 \rho^2 \sin\phi \, d\rho \, d\phi \, d\theta = \int_0^{2\pi} d\theta \int_{\pi/2}^{\pi} \sin\phi \, d\phi \int_1^2 \rho^2 \, d\rho$$

$$= \left[\theta\right]_0^{2\pi} \left[-\cos\phi\right]_{\pi/2}^{\pi} \left[\tfrac{1}{3}\rho^3\right]_1^2$$

$$= 2\pi(1)\left(\tfrac{7}{3}\right) = \tfrac{14\pi}{3}$$

19. The solid E is most conveniently described if we use cylindrical coordinates:
$E = \{(r, \theta, z) \mid 0 \le \theta \le \frac{\pi}{2}, 0 \le r \le 3, 0 \le z \le 2\}$. Then

$$\iiint_E f(x, y, z) \, dV = \int_0^{\pi/2} \int_0^3 \int_0^2 f(r\cos\theta, r\sin\theta, z) \, r \, dz \, dr \, d\theta.$$

20. The solid E is most conveniently described if we use spherical coordinates:
$E = \{(\rho, \theta, \phi) \mid 1 \le \rho \le 2, \frac{\pi}{2} \le \theta \le 2\pi, 0 \le \phi \le \frac{\pi}{2}\}$. Then

$$\iiint_E f(x, y, z) \, dV = \int_0^{\pi/2} \int_{\pi/2}^{2\pi} \int_1^2 f(\rho\sin\phi\cos\theta, \rho\sin\phi\sin\theta, \rho\cos\phi) \, \rho^2 \sin\phi \, d\rho \, d\theta \, d\phi.$$

21. In spherical coordinates, B is represented by $\{(\rho, \theta, \phi) \mid 0 \le \rho \le 5, 0 \le \theta \le 2\pi, 0 \le \phi \le \pi\}$. Thus

$$\iiint_B (x^2 + y^2 + z^2)^2 \, dV = \int_0^{\pi} \int_0^{2\pi} \int_0^5 (\rho^2)^2 \rho^2 \sin\phi \, d\rho \, d\theta \, d\phi = \int_0^{\pi} \sin\phi \, d\phi \int_0^{2\pi} d\theta \int_0^5 \rho^6 \, d\rho$$

$$= \left[-\cos\phi\right]_0^{\pi} \left[\theta\right]_0^{2\pi} \left[\tfrac{1}{7}\rho^7\right]_0^5 = (2)(2\pi)\left(\tfrac{78,125}{7}\right)$$

$$= \tfrac{312,500}{7}\pi \approx 140,249.7$$

22. In spherical coordinates, H is represented by $\left\{(\rho, \theta, \phi) \mid 0 \le \rho \le 3, 0 \le \theta \le 2\pi, 0 \le \phi \le \frac{\pi}{2}\right\}$. Thus

$$\iiint_H (9 - x^2 - y^2) \, dV = \int_0^{\pi/2} \int_0^{2\pi} \int_0^3 \left[9 - (\rho^2 \sin^2\phi \cos^2\theta + \rho^2 \sin^2\phi \sin^2\theta)\right] \rho^2 \sin\phi \, d\rho \, d\theta \, d\phi$$

$$= \int_0^{\pi/2} \int_0^{2\pi} \int_0^3 (9 - \rho^2 \sin^2\phi) \rho^2 \sin\phi \, d\rho \, d\theta \, d\phi$$

$$= \int_0^{\pi/2} \int_0^{2\pi} \left[3\rho^3 - \tfrac{1}{5}\rho^5 \sin^2\phi\right]_{\rho=0}^{\rho=3} \sin\phi \, d\theta \, d\phi$$

$$= \int_0^{\pi/2} \int_0^{2\pi} \left(81\sin\phi - \tfrac{243}{5}\sin^3\phi\right) d\theta \, d\phi$$

$$= \int_0^{2\pi} d\theta \int_0^{\pi/2} \left[81\sin\phi - \tfrac{243}{5}(1 - \cos^2\phi)\sin\phi\right] d\phi$$

$$= 2\pi\left[-81\cos\phi - \tfrac{243}{5}\left(\tfrac{1}{3}\cos^3\phi - \cos\phi\right)\right]_0^{\pi/2}$$

$$= 2\pi\left[0 + 81 + \tfrac{243}{5}\left(-\tfrac{2}{3}\right)\right] = \tfrac{486}{5}\pi$$

23. In spherical coordinates, E is represented by $\{(\rho, \theta, \phi) \mid 1 \le \rho \le 2, 0 \le \theta \le \frac{\pi}{2}, 0 \le \phi \le \frac{\pi}{2}\}$. Thus

$$\iiint_E z \, dV = \int_0^{\pi/2} \int_0^{\pi/2} \int_1^2 (\rho \cos \phi) \, \rho^2 \sin \phi \, d\rho \, d\theta \, d\phi = \int_0^{\pi/2} \cos \phi \sin \phi \, d\phi \int_0^{\pi/2} d\theta \int_1^2 \rho^3 \, d\rho$$

$$= \left[\tfrac{1}{2} \sin^2 \phi \right]_0^{\pi/2} \left[\theta \right]_0^{\pi/2} \left[\tfrac{1}{4} \rho^4 \right]_1^2 = \left(\tfrac{1}{2} \right) \left(\tfrac{\pi}{2} \right) \left(\tfrac{15}{4} \right) = \tfrac{15\pi}{16}$$

24. $\iiint_E e^{\sqrt{x^2+y^2+z^2}} \, dV = \int_0^{\pi/2} \int_0^{\pi/2} \int_0^3 e^{\sqrt{\rho^2}} \rho^2 \sin \phi \, d\rho \, d\phi \, d\theta = \int_0^{\pi/2} \int_0^{\pi/2} \int_0^3 \rho^2 e^{\rho} \sin \phi \, d\rho \, d\phi \, d\theta$

$$= \int_0^{\pi/2} d\theta \int_0^{\pi/2} \sin \phi \, d\phi \int_0^3 \rho^2 e^{\rho} \, d\rho = \left[\theta \right]_0^{\pi/2} \left[-\cos \phi \right]_0^{\pi/2} \left[(\rho^2 - 2\rho + 2) e^{\rho} \right]_0^3 \quad \begin{bmatrix} \text{integrate by} \\ \text{parts twice} \end{bmatrix}$$

$$= \tfrac{\pi}{2}(0+1)(5e^3 - 2) = \tfrac{\pi}{2}(5e^3 - 2)$$

25. $\iiint_E x^2 \, dV = \int_0^\pi \int_0^\pi \int_3^4 (\rho \sin \phi \cos \theta)^2 \, \rho^2 \sin \phi \, d\rho \, d\phi \, d\theta = \int_0^\pi \cos^2 \theta \, d\theta \int_0^\pi \sin^3 \phi \, d\phi \int_3^4 \rho^4 \, d\rho$

$$= \left[\tfrac{1}{2}\theta + \tfrac{1}{4} \sin 2\theta \right]_0^\pi \left[-\tfrac{1}{3}(2 + \sin^2 \phi) \cos \phi \right]_0^\pi \left[\tfrac{1}{5} \rho^5 \right]_3^4 = \left(\tfrac{\pi}{2} \right) \left(\tfrac{2}{3} + \tfrac{2}{3} \right) \tfrac{1}{5}(4^5 - 3^5) = \tfrac{1562}{15} \pi$$

26. $\iiint_E xyz \, dV = \int_0^{\pi/3} \int_0^{2\pi} \int_2^4 (\rho \sin \phi \cos \theta)(\rho \sin \phi \sin \theta)(\rho \cos \phi) \, \rho^2 \sin \phi \, d\rho \, d\theta \, d\phi$

$$= \int_0^{\pi/3} \sin^3 \phi \cos \phi \, d\phi \int_0^{2\pi} \sin \theta \cos \theta \, d\theta \int_2^4 \rho^5 \, d\rho = \left[\tfrac{1}{4} \sin^4 \phi \right]_0^{\pi/3} \left[\tfrac{1}{2} \sin^2 \theta \right]_0^{2\pi} \left[\tfrac{1}{6} \rho^6 \right]_2^4 = 0$$

27. The solid region is given by $E = \{(\rho, \theta, \phi) \mid 0 \le \rho \le a, 0 \le \theta \le 2\pi, \frac{\pi}{6} \le \phi \le \frac{\pi}{3}\}$ and its volume is

$$V = \iiint_E dV = \int_{\pi/6}^{\pi/3} \int_0^{2\pi} \int_0^a \rho^2 \sin \phi \, d\rho \, d\theta \, d\phi = \int_{\pi/6}^{\pi/3} \sin \phi \, d\phi \int_0^{2\pi} d\theta \int_0^a \rho^2 \, d\rho$$

$$= \left[-\cos \phi \right]_{\pi/6}^{\pi/3} \left[\theta \right]_0^{2\pi} \left[\tfrac{1}{3} \rho^3 \right]_0^a = \left(-\tfrac{1}{2} + \tfrac{\sqrt{3}}{2} \right) (2\pi) \left(\tfrac{1}{3} a^3 \right) = \tfrac{\sqrt{3}-1}{3} \pi a^3$$

28. If we center the ball at the origin, then the ball is given by

$B = \{(\rho, \theta, \phi) \mid 0 \le \rho \le a, 0 \le \theta \le 2\pi, 0 \le \phi \le \pi\}$ and the distance from any point (x, y, z) in the ball to the center $(0, 0, 0)$ is $\sqrt{x^2 + y^2 + z^2} = \rho$. Thus the average distance is

$$\frac{1}{V(B)} \iiint_B \rho \, dV = \frac{1}{\frac{4}{3}\pi a^3} \int_0^\pi \int_0^{2\pi} \int_0^a \rho \cdot \rho^2 \sin \phi \, d\rho \, d\theta \, d\phi = \frac{3}{4\pi a^3} \int_0^\pi \sin \phi \, d\phi \int_0^{2\pi} d\theta \int_0^a \rho^3 \, d\rho$$

$$= \frac{3}{4\pi a^3} \left[-\cos \phi \right]_0^\pi \left[\theta \right]_0^{2\pi} \left[\tfrac{1}{4} \rho^4 \right]_0^a = \frac{3}{4\pi a^3} (2)(2\pi) \left(\tfrac{1}{4} a^4 \right) = \tfrac{3}{4} a$$

29. (a) Since $\rho = 4 \cos \phi$ implies $\rho^2 = 4\rho \cos \phi$, the equation is that of a sphere of radius 2 with center at $(0, 0, 2)$. Thus

$$V = \int_0^{2\pi} \int_0^{\pi/3} \int_0^{4\cos\phi} \rho^2 \sin \phi \, d\rho \, d\phi \, d\theta = \int_0^{2\pi} \int_0^{\pi/3} \left[\tfrac{1}{3} \rho^3 \right]_{\rho=0}^{\rho=4\cos\phi} \sin \phi \, d\phi \, d\theta = \int_0^{2\pi} \int_0^{\pi/3} \left(\tfrac{64}{3} \cos^3 \phi \right) \sin \phi \, d\phi \, d\theta$$

$$= \int_0^{2\pi} \left[-\tfrac{16}{3} \cos^4 \phi \right]_{\phi=0}^{\phi=\pi/3} d\theta = \int_0^{2\pi} -\tfrac{16}{3} \left(\tfrac{1}{16} - 1 \right) d\theta = 5\theta \Big]_0^{2\pi} = 10\pi$$

(b) By the symmetry of the problem $M_{yz} = M_{xz} = 0$. Then

$$M_{xy} = \int_0^{2\pi} \int_0^{\pi/3} \int_0^{4\cos\phi} \rho^3 \cos \phi \sin \phi \, d\rho \, d\phi \, d\theta = \int_0^{2\pi} \int_0^{\pi/3} \cos \phi \sin \phi \left(64 \cos^4 \phi \right) d\phi \, d\theta$$

$$= \int_0^{2\pi} 64 \left[-\tfrac{1}{6} \cos^6 \phi \right]_{\phi=0}^{\phi=\pi/3} d\theta = \int_0^{2\pi} \tfrac{21}{2} \, d\theta = 21\pi$$

Hence $(\overline{x}, \overline{y}, \overline{z}) = (0, 0, 2.1)$.

30. In spherical coordinates, the sphere $x^2 + y^2 + z^2 = 4$ is equivalent to $\rho = 2$ and the cone $z = \sqrt{x^2 + y^2}$ is represented

by $\phi = \frac{\pi}{4}$. Thus, the solid is given by $\left\{ (\rho, \theta, \phi) \mid 0 \le \rho \le 2, 0 \le \theta \le 2\pi, \frac{\pi}{4} \le \phi \le \frac{\pi}{2} \right\}$ and

$$V = \int_{\pi/4}^{\pi/2} \int_0^{2\pi} \int_0^2 \rho^2 \sin \phi \, d\rho \, d\theta \, d\phi = \int_{\pi/4}^{\pi/2} \sin \phi \, d\phi \int_0^{2\pi} d\theta \int_0^2 \rho^2 \, d\rho$$

$$= \left[-\cos \phi \right]_{\pi/4}^{\pi/2} \left[\theta \right]_0^{2\pi} \left[\tfrac{1}{3}\rho^3 \right]_0^2 = \left(\tfrac{\sqrt{2}}{2} \right)(2\pi)\left(\tfrac{8}{3} \right) = \tfrac{8\sqrt{2}\,\pi}{3}$$

31. By the symmetry of the region, $M_{xy} = 0$ and $M_{yz} = 0$. Assuming constant density K,

$$m = \iiint_E KV = K \int_0^\pi \int_0^\pi \int_3^4 \rho^2 \sin \phi \, d\rho \, d\phi \, d\theta = K \int_0^\pi d\theta \int_0^\pi \sin \phi \, d\phi \int_3^4 \rho^2 \, d\rho$$

$$= K\pi \left[-\cos \phi \right]_0^\pi \left[\tfrac{1}{3}\rho^3 \right]_3^4 = 2K\pi \cdot \tfrac{37}{3} = \tfrac{74}{3}\pi K$$

and $\quad M_{xz} = \iiint_E y K \, dV = K \int_0^\pi \int_0^\pi \int_3^4 (\rho \sin \phi \sin \theta) \, \rho^2 \sin \phi \, d\rho \, d\phi \, d\theta = K \int_0^\pi \sin \theta \, d\theta \int_0^\pi \sin^2 \phi \, d\phi \int_3^4 \rho^3 \, d\rho$

$$= K \left[-\cos \theta \right]_0^\pi \left[\tfrac{1}{2}\phi - \tfrac{1}{4}\sin 2\phi \right]_0^\pi \left[\tfrac{1}{4}\rho^4 \right]_3^4 = K(2)\left(\tfrac{\pi}{2} \right)\tfrac{1}{4}(256 - 81) = \tfrac{175}{4}\pi K$$

Thus the centroid is $(\overline{x}, \overline{y}, \overline{z}) = \left(\dfrac{M_{yz}}{m}, \dfrac{M_{xz}}{m}, \dfrac{M_{xy}}{m} \right) = \left(0, \dfrac{175\pi K/4}{74\pi K/3}, 0 \right) = \left(0, \tfrac{525}{296}, 0 \right)$.

32. (a) Placing the center of the base at $(0, 0, 0)$, $\rho(x, y, z) = K \sqrt{x^2 + y^2 + z^2}$ is the density function. So

$$m = \int_0^{2\pi} \int_0^{\pi/2} \int_0^a K\rho^3 \sin \phi \, d\rho \, d\phi \, d\theta = K \int_0^{2\pi} d\theta \int_0^{\pi/2} \sin \phi \, d\phi \int_0^a \rho^3 \, d\rho$$

$$= K \left[\theta \right]_0^{2\pi} \left[-\cos \phi \right]_0^{\pi/2} \left[\tfrac{1}{4}\rho^4 \right]_0^a = K(2\pi)(1)\left(\tfrac{1}{4}a^4 \right) = \tfrac{1}{2}\pi K a^4$$

(b) By the symmetry of the problem $M_{yz} = M_{xz} = 0$. Then

$$M_{xy} = \int_0^{2\pi} \int_0^{\pi/2} \int_0^a K\rho^4 \sin \phi \cos \phi \, d\rho \, d\phi \, d\theta = K \int_0^{2\pi} d\theta \int_0^{\pi/2} \sin \phi \cos \phi \, d\phi \int_0^a \rho^4 \, d\rho$$

$$= K \left[\theta \right]_0^{2\pi} \left[\tfrac{1}{2} \sin^2 \phi \right]_0^{\pi/2} \left[\tfrac{1}{5}\rho^5 \right]_0^a = K(2\pi)\left(\tfrac{1}{2} \right)\left(\tfrac{1}{5}a^5 \right) = \tfrac{1}{5}\pi K a^5$$

Hence $(\overline{x}, \overline{y}, \overline{z}) = \left(0, 0, \tfrac{2}{5}a \right)$.

(c) $I_z = \int_0^{2\pi} \int_0^{\pi/2} \int_0^a (K\rho^3 \sin \phi)(\rho^2 \sin^2 \phi) \, d\rho \, d\phi \, d\theta = K \int_0^{2\pi} d\theta \int_0^{\pi/2} \sin^3 \phi \, d\phi \int_0^a \rho^5 \, d\rho$

$$= K \left[\theta \right]_0^{2\pi} \left[-\cos \phi + \tfrac{1}{3} \cos^3 \phi \right]_0^{\pi/2} \left[\tfrac{1}{6}\rho^6 \right]_0^a = K(2\pi)\left(\tfrac{2}{3} \right)\left(\tfrac{1}{6}a^6 \right) = \tfrac{2}{9}\pi K a^6$$

33. (a) The density function is $\rho(x, y, z) = K$, a constant, and by the symmetry of the problem $M_{xz} = M_{yz} = 0$. Then

$M_{xy} = \int_0^{2\pi} \int_0^{\pi/2} \int_0^a K\rho^3 \sin \phi \, \cos \phi \, d\rho \, d\phi \, d\theta = \tfrac{1}{2}\pi K a^4 \int_0^{\pi/2} \sin \phi \, \cos \phi \, d\phi = \tfrac{1}{8}\pi K a^4$. But the mass is K(volume of

the hemisphere) $= \tfrac{2}{3}\pi K a^3$, so the centroid is $\left(0, 0, \tfrac{3}{8}a \right)$.

(b) Place the center of the base at $(0, 0, 0)$; the density function is $\rho(x, y, z) = K$. By symmetry, the moments of inertia about

any two such diameters will be equal, so we just need to find I_x:

$$I_x = \int_0^{2\pi} \int_0^{\pi/2} \int_0^a (K\rho^2 \sin \phi) \, \rho^2 \, (\sin^2 \phi \, \sin^2 \theta + \cos^2 \phi) \, d\rho \, d\phi \, d\theta$$

$$= K \int_0^{2\pi} \int_0^{\pi/2} (\sin^3 \phi \, \sin^2 \theta + \sin \phi \cos^2 \phi)\left(\tfrac{1}{5}a^5 \right) d\phi \, d\theta$$

$$= \tfrac{1}{5}K a^5 \int_0^{2\pi} \left[\sin^2 \theta \left(-\cos \phi + \tfrac{1}{3} \cos^3 \phi \right) + \left(-\tfrac{1}{3} \cos^3 \phi \right) \right]_{\phi=0}^{\phi=\pi/2} d\theta = \tfrac{1}{5}K a^5 \int_0^{2\pi} \left[\tfrac{2}{3} \sin^2 \theta + \tfrac{1}{3} \right] d\theta$$

$$= \tfrac{1}{5}K a^5 \left[\tfrac{2}{3}\left(\tfrac{1}{2}\theta - \tfrac{1}{4} \sin 2\theta \right) + \tfrac{1}{3}\theta \right]_0^{2\pi} = \tfrac{1}{5}K a^5 \left[\tfrac{2}{3}(\pi - 0) + \tfrac{1}{3}(2\pi - 0) \right] = \tfrac{4}{15}K a^5 \pi$$

34. Place the center of the base at $(0, 0, 0)$, then the density is $\rho(x, y, z) = Kz$, K a constant. Then

$$m = \int_0^{2\pi} \int_0^{\pi/2} \int_0^a (K\rho\cos\phi)\,\rho^2 \sin\phi\, d\rho\, d\phi\, d\theta = 2\pi K \int_0^{\pi/2} \cos\phi\sin\phi \cdot \tfrac{1}{4}a^4\, d\phi = \tfrac{1}{2}\pi K a^4 \left[-\tfrac{1}{4}\cos 2\phi\right]_0^{\pi/2} = \tfrac{\pi}{4}K a^4.$$

By the symmetry of the problem $M_{xz} = M_{yz} = 0$, and

$$M_{xy} = \int_0^{2\pi} \int_0^{\pi/2} \int_0^a K\rho^4 \cos^2\phi\sin\phi\, d\rho\, d\phi\, d\theta = \tfrac{2}{5}\pi K a^5 \int_0^{\pi/2} \cos^2\phi\sin\phi\, d\phi = \tfrac{2}{5}\pi K a^5 \left[-\tfrac{1}{3}\cos^3\theta\right]_0^{\pi/2} = \tfrac{2}{15}\pi K a^5.$$

Hence $(\overline{x}, \overline{y}, \overline{z}) = \left(0, 0, \tfrac{8}{15}a\right)$.

35. In spherical coordinates $z = \sqrt{x^2 + y^2}$ becomes $\cos\phi = \sin\phi$ or $\phi = \tfrac{\pi}{4}$. Then

$$V = \int_0^{2\pi} \int_0^{\pi/4} \int_0^1 \rho^2 \sin\phi\, d\rho\, d\phi\, d\theta = \int_0^{2\pi} d\theta \int_0^{\pi/4} \sin\phi\, d\phi \int_0^1 \rho^2\, d\rho = \tfrac{1}{3}\pi\left(2 - \sqrt{2}\right),$$

$$M_{xy} = \int_0^{2\pi} \int_0^{\pi/4} \int_0^1 \rho^3 \sin\phi\cos\phi\, d\rho\, d\phi\, d\theta = 2\pi\left[-\tfrac{1}{4}\cos 2\phi\right]_0^{\pi/4}\left(\tfrac{1}{4}\right) = \tfrac{\pi}{8}$$ and by symmetry $M_{yz} = M_{xz} = 0$.

Hence $(\overline{x}, \overline{y}, \overline{z}) = \left(0, 0, \dfrac{3}{8\left(2 - \sqrt{2}\right)}\right)$.

36. Place the center of the sphere at $(0, 0, 0)$, let the diameter of intersection be along the z-axis, one of the planes be the xz-plane and the other be the plane whose angle with the xz-plane is $\theta = \tfrac{\pi}{6}$. Then in spherical coordinates the volume is given by

$$V = \int_0^{\pi/6} \int_0^{\pi} \int_0^a \rho^2 \sin\phi\, d\rho\, d\phi\, d\theta = \int_0^{\pi/6} d\theta \int_0^{\pi} \sin\phi\, d\phi \int_0^a \rho^2\, d\rho = \tfrac{\pi}{6}(2)\left(\tfrac{1}{3}a^3\right) = \tfrac{1}{9}\pi a^3.$$

37. In cylindrical coordinates the paraboloid is given by $z = r^2$ and the plane by $z = 2r\sin\theta$ and they intersect in the circle $r = 2\sin\theta$. Then $\iiint_E z\, dV = \int_0^{\pi} \int_0^{2\sin\theta} \int_{r^2}^{2r\sin\theta} rz\, dz\, dr\, d\theta = \tfrac{5\pi}{6}$ [using a CAS].

38. (a) The region enclosed by the torus is $\{(\rho, \theta, \phi) \mid 0 \le \theta \le 2\pi, 0 \le \phi \le \pi, 0 \le \rho \le \sin\phi\}$, so its volume is

$$V = \int_0^{2\pi} \int_0^{\pi} \int_0^{\sin\phi} \rho^2 \sin\phi\, d\rho\, d\phi\, d\theta = 2\pi \int_0^{\pi} \tfrac{1}{3}\sin^4\phi\, d\phi = \tfrac{2}{3}\pi\left[\tfrac{3}{8}\phi - \tfrac{1}{4}\sin 2\phi + \tfrac{1}{16}\sin 4\phi\right]_0^{\pi} = \tfrac{1}{4}\pi^2.$$

(b) In Maple, we can plot the torus using the

`coords=spherical` option in a regular `plot3d`

command. In Mathematica, use `ParametricPlot3D`.

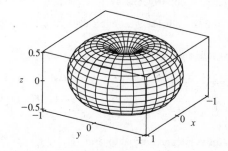

39. The region E of integration is the region above the cone $z = \sqrt{x^2 + y^2}$ and below the sphere $x^2 + y^2 + z^2 = 2$ in the first octant. Because E is in the first octant we have $0 \le \theta \le \tfrac{\pi}{2}$. The cone has equation $\phi = \tfrac{\pi}{4}$ (as in Example 4), so $0 \le \phi \le \tfrac{\pi}{4}$, and $0 \le \rho \le \sqrt{2}$. So the integral becomes

$$\int_0^{\pi/4} \int_0^{\pi/2} \int_0^{\sqrt{2}} (\rho\sin\phi\cos\theta)(\rho\sin\phi\sin\theta)\,\rho^2 \sin\phi\, d\rho\, d\theta\, d\phi$$

$$= \int_0^{\pi/4} \sin^3\phi\, d\phi \int_0^{\pi/2} \sin\theta\cos\theta\, d\theta \int_0^{\sqrt{2}} \rho^4\, d\rho = \left(\int_0^{\pi/4} (1 - \cos^2\phi)\sin\phi\, d\phi\right)\left[\tfrac{1}{2}\sin^2\theta\right]_0^{\pi/2}\left[\tfrac{1}{5}\rho^5\right]_0^{\sqrt{2}}$$

$$= \left[\tfrac{1}{3}\cos^3\phi - \cos\phi\right]_0^{\pi/4} \cdot \tfrac{1}{2} \cdot \tfrac{1}{5}\left(\sqrt{2}\right)^5 = \left[\tfrac{\sqrt{2}}{12} - \tfrac{\sqrt{2}}{2} - \left(\tfrac{1}{3} - 1\right)\right] \cdot \tfrac{2\sqrt{2}}{5} = \tfrac{4\sqrt{2} - 5}{15}$$

40. The region of integration is the solid sphere $x^2 + y^2 + z^2 \le a^2$, so $0 \le \theta \le 2\pi$, $0 \le \phi \le \pi$, and $0 \le \rho \le a$. Also

$x^2 z + y^2 z + z^3 = (x^2 + y^2 + z^2)z = \rho^2 z = \rho^3 \cos\phi$, so the integral becomes

$$\int_0^\pi \int_0^{2\pi} \int_0^a \left(\rho^3 \cos\phi\right) \rho^2 \sin\phi \, d\rho \, d\theta \, d\phi = \int_0^\pi \sin\phi\cos\phi \, d\phi \int_0^{2\pi} d\theta \int_0^a \rho^5 \, d\rho = \left[\tfrac{1}{2}\sin^2\phi\right]_0^\pi \left[\theta\right]_0^{2\pi} \left[\tfrac{1}{6}\rho^6\right]_0^a = 0$$

41. In cylindrical coordinates, the equation of the cylinder is $r = 3$, $0 \le z \le 10$.

The hemisphere is the upper part of the sphere radius 3, center $(0, 0, 10)$, equation

$r^2 + (z - 10)^2 = 3^2$, $z \ge 10$. In Maple, we can use the `coords=cylindrical` option

in a regular `plot3d` command. In Mathematica, we can use `ParametricPlot3D`.

42. We begin by finding the positions of Los Angeles and Montréal in spherical coordinates, using the method described in the exercise:

Montréal	Los Angeles
$\rho = 3960$ mi	$\rho = 3960$ mi
$\theta = 360° - 73.60° = 286.40°$	$\theta = 360° - 118.25° = 241.75°$
$\phi = 90° - 45.50° = 44.50°$	$\phi = 90° - 34.06° = 55.94°$

Now we change the above to Cartesian coordinates using $x = \rho\cos\theta\sin\phi$, $y = \rho\sin\theta\sin\phi$ and $z = \rho\cos\phi$ to get two

position vectors of length 3960 mi (since both cities must lie on the surface of the Earth). In particular:

Montréal: $\langle 783.67, -2662.67, 2824.47 \rangle$ Los Angeles: $\langle -1552.80, -2889.91, 2217.84 \rangle$

To find the angle α between these two vectors we use the dot product:

$\langle 783.67, -2662.67, 2824.47 \rangle \cdot \langle -1552.80, -2889.91, 2217.84 \rangle = (3960)^2 \cos\alpha \quad \Rightarrow \quad \cos\alpha \approx 0.8126 \quad \Rightarrow$

$\alpha \approx 0.6223$ rad. The great circle distance between the cities is $s = \rho\theta \approx 3960(0.6223) \approx 2464$ mi.

43. If E is the solid enclosed by the surface $\rho = 1 + \frac{1}{5}\sin 6\theta \sin 5\phi$, it can be described in spherical coordinates as

$E = \left\{(\rho, \theta, \phi) \mid 0 \le \rho \le 1 + \frac{1}{5}\sin 6\theta \sin 5\phi, 0 \le \theta \le 2\pi, 0 \le \phi \le \pi\right\}$. Its volume is given by

$V(E) = \iiint_E dV = \int_0^\pi \int_0^{2\pi} \int_0^{1+(\sin 6\theta \sin 5\phi)/5} \rho^2 \sin\phi \, d\rho \, d\theta \, d\phi = \frac{136\pi}{99}$ [using a CAS].

44. The given integral is equal to $\displaystyle\lim_{R\to\infty} \int_0^{2\pi} \int_0^\pi \int_0^R \rho e^{-\rho^2} \rho^2 \sin\phi \, d\rho \, d\phi \, d\theta = \lim_{R\to\infty}\left(\int_0^{2\pi} d\theta\right)\left(\int_0^\pi \sin\phi \, d\phi\right)\left(\int_0^R \rho^3 e^{-\rho^2} \, d\rho\right)$.

Now use integration by parts with $u = \rho^2$, $dv = \rho e^{-\rho^2} \, d\rho$ to get

$$\lim_{R\to\infty} 2\pi(2)\left(\rho^2\left(-\tfrac{1}{2}\right)e^{-\rho^2}\Big]_0^R - \int_0^R 2\rho\left(-\tfrac{1}{2}\right)e^{-\rho^2} \, d\rho\right) = \lim_{R\to\infty} 4\pi\left(-\tfrac{1}{2}R^2 e^{-R^2} + \left[-\tfrac{1}{2}e^{-\rho^2}\right]_0^R\right)$$

$$= 4\pi \lim_{R\to\infty}\left[-\tfrac{1}{2}R^2 e^{-R^2} - \tfrac{1}{2}e^{-R^2} + \tfrac{1}{2}\right] = 4\pi\left(\tfrac{1}{2}\right) = 2\pi$$

(Note that $R^2 e^{-R^2} \to 0$ as $R \to \infty$ by l'Hospital's Rule.)

45. (a) From the diagram, $z = r \cot \phi_0$ to $z = \sqrt{a^2 - r^2}$, $r = 0$

to $r = a \sin \phi_0$ (or use $a^2 - r^2 = r^2 \cot^2 \phi_0$). Thus

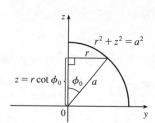

$$V = \int_0^{2\pi} \int_0^{a \sin \phi_0} \int_{r \cot \phi_0}^{\sqrt{a^2 - r^2}} r \, dz \, dr \, d\theta$$

$$= 2\pi \int_0^{a \sin \phi_0} \left(r \sqrt{a^2 - r^2} - r^2 \cot \phi_0 \right) dr$$

$$= \frac{2\pi}{3} \left[-(a^2 - r^2)^{3/2} - r^3 \cot \phi_0 \right]_0^{a \sin \phi_0}$$

$$= \frac{2\pi}{3} \left[-\left(a^2 - a^2 \sin^2 \phi_0\right)^{3/2} - a^3 \sin^3 \phi_0 \cot \phi_0 + a^3 \right]$$

$$= \frac{2}{3} \pi a^3 \left[1 - \left(\cos^3 \phi_0 + \sin^2 \phi_0 \cos \phi_0\right) \right] = \frac{2}{3} \pi a^3 (1 - \cos \phi_0)$$

(b) The wedge in question is the shaded area rotated from $\theta = \theta_1$ to $\theta = \theta_2$.
Letting

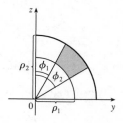

V_{ij} = volume of the region bounded by the sphere of radius ρ_i

and the cone with angle ϕ_j $(\theta = \theta_1$ to $\theta_2)$

and letting V be the volume of the wedge, we have

$$V = (V_{22} - V_{21}) - (V_{12} - V_{11})$$

$$= \frac{1}{3}(\theta_2 - \theta_1)\left[\rho_2^3(1 - \cos \phi_2) - \rho_2^3(1 - \cos \phi_1) - \rho_1^3(1 - \cos \phi_2) + \rho_1^3(1 - \cos \phi_1)\right]$$

$$= \frac{1}{3}(\theta_2 - \theta_1)\left[(\rho_2^3 - \rho_1^3)(1 - \cos \phi_2) - (\rho_2^3 - \rho_1^3)(1 - \cos \phi_1)\right] = \frac{1}{3}(\theta_2 - \theta_1)\left[(\rho_2^3 - \rho_1^3)(\cos \phi_1 - \cos \phi_2)\right]$$

Or: Show that $V = \displaystyle\int_{\theta_1}^{\theta_2} \int_{\rho_1 \sin \phi_1}^{\rho_2 \sin \phi_2} \int_{r \cot \phi_2}^{r \cot \phi_1} r \, dz \, dr \, d\theta$.

(c) By the Mean Value Theorem with $f(\rho) = \rho^3$ there exists some $\tilde{\rho}$ with $\rho_1 \leq \tilde{\rho} \leq \rho_2$ such that

$$f(\rho_2) - f(\rho_1) = f'(\tilde{\rho})(\rho_2 - \rho_1) \text{ or } \rho_1^3 - \rho_2^3 = 3\tilde{\rho}^2 \Delta\rho.$$ Similarly there exists ϕ with $\phi_1 \leq \tilde{\phi} \leq \phi_2$

such that $\cos \phi_2 - \cos \phi_1 = \left(-\sin \tilde{\phi}\right) \Delta\phi$. Substituting into the result from (b) gives

$$\Delta V = (\tilde{\rho}^2 \, \Delta\rho)(\theta_2 - \theta_1)(\sin \tilde{\phi}) \, \Delta\phi = \tilde{\rho}^2 \sin \tilde{\phi} \, \Delta\rho \, \Delta\phi \, \Delta\theta.$$

APPLIED PROJECT   Roller Derby

1. $mgh = \frac{1}{2}mv^2 + \frac{1}{2}I\omega^2 = \frac{1}{2}(m + I/r^2)v^2$, so $v^2 = \dfrac{2mgh}{m + I/r^2} = \dfrac{2gh}{1 + I^*}$.

2. The vertical component of the speed is $v \sin \alpha$, so

$$\frac{dy}{dt} = \sqrt{\frac{2gy}{1 + I^*}} \sin \alpha = \sqrt{\frac{2g}{1 + I^*}} \sin \alpha \, \sqrt{y}.$$

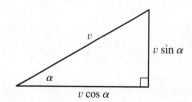

3. Solving the separable differential equation, we get $\dfrac{dy}{\sqrt{y}} = \sqrt{\dfrac{2g}{1+I^*}}\,\sin\alpha\,dt \ \Rightarrow \ 2\sqrt{y} = \sqrt{\dfrac{2g}{1+I^*}}\,(\sin\alpha)t + C.$

But $y = 0$ when $t = 0$, so $C = 0$ and we have $2\sqrt{y} = \sqrt{\dfrac{2g}{1+I^*}}\,(\sin\alpha)t$. Solving for t when $y = h$ gives

$$T = \frac{2\sqrt{h}}{\sin\alpha}\sqrt{\frac{1+I^*}{2g}} = \sqrt{\frac{2h(1+I^*)}{g\sin^2\alpha}}.$$

4. Assume that the length of each cylinder is ℓ. Then the density of the solid cylinder is $\dfrac{m}{\pi r^2\ell}$, and from Formulas 16.6.16

[ET 15.6.16], its moment of inertia (using cylindrical coordinates) is

$$I_z = \iiint \frac{m}{\pi r^2\ell}(x^2+y^2)\,dV = \int_0^\ell\int_0^{2\pi}\int_0^r \frac{m}{\pi r^2\ell}R^2\,R\,dR\,d\theta\,dz = \frac{m}{\pi r^2\ell}\,2\pi\ell\big[\tfrac{1}{4}R^4\big]_0^r = \frac{mr^2}{2}$$

and so $I^* = \dfrac{I_z}{mr^2} = \dfrac{1}{2}.$

For the hollow cylinder, we consider its entire mass to lie a distance r from the axis of rotation, so $x^2 + y^2 = r^2$ is a

constant. We express the density in terms of mass per unit area as $\rho = \dfrac{m}{2\pi r\ell}$, and then the moment of inertia is calculated as a

double integral: $I_z = \iint (x^2+y^2)\dfrac{m}{2\pi r\ell}\,dA = \dfrac{mr^2}{2\pi r\ell}\iint dA = mr^2$, so $I^* = \dfrac{I_z}{mr^2} = 1.$

5. The volume of such a ball is $\tfrac{4}{3}\pi(r^3 - a^3) = \tfrac{4}{3}\pi r(1-b^3)$, and so its density is $\dfrac{m}{\tfrac{4}{3}\pi r^3(1-b^3)}$. Using Formula 16.8.3

[ET 15.8.3], we get

$$I_z = \iiint (x^2+y^2)\frac{m}{\tfrac{4}{3}\pi r^3(1-b^3)}\,dV$$

$$= \frac{m}{\tfrac{4}{3}\pi r^3(1-b^3)}\int_a^r\int_0^{2\pi}\int_0^\pi (\rho^2\sin^2\phi)(\rho^2\sin\phi)\,d\phi\,d\theta\,d\rho$$

$$= \frac{m}{\tfrac{4}{3}\pi r^3(1-b^3)}\cdot 2\pi\left[-\frac{(2+\sin^2\phi)\cos\phi}{3}\right]_0^\pi\left[\frac{\rho^5}{5}\right]_a^r \qquad \text{[from the Table of Integrals]}$$

$$= \frac{m}{\tfrac{4}{3}\pi r^3(1-b^3)}\cdot 2\pi\cdot\frac{4}{3}\cdot\frac{r^5-a^5}{5} = \frac{2mr^5(1-b^5)}{5r^3(1-b^3)} = \frac{2(1-b^5)mr^2}{5(1-b^3)}$$

Therefore $I^* = \dfrac{2(1-b^5)}{5(1-b^3)}$. Since a represents the inner radius, $a \to 0$ corresponds to a solid ball, and $a \to r$ corresponds to

a hollow ball.

6. For a solid ball, $a \to 0 \ \Rightarrow \ b \to 0$, so $I^* = \lim\limits_{b\to 0}\dfrac{2(1-b^5)}{5(1-b^3)} = \dfrac{2}{5}$. For a hollow ball, $a \to r \ \Rightarrow \ b \to 1$, so

$$I^* = \lim_{b\to 1}\frac{2(1-b^5)}{5(1-b^3)} = \frac{2}{5}\lim_{b\to 1}\frac{-5b^4}{-3b^2} = \frac{2}{5}\left(\frac{5}{3}\right) = \frac{2}{3} \qquad \text{[by l'Hospital's Rule]}$$

Note: We could instead have calculated $I^* = \lim\limits_{b\to 1}\dfrac{2(1-b)(1+b+b^2+b^3+b^4)}{5(1-b)(1+b+b^2)} = \dfrac{2\cdot 5}{5\cdot 3} = \dfrac{2}{3}.$

Thus the objects finish in the following order: solid ball $\left(I^* = \tfrac{2}{5}\right)$, solid cylinder $\left(I^* = \tfrac{1}{2}\right)$, hollow ball $\left(I^* = \tfrac{2}{3}\right)$, hollow

cylinder $\left(I^* = 1\right)$.

16.9 Change of Variables in Multiple Integrals

1. $x = 5u - v,\ y = u + 3v$.

The Jacobian is $\dfrac{\partial(x, y)}{\partial(u, v)} = \begin{vmatrix} \partial x/\partial u & \partial x/\partial v \\ \partial y/\partial u & \partial y/\partial v \end{vmatrix} = \begin{vmatrix} 5 & -1 \\ 1 & 3 \end{vmatrix} = 5(3) - (-1)(1) = 16.$

2. $x = uv,\ y = u/v$.

$\dfrac{\partial(x, y)}{\partial(u, v)} = \begin{vmatrix} \partial x/\partial u & \partial x/\partial v \\ \partial y/\partial u & \partial y/\partial v \end{vmatrix} = \begin{vmatrix} v & u \\ 1/v & -u/v^2 \end{vmatrix} = v\left(-\dfrac{u}{v^2}\right) - u\left(\dfrac{1}{v}\right) = -\dfrac{u}{v} - \dfrac{u}{v} = -\dfrac{2u}{v}$

3. $x = e^{-r}\sin\theta,\ y = e^r\cos\theta$.

$\dfrac{\partial(x, y)}{\partial(r, \theta)} = \begin{vmatrix} \partial x/\partial r & \partial x/\partial\theta \\ \partial y/\partial r & \partial y/\partial\theta \end{vmatrix} = \begin{vmatrix} -e^{-r}\sin\theta & e^{-r}\cos\theta \\ e^r\cos\theta & -e^r\sin\theta \end{vmatrix} = e^{-r}e^r\sin^2\theta - e^{-r}e^r\cos^2\theta = \sin^2\theta - \cos^2\theta \text{ or } -\cos 2\theta$

4. $x = e^{s+t},\ y = e^{s-t}$.

$\dfrac{\partial(x, y)}{\partial(s, t)} = \begin{vmatrix} \partial x/\partial s & \partial x/\partial t \\ \partial y/\partial s & \partial y/\partial t \end{vmatrix} = \begin{vmatrix} e^{s+t} & e^{s+t} \\ e^{s-t} & -e^{s-t} \end{vmatrix} = -e^{s+t}e^{s-t} - e^{s+t}e^{s-t} = -2e^{2s}$

5. $x = u/v,\ y = v/w,\ z = w/u$.

$\dfrac{\partial(x, y, z)}{\partial(u, v, w)} = \begin{vmatrix} \partial x/\partial u & \partial x/\partial v & \partial x/\partial w \\ \partial y/\partial u & \partial y/\partial v & \partial y/\partial w \\ \partial z/\partial u & \partial z/\partial v & \partial z/\partial w \end{vmatrix} = \begin{vmatrix} 1/v & -u/v^2 & 0 \\ 0 & 1/w & -v/w^2 \\ -w/u^2 & 0 & 1/u \end{vmatrix}$

$= \dfrac{1}{v}\begin{vmatrix} 1/w & -v/w^2 \\ 0 & 1/u \end{vmatrix} - \left(-\dfrac{u}{v^2}\right)\begin{vmatrix} 0 & -v/w^2 \\ -w/u^2 & 1/u \end{vmatrix} + 0\begin{vmatrix} 0 & 1/w \\ -w/u^2 & 0 \end{vmatrix}$

$= \dfrac{1}{v}\left(\dfrac{1}{uw} - 0\right) + \dfrac{u}{v^2}\left(0 - \dfrac{v}{u^2 w}\right) + 0 = \dfrac{1}{uvw} - \dfrac{1}{uvw} = 0$

6. $x = v + w^2,\ y = w + u^2,\ z = u + v^2$.

$\dfrac{\partial(x, y, z)}{\partial(u, v, w)} = \begin{vmatrix} 0 & 1 & 2w \\ 2u & 0 & 1 \\ 1 & 2v & 0 \end{vmatrix} = 0\begin{vmatrix} 0 & 1 \\ 2v & 0 \end{vmatrix} - 1\begin{vmatrix} 2u & 1 \\ 1 & 0 \end{vmatrix} + 2w\begin{vmatrix} 2u & 0 \\ 1 & 2v \end{vmatrix} = 0 - (0 - 1) + 2w(4uv - 0) = 1 + 8uvw$

7. The transformation maps the boundary of S to the boundary of the image R, so we first look at side S_1 in the uv-plane. S_1 is described by $v = 0$ $[0 \le u \le 3]$, so $x = 2u + 3v = 2u$ and $y = u - v = u$. Eliminating u, we have $x = 2y,\ 0 \le x \le 6$. S_2 is the line segment $u = 3,\ 0 \le v \le 2$, so $x = 6 + 3v$ and $y = 3 - v$. Then $v = 3 - y \Rightarrow x = 6 + 3(3 - y) = 15 - 3y$, $6 \le x \le 12$. S_3 is the line segment $v = 2,\ 0 \le u \le 3$, so $x = 2u + 6$ and $y = u - 2$, giving $u = y + 2 \Rightarrow x = 2y + 10$, $6 \le x \le 12$. Finally, S_4 is the segment $u = 0,\ 0 \le v \le 2$, so $x = 3v$ and $y = -v \Rightarrow x = -3y,\ 0 \le x \le 6$. The image of set S is the region R shown in the xy-plane, a parallelogram bounded by these four segments.

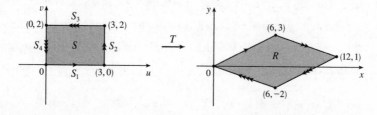

8. S_1 is the line segment $v = 0, 0 \le u \le 1$, so $x = v = 0$ and $y = u(1 + v^2) = u$. Since $0 \le u \le 1$, the image is the line

segment $x = 0, 0 \le y \le 1$. S_2 is the segment $u = 1, 0 \le v \le 1$, so $x = v$ and $y = u(1 + v^2) = 1 + x^2$. Thus the image is

the portion of the parabola $y = 1 + x^2$ for $0 \le x \le 1$. S_3 is the segment $v = 1, 0 \le u \le 1$, so $x = 1$ and $y = 2u$. The image

is the segment $x = 1, 0 \le y \le 2$. S_4 is described by $u = 0, 0 \le v \le 1$, so $0 \le x = v \le 1$ and $y = u(1 + v^2) = 0$. The

image is the line segment $y = 0, 0 \le x \le 1$. Thus, the image of S is the region R bounded by the parabola $y = 1 + x^2$, the

x-axis, and the lines $x = 0, x = 1$.

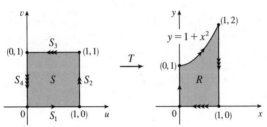

9. S_1 is the line segment $u = v, 0 \le u \le 1$, so $y = v = u$ and $x = u^2 = y^2$. Since $0 \le u \le 1$, the image is the portion of the

parabola $x = y^2, 0 \le y \le 1$. S_2 is the segment $v = 1, 0 \le u \le 1$, thus $y = v = 1$ and $x = u^2$, so $0 \le x \le 1$. The image is

the line segment $y = 1, 0 \le x \le 1$. S_3 is the segment $u = 0, 0 \le v \le 1$, so $x = u^2 = 0$ and $y = v \Rightarrow 0 \le y \le 1$. The

image is the segment $x = 0, 0 \le y \le 1$. Thus, the image of S is the region R in the first quadrant bounded by the parabola

$x = y^2$, the y-axis, and the line $y = 1$.

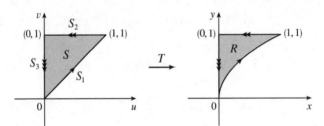

10. Substituting $u = \dfrac{x}{a}, v = \dfrac{y}{b}$ into $u^2 + v^2 \le 1$ gives

$\dfrac{x^2}{a^2} + \dfrac{y^2}{b^2} \le 1$, so the image of $u^2 + v^2 \le 1$ is the

elliptical region $\dfrac{x^2}{a^2} + \dfrac{y^2}{b^2} \le 1$.

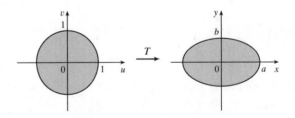

11. $\dfrac{\partial(x, y)}{\partial(u, v)} = \begin{vmatrix} 2 & 1 \\ 1 & 2 \end{vmatrix} = 3$ and $x - 3y = (2u + v) - 3(u + 2v) = -u - 5v$. To find the region S in the uv-plane that

corresponds to R we first find the corresponding boundary under the given transformation. The line through $(0, 0)$ and $(2, 1)$ is

$y = \frac{1}{2}x$ which is the image of $u + 2v = \frac{1}{2}(2u + v) \Rightarrow v = 0$; the line through $(2, 1)$ and $(1, 2)$ is $x + y = 3$ which is the

image of $(2u + v) + (u + 2v) = 3 \Rightarrow u + v = 1$; the line through $(0, 0)$ and $(1, 2)$ is $y = 2x$ which is the image of

$u + 2v = 2(2u + v) \Rightarrow u = 0$. Thus S is the triangle $0 \le v \le 1 - u, 0 \le u \le 1$ in the uv-plane and

$$\iint_R (x - 3y)\, dA = \int_0^1 \int_0^{1-u} (-u - 5v)\, |3|\, dv\, du = -3 \int_0^1 \left[uv + \tfrac{5}{2}v^2 \right]_{v=0}^{v=1-u}\, du$$

$$= -3 \int_0^1 \left(u - u^2 + \tfrac{5}{2}(1 - u)^2 \right) du = -3 \left[\tfrac{1}{2}u^2 - \tfrac{1}{3}u^3 - \tfrac{5}{6}(1 - u)^3 \right]_0^1 = -3\left(\tfrac{1}{2} - \tfrac{1}{3} + \tfrac{5}{6} \right) = -3$$

12. $\dfrac{\partial(x,y)}{\partial(u,v)} = \begin{vmatrix} 1/4 & 1/4 \\ -3/4 & 1/4 \end{vmatrix} = \dfrac{1}{4}$, $4x + 8y = 4 \cdot \frac{1}{4}(u+v) + 8 \cdot \frac{1}{4}(v-3u) = 3v - 5u$. R is a parallelogram bounded by the

lines $x - y = -4$, $x - y = 4$, $3x + y = 0$, $3x + y = 8$. Since $u = x - y$ and $v = 3x + y$, R is the image of the rectangle

enclosed by the lines $u = -4$, $u = 4$, $v = 0$, and $v = 8$. Thus

$$\iint_R (4x + 8y)\,dA = \int_{-4}^4 \int_0^8 (3v - 5u)\left|\tfrac{1}{4}\right|\,dv\,du = \tfrac{1}{4}\int_{-4}^4 \left[\tfrac{3}{2}v^2 - 5uv\right]_{v=0}^{v=8} du$$

$$= \tfrac{1}{4}\int_{-4}^4 (96 - 40u)\,du = \tfrac{1}{4}\left[96u - 20u^2\right]_{-4}^4 = 192$$

13. $\dfrac{\partial(x,y)}{\partial(u,v)} = \begin{vmatrix} 2 & 0 \\ 0 & 3 \end{vmatrix} = 6$, $x^2 = 4u^2$ and the planar ellipse $9x^2 + 4y^2 \le 36$ is the image of the disk $u^2 + v^2 \le 1$. Thus

$$\iint_R x^2\,dA = \iint_{u^2+v^2\le 1} (4u^2)(6)\,du\,dv = \int_0^{2\pi}\int_0^1 (24r^2 \cos^2 \theta)\,r\,dr\,d\theta = 24\int_0^{2\pi} \cos^2 \theta\,d\theta \int_0^1 r^3\,dr$$

$$= 24\left[\tfrac{1}{2}x + \tfrac{1}{4}\sin 2x\right]_0^{2\pi} \left[\tfrac{1}{4}r^4\right]_0^1 = 24(\pi)\left(\tfrac{1}{4}\right) = 6\pi$$

14. $\dfrac{\partial(x,y)}{\partial(u,v)} = \begin{vmatrix} \sqrt{2} & -\sqrt{2/3} \\ \sqrt{2} & \sqrt{2/3} \end{vmatrix} = \dfrac{4}{\sqrt{3}}$, $x^2 - xy + y^2 = 2u^2 + 2v^2$ and the planar ellipse $x^2 - xy + y^2 \le 2$

is the image of the disk $u^2 + v^2 \le 1$. Thus

$$\iint_R (x^2 - xy + y^2)\,dA = \iint_{u^2+v^2\le 1} (2u^2 + 2v^2)\left(\tfrac{4}{\sqrt{3}}\,du\,dv\right) = \int_0^{2\pi}\int_0^1 \tfrac{8}{\sqrt{3}}r^3\,dr\,d\theta = \tfrac{4\pi}{\sqrt{3}}$$

15. $\dfrac{\partial(x,y)}{\partial(u,v)} = \begin{vmatrix} 1/v & -u/v^2 \\ 0 & 1 \end{vmatrix} = \dfrac{1}{v}$, $xy = u$, $y = x$ is the image of the parabola $v^2 = u$, $y = 3x$ is the image of the parabola

$v^2 = 3u$, and the hyperbolas $xy = 1$, $xy = 3$ are the images of the lines $u = 1$ and $u = 3$ respectively. Thus

$$\iint_R xy\,dA = \int_1^3 \int_{\sqrt{u}}^{\sqrt{3u}} u\left(\tfrac{1}{v}\right)dv\,du = \int_1^3 u\left(\ln\sqrt{3u} - \ln\sqrt{u}\right)du = \int_1^3 u\ln\sqrt{3}\,du = 4\ln\sqrt{3} = 2\ln 3.$$

16. Here $y = \dfrac{v}{u}$, $x = \dfrac{u^2}{v}$ so $\dfrac{\partial(x,y)}{\partial(u,v)} = \begin{vmatrix} 2u/v & -u^2/v^2 \\ -v/u^2 & 1/u \end{vmatrix} = \dfrac{1}{v}$ and R is the

image of the square with vertices $(1,1)$, $(2,1)$, $(2,2)$, and $(1,2)$. So

$$\iint_R y^2\,dA = \int_1^2 \int_1^2 \dfrac{v^2}{u^2}\left(\tfrac{1}{v}\right)du\,dv = \int_1^2 \dfrac{v}{2}\,dv = \dfrac{3}{4}$$

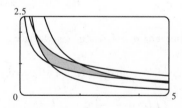

17. (a) $\dfrac{\partial(x,y,z)}{\partial(u,v,w)} = \begin{vmatrix} a & 0 & 0 \\ 0 & b & 0 \\ 0 & 0 & c \end{vmatrix} = abc$ and since $u = \dfrac{x}{a}$, $v = \dfrac{y}{b}$, $w = \dfrac{z}{c}$ the solid enclosed by the ellipsoid is the image of the

ball $u^2 + v^2 + w^2 \le 1$. So

$$\iiint_E dV = \iiint_{u^2+v^2+w^2\le 1} abc\,du\,dv\,dw = (abc)(\text{volume of the ball}) = \tfrac{4}{3}\pi abc$$

(b) If we approximate the surface of the earth by the ellipsoid $\dfrac{x^2}{6378^2} + \dfrac{y^2}{6378^2} + \dfrac{z^2}{6356^2} = 1$, then we can estimate

the volume of the earth by finding the volume of the solid E enclosed by the ellipsoid. From part (a), this is

$$\iiint_E dV = \tfrac{4}{3}\pi(6378)(6378)(6356) \approx 1.083 \times 10^{12}\ \text{km}^3.$$

18. The moment of intertia about the z-axis is $I_z = \iiint_E (x^2 + y^2)\, \rho(x, y, z)\, dV$, where E is the solid enclosed by

$\dfrac{x^2}{a^2} + \dfrac{y^2}{b^2} + \dfrac{z^2}{c^2} = 1$. As in Exercise 17(a), we use the transformation $x = au$, $y = bv$, $z = cw$, so $\left| \dfrac{\partial(x, y, z)}{\partial(u, v, w)} \right| = abc$ and

$$I_z = \iiint_E (x^2 + y^2)\, k\, dV = \iiint_{u^2+v^2+w^2 \le 1} k(a^2 u^2 + b^2 v^2)(abc)\, du\, dv\, dw$$

$$= abck \int_0^\pi \int_0^{2\pi} \int_0^1 (a^2 \rho^2 \sin^2\phi \cos^2\theta + b^2 \rho^2 \sin^2\phi \sin^2\theta)\, \rho^2 \sin\phi\, d\rho\, d\theta\, d\phi$$

$$= abck \left[a^2 \int_0^\pi \int_0^{2\pi} \int_0^1 (\rho^2 \sin^2\phi \cos^2\theta)\, \rho^2 \sin\phi\, d\rho\, d\theta\, d\phi + b^2 \int_0^\pi \int_0^{2\pi} \int_0^1 (\rho^2 \sin^2\phi \sin^2\theta)\, \rho^2 \sin\phi\, d\rho\, d\theta\, d\phi \right]$$

$$= a^3 bck \int_0^\pi \sin^3\phi\, d\phi \int_0^{2\pi} \cos^2\theta\, d\theta \int_0^1 \rho^4\, d\rho + ab^3 ck \int_0^\pi \sin^3\phi\, d\phi \int_0^{2\pi} \sin^2\theta\, d\theta \int_0^1 \rho^4\, d\rho$$

$$= a^3 bck \left[\tfrac{1}{3}\cos^3\phi - \cos\phi \right]_0^\pi \left[\tfrac{1}{2}\theta + \tfrac{1}{4}\sin 2\theta \right]_0^{2\pi} \left[\tfrac{1}{5}\rho^5 \right]_0^1 + ab^3 ck \left[\tfrac{1}{3}\cos^3\phi - \cos\phi \right]_0^\pi \left[\tfrac{1}{2}\theta - \tfrac{1}{4}\sin 2\theta \right]_0^{2\pi} \left[\tfrac{1}{5}\rho^5 \right]_0^1$$

$$= a^3 bck \left(\tfrac{4}{3} \right) (\pi) \left(\tfrac{1}{5} \right) + ab^3 ck \left(\tfrac{4}{3} \right) (\pi) \left(\tfrac{1}{5} \right) = \tfrac{4}{15}\pi \left(a^2 + b^2 \right) abck$$

19. Letting $u = x - 2y$ and $v = 3x - y$, we have $x = \tfrac{1}{5}(2v - u)$ and $y = \tfrac{1}{5}(v - 3u)$. Then $\dfrac{\partial(x, y)}{\partial(u, v)} = \begin{vmatrix} -1/5 & 2/5 \\ -3/5 & 1/5 \end{vmatrix} = \dfrac{1}{5}$

and R is the image of the rectangle enclosed by the lines $u = 0$, $u = 4$, $v = 1$, and $v = 8$. Thus

$$\iint_R \frac{x - 2y}{3x - y}\, dA = \int_0^4 \int_1^8 \frac{u}{v} \left| \frac{1}{5} \right| dv\, du = \frac{1}{5} \int_0^4 u\, du \int_1^8 \frac{1}{v}\, dv = \tfrac{1}{5} \left[\tfrac{1}{2}u^2 \right]_0^4 \left[\ln|v| \right]_1^8 = \tfrac{8}{5}\ln 8.$$

20. Letting $u = x + y$ and $v = x - y$, we have $x = \tfrac{1}{2}(u + v)$ and $y = \tfrac{1}{2}(u - v)$. Then $\dfrac{\partial(x, y)}{\partial(u, v)} = \begin{vmatrix} 1/2 & 1/2 \\ 1/2 & -1/2 \end{vmatrix} = -\dfrac{1}{2}$ and R is

the image of the rectangle enclosed by the lines $u = 0$, $u = 3$, $v = 0$, and $v = 2$. Thus

$$\iint_R (x + y)\, e^{x^2 - y^2}\, dA = \int_0^3 \int_0^2 u e^{uv} \left| -\tfrac{1}{2} \right| dv\, du = \tfrac{1}{2} \int_0^3 \left[e^{uv} \right]_{v=0}^{v=2} du = \tfrac{1}{2} \int_0^3 (e^{2u} - 1)\, du$$

$$= \tfrac{1}{2} \left[\tfrac{1}{2}e^{2u} - u \right]_0^3 = \tfrac{1}{2} \left(\tfrac{1}{2}e^6 - 3 - \tfrac{1}{2} \right) = \tfrac{1}{4}(e^6 - 7)$$

21. Letting $u = y - x$, $v = y + x$, we have $y = \tfrac{1}{2}(u + v)$, $x = \tfrac{1}{2}(v - u)$. Then $\dfrac{\partial(x, y)}{\partial(u, v)} = \begin{vmatrix} -1/2 & 1/2 \\ 1/2 & 1/2 \end{vmatrix} = -\dfrac{1}{2}$ and R is the

image of the trapezoidal region with vertices $(-1, 1)$, $(-2, 2)$, $(2, 2)$, and $(1, 1)$. Thus

$$\iint_R \cos \frac{y - x}{y + x}\, dA = \int_1^2 \int_{-v}^v \cos \frac{u}{v} \left| -\tfrac{1}{2} \right| du\, dv = \frac{1}{2} \int_1^2 \left[v \sin \frac{u}{v} \right]_{u=-v}^{u=v} dv = \frac{1}{2} \int_1^2 2v \sin(1)\, dv = \tfrac{3}{2}\sin 1$$

22. Letting $u = 3x$, $v = 2y$, we have $9x^2 + 4y^2 = u^2 + v^2$, $x = \tfrac{1}{3}u$, and $y = \tfrac{1}{2}v$. Then $\dfrac{\partial(x, y)}{\partial(u, v)} = \dfrac{1}{6}$ and R is the image of the

quarter-disk D given by $u^2 + v^2 \le 1$, $u \ge 0$, $v \ge 0$. Thus

$$\iint_R \sin(9x^2 + 4y^2)\, dA = \iint_D \tfrac{1}{6}\sin(u^2 + v^2)\, du\, dv = \int_0^{\pi/2} \int_0^1 \tfrac{1}{6}\sin(r^2)\, r\, dr\, d\theta = \tfrac{\pi}{12} \left[-\tfrac{1}{2}\cos r^2 \right]_0^1 = \tfrac{\pi}{24}(1 - \cos 1)$$

23. Let $u = x + y$ and $v = -x + y$. Then $u + v = 2y \;\Rightarrow\; y = \tfrac{1}{2}(u + v)$ and $u - v = 2x \;\Rightarrow\; x = \tfrac{1}{2}(u - v)$.

$\dfrac{\partial(x, y)}{\partial(u, v)} = \begin{vmatrix} 1/2 & -1/2 \\ 1/2 & 1/2 \end{vmatrix} = \dfrac{1}{2}$. Now $|u| = |x + y| \le |x| + |y| \le 1 \;\Rightarrow\; -1 \le u \le 1$, and

$|v| = |-x + y| \le |x| + |y| \le 1 \Rightarrow -1 \le v \le 1$. R is the image of the square

region with vertices $(1, 1)$, $(1, -1)$, $(-1, -1)$, and $(-1, 1)$.

So $\iint_R e^{x+y} \, dA = \frac{1}{2} \int_{-1}^{1} \int_{-1}^{1} e^u \, du \, dv = \frac{1}{2} [e^u]_{-1}^{1} [v]_{-1}^{1} = e - e^{-1}$.

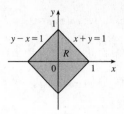

24. Let $u = x + y$ and $v = y$, then $x = u - v$, $y = v$, $\dfrac{\partial(x, y)}{\partial(u, v)} = 1$ and R is the image under T of the triangular region with

vertices $(0, 0)$, $(1, 0)$ and $(1, 1)$. Thus

$$\iint_R f(x + y) \, dA = \int_0^1 \int_0^u (1) \, f(u) \, dv \, du = \int_0^1 f(u) \left[v \right]_{v=0}^{v=u} du = \int_0^1 u f(u) \, du \quad \text{as desired.}$$

16 Review   ET 15

CONCEPT CHECK

1. (a) A double Riemann sum of f is $\displaystyle\sum_{i=1}^{m} \sum_{j=1}^{n} f(x_{ij}^*, y_{ij}^*) \, \Delta A$, where ΔA is the area of each subrectangle and (x_{ij}^*, y_{ij}^*) is a

sample point in each subrectangle. If $f(x, y) \ge 0$, this sum represents an approximation to the volume of the solid that lies

above the rectangle R and below the graph of f.

(b) $\iint_R f(x, y) \, dA = \displaystyle\lim_{m,n \to \infty} \sum_{i=1}^{m} \sum_{j=1}^{n} f(x_{ij}^*, y_{ij}^*) \, \Delta A$

(c) If $f(x, y) \ge 0$, $\iint_R f(x, y) \, dA$ represents the volume of the solid that lies above the rectangle R and below the surface

$z = f(x, y)$. If f takes on both positive and negative values, $\iint_R f(x, y) \, dA$ is the difference of the volume above R but

below the surface $z = f(x, y)$ and the volume below R but above the surface $z = f(x, y)$.

(d) We usually evaluate $\iint_R f(x, y) \, dA$ as an iterated integral according to Fubini's Theorem (see Theorem 16.2.4

[ET 15.2.4]).

(e) The Midpoint Rule for Double Integrals says that we approximate the double integral $\iint_R f(x, y) \, dA$ by the double

Riemann sum $\displaystyle\sum_{i=1}^{m} \sum_{j=1}^{n} f(\overline{x}_i, \overline{y}_j) \, \Delta A$ where the sample points $(\overline{x}_i, \overline{y}_j)$ are the centers of the subrectangles.

(f) $f_{\text{ave}} = \dfrac{1}{A(R)} \displaystyle\iint_R f(x, y) \, dA$ where $A(R)$ is the area of R.

2. (a) See (1) and (2) and the accompanying discussion in Section 16.3 [ET 15.3].

(b) See (3) and the accompanying discussion in Section 16.3 [ET 15.3].

(c) See (5) and the preceding discussion in Section 16.3 [ET 15.3].

(d) See (6)–(11) in Section 16.3 [ET 15.3].

3. We may want to change from rectangular to polar coordinates in a double integral if the region R of integration is more easily

described in polar coordinates. To accomplish this, we use $\iint_R f(x, y) \, dA = \int_\alpha^\beta \int_a^b f(r \cos \theta, r \sin \theta) \, r \, dr \, d\theta$ where R is

given by $0 \le a \le r \le b$, $\alpha \le \theta \le \beta$.

4. (a) $m = \iint_D \rho(x, y)\, dA$

(b) $M_x = \iint_D y\rho(x, y)\, dA$, $M_y = \iint_D x\rho(x, y)\, dA$

(c) The center of mass is $(\overline{x}, \overline{y})$ where $\overline{x} = \dfrac{M_y}{m}$ and $\overline{y} = \dfrac{M_x}{m}$.

(d) $I_x = \iint_D y^2 \rho(x, y)\, dA$, $I_y = \iint_D x^2 \rho(x, y)\, dA$, $I_0 = \iint_D (x^2 + y^2)\rho(x, y)\, dA$

5. (a) $P(a \le X \le b, c \le Y \le d) = \int_a^b \int_c^d f(x, y)\, dy\, dx$

(b) $f(x, y) \ge 0$ and $\iint_{\mathbb{R}^2} f(x, y)\, dA = 1$.

(c) The expected value of X is $\mu_1 = \iint_{\mathbb{R}^2} x f(x, y)\, dA$; the expected value of Y is $\mu_2 = \iint_{\mathbb{R}^2} y f(x, y)\, dA$.

6. (a) $\iiint_B f(x, y, z)\, dV = \displaystyle\lim_{l,m,n \to \infty} \sum_{i=1}^{l} \sum_{j=1}^{m} \sum_{k=1}^{n} f\left(x_{ijk}^*, y_{ijk}^*, z_{ijk}^*\right) \Delta V$

(b) We usually evaluate $\iiint_B f(x, y, z)\, dV$ as an iterated integral according to Fubini's Theorem for Triple Integrals (see Theorem 16.6.4 [ET 15.6.4]).

(c) See the paragraph following Example 16.6.1 [ET 15.6.1].

(d) See (5) and (6) and the accompanying discussion in Section 16.6 [ET 15.6].

(e) See (10) and the accompanying discussion in Section 16.6 [ET 15.6].

(f) See (11) and the preceding discussion in Section 16.6 [ET 15.6].

7. (a) $m = \iiint_E \rho(x, y, z)\, dV$

(b) $M_{yz} = \iiint_E x\rho(x, y, z)\, dV$, $M_{xz} = \iiint_E y\rho(x, y, z)\, dV$, $M_{xy} = \iiint_E z\rho(x, y, z)\, dV$.

(c) The center of mass is $(\overline{x}, \overline{y}, \overline{z})$ where $\overline{x} = \dfrac{M_{yz}}{m}$, $\overline{y} = \dfrac{M_{xz}}{m}$, and $\overline{z} = \dfrac{M_{xy}}{m}$.

(d) $I_x = \iiint_E (y^2 + z^2)\rho(x, y, z)\, dV$, $I_y = \iiint_E (x^2 + z^2)\rho(x, y, z)\, dV$, $I_z = \iiint_E (x^2 + y^2)\rho(x, y, z)\, dV$.

8. (a) See Formula 16.7.4 [ET 15.7.4] and the accompanying discussion.

(b) See Formula 16.8.3 [ET 15.8.3] and the accompanying discussion.

(c) We may want to change from rectangular to cylindrical or spherical coordinates in a triple integral if the region E of integration is more easily described in cylindrical or spherical coordinates or if the triple integral is easier to evaluate using cylindrical or spherical coordinates.

9. (a) $\dfrac{\partial(x, y)}{\partial(u, v)} = \begin{vmatrix} \partial x/\partial u & \partial x/\partial v \\ \partial y/\partial u & \partial y/\partial v \end{vmatrix} = \dfrac{\partial x}{\partial u}\dfrac{\partial y}{\partial v} - \dfrac{\partial x}{\partial v}\dfrac{\partial y}{\partial u}$

(b) See (9) and the accompanying discussion in Section 16.9 [ET 15.9].

(c) See (13) and the accompanying discussion in Section 16.9 [ET 15.9].

TRUE-FALSE QUIZ

1. This is true by Fubini's Theorem.

2. False. $\int_0^1 \int_0^x \sqrt{x + y^2} \, dy \, dx$ describes the region of integration as a Type I region. To reverse the order of integration, we must consider the region as a Type II region: $\int_0^1 \int_y^1 \sqrt{x + y^2} \, dx \, dy$.

3. True by Equation 16.2.5 [ET 15.2.5].

4. $\int_{-1}^1 \int_0^1 e^{x^2 + y^2} \sin y \, dx \, dy = \left(\int_0^1 e^{x^2} \, dx \right) \left(\int_{-1}^1 e^{y^2} \sin y \, dy \right) = \left(\int_0^1 e^{x^2} \, dx \right)(0) = 0$, since $e^{y^2} \sin y$ is an odd function.

Therefore the statement is true.

5. True: $\iint_D \sqrt{4 - x^2 - y^2} \, dA =$ the volume under the surface $x^2 + y^2 + z^2 = 4$ and above the xy-plane

$= \frac{1}{2}$ (the volume of the sphere $x^2 + y^2 + z^2 = 4$) $= \frac{1}{2} \cdot \frac{4}{3}\pi(2)^3 = \frac{16}{3}\pi$

6. This statement is true because in the given region, $(x^2 + \sqrt{y}) \sin(x^2 y^2) \le (1 + 2)(1) = 3$, so

$\int_1^4 \int_0^1 (x^2 + \sqrt{y}) \sin(x^2 y^2) \, dx \, dy \le \int_1^4 \int_0^1 3 \, dA = 3A(D) = 3(3) = 9$.

7. The volume enclosed by the cone $z = \sqrt{x^2 + y^2}$ and the plane $z = 2$ is, in cylindrical coordinates,

$V = \int_0^{2\pi} \int_0^2 \int_r^2 r \, dz \, dr \, d\theta \ne \int_0^{2\pi} \int_0^2 \int_r^2 dz \, dr \, d\theta$, so the assertion is false.

8. True. The moment of inertia about the z-axis of a solid E with constant density k is

$I_z = \iiint_E (x^2 + y^2)\rho(x, y, z) \, dV = \iiint_E (kr^2) \, r \, dz \, dr \, d\theta = \iiint_E kr^3 \, dz \, dr \, d\theta$.

EXERCISES

1. As shown in the contour map, we divide R into 9 equally sized subsquares, each with area $\Delta A = 1$. Then we approximate $\iint_R f(x, y) \, dA$ by a Riemann sum with $m = n = 3$ and the sample points the upper right corners of each square, so

$$\iint_R f(x, y) \, dA \approx \sum_{i=1}^3 \sum_{j=1}^3 f(x_i, y_j) \, \Delta A$$

$$= \Delta A \left[f(1, 1) + f(1, 2) + f(1, 3) + f(2, 1) + f(2, 2) + f(2, 3) + f(3, 1) + f(3, 2) + f(3, 3) \right]$$

Using the contour lines to estimate the function values, we have

$$\iint_R f(x, y) \, dA \approx 1[2.7 + 4.7 + 8.0 + 4.7 + 6.7 + 10.0 + 6.7 + 8.6 + 11.9] \approx 64.0$$

2. As in Exercise 1, we have $m = n = 3$ and $\Delta A = 1$. Using the contour map to estimate the value of f at the center of each subsquare, we have

$$\iint_R f(x, y) \, dA \approx \sum_{i=1}^3 \sum_{j=1}^3 f(\overline{x}_i, \overline{y}_j) \, \Delta A$$

$$= \Delta A \left[f(0.5, 0.5) + (0.5, 1.5) + (0.5, 2.5) + (1.5, 0.5) + f(1.5, 1.5) \right.$$

$$\left. + f(1.5, 2.5) + (2.5, 0.5) + f(2.5, 1.5) + f(2.5, 2.5) \right]$$

$$\approx 1[1.2 + 2.5 + 5.0 + 3.2 + 4.5 + 7.1 + 5.2 + 6.5 + 9.0] = 44.2$$

3. $\int_1^2 \int_0^2 (y + 2xe^y)\, dx\, dy = \int_1^2 \left[xy + x^2 e^y \right]_{x=0}^{x=2} dy = \int_1^2 (2y + 4e^y)\, dy = \left[y^2 + 4e^y \right]_1^2$

$$= 4 + 4e^2 - 1 - 4e = 4e^2 - 4e + 3$$

4. $\int_0^1 \int_0^1 ye^{xy}\, dx\, dy = \int_0^1 \left[e^{xy} \right]_{x=0}^{x=1} dy = \int_0^1 (e^y - 1)\, dy = \left[e^y - y \right]_0^1 = e - 2$

5. $\int_0^1 \int_0^x \cos(x^2)\, dy\, dx = \int_0^1 \left[\cos(x^2)y \right]_{y=0}^{y=x} dx = \int_0^1 x\cos(x^2)\, dx = \frac{1}{2}\sin(x^2)\big]_0^1 = \frac{1}{2}\sin 1$

6. $\int_0^1 \int_x^{e^x} 3xy^2\, dy\, dx = \int_0^1 \left[xy^3 \right]_{y=x}^{y=e^x} dx = \int_0^1 (xe^{3x} - x^4)\, dx = \frac{1}{3}xe^{3x}\big]_0^1 - \int_0^1 \frac{1}{3}e^{3x}\, dx - \left[\frac{1}{5}x^5 \right]_0^1$ $\quad\begin{bmatrix} \text{integrate by parts} \\ \text{in the first term} \end{bmatrix}$

$$= \frac{1}{3}e^3 - \left[\frac{1}{9}e^{3x} \right]_0^1 - \frac{1}{5} = \frac{2}{9}e^3 - \frac{4}{45}$$

7. $\int_0^\pi \int_0^1 \int_0^{\sqrt{1-y^2}} y\sin x\, dz\, dy\, dx = \int_0^\pi \int_0^1 \left[(y\sin x)z \right]_{z=0}^{z=\sqrt{1-y^2}} dy\, dx = \int_0^\pi \int_0^1 y\sqrt{1-y^2}\sin x\, dy\, dx$

$$= \int_0^\pi \left[-\frac{1}{3}(1-y^2)^{3/2}\sin x \right]_{y=0}^{y=1} dx = \int_0^\pi \frac{1}{3}\sin x\, dx = -\frac{1}{3}\cos x\big]_0^\pi = \frac{2}{3}$$

8. $\int_0^1 \int_0^y \int_x^1 6xyz\, dz\, dx\, dy = \int_0^1 \int_0^y \left[3xyz^2 \right]_{z=x}^{z=1} dx\, dy = \int_0^1 \int_0^y (3xy - 3x^3 y)\, dx\, dy$

$$= \int_0^1 \left[\frac{3}{2}x^2 y - \frac{3}{4}x^4 y \right]_{x=0}^{x=y} dy = \int_0^1 \left(\frac{3}{2}y^3 - \frac{3}{4}y^5 \right) dy = \left[\frac{3}{8}y^4 - \frac{1}{8}y^6 \right]_0^1 = \frac{1}{4}$$

9. The region R is more easily described by polar coordinates: $R = \{(r, \theta) \mid 2 \le r \le 4, 0 \le \theta \le \pi\}$. Thus

$$\iint_R f(x, y)\, dA = \int_0^\pi \int_2^4 f(r\cos\theta, r\sin\theta)\, r\, dr\, d\theta.$$

10. The region R is a type II region that can be described as the region enclosed by the lines $y = 4 - x$, $y = 4 + x$, and the x-axis. So using rectangular coordinates, we can say $R = \{(x, y) \mid y - 4 \le x \le 4 - y, 0 \le y \le 4\}$

and $\iint_R f(x, y)\, dA = \int_0^4 \int_{y-4}^{4-y} f(x, y)\, dx\, dy.$

11.

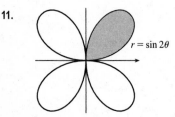

$r = \sin 2\theta$

The region whose area is given by $\int_0^{\pi/2} \int_0^{\sin 2\theta} r\, dr\, d\theta$ is

$\{(r, \theta) \mid 0 \le \theta \le \frac{\pi}{2}, 0 \le r \le \sin 2\theta\}$, which is the region contained in the loop in the first quadrant of the four-leaved rose $r = \sin 2\theta$.

12. The solid is $\{(\rho, \theta, \phi) \mid 1 \le \rho \le 2, 0 \le \theta \le \frac{\pi}{2}, 0 \le \phi \le \frac{\pi}{2}\}$ which is the region in the first octant on or between the two spheres $\rho = 1$ and $\rho = 2$.

13.

$\int_0^1 \int_x^1 \cos(y^2)\, dy\, dx = \int_0^1 \int_0^y \cos(y^2)\, dx\, dy$

$$= \int_0^1 \cos(y^2)\left[x \right]_{x=0}^{x=y} dy = \int_0^1 y\cos(y^2)\, dy$$

$$= \left[\frac{1}{2}\sin(y^2) \right]_0^1 = \frac{1}{2}\sin 1$$

(1, 1)

$y = x$

0

14.

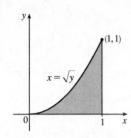

$$\int_0^1 \int_{\sqrt{y}}^1 \frac{ye^{x^2}}{x^3}\, dx\, dy = \int_0^1 \int_0^{x^2} \frac{ye^{x^2}}{x^3}\, dy\, dx = \int_0^1 \frac{e^{x^2}}{x^3} \left[\tfrac{1}{2}y^2\right]_{y=0}^{y=x^2} dx$$

$$= \int_0^1 \tfrac{1}{2}xe^{x^2}\, dx = \tfrac{1}{4}e^{x^2}\Big]_0^1 = \tfrac{1}{4}(e-1)$$

15. $\iint_R ye^{xy}\, dA = \int_0^3 \int_0^2 ye^{xy}\, dx\, dy = \int_0^3 \left[e^{xy}\right]_{x=0}^{x=2} dy = \int_0^3 (e^{2y}-1)\, dy = \left[\tfrac{1}{2}e^{2y}-y\right]_0^3 = \tfrac{1}{2}e^6 - 3 - \tfrac{1}{2} = \tfrac{1}{2}e^6 - \tfrac{7}{2}$

16. $\iint_D xy\, dA = \int_0^1 \int_{y^2}^{y+2} xy\, dx\, dy = \int_0^1 y\left[\tfrac{1}{2}x^2\right]_{x=y^2}^{x=y+2} dy = \tfrac{1}{2}\int_0^1 y((y+2)^2 - y^4)\, dy$

$$= \tfrac{1}{2}\int_0^1 (y^3 + 4y^2 + 4y - y^5)\, dy = \tfrac{1}{2}\left[\tfrac{1}{4}y^4 + \tfrac{4}{3}y^3 + 2y^2 - \tfrac{1}{6}y^6\right]_0^1 = \tfrac{41}{24}$$

17.

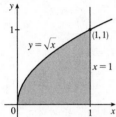

$$\iint_D \frac{y}{1+x^2}\, dA = \int_0^1 \int_0^{\sqrt{x}} \frac{y}{1+x^2}\, dy\, dx = \int_0^1 \frac{1}{1+x^2}\left[\tfrac{1}{2}y^2\right]_{y=0}^{y=\sqrt{x}} dx$$

$$= \tfrac{1}{2}\int_0^1 \frac{x}{1+x^2}\, dx = \left[\tfrac{1}{4}\ln(1+x^2)\right]_0^1 = \tfrac{1}{4}\ln 2$$

18. $\iint_D \frac{1}{1+x^2}\, dA = \int_0^1 \int_x^1 \frac{1}{1+x^2}\, dy\, dx = \int_0^1 \frac{1}{1+x^2}\left[y\right]_{y=x}^{y=1} dx = \int_0^1 \frac{1-x}{1+x^2}\, dx = \int_0^1 \left(\frac{1}{1+x^2} - \frac{x}{1+x^2}\right) dx$

$$= \left[\tan^{-1} x - \tfrac{1}{2}\ln(1+x^2)\right]_0^1 = \tan^{-1} 1 - \tfrac{1}{2}\ln 2 - \left(\tan^{-1} 0 - \tfrac{1}{2}\ln 1\right) = \tfrac{\pi}{4} - \tfrac{1}{2}\ln 2$$

19.

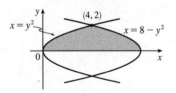

$$\iint_D y\, dA = \int_0^2 \int_{y^2}^{8-y^2} y\, dx\, dy$$

$$= \int_0^2 y\left[x\right]_{x=y^2}^{x=8-y^2} dy = \int_0^2 y(8 - y^2 - y^2)\, dy$$

$$= \int_0^2 (8y - 2y^3)\, dy = \left[4y^2 - \tfrac{1}{2}y^4\right]_0^2 = 8$$

20.

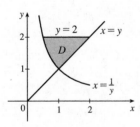

$$\iint_D y\, dA = \int_1^2 \int_{1/y}^y y\, dx\, dy = \int_1^2 y\left(y - \frac{1}{y}\right) dy$$

$$= \int_1^2 (y^2 - 1)\, dy = \left[\tfrac{1}{3}y^3 - y\right]_1^2$$

$$= \left(\tfrac{8}{3} - 2\right) - \left(\tfrac{1}{3} - 1\right) = \tfrac{4}{3}$$

21.

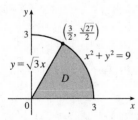

$$\iint_D (x^2+y^2)^{3/2}\, dA = \int_0^{\pi/3} \int_0^3 (r^2)^{3/2} r\, dr\, d\theta$$

$$= \int_0^{\pi/3} d\theta \int_0^3 r^4\, dr = \left[\theta\right]_0^{\pi/3} \left[\tfrac{1}{5}r^5\right]_0^3$$

$$= \frac{\pi}{3}\frac{3^5}{5} = \frac{81\pi}{5}$$

22. $\iint_D x\, dA = \int_0^{\pi/2} \int_1^{\sqrt{2}} (r\cos\theta)\, r\, dr\, d\theta = \int_0^{\pi/2} \cos\theta\, d\theta \int_1^{\sqrt{2}} r^2\, dr = \left[\sin\theta\right]_0^{\pi/2} \left[\tfrac{1}{3}r^3\right]_1^{\sqrt{2}}$

$$= 1 \cdot \tfrac{1}{3}(2^{3/2} - 1) = \tfrac{1}{3}(2^{3/2} - 1)$$

23. $\iiint_E xy \, dV = \int_0^3 \int_0^x \int_0^{x+y} xy \, dz \, dy \, dx = \int_0^3 \int_0^x xy \left[z \right]_{z=0}^{z=x+y} dy \, dx = \int_0^3 \int_0^x xy(x+y) \, dy \, dx$

$\qquad = \int_0^3 \int_0^x (x^2y + xy^2) \, dy \, dx = \int_0^3 \left[\frac{1}{2}x^2y^2 + \frac{1}{3}xy^3 \right]_{y=0}^{y=x} dx = \int_0^3 \left(\frac{1}{2}x^4 + \frac{1}{3}x^4 \right) dx$

$\qquad = \frac{5}{6} \int_0^3 x^4 \, dx = \left[\frac{1}{6}x^5 \right]_0^3 = \frac{81}{2} = 40.5$

24. $\iiint_T xy \, dV = \int_0^{1/3} \int_0^{1-3x} \int_0^{1-3x-y} xy \, dz \, dy \, dx = \int_0^{1/3} \int_0^{1-3x} xy(1 - 3x - y) \, dy \, dx$

$\qquad = \int_0^{1/3} \int_0^{1-3x} (xy - 3x^2y - xy^2) \, dy \, dx$

$\qquad = \int_0^{1/3} \left[\frac{1}{2}xy^2 - \frac{3}{2}x^2y^2 - \frac{1}{3}xy^3 \right]_{y=0}^{y=1-3x} dx$

$\qquad = \int_0^{1/3} \left[\frac{1}{2}x(1 - 3x)^2 - \frac{3}{2}x^2(1 - 3x)^2 - \frac{1}{3}x(1 - 3x)^3 \right] dx$

$\qquad = \int_0^{1/3} \left(\frac{1}{6}x - \frac{3}{2}x^2 + \frac{9}{2}x^3 - \frac{9}{2}x^4 \right) dx$

$\qquad = \frac{1}{12}x^2 - \frac{1}{2}x^3 + \frac{9}{8}x^4 - \frac{9}{10}x^5 \Big]_0^{1/3} = \frac{1}{1080}$

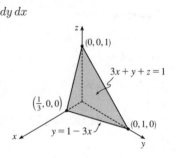

25. $\iiint_E y^2z^2 \, dV = \int_{-1}^1 \int_{-\sqrt{1-y^2}}^{\sqrt{1-y^2}} \int_0^{1-y^2-z^2} y^2z^2 \, dx \, dz \, dy = \int_{-1}^1 \int_{-\sqrt{1-y^2}}^{\sqrt{1-y^2}} y^2z^2(1 - y^2 - z^2) \, dz \, dy$

$\qquad = \int_0^{2\pi} \int_0^1 (r^2 \cos^2\theta)(r^2 \sin^2\theta)(1 - r^2) \, r \, dr \, d\theta = \int_0^{2\pi} \int_0^1 \frac{1}{4} \sin^2 2\theta (r^5 - r^7) \, dr \, d\theta$

$\qquad = \int_0^{2\pi} \frac{1}{8}(1 - \cos 4\theta) \left[\frac{1}{6}r^6 - \frac{1}{8}r^8 \right]_{r=0}^{r=1} d\theta = \frac{1}{192} \left[\theta - \frac{1}{4} \sin 4\theta \right]_0^{2\pi} = \frac{2\pi}{192} = \frac{\pi}{96}$

26. $\iiint_E z \, dV = \int_0^1 \int_0^{\sqrt{1-y^2}} \int_0^{2-y} z \, dx \, dz \, dy = \int_0^1 \int_0^{\sqrt{1-y^2}} (2-y)z \, dz \, dy = \int_0^1 \frac{1}{2}(2-y)(1-y^2) \, dy$

$\qquad = \int_0^1 \frac{1}{2}(2 - y - 2y^2 + y^3) \, dy = \frac{13}{24}$

27. $\iiint_E yz \, dV = \int_{-2}^2 \int_0^{\sqrt{4-x^2}} \int_0^y yz \, dz \, dy \, dx = \int_{-2}^2 \int_0^{\sqrt{4-x^2}} \frac{1}{2}y^3 \, dy \, dx = \int_0^\pi \int_0^2 \frac{1}{2}r^3(\sin^3\theta) \, r \, dr \, d\theta$

$\qquad = \frac{16}{5} \int_0^\pi \sin^3\theta \, d\theta = \frac{16}{5} \left[-\cos\theta + \frac{1}{3}\cos^3\theta \right]_0^\pi = \frac{64}{15}$

28. $\iiint_H z^3 \sqrt{x^2 + y^2 + z^2} \, dV = \int_0^{2\pi} \int_0^{\pi/2} \int_0^1 (\rho^3 \cos^3\phi)\rho(\rho^2 \sin\phi) \, d\rho \, d\phi \, d\theta$

$\qquad = \int_0^{2\pi} d\theta \int_0^{\pi/2} \cos^3\phi \sin\phi \, d\phi \int_0^1 \rho^6 \, d\rho = 2\pi \left[-\frac{1}{4}\cos^4\phi \right]_0^{\pi/2} \left(\frac{1}{7} \right) = \frac{\pi}{14}$

29. $V = \int_0^2 \int_1^4 (x^2 + 4y^2) \, dy \, dx = \int_0^2 \left[x^2y + \frac{4}{3}y^3 \right]_{y=1}^{y=4} dx = \int_0^2 (3x^2 + 84) \, dx = 176$

30.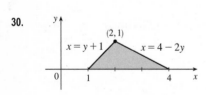

$\qquad V = \int_0^1 \int_{y+1}^{4-2y} \int_0^{x^2y} dz \, dx \, dy = \int_0^1 \int_{y+1}^{4-2y} x^2y \, dx \, dy$

$\qquad = \int_0^1 \frac{1}{3} \left[(4 - 2y)^3 y - (y+1)^3 y \right] dy$

$\qquad = \int_0^1 3(-y^4 + 5y^3 - 11y^2 + 7y) \, dy = 3\left(-\frac{1}{5} + \frac{5}{4} - \frac{11}{3} + \frac{7}{2} \right) = \frac{53}{20}$

31.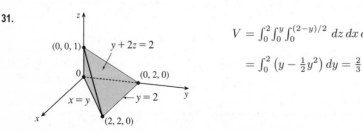

$\qquad V = \int_0^2 \int_0^y \int_0^{(2-y)/2} dz \, dx \, dy = \int_0^2 \int_0^y \left(1 - \frac{1}{2}y \right) dx \, dy$

$\qquad = \int_0^2 \left(y - \frac{1}{2}y^2 \right) dy = \frac{2}{3}$

32. $V = \int_0^{2\pi} \int_0^2 \int_0^{3-r\sin\theta} r\, dz\, dr\, d\theta = \int_0^{2\pi} \int_0^2 (3r - r^2 \sin\theta)\, dr\, d\theta = \int_0^{2\pi} \left[6 - \frac{8}{3}\sin\theta \right] d\theta = 6\theta\big]_0^{2\pi} + 0 = 12\pi$

33. Using the wedge above the plane $z = 0$ and below the plane $z = mx$ and noting that we have the same volume for $m < 0$ as

for $m > 0$ (so use $m > 0$), we have

$V = 2\int_0^{a/3} \int_0^{\sqrt{a^2 - 9y^2}} mx\, dx\, dy = 2\int_0^{a/3} \frac{1}{2}m(a^2 - 9y^2)\, dy = m\left[a^2 y - 3y^3 \right]_0^{a/3} = m\left(\frac{1}{3}a^3 - \frac{1}{9}a^3 \right) = \frac{2}{9}ma^3.$

34. The paraboloid and the half-cone intersect when $x^2 + y^2 = \sqrt{x^2 + y^2}$, that is when $x^2 + y^2 = 1$ or 0. So

$V = \iint\limits_{x^2+y^2 \leq 1} \int_{x^2+y^2}^{\sqrt{x^2+y^2}} dz\, dA = \int_0^{2\pi} \int_0^1 \int_{r^2}^r r\, dz\, dr\, d\theta = \int_0^{2\pi} \int_0^1 (r^2 - r^3)\, dr\, d\theta = \int_0^{2\pi} \left(\frac{1}{3} - \frac{1}{4} \right) d\theta = \frac{1}{12}(2\pi) = \frac{\pi}{6}.$

35. (a) $m = \int_0^1 \int_0^{1-y^2} y\, dx\, dy = \int_0^1 (y - y^3)\, dy = \frac{1}{2} - \frac{1}{4} = \frac{1}{4}$

(b) $M_y = \int_0^1 \int_0^{1-y^2} xy\, dx\, dy = \int_0^1 \frac{1}{2}y(1-y^2)^2\, dy = -\frac{1}{12}(1-y^2)^3 \big]_0^1 = \frac{1}{12}$,

$M_x = \int_0^1 \int_0^{1-y^2} y^2\, dx\, dy = \int_0^1 (y^2 - y^4)\, dy = \frac{2}{15}$. Hence $(\overline{x}, \overline{y}) = \left(\frac{1}{3}, \frac{8}{15} \right)$.

(c) $I_x = \int_0^1 \int_0^{1-y^2} y^3\, dx\, dy = \int_0^1 (y^3 - y^5)\, dy = \frac{1}{12}$,

$I_y = \int_0^1 \int_0^{1-y^2} yx^2\, dx\, dy = \int_0^1 \frac{1}{3}y(1-y^2)^3\, dy = -\frac{1}{24}(1-y^2)^4 \big]_0^1 = \frac{1}{24}$,

$I_0 = I_x + I_y = \frac{1}{8}$, $\overline{\overline{y}}^2 = \frac{1/12}{1/4} = \frac{1}{3} \Rightarrow \overline{\overline{y}} = \frac{1}{\sqrt{3}}$, and $\overline{\overline{x}}^2 = \frac{1/24}{1/4} = \frac{1}{6} \Rightarrow \overline{\overline{x}} = \frac{1}{\sqrt{6}}$.

36. (a) $m = \frac{1}{4}\pi K a^2$ where K is constant,

$M_y = \iint\limits_{x^2+y^2 \leq a^2} Kx\, dA = K\int_0^{\pi/2} \int_0^a r^2 \cos\theta\, dr\, d\theta = \frac{1}{3}Ka^3 \int_0^{\pi/2} \cos\theta\, d\theta = \frac{1}{3}a^3 K$, and

$M_x = K\int_0^{\pi/2} \int_0^a r^2 \sin\theta\, dr\, d\theta = \frac{1}{3}a^3 K$ [by symmetry $M_y = M_x$].

Hence the centroid is $(\overline{x}, \overline{y}) = \left(\frac{4}{3\pi}a, \frac{4}{3\pi}a \right)$.

(b) $m = \int_0^{\pi/2} \int_0^a r^4 \cos\theta \sin^2\theta\, dr\, d\theta = \left[\frac{1}{3}\sin^3\theta \right]_0^{\pi/2} \left(\frac{1}{5}a^5 \right) = \frac{1}{15}a^5$,

$M_y = \int_0^{\pi/2} \int_0^a r^5 \cos^2\theta \sin^2\theta\, dr\, d\theta = \frac{1}{8}\left[\theta - \frac{1}{4}\sin 4\theta \right]_0^{\pi/2} \left(\frac{1}{6}a^6 \right) = \frac{1}{96}\pi a^6$, and

$M_x = \int_0^{\pi/2} \int_0^a r^5 \cos\theta \sin^3\theta\, dr\, d\theta = \left[\frac{1}{4}\sin^4\theta \right]_0^{\pi/2} \left(\frac{1}{6}a^6 \right) = \frac{1}{24}a^6$. Hence $(\overline{x}, \overline{y}) = \left(\frac{5}{32}\pi a, \frac{5}{8}a \right)$.

37. The equation of the cone with the suggested orientation is $(h - z) = \frac{h}{a}\sqrt{x^2 + y^2}, 0 \leq z \leq h$. Then $V = \frac{1}{3}\pi a^2 h$ is the

volume of one frustum of a cone; by symmetry $M_{yz} = M_{xz} = 0$; and

$M_{xy} = \iint\limits_{x^2+y^2 \leq a^2} \int_0^{h-(h/a)\sqrt{x^2+y^2}} z\, dz\, dA = \int_0^{2\pi} \int_0^a \int_0^{(h/a)(a-r)} rz\, dz\, dr\, d\theta = \pi \int_0^a r\frac{h^2}{a^2}(a-r)^2\, dr$

$= \frac{\pi h^2}{a^2} \int_0^a (a^2 r - 2ar^2 + r^3)\, dr = \frac{\pi h^2}{a^2}\left(\frac{a^4}{2} - \frac{2a^4}{3} + \frac{a^4}{4} \right) = \frac{\pi h^2 a^2}{12}$

Hence the centroid is $(\overline{x}, \overline{y}, \overline{z}) = \left(0, 0, \frac{1}{4}h \right)$.

38. $I_z = \int_0^{2\pi} \int_0^a \int_0^{(h/a)(a-r)} r^3\, dz\, dr\, d\theta = 2\pi \int_0^a \frac{h}{a}(ar^3 - r^4)\, dr = \frac{2\pi h}{a}\left(\frac{a^5}{4} - \frac{a^5}{5} \right) = \frac{\pi a^4 h}{10}$

39.

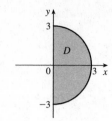

$$\int_0^3 \int_{-\sqrt{9-x^2}}^{\sqrt{9-x^2}} (x^3 + xy^2)\, dy\, dx = \int_0^3 \int_{-\sqrt{9-x^2}}^{\sqrt{9-x^2}} x(x^2 + y^2)\, dy\, dx$$

$$= \int_{-\pi/2}^{\pi/2} \int_0^3 (r\cos\theta)(r^2)\, r\, dr\, d\theta$$

$$= \int_{-\pi/2}^{\pi/2} \cos\theta\, d\theta \int_0^3 r^4\, dr$$

$$= \left[\sin\theta\right]_{-\pi/2}^{\pi/2} \left[\tfrac{1}{5}r^5\right]_0^3 = 2\cdot\tfrac{1}{5}(243) = \tfrac{486}{5} = 97.2$$

40. The region of integration is the solid hemisphere $x^2 + y^2 + z^2 \le 4$, $x \ge 0$.

$$\int_{-2}^2 \int_0^{\sqrt{4-y^2}} \int_{-\sqrt{4-x^2-y^2}}^{\sqrt{4-x^2-y^2}} y^2\,\sqrt{x^2+y^2+z^2}\, dz\, dx\, dy$$

$$= \int_{-\pi/2}^{\pi/2} \int_0^\pi \int_0^2 (\rho\sin\phi\sin\theta)^2 \left(\sqrt{\rho^2}\right)\rho^2 \sin\phi\, d\rho\, d\phi\, d\theta = \int_{-\pi/2}^{\pi/2} \sin^2\theta\, d\theta \int_0^\pi \sin^3\phi\, d\phi \int_0^2 \rho^5\, d\rho$$

$$= \left[\tfrac{1}{2}\theta - \tfrac{1}{4}\sin 2\theta\right]_{-\pi/2}^{\pi/2} \left[-\tfrac{1}{3}(2+\sin^2\phi)\cos\phi\right]_0^\pi \left[\tfrac{1}{6}\rho^6\right]_0^2 = \left(\tfrac{\pi}{2}\right)\left(\tfrac{2}{3} + \tfrac{2}{3}\right)\left(\tfrac{32}{3}\right) = \tfrac{64}{9}\pi$$

41. From the graph, it appears that $1 - x^2 = e^x$ at $x \approx -0.71$ and at

$x = 0$, with $1 - x^2 > e^x$ on $(-0.71, 0)$. So the desired integral is

$$\iint_D y^2\, dA \approx \int_{-0.71}^0 \int_{e^x}^{1-x^2} y^2\, dy\, dx$$

$$= \tfrac{1}{3} \int_{-0.71}^0 [(1-x^2)^3 - e^{3x}]\, dx$$

$$= \tfrac{1}{3}\left[x - x^3 + \tfrac{3}{5}x^5 - \tfrac{1}{7}x^7 - \tfrac{1}{3}e^{3x}\right]_{-0.71}^0 \approx 0.0512$$

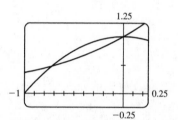

42. Let the tetrahedron be called T. The front face of T is given by the plane $x + \tfrac{1}{2}y + \tfrac{1}{3}z = 1$, or $z = 3 - 3x - \tfrac{3}{2}y$,

which intersects the xy-plane in the line $y = 2 - 2x$. So the total mass is

$$m = \iiint_T \rho(x,y,z)\, dV = \int_0^1 \int_0^{2-2x} \int_0^{3-3x-3y/2} (x^2 + y^2 + z^2)\, dz\, dy\, dx = \tfrac{7}{5}.$$ The center of mass is

$$(\overline{x},\overline{y},\overline{z}) = \left(m^{-1}\iiint_T x\rho(x,y,z)\, dV,\, m^{-1}\iiint_T y\rho(x,y,z)\, dV,\, m^{-1}\iiint_T z\rho(x,y,z)\, dV\right) = \left(\tfrac{4}{21}, \tfrac{11}{21}, \tfrac{8}{7}\right).$$

43. (a) $f(x,y)$ is a joint density function, so we know that $\iint_{\mathbb{R}^2} f(x,y)\, dA = 1$. Since $f(x,y) = 0$ outside the rectangle

$[0,3] \times [0,2]$, we can say

$$\iint_{\mathbb{R}^2} f(x,y)\, dA = \int_{-\infty}^\infty \int_{-\infty}^\infty f(x,y)\, dy\, dx = \int_0^3 \int_0^2 C(x+y)\, dy\, dx$$

$$= C\int_0^3 \left[xy + \tfrac{1}{2}y^2\right]_{y=0}^{y=2} dx = C\int_0^3 (2x+2)\, dx = C\left[x^2 + 2x\right]_0^3 = 15C$$

Then $15C = 1 \;\Rightarrow\; C = \tfrac{1}{15}$.

(b) $P(X \le 2, Y \ge 1) = \int_{-\infty}^2 \int_1^\infty f(x,y)\, dy\, dx = \int_0^2 \int_1^2 \tfrac{1}{15}(x,y)\, dy\, dx = \tfrac{1}{15}\int_0^2 \left[xy + \tfrac{1}{2}y^2\right]_{y=1}^{y=2} dx$

$$= \tfrac{1}{15}\int_0^2 \left(x + \tfrac{3}{2}\right) dx = \tfrac{1}{15}\left[\tfrac{1}{2}x^2 + \tfrac{3}{2}x\right]_0^2 = \tfrac{1}{3}$$

(c) $P(X + Y \le 1) = P((X,Y) \in D)$ where D is the triangular region shown in the figure. Thus

$$P(X + Y \le 1) = \iint_D f(x,y)\,dA = \int_0^1 \int_0^{1-x} \tfrac{1}{15}(x+y)\,dy\,dx$$

$$= \tfrac{1}{15} \int_0^1 \left[xy + \tfrac{1}{2}y^2 \right]_{y=0}^{y=1-x}\,dx$$

$$= \tfrac{1}{15} \int_0^1 \left[x(1-x) + \tfrac{1}{2}(1-x)^2 \right]\,dx$$

$$= \tfrac{1}{30} \int_0^1 (1-x^2)\,dx = \tfrac{1}{30}\left[x - \tfrac{1}{3}x^3 \right]_0^1 = \tfrac{1}{45}$$

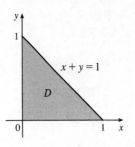

44. Each lamp has exponential density function

$$f(t) = \begin{cases} 0 & \text{if } t < 0 \\ \frac{1}{800} e^{-t/800} & \text{if } t \ge 0 \end{cases}$$

If X, Y, and Z are the lifetimes of the individual bulbs, then X, Y, and Z are independent, so the joint density function is the product of the individual density functions:

$$f(x,y,z) = \begin{cases} \frac{1}{800^3} e^{-(x+y+z)/800} & \text{if } x \ge 0, y \ge 0, z \ge 0 \\ 0 & \text{otherwise} \end{cases}$$

The probability that all three bulbs fail within a total of 1000 hours is $P(X + Y + Z \le 1000)$, or equivalently $P((X,Y,Z) \in E)$ where E is the solid region in the first octant bounded by the coordinate planes and the plane $x + y + z = 1000$. The plane $x + y + z = 1000$ meets the xy-plane in the line $x + y = 1000$, so we have

$$P(X + Y + Z \le 1000) = \iiint_E f(x,y,z)\,dV = \int_0^{1000} \int_0^{1000-x} \int_0^{1000-x-y} \frac{1}{800^3} e^{-(x+y+z)/800}\,dz\,dy\,dx$$

$$= \frac{1}{800^3} \int_0^{1000} \int_0^{1000-x} -800 \left[e^{-(x+y+z)/800} \right]_{z=0}^{z=1000-x-y}\,dy\,dx$$

$$= \frac{-1}{800^2} \int_0^{1000} \int_0^{1000-x} \left[e^{-5/4} - e^{-(x+y)/800} \right]\,dy\,dx$$

$$= \frac{-1}{800^2} \int_0^{1000} \left[e^{-5/4}y + 800e^{-(x+y)/800} \right]_{y=0}^{y=1000-x}\,dx$$

$$= \frac{-1}{800^2} \int_0^{1000} \left[e^{-5/4}(1800 - x) - 800e^{-x/800} \right]\,dx$$

$$= \frac{-1}{800^2} \left[-\tfrac{1}{2}e^{-5/4}(1800 - x)^2 + 800^2 e^{-x/800} \right]_0^{1000}$$

$$= \frac{-1}{800^2} \left[-\tfrac{1}{2}e^{-5/4}(800)^2 + 800^2 e^{-5/4} + \tfrac{1}{2}e^{-5/4}(1800)^2 - 800^2 \right]$$

$$= 1 - \tfrac{97}{32}e^{-5/4} \approx 0.1315$$

45.

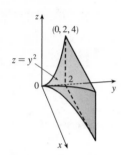
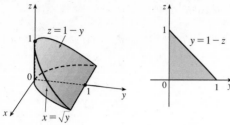
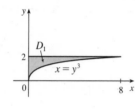

$$\int_{-1}^{1} \int_{x^2}^{1} \int_{0}^{1-y} f(x,y,z)\,dz\,dy\,dx = \int_{0}^{1} \int_{0}^{1-z} \int_{-\sqrt{y}}^{\sqrt{y}} f(x,y,z)\,dx\,dy\,dz$$

46.

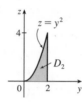
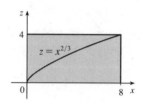

$$\int_{0}^{2} \int_{0}^{y^3} \int_{0}^{y^2} f(x,y,z)\,dz\,dx\,dy = \iiint_{E} f(x,y,z)\,dV \text{ where } E = \left\{(x,y,z) \mid 0 \le y \le 2, 0 \le x \le y^3, 0 \le z \le y^2\right\}.$$

If D_1, D_2, and D_3 are the projections of E on the xy-, yz-, and xz-planes, then

$$D_1 = \left\{(x,y) \mid 0 \le y \le 2, 0 \le x \le y^3\right\} = \left\{(x,y) \mid 0 \le x \le 8, \sqrt[3]{x} \le y \le 2\right\},$$

$$D_2 = \left\{(y,z) \mid 0 \le z \le 4, \sqrt{z} \le y \le 2\right\} = \left\{(y,z) \mid 0 \le y \le 2, 0 \le z \le y^2\right\}, D_3 = \left\{(x,z) \mid 0 \le x \le 8, 0 \le z \le 4\right\}.$$

Therefore we have

$$\int_{0}^{2} \int_{0}^{y^3} \int_{0}^{y^2} f(x,y,z)\,dz\,dx\,dy = \int_{0}^{8} \int_{\sqrt[3]{x}}^{2} \int_{0}^{y^2} f(x,y,z)\,dz\,dy\,dx = \int_{0}^{4} \int_{\sqrt{z}}^{2} \int_{0}^{y^3} f(x,y,z)\,dx\,dy\,dz$$

$$= \int_{0}^{2} \int_{0}^{y^2} \int_{0}^{y^3} f(x,y,z)\,dx\,dz\,dy$$

$$= \int_{0}^{8} \int_{0}^{x^{2/3}} \int_{\sqrt[3]{x}}^{2} f(x,y,z)\,dy\,dz\,dx + \int_{0}^{8} \int_{x^{2/3}}^{4} \int_{\sqrt{z}}^{2} f(x,y,z)\,dy\,dz\,dx$$

$$= \int_{0}^{4} \int_{0}^{z^{3/2}} \int_{\sqrt{z}}^{2} f(x,y,z)\,dy\,dx\,dz + \int_{0}^{4} \int_{z^{3/2}}^{8} \int_{\sqrt[3]{x}}^{2} f(x,y,z)\,dy\,dx\,dz$$

47. Since $u = x - y$ and $v = x + y$, $x = \frac{1}{2}(u+v)$ and $y = \frac{1}{2}(v-u)$.

Thus $\dfrac{\partial(x,y)}{\partial(u,v)} = \begin{vmatrix} 1/2 & 1/2 \\ -1/2 & 1/2 \end{vmatrix} = \dfrac{1}{2}$ and $\iint_{R} \dfrac{x-y}{x+y}\,dA = \int_{2}^{4} \int_{-2}^{0} \dfrac{u}{v}\left(\dfrac{1}{2}\right)\,du\,dv = -\int_{2}^{4} \dfrac{dv}{v} = -\ln 2.$

48. $\dfrac{\partial(x,y,z)}{\partial(u,v,w)} = \begin{vmatrix} 2u & 0 & 0 \\ 0 & 2v & 0 \\ 0 & 0 & 2w \end{vmatrix} = 8uvw$, so

$$V = \iiint_{E} dV = \int_{0}^{1} \int_{0}^{1-u} \int_{0}^{1-u-v} 8uvw\,dw\,dv\,du = \int_{0}^{1} \int_{0}^{1-u} 4uv(1-u-v)^2\,du$$

$$= \int_{0}^{1} \int_{0}^{1-u} \left[4u(1-u)^2 v - 8u(1-u)v^2 + 4uv^3\right]\,dv\,du$$

$$= \int_{0}^{1} \left[2u(1-u)^4 - \tfrac{8}{3}u(1-u)^4 + u(1-u)^4\right]\,du = \int_{0}^{1} \tfrac{1}{3}u(1-u)^4\,du$$

$$= \int_{0}^{1} \tfrac{1}{3}\left[(1-u)^4 - (1-u)^5\right]\,du = \tfrac{1}{3}\left[-\tfrac{1}{5}(1-u)^5 + \tfrac{1}{6}(1-u)^6\right]_{0}^{1} = \tfrac{1}{3}\left(-\tfrac{1}{6} + \tfrac{1}{5}\right) = \tfrac{1}{90}$$

49. Let $u = y - x$ and $v = y + x$ so $x = y - u = (v - x) - u \;\Rightarrow\; x = \frac{1}{2}(v - u)$ and $y = v - \frac{1}{2}(v - u) = \frac{1}{2}(v + u)$.

$\left| \dfrac{\partial(x,y)}{\partial(u,v)} \right| = \left| \dfrac{\partial x}{\partial u}\dfrac{\partial y}{\partial v} - \dfrac{\partial x}{\partial v}\dfrac{\partial y}{\partial u} \right| = \left| -\frac{1}{2}\left(\frac{1}{2}\right) - \frac{1}{2}\left(\frac{1}{2}\right) \right| = \left| -\frac{1}{2} \right| = \frac{1}{2}$. R is the image under this transformation of the square

with vertices $(u, v) = (0, 0)$, $(-2, 0)$, $(0, 2)$, and $(-2, 2)$. So

$$\iint_R xy\,dA = \int_0^2 \int_{-2}^0 \frac{v^2 - u^2}{4}\left(\frac{1}{2}\right) du\,dv = \frac{1}{8}\int_0^2 \left[v^2 u - \frac{1}{3}u^3\right]_{u=-2}^{u=0} dv = \frac{1}{8}\int_0^2 \left(2v^2 - \frac{8}{3}\right) dv = \frac{1}{8}\left[\frac{2}{3}v^3 - \frac{8}{3}v\right]_0^2 = 0$$

This result could have been anticipated by symmetry, since the integrand is an odd function of y and R is symmetric about the x-axis.

50. By the Extreme Value Theorem (15.7.8 [ET 14.7.8]), f has an absolute minimum value m and an absolute maximum value M in D. Then by Property 16.3.11 [ET 15.3.11], $mA(D) \le \iint_D f(x, y)\,dA \le MA(D)$. Dividing through by the positive

number $A(D)$, we get $m \le \dfrac{1}{A(D)}\iint_D f(x, y)\,dA \le M$. This says that the average value of f over D lies between m and

M. But f is continuous on D and takes on the values m and M, and so by the Intermediate Value Theorem must take on all

values between m and M. Specifically, there exists a point (x_0, y_0) in D such that $f(x_0, y_0) = \dfrac{1}{A(D)}\iint_D f(x, y)\,dA$ or

equivalently $\iint_D f(x, y)\,dA = f(x_0, y_0)\,A(D)$.

51. For each r such that D_r lies within the domain, $A(D_r) = \pi r^2$, and by the Mean Value Theorem for Double Integrals there

exists (x_r, y_r) in D_r such that $f(x_r, y_r) = \dfrac{1}{\pi r^2}\iint_{D_r} f(x, y)\,dA$. But $\lim\limits_{r \to 0^+}(x_r, y_r) = (a, b)$,

so $\lim\limits_{r \to 0^+} \dfrac{1}{\pi r^2}\iint_{D_r} f(x, y)\,dA = \lim\limits_{r \to 0^+} f(x_r, y_r) = f(a, b)$ by the continuity of f.

52. (a) $\displaystyle\iint_D \frac{1}{(x^2 + y^2)^{n/2}}\,dA = \int_0^{2\pi}\int_r^R \frac{1}{(t^2)^{n/2}}t\,dt\,d\theta = 2\pi\int_r^R t^{1-n}\,dt$

$$= \begin{cases} \dfrac{2\pi}{2-n}t^{2-n}\Big]_r^R = \dfrac{2\pi}{2-n}\left(R^{2-n} - r^{2-n}\right) & \text{if } n \ne 2 \\[2mm] 2\pi \ln(R/r) & \text{if } n = 2 \end{cases}$$

(b) The integral in part (a) has a limit as $r \to 0^+$ for all values of n such that $2 - n > 0 \;\Leftrightarrow\; n < 2$.

(c) $\displaystyle\iiint_E \frac{1}{(x^2 + y^2 + z^2)^{n/2}}\,dV = \int_r^R\int_0^\pi\int_0^{2\pi} \frac{1}{(\rho^2)^{n/2}}\rho^2 \sin\phi\,d\theta\,d\phi\,d\rho = 2\pi\int_r^R\int_0^\pi \rho^{2-n}\sin\phi\,d\phi\,d\rho$

$$= \begin{cases} \dfrac{4\pi}{3-n}\rho^{3-n}\Big]_r^R = \dfrac{4\pi}{3-n}\left(R^{3-n} - r^{3-n}\right) & \text{if } n \ne 3 \\[2mm] 4\pi \ln(R/r) & \text{if } n = 3 \end{cases}$$

(d) As $r \to 0^+$, the above integral has a limit, provided that $3 - n > 0 \;\Leftrightarrow\; n < 3$.

□ PROBLEMS PLUS

1.

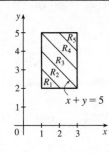

Let $R = \bigcup_{i=1}^{5} R_i$, where

$R_i = \{(x,y) \mid x+y \geq i+2, x+y < i+3, 1 \leq x \leq 3, 2 \leq y \leq 5\}$.

$\iint_R [\![x+y]\!] \, dA = \sum_{i=1}^{5} \iint_{R_i} [\![x+y]\!] \, dA = \sum_{i=1}^{5} [\![x+y]\!] \iint_{R_i} dA$, since

$[\![x+y]\!] = \text{constant} = i+2$ for $(x,y) \in R_i$. Therefore

$\iint_R [\![x+y]\!] \, dA = \sum_{i=1}^{5} (i+2) [A(R_i)]$
$= 3A(R_1) + 4A(R_2) + 5A(R_3) + 6A(R_4) + 7A(R_5)$
$= 3\left(\frac{1}{2}\right) + 4\left(\frac{3}{2}\right) + 5(2) + 6\left(\frac{3}{2}\right) + 7\left(\frac{1}{2}\right) = 30$

2.

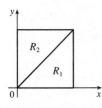

Let $R = \{(x,y) \mid 0 \leq x, y \leq 1\}$. For $x, y \in R$, $\max\{x^2, y^2\} = x^2$ if $x \geq y$,

and $\max\{x^2, y^2\} = y^2$ if $x \leq y$. Therefore we divide R into two regions:

$R = R_1 \cup R_2$, where $R_1 = \{(x,y) \mid 0 \leq x \leq 1, 0 \leq y \leq x\}$ and

$R_2 = \{(x,y) \mid 0 \leq y \leq 1, 0 \leq x \leq y\}$. Now $\max\{x^2, y^2\} = x^2$ for

$(x,y) \in R_1$, and $\max\{x^2, y^2\} = y^2$ for $(x,y) \in R_2 \Rightarrow$

$\int_0^1 \int_0^1 e^{\max\{x^2, y^2\}} \, dy \, dx = \iint_R e^{\max\{x^2, y^2\}} \, dA = \iint_{R_1} e^{\max\{x^2, y^2\}} \, dA + \iint_{R_2} e^{\max\{x^2, y^2\}} \, dA$

$= \int_0^1 \int_0^x e^{x^2} \, dy \, dx + \int_0^1 \int_0^y e^{y^2} \, dx \, dy = \int_0^1 x e^{x^2} \, dx + \int_0^1 y e^{y^2} \, dy = e^{x^2}\Big]_0^1 = e - 1$

3. $f_{\text{ave}} = \dfrac{1}{b-a} \displaystyle\int_a^b f(x) \, dx = \dfrac{1}{1-0} \int_0^1 \left[\int_x^1 \cos(t^2) \, dt \right] dx$

$= \int_0^1 \int_x^1 \cos(t^2) \, dt \, dx = \int_0^1 \int_0^t \cos(t^2) \, dx \, dt$ [changing the order of integration]

$= \int_0^1 t \cos(t^2) \, dt = \frac{1}{2} \sin(t^2)\Big]_0^1 = \frac{1}{2} \sin 1$

4. Let $u = \mathbf{a} \cdot \mathbf{r}$, $v = \mathbf{b} \cdot \mathbf{r}$, $w = \mathbf{c} \cdot \mathbf{r}$, where $\mathbf{a} = \langle a_1, a_2, a_3 \rangle$, $\mathbf{b} = \langle b_1, b_2, b_3 \rangle$, $\mathbf{c} = \langle c_1, c_2, c_3 \rangle$. Under this change of variables,

E corresponds to the rectangular box $0 \leq u \leq \alpha$, $0 \leq v \leq \beta$, $0 \leq w \leq \gamma$. So, by Formula 16.9.13 [ET 15.9.13],

$\displaystyle\int_0^\gamma \int_0^\beta \int_0^\alpha uvw \, du \, dv \, dw = \iiint_E (\mathbf{a} \cdot \mathbf{r})(\mathbf{b} \cdot \mathbf{r})(\mathbf{c} \cdot \mathbf{r}) \left| \dfrac{\partial(u,v,w)}{\partial(x,y,z)} \right| dV$. But

$\left| \dfrac{\partial(u,v,w)}{\partial(x,y,z)} \right| = \begin{vmatrix} a_1 & a_2 & a_3 \\ b_1 & b_2 & b_3 \\ c_1 & c_2 & c_3 \end{vmatrix} = |\mathbf{a} \cdot \mathbf{b} \times \mathbf{c}| \Rightarrow$

$\iiint_E (\mathbf{a} \cdot \mathbf{r})(\mathbf{b} \cdot \mathbf{r})(\mathbf{c} \cdot \mathbf{r}) \, dV = \dfrac{1}{|\mathbf{a} \cdot \mathbf{b} \times \mathbf{c}|} \int_0^\gamma \int_0^\beta \int_0^\alpha uvw \, du \, dv \, dw$

$= \dfrac{1}{|\mathbf{a} \cdot \mathbf{b} \times \mathbf{c}|} \left(\dfrac{\alpha^2}{2}\right)\left(\dfrac{\beta^2}{2}\right)\left(\dfrac{\gamma^2}{2}\right) = \dfrac{(\alpha\beta\gamma)^2}{8\,|\mathbf{a} \cdot \mathbf{b} \times \mathbf{c}|}$

5. Since $|xy| < 1$, except at $(1, 1)$, the formula for the sum of a geometric series gives $\dfrac{1}{1 - xy} = \sum\limits_{n=0}^{\infty} (xy)^n$, so

$$\int_0^1 \int_0^1 \frac{1}{1-xy}\, dx\, dy = \int_0^1 \int_0^1 \sum_{n=0}^{\infty} (xy)^n\, dx\, dy = \sum_{n=0}^{\infty} \int_0^1 \int_0^1 (xy)^n\, dx\, dy = \sum_{n=0}^{\infty} \left[\int_0^1 x^n\, dx\right]\left[\int_0^1 y^n\, dy\right]$$

$$= \sum_{n=0}^{\infty} \frac{1}{n+1} \cdot \frac{1}{n+1} = \sum_{n=0}^{\infty} \frac{1}{(n+1)^2} = \frac{1}{1^2} + \frac{1}{2^2} + \frac{1}{3^2} + \cdots = \sum_{n=1}^{\infty} \frac{1}{n^2}$$

6. Let $x = \dfrac{u - v}{\sqrt{2}}$ and $y = \dfrac{u + v}{\sqrt{2}}$. We know the region of integration in the xy-plane, so to find its image in the uv-plane we get

u and v in terms of x and y, and then use the methods of Section 16.9 [ET 15.9]. $x + y = \dfrac{u - v}{\sqrt{2}} + \dfrac{u + v}{\sqrt{2}} = \sqrt{2}\,u$, so

$u = \dfrac{x + y}{\sqrt{2}}$, and similarly $v = \dfrac{y - x}{\sqrt{2}}$. S_1 is given by $y = 0$, $0 \le x \le 1$, so from the equations derived above, the image of S_1

is S_1': $u = \frac{1}{\sqrt{2}} x$, $v = -\frac{1}{\sqrt{2}} x$, $0 \le x \le 1$, that is, $v = -u$, $0 \le u \le \frac{1}{\sqrt{2}}$. Similarly, the image of S_2 is S_2': $v = u - \sqrt{2}$,

$\frac{1}{\sqrt{2}} \le u \le \sqrt{2}$, the image of S_3 is S_3': $v = \sqrt{2} - u$, $\frac{1}{\sqrt{2}} \le u \le \sqrt{2}$, and the image of S_4 is S_4': $v = u$, $0 \le u \le \frac{1}{\sqrt{2}}$.

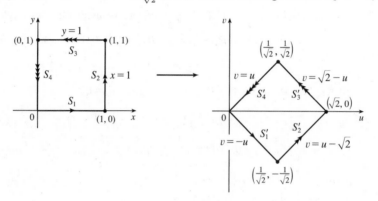

The Jacobian of the transformation is $\dfrac{\partial(x, y)}{\partial(u, v)} = \begin{vmatrix} \partial x/\partial u & \partial x/\partial v \\ \partial y/\partial u & \partial y/\partial v \end{vmatrix} = \begin{vmatrix} \frac{1}{\sqrt{2}} & -\frac{1}{\sqrt{2}} \\ \frac{1}{\sqrt{2}} & \frac{1}{\sqrt{2}} \end{vmatrix} = 1$. From the diagram,

we see that we must evaluate two integrals: one over the region $\left\{(u, v) \mid 0 \le u \le \frac{1}{\sqrt{2}},\ -u \le v \le u\right\}$ and the other

over $\left\{(u, v) \mid \frac{1}{\sqrt{2}} \le u \le \sqrt{2},\ -\sqrt{2} + u \le v \le \sqrt{2} - u\right\}$. So

$$\int_0^1 \int_0^1 \frac{dx\, dy}{1 - xy} = \int_0^{\sqrt{2}/2} \int_{-u}^{u} \frac{dv\, du}{1 - \left[\frac{1}{\sqrt{2}}(u + v)\right]\left[\frac{1}{\sqrt{2}}(u - v)\right]} + \int_{\sqrt{2}/2}^{\sqrt{2}} \int_{-\sqrt{2}+u}^{\sqrt{2}-u} \frac{dv\, du}{1 - \left[\frac{1}{\sqrt{2}}(u + v)\right]\left[\frac{1}{\sqrt{2}}(u - v)\right]}$$

$$= \int_0^{\sqrt{2}/2} \int_{-u}^{u} \frac{2\, dv\, du}{2 - u^2 + v^2} + \int_{\sqrt{2}/2}^{\sqrt{2}} \int_{-\sqrt{2}+u}^{\sqrt{2}-u} \frac{2\, dv\, du}{2 - u^2 + v^2}$$

$$= 2\left[\int_0^{\sqrt{2}/2} \frac{1}{\sqrt{2 - u^2}} \left[\arctan \frac{v}{\sqrt{2 - u^2}}\right]_{-u}^{u} du + \int_{\sqrt{2}/2}^{\sqrt{2}} \frac{1}{\sqrt{2 - u^2}} \left[\arctan \frac{v}{\sqrt{2 - u^2}}\right]_{-\sqrt{2}+u}^{\sqrt{2}-u} du\right]$$

$$= 4\left[\int_0^{\sqrt{2}/2} \frac{1}{\sqrt{2 - u^2}} \arctan \frac{u}{\sqrt{2 - u^2}}\, du + \int_{\sqrt{2}/2}^{\sqrt{2}} \frac{1}{\sqrt{2 - u^2}} \arctan \frac{\sqrt{2} - u}{\sqrt{2 - u^2}}\, du\right]$$

Now let $u = \sqrt{2}\sin\theta$, so $du = \sqrt{2}\cos\theta\, d\theta$ and the limits change to 0 and $\frac{\pi}{6}$ (in the first integral) and $\frac{\pi}{6}$ and $\frac{\pi}{2}$ (in the second integral). Continuing:

$$\int_0^1 \int_0^1 \frac{dx\,dy}{1-xy} = 4\left[\int_0^{\pi/6} \frac{1}{\sqrt{2-2\sin^2\theta}} \arctan\left(\frac{\sqrt{2}\sin\theta}{\sqrt{2-2\sin^2\theta}}\right)\left(\sqrt{2}\cos\theta\,d\theta\right)\right.$$

$$\left. + \int_{\pi/6}^{\pi/2} \frac{1}{\sqrt{2-2\sin^2\theta}} \arctan\left(\frac{\sqrt{2}-\sqrt{2}\sin\theta}{\sqrt{2-2\sin^2\theta}}\right)\left(\sqrt{2}\cos\theta\,d\theta\right)\right]$$

$$= 4\left[\int_0^{\pi/6} \frac{\sqrt{2}\cos\theta}{\sqrt{2}\cos\theta} \arctan\left(\frac{\sqrt{2}\sin\theta}{\sqrt{2}\cos\theta}\right)d\theta + \int_{\pi/6}^{\pi/2} \frac{\sqrt{2}\cos\theta}{\sqrt{2}\cos\theta} \arctan\left(\frac{\sqrt{2}(1-\sin\theta)}{\sqrt{2}\cos\theta}\right)d\theta\right]$$

$$= 4\left[\int_0^{\pi/6} \arctan(\tan\theta)\,d\theta + \int_{\pi/6}^{\pi/2} \arctan\left(\frac{1-\sin\theta}{\cos\theta}\right)d\theta\right]$$

But (following the hint)

$$\frac{1-\sin\theta}{\cos\theta} = \frac{1-\cos\left(\frac{\pi}{2}-\theta\right)}{\sin\left(\frac{\pi}{2}-\theta\right)} = \frac{1-\left[1-2\sin^2\left(\frac{1}{2}\left(\frac{\pi}{2}-\theta\right)\right)\right]}{2\sin\left(\frac{1}{2}\left(\frac{\pi}{2}-\theta\right)\right)\cos\left(\frac{1}{2}\left(\frac{\pi}{2}-\theta\right)\right)} \quad \text{[half-angle formulas]}$$

$$= \frac{2\sin^2\left(\frac{1}{2}\left(\frac{\pi}{2}-\theta\right)\right)}{2\sin\left(\frac{1}{2}\left(\frac{\pi}{2}-\theta\right)\right)\cos\left(\frac{1}{2}\left(\frac{\pi}{2}-\theta\right)\right)} = \tan\left(\frac{1}{2}\left(\frac{\pi}{2}-\theta\right)\right)$$

Continuing:

$$\int_0^1 \int_0^1 \frac{dx\,dy}{1-xy} = 4\left[\int_0^{\pi/6} \arctan(\tan\theta)\,d\theta + \int_{\pi/6}^{\pi/2} \arctan\left(\tan\left(\tfrac{1}{2}\left(\tfrac{\pi}{2}-\theta\right)\right)\right)d\theta\right]$$

$$= 4\left[\int_0^{\pi/6} \theta\,d\theta + \int_{\pi/6}^{\pi/2} \left[\frac{1}{2}\left(\frac{\pi}{2}-\theta\right)\right]d\theta\right] = 4\left(\left[\frac{\theta^2}{2}\right]_0^{\pi/6} + \left[\frac{\pi\theta}{4} - \frac{\theta^2}{4}\right]_{\pi/6}^{\pi/2}\right) = 4\left(\frac{3\pi^2}{72}\right) = \frac{\pi^2}{6}$$

7. (a) Since $|xyz| < 1$ except at $(1,1,1)$, the formula for the sum of a geometric series gives $\dfrac{1}{1-xyz} = \displaystyle\sum_{n=0}^{\infty} (xyz)^n$, so

$$\int_0^1 \int_0^1 \int_0^1 \frac{1}{1-xyz}\,dx\,dy\,dz = \int_0^1 \int_0^1 \int_0^1 \sum_{n=0}^{\infty} (xyz)^n\,dx\,dy\,dz = \sum_{n=0}^{\infty} \int_0^1 \int_0^1 \int_0^1 (xyz)^n\,dx\,dy\,dz$$

$$= \sum_{n=0}^{\infty} \left[\int_0^1 x^n\,dx\right]\left[\int_0^1 y^n\,dy\right]\left[\int_0^1 z^n\,dz\right] = \sum_{n=0}^{\infty} \frac{1}{n+1}\cdot\frac{1}{n+1}\cdot\frac{1}{n+1}$$

$$= \sum_{n=0}^{\infty} \frac{1}{(n+1)^3} = \frac{1}{1^3} + \frac{1}{2^3} + \frac{1}{3^3} + \cdots = \sum_{n=1}^{\infty} \frac{1}{n^3}$$

(b) Since $|-xyz| < 1$, except at $(1,1,1)$, the formula for the sum of a geometric series gives $\dfrac{1}{1+xyz} = \displaystyle\sum_{n=0}^{\infty} (-xyz)^n$, so

$$\int_0^1 \int_0^1 \int_0^1 \frac{1}{1+xyz}\,dx\,dy\,dz = \int_0^1 \int_0^1 \int_0^1 \frac{1}{1+xyz} \sum_{n=0}^{\infty} (-xyz)^n\,dx\,dy\,dz = \sum_{n=0}^{\infty} \int_0^1 \int_0^1 \int_0^1 (-xyz)^n\,dx\,dy\,dz$$

$$= \sum_{n=0}^{\infty} (-1)^n \left[\int_0^1 x^n\,dx\right]\left[\int_0^1 y^n\,dy\right]\left[\int_0^1 z^n\,dz\right] = \sum_{n=0}^{\infty} (-1)^n \frac{1}{n+1}\cdot\frac{1}{n+1}\cdot\frac{1}{n+1}$$

$$= \sum_{n=0}^{\infty} \frac{(-1)^n}{(n+1)^3} = \frac{1}{1^3} - \frac{1}{2^3} + \frac{1}{3^3} - \cdots = \sum_{n=0}^{\infty} \frac{(-1)^{n-1}}{n^3}$$

To evaluate this sum, we first write out a few terms: $s = 1 - \dfrac{1}{2^3} + \dfrac{1}{3^3} - \dfrac{1}{4^3} + \dfrac{1}{5^3} - \dfrac{1}{6^3} \approx 0.8998$. Notice that

$a_7 = \dfrac{1}{7^3} < 0.003$. By the Alternating Series Estimation Theorem from Section 12.5 [ET 11.5], we have

$|s - s_6| \le a_7 < 0.003$. This error of 0.003 will not affect the second decimal place, so we have $s \approx 0.90$.

8. $\displaystyle\int_0^\infty \frac{\arctan \pi x - \arctan x}{x}\,dx = \int_0^\infty \left[\frac{\arctan yx}{x}\right]_{y=1}^{y=\pi}\,dx = \int_0^\infty \int_1^\pi \frac{1}{1+y^2x^2}\,dy\,dx = \int_1^\pi \int_0^\infty \frac{1}{1+y^2x^2}\,dx\,dy$

$$= \int_1^\pi \lim_{t\to\infty}\left[\frac{\arctan yx}{y}\right]_{x=0}^{x=t}\,dy = \int_1^\pi \frac{\pi}{2y}\,dy = \frac{\pi}{2}\left[\ln y\right]_1^\pi = \frac{\pi}{2}\ln \pi$$

9. (a) $x = r\cos\theta,\ y = r\sin\theta,\ z = z.$ Then $\dfrac{\partial u}{\partial r} = \dfrac{\partial u}{\partial x}\dfrac{\partial x}{\partial r} + \dfrac{\partial u}{\partial y}\dfrac{\partial y}{\partial r} + \dfrac{\partial u}{\partial z}\dfrac{\partial z}{\partial r} = \dfrac{\partial u}{\partial x}\cos\theta + \dfrac{\partial u}{\partial y}\sin\theta$ and

$$\frac{\partial^2 u}{\partial r^2} = \cos\theta\left[\frac{\partial^2 u}{\partial x^2}\frac{\partial x}{\partial r} + \frac{\partial^2 u}{\partial y\,\partial x}\frac{\partial y}{\partial r} + \frac{\partial^2 u}{\partial z\,\partial x}\frac{\partial z}{\partial r}\right] + \sin\theta\left[\frac{\partial^2 u}{\partial y^2}\frac{\partial y}{\partial r} + \frac{\partial^2 u}{\partial x\,\partial y}\frac{\partial x}{\partial r} + \frac{\partial^2 u}{\partial z\,\partial y}\frac{\partial z}{\partial r}\right]$$

$$= \frac{\partial^2 u}{\partial x^2}\cos^2\theta + \frac{\partial^2 u}{\partial y^2}\sin^2\theta + 2\frac{\partial^2 u}{\partial y\,\partial x}\cos\theta\,\sin\theta$$

Similarly $\dfrac{\partial u}{\partial \theta} = -\dfrac{\partial u}{\partial x}r\sin\theta + \dfrac{\partial u}{\partial y}r\cos\theta$ and

$$\frac{\partial^2 u}{\partial \theta^2} = \frac{\partial^2 u}{\partial x^2}r^2\sin^2\theta + \frac{\partial^2 u}{\partial y^2}r^2\cos^2\theta - 2\frac{\partial^2 u}{\partial y\,\partial x}r^2\sin\theta\cos\theta - \frac{\partial u}{\partial x}r\cos\theta - \frac{\partial u}{\partial y}r\sin\theta. \text{ So}$$

$$\frac{\partial^2 u}{\partial r^2} + \frac{1}{r}\frac{\partial u}{\partial r} + \frac{1}{r^2}\frac{\partial^2 u}{\partial \theta^2} + \frac{\partial^2 u}{\partial z^2} = \frac{\partial^2 u}{\partial x^2}\cos^2\theta + \frac{\partial^2 u}{\partial y^2}\sin^2\theta + 2\frac{\partial^2 u}{\partial y\,\partial x}\cos\theta\,\sin\theta + \frac{\partial u}{\partial x}\frac{\cos\theta}{r} + \frac{\partial u}{\partial y}\frac{\sin\theta}{r}$$

$$+ \frac{\partial^2 u}{\partial x^2}\sin^2\theta + \frac{\partial^2 u}{\partial y^2}\cos^2\theta - 2\frac{\partial^2 u}{\partial y\,\partial x}\sin\theta\,\cos\theta$$

$$- \frac{\partial u}{\partial x}\frac{\cos\theta}{r} - \frac{\partial u}{\partial y}\frac{\sin\theta}{r} + \frac{\partial^2 u}{\partial z^2}$$

$$= \frac{\partial^2 u}{\partial x^2} + \frac{\partial^2 u}{\partial y^2} + \frac{\partial^2 u}{\partial z^2}$$

(b) $x = \rho\sin\phi\cos\theta,\ y = \rho\sin\phi\sin\theta,\ z = \rho\cos\phi.$ Then

$$\frac{\partial u}{\partial \rho} = \frac{\partial u}{\partial x}\frac{\partial x}{\partial \rho} + \frac{\partial u}{\partial y}\frac{\partial y}{\partial \rho} + \frac{\partial u}{\partial z}\frac{\partial z}{\partial \rho} = \frac{\partial u}{\partial x}\sin\phi\,\cos\theta + \frac{\partial u}{\partial y}\sin\phi\,\sin\theta + \frac{\partial u}{\partial z}\cos\phi,\ \text{and}$$

$$\frac{\partial^2 u}{\partial \rho^2} = \sin\phi\,\cos\theta\left[\frac{\partial^2 u}{\partial x^2}\frac{\partial x}{\partial \rho} + \frac{\partial^2 u}{\partial y\,\partial x}\frac{\partial y}{\partial \rho} + \frac{\partial^2 u}{\partial z\,\partial x}\frac{\partial z}{\partial \rho}\right]$$

$$+ \sin\phi\,\sin\theta\left[\frac{\partial^2 u}{\partial y^2}\frac{\partial y}{\partial \rho} + \frac{\partial^2 u}{\partial x\,\partial y}\frac{\partial x}{\partial \rho} + \frac{\partial^2 u}{\partial z\,\partial y}\frac{\partial z}{\partial \rho}\right]$$

$$+ \cos\phi\left[\frac{\partial^2 u}{\partial z^2}\frac{\partial z}{\partial \rho} + \frac{\partial^2 u}{\partial x\,\partial z}\frac{\partial x}{\partial \rho} + \frac{\partial^2 u}{\partial y\,\partial z}\frac{\partial y}{\partial \rho}\right]$$

$$= 2\frac{\partial^2 u}{\partial y\,\partial x}\sin^2\phi\,\sin\theta\,\cos\theta + 2\frac{\partial^2 u}{\partial z\,\partial x}\sin\phi\,\cos\phi\,\cos\theta + 2\frac{\partial^2 u}{\partial y\,\partial z}\sin\phi\,\cos\phi\,\sin\theta$$

$$+ \frac{\partial^2 u}{\partial x^2}\sin^2\phi\cos^2\theta + \frac{\partial^2 u}{\partial y^2}\sin^2\phi\sin^2\theta + \frac{\partial^2 u}{\partial z^2}\cos^2\phi$$

Similarly $\dfrac{\partial u}{\partial \phi} = \dfrac{\partial u}{\partial x}\rho\cos\phi\cos\theta + \dfrac{\partial u}{\partial y}\rho\cos\phi\sin\theta - \dfrac{\partial u}{\partial z}\rho\sin\phi,$ and

$$\frac{\partial^2 u}{\partial \phi^2} = 2\frac{\partial^2 u}{\partial y\,\partial x}\rho^2\cos^2\phi\,\sin\theta\,\cos\theta - 2\frac{\partial^2 u}{\partial x\,\partial z}\rho^2\sin\phi\,\cos\phi\,\cos\theta$$

$$- 2\frac{\partial^2 u}{\partial y\,\partial z}\rho^2\sin\phi\,\cos\phi\,\sin\theta + \frac{\partial^2 u}{\partial x^2}\rho^2\cos^2\phi\,\cos^2\theta + \frac{\partial^2 u}{\partial y^2}\rho^2\cos^2\phi\,\sin^2\theta$$

$$+ \frac{\partial^2 u}{\partial z^2}\rho^2\sin^2\phi - \frac{\partial u}{\partial x}\rho\sin\phi\,\cos\theta - \frac{\partial u}{\partial y}\rho\sin\phi\,\sin\theta - \frac{\partial u}{\partial z}\rho\cos\phi$$

And $\dfrac{\partial u}{\partial \theta} = -\dfrac{\partial u}{\partial x}\,\rho \sin \phi \sin \theta + \dfrac{\partial u}{\partial y}\,\rho \sin \phi \cos \theta$, while

$$\dfrac{\partial^2 u}{\partial \theta^2} = -2\,\dfrac{\partial^2 u}{\partial y\,\partial x}\,\rho^2 \sin^2 \phi \,\cos \theta \,\sin \theta + \dfrac{\partial^2 u}{\partial x^2}\,\rho^2 \sin^2 \phi \,\sin^2 \theta$$

$$+ \dfrac{\partial^2 u}{\partial y^2}\,\rho^2 \sin^2 \phi \cos^2 \theta - \dfrac{\partial u}{\partial x}\,\rho \sin \phi \,\cos \theta - \dfrac{\partial u}{\partial y}\,\rho \sin \phi \,\sin \theta$$

Therefore

$$\dfrac{\partial^2 u}{\partial \rho^2} + \dfrac{2}{\rho}\dfrac{\partial u}{\partial \rho} + \dfrac{\cot \phi}{\rho^2}\dfrac{\partial u}{\partial \phi} + \dfrac{1}{\rho^2}\dfrac{\partial^2 u}{\partial \phi^2} + \dfrac{1}{\rho^2 \sin^2 \phi}\dfrac{\partial^2 u}{\partial \theta^2}$$

$$= \dfrac{\partial^2 u}{\partial x^2}\left[(\sin^2 \phi \,\cos^2 \theta) + (\cos^2 \phi \,\cos^2 \theta) + \sin^2 \theta\right]$$

$$+ \dfrac{\partial^2 u}{\partial y^2}\left[(\sin^2 \phi \,\sin^2 \theta) + (\cos^2 \phi \,\sin^2 \theta) + \cos^2 \theta\right] + \dfrac{\partial^2 u}{\partial z^2}\left[\cos^2 \phi + \sin^2 \phi\right]$$

$$+ \dfrac{\partial u}{\partial x}\left[\dfrac{2 \sin^2 \phi \,\cos \theta + \cos^2 \phi \,\cos \theta - \sin^2 \phi \,\cos \theta - \cos \theta}{\rho \sin \phi}\right]$$

$$+ \dfrac{\partial u}{\partial y}\left[\dfrac{2 \sin^2 \phi \,\sin \theta + \cos^2 \phi \,\sin \theta - \sin^2 \phi \,\sin \theta - \sin \theta}{\rho \sin \phi}\right]$$

But $2 \sin^2 \phi \cos \theta + \cos^2 \phi \cos \theta - \sin^2 \phi \cos \theta - \cos \theta = (\sin^2 \phi + \cos^2 \phi - 1)\cos \theta = 0$ and similarly the coefficient of $\partial u/\partial y$ is 0. Also $\sin^2 \phi \cos^2 \theta + \cos^2 \phi \cos^2 \theta + \sin^2 \theta = \cos^2 \theta\,(\sin^2 \phi + \cos^2 \phi) + \sin^2 \theta = 1$, and similarly the coefficient of $\partial^2 u/\partial y^2$ is 1. So Laplace's Equation in spherical coordinates is as stated.

10. (a) Consider a polar division of the disk, similar to that in Figure 16.4.4 [ET 15.4.4], where

$0 = \theta_0 < \theta_1 < \theta_2 < \cdots < \theta_n = 2\pi,\ 0 = r_1 < r_2 < \cdots < r_m = R$, and where the polar subrectangle R_{ij}, as well as $r_i^*,\theta_j^*,\Delta r$ and $\Delta \theta$ are the same as in that figure. Thus $\Delta A_i = r_i^*\,\Delta r\,\Delta \theta$. The mass of R_{ij} is $\rho\,\Delta A_i$, and its distance from m is $s_{ij} \approx \sqrt{(r_i^*)^2 + d^2}$. According to Newton's Law of Gravitation, the force of attraction experienced by m due to this polar subrectangle is in the direction from m towards R_{ij} and has magnitude $\dfrac{Gm\rho\,\Delta A_i}{s_{ij}^2}$. The symmetry of the lamina with respect to the x- and y-axes and the position of m are such that all horizontal components of the gravitational force cancel, so that the total force is simply in the z-direction. Thus, we need only be concerned with the components of this vertical force; that is, $\dfrac{Gm\rho\,\Delta A_i}{s_{ij}^2}\sin \alpha$, where α is the angle between the origin, r_i^* and the mass m. Thus $\sin \alpha = \dfrac{d}{s_{ij}}$ and the previous result becomes $\dfrac{Gm\rho d\,\Delta A_i}{s_{ij}^3}$. The total attractive force is just the Riemann sum

$$\sum_{i=1}^{m}\sum_{j=1}^{n}\dfrac{Gm\rho d\,\Delta A_i}{s_{ij}^3} = \sum_{i=1}^{m}\sum_{j=1}^{n}\dfrac{Gm\rho d(r_i^*)\,\Delta r\,\Delta \theta}{[(r_i^*)^2 + d^2]^{3/2}}\quad\text{which becomes}\quad \int_0^R\int_0^{2\pi}\dfrac{Gm\rho d}{(r^2 + d^2)^{3/2}}\,r\,d\theta\,dr\quad\text{as } m \to \infty \text{ and}$$

$n \to \infty$. Therefore,

$$F = 2\pi Gm\rho d\int_0^R \dfrac{r}{(r^2 + d^2)^{3/2}}\,dr = 2\pi Gm\rho d\left[-\dfrac{1}{\sqrt{r^2 + d^2}}\right]_0^R = 2\pi Gm\rho d\left(\dfrac{1}{d} - \dfrac{1}{\sqrt{R^2 + d^2}}\right)$$

(b) This is just the result of part (a) in the limit as $R \to \infty$. In this case $\dfrac{1}{\sqrt{R^2 + d^2}} \to 0$, and we are left with

$$F = 2\pi Gm\rho d\left(\dfrac{1}{d} - 0\right) = 2\pi Gm\rho.$$

11. $\int_0^x \int_0^y \int_0^z f(t)\, dt\, dz\, dy = \iiint_E f(t)\, dV$, where

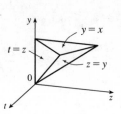

$E = \{(t, z, y) \mid 0 \le t \le z, 0 \le z \le y, 0 \le y \le x\}$.

If we let D be the projection of E on the yt-plane then

$D = \{(y, t) \mid 0 \le t \le x, t \le y \le x\}$. And we see from the diagram

that $E = \{(t, z, y) \mid t \le z \le y, t \le y \le x, 0 \le t \le x\}$. So

$$\int_0^x \int_0^y \int_0^z f(t)\, dt\, dz\, dy = \int_0^x \int_t^x \int_t^y f(t)\, dz\, dy\, dt = \int_0^x \left[\int_t^x (y - t) f(t)\, dy \right] dt$$

$$= \int_0^x \left[\left(\tfrac{1}{2}y^2 - ty \right) f(t) \right]_{y=t}^{y=x} dt = \int_0^x \left[\tfrac{1}{2}x^2 - tx - \tfrac{1}{2}t^2 + t^2 \right] f(t)\, dt$$

$$= \int_0^x \left[\tfrac{1}{2}x^2 - tx + \tfrac{1}{2}t^2 \right] f(t)\, dt = \int_0^x \left(\tfrac{1}{2}x^2 - 2tx + t^2 \right) f(t)\, dt$$

$$= \tfrac{1}{2} \int_0^x (x - t)^2 f(t)\, dt$$

17.1 Vector Fields

1. $\mathbf{F}(x, y) = \frac{1}{2}(\mathbf{i} + \mathbf{j})$

All vectors in this field are identical, with length $\frac{1}{\sqrt{2}}$ and direction parallel to the line $y = x$.

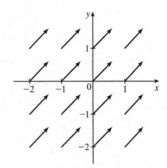

2. $\mathbf{F}(x, y) = \mathbf{i} + x\,\mathbf{j}$

The length of the vector $\mathbf{i} + x\,\mathbf{j}$ is $\sqrt{1 + x^2}$. Vectors are tangent to parabolas opening about the y-axis.

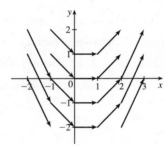

3. $\mathbf{F}(x, y) = y\,\mathbf{i} + \frac{1}{2}\,\mathbf{j}$

The length of the vector $y\,\mathbf{i} + \frac{1}{2}\,\mathbf{j}$ is $\sqrt{y^2 + \frac{1}{4}}$. Vectors are tangent to parabolas opening about the x-axis.

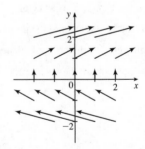

4. $\mathbf{F}(x, y) = (x - y)\,\mathbf{i} + x\,\mathbf{j}$

The length of the vector $(x - y)\,\mathbf{i} + x\,\mathbf{j}$ is $\sqrt{(x - y)^2 + x^2}$. Vectors along the line $y = x$ are vertical.

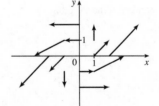

5. $\mathbf{F}(x, y) = \dfrac{y\,\mathbf{i} + x\,\mathbf{j}}{\sqrt{x^2 + y^2}}$

The length of the vector $\dfrac{y\,\mathbf{i} + x\,\mathbf{j}}{\sqrt{x^2 + y^2}}$ is 1.

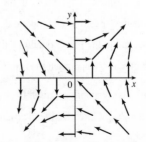

6. $\mathbf{F}(x, y) = \dfrac{y\,\mathbf{i} - x\,\mathbf{j}}{\sqrt{x^2 + y^2}}$

All the vectors $\mathbf{F}(x, y)$ are unit vectors tangent to circles

centered at the origin with radius $\sqrt{x^2 + y^2}$.

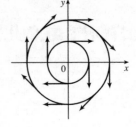

7. $\mathbf{F}(x, y, z) = \mathbf{k}$

All vectors in this field are parallel to the z-axis and have

length 1.

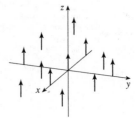

8. $\mathbf{F}(x, y, z) = -y\,\mathbf{k}$

At each point (x, y, z), $\mathbf{F}(x, y, z)$ is a vector of length $|y|$.

For $y > 0$, all point in the direction of the negative z-axis,

while for $y < 0$, all are in the direction of the positive

z-axis. In each plane $y = k$, all the vectors are identical.

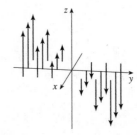

9. $\mathbf{F}(x, y, z) = x\,\mathbf{k}$

At each point (x, y, z), $\mathbf{F}(x, y, z)$ is a vector of length $|x|$.

For $x > 0$, all point in the direction of the positive z-axis,

while for $x < 0$, all are in the direction of the negative

z-axis. In each plane $x = k$, all the vectors are identical.

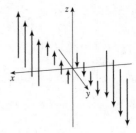

10. $\mathbf{F}(x, y, z) = \mathbf{j} - \mathbf{i}$

All vectors in this field have length $\sqrt{2}$ and point in the

same direction, parallel to the xy-plane.

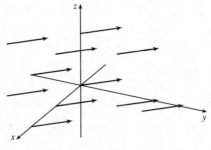

11. $\mathbf{F}(x, y) = \langle y, x \rangle$ corresponds to graph II. In the first quadrant all the vectors have positive x- and y-components, in the second
quadrant all vectors have positive x-components and negative y-components, in the third quadrant all vectors have negative x-
and y-components, and in the fourth quadrant all vectors have negative x-components and positive y-components. In addition,
the vectors get shorter as we approach the origin.

12. $\mathbf{F}(x, y) = \langle 1, \sin y \rangle$ corresponds to graph IV since the x-component of each vector is constant, the vectors are independent of
x (vectors along horizontal lines are identical), and the vector field appears to repeat the same pattern vertically.

13. $\mathbf{F}(x, y) = \langle x - 2, x + 1 \rangle$ corresponds to graph I since the vectors are independent of y (vectors along vertical lines are identical) and, as we move to the right, both the x- and the y-components get larger.

14. $\mathbf{F}(x, y) = \langle y, 1/x \rangle$ corresponds to graph III. As in Exercise 11, all the vectors in the first quadrant have positive x- and y-components, in the second quadrant all vectors have positive x-components and negative y-components, in the third quadrant all vectors have negative x- and y-components, and in the fourth quadrant all vectors have negative x-components and positive y-components. Also, the vectors become longer as we approach the y-axis.

15. $\mathbf{F}(x, y, z) = \mathbf{i} + 2\mathbf{j} + 3\mathbf{k}$ corresponds to graph IV, since all vectors have identical length and direction.

16. $\mathbf{F}(x, y, z) = \mathbf{i} + 2\mathbf{j} + z\mathbf{k}$ corresponds to graph I, since the horizontal vector components remain constant, but the vectors above the xy-plane point generally upward while the vectors below the xy-plane point generally downward.

17. $\mathbf{F}(x, y, z) = x\mathbf{i} + y\mathbf{j} + 3\mathbf{k}$ corresponds to graph III; the projection of each vector onto the xy-plane is $x\mathbf{i} + y\mathbf{j}$, which points away from the origin, and the vectors point generally upward because their z-components are all 3.

18. $\mathbf{F}(x, y, z) = x\mathbf{i} + y\mathbf{j} + z\mathbf{k}$ corresponds to graph II; each vector $\mathbf{F}(x, y, z)$ has the same length and direction as the position vector of the point (x, y, z), and therefore the vectors all point directly away from the origin.

19.

The vector field seems to have very short vectors near the line $y = 2x$.

For $\mathbf{F}(x, y) = \langle 0, 0 \rangle$ we must have $y^2 - 2xy = 0$ and $3xy - 6x^2 = 0$. The first equation holds if $y = 0$ or $y = 2x$, and the second holds if $x = 0$ or $y = 2x$. So both equations hold [and thus $\mathbf{F}(x, y) = \mathbf{0}$] along the line $y = 2x$.

20.

From the graph, it appears that all of the vectors in the field lie on lines through the origin, and that the vectors have very small magnitudes near the circle $|\mathbf{x}| = 2$ and near the origin. Note that $\mathbf{F}(\mathbf{x}) = \mathbf{0}$ ⇔ $r(r - 2) = 0$ ⇔ $r = 0$ or 2, so as we suspected, $\mathbf{F}(\mathbf{x}) = \mathbf{0}$ for $|\mathbf{x}| = 2$ and for $|\mathbf{x}| = 0$. Note that where $r^2 - r < 0$, the vectors point towards the origin, and where $r^2 - r > 0$, they point away from the origin.

21. $f(x, y) = xe^{xy}$ ⇒

$\nabla f(x, y) = f_x(x, y)\,\mathbf{i} + f_y(x, y)\,\mathbf{j} = (xe^{xy} \cdot y + e^{xy})\,\mathbf{i} + (xe^{xy} \cdot x)\mathbf{j} = (xy + 1)e^{xy}\,\mathbf{i} + x^2 e^{xy}\,\mathbf{j}$

22. $f(x, y) = \tan(3x - 4y)$ ⇒

$\nabla f(x, y) = f_x(x, y)\,\mathbf{i} + f_y(x, y)\,\mathbf{j} = \left[\sec^2(3x - 4y) \cdot 3\right]\mathbf{i} + \left[\sec^2(3x - 4y) \cdot (-4)\right]\mathbf{j}$

$\qquad = 3\sec^2(3x - 4y)\,\mathbf{i} - 4\sec^2(3x - 4y)\,\mathbf{j}$

23. $\nabla f(x, y, z) = f_x(x, y, z)\,\mathbf{i} + f_y(x, y, z)\,\mathbf{j} + f_z(x, y, z)\,\mathbf{k} = \dfrac{x}{\sqrt{x^2 + y^2 + z^2}}\,\mathbf{i} + \dfrac{y}{\sqrt{x^2 + y^2 + z^2}}\,\mathbf{j} + \dfrac{z}{\sqrt{x^2 + y^2 + z^2}}\,\mathbf{k}$

24. $\nabla f(x, y, z) = f_x(x, y, z)\,\mathbf{i} + f_y(x, y, z)\,\mathbf{j} + f_z(x, y, z)\,\mathbf{k} = \left(\cos\dfrac{y}{z}\right)\mathbf{i} - x\left(\sin\dfrac{y}{z}\right)\left(\dfrac{1}{z}\right)\mathbf{j} - x\left(\sin\dfrac{y}{z}\right)\left(-\dfrac{y}{z^2}\right)\mathbf{k}$

$\qquad = \left(\cos\dfrac{y}{z}\right)\mathbf{i} - \dfrac{x}{z}\left(\sin\dfrac{y}{z}\right)\mathbf{j} + \dfrac{xy}{z^2}\left(\sin\dfrac{y}{z}\right)\mathbf{k}$

25. $f(x, y) = x^2 - y \;\Rightarrow\; \nabla f(x, y) = 2x\,\mathbf{i} - \mathbf{j}$.
The length of $\nabla f(x, y)$ is $\sqrt{4x^2 + 1}$. When
$x \neq 0$, the vectors point away from the y-axis in
a slightly downward direction with length that
increases as the distance from the y-axis
increases.

26. $f(x, y) = \sqrt{x^2 + y^2} \;\Rightarrow$
$\qquad \nabla f(x, y) = \frac{1}{2}(x^2 + y^2)^{-1/2}(2x)\,\mathbf{i} + \frac{1}{2}(x^2 + y^2)^{-1/2}(2y)\,\mathbf{j}$
$\qquad = \dfrac{x}{\sqrt{x^2 + y^2}}\,\mathbf{i} + \dfrac{y}{\sqrt{x^2 + y^2}}\,\mathbf{j}$ or $\dfrac{1}{\sqrt{x^2 + y^2}}\,(x\,\mathbf{i} + y\,\mathbf{j})$.

$\nabla f(x, y)$ is not defined at the origin, but elsewhere all vectors have
length 1 and point away from the origin.

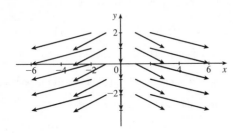

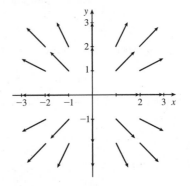

27. We graph ∇f along with a contour map of f.

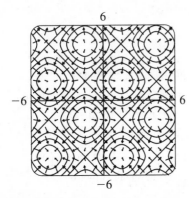

The graph shows that the gradient vectors are
perpendicular to the level curves. Also, the
gradient vectors point in the direction in which
f is increasing and are longer where the level
curves are closer together.

28. We graph ∇f along with a contour map of f.

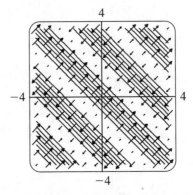

The graph shows that the gradient vectors are perpendicular to the
level curves. Also, the gradient vectors point in the direction in
which f is increasing and are longer where the level curves are
closer together.

29. $f(x, y) = x^2 + y^2 \;\Rightarrow\; \nabla f(x, y) = 2x\,\mathbf{i} + 2y\,\mathbf{j}$. Thus, each vector $\nabla f(x, y)$ has the same direction and twice the length of
the position vector of the point (x, y), so the vectors all point directly away from the origin and their lengths increase as we
move away from the origin. Hence, ∇f is graph II.

30. $f(x, y) = x(x + y) = x^2 + xy$ $\Rightarrow$ $\nabla f(x, y) = (2x + y)\,\mathbf{i} + x\,\mathbf{j}$. The y-component of each vector is x, so the vectors point upward in quadrants I and IV and downward in quadrants II and III. Also, the x-component of each vector is 0 along the line $y = -2x$ so the vectors are vertical there. Thus, ∇f is graph IV.

31. $f(x, y) = (x + y)^2$ $\Rightarrow$ $\nabla f(x, y) = 2(x + y)\,\mathbf{i} + 2(x + y)\,\mathbf{j}$. The x- and y-components of each vector are equal, so all vectors are parallel to the line $y = x$. The vectors are $\mathbf{0}$ along the line $y = -x$ and their length increases as the distance from this line increases. Thus, ∇f is graph II.

32. $f(x, y) = \sin\sqrt{x^2 + y^2}$ $\Rightarrow$

$$\nabla f(x, y) = \left[\cos\sqrt{x^2 + y^2} \cdot \tfrac{1}{2}(x^2 + y^2)^{-1/2}(2x)\right]\mathbf{i} + \left[\cos\sqrt{x^2 + y^2} \cdot \tfrac{1}{2}(x^2 + y^2)^{-1/2}(2y)\right]\mathbf{j}$$

$$= \frac{\cos\sqrt{x^2 + y^2}}{\sqrt{x^2 + y^2}}\,x\,\mathbf{i} + \frac{\cos\sqrt{x^2 + y^2}}{\sqrt{x^2 + y^2}}\,y\,\mathbf{j} \quad \text{or} \quad \frac{\cos\sqrt{x^2 + y^2}}{\sqrt{x^2 + y^2}}\,(x\,\mathbf{i} + y\,\mathbf{j})$$

Thus each vector is a scalar multiple of its position vector, so the vectors point toward or away from the origin with length that changes in a periodic fashion as we move away from the origin. ∇f is graph I.

33. At $t = 3$ the particle is at $(2, 1)$ so its velocity is $\mathbf{V}(2, 1) = \langle 4, 3\rangle$. After 0.01 units of time, the particle's change in location should be approximately $0.01\,\mathbf{V}(2, 1) = 0.01\,\langle 4, 3\rangle = \langle 0.04, 0.03\rangle$, so the particle should be approximately at the point $(2.04, 1.03)$.

34. At $t = 1$ the particle is at $(1, 3)$ so its velocity is $\mathbf{F}(1, 3) = \langle 1, -1\rangle$. After 0.05 units of time, the particle's change in location should be approximately $0.05\,\mathbf{F}(1, 3) = 0.05\,\langle 1, -1\rangle = \langle 0.05, -0.05\rangle$, so the particle should be approximately at the point $(1.05, 2.95)$.

35. (a) We sketch the vector field $\mathbf{F}(x, y) = x\,\mathbf{i} - y\,\mathbf{j}$ along with several approximate flow lines. The flow lines appear to be hyperbolas with shape similar to the graph of $y = \pm 1/x$, so we might guess that the flow lines have equations

$$y = C/x.$$

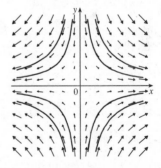

(b) If $x = x(t)$ and $y = y(t)$ are parametric equations of a flow line, then the velocity vector of the flow line at the point (x, y) is $x'(t)\,\mathbf{i} + y'(t)\,\mathbf{j}$. Since the velocity vectors coincide with the vectors in the vector field, we have

$x'(t)\,\mathbf{i} + y'(t)\,\mathbf{j} = x\,\mathbf{i} - y\,\mathbf{j}$ $\Rightarrow$ $dx/dt = x$, $dy/dt = -y$. To solve these differential equations, we know

$dx/dt = x$ $\Rightarrow$ $dx/x = dt$ $\Rightarrow$ $\ln|x| = t + C$ $\Rightarrow$ $x = \pm e^{t+C} = Ae^t$ for some constant A, and

$dy/dt = -y$ $\Rightarrow$ $dy/y = -dt$ $\Rightarrow$ $\ln|y| = -t + K$ $\Rightarrow$ $y = \pm e^{-t+K} = Be^{-t}$ for some constant B. Therefore

$xy = Ae^t Be^{-t} = AB = $ constant. If the flow line passes through $(1, 1)$ then $(1)(1) = $ constant $= 1$ $\Rightarrow$ $xy = 1$ $\Rightarrow$

$y = 1/x$, $x > 0$.

36. (a) We sketch the vector field $\mathbf{F}(x, y) = \mathbf{i} + x\mathbf{j}$ along with several approximate flow lines. The flow lines appear to be parabolas.

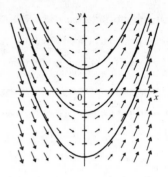

(b) If $x = x(t)$ and $y = y(t)$ are parametric equations of a flow line, then the velocity vector of the flow line at the point (x, y) is $x'(t)\mathbf{i} + y'(t)\mathbf{j}$. Since the velocity vectors coincide with the vectors in the vector field, we have

$$x'(t)\mathbf{i} + y'(t)\mathbf{j} = \mathbf{i} + x\mathbf{j} \quad \Rightarrow \quad \frac{dx}{dt} = 1, \frac{dy}{dt} = x. \text{ Thus } \frac{dy}{dx} = \frac{dy/dt}{dx/dt} = \frac{x}{1} = x.$$

(c) From part (b), $dy/dx = x$. Integrating, we have $y = \frac{1}{2}x^2 + c$. Since the particle starts at the origin, we know $(0,0)$ is on the curve, so $0 = 0 + c \quad \Rightarrow \quad c = 0$ and the path the particle follows is $y = \frac{1}{2}x^2$.

17.2 Line Integrals ET 16.2

1. $x = t^3$ and $y = t$, $0 \le t \le 2$, so by Formula 3

$$\int_C y^3 \, ds = \int_0^2 t^3 \sqrt{\left(\frac{dx}{dt}\right)^2 + \left(\frac{dy}{dt}\right)^2} \, dt = \int_0^2 t^3 \sqrt{(3t^2)^2 + (1)^2} \, dt = \int_0^2 t^3 \sqrt{9t^4 + 1} \, dt$$

$$= \frac{1}{36} \cdot \frac{2}{3} \left(9t^4 + 1\right)^{3/2} \Big]_0^2 = \frac{1}{54}(145^{3/2} - 1) \text{ or } \frac{1}{54}\left(145\sqrt{145} - 1\right)$$

2. $\int_C xy \, ds = \int_0^1 (t^2)(2t)\sqrt{(2t)^2 + (2)^2} \, dt = \int_0^1 2t^3 \sqrt{4t^2 + 4} \, dt = \int_0^1 4t^3 \sqrt{t^2 + 1} \, dt$ $\quad \begin{bmatrix} \text{Substitute } u = t^2 + 1 & \Rightarrow \\ t^2 = u - 1, du = 2t \, dt \end{bmatrix}$

$$= \int_1^2 2(u-1)\sqrt{u} \, du = 2\int_1^2 (u^{3/2} - u^{1/2}) \, du = 2\left[\frac{2}{5}u^{5/2} - \frac{2}{3}u^{3/2}\right]_1^2$$

$$= 2\left(\frac{8}{5}\sqrt{2} - \frac{4}{3}\sqrt{2} - \frac{2}{5} + \frac{2}{3}\right) = \frac{8}{15}\left(\sqrt{2} + 1\right)$$

3. Parametric equations for C are $x = 4\cos t$, $y = 4\sin t$, $-\frac{\pi}{2} \le t \le \frac{\pi}{2}$. Then

$$\int_C xy^4 \, ds = \int_{-\pi/2}^{\pi/2} (4\cos t)(4\sin t)^4 \sqrt{(-4\sin t)^2 + (4\cos t)^2} \, dt = \int_{-\pi/2}^{\pi/2} 4^5 \cos t \, \sin^4 t \, \sqrt{16(\sin^2 t + \cos^2 t)} \, dt$$

$$= 4^5 \int_{-\pi/2}^{\pi/2} (\sin^4 t \cos t)(4) \, dt = (4)^6 \left[\frac{1}{5}\sin^5 t\right]_{-\pi/2}^{\pi/2} = \frac{2 \cdot 4^6}{5} = 1638.4$$

4. Parametric equations for C are $x = 4t$, $y = 3 + 3t$, $0 \le t \le 1$. Then

$$\int_C x\sin y \, ds = \int_0^1 (4t)\sin(3 + 3t)\sqrt{4^2 + 3^2} \, dt = 20\int_0^1 t\sin(3 + 3t) \, dt$$

Integrating by parts with $u = t \quad \Rightarrow \quad du = dt, dv = \sin(3 + 3t) \, dt \quad \Rightarrow \quad v = -\frac{1}{3}\cos(3 + 3t)$ gives

$$\int_C x\sin y \, ds = 20\left[-\frac{1}{3}t\cos(3 + 3t) + \frac{1}{9}\sin(3 + 3t)\right]_0^1 = 20\left[-\frac{1}{3}\cos 6 + \frac{1}{9}\sin 6 + 0 - \frac{1}{9}\sin 3\right]$$

$$= \frac{20}{9}(\sin 6 - 3\cos 6 - \sin 3)$$

5. If we choose x as the parameter, parametric equations for C are $x = x$, $y = \sqrt{x}$ for $1 \le x \le 4$ and

$$\int_C \left(x^2 y^3 - \sqrt{x} \right) dy = \int_1^4 \left[x^2 \cdot (\sqrt{x})^3 - \sqrt{x} \right] \frac{1}{2\sqrt{x}} \, dx = \tfrac{1}{2} \int_1^4 \left(x^3 - 1 \right) dx$$

$$= \tfrac{1}{2} \left[\tfrac{1}{4} x^4 - x \right]_1^4 = \tfrac{1}{2} \left(64 - 4 - \tfrac{1}{4} + 1 \right) = \tfrac{243}{8}$$

6. Choosing y as the parameter, we have $x = e^y$, $y = y$, $0 \le y \le 1$. Then

$$\int_c x e^y \, dx = \int_0^1 e^y (e^y) e^y \, dy = \int_0^1 e^{3y} \, dy = \tfrac{1}{3} e^{3y} \big]_0^1 = \tfrac{1}{3} (e^3 - 1).$$

7.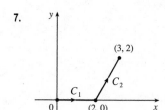

$C = C_1 + C_2$

On C_1: $x = x, y = 0 \implies dy = 0 \, dx$, $0 \le x \le 2$.

On C_2: $x = x, y = 2x - 4 \implies dy = 2 \, dx$, $2 \le x \le 3$.

Then

$$\int_C xy \, dx + (x - y) \, dy = \int_{C_1} xy \, dx + (x - y) \, dy + \int_{C_2} xy \, dx + (x - y) \, dy$$

$$= \int_0^2 (0 + 0) \, dx + \int_2^3 \left[(2x^2 - 4x) + (-x + 4)(2) \right] dx$$

$$= \int_2^3 (2x^2 - 6x + 8) \, dx = \tfrac{17}{3}$$

8.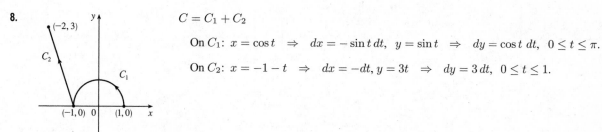

$C = C_1 + C_2$

On C_1: $x = \cos t \implies dx = -\sin t \, dt$, $y = \sin t \implies dy = \cos t \, dt$, $0 \le t \le \pi$.

On C_2: $x = -1 - t \implies dx = -dt, y = 3t \implies dy = 3 \, dt$, $0 \le t \le 1$.

Then

$$\int_C \sin x \, dx + \cos y \, dy = \int_{C_1} \sin x \, dx + \cos y \, dy + \int_{C_2} \sin x \, dx + \cos y \, dy$$

$$= \int_0^\pi \sin(\cos t)(-\sin t \, dt) + \cos(\sin t) \cos t \, dt + \int_0^1 \sin(-1 - t)(-dt) + \cos(3t)(3 \, dt)$$

$$= \left[-\cos(\cos t) + \sin(\sin t) \right]_0^\pi + \left[-\cos(-1 - t) + \sin(3t) \right]_0^1$$

$$= -\cos(\cos \pi) + \sin(\sin \pi) + \cos(\cos 0) - \sin(\sin 0) - \cos(-2) + \sin(3) + \cos(-1) - \sin(0)$$

$$= -\cos(-1) + \sin 0 + \cos(1) - \sin 0 - \cos(-2) + \sin 3 + \cos(-1)$$

$$= -\cos 1 + \cos 1 - \cos 2 + \sin 3 + \cos 1 = \cos 1 - \cos 2 + \sin 3$$

where we have used the identity $\cos(-\theta) = \cos \theta$.

9. $x = 2 \sin t$, $y = t$, $z = -2 \cos t$, $0 \le t \le \pi$. Then by Formula 9,

$$\int_C xyz \, ds = \int_0^\pi (2 \sin t)(t)(-2 \cos t) \sqrt{\left(\tfrac{dx}{dt} \right)^2 + \left(\tfrac{dy}{dt} \right)^2 + \left(\tfrac{dz}{dt} \right)^2} \, dt$$

$$= \int_0^\pi -4t \sin t \cos t \sqrt{(2 \cos t)^2 + (1)^2 + (2 \sin t)^2} \, dt = \int_0^\pi -2t \sin 2t \sqrt{4(\cos^2 t + \sin^2 t) + 1} \, dt$$

$$= -2\sqrt{5} \int_0^\pi t \sin 2t \, dt = -2\sqrt{5} \left[-\tfrac{1}{2} t \cos 2t + \tfrac{1}{4} \sin 2t \right]_0^\pi \qquad \left[\begin{array}{l} \text{integrate by parts with} \\ u = t, dv = \sin 2t \, dt \end{array} \right]$$

$$= -2\sqrt{5} \left(-\tfrac{\pi}{2} - 0 \right) = \sqrt{5}\,\pi$$

10. Parametric equations for C are $x = -1 + 2t$, $y = 5 + t$, $z = 4t$, $0 \le t \le 1$. Then

$$\int_C xyz^2 \, ds = \int_0^1 (-1 + 2t)(5 + t)(4t)^2 \sqrt{2^2 + 1^2 + 4^2} \, dt = \sqrt{21} \int_0^1 (32t^4 + 144t^3 - 80t^2) \, dt$$

$$= \sqrt{21} \left[32 \cdot \frac{t^5}{5} + 144 \cdot \frac{t^4}{4} - 80 \cdot \frac{t^3}{3} \right]_0^1 = \sqrt{21} \left(\frac{32}{5} + 36 - \frac{80}{3} \right) = \frac{236}{15} \sqrt{21}$$

11. Parametric equations for C are $x = t$, $y = 2t$, $z = 3t$, $0 \le t \le 1$. Then

$$\int_C xe^{yz} \, ds = \int_0^1 te^{(2t)(3t)} \sqrt{1^2 + 2^2 + 3^2} \, dt = \sqrt{14} \int_0^1 te^{6t^2} \, dt = \sqrt{14} \left[\frac{1}{12} e^{6t^2} \right]_0^1 = \frac{\sqrt{14}}{12} (e^6 - 1).$$

12. $\sqrt{(dx/dt)^2 + (dy/dt)^2 + (dz/dt)^2} = \sqrt{1^2 + (2t)^2 + (3t^2)^2} = \sqrt{1 + 4t^2 + 9t^4}$. Then

$$\int_C (2x + 9z) \, ds = \int_0^1 (2t + 9t^3) \sqrt{1 + 4t^2 + 9t^4} \, dt \qquad [\text{let } u = 1 + 4t^2 + 9t^4 \ \Rightarrow \ \tfrac{1}{4} \, du = (2t + 9t^3) \, dt]$$

$$= \int_1^{14} \tfrac{1}{4} \sqrt{u} \, du = \tfrac{1}{6} u^{3/2} \Big]_1^{14} = \tfrac{1}{6} (14^{3/2} - 1)$$

13. $\int_C x^2 y \sqrt{z} \, dz = \int_0^1 (t^3)^2 (t) \sqrt{t^2} \cdot 2t \, dt = \int_0^1 2t^9 \, dt = \tfrac{1}{5} t^{10} \Big]_0^1 = \tfrac{1}{5}$

14. $\int_C z \, dx + x \, dy + y \, dz = \int_0^1 t^2 \cdot 2t \, dt + t^2 \cdot 3t^2 \, dt + t^3 \cdot 2t \, dt = \int_0^1 (2t^3 + 5t^4) \, dt = \left[\tfrac{1}{2} t^4 + t^5 \right]_0^1 = \tfrac{1}{2} + 1 = \tfrac{3}{2}$

15.

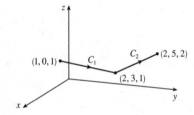

On C_1: $x = 1 + t \ \Rightarrow \ dx = dt$, $y = 3t \ \Rightarrow$

$dy = 3 \, dt$, $z = 1 \ \Rightarrow \ dz = 0 \, dt$, $0 \le t \le 1$.

On C_2: $x = 2 \ \Rightarrow \ dx = 0 \, dt$, $y = 3 + 2t \ \Rightarrow$

$dy = 2 \, dt$, $z = 1 + t \ \Rightarrow \ dz = dt$, $0 \le t \le 1$.

Then

$$\int_C (x + yz) \, dx + 2x \, dy + xyz \, dz$$

$$= \int_{C_1} (x + yz) \, dx + 2x \, dy + xyz \, dz + \int_{C_2} (x + yz) \, dx + 2x \, dy + xyz \, dz$$

$$= \int_0^1 (1 + t + (3t)(1)) \, dt + 2(1 + t) \cdot 3 \, dt + (1 + t)(3t)(1) \cdot 0 \, dt$$

$$\qquad + \int_0^1 (2 + (3 + 2t)(1 + t)) \cdot 0 \, dt + 2(2) \cdot 2 \, dt + (2)(3 + 2t)(1 + t) \, dt$$

$$= \int_0^1 (10t + 7) \, dt + \int_0^1 (4t^2 + 10t + 14) \, dt = \left[5t^2 + 7t \right]_0^1 + \left[\tfrac{4}{3} t^3 + 5t^2 + 14t \right]_0^1 = 12 + \tfrac{61}{3} = \tfrac{97}{3}$$

16.

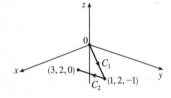

On C_1: $x = t \ \Rightarrow \ dx = dt$, $y = 2t \ \Rightarrow$

$dy = 2 \, dt$, $z = -t \ \Rightarrow \ dz = -dt$, $0 \le t \le 1$.

On C_2: $= 1 + 2t \ \Rightarrow \ dx = 2 \, dt$, $y = 2 \ \Rightarrow$

$dy = 0 \, dt$, $z = -1 + t \ \Rightarrow \ dz = dt$, $0 \le t \le 1$.

Then

$$\int_C x^2 \, dx + y^2 \, dy + z^2 \, dz = \int_{C_1} x^2 \, dx + y^2 \, dy + z^2 \, dz + \int_{C_2} x^2 \, dx + y^2 \, dy + z^2 \, dz$$

$$= \int_0^1 t^2 \, dt + (2t)^2 \cdot 2 \, dt + (-t)^2 (-dt) + \int_0^1 (1 + 2t)^2 \cdot 2 \, dt + 2^2 \cdot 0 \, dt + (-1 + t)^2 \, dt$$

$$= \int_0^1 8t^2 \, dt + \int_0^1 (9t^2 + 6t + 3) \, dt = \left[\tfrac{8}{3} t^3 \right]_0^1 + \left[3t^3 + 3t^2 + 3t \right]_0^1 = \tfrac{35}{3}$$

17. (a) Along the line $x = -3$, the vectors of $\mathbf{F}$ have positive y-components, so since the path goes upward, the integrand $\mathbf{F} \cdot \mathbf{T}$ is always positive. Therefore $\int_{C_1} \mathbf{F} \cdot d\mathbf{r} = \int_{C_1} \mathbf{F} \cdot \mathbf{T} \, ds$ is positive.

(b) All of the (nonzero) field vectors along the circle with radius 3 are pointed in the clockwise direction, that is, opposite the direction to the path. So $\mathbf{F} \cdot \mathbf{T}$ is negative, and therefore $\int_{C_2} \mathbf{F} \cdot d\mathbf{r} = \int_{C_2} \mathbf{F} \cdot \mathbf{T} \, ds$ is negative.

18. Vectors starting on C_1 point in roughly the same direction as C_1, so the tangential component $\mathbf{F} \cdot \mathbf{T}$ is positive. Then $\int_{C_1} \mathbf{F} \cdot d\mathbf{r} = \int_{C_1} \mathbf{F} \cdot \mathbf{T} \, ds$ is positive. On the other hand, no vectors starting on C_2 point in the same direction as C_2, while some vectors point in roughly the opposite direction, so we would expect $\int_{C_2} \mathbf{F} \cdot d\mathbf{r} = \int_{C_2} \mathbf{F} \cdot \mathbf{T} \, ds$ to be negative.

19. $\mathbf{r}(t) = 11t^4 \, \mathbf{i} + t^3 \, \mathbf{j}$, so $\mathbf{F}(\mathbf{r}(t)) = (11t^4)(t^3) \, \mathbf{i} + 3(t^3)^2 \, \mathbf{j} = 11t^7 \, \mathbf{i} + 3t^6 \, \mathbf{j}$ and $\mathbf{r}'(t) = 44t^3 \, \mathbf{i} + 3t^2 \, \mathbf{j}$. Then

$$\int_C \mathbf{F} \cdot d\mathbf{r} = \int_0^1 \mathbf{F}(\mathbf{r}(t)) \cdot \mathbf{r}'(t) \, dt = \int_0^1 (11t^7 \cdot 44t^3 + 3t^6 \cdot 3t^2) \, dt = \int_0^1 (484t^{10} + 9t^8) \, dt = \left[44t^{11} + t^9 \right]_0^1 = 45.$$

20. $\mathbf{F}(\mathbf{r}(t)) = (t^2 + t^3) \, \mathbf{i} + (t^3 - t^2) \, \mathbf{j} + (t^2)^2 \, \mathbf{k} = (t^2 + t^3) \, \mathbf{i} + (t^3 - t^2) \, \mathbf{j} + t^4 \, \mathbf{k}$, $\mathbf{r}'(t) = 2t \, \mathbf{i} + 3t^2 \, \mathbf{j} + 2t \, \mathbf{k}$. Then

$$\int_C \mathbf{F} \cdot d\mathbf{r} = \int_0^1 \mathbf{F}(\mathbf{r}(t)) \cdot \mathbf{r}'(t) \, dt = \int_0^1 (2t^3 + 2t^4 + 3t^5 - 3t^4 + 2t^5) \, dt = \int_0^1 (5t^5 - t^4 + 2t^3) \, dt$$

$$= \left[\tfrac{5}{6} t^6 - \tfrac{1}{5} t^5 + \tfrac{1}{2} t^4 \right]_0^1 = \tfrac{5}{6} - \tfrac{1}{5} + \tfrac{1}{2} = \tfrac{17}{15}.$$

21. $\int_C \mathbf{F} \cdot d\mathbf{r} = \int_0^1 \left\langle \sin t^3, \cos(-t^2), t^4 \right\rangle \cdot \left\langle 3t^2, -2t, 1 \right\rangle dt$

$$= \int_0^1 (3t^2 \sin t^3 - 2t \cos t^2 + t^4) \, dt = \left[-\cos t^3 - \sin t^2 + \tfrac{1}{5} t^5 \right]_0^1 = \tfrac{6}{5} - \cos 1 - \sin 1$$

22. $\int_C \mathbf{F} \cdot d\mathbf{r} = \int_0^\pi \left\langle \cos t, \sin t, -t \right\rangle \cdot \left\langle 1, \cos t, -\sin t \right\rangle dt = \int_0^\pi (\cos t + \sin t \, \cos t + t \sin t) \, dt$

$$= \left[\sin t + \tfrac{1}{2} \sin^2 t + (\sin t - t \cos t) \right]_0^\pi = \pi$$

23. $\mathbf{F}(\mathbf{r}(t)) = (e^t)\left(e^{-t^2}\right) \mathbf{i} + \sin\left(e^{-t^2}\right) \mathbf{j} = e^{t-t^2} \, \mathbf{i} + \sin\left(e^{-t^2}\right) \mathbf{j}$, $\mathbf{r}'(t) = e^t \, \mathbf{i} - 2te^{-t^2} \, \mathbf{j}$. Then

$$\int_C \mathbf{F} \cdot d\mathbf{r} = \int_1^2 \mathbf{F}(\mathbf{r}(t)) \cdot \mathbf{r}'(t) \, dt = \int_1^2 \left[e^{t-t^2} e^t + \sin\left(e^{-t^2}\right) \cdot \left(-2te^{-t^2}\right) \right] dt$$

$$= \int_1^2 \left[e^{2t-t^2} - 2te^{-t^2} \sin\left(e^{-t^2}\right) \right] dt \approx 1.9633$$

24. $\mathbf{F}(\mathbf{r}(t)) = (\sin t) \sin(\sin 5t) \, \mathbf{i} + (\sin 5t) \sin(\cos t) \, \mathbf{j} + (\cos t) \sin(\sin t) \, \mathbf{k}$, $\mathbf{r}'(t) = -\sin t \, \mathbf{i} + \cos t \, \mathbf{j} + 5 \cos 5t \, \mathbf{k}$. Then

$$\int_C \mathbf{F} \cdot d\mathbf{r} = \int_0^\pi \mathbf{F}(\mathbf{r}(t)) \cdot \mathbf{r}'(t) \, dt$$

$$= \int_0^\pi [-\sin^2 t \, \sin(\sin 5t) + \cos t \, \sin 5t \, \sin(\cos t) + 5 \cos t \, \cos 5t \, \sin(\sin t)] \, dt \approx -0.1363$$

25. $x = t^2$, $y = t^3$, $z = t^4$ so by Formula 9,

$$\int_C x \sin(y + z) \, ds = \int_0^5 (t^2) \sin(t^3 + t^4) \sqrt{(2t)^2 + (3t^2)^2 + (4t^3)^2} \, dt$$

$$= \int_0^5 t^2 \sin(t^3 + t^4) \sqrt{4t^2 + 9t^4 + 16t^6} \, dt \approx 15.0074$$

26. $\int_C ze^{-xy} \, ds = \int_0^1 (e^{-t}) e^{-t \cdot t^2} \sqrt{(1)^2 + (2t)^2 + (-e^{-t})^2} \, dt = \int_0^1 e^{-t-t^3} \sqrt{1 + 4t^2 + e^{-2t}} \, dt \approx 0.8208$

27. We graph $\mathbf{F}(x, y) = (x - y) \, \mathbf{i} + xy \, \mathbf{j}$ and the curve C. We see that most of the vectors starting on C point in roughly the same direction as C, so for these portions of C the tangential component $\mathbf{F} \cdot \mathbf{T}$ is positive. Although some vectors in the third quadrant which start on C point in roughly the opposite direction, and hence give negative tangential components, it seems reasonable that the effect of these portions of C is outweighed by the positive tangential components. Thus, we would expect $\int_C \mathbf{F} \cdot d\mathbf{r} = \int_C \mathbf{F} \cdot \mathbf{T}\, ds$ to be positive.

To verify, we evaluate $\int_C \mathbf{F} \cdot d\mathbf{r}$. The curve C can be represented by $\mathbf{r}(t) = 2\cos t\, \mathbf{i} + 2\sin t\, \mathbf{j}$, $0 \le t \le \frac{3\pi}{2}$, so $\mathbf{F}(\mathbf{r}(t)) = (2\cos t - 2\sin t)\, \mathbf{i} + 4\cos t \sin t\, \mathbf{j}$ and $\mathbf{r}'(t) = -2\sin t\, \mathbf{i} + 2\cos t\, \mathbf{j}$. Then

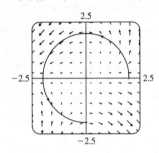

$$\int_C \mathbf{F} \cdot d\mathbf{r} = \int_0^{3\pi/2} \mathbf{F}(\mathbf{r}(t)) \cdot \mathbf{r}'(t)\, dt$$
$$= \int_0^{3\pi/2} [-2\sin t(2\cos t - 2\sin t) + 2\cos t(4\cos t \sin t)]\, dt$$
$$= 4 \int_0^{3\pi/2} (\sin^2 t - \sin t \cos t + 2\sin t \cos^2 t)\, dt$$
$$= 3\pi + \tfrac{2}{3} \qquad \text{[using a CAS]}$$

28. We graph $\mathbf{F}(x, y) = \dfrac{x}{\sqrt{x^2 + y^2}}\, \mathbf{i} + \dfrac{y}{\sqrt{x^2 + y^2}}\, \mathbf{j}$ and the curve C. In the

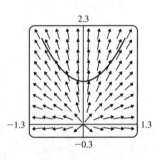

first quadrant, each vector starting on C points in roughly the same direction as C, so the tangential component $\mathbf{F} \cdot \mathbf{T}$ is positive. In the second quadrant, each vector starting on C points in roughly the direction opposite to C, so $\mathbf{F} \cdot \mathbf{T}$ is negative. Here, it appears that the tangential components in the first and second quadrants counteract each other, so it seems reasonable to guess that $\int_C \mathbf{F} \cdot d\mathbf{r} = \int_C \mathbf{F} \cdot \mathbf{T}\, ds$ is zero. To verify, we evaluate $\int_C \mathbf{F} \cdot d\mathbf{r}$. The curve C can be represented by

$\mathbf{r}(t) = t\, \mathbf{i} + (1 + t^2)\, \mathbf{j}$, $-1 \le t \le 1$, so $\mathbf{F}(\mathbf{r}(t)) = \dfrac{t}{\sqrt{t^2 + (1 + t^2)^2}}\, \mathbf{i} + \dfrac{1 + t^2}{\sqrt{t^2 + (1 + t^2)^2}}\, \mathbf{j}$ and $\mathbf{r}'(t) = \mathbf{i} + 2t\, \mathbf{j}$. Then

$$\int_C \mathbf{F} \cdot d\mathbf{r} = \int_{-1}^1 \mathbf{F}(\mathbf{r}(t)) \cdot \mathbf{r}'(t)\, dt = \int_{-1}^1 \left(\frac{t}{\sqrt{t^2 + (1 + t^2)^2}} + \frac{2t(1 + t^2)}{\sqrt{t^2 + (1 + t^2)^2}} \right) dt$$
$$= \int_{-1}^1 \frac{t(3 + 2t^2)}{\sqrt{t^4 + 3t^2 + 1}}\, dt = 0 \qquad \text{[since the integrand is an odd function]}$$

29. (a) $\int_C \mathbf{F} \cdot d\mathbf{r} = \int_0^1 \left\langle e^{t^2 - 1}, t^5 \right\rangle \cdot \left\langle 2t, 3t^2 \right\rangle dt = \int_0^1 \left(2t e^{t^2 - 1} + 3t^7 \right) dt = \left[e^{t^2 - 1} + \frac{3}{8}t^8 \right]_0^1 = \frac{11}{8} - 1/e$

(b) $\mathbf{r}(0) = \mathbf{0}$, $\mathbf{F}(\mathbf{r}(0)) = \left\langle e^{-1}, 0 \right\rangle$;

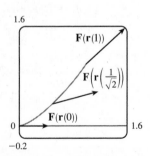

$\mathbf{r}\left(\frac{1}{\sqrt{2}}\right) = \left\langle \frac{1}{2}, \frac{1}{2\sqrt{2}} \right\rangle$, $\mathbf{F}\left(\mathbf{r}\left(\frac{1}{\sqrt{2}}\right)\right) = \left\langle e^{-1/2}, \frac{1}{4\sqrt{2}} \right\rangle$;

$\mathbf{r}(1) = \langle 1, 1 \rangle$, $\mathbf{F}(\mathbf{r}(1)) = \langle 1, 1 \rangle$.

In order to generate the graph with Maple, we use the PLOT command

(not to be confused with the `plot` command) to define each of the vectors.

For example,

```
v1:=PLOT(CURVES([[0,0],[evalf(1/exp(1)),0]]));
```

generates the vector from the vector field at the point $(0, 0)$ (but without an arrowhead) and gives it the name `v1`. To show everything on the same screen, we use the `display` command. In Mathematica, we use `ListPlot` (with the `PlotJoined -> True` option) to generate the vectors, and then `Show` to show everything on the same screen.

30. (a) $\int_C \mathbf{F} \cdot d\mathbf{r} = \int_{-1}^{1} \langle 2t, t^2, 3t \rangle \cdot \langle 2, 3, -2t \rangle \, dt = \int_{-1}^{1} (4t + 3t^2 - 6t^2) \, dt = \left[2t^2 - t^3 \right]_{-1}^{1} = -2$

(b) Now $\mathbf{F}(\mathbf{r}(t)) = \langle 2t, t^2, 3t \rangle$, so $\mathbf{F}(\mathbf{r}(-1)) = \langle -2, 1, -3 \rangle$, $\mathbf{F}\left(\mathbf{r}\left(-\frac{1}{2}\right)\right) = \langle -1, \frac{1}{4}, -\frac{3}{2} \rangle$, $\mathbf{F}\left(\mathbf{r}\left(\frac{1}{2}\right)\right) = \langle 1, \frac{1}{4}, \frac{3}{2} \rangle$, and $\mathbf{F}(\mathbf{r}(1)) = \langle 2, 1, 3 \rangle$.

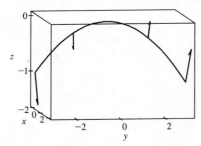 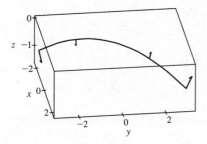

31. $x = e^{-t} \cos 4t, \;\; y = e^{-t} \sin 4t, \;\; z = e^{-t}, \;\; 0 \le t \le 2\pi$.

Then $\dfrac{dx}{dt} = e^{-t}(-\sin 4t)(4) - e^{-t} \cos 4t = -e^{-t}(4 \sin 4t + \cos 4t)$,

$\dfrac{dy}{dt} = e^{-t}(\cos 4t)(4) - e^{-t} \sin 4t = -e^{-t}(-4 \cos 4t + \sin 4t)$, and $\dfrac{dz}{dt} = -e^{-t}$, so

$$\sqrt{\left(\frac{dx}{dt}\right)^2 + \left(\frac{dy}{dt}\right)^2 + \left(\frac{dz}{dt}\right)^2} = \sqrt{(-e^{-t})^2 [(4 \sin 4t + \cos 4t)^2 + (-4 \cos 4t + \sin 4t)^2 + 1]}$$

$$= e^{-t} \sqrt{16(\sin^2 4t + \cos^2 4t) + \sin^2 4t + \cos^2 4t + 1} = 3\sqrt{2}\, e^{-t}$$

Therefore $\int_C x^3 y^2 z \, ds = \int_0^{2\pi} (e^{-t} \cos 4t)^3 (e^{-t} \sin 4t)^2 (e^{-t})(3\sqrt{2}\, e^{-t}) \, dt$

$$= \int_0^{2\pi} 3\sqrt{2}\, e^{-7t} \cos^3 4t \sin^2 4t \, dt = \tfrac{172{,}704}{5{,}632{,}705} \sqrt{2}\, (1 - e^{-14\pi})$$

32. (a) We parametrize the circle C as $\mathbf{r}(t) = 2 \cos t \, \mathbf{i} + 2 \sin t \, \mathbf{j}, \;\; 0 \le t \le 2\pi$. So $\mathbf{F}(\mathbf{r}(t)) = \langle 4 \cos^2 t, 4 \cos t \sin t \rangle$,

$\mathbf{r}'(t) = \langle -2 \sin t, 2 \cos t \rangle$, and $W = \int_C \mathbf{F} \cdot d\mathbf{r} = \int_0^{2\pi} (-8 \cos^2 t \sin t + 8 \cos^2 t \sin t) \, dt = 0$.

(b)

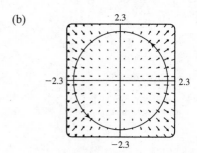

From the graph, we see that all of the vectors in the field are perpendicular to the path. This indicates that the field does no work on the particle, since the field never pulls the particle in the direction in which it is going. In other words, at any point along C, $\mathbf{F} \cdot \mathbf{T} = 0$, and so certainly $\int_C \mathbf{F} \cdot d\mathbf{r} = 0$.

33. We use the parametrization $x = 2 \cos t, \; y = 2 \sin t, \; -\frac{\pi}{2} \le t \le \frac{\pi}{2}$. Then

$ds = \sqrt{\left(\frac{dx}{dt}\right)^2 + \left(\frac{dy}{dt}\right)^2} \, dt = \sqrt{(-2 \sin t)^2 + (2 \cos t)^2} \, dt = 2 \, dt$, so $m = \int_C k \, ds = 2k \int_{-\pi/2}^{\pi/2} dt = 2k(\pi)$,

$\bar{x} = \frac{1}{2\pi k} \int_C xk \, ds = \frac{1}{2\pi} \int_{-\pi/2}^{\pi/2} (2 \cos t) 2 \, dt = \frac{1}{2\pi} \left[4 \sin t \right]_{-\pi/2}^{\pi/2} = \frac{4}{\pi}, \; \bar{y} = \frac{1}{2\pi k} \int_C yk \, ds = \frac{1}{2\pi} \int_{-\pi/2}^{\pi/2} (2 \sin t) 2 \, dt = 0$.

Hence $(\bar{x}, \bar{y}) = \left(\frac{4}{\pi}, 0 \right)$.

34. We use the parametrization $x = a\cos t$, $y = a\sin t$, $0 \le t \le \frac{\pi}{2}$. Then

$$ds = \sqrt{\left(\frac{dx}{dt}\right)^2 + \left(\frac{dy}{dt}\right)^2}\, dt = \sqrt{(-a\sin t)^2 + (a\cos t)^2}\, dt = a\, dt, \text{ so}$$

$$m = \int_C \rho(x,y)\, ds = \int_C kxy\, ds = \int_0^{\pi/2} k(a\cos t)(a\sin t)\, a\, dt = ka^3 \int_0^{\pi/2} \cos t \sin t\, dt = ka^3 \left[\tfrac{1}{2}\sin^2 t\right]_0^{\pi/2} = \tfrac{1}{2}ka^3,$$

$$\overline{x} = \frac{1}{ka^3/2}\int_C x(kxy)\, ds = \frac{2}{ka^3}\int_0^{\pi/2} k(a\cos t)^2(a\sin t)a\, dt = \frac{2}{ka^3}\cdot ka^4 \int_0^{\pi/2} \cos^2 t \sin t\, dt$$

$$= 2a\left[-\tfrac{1}{3}\cos^3 t\right]_0^{\pi/2} = 2a\left(0 + \tfrac{1}{3}\right) = \tfrac{2}{3}a, \text{ and}$$

$$\overline{y} = \frac{1}{ka^3/2}\int_C y(kxy)\, ds = \frac{2}{ka^3}\int_0^{\pi/2} k(a\cos t)(a\sin t)^2 a\, dt = \frac{2}{ka^3}\cdot ka^4 \int_0^{\pi/2} \sin^2 t \cos t\, dt$$

$$= 2a\left[\tfrac{1}{3}\sin^3 t\right]_0^{\pi/2} = 2a\left(\tfrac{1}{3} - 0\right) = \tfrac{2}{3}a.$$

Therefore the mass is $\frac{1}{2}ka^3$ and the center of mass is $(\overline{x}, \overline{y}) = \left(\frac{2}{3}a, \frac{2}{3}a\right)$.

35. (a) $\overline{x} = \dfrac{1}{m}\displaystyle\int_C x\rho(x,y,z)\, ds$, $\overline{y} = \dfrac{1}{m}\displaystyle\int_C y\rho(x,y,z)\, ds$, $\overline{z} = \dfrac{1}{m}\displaystyle\int_C z\rho(x,y,z)\, ds$ where $m = \int_C \rho(x,y,z)\, ds$.

(b) $m = \int_C k\, ds = k\int_0^{2\pi} \sqrt{4\sin^2 t + 4\cos^2 t + 9}\, dt = k\sqrt{13}\int_0^{2\pi} dt = 2\pi k\sqrt{13}$,

$$\overline{x} = \frac{1}{2\pi k\sqrt{13}}\int_0^{2\pi} 2k\sqrt{13}\sin t\, dt = 0, \quad \overline{y} = \frac{1}{2\pi k\sqrt{13}}\int_0^{2\pi} 2k\sqrt{13}\cos t\, dt = 0,$$

$$\overline{z} = \frac{1}{2\pi k\sqrt{13}}\int_0^{2\pi}\left(k\sqrt{13}\right)(3t)\, dt = \frac{3}{2\pi}\left(2\pi^2\right) = 3\pi. \text{ Hence } (\overline{x}, \overline{y}, \overline{z}) = (0, 0, 3\pi).$$

36. $m = \int_C (x^2 + y^2 + z^2)\, ds = \int_0^{2\pi} (t^2 + 1)\sqrt{(1)^2 + (-\sin t)^2 + (\cos t)^2}\, dt = \int_0^{2\pi} (t^2 + 1)\sqrt{2}\, dt = \sqrt{2}\left(\frac{8}{3}\pi^3 + 2\pi\right)$,

$$\overline{x} = \frac{1}{\sqrt{2}\left(\frac{8}{3}\pi^3 + 2\pi\right)}\int_0^{2\pi} \sqrt{2}(t^3 + t)\, dt = \frac{4\pi^4 + 2\pi^2}{\frac{8}{3}\pi^3 + 2\pi} = \frac{3\pi(2\pi^2 + 1)}{4\pi^2 + 3},$$

$$\overline{y} = \frac{3}{2\sqrt{2}\,\pi(4\pi^2 + 3)}\int_0^{2\pi}\left(\sqrt{2}\cos t\right)(t^2 + 1)\, dt = 0, \text{ and}$$

$$\overline{z} = \frac{3}{2\sqrt{2}\,\pi(4\pi^2 + 3)}\int_0^{2\pi}\left(\sqrt{2}\sin t\right)(t^2 + 1)\, dt = 0. \text{ Hence } (\overline{x}, \overline{y}, \overline{z}) = \left(\frac{3\pi(2\pi^2 + 1)}{4\pi^2 + 3}, 0, 0\right).$$

37. From Example 3, $\rho(x,y) = k(1 - y)$, $x = \cos t$, $y = \sin t$, and $ds = dt$, $0 \le t \le \pi$ $\Rightarrow$

$$I_x = \int_C y^2\rho(x,y)\, ds = \int_0^\pi \sin^2 t\left[k(1 - \sin t)\right]dt = k\int_0^\pi (\sin^2 t - \sin^3 t)\, dt$$

$$= \tfrac{1}{2}k\int_0^\pi (1 - \cos 2t)\, dt - k\int_0^\pi (1 - \cos^2 t)\sin t\, dt \qquad \left[\begin{array}{c}\text{Let } u = \cos t,\, du = -\sin t\, dt \\ \text{in the second integral}\end{array}\right]$$

$$= k\left[\tfrac{\pi}{2} + \int_1^{-1}(1 - u^2)\, du\right] = k\left(\tfrac{\pi}{2} - \tfrac{4}{3}\right)$$

$$I_y = \int_C x^2\rho(x,y)\, ds = k\int_0^\pi \cos^2 t\,(1 - \sin t)\, dt = \tfrac{k}{2}\int_0^\pi (1 + \cos 2t)\, dt - k\int_0^\pi \cos^2 t \sin t\, dt$$

$$= k\left(\tfrac{\pi}{2} - \tfrac{2}{3}\right), \text{ using the same substitution as above.}$$

38. The wire is given as $x = 2\sin t$, $y = 2\cos t$, $z = 3t$, $0 \le t \le 2\pi$ with $\rho(x,y,z) = k$. Then

$$ds = \sqrt{(2\cos t)^2 + (-2\sin t)^2 + 3^2}\, dt = \sqrt{4(\cos^2 t + \sin^2 t) + 9}\, dt = \sqrt{13}\, dt \text{ and}$$

$$I_x = \int_C (y^2 + z^2)\rho(x,y,z)\, ds = \int_0^{2\pi} (4\cos^2 t + 9t^2)(k)\sqrt{13}\, dt = \sqrt{13}\, k\left[4\left(\tfrac{1}{2}t + \tfrac{1}{4}\sin 2t\right) + 3t^3\right]_0^{2\pi}$$
$$= \sqrt{13}\, k(4\pi + 24\pi^3) = 4\sqrt{13}\,\pi k(1 + 6\pi^2)$$

$$I_y = \int_C (x^2 + z^2)\rho(x,y,z)\, ds = \int_0^{2\pi} \left(4\sin^2 t + 9t^2\right)(k)\sqrt{13}\, dt = \sqrt{13}\, k\left[4\left(\tfrac{1}{2}t - \tfrac{1}{4}\sin 2t\right) + 3t^3\right]_0^{2\pi}$$
$$= \sqrt{13}\, k(4\pi + 24\pi^3) = 4\sqrt{13}\,\pi k(1 + 6\pi^2)$$

$$I_z = \int_C (x^2 + y^2)\rho(x,y,z)\, ds = \int_0^{2\pi} (4\sin^2 t + 4\cos^2 t)(k)\sqrt{13}\, dt = 4\sqrt{13}\, k\int_0^{2\pi} dt = 8\pi\sqrt{13}\, k$$

39. $W = \int_C \mathbf{F} \cdot d\mathbf{r} = \int_0^{2\pi} \langle t - \sin t, 3 - \cos t\rangle \cdot \langle 1 - \cos t, \sin t\rangle\, dt$

$$= \int_0^{2\pi} (t - t\cos t - \sin t + \sin t\cos t + 3\sin t - \sin t\cos t)\, dt$$

$$= \int_0^{2\pi} (t - t\cos t + 2\sin t)\, dt = \left[\tfrac{1}{2}t^2 - (t\sin t + \cos t) - 2\cos t\right]_0^{2\pi} \quad \begin{bmatrix} \text{integrate by parts} \\ \text{in the second term} \end{bmatrix}$$

$$= 2\pi^2$$

40. $x = x$, $y = x^2$, $-1 \le x \le 2$,

$$W = \int_{-1}^2 \left\langle x\sin x^2, x^2\right\rangle \cdot \langle 1, 2x\rangle\, dx = \int_{-1}^2 (x\sin x^2 + 2x^3)\, dx = \left[-\tfrac{1}{2}\cos x^2 + \tfrac{1}{2}x^4\right]_{-1}^2 = \tfrac{1}{2}(15 + \cos 1 - \cos 4).$$

41. $\mathbf{r}(t) = \langle 1 + 2t, 4t, 2t\rangle$, $0 \le t \le 1$,

$$W = \int_C \mathbf{F} \cdot d\mathbf{r} = \int_0^1 \langle 6t, 1 + 4t, 1 + 6t\rangle \cdot \langle 2, 4, 2\rangle\, dt = \int_0^1 \left(12t + 4(1 + 4t) + 2(1 + 6t)\right) dt$$

$$= \int_0^1 (40t + 6)\, dt = \left[20t^2 + 6t\right]_0^1 = 26$$

42. $\mathbf{r}(t) = 2\mathbf{i} + t\mathbf{j} + 5t\mathbf{k}$, $0 \le t \le 1$. Therefore

$$W = \int_C \mathbf{F} \cdot d\mathbf{r} = \int_0^1 \frac{K\langle 2, t, 5t\rangle}{(4 + 26t^2)^{3/2}} \cdot \langle 0, 1, 5\rangle\, dt = K\int_0^1 \frac{26t}{(4 + 26t^2)^{3/2}}\, dt = K\left[-(4 + 26t^2)^{-1/2}\right]_0^1 = K\left(\tfrac{1}{2} - \tfrac{1}{\sqrt{30}}\right).$$

43. Let $\mathbf{F} = 185\,\mathbf{k}$. To parametrize the staircase, let $x = 20\cos t$, $y = 20\sin t$, $z = \frac{90}{6\pi}t = \frac{15}{\pi}t$, $0 \le t \le 6\pi \Rightarrow$

$$W = \int_C \mathbf{F} \cdot d\mathbf{r} = \int_0^{6\pi} \langle 0, 0, 185\rangle \cdot \left\langle -20\sin t, 20\cos t, \tfrac{15}{\pi}\right\rangle dt = (185)\tfrac{15}{\pi}\int_0^{6\pi} dt = (185)(90) \approx 1.67 \times 10^4 \text{ ft-lb}$$

44. This time m is a function of t: $m = 185 - \frac{9}{6\pi}t = 185 - \frac{3}{2\pi}t$. So let $\mathbf{F} = \left(185 - \frac{3}{2\pi}t\right)\mathbf{k}$. To parametrize the staircase,

let $x = 20\cos t$, $y = 20\sin t$, $z = \frac{90}{6\pi}t = \frac{15}{\pi}t$, $0 \le t \le 6\pi$. Therefore

$$W = \int_C \mathbf{F} \cdot d\mathbf{r} = \int_0^{6\pi} \left\langle 0, 0, 185 - \tfrac{3}{2\pi}t\right\rangle \cdot \left\langle -20\sin t, 20\cos t, \tfrac{15}{\pi}\right\rangle dt = \tfrac{15}{\pi}\int_0^{6\pi} \left(185 - \tfrac{3}{2\pi}t\right) dt$$

$$= \tfrac{15}{\pi}\left[185t - \tfrac{3}{4\pi}t^2\right]_0^{6\pi} = 90\left(185 - \tfrac{9}{2}\right) \approx 1.62 \times 10^4 \text{ ft-lb}$$

45. (a) $\mathbf{r}(t) = \langle \cos t, \sin t\rangle$, $0 \le t \le 2\pi$, and let $\mathbf{F} = \langle a, b\rangle$. Then

$$W = \int_C \mathbf{F} \cdot d\mathbf{r} = \int_0^{2\pi} \langle a, b\rangle \cdot \langle -\sin t, \cos t\rangle\, dt = \int_0^{2\pi} (-a\sin t + b\cos t)\, dt = \left[a\cos t + b\sin t\right]_0^{2\pi}$$

$$= a + 0 - a + 0 = 0$$

(b) Yes. $\mathbf{F}(x,y) = k\,\mathbf{x} = \langle kx, ky\rangle$ and

$$W = \int_C \mathbf{F} \cdot d\mathbf{r} = \int_0^{2\pi} \langle k\cos t, k\sin t\rangle \cdot \langle -\sin t, \cos t\rangle\, dt = \int_0^{2\pi} (-k\sin t\cos t + k\sin t\cos t)\, dt = \int_0^{2\pi} 0\, dt = 0.$$

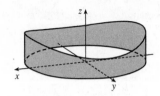

46. Consider the base of the fence in the xy-plane, centered at the origin, with the height given by $z = h(x, y)$. The fence can be graphed using the parametric equations $x = 10 \cos u$, $y = 10 \sin u$,

$$z = v\left[4 + 0.01((10\cos u)^2 - (10\sin u)^2)\right] = v(4 + \cos^2 u - \sin^2 u)$$
$$= v(4 + \cos 2u), \ 0 \le u \le 2\pi, \ 0 \le v \le 1.$$

The area of the fence is $\int_C h(x, y)\, ds$ where C, the base of the fence, is given by $x = 10\cos t$, $y = 10\sin t$, $0 \le t \le 2\pi$. Then

$$\int_C h(x, y)\, ds = \int_0^{2\pi} \left[4 + 0.01((10\cos t)^2 - (10\sin t)^2)\right] \sqrt{(-10\sin t)^2 + (10\cos t)^2}\, dt$$
$$= \int_0^{2\pi} (4 + \cos 2t)\sqrt{100}\, dt = 10\left[4t + \tfrac{1}{2}\sin 2t\right]_0^{2\pi} = 10(8\pi) = 80\pi \text{ m}^2$$

If we paint both sides of the fence, the total surface area to cover is 160π m^2, and since 1 L of paint covers 100 m^2, we require $\frac{160\pi}{100} = 1.6\pi \approx 5.03$ L of paint.

47. The work done in moving the object is $\int_C \mathbf{F} \cdot d\mathbf{r} = \int_C \mathbf{F} \cdot \mathbf{T}\, ds$. We can approximate this integral by dividing C into 7 segments of equal length $\Delta s = 2$ and approximating $\mathbf{F} \cdot \mathbf{T}$, that is, the tangential component of force, at a point (x_i^*, y_i^*) on each segment. Since C is composed of straight line segments, $\mathbf{F} \cdot \mathbf{T}$ is the scalar projection of each force vector onto C. If we choose (x_i^*, y_i^*) to be the point on the segment closest to the origin, then the work done is

$$\int_C \mathbf{F} \cdot \mathbf{T}\, ds \approx \sum_{i=1}^{7} \left[\mathbf{F}(x_i^*, y_i^*) \cdot \mathbf{T}(x_i^*, y_i^*)\right] \Delta s = [2 + 2 + 2 + 2 + 1 + 1 + 1](2) = 22.$$ Thus, we estimate the work done to

be approximately 22 J.

48. Use the orientation pictured in the figure. Then since $\mathbf{B}$ is tangent to any circle that lies in the plane perpendicular to the wire, $\mathbf{B} = |\mathbf{B}|\, \mathbf{T}$ where $\mathbf{T}$ is the unit tangent to the circle $C: x = r\cos\theta$, $y = r\sin\theta$. Thus $\mathbf{B} = |\mathbf{B}|\langle -\sin\theta, \cos\theta\rangle$. Then

$$\int_C \mathbf{B} \cdot d\mathbf{r} = \int_0^{2\pi} |\mathbf{B}|\langle -\sin\theta, \cos\theta\rangle \cdot \langle -r\sin\theta, r\cos\theta\rangle\, d\theta = \int_0^{2\pi} |\mathbf{B}|\, r\, d\theta = 2\pi r\, |\mathbf{B}|.$$ (Note that $|\mathbf{B}|$ here is the magnitude

of the field at a distance r from the wire's center.) But by Ampere's Law $\int_C \mathbf{B} \cdot d\mathbf{r} = \mu_0 I$. Hence $|\mathbf{B}| = \mu_0 I/(2\pi r)$.

17.3   The Fundamental Theorem for Line Integrals   ET 16.3

1. C appears to be a smooth curve, and since ∇f is continuous, we know f is differentiable. Then Theorem 2 says that the value of $\int_C \nabla f \cdot d\mathbf{r}$ is simply the difference of the values of f at the terminal and initial points of C. From the graph, this is $50 - 10 = 40$.

2. C is represented by the vector function $\mathbf{r}(t) = (t^2 + 1)\mathbf{i} + (t^3 + t)\mathbf{j}$, $0 \le t \le 1$, so $\mathbf{r}'(t) = 2t\,\mathbf{i} + (3t^2 + 1)\,\mathbf{j}$. Since $3t^2 + 1 \ne 0$, we have $\mathbf{r}'(t) \ne \mathbf{0}$, thus C is a smooth curve. ∇f is continuous, and hence f is differentiable, so by Theorem 2 we have $\int_C \nabla f \cdot d\mathbf{r} = f(\mathbf{r}(1)) - f(\mathbf{r}(0)) = f(2, 2) - f(1, 0) = 9 - 3 = 6$.

3. $\partial(2x - 3y)/\partial y = -3 = \partial(-3x + 4y - 8)/\partial x$ and the domain of $\mathbf{F}$ is $\mathbb{R}^2$ which is open and simply-connected, so by Theorem 6 $\mathbf{F}$ is conservative. Thus, there exists a function f such that $\nabla f = \mathbf{F}$, that is, $f_x(x, y) = 2x - 3y$ and $f_y(x, y) = -3x + 4y - 8$. But $f_x(x, y) = 2x - 3y$ implies $f(x, y) = x^2 - 3xy + g(y)$ and differentiating both sides of this equation with respect to y gives $f_y(x, y) = -3x + g'(y)$. Thus $-3x + 4y - 8 = -3x + g'(y)$ so $g'(y) = 4y - 8$ and

$g(y) = 2y^2 - 8y + K$ where K is a constant. Hence $f(x, y) = x^2 - 3xy + 2y^2 - 8y + K$ is a potential function for $\mathbf{F}$.

4. $\partial(e^x \cos y)/\partial y = -e^x \sin y$, $\partial(e^x \sin y)/\partial x = e^x \sin y$. Since these are not equal, $\mathbf{F}$ is not conservative.

5. $\partial(e^x \sin y)/\partial y = e^x \cos y = \partial(e^x \cos y)/\partial x$ and the domain of $\mathbf{F}$ is $\mathbb{R}^2$. Hence $\mathbf{F}$ is conservative so there exists a function f

such that $\nabla f = \mathbf{F}$. Then $f_x(x, y) = e^x \sin y$ implies $f(x, y) = e^x \sin y + g(y)$ and $f_y(x, y) = e^x \cos y + g'(y)$. But

$f_y(x, y) = e^x \cos y$ so $g'(y) = 0 \Rightarrow g(y) = K$. Then $f(x, y) = e^x \sin y + K$ is a potential function for $\mathbf{F}$.

6. $\partial(3x^2 - 2y^2)/\partial y = -4y$, $\partial(4xy + 3)/\partial x = 4y$. Since these are not equal, $\mathbf{F}$ is not conservative.

7. $\partial(ye^x + \sin y)/\partial y = e^x + \cos y = \partial(e^x + x \cos y)/\partial x$ and the domain of $\mathbf{F}$ is $\mathbb{R}^2$. Hence $\mathbf{F}$ is conservative so there

exists a function f such that $\nabla f = \mathbf{F}$. Then $f_x(x, y) = ye^x + \sin y$ implies $f(x, y) = ye^x + x \sin y + g(y)$ and

$f_y(x, y) = e^x + x \cos y + g'(y)$. But $f_y(x, y) = e^x + x \cos y$ so $g(y) = K$ and $f(x, y) = ye^x + x \sin y + K$ is a potential

function for $\mathbf{F}$.

8. $\partial(xy \cos xy + \sin xy)/\partial y = -x^2 y \sin xy + 2x \cos xy = \partial(x^2 \cos xy)/\partial x$ and the domain of $\mathbf{F}$ is $\mathbb{R}^2$. Hence $\mathbf{F}$ is

conservative, so there exists a function f such that $\nabla f = \mathbf{F}$. Then $f_y(x, y) = x^2 \cos xy$ implies $f(x, y) = x \sin xy + g(x)$

and $f_x(x, y) = xy \cos xy + \sin xy + g'(x)$. But $f_x(x, y) = xy \cos xy + \sin xy$ so $g(x) = K$ and $f(x, y) = x \sin xy + K$ is

a potential function for $\mathbf{F}$.

9. $\partial(\ln y + 2xy^3)/\partial y = 1/y + 6xy^2 = \partial(3x^2 y^2 + x/y)/\partial x$ and the domain of $\mathbf{F}$ is $\{(x, y) \mid y > 0\}$ which is open and simply

connected. Hence $\mathbf{F}$ is conservative so there exists a function f such that $\nabla f = \mathbf{F}$. Then $f_x(x, y) = \ln y + 2xy^3$ implies

$f(x, y) = x \ln y + x^2 y^3 + g(y)$ and $f_y(x, y) = x/y + 3x^2 y^2 + g'(y)$. But $f_y(x, y) = 3x^2 y^2 + x/y$ so $g'(y) = 0 \Rightarrow$

$g(y) = K$ and $f(x, y) = x \ln y + x^2 y^3 + K$ is a potential function for $\mathbf{F}$.

10. $\dfrac{\partial(xy \cosh xy + \sinh xy)}{\partial y} = x^2 y \sinh xy + x \cosh xy + x \cosh xy = x^2 y \sinh xy + 2x \cosh xy = \dfrac{\partial(x^2 \cosh xy)}{\partial x}$

and the domain of $\mathbf{F}$ is $\mathbb{R}^2$. Thus $\mathbf{F}$ is conservative, so there exists a function f such that $\nabla f = \mathbf{F}$. Then

$f_x(x, y) = xy \cosh xy + \sinh xy$ implies $f(x, y) = x \sinh xy + g(y) \Rightarrow f_y(x, y) = x^2 \cosh xy + g'(y)$. But

$f_y(x, y) = x^2 \cosh xy$ so $g(y) = K$ and $f(x, y) = x \sinh xy + K$ is a potential function for $\mathbf{F}$.

11. (a) $\mathbf{F}$ has continuous first-order partial derivatives and $\dfrac{\partial}{\partial y} 2xy = 2x = \dfrac{\partial}{\partial x}(x^2)$ on $\mathbb{R}^2$, which is open and simply-connected.

Thus, $\mathbf{F}$ is conservative by Theorem 6. Then we know that the line integral of $\mathbf{F}$ is independent of path; in particular, the

value of $\int_C \mathbf{F} \cdot d\mathbf{r}$ depends only on the endpoints of C. Since all three curves have the same initial and terminal points,

$\int_C \mathbf{F} \cdot d\mathbf{r}$ will have the same value for each curve.

(b) We first find a potential function f, so that $\nabla f = \mathbf{F}$. We know $f_x(x, y) = 2xy$ and $f_y(x, y) = x^2$. Integrating

$f_x(x, y)$ with respect to x, we have $f(x, y) = x^2 y + g(y)$. Differentiating both sides with respect to y gives

$f_y(x, y) = x^2 + g'(y)$, so we must have $x^2 + g'(y) = x^2 \Rightarrow g'(y) = 0 \Rightarrow g(y) = K$, a constant.

Thus $f(x, y) = x^2 y + K$. All three curves start at $(1, 2)$ and end at $(3, 2)$, so by Theorem 2,

$\int_C \mathbf{F} \cdot d\mathbf{r} = f(3, 2) - f(1, 2) = 18 - 2 = 16$ for each curve.

12. (a) $f_x(x,y) = x^2$ implies $f(x,y) = \frac{1}{3}x^3 + g(y)$ and $f_y(x,y) = 0 + g'(y)$. But $f_y(x,y) = y^2$ so

$g'(y) = y^2 \Rightarrow g(y) = \frac{1}{3}y^3 + K$. We can take $K = 0$, so $f(x,y) = \frac{1}{3}x^3 + \frac{1}{3}y^3$.

(b) $\int_C \mathbf{F} \cdot d\mathbf{r} = f(2,8) - f(-1,2) = \left(\frac{8}{3} + \frac{512}{3}\right) - \left(-\frac{1}{3} + \frac{8}{3}\right) = 171$.

13. (a) $f_x(x,y) = xy^2$ implies $f(x,y) = \frac{1}{2}x^2y^2 + g(y)$ and $f_y(x,y) = x^2y + g'(y)$. But $f_y(x,y) = x^2y$ so $g'(y) = 0 \Rightarrow$

$g(y) = K$, a constant. We can take $K = 0$, so $f(x,y) = \frac{1}{2}x^2y^2$.

(b) The initial point of C is $\mathbf{r}(0) = (0,1)$ and the terminal point is $\mathbf{r}(1) = (2,1)$, so

$\int_C \mathbf{F} \cdot d\mathbf{r} = f(2,1) - f(0,1) = 2 - 0 = 2$.

14. (a) $f_x(x,y) = y^2/(1+x^2)$ implies $f(x,y) = y^2 \arctan x + g(y) \Rightarrow f_y(x,y) = 2y \arctan x + g'(y)$. But

$f_y(x,y) = 2y \arctan x$ so $g'(y) = 0 \Rightarrow g(y) = K$. We can take $K = 0$, so $f(x,y) = y^2 \arctan x$.

(b) The initial point of C is $\mathbf{r}(0) = (0,0)$ and the terminal point is $\mathbf{r}(1) = (1,2)$, so

$\int_C \mathbf{F} \cdot d\mathbf{r} = f(1,2) - f(0,0) = 4 \arctan 1 - 0 = 4 \cdot \frac{\pi}{4} = \pi$.

15. (a) $f_x(x,y,z) = yz$ implies $f(x,y,z) = xyz + g(y,z)$ and so $f_y(x,y,z) = xz + g_y(y,z)$. But $f_y(x,y,z) = xz$ so

$g_y(y,z) = 0 \Rightarrow g(y,z) = h(z)$. Thus $f(x,y,z) = xyz + h(z)$ and $f_z(x,y,z) = xy + h'(z)$. But

$f_z(x,y,z) = xy + 2z$, so $h'(z) = 2z \Rightarrow h(z) = z^2 + K$. Hence $f(x,y,z) = xyz + z^2$ (taking $K = 0$).

(b) $\int_C \mathbf{F} \cdot d\mathbf{r} = f(4,6,3) - f(1,0,-2) = 81 - 4 = 77$.

16. (a) $f_x(x,y,z) = 2xz + y^2$ implies $f(x,y,z) = x^2z + xy^2 + g(y,z)$ and so $f_y(x,y,z) = 2xy + g_y(y,z)$. But

$f_y(x,y,z) = 2xy$ so $g_y(y,z) = 0 \Rightarrow g(y,z) = h(z)$. Thus $f(x,y,z) = x^2z + xy^2 + h(z)$ and

$f_z(x,y,z) = x^2 + h'(z)$. But $f_z(x,y,z) = x^2 + 3z^2$, so $h'(z) = 3z^2 \Rightarrow h(z) = z^3 + K$. Hence

$f(x,y,z) = x^2z + xy^2 + z^3$ (taking $K = 0$).

(b) $t = 0$ corresponds to the point $(0,1,-1)$ and $t = 1$ corresponds to $(1,2,1)$, so

$\int_C \mathbf{F} \cdot d\mathbf{r} = f(1,2,1) - f(0,1,-1) = 6 - (-1) = 7$.

17. (a) $f_x(x,y,z) = y^2 \cos z$ implies $f(x,y,z) = xy^2 \cos z + g(y,z)$ and so $f_y(x,y,z) = 2xy \cos z + g_y(y,z)$. But

$f_y(x,y,z) = 2xy \cos z$ so $g_y(y,z) = 0 \Rightarrow g(y,z) = h(z)$. Thus $f(x,y,z) = xy^2 \cos z + h(z)$ and

$f_z(x,y,z) = -xy^2 \sin z + h'(z)$. But $f_z(x,y,z) = -xy^2 \sin z$, so $h'(z) = 0 \Rightarrow h(z) = K$. Hence

$f(x,y,z) = xy^2 \cos z$ (taking $K = 0$).

(b) $\mathbf{r}(0) = \langle 0,0,0 \rangle$, $\mathbf{r}(\pi) = \langle \pi^2, 0, \pi \rangle$ so $\int_C \mathbf{F} \cdot d\mathbf{r} = f(\pi^2, 0, \pi) - f(0,0,0) = 0 - 0 = 0$.

18. (a) $f_x(x,y,z) = e^y$ implies $f(x,y,z) = xe^y + g(y,z)$ and so $f_y(x,y,z) = xe^y + g_y(y,z)$. But $f_y(x,y,z) = xe^y$ so

$g_y(y,z) = 0 \Rightarrow g(y,z) = h(z)$. Thus $f(x,y,z) = xe^y + h(z)$ and $f_z(x,y,z) = 0 + h'(z)$. But

$f_z(x,y,z) = (z+1)e^z$, so $h'(z) = (z+1)e^z \Rightarrow h(z) = ze^z + K$ (using integration by parts). Hence

$f(x,y,z) = xe^y + ze^z$ (taking $K = 0$).

(b) $\mathbf{r}(0) = \langle 0,0,0 \rangle$, $\mathbf{r}(1) = \langle 1,1,1 \rangle$ so $\int_C \mathbf{F} \cdot d\mathbf{r} = f(1,1,1) - f(0,0,0) = 2e - 0 = 2e$.

19. Here $\mathbf{F}(x, y) = \tan y \, \mathbf{i} + x \sec^2 y \, \mathbf{j}$. Then $f(x, y) = x \tan y$ is a potential function for $\mathbf{F}$, that is, $\nabla f = \mathbf{F}$ so

$\mathbf{F}$ is conservative and thus its line integral is independent of path. Hence

$\int_C \tan y \, dx + x \sec^2 y \, dy = \int_C \mathbf{F} \cdot d\mathbf{r} = f\left(2, \frac{\pi}{4}\right) - f(1, 0) = 2 \tan \frac{\pi}{4} - \tan 0 = 2.$

20. Here $\mathbf{F}(x, y) = (1 - ye^{-x}) \, \mathbf{i} + e^{-x} \, \mathbf{j}$. Then $f(x, y) = x + ye^{-x}$ is a potential function for $\mathbf{F}$, that is, $\nabla f = \mathbf{F}$ so

$\mathbf{F}$ is conservative and thus its line integral is independent of path. Hence

$\int_C (1 - ye^{-x}) \, dx + e^{-x} \, dy = \int_C \mathbf{F} \cdot d\mathbf{r} = f(1, 2) - f(0, 1) = (1 + 2e^{-1}) - 1 = 2/e.$

21. $\mathbf{F}(x, y) = 2y^{3/2} \, \mathbf{i} + 3x \sqrt{y} \, \mathbf{j}$, $W = \int_C \mathbf{F} \cdot d\mathbf{r}$. Since $\partial(2y^{3/2})/\partial y = 3\sqrt{y} = \partial(3x\sqrt{y})/\partial x$, there exists a function f such

that $\nabla f = \mathbf{F}$. In fact, $f_x(x, y) = 2y^{3/2} \;\Rightarrow\; f(x, y) = 2xy^{3/2} + g(y) \;\Rightarrow\; f_y(x, y) = 3xy^{1/2} + g'(y)$. But

$f_y(x, y) = 3x\sqrt{y}$ so $g'(y) = 0$ or $g(y) = K$. We can take $K = 0 \;\Rightarrow\; f(x, y) = 2xy^{3/2}$. Thus

$W = \int_C \mathbf{F} \cdot d\mathbf{r} = f(2, 4) - f(1, 1) = 2(2)(8) - 2(1) = 30.$

22. $\mathbf{F}(x, y) = e^{-y} \, \mathbf{i} - xe^{-y} \, \mathbf{j}$, $W = \int_C \mathbf{F} \cdot d\mathbf{r}$. Since $\dfrac{\partial}{\partial y}(e^{-y}) = -e^{-y} = \dfrac{\partial}{\partial x}(-xe^{-y})$, there exists a function f such that

$\nabla f = \mathbf{F}$. In fact, $f_x = e^{-y} \;\Rightarrow\; f(x, y) = xe^{-y} + g(y) \;\Rightarrow\; f_y = -xe^{-y} + g'(y) \;\Rightarrow\; g'(y) = 0$, so we can take

$f(x, y) = xe^{-y}$ as a potential function for $\mathbf{F}$. Thus $W = \int_C \mathbf{F} \cdot d\mathbf{r} = f(2, 0) - f(0, 1) = 2 - 0 = 2.$

23. We know that if the vector field (call it $\mathbf{F}$) is conservative, then around any closed path C, $\int_C \mathbf{F} \cdot d\mathbf{r} = 0$. But take C to be a

circle centered at the origin, oriented counterclockwise. All of the field vectors that start on C are roughly in the direction of

motion along C, so the integral around C will be positive. Therefore the field is not conservative.

24. If a vector field $\mathbf{F}$ is conservative, then around any closed path C, $\int_C \mathbf{F} \cdot d\mathbf{r} = 0$. For any closed path we draw in the field, it

appears that some vectors on the curve point in approximately the same direction as the curve and a similar number point in

roughly the opposite direction. (Some appear perpendicular to the curve as well.) Therefore it is plausible that $\int_C \mathbf{F} \cdot d\mathbf{r} = 0$

for every closed curve C which means $\mathbf{F}$ is conservative.

25.

From the graph, it appears that $\mathbf{F}$ is conservative, since around all closed

paths, the number and size of the field vectors pointing in directions similar

to that of the path seem to be roughly the same as the number and size of the

vectors pointing in the opposite direction. To check, we calculate

$\dfrac{\partial}{\partial y}(\sin y) = \cos y = \dfrac{\partial}{\partial x}(1 + x \cos y)$. Thus $\mathbf{F}$ is conservative, by

Theorem 6.

26. $\nabla f(x, y) = \cos(x - 2y) \, \mathbf{i} - 2\cos(x - 2y) \, \mathbf{j}$

(a) We use Theorem 2: $\int_{C_1} \mathbf{F} \cdot d\mathbf{r} = \int_{C_1} \nabla f \cdot d\mathbf{r} = f(\mathbf{r}(b)) - f(\mathbf{r}(a))$ where C_1 starts at $t = a$ and ends at $t = b$. So

because $f(0, 0) = \sin 0 = 0$ and $f(\pi, \pi) = \sin(\pi - 2\pi) = 0$, one possible curve C_1 is the straight line from $(0, 0)$ to

(π, π); that is, $\mathbf{r}(t) = \pi t \, \mathbf{i} + \pi t \, \mathbf{j}$, $0 \le t \le 1$.

(b) From (a), $\int_{C_2} \mathbf{F} \cdot d\mathbf{r} = f(\mathbf{r}(b)) - f(\mathbf{r}(a))$. So because $f(0,0) = \sin 0 = 0$ and $f\left(\frac{\pi}{2}, 0\right) = 1$, one possible curve C_2 is

$\mathbf{r}(t) = \frac{\pi}{2} t\,\mathbf{i}$, $0 \le t \le 1$, the straight line from $(0,0)$ to $\left(\frac{\pi}{2}, 0\right)$.

27. Since $\mathbf{F}$ is conservative, there exists a function f such that $\mathbf{F} = \nabla f$, that is, $P = f_x$, $Q = f_y$, and $R = f_z$. Since P,

Q and R have continuous first order partial derivatives, Clairaut's Theorem says that $\partial P/\partial y = f_{xy} = f_{yx} = \partial Q/\partial x$,

$\partial P/\partial z = f_{xz} = f_{zx} = \partial R/\partial x$, and $\partial Q/\partial z = f_{yz} = f_{zy} = \partial R/\partial y$.

28. Here $\mathbf{F}(x, y, z) = y\,\mathbf{i} + x\,\mathbf{j} + xyz\,\mathbf{k}$. Then using the notation of Exercise 27, $\partial P/\partial z = 0$ while $\partial R/\partial x = yz$. Since these

aren't equal, $\mathbf{F}$ is not conservative. Thus by Theorem 4, the line integral of $\mathbf{F}$ is not independent of path.

29. $D = \{(x, y) \mid x > 0, y > 0\} = $ the first quadrant (excluding the axes).

(a) D is open because around every point in D we can put a disk that lies in D.

(b) D is connected because the straight line segment joining any two points in D lies in D.

(c) D is simply-connected because it's connected and has no holes.

30. $D = \{(x, y) \mid x \ne 0\}$ consists of all points in the xy-plane except for those on the y-axis.

(a) D is open.

(b) Points on opposite sides of the y-axis cannot be joined by a path that lies in D, so D is not connected.

(c) D is not simply-connected because it is not connected.

31. $D = \{(x, y) \mid 1 < x^2 + y^2 < 4\} = $ the annular region between the circles with center $(0, 0)$ and radii 1 and 2.

(a) D is open.

(b) D is connected.

(c) D is not simply-connected. For example, $x^2 + y^2 = (1.5)^2$ is simple and closed and lies within D but encloses points that

are not in D. (Or we can say, D has a hole, so is not simply-connected.)

32. $D = \{(x, y) \mid x^2 + y^2 \le 1 \text{ or } 4 \le x^2 + y^2 \le 9\} = $ the points on or inside the circle $x^2 + y^2 = 1$, together with the points

on or between the circles $x^2 + y^2 = 4$ and $x^2 + y^2 = 9$.

(a) D is not open because, for instance, no disk with center $(0, 2)$ lies entirely within D.

(b) D is not connected because, for example, $(0, 0)$ and $(0, 2.5)$ lie in D but cannot be joined by a path that lies entirely in D.

(c) D is not simply-connected because, for example, $x^2 + y^2 = 9$ is a simple closed curve in D but encloses points that are

not in D.

33. (a) $P = -\dfrac{y}{x^2 + y^2}$, $\dfrac{\partial P}{\partial y} = \dfrac{y^2 - x^2}{(x^2 + y^2)^2}$ and $Q = \dfrac{x}{x^2 + y^2}$, $\dfrac{\partial Q}{\partial x} = \dfrac{y^2 - x^2}{(x^2 + y^2)^2}$. Thus $\dfrac{\partial P}{\partial y} = \dfrac{\partial Q}{\partial x}$.

(b) C_1: $x = \cos t$, $y = \sin t$, $0 \le t \le \pi$, C_2: $x = \cos t$, $y = \sin t$, $t = 2\pi$ to $t = \pi$. Then

$$\int_{C_1} \mathbf{F} \cdot d\mathbf{r} = \int_0^\pi \frac{(-\sin t)(-\sin t) + (\cos t)(\cos t)}{\cos^2 t + \sin^2 t}\, dt = \int_0^\pi dt = \pi \text{ and } \int_{C_2} \mathbf{F} \cdot d\mathbf{r} = \int_{2\pi}^\pi dt = -\pi$$

Since these aren't equal, the line integral of $\mathbf{F}$ isn't independent of path. (Or notice that $\int_{C_3} \mathbf{F} \cdot d\mathbf{r} = \int_0^{2\pi} dt = 2\pi$ where

C_3 is the circle $x^2 + y^2 = 1$, and apply the contrapositive of Theorem 3.) This doesn't contradict Theorem 6, since the domain of $\mathbf{F}$, which is $\mathbb{R}^2$ except the origin, isn't simply-connected.

34. (a) Here $\mathbf{F}(\mathbf{r}) = c\mathbf{r}/|\mathbf{r}|^3$ and $\mathbf{r} = x\,\mathbf{i} + y\,\mathbf{j} + z\,\mathbf{k}$. Then $f(\mathbf{r}) = -c/|\mathbf{r}|$ is a potential function for $\mathbf{F}$, that is, $\nabla f = \mathbf{F}$. (See the discussion of gradient fields in Section 17.1 [ET 16.1].) Hence $\mathbf{F}$ is conservative and its line integral is independent of path. Let $P_1 = (x_1, y_1, z_1)$ and $P_2 = (x_2, y_2, z_2)$.

$$W = \int_C \mathbf{F} \cdot d\mathbf{r} = f(P_2) - f(P_1) = -\frac{c}{(x_2^2 + y_2^2 + z_2^2)^{1/2}} + \frac{c}{(x_1^2 + y_1^2 + z_1^2)^{1/2}} = c\left(\frac{1}{d_1} - \frac{1}{d_2}\right).$$

(b) In this case, $c = -(mMG) \Rightarrow$

$$W = -mMG\left(\frac{1}{1.52 \times 10^{11}} - \frac{1}{1.47 \times 10^{11}}\right)$$

$$= -(5.97 \times 10^{24})(1.99 \times 10^{30})(6.67 \times 10^{-11})(-2.2377 \times 10^{-13}) \approx 1.77 \times 10^{32} \text{ J}$$

(c) In this case, $c = \epsilon qQ \Rightarrow$

$$W = \epsilon qQ\left(\frac{1}{10^{-12}} - \frac{1}{5 \times 10^{-13}}\right) = (8.985 \times 10^9)(1)(-1.6 \times 10^{-19})(-10^{12}) \approx 1400 \text{ J}.$$

17.4 Green's Theorem

1. (a) Parametric equations for C are $x = 2\cos t$, $y = 2\sin t$, $0 \le t \le 2\pi$. Then

$$\oint_C (x - y)\,dx + (x + y)\,dy = \int_0^{2\pi} [(2\cos t - 2\sin t)(-2\sin t) + (2\cos t + 2\sin t)(2\cos t)]\,dt$$

$$= \int_0^{2\pi} (4\sin^2 t + 4\cos^2 t)\,dt = \int_0^{2\pi} 4\,dt = 4t\Big]_0^{2\pi} = 8\pi$$

(b) Note that C as given in part (a) is a positively oriented, smooth, simple closed curve. Then by Green's Theorem,

$$\oint_C (x - y)\,dx + (x + y)\,dy = \iint_D \left[\frac{\partial}{\partial x}(x + y) - \frac{\partial}{\partial y}(x - y)\right] dA = \iint_D [1 - (-1)]\,dA = 2\iint_D dA$$

$$= 2A(D) = 2\pi(2)^2 = 8\pi$$

2. (a)

C_1: $x = t \Rightarrow dx = dt$, $y = 0 \Rightarrow dy = 0\,dt$, $0 \le t \le 3$.

C_2: $x = 3 \Rightarrow dx = 0\,dt$, $y = t \Rightarrow dy = dt$, $0 \le t \le 1$.

C_3: $x = 3 - t \Rightarrow dx = -dt$, $y = 1 \Rightarrow dy = 0\,dt$, $0 \le t \le 3$.

C_4: $x = 0 \Rightarrow dx = 0\,dt$, $y = 1 - t \Rightarrow dy = -dt$, $0 \le t \le 1$

Thus $\displaystyle\oint_C xy\,dx + x^2\,dy = \oint_{C_1 + C_2 + C_3 + C_4} xy\,dx + x^2\,dy = \int_0^3 0\,dt + \int_0^1 9\,dt + \int_0^3 (3 - t)(-1)\,dt + \int_0^1 0\,dt$

$$= [9t]_0^1 + \left[\tfrac{1}{2}t^2 - 3t\right]_0^3 = 9 + \tfrac{9}{2} - 9 = \tfrac{9}{2}$$

(b) $\displaystyle\oint_C xy\,dx + x^2\,dy = \iint_D \left[\frac{\partial}{\partial x}(x^2) - \frac{\partial}{\partial y}(xy)\right] dA = \int_0^3 \int_0^1 (2x - x)\,dy\,dx = \int_0^3 x\,dx \int_0^1 dy = \left[\tfrac{1}{2}x^2\right]_0^3 \cdot 1 = \tfrac{9}{2}$

3. (a)

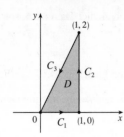

C_1: $x = t$ $\Rightarrow$ $dx = dt$, $y = 0$ $\Rightarrow$ $dy = 0\,dt$, $0 \le t \le 1$.

C_2: $x = 1$ $\Rightarrow$ $dx = 0\,dt$, $y = t$ $\Rightarrow$ $dy = dt$, $0 \le t \le 2$.

C_3: $x = 1 - t$ $\Rightarrow$ $dx = -dt$, $y = 2 - 2t$ $\Rightarrow$ $dy = -2\,dt$, $0 \le t \le 1$.

Thus
$$\oint_C xy\,dx + x^2 y^3 \,dy = \oint_{C_1 + C_2 + C_3} xy\,dx + x^2 y^3\,dy$$

$$= \int_0^1 0\,dt + \int_0^2 t^3\,dt + \int_0^1 \left[-(1-t)(2-2t) - 2(1-t)^2(2-2t)^3\right]dt$$

$$= 0 + \left[\tfrac{1}{4}t^4\right]_0^2 + \left[\tfrac{2}{3}(1-t)^3 + \tfrac{8}{3}(1-t)^6\right]_0^1 = 4 - \tfrac{10}{3} = \tfrac{2}{3}$$

(b) $\oint_C xy\,dx + x^2 y^3\,dy = \iint_D \left[\frac{\partial}{\partial x}(x^2 y^3) - \frac{\partial}{\partial y}(xy)\right]dA = \int_0^1 \int_0^{2x}(2xy^3 - x)\,dy\,dx$

$$= \int_0^1 \left[\tfrac{1}{2}xy^4 - xy\right]_{y=0}^{y=2x} dx = \int_0^1 (8x^5 - 2x^2)\,dx = \tfrac{4}{3} - \tfrac{2}{3} = \tfrac{2}{3}$$

4. (a) C_1: $x = 0$ $\Rightarrow$ $dx = 0\,dt$, $y = 1 - t$ $\Rightarrow$ $dy = -dt$, $0 \le t \le 1$

C_2: $x = t$ $\Rightarrow$ $dx = dt$, $y = 0$ $\Rightarrow$ $dy = 0\,dt$, $0 \le t \le 1$

C_3: $x = 1 - t$ $\Rightarrow$ $dx = -dt$, $y = 1 - (1-t)^2 = 2t - t^2$ $\Rightarrow$

$\qquad dy = (2 - 2t)\,dt$, $0 \le t \le 1$

Thus

$$\oint_C x\,dx + y\,dy = \oint_{C_1 + C_2 + C_3} x\,dx + y\,dy$$

$$= \int_0^1 (0\,dt + (1-t)(-dt)) + \int_0^1 (t\,dt + 0\,dt) + \int_0^1 ((1-t)(-dt) + (2t - t^2)(2 - 2t)\,dt)$$

$$= \left[\tfrac{1}{2}t^2 - t\right]_0^1 + \left[\tfrac{1}{2}t^2\right]_0^1 + \left[\tfrac{1}{2}t^4 - 2t^3 + \tfrac{5}{2}t^2 - t\right]_0^1 = -\tfrac{1}{2} + \tfrac{1}{2} + \left(\tfrac{1}{2} - 2 + \tfrac{5}{2} - 1\right) = 0$$

(b) $\oint_C x\,dx + y\,dy = \iint_D \left[\frac{\partial}{\partial x}(y) - \frac{\partial}{\partial y}(x)\right]dA = \iint_D 0\,dA = 0$

5.

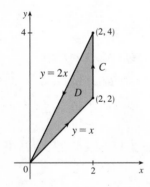

The region D enclosed by C is given by $\{(x, y) \mid 0 \le x \le 2,\ x \le y \le 2x\}$, so

$$\int_C xy^2\,dx + 2x^2 y\,dy = \iint_D \left[\frac{\partial}{\partial x}(2x^2 y) - \frac{\partial}{\partial y}(xy^2)\right]dA$$

$$= \int_0^2 \int_x^{2x}(4xy - 2xy)\,dy\,dx$$

$$= \int_0^2 \left[xy^2\right]_{y=x}^{y=2x} dx$$

$$= \int_0^2 3x^3\,dx = \tfrac{3}{4}x^4 \Big]_0^2 = 12$$

6. The region D enclosed by C is $[0, 5] \times [0, 2]$, so

$$\int_C \cos y\,dx + x^2 \sin y\,dy = \iint_D \left[\frac{\partial}{\partial x}(x^2 \sin y) - \frac{\partial}{\partial y}(\cos y)\right]dA = \int_0^5 \int_0^2 [2x \sin y - (-\sin y)]\,dy\,dx$$

$$= \int_0^5 (2x + 1)\,dx \int_0^2 \sin y\,dy = \left[x^2 + x\right]_0^5 \left[-\cos y\right]_0^2 = 30(1 - \cos 2)$$

7. $\int_C \left(y + e^{\sqrt{x}}\right) dx + (2x + \cos y^2) \, dy = \iint_D \left[\frac{\partial}{\partial x} (2x + \cos y^2) - \frac{\partial}{\partial y} \left(y + e^{\sqrt{x}}\right)\right] dA$

$$= \int_0^1 \int_{y^2}^{\sqrt{y}} (2 - 1) \, dx \, dy = \int_0^1 (y^{1/2} - y^2) \, dy = \frac{1}{3}$$

8. $\int_C xe^{-2x} \, dx + (x^4 + 2x^2 y^2) \, dy = \iint_D \left[\frac{\partial}{\partial x} (x^4 + 2x^2 y^2) - \frac{\partial}{\partial y} (xe^{-2x})\right] dA = \iint_D (4x^3 + 4xy^2 - 0) \, dA$

$$= 4 \iint_D x(x^2 + y^2) \, dA = 4 \int_0^{2\pi} \int_1^2 (r \cos \theta)(r^2) \, r \, dr \, d\theta$$

$$= 4 \int_0^{2\pi} \cos \theta \, d\theta \int_1^2 r^4 \, dr = 4 \left[\sin \theta\right]_0^{2\pi} \left[\tfrac{1}{5} r^5\right]_1^2 = 0$$

9. $\int_C y^3 \, dx - x^3 \, dy = \iint_D \left[\frac{\partial}{\partial x} (-x^3) - \frac{\partial}{\partial y} (y^3)\right] dA = \iint_D (-3x^2 - 3y^2) \, dA = \int_0^{2\pi} \int_0^2 (-3r^2) \, r \, dr \, d\theta$

$$= -3 \int_0^{2\pi} d\theta \int_0^2 r^3 \, dr = -3(2\pi)(4) = -24\pi$$

10. $\int_C \sin y \, dx + x \cos y \, dy = \iint_D \left[\frac{\partial}{\partial x} (x \cos y) - \frac{\partial}{\partial y} (\sin y)\right] dA = \iint_D (\cos y - \cos y) \, dA = \iint_D 0 \, dA = 0$

11. $\mathbf{F}(x, y) = \left\langle \sqrt{x} + y^3, x^2 + \sqrt{y} \right\rangle$ and the region D enclosed by C is given by $\{(x, y) \mid 0 \le x \le \pi, 0 \le y \le \sin x\}$.

C is traversed clockwise, so $-C$ gives the positive orientation.

$$\int_C \mathbf{F} \cdot d\mathbf{r} = -\int_{-C} \left(\sqrt{x} + y^3\right) dx + \left(x^2 + \sqrt{y}\right) dy = -\iint_D \left[\frac{\partial}{\partial x} \left(x^2 + \sqrt{y}\right) - \frac{\partial}{\partial y} \left(\sqrt{x} + y^3\right)\right] dA$$

$$= -\int_0^{\pi} \int_0^{\sin x} (2x - 3y^2) \, dy \, dx = -\int_0^{\pi} \left[2xy - y^3\right]_{y=0}^{y=\sin x} dx$$

$$= -\int_0^{\pi} (2x \sin x - \sin^3 x) \, dx = -\int_0^{\pi} (2x \sin x - (1 - \cos^2 x) \sin x) \, dx$$

$$= -\left[2 \sin x - 2x \cos x + \cos x - \tfrac{1}{3} \cos^3 x\right]_0^{\pi} \quad \text{[integrate by parts in the first term]}$$

$$= -\left(2\pi - 2 + \tfrac{2}{3}\right) = \tfrac{4}{3} - 2\pi$$

12. $\mathbf{F}(x, y) = \left\langle y^2 \cos x, x^2 + 2y \sin x \right\rangle$ and the region D enclosed by C is given by $\{(x, y) \mid 0 \le x \le 2, 0 \le y \le 3x\}$.

C is traversed clockwise, so $-C$ gives the positive orientation.

$$\int_C \mathbf{F} \cdot d\mathbf{r} = -\int_{-C} (y^2 \cos x) \, dx + (x^2 + 2y \sin x) \, dy = -\iint_D \left[\frac{\partial}{\partial x} (x^2 + 2y \sin x) - \frac{\partial}{\partial y} (y^2 \cos x)\right] dA$$

$$= -\iint_D (2x + 2y \cos x - 2y \cos x) \, dA = -\int_0^2 \int_0^{3x} 2x \, dy \, dx$$

$$= -\int_0^2 2x \left[y\right]_{y=0}^{y=3x} dx = -\int_0^2 6x^2 \, dx = -2x^3 \big]_0^2 = -16$$

13. $\mathbf{F}(x, y) = \left\langle e^x + x^2 y, e^y - xy^2 \right\rangle$ and the region D enclosed by C is the disk $x^2 + y^2 \le 25$.

C is traversed clockwise, so $-C$ gives the positive orientation.

$$\int_C \mathbf{F} \cdot d\mathbf{r} = -\int_{-C} (e^x + x^2 y) \, dx + (e^y - xy^2) \, dy = -\iint_D \left[\frac{\partial}{\partial x} (e^y - xy^2) - \frac{\partial}{\partial y} (e^x + x^2 y)\right] dA$$

$$= -\iint_D (-y^2 - x^2) \, dA = \iint_D (x^2 + y^2) \, dA = \int_0^{2\pi} \int_0^5 (r^2) \, r \, dr \, d\theta$$

$$= \int_0^{2\pi} d\theta \int_0^5 r^3 \, dr = 2\pi \left[\tfrac{1}{4} r^4\right]_0^5 = \tfrac{625}{2} \pi$$

14. $\mathbf{F}(x, y) = \left\langle y - \ln(x^2 + y^2), 2\tan^{-1}\left(\frac{y}{x}\right)\right\rangle$ and the region D enclosed by C is the disk with radius 1 centered at $(2, 3)$.

C is oriented positively, so

$$\int_C \mathbf{F} \cdot d\mathbf{r} = \int_C (y - \ln(x^2 + y^2))\, dx + \left(2\tan^{-1}\left(\frac{y}{x}\right)\right) dy = \iint_D \left[\frac{\partial}{\partial x}\left(2\tan^{-1}\left(\frac{y}{x}\right)\right) - \frac{\partial}{\partial y}\left(y - \ln(x^2 + y^2)\right)\right] dA$$

$$= \iint_D \left[2\left(\frac{-yx^{-2}}{1 + (y/x)^2}\right) - \left(1 - \frac{2y}{x^2 + y^2}\right)\right] dA = \iint_D \left[-\frac{2y}{x^2 + y^2} - 1 + \frac{2y}{x^2 + y^2}\right] dA$$

$$= -\iint_D dA = -(\text{area of } D) = -\pi$$

15. Here $C = C_1 + C_2$ where

C_1 can be parametrized as $x = t,\ y = 1,\ -1 \le t \le 1$, and

C_2 is given by $x = -t,\ y = 2 - t^2,\ -1 \le t \le 1$.

Then the line integral is

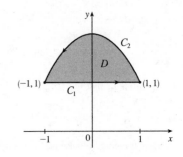

$$\oint_{C_1 + C_2} y^2 e^x\, dx + x^2 e^y\, dy = \int_{-1}^1 [1 \cdot e^t + t^2 e \cdot 0]\, dt$$

$$+ \int_{-1}^1 [(2 - t^2)^2 e^{-t}(-1) + (-t)^2 e^{2-t^2}(-2t)]\, dt$$

$$= \int_{-1}^1 [e^t - (2 - t^2)^2 e^{-t} - 2t^3 e^{2-t^2}]\, dt = -8e + 48e^{-1}$$

according to a CAS. The double integral is

$$\iint_D \left(\frac{\partial Q}{\partial x} - \frac{\partial P}{\partial y}\right) dA = \int_{-1}^1 \int_1^{2 - x^2} (2xe^y - 2ye^x)\, dy\, dx = -8e + 48e^{-1},\ \text{verifying Green's Theorem in this case.}$$

16. We can parametrize C as $x = \cos\theta,\ y = 2\sin\theta,\ 0 \le \theta \le 2\pi$. Then the line integral is

$$\oint_C P\, dx + Q\, dy = \int_0^{2\pi} \left[2\cos\theta - (\cos\theta)^3(2\sin\theta)^5\right](-\sin\theta)\, d\theta + \int_0^{2\pi} (\cos\theta)^3(2\sin\theta)^8 \cdot 2\cos\theta\, d\theta$$

$$= \int_0^{2\pi} [-2\cos\theta\sin\theta + 32\cos^3\theta\sin^6\theta + 512\cos^4\theta\sin^8\theta]\, d\theta = 7\pi,$$

according to a CAS. The double integral is $\iint_D \left(\frac{\partial Q}{\partial x} - \frac{\partial P}{\partial y}\right) dA = \int_{-1}^1 \int_{-\sqrt{4 - 4x^2}}^{\sqrt{4 - 4x^2}} (3x^2 y^8 + 5x^3 y^4)\, dy\, dx = 7\pi$.

17. By Green's Theorem, $W = \int_C \mathbf{F} \cdot d\mathbf{r} = \int_C x(x + y)\, dx + xy^2\, dy = \iint_D (y^2 - x)\, dy\, dx$ where C is the path described in the question and D is the triangle bounded by C. So

$$W = \int_0^1 \int_0^{1 - x} (y^2 - x)\, dy\, dx = \int_0^1 \left[\frac{1}{3}y^3 - xy\right]_{y=0}^{y=1-x} dx = \int_0^1 \left(\frac{1}{3}(1 - x)^3 - x(1 - x)\right) dx$$

$$= \left[-\frac{1}{12}(1 - x)^4 - \frac{1}{2}x^2 + \frac{1}{3}x^3\right]_0^1 = \left(-\frac{1}{2} + \frac{1}{3}\right) - \left(-\frac{1}{12}\right) = -\frac{1}{12}$$

18. By Green's Theorem, $W = \int_C \mathbf{F} \cdot d\mathbf{r} = \int_C x\, dx + (x^3 + 3xy^2)\, dy = \iint_D (3x^2 + 3y^2 - 0)\, dA$, where D is the semicircular region bounded by C. Converting to polar coordinates, we have $W = 3\int_0^2 \int_0^\pi r^2 \cdot r\, d\theta\, dr = 3\pi \left[\frac{1}{4}r^4\right]_0^2 = 12\pi$.

19. Let C_1 be the arch of the cycloid from $(0, 0)$ to $(2\pi, 0)$, which corresponds to $0 \le t \le 2\pi$, and let C_2 be the segment from $(2\pi, 0)$ to $(0, 0)$, so C_2 is given by $x = 2\pi - t, y = 0, 0 \le t \le 2\pi$. Then $C = C_1 \cup C_2$ is traversed clockwise, so $-C$ is oriented positively. Thus $-C$ encloses the area under one arch of the cycloid and from (5) we have

$$A = -\oint_{-C} y\, dx = \int_{C_1} y\, dx + \int_{C_2} y\, dx = \int_0^{2\pi} (1 - \cos t)(1 - \cos t)\, dt + \int_0^{2\pi} 0\, (-dt)$$

$$= \int_0^{2\pi} (1 - 2\cos t + \cos^2 t)\, dt + 0 = \left[t - 2\sin t + \frac{1}{2}t + \frac{1}{4}\sin 2t\right]_0^{2\pi} = 3\pi$$

20.

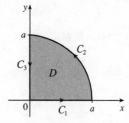

$$A = \oint_C x \, dy = \int_0^{2\pi} (5\cos t - \cos 5t)(5\cos t - 5\cos 5t) \, dt$$

$$= \int_0^{2\pi} (25\cos^2 t - 30\cos t \cos 5t + 5\cos^2 5t) \, dt$$

$$= \left[25\left(\tfrac{1}{2}t + \tfrac{1}{4}\sin 2t\right) - 30\left(\tfrac{1}{8}\sin 4t + \tfrac{1}{12}\sin 6t\right) + 5\left(\tfrac{1}{2}t + \tfrac{1}{20}\sin 10t\right)\right]_0^{2\pi}$$

[Use Formula 80 in the Table of Integrals]

$$= 30\pi$$

21. (a) Using Equation 17.2.8 [ET 16.2.8], we write parametric equations of the line segment as $x = (1-t)x_1 + tx_2$, $y = (1-t)y_1 + ty_2$, $0 \le t \le 1$. Then $dx = (x_2 - x_1) \, dt$ and $dy = (y_2 - y_1) \, dt$, so

$$\int_C x \, dy - y \, dx = \int_0^1 [(1-t)x_1 + tx_2](y_2 - y_1) \, dt + [(1-t)y_1 + ty_2](x_2 - x_1) \, dt$$

$$= \int_0^1 (x_1(y_2 - y_1) - y_1(x_2 - x_1) + t[(y_2 - y_1)(x_2 - x_1) - (x_2 - x_1)(y_2 - y_1)]) \, dt$$

$$= \int_0^1 (x_1 y_2 - x_2 y_1) \, dt = x_1 y_2 - x_2 y_1$$

(b) We apply Green's Theorem to the path $C = C_1 \cup C_2 \cup \cdots \cup C_n$, where C_i is the line segment that joins (x_i, y_i) to (x_{i+1}, y_{i+1}) for $i = 1, 2, \ldots, n-1$, and C_n is the line segment that joins (x_n, y_n) to (x_1, y_1). From (5),
$\frac{1}{2} \int_C x \, dy - y \, dx = \iint_D dA$, where D is the polygon bounded by C. Therefore

$$\text{area of polygon} = A(D) = \iint_D dA = \tfrac{1}{2} \int_C x \, dy - y \, dx$$

$$= \tfrac{1}{2}\left(\int_{C_1} x \, dy - y \, dx + \int_{C_2} x \, dy - y \, dx + \cdots + \int_{C_{n-1}} x \, dy - y \, dx + \int_{C_n} x \, dy - y \, dx\right)$$

To evaluate these integrals we use the formula from (a) to get

$$A(D) = \tfrac{1}{2}[(x_1 y_2 - x_2 y_1) + (x_2 y_3 - x_3 y_2) + \cdots + (x_{n-1} y_n - x_n y_{n-1}) + (x_n y_1 - x_1 y_n)].$$

(c) $A = \tfrac{1}{2}[(0 \cdot 1 - 2 \cdot 0) + (2 \cdot 3 - 1 \cdot 1) + (1 \cdot 2 - 0 \cdot 3) + (0 \cdot 1 - (-1) \cdot 2) + (-1 \cdot 0 - 0 \cdot 1)]$

$$= \tfrac{1}{2}(0 + 5 + 2 + 2) = \tfrac{9}{2}$$

22. By Green's Theorem, $\frac{1}{2A} \oint_C x^2 \, dy = \frac{1}{2A} \iint_D 2x \, dA = \frac{1}{A} \iint_D x \, dA = \overline{x}$ and

$$-\tfrac{1}{2A} \oint_C y^2 \, dx = -\tfrac{1}{2A} \iint_D (-2y) \, dA = \tfrac{1}{A} \iint_D y \, dA = \overline{y}.$$

23. We orient the quarter-circular region as shown in the figure.

$$A = \tfrac{1}{4}\pi a^2 \text{ so } \overline{x} = \frac{1}{\pi a^2 / 2} \oint_C x^2 \, dy \text{ and } \overline{y} = -\frac{1}{\pi a^2 / 2} \oint_C y^2 \, dx.$$

Here $C = C_1 + C_2 + C_3$ where C_1: $x = t$, $y = 0$, $0 \le t \le a$;

C_2: $x = a\cos t$, $y = a\sin t$, $0 \le t \le \tfrac{\pi}{2}$; and

C_3: $x = 0, y = a - t, 0 \le t \le a$. Then

$$\oint_C x^2 \, dy = \int_{C_1} x^2 \, dy + \int_{C_2} x^2 \, dy + \int_{C_3} x^2 \, dy = \int_0^a 0 \, dt + \int_0^{\pi/2} (a\cos t)^2 (a\cos t) \, dt + \int_0^a 0 \, dt$$

$$= \int_0^{\pi/2} a^3 \cos^3 t \, dt = a^3 \int_0^{\pi/2} (1 - \sin^2 t) \cos t \, dt = a^3 \left[\sin t - \tfrac{1}{3}\sin^3 t\right]_0^{\pi/2} = \tfrac{2}{3}a^3$$

$$\text{so } \overline{x} = \frac{1}{\pi a^2 / 2} \oint_C x^2 \, dy = \frac{4a}{3\pi}.$$

$\oint_C y^2 dx = \int_{C_1} y^2 dx + \int_{C_2} y^2 dx + \int_{C_3} y^2 dx = \int_0^a 0\, dt + \int_0^{\pi/2} (a\sin t)^2 (-a\sin t)\, dt + \int_0^a 0\, dt$

$= \int_0^{\pi/2} (-a^3 \sin^3 t)\, dt = -a^3 \int_0^{\pi/2} (1 - \cos^2 t)\sin t\, dt = -a^3 \left[\tfrac{1}{3}\cos^3 t - \cos t\right]_0^{\pi/2} = -\tfrac{2}{3} a^3,$

so $\overline{y} = -\dfrac{1}{\pi a^2/2} \oint_C y^2 dx = \dfrac{4a}{3\pi}$. Thus $(\overline{x}, \overline{y}) = \left(\dfrac{4a}{3\pi}, \dfrac{4a}{3\pi}\right)$.

24. Here $A = \tfrac{1}{2}ab$ and $C = C_1 + C_2 + C_3$, where $C_1: x = x, y = 0, 0 \le x \le a$;

$C_2: x = a, y = y, 0 \le y \le b$; and $C_3: x = x, y = \tfrac{b}{a}x, x = a$ to $x = 0$. Then

$\oint_C x^2 dy = \int_{C_1} x^2 dy + \int_{C_2} x^2 dy + \int_{C_3} x^2 dy = 0 + \int_0^b a^2 dy + \int_a^0 (x^2)\left(\tfrac{b}{a} dx\right)$

$= a^2 b + \tfrac{b}{a}\left[\tfrac{1}{3}x^3\right]_a^0 = a^2 b - \tfrac{1}{3}a^2 b = \tfrac{2}{3}a^2 b.$

Similarly, $\oint_C y^2 dx = \int_{C_1} y^2 dx + \int_{C_2} y^2 dx + \int_{C_3} y^2 dx = 0 + 0 + \int_a^0 \left(\tfrac{b}{a}x\right)^2 dx = \tfrac{b^2}{a^2} \cdot \tfrac{1}{3}x^3 \Big]_a^0 = -\tfrac{1}{3}ab^2$. Thus

$\overline{x} = \tfrac{1}{2A} \oint_C x^2 dy = \tfrac{1}{ab} \cdot \tfrac{2}{3}a^2 b = \tfrac{2}{3}a$ and $\overline{y} = -\tfrac{1}{2A} \oint_C y^2 dx = -\tfrac{1}{ab}\left(-\tfrac{1}{3}ab^2\right) = \tfrac{1}{3}b$, so $(\overline{x}, \overline{y}) = \left(\tfrac{2}{3}a, \tfrac{1}{3}b\right)$.

25. By Green's Theorem, $-\tfrac{1}{3}\rho \oint_C y^3 dx = -\tfrac{1}{3}\rho \iint_D (-3y^2)\, dA = \iint_D y^2 \rho\, dA = I_x$ and

$\tfrac{1}{3}\rho \oint_C x^3 dy = \tfrac{1}{3}\rho \iint_D (3x^2)\, dA = \iint_D x^2 \rho\, dA = I_y.$

26. By symmetry the moments of inertia about any two diameters are equal. Centering the disk at the origin, the moment of inertia about a diameter equals

$I_y = \tfrac{1}{3}\rho \oint_C x^3 dy = \tfrac{1}{3}\rho \int_0^{2\pi} (a^4 \cos^4 t)\, dt = \tfrac{1}{3}a^4 \rho \int_0^{2\pi} \left[\tfrac{3}{8} + \tfrac{1}{2}\cos 2t + \tfrac{1}{8}\cos 4t\right] dt = \tfrac{1}{3}a^4 \rho \cdot \tfrac{3(2\pi)}{8} = \tfrac{1}{4}\pi a^4 \rho$

27. Since C is a simple closed path which doesn't pass through or enclose the origin, there exists an open region that doesn't contain the origin but does contain D. Thus $P = -y/(x^2 + y^2)$ and $Q = x/(x^2 + y^2)$ have continuous partial derivatives on this open region containing D and we can apply Green's Theorem. But by Exercise 17.3.33(a) [ET 16.3.33(a)],

$\partial P/\partial y = \partial Q/\partial x$, so $\oint_C \mathbf{F} \cdot d\mathbf{r} = \iint_D 0\, dA = 0.$

28. We express D as a type II region: $D = \{(x, y) \mid f_1(y) \le x \le f_2(y), c \le y \le d\}$ where f_1 and f_2 are continuous functions.

Then $\displaystyle\iint_D \frac{\partial Q}{\partial x}\, dA = \int_c^d \int_{f_1(y)}^{f_2(y)} \frac{\partial Q}{\partial x}\, dx\, dy = \int_c^d [Q(f_2(y), y) - Q(f_1(y), y)]\, dy$ by the Fundamental Theorem of

Calculus. But referring to the figure, $\displaystyle\oint_C Q\, dy = \oint_{C_1 + C_2 + C_3 + C_4} Q\, dy.$

Then $\int_{C_1} Q\, dy = \int_d^c Q(f_1(y), y)\, dy$, $\int_{C_2} Q\, dy = \int_{C_4} Q\, dy = 0$,

and $\int_{C_3} Q\, dy = \int_c^d Q(f_2(y), y)\, dy$. Hence

$\oint_C Q\, dy = \int_c^d [Q(f_2(y), y) - Q(f_1(y), y)]\, dy = \iint_D (\partial Q/\partial x)\, dA.$

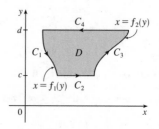

29. Using the first part of (5), we have that $\iint_R dx\, dy = A(R) = \int_{\partial R} x\, dy$. But $x = g(u, v)$, and $dy = \dfrac{\partial h}{\partial u}\, du + \dfrac{\partial h}{\partial v}\, dv$,

and we orient ∂S by taking the positive direction to be that which corresponds, under the mapping, to the positive direction along ∂R, so

$$\int_{\partial R} x\,dy = \int_{\partial S} g(u,v)\left(\frac{\partial h}{\partial u}\,du + \frac{\partial h}{\partial v}\,dv\right) = \int_{\partial S} g(u,v)\frac{\partial h}{\partial u}\,du + g(u,v)\frac{\partial h}{\partial v}\,dv$$

$$= \pm \iint_S \left[\frac{\partial}{\partial u}\left(g(u,v)\frac{\partial h}{\partial v}\right) - \frac{\partial}{\partial v}\left(g(u,v)\frac{\partial h}{\partial u}\right)\right]dA \qquad \text{[using Green's Theorem in the } uv\text{-plane]}$$

$$= \pm \iint_S \left(\frac{\partial g}{\partial u}\frac{\partial h}{\partial v} + g(u,v)\frac{\partial^2 h}{\partial u\,\partial v} - \frac{\partial g}{\partial v}\frac{\partial h}{\partial u} - g(u,v)\frac{\partial^2 h}{\partial v\,\partial u}\right)dA \qquad \text{[using the Chain Rule]}$$

$$= \pm \iint_S \left(\frac{\partial x}{\partial u}\frac{\partial y}{\partial v} - \frac{\partial x}{\partial v}\frac{\partial y}{\partial u}\right)dA \quad \text{[by the equality of mixed partials]} \quad = \pm \iint_S \frac{\partial(x,y)}{\partial(u,v)}\,du\,dv$$

The sign is chosen to be positive if the orientation that we gave to ∂S corresponds to the usual positive orientation, and it is negative otherwise. In either case, since $A(R)$ is positive, the sign chosen must be the same as the sign of $\dfrac{\partial(x,y)}{\partial(u,v)}$.

Therefore $A(R) = \displaystyle\iint_R dx\,dy = \iint_S \left|\frac{\partial(x,y)}{\partial(u,v)}\right|du\,dv$.

17.5 Curl and Divergence

ET 16.5

1. (a) $\operatorname{curl}\mathbf{F} = \nabla \times \mathbf{F} = \begin{vmatrix} \mathbf{i} & \mathbf{j} & \mathbf{k} \\ \partial/\partial x & \partial/\partial y & \partial/\partial z \\ xyz & 0 & -x^2y \end{vmatrix} = (-x^2 - 0)\,\mathbf{i} - (-2xy - xy)\,\mathbf{j} + (0 - xz)\,\mathbf{k}$

$\qquad = -x^2\,\mathbf{i} + 3xy\,\mathbf{j} - xz\,\mathbf{k}$

(b) $\operatorname{div}\mathbf{F} = \nabla \cdot \mathbf{F} = \dfrac{\partial}{\partial x}(xyz) + \dfrac{\partial}{\partial y}(0) + \dfrac{\partial}{\partial z}(-x^2y) = yz + 0 + 0 = yz$

2. (a) $\operatorname{curl}\mathbf{F} = \nabla \times \mathbf{F} = \begin{vmatrix} \mathbf{i} & \mathbf{j} & \mathbf{k} \\ \partial/\partial x & \partial/\partial y & \partial/\partial z \\ x^2yz & xy^2z & xyz^2 \end{vmatrix} = (xz^2 - xy^2)\,\mathbf{i} - (yz^2 - x^2y)\,\mathbf{j} + (y^2z - x^2z)\,\mathbf{k}$

$\qquad = x(z^2 - y^2)\,\mathbf{i} + y(x^2 - z^2)\,\mathbf{j} + z(y^2 - x^2)\,\mathbf{k}$

(b) $\operatorname{div}\mathbf{F} = \nabla \cdot \mathbf{F} = \dfrac{\partial}{\partial x}(x^2yz) + \dfrac{\partial}{\partial y}(xy^2z) + \dfrac{\partial}{\partial z}(xyz^2) = 2xyz + 2xyz + 2xyz = 6xyz$

3. (a) $\operatorname{curl}\mathbf{F} = \nabla \times \mathbf{F} = \begin{vmatrix} \mathbf{i} & \mathbf{j} & \mathbf{k} \\ \partial/\partial x & \partial/\partial y & \partial/\partial z \\ 1 & x+yz & xy-\sqrt{z} \end{vmatrix} = (x - y)\,\mathbf{i} - (y - 0)\,\mathbf{j} + (1 - 0)\,\mathbf{k}$

$\qquad = (x - y)\,\mathbf{i} - y\,\mathbf{j} + \mathbf{k}$

(b) $\operatorname{div}\mathbf{F} = \nabla \cdot \mathbf{F} = \dfrac{\partial}{\partial x}(1) + \dfrac{\partial}{\partial y}(x + yz) + \dfrac{\partial}{\partial z}\left(xy - \sqrt{z}\right) = z - \dfrac{1}{2\sqrt{z}}$

4. (a) $\operatorname{curl}\mathbf{F} = \nabla \times \mathbf{F} = \begin{vmatrix} \mathbf{i} & \mathbf{j} & \mathbf{k} \\ \partial/\partial x & \partial/\partial y & \partial/\partial z \\ 0 & \cos xz & -\sin xy \end{vmatrix} = (-x\cos xy + x\sin xz)\,\mathbf{i} - (-y\cos xy - 0)\,\mathbf{j} + (-z\sin xz - 0)\,\mathbf{k}$

$\qquad = x(\sin xz - \cos xy)\,\mathbf{i} + y\cos xy\,\mathbf{j} - z\sin xz\,\mathbf{k}$

(b) $\operatorname{div}\mathbf{F} = \nabla \cdot \mathbf{F} = \dfrac{\partial}{\partial x}(0) + \dfrac{\partial}{\partial y}(\cos xz) + \dfrac{\partial}{\partial z}(-\sin xy) = 0 + 0 + 0 = 0$

5. (a) $\text{curl } \mathbf{F} = \nabla \times \mathbf{F} = \begin{vmatrix} \mathbf{i} & \mathbf{j} & \mathbf{k} \\ \partial/\partial x & \partial/\partial y & \partial/\partial z \\ \dfrac{x}{\sqrt{x^2+y^2+z^2}} & \dfrac{y}{\sqrt{x^2+y^2+z^2}} & \dfrac{z}{\sqrt{x^2+y^2+z^2}} \end{vmatrix}$

$$= \frac{1}{(x^2+y^2+z^2)^{3/2}}\left[(-yz+yz)\,\mathbf{i} - (-xz+xz)\,\mathbf{j} + (-xy+xy)\,\mathbf{k}\right] = \mathbf{0}$$

(b) $\text{div } \mathbf{F} = \nabla \cdot \mathbf{F} = \dfrac{\partial}{\partial x}\left(\dfrac{x}{\sqrt{x^2+y^2+z^2}}\right) + \dfrac{\partial}{\partial y}\left(\dfrac{y}{\sqrt{x^2+y^2+z^2}}\right) + \dfrac{\partial}{\partial z}\left(\dfrac{z}{\sqrt{x^2+y^2+z^2}}\right)$

$$= \frac{x^2+y^2+z^2-x^2}{(x^2+y^2+z^2)^{3/2}} + \frac{x^2+y^2+z^2-y^2}{(x^2+y^2+z^2)^{3/2}} + \frac{x^2+y^2+z^2-z^2}{(x^2+y^2+z^2)^{3/2}} = \frac{2x^2+2y^2+2z^2}{(x^2+y^2+z^2)^{3/2}} = \frac{2}{\sqrt{x^2+y^2+z^2}}$$

6. (a) $\text{curl } \mathbf{F} = \nabla \times \mathbf{F} = \begin{vmatrix} \mathbf{i} & \mathbf{j} & \mathbf{k} \\ \partial/\partial x & \partial/\partial y & \partial/\partial z \\ 0 & e^{xy}\sin z & y\tan^{-1}(x/z) \end{vmatrix}$

$$= \left[\tan^{-1}(x/z) - e^{xy}\cos z\right]\mathbf{i} - \left(y \cdot \frac{1}{1+(x/z)^2} \cdot \frac{1}{z} - 0\right)\mathbf{j} + \left(ye^{xy}\sin z - 0\right)\mathbf{k}$$

$$= \left[\tan^{-1}(x/z) - e^{xy}\cos z\right]\mathbf{i} - \frac{yz}{x^2+z^2}\,\mathbf{j} + ye^{xy}\sin z\,\mathbf{k}$$

(b) $\text{div } \mathbf{F} = \nabla \cdot \mathbf{F} = \dfrac{\partial}{\partial x}(0) + \dfrac{\partial}{\partial y}(e^{xy}\sin z) + \dfrac{\partial}{\partial z}\left[y\tan^{-1}(x/z)\right]$

$$= 0 + xe^{xy}\sin z + y \cdot \frac{1}{1+(x/z)^2}\left(-\frac{x}{z^2}\right) = xe^{xy}\sin z - \frac{xy}{x^2+z^2}$$

7. (a) $\text{curl } \mathbf{F} = \nabla \times \mathbf{F} = \begin{vmatrix} \mathbf{i} & \mathbf{j} & \mathbf{k} \\ \partial/\partial x & \partial/\partial y & \partial/\partial z \\ \ln x & \ln(xy) & \ln(xyz) \end{vmatrix} = \left(\dfrac{xz}{xyz} - 0\right)\mathbf{i} - \left(\dfrac{yz}{xyz} - 0\right)\mathbf{j} + \left(\dfrac{y}{xy} - 0\right)\mathbf{k} = \left\langle \dfrac{1}{y}, -\dfrac{1}{x}, \dfrac{1}{x}\right\rangle$

(b) $\text{div } \mathbf{F} = \nabla \cdot \mathbf{F} = \dfrac{\partial}{\partial x}(\ln x) + \dfrac{\partial}{\partial y}(\ln(xy)) + \dfrac{\partial}{\partial z}(\ln(xyz)) = \dfrac{1}{x} + \dfrac{x}{xy} + \dfrac{xy}{xyz} = \dfrac{1}{x} + \dfrac{1}{y} + \dfrac{1}{z}$

8. (a) $\text{curl } \mathbf{F} = \nabla \times \mathbf{F} = \begin{vmatrix} \mathbf{i} & \mathbf{j} & \mathbf{k} \\ \partial/\partial x & \partial/\partial y & \partial/\partial z \\ e^x & e^{xy} & e^{xyz} \end{vmatrix} = (xze^{xyz} - 0)\,\mathbf{i} - (yze^{xyz} - 0)\,\mathbf{j} + (ye^{xy} - 0)\,\mathbf{k}$

$$= \langle xze^{xyz}, -yze^{xyz}, ye^{xy}\rangle$$

(b) $\text{div } \mathbf{F} = \nabla \cdot \mathbf{F} = \dfrac{\partial}{\partial x}(e^x) + \dfrac{\partial}{\partial y}(e^{xy}) + \dfrac{\partial}{\partial z}(e^{xyz}) = e^x + xe^{xy} + xye^{xyz}$

9. If the vector field is $\mathbf{F} = P\mathbf{i} + Q\mathbf{j} + R\mathbf{k}$, then we know $R = 0$. In addition, the x-component of each vector of $\mathbf{F}$ is 0, so

$P = 0$, hence $\dfrac{\partial P}{\partial x} = \dfrac{\partial P}{\partial y} = \dfrac{\partial P}{\partial z} = \dfrac{\partial R}{\partial x} = \dfrac{\partial R}{\partial y} = \dfrac{\partial R}{\partial z} = 0$. Q decreases as y increases, so $\dfrac{\partial Q}{\partial y} < 0$, but Q doesn't change

in the x- or z-directions, so $\dfrac{\partial Q}{\partial x} = \dfrac{\partial Q}{\partial z} = 0$.

(a) $\text{div } \mathbf{F} = \dfrac{\partial P}{\partial x} + \dfrac{\partial Q}{\partial y} + \dfrac{\partial R}{\partial z} = 0 + \dfrac{\partial Q}{\partial y} + 0 < 0$

(b) $\text{curl } \mathbf{F} = \left(\dfrac{\partial R}{\partial y} - \dfrac{\partial Q}{\partial z}\right)\mathbf{i} + \left(\dfrac{\partial P}{\partial z} - \dfrac{\partial R}{\partial x}\right)\mathbf{j} + \left(\dfrac{\partial Q}{\partial x} - \dfrac{\partial P}{\partial y}\right)\mathbf{k} = (0-0)\,\mathbf{i} + (0-0)\,\mathbf{j} + (0-0)\,\mathbf{k} = \mathbf{0}$

10. If the vector field is $\mathbf{F} = P\mathbf{i} + Q\mathbf{j} + R\mathbf{k}$, then we know $R = 0$. In addition, P and Q don't vary in the z-direction, so

$\dfrac{\partial R}{\partial x} = \dfrac{\partial R}{\partial y} = \dfrac{\partial R}{\partial z} = \dfrac{\partial P}{\partial z} = \dfrac{\partial Q}{\partial z} = 0$. As x increases, the x-component of each vector of $\mathbf{F}$ increases while the y-component

remains constant, so $\dfrac{\partial P}{\partial x} > 0$ and $\dfrac{\partial Q}{\partial x} = 0$. Similarly, as y increases, the y-component of each vector increases while the

x-component remains constant, so $\dfrac{\partial Q}{\partial y} > 0$ and $\dfrac{\partial P}{\partial y} = 0$.

(a) $\operatorname{div}\mathbf{F} = \dfrac{\partial P}{\partial x} + \dfrac{\partial Q}{\partial y} + \dfrac{\partial R}{\partial z} = \dfrac{\partial P}{\partial x} + \dfrac{\partial Q}{\partial y} + 0 > 0$

(b) $\operatorname{curl}\mathbf{F} = \left(\dfrac{\partial R}{\partial y} - \dfrac{\partial Q}{\partial z}\right)\mathbf{i} + \left(\dfrac{\partial P}{\partial z} - \dfrac{\partial R}{\partial x}\right)\mathbf{j} + \left(\dfrac{\partial Q}{\partial x} - \dfrac{\partial P}{\partial y}\right)\mathbf{k} = (0-0)\mathbf{i} + (0-0)\mathbf{j} + (0-0)\mathbf{k} = \mathbf{0}$

11. If the vector field is $\mathbf{F} = P\mathbf{i} + Q\mathbf{j} + R\mathbf{k}$, then we know $R = 0$. In addition, the y-component of each vector of $\mathbf{F}$ is 0, so

$Q = 0$, hence $\dfrac{\partial Q}{\partial x} = \dfrac{\partial Q}{\partial y} = \dfrac{\partial Q}{\partial z} = \dfrac{\partial R}{\partial x} = \dfrac{\partial R}{\partial y} = \dfrac{\partial R}{\partial z} = 0$. P increases as y increases, so $\dfrac{\partial P}{\partial y} > 0$, but P doesn't change in

the x- or z-directions, so $\dfrac{\partial P}{\partial x} = \dfrac{\partial P}{\partial z} = 0$.

(a) $\operatorname{div}\mathbf{F} = \dfrac{\partial P}{\partial x} + \dfrac{\partial Q}{\partial y} + \dfrac{\partial R}{\partial z} = 0 + 0 + 0 = 0$

(b) $\operatorname{curl}\mathbf{F} = \left(\dfrac{\partial R}{\partial y} - \dfrac{\partial Q}{\partial z}\right)\mathbf{i} + \left(\dfrac{\partial P}{\partial z} - \dfrac{\partial R}{\partial x}\right)\mathbf{j} + \left(\dfrac{\partial Q}{\partial x} - \dfrac{\partial P}{\partial y}\right)\mathbf{k} = (0-0)\mathbf{i} + (0-0)\mathbf{j} + \left(0 - \dfrac{\partial P}{\partial y}\right)\mathbf{k} = -\dfrac{\partial P}{\partial y}\mathbf{k}$

Since $\dfrac{\partial P}{\partial y} > 0$, $-\dfrac{\partial P}{\partial y}\mathbf{k}$ is a vector pointing in the negative z-direction.

12. (a) $\operatorname{curl} f = \nabla \times f$ is meaningless because f is a scalar field.

(b) $\operatorname{grad} f$ is a vector field.

(c) $\operatorname{div}\mathbf{F}$ is a scalar field.

(d) $\operatorname{curl}(\operatorname{grad} f)$ is a vector field.

(e) $\operatorname{grad}\mathbf{F}$ is meaningless because $\mathbf{F}$ is not a scalar field.

(f) $\operatorname{grad}(\operatorname{div}\mathbf{F})$ is a vector field.

(g) $\operatorname{div}(\operatorname{grad} f)$ is a scalar field.

(h) $\operatorname{grad}(\operatorname{div} f)$ is meaningless because f is a scalar field.

(i) $\operatorname{curl}(\operatorname{curl}\mathbf{F})$ is a vector field.

(j) $\operatorname{div}(\operatorname{div}\mathbf{F})$ is meaningless because $\operatorname{div}\mathbf{F}$ is a scalar field.

(k) $(\operatorname{grad} f) \times (\operatorname{div}\mathbf{F})$ is meaningless because $\operatorname{div}\mathbf{F}$ is a scalar field.

(l) $\operatorname{div}(\operatorname{curl}(\operatorname{grad} f))$ is a scalar field.

13. $\operatorname{curl}\mathbf{F} = \nabla \times \mathbf{F} = \begin{vmatrix} \mathbf{i} & \mathbf{j} & \mathbf{k} \\ \partial/\partial x & \partial/\partial y & \partial/\partial z \\ y^2 z^3 & 2xyz^3 & 3xy^2 z^2 \end{vmatrix} = (6xyz^2 - 6xyz^2)\mathbf{i} - (3y^2 z^2 - 3y^2 z^2)\mathbf{j} + (2yz^3 - 2yz^3)\mathbf{k} = \mathbf{0}$

and $\mathbf{F}$ is defined on all of $\mathbb{R}^3$ with component functions which have continuous partial derivatives, so by Theorem 4,

F is conservative. Thus, there exists a function f such that $\mathbf{F} = \nabla f$. Then $f_x(x, y, z) = y^2 z^3$ implies

$f(x, y, z) = xy^2 z^3 + g(y, z)$ and $f_y(x, y, z) = 2xyz^3 + g_y(y, z)$. But $f_y(x, y, z) = 2xyz^3$, so $g(y, z) = h(z)$ and

$f(x, y, z) = xy^2 z^3 + h(z)$. Thus $f_z(x, y, z) = 3xy^2 z^2 + h'(z)$ but $f_z(x, y, z) = 3xy^2 z^2$ so $h(z) = K$, a constant.

Hence a potential function for **F** is $f(x, y, z) = xy^2 z^3 + K$.

14. $\operatorname{curl} \mathbf{F} = \nabla \times \mathbf{F} = \begin{vmatrix} \mathbf{i} & \mathbf{j} & \mathbf{k} \\ \partial/\partial x & \partial/\partial y & \partial/\partial z \\ xyz^2 & x^2 yz^2 & x^2 y^2 z \end{vmatrix} = (2x^2 yz - 2x^2 yz)\,\mathbf{i} - (2xy^2 z - 2xyz)\,\mathbf{j} + (2xyz^2 - xz^2)\,\mathbf{k} \neq \mathbf{0}$,

so **F** is not conservative.

15. $\operatorname{curl} \mathbf{F} = \nabla \times \mathbf{F} = \begin{vmatrix} \mathbf{i} & \mathbf{j} & \mathbf{k} \\ \partial/\partial x & \partial/\partial y & \partial/\partial z \\ 2xy & x^2 + 2yz & y^2 \end{vmatrix} = (2y - 2y)\,\mathbf{i} - (0 - 0)\,\mathbf{j} + (2x - 2x)\,\mathbf{k} = \mathbf{0}$, **F** is defined on all of $\mathbb{R}^3$,

and the partial derivatives of the component functions are continuous, so **F** is conservative. Thus there exists a function f

such that $\nabla f = \mathbf{F}$. Then $f_x(x, y, z) = 2xy$ implies $f(x, y, z) = x^2 y + g(y, z)$ and $f_y(x, y, z) = x^2 + g_y(y, z)$. But

$f_y(x, y, z) = x^2 + 2yz$, so $g(y, z) = y^2 z + h(z)$ and $f(x, y, z) = x^2 y + y^2 z + h(z)$. Thus $f_z(x, y, z) = y^2 + h'(z)$ but

$f_z(x, y, z) = y^2$ so $h(z) = K$ and $f(x, y, z) = x^2 y + y^2 z + K$.

16. $\operatorname{curl} \mathbf{F} = \nabla \times \mathbf{F} = \begin{vmatrix} \mathbf{i} & \mathbf{j} & \mathbf{k} \\ \partial/\partial x & \partial/\partial y & \partial/\partial z \\ e^z & 1 & xe^z \end{vmatrix} = (0 - 0)\,\mathbf{i} - (e^z - e^z)\,\mathbf{j} + (0 - 0)\,\mathbf{k} = \mathbf{0}$ and **F** is defined on all of $\mathbb{R}^3$ with

component functions that have continuous partial deriatives, so **F** is conservative. Thus there exists a function f such that

$\nabla f = \mathbf{F}$. Then $f_x(x, y, z) = e^z$ implies $f(x, y, z) = xe^z + g(y, z) \;\Rightarrow\; f_y(x, y, z) = g_y(y, z)$. But $f_y(x, y, z) = 1$, so

$g(y, z) = y + h(z)$ and $f(x, y, z) = xe^z + y + h(z)$. Thus $f_z(x, y, z) = xe^z + h'(z)$ but $f_z(x, y, z) = xe^z$, so $h(z) = K$,

a constant. Hence a potential function for **F** is $f(x, y, z) = xe^z + y + K$.

17. $\operatorname{curl} \mathbf{F} = \nabla \times \mathbf{F} = \begin{vmatrix} \mathbf{i} & \mathbf{j} & \mathbf{k} \\ \partial/\partial x & \partial/\partial y & \partial/\partial z \\ ye^{-x} & e^{-x} & 2z \end{vmatrix} = (0 - 0)\,\mathbf{i} - (0 - 0)\,\mathbf{j} + (-e^{-x} - e^{-x})\,\mathbf{k} = -2e^{-x}\,\mathbf{k} \neq \mathbf{0}$,

so **F** is not conservative.

18. $\operatorname{curl} \mathbf{F} = \nabla \times \mathbf{F} = \begin{vmatrix} \mathbf{i} & \mathbf{j} & \mathbf{k} \\ \partial/\partial x & \partial/\partial y & \partial/\partial z \\ y \cos xy & x \cos xy & -\sin z \end{vmatrix}$

$\qquad = (0 - 0)\,\mathbf{i} - (0 - 0)\,\mathbf{j} + [(-xy \sin xy + \cos xy) - (-xy \sin xy + \cos xy)]\,\mathbf{k} = \mathbf{0}$

F is defined on all of $\mathbb{R}^3$, and the partial derivatives of the component functions are continuous, so **F** is conservative. Thus

there exists a function f such that $\nabla f = \mathbf{F}$. Then $f_x(x, y, z) = y \cos xy$ implies $f(x, y, z) = \sin xy + g(y, z) \;\Rightarrow\;$

$f_y(x, y, z) = x \cos xy + g_y(y, z)$. But $f_y(x, y, z) = x \cos xy$, so $g(y, z) = h(z)$ and $f(x, y, z) = \sin xy + h(z)$.

Thus $f_z(x, y, z) = h'(z)$ but $f_z(x, y, z) = -\sin z$ so $h(z) = \cos z + K$ and a potential function for **F** is

$f(x, y, z) = \sin xy + \cos z + K$.

19. No. Assume there is such a **G**. Then $\text{div}(\text{curl}\,\mathbf{G}) = \dfrac{\partial}{\partial x}\,(x\sin y) + \dfrac{\partial}{\partial y}\,(\cos y) + \dfrac{\partial}{\partial z}\,(z - xy) = \sin y - \sin y + 1 \neq 0$,

which contradicts Theorem 11.

20. No. Assume there is such a **G**. Then $\text{div}(\text{curl}\,\mathbf{G}) = yz - 2yz + 2yz = yz \neq 0$ which contradicts Theorem 11.

21. $\text{curl}\,\mathbf{F} = \begin{vmatrix} \mathbf{i} & \mathbf{j} & \mathbf{k} \\ \partial/\partial x & \partial/\partial y & \partial/\partial z \\ f(x) & g(y) & h(z) \end{vmatrix} = (0 - 0)\,\mathbf{i} + (0 - 0)\,\mathbf{j} + (0 - 0)\,\mathbf{k} = \mathbf{0}$. Hence $\mathbf{F} = f(x)\,\mathbf{i} + g(y)\,\mathbf{j} + h(z)\,\mathbf{k}$

is irrotational.

22. $\text{div}\,\mathbf{F} = \dfrac{\partial(f(y, z))}{\partial x} + \dfrac{\partial(g(x, z))}{\partial y} + \dfrac{\partial(h(x, y))}{\partial z} = 0$ so **F** is incompressible.

For Exercises 23–29, let $\mathbf{F}(x, y, z) = P_1\,\mathbf{i} + Q_1\,\mathbf{j} + R_1\,\mathbf{k}$ and $\mathbf{G}(x, y, z) = P_2\,\mathbf{i} + Q_2\,\mathbf{j} + R_2\,\mathbf{k}$.

23. $\text{div}(\mathbf{F} + \mathbf{G}) = \text{div}\langle P_1 + P_2, Q_1 + Q_2, R_1 + R_2\rangle = \dfrac{\partial(P_1 + P_2)}{\partial x} + \dfrac{\partial(Q_1 + Q_2)}{\partial y} + \dfrac{\partial(R_1 + R_2)}{\partial z}$

$\quad = \dfrac{\partial P_1}{\partial x} + \dfrac{\partial P_2}{\partial x} + \dfrac{\partial Q_1}{\partial y} + \dfrac{\partial Q_2}{\partial y} + \dfrac{\partial R_1}{\partial z} + \dfrac{\partial R_2}{\partial z} = \left(\dfrac{\partial P_1}{\partial x} + \dfrac{\partial Q_1}{\partial y} + \dfrac{\partial R_1}{\partial z}\right) + \left(\dfrac{\partial P_2}{\partial x} + \dfrac{\partial Q_2}{\partial y} + \dfrac{\partial R_2}{\partial z}\right)$

$\quad = \text{div}\langle P_1, Q_1, R_1\rangle + \text{div}\langle P_2, Q_2, R_2\rangle = \text{div}\,\mathbf{F} + \text{div}\,\mathbf{G}$

24. $\text{curl}\,\mathbf{F} + \text{curl}\,\mathbf{G} = \left[\left(\dfrac{\partial R_1}{\partial y} - \dfrac{\partial Q_1}{\partial z}\right)\mathbf{i} + \left(\dfrac{\partial P_1}{\partial z} - \dfrac{\partial R_1}{\partial x}\right)\mathbf{j} + \left(\dfrac{\partial Q_1}{\partial x} - \dfrac{\partial P_1}{\partial y}\right)\mathbf{k}\right]$

$\qquad\qquad\qquad + \left[\left(\dfrac{\partial R_2}{\partial y} - \dfrac{\partial Q_2}{\partial z}\right)\mathbf{i} + \left(\dfrac{\partial P_2}{\partial z} - \dfrac{\partial R_2}{\partial x}\right)\mathbf{j} + \left(\dfrac{\partial Q_2}{\partial x} - \dfrac{\partial P_2}{\partial y}\right)\mathbf{k}\right]$

$\qquad = \left[\dfrac{\partial(R_1 + R_2)}{\partial y} - \dfrac{\partial(Q_1 + Q_2)}{\partial z}\right]\mathbf{i} + \left[\dfrac{\partial(P_1 + P_2)}{\partial z} - \dfrac{\partial(R_1 + R_2)}{\partial x}\right]\mathbf{j}$

$\qquad\qquad + \left[\dfrac{\partial(Q_1 + Q_2)}{\partial x} - \dfrac{\partial(P_1 + P_2)}{\partial y}\right]\mathbf{k} = \text{curl}(\mathbf{F} + \mathbf{G})$

25. $\text{div}(f\mathbf{F}) = \text{div}(f\langle P_1, Q_1, R_1\rangle) = \text{div}\langle fP_1, fQ_1, fR_1\rangle = \dfrac{\partial(fP_1)}{\partial x} + \dfrac{\partial(fQ_1)}{\partial y} + \dfrac{\partial(fR_1)}{\partial z}$

$\quad = \left(f\dfrac{\partial P_1}{\partial x} + P_1\dfrac{\partial f}{\partial x}\right) + \left(f\dfrac{\partial Q_1}{\partial y} + Q_1\dfrac{\partial f}{\partial y}\right) + \left(f\dfrac{\partial R_1}{\partial z} + R_1\dfrac{\partial f}{\partial z}\right)$

$\quad = f\left(\dfrac{\partial P_1}{\partial x} + \dfrac{\partial Q_1}{\partial y} + \dfrac{\partial R_1}{\partial z}\right) + \langle P_1, Q_1, R_1\rangle \cdot \left\langle \dfrac{\partial f}{\partial x}, \dfrac{\partial f}{\partial y}, \dfrac{\partial f}{\partial z}\right\rangle = f\,\text{div}\,\mathbf{F} + \mathbf{F} \cdot \nabla f$

26. $\text{curl}(f\mathbf{F}) = \left[\dfrac{\partial(fR_1)}{\partial y} - \dfrac{\partial(fQ_1)}{\partial z}\right]\mathbf{i} + \left[\dfrac{\partial(fP_1)}{\partial z} - \dfrac{\partial(fR_1)}{\partial x}\right]\mathbf{j} + \left[\dfrac{\partial(fQ_1)}{\partial x} - \dfrac{\partial(fP_1)}{\partial y}\right]\mathbf{k}$

$\quad = \left[f\dfrac{\partial R_1}{\partial y} + R_1\dfrac{\partial f}{\partial y} - f\dfrac{\partial Q_1}{\partial z} - Q_1\dfrac{\partial f}{\partial z}\right]\mathbf{i} + \left[f\dfrac{\partial P_1}{\partial z} + P_1\dfrac{\partial f}{\partial z} - f\dfrac{\partial R_1}{\partial x} - R_1\dfrac{\partial f}{\partial x}\right]\mathbf{j}$

$\qquad\qquad + \left[f\dfrac{\partial Q_1}{\partial x} + Q_1\dfrac{\partial f}{\partial x} - f\dfrac{\partial P_1}{\partial y} - P_1\dfrac{\partial f}{\partial y}\right]\mathbf{k}$

$\quad = f\left[\dfrac{\partial R_1}{\partial y} - \dfrac{\partial Q_1}{\partial z}\right]\mathbf{i} + f\left[\dfrac{\partial P_1}{\partial z} - \dfrac{\partial R_1}{\partial x}\right]\mathbf{j} + f\left[\dfrac{\partial Q_1}{\partial x} - \dfrac{\partial P_1}{\partial y}\right]\mathbf{k}$

$\qquad\qquad + \left[R_1\dfrac{\partial f}{\partial y} - Q_1\dfrac{\partial f}{\partial z}\right]\mathbf{i} + \left[P_1\dfrac{\partial f}{\partial z} - R_1\dfrac{\partial f}{\partial x}\right]\mathbf{j} + \left[Q_1\dfrac{\partial f}{\partial x} - P_1\dfrac{\partial f}{\partial y}\right]\mathbf{k}$

$\quad = f\,\text{curl}\,\mathbf{F} + (\nabla f) \times \mathbf{F}$

27. $\operatorname{div}(\mathbf{F} \times \mathbf{G}) = \nabla \cdot (\mathbf{F} \times \mathbf{G}) = \begin{vmatrix} \partial/\partial x & \partial/\partial y & \partial/\partial z \\ P_1 & Q_1 & R_1 \\ P_2 & Q_2 & R_2 \end{vmatrix} = \frac{\partial}{\partial x}\begin{vmatrix} Q_1 & R_1 \\ Q_2 & R_2 \end{vmatrix} - \frac{\partial}{\partial y}\begin{vmatrix} P_1 & R_1 \\ P_2 & R_2 \end{vmatrix} + \frac{\partial}{\partial z}\begin{vmatrix} P_1 & Q_1 \\ P_2 & Q_2 \end{vmatrix}$

$$= \left[Q_1 \frac{\partial R_2}{\partial x} + R_2 \frac{\partial Q_1}{\partial x} - Q_2 \frac{\partial R_1}{\partial x} - R_1 \frac{\partial Q_2}{\partial x} \right] - \left[P_1 \frac{\partial R_2}{\partial y} + R_2 \frac{\partial P_1}{\partial y} - P_2 \frac{\partial R_1}{\partial y} - R_1 \frac{\partial P_2}{\partial y} \right]$$

$$+ \left[P_1 \frac{\partial Q_2}{\partial z} + Q_2 \frac{\partial P_1}{\partial z} - P_2 \frac{\partial Q_1}{\partial z} - Q_1 \frac{\partial P_2}{\partial z} \right]$$

$$= \left[P_2 \left(\frac{\partial R_1}{\partial y} - \frac{\partial Q_1}{\partial z} \right) + Q_2 \left(\frac{\partial P_1}{\partial z} - \frac{\partial R_1}{\partial x} \right) + R_2 \left(\frac{\partial Q_1}{\partial x} - \frac{\partial P_1}{\partial y} \right) \right]$$

$$- \left[P_1 \left(\frac{\partial R_2}{\partial y} - \frac{\partial Q_2}{\partial z} \right) + Q_1 \left(\frac{\partial P_2}{\partial z} - \frac{\partial R_2}{\partial x} \right) + R_1 \left(\frac{\partial Q_2}{\partial x} - \frac{\partial P_2}{\partial y} \right) \right]$$

$$= \mathbf{G} \cdot \operatorname{curl} \mathbf{F} - \mathbf{F} \cdot \operatorname{curl} \mathbf{G}$$

28. $\operatorname{div}(\nabla f \times \nabla g) = \nabla g \cdot \operatorname{curl}(\nabla f) - \nabla f \cdot \operatorname{curl}(\nabla g)$ [by Exercise 27] $= 0$ [by Theorem 3]

29. $\operatorname{curl}(\operatorname{curl} \mathbf{F}) = \nabla \times (\nabla \times \mathbf{F}) = \begin{vmatrix} \mathbf{i} & \mathbf{j} & \mathbf{k} \\ \partial/\partial x & \partial/\partial y & \partial/\partial z \\ \partial R_1/\partial y - \partial Q_1/\partial z & \partial P_1/\partial z - \partial R_1/\partial x & \partial Q_1/\partial x - \partial P_1/\partial y \end{vmatrix}$

$$= \left(\frac{\partial^2 Q_1}{\partial y \partial x} - \frac{\partial^2 P_1}{\partial y^2} - \frac{\partial^2 P_1}{\partial z^2} + \frac{\partial^2 R_1}{\partial z \partial x} \right) \mathbf{i} + \left(\frac{\partial^2 R_1}{\partial z \partial y} - \frac{\partial^2 Q_1}{\partial z^2} - \frac{\partial^2 Q_1}{\partial x^2} + \frac{\partial^2 P_1}{\partial x \partial y} \right) \mathbf{j}$$

$$+ \left(\frac{\partial^2 P_1}{\partial x \partial z} - \frac{\partial^2 R_1}{\partial x^2} - \frac{\partial^2 R_1}{\partial y^2} + \frac{\partial^2 Q_1}{\partial y \partial z} \right) \mathbf{k}$$

Now let's consider $\operatorname{grad}(\operatorname{div} \mathbf{F}) - \nabla^2 \mathbf{F}$ and compare with the above.

(Note that $\nabla^2 \mathbf{F}$ is defined on page 1102 [ET 1066].)

$$\operatorname{grad}(\operatorname{div} \mathbf{F}) - \nabla^2 \mathbf{F} = \left[\left(\frac{\partial^2 P_1}{\partial x^2} + \frac{\partial^2 Q_1}{\partial x \partial y} + \frac{\partial^2 R_1}{\partial x \partial z} \right) \mathbf{i} + \left(\frac{\partial^2 P_1}{\partial y \partial x} + \frac{\partial^2 Q_1}{\partial y^2} + \frac{\partial^2 R_1}{\partial y \partial z} \right) \mathbf{j} + \left(\frac{\partial^2 P_1}{\partial z \partial x} + \frac{\partial^2 Q_1}{\partial z \partial y} + \frac{\partial^2 R_1}{\partial z^2} \right) \mathbf{k} \right]$$

$$- \left[\left(\frac{\partial^2 P_1}{\partial x^2} + \frac{\partial^2 P_1}{\partial y^2} + \frac{\partial^2 P_1}{\partial z^2} \right) \mathbf{i} + \left(\frac{\partial^2 Q_1}{\partial x^2} + \frac{\partial^2 Q_1}{\partial y^2} + \frac{\partial^2 Q_1}{\partial z^2} \right) \mathbf{j} \right.$$

$$\left. + \left(\frac{\partial^2 R_1}{\partial x^2} + \frac{\partial^2 R_1}{\partial y^2} + \frac{\partial^2 R_1}{\partial z^2} \right) \mathbf{k} \right]$$

$$= \left(\frac{\partial^2 Q_1}{\partial x \partial y} + \frac{\partial^2 R_1}{\partial x \partial z} - \frac{\partial^2 P_1}{\partial y^2} - \frac{\partial^2 P_1}{\partial z^2} \right) \mathbf{i} + \left(\frac{\partial^2 P_1}{\partial y \partial x} + \frac{\partial^2 R_1}{\partial y \partial z} - \frac{\partial^2 Q_1}{\partial x^2} - \frac{\partial^2 Q_1}{\partial z^2} \right) \mathbf{j}$$

$$+ \left(\frac{\partial^2 P_1}{\partial z \partial x} + \frac{\partial^2 Q_1}{\partial z \partial y} - \frac{\partial^2 R_1}{\partial x^2} - \frac{\partial^2 R_2}{\partial y^2} \right) \mathbf{k}$$

Then applying Clairaut's Theorem to reverse the order of differentiation in the second partial derivatives as needed and comparing, we have $\operatorname{curl} \operatorname{curl} \mathbf{F} = \operatorname{grad} \operatorname{div} \mathbf{F} - \nabla^2 \mathbf{F}$ as desired.

30. (a) $\nabla \cdot \mathbf{r} = \left(\dfrac{\partial}{\partial x} \mathbf{i} + \dfrac{\partial}{\partial y} \mathbf{j} + \dfrac{\partial}{\partial z} \mathbf{k} \right) \cdot (x\,\mathbf{i} + y\,\mathbf{j} + z\,\mathbf{k}) = 1 + 1 + 1 = 3$

(b) $\nabla \cdot (r\mathbf{r}) = \nabla \cdot \sqrt{x^2 + y^2 + z^2}\,(x\,\mathbf{i} + y\,\mathbf{j} + z\,\mathbf{k})$

$$= \left(\dfrac{x^2}{\sqrt{x^2 + y^2 + z^2}} + \sqrt{x^2 + y^2 + z^2} \right) + \left(\dfrac{y^2}{\sqrt{x^2 + y^2 + z^2}} + \sqrt{x^2 + y^2 + z^2} \right)$$

$$+ \left(\dfrac{z^2}{\sqrt{x^2 + y^2 + z^2}} + \sqrt{x^2 + y^2 + z^2} \right)$$

$$= \dfrac{1}{\sqrt{x^2 + y^2 + z^2}} (4x^2 + 4y^2 + 4z^2) = 4\sqrt{x^2 + y^2 + z^2} = 4r$$

Another method:

By Exercise 25, $\nabla \cdot (r\mathbf{r}) = \operatorname{div}(r\mathbf{r}) = r\operatorname{div}\mathbf{r} + \mathbf{r} \cdot \nabla r = 3r + \mathbf{r} \cdot \dfrac{\mathbf{r}}{r}$ [see Exercise 31(a) below] $= 4r$.

(c) $\nabla^2 r^3 = \nabla^2 \left(x^2 + y^2 + z^2 \right)^{3/2}$

$$= \dfrac{\partial}{\partial x}\left[\dfrac{3}{2}(x^2 + y^2 + z^2)^{1/2}(2x) \right] + \dfrac{\partial}{\partial y}\left[\dfrac{3}{2}(x^2 + y^2 + z^2)^{1/2}(2y) \right] + \dfrac{\partial}{\partial z}\left[\dfrac{3}{2}(x^2 + y^2 + z^2)^{1/2}(2z) \right]$$

$$= 3\left[\dfrac{1}{2}(x^2 + y^2 + z^2)^{-1/2}(2x)(x) + (x^2 + y^2 + z^2)^{1/2} \right]$$

$$+ 3\left[\dfrac{1}{2}(x^2 + y^2 + z^2)^{-1/2}(2y)(y) + (x^2 + y^2 + z^2)^{1/2} \right]$$

$$+ 3\left[\dfrac{1}{2}(x^2 + y^2 + z^2)^{-1/2}(2z)(z) + (x^2 + y^2 + z^2)^{1/2} \right]$$

$$= 3(x^2 + y^2 + z^2)^{-1/2}(4x^2 + 4y^2 + 4z^2) = 12(x^2 + y^2 + z^2)^{1/2} = 12r$$

Another method: $\dfrac{\partial}{\partial x}\left(x^2 + y^2 + z^2 \right)^{3/2} = 3x\sqrt{x^2 + y^2 + z^2} \;\Rightarrow\; \nabla r^3 = 3r(x\,\mathbf{i} + y\,\mathbf{j} + z\,\mathbf{k}) = 3r\,\mathbf{r}$,

so $\nabla^2 r^3 = \nabla \cdot \nabla r^3 = \nabla \cdot (3r\,\mathbf{r}) = 3(4r) = 12r$ by part (b).

31. (a) $\nabla r = \nabla \sqrt{x^2 + y^2 + z^2} = \dfrac{x}{\sqrt{x^2 + y^2 + z^2}}\,\mathbf{i} + \dfrac{y}{\sqrt{x^2 + y^2 + z^2}}\,\mathbf{j} + \dfrac{z}{\sqrt{x^2 + y^2 + z^2}}\,\mathbf{k} = \dfrac{x\,\mathbf{i} + y\,\mathbf{j} + z\,\mathbf{k}}{\sqrt{x^2 + y^2 + z^2}} = \dfrac{\mathbf{r}}{r}$

(b) $\nabla \times \mathbf{r} = \begin{vmatrix} \mathbf{i} & \mathbf{j} & \mathbf{k} \\ \dfrac{\partial}{\partial x} & \dfrac{\partial}{\partial y} & \dfrac{\partial}{\partial z} \\ x & y & z \end{vmatrix} = \left[\dfrac{\partial}{\partial y}(z) - \dfrac{\partial}{\partial z}(y) \right]\mathbf{i} + \left[\dfrac{\partial}{\partial z}(x) - \dfrac{\partial}{\partial x}(z) \right]\mathbf{j} + \left[\dfrac{\partial}{\partial x}(y) - \dfrac{\partial}{\partial y}(x) \right]\mathbf{k} = \mathbf{0}$

(c) $\nabla \left(\dfrac{1}{r} \right) = \nabla \left(\dfrac{1}{\sqrt{x^2 + y^2 + z^2}} \right)$

$$= \dfrac{-\dfrac{1}{2\sqrt{x^2 + y^2 + z^2}}(2x)}{x^2 + y^2 + z^2}\,\mathbf{i} - \dfrac{\dfrac{1}{2\sqrt{x^2 + y^2 + z^2}}(2y)}{x^2 + y^2 + z^2}\,\mathbf{j} - \dfrac{\dfrac{1}{2\sqrt{x^2 + y^2 + z^2}}(2z)}{x^2 + y^2 + z^2}\,\mathbf{k}$$

$$= -\dfrac{x\,\mathbf{i} + y\,\mathbf{j} + z\,\mathbf{k}}{(x^2 + y^2 + z^2)^{3/2}} = -\dfrac{\mathbf{r}}{r^3}$$

(d) $\nabla \ln r = \nabla \ln(x^2 + y^2 + z^2)^{1/2} = \dfrac{1}{2}\nabla \ln(x^2 + y^2 + z^2)$

$$= \dfrac{x}{x^2 + y^2 + z^2}\,\mathbf{i} + \dfrac{y}{x^2 + y^2 + z^2}\,\mathbf{j} + \dfrac{z}{x^2 + y^2 + z^2}\,\mathbf{k} = \dfrac{x\,\mathbf{i} + y\,\mathbf{j} + z\,\mathbf{k}}{x^2 + y^2 + z^2} = \dfrac{\mathbf{r}}{r^2}$$

32. $\mathbf{r} = x\mathbf{i} + y\mathbf{j} + z\mathbf{k} \Rightarrow r = |\mathbf{r}| = \sqrt{x^2 + y^2 + z^2}$, so

$$\mathbf{F} = \frac{\mathbf{r}}{r^p} = \frac{x}{(x^2 + y^2 + z^2)^{p/2}}\mathbf{i} + \frac{y}{(x^2 + y^2 + z^2)^{p/2}}\mathbf{j} + \frac{z}{(x^2 + y^2 + z^2)^{p/2}}\mathbf{k}$$

Then $\dfrac{\partial}{\partial x}\dfrac{x}{(x^2 + y^2 + z^2)^{p/2}} = \dfrac{(x^2 + y^2 + z^2) - px^2}{(x^2 + y^2 + z^2)^{1 + p/2}} = \dfrac{r^2 - px^2}{r^{p+2}}$. Similarly,

$\dfrac{\partial}{\partial y}\dfrac{y}{(x^2 + y^2 + z^2)^{p/2}} = \dfrac{r^2 - py^2}{r^{p+2}}$ and $\dfrac{\partial}{\partial z}\dfrac{z}{(x^2 + y^2 + z^2)^{p/2}} = \dfrac{r^2 - pz^2}{r^{p+2}}$. Thus

$$\text{div } \mathbf{F} = \nabla \cdot \mathbf{F} = \frac{r^2 - px^2}{r^{p+2}} + \frac{r^2 - py^2}{r^{p+2}} + \frac{r^2 - pz^2}{r^{p+2}} = \frac{3r^2 - px^2 - py^2 - pz^2}{r^{p+2}}$$

$$= \frac{3r^2 - p(x^2 + y^2 + z^2)}{r^{p+2}} = \frac{3r^2 - pr^2}{r^{p+2}} = \frac{3 - p}{r^p}$$

Consequently, if $p = 3$ we have div $\mathbf{F} = 0$.

33. By (13), $\oint_C f(\nabla g) \cdot \mathbf{n}\, ds = \iint_D \text{div}(f\nabla g)\, dA = \iint_D [f \,\text{div}(\nabla g) + \nabla g \cdot \nabla f]\, dA$ by Exercise 25. But $\text{div}(\nabla g) = \nabla^2 g$.

Hence $\iint_D f\nabla^2 g\, dA = \oint_C f(\nabla g) \cdot \mathbf{n}\, ds - \iint_D \nabla g \cdot \nabla f\, dA$.

34. By Exercise 33, $\iint_D f\nabla^2 g\, dA = \oint_C f(\nabla g) \cdot \mathbf{n}\, ds - \iint_D \nabla g \cdot \nabla f\, dA$ and

$\iint_D g\nabla^2 f\, dA = \oint_C g(\nabla f) \cdot \mathbf{n}\, ds - \iint_D \nabla f \cdot \nabla g\, dA$. Hence

$\iint_D (f\nabla^2 g - g\nabla^2 f)\, dA = \oint_C [f(\nabla g) \cdot \mathbf{n} - g(\nabla f) \cdot \mathbf{n}]\, ds + \iint_D (\nabla f \cdot \nabla g - \nabla g \cdot \nabla f)\, dA = \oint_C [f\nabla g - g\nabla f] \cdot \mathbf{n}\, ds$.

35. Let $f(x, y) = 1$. Then $\nabla f = \mathbf{0}$ and Green's first identity (see Exercise 33) says

$\iint_D \nabla^2 g\, dA = \oint_C (\nabla g) \cdot \mathbf{n}\, ds - \iint_D \mathbf{0} \cdot \nabla g\, dA \Rightarrow \iint_D \nabla^2 g\, dA = \oint_C \nabla g \cdot \mathbf{n}\, ds$. But g is harmonic on D, so

$\nabla^2 g = 0 \Rightarrow \oint_C \nabla g \cdot \mathbf{n}\, ds = 0$ and $\oint_C D_{\mathbf{n}} g\, ds = \oint_C (\nabla g \cdot \mathbf{n})\, ds = 0$.

36. Let $g = f$. Then Green's first identity (see Exercise 33) says $\iint_D f\nabla^2 f\, dA = \oint_C (f)(\nabla f) \cdot \mathbf{n}\, ds - \iint_D \nabla f \cdot \nabla f\, dA$.

But f is harmonic, so $\nabla^2 f = 0$, and $\nabla f \cdot \nabla f = |\nabla f|^2$, so we have $0 = \oint_C (f)(\nabla f) \cdot \mathbf{n}\, ds - \iint_D |\nabla f|^2\, dA \Rightarrow$

$\iint_D |\nabla f|^2\, dA = \oint_C (f)(\nabla f) \cdot \mathbf{n}\, ds = 0$ since $f(x, y) = 0$ on C.

37. (a) We know that $\omega = v/d$, and from the diagram $\sin\theta = d/r \Rightarrow v = d\omega = (\sin\theta)r\omega = |\mathbf{w} \times \mathbf{r}|$. But $\mathbf{v}$ is perpendicular to both $\mathbf{w}$ and $\mathbf{r}$, so that $\mathbf{v} = \mathbf{w} \times \mathbf{r}$.

(b) From (a), $\mathbf{v} = \mathbf{w} \times \mathbf{r} = \begin{vmatrix} \mathbf{i} & \mathbf{j} & \mathbf{k} \\ 0 & 0 & \omega \\ x & y & z \end{vmatrix} = (0 \cdot z - \omega y)\mathbf{i} + (\omega x - 0 \cdot z)\mathbf{j} + (0 \cdot y - x \cdot 0)\mathbf{k} = -\omega y\mathbf{i} + \omega x\mathbf{j}$

(c) $\text{curl } \mathbf{v} = \nabla \times \mathbf{v} = \begin{vmatrix} \mathbf{i} & \mathbf{j} & \mathbf{k} \\ \partial/\partial x & \partial/\partial y & \partial/\partial z \\ -\omega y & \omega x & 0 \end{vmatrix}$

$= \left[\dfrac{\partial}{\partial y}(0) - \dfrac{\partial}{\partial z}(\omega x)\right]\mathbf{i} + \left[\dfrac{\partial}{\partial z}(-\omega y) - \dfrac{\partial}{\partial x}(0)\right]\mathbf{j} + \left[\dfrac{\partial}{\partial x}(\omega x) - \dfrac{\partial}{\partial y}(-\omega y)\right]\mathbf{k}$

$= [\omega - (-\omega)]\mathbf{k} = 2\omega\mathbf{k} = 2\mathbf{w}$

38. Let $\mathbf{H} = \langle h_1, h_2, h_3 \rangle$ and $\mathbf{E} = \langle E_1, E_2, E_3 \rangle$.

(a) $\nabla \times (\nabla \times \mathbf{E}) = \nabla \times (\operatorname{curl} \mathbf{E}) = \nabla \times \left(-\frac{1}{c}\frac{\partial \mathbf{H}}{\partial t}\right) = -\frac{1}{c}\begin{vmatrix} \mathbf{i} & \mathbf{j} & \mathbf{k} \\ \partial/\partial x & \partial/\partial y & \partial/\partial z \\ \partial h_1/\partial t & \partial h_2/\partial t & \partial h_3/\partial t \end{vmatrix}$

$$= -\frac{1}{c}\left[\left(\frac{\partial^2 h_3}{\partial y\,\partial t} - \frac{\partial^2 h_2}{\partial z\,\partial t}\right)\mathbf{i} + \left(\frac{\partial^2 h_1}{\partial z\,\partial t} - \frac{\partial^2 h_3}{\partial x\,\partial t}\right)\mathbf{j} + \left(\frac{\partial^2 h_2}{\partial x\,\partial t} - \frac{\partial^2 h_1}{\partial y\,\partial t}\right)\mathbf{k}\right]$$

$$= -\frac{1}{c}\frac{\partial}{\partial t}\left[\left(\frac{\partial h_3}{\partial y} - \frac{\partial h_2}{\partial z}\right)\mathbf{i} + \left(\frac{\partial h_1}{\partial z} - \frac{\partial h_3}{\partial x}\right)\mathbf{j} + \left(\frac{\partial h_2}{\partial x} - \frac{\partial h_1}{\partial y}\right)\mathbf{k}\right]$$

[assuming that the partial derivatives are continuous so that the order of differentiation does not matter]

$$= -\frac{1}{c}\frac{\partial}{\partial t}\operatorname{curl} \mathbf{H} = -\frac{1}{c}\frac{\partial}{\partial t}\left(\frac{1}{c}\frac{\partial \mathbf{E}}{\partial t}\right) = -\frac{1}{c^2}\frac{\partial^2 \mathbf{E}}{\partial t^2}$$

(b) $\nabla \times (\nabla \times \mathbf{H}) = \nabla \times (\operatorname{curl} \mathbf{H}) = \nabla \times \left(\frac{1}{c}\frac{\partial \mathbf{E}}{\partial t}\right) = \frac{1}{c}\begin{vmatrix} \mathbf{i} & \mathbf{j} & \mathbf{k} \\ \partial/\partial x & \partial/\partial y & \partial/\partial z \\ \partial E_1/\partial t & \partial E_2/\partial t & \partial E_3/\partial t \end{vmatrix}$

$$= \frac{1}{c}\left[\left(\frac{\partial^2 E_3}{\partial y\,\partial t} - \frac{\partial^2 E_2}{\partial z\,\partial t}\right)\mathbf{i} + \left(\frac{\partial^2 E_1}{\partial z\,\partial t} - \frac{\partial^2 E_3}{\partial x\,\partial t}\right)\mathbf{j} + \left(\frac{\partial^2 E_2}{\partial x\,\partial t} - \frac{\partial^2 E_1}{\partial y\,\partial t}\right)\mathbf{k}\right]$$

$$= \frac{1}{c}\frac{\partial}{\partial t}\left[\left(\frac{\partial E_3}{\partial y} - \frac{\partial E_2}{\partial z}\right)\mathbf{i} + \left(\frac{\partial E_1}{\partial z} - \frac{\partial E_3}{\partial x}\right)\mathbf{j} + \left(\frac{\partial E_2}{\partial x} - \frac{\partial E_1}{\partial y}\right)\mathbf{k}\right]$$

[assuming that the partial derivatives are continuous so that the order of differentiation does not matter]

$$= \frac{1}{c}\frac{\partial}{\partial t}\operatorname{curl} \mathbf{E} = \frac{1}{c}\frac{\partial}{\partial t}\left(-\frac{1}{c}\frac{\partial \mathbf{H}}{\partial t}\right) = -\frac{1}{c^2}\frac{\partial^2 \mathbf{H}}{\partial t^2}$$

(c) Using Exercise 29, we have that $\operatorname{curl}\operatorname{curl}\mathbf{E} = \operatorname{grad}\operatorname{div}\mathbf{E} - \nabla^2\mathbf{E} \;\Rightarrow$

$$\nabla^2\mathbf{E} = \operatorname{grad}\operatorname{div}\mathbf{E} - \operatorname{curl}\operatorname{curl}\mathbf{E} = \operatorname{grad} 0 + \frac{1}{c^2}\frac{\partial^2 \mathbf{E}}{\partial t^2} \quad \text{[from part (a)]} \;\; = \frac{1}{c^2}\frac{\partial^2 \mathbf{E}}{\partial t^2}.$$

(d) As in part (c), $\nabla^2\mathbf{H} = \operatorname{grad}\operatorname{div}\mathbf{H} - \operatorname{curl}\operatorname{curl}\mathbf{H} = \operatorname{grad} 0 + \frac{1}{c^2}\frac{\partial^2 \mathbf{H}}{\partial t^2} \quad \text{[using part (b)]} \;\; = \frac{1}{c^2}\frac{\partial^2 \mathbf{H}}{\partial t^2}.$

39. For any continuous function f on $\mathbb{R}^3$, define a vector field $\mathbf{G}(x, y, z) = \langle g(x, y, z), 0, 0\rangle$ where $g(x, y, z) = \int_0^x f(t, y, z)\,dt$.

Then $\operatorname{div}\mathbf{G} = \frac{\partial}{\partial x}\left(g(x, y, z)\right) + \frac{\partial}{\partial y}(0) + \frac{\partial}{\partial z}(0) = \frac{\partial}{\partial x}\int_0^x f(t, y, z)\,dt = f(x, y, z)$ by the Fundamental Theorem of

Calculus. Thus every continuous function f on $\mathbb{R}^3$ is the divergence of some vector field.

17.6   Parametric Surfaces and Their Areas

ET 16.6

1. $P(7, 10, 4)$ lies on the parametric surface $\mathbf{r}(u, v) = \langle 2u + 3v, 1 + 5u - v, 2 + u + v\rangle$ if and only if there are values for u and v where $2u + 3v = 7$, $1 + 5u - v = 10$, and $2 + u + v = 4$. But solving the first two equations simultaneously gives $u = 2$, $v = 1$ and these values do not satisfy the third equation, so P does not lie on the surface.

$Q(5, 22, 5)$ lies on the surface if $2u + 3v = 5$, $1 + 5u - v = 22$, and $2 + u + v = 5$ for some values of u and v. Solving the first two equations simultaneously gives $u = 4$, $v = -1$ and these values satisfy the third equation, so Q lies on the surface.

2. $P(3, -1, 5)$ lies on the parametric surface $\mathbf{r}(u, v) = \langle u + v, u^2 - v, u + v^2 \rangle$ if and only if there are values for u and v where

$u + v = 3$, $u^2 - v = -1$, and $u + v^2 = 5$. From the first equation we have $v = 3 - u$ and substituting into the second

equation gives $u^2 - 3 + u = -1$ $\Leftrightarrow$ $u^2 + u - 2 = 0$ $\Leftrightarrow$ $(u + 2)(u - 1) = 0$, so $u = -2$ $\Rightarrow$ $v = 5$ or $u = 1$ $\Rightarrow$

$v = 2$. The third equation is satisified by $u = 1$, $v = 2$ so P does lie on the surface.

$Q(-1, 3, 4)$ lies on $\mathbf{r}(u, v)$ if and only if $u + v = -1$, $u^2 - v = 3$, and $u + v^2 = 4$, but substituting the first equation into the

second gives $u = -2$, $v = 1$ or $u = 1$, $v = -2$, and neither of these pairs satisfies the third equation. Thus, Q does not lie on

the surface.

3. $\mathbf{r}(u, v) = (u + v)\mathbf{i} + (3 - v)\mathbf{j} + (1 + 4u + 5v)\mathbf{k} = \langle 0, 3, 1 \rangle + u\langle 1, 0, 4 \rangle + v\langle 1, -1, 5 \rangle$. From Example 3, we recognize

this as a vector equation of a plane through the point $(0, 3, 1)$ and containing vectors $\mathbf{a} = \langle 1, 0, 4 \rangle$ and $\mathbf{b} = \langle 1, -1, 5 \rangle$. If we

wish to find a more conventional equation for the plane, a normal vector to the plane is $\mathbf{a} \times \mathbf{b} = \begin{vmatrix} \mathbf{i} & \mathbf{j} & \mathbf{k} \\ 1 & 0 & 4 \\ 1 & -1 & 5 \end{vmatrix} = 4\mathbf{i} - \mathbf{j} - \mathbf{k}$

and an equation of the plane is $4(x - 0) - (y - 3) - (z - 1) = 0$ or $4x - y - z = -4$.

4. $\mathbf{r}(u, v) = 2\sin u\,\mathbf{i} + 3\cos u\,\mathbf{j} + v\,\mathbf{k}$, so the corresponding parametric equations for the surface are $x = 2\sin u$, $y = 3\cos u$,

$z = v$. For any point (x, y, z) on the surface, we have $(x/2)^2 + (y/3)^2 = \sin^2 u + \cos^2 u = 1$, so cross-sections parallel to

the yz-plane are all ellipses. Since $z = v$ with $0 \le v \le 2$, the surface is the portion of the elliptical cylinder $x^2/4 + y^2/9 = 1$

for $0 \le z \le 2$.

5. $\mathbf{r}(s, t) = \langle s, t, t^2 - s^2 \rangle$, so the corresponding parametric equations for the surface are $x = s$, $y = t$, $z = t^2 - s^2$. For any

point (x, y, z) on the surface, we have $z = y^2 - x^2$. With no restrictions on the parameters, the surface is $z = y^2 - x^2$, which

we recognize as a hyperbolic paraboloid.

6. $\mathbf{r}(s, t) = s\sin 2t\,\mathbf{i} + s^2\,\mathbf{j} + s\cos 2t\,\mathbf{k}$, so the corresponding parametric equations for the surface are $x = s\sin 2t$, $y = s^2$,

$z = s\cos 2t$. For any point (x, y, z) on the surface, we have $x^2 + z^2 = s^2\sin^2 2t + s^2\cos^2 2t = s^2 = y$. Since no

restrictions are placed on the parameters, the surface is $y = x^2 + z^2$, which we recognize as a circular paraboloid whose axis

is the y-axis.

7. $\mathbf{r}(u, v) = \langle u^2 + 1, v^3 + 1, u + v \rangle$, $-1 \le u \le 1$, $-1 \le v \le 1$.

The surface has parametric equations $x = u^2 + 1$, $y = v^3 + 1$, $z = u + v$,

$-1 \le u \le 1$, $-1 \le v \le 1$. In Maple, the surface can be graphed by entering

`plot3d([u^2+1,v^3+1,u+v],u=-1..1,v=-1..1);`. In

Mathematica we use the `ParametricPlot3D` command. If we keep u

constant at u_0, $x = u_0^2 + 1$, a constant, so the corresponding grid curves must

be the curves parallel to the yz-plane. If v is constant, we have $y = v_0^3 + 1$,

a constant, so these grid curves are the curves parallel to the xz-plane.

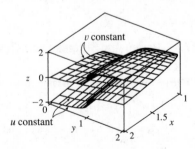

8. $\mathbf{r}(u,v) = \langle u+v, u^2, v^2 \rangle$, $-1 \le u \le 1$, $-1 \le v \le 1$.

The surface has parametric equations $x = u+v$, $y = u^2$, $z = v^2$,

$-1 \le u \le 1$, $-1 \le v \le 1$. If $u = u_0$ is constant, $y = u_0^2 = $ constant, so

the corresponding grid curves are the curves parallel to the xz-plane. If

$v = v_0$ is constant, $z = v_0^2 = $ constant, so the corresponding grid curves

are the curves parallel to the xy-plane.

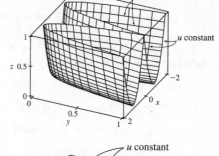

9. $\mathbf{r}(u,v) = \langle u\cos v, u\sin v, u^5 \rangle$.

The surface has parametric equations $x = u\cos v$, $y = u\sin v$,

$z = u^5$, $-1 \le u \le 1$, $0 \le v \le 2\pi$. Note that if $u = u_0$ is constant

then $z = u_0^5$ is constant and $x = u_0\cos v$, $y = u_0\sin v$ describe a

circle in x, y of radius $|u_0|$, so the corresponding grid curves are

circles parallel to the xy-plane. If $v = v_0$, a constant, the parametric

equations become $x = u\cos v_0$, $y = u\sin v_0$, $z = u^5$. Then

$y = (\tan v_0)x$, so these are the grid curves we see that lie in vertical

planes $y = kx$ through the z-axis.

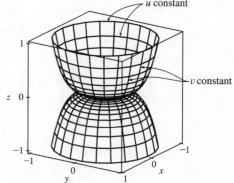

10. $\mathbf{r}(u,v) = \langle \cos u\sin v, \sin u\sin v, \cos v + \ln\tan(v/2) \rangle$.

The surface has parametric equations $x = \cos u\sin v$, $y = \sin u\sin v$,

$z = \cos v + \ln\tan(v/2)$, $0 \le u \le 2\pi$, $0.1 \le v \le 6.2$. Note that if

$v = v_0$ is constant, the parametric equations become $x = \cos u\sin v_0$,

$y = \sin u\sin v_0$, $z = \cos v_0 + \ln\tan(v_0/2)$ which represent a circle of

radius $\sin v_0$ in the plane $z = \cos v_0 + \ln\tan(v_0/2)$. So the circular grid

curves we see lying horizontally are the grid curves with v constant. The

vertically oriented grid curves correspond to $u = u_0$ being held constant,

giving $x = \cos u_0\sin v$, $y = \sin u_0\sin v$, $z = \cos v + \ln\tan(v/2)$. These

curves lie in vertical planes that contain the z-axis.

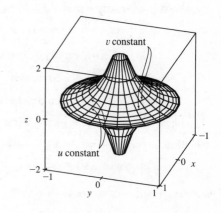

11. $x = \sin v$, $y = \cos u\sin 4v$, $z = \sin 2u\sin 4v$, $0 \le u \le 2\pi$, $-\frac{\pi}{2} \le v \le \frac{\pi}{2}$.

Note that if $v = v_0$ is constant, then $x = \sin v_0$ is constant, so the

corresponding grid curves must be parallel to the yz-plane. These

are the vertically oriented grid curves we see, each shaped like a

"figure-eight." When $u = u_0$ is held constant, the parametric

equations become $x = \sin v$, $y = \cos u_0\sin 4v$,

$z = \sin 2u_0\sin 4v$. Since z is a constant multiple of y, the

corresponding grid curves are the curves contained in planes

$z = ky$ that pass through the x-axis.

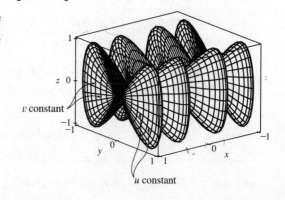

12. $x = u \sin u \cos v$, $y = u \cos u \cos v$, $z = u \sin v$.

We graph the portion of the surface with parametric domain

$0 \leq u \leq 4\pi$, $0 \leq v \leq 2\pi$. Note that if $v = v_0$ is constant, the parametric

equations become $x = u \sin u \cos v_0$, $y = u \cos u \cos v_0$, $z = u \sin v_0$.

The equations for x and y show that the projections onto the xy-plane give

a spiral shape, so the corresponding grid curves are the almost-horizontal

spiral curves we see. The vertical grid curves, which look approximately

circular, correspond to $u = u_0$ being held constant, giving

$x = u_0 \sin u_0 \ \cos v$, $y = u_0 \cos u_0 \ \cos v$, $z = u_0 \sin v$.

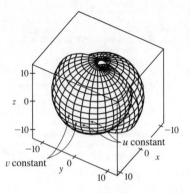

13. $\mathbf{r}(u, v) = u \cos v \, \mathbf{i} + u \sin v \, \mathbf{j} + v \, \mathbf{k}$. The parametric equations for the surface are $x = u \cos v$, $y = u \sin v$, $z = v$. We look at the grid curves first; if we fix v, then x and y parametrize a straight line in the plane $z = v$ which intersects the z-axis. If u is held constant, the projection onto the xy-plane is circular; with $z = v$, each grid curve is a helix. The surface is a spiraling ramp, graph I.

14. $\mathbf{r}(u, v) = u \cos v \, \mathbf{i} + u \sin v \, \mathbf{j} + \sin u \, \mathbf{k}$. The corresponding parametric equations for the surface are $x = u \cos v$, $y = u \sin v$, $z = \sin u$, $-\pi \leq u \leq \pi$. If $u = u_0$ is held constant, then $x = u_0 \cos v$, $y = u_0 \sin v$ so each grid curve is a circle of radius $|u_0|$ in the horizontal plane $z = \sin u_0$. If $v = v_0$ is constant, then $x = u \cos v_0$, $y = u \sin v_0$ $\Rightarrow$ $y = (\tan v_0)x$, so the grid curves lie in vertical planes $y = kx$ through the z-axis. In fact, since x and y are constant multiples of u and $z = \sin u$, each of these traces is a sine wave. The surface is graph I.

15. $\mathbf{r}(u, v) = \sin v \, \mathbf{i} + \cos u \sin 2v \, \mathbf{j} + \sin u \sin 2v \, \mathbf{k}$. Parametric equations for the surface are $x = \sin v$, $y = \cos u \sin 2v$, $z = \sin u \sin 2v$. If $v = v_0$ is fixed, then $x = \sin v_0$ is constant, and $y = (\sin 2v_0) \cos u$ and $z = (\sin 2v_0) \sin u$ describe a circle of radius $|\sin 2v_0|$, so each corresponding grid curve is a circle contained in the vertical plane $x = \sin v_0$ parallel to the yz-plane. The only possible surface is graph II. The grid curves we see running lengthwise along the surface correspond to holding u constant, in which case $y = (\cos u_0) \sin 2v$, $z = (\sin u_0) \sin 2v$ $\Rightarrow$ $z = (\tan u_0)y$, so each grid curve lies in a plane $z = ky$ that includes the x-axis.

16. $x = (1 - u)(3 + \cos v) \cos 4\pi u$, $y = (1 - u)(3 + \cos v) \sin 4\pi u$, $z = 3u + (1 - u) \sin v$. These equations correspond to graph VI: when $u = 0$, then $x = 3 + \cos v$, $y = 0$, and $z = \sin v$, which are equations of a circle with radius 1 in the xz-plane centered at $(3, 0, 0)$. When $u = \frac{1}{2}$, then $x = \frac{3}{2} + \frac{1}{2} \cos v$, $y = 0$, and $z = \frac{3}{2} + \frac{1}{2} \sin v$, which are equations of a circle with radius $\frac{1}{2}$ in the xz-plane centered at $\left(\frac{3}{2}, 0, \frac{3}{2}\right)$. When $u = 1$, then $x = y = 0$ and $z = 3$, giving the topmost point shown in the graph. This suggests that the grid curves with u constant are the vertically oriented circles visible on the surface. The spiralling grid curves correspond to keeping v constant.

17. $x = \cos^3 u \cos^3 v$, $y = \sin^3 u \cos^3 v$, $z = \sin^3 v$. If $v = v_0$ is held constant then $z = \sin^3 v_0$ is constant, so the corresponding grid curve lies in a horizontal plane. Several of the graphs exhibit horizontal grid curves, but the curves for this surface are neither circles nor straight lines, so graph III is the only possibility. (In fact, the horizontal grid curves here are members of the family $x = a \cos^3 u$, $y = a \sin^3 u$ and are called astroids.) The vertical grid curves we see on the surface correspond to $u = u_0$ held constant, as then we have $x = \cos^3 u_0 \cos^3 v$, $y = \sin^3 u_0 \ \cos^3 v$ so the corresponding grid curve lies in the vertical plane $y = (\tan^3 u_0)x$ through the z-axis.

18. $x = (1 - |u|)\cos v$, $y = (1 - |u|)\sin v$, $z = u$. Then $x^2 + y^2 = (1 - |u|)^2 \cos^2 v + (1 - |u|)^2 \sin^2 v = (1 - |u|)^2$, so if u is held constant, each grid curve is a circle of radius $(1 - |u|)$ in the horizontal plane $z = u$. The graph then must be graph VI. If v is held constant, so $v = v_0$, we have $x = (1 - |u|)\cos v_0$ and $y = (1 - |u|)\sin v_0$. Then $y = (\tan v_0)x$, so the grid curves we see running vertically along the surface in the planes $y = kx$ correspond to keeping v constant.

19. From Example 3, parametric equations for the plane through the point $(1, 2, -3)$ that contains the vectors $\mathbf{a} = \langle 1, 1, -1 \rangle$ and $\mathbf{b} = \langle 1, -1, 1 \rangle$ are $x = 1 + u(1) + v(1) = 1 + u + v$, $y = 2 + u(1) + v(-1) = 2 + u - v$, $z = -3 + u(-1) + v(1) = -3 - u + v$.

20. Solving the equation for z gives $z^2 = 1 - 2x^2 - 4y^2 \ \Rightarrow \ z = -\sqrt{1 - 2x^2 - 4y^2}$ (since we want the lower half of the ellipsoid). If we let x and y be the parameters, parametric equations are $x = x$, $y = y$, $z = -\sqrt{1 - 2x^2 - 4y^2}$.

Alternate solution: The equation can be rewritten as $\dfrac{x^2}{\left(1/\sqrt{2}\right)^2} + \dfrac{y^2}{(1/2)^2} + z^2 = 1$, and if we let $x = \dfrac{1}{\sqrt{2}}\, u\cos v$ and

$y = \frac{1}{2}u\sin v$, then $z = -\sqrt{1 - 2x^2 - 4y^2} = -\sqrt{1 - u^2\cos^2 v - u^2\sin^2 v} = -\sqrt{1 - u^2}$, where $0 \le u \le 1$ and $0 \le v \le 2\pi$.

21. Solving the equation for y gives $y^2 = 1 - x^2 + z^2 \ \Rightarrow \ y = \sqrt{1 - x^2 + z^2}$. (We choose the positive root since we want the part of the hyperboloid that corresponds to $y \ge 0$.) If we let x and z be the parameters, parametric equations are $x = x$, $z = z$, $y = \sqrt{1 - x^2 + z^2}$.

22. $x = 4 - y^2 - 2z^2$, $y = y$, $z = z$ where $y^2 + 2z^2 \le 4$ since $x \ge 0$. Then the associated vector equation is $\mathbf{r}(y, z) = (4 - y^2 - 2z^2)\,\mathbf{i} + y\,\mathbf{j} + z\,\mathbf{k}$.

23. Since the cone intersects the sphere in the circle $x^2 + y^2 = 2$, $z = \sqrt{2}$ and we want the portion of the sphere above this, we can parametrize the surface as $x = x$, $y = y$, $z = \sqrt{4 - x^2 - y^2}$ where $x^2 + y^2 \le 2$.
Alternate solution: Using spherical coordinates, $x = 2\sin\phi\cos\theta$, $y = 2\sin\phi\sin\theta$, $z = 2\cos\phi$ where $0 \le \phi \le \frac{\pi}{4}$ and $0 \le \theta \le 2\pi$.

24. In spherical coordinates, parametric equations are $x = 4\sin\phi\cos\theta$, $y = 4\sin\phi\sin\theta$, $z = 4\cos\phi$. The intersection of the sphere with the plane $z = 2$ corresponds to $z = 4\cos\phi = 2 \ \Rightarrow \ \cos\phi = \frac{1}{2} \ \Rightarrow \ \phi = \frac{\pi}{3}$. By symmetry, the intersection of the sphere with the plane $z = -2$ corresponds to $\phi = \pi - \frac{\pi}{3} = \frac{2\pi}{3}$. Thus the surface is described by $0 \le \theta \le 2\pi$, $\frac{\pi}{3} \le \phi \le \frac{2\pi}{3}$.

25. Parametric equations are $x = x$, $y = 4\cos\theta$, $z = 4\sin\theta$, $0 \le x \le 5$, $0 \le \theta \le 2\pi$.

26. Using x and y as the parameters, $x = x$, $y = y$, $z = x + 3$ where $0 \le x^2 + y^2 \le 1$. Also, since the plane intersects the cylinder in an ellipse, the surface is a planar ellipse in the plane $z = x + 3$. Thus, parametrizing with respect to s and θ, we have $x = s\cos\theta$, $y = s\sin\theta$, $z = 3 + s\cos\theta$ where $0 \le s \le 1$ and $0 \le \theta \le 2\pi$.

27. The surface appears to be a portion of a circular cylinder of radius 3 with axis the x-axis. An equation of the cylinder is $y^2 + z^2 = 9$, and we can impose the restrictions $0 \le x \le 5$, $y \le 0$ to obtain the portion shown. To graph the surface on a CAS, we can use parametric equations $x = u$, $y = 3\cos v$, $z = 3\sin v$ with the parameter domain $0 \le u \le 5$, $\frac{\pi}{2} \le v \le \frac{3\pi}{2}$. Alternatively, we can regard x and z as parameters. Then parametric equations are $x = x$, $z = z$, $y = -\sqrt{9 - z^2}$, where $0 \le x \le 5$ and $-3 \le z \le 3$.

28. The surface appears to be a portion of a sphere of radius 1 centered at the origin. In spherical coordinates, the sphere has equation $\rho = 1$, and imposing the restrictions $\frac{\pi}{2} \le \theta \le 2\pi$, $\frac{\pi}{4} \le \phi \le \pi$ will give only the portion of the sphere shown. Thus, to graph the surface on a CAS we can either use spherical coordinates with the stated restrictions, or we can use parametric equations: $x = \sin\phi\cos\theta$, $y = \sin\phi\sin\theta$, $z = \cos\phi$, $\frac{\pi}{2} \le \theta \le 2\pi$, $\frac{\pi}{4} \le \phi \le \pi$.

29. Using Equations 3, we have the parametrization $x = x$, $y = e^{-x}\cos\theta$, $z = e^{-x}\sin\theta$, $0 \le x \le 3$, $0 \le \theta \le 2\pi$.

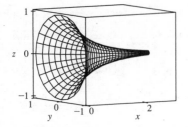

30. Letting θ be the angle of rotation about the y-axis, we have the parametrization $x = (4y^2 - y^4)\cos\theta$, $y = y$, $z = (4y^2 - y^4)\sin\theta$, $-2 \le y \le 2$, $0 \le \theta \le 2\pi$.

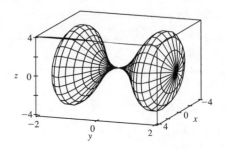

31. (a) Replacing $\cos u$ by $\sin u$ and $\sin u$ by $\cos u$ gives parametric equations $x = (2 + \sin v)\sin u$, $y = (2 + \sin v)\cos u$, $z = u + \cos v$. From the graph, it appears that the direction of the spiral is reversed. We can verify this observation by noting that the projection of the spiral grid curves onto the xy-plane, given by $x = (2 + \sin v)\sin u$, $y = (2 + \sin v)\cos u$, $z = 0$, draws a circle in the clockwise direction for each value of v. The original equations, on the other hand, give circular projections drawn in the counterclockwise direction. The equation for z is identical in both surfaces, so as z increases, these grid curves spiral up in opposite directions for the two surfaces.

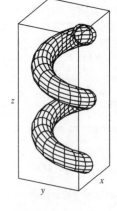

(b) Replacing $\cos u$ by $\cos 2u$ and $\sin u$ by $\sin 2u$ gives parametric equations $x = (2 + \sin v)\cos 2u$, $y = (2 + \sin v)\sin 2u$, $z = u + \cos v$. From the graph, it appears that the number of coils in the surface doubles within the same parametric domain. We can verify this observation by noting that the projection of the spiral grid curves onto the xy-plane, given by $x = (2 + \sin v)\cos 2u$, $y = (2 + \sin v)\sin 2u$, $z = 0$ (where v is constant), complete circular revolutions for $0 \le u \le \pi$ while the original surface requires $0 \le u \le 2\pi$ for a complete revolution. Thus, the new surface winds around twice as fast as the original surface, and since the equation for z is identical in both surfaces, we observe twice as many circular coils in the same z-interval.

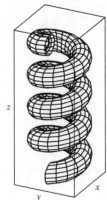

32. First we graph the surface as viewed from the front, then from two additional viewpoints.

  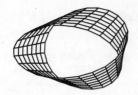

The surface appears as a twisted sheet, and is unusual because it has only one side. (The Möbius strip is discussed in more detail in Section 17.7 [ET 16.7].)

33. $\mathbf{r}(u, v) = (u + v)\,\mathbf{i} + 3u^2\,\mathbf{j} + (u - v)\,\mathbf{k}$.

$\mathbf{r}_u = \mathbf{i} + 6u\,\mathbf{j} + \mathbf{k}$ and $\mathbf{r}_v = \mathbf{i} - \mathbf{k}$, so $\mathbf{r}_u \times \mathbf{r}_v = -6u\,\mathbf{i} + 2\,\mathbf{j} - 6u\,\mathbf{k}$.
Since the point $(2, 3, 0)$ corresponds to $u = 1$, $v = 1$, a normal vector
to the surface at $(2, 3, 0)$ is $-6\,\mathbf{i} + 2\,\mathbf{j} - 6\,\mathbf{k}$, and an equation of the
tangent plane is $-6x + 2y - 6z = -6$ or $3x - y + 3z = 3$.

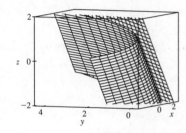

34. $\mathbf{r}(u, v) = u^2\,\mathbf{i} + v^2\,\mathbf{j} + uv\,\mathbf{k} \quad\Rightarrow\quad \mathbf{r}(1, 1) = (1, 1, 1)$.

$\mathbf{r}_u = 2u\,\mathbf{i} + v\,\mathbf{k}$ and $\mathbf{r}_v = 2v\,\mathbf{j} + u\,\mathbf{k}$, so a normal vector to the
surface at the point $(1, 1, 1)$ is

$\mathbf{r}_u(1, 1) \times \mathbf{r}_v(1, 1) = (2\,\mathbf{i} + \mathbf{k}) \times (2\,\mathbf{j} + \mathbf{k}) = -2\,\mathbf{i} - 2\,\mathbf{j} + 4\,\mathbf{k}$.
Thus an equation of the tangent plane at the point $(1, 1, 1)$ is
$-2(x - 1) - 2(y - 1) + 4(z - 1) = 0$ or $x + y - 2z = 0$.

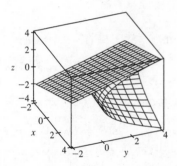

35. $\mathbf{r}(u, v) = u^2\,\mathbf{i} + 2u \sin v\,\mathbf{j} + u \cos v\,\mathbf{k} \quad\Rightarrow\quad \mathbf{r}(1, 0) = (1, 0, 1)$.

$\mathbf{r}_u = 2u\,\mathbf{i} + 2 \sin v\,\mathbf{j} + \cos v\,\mathbf{k}$ and $\mathbf{r}_v = 2u \cos v\,\mathbf{j} - u \sin v\,\mathbf{k}$,
so a normal vector to the surface at the point $(1, 0, 1)$ is

$\mathbf{r}_u(1, 0) \times \mathbf{r}_v(1, 0) = (2\,\mathbf{i} + \mathbf{k}) \times (2\,\mathbf{j}) = -2\,\mathbf{i} + 4\,\mathbf{k}$.
Thus an equation of the tangent plane at $(1, 0, 1)$ is
$-2(x - 1) + 0(y - 0) + 4(z - 1) = 0$ or $-x + 2z = 1$.

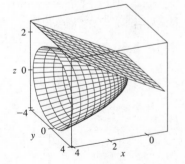

36. $\mathbf{r}(u, v) = uv\,\mathbf{i} + u \sin v\,\mathbf{j} + v \cos u\,\mathbf{k} \quad\Rightarrow\quad \mathbf{r}(0, \pi) = (0, 0, \pi)$.

$\mathbf{r}_u = v\,\mathbf{i} + \sin v\,\mathbf{j} - v \sin u\,\mathbf{k}$ and $\mathbf{r}_v = u\,\mathbf{i} + u \cos v\,\mathbf{j} + \cos u\ \mathbf{k}$, so
a normal vector to the surface at the point $(0, 0, \pi)$ is

$\mathbf{r}_u(0, \pi) \times \mathbf{r}_v(0, \pi) = (\pi\,\mathbf{i}) \times (\mathbf{k}) = -\pi\,\mathbf{j}$. Thus an equation of the
tangent plane is $-\pi(y - 0) = 0$ or $y = 0$.

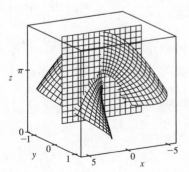

37. The surface S is given by $z = f(x, y) = 6 - 3x - 2y$ which intersects the xy-plane in the line $3x + 2y = 6$, so D is the

triangular region given by $\{(x, y) \mid 0 \le x \le 2, 0 \le y \le 3 - \frac{3}{2}x\}$. By Formula 9, the surface area of S is

$$A(S) = \iint_D \sqrt{1 + \left(\frac{\partial z}{\partial x}\right)^2 + \left(\frac{\partial z}{\partial y}\right)^2}\, dA$$

$$= \iint_D \sqrt{1 + (-3)^2 + (-2)^2}\, dA = \sqrt{14} \iint_D dA = \sqrt{14}\, A(D) = \sqrt{14}\left(\frac{1}{2} \cdot 2 \cdot 3\right) = 3\sqrt{14}.$$

38. $z = f(x, y) = 10 - 2x - 5y$ and D is the disk $x^2 + y^2 \le 9$, so by Formula 9,

$$A(S) = \iint_D \sqrt{1 + (-2)^2 + (-5)^2}\, dA = \sqrt{30} \iint_D dA = \sqrt{30}\, A(D) = \sqrt{30}\left(\pi \cdot 3^2\right) = 9\sqrt{30}\,\pi.$$

39. $z = f(x, y) = \frac{2}{3}(x^{3/2} + y^{3/2})$ and $D = \{(x, y) \mid 0 \le x \le 1, 0 \le y \le 1\}$. Then $f_x = x^{1/2}$, $f_y = y^{1/2}$ and

$$A(S) = \iint_D \sqrt{1 + (\sqrt{x})^2 + (\sqrt{y})^2}\, dA = \int_0^1 \int_0^1 \sqrt{1 + x + y}\, dy\, dx$$

$$= \int_0^1 \left[\frac{2}{3}(x + y + 1)^{3/2}\right]_{y=0}^{y=1} dx = \frac{2}{3} \int_0^1 \left[(x + 2)^{3/2} - (x + 1)^{3/2}\right] dx$$

$$= \frac{2}{3}\left[\frac{2}{5}(x + 2)^{5/2} - \frac{2}{5}(x + 1)^{5/2}\right]_0^1 = \frac{4}{15}(3^{5/2} - 2^{5/2} - 2^{5/2} + 1) = \frac{4}{15}(3^{5/2} - 2^{7/2} + 1)$$

40. $\mathbf{r}_u = \langle 0, 1, -5\rangle$, $\mathbf{r}_v = \langle 1, -2, 1\rangle$, and $\mathbf{r}_u \times \mathbf{r}_v = \langle -9, -5, -1\rangle$. Then by Definition 6,

$$A(S) = \iint_D |\mathbf{r}_u \times \mathbf{r}_v|\, dA = \int_0^1 \int_0^1 |\langle -9, -5, -1\rangle|\, du\, dv = \sqrt{107} \int_0^1 du \int_0^1 dv = \sqrt{107}$$

41. $z = f(x, y) = xy$ with $0 \le x^2 + y^2 \le 1$, so $f_x = y$, $f_y = x$ $\Rightarrow$

$$A(S) = \iint_D \sqrt{1 + y^2 + x^2}\, dA = \int_0^{2\pi} \int_0^1 \sqrt{r^2 + 1}\, r\, dr\, d\theta = \int_0^{2\pi} \left[\frac{1}{3}(r^2 + 1)^{3/2}\right]_{r=0}^{r=1} d\theta$$

$$= \int_0^{2\pi} \frac{1}{3}(2\sqrt{2} - 1)\, d\theta = \frac{2\pi}{3}(2\sqrt{2} - 1)$$

42. $z = f(x, y) = 1 + 3x + 2y^2$ with $0 \le x \le 2y$, $0 \le y \le 1$. Thus, by Formula 9,

$$A(S) = \iint_D \sqrt{1 + 3^2 + (4y)^2}\, dA = \int_0^1 \int_0^{2y} \sqrt{10 + 16y^2}\, dx\, dy = \int_0^1 2y\sqrt{10 + 16y^2}\, dy$$

$$= \frac{1}{16} \cdot \frac{2}{3}(10 + 16y^2)^{3/2}\Big]_0^1 = \frac{1}{24}(26^{3/2} - 10^{3/2})$$

43. $z = f(x, y) = y^2 - x^2$ with $1 \le x^2 + y^2 \le 4$. Then

$$A(S) = \iint_D \sqrt{1 + 4x^2 + 4y^2}\, dA = \int_0^{2\pi} \int_1^2 \sqrt{1 + 4r^2}\, r\, dr\, d\theta = \int_0^{2\pi} d\theta \int_1^2 r\sqrt{1 + 4r^2}\, dr$$

$$= \left[\theta\right]_0^{2\pi} \left[\frac{1}{12}(1 + 4r^2)^{3/2}\right]_1^2 = \frac{\pi}{6}\left(17\sqrt{17} - 5\sqrt{5}\right)$$

44. A parametric representation of the surface is $x = y^2 + z^2$, $y = y$, $z = z$ with $0 \le y^2 + z^2 \le 9$.

Hence $\mathbf{r}_y \times \mathbf{r}_z = (2y\,\mathbf{i} + \mathbf{j}) \times (2z\,\mathbf{i} + \mathbf{k}) = \mathbf{i} - 2y\,\mathbf{j} - 2z\,\mathbf{k}$.

Note: In general, if $x = f(y, z)$ then $\mathbf{r}_y \times \mathbf{r}_z = \mathbf{i} - \dfrac{\partial f}{\partial y}\,\mathbf{j} - \dfrac{\partial f}{\partial z}\,\mathbf{k}$, and $A(S) = \iint_D \sqrt{1 + \left(\dfrac{\partial f}{\partial y}\right)^2 + \left(\dfrac{\partial f}{\partial z}\right)^2}\, dA$. Then

$$A(S) = \iint_{0 \le y^2 + z^2 \le 9} \sqrt{1 + 4y^2 + 4z^2}\, dA = \int_0^{2\pi} \int_0^3 \sqrt{1 + 4r^2}\, r\, dr\, d\theta$$

$$= \int_0^{2\pi} d\theta \int_0^3 r\sqrt{1 + 4r^2}\, dr = 2\pi\left[\frac{1}{12}(1 + 4r^2)^{3/2}\right]_0^3 = \frac{\pi}{6}\left(37\sqrt{37} - 1\right)$$

45. A parametric representation of the surface is $x = x$, $y = 4x + z^2$, $z = z$ with $0 \leq x \leq 1, 0 \leq z \leq 1$.

Hence $\mathbf{r}_x \times \mathbf{r}_z = (\mathbf{i} + 4\mathbf{j}) \times (2z\,\mathbf{j} + \mathbf{k}) = 4\,\mathbf{i} - \mathbf{j} + 2z\,\mathbf{k}$.

Note: In general, if $y = f(x, z)$ then $\mathbf{r}_x \times \mathbf{r}_z = \dfrac{\partial f}{\partial x}\,\mathbf{i} - \mathbf{j} + \dfrac{\partial f}{\partial z}\,\mathbf{k}$ and $A(S) = \displaystyle\iint_D \sqrt{1 + \left(\dfrac{\partial f}{\partial x}\right)^2 + \left(\dfrac{\partial f}{\partial z}\right)^2}\,dA$. Then

$$A(S) = \int_0^1 \int_0^1 \sqrt{17 + 4z^2}\,dx\,dz = \int_0^1 \sqrt{17 + 4z^2}\,dz$$

$$= \tfrac{1}{2}\left(z\sqrt{17 + 4z^2} + \tfrac{17}{2}\ln\left|2z + \sqrt{4z^2 + 17}\right|\right)\Big]_0^1 = \tfrac{\sqrt{21}}{2} + \tfrac{17}{4}\left[\ln\left(2 + \sqrt{21}\right) - \ln\sqrt{17}\right]$$

46. $\mathbf{r}_u = \langle \cos v, \sin v, 0 \rangle$, $\mathbf{r}_v = \langle -u\sin v, u\cos v, 1 \rangle$, and $\mathbf{r}_u \times \mathbf{r}_v = \langle \sin v, -\cos v, u \rangle$. Then

$$A(S) = \int_0^\pi \int_0^1 \sqrt{1 + u^2}\,du\,dv = \int_0^\pi dv \int_0^1 \sqrt{1 + u^2}\,du$$

$$= \pi\left[\tfrac{u}{2}\sqrt{u^2 + 1} + \tfrac{1}{2}\ln\left|u + \sqrt{u^2 + 1}\right|\right]\Big|_0^1 = \tfrac{\pi}{2}\left[\sqrt{2} + \ln\left(1 + \sqrt{2}\right)\right]$$

47. $\mathbf{r}_u = \langle 2u, v, 0 \rangle$, $\mathbf{r}_v = \langle 0, u, v \rangle$, and $\mathbf{r}_u \times \mathbf{r}_v = \langle v^2, -2uv, 2u^2 \rangle$. Then

$$A(S) = \iint_D |\mathbf{r}_u \times \mathbf{r}_v|\,dA = \int_0^1 \int_0^2 \sqrt{v^4 + 4u^2v^2 + 4u^4}\,dv\,du = \int_0^1 \int_0^2 \sqrt{(v^2 + 2u^2)^2}\,dv\,du$$

$$= \int_0^1 \int_0^2 (v^2 + 2u^2)\,dv\,du = \int_0^1 \left[\tfrac{1}{3}v^3 + 2u^2 v\right]_{v=0}^{v=2} du = \int_0^1 \left(\tfrac{8}{3} + 4u^2\right)du = \left[\tfrac{8}{3}u + \tfrac{4}{3}u^3\right]_0^1 = 4$$

48. $z = f(x, y) = \cos(x^2 + y^2)$ with $x^2 + y^2 \leq 1$.

$$A(S) = \iint_D \sqrt{1 + (-2x\sin(x^2 + y^2))^2 + (-2y\sin(x^2 + y^2))^2}\,dA$$

$$= \iint_D \sqrt{1 + 4x^2\sin^2(x^2 + y^2) + 4y^2\sin^2(x^2 + y^2)}\,dA = \iint_D \sqrt{1 + 4(x^2 + y^2)\sin^2(x^2 + y^2)}\,dA$$

$$= \int_0^{2\pi} \int_0^1 \sqrt{1 + 4r^2\sin^2(r^2)}\,r\,dr\,d\theta = \int_0^{2\pi} d\theta \int_0^1 r\sqrt{1 + 4r^2\sin^2(r^2)}\,dr$$

$$= 2\pi \int_0^1 r\sqrt{1 + 4r^2\sin^2(r^2)}\,dr \approx 4.1073$$

49. $z = f(x, y) = e^{-x^2 - y^2}$ with $x^2 + y^2 \leq 4$.

$$A(S) = \iint_D \sqrt{1 + \left(-2xe^{-x^2 - y^2}\right)^2 + \left(-2ye^{-x^2 - y^2}\right)^2}\,dA = \iint_D \sqrt{1 + 4(x^2 + y^2)e^{-2(x^2 + y^2)}}\,dA$$

$$= \int_0^{2\pi} \int_0^2 \sqrt{1 + 4r^2 e^{-2r^2}}\,r\,dr\,d\theta = \int_0^{2\pi} d\theta \int_0^2 r\sqrt{1 + 4r^2 e^{-2r^2}}\,dr = 2\pi \int_0^2 r\sqrt{1 + 4r^2 e^{-2r^2}}\,dr \approx 13.9783$$

50. Let $f(x, y) = \dfrac{1 + x^2}{1 + y^2}$. Then $f_x = \dfrac{2x}{1 + y^2}$,

$$f_y = (1 + x^2)\left[-\dfrac{2y}{(1 + y^2)^2}\right] = -\dfrac{2y(1 + x^2)}{(1 + y^2)^2}.$$

We use a CAS to estimate

$\int_{-1}^1 \int_{-(1 - |x|)}^{1 - |x|} \sqrt{1 + f_x^2 + f_y^2}\,dy\,dx \approx 2.6959$.

In order to graph only the part of the surface above the square, we

use $-(1 - |x|) \leq y \leq 1 - |x|$ as the y-range in our plot command.

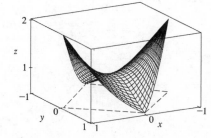

51. (a) $A(S) = \displaystyle\iint_D \sqrt{1 + \left(\dfrac{\partial z}{\partial x}\right)^2 + \left(\dfrac{\partial z}{\partial y}\right)^2}\,dA = \int_0^6 \int_0^4 \sqrt{1 + \dfrac{4x^2 + 4y^2}{(1 + x^2 + y^2)^4}}\,dy\,dx$.

Using the Midpoint Rule with $f(x, y) = \sqrt{1 + \dfrac{4x^2 + 4y^2}{(1 + x^2 + y^2)^4}}$, $m = 3$, $n = 2$ we have

$$A(S) \approx \sum_{i=1}^3 \sum_{j=1}^2 f(\overline{x}_i, \overline{y}_j)\,\Delta A = 4\left[f(1, 1) + f(1, 3) + f(3, 1) + f(3, 3) + f(5, 1) + f(5, 3)\right] \approx 24.2055$$

(b) Using a CAS we have $A(S) = \int_0^6 \int_0^4 \sqrt{1 + \dfrac{4x^2 + 4y^2}{(1 + x^2 + y^2)^4}}\, dy\, dx \approx 24.2476$. This agrees with the estimate in part (a)

to the first decimal place.

52. $\mathbf{r}(u, v) = \langle \cos^3 u \cos^3 v, \sin^3 u \cos^3 v, \sin^3 v \rangle$, so $\mathbf{r}_u = \langle -3 \cos^2 u \sin u \cos^3 v, 3 \sin^2 u \cos u \cos^3 v, 0 \rangle$,

$\mathbf{r}_v = \langle -3 \cos^3 u \cos^2 v \sin v, -3 \sin^3 u \cos^2 v \sin v, 3 \sin^2 v \cos v \rangle$, and

$\mathbf{r}_u \times \mathbf{r}_v = \langle 9 \cos u \sin^2 u \cos^4 v \sin^2 v, 9 \cos^2 u \sin u \cos^4 v \sin^2 v, 9 \cos^2 u \sin^2 u \cos^5 v \sin v \rangle$. Then

$$|\mathbf{r}_u \times \mathbf{r}_v| = 9 \sqrt{\cos^2 u \sin^4 u \cos^8 v \sin^4 v + \cos^4 u \sin^2 u \cos^8 v \sin^4 v + \cos^4 u \sin^4 u \cos^{10} v \sin^2 v}$$

$$= 9 \sqrt{\cos^2 u \sin^2 u \cos^8 v \sin^2 v \left(\sin^2 v + \cos^2 u \sin^2 u \cos^2 v\right)}$$

$$= 9 \cos^4 v \left|\cos u \sin u \sin v\right| \sqrt{\sin^2 v + \cos^2 u \sin^2 u \cos^2 v}$$

Using a CAS, we have $A(S) = \int_0^\pi \int_0^{2\pi} 9 \cos^4 v \left|\cos u \sin u \sin v\right| \sqrt{\sin^2 v + \cos^2 u \sin^2 u \cos^2 v}\, dv\, du \approx 4.4506$.

53. $z = 1 + 2x + 3y + 4y^2$, so

$$A(S) = \iint_D \sqrt{1 + \left(\frac{\partial z}{\partial x}\right)^2 + \left(\frac{\partial z}{\partial y}\right)^2}\, dA = \int_1^4 \int_0^1 \sqrt{1 + 4 + (3 + 8y)^2}\, dy\, dx = \int_1^4 \int_0^1 \sqrt{14 + 48y + 64y^2}\, dy\, dx.$$

Using a CAS, we have

$\int_1^4 \int_0^1 \sqrt{14 + 48y + 64y^2}\, dy\, dx = \frac{45}{8} \sqrt{14} + \frac{15}{16} \ln\left(11 \sqrt{5} + 3 \sqrt{14}\, \sqrt{5}\right) - \frac{15}{16} \ln\left(3 \sqrt{5} + \sqrt{14}\, \sqrt{5}\right)$

or $\frac{45}{8} \sqrt{14} + \frac{15}{16} \ln \frac{11 \sqrt{5} + 3 \sqrt{70}}{3 \sqrt{5} + \sqrt{70}}$.

54. (a) $\mathbf{r}_u = a \cos v\, \mathbf{i} + b \sin v\, \mathbf{j} + 2u\, \mathbf{k}$, $\mathbf{r}_v = -au \sin v\, \mathbf{i} + bu \cos v\, \mathbf{j} + 0\, \mathbf{k}$, and

$\mathbf{r}_u \times \mathbf{r}_v = -2bu^2 \cos v\, \mathbf{i} - 2au^2 \sin v\, \mathbf{j} + abu\, \mathbf{k}$.

$A(S) = \int_0^{2\pi} \int_0^2 |\mathbf{r}_u \times \mathbf{r}_v|\, du\, dv = \int_0^{2\pi} \int_0^2 \sqrt{4b^2 u^4 \cos^2 v + 4a^2 u^4 \sin^2 v + a^2 b^2 u^2}\, du\, dv$

(b) $x^2 = a^2 u^2 \cos^2 v$, $y^2 = b^2 u^2 \sin^2 v$, $z = u^2$ $\Rightarrow$ $x^2/a^2 + y^2/b^2 = u^2 = z$ which is an elliptic paraboloid. To find D,

notice that $0 \le u \le 2$ $\Rightarrow$ $0 \le z \le 4$ $\Rightarrow$ $0 \le x^2/a^2 + y^2/b^2 \le 4$. Therefore, using Formula 9, we have

$$A(S) = \int_{-2a}^{2a} \int_{-b\sqrt{4 - (x^2/a^2)}}^{b\sqrt{4 - (x^2/a^2)}} \sqrt{1 + (2x/a^2)^2 + (2y/b^2)^2}\, dy\, dx.$$

(c)

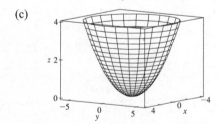

(d) We substitute $a = 2$, $b = 3$ in the integral in part (a) to get

$A(S) = \int_0^{2\pi} \int_0^2 2u \sqrt{9u^2 \cos^2 v + 4u^2 \sin^2 v + 9}\, du\, dv$. We use a CAS

to estimate the integral accurate to four decimal places. To speed up the

calculation, we can set `Digits:=7;` (in Maple) or use the approximation

command `N` (in Mathematica). We find that $A(S) \approx 115.6596$.

55. (a) $x = a \sin u \cos v$, $y = b \sin u \sin v$, $z = c \cos u$ $\Rightarrow$

$$\frac{x^2}{a^2} + \frac{y^2}{b^2} + \frac{z^2}{c^2} = (\sin u \cos v)^2 + (\sin u \sin v)^2 + (\cos u)^2$$

$$= \sin^2 u + \cos^2 u = 1$$

and since the ranges of u and v are sufficient to generate the entire graph,

the parametric equations represent an ellipsoid.

(b)

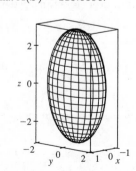

(c) From the parametric equations (with $a = 1$, $b = 2$, and $c = 3$), we calculate

$\mathbf{r}_u = \cos u \, \cos v \, \mathbf{i} + 2 \cos u \, \sin v \, \mathbf{j} - 3 \sin u \, \mathbf{k}$ and $\mathbf{r}_v = -\sin u \, \sin v \, \mathbf{i} + 2 \sin u \, \cos v \, \mathbf{j}$. So

$\mathbf{r}_u \times \mathbf{r}_v = 6 \sin^2 u \, \cos v \, \mathbf{i} + 3 \sin^2 u \, \sin v \, \mathbf{j} + 2 \sin u \, \cos u \, \mathbf{k}$, and the surface area is given by

$$A(S) = \int_0^{2\pi} \int_0^\pi |\mathbf{r}_u \times \mathbf{r}_v| \, du \, dv = \int_0^{2\pi} \int_0^\pi \sqrt{36 \sin^4 u \, \cos^2 v + 9 \sin^4 u \, \sin^2 v + 4 \cos^2 u \, \sin^2 u} \, du \, dv$$

56. (a) $x = a \cosh u \, \cos v$, $y = b \cosh u \, \sin v$, $z = c \sinh u$ $\Rightarrow$

$$\frac{x^2}{a^2} + \frac{y^2}{b^2} - \frac{z^2}{c^2} = \cosh^2 u \, \cos^2 v + \cosh^2 u \, \sin^2 v - \sinh^2 u$$

$$= \cosh^2 u - \sinh^2 u = 1$$

and the parametric equations represent a hyperboloid of

one sheet.

(b)

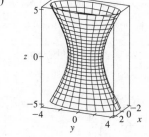

(c) $\mathbf{r}_u = \sinh u \, \cos v \, \mathbf{i} + 2 \sinh u \, \sin v \, \mathbf{j} + 3 \cosh u \, \mathbf{k}$ and

$\mathbf{r}_v = -\cosh u \, \sin v \, \mathbf{i} + 2 \cosh u \, \cos v \, \mathbf{j}$, so $\mathbf{r}_u \times \mathbf{r}_v = -6 \cosh^2 u \, \cos v \, \mathbf{i} - 3 \cosh^2 u \, \sin v \, \mathbf{j} + 2 \cosh u \, \sinh u \, \mathbf{k}$.

We integrate between $u = \sinh^{-1}(-1) = -\ln(1 + \sqrt{2})$ and $u = \sinh^{-1} 1 = \ln(1 + \sqrt{2})$, since then z varies between

-3 and 3, as desired. So the surface area is

$$A(S) = \int_0^{2\pi} \int_{-\ln(1+\sqrt{2})}^{\ln(1+\sqrt{2})} |\mathbf{r}_u \times \mathbf{r}_v| \, du \, dv$$

$$= \int_0^{2\pi} \int_{-\ln(1+\sqrt{2})}^{\ln(1+\sqrt{2})} \sqrt{36 \cosh^4 u \, \cos^2 v + 9 \cosh^4 u \, \sin^2 v + 4 \cosh^2 u \, \sinh^2 u} \, du \, dv$$

57. To find the region D: $z = x^2 + y^2$ implies $z + z^2 = 4z$ or $z^2 - 3z = 0$. Thus $z = 0$ or $z = 3$ are the planes where the surfaces intersect. But $x^2 + y^2 + z^2 = 4z$ implies $x^2 + y^2 + (z - 2)^2 = 4$, so $z = 3$ intersects the upper hemisphere.

Thus $(z - 2)^2 = 4 - x^2 - y^2$ or $z = 2 + \sqrt{4 - x^2 - y^2}$. Therefore D is the region inside the circle $x^2 + y^2 + (3 - 2)^2 = 4$,

that is, $D = \{(x, y) \mid x^2 + y^2 \le 3\}$.

$$A(S) = \iint_D \sqrt{1 + [(-x)(4 - x^2 - y^2)^{-1/2}]^2 + [(-y)(4 - x^2 - y^2)^{-1/2}]^2} \, dA$$

$$= \int_0^{2\pi} \int_0^{\sqrt{3}} \sqrt{1 + \frac{r^2}{4 - r^2}} \, r \, dr \, d\theta = \int_0^{2\pi} \int_0^{\sqrt{3}} \frac{2r \, dr}{\sqrt{4 - r^2}} \, d\theta = \int_0^{2\pi} \left[-2(4 - r^2)^{1/2} \right]_{r=0}^{r=\sqrt{3}} d\theta$$

$$= \int_0^{2\pi} (-2 + 4) \, d\theta = 2\theta \Big]_0^{2\pi} = 4\pi$$

58. We first find the area of the face of the surface that intersects the positive y-axis. A parametric representation of the surface is

$x = x$, $y = \sqrt{1 - z^2}$, $z = z$ with $x^2 + z^2 \le 1$. Then $\mathbf{r}(x, z) = \langle x, \sqrt{1 - z^2}, z \rangle$ $\Rightarrow$ $\mathbf{r}_x = \langle 1, 0, 0 \rangle$,

$\mathbf{r}_z = \langle 0, -z/\sqrt{1 - z^2}, 1 \rangle$ and $\mathbf{r}_x \times \mathbf{r}_z = \langle 0, -1, -z/\sqrt{1 - z^2} \rangle$ $\Rightarrow$ $|\mathbf{r}_x \times \mathbf{r}_z| = \sqrt{1 + \frac{z^2}{1 - z^2}} = \frac{1}{\sqrt{1 - z^2}}$.

$$A(S) = \iint_{x^2 + z^2 \le 1} |\mathbf{r}_x \times \mathbf{r}_z| \, dA = \int_{-1}^1 \int_{-\sqrt{1-z^2}}^{\sqrt{1-z^2}} \frac{1}{\sqrt{1 - z^2}} \, dx \, dz = 4 \int_0^1 \int_0^{\sqrt{1-z^2}} \frac{1}{\sqrt{1 - z^2}} \, dx \, dz \quad \begin{bmatrix} \text{by the symmetry} \\ \text{of the surface} \end{bmatrix}$$

This integral is improper [when $z = 1$], so

$$A(S) = \lim_{t \to 1^-} 4 \int_0^t \int_0^{\sqrt{1-z^2}} \frac{1}{\sqrt{1-z^2}} \, dx \, dz = \lim_{t \to 1^-} 4 \int_0^t \frac{\sqrt{1-z^2}}{\sqrt{1-z^2}} \, dz = \lim_{t \to 1^-} 4 \int_0^t dz = \lim_{t \to 1^-} 4t = 4$$

Since the complete surface consists of four congruent faces, the total surface area is $4(4) = 16$.

Alternate solution: The face of the surface that intersects the positive y-axis can also be parametrized as

$\mathbf{r}(x, \theta) = \langle x, \cos\theta, \sin\theta \rangle$ for $-\frac{\pi}{2} \le \theta \le \frac{\pi}{2}$ and $x^2 + z^2 \le 1 \iff x^2 + \sin^2\theta \le 1 \iff$

$-\sqrt{1-\sin^2\theta} \le x \le \sqrt{1-\sin^2\theta} \iff -\cos\theta \le x \le \cos\theta$. Then $\mathbf{r}_x = \langle 1, 0, 0 \rangle$, $\mathbf{r}_\theta = \langle 0, -\sin\theta, \cos\theta \rangle$ and

$\mathbf{r}_x \times \mathbf{r}_\theta = \langle 0, -\cos\theta, -\sin\theta \rangle \Rightarrow |\mathbf{r}_x \times \mathbf{r}_\theta| = 1$, so

$A(S) = \int_{-\pi/2}^{\pi/2} \int_{-\cos\theta}^{\cos\theta} 1 \, dx \, d\theta = \int_{-\pi/2}^{\pi/2} 2\cos\theta \, d\theta = 2\sin\theta \Big]_{-\pi/2}^{\pi/2} = 4$. Again, the area of the complete surface

is $4(4) = 16$.

59. Let $A(S_1)$ be the surface area of that portion of the surface which lies above the plane $z = 0$. Then $A(S) = 2A(S_1)$.

Following Example 10, a parametric representation of S_1 is $x = a\sin\phi\cos\theta$, $y = a\sin\phi\sin\theta$,

$z = a\cos\phi$ and $|\mathbf{r}_\phi \times \mathbf{r}_\theta| = a^2 \sin\phi$. For D, $0 \le \phi \le \frac{\pi}{2}$ and for each fixed ϕ, $\left(x - \frac{1}{2}a\right)^2 + y^2 \le \left(\frac{1}{2}a\right)^2$ or

$\left[a\sin\phi\cos\theta - \frac{1}{2}a\right]^2 + a^2\sin^2\phi\sin^2\theta \le (a/2)^2$ implies $a^2\sin^2\phi - a^2\sin\phi\cos\theta \le 0$ or

$\sin\phi(\sin\phi - \cos\theta) \le 0$. But $0 \le \phi \le \frac{\pi}{2}$, so $\cos\theta \ge \sin\phi$ or $\sin\left(\frac{\pi}{2} + \theta\right) \ge \sin\phi$ or $\phi - \frac{\pi}{2} \le \theta \le \frac{\pi}{2} - \phi$.

Hence $D = \left\{(\phi, \theta) \mid 0 \le \phi \le \frac{\pi}{2}, \phi - \frac{\pi}{2} \le \theta \le \frac{\pi}{2} - \phi\right\}$. Then

$$A(S_1) = \int_0^{\pi/2} \int_{\phi - (\pi/2)}^{(\pi/2) - \phi} a^2 \sin\phi \, d\theta \, d\phi = a^2 \int_0^{\pi/2} (\pi - 2\phi)\sin\phi \, d\phi$$

$$= a^2 \left[(-\pi\cos\phi) - 2(-\phi\cos\phi + \sin\phi)\right]_0^{\pi/2} = a^2(\pi - 2)$$

Thus $A(S) = 2a^2(\pi - 2)$.

Alternate solution: Working on S_1 we could parametrize the portion of the sphere by $x = x$, $y = y$, $z = \sqrt{a^2 - x^2 - y^2}$.

Then $|\mathbf{r}_x \times \mathbf{r}_y| = \sqrt{1 + \dfrac{x^2}{a^2 - x^2 - y^2} + \dfrac{y^2}{a^2 - x^2 - y^2}} = \dfrac{a}{\sqrt{a^2 - x^2 - y^2}}$ and

$$A(S_1) = \iint_{0 \le (x - (a/2))^2 + y^2 \le (a/2)^2} \frac{a}{\sqrt{a^2 - x^2 - y^2}} \, dA = \int_{-\pi/2}^{\pi/2} \int_0^{a\cos\theta} \frac{a}{\sqrt{a^2 - r^2}} r \, dr \, d\theta$$

$$= \int_{-\pi/2}^{\pi/2} -a(a^2 - r^2)^{1/2} \Big]_{r=0}^{r = a\cos\theta} d\theta = \int_{-\pi/2}^{\pi/2} a^2[1 - (1 - \cos^2\theta)^{1/2}] \, d\theta$$

$$= \int_{-\pi/2}^{\pi/2} a^2(1 - |\sin\theta|) \, d\theta = 2a^2 \int_0^{\pi/2} (1 - \sin\theta) \, d\theta = 2a^2\left(\frac{\pi}{2} - 1\right)$$

Thus $A(S) = 4a^2\left(\frac{\pi}{2} - 1\right) = 2a^2(\pi - 2)$.

Notes:

(1) Perhaps working in spherical coordinates is the most obvious approach here. However, you must be careful in setting up D.

(2) In the alternate solution, you can avoid having to use $|\sin\theta|$ by working in the first octant and then multiplying by 4. However, if you set up S_1 as above and arrived at $A(S_1) = a^2\pi$, you now see your error.

60. (a) Here $z = a \sin \alpha$, $y = |AB|$, and $x = |OA|$. But

$$|OB| = |OC| + |CB| = b + a \cos \alpha \text{ and } \sin \theta = \frac{|AB|}{|OB|} \text{ so that}$$

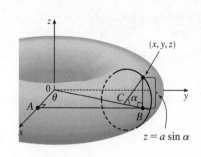

$$y = |OB| \sin \theta = (b + a \cos \alpha) \sin \theta. \text{ Similarly } \cos \theta = \frac{|OA|}{|OB|} \text{ so}$$

$x = (b + a \cos \alpha) \cos \theta$. Hence a parametric representation for the

torus is $x = b \cos \theta + a \cos \alpha \cos \theta$, $y = b \sin \theta + a \cos \alpha \sin \theta$,

$z = a \sin \alpha$, where $0 \le \alpha \le 2\pi$, $0 \le \theta \le 2\pi$.

(b)

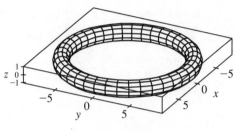

$a = 1, b = 8$

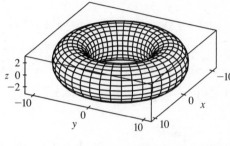

$a = 3, b = 8$

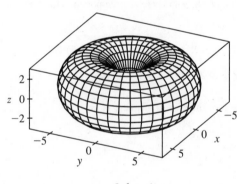

$a = 3, b = 4$

(c) $x = b \cos \theta + a \cos \alpha \cos \theta$, $y = b \sin \theta + a \cos \alpha \sin \theta$, $z = a \sin \alpha$, so $\mathbf{r}_\alpha = \langle -a \sin \alpha \cos \theta, -a \sin \alpha \sin \theta, a \cos \alpha \rangle$,

$\mathbf{r}_\theta = \langle -(b + a \cos \alpha) \sin \theta, (b + a \cos \alpha) \cos \theta, 0 \rangle$ and

$$\mathbf{r}_\alpha \times \mathbf{r}_\theta = \left(-ab \cos \alpha \cos \theta - a^2 \cos \alpha \cos^2 \theta \right) \mathbf{i} + \left(-ab \sin \alpha \cos \theta - a^2 \sin \alpha \cos^2 \theta \right) \mathbf{j}$$
$$+ \left(-ab \cos^2 \alpha \sin \theta - a^2 \cos^2 \alpha \sin \theta \cos \theta - ab \sin^2 \alpha \sin \theta - a^2 \sin^2 \alpha \sin \theta \cos \theta \right) \mathbf{k}$$
$$= -a (b + a \cos \alpha) \left[(\cos \theta \cos \alpha) \mathbf{i} + (\sin \theta \cos \alpha) \mathbf{j} + (\sin \alpha) \mathbf{k} \right]$$

Then $|\mathbf{r}_\alpha \times \mathbf{r}_\theta| = a(b + a \cos \alpha) \sqrt{\cos^2 \theta \cos^2 \alpha + \sin^2 \theta \cos^2 \alpha + \sin^2 \alpha} = a(b + a \cos \alpha)$.

Note: $b > a$, $-1 \le \cos \alpha \le 1$ so $|b + a \cos \alpha| = b + a \cos \alpha$. Hence

$A(S) = \int_0^{2\pi} \int_0^{2\pi} a(b + a \cos \alpha) \, d\alpha \, d\theta = 2\pi \left[ab\alpha + a^2 \sin \alpha \right]_0^{2\pi} = 4\pi^2 ab$.

17.7 Surface Integrals

1. The faces of the box in the planes $x = 0$ and $x = 2$ have surface area 24 and centers $(0, 2, 3)$, $(2, 2, 3)$. The faces in $y = 0$ and $y = 4$ have surface area 12 and centers $(1, 0, 3)$, $(1, 4, 3)$, and the faces in $z = 0$ and $z = 6$ have area 8 and centers $(1, 2, 0)$, $(1, 2, 6)$. For each face we take the point P_{ij}^* to be the center of the face and $f(x, y, z) = e^{-0.1(x+y+z)}$, so by Definition 1,

$$\iint_S f(x, y, z)\, dS \approx [f(0, 2, 3)](24) + [f(2, 2, 3)](24) + [f(1, 0, 3)](12)$$
$$+ [f(1, 4, 3)](12) + [f(1, 2, 0)](8) + [f(1, 2, 6)](8)$$
$$= 24(e^{-0.5} + e^{-0.7}) + 12(e^{-0.4} + e^{-0.8}) + 8(e^{-0.3} + e^{-0.9}) \approx 49.09$$

2. Each quarter-cylinder has surface area $\frac{1}{4}[2\pi(1)(2)] = \pi$, and the top and bottom disks have surface area $\pi(1)^2 = \pi$. We can take $(0, 0, 1)$ as a sample point in the top disk, $(0, 0, -1)$ in the bottom disk, and $(\pm 1, 0, 0)$, $(0, \pm 1, 0)$ in the four quarter-cylinders. Then $\iint_S f(x, y, z)\, dS$ can be approximated by the Riemann sum

$$f(1, 0, 0)(\pi) + f(-1, 0, 0)(\pi) + f(0, 1, 0)\,(\pi) + f(0, -1, 0)(\pi) + f(0, 0, 1)(\pi) + f(0, 0, -1)(\pi)$$
$$= (2 + 2 + 3 + 3 + 4 + 4)\pi = 18\pi \approx 56.5.$$

3. We can use the xz- and yz-planes to divide H into four patches of equal size, each with surface area equal to $\frac{1}{8}$ the surface area of a sphere with radius $\sqrt{50}$, so $\Delta S = \frac{1}{8}(4)\pi\left(\sqrt{50}\right)^2 = 25\pi$. Then $(\pm 3, \pm 4, 5)$ are sample points in the four patches, and using a Riemann sum as in Definition 1, we have

$$\iint_H f(x, y, z)\, dS \approx f(3, 4, 5)\,\Delta S + f(3, -4, 5)\,\Delta S + f(-3, 4, 5)\,\Delta S + f(-3, -4, 5)\,\Delta S$$
$$= (7 + 8 + 9 + 12)(25\pi) = 900\pi \approx 2827$$

4. On the surface, $f(x, y, z) = g\left(\sqrt{x^2 + y^2 + z^2}\right) = g(2) = -5$. So since the area of a sphere is $4\pi r^2$,

$$\iint_S f(x, y, z)\, dS = \iint_S g(2)\, dS = -5 \iint_S dS = -5[4\pi(2)^2] = -80\pi.$$

5. $z = 1 + 2x + 3y$ so $\dfrac{\partial z}{\partial x} = 2$ and $\dfrac{\partial z}{\partial y} = 3$. Then by Formula 4,

$$\iint_S x^2 yz\, dS = \iint_D x^2 yz \sqrt{\left(\frac{\partial z}{\partial x}\right)^2 + \left(\frac{\partial z}{\partial y}\right)^2 + 1}\, dA = \int_0^3 \int_0^2 x^2 y(1 + 2x + 3y)\sqrt{4 + 9 + 1}\, dy\, dx$$

$$= \sqrt{14} \int_0^3 \int_0^2 (x^2 y + 2x^3 y + 3x^2 y^2)\, dy\, dx = \sqrt{14} \int_0^3 \left[\frac{1}{2}x^2 y^2 + x^3 y^2 + x^2 y^3\right]_{y=0}^{y=2} dx$$

$$= \sqrt{14} \int_0^3 (10x^2 + 4x^3)\, dx = \sqrt{14}\left[\frac{10}{3}x^3 + x^4\right]_0^3 = 171\sqrt{14}$$

6. S is the region in the plane $2x + y + z = 2$ or $z = 2 - 2x - y$ over $D = \{(x, y) \mid 0 \le x \le 1, 0 \le y \le 2 - 2x\}$. Thus by Formula 4,

$$\iint_S xy\, dS = \iint_D xy \sqrt{(-2)^2 + (-1)^2 + 1}\, dA = \sqrt{6} \int_0^1 \int_0^{2-2x} xy\, dy\, dx = \sqrt{6} \int_0^1 \left[\frac{1}{2}xy^2\right]_{y=0}^{y=2-2x} dx$$

$$= \frac{\sqrt{6}}{2} \int_0^1 (4x - 8x^2 + 4x^3)\, dx = \frac{\sqrt{6}}{2}\left(2 - \frac{8}{3} + 1\right) = \frac{\sqrt{6}}{6}$$

7. S is the part of the plane $z = 1 - x - y$ over the region $D = \{(x, y) \mid 0 \le x \le 1, 0 \le y \le 1 - x\}$. Thus

$$\iint_S yz\, dS = \iint_D y(1 - x - y)\sqrt{(-1)^2 + (-1)^2 + 1}\, dA = \sqrt{3} \int_0^1 \int_0^{1-x} (y - xy - y^2)\, dy\, dx$$

$$= \sqrt{3} \int_0^1 \left[\frac{1}{2}y^2 - \frac{1}{2}xy^2 - \frac{1}{3}y^3\right]_{y=0}^{y=1-x} dx = \sqrt{3} \int_0^1 \frac{1}{6}(1 - x)^3\, dx = -\frac{\sqrt{3}}{24}(1 - x)^4\Big|_0^1 = \frac{\sqrt{3}}{24}$$

8. $z = \frac{2}{3}(x^{3/2} + y^{3/2})$ and

$$\iint_S y\, dS = \iint_D y\sqrt{(\sqrt{x})^2 + (\sqrt{y})^2 + 1}\, dA = \int_0^1 \int_0^1 y\sqrt{x + y + 1}\, dx\, dy$$

$$= \int_0^1 y\left[\frac{2}{3}(x + y + 1)^{3/2}\right]_{x=0}^{x=1} dy = \int_0^1 \frac{2}{3}y\left[(y+2)^{3/2} - (y+1)^{3/2}\right] dy$$

Substituting $u = y + 2$ in the first term and $t = y + 1$ in the second, we have

$$\iint_S y\, dS = \frac{2}{3}\int_2^3 (u - 2)u^{3/2}\, du - \frac{2}{3}\int_1^2 (t - 1)t^{3/2}\, dt = \frac{2}{3}\left[\frac{2}{7}u^{7/2} - \frac{4}{5}u^{5/2}\right]_2^3 - \frac{2}{3}\left[\frac{2}{7}t^{7/2} - \frac{2}{5}t^{5/2}\right]_1^2$$

$$= \frac{2}{3}\left[\frac{2}{7}(3^{7/2} - 2^{7/2}) - \frac{4}{5}(3^{5/2} - 2^{5/2}) - \frac{2}{7}(2^{7/2} - 1) + \frac{2}{5}(2^{5/2} - 1)\right]$$

$$= \frac{2}{3}\left(\frac{18}{35}\sqrt{3} + \frac{8}{35}\sqrt{2} - \frac{4}{35}\right) = \frac{4}{105}\left(9\sqrt{3} + 4\sqrt{2} - 2\right)$$

9. $\mathbf{r}(u, v) = u^2\,\mathbf{i} + u\sin v\,\mathbf{j} + u\cos v\,\mathbf{k},\ 0 \le u \le 1,\ 0 \le v \le \pi/2$, so

$\mathbf{r}_u \times \mathbf{r}_v = (2u\,\mathbf{i} + \sin v\,\mathbf{j} + \cos v\,\mathbf{k}) \times (u\cos v\,\mathbf{j} - u\sin v\,\mathbf{k}) = -u\,\mathbf{i} + 2u^2\sin v\,\mathbf{j} + 2u^2\cos v\,\mathbf{k}$ and

$|\mathbf{r}_u \times \mathbf{r}_v| = \sqrt{u^2 + 4u^4\sin^2 v + 4u^4\cos^2 v} = \sqrt{u^2 + 4u^4(\sin^2 v + \cos^2 v)} = u\sqrt{1 + 4u^2}$ (since $u \ge 0$). Then by Formula 2,

$$\iint_S yz\, dS = \iint_D (u\sin v)(u\cos v)\,|\mathbf{r}_u \times \mathbf{r}_v|\, dA = \int_0^{\pi/2}\int_0^1 (u\sin v)(u\cos v)\cdot u\sqrt{1 + 4u^2}\, du\, dv$$

$$= \int_0^1 u^3\sqrt{1 + 4u^2}\, du \int_0^{\pi/2}\sin v\cos v\, dv \qquad \left[\text{let } t = 1 + 4u^2 \ \Rightarrow\ u^2 = \frac{1}{4}(t - 1) \text{ and } \frac{1}{8}dt = u\, du\right]$$

$$= \int_1^5 \frac{1}{8}\cdot\frac{1}{4}(t - 1)\sqrt{t}\, dt \int_0^{\pi/2}\sin v\cos v\, dv = \frac{1}{32}\int_1^5 \left(t^{3/2} - \sqrt{t}\right) dt \int_0^{\pi/2}\sin v\cos v\, dv$$

$$= \frac{1}{32}\left[\frac{2}{5}t^{5/2} - \frac{2}{3}t^{3/2}\right]_1^5 \left[\frac{1}{2}\sin^2 v\right]_0^{\pi/2} = \frac{1}{32}\left(\frac{2}{5}(5)^{5/2} - \frac{2}{3}(5)^{3/2} - \frac{2}{5} + \frac{2}{3}\right)\cdot\frac{1}{2}(1 - 0) = \frac{5}{48}\sqrt{5} + \frac{1}{240}$$

10. $\mathbf{r}_u = \cos v\,\mathbf{i} + \sin v\,\mathbf{j},\ \mathbf{r}_v = -u\sin v\,\mathbf{i} + u\cos v\,\mathbf{j} + \mathbf{k}\ \Rightarrow\ \mathbf{r}_u \times \mathbf{r}_v = \sin v\,\mathbf{i} - \cos v\,\mathbf{j} + u\,\mathbf{k}\ \Rightarrow\ |\mathbf{r}_u \times \mathbf{r}_v| = \sqrt{1 + u^2}$,

so $\iint_S \sqrt{1 + x^2 + y^2}\, dS = \int_0^\pi\int_0^1 \sqrt{1 + u^2}\sqrt{1 + u^2}\, du\, dv = \frac{4}{3}\pi$.

11. S is the portion of the cone $z^2 = x^2 + y^2$ for $1 \le z \le 3$, or equivalently, S is the part of the surface $z = \sqrt{x^2 + y^2}$ over the region $D = \{(x, y) \mid 1 \le x^2 + y^2 \le 9\}$. Thus

$$\iint_S x^2 z^2\, dS = \iint_D x^2(x^2 + y^2)\sqrt{\left(\frac{x}{\sqrt{x^2 + y^2}}\right)^2 + \left(\frac{y}{\sqrt{x^2 + y^2}}\right)^2 + 1}\, dA$$

$$= \iint_D x^2(x^2 + y^2)\sqrt{\frac{x^2 + y^2}{x^2 + y^2} + 1}\, dA = \iint_D \sqrt{2}\,x^2(x^2 + y^2)\, dA = \sqrt{2}\int_0^{2\pi}\int_1^3 (r\cos\theta)^2(r^2)\,r\, dr\, d\theta$$

$$= \sqrt{2}\int_0^{2\pi}\cos^2\theta\, d\theta \int_1^3 r^5\, dr = \sqrt{2}\left[\frac{1}{2}\theta + \frac{1}{4}\sin 2\theta\right]_0^{2\pi}\left[\frac{1}{6}r^6\right]_1^3 = \sqrt{2}\,(\pi)\cdot\frac{1}{6}(3^6 - 1) = \frac{364\sqrt{2}}{3}\pi$$

12. Using y and z as parameters, we have $\mathbf{r}(y, z) = (y + 2z^2)\,\mathbf{i} + y\,\mathbf{j} + z\,\mathbf{k},\ 0 \le y \le 1,\ 0 \le z \le 1$.

Then $\mathbf{r}_y \times \mathbf{r}_z = (\mathbf{i} + \mathbf{j}) \times (4z\,\mathbf{i} + \mathbf{k}) = \mathbf{i} - \mathbf{j} - 4z\,\mathbf{k}$ and $|\mathbf{r}_y \times \mathbf{r}_z| = \sqrt{2 + 16z^2}$. Thus

$$\iint_S z\, dS = \int_0^1 \int_0^1 z\sqrt{2 + 16z^2}\, dy\, dz = \int_0^1 z\sqrt{2 + 16z^2}\, dz = \left[\frac{1}{32}\cdot\frac{2}{3}(2 + 16z^2)^{3/2}\right]_0^1 = \frac{1}{48}(18^{3/2} - 2^{3/2}) = \frac{13}{12}\sqrt{2}.$$

13. Using x and z as parameters, we have $\mathbf{r}(x, z) = x\,\mathbf{i} + (x^2 + z^2)\,\mathbf{j} + z\,\mathbf{k},\ x^2 + z^2 \le 4$. Then

$\mathbf{r}_x \times \mathbf{r}_z = (\mathbf{i} + 2x\,\mathbf{j}) \times (2z\,\mathbf{j} + \mathbf{k}) = 2x\,\mathbf{i} - \mathbf{j} + 2z\,\mathbf{k}$ and $|\mathbf{r}_x \times \mathbf{r}_z| = \sqrt{4x^2 + 1 + 4z^2} = \sqrt{1 + 4(x^2 + z^2)}$. Thus

$$\iint_S y \, dS = \iint_{x^2+z^2\le 4} (x^2+z^2)\sqrt{1+4(x^2+z^2)} \, dA = \int_0^{2\pi} \int_0^2 r^2\sqrt{1+4r^2} \, r \, dr \, d\theta = \int_0^{2\pi} d\theta \int_0^2 r^2\sqrt{1+4r^2} \, r \, dr$$

$$= 2\pi \int_0^2 r^2\sqrt{1+4r^2} \, r \, dr \qquad \left[\text{let } u = 1+4r^2 \;\Rightarrow\; r^2 = \tfrac{1}{4}(u-1) \text{ and } \tfrac{1}{8}du = r\,dr\right]$$

$$= 2\pi \int_1^{17} \tfrac{1}{4}(u-1)\sqrt{u}\cdot\tfrac{1}{8}du = \tfrac{1}{16}\pi\int_1^{17}(u^{3/2}-u^{1/2})\,du$$

$$= \tfrac{1}{16}\pi\left[\tfrac{2}{5}u^{5/2} - \tfrac{2}{3}u^{3/2}\right]_1^{17} = \tfrac{1}{16}\pi\left[\tfrac{2}{5}(17)^{5/2} - \tfrac{2}{3}(17)^{3/2} - \tfrac{2}{5} + \tfrac{2}{3}\right] = \frac{\pi}{60}\left(391\sqrt{17}+1\right)$$

14. The sphere intersects the cylinder in the circle $x^2+y^2=1$, $z=\sqrt{3}$, so S is the portion of the sphere where $z \ge \sqrt{3}$.

Using spherical coordinates to parametrize the sphere we have $\mathbf{r}(\phi,\theta) = 2\sin\phi\cos\theta\,\mathbf{i} + 2\sin\phi\sin\theta\,\mathbf{j} + 2\cos\phi\,\mathbf{k}$, and

$|\mathbf{r}_\phi \times \mathbf{r}_\theta| = 4\sin\phi$ (see Example 17.6.10 [ET 16.6.10]). The portion where $z \ge \sqrt{3}$ corresponds to $0 \le \phi \le \frac{\pi}{6}, 0 \le \theta \le 2\pi$

so

$$\iint_S y^2 \, dS = \int_0^{2\pi}\int_0^{\pi/6}(2\sin\phi\sin\theta)^2(4\sin\phi)\,d\phi\,d\theta = 16\int_0^{2\pi}\sin^2\theta\,d\theta\int_0^{\pi/6}\sin^3\phi\,d\phi$$

$$= 16\left[\tfrac{1}{2}\theta - \tfrac{1}{4}\sin 2\theta\right]_0^{2\pi}\left[\tfrac{1}{3}\cos^3\phi - \cos\phi\right]_0^{\pi/6} = 16(\pi)\left(\frac{\sqrt{3}}{8} - \frac{\sqrt{3}}{2} - \tfrac{1}{3} + 1\right) = \left(\tfrac{32}{3} - 6\sqrt{3}\right)\pi$$

15. Using spherical coordinates and Example 17.6.10 [ET 16.6.10] we have $\mathbf{r}(\phi,\theta) = 2\sin\phi\cos\theta\,\mathbf{i} + 2\sin\phi\sin\theta\,\mathbf{j} + 2\cos\phi\,\mathbf{k}$

and $|\mathbf{r}_\phi \times \mathbf{r}_\theta| = 4\sin\phi$. Then $\iint_S(x^2 z + y^2 z)\,dS = \int_0^{2\pi}\int_0^{\pi/2}(4\sin^2\phi)(2\cos\phi)(4\sin\phi)\,d\phi\,d\theta = 16\pi\sin^4\phi\big]_0^{\pi/2} = 16\pi.$

16. Here S consists of three surfaces: S_1, the lateral surface of the cylinder; S_2, the front formed by the plane $x+y=5$;

and the back, S_3, in the plane $x=0$.

On S_1: the surface is given by $\mathbf{r}(u,v) = u\,\mathbf{i} + 3\cos v\,\mathbf{j} + 3\sin v\,\mathbf{k}, 0 \le v \le 2\pi$, and $0 \le x \le 5-y \;\Rightarrow$

$0 \le u \le 5-3\cos v$. Then $\mathbf{r}_u \times \mathbf{r}_v = -3\cos v\,\mathbf{j} - 3\sin v\,\mathbf{k}$ and $|\mathbf{r}_u \times \mathbf{r}_v| = \sqrt{9\cos^2 v + 9\sin^2 v} = 3$, so

$$\iint_{S_1} xz \, dS = \int_0^{2\pi}\int_0^{5-3\cos v} u(3\sin v)(3)\,du\,dv = 9\int_0^{2\pi}\left[\tfrac{1}{2}u^2\right]_{u=0}^{u=5-3\cos v}\sin v\,dv$$

$$= \tfrac{9}{2}\int_0^{2\pi}(5-3\cos v)^2\sin v\,dv = \tfrac{9}{2}\left[\tfrac{1}{9}(5-3\cos v)^3\right]_0^{2\pi} = 0.$$

On S_2: $\mathbf{r}(y,z) = (5-y)\,\mathbf{i} + y\,\mathbf{j} + z\,\mathbf{k}$ and $|\mathbf{r}_y \times \mathbf{r}_z| = |\mathbf{i}+\mathbf{j}| = \sqrt{2}$, where $y^2+z^2 \le 9$ and

$$\iint_{S_2} xz \, dS = \iint_{y^2+z^2\le 9}(5-y)z\sqrt{2}\,dA = \sqrt{2}\int_0^{2\pi}\int_0^3(5-r\cos\theta)(r\sin\theta)\,r\,dr\,d\theta$$

$$= \sqrt{2}\int_0^{2\pi}\int_0^3(5r^2 - r^3\cos\theta)(\sin\theta)\,dr\,d\theta = \sqrt{2}\int_0^{2\pi}\left[\tfrac{5}{3}r^3 - \tfrac{1}{4}r^4\cos\theta\right]_{r=0}^{r=3}\sin\theta\,d\theta$$

$$= \sqrt{2}\int_0^{2\pi}\left(45 - \tfrac{81}{4}\cos\theta\right)\sin\theta\,d\theta = \sqrt{2}\left(\tfrac{4}{81}\right)\cdot\tfrac{1}{2}\left(45 - \tfrac{81}{4}\cos\theta\right)^2\Big]_0^{2\pi} = 0$$

On S_3: $x=0$ so $\iint_{S_3} xz \, dS = 0$. Hence $\iint_S xz \, dS = 0 + 0 + 0 = 0.$

17. S is given by $\mathbf{r}(u,v) = u\,\mathbf{i} + \cos v\,\mathbf{j} + \sin v\,\mathbf{k}, 0 \le u \le 3, 0 \le v \le \pi/2$. Then

$$\mathbf{r}_u \times \mathbf{r}_v = \mathbf{i} \times (-\sin v\,\mathbf{j} + \cos v\,\mathbf{k}) = -\cos v\,\mathbf{j} - \sin v\,\mathbf{k} \text{ and } |\mathbf{r}_u \times \mathbf{r}_v| = \sqrt{\cos^2 v + \sin^2 v} = 1, \text{ so}$$

$$\iint_S(z + x^2 y)\,dS = \int_0^{\pi/2}\int_0^3(\sin v + u^2\cos v)(1)\,du\,dv = \int_0^{\pi/2}(3\sin v + 9\cos v)\,dv$$

$$= [-3\cos v + 9\sin v]_0^{\pi/2} = 0 + 9 + 3 - 0 = 12$$

18. Let S_1 be the lateral surface, S_2 the top disk, and S_3 the bottom disk.

On S_1: $\mathbf{r}(\theta,z) = 3\cos\theta\,\mathbf{i} + 3\sin\theta\,\mathbf{j} + z\,\mathbf{k}, 0 \le \theta \le 2\pi, 0 \le z \le 2, |\mathbf{r}_\theta \times \mathbf{r}_z| = 3,$

$\iint_{S_1}(x^2+y^2+z^2)\,dS = \int_0^{2\pi}\int_0^2(9+z^2)\,3\,dz\,d\theta = 2\pi(54+8) = 124\pi.$

On S_2: $\mathbf{r}(\theta, r) = r\cos\theta\,\mathbf{i} + r\sin\theta\,\mathbf{j} + 2\,\mathbf{k}$, $0 \le r \le 3$, $0 \le \theta \le 2\pi$, $|\mathbf{r}_\theta \times \mathbf{r}_r| = r$,

$\iint_{S_2}(x^2 + y^2 + z^2)\,dS = \int_0^{2\pi}\int_0^3 (r^2 + 4)\,r\,dr\,d\theta = 2\pi\left(\frac{81}{4} + 18\right) = \frac{153}{2}\pi$.

On S_3: $\mathbf{r}(\theta, r) = r\cos\theta\,\mathbf{i} + r\sin\theta\,\mathbf{j}$, $0 \le r \le 3$, $0 \le \theta \le 2\pi$, $|\mathbf{r}_\theta \times \mathbf{r}_r| = r$,

$\iint_{S_3}(x^2 + y^2 + z^2)\,dS = \int_0^{2\pi}\int_0^3 (r^2 + 0)\,r\,dr\,d\theta = 2\pi\left(\frac{81}{4}\right) = \frac{81}{2}\pi$.

Hence $\iint_S (x^2 + y^2 + z^2)\,dS = 124\pi + \frac{153}{2}\pi + \frac{81}{2}\pi = 241\pi$.

19. $\mathbf{F}(x, y, z) = xy\,\mathbf{i} + yz\,\mathbf{j} + zx\,\mathbf{k}$, $z = g(x, y) = 4 - x^2 - y^2$, and D is the square $[0, 1] \times [0, 1]$, so by Equation 10

$$\iint_S \mathbf{F} \cdot d\mathbf{S} = \iint_D [-xy(-2x) - yz(-2y) + zx]\,dA = \int_0^1 \int_0^1 [2x^2y + 2y^2(4 - x^2 - y^2) + x(4 - x^2 - y^2)]\,dy\,dx$$

$$= \int_0^1 \left(\tfrac{1}{3}x^2 + \tfrac{11}{3}x - x^3 + \tfrac{34}{15}\right)dx = \tfrac{713}{180}$$

20. $\mathbf{r}_u = \cos v\,\mathbf{i} + \sin v\,\mathbf{j}$, $\mathbf{r}_v = -u\sin v\,\mathbf{i} + u\cos v\,\mathbf{j} + \mathbf{k}$ $\Rightarrow$ $\mathbf{r}_u \times \mathbf{r}_v = \sin v\,\mathbf{i} - \cos v\,\mathbf{j} + u\,\mathbf{k}$ and

$\mathbf{F}(\mathbf{r}(u, v)) = u\sin v\,\mathbf{i} + u\cos v\,\mathbf{j} + v^2\,\mathbf{k}$. Then by Formula 9,

$$\iint_S \mathbf{F} \cdot d\mathbf{S} = \iint_D \mathbf{F} \cdot (\mathbf{r}_u \times \mathbf{r}_v)\,dA = \int_0^\pi \int_0^1 (u\sin^2 v - u\cos^2 v + uv^2)\,du\,dv$$

$$= \int_0^\pi \int_0^1 (-u\cos 2v + uv^2)\,du\,dv = \int_0^\pi \left[-\tfrac{1}{2}\cos 2v + \tfrac{1}{2}v^2\right]dv = \tfrac{1}{6}\pi^3$$

21. $\mathbf{F}(x, y, z) = xze^y\,\mathbf{i} - xze^y\,\mathbf{j} + z\,\mathbf{k}$, $z = g(x, y) = 1 - x - y$, and $D = \{(x, y) \mid 0 \le x \le 1, 0 \le y \le 1 - x\}$. Since S has downward orientation, we have

$$\iint_S \mathbf{F} \cdot d\mathbf{S} = -\iint_D [-xze^y(-1) - (-xze^y)(-1) + z]\,dA = -\int_0^1 \int_0^{1-x} (1 - x - y)\,dy\,dx$$

$$= -\int_0^1 \left(\tfrac{1}{2}x^2 - x + \tfrac{1}{2}\right)dx = -\tfrac{1}{6}$$

22. $\mathbf{F}(x, y, z) = x\,\mathbf{i} + y\,\mathbf{j} + z^4\,\mathbf{k}$, $z = g(x, y) = \sqrt{x^2 + y^2}$, and D is the disk $\{(x, y) \mid x^2 + y^2 \le 1\}$. Since S has downward orientation, we have

$$\iint_S \mathbf{F} \cdot d\mathbf{S} = -\iint_D \left[-x\left(\frac{x}{\sqrt{x^2 + y^2}}\right) - y\left(\frac{y}{\sqrt{x^2 + y^2}}\right) + z^4\right]dA = -\iint_D \left[\frac{-x^2 - y^2}{\sqrt{x^2 + y^2}} + \left(\sqrt{x^2 + y^2}\right)^4\right]dA$$

$$= -\int_0^{2\pi}\int_0^1 \left(\frac{-r^2}{r} + r^4\right)r\,dr\,d\theta = -\int_0^{2\pi}d\theta \int_0^1 (r^5 - r^2)\,dr = -2\pi\left(\tfrac{1}{6} - \tfrac{1}{3}\right) = \frac{\pi}{3}$$

23. $\mathbf{F}(x, y, z) = x\,\mathbf{i} - z\,\mathbf{j} + y\,\mathbf{k}$, $z = g(x, y) = \sqrt{4 - x^2 - y^2}$ and D is the quarter disk $\{(x, y) \mid 0 \le x \le 2, 0 \le y \le \sqrt{4 - x^2}\}$. S has downward orientation, so by Formula 10,

$$\iint_S \mathbf{F} \cdot d\mathbf{S} = -\iint_D \left[-x \cdot \tfrac{1}{2}(4 - x^2 - y^2)^{-1/2}(-2x) - (-z) \cdot \tfrac{1}{2}(4 - x^2 - y^2)^{-1/2}(-2y) + y\right]dA$$

$$= -\iint_D \left(\frac{x^2}{\sqrt{4 - x^2 - y^2}} - \sqrt{4 - x^2 - y^2} \cdot \frac{y}{\sqrt{4 - x^2 - y^2}} + y\right)dA$$

$$= -\iint_D x^2(4 - (x^2 + y^2))^{-1/2}\,dA = -\int_0^{\pi/2}\int_0^2 (r\cos\theta)^2(4 - r^2)^{-1/2}\,r\,dr\,d\theta$$

$$= -\int_0^{\pi/2}\cos^2\theta\,d\theta \int_0^2 r^3(4 - r^2)^{-1/2}\,dr \qquad [\text{let } u = 4 - r^2 \;\Rightarrow\; r^2 = 4 - u \text{ and } -\tfrac{1}{2}\,du = r\,dr]$$

$$= -\int_0^{\pi/2}\left(\tfrac{1}{2} + \tfrac{1}{2}\cos 2\theta\right)d\theta \int_4^0 -\tfrac{1}{2}(4 - u)(u)^{-1/2}\,du$$

$$= -\left[\tfrac{1}{2}\theta + \tfrac{1}{4}\sin 2\theta\right]_0^{\pi/2} \left(-\tfrac{1}{2}\right)\left[8\sqrt{u} - \tfrac{2}{3}u^{3/2}\right]_4^0 = -\tfrac{\pi}{4}\left(-\tfrac{1}{2}\right)\left(-16 + \tfrac{16}{3}\right) = -\tfrac{4}{3}\pi$$

24. $\mathbf{F}(x, y, z) = xz\,\mathbf{i} + x\,\mathbf{j} + y\,\mathbf{k}$

Using spherical coordinates, S is given by $x = 5\sin\phi\cos\theta$, $y = 5\sin\phi\sin\theta$, $z = 5\cos\phi$, $0 \le \theta \le \pi$,

$0 \le \phi \le \pi$. $\mathbf{F}(\mathbf{r}(\phi, \theta)) = (5\sin\phi\cos\theta)(5\cos\phi)\,\mathbf{i} + (5\sin\phi\cos\theta)\,\mathbf{j} + (5\sin\phi\sin\theta)\,\mathbf{k}$ and

$\mathbf{r}_\phi \times \mathbf{r}_\theta = 25\sin^2\phi\cos\theta\,\mathbf{i} + 25\sin^2\phi\sin\theta\,\mathbf{j} + 25\cos\phi\sin\phi\,\mathbf{k}$, so

$$\mathbf{F}(\mathbf{r}(\phi, \theta)) \cdot (\mathbf{r}_\phi \times \mathbf{r}_\theta) = 625\sin^3\phi\cos\phi\cos^2\theta + 125\sin^3\phi\cos\theta\sin\theta + 125\sin^2\phi\cos\phi\sin\theta$$

Then

$$\iint_S \mathbf{F} \cdot d\mathbf{S} = \iint_D [\mathbf{F}(\mathbf{r}(\phi, \theta)) \cdot (\mathbf{r}_\phi \times \mathbf{r}_\theta)]\, dA$$

$$= \int_0^\pi \int_0^\pi (625\sin^3\phi\cos\phi\cos^2\theta + 125\sin^3\phi\cos\theta\sin\theta + 125\sin^2\phi\cos\phi\sin\theta)\, d\theta\, d\phi$$

$$= 125\int_0^\pi \left[5\sin^3\phi\cos\phi\left(\tfrac{1}{2}\theta + \tfrac{1}{4}\sin 2\theta\right) + \sin^3\phi\left(\tfrac{1}{2}\sin^2\theta\right) + \sin^2\phi\cos\phi(-\cos\theta)\right]_{\theta=0}^{\theta=\pi}\, d\phi$$

$$= 125\int_0^\pi \left(\tfrac{5}{2}\pi\sin^3\phi\cos\phi + 2\sin^2\phi\cos\phi\right)\, d\phi = 125\left[\tfrac{5}{2}\pi \cdot \tfrac{1}{4}\sin^4\phi + 2 \cdot \tfrac{1}{3}\sin^3\phi\right]_0^\pi = 0$$

25. Let S_1 be the paraboloid $y = x^2 + z^2$, $0 \le y \le 1$ and S_2 the disk $x^2 + z^2 \le 1$, $y = 1$. Since S is a closed

surface, we use the outward orientation.

On S_1: $\mathbf{F}(\mathbf{r}(x, z)) = (x^2 + z^2)\,\mathbf{j} - z\,\mathbf{k}$ and $\mathbf{r}_x \times \mathbf{r}_z = 2x\,\mathbf{i} - \mathbf{j} + 2z\,\mathbf{k}$ (since the $\mathbf{j}$-component must be negative on S_1). Then

$$\iint_{S_1} \mathbf{F} \cdot d\mathbf{S} = \iint_{x^2 + z^2 \le 1} [-(x^2 + z^2) - 2z^2]\, dA = -\int_0^{2\pi}\int_0^1 (r^2 + 2r^2\cos^2\theta)\, r\, dr\, d\theta$$

$$= -\int_0^{2\pi} \tfrac{1}{4}(1 + 2\cos^2\theta)\, d\theta = -\left(\tfrac{\pi}{2} + \tfrac{\pi}{2}\right) = -\pi$$

On S_2: $\mathbf{F}(\mathbf{r}(x, z)) = \mathbf{j} - z\,\mathbf{k}$ and $\mathbf{r}_z \times \mathbf{r}_x = \mathbf{j}$. Then $\iint_{S_2} \mathbf{F} \cdot d\mathbf{S} = \iint_{x^2 + z^2 \le 1} (1)\, dA = \pi$.

Hence $\iint_S \mathbf{F} \cdot d\mathbf{S} = -\pi + \pi = 0$.

26. $\mathbf{F}(x, y, z) = xy\,\mathbf{i} + 4x^2\,\mathbf{j} + yz\,\mathbf{k}$, $z = g(x, y) = xe^y$, and D is the square $[0, 1] \times [0, 1]$, so by Equation 10

$$\iint_S \mathbf{F} \cdot d\mathbf{S} = \iint_D [-xy(e^y) - 4x^2(xe^y) + yz]\, dA = \int_0^1\int_0^1 (-xye^y - 4x^3e^y + xye^y)\, dy\, dx$$

$$= \int_0^1 \left[-4x^3e^y\right]_{y=0}^{y=1}\, dx = (e - 1)\int_0^1 (-4x^3)\, dx = 1 - e$$

27. Here S consists of the six faces of the cube as labeled in the figure. On S_1:

$\mathbf{F} = \mathbf{i} + 2y\,\mathbf{j} + 3z\,\mathbf{k}$, $\mathbf{r}_y \times \mathbf{r}_z = \mathbf{i}$ and $\iint_{S_1} \mathbf{F} \cdot d\mathbf{S} = \int_{-1}^1\int_{-1}^1 dy\, dz = 4$;

S_2: $\mathbf{F} = x\,\mathbf{i} + 2\,\mathbf{j} + 3z\,\mathbf{k}$, $\mathbf{r}_z \times \mathbf{r}_x = \mathbf{j}$ and $\iint_{S_2} \mathbf{F} \cdot d\mathbf{S} = \int_{-1}^1\int_{-1}^1 2\, dx\, dz = 8$;

S_3: $\mathbf{F} = x\,\mathbf{i} + 2y\,\mathbf{j} + 3\,\mathbf{k}$, $\mathbf{r}_x \times \mathbf{r}_y = \mathbf{k}$ and $\iint_{S_3} \mathbf{F} \cdot d\mathbf{S} = \int_{-1}^1\int_{-1}^1 3\, dx\, dy = 12$;

S_4: $\mathbf{F} = -\mathbf{i} + 2y\,\mathbf{j} + 3z\,\mathbf{k}$, $\mathbf{r}_z \times \mathbf{r}_y = -\mathbf{i}$ and $\iint_{S_4} \mathbf{F} \cdot d\mathbf{S} = 4$;

S_5: $\mathbf{F} = x\,\mathbf{i} - 2\,\mathbf{j} + 3z\,\mathbf{k}$, $\mathbf{r}_x \times \mathbf{r}_z = -\mathbf{j}$ and $\iint_{S_5} \mathbf{F} \cdot d\mathbf{S} = 8$;

S_6: $\mathbf{F} = x\,\mathbf{i} + 2y\,\mathbf{j} - 3\,\mathbf{k}$, $\mathbf{r}_y \times \mathbf{r}_x = -\mathbf{k}$ and $\iint_{S_6} \mathbf{F} \cdot d\mathbf{S} = \int_{-1}^1\int_{-1}^1 3\, dx\, dy = 12$.

Hence $\iint_S \mathbf{F} \cdot d\mathbf{S} = \sum_{i=1}^6 \iint_{S_i} \mathbf{F} \cdot d\mathbf{S} = 48$.

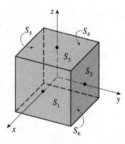

28. Here S consists of three surfaces: S_1, the lateral surface of the cylinder; S_2, the front formed by the plane $x + y = 2$; and the

back, S_3, in the plane $y = 0$.

On S_1: $\mathbf{F}(\mathbf{r}(\theta, y)) = \sin\theta\,\mathbf{i} + y\,\mathbf{j} + 5\,\mathbf{k}$ and $\mathbf{r}_\theta \times \mathbf{r}_y = \sin\theta\,\mathbf{i} + \cos\theta\,\mathbf{k}$ $\Rightarrow$

$$\iint_{S_1} \mathbf{F} \cdot d\mathbf{S} = \int_0^{2\pi} \int_0^{2-\sin\theta} (\sin^2\theta + 5\cos\theta)\, dy\, d\theta$$

$$= \int_0^{2\pi} (2\sin^2\theta + 10\cos\theta - \sin^3\theta - 5\sin\theta\,\cos\theta)\, d\theta = 2\pi$$

On S_2: $\mathbf{F}(\mathbf{r}(x,z)) = x\,\mathbf{i} + (2-x)\,\mathbf{j} + 5\,\mathbf{k}$ and $\mathbf{r}_z \times \mathbf{r}_x = \mathbf{i} + \mathbf{j}$.

$$\iint_{S_2} \mathbf{F} \cdot d\mathbf{S} = \iint_{x^2+z^2 \le 1} [x + (2-x)]\, dA = 2\pi$$

On S_3: $\mathbf{F}(\mathbf{r}(x,z)) = x\,\mathbf{i} + 5\,\mathbf{k}$ and $\mathbf{r}_x \times \mathbf{r}_z = -\mathbf{j}$ so $\iint_{S_3} \mathbf{F} \cdot d\mathbf{S} = 0$. Hence $\iint_S \mathbf{F} \cdot d\mathbf{S} = 4\pi$.

29. Here S consists of four surfaces: S_1, the top surface (a portion of the circular cylinder $y^2 + z^2 = 1$); S_2, the bottom surface (a portion of the xy-plane); S_3, the front half-disk in the plane $x = 2$, and S_4, the back half-disk in the plane $x = 0$.

On S_1: The surface is $z = \sqrt{1-y^2}$ for $0 \le x \le 2$, $-1 \le y \le 1$ with upward orientation, so

$$\iint_{S_1} \mathbf{F} \cdot d\mathbf{S} = \int_0^2 \int_{-1}^1 \left[-x^2\,(0) - y^2\left(-\frac{y}{\sqrt{1-y^2}}\right) + z^2 \right] dy\, dx = \int_0^2 \int_{-1}^1 \left(\frac{y^3}{\sqrt{1-y^2}} + 1 - y^2 \right) dy\, dx$$

$$= \int_0^2 \left[-\sqrt{1-y^2} + \tfrac{1}{3}(1-y^2)^{3/2} + y - \tfrac{1}{3}y^3 \right]_{y=-1}^{y=1} dx = \int_0^2 \tfrac{4}{3}\, dx = \tfrac{8}{3}$$

On S_2: The surface is $z = 0$ with downward orientation, so

$$\iint_{S_2} \mathbf{F} \cdot d\mathbf{S} = \int_0^2 \int_{-1}^1 (-z^2)\, dy\, dx = \int_0^2 \int_{-1}^1 (0)\, dy\, dx = 0$$

On S_3: The surface is $x = 2$ for $-1 \le y \le 1$, $0 \le z \le \sqrt{1-y^2}$, oriented in the positive x-direction. Regarding y and z as parameters, we have $\mathbf{r}_y \times \mathbf{r}_z = \mathbf{i}$ and

$$\iint_{S_3} \mathbf{F} \cdot d\mathbf{S} = \int_{-1}^1 \int_0^{\sqrt{1-y^2}} x^2\, dz\, dy = \int_{-1}^1 \int_0^{\sqrt{1-y^2}} 4\, dz\, dy = 4A(S_3) = 2\pi$$

On S_4: The surface is $x = 0$ for $-1 \le y \le 1$, $0 \le z \le \sqrt{1-y^2}$, oriented in the negative x-direction. Regarding y and z as parameters, we use $-(\mathbf{r}_y \times \mathbf{r}_z) = -\mathbf{i}$ and

$$\iint_{S_4} \mathbf{F} \cdot d\mathbf{S} = \int_{-1}^1 \int_0^{\sqrt{1-y^2}} x^2\, dz\, dy = \int_{-1}^1 \int_0^{\sqrt{1-y^2}} (0)\, dz\, dy = 0$$

Thus $\iint_S \mathbf{F} \cdot d\mathbf{S} = \tfrac{8}{3} + 0 + 2\pi + 0 = 2\pi + \tfrac{8}{3}$.

30. Here S consists of four surfaces: S_1, the triangular face with vertices $(1,0,0)$, $(0,1,0)$, and $(0,0,1)$; S_2, the face of the tetrahedron in the xy-plane; S_3, the face in the xz-plane; and S_4, the face in the yz-plane.

On S_1: The face is the portion of the plane $z = 1 - x - y$ for $0 \le x \le 1$, $0 \le y \le 1 - x$ with upward orientation, so

$$\iint_{S_1} \mathbf{F} \cdot d\mathbf{S} = \int_0^1 \int_0^{1-x} [-y\,(-1) - (z-y)\,(-1) + x]\, dy\, dx = \int_0^1 \int_0^{1-x} (z+x)\, dy\, dx = \int_0^1 \int_0^{1-x} (1-y)\, dy\, dx$$

$$= \int_0^1 \left[y - \tfrac{1}{2}y^2 \right]_{y=0}^{y=1-x} dx = \tfrac{1}{2} \int_0^1 (1-x^2)\, dx = \tfrac{1}{2} \left[x - \tfrac{1}{3}x^3 \right]_0^1 = \tfrac{1}{3}$$

On S_2: The surface is $z = 0$ with downward orientation, so

$$\iint_{S_2} \mathbf{F} \cdot d\mathbf{S} = \int_0^1 \int_0^{1-x} (-x)\, dy\, dx = -\int_0^1 x\,(1-x)\, dx = -\left[\tfrac{1}{2}x^2 - \tfrac{1}{3}x^3 \right]_0^1 = -\tfrac{1}{6}$$

On S_3: The surface is $y = 0$ for $0 \le x \le 1$, $0 \le z \le 1 - x$, oriented in the negative y-direction. Regarding x and z as parameters, we have $\mathbf{r}_x \times \mathbf{r}_z = -\mathbf{j}$ and

$$\iint_{S_3} \mathbf{F} \cdot d\mathbf{S} = \int_0^1 \int_0^{1-x} -(z-y)\, dz\, dx = -\int_0^1 \int_0^{1-x} z\, dz\, dx = -\int_0^1 \left[\tfrac{1}{2}z^2 \right]_{z=0}^{z=1-x} dx$$

$$= -\tfrac{1}{2} \int_0^1 (1-x)^2\, dx = \tfrac{1}{6} \left[(1-x)^3 \right]_0^1 = -\tfrac{1}{6}$$

On S_4: The surface is $x = 0$ for $0 \le y \le 1$, $0 \le z \le 1 - y$, oriented in the negative x-direction. Regarding y and z as parameters, we have $\mathbf{r}_y \times \mathbf{r}_z = \mathbf{i}$ so we use $-(\mathbf{r}_y \times \mathbf{r}_z) = -\mathbf{i}$ and

$$\iint_{S_4} \mathbf{F} \cdot d\mathbf{S} = \int_0^1 \int_0^{1-y} (-y)\, dz\, dy = -\int_0^1 y(1-y)\, dy = -\left[\tfrac{1}{2}y^2 - \tfrac{1}{3}y^3\right]_0^1 = -\tfrac{1}{6}$$

Thus $\iint_S \mathbf{F} \cdot d\mathbf{S} = \tfrac{1}{3} - \tfrac{1}{6} - \tfrac{1}{6} - \tfrac{1}{6} = -\tfrac{1}{6}$.

31. $z = xy \Rightarrow \partial z/\partial x = y$, $\partial z/\partial y = x$, so by Formula 4, a CAS gives

$\iint_S xyz\, dS = \int_0^1 \int_0^1 xy(xy)\sqrt{y^2 + x^2 + 1}\, dx\, dy \approx 0.1642$.

32. As in Exercise 31, we use a CAS to calculate

$$\iint_S x^2 yz\, dS = \int_0^1 \int_0^1 x^2 y(xy)\sqrt{y^2 + x^2 + 1}\, dx\, dy$$
$$= \tfrac{1}{60}\sqrt{3} - \tfrac{1}{12}\ln(1 + \sqrt{3}) - \tfrac{317}{192}\ln(\sqrt{2} + 1) + \tfrac{317}{2880}\sqrt{2} + \tfrac{1}{24}\ln 2$$

33. We use Formula 4 with $z = 3 - 2x^2 - y^2 \Rightarrow \partial z/\partial x = -4x$, $\partial z/\partial y = -2y$. The boundaries of the region

$3 - 2x^2 - y^2 \ge 0$ are $-\sqrt{\tfrac{3}{2}} \le x \le \sqrt{\tfrac{3}{2}}$ and $-\sqrt{3 - 2x^2} \le y \le \sqrt{3 - 2x^2}$, so we use a CAS (with precision reduced to

seven or fewer digits; otherwise the calculation may take a long time) to calculate

$$\iint_S x^2 y^2 z^2\, dS = \int_{-\sqrt{3/2}}^{\sqrt{3/2}} \int_{-\sqrt{3-2x^2}}^{\sqrt{3-2x^2}} x^2 y^2 (3 - 2x^2 - y^2)^2 \sqrt{16x^2 + 4y^2 + 1}\, dy\, dx \approx 3.4895$$

34. The flux of $\mathbf{F}$ across S is given by $\iint_S \mathbf{F} \cdot d\mathbf{S} = \iint_S \mathbf{F} \cdot \mathbf{n}\, dS$. Now on S, $z = g(x, y) = 2\sqrt{1 - y^2}$, so $\partial g/\partial x = 0$ and

$\partial g/\partial y = -2y(1 - y^2)^{-1/2}$. Therefore, by (10),

$$\iint_S \mathbf{F} \cdot d\mathbf{S} = \int_{-2}^2 \int_{-1}^1 \left(-x^2 y \left[-2y(1 - y^2)^{-1/2} \right] + \left[2\sqrt{1 - y^2} \right]^2 e^{x/5} \right) dy\, dx = \tfrac{1}{3}(16\pi + 80e^{2/5} - 80e^{-2/5})$$

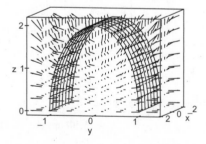 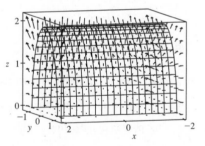

35. If S is given by $y = h(x, z)$, then S is also the level surface $f(x, y, z) = y - h(x, z) = 0$.

$\mathbf{n} = \dfrac{\nabla f(x, y, z)}{|\nabla f(x, y, z)|} = \dfrac{-h_x\, \mathbf{i} + \mathbf{j} - h_z\, \mathbf{k}}{\sqrt{h_x^2 + 1 + h_z^2}}$, and $-\mathbf{n}$ is the unit normal that points to the left. Now we proceed as in the

derivation of (10), using Formula 4 to evaluate

$$\iint_S \mathbf{F} \cdot d\mathbf{S} = \iint_S \mathbf{F} \cdot \mathbf{n}\, dS = \iint_D (P\mathbf{i} + Q\mathbf{j} + R\mathbf{k}) \frac{\dfrac{\partial h}{\partial x}\mathbf{i} - \mathbf{j} + \dfrac{\partial h}{\partial z}\mathbf{k}}{\sqrt{\left(\dfrac{\partial h}{\partial x}\right)^2 + 1 + \left(\dfrac{\partial h}{\partial z}\right)^2}} \sqrt{\left(\dfrac{\partial h}{\partial x}\right)^2 + 1 + \left(\dfrac{\partial h}{\partial z}\right)^2}\, dA$$

where D is the projection of S onto the xz-plane. Therefore $\displaystyle\iint_S \mathbf{F} \cdot d\mathbf{S} = \iint_D \left(P\frac{\partial h}{\partial x} - Q + R\frac{\partial h}{\partial z} \right) dA$.

36. If S is given by $x = k(y, z)$, then S is also the level surface $f(x, y, z) = x - k(y, z) = 0$.

$\mathbf{n} = \dfrac{\nabla f(x, y, z)}{|\nabla f(x, y, z)|} = \dfrac{\mathbf{i} - k_y\mathbf{j} - k_z\mathbf{k}}{\sqrt{1 + k_y^2 + k_z^2}}$, and since the x-component is positive this is the unit normal that points forward.

Now we proceed as in the derivation of (10), using Formula 4 for

$$\iint_S \mathbf{F} \cdot d\mathbf{S} = \iint_S \mathbf{F} \cdot \mathbf{n}\,dS = \iint_D (P\mathbf{i} + Q\mathbf{j} + R\mathbf{k}) \dfrac{\mathbf{i} - \dfrac{\partial k}{\partial y}\mathbf{j} - \dfrac{\partial k}{\partial z}\mathbf{k}}{\sqrt{1 + \left(\dfrac{\partial k}{\partial y}\right)^2 + \left(\dfrac{\partial k}{\partial z}\right)^2}} \sqrt{1 + \left(\dfrac{\partial k}{\partial y}\right)^2 + \left(\dfrac{\partial k}{\partial z}\right)^2}\,dA$$

where D is the projection of S onto the yz-plane. Therefore $\displaystyle\iint_S \mathbf{F} \cdot d\mathbf{S} = \iint_D \left(P - Q\dfrac{\partial k}{\partial y} - R\dfrac{\partial k}{\partial z}\right) dA$.

37. $m = \iint_S K\,dS = K \cdot 4\pi\left(\tfrac{1}{2}a^2\right) = 2\pi a^2 K$; by symmetry $M_{xz} = M_{yz} = 0$, and

$M_{xy} = \iint_S zK\,dS = K\int_0^{2\pi}\int_0^{\pi/2}(a\cos\phi)(a^2\sin\phi)\,d\phi\,d\theta = 2\pi Ka^3\left[-\tfrac{1}{4}\cos 2\phi\right]_0^{\pi/2} = \pi Ka^3$.

Hence $(\overline{x}, \overline{y}, \overline{z}) = \left(0, 0, \tfrac{1}{2}a\right)$.

38. S is given by $\mathbf{r}(x, y) = x\mathbf{i} + y\mathbf{j} + \sqrt{x^2 + y^2}\,\mathbf{k}$, $|\mathbf{r}_x \times \mathbf{r}_y| = \sqrt{1 + \dfrac{x^2 + y^2}{x^2 + y^2}} = \sqrt{2}$ so

$$m = \iint_S \left(10 - \sqrt{x^2 + y^2}\right) dS = \iint_{1 \le x^2 + y^2 \le 16} \left(10 - \sqrt{x^2 + y^2}\right)\sqrt{2}\,dA$$

$$= \int_0^{2\pi}\int_1^4 \sqrt{2}\,(10 - r)\,r\,dr\,d\theta = 2\pi\sqrt{2}\left[5r^2 - \tfrac{1}{3}r^3\right]_1^4 = 108\sqrt{2}\,\pi$$

39. (a) $I_z = \iint_S (x^2 + y^2)\rho(x, y, z)\,dS$

(b) $I_z = \iint_S (x^2 + y^2)\left(10 - \sqrt{x^2 + y^2}\right) dS = \iint_{1 \le x^2 + y^2 \le 16} (x^2 + y^2)\left(10 - \sqrt{x^2 + y^2}\right)\sqrt{2}\,dA$

$$= \int_0^{2\pi}\int_1^4 \sqrt{2}\,(10r^3 - r^4)\,dr\,d\theta = 2\sqrt{2}\,\pi\left(\tfrac{4329}{10}\right) = \tfrac{4329}{5}\sqrt{2}\,\pi$$

40. Using spherical coordinates to parametrize the sphere we have $\mathbf{r}(\phi, \theta) = 5\sin\phi\cos\theta\,\mathbf{i} + 5\sin\phi\sin\theta\,\mathbf{j} + 5\cos\phi\,\mathbf{k}$, and

$|\mathbf{r}_\phi \times \mathbf{r}_\theta| = 25\sin\phi$ (see Example 17.6.10 [ET 16.6.10]). S is the portion of the sphere where $z \ge 4$, so

$0 \le \phi \le \tan^{-1}(3/4)$ and $0 \le \theta \le 2\pi$.

(a) $m = \iint_S \rho(x, y, z)\,dS = \int_0^{2\pi}\int_0^{\tan^{-1}(3/4)} k(25\sin\phi)\,d\phi\,d\theta = 25k\int_0^{2\pi} d\theta \int_0^{\tan^{-1}(3/4)} \sin\phi\,d\phi$

$= 25k(2\pi)\left[-\cos\left(\tan^{-1}\tfrac{3}{4}\right) + 1\right] = 50\pi k\left(-\tfrac{4}{5} + 1\right) = 10\pi k.$

Because S has constant density, $\overline{x} = \overline{y} = 0$ by symmetry, and

$\overline{z} = \dfrac{1}{m}\iint_S z\rho(x, y, z)\,dS = \dfrac{1}{10\pi k}\int_0^{2\pi}\int_0^{\tan^{-1}(3/4)} k(5\cos\phi)(25\sin\phi)\,d\phi\,d\theta$

$= \dfrac{1}{10\pi k}\,(125k)\int_0^{2\pi} d\theta \int_0^{\tan^{-1}(3/4)} \sin\phi\cos\phi\,d\phi = \dfrac{1}{10\pi k}\,(125k)\,(2\pi)\left[\tfrac{1}{2}\sin^2\phi\right]_0^{\tan^{-1}(3/4)} = 25 \cdot \tfrac{1}{2}\left(\tfrac{3}{5}\right)^2 = \tfrac{9}{2},$

so the center of mass is $(\overline{x}, \overline{y}, \overline{z}) = \left(0, 0, \tfrac{9}{2}\right)$.

(b) $I_z = \iint_S (x^2 + y^2)\rho(x, y, z)\,dS = \int_0^{2\pi}\int_0^{\tan^{-1}(3/4)} k(25\sin^2\phi)(25\sin\phi)\,d\phi\,d\theta$

$= 625k\int_0^{2\pi} d\theta \int_0^{\tan^{-1}(3/4)} \sin^3\phi\,d\phi = 625k(2\pi)\left[\tfrac{1}{3}\cos^3\phi - \cos\phi\right]_0^{\tan^{-1}(3/4)}$

$= 1250\pi k\left[\tfrac{1}{3}\left(\tfrac{4}{5}\right)^3 - \tfrac{4}{5} - \tfrac{1}{3} + 1\right] = 1250\pi k\left(\tfrac{14}{375}\right) = \tfrac{140}{3}\pi k$

41. The rate of flow through the cylinder is the flux $\iint_S \rho\mathbf{v}\cdot\mathbf{n}\,dS = \iint_S \rho\mathbf{v}\cdot d\mathbf{S}$. We use the parametric representation

$\mathbf{r}(u,v) = 2\cos u\,\mathbf{i} + 2\sin u\,\mathbf{j} + v\,\mathbf{k}$ for S, where $0 \le u \le 2\pi$, $0 \le v \le 1$, so $\mathbf{r}_u = -2\sin u\,\mathbf{i} + 2\cos u\,\mathbf{j}$, $\mathbf{r}_v = \mathbf{k}$, and the

outward orientation is given by $\mathbf{r}_u \times \mathbf{r}_v = 2\cos u\,\mathbf{i} + 2\sin u\,\mathbf{j}$. Then

$$\iint_S \rho\mathbf{v}\cdot d\mathbf{S} = \rho\int_0^{2\pi}\int_0^1 \left(v\,\mathbf{i} + 4\sin^2 u\,\mathbf{j} + 4\cos^2 u\,\mathbf{k}\right)\cdot(2\cos u\,\mathbf{i} + 2\sin u\,\mathbf{j})\,dv\,du$$

$$= \rho\int_0^{2\pi}\int_0^1 \left(2v\cos u + 8\sin^3 u\right)dv\,du = \rho\int_0^{2\pi}\left(\cos u + 8\sin^3 u\right)du$$

$$= \rho\Big[\sin u + 8\big(-\tfrac{1}{3}\big)\big(2 + \sin^2 u\big)\cos u\Big]_0^{2\pi} = 0 \text{ kg/s}$$

42. A parametric representation for the hemisphere S is $\mathbf{r}(\phi,\theta) = 3\sin\phi\,\cos\theta\,\mathbf{i} + 3\sin\phi\,\sin\theta\,\mathbf{j} + 3\cos\phi\,\mathbf{k}$, $0 \le \phi \le \pi/2$,

$0 \le \theta \le 2\pi$. Then $\mathbf{r}_\phi = 3\cos\phi\,\cos\theta\,\mathbf{i} + 3\cos\phi\,\sin\theta\,\mathbf{j} - 3\sin\phi\,\mathbf{k}$, $\mathbf{r}_\theta = -3\sin\phi\,\sin\theta\,\mathbf{i} + 3\sin\phi\,\cos\theta\,\mathbf{j}$, and the outward

orientation is given by $\mathbf{r}_\phi \times \mathbf{r}_\theta = 9\sin^2\phi\,\cos\theta\,\mathbf{i} + 9\sin^2\phi\,\sin\theta\,\mathbf{j} + 9\sin\phi\,\cos\phi\,\mathbf{k}$. The rate of flow through S is

$$\iint_S \rho\mathbf{v}\cdot d\mathbf{S} = \rho\int_0^{\pi/2}\int_0^{2\pi}\left(3\sin\phi\,\sin\theta\,\mathbf{i} + 3\sin\phi\,\cos\theta\,\mathbf{j}\right)\cdot\left(9\sin^2\phi\,\cos\theta\,\mathbf{i} + 9\sin^2\phi\,\sin\theta\,\mathbf{j} + 9\sin\phi\,\cos\phi\,\mathbf{k}\right)d\theta\,d\phi$$

$$= 27\rho\int_0^{\pi/2}\int_0^{2\pi}\left(\sin^3\phi\,\sin\theta\,\cos\theta + \sin^3\phi\,\sin\theta\,\cos\theta\right)d\theta\,d\phi = 54\rho\int_0^{\pi/2}\sin^3\phi\,d\phi\int_0^{2\pi}\sin\theta\,\cos\theta\,d\theta$$

$$= 54\rho\Big[-\tfrac{1}{3}(2 + \sin^2\phi)\cos\phi\Big]_0^{\pi/2}\Big[\tfrac{1}{2}\sin^2\theta\Big]_0^{2\pi} = 0 \text{ kg/s}$$

43. S consists of the hemisphere S_1 given by $z = \sqrt{a^2 - x^2 - y^2}$ and the disk S_2 given by $0 \le x^2 + y^2 \le a^2$, $z = 0$.

On S_1: $\mathbf{E} = a\sin\phi\,\cos\theta\,\mathbf{i} + a\sin\phi\,\sin\theta\,\mathbf{j} + 2a\cos\phi\,\mathbf{k}$,

$\mathbf{T}_\phi \times \mathbf{T}_\theta = a^2\sin^2\phi\,\cos\theta\,\mathbf{i} + a^2\sin^2\phi\,\sin\theta\,\mathbf{j} + a^2\sin\phi\,\cos\phi\,\mathbf{k}$. Thus

$$\iint_{S_1} \mathbf{E}\cdot d\mathbf{S} = \int_0^{2\pi}\int_0^{\pi/2}(a^3\sin^3\phi + 2a^3\sin\phi\,\cos^2\phi)\,d\phi\,d\theta$$

$$= \int_0^{2\pi}\int_0^{\pi/2}(a^3\sin\phi + a^3\sin\phi\,\cos^2\phi)\,d\phi\,d\theta = (2\pi)a^3\big(1 + \tfrac{1}{3}\big) = \tfrac{8}{3}\pi a^3$$

On S_2: $\mathbf{E} = x\,\mathbf{i} + y\,\mathbf{j}$, and $\mathbf{r}_y \times \mathbf{r}_x = -\mathbf{k}$ so $\iint_{S_2} \mathbf{E}\cdot d\mathbf{S} = 0$. Hence the total charge is $q = \varepsilon_0\iint_S \mathbf{E}\cdot d\mathbf{S} = \tfrac{8}{3}\pi a^3\varepsilon_0$.

44. Referring to the figure, on

S_1: $\mathbf{E} = \mathbf{i} + y\,\mathbf{j} + z\,\mathbf{k}$, $\mathbf{r}_y \times \mathbf{r}_z = \mathbf{i}$ and $\iint_{S_1} \mathbf{E}\cdot d\mathbf{S} = \int_{-1}^1\int_{-1}^1 dy\,dz = 4$;

S_2: $\mathbf{E} = x\,\mathbf{i} + \mathbf{j} + z\,\mathbf{k}$, $\mathbf{r}_z \times \mathbf{r}_x = \mathbf{j}$ and $\iint_{S_2} \mathbf{E}\cdot d\mathbf{S} = \int_{-1}^1\int_{-1}^1 dx\,dz = 4$;

S_3: $\mathbf{E} = x\,\mathbf{i} + y\,\mathbf{j} + \mathbf{k}$, $\mathbf{r}_x \times \mathbf{r}_y = \mathbf{k}$ and $\iint_{S_3} \mathbf{E}\cdot d\mathbf{S} = \int_{-1}^1\int_{-1}^1 dx\,dy = 4$;

S_4: $\mathbf{E} = -\mathbf{i} + y\,\mathbf{j} + z\,\mathbf{k}$, $\mathbf{r}_z \times \mathbf{r}_y = -\mathbf{i}$ and $\iint_{S_4} \mathbf{E}\cdot d\mathbf{S} = 4$.

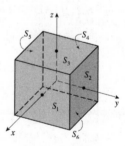

Similarly $\iint_{S_5} \mathbf{E}\cdot d\mathbf{S} = \iint_{S_6} \mathbf{E}\cdot d\mathbf{S} = 4$. Hence $q = \varepsilon_0\iint_S \mathbf{E}\cdot d\mathbf{S} = \varepsilon_0\sum_{i=1}^6 \iint_{S_i} \mathbf{E}\cdot d\mathbf{S} = 24\varepsilon_0$.

45. $K\nabla u = 6.5(4y\,\mathbf{j} + 4z\,\mathbf{k})$. S is given by $\mathbf{r}(x,\theta) = x\,\mathbf{i} + \sqrt{6}\,\cos\theta\,\mathbf{j} + \sqrt{6}\,\sin\theta\,\mathbf{k}$ and since we want the inward heat flow, we

use $\mathbf{r}_x \times \mathbf{r}_\theta = -\sqrt{6}\,\cos\theta\,\mathbf{j} - \sqrt{6}\,\sin\theta\,\mathbf{k}$. Then the rate of heat flow inward is given by

$$\iint_S (-K\,\nabla u)\cdot d\mathbf{S} = \int_0^{2\pi}\int_0^4 -(6.5)(-24)\,dx\,d\theta = (2\pi)(156)(4) = 1248\pi.$$

46. $u(x, y, z) = c/\sqrt{x^2 + y^2 + z^2}$,

$$\mathbf{F} = -K \nabla u = -K \left[-\frac{cx}{(x^2 + y^2 + z^2)^{3/2}} \mathbf{i} - \frac{cy}{(x^2 + y^2 + z^2)^{3/2}} \mathbf{j} - \frac{cz}{(x^2 + y^2 + z^2)^{3/2}} \mathbf{k} \right]$$

$$= \frac{cK}{(x^2 + y^2 + z^2)^{3/2}} (x\mathbf{i} + y\mathbf{j} + z\mathbf{k})$$

and the outward unit normal is $\mathbf{n} = \dfrac{1}{a} (x\mathbf{i} + y\mathbf{j} + z\mathbf{k})$.

Thus $\mathbf{F} \cdot \mathbf{n} = \dfrac{cK}{a(x^2 + y^2 + z^2)^{3/2}} (x^2 + y^2 + z^2)$, but on S, $x^2 + y^2 + z^2 = a^2$ so $\mathbf{F} \cdot \mathbf{n} = \dfrac{cK}{a^2}$. Hence the rate of heat flow

across S is $\displaystyle\iint_S \mathbf{F} \cdot d\mathbf{S} = \frac{cK}{a^2} \iint_S dS = \frac{cK}{a^2} (4\pi a^2) = 4\pi Kc$.

47. Let S be a sphere of radius a centered at the origin. Then $|\mathbf{r}| = a$ and $\mathbf{F(r)} = c\mathbf{r}/|\mathbf{r}|^3 = (c/a^3)(x\mathbf{i} + y\mathbf{j} + z\mathbf{k})$. A

parametric representation for S is $\mathbf{r}(\phi, \theta) = a\sin\phi\cos\theta\,\mathbf{i} + a\sin\phi\sin\theta\,\mathbf{j} + a\cos\phi\,\mathbf{k}$, $0 \le \phi \le \pi$, $0 \le \theta \le 2\pi$. Then

$\mathbf{r}_\phi = a\cos\phi\cos\theta\,\mathbf{i} + a\cos\phi\sin\theta\,\mathbf{j} - a\sin\phi\,\mathbf{k}$, $\mathbf{r}_\theta = -a\sin\phi\sin\theta\,\mathbf{i} + a\sin\phi\cos\theta\,\mathbf{j}$, and the outward orientation is given

by $\mathbf{r}_\phi \times \mathbf{r}_\theta = a^2\sin^2\phi\cos\theta\,\mathbf{i} + a^2\sin^2\phi\sin\theta\,\mathbf{j} + a^2\sin\phi\cos\phi\,\mathbf{k}$. The flux of $\mathbf{F}$ across S is

$$\iint_S \mathbf{F} \cdot d\mathbf{S} = \int_0^\pi \int_0^{2\pi} \frac{c}{a^3} (a\sin\phi\cos\theta\,\mathbf{i} + a\sin\phi\sin\theta\,\mathbf{j} + a\cos\phi\,\mathbf{k})$$

$$\cdot \left(a^2\sin^2\phi\cos\theta\,\mathbf{i} + a^2\sin^2\phi\sin\theta\,\mathbf{j} + a^2\sin\phi\cos\phi\,\mathbf{k} \right) d\theta\,d\phi$$

$$= \frac{c}{a^3} \int_0^\pi \int_0^{2\pi} a^3 \left(\sin^3\phi + \sin\phi\cos^2\phi \right) d\theta\,d\phi = c \int_0^\pi \int_0^{2\pi} \sin\phi\,d\theta\,d\phi = 4\pi c$$

Thus the flux does not depend on the radius a.

17.8 Stokes' Theorem

ET 16.8

1. Both H and P are oriented piecewise-smooth surfaces that are bounded by the simple, closed, smooth curve $x^2 + y^2 = 4$,

$z = 0$ (which we can take to be oriented positively for both surfaces). Then H and P satisfy the hypotheses of Stokes'

Theorem, so by (3) we know $\iint_H \text{curl } \mathbf{F} \cdot d\mathbf{S} = \int_C \mathbf{F} \cdot d\mathbf{r} = \iint_P \text{curl } \mathbf{F} \cdot d\mathbf{S}$ (where C is the boundary curve).

2. The boundary curve C is the circle $x^2 + y^2 = 9$, $z = 0$ oriented in the counterclockwise direction when viewed from above.

A vector equation of C is $\mathbf{r}(t) = 3\cos t\,\mathbf{i} + 3\sin t\,\mathbf{j}$, $0 \le t \le 2\pi$, so $\mathbf{r}'(t) = -3\sin t\,\mathbf{i} + 3\cos t\,\mathbf{j}$ and

$\mathbf{F}(\mathbf{r}(t)) = 2(3\sin t)(\cos 0)\,\mathbf{i} + e^{3\cos t}(\sin 0)\,\mathbf{j} + (3\cos t)e^{3\sin t}\,\mathbf{k} = 6\sin t\,\mathbf{i} + (3\cos t)e^{3\sin t}\,\mathbf{k}$. Then, by Stokes' Theorem,

$\displaystyle\iint_S \text{curl } \mathbf{F} \cdot d\mathbf{S} = \int_C \mathbf{F} \cdot d\mathbf{r} = \int_0^{2\pi} \mathbf{F}(\mathbf{r}(t)) \cdot \mathbf{r}'(t)\,dt = \int_0^{2\pi} (-18\sin^2 t + 0 + 0)\,dt = -18\left[\frac{1}{2}t - \frac{1}{4}\sin 2t \right]_0^{2\pi} = -18\pi$.

3. The paraboloid $z = x^2 + y^2$ intersects the cylinder $x^2 + y^2 = 4$ in the circle $x^2 + y^2 = 4$, $z = 4$. This boundary curve C

should be oriented in the counterclockwise direction when viewed from above, so a vector equation of C is

$\mathbf{r}(t) = 2\cos t\,\mathbf{i} + 2\sin t\,\mathbf{j} + 4\,\mathbf{k}$, $0 \le t \le 2\pi$. Then $\mathbf{r}'(t) = -2\sin t\,\mathbf{i} + 2\cos t\,\mathbf{j}$,

$$\mathbf{F}(\mathbf{r}(t)) = (4\cos^2 t)(16)\,\mathbf{i} + (4\sin^2 t)(16)\,\mathbf{j} + (2\cos t)(2\sin t)(4)\,\mathbf{k} = 64\cos^2 t\,\mathbf{i} + 64\sin^2 t\,\mathbf{j} + 16\sin t\,\cos t\,\mathbf{k},$$

and by Stokes' Theorem,

$$\iint_S \text{curl } \mathbf{F} \cdot d\mathbf{S} = \int_C \mathbf{F} \cdot d\mathbf{r} = \int_0^{2\pi} \mathbf{F}(\mathbf{r}(t)) \cdot \mathbf{r}'(t)\,dt = \int_0^{2\pi} (-128\cos^2 t\,\sin t + 128\sin^2 t\,\cos t + 0)\,dt$$

$$= 128\left[\frac{1}{3}\cos^3 t + \frac{1}{3}\sin^3 t \right]_0^{2\pi} = 0$$

4. The boundary curve C is the circle $x^2 + z^2 = 9$, $y = 3$ with vector equation $\mathbf{r}(t) = 3\sin t\,\mathbf{i} + 3\mathbf{j} + 3\cos t\,\mathbf{k}$, $0 \le t \le 2\pi$

which gives the positive orientation. Then $\mathbf{F}(\mathbf{r}(t)) = 729\sin^2 t\cos t\,\mathbf{i} + \sin(27\sin t\cos t)\,\mathbf{j} + 27\sin t\cos t\,\mathbf{k}$ and

$\mathbf{F}(\mathbf{r}(t)) \cdot \mathbf{r}'(t) = 2187\sin^2 t\cos^2 t - 81\sin^2 t\cos t$. Thus

$$\iint_S \operatorname{curl}\mathbf{F} \cdot d\mathbf{S} = \oint_C \mathbf{F} \cdot d\mathbf{r} = \int_0^{2\pi} \mathbf{F}(\mathbf{r}(t)) \cdot \mathbf{r}'(t)\,dt = \int_0^{2\pi} (2187\sin^2 t\cos^2 t - 81\sin^2 t\cos t)\,dt$$

$$= \int_0^{2\pi} \left(2187\left(\tfrac{1}{2}\sin 2t\right)^2 - 81\sin^2 t\cos t\right) dt = \left[\tfrac{2187}{4}\left(\tfrac{1}{2}t - \tfrac{1}{8}\sin 4t\right) - 81 \cdot \tfrac{1}{3}\sin^3 t\right]_0^{2\pi}$$

$$= \tfrac{2187}{4}(\pi) - 0 = \tfrac{2187}{4}\pi$$

5. C is the square in the plane $z = -1$. By (3), $\iint_{S_1} \operatorname{curl}\mathbf{F} \cdot d\mathbf{S} = \oint_C \mathbf{F} \cdot d\mathbf{r} = \iint_{S_2} \operatorname{curl}\mathbf{F} \cdot d\mathbf{S}$ where S_1 is the original cube

without the bottom and S_2 is the bottom face of the cube. $\operatorname{curl}\mathbf{F} = x^2 z\,\mathbf{i} + (xy - 2xyz)\,\mathbf{j} + (y - xz)\,\mathbf{k}$. For S_2, we choose

$\mathbf{n} = \mathbf{k}$ so that C has the same orientation for both surfaces. Then $\operatorname{curl}\mathbf{F} \cdot \mathbf{n} = y - xz = x + y$ on S_2, where $z = -1$. Thus

$\iint_{S_2} \operatorname{curl}\mathbf{F} \cdot d\mathbf{S} = \int_{-1}^{1}\int_{-1}^{1}(x + y)\,dx\,dy = 0$ so $\iint_{S_1} \operatorname{curl}\mathbf{F} \cdot d\mathbf{S} = 0$.

6. The boundary curve C is the unit circle in the yz-plane. By Equation 3, $\iint_{S_1} \operatorname{curl}\mathbf{F} \cdot d\mathbf{S} = \oint_C \mathbf{F} \cdot d\mathbf{r} = \iint_{S_2} \operatorname{curl}\mathbf{F} \cdot d\mathbf{S}$

where S_1 is the original hemisphere and S_2 is the disk $y^2 + z^2 \le 1$, $x = 0$.

$\operatorname{curl}\mathbf{F} = (x - x^2)\,\mathbf{i} - (y + e^{xy}\sin z)\,\mathbf{j} + (2xz - xe^{xy}\cos z)\,\mathbf{k}$, and for S_2 we choose $\mathbf{n} = \mathbf{i}$ so that C has the

same orientation for both surfaces. Then $\operatorname{curl}\mathbf{F} \cdot \mathbf{n} = x - x^2$ on S_2, where $x = 0$. Thus

$$\iint_{S_2} \operatorname{curl}\mathbf{F} \cdot d\mathbf{S} = \iint_{y^2+z^2\le 1} (x - x^2)\,dA = \iint_{y^2+z^2\le 1} 0\,dA = 0.$$

Alternatively, we can evaluate $\oint_C \mathbf{F} \cdot d\mathbf{r}$: C with positive orientation is given by $\mathbf{r}(t) = \langle 0, \cos t, \sin t \rangle$, $0 \le t \le 2\pi$, and

$$\iint_S \operatorname{curl}\mathbf{F} \cdot d\mathbf{S} = \oint_C \mathbf{F} \cdot d\mathbf{r} = \int_0^{2\pi} \left\langle e^{0(\cos t)}\cos(\sin t), (0)^2(\sin t), (0)(\cos t) \right\rangle \cdot \langle 0, -\sin t, \cos t \rangle\,dt = \int_0^{2\pi} 0\,dt = 0.$$

7. $\operatorname{curl}\mathbf{F} = -2z\,\mathbf{i} - 2x\,\mathbf{j} - 2y\,\mathbf{k}$ and we take the surface S to be the planar region enclosed by C, so S is the portion of the plane

$x + y + z = 1$ over $D = \{(x, y) \mid 0 \le x \le 1, 0 \le y \le 1 - x\}$. Since C is oriented counterclockwise, we orient S upward.

Using Equation 17.7.10 [ET 16.7.10], we have $z = g(x, y) = 1 - x - y$, $P = -2z$, $Q = -2x$, $R = -2y$, and

$$\int_C \mathbf{F} \cdot d\mathbf{r} = \iint_S \operatorname{curl}\mathbf{F} \cdot d\mathbf{S} = \iint_D [-(-2z)(-1) - (-2x)(-1) + (-2y)]\,dA$$
$$= \int_0^1 \int_0^{1-x}(-2)\,dy\,dx = -2\int_0^1(1 - x)\,dx = -1$$

8. $\operatorname{curl}\mathbf{F} = e^x\,\mathbf{k}$ and S is the portion of the plane $2x + y + 2z = 2$ over $D = \{(x, y) \mid 0 \le x \le 1, 0 \le y \le 2 - 2x\}$. We orient

S upward and use Equation 17.7.10 [ET 16.7.10] with $z = g(x, y) = 1 - x - \tfrac{1}{2}y$:

$$\int_C \mathbf{F} \cdot d\mathbf{r} = \iint_S \operatorname{curl}\mathbf{F} \cdot d\mathbf{S} = \iint_D(0 + 0 + e^x)\,dA = \int_0^1\int_0^{2-2x} e^x\,dy\,dx = \int_0^1(2 - 2x)e^x\,dx$$
$$= \left[(2 - 2x)e^x + 2e^x\right]_0^1 \quad \text{[integrating by parts]} \quad = 2e - 4$$

9. $\operatorname{curl}\mathbf{F} = (xe^{xy} - 2x)\,\mathbf{i} - (ye^{xy} - y)\,\mathbf{j} + (2z - z)\,\mathbf{k}$ and we take S to be the disk $x^2 + y^2 \le 16$, $z = 5$. Since C is oriented

counterclockwise (from above), we orient S upward. Then $\mathbf{n} = \mathbf{k}$ and $\operatorname{curl}\mathbf{F} \cdot \mathbf{n} = 2z - z$ on S, where $z = 5$. Thus

$$\oint \mathbf{F} \cdot d\mathbf{r} = \iint_S \operatorname{curl}\mathbf{F} \cdot \mathbf{n}\,dS = \iint_S(2z - z)\,dS = \iint_S(10 - 5)\,dS = 5(\text{area of }S) = 5(\pi \cdot 4^2) = 80\pi$$

10. The curve of intersection is an ellipse in the plane $z = 5 - x$. $\operatorname{curl}\mathbf{F} = \mathbf{i} - x\,\mathbf{k}$ and we take the surface S to be the planar region enclosed by C with upward orientation, so

$$\oint_C \mathbf{F} \cdot d\mathbf{r} = \iint_S \operatorname{curl}\mathbf{F} \cdot d\mathbf{S} = \iint_{x^2+y^2\leq 9} [-1(-1) - 0 + (-x)]\, dA = \int_0^{2\pi} \int_0^3 (1 - r\cos\theta)\, r\, dr\, d\theta$$

$$= \int_0^{2\pi} \int_0^3 (r - r^2\cos\theta)\, dr\, d\theta = \int_0^{2\pi} \left(\tfrac{9}{2} - 9\cos\theta\right) d\theta = \left[\tfrac{9}{2}\theta - 9\sin\theta\right]_0^{2\pi} = 9\pi$$

11. (a) The curve of intersection is an ellipse in the plane $x + y + z = 1$ with unit normal $\mathbf{n} = \tfrac{1}{\sqrt{3}}(\mathbf{i} + \mathbf{j} + \mathbf{k})$,

$\operatorname{curl}\mathbf{F} = x^2\,\mathbf{j} + y^2\,\mathbf{k}$, and $\operatorname{curl}\mathbf{F} \cdot \mathbf{n} = \tfrac{1}{\sqrt{3}}(x^2 + y^2)$. Then

$$\oint_C \mathbf{F} \cdot d\mathbf{r} = \iint_S \tfrac{1}{\sqrt{3}}(x^2 + y^2)\, dS = \iint_{x^2+y^2\leq 9} (x^2 + y^2)\, dx\, dy = \int_0^{2\pi} \int_0^3 r^3\, dr\, d\theta = 2\pi\left(\tfrac{81}{4}\right) = \tfrac{81\pi}{2}$$

(b)

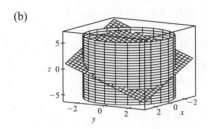

(c) One possible parametrization is $x = 3\cos t$, $y = 3\sin t$, $z = 1 - 3\cos t - 3\sin t$, $0 \leq t \leq 2\pi$.

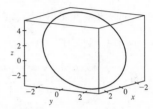

12. (a) S is the part of the surface $z = y^2 - x^2$ that lies above the unit disk D. $\operatorname{curl}\mathbf{F} = x\,\mathbf{i} - y\,\mathbf{j} + (x^2 - x^2)\,\mathbf{k} = x\,\mathbf{i} - y\,\mathbf{j}$.

Using Equation 17.7.10 [ET 16.7.10] with $g(x, y) = y^2 - x^2$, $P = x$, $Q = -y$, we have

$$\int_C \mathbf{F} \cdot d\mathbf{r} = \iint_S \operatorname{curl}\mathbf{F} \cdot d\mathbf{S} = \iint_D [-x(-2x) - (-y)(2y)]\, dA = 2\iint_D (x^2 + y^2)\, dA$$

$$= 2\int_0^{2\pi} \int_0^1 r^2 r\, dr\, d\theta = 2(2\pi)\left[\tfrac{1}{4}r^4\right]_0^1 = \pi$$

(b)

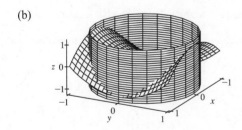

(c) One possible set of parametric equations is $x = \cos t$, $y = \sin t$, $z = \sin^2 t - \cos^2 t$, $0 \leq t \leq 2\pi$.

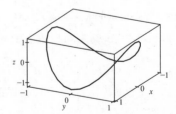

13. The boundary curve C is the circle $x^2 + y^2 = 1$, $z = 1$ oriented in the counterclockwise direction as viewed from above.

We can parametrize C by $\mathbf{r}(t) = \cos t\,\mathbf{i} + \sin t\,\mathbf{j} + \mathbf{k}$, $0 \leq t \leq 2\pi$, and then $\mathbf{r}'(t) = -\sin t\,\mathbf{i} + \cos t\,\mathbf{j}$. Thus

$\mathbf{F}(\mathbf{r}(t)) = \sin^2 t\,\mathbf{i} + \cos t\,\mathbf{j} + \mathbf{k}$, $\mathbf{F}(\mathbf{r}(t)) \cdot \mathbf{r}'(t) = \cos^2 t - \sin^3 t$, and

$$\int_C \mathbf{F} \cdot d\mathbf{r} = \int_0^{2\pi} (\cos^2 t - \sin^3 t)\, dt = \int_0^{2\pi} \tfrac{1}{2}(1 + \cos 2t)\, dt - \int_0^{2\pi} (1 - \cos^2 t)\sin t\, dt$$

$$= \tfrac{1}{2}\left[t + \tfrac{1}{2}\sin 2t\right]_0^{2\pi} - \left[-\cos t + \tfrac{1}{3}\cos^3 t\right]_0^{2\pi} = \pi$$

Now $\operatorname{curl}\mathbf{F} = (1 - 2y)\,\mathbf{k}$, and the projection D of S on the xy-plane is the disk $x^2 + y^2 \leq 1$, so by Equation 17.7.10

[ET 16.7.10] with $z = g(x, y) = x^2 + y^2$ we have

$$\iint_S \operatorname{curl}\mathbf{F} \cdot d\mathbf{S} = \iint_D (1 - 2y)\, dA = \int_0^{2\pi} \int_0^1 (1 - 2r\sin\theta)\, r\, dr\, d\theta = \int_0^{2\pi} \left(\tfrac{1}{2} - \tfrac{2}{3}\sin\theta\right) d\theta = \pi.$$

14. The plane intersects the coordinate axes at $x = 1$, $y = z = 2$ so the boundary curve C consists of the three line segments

C_1: $\mathbf{r}_1(t) = (1-t)\,\mathbf{i} + 2t\,\mathbf{j}$, $0 \le t \le 1$, C_2: $\mathbf{r}_2(t) = (2-2t)\,\mathbf{j} + 2t\,\mathbf{k}$, $0 \le t \le 1$, C_3: $\mathbf{r}_3(t) = t\,\mathbf{i} + (2-2t)\,\mathbf{k}$,

$0 \le t \le 1$. Then

$$\oint_C \mathbf{F} \cdot d\mathbf{r} = \int_0^1 [(1-t)\,\mathbf{i} + 2t\,\mathbf{j}] \cdot (-\mathbf{i} + 2\,\mathbf{j})\,dt + \int_0^1 [(2-2t)\,\mathbf{j}] \cdot (-2\,\mathbf{j} + 2\,\mathbf{k})\,dt + \int_0^1 (t\,\mathbf{i}) \cdot (\mathbf{i} - 2\,\mathbf{k})\,dt$$

$$= \int_0^1 (5t - 1)\,dt + \int_0^1 (4t - 4)\,dt + \int_0^1 t\,dt = \tfrac{3}{2} - 2 + \tfrac{1}{2} = 0$$

Now curl $\mathbf{F} = xz\,\mathbf{i} - yz\,\mathbf{j}$, so by Equation 17.7.10 [ET 16.7.10] with $z = g(x, y) = 2 - 2x - y$ we have

$$\iint_S \operatorname{curl}\mathbf{F} \cdot d\mathbf{S} = \iint_D [-x(2 - 2x - y)(-2) + y(2 - 2x - y)(-1)]\,dA = \int_0^1 \int_0^{2-2x} (4x - 4x^2 - 2y + y^2)\,dy\,dx$$

$$= \int_0^1 \left[4x(2 - 2x) - 4x^2(2 - 2x) - (2 - 2x)^2 + \tfrac{1}{3}(2 - 2x)^3\right] dx$$

$$= \int_0^1 \left(\tfrac{16}{3}x^3 - 12x^2 + 8x - \tfrac{4}{3}\right) dx = \left[\tfrac{4}{3}x^4 - 4x^3 + 4x^2 - \tfrac{4}{3}x\right]_0^1 = 0$$

15. The boundary curve C is the circle $x^2 + z^2 = 1$, $y = 0$ oriented in the counterclockwise direction as viewed from the positive y-axis. Then C can be described by $\mathbf{r}(t) = \cos t\,\mathbf{i} - \sin t\,\mathbf{k}$, $0 \le t \le 2\pi$, and $\mathbf{r}'(t) = -\sin t\,\mathbf{i} - \cos t\,\mathbf{k}$. Thus

$\mathbf{F}(\mathbf{r}(t)) = -\sin t\,\mathbf{j} + \cos t\,\mathbf{k}$, $\mathbf{F}(\mathbf{r}(t)) \cdot \mathbf{r}'(t) = -\cos^2 t$, and $\oint_C \mathbf{F} \cdot d\mathbf{r} = \int_0^{2\pi} -\cos^2 t\,dt = -\tfrac{1}{2}t - \tfrac{1}{4}\sin 2t\big]_0^{2\pi} = -\pi$.

Now curl $\mathbf{F} = -\mathbf{i} - \mathbf{j} - \mathbf{k}$, and S can be parametrized (see Example 17.6.10 [ET 16.6.10]) by

$\mathbf{r}(\phi, \theta) = \sin\phi\,\cos\theta\,\mathbf{i} + \sin\phi\,\sin\theta\,\mathbf{j} + \cos\phi\,\mathbf{k}$, $0 \le \theta \le \pi$, $0 \le \phi \le \pi$. Then

$\mathbf{r}_\phi \times \mathbf{r}_\theta = \sin^2\phi\,\cos\theta\,\mathbf{i} + \sin^2\phi\,\sin\theta\,\mathbf{j} + \sin\phi\,\cos\phi\,\mathbf{k}$ and

$$\iint_S \operatorname{curl}\mathbf{F} \cdot d\mathbf{S} = \iint_{x^2+z^2 \le 1} \operatorname{curl}\mathbf{F} \cdot (\mathbf{r}_\phi \times \mathbf{r}_\theta)\,dA = \int_0^\pi \int_0^\pi (-\sin^2\phi\,\cos\theta - \sin^2\phi\,\sin\theta - \sin\phi\,\cos\phi)\,d\theta\,d\phi$$

$$= \int_0^\pi (-2\sin^2\phi - \pi\sin\phi\,\cos\phi)\,d\phi = \left[\tfrac{1}{2}\sin 2\phi - \phi - \tfrac{\pi}{2}\sin^2\phi\right]_0^\pi = -\pi$$

16. Let S be the surface in the plane $x + y + z = 1$ with upward orientation enclosed by C. Then an upward unit normal vector for S is $\mathbf{n} = \tfrac{1}{\sqrt{3}}(\mathbf{i} + \mathbf{j} + \mathbf{k})$. Orient C in the counterclockwise direction, as viewed from above. $\int_C z\,dx - 2x\,dy + 3y\,dz$ is equivalent to $\int_C \mathbf{F} \cdot d\mathbf{r}$ for $\mathbf{F}(x, y, z) = z\,\mathbf{i} - 2x\,\mathbf{j} + 3y\,\mathbf{k}$, and the components of $\mathbf{F}$ are polynomials, which have continuous partial derivatives throughout $\mathbb{R}^3$. We have curl $\mathbf{F} = 3\,\mathbf{i} + \mathbf{j} - 2\,\mathbf{k}$, so by Stokes' Theorem,

$$\int_C z\,dx - 2x\,dy + 3y\,dz = \int_C \mathbf{F} \cdot d\mathbf{r} = \iint_S \operatorname{curl}\mathbf{F} \cdot \mathbf{n}\,dS = \iint_S (3\,\mathbf{i} + \mathbf{j} - 2\,\mathbf{k}) \cdot \tfrac{1}{\sqrt{3}}(\mathbf{i} + \mathbf{j} + \mathbf{k})\,dS$$

$$= \tfrac{2}{\sqrt{3}} \iint_S dS = \tfrac{2}{\sqrt{3}} \text{ (surface area of } S\text{)}$$

Thus the value of $\int_C z\,dx - 2x\,dy + 3y\,dz$ is always $\tfrac{2}{\sqrt{3}}$ times the area of the region enclosed by C, regardless of its shape or location. [Notice that because $\mathbf{n}$ is normal to a plane, it is constant. But curl $\mathbf{F}$ is also constant, so the dot product curl $\mathbf{F} \cdot \mathbf{n}$ is constant and we could have simply argued that $\iint_S \operatorname{curl}\mathbf{F} \cdot \mathbf{n}\,dS$ is a constant multple of $\iint_S dS$, the surface area of S.]

17. It is easier to use Stokes' Theorem than to compute the work directly. Let S be the planar region enclosed by the path of the particle, so S is the portion of the plane $z = \tfrac{1}{2}y$ for $0 \le x \le 1$, $0 \le y \le 2$, with upward orientation.

curl $\mathbf{F} = 8y\,\mathbf{i} + 2z\,\mathbf{j} + 2y\,\mathbf{k}$ and

$$\oint_C \mathbf{F} \cdot d\mathbf{r} = \iint_S \operatorname{curl}\mathbf{F} \cdot d\mathbf{S} = \iint_D \left[-8y\,(0) - 2z\left(\tfrac{1}{2}\right) + 2y\right] dA = \int_0^1 \int_0^2 \left(2y - \tfrac{1}{2}y\right) dy\,dx$$

$$= \int_0^1 \int_0^2 \tfrac{3}{2}y\,dy\,dx = \int_0^1 \left[\tfrac{3}{4}y^2\right]_{y=0}^{y=2} dx = \int_0^1 3\,dx = 3$$

18. $\int_C (y + \sin x)\,dx + (z^2 + \cos y)\,dy + x^3\,dz = \int_C \mathbf{F} \cdot d\mathbf{r}$, where $\mathbf{F}(x, y, z) = (y + \sin x)\,\mathbf{i} + (z^2 + \cos y)\,\mathbf{j} + x^3\,\mathbf{k}$ $\Rightarrow$
curl $\mathbf{F} = -2z\,\mathbf{i} - 3x^2\,\mathbf{j} - \mathbf{k}$. Since $\sin 2t = 2\sin t\,\cos t$, C lies on the surface $z = 2xy$. Let S be the part of this surface that
is bounded by C. Then the projection of S onto the xy-plane is the unit disk D $[x^2 + y^2 \le 1]$. C is traversed clockwise
(when viewed from above) so S is oriented downward. Using Equation 17.7.10 [ET 16.7.10] with $g(x, y) = 2xy$,
$P = -2z = -2(2xy) = -4xy$, $Q = -3x^2$, $R = -1$ and multiplying by -1 for the downward orientation, we have

$$\int_C \mathbf{F} \cdot d\mathbf{r} = -\iint_S \text{curl}\,\mathbf{F} \cdot d\mathbf{S} = -\iint_D \left[-(-4xy)(2y) - (-3x^2)(2x) - 1\right] dA$$

$$= -\iint_D (8xy^2 + 6x^3 - 1)\,dA = -\int_0^{2\pi} \int_0^1 (8r^3 \cos\theta \sin^2\theta + 6r^3 \cos^3\theta - 1)\,r\,dr\,d\theta$$

$$= -\int_0^{2\pi} \left(\tfrac{8}{5} \cos\theta \sin^2\theta + \tfrac{6}{5}\cos^3\theta - \tfrac{1}{2}\right) d\theta = -\left[\tfrac{8}{15}\sin^3\theta + \tfrac{6}{5}\left(\sin\theta - \tfrac{1}{3}\sin^3\theta\right) - \tfrac{1}{2}\theta\right]_0^{2\pi} = \pi$$

19. Assume S is centered at the origin with radius a and let H_1 and H_2 be the upper and lower hemispheres, respectively, of S.
Then $\iint_S \text{curl}\,\mathbf{F} \cdot d\mathbf{S} = \iint_{H_1} \text{curl}\,\mathbf{F} \cdot d\mathbf{S} + \iint_{H_2} \text{curl}\,\mathbf{F} \cdot d\mathbf{S} = \oint_{C_1} \mathbf{F} \cdot d\mathbf{r} + \oint_{C_2} \mathbf{F} \cdot d\mathbf{r}$ by Stokes' Theorem. But C_1 is the
circle $x^2 + y^2 = a^2$ oriented in the counterclockwise direction while C_2 is the same circle oriented in the clockwise direction.
Hence $\oint_{C_2} \mathbf{F} \cdot d\mathbf{r} = -\oint_{C_1} \mathbf{F} \cdot d\mathbf{r}$ so $\iint_S \text{curl}\,\mathbf{F} \cdot d\mathbf{S} = 0$ as desired.

20. (a) By Exercise 17.5.26 [ET 16.5.26], $\text{curl}(f\nabla g) = f\,\text{curl}(\nabla g) + \nabla f \times \nabla g = \nabla f \times \nabla g$ since $\text{curl}(\nabla g) = \mathbf{0}$. Hence by
Stokes' Theorem $\int_C (f\nabla g) \cdot d\mathbf{r} = \iint_S (\nabla f \times \nabla g) \cdot d\mathbf{S}$.

(b) As in (a), $\text{curl}(f\nabla f) = \nabla f \times \nabla f = \mathbf{0}$, so by Stokes' Theorem, $\int_C (f\nabla f) \cdot d\mathbf{r} = \iint_S [\text{curl}(f\nabla f)] \cdot d\mathbf{S} = 0$.

(c) As in part (a),

$$\text{curl}(f\nabla g + g\nabla f) = \text{curl}(f\nabla g) + \text{curl}(g\nabla f) \quad \text{[by Exercise 17.5.24 [ET 16.5.24]]}$$

$$= (\nabla f \times \nabla g) + (\nabla g \times \nabla f) = \mathbf{0} \quad \text{[since } \mathbf{u} \times \mathbf{v} = -(\mathbf{v} \times \mathbf{u})\text{]}$$

Hence by Stokes' Theorem, $\int_C (f\nabla g + g\nabla f) \cdot d\mathbf{r} = \iint_S \text{curl}(f\nabla g + g\nabla f) \cdot d\mathbf{S} = 0$.

17.9 The Divergence Theorem   ET 16.9

1. div $\mathbf{F} = 3 + x + 2x = 3 + 3x$, so

$\iiint_E \text{div}\,\mathbf{F}\,dV = \int_0^1 \int_0^1 \int_0^1 (3x + 3)\,dx\,dy\,dz = \tfrac{9}{2}$ (notice the triple integral is

three times the volume of the cube plus three times $\bar{x}$).

To compute $\iint_S \mathbf{F} \cdot d\mathbf{S}$, on

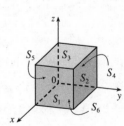

S_1: $\mathbf{n} = \mathbf{i}$, $\mathbf{F} = 3\,\mathbf{i} + y\,\mathbf{j} + 2z\,\mathbf{k}$, and $\iint_{S_1} \mathbf{F} \cdot d\mathbf{S} = \iint_{S_1} 3\,dS = 3$;

S_2: $\mathbf{F} = 3x\,\mathbf{i} + x\,\mathbf{j} + 2xz\,\mathbf{k}$, $\mathbf{n} = \mathbf{j}$ and $\iint_{S_2} \mathbf{F} \cdot d\mathbf{S} = \iint_{S_2} x\,dS = \tfrac{1}{2}$;

S_3: $\mathbf{F} = 3x\,\mathbf{i} + xy\,\mathbf{j} + 2x\,\mathbf{k}$, $\mathbf{n} = \mathbf{k}$ and $\iint_{S_3} \mathbf{F} \cdot d\mathbf{S} = \iint_{S_3} 2x\,dS = 1$;

S_4: $\mathbf{F} = 0$, $\iint_{S_4} \mathbf{F} \cdot d\mathbf{S} = 0$; S_5: $\mathbf{F} = 3x\,\mathbf{i} + 2x\,\mathbf{k}$, $\mathbf{n} = -\mathbf{j}$ and $\iint_{S_5} \mathbf{F} \cdot d\mathbf{S} = \iint_{S_5} 0\,dS = 0$;

S_6: $\mathbf{F} = 3x\,\mathbf{i} + xy\,\mathbf{j}$, $\mathbf{n} = -\mathbf{k}$ and $\iint_{S_6} \mathbf{F} \cdot d\mathbf{S} = \iint_{S_6} 0\,dS = 0$. Thus $\iint_S \mathbf{F} \cdot d\mathbf{S} = \tfrac{9}{2}$.

2. div $\mathbf{F} = 2x + x + 1 = 3x + 1$ so

$$\iiint_E \text{div } \mathbf{F} \, dV = \iiint_E (3x + 1) \, dV = \int_0^{2\pi} \int_0^2 \int_0^{4-r^2} (3r \cos\theta + 1) \, r \, dz \, dr \, d\theta$$

$$= \int_0^2 \int_0^{2\pi} r(3r \cos\theta + 1)(4 - r^2) \, d\theta \, dr$$

$$= \int_0^{2\pi} r(4 - r^2) \left[3r \sin\theta + \theta \right]_{\theta=0}^{\theta=2\pi} \, dr$$

$$= 2\pi \int_0^2 (4r - r^3) \, dr = 2\pi \left[2r^2 - \tfrac{1}{4}r^4 \right]_0^2$$

$$= 2\pi(8 - 4) = 8\pi$$

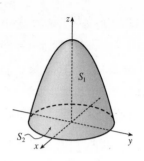

On S_1: The surface is $z = 4 - x^2 - y^2$, $x^2 + y^2 \le 4$, with upward orientation, and $\mathbf{F} = x^2 \, \mathbf{i} + xy \, \mathbf{j} + (4 - x^2 - y^2) \, \mathbf{k}$. Then

$$\iint_{S_1} \mathbf{F} \cdot d\mathbf{S} = \iint_D [-(x^2)(-2x) - (xy)(-2y) + (4 - x^2 - y^2)] \, dA$$

$$= \iint_D \left[2x(x^2 + y^2) + 4 - (x^2 + y^2) \right] \, dA = \int_0^{2\pi} \int_0^2 (2r \cos\theta \cdot r^2 + 4 - r^2) \, r \, dr \, d\theta$$

$$= \int_0^{2\pi} \left[\tfrac{2}{5} r^5 \cos\theta + 2r^2 - \tfrac{1}{4} r^4 \right]_{r=0}^{r=2} \, d\theta = \int_0^{2\pi} \left(\tfrac{64}{5} \cos\theta + 4 \right) \, d\theta = \left[\tfrac{64}{5} \sin\theta + 4\theta \right]_0^{2\pi} = 8\pi$$

On S_2: The surface is $z = 0$ with downward orientation, so $\mathbf{F} = x^2 \, \mathbf{i} + xy \, \mathbf{j}$, $\mathbf{n} = -\mathbf{k}$ and $\iint_{S_2} \mathbf{F} \cdot \mathbf{n} \, dS = \iint_{S_2} 0 \, dS = 0$.

Thus $\iint_S \mathbf{F} \cdot d\mathbf{S} = \iint_{S_1} \mathbf{F} \cdot d\mathbf{S} + \iint_{S_2} \mathbf{F} \cdot d\mathbf{S} = 8\pi$.

3. div $\mathbf{F} = x + y + z$, so

$$\iiint_E \text{div } \mathbf{F} \, dV = \int_0^{2\pi} \int_0^1 \int_0^1 (r \cos\theta + r \sin\theta + z) \, r \, dz \, dr \, d\theta = \int_0^{2\pi} \int_0^1 \left(r^2 \cos\theta + r^2 \sin\theta + \tfrac{1}{2} r \right) \, dr \, d\theta$$

$$= \int_0^{2\pi} \left(\tfrac{1}{3} \cos\theta + \tfrac{1}{3} \sin\theta + \tfrac{1}{4} \right) \, d\theta = \tfrac{1}{4}(2\pi) = \tfrac{\pi}{2}$$

Let S_1 be the top of the cylinder, S_2 the bottom, and S_3 the vertical edge. On S_1, $z = 1$, $\mathbf{n} = \mathbf{k}$, and $\mathbf{F} = xy \, \mathbf{i} + y \, \mathbf{j} + x \, \mathbf{k}$, so

$$\iint_{S_1} \mathbf{F} \cdot d\mathbf{S} = \iint_{S_1} \mathbf{F} \cdot \mathbf{n} \, dS = \iint_{S_1} x \, dS = \int_0^{2\pi} \int_0^1 (r \cos\theta) \, r \, dr \, d\theta = \left[\sin\theta \right]_0^{2\pi} \left[\tfrac{1}{3} r^3 \right]_0^1 = 0.$$

On S_2, $z = 0$, $\mathbf{n} = -\mathbf{k}$, and $\mathbf{F} = xy \, \mathbf{i}$ so $\iint_{S_2} \mathbf{F} \cdot d\mathbf{S} = \iint_{S_2} 0 \, dS = 0$.

S_3 is given by $\mathbf{r}(\theta, z) = \cos\theta \, \mathbf{i} + \sin\theta \, \mathbf{j} + z \, \mathbf{k}$, $0 \le \theta \le 2\pi$, $0 \le z \le 1$. Then $\mathbf{r}_\theta \times \mathbf{r}_z = \cos\theta \, \mathbf{i} + \sin\theta \, \mathbf{j}$ and

$$\iint_{S_3} \mathbf{F} \cdot d\mathbf{S} = \iint_D \mathbf{F} \cdot (\mathbf{r}_\theta \times \mathbf{r}_z) \, dA = \int_0^{2\pi} \int_0^1 (\cos^2\theta \sin\theta + z \sin^2\theta) \, dz \, d\theta$$

$$= \int_0^{2\pi} \left(\cos^2\theta \sin\theta + \tfrac{1}{2} \sin^2\theta \right) \, d\theta = \left[-\tfrac{1}{3} \cos^3\theta + \tfrac{1}{4} \left(\theta - \tfrac{1}{2} \sin 2\theta \right) \right]_0^{2\pi} = \tfrac{\pi}{2}$$

Thus $\iint_S \mathbf{F} \cdot d\mathbf{S} = 0 + 0 + \tfrac{\pi}{2} = \tfrac{\pi}{2}$.

4. div $\mathbf{F} = 1 + 1 + 1 = 3$, so $\iiint_E \text{div } \mathbf{F} \, dV = \iiint_E 3 \, dV = 3(\text{volume of ball}) = 3\left(\tfrac{4}{3}\pi\right) = 4\pi$. To find $\iint_S \mathbf{F} \cdot d\mathbf{S}$ we use spherical coordinates. S is the unit sphere, represented by $\mathbf{r}(\phi, \theta) = \sin\phi \cos\theta \, \mathbf{i} + \sin\phi \sin\theta \, \mathbf{j} + \cos\phi \, \mathbf{k}$, $0 \le \phi \le \pi$, $0 \le \theta \le 2\pi$. Then $\mathbf{r}_\phi \times \mathbf{r}_\theta = \sin^2\phi \cos\theta \, \mathbf{i} + \sin^2\phi \sin\theta \, \mathbf{j} + \sin\phi \cos\phi \, \mathbf{k}$ (see Example 17.6.10 [ET 16.6.10]) and $\mathbf{F}(\mathbf{r}(\phi, \theta)) = \sin\phi \cos\theta \, \mathbf{i} + \sin\phi \sin\theta \, \mathbf{j} + \cos\phi \, \mathbf{k}$. Thus

$$\iint_S \mathbf{F} \cdot d\mathbf{S} = \iint_D \mathbf{F} \cdot (\mathbf{r}_\phi \times \mathbf{r}_\theta) \, dA = \int_0^{2\pi} \int_0^\pi \left(\sin^3\phi \cos^2\theta + \sin^3\phi \sin^2\theta + \sin\phi \cos^2\phi \right) \, d\phi \, d\theta$$

$$= \int_0^{2\pi} d\theta \int_0^\pi \sin\phi \, d\phi = (2\pi)(2) = 4\pi$$

5. div $\mathbf{F} = \frac{\partial}{\partial x} (e^x \sin y) + \frac{\partial}{\partial y} (e^x \cos y) + \frac{\partial}{\partial z} (yz^2) = e^x \sin y - e^x \sin y + 2yz = 2yz$, so by the Divergence Theorem,

$$\iint_S \mathbf{F} \cdot d\mathbf{S} = \iiint_E \text{div } \mathbf{F} \, dV = \int_0^1 \int_0^1 \int_0^2 2yz \, dz \, dy \, dx = 2 \int_0^1 dx \int_0^1 y \, dy \int_0^1 z \, dz = 2 \left[x \right]_0^1 \left[\tfrac{1}{2} y^2 \right]_0^1 \left[\tfrac{1}{2} z^2 \right]_0^2 = 2.$$

6. $\operatorname{div} \mathbf{F} = \frac{\partial}{\partial x}\left(x^2 z^3\right) + \frac{\partial}{\partial y}\left(2xyz^3\right) + \frac{\partial}{\partial z}\left(xz^4\right) = 2xz^3 + 2xz^3 + 4xz^3 = 8xz^3$, so by the Divergence Theorem,

$$\iint_S \mathbf{F} \cdot d\mathbf{S} = \iiint_E \operatorname{div} \mathbf{F}\, dV = \int_{-1}^{1}\int_{-2}^{2}\int_{-3}^{3} 8xz^3\, dz\, dy\, dx = 8\int_{-1}^{1} x\, dx \int_{-2}^{2} dy \int_{-3}^{3} z^3\, dz$$

$$= 8\left[\tfrac{1}{2}x^2\right]_{-1}^{1}\left[y\right]_{-2}^{2}\left[\tfrac{1}{4}z^4\right]_{-3}^{3} = 0$$

7. $\operatorname{div} \mathbf{F} = 3y^2 + 0 + 3z^2$, so using cylindrical coordinates with $y = r\cos\theta$, $z = r\sin\theta$, $x = x$ we have

$$\iint_S \mathbf{F} \cdot d\mathbf{S} = \iiint_E (3y^2 + 3z^2)\, dV = \int_0^{2\pi}\int_0^1\int_{-1}^{2} (3r^2\cos^2\theta + 3r^2\sin^2\theta)\, r\, dx\, dr\, d\theta$$

$$= 3\int_0^{2\pi} d\theta \int_0^1 r^3\, dr \int_{-1}^{2} dx = 3(2\pi)\left(\tfrac{1}{4}\right)(3) = \tfrac{9\pi}{2}$$

8. $\operatorname{div} \mathbf{F} = 3x^2 y - 2x^2 y - x^2 y = 0$, so $\iint_S \mathbf{F} \cdot d\mathbf{S} = \iiint_E 0\, dV = 0.$

9. $\operatorname{div} \mathbf{F} = y\sin z + 0 - y\sin z = 0$, so by the Divergence Theorem, $\iint_S \mathbf{F} \cdot d\mathbf{S} = \iiint_E 0\, dV = 0.$

10. $\operatorname{div} \mathbf{F} = 2xy + 2xy + 2xy = 6xy$, so

$$\iint_S \mathbf{F} \cdot d\mathbf{S} = \iiint_E 6xy\, dV = \int_0^1\int_0^{2-2y}\int_0^{2-x-2y} 6xy\, dz\, dx\, dy = \int_0^1\int_0^{2-2y} 6xy(2 - x - 2y)\, dx\, dy$$

$$= \int_0^1\int_0^{2-2y} (12xy - 6x^2 y - 12xy^2)\, dx\, dy = \int_0^1 \left[6x^2 y - 2x^3 y - 6x^2 y^2\right]_{x=0}^{x=2-2y}\, dy$$

$$= \int_0^1 y(2 - 2y)^3\, dy = \left[-\tfrac{8}{5}y^5 + 6y^4 - 8y^3 + 4y^2\right]_0^1 = \tfrac{2}{5}$$

11. $\operatorname{div} \mathbf{F} = y^2 + 0 + x^2 = x^2 + y^2$ so

$$\iint_S \mathbf{F} \cdot d\mathbf{S} = \iiint_E (x^2 + y^2)\, dV = \int_0^{2\pi}\int_0^2\int_{r^2}^{4} r^2 \cdot r\, dz\, dr\, d\theta = \int_0^{2\pi}\int_0^2 r^3(4 - r^2)\, dr\, d\theta$$

$$= \int_0^{2\pi} d\theta \int_0^2 (4r^3 - r^5)\, dr = 2\pi\left[r^4 - \tfrac{1}{6}r^6\right]_0^2 = \tfrac{32}{3}\pi$$

12. $\operatorname{div} \mathbf{F} = 4x^3 + 4xy^2$ so

$$\iint_S \mathbf{F} \cdot d\mathbf{S} = \iiint_E 4x(x^2 + y^2)\, dV = \int_0^{2\pi}\int_0^1\int_0^{r\cos\theta + 2} (4r^3\cos\theta)\, r\, dz\, dr\, d\theta$$

$$= \int_0^{2\pi}\int_0^1 (4r^5\cos^2\theta + 8r^4\cos\theta)\, dr\, d\theta = \int_0^{2\pi}\left(\tfrac{2}{3}\cos^2\theta + \tfrac{8}{5}\cos\theta\right) d\theta = \tfrac{2}{3}\pi$$

13. $\operatorname{div} \mathbf{F} = 12x^2 z + 12y^2 z + 12z^3$ so

$$\iint_S \mathbf{F} \cdot d\mathbf{S} = \iiint_E 12z(x^2 + y^2 + z^2)\, dV = \int_0^{2\pi}\int_0^{\pi}\int_0^{R} 12(\rho\cos\phi)(\rho^2)\rho^2\sin\phi\, d\rho\, d\phi\, d\theta$$

$$= 12\int_0^{2\pi} d\theta \int_0^{\pi} \sin\phi\cos\phi\, d\phi \int_0^{R} \rho^5\, d\rho = 12(2\pi)\left[\tfrac{1}{2}\sin^2\phi\right]_0^{\pi}\left[\tfrac{1}{6}\rho^6\right]_0^{R} = 0$$

14. $\mathbf{F}(x, y, z) = \dfrac{x}{\sqrt{x^2 + y^2 + z^2}}\,\mathbf{i} + \dfrac{y}{\sqrt{x^2 + y^2 + z^2}}\,\mathbf{j} + \dfrac{z}{\sqrt{x^2 + y^2 + z^2}}\,\mathbf{k}$, so

$$\operatorname{div}\mathbf{F} = \frac{\sqrt{x^2 + y^2 + z^2} - x^2/\sqrt{x^2 + y^2 + z^2}}{x^2 + y^2 + z^2} + \frac{\sqrt{x^2 + y^2 + z^2} - y^2/\sqrt{x^2 + y^2 + z^2}}{x^2 + y^2 + z^2}$$

$$+ \frac{\sqrt{x^2 + y^2 + z^2} - z^2/\sqrt{x^2 + y^2 + z^2}}{x^2 + y^2 + z^2}$$

$$= \frac{x^2 + y^2 + z^2 - x^2}{(x^2 + y^2 + z^2)^{3/2}} + \frac{x^2 + y^2 + z^2 - y^2}{(x^2 + y^2 + z^2)^{3/2}} + \frac{x^2 + y^2 + z^2 - z^2}{(x^2 + y^2 + z^2)^{3/2}} = \frac{2(x^2 + y^2 + z^2)}{(x^2 + y^2 + z^2)^{3/2}} = \frac{2}{\sqrt{x^2 + y^2 + z^2}}.$$

Then

$$\iint_S \mathbf{F} \cdot d\mathbf{S} = \iiint_E \frac{2}{\sqrt{x^2 + y^2 + z^2}}\, dV = \int_0^{\pi/2}\int_0^{2\pi}\int_0^1 \frac{2}{\rho}\rho^2\sin\phi\, d\rho\, d\theta\, d\phi$$

$$= 2\int_0^{\pi/2} \sin\phi\, d\phi \int_0^{2\pi} d\theta \int_0^1 \rho\, d\rho = 2\,(1)\,(2\pi)\left(\tfrac{1}{2}\right) = 2\pi$$

15. $\iint_S \mathbf{F} \cdot d\mathbf{S} = \iiint_E \sqrt{3 - x^2}\, dV = \int_{-1}^1 \int_{-1}^1 \int_0^{2 - x^4 - y^4} \sqrt{3 - x^2}\, dz\, dy\, dx = \frac{341}{60}\sqrt{2} + \frac{81}{20}\sin^{-1}\left(\frac{\sqrt{3}}{3}\right)$

16.

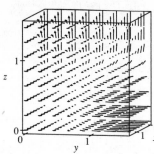 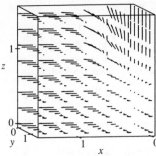

By the Divergence Theorem, the flux of $\mathbf{F}$ across the surface of the cube is

$$\iint_S \mathbf{F} \cdot d\mathbf{S} = \int_0^{\pi/2}\int_0^{\pi/2}\int_0^{\pi/2}\left[\cos x\,\cos^2 y + 3\sin^2 y\,\cos y\,\cos^4 z + 5\sin^4 z\,\cos z\,\cos^6 x\right] dz\, dy\, dx = \frac{19}{64}\pi^2.$$

17. For S_1 we have $\mathbf{n} = -\mathbf{k}$, so $\mathbf{F} \cdot \mathbf{n} = \mathbf{F} \cdot (-\mathbf{k}) = -x^2 z - y^2 = -y^2$ (since $z = 0$ on S_1). So if D is the unit disk, we get

$$\iint_{S_1} \mathbf{F} \cdot d\mathbf{S} = \iint_{S_1} \mathbf{F} \cdot \mathbf{n}\, dS = \iint_D (-y^2)\, dA = -\int_0^{2\pi}\int_0^1 r^2 (\sin^2\theta)\, r\, dr\, d\theta = -\frac{1}{4}\pi.$$ Now since S_2 is closed, we can use

the Divergence Theorem. Since $\operatorname{div}\mathbf{F} = \frac{\partial}{\partial x}\left(z^2 x\right) + \frac{\partial}{\partial y}\left(\frac{1}{3}y^3 + \tan z\right) + \frac{\partial}{\partial z}\left(x^2 z + y^2\right) = z^2 + y^2 + x^2$, we use spherical

coordinates to get $\iint_{S_2} \mathbf{F} \cdot d\mathbf{S} = \iiint_E \operatorname{div}\mathbf{F}\, dV = \int_0^{2\pi}\int_0^{\pi/2}\int_0^1 \rho^2 \cdot \rho^2 \sin\phi\, d\rho\, d\phi\, d\theta = \frac{2}{5}\pi$. Finally

$$\iint_S \mathbf{F} \cdot d\mathbf{S} = \iint_{S_2} \mathbf{F} \cdot d\mathbf{S} - \iint_{S_1} \mathbf{F} \cdot d\mathbf{S} = \frac{2}{5}\pi - \left(-\frac{1}{4}\pi\right) = \frac{13}{20}\pi.$$

18. As in the hint to Exercise 19, we create a closed surface $S_2 = S \cup S_1$, where S is the part of the paraboloid $x^2 + y^2 + z = 2$

that lies above the plane $z = 1$, and S_1 is the disk $x^2 + y^2 = 1$ on the plane $z = 1$ oriented downward, and we then apply the

Divergence Theorem. Since the disk S_1 is oriented downward, its unit normal vector is $\mathbf{n} = -\mathbf{k}$ and $\mathbf{F} \cdot (-\mathbf{k}) = -z = -1$ on

S_1. So $\iint_{S_1} \mathbf{F} \cdot d\mathbf{S} = \iint_{S_1} \mathbf{F} \cdot \mathbf{n}\, dS = \iint_{S_1}(-1)\, dS = -A(S_1) = -\pi$. Let E be the region bounded by S_2. Then

$$\iint_{S_2} \mathbf{F} \cdot d\mathbf{S} = \iiint_E \operatorname{div}\mathbf{F}\, dV = \iiint_E 1\, dV = \int_0^1\int_0^{2\pi}\int_1^{2-r^2} r\, dz\, d\theta\, dr = \int_0^1\int_0^{2\pi}(r - r^3)\, d\theta\, dr = (2\pi)\frac{1}{4} = \frac{\pi}{2}.$$ Thus the

flux of $\mathbf{F}$ across S is $\iint_S \mathbf{F} \cdot d\mathbf{S} = \iint_{S_2} \mathbf{F} \cdot d\mathbf{S} - \iint_{S_1} \mathbf{F} \cdot d\mathbf{S} = \frac{\pi}{2} - (-\pi) = \frac{3\pi}{2}.$

19. The vectors that end near P_1 are longer than the vectors that start near P_1, so the net flow is inward near P_1 and $\operatorname{div}\mathbf{F}(P_1)$ is

negative. The vectors that end near P_2 are shorter than the vectors that start near P_2, so the net flow is outward near P_2 and

$\operatorname{div}\mathbf{F}(P_2)$ is positive.

20. (a) The vectors that end near P_1 are shorter than the vectors that start near P_1, so the net flow is outward and P_1 is a source.

The vectors that end near P_2 are longer than the vectors that start near P_2, so the net flow is inward and P_2 is a sink.

(b) $\mathbf{F}(x, y) = \langle x, y^2 \rangle \Rightarrow \operatorname{div}\mathbf{F} = \nabla \cdot \mathbf{F} = 1 + 2y$. The y-value at P_1 is positive, so $\operatorname{div}\mathbf{F} = 1 + 2y$ is positive, thus P_1

is a source. At P_2, $y < -1$, so $\operatorname{div}\mathbf{F} = 1 + 2y$ is negative, and P_2 is a sink.

21.

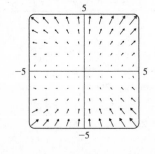

From the graph it appears that for points above the x-axis, vectors starting near a

particular point are longer than vectors ending there, so divergence is positive.

The opposite is true at points below the x-axis, where divergence is negative.

$\mathbf{F}(x, y) = \langle xy, x + y^2 \rangle \Rightarrow \operatorname{div}\mathbf{F} = \frac{\partial}{\partial x}(xy) + \frac{\partial}{\partial y}(x + y^2) = y + 2y = 3y.$

Thus $\operatorname{div}\mathbf{F} > 0$ for $y > 0$, and $\operatorname{div}\mathbf{F} < 0$ for $y < 0$.

22.

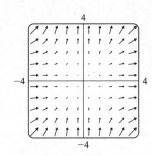

From the graph it appears that for points above the line $y = -x$, vectors starting near a particular point are longer than vectors ending there, so divergence is positive. The opposite is true at points below the line $y = -x$, where divergence is negative.

$\mathbf{F}(x, y) = \langle x^2, y^2 \rangle \;\Rightarrow\; \operatorname{div} \mathbf{F} = \frac{\partial}{\partial x}(x^2) + \frac{\partial}{\partial y}(y^2) = 2x + 2y$. Then $\operatorname{div} \mathbf{F} > 0$ for $2x + 2y > 0 \;\Rightarrow\; y > -x$, and $\operatorname{div} \mathbf{F} < 0$ for $y < -x$.

23. Since $\dfrac{\mathbf{x}}{|\mathbf{x}|^3} = \dfrac{x\,\mathbf{i} + y\,\mathbf{j} + z\,\mathbf{k}}{(x^2 + y^2 + z^2)^{3/2}}$ and $\dfrac{\partial}{\partial x}\left(\dfrac{x}{(x^2 + y^2 + z^2)^{3/2}}\right) = \dfrac{(x^2 + y^2 + z^2) - 3x^2}{(x^2 + y^2 + z^2)^{5/2}}$ with similar expressions

for $\dfrac{\partial}{\partial y}\left(\dfrac{y}{(x^2 + y^2 + z^2)^{3/2}}\right)$ and $\dfrac{\partial}{\partial z}\left(\dfrac{z}{(x^2 + y^2 + z^2)^{3/2}}\right)$, we have

$\operatorname{div}\left(\dfrac{\mathbf{x}}{|\mathbf{x}|^3}\right) = \dfrac{3(x^2 + y^2 + z^2) - 3(x^2 + y^2 + z^2)}{(x^2 + y^2 + z^2)^{5/2}} = 0$, except at $(0, 0, 0)$ where it is undefined.

24. We first need to find $\mathbf{F}$ so that $\iint_S \mathbf{F} \cdot \mathbf{n}\,dS = \iint_S (2x + 2y + z^2)\,dS$, so $\mathbf{F} \cdot \mathbf{n} = 2x + 2y + z^2$. But for S,

$\mathbf{n} = \dfrac{x\,\mathbf{i} + y\,\mathbf{j} + z\,\mathbf{k}}{\sqrt{x^2 + y^2 + z^2}} = x\,\mathbf{i} + y\,\mathbf{j} + z\,\mathbf{k}$. Thus $\mathbf{F} = 2\,\mathbf{i} + 2\,\mathbf{j} + z\,\mathbf{k}$ and $\operatorname{div} \mathbf{F} = 1$.

If $B = \{(x, y, z) \mid x^2 + y^2 + z^2 \le 1\}$, then $\iint_S (2x + 2y + z^2)\,dS = \iiint_B dV = V(B) = \frac{4}{3}\pi(1)^3 = \frac{4}{3}\pi$.

25. $\iint_S \mathbf{a} \cdot \mathbf{n}\,dS = \iiint_E \operatorname{div} \mathbf{a}\,dV = 0$ since $\operatorname{div} \mathbf{a} = 0$.

26. $\frac{1}{3}\iint_S \mathbf{F} \cdot d\mathbf{S} = \frac{1}{3}\iiint_E \operatorname{div} \mathbf{F}\,dV = \frac{1}{3}\iiint_E 3\,dV = V(E)$

27. $\iint_S \operatorname{curl} \mathbf{F} \cdot d\mathbf{S} = \iiint_E \operatorname{div}(\operatorname{curl} \mathbf{F})\,dV = 0$ by Theorem 17.5.11 [ET 16.5.11].

28. $\iint_S D_{\mathbf{n}} f\,dS = \iint_S (\nabla f \cdot \mathbf{n})\,dS = \iiint_E \operatorname{div}(\nabla f)\,dV = \iiint_E \nabla^2 f\,dV$

29. $\iint_S (f\nabla g) \cdot \mathbf{n}\,dS = \iiint_E \operatorname{div}(f\nabla g)\,dV = \iiint_E (f\nabla^2 g + \nabla g \cdot \nabla f)\,dV$ by Exercise 17.5.25 [ET 16.5.25].

30. $\iint_S (f\nabla g - g\nabla f) \cdot \mathbf{n}\,dS = \iiint_E \left[(f\nabla^2 g + \nabla g \cdot \nabla f) - (g\nabla^2 f + \nabla g \cdot \nabla f)\right] dV$ [by Exercise 29].

But $\nabla g \cdot \nabla f = \nabla f \cdot \nabla g$, so that $\iint_S (f\nabla g - g\nabla f) \cdot \mathbf{n}\,dS = \iiint_E (f\nabla^2 g - g\nabla^2 f)\,dV$.

31. If $\mathbf{c} = c_1\,\mathbf{i} + c_2\,\mathbf{j} + c_3\,\mathbf{k}$ is an arbitrary constant vector, we define $\mathbf{F} = f\mathbf{c} = fc_1\,\mathbf{i} + fc_2\,\mathbf{j} + fc_3\,\mathbf{k}$. Then

$\operatorname{div} \mathbf{F} = \operatorname{div} f\mathbf{c} = \dfrac{\partial f}{\partial x}c_1 + \dfrac{\partial f}{\partial y}c_2 + \dfrac{\partial f}{\partial z}c_3 = \nabla f \cdot \mathbf{c}$ and the Divergence Theorem says $\iint_S \mathbf{F} \cdot d\mathbf{S} = \iiint_E \operatorname{div} \mathbf{F}\,dV \;\Rightarrow$

$\iint_S \mathbf{F} \cdot \mathbf{n}\,dS = \iiint_E \nabla f \cdot \mathbf{c}\,dV$. In particular, if $\mathbf{c} = \mathbf{i}$ then $\iint_S f\mathbf{i} \cdot \mathbf{n}\,dS = \iiint_E \nabla f \cdot \mathbf{i}\,dV \;\Rightarrow$

$\iint_S fn_1\,dS = \iiint_E \dfrac{\partial f}{\partial x}\,dV$ (where $\mathbf{n} = n_1\,\mathbf{i} + n_2\,\mathbf{j} + n_3\,\mathbf{k}$). Similarly, if $\mathbf{c} = \mathbf{j}$ we have $\iint_S fn_2\,dS = \iiint_E \dfrac{\partial f}{\partial y}\,dV$,

and $\mathbf{c} = \mathbf{k}$ gives $\iint_S fn_3\,dS = \iiint_E \dfrac{\partial f}{\partial z}\,dV$. Then

$\iint_S f\mathbf{n}\,dS = \left(\iint_S fn_1\,dS\right)\mathbf{i} + \left(\iint_S fn_2\,dS\right)\mathbf{j} + \left(\iint_S fn_3\,dS\right)\mathbf{k}$

$\qquad = \left(\iiint_E \dfrac{\partial f}{\partial x}\,dV\right)\mathbf{i} + \left(\iiint_E \dfrac{\partial f}{\partial y}\,dV\right)\mathbf{j} + \left(\iiint_E \dfrac{\partial f}{\partial z}\,dV\right)\mathbf{k} = \iiint_E \left(\dfrac{\partial f}{\partial x}\mathbf{i} + \dfrac{\partial f}{\partial y}\mathbf{j} + \dfrac{\partial f}{\partial z}\mathbf{k}\right)dV$

$\qquad = \iiint_E \nabla f\,dV$ as desired.

32. By Exercise 31, $\iint_S p\mathbf{n}\,dS = \iiint_E \nabla p\,dV$, so

$$\mathbf{F} = -\iint_S p\mathbf{n}\,dS = -\iiint_E \nabla p\,dV = -\iiint_E \nabla(\rho g z)\,dV = -\iiint_E (\rho g\,\mathbf{k})\,dV = -\rho g\left(\iiint_E dV\right)\mathbf{k} = -\rho g V(E)\,\mathbf{k}.$$

But the weight of the displaced liquid is volume $\times$ density $\times$ $g = \rho g V(E)$, thus $\mathbf{F} = -W\mathbf{k}$ as desired.

17 Review

ET 16

CONCEPT CHECK

---

1. See Definitions 1 and 2 in Section 17.1 [ET 16.1]. A vector field can represent, for example, the wind velocity at any location in space, the speed and direction of the ocean current at any location, or the force vectors of Earth's gravitational field at a location in space.

2. (a) A conservative vector field $\mathbf{F}$ is a vector field which is the gradient of some scalar function f.

(b) The function f in part (a) is called a potential function for $\mathbf{F}$, that is, $\mathbf{F} = \nabla f$.

3. (a) See Definition 17.2.2 [ET 16.2.2].

(b) We normally evaluate the line integral using Formula 17.2.3 [ET 16.2.3].

(c) The mass is $m = \int_C \rho(x,y)\,ds$, and the center of mass is $(\overline{x},\overline{y})$ where $\overline{x} = \frac{1}{m}\int_C x\rho(x,y)\,ds$, $\overline{y} = \frac{1}{m}\int_C y\rho(x,y)\,ds$.

(d) See (5) and (6) in Section 17.2 [ET 16.2] for plane curves; we have similar definitions when C is a space curve
(see the equation preceding (10) in Section 17.2 [ET 16.2]).

(e) For plane curves, see Equations 17.2.7 [ET 16.2.7]. We have similar results for space curves
(see the equation preceding (10) in Section 17.2 [ET 16.2]).

4. (a) See Definition 17.2.13 [ET 16.2.13].

(b) If $\mathbf{F}$ is a force field, $\int_C \mathbf{F}\cdot d\mathbf{r}$ represents the work done by $\mathbf{F}$ in moving a particle along the curve C.

(c) $\int_C \mathbf{F}\cdot d\mathbf{r} = \int_C P\,dx + Q\,dy + R\,dz$

5. See Theorem 17.3.2 [ET 16.3.2].

6. (a) $\int_C \mathbf{F}\cdot d\mathbf{r}$ is independent of path if the line integral has the same value for any two curves that have the same initial and terminal points.

(b) See Theorem 17.3.4 [ET 16.3.4].

7. See the statement of Green's Theorem on page 1091 [ET 1055].

8. See Equations 17.4.5 [ET 16.4.5].

9. (a) $\operatorname{curl}\mathbf{F} = \left(\dfrac{\partial R}{\partial y} - \dfrac{\partial Q}{\partial z}\right)\mathbf{i} + \left(\dfrac{\partial P}{\partial z} - \dfrac{\partial R}{\partial x}\right)\mathbf{j} + \left(\dfrac{\partial Q}{\partial x} - \dfrac{\partial P}{\partial y}\right)\mathbf{k} = \nabla\times\mathbf{F}$

(b) $\operatorname{div}\mathbf{F} = \dfrac{\partial P}{\partial x} + \dfrac{\partial Q}{\partial y} + \dfrac{\partial R}{\partial z} = \nabla\cdot\mathbf{F}$

(c) For curl $\mathbf{F}$, see the discussion accompanying Figure 1 on page 1100 [ET 1064] as well as Figure 6 and the accompanying discussion on page 1132 [ET 1096]. For div $\mathbf{F}$, see the discussion following Example 5 on page 1102 [ET 1066] as well as the discussion preceding (8) on page 1139 [ET 1103].

10. See Theorem 17.3.6 [ET 16.3.6]; see Theorem 17.5.4 [ET 16.5.4].

11. (a) See (1) and (2) and the accompanying discussion in Section 17.6 [ET 16.6] ; See Figure 4 and the accompanying discussion on page 1107 [ET 1071] .

(b) See Definition 17.6.6 [ET 16.6.6].

(c) See Equation 17.6.9 [ET 16.6.9].

12. (a) See (1) in Section 17.7 [ET 16.7].

(b) We normally evaluate the surface integral using Formula 17.7.2 [ET 16.7.2].

(c) See Formula 17.7.4 [ET 16.7.4].

(d) The mass is $m = \iint_S \rho(x, y, z) \, dS$ and the center of mass is $(\overline{x}, \overline{y}, \overline{z})$ where $\overline{x} = \frac{1}{m} \iint_S x\rho(x, y, z) \, dS$, $\overline{y} = \frac{1}{m} \iint_S y\rho(x, y, z) \, dS, \overline{z} = \frac{1}{m} \iint_S z\rho(x, y, z) \, dS$.

13. (a) See Figures 6 and 7 and the accompanying discussion in Section 17.7 [ET 16.7]. A Möbius strip is a nonorientable surface; see Figures 4 and 5 and the accompanying discussion on page 1121 [ET 1085].

(b) See Definition 17.7.8 [ET 16.7.8].

(c) See Formula 17.7.9 [ET 16.7.9].

(d) See Formula 17.7.10 [ET 16.7.10].

14. See the statement of Stokes' Theorem on page 1129 [ET 1093.].

15. See the statement of the Divergence Theorem on page 1135 [ET 1099].

16. In each theorem, we have an integral of a "derivative" over a region on the left side, while the right side involves the values of the original function only on the boundary of the region.

TRUE-FALSE QUIZ

1. False; div **F** is a scalar field.

2. True. (See Definition 17.5.1 [ET 16.5.1].)

3. True, by Theorem 17.5.3 [ET 16.5.3] and the fact that div **0** = 0.

4. True, by Theorem 17.3.2 [ET 16.3.2].

5. False. See Exercise 17.3.33 [ET 16.3.33]. (But the assertion is true if D is simply-connected; see Theorem 17.3.6 [ET 16.3.6].)

6. False. See the discussion accompanying Figure 8 on page 1075 [ET 1039].

7. True. Apply the Divergence Theorem and use the fact that div **F** = 0.

8. False by Theorem 17.5.11 [ET 16.5.11], because if it were true, then div curl **F** = $3 \neq 0$.

EXERCISES

1. (a) Vectors starting on C point in roughly the direction opposite to C, so the tangential component $\mathbf{F} \cdot \mathbf{T}$ is negative.

   Thus $\int_C \mathbf{F} \cdot d\mathbf{r} = \int_C \mathbf{F} \cdot \mathbf{T} \, ds$ is negative.

   (b) The vectors that end near P are shorter than the vectors that start near P, so the net flow is outward near P and

   div $\mathbf{F}(P)$ is positive.

2. We can parametrize C by $x = x$, $y = x^2$, $0 \leq x \leq 1$ so

   $\int_C x \, ds = \int_0^1 x \sqrt{1 + (2x)^2} \, dx = \frac{1}{12}(1 + 4x^2)^{3/2}\Big]_0^1 = \frac{1}{12}(5\sqrt{5} - 1)$.

3. $\int_C yz \cos x \, ds = \int_0^\pi (3 \cos t)(3 \sin t) \cos t \sqrt{(1)^2 + (-3 \sin t)^2 + (3 \cos t)^2} \, dt = \int_0^\pi (9 \cos^2 t \sin t)\sqrt{10} \, dt$

   $= 9\sqrt{10}\left(-\frac{1}{3}\cos^3 t\right)\Big]_0^\pi = -3\sqrt{10}(-2) = 6\sqrt{10}$

4. $x = 3 \cos t \implies dx = -3 \sin t \, dt$, $y = 2 \sin t \implies dy = 2 \cos t \, dt$, $0 \leq t \leq 2\pi$, so

   $\int_C y \, dx + (x + y^2) \, dy = \int_0^{2\pi} \left[(2 \sin t)(-3 \sin t) + (3 \cos t + 4 \sin^2 t)(2 \cos t)\right] dt$

   $= \int_0^{2\pi} (-6 \sin^2 t + 6 \cos^2 t + 8 \sin^2 t \cos t) \, dt = \int_0^{2\pi} \left[6(\cos^2 t - \sin^2 t) + 8 \sin^2 t \cos t\right] dt$

   $= \int_0^{2\pi} (6 \cos 2t + 8 \sin^2 t \cos t) \, dt = 3 \sin 2t + \frac{8}{3} \sin^3 t\Big]_0^{2\pi} = 0$

   Or: Notice that $\frac{\partial}{\partial y}(y) = 1 = \frac{\partial}{\partial x}(x + y^2)$, so $\mathbf{F}(x, y) = \langle y, x + y^2 \rangle$ is a conservative vector field. Since C is a closed

   curve, $\int_C \mathbf{F} \cdot d\mathbf{r} = \int_C y \, dx + (x + y^2) \, dy = 0$.

5. $\int_C y^3 \, dx + x^2 \, dy = \int_{-1}^1 \left[y^3(-2y) + (1 - y^2)^2\right] dy = \int_{-1}^1 (-y^4 - 2y^2 + 1) \, dy$

   $= \left[-\frac{1}{5}y^5 - \frac{2}{3}y^3 + y\right]_{-1}^1 = -\frac{1}{5} - \frac{2}{3} + 1 - \frac{1}{5} - \frac{2}{3} + 1 = \frac{4}{15}$

6. $\int_C \sqrt{xy} \, dx + e^y \, dy + xz \, dz = \int_0^1 \left(\sqrt{t^4 \cdot t^2} \cdot 4t^3 + e^{t^2} \cdot 2t + t^4 \cdot t^3 \cdot 3t^2\right) dt = \int_0^1 (4t^6 + 2te^{t^2} + 3t^9) \, dt$

   $= \left[\frac{4}{7}t^7 + e^{t^2} + \frac{3}{10}t^{10}\right]_0^1 = e - \frac{9}{70}$

7. C: $x = 1 + 2t \implies dx = 2 \, dt$, $y = 4t \implies dy = 4 \, dt$, $z = -1 + 3t \implies dz = 3 \, dt$, $0 \leq t \leq 1$.

   $\int_C xy \, dx + y^2 \, dy + yz \, dz = \int_0^1 [(1 + 2t)(4t)(2) + (4t)^2(4) + (4t)(-1 + 3t)(3)] \, dt$

   $= \int_0^1 (116t^2 - 4t) \, dt = \left[\frac{116}{3}t^3 - 2t^2\right]_0^1 = \frac{116}{3} - 2 = \frac{110}{3}$

8. $\mathbf{F}(\mathbf{r}(t)) = (\sin t)(1 + t)\mathbf{i} + (\sin^2 t)\mathbf{j}$, $\mathbf{r}'(t) = \cos t \, \mathbf{i} + \mathbf{j}$ and

   $\int_C \mathbf{F} \cdot d\mathbf{r} = \int_0^\pi ((1 + t) \sin t \cos t + \sin^2 t) \, dt = \int_0^\pi \left(\frac{1}{2}(1 + t) \sin 2t + \sin^2 t\right) dt$

   $= \left[\frac{1}{2}\left((1 + t)\left(-\frac{1}{2}\cos 2t\right) + \frac{1}{4}\sin 2t\right) + \frac{1}{2}t - \frac{1}{4}\sin 2t\right]_0^\pi = \frac{\pi}{4}$

9. $\mathbf{F}(\mathbf{r}(t)) = e^{-t}\mathbf{i} + t^2(-t)\mathbf{j} + (t^2 + t^3)\mathbf{k}$, $\mathbf{r}'(t) = 2t \, \mathbf{i} + 3t^2 \, \mathbf{j} - \mathbf{k}$ and

   $\int_C \mathbf{F} \cdot d\mathbf{r} = \int_0^1 (2te^{-t} - 3t^5 - (t^2 + t^3)) \, dt = \left[-2te^{-t} - 2e^{-t} - \frac{1}{2}t^6 - \frac{1}{3}t^3 - \frac{1}{4}t^4\right]_0^1 = \frac{11}{12} - \frac{4}{e}$.

10. (a) C: $x = 3 - 3t$, $y = \frac{\pi}{2}t$, $z = 3t$, $0 \le t \le 1$. Then

$$W = \int_C \mathbf{F} \cdot d\mathbf{r} = \int_0^1 \left[3t\,\mathbf{i} + (3 - 3t)\,\mathbf{j} + \tfrac{\pi}{2}t\,\mathbf{k}\right] \cdot \left[-3\,\mathbf{i} + \tfrac{\pi}{2}\,\mathbf{j} + 3\,\mathbf{k}\right] dt = \int_0^1 \left[-9t + \tfrac{3\pi}{2}\right] dt = \tfrac{1}{2}(3\pi - 9).$$

(b) $W = \int_C \mathbf{F} \cdot d\mathbf{r} = \int_0^{\pi/2} (3\sin t\,\mathbf{i} + 3\cos t\,\mathbf{j} + t\,\mathbf{k}) \cdot (-3\sin t\,\mathbf{i} + \mathbf{j} + 3\cos t\,\mathbf{k})\,dt$

$$= \int_0^{\pi/2} (-9\sin^2 t + 3\cos t + 3t\cos t)\,dt = \left[-\tfrac{9}{2}(t - \sin t\cos t) + 3\sin t + 3(t\sin t + \cos t)\right]_0^{\pi/2}$$

$$= -\tfrac{9\pi}{4} + 3 + \tfrac{3\pi}{2} - 3 = -\tfrac{3\pi}{4}$$

11. $\frac{\partial}{\partial y}\left[(1 + xy)e^{xy}\right] = 2xe^{xy} + x^2 ye^{xy} = \frac{\partial}{\partial x}\left[e^y + x^2 e^{xy}\right]$ and the domain of $\mathbf{F}$ is $\mathbb{R}^2$, so $\mathbf{F}$ is conservative. Thus there

exists a function f such that $\mathbf{F} = \nabla f$. Then $f_y(x, y) = e^y + x^2 e^{xy}$ implies $f(x, y) = e^y + xe^{xy} + g(x)$ and then

$f_x(x, y) = xye^{xy} + e^{xy} + g'(x) = (1 + xy)e^{xy} + g'(x)$. But $f_x(x, y) = (1 + xy)e^{xy}$, so $g'(x) = 0 \;\Rightarrow\; g(x) = K$.

Thus $f(x, y) = e^y + xe^{xy} + K$ is a potential function for $\mathbf{F}$.

12. $\mathbf{F}$ is defined on all of $\mathbb{R}^3$, its components have continuous partial derivatives, and

curl $\mathbf{F} = (0 - 0)\,\mathbf{i} - (0 - 0)\,\mathbf{j} + (\cos y - \cos y)\,\mathbf{k} = \mathbf{0}$, so $\mathbf{F}$ is conservative by Theorem 17.5.4 [ET 16.5.4]. Thus there

exists a function f such that $\nabla f = \mathbf{F}$. Then $f_x(x, y, z) = \sin y$ implies $f(x, y, z) = x\sin y + g(y, z)$ and then

$f_y(x, y, z) = x\cos y + g_y(y, z)$. But $f_y(x, y, z) = x\cos y$, so $g_y(y, z) = 0 \;\Rightarrow\; g(y, z) = h(z)$. Then

$f(x, y, z) = x\sin y + h(z)$ implies $f_z(x, y, z) = h'(z)$. But $f_z(x, y, z) = -\sin z$, so $h(z) = \cos z + K$. Thus a potential

function for $\mathbf{F}$ is $f(x, y, z) = x\sin y + \cos z + K$.

13. Since $\frac{\partial}{\partial y}(4x^3 y^2 - 2xy^3) = 8x^3 y - 6xy^2 = \frac{\partial}{\partial x}(2x^4 y - 3x^2 y^2 + 4y^3)$ and the domain of $\mathbf{F}$ is $\mathbb{R}^2$, $\mathbf{F}$ is conservative.

Furthermore $f(x, y) = x^4 y^2 - x^2 y^3 + y^4$ is a potential function for $\mathbf{F}$. $t = 0$ corresponds to the point $(0, 1)$ and $t = 1$

corresponds to $(1, 1)$, so $\int_C \mathbf{F} \cdot d\mathbf{r} = f(1, 1) - f(0, 1) = 1 - 1 = 0$.

14. Here curl $\mathbf{F} = \mathbf{0}$, the domain of $\mathbf{F}$ is $\mathbb{R}^3$, and the components of $\mathbf{F}$ have continuous partial derivatives, so $\mathbf{F}$ is conservative.

Furthermore $f(x, y, z) = xe^y + ye^z$ is a potential function for $\mathbf{F}$. Then $\int_C \mathbf{F} \cdot d\mathbf{r} = f(4, 0, 3) - f(0, 2, 0) = 4 - 2 = 2$.

15. C_1: $\mathbf{r}(t) = t\,\mathbf{i} + t^2\,\mathbf{j}$, $-1 \le t \le 1$;

C_2: $\mathbf{r}(t) = -t\,\mathbf{i} + \mathbf{j}$, $-1 \le t \le 1$.

Then

$$\int_C xy^2\,dx - x^2 y\,dy = \int_{-1}^1 (t^5 - 2t^5)\,dt + \int_{-1}^1 t\,dt$$

$$= \left[-\tfrac{1}{6}t^6\right]_{-1}^1 + \left[\tfrac{1}{2}t^2\right]_{-1}^1 = 0$$

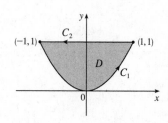

Using Green's Theorem, we have

$$\int_C xy^2\,dx - x^2 y\,dy = \iint_D \left[\frac{\partial}{\partial x}(-x^2 y) - \frac{\partial}{\partial y}(xy^2)\right] dA = \iint_D (-2xy - 2xy)\,dA = \int_{-1}^1 \int_{x^2}^1 -4xy\,dy\,dx$$

$$= \int_{-1}^1 \left[-2xy^2\right]_{y=x^2}^{y=1} dx = \int_{-1}^1 (2x^5 - 2x)\,dx = \left[\tfrac{1}{3}x^6 - x^2\right]_{-1}^1 = 0$$

16. $\int_C \sqrt{1 + x^3}\,dx + 2xy\,dy = \iint_D \left[\frac{\partial}{\partial x}(2xy) - \frac{\partial}{\partial y}\left(\sqrt{1 + x^3}\right)\right] dA = \int_0^1 \int_0^{3x}(2y - 0)\,dy\,dx = \int_0^1 9x^2\,dx = 3x^3\big|_0^1 = 3$

17. $\int_C x^2 y\,dx - xy^2\,dy = \iint_{x^2 + y^2 \le 4} \left[\frac{\partial}{\partial x}(-xy^2) - \frac{\partial}{\partial y}(x^2 y)\right] dA = \iint_{x^2 + y^2 \le 4} (-y^2 - x^2)\,dA = -\int_0^{2\pi} \int_0^2 r^3\,dr\,d\theta = -8\pi$

18. $\operatorname{curl}\mathbf{F} = (0 - e^{-y}\cos z)\,\mathbf{i} - (e^{-z}\cos x - 0)\,\mathbf{j} + (0 - e^{-x}\cos y)\,\mathbf{k} = -e^{-y}\cos z\,\mathbf{i} - e^{-z}\cos x\,\mathbf{j} - e^{-x}\cos y\,\mathbf{k}$,

$\operatorname{div}\mathbf{F} = -e^{-x}\sin y - e^{-y}\sin z - e^{-z}\sin x$

19. If we assume there is such a vector field $\mathbf{G}$, then $\operatorname{div}(\operatorname{curl}\mathbf{G}) = 2 + 3z - 2xz$. But $\operatorname{div}(\operatorname{curl}\mathbf{F}) = 0$ for all vector fields $\mathbf{F}$. Thus such a $\mathbf{G}$ cannot exist.

20. Let $\mathbf{F} = P_1\,\mathbf{i} + Q_1\,\mathbf{j} + R_1\,\mathbf{k}$ and $\mathbf{G} = P_2\,\mathbf{i} + Q_2\,\mathbf{j} + R_2\,\mathbf{k}$ be vector fields whose first partials exist and are continuous. Then

$$\mathbf{F}\operatorname{div}\mathbf{G} - \mathbf{G}\operatorname{div}\mathbf{F} = \left[P_1\left(\frac{\partial P_2}{\partial x} + \frac{\partial Q_2}{\partial y} + \frac{\partial R_2}{\partial z}\right)\mathbf{i} + Q_1\left(\frac{\partial P_2}{\partial x} + \frac{\partial Q_2}{\partial y} + \frac{\partial R_2}{\partial z}\right)\mathbf{j} + R_1\left(\frac{\partial P_2}{\partial x} + \frac{\partial Q_2}{\partial y} + \frac{\partial R_2}{\partial z}\right)\mathbf{k}\right]$$

$$- \left[P_2\left(\frac{\partial P_1}{\partial x} + \frac{\partial Q_1}{\partial y} + \frac{\partial R_1}{\partial z}\right)\mathbf{i} + Q_2\left(\frac{\partial P_1}{\partial x} + \frac{\partial Q_1}{\partial y} + \frac{\partial R_1}{\partial z}\right)\mathbf{j}\right.$$

$$\left.+ R_2\left(\frac{\partial P_1}{\partial x} + \frac{\partial Q}{\partial y} + \frac{\partial R_1}{\partial z}\right)\mathbf{k}\right]$$

and

$$(\mathbf{G}\cdot\nabla)\mathbf{F} - (\mathbf{F}\cdot\nabla)\mathbf{G} = \left[\left(P_2\frac{\partial P_1}{\partial x} + Q_2\frac{\partial P_1}{\partial y} + R_2\frac{\partial P_1}{\partial z}\right)\mathbf{i} + \left(P_2\frac{\partial Q_1}{\partial x} + Q_2\frac{\partial Q_1}{\partial y} + R_2\frac{\partial Q_1}{\partial z}\right)\mathbf{j}\right.$$

$$+ \left(P_2\frac{\partial R_1}{\partial x} + Q_2\frac{\partial R_1}{\partial y} + R_2\frac{\partial R_1}{\partial z}\right)\mathbf{k}\bigg]$$

$$- \left[\left(P_1\frac{\partial P_2}{\partial x} + Q_1\frac{\partial P_2}{\partial y} + R_1\frac{\partial P_2}{\partial z}\right)\mathbf{i} + \left(P_1\frac{\partial Q_2}{\partial x} + Q_1\frac{\partial Q_2}{\partial y} + R_1\frac{\partial Q_2}{\partial z}\right)\mathbf{j}\right.$$

$$\left.+ \left(P_1\frac{\partial R_2}{\partial x} + Q_1\frac{\partial R_2}{\partial y} + R_1\frac{\partial R_2}{\partial z}\right)\mathbf{k}\right]$$

Hence

$$\mathbf{F}\operatorname{div}\mathbf{G} - \mathbf{G}\operatorname{div}\mathbf{F} + (\mathbf{G}\cdot\nabla)\mathbf{F} - (\mathbf{F}\cdot\nabla)\mathbf{G}$$

$$= \left[\left(P_1\frac{\partial Q_2}{\partial y} + Q_2\frac{\partial P_1}{\partial x}\right) - \left(P_2\frac{\partial Q_1}{\partial y} + Q_1\frac{\partial P_2}{\partial y}\right)\right.$$

$$\left.- \left(P_2\frac{\partial R_1}{\partial z} + R_1\frac{\partial P_2}{\partial z}\right) + \left(P_1\frac{\partial R_2}{\partial z} + R_2\frac{\partial P_1}{\partial z}\right)\right]\mathbf{i}$$

$$+ \left[\left(Q_1\frac{\partial R_2}{\partial z} + R_2\frac{\partial Q_1}{\partial z}\right) - \left(Q_2\frac{\partial R_1}{\partial z} + R_1\frac{\partial Q_2}{\partial z}\right)\right.$$

$$\left.- \left(P_1\frac{\partial Q_2}{\partial x} + Q_2\frac{\partial P_1}{\partial x}\right) + \left(P_2\frac{\partial Q_1}{\partial x} + Q_1\frac{\partial P_2}{\partial x}\right)\right]\mathbf{j}$$

$$+ \left[\left(P_2\frac{\partial R_1}{\partial x} + R_1\frac{\partial P_2}{\partial x}\right) - \left(P_1\frac{\partial R_2}{\partial x} + R_2\frac{\partial P_1}{\partial x}\right)\right.$$

$$\left.- \left(Q_1\frac{\partial R_2}{\partial y} + R_2\frac{\partial Q_1}{\partial y}\right) + \left(Q_2\frac{\partial R_1}{\partial y} + R_1\frac{\partial Q_2}{\partial y}\right)\right]\mathbf{k}$$

$$= \left[\frac{\partial}{\partial y}(P_1 Q_2 - P_2 Q_1) - \frac{\partial}{\partial z}(P_2 R_1 - P_1 R_2)\right]\mathbf{i}$$

$$+ \left[\frac{\partial}{\partial z}(Q_1 R_2 - Q_2 R_1) - \frac{\partial}{\partial x}(P_1 Q_2 - P_2 Q_1)\right]\mathbf{j}$$

$$+ \left[\frac{\partial}{\partial x}(P_2 R_1 - P_1 R_2) - \frac{\partial}{\partial y}(Q_1 R_2 - Q_2 R_1)\right]\mathbf{k}$$

$$= \operatorname{curl}(\mathbf{F}\times\mathbf{G})$$

21. For any piecewise-smooth simple closed plane curve C bounding a region D, we can apply Green's Theorem to

$\mathbf{F}(x, y) = f(x)\,\mathbf{i} + g(y)\,\mathbf{j}$ to get $\int_C f(x)\,dx + g(y)\,dy = \iint_D \left[\frac{\partial}{\partial x}\,g(y) - \frac{\partial}{\partial y}\,f(x)\right] dA = \iint_D 0\,dA = 0.$

22. $\nabla^2(fg) = \dfrac{\partial^2(fg)}{\partial x^2} + \dfrac{\partial^2(fg)}{\partial y^2} + \dfrac{\partial^2(fg)}{\partial z^2}$

$= \dfrac{\partial}{\partial x}\left(\dfrac{\partial f}{\partial x}\,g + f\,\dfrac{\partial g}{\partial x}\right) + \dfrac{\partial}{\partial y}\left(\dfrac{\partial f}{\partial y}\,g + f\,\dfrac{\partial g}{\partial y}\right) + \dfrac{\partial}{\partial z}\left(\dfrac{\partial f}{\partial z}\,g + f\,\dfrac{\partial g}{\partial z}\right)$ [Product Rule]

$= \dfrac{\partial^2 f}{\partial x^2}\,g + 2\,\dfrac{\partial f}{\partial x}\,\dfrac{\partial g}{\partial x} + f\,\dfrac{\partial^2 g}{\partial x^2} + \dfrac{\partial^2 f}{\partial y^2}\,g + 2\,\dfrac{\partial f}{\partial y}\,\dfrac{\partial g}{\partial y}$

$\qquad\qquad\qquad + f\,\dfrac{\partial^2 g}{\partial y^2} + \dfrac{\partial^2 f}{\partial z^2}\,g + 2\,\dfrac{\partial f}{\partial z}\,\dfrac{\partial g}{\partial z} + f\,\dfrac{\partial^2 g}{\partial z^2}$ [Product Rule]

$= f\left(\dfrac{\partial^2 g}{\partial x^2} + \dfrac{\partial^2 g}{\partial y^2} + \dfrac{\partial^2 g}{\partial z^2}\right) + g\left(\dfrac{\partial^2 f}{\partial x^2} + \dfrac{\partial^2 f}{\partial y^2} + \dfrac{\partial^2 f}{\partial z^2}\right) + 2\left\langle \dfrac{\partial f}{\partial x}, \dfrac{\partial f}{\partial y}, \dfrac{\partial f}{\partial z}\right\rangle \cdot \left\langle \dfrac{\partial g}{\partial x}, \dfrac{\partial g}{\partial y}, \dfrac{\partial g}{\partial z}\right\rangle$

$= f\nabla^2 g + g\nabla^2 f + 2\nabla f \cdot \nabla g$

Another method: Using the rules in Exercises 15.6.37(b) [ET 14.6.37(b)] and 17.5.25 [ET 16.5.25], we have

$$\nabla^2(fg) = \nabla \cdot \nabla(fg) = \nabla \cdot (g\,\nabla f + f\,\nabla g) = \nabla g \cdot \nabla f + g\nabla \cdot \nabla f + \nabla f \cdot \nabla g + f\nabla \cdot \nabla g$$

$$= g\,\nabla^2 f + f\,\nabla^2 g + 2\nabla f \cdot \nabla g$$

23. $\nabla^2 f = 0$ means that $\dfrac{\partial^2 f}{\partial x^2} + \dfrac{\partial^2 f}{\partial y^2} = 0$. Now if $\mathbf{F} = f_y\,\mathbf{i} - f_x\,\mathbf{j}$ and C is any closed path in D, then applying Green's

Theorem, we get

$$\int_C \mathbf{F} \cdot d\mathbf{r} = \int_C f_y\,dx - f_x\,dy = \iint_D \left[\dfrac{\partial}{\partial x}\,(-f_x) - \dfrac{\partial}{\partial y}\,(f_y)\right] dA = -\iint_D (f_{xx} + f_{yy})\,dA = -\iint_D 0\,dA = 0$$

Therefore the line integral is independent of path, by Theorem 17.3.3 [ET 16.3.3].

24. (a) $x^2 + y^2 = \cos^2 t + \sin^2 t = 1$, so C lies on the circular cylinder $x^2 + y^2 = 1$.

But also $y = z$, so C lies on the plane $y = z$. Thus C is the intersection of the

plane $y = z$ and the cylinder $x^2 + y^2 = 1$.

(b) Apply Stokes' Theorem, $\int_C \mathbf{F} \cdot d\mathbf{r} = \iint_S \text{curl } \mathbf{F} \cdot d\mathbf{S}$:

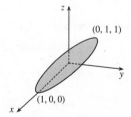

$$\text{curl } \mathbf{F} = \begin{vmatrix} \mathbf{i} & \mathbf{j} & \mathbf{k} \\ \partial/\partial x & \partial/\partial y & \partial/\partial z \\ 2xe^{2y} & 2x^2 e^{2y} + 2y\cot z & -y^2\csc^2 z \end{vmatrix} = \left\langle -2y\csc^2 z - (-2y\csc^2 z), 0, 4xe^{2y} - 4xe^{2y}\right\rangle = \mathbf{0}$$

Therefore $\int_C \mathbf{F} \cdot d\mathbf{r} = \iint_S \mathbf{0} \cdot d\mathbf{S} = 0$.

25. $z = f(x, y) = x^2 + 2y$ with $0 \le x \le 1$, $0 \le y \le 2x$. Thus

$A(S) = \iint_D \sqrt{1 + 4x^2 + 4}\,dA = \int_0^1 \int_0^{2x} \sqrt{5 + 4x^2}\,dy\,dx = \int_0^1 2x\sqrt{5 + 4x^2}\,dx = \frac{1}{6}(5 + 4x^2)^{3/2}\Big]_0^1 = \frac{1}{6}\left(27 - 5\sqrt{5}\right).$

26. (a) $\mathbf{r}_u = -v\,\mathbf{j} + 2u\,\mathbf{k}$, $\mathbf{r}_v = 2v\,\mathbf{i} - u\,\mathbf{j}$ and

$\mathbf{r}_u \times \mathbf{r}_v = 2u^2\,\mathbf{i} + 4uv\,\mathbf{j} + 2v^2\,\mathbf{k}$. Since the point $(4, -2, 1)$

corresponds to $u = 1$, $v = 2$ (or $u = -1$, $v = -2$ but $\mathbf{r}_u \times \mathbf{r}_v$

is the same for both), a normal vector to the surface at $(4, -2, 1)$

is $2\,\mathbf{i} + 8\,\mathbf{j} + 8\,\mathbf{k}$ and an equation of the tangent plane is

$2x + 8y + 8z = 0$ or $x + 4y + 4z = 0$.

(b)

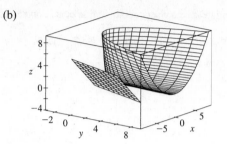

(c) By Definition 17.6.6 [ET 16.6.6] , the area of S is given by

$$A(S) = \int_0^3 \int_{-3}^3 \sqrt{(2u^2)^2 + (4uv)^2 + (2v^2)^2}\, dv\, du = 2\int_0^3 \int_{-3}^3 \sqrt{u^4 + 4u^2v^2 + v^4}\, dv\, du.$$

(d) By Equation 17.7.9 [ET 16.7.9], the surface integral is

$$\iint_S \mathbf{F} \cdot d\mathbf{S} = \int_0^3 \int_{-3}^3 \left\langle \frac{(u^2)^2}{1 + (v^2)^2}, \frac{(v^2)^2}{1 + (-uv)^2}, \frac{(-uv)^2}{1 + (u^2)^2} \right\rangle \cdot \left\langle 2u^2, 4uv, 2v^2 \right\rangle\, dv\, du$$

$$= \int_0^3 \int_{-3}^3 \left(\frac{2u^6}{1 + v^4} + \frac{4uv^5}{1 + u^2v^2} + \frac{2u^2v^4}{1 + u^4} \right)\, dv\, du \approx 1524.0190$$

27. $z = f(x, y) = x^2 + y^2$ with $0 \le x^2 + y^2 \le 4$ so $\mathbf{r}_x \times \mathbf{r}_y = -2x\,\mathbf{i} - 2y\,\mathbf{j} + \mathbf{k}$ (using upward orientation). Then

$$\iint_S z\, dS = \iint_{x^2 + y^2 \le 4} (x^2 + y^2)\sqrt{4x^2 + 4y^2 + 1}\, dA = \int_0^{2\pi} \int_0^2 r^3\sqrt{1 + 4r^2}\, dr\, d\theta = \tfrac{1}{60}\pi\left(391\sqrt{17} + 1\right)$$

(Substitute $u = 1 + 4r^2$ and use tables.)

28. $z = f(x, y) = 4 + x + y$ with $0 \le x^2 + y^2 \le 4$ so $\mathbf{r}_x \times \mathbf{r}_y = -\mathbf{i} - \mathbf{j} + \mathbf{k}$. Then

$$\iint_S (x^2 z + y^2 z)\, dS = \iint_{x^2 + y^2 \le 4} (x^2 + y^2)(4 + x + y)\sqrt{3}\, dA$$

$$= \int_0^2 \int_0^{2\pi} \sqrt{3}\, r^3(4 + r\cos\theta + r\sin\theta)\, d\theta\, dr = \int_0^2 8\pi\sqrt{3}\, r^3\, dr = 32\pi\sqrt{3}$$

29. Since the sphere bounds a simple solid region, the Divergence Theorem applies and

$$\iint_S \mathbf{F} \cdot d\mathbf{S} = \iiint_E (z - 2)\, dV = \iiint_E z\, dV - 2\iiint_E dV = m\bar{z} - 2\left(\tfrac{4}{3}\pi 2^3\right) = -\tfrac{64}{3}\pi.$$

Alternate solution: $\mathbf{F}(\mathbf{r}(\phi, \theta)) = 4\sin\phi\,\cos\theta\,\cos\phi\,\mathbf{i} - 4\sin\phi\,\sin\theta\,\mathbf{j} + 6\sin\phi\,\cos\theta\,\mathbf{k}$,

$\mathbf{r}_\phi \times \mathbf{r}_\theta = 4\sin^2\phi\,\cos\theta\,\mathbf{i} + 4\sin^2\phi\,\sin\theta\,\mathbf{j} + 4\sin\phi\,\cos\phi\,\mathbf{k}$, and

$\mathbf{F} \cdot (\mathbf{r}_\phi \times \mathbf{r}_\theta) = 16\sin^3\phi\,\cos^2\theta\,\cos\phi - 16\sin^3\phi\,\sin^2\theta + 24\sin^2\phi\,\cos\phi\,\cos\theta$. Then

$$\iint_S \mathbf{F} \cdot d\mathbf{S} = \int_0^{2\pi} \int_0^\pi \left(16\sin^3\phi\,\cos\phi\,\cos^2\theta - 16\sin^3\phi\,\sin^2\theta + 24\sin^2\phi\,\cos\phi\,\cos\theta\right)\, d\phi\, d\theta$$

$$= \int_0^{2\pi} \tfrac{4}{3}(-16\sin^2\theta)\, d\theta = -\tfrac{64}{3}\pi$$

30. $z = f(x, y) = x^2 + y^2$, $\mathbf{r}_x \times \mathbf{r}_y = -2x\,\mathbf{i} - 2y\,\mathbf{j} + \mathbf{k}$ (because of upward orientation) and

$\mathbf{F}(\mathbf{r}(x, y)) \cdot (\mathbf{r}_x \times \mathbf{r}_y) = -2x^3 - 2xy^2 + x^2 + y^2$. Then

$$\iint_S \mathbf{F} \cdot d\mathbf{S} = \iint_{x^2 + y^2 \le 1} \left(-2x^3 - 2xy^2 + x^2 + y^2\right)\, dA$$

$$= \int_0^1 \int_0^{2\pi} \left(-2r^3\cos^3\theta - 2r^3\cos\theta\,\sin^2\theta + r^2\right) r\, dr\, d\theta = \int_0^1 r^3(2\pi)\, dr = \tfrac{\pi}{2}$$

31. Since $\text{curl}\,\mathbf{F} = \mathbf{0}$, $\iint_S (\text{curl}\,\mathbf{F}) \cdot d\mathbf{S} = 0$. We parametrize C: $\mathbf{r}(t) = \cos t\,\mathbf{i} + \sin t\,\mathbf{j}$, $0 \le t \le 2\pi$ and

$$\oint_C \mathbf{F} \cdot d\mathbf{r} = \int_0^{2\pi} \left(-\cos^2 t\,\sin t + \sin^2 t\,\cos t\right)\, dt = \tfrac{1}{3}\cos^3 t + \tfrac{1}{3}\sin^3 t\,\Big]_0^{2\pi} = 0.$$

32. $\iint_S \text{curl}\,\mathbf{F} \cdot d\mathbf{S} = \oint_C \mathbf{F} \cdot d\mathbf{r}$ where C: $\mathbf{r}(t) = 2\cos t\,\mathbf{i} + 2\sin t\,\mathbf{j} + \mathbf{k}$, $0 \leq t \leq 2\pi$, so $\mathbf{r}'(t) = -2\sin t\,\mathbf{i} + 2\cos t\,\mathbf{j}$,

$\mathbf{F}(\mathbf{r}(t)) = 8\cos^2 t\,\sin t\,\mathbf{i} + 2\sin t\,\mathbf{j} + e^{4\cos t\,\sin t}\,\mathbf{k}$, and $\mathbf{F}(\mathbf{r}(t)) \cdot \mathbf{r}'(t) = -16\cos^2 t\sin^2 t + 4\sin t\cos t$. Thus

$\oint_C \mathbf{F} \cdot d\mathbf{r} = \int_0^{2\pi}(-16\cos^2 t\,\sin^2 t + 4\sin t\,\cos t)\,dt = \left[-16\left(-\tfrac{1}{4}\sin t\cos^3 t + \tfrac{1}{16}\sin 2t + \tfrac{1}{8}t\right) + 2\sin^2 t\right]_0^{2\pi} = -4\pi$.

33. The surface is given by $x + y + z = 1$ or $z = 1 - x - y$, $0 \leq x \leq 1$, $0 \leq y \leq 1 - x$ and $\mathbf{r}_x \times \mathbf{r}_y = \mathbf{i} + \mathbf{j} + \mathbf{k}$. Then

$\oint_C \mathbf{F} \cdot d\mathbf{r} = \iint_S \text{curl}\,\mathbf{F} \cdot d\mathbf{S} = \iint_D(-y\,\mathbf{i} - z\,\mathbf{j} - x\,\mathbf{k}) \cdot (\mathbf{i} + \mathbf{j} + \mathbf{k})\,dA = \iint_D(-1)\,dA = -(\text{area of }D) = -\tfrac{1}{2}$.

34. $\iint_S \mathbf{F} \cdot d\mathbf{S} = \iiint_E 3(x^2 + y^2 + z^2)\,dV = \int_0^{2\pi}\int_0^1\int_0^2(3r^2 + 3z^2)\,r\,dz\,dr\,d\theta = 2\pi\int_0^1(6r^3 + 8r)\,dr = 11\pi$

35. $\iiint_E \text{div}\,\mathbf{F}\,dV = \iiint_{x^2+y^2+z^2\leq 1} 3\,dV = 3(\text{volume of sphere}) = 4\pi$. Then

$\mathbf{F}(\mathbf{r}(\phi,\theta)) \cdot (\mathbf{r}_\phi \times \mathbf{r}_\theta) = \sin^3\phi\cos^2\theta + \sin^3\phi\sin^2\theta + \sin\phi\cos^2\phi = \sin\phi$ and

$\iint_S \mathbf{F} \cdot d\mathbf{S} = \int_0^{2\pi}\int_0^\pi \sin\phi\,d\phi\,d\theta = (2\pi)(2) = 4\pi$.

36. Here we must use Equation 17.9.7 [ET 16.9.7] since $\mathbf{F}$ is not defined at the origin. Let S_1 be the sphere of radius 1 with center at the origin and outer unit normal $\mathbf{n}_1$. Let S_2 be the surface of the ellipsoid with outer unit normal $\mathbf{n}_2$ and let E be the solid region between S_1 and S_2. Then the outward flux of $\mathbf{F}$ through the ellipsoid is given by

$\iint_{S_2} \mathbf{F} \cdot \mathbf{n}_2\,dS = -\iint_{S_1} \mathbf{F} \cdot (-\mathbf{n}_1)\,dS + \iiint_E \text{div}\,\mathbf{F}\,dV$. But $\mathbf{F} = \mathbf{r}/|\mathbf{r}|^3$, so

$\text{div}\,\mathbf{F} = \nabla \cdot (|\mathbf{r}|^{-3}\,\mathbf{r}) = |\mathbf{r}|^{-3}(\nabla \cdot \mathbf{r}) + \mathbf{r} \cdot (\nabla|\mathbf{r}|^{-3}) = |\mathbf{r}|^{-3}(3) + \mathbf{r} \cdot (-3|\mathbf{r}|^{-4})(\mathbf{r}|\mathbf{r}|^{-1}) = 0$. [Here we have used

Exercises 17.5.30(a) [ET 16.5.30(a)] and 17.5.31(a) [ET 16.5.31(a)].] And $\mathbf{F} \cdot \mathbf{n}_1 = \dfrac{\mathbf{r}}{|\mathbf{r}|^3} \cdot \dfrac{\mathbf{r}}{|\mathbf{r}|} = |\mathbf{r}|^{-2} = 1$ on S_1.

Thus $\iint_{S_2} \mathbf{F} \cdot \mathbf{n}_2\,dS = \iint_{S_1} dS + \iiint_E 0\,dV = (\text{surface area of the unit sphere}) = 4\pi(1)^2 = 4\pi$.

37. Because $\text{curl}\,\mathbf{F} = \mathbf{0}$, $\mathbf{F}$ is conservative, and if $f(x, y, z) = x^3yz - 3xy + z^2$, then $\nabla f = \mathbf{F}$. Hence

$\int_C \mathbf{F} \cdot d\mathbf{r} = \int_C \nabla f \cdot d\mathbf{r} = f(0, 3, 0) - f(0, 0, 2) = 0 - 4 = -4$.

38. Let C' be the circle with center at the origin and radius a as in the figure.

Let D be the region bounded by C and C'. Then D's positively oriented

boundary is $C \cup (-C')$. Hence by Green's Theorem

$\displaystyle\int_C \mathbf{F} \cdot d\mathbf{r} + \int_{-C'} \mathbf{F} \cdot d\mathbf{r} = \iint_D \left(\frac{\partial Q}{\partial x} - \frac{\partial P}{\partial y}\right)dA = 0$, so

$\int_C \mathbf{F} \cdot d\mathbf{r} = -\int_{-C'} \mathbf{F} \cdot d\mathbf{r} = \int_{C'} \mathbf{F} \cdot d\mathbf{r} = \int_0^{2\pi} \mathbf{F}(\mathbf{r}(t)) \cdot \mathbf{r}'(t)\,dt$

$= \displaystyle\int_0^{2\pi}\left[\frac{2a^3\cos^3 t + 2a^3\cos t\,\sin^2 t - 2a\sin t}{a^2}(-a\sin t) + \frac{2a^3\sin^3 t + 2a^3\cos^2 t\,\sin t + 2a\cos t}{a^2}(a\cos t)\right]dt$

$= \displaystyle\int_0^{2\pi} \frac{2a^2}{a^2}\,dt = 4\pi$

39. By the Divergence Theorem, $\iint_S \mathbf{F} \cdot \mathbf{n}\,dS = \iiint_E \text{div}\,\mathbf{F}\,dV = 3(\text{volume of }E) = 3(8 - 1) = 21$.

40. The stated conditions allow us to use the Divergence Theorem. Hence $\iint_S \text{curl}\,\mathbf{F} \cdot d\mathbf{S} = \iiint_E \text{div}(\text{curl}\,\mathbf{F})\,dV = 0$ since $\text{div}(\text{curl}\,\mathbf{F}) = 0$.

41. Let $\mathbf{F} = \mathbf{a} \times \mathbf{r} = \langle a_1, a_2, a_3 \rangle \times \langle x, y, z \rangle = \langle a_2 z - a_3 y, a_3 x - a_1 z, a_1 y - a_2 x \rangle$. Then $\text{curl}\,\mathbf{F} = \langle 2a_1, 2a_2, 2a_3 \rangle = 2\mathbf{a}$, and $\iint_S 2\mathbf{a} \cdot d\mathbf{S} = \iint_S \text{curl}\,\mathbf{F} \cdot d\mathbf{S} = \int_C \mathbf{F} \cdot d\mathbf{r} = \int_C(\mathbf{a} \times \mathbf{r}) \cdot d\mathbf{r}$ by Stokes' Theorem.

□ PROBLEMS PLUS

1. Let S_1 be the portion of $\Omega(S)$ between $S(a)$ and S, and let ∂S_1 be its boundary. Also let S_L be the lateral surface of S_1 [that is, the surface of S_1 except S and $S(a)$]. Applying the Divergence Theorem we have $\iint_{\partial S_1} \frac{\mathbf{r} \cdot \mathbf{n}}{r^3} \, dS = \iiint_{S_1} \nabla \cdot \frac{\mathbf{r}}{r^3} \, dV$.

But

$$\nabla \cdot \frac{\mathbf{r}}{r^3} = \left\langle \frac{\partial}{\partial x}, \frac{\partial}{\partial y}, \frac{\partial}{\partial z} \right\rangle \cdot \left\langle \frac{x}{(x^2 + y^2 + z^2)^{3/2}}, \frac{y}{(x^2 + y^2 + z^2)^{3/2}}, \frac{z}{(x^2 + y^2 + z^2)^{3/2}} \right\rangle$$

$$= \frac{(x^2 + y^2 + z^2 - 3x^2) + (x^2 + y^2 + z^2 - 3y^2) + (x^2 + y^2 + z^2 - 3z^2)}{(x^2 + y^2 + z^2)^{5/2}} = 0$$

$\Rightarrow \iint_{\partial S_1} \frac{\mathbf{r} \cdot \mathbf{n}}{r^3} \, dS = \iiint_{S_1} 0 \, dV = 0$. On the other hand, notice that for the surfaces of ∂S_1 other than $S(a)$ and S,

$\mathbf{r} \cdot \mathbf{n} = 0 \Rightarrow$

$$0 = \iint_{\partial S_1} \frac{\mathbf{r} \cdot \mathbf{n}}{r^3} \, dS = \iint_S \frac{\mathbf{r} \cdot \mathbf{n}}{r^3} \, dS + \iint_{S(a)} \frac{\mathbf{r} \cdot \mathbf{n}}{r^3} \, dS + \iint_{S_L} \frac{\mathbf{r} \cdot \mathbf{n}}{r^3} \, dS = \iint_S \frac{\mathbf{r} \cdot \mathbf{n}}{r^3} \, dS + \iint_{S(a)} \frac{\mathbf{r} \cdot \mathbf{n}}{r^3} \, dS \Rightarrow$$

$\iint_S \frac{\mathbf{r} \cdot \mathbf{n}}{r^3} \, dS = -\iint_{S(a)} \frac{\mathbf{r} \cdot \mathbf{n}}{r^3} \, dS$. Notice that on $S(a)$, $r = a \Rightarrow \mathbf{n} = -\frac{\mathbf{r}}{r} = -\frac{\mathbf{r}}{a}$ and $\mathbf{r} \cdot \mathbf{r} = r^2 = a^2$, so

that $-\iint_{S(a)} \frac{\mathbf{r} \cdot \mathbf{n}}{r^3} \, dS = \iint_{S(a)} \frac{\mathbf{r} \cdot \mathbf{r}}{a^4} \, dS = \iint_{S(a)} \frac{a^2}{a^4} \, dS = \frac{1}{a^2} \iint_{S(a)} dS = \frac{\text{area of } S(a)}{a^2} = |\Omega(S)|$.

Therefore $|\Omega(S)| = \iint_S \frac{\mathbf{r} \cdot \mathbf{n}}{r^3} \, dS$.

2. By Green's Theorem

$$\int_C (y^3 - y) \, dx - 2x^3 \, dy = \iint_D \left[\frac{\partial(-2x^3)}{\partial x} - \frac{\partial(y^3 - y)}{\partial y} \right] dA = \iint_D (1 - 6x^2 - 3y^2) \, dA$$

Notice that for $6x^2 + 3y^2 > 1$, the integrand is negative. The integral has maximum value if it is evaluated only in the region where the integrand is positive, which is within the ellipse $6x^2 + 3y^2 = 1$. So the simple closed curve that gives a maximum value for the line integral is the ellipse $6x^2 + 3y^2 = 1$.

3. The given line integral $\frac{1}{2} \int_C (bz - cy) \, dx + (cx - az) \, dy + (ay - bx) \, dz$ can be expressed as $\int_C \mathbf{F} \cdot d\mathbf{r}$ if we define the vector field $\mathbf{F}$ by $\mathbf{F}(x, y, z) = P\mathbf{i} + Q\mathbf{j} + R\mathbf{k} = \frac{1}{2}(bz - cy)\mathbf{i} + \frac{1}{2}(cx - az)\mathbf{j} + \frac{1}{2}(ay - bx)\mathbf{k}$. Then define S to be the planar interior of C, so S is an oriented, smooth surface. Stokes' Theorem says $\int_C \mathbf{F} \cdot d\mathbf{r} = \iint_S \text{curl } \mathbf{F} \cdot d\mathbf{S} = \iint_S \text{curl } \mathbf{F} \cdot \mathbf{n} \, dS$. Now

$$\text{curl } \mathbf{F} = \left(\frac{\partial R}{\partial y} - \frac{\partial Q}{\partial z} \right)\mathbf{i} + \left(\frac{\partial P}{\partial z} - \frac{\partial R}{\partial x} \right)\mathbf{j} + \left(\frac{\partial Q}{\partial x} - \frac{\partial P}{\partial y} \right)\mathbf{k}$$

$$= \left(\tfrac{1}{2}a + \tfrac{1}{2}a \right)\mathbf{i} + \left(\tfrac{1}{2}b + \tfrac{1}{2}b \right)\mathbf{j} + \left(\tfrac{1}{2}c + \tfrac{1}{2}c \right)\mathbf{k} = a\mathbf{i} + b\mathbf{j} + c\mathbf{k} = \mathbf{n}$$

so $\text{curl } \mathbf{F} \cdot \mathbf{n} = \mathbf{n} \cdot \mathbf{n} = |\mathbf{n}|^2 = 1$, hence $\iint_S \text{curl } \mathbf{F} \cdot \mathbf{n} \, dS = \iint_S dS$ which is simply the surface area of S. Thus, $\int_C \mathbf{F} \cdot d\mathbf{r} = \frac{1}{2} \int_C (bz - cy) \, dx + (cx - az) \, dy + (ay - bx) \, dz$ is the plane area enclosed by C.

4. The surface given by $x = \sin u$, $y = \sin v$, $z = \sin(u + v)$ is difficult to visualize, so we first graph the surface from three different points of view.

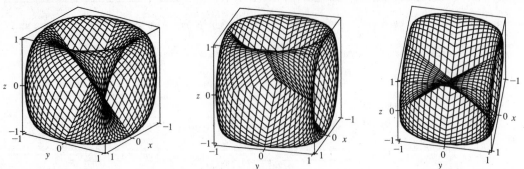

The trace in the horizontal plane $z = 0$ is given by $z = \sin(u + v) = 0 \;\Rightarrow\; u + v = k\pi$ [k an integer]. Then

we can write $v = k\pi - u$, and the trace is given by the parametric equations $x = \sin u$,

$y = \sin v = \sin(k\pi - u) = \sin k\pi \cos u - \cos k\pi \sin u = \pm \sin u$, and since $\sin u = x$, the trace consists of the two lines

$y = \pm x$.

If $z = 1$, $z = \sin(u + v) = 1 \;\Rightarrow\; u + v = \frac{\pi}{2} + 2k\pi$. So $v = \left(\frac{\pi}{2} + 2k\pi\right) - u$ and the trace in $z = 1$ is given by the

parametric equations $x = \sin u$, $y = \sin v = \sin\left(\left(\frac{\pi}{2} + 2k\pi\right) - u\right) = \sin\left(\frac{\pi}{2} + 2k\pi\right)\cos u - \cos\left(\frac{\pi}{2} + 2k\pi\right)\sin u = \cos u$.

This curve is equivalent to $x^2 + y^2 = 1$, $z = 1$, a circle of radius 1. Similarly, in $z = -1$ we have $z = \sin(u + v) = -1 \;\Rightarrow\;$

$u + v = \frac{3\pi}{2} + 2k\pi \;\Rightarrow\; v = \left(\frac{3\pi}{2} + 2k\pi\right) - u$, so the trace is given by the parametric equations $x = \sin u$,

$y = \sin v = \sin\left(\left(\frac{3\pi}{2} + 2k\pi\right) - u\right) = \sin\left(\frac{3\pi}{2} + 2k\pi\right)\cos u - \cos\left(\frac{3\pi}{2} + 2k\pi\right)\sin u = -\cos u$, which again is a circle,

$x^2 + y^2 = 1$, $z = -1$.

If $z = \frac{1}{2}$, $z = \sin(u + v) = \frac{1}{2} \;\Rightarrow\; u + v = \alpha + 2k\pi$ where $\alpha = \frac{\pi}{6}$ or $\frac{5\pi}{6}$. Then

$v = (\alpha + 2k\pi) - u$ and the trace in $z = \frac{1}{2}$ is given by the parametric equations $x = \sin u$,

$y = \sin v = \sin[(\alpha + 2k\pi) - u] = \sin(\alpha + 2k\pi)\cos u - \cos(\alpha + 2k\pi)\sin u = \frac{1}{2}\cos u \pm \frac{\sqrt{3}}{2}\sin u$. In rectangular

coordinates, $x = \sin u$ so $y = \frac{1}{2}\cos u \pm \frac{\sqrt{3}}{2}x \;\Rightarrow\; y \pm \frac{\sqrt{3}}{2}x = \frac{1}{2}\cos u \;\Rightarrow\; 2y \pm \sqrt{3}x = \cos u$. But then

$x^2 + \left(2y \pm \sqrt{3}x\right)^2 = \sin^2 u + \cos^2 u = 1 \;\Rightarrow\; x^2 + 4y^2 \pm 4\sqrt{3}xy + 3x^2 = 1 \;\Rightarrow\; 4x^2 \pm 4\sqrt{3}xy + 4y^2 = 1$, which

may be recognized as a conic section. In particular, each equation is an ellipse rotated $\pm 45°$ from the standard orientation (see

the graph below). The trace in $z = -\frac{1}{2}$ is similar: $z = \sin(u + v) = -\frac{1}{2} \;\Rightarrow\; u + v = \beta + 2k\pi$ where $\beta = \frac{7\pi}{6}$ or $\frac{11\pi}{6}$.

Then $v = (\beta + 2k\pi) - u$ and the trace is given by the parametric equations $x = \sin u$,

$y = \sin v = \sin[(\beta + 2k\pi) - u] = \sin(\beta + 2k\pi)\cos u - \cos(\beta + 2k\pi)\sin u = -\frac{1}{2}\cos u \pm \frac{\sqrt{3}}{2}\sin u$. If we convert to

rectangular coordinates, we arrive at the same pair of equations, $4x^2 \pm 4\sqrt{3}xy + 4y^2 = 1$, so the trace is identical to the trace

in $z = \frac{1}{2}$.

Graphing each of these, we have the following 5 traces.

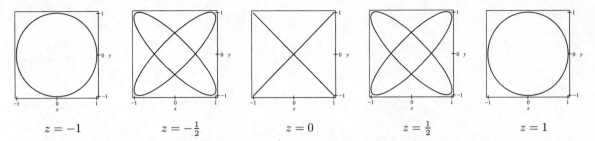

$z = -1$ $\qquad z = -\frac{1}{2}$ $\qquad z = 0$ $\qquad z = \frac{1}{2}$ $\qquad z = 1$

Visualizing these traces on the surface reveals that horizontal cross sections are pairs of intersecting ellipses whose major axes are perpendicular to each other. At the bottom of the surface, $z = -1$, the ellipses coincide as circles of radius 1. As we move up the surface, the ellipses become narrower until at $z = 0$ they collapse into line segments, after which the process is reversed, and the ellipses widen to again coincide as circles at $z = 1$.

5. $(\mathbf{F} \cdot \nabla) \mathbf{G} = \left(P_1 \dfrac{\partial}{\partial x} + Q_1 \dfrac{\partial}{\partial y} + R_1 \dfrac{\partial}{\partial z} \right) (P_2 \,\mathbf{i} + Q_2 \,\mathbf{j} + R_2 \,\mathbf{k})$

$$= \left(P_1 \frac{\partial P_2}{\partial x} + Q_1 \frac{\partial P_2}{\partial y} + R_1 \frac{\partial P_2}{\partial z} \right) \mathbf{i} + \left(P_1 \frac{\partial Q_2}{\partial x} + Q_1 \frac{\partial Q_2}{\partial y} + R_1 \frac{\partial Q_2}{\partial z} \right) \mathbf{j}$$

$$+ \left(P_1 \frac{\partial R_2}{\partial x} + Q_1 \frac{\partial R_2}{\partial y} + R_1 \frac{\partial R_2}{\partial z} \right) \mathbf{k}$$

$$= (\mathbf{F} \cdot \nabla P_2) \,\mathbf{i} + (\mathbf{F} \cdot \nabla Q_2) \,\mathbf{j} + (\mathbf{F} \cdot \nabla R_2) \,\mathbf{k}.$$

Similarly, $(\mathbf{G} \cdot \nabla) \mathbf{F} = (\mathbf{G} \cdot \nabla P_1) \,\mathbf{i} + (\mathbf{G} \cdot \nabla Q_1) \,\mathbf{j} + (\mathbf{G} \cdot \nabla R_1) \,\mathbf{k}$. Then

$$\mathbf{F} \times \operatorname{curl} \mathbf{G} = \begin{vmatrix} \mathbf{i} & \mathbf{j} & \mathbf{k} \\ P_1 & Q_1 & R_1 \\ \partial R_2/\partial y - \partial Q_2/\partial z & \partial P_2/\partial z - \partial R_2/\partial x & \partial Q_2/\partial x - \partial P_2/\partial y \end{vmatrix}$$

$$= \left(Q_1 \frac{\partial Q_2}{\partial x} - Q_1 \frac{\partial P_2}{\partial y} - R_1 \frac{\partial P_2}{\partial z} + R_1 \frac{\partial R_2}{\partial x} \right) \mathbf{i} + \left(R_1 \frac{\partial R_2}{\partial y} - R_1 \frac{\partial Q_2}{\partial z} - P_1 \frac{\partial Q_2}{\partial x} + P_1 \frac{\partial P_2}{\partial y} \right) \mathbf{j}$$

$$+ \left(P_1 \frac{\partial P_2}{\partial z} - P_1 \frac{\partial R_2}{\partial x} - Q_1 \frac{\partial R_2}{\partial y} + Q_1 \frac{\partial Q_2}{\partial z} \right) \mathbf{k}$$

and

$$\mathbf{G} \times \operatorname{curl} \mathbf{F} = \left(Q_2 \frac{\partial Q_1}{\partial x} - Q_2 \frac{\partial P_1}{\partial y} - R_2 \frac{\partial P_1}{\partial z} + R_2 \frac{\partial R_1}{\partial x} \right) \mathbf{i} + \left(R_2 \frac{\partial R_1}{\partial y} - R_2 \frac{\partial Q_1}{\partial z} - P_2 \frac{\partial Q_1}{\partial x} + P_2 \frac{\partial P_1}{\partial y} \right) \mathbf{j}$$

$$+ \left(P_2 \frac{\partial P_1}{\partial z} - P_2 \frac{\partial R_1}{\partial x} - Q_2 \frac{\partial R_1}{\partial y} + Q_2 \frac{\partial Q_1}{\partial z} \right) \mathbf{k}.$$

Then

$$(\mathbf{F} \cdot \nabla) \mathbf{G} + \mathbf{F} \times \operatorname{curl} \mathbf{G} = \left(P_1 \frac{\partial P_2}{\partial x} + Q_1 \frac{\partial Q_2}{\partial x} + R_1 \frac{\partial R_2}{\partial x} \right) \mathbf{i} + \left(P_1 \frac{\partial P_2}{\partial y} + Q_1 \frac{\partial Q_2}{\partial y} + R_1 \frac{\partial R_2}{\partial y} \right) \mathbf{j}$$

$$+ \left(P_1 \frac{\partial P_2}{\partial z} + Q_1 \frac{\partial Q_2}{\partial z} + R_1 \frac{\partial R_2}{\partial z} \right) \mathbf{k}$$

and

$$(\mathbf{G} \cdot \nabla)\mathbf{F} + \mathbf{G} \times \text{curl}\,\mathbf{F} = \left(P_2\frac{\partial P_1}{\partial x} + Q_2\frac{\partial Q_1}{\partial x} + R_2\frac{\partial R_1}{\partial x}\right)\mathbf{i} + \left(P_2\frac{\partial P_1}{\partial y} + Q_2\frac{\partial Q_1}{\partial y} + R_2\frac{\partial R_1}{\partial y}\right)\mathbf{j}$$

$$+ \left(P_2\frac{\partial P_1}{\partial z} + Q_2\frac{\partial Q_1}{\partial z} + R_2\frac{\partial R_1}{\partial z}\right)\mathbf{k}.$$

Hence

$$(\mathbf{F} \cdot \nabla)\mathbf{G} + \mathbf{F} \times \text{curl}\,\mathbf{G} + (\mathbf{G} \cdot \nabla)\mathbf{F} + \mathbf{G} \times \text{curl}\,\mathbf{F}$$

$$= \left[\left(P_1\frac{\partial P_2}{\partial x} + P_2\frac{\partial P_1}{\partial x}\right) + \left(Q_1\frac{\partial Q_2}{\partial x} + Q_2\frac{\partial Q_1}{\partial y}\right) + \left(R_1\frac{\partial R_2}{\partial x} + R_2\frac{\partial R_1}{\partial x}\right)\right]\mathbf{i}$$

$$+ \left[\left(P_1\frac{\partial P_2}{\partial y} + P_2\frac{\partial P_1}{\partial y}\right) + \left(Q_1\frac{\partial Q_2}{\partial y} + Q_2\frac{\partial Q_1}{\partial y}\right) + \left(R_1\frac{\partial R_2}{\partial y} + R_2\frac{\partial R_1}{\partial y}\right)\right]\mathbf{j}$$

$$+ \left[\left(P_1\frac{\partial P_2}{\partial z} + P_2\frac{\partial P_1}{\partial z}\right) + \left(Q_1\frac{\partial Q_2}{\partial z} + Q_2\frac{\partial Q_1}{\partial z}\right) + \left(R_1\frac{\partial R_2}{\partial z} + R_2\frac{\partial R_1}{\partial z}\right)\right]\mathbf{k}$$

$$= \nabla(P_1 P_2 + Q_1 Q_2 + R_1 R_2) = \nabla(\mathbf{F} \cdot \mathbf{G}).$$

6. (a) First we place the piston on coordinate axes so the top of the cylinder is at the origin and $x(t) \geq 0$ is the distance from the top of the cylinder to the piston at time t. Let C_1 be the curve traced out by the piston during one four-stroke cycle, so C_1 is given by $\mathbf{r}(t) = x(t)\,\mathbf{i}$, $a \leq t \leq b$. (Thus, the curve lies on the positive x-axis and reverses direction several times.) The force on the piston is $AP(t)\,\mathbf{i}$, where A is the area of the top of the piston and $P(t)$ is the pressure in the cylinder at time t. As in Section 17.2 [ET 16.2], the work done on the piston is $\int_{C_1} \mathbf{F} \cdot d\mathbf{r} = \int_a^b AP(t)\,\mathbf{i} \cdot x'(t)\,\mathbf{i}\,dt = \int_a^b AP(t)\,x'(t)\,dt$. Here, the volume of the cylinder at time t is $V(t) = Ax(t) \Rightarrow V'(t) = Ax'(t) \Rightarrow$ $\int_a^b AP(t)\,x'(t)\,dt = \int_a^b P(t)\,V'(t)\,dt$. Since the curve C in the PV-plane corresponds to the values of P and V at time t, $a \leq t \leq b$, we have

$$W = \int_a^b AP(t)\,x'(t)\,dt = \int_a^b P(t)\,V'(t)\,dt = \int_C P\,dV$$

Another method: If we divide the time interval $[a, b]$ into n subintervals of equal length Δt, the amount of work done on the piston in the ith time interval is approximately $AP(t_i)[x(t_i) - x(t_{i-1})]$. Thus we estimate the total work done during one cycle to be $\sum_{i=1}^{n} AP(t_i)[x(t_i) - x(t_{i-1})]$. If we allow $n \to \infty$, we have

$$W = \lim_{n\to\infty} \sum_{i=1}^{n} AP(t_i)[x(t_i) - x(t_{i-1})] = \lim_{n\to\infty} \sum_{i=1}^{n} P(t_i)[Ax(t_i) - Ax(t_{i-1})] = \lim_{n\to\infty} \sum_{i=1}^{n} P(t_i)[V(t_i) - V(t_{i-1})]$$

$$= \int_C P\,dV$$

(b) Let C_L be the lower loop of the curve C and C_U the upper loop. Then $C = C_L \cup C_U$. C_L is positively oriented, so from Formula 17.4.5 [ET 16.4.5] we know the area of the lower loop in the PV-plane is given by $-\oint_{C_L} P\,dV$. C_U is negatively oriented, so the area of the upper loop is given by $-\left(-\oint_{C_U} P\,dV\right) = \oint_{C_U} P\,dV$. From part (a),

$$W = \int_C P\,dV = \int_{C_L \cup C_U} P\,dV = \oint_{C_L} P\,dV + \oint_{C_U} P\,dV = \oint_{C_U} P\,dV - \left(-\oint_{C_L} P\,dV\right),$$

the difference of the areas enclosed by the two loops of C.

18.1 Second-Order Linear Equations

1. The auxiliary equation is $r^2 - r - 6 = 0 \;\Rightarrow\; (r-3)(r+2) = 0 \;\Rightarrow\; r = 3, r = -2$. Then by (8) the general solution is

$y = c_1 e^{3x} + c_2 e^{-2x}$.

2. The auxiliary equation is $r^2 + 4r + 4 = 0 \;\Rightarrow\; (r+2)^2 = 0 \;\Rightarrow\; r = -2$. Then by (10), the general solution is

$y = c_1 e^{-2x} + c_2 x e^{-2x}$.

3. The auxiliary equation is $r^2 + 16 = 0 \;\Rightarrow\; r = \pm 4i$. Then by (11) the general solution is

$y = e^{0x}(c_1 \cos 4x + c_2 \sin 4x) = c_1 \cos 4x + c_2 \sin 4x$.

4. The auxiliary equation is $r^2 - 8r + 12 = (r-6)(r-2) = 0 \;\Rightarrow\; r = 6, r = 2$. Then the general solution is

$y = c_1 e^{6x} + c_2 e^{2x}$.

5. The auxiliary equation is $9r^2 - 12r + 4 = 0 \;\Rightarrow\; (3r-2)^2 = 0 \;\Rightarrow\; r = \frac{2}{3}$. Then by (10), the general solution is

$y = c_1 e^{2x/3} + c_2 x e^{2x/3}$.

6. The auxiliary equation is $25r^2 + 9 = 0 \;\Rightarrow\; r^2 = -\frac{9}{25} \;\Rightarrow\; r = \pm\frac{3}{5}i$, so the general solution is

$y = e^{0x}\left[c_1 \cos\left(\frac{3}{5}x\right) + c_2 \sin\left(\frac{3}{5}x\right)\right] = c_1 \cos\left(\frac{3}{5}x\right) + c_2 \sin\left(\frac{3}{5}x\right)$.

7. The auxiliary equation is $2r^2 - r = r(2r-1) = 0 \;\Rightarrow\; r = 0, r = \frac{1}{2}$, so $y = c_1 e^{0x} + c_2 e^{x/2} = c_1 + c_2 e^{x/2}$.

8. The auxiliary equation is $r^2 - 4r + 1 = 0 \;\Rightarrow\; r = \dfrac{4 \pm \sqrt{12}}{2} = 2 \pm \sqrt{3}$, so $y = c_1 e^{(2+\sqrt{3})x} + c_2 e^{(2-\sqrt{3})x}$.

9. The auxiliary equation is $r^2 - 4r + 13 = 0 \;\Rightarrow\; r = \dfrac{4 \pm \sqrt{-36}}{2} = 2 \pm 3i$, so $y = e^{2x}(c_1 \cos 3x + c_2 \sin 3x)$.

10. The auxiliary equation is $r^2 + 3r = r(r+3) = 0 \;\Rightarrow\; r = 0, r = -3$, so $y = c_1 + c_2 e^{-3x}$.

11. The auxiliary equation is $2r^2 + 2r - 1 = 0 \;\Rightarrow\; r = \dfrac{-2 \pm \sqrt{12}}{4} = -\dfrac{1}{2} \pm \dfrac{\sqrt{3}}{2}$, so

$y = c_1 e^{\left(-1/2 + \sqrt{3}/2\right)t} + c_2 e^{\left(-1/2 - \sqrt{3}/2\right)t}$.

12. The auxiliary equation is $8r^2 + 12r + 5 = 0 \;\Rightarrow\; r = \dfrac{-12 \pm \sqrt{-16}}{16} = -\dfrac{3}{4} \pm \dfrac{1}{4}i$, so

$y = e^{-3t/4}\left[c_1 \cos\left(\frac{1}{4}t\right) + c_2 \sin\left(\frac{1}{4}t\right)\right]$.

13. The auxiliary equation is $100r^2 + 200r + 101 = 0 \;\Rightarrow\; r = \dfrac{-200 \pm \sqrt{-400}}{200} = -1 \pm \dfrac{1}{10}i$, so

$P = e^{-t}\left[c_1 \cos\left(\frac{1}{10}t\right) + c_2 \sin\left(\frac{1}{10}t\right)\right]$.

14. The auxiliary equation is $r^2 + 4r + 20 = 0 \Rightarrow r = \dfrac{-4 \pm \sqrt{-64}}{2} = -2 \pm 4i$,

so the general solution is $y = e^{-2x}(c_1 \cos 4x + c_2 \sin 4x)$. We graph the basic

solutions $f(x) = e^{-2x} \cos 4x$, $g(x) = e^{-2x} \sin 4x$ as well as

$y = e^{-2x}(\cos 4x - \sin 4x)$ and $y = e^{-2x}(-2 \cos 4x + 2 \sin 4x)$. All the

solutions oscillate with amplitudes that become arbitrarily large as $x \to -\infty$ and

the solutions are asymptotic to the x-axis as $x \to \infty$.

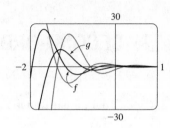

15. The auxiliary equation is $5r^2 - 2r - 3 = (5r + 3)(r - 1) = 0 \Rightarrow r = -\frac{3}{5}$,

$r = 1$, so the general solution is $y = c_1 e^{-3x/5} + c_2 e^x$. We graph the basic

solutions $f(x) = e^{-3x/5}$, $g(x) = e^x$ as well as $y = e^{-3x/5} + 2e^x$,

$y = e^{-3x/5} - e^x$, and $y = -2e^{-3x/5} - e^x$. Each solution consists of a single

continuous curve that approaches either 0 or $\pm\infty$ as $x \to \pm\infty$.

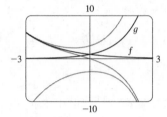

16. The auxiliary equation is $9r^2 + 6r + 1 = (3r + 1)^2 = 0 \Rightarrow r = -\frac{1}{3}$, so the

general solution is $y = c_1 e^{-x/3} + c_2 x e^{-x/3}$. We graph the basic solutions

$f(x) = e^{-x/3}$, $g(x) = x e^{-x/3}$ as well as $y = 3e^{-x/3} + 2x e^{-x/3}$,

$y = -e^{-x/3} - 2x e^{-x/3}$, and $y = -4e^{-x/3} + 3x e^{-x/3}$. The graphs are all

asymptotic to the x-axis as $x \to \infty$, and as $x \to -\infty$ the solutions approach $\pm\infty$.

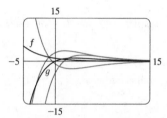

17. $2r^2 + 5r + 3 = (2r + 3)(r + 1) = 0$, so $r = -\frac{3}{2}$, $r = -1$ and the general solution is $y = c_1 e^{-3x/2} + c_2 e^{-x}$.

Then $y(0) = 3 \Rightarrow c_1 + c_2 = 3$ and $y'(0) = -4 \Rightarrow -\frac{3}{2}c_1 - c_2 = -4$, so $c_1 = 2$ and $c_2 = 1$. Thus the solution to the

initial-value problem is $y = 2e^{-3x/2} + e^{-x}$.

18. $r^2 + 3 = 0 \Rightarrow r = \pm\sqrt{3}\,i$ and the general solution is

$y = e^{0x}\left(c_1 \cos(\sqrt{3}\,x) + c_2 \sin(\sqrt{3}\,x)\right) = c_1 \cos(\sqrt{3}\,x) + c_2 \sin(\sqrt{3}\,x)$. Then $y(0) = 1 \Rightarrow c_1 = 1$ and $y'(0) = 3 \Rightarrow$

$c_2 = \sqrt{3}$, so the solution to the initial-value problem is $y = \cos(\sqrt{3}\,x) + \sqrt{3} \sin(\sqrt{3}\,x)$.

19. $4r^2 - 4r + 1 = (2r - 1)^2 = 0 \Rightarrow r = \frac{1}{2}$ and the general solution is $y = c_1 e^{x/2} + c_2 x e^{x/2}$. Then $y(0) = 1 \Rightarrow c_1 = 1$

and $y'(0) = -1.5 \Rightarrow \frac{1}{2}c_1 + c_2 = -1.5$, so $c_2 = -2$ and the solution to the initial-value problem is $y = e^{x/2} - 2x e^{x/2}$.

20. $2r^2 + 5r - 3 = (2r - 1)(r + 3) = 0 \Rightarrow r = \frac{1}{2}$, $r = -3$ and the general solution is $y = c_1 e^{x/2} + c_2 e^{-3x}$. Then

$1 = y(0) = c_1 + c_2$ and $4 = y'(0) = \frac{1}{2}c_1 - 3c_2$ so $c_1 = 2$, $c_2 = -1$ and the solution to the initial-value problem is

$y = 2e^{x/2} - e^{-3x}$.

21. $r^2 + 16 = 0 \Rightarrow r = \pm 4i$ and the general solution is $y = e^{0x}(c_1 \cos 4x + c_2 \sin 4x) = c_1 \cos 4x + c_2 \sin 4x$. Then

$y\left(\frac{\pi}{4}\right) = -3 \Rightarrow -c_1 = -3 \Rightarrow c_1 = 3$ and $y'\left(\frac{\pi}{4}\right) = 4 \Rightarrow -4c_2 = 4 \Rightarrow c_2 = -1$, so the solution to the

initial-value problem is $y = 3 \cos 4x - \sin 4x$.

22. $r^2 - 2r + 5 = 0 \;\Rightarrow\; r = 1 \pm 2i$ and the general solution is $y = e^x(c_1 \cos 2x + c_2 \sin 2x)$. Then

$0 = y(\pi) = e^\pi(c_1 + 0) \;\Rightarrow\; c_1 = 0$ and $2 = y'(\pi) = (c_1 + 2c_2)e^\pi \;\Rightarrow\; c_2 = 1/e^\pi$ and the solution to the

initial-value problem is $y = \dfrac{e^x}{e^\pi} \sin 2x = e^{x-\pi} \sin 2x$.

23. $r^2 + 2r + 2 = 0 \;\Rightarrow\; r = -1 \pm i$ and the general solution is $y = e^{-x}(c_1 \cos x + c_2 \sin x)$. Then $2 = y(0) = c_1$ and

$1 = y'(0) = c_2 - c_1 \;\Rightarrow\; c_2 = 3$ and the solution to the initial-value problem is $y = e^{-x}(2 \cos x + 3 \sin x)$.

24. $r^2 + 12r + 36 = (r + 6)^2 = 0 \;\Rightarrow\; r = -6$ and the general solution is $y = c_1 e^{-6x} + c_2 x e^{-6x}$. Then

$0 = y(1) = c_1 e^{-6} + c_2 e^{-6} \;\Rightarrow\; c_1 + c_2 = 0$ and $1 = y'(1) = -6c_1 e^{-6} - 5c_2 e^{-6} \;\Rightarrow\; 6c_1 + 5c_2 = -e^6$, so

$c_1 = -e^6$ and $c_2 = e^6$. The solution to the initial-value problem is $y = -e^6 e^{-6x} + e^6 x e^{-6x} = (x - 1)e^{6-6x}$.

25. $4r^2 + 1 = 0 \;\Rightarrow\; r = \pm\frac{1}{2}i$ and the general solution is $y = c_1 \cos\left(\frac{1}{2}x\right) + c_2 \sin\left(\frac{1}{2}x\right)$. Then $3 = y(0) = c_1$ and

$-4 = y(\pi) = c_2$, so the solution of the boundary-value problem is $y = 3\cos\left(\frac{1}{2}x\right) - 4\sin\left(\frac{1}{2}x\right)$.

26. $r^2 + 2r = r(2 + r) = 0 \;\Rightarrow\; r = 0, r = -2$ and the general solution is $y = c_1 + c_2 e^{-2x}$. Then $1 = y(0) = c_1 + c_2$

and $2 = y(1) = c_1 + c_2 e^{-2}$ so $c_2 = \dfrac{e^2}{1 - e^2}, c_1 = \dfrac{1 - 2e^2}{1 - e^2}$. The solution of the boundary-value problem is

$y = \dfrac{1 - 2e^2}{1 - e^2} + \dfrac{e^2}{1 - e^2} \cdot e^{-2x}$.

27. $r^2 - 3r + 2 = (r - 2)(r - 1) = 0 \;\Rightarrow\; r = 1, r = 2$ and the general solution is $y = c_1 e^x + c_2 e^{2x}$. Then

$1 = y(0) = c_1 + c_2$ and $0 = y(3) = c_1 e^3 + c_2 e^6$ so $c_2 = 1/(1 - e^3)$ and $c_1 = e^3/(e^3 - 1)$. The solution of the

boundary-value problem is $y = \dfrac{e^{x+3}}{e^3 - 1} + \dfrac{e^{2x}}{1 - e^3}$.

28. $r^2 + 100 = 0 \;\Rightarrow\; r = \pm 10i$ and the general solution is $y = c_1 \cos 10x + c_2 \sin 10x$. But $2 = y(0) = c_1$ and

$5 = y(\pi) = c_1$, so there is no solution.

29. $r^2 - 6r + 25 = 0 \;\Rightarrow\; r = 3 \pm 4i$ and the general solution is $y = e^{3x}(c_1 \cos 4x + c_2 \sin 4x)$. But $1 = y(0) = c_1$ and

$2 = y(\pi) = c_1 e^{3\pi} \;\Rightarrow\; c_1 = 2/e^{3\pi}$, so there is no solution.

30. $r^2 - 6r + 9 = (r - 3)^2 = 0 \;\Rightarrow\; r = 3$ and the general solution is $y = c_1 e^{3x} + c_2 x e^{3x}$. Then $1 = y(0) = c_1$ and

$0 = y(1) = c_1 e^3 + c_2 e^3 \;\Rightarrow\; c_2 = -1$. The solution of the boundary-value problem is $y = e^{3x} - x e^{3x}$.

31. $r^2 + 4r + 13 = 0 \;\Rightarrow\; r = -2 \pm 3i$ and the general solution is $y = e^{-2x}(c_1 \cos 3x + c_2 \sin 3x)$. But $2 = y(0) = c_1$

and $1 = y\left(\frac{\pi}{2}\right) = e^{-\pi}(-c_2)$, so the solution to the boundary-value problem is $y = e^{-2x}(2 \cos 3x - e^\pi \sin 3x)$.

32. $9r^2 - 18r + 10 = 0 \;\Rightarrow\; r = 1 \pm \frac{1}{3}i$ and the general solution is $y = e^x\left(c_1 \cos \frac{x}{3} + c_2 \sin \frac{x}{3}\right)$. Then $0 = y(0) = c_1$

and $1 = y(\pi) = e^\pi\left(\frac{1}{2}c_1 + \frac{\sqrt{3}}{2}c_2\right) \;\Rightarrow\; c_2 = \dfrac{2}{\sqrt{3}\,e^\pi}$. The solution of the boundary-value problem is

$y = \dfrac{2e^x}{\sqrt{3}\,e^\pi} \sin\left(\dfrac{x}{3}\right) = \dfrac{2}{\sqrt{3}} e^{x-\pi} \sin\left(\dfrac{x}{3}\right)$.

33. (a) *Case 1* ($\lambda = 0$): $y'' + \lambda y = 0 \ \Rightarrow \ y'' = 0$ which has an auxiliary equation $r^2 = 0 \ \Rightarrow \ r = 0 \ \Rightarrow \ y = c_1 + c_2 x$

where $y(0) = 0$ and $y(L) = 0$. Thus, $0 = y(0) = c_1$ and $0 = y(L) = c_2 L \ \Rightarrow \ c_1 = c_2 = 0$. Thus $y = 0$.

Case 2 ($\lambda < 0$): $y'' + \lambda y = 0$ has auxiliary equation $r^2 = -\lambda \ \Rightarrow \ r = \pm\sqrt{-\lambda}$ [distinct and real since $\lambda < 0$] $\ \Rightarrow$

$y = c_1 e^{\sqrt{-\lambda} x} + c_2 e^{-\sqrt{-\lambda} x}$ where $y(0) = 0$ and $y(L) = 0$. Thus $0 = y(0) = c_1 + c_2$ (∗) and

$0 = y(L) = c_1 e^{\sqrt{-\lambda} L} + c_2 e^{-\sqrt{-\lambda} L}$ (†).

Multiplying (∗) by $e^{\sqrt{-\lambda} L}$ and subtracting (†) gives $c_2 \left(e^{\sqrt{-\lambda} L} - e^{-\sqrt{-\lambda} L} \right) = 0 \ \Rightarrow \ c_2 = 0$ and thus $c_1 = 0$ from (∗).

Thus $y = 0$ for the cases $\lambda = 0$ and $\lambda < 0$.

(b) $y'' + \lambda y = 0$ has an auxiliary equation $r^2 + \lambda = 0 \ \Rightarrow \ r = \pm i \sqrt{\lambda} \ \Rightarrow \ y = c_1 \cos \sqrt{\lambda} \, x + c_2 \sin \sqrt{\lambda} \, x$ where

$y(0) = 0$ and $y(L) = 0$. Thus, $0 = y(0) = c_1$ and $0 = y(L) = c_2 \sin \sqrt{\lambda} L$ since $c_1 = 0$. Since we cannot have a trivial

solution, $c_2 \neq 0$ and thus $\sin \sqrt{\lambda} \, L = 0 \ \Rightarrow \ \sqrt{\lambda} \, L = n\pi$ where n is an integer $\ \Rightarrow \ \lambda = n^2 \pi^2 / L^2$ and

$y = c_2 \sin(n\pi x / L)$ where n is an integer.

34. The auxiliary equation is $ar^2 + br + c = 0$. If $b^2 - 4ac > 0$, then any solution is of the form $y(x) = c_1 e^{r_1 x} + c_2 e^{r_2 x}$ where

$r_1 = \dfrac{-b + \sqrt{b^2 - 4ac}}{2a}$ and $r_2 = \dfrac{-b - \sqrt{b^2 - 4ac}}{2a}$. But a, b, and c are all positive so both r_1 and r_2 are negative and

$\lim_{x \to \infty} y(x) = 0$. If $b^2 - 4ac = 0$, then any solution is of the form $y(x) = c_1 e^{rx} + c_2 x e^{rx}$ where $r = -b/(2a) < 0$

since a, b are positive. Hence $\lim_{x \to \infty} y(x) = 0$. Finally if $b^2 - 4ac < 0$, then any solution is of the form

$y(x) = e^{\alpha x}(c_1 \cos \beta x + c_2 \sin \beta x)$ where $\alpha = -b/(2a) < 0$ since a and b are positive. Thus $\lim_{x \to \infty} y(x) = 0$.

18.2 Nonhomogeneous Linear Equations

1. The auxiliary equation is $r^2 + 3r + 2 = (r + 2)(r + 1) = 0$, so the complementary solution is $y_c(x) = c_1 e^{-2x} + c_2 e^{-x}$.

We try the particular solution $y_p(x) = Ax^2 + Bx + C$, so $y_p' = 2Ax + B$ and $y_p'' = 2A$. Substituting into the differential

equation, we have $(2A) + 3(2Ax + B) + 2(Ax^2 + Bx + C) = x^2$ or $2Ax^2 + (6A + 2B)x + (2A + 3B + 2C) = x^2$.

Comparing coefficients gives $2A = 1$, $6A + 2B = 0$, and $2A + 3B + 2C = 0$, so $A = \frac{1}{2}$, $B = -\frac{3}{2}$, and $C = \frac{7}{4}$. Thus the

general solution is $y(x) = y_c(x) + y_p(x) = c_1 e^{-2x} + c_2 e^{-x} + \frac{1}{2} x^2 - \frac{3}{2} x + \frac{7}{4}$.

2. The auxiliary equation is $r^2 + 9 = 0$ with roots $r = \pm 3i$, so the complementary solution is $y_c(x) = c_1 \cos(3x) + c_2 \sin(3x)$.

Try the particular solution $y_p(x) = Ae^{3x}$, so $y_p' = 3Ae^{3x}$ and $y_p'' = 9Ae^{3x}$. Substitution into the differential equation

gives $9Ae^{3x} + 9(Ae^{3x}) = e^{3x}$ or $18Ae^{3x} = e^{3x}$. Thus $A = \frac{1}{18}$ and the general solution is

$y(x) = y_c(x) + y_p(x) = c_1 \cos(3x) + c_2 \sin(3x) + \frac{1}{18} e^{3x}$.

3. The auxiliary equation is $r^2 - 2r = r(r - 2) = 0$, so the complementary solution is $y_c(x) = c_1 + c_2 e^{2x}$. Try the particular

solution $y_p(x) = A \cos 4x + B \sin 4x$, so $y_p' = -4A \sin 4x + 4B \cos 4x$ and $y_p'' = -16A \cos 4x - 16B \sin 4x$. Substitution

into the differential equation gives $(-16A \cos 4x - 16B \sin 4x) - 2(-4A \sin 4x + 4B \cos 4x) = \sin 4x \ \Rightarrow$

$(-16A - 8B)\cos 4x + (8A - 16B)\sin 4x = \sin 4x$. Then $-16A - 8B = 0$ and $8A - 16B = 1 \Rightarrow A = \frac{1}{40}$ and

$B = -\frac{1}{20}$. Thus the general solution is $y(x) = y_c(x) + y_p(x) = c_1 + c_2 e^{2x} + \frac{1}{40}\cos 4x - \frac{1}{20}\sin 4x$.

4. The auxiliary equation is $r^2 + 6r + 9 = (r + 3)^2 = 0$, so the complementary solution is $y_c(x) = c_1 e^{-3x} + c_2 x e^{-3x}$.

Try the particular solution $y_p(x) = Ax + B$, so $y_p' = A$ and $y_p'' = 0$. Substitution into the differential equation gives

$0 + 6A + 9(Ax + B) = 1 + x$ or $(9A)x + (6A + 9B) = 1 + x$. Comparing coefficients, we have $9A = 1$ and

$6A + 9B = 1$, so $A = \frac{1}{9}$ and $B = \frac{1}{27}$. Thus the general solution is $y(x) = c_1 e^{-3x} + c_2 x e^{-3x} + \frac{1}{9}x + \frac{1}{27}$.

5. The auxiliary equation is $r^2 - 4r + 5 = 0$ with roots $r = 2 \pm i$, so the complementary solution is

$y_c(x) = e^{2x}(c_1 \cos x + c_2 \sin x)$. Try $y_p(x) = Ae^{-x}$, so $y_p' = -Ae^{-x}$ and $y_p'' = Ae^{-x}$. Substitution gives

$Ae^{-x} - 4(-Ae^{-x}) + 5(Ae^{-x}) = e^{-x} \Rightarrow 10Ae^{-x} = e^{-x} \Rightarrow A = \frac{1}{10}$. Thus the general solution is

$y(x) = e^{2x}(c_1 \cos x + c_2 \sin x) + \frac{1}{10}e^{-x}$.

6. $y_c(x) = e^{-x}(c_1 x + c_2)$. Try $y_p(x) = x^2(Ax + B)e^{-x}$ so that no term in y_p is a solution of the complementary equation.

Then $y_p' = [-Ax^3 + (3A - B)x^2 + 2Bx]e^{-x}$, $y_p'' = [Ax^3 + (B - 6A)x^2 + (6A - 4B)x + 2B]e^{-x}$ and substitution gives

$[Ax^3 + (B - 6A)x^2 + (6A - 4B)x + 2B] + 2[-Ax^3 + (3A - B)x^2 + 2Bx] + (Ax^3 + Bx^2) = x \Rightarrow 6Ax + 2B = x$.

So $y_p(x) = x^2\left(\frac{1}{6}x\right)e^{-x}$ and the general solution is $y(x) = e^{-x}(c_1 x + c_2) + \frac{1}{6}x^3 e^{-x}$.

7. The auxiliary equation is $r^2 + 1 = 0$ with roots $r = \pm i$, so the complementary solution is $y_c(x) = c_1 \cos x + c_2 \sin x$.

For $y'' + y = e^x$ try $y_{p_1}(x) = Ae^x$. Then $y_{p_1}' = y_{p_1}'' = Ae^x$ and substitution gives $Ae^x + Ae^x = e^x \Rightarrow A = \frac{1}{2}$,

so $y_{p_1}(x) = \frac{1}{2}e^x$. For $y'' + y = x^3$ try $y_{p_2}(x) = Ax^3 + Bx^2 + Cx + D$. Then $y_{p_2}' = 3Ax^2 + 2Bx + C$ and

$y_{p_2}'' = 6Ax + 2B$. Substituting, we have $6Ax + 2B + Ax^3 + Bx^2 + Cx + D = x^3$, so $A = 1$, $B = 0$,

$6A + C = 0 \Rightarrow C = -6$, and $2B + D = 0 \Rightarrow D = 0$. Thus $y_{p_2}(x) = x^3 - 6x$ and the general solution is

$y(x) = y_c(x) + y_{p_1}(x) + y_{p_2}(x) = c_1 \cos x + c_2 \sin x + \frac{1}{2}e^x + x^3 - 6x$. But $2 = y(0) = c_1 + \frac{1}{2} \Rightarrow$

$c_1 = \frac{3}{2}$ and $0 = y'(0) = c_2 + \frac{1}{2} - 6 \Rightarrow c_2 = \frac{11}{2}$. Thus the solution to the initial-value problem is

$y(x) = \frac{3}{2}\cos x + \frac{11}{2}\sin x + \frac{1}{2}e^x + x^3 - 6x$.

8. The auxiliary equation is $r^2 - 4 = 0$ with roots $r = \pm 2$, so the complementary solution is $y_c(x) = c_1 e^{2x} + c_2 e^{-2x}$.

Try $y_p(x) = e^x(A\cos x + B\sin x)$, so $y_p' = e^x(A\cos x + B\sin x + B\cos x - A\sin x)$ and

$y_p'' = e^x(2B\cos x - 2A\sin x)$. Substitution gives $e^x(2B\cos x - 2A\sin x) - 4e^x(A\cos x + B\sin x) = e^x \cos x \Rightarrow$

$(2B - 4A)e^x \cos x + (-2A - 4B)e^x \sin x = e^x \cos x \Rightarrow A = -\frac{1}{5}$, $B = \frac{1}{10}$. Thus the general solution is

$y(x) = c_1 e^{2x} + c_2 e^{-2x} + e^x\left(-\frac{1}{5}\cos x + \frac{1}{10}\sin x\right)$. But $1 = y(0) = c_1 + c_2 - \frac{1}{5}$ and $2 = y'(0) = 2c_1 - 2c_2 - \frac{1}{10}$. Then

$c_1 = \frac{9}{8}$, $c_2 = \frac{3}{40}$, and the solution to the initial-value problem is $y(x) = \frac{9}{8}e^{2x} + \frac{3}{40}e^{-2x} + e^x\left(-\frac{1}{5}\cos x + \frac{1}{10}\sin x\right)$.

9. The auxiliary equation is $r^2 - r = 0$ with roots $r = 0$, $r = 1$ so the complementary solution is $y_c(x) = c_1 + c_2 e^x$.

Try $y_p(x) = x(Ax + B)e^x$ so that no term in y_p is a solution of the complementary equation. Then

$y_p' = (Ax^2 + (2A + B)x + B)e^x$ and $y_p'' = (Ax^2 + (4A + B)x + (2A + 2B))e^x$. Substitution into the differential equation gives $(Ax^2 + (4A + B)x + (2A + 2B))e^x - (Ax^2 + (2A + B)x + B)e^x = xe^x$ $\Rightarrow$ $(2Ax + (2A + B))e^x = xe^x$ $\Rightarrow$

$A = \frac{1}{2}, B = -1$. Thus $y_p(x) = \left(\frac{1}{2}x^2 - x\right)e^x$ and the general solution is $y(x) = c_1 + c_2 e^x + \left(\frac{1}{2}x^2 - x\right)e^x$. But

$2 = y(0) = c_1 + c_2$ and $1 = y'(0) = c_2 - 1$, so $c_2 = 2$ and $c_1 = 0$. The solution to the initial-value problem is

$y(x) = 2e^x + \left(\frac{1}{2}x^2 - x\right)e^x = e^x\left(\frac{1}{2}x^2 - x + 2\right)$.

10. $y_c(x) = c_1 e^x + c_2 e^{-2x}$. For $y'' + y' - 2y = x$ try $y_{p_1}(x) = Ax + B$. Then $y'_{p_1} = A$, $y''_{p_1} = 0$, and substitution gives

$0 + A - 2(Ax + B) = x$ $\Rightarrow$ $A = -\frac{1}{2}, B = -\frac{1}{4}$, so $y_{p_1}(x) = -\frac{1}{2}x - \frac{1}{4}$. For $y'' + y' - 2y = \sin 2x$ try

$y_{p_2}(x) = A\cos 2x + B\sin 2x$. Then $y'_{p_2} = -2A\sin 2x + 2B\cos 2x$, $y''_{p_2} = -4A\cos 2x - 4B\sin 2x$, and substitution

gives $(-4A\cos 2x - 4B\sin 2x) + (-2A\sin 2x + 2B\cos 2x) - 2(A\cos 2x + B\sin 2x) = \sin 2x$ $\Rightarrow$ $A = -\frac{1}{20}$,

$B = -\frac{3}{20}$. Thus $y_{p_2}(x) = -\frac{1}{20}\cos 2x + -\frac{3}{20}\sin 2x$ and the general solution is

$y(x) = c_1 e^x + c_2 e^{-2x} - \frac{1}{2}x - \frac{1}{4} - \frac{1}{20}\cos 2x - \frac{3}{20}\sin 2x$. But $1 = y(0) = c_1 + c_2 - \frac{1}{4} - \frac{1}{20}$ and

$0 = y'(0) = c_1 - 2c_2 - \frac{1}{2} - \frac{3}{10}$ $\Rightarrow$ $c_1 = \frac{17}{15}$ and $c_2 = \frac{1}{6}$. Thus the solution to the initial-value problem is

$y(x) = \frac{17}{15}e^x + \frac{1}{6}e^{-2x} - \frac{1}{2}x - \frac{1}{4} - \frac{1}{20}\cos 2x - \frac{3}{20}\sin 2x$.

11. The auxiliary equation is $r^2 + 3r + 2 = (r + 1)(r + 2) = 0$, so $r = -1, r = -2$ and $y_c(x) = c_1 e^{-x} + c_2 e^{-2x}$.

Try $y_p = A\cos x + B\sin x$ $\Rightarrow$ $y'_p = -A\sin x + B\cos x$, $y''_p = -A\cos x - B\sin x$. Substituting into the differential

equation gives $(-A\cos x - B\sin x) + 3(-A\sin x + B\cos x) + 2(A\cos x + B\sin x) = \cos x$ or

$(A + 3B)\cos x + (-3A + B)\sin x = \cos x$. Then solving the equations

$A + 3B = 1, -3A + B = 0$ gives $A = \frac{1}{10}, B = \frac{3}{10}$ and the general

solution is $y(x) = c_1 e^{-x} + c_2 e^{-2x} + \frac{1}{10}\cos x + \frac{3}{10}\sin x$. The graph

shows y_p and several other solutions. Notice that all solutions are

asymptotic to y_p as $x \to \infty$. Except for y_p, all solutions approach either ∞

or $-\infty$ as $x \to -\infty$.

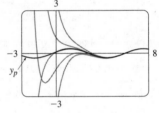

12. The auxiliary equation is $r^2 + 4 = 0$ $\Rightarrow$ $r = \pm 2i$, so $y_c(x) = c_1\cos 2x + c_2\sin 2x$. Try $y_p = Ae^{-x}$ $\Rightarrow$

$y'_p = -Ae^{-x}$, $y''_p = Ae^{-x}$. Substituting into the differential equation gives $Ae^{-x} + 4Ae^{-x} = e^{-x}$ $\Rightarrow$

$5A = 1$ $\Rightarrow$ $A = \frac{1}{5}$, so $y_p = \frac{1}{5}e^{-x}$ and the general solution is

$y(x) = c_1\cos 2x + c_2\sin 2x + \frac{1}{5}e^{-x}$. We graph y_p along with several

other solutions. All of the solutions except y_p oscillate around $y_p = \frac{1}{5}e^{-x}$,

and all solutions approach ∞ as $x \to -\infty$.

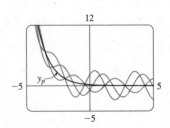

13. Here $y_c(x) = c_1\cos 3x + c_2\sin 3x$. For $y'' + 9y = e^{2x}$ try $y_{p_1}(x) = Ae^{2x}$ and for $y'' + 9y = x^2\sin x$

try $y_{p_2}(x) = (Bx^2 + Cx + D)\cos x + (Ex^2 + Fx + G)\sin x$. Thus a trial solution is

$y_p(x) = y_{p_1}(x) + y_{p_2}(x) = Ae^{2x} + (Bx^2 + Cx + D)\cos x + (Ex^2 + Fx + G)\sin x$.

14. Since $y_c(x) = c_1 + c_2 e^{-9x}$, try $y_p(x) = (Ax + B)e^{-x}\cos \pi x + (Cx + D)e^{-x}\sin \pi x$.

15. Here $y_c(x) = c_1 + c_2 e^{-9x}$. For $y'' + 9y' = 1$ try $y_{p_1}(x) = Ax$ (since $y = A$ is a solution to the complementary equation) and for $y'' + 9y' = xe^{9x}$ try $y_{p_2}(x) = (Bx + C)e^{9x}$.

16. Since $y_c(x) = c_1 e^x + c_2 e^{-4x}$ try $y_p(x) = x(Ax^3 + Bx^2 + Cx + D)e^x$ so that no term of $y_p(x)$ satisfies the complementary equation.

17. Since $y_c(x) = e^{-x}(c_1 \cos 3x + c_2 \sin 3x)$ we try $y_p(x) = x(Ax^2 + Bx + C)e^{-x} \cos 3x + x(Dx^2 + Ex + F)e^{-x} \sin 3x$ (so that no term of y_p is a solution of the complementary equation).

18. Here $y_c(x) = c_1 \cos 2x + c_2 \sin 2x$. For $y'' + 4y = e^{3x}$ try $y_{p_1}(x) = Ae^{3x}$ and for $y'' + 4y = x \sin 2x$ try $y_{p_2}(x) = x(Bx + C) \cos 2x + x(Dx + E) \sin 2x$ (so that no term of y_{p_2} is a solution of the complementary equation).

Note: Solving Equations (7) and (9) in The Method of Variation of Parameters gives

$$u_1' = -\frac{Gy_2}{a(y_1 y_2' - y_2 y_1')} \qquad \text{and} \qquad u_2' = \frac{Gy_1}{a(y_1 y_2' - y_2 y_1')}$$

We will use these equations rather than resolving the system in each of the remaining exercises in this section.

19. (a) Here $4r^2 + 1 = 0 \implies r = \pm \frac{1}{2}i$ and $y_c(x) = c_1 \cos\left(\frac{1}{2}x\right) + c_2 \sin\left(\frac{1}{2}x\right)$. We try a particular solution of the form

$y_p(x) = A \cos x + B \sin x \implies y_p' = -A \sin x + B \cos x$ and $y_p'' = -A \cos x - B \sin x$. Then the equation

$4y'' + y = \cos x$ becomes $4(-A \cos x - B \sin x) + (A \cos x + B \sin x) = \cos x$ or

$-3A \cos x - 3B \sin x = \cos x \implies A = -\frac{1}{3}, B = 0$. Thus, $y_p(x) = -\frac{1}{3} \cos x$ and the general solution is

$y(x) = y_c(x) + y_p(x) = c_1 \cos\left(\frac{1}{2}x\right) + c_2 \sin\left(\frac{1}{2}x\right) - \frac{1}{3} \cos x$.

(b) From (a) we know that $y_c(x) = c_1 \cos\frac{x}{2} + c_2 \sin\frac{x}{2}$. Setting $y_1 = \cos\frac{x}{2}$, $y_2 = \sin\frac{x}{2}$, we have

$y_1 y_2' - y_2 y_1' = \frac{1}{2} \cos^2\frac{x}{2} + \frac{1}{2} \sin^2\frac{x}{2} = \frac{1}{2}$. Thus $u_1' = -\dfrac{\cos x \sin\frac{x}{2}}{4 \cdot \frac{1}{2}} = -\frac{1}{2} \cos\left(2 \cdot \frac{x}{2}\right) \sin\frac{x}{2} = -\frac{1}{2}\left(2 \cos^2\frac{x}{2} - 1\right) \sin\frac{x}{2}$

and $u_2' = \dfrac{\cos x \cos\frac{x}{2}}{4 \cdot \frac{1}{2}} = \frac{1}{2} \cos\left(2 \cdot \frac{x}{2}\right) \cos\frac{x}{2} = \frac{1}{2}\left(1 - 2 \sin^2\frac{x}{2}\right) \cos\frac{x}{2}$. Then

$u_1(x) = \int \left(\frac{1}{2} \sin\frac{x}{2} - \cos^2\frac{x}{2} \sin\frac{x}{2}\right) dx = -\cos\frac{x}{2} + \frac{2}{3} \cos^3\frac{x}{2}$ and

$u_2(x) = \int \left(\frac{1}{2} \cos\frac{x}{2} - \sin^2\frac{x}{2} \cos\frac{x}{2}\right) dx = \sin\frac{x}{2} - \frac{2}{3} \sin^3\frac{x}{2}$. Thus

$$y_p(x) = \left(-\cos\frac{x}{2} + \frac{2}{3} \cos^3\frac{x}{2}\right) \cos\frac{x}{2} + \left(\sin\frac{x}{2} - \frac{2}{3} \sin^3\frac{x}{2}\right) \sin\frac{x}{2} = -\left(\cos^2\frac{x}{2} - \sin^2\frac{x}{2}\right) + \frac{2}{3}\left(\cos^4\frac{x}{2} - \sin^4\frac{x}{2}\right)$$

$$= -\cos\left(2 \cdot \frac{x}{2}\right) + \frac{2}{3}\left(\cos^2\frac{x}{2} + \sin^2\frac{x}{2}\right)\left(\cos^2\frac{x}{2} - \sin^2\frac{x}{2}\right) = -\cos x + \frac{2}{3} \cos x = -\frac{1}{3} \cos x$$

and the general solution is $y(x) = y_c(x) + y_p(x) = c_1 \cos\frac{x}{2} + c_2 \sin\frac{x}{2} - \frac{1}{3} \cos x$.

20. (a) Here $r^2 - 2r - 3 = (r - 3)(r + 1) = 0 \implies r = 3, r = -1$ and the complementary solution is

$y_c(x) = c_1 e^{3x} + c_2 e^{-x}$. A particular solution is of the form $y_p(x) = Ax + B \implies y_p' = A, y_p'' = 0$, and

substituting into the differential equation gives $0 - 2A - 3(Ax + B) = x + 2$ or $-3Ax + (-2A - 3B) = x + 2$,

so $A = -\frac{1}{3}$ and $-2A - 3B = 2 \implies B = -\frac{4}{9}$. Thus $y_p(x) = -\frac{1}{3}x - \frac{4}{9}$ and the general solution is

$y(x) = y_c(x) + y_p(x) = c_1 e^{3x} + c_2 e^{-x} - \frac{1}{3}x - \frac{4}{9}$.

(b) In (a), $y_c(x) = c_1 e^{3x} + c_2 e^{-x}$, so set $y_1 = e^{3x}$, $y_2 = e^{-x}$. Then $y_1 y_2' - y_2 y_1' = -e^{3x} e^{-x} - 3e^{3x} e^{-x} = -4e^{2x}$ so

$$u_1' = -\frac{(x+2)e^{-x}}{-4e^{2x}} = \tfrac{1}{4}(x+2)e^{-3x} \;\Rightarrow\; u_1(x) = \tfrac{1}{4}\int (x+2)e^{-3x}\,dx = \tfrac{1}{4}\left[-\tfrac{1}{3}(x+2)e^{-3x} - \tfrac{1}{9}e^{-3x}\right] \quad \text{[by parts]}$$

and $u_2' = \dfrac{(x+2)e^{3x}}{-4e^{2x}} = -\tfrac{1}{4}(x+2)e^{x} \;\Rightarrow\; u_2(x) = -\tfrac{1}{4}\int(x+2)e^{x}\,dx = -\tfrac{1}{4}[(x+2)e^{x} - e^{x}]$ [by parts].

Hence $y_p(x) = \tfrac{1}{4}\left[\left(-\tfrac{1}{3}x - \tfrac{7}{9}\right)e^{-3x}\right]e^{3x} - \tfrac{1}{4}[(x+1)e^{x}]e^{-x} = -\tfrac{1}{3}x - \tfrac{4}{9}$ and

$y(x) = y_c(x) + y_p(x) = c_1 e^{3x} + c_2 e^{-x} - \tfrac{1}{3}x - \tfrac{4}{9}$.

21. (a) $r^2 - 2r + 1 = (r-1)^2 = 0 \;\Rightarrow\; r = 1$, so the complementary solution is $y_c(x) = c_1 e^{x} + c_2 x e^{x}$. A particular solution

is of the form $y_p(x) = Ae^{2x}$. Thus $4Ae^{2x} - 4Ae^{2x} + Ae^{2x} = e^{2x} \;\Rightarrow\; Ae^{2x} = e^{2x} \;\Rightarrow\; A = 1 \;\Rightarrow\; y_p(x) = e^{2x}$.

So a general solution is $y(x) = y_c(x) + y_p(x) = c_1 e^{x} + c_2 x e^{x} + e^{2x}$.

(b) From (a), $y_c(x) = c_1 e^{x} + c_2 x e^{x}$, so set $y_1 = e^{x}$, $y_2 = x e^{x}$. Then, $y_1 y_2' - y_2 y_1' = e^{2x}(1+x) - x e^{2x} = e^{2x}$ and so

$u_1' = -x e^{x} \;\Rightarrow\; u_1(x) = -\int x e^{x}\,dx = -(x-1)e^{x}$ [by parts] and $u_2' = e^{x} \;\Rightarrow\; u_2(x) = \int e^{x}\,dx = e^{x}$. Hence

$y_p(x) = (1-x)e^{2x} + x e^{2x} = e^{2x}$ and the general solution is $y(x) = y_c(x) + y_p(x) = c_1 e^{x} + c_2 x e^{x} + e^{2x}$.

22. (a) Here $r^2 - r = r(r-1) = 0 \;\Rightarrow\; r = 0, 1$ and $y_c(x) = c_1 + c_2 e^{x}$ and so we try a particular solution of the form

$y_p(x) = Ax e^{x}$. Thus, after calculating the necessary derivatives, we get $y'' - y' = e^{x} \;\Rightarrow$

$Ae^{x}(2+x) - Ae^{x}(1+x) = e^{x} \;\Rightarrow\; A = 1$. Thus $y_p(x) = x e^{x}$ and the general solution is $y(x) = c_1 + c_2 e^{x} + x e^{x}$.

(b) From (a) we know that $y_c(x) = c_1 + c_2 e^{x}$, so setting $y_1 = 1$, $y_2 = e^{x}$, then $y_1 y_2' - y_2 y_1' = e^{x} - 0 = e^{x}$. Thus

$u_1' = -e^{2x}/e^{x} = -e^{x}$ and $u_2' = e^{x}/e^{x} = 1$. Then $u_1(x) = -\int e^{x}\,dx = -e^{x}$ and $u_2(x) = x$. Thus

$y_p(x) = -e^{x} + x e^{x}$ and the general solution is $y(x) = c_1 + c_2 e^{x} - e^{x} + x e^{x} = c_1 + c_3 e^{x} + x e^{x}$.

23. As in Example 5, $y_c(x) = c_1 \sin x + c_2 \cos x$, so set $y_1 = \sin x$, $y_2 = \cos x$. Then $y_1 y_2' - y_2 y_1' = -\sin^2 x - \cos^2 x = -1$,

so $u_1' = -\dfrac{\sec^2 x \cos x}{-1} = \sec x \;\Rightarrow\; u_1(x) = \int \sec x\,dx = \ln(\sec x + \tan x)$ for $0 < x < \tfrac{\pi}{2}$,

and $u_2' = \dfrac{\sec^2 x \sin x}{-1} = -\sec x \tan x \;\Rightarrow\; u_2(x) = -\sec x$. Hence

$y_p(x) = \ln(\sec x + \tan x) \cdot \sin x - \sec x \cdot \cos x = \sin x \ln(\sec x + \tan x) - 1$ and the general solution is

$y(x) = c_1 \sin x + c_2 \cos x + \sin x \ln(\sec x + \tan x) - 1$.

24. As in Exercise 23, $y_c(x) = c_1 \sin x + c_2 \cos x$, $y_1 = \sin x$, $y_2 = \cos x$, and $y_1 y_2' - y_2 y_1' = -1$. Then

$u_1' = -\dfrac{\sec^3 x \cos x}{-1} = \sec^2 x \;\Rightarrow\; u_1(x) = \tan x$ and $u_2' = \dfrac{\sec^3 x \sin x}{-1} = -\sec^2 x \tan x \;\Rightarrow$

$u_2(x) = -\int \tan x \sec^2 x\,dx = -\tfrac{1}{2}\tan^2 x$. Hence

$y_p(x) = \tan x \sin x - \tfrac{1}{2}\tan^2 x \cos x = \tan x \sin x - \tfrac{1}{2}\tan x \sin x = \tfrac{1}{2}\tan x \sin x$ and the general solution

is $y(x) = c_1 \sin x + c_2 \cos x + \tfrac{1}{2}\tan x \sin x$.

25. $y_1 = e^x$, $y_2 = e^{2x}$ and $y_1 y_2' - y_2 y_1' = e^{3x}$. So $u_1' = \dfrac{-e^{2x}}{(1 + e^{-x})e^{3x}} = -\dfrac{e^{-x}}{1 + e^{-x}}$ and

$u_1(x) = \displaystyle\int -\dfrac{e^{-x}}{1 + e^{-x}}\, dx = \ln(1 + e^{-x})$. $u_2' = \dfrac{e^x}{(1 + e^{-x})e^{3x}} = \dfrac{e^x}{e^{3x} + e^{2x}}$ so

$u_2(x) = \displaystyle\int \dfrac{e^x}{e^{3x} + e^{2x}}\, dx = \ln\left(\dfrac{e^x + 1}{e^x}\right) - e^{-x} = \ln(1 + e^{-x}) - e^{-x}$. Hence

$y_p(x) = e^x \ln(1 + e^{-x}) + e^{2x}[\ln(1 + e^{-x}) - e^{-x}]$ and the general solution is

$y(x) = [c_1 + \ln(1 + e^{-x})]e^x + [c_2 - e^{-x} + \ln(1 + e^{-x})]e^{2x}$.

26. $y_1 = e^{-x}$, $y_2 = e^{-2x}$ and $y_1 y_2' - y_2 y_1' = -e^{-3x}$. So $u_1' = -\dfrac{(\sin e^x)e^{-2x}}{-e^{-3x}} = e^x \sin e^x$

and $u_2' = \dfrac{(\sin e^x)e^{-x}}{-e^{-3x}} = -e^{2x} \sin e^x$. Hence $u_1(x) = \int e^x \sin e^x dx = -\cos e^x$ and

$u_2(x) = \int -e^{2x} \sin e^x dx = e^x \cos e^x - \sin e^x$. Then $y_p(x) = -e^{-x} \cos e^x - e^{-2x}[\sin e^x - e^x \cos e^x]$

and the general solution is $y(x) = (c_1 - \cos e^x)e^{-x} + [c_2 - \sin e^x + e^x \cos e^x]e^{-2x}$.

27. $r^2 - 2r + 1 = (r - 1)^2 = 0 \;\Rightarrow\; r = 1$ so $y_c(x) = c_1 e^x + c_2 x e^x$. Thus $y_1 = e^x$, $y_2 = x e^x$ and

$y_1 y_2' - y_2 y_1' = e^x(x + 1)e^x - x e^x e^x = e^{2x}$. So $u_1' = -\dfrac{x e^x \cdot e^x/(1 + x^2)}{e^{2x}} = -\dfrac{x}{1 + x^2} \;\Rightarrow\;$

$u_1 = -\int \dfrac{x}{1 + x^2}\, dx = -\tfrac{1}{2}\ln(1 + x^2)$, $u_2' = \dfrac{e^x \cdot e^x/(1 + x^2)}{e^{2x}} = \dfrac{1}{1 + x^2} \;\Rightarrow\; u_2 = \int \dfrac{1}{1 + x^2}\, dx = \tan^{-1} x$ and

$y_p(x) = -\tfrac{1}{2}e^x \ln(1 + x^2) + x e^x \tan^{-1} x$. Hence the general solution is $y(x) = e^x \left[c_1 + c_2 x - \tfrac{1}{2}\ln(1 + x^2) + x \tan^{-1} x\right]$.

28. $y_1 = e^{-2x}$, $y_2 = x e^{-2x}$ and $y_1 y_2' - y_2 y_1' = e^{-4x}$. Then $u_1' = \dfrac{-e^{-2x} x e^{-2x}}{x^3 e^{-4x}} = -\dfrac{1}{x^2}$ so $u_1(x) = x^{-1}$ and

$u_2' = \dfrac{e^{-2x} e^{-2x}}{x^3 e^{-4x}} = \dfrac{1}{x^3}$ so $u_2(x) = -\dfrac{1}{2x^2}$. Thus $y_p(x) = \dfrac{e^{-2x}}{x} - \dfrac{x e^{-2x}}{2x^2} = \dfrac{e^{-2x}}{2x}$ and the general solution is

$y(x) = e^{-2x}[c_1 + c_2 x + 1/(2x)]$.

18.3 Applications of Second-Order Differential Equations ET 17.3

1. By Hooke's Law $k(0.25) = 25$ so $k = 100$ is the spring constant and the differential equation is $5x'' + 100x = 0$.

The auxiliary equation is $5r^2 + 100 = 0$ with roots $r = \pm 2\sqrt{5}\, i$, so the general solution to the differential equation is

$x(t) = c_1 \cos(2\sqrt{5}\, t) + c_2 \sin(2\sqrt{5}\, t)$. We are given that $x(0) = 0.35 \;\Rightarrow\; c_1 = 0.35$ and $x'(0) = 0 \;\Rightarrow\;$

$2\sqrt{5}\, c_2 = 0 \;\Rightarrow\; c_2 = 0$, so the position of the mass after t seconds is $x(t) = 0.35 \cos(2\sqrt{5}\, t)$.

2. By Hooke's Law $k(0.4) = 32$ so $k = \dfrac{32}{0.4} = 80$ is the spring constant and the differential equation is $8x'' + 80x = 0$.

The general solution is $x(t) = c_1 \cos(\sqrt{10}\, t) + c_2 \sin(\sqrt{10}\, t)$. But $0 = x(0) = c_1$ and $1 = x'(0) = \sqrt{10}\, c_2 \;\Rightarrow\;$

$c_2 = \dfrac{1}{\sqrt{10}}$, so the position of the mass after t seconds is $x(t) = \dfrac{1}{\sqrt{10}} \sin(\sqrt{10}\, t)$.

3. $k(0.5) = 6$ or $k = 12$ is the spring constant, so the initial-value problem is $2x'' + 14x' + 12x = 0$, $x(0) = 1$, $x'(0) = 0$.

The general solution is $x(t) = c_1 e^{-6t} + c_2 e^{-t}$. But $1 = x(0) = c_1 + c_2$ and $0 = x'(0) = -6c_1 - c_2$. Thus the position is

given by $x(t) = -\frac{1}{5}e^{-6t} + \frac{6}{5}e^{-t}$.

4. (a) $k(0.25) = 13 \;\Rightarrow\; k = 52$, so the differential equation is

   $2x'' + 8x' + 52x = 0$ with general solution

   $x(t) = e^{-2t}\left[c_1 \cos\left(\sqrt{22}\,t\right) + c_2 \sin\left(\sqrt{22}\,t\right)\right]$. Then $0 = x(0) = c_1$

   and $0.5 = x'(0) = \sqrt{22}\,c_2 \;\Rightarrow\; c_2 = \frac{1}{2\sqrt{22}}$, so the position is

   given by $x(t) = \frac{1}{2\sqrt{22}}e^{-2t}\sin\left(\sqrt{22}\,t\right)$.

(b)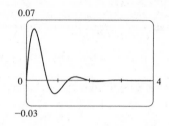

5. For critical damping we need $c^2 - 4mk = 0$ or $m = c^2/(4k) = 14^2/(4 \cdot 12) = \frac{49}{12}$ kg.

6. For critical damping we need $c^2 = 4mk$ or $c = 2\sqrt{mk} = 2\sqrt{2 \cdot 52} = 4\sqrt{26}$.

7. We are given $m = 1$, $k = 100$, $x(0) = -0.1$ and $x'(0) = 0$. From (3), the differential equation is $\dfrac{d^2x}{dt^2} + c\dfrac{dx}{dt} + 100x = 0$

with auxiliary equation $r^2 + cr + 100 = 0$.

If $c = 10$, we have two complex roots $r = -5 \pm 5\sqrt{3}\,i$, so the motion is underdamped and the solution is

$x = e^{-5t}\left[c_1 \cos\left(5\sqrt{3}\,t\right) + c_2 \sin\left(5\sqrt{3}\,t\right)\right]$. Then $-0.1 = x(0) = c_1$ and $0 = x'(0) = 5\sqrt{3}\,c_2 - 5c_1 \;\Rightarrow\; c_2 = -\frac{1}{10\sqrt{3}}$,

so $x = e^{-5t}\left[-0.1\cos\left(5\sqrt{3}\,t\right) - \frac{1}{10\sqrt{3}}\sin\left(5\sqrt{3}\,t\right)\right]$.

If $c = 15$, we again have underdamping since the auxiliary equation has roots $r = -\frac{15}{2} \pm \frac{5\sqrt{7}}{2}i$. The general solution is

$x = e^{-15t/2}\left[c_1 \cos\left(\frac{5\sqrt{7}}{2}t\right) + c_2 \sin\left(\frac{5\sqrt{7}}{2}t\right)\right]$, so $-0.1 = x(0) = c_1$ and $0 = x'(0) = \frac{5\sqrt{7}}{2}c_2 - \frac{15}{2}c_1 \;\Rightarrow\; c_2 = -\frac{3}{10\sqrt{7}}$.

Thus $x = e^{-15t/2}\left[-0.1\cos\left(\frac{5\sqrt{7}}{2}t\right) - \frac{3}{10\sqrt{7}}\sin\left(\frac{5\sqrt{7}}{2}t\right)\right]$.

For $c = 20$, we have equal roots $r_1 = r_2 = -10$, so the oscillation is critically damped and the solution is

$x = (c_1 + c_2 t)e^{-10t}$. Then $-0.1 = x(0) = c_1$ and $0 = x'(0) = -10c_1 + c_2 \;\Rightarrow\; c_2 = -1$, so $x = (-0.1 - t)e^{-10t}$.

If $c = 25$ the auxiliary equation has roots $r_1 = -5$, $r_2 = -20$, so we have overdamping and the solution is

$x = c_1 e^{-5t} + c_2 e^{-20t}$. Then $-0.1 = x(0) = c_1 + c_2$ and $0 = x'(0) = -5c_1 - 20c_2 \;\Rightarrow\; c_1 = -\frac{2}{15}$ and $c_2 = \frac{1}{30}$,

so $x = -\frac{2}{15}e^{-5t} + \frac{1}{30}e^{-20t}$.

If $c = 30$ we have roots $r = -15 \pm 5\sqrt{5}$, so the motion is

overdamped and the solution is $x = c_1 e^{\left(-15 + 5\sqrt{5}\right)t} + c_2 e^{\left(-15 - 5\sqrt{5}\right)t}$.

Then $-0.1 = x(0) = c_1 + c_2$ and

$0 = x'(0) = \left(-15 + 5\sqrt{5}\right)c_1 + \left(-15 - 5\sqrt{5}\right)c_2 \;\Rightarrow\;$

$c_1 = \frac{-5 - 3\sqrt{5}}{100}$ and $c_2 = \frac{-5 + 3\sqrt{5}}{100}$, so

$x = \left(\frac{-5 - 3\sqrt{5}}{100}\right)e^{\left(-15 + 5\sqrt{5}\right)t} + \left(\frac{-5 + 3\sqrt{5}}{100}\right)e^{\left(-15 - 5\sqrt{5}\right)t}$.

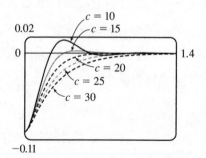

8. We are given $m = 1$, $c = 10$, $x(0) = 0$ and $x'(0) = 1$. The differential equation is $\dfrac{d^2 x}{dt^2} + 10\dfrac{dx}{dt} + kx = 0$ with auxiliary

equation $r^2 + 10r + k = 0$. $k = 10$: the auxiliary equation has roots $r = -5 \pm \sqrt{15}$ so we have overdamping and the

solution is $x = c_1 e^{(-5 + \sqrt{15})t} + c_2 e^{(-5 - \sqrt{15})t}$. Entering the initial conditions gives $c_1 = \dfrac{1}{2\sqrt{15}}$ and $c_2 = -\dfrac{1}{2\sqrt{15}}$, so

$x = \dfrac{1}{2\sqrt{15}} e^{(-5 + \sqrt{15})t} - \dfrac{1}{2\sqrt{15}} e^{(-5 - \sqrt{15})t}$.

$k = 20$: $r = -5 \pm \sqrt{5}$ and the solution is $x = c_1 e^{(-5 + \sqrt{5})t} + c_2 e^{(-5 - \sqrt{5})t}$ so again the motion is overdamped.

The initial conditions give $c_1 = \dfrac{1}{2\sqrt{5}}$ and $c_2 = -\dfrac{1}{2\sqrt{5}}$, so $x = \dfrac{1}{2\sqrt{5}} e^{(-5 + \sqrt{5})t} - \dfrac{1}{2\sqrt{5}} e^{(-5 - \sqrt{5})t}$.

$k = 25$: we have equal roots $r_1 = r_2 = -5$, so the motion is critically damped and the solution is $x = (c_1 + c_2 t)e^{-5t}$.

The initial conditions give $c_1 = 0$ and $c_2 = 1$, so $x = te^{-5t}$.

$k = 30$: $r = -5 \pm \sqrt{5}\, i$ so the motion is underdamped and the solution is $x = e^{-5t}\left[c_1 \cos\left(\sqrt{5}\,t\right) + c_2 \sin\left(\sqrt{5}\,t\right)\right]$.

The initial conditions give $c_1 = 0$ and $c_2 = \dfrac{1}{\sqrt{5}}$, so $x = \dfrac{1}{\sqrt{5}} e^{-5t} \sin\left(\sqrt{5}\,t\right)$.

$k = 40$: $r = -5 \pm \sqrt{15}\, i$ so we again have underdamping.

The solution is $x = e^{-5t}\left[c_1 \cos\left(\sqrt{15}\,t\right) + c_2 \sin\left(\sqrt{15}\,t\right)\right]$,

and the initial conditions give $c_1 = 0$ and $c_2 = \dfrac{1}{\sqrt{15}}$.

Thus $x = \dfrac{1}{\sqrt{15}} e^{-5t} \sin\left(\sqrt{15}\,t\right)$.

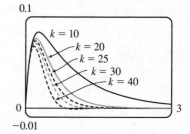

9. The differential equation is $mx'' + kx = F_0 \cos \omega_0 t$ and $\omega_0 \neq \omega = \sqrt{k/m}$. Here the auxiliary equation is $mr^2 + k = 0$

with roots $\pm\sqrt{k/m}\, i = \pm\omega i$ so $x_c(t) = c_1 \cos \omega t + c_2 \sin \omega t$. Since $\omega_0 \neq \omega$, try $x_p(t) = A \cos \omega_0 t + B \sin \omega_0 t$.

Then we need $(m)(-\omega_0^2)(A \cos \omega_0 t + B \sin \omega_0 t) + k(A \cos \omega_0 t + B \sin \omega_0 t) = F_0 \cos \omega_0 t$ or $A(k - m\omega_0^2) = F_0$ and

$B(k - m\omega_0^2) = 0$. Hence $B = 0$ and $A = \dfrac{F_0}{k - m\omega_0^2} = \dfrac{F_0}{m(\omega^2 - \omega_0^2)}$ since $\omega^2 = \dfrac{k}{m}$. Thus the motion of the mass is given

by $x(t) = c_1 \cos \omega t + c_2 \sin \omega t + \dfrac{F_0}{m(\omega^2 - \omega_0^2)} \cos \omega_0 t$.

10. As in Exercise 9, $x_c(t) = c_1 \cos \omega t + c_2 \sin \omega t$. But the natural frequency of the system equals the frequency of the

external force, so try $x_p(t) = t(A \cos \omega t + B \sin \omega t)$. Then we need

$m(2\omega B - \omega^2 At) \cos \omega t - m(2\omega A + \omega^2 Bt) \sin \omega t + kAt \cos \omega t + kBt \sin \omega t = F_0 \cos \omega t$ or $2m\omega B = F_0$ and

$-2m\omega A = 0$ [noting $-m\omega^2 A + kA = 0$ and $-m\omega^2 B + kB = 0$ since $\omega^2 = k/m$]. Hence the general solution is

$x(t) = c_1 \cos \omega t + c_2 \sin \omega t + [F_0 t/(2m\omega)] \sin \omega t$.

11. From Equation 6, $x(t) = f(t) + g(t)$ where $f(t) = c_1 \cos \omega t + c_2 \sin \omega t$ and $g(t) = \dfrac{F_0}{m(\omega^2 - \omega_0^2)} \cos \omega_0 t$. Then f

is periodic, with period $\dfrac{2\pi}{\omega}$, and if $\omega \neq \omega_0$, g is periodic with period $\dfrac{2\pi}{\omega_0}$. If $\dfrac{\omega}{\omega_0}$ is a rational number, then we can say

$\dfrac{\omega}{\omega_0} = \dfrac{a}{b} \;\Rightarrow\; a = \dfrac{b\omega}{\omega_0}$ where a and b are non-zero integers. Then

$x\left(t + a \cdot \dfrac{2\pi}{\omega}\right) = f\left(t + a \cdot \dfrac{2\pi}{\omega}\right) + g\left(t + a \cdot \dfrac{2\pi}{\omega}\right) = f(t) + g\left(t + \dfrac{b\omega}{\omega_0} \cdot \dfrac{2\pi}{\omega}\right) = f(t) + g\left(t + b \cdot \dfrac{2\pi}{\omega_0}\right) = f(t) + g(t) = x(t)$

so $x(t)$ is periodic.

12. (a) The graph of $x = c_1 e^{rt} + c_2 t e^{rt}$ has a t-intercept when $c_1 e^{rt} + c_2 t e^{rt} = 0 \Leftrightarrow e^{rt}(c_1 + c_2 t) = 0 \Leftrightarrow c_1 = -c_2 t$.

Since $t > 0$, x has a t-intercept if and only if c_1 and c_2 have opposite signs.

(b) For $t > 0$, the graph of x crosses the t-axis when $c_1 e^{r_1 t} + c_2 e^{r_2 t} = 0 \Leftrightarrow c_2 e^{r_2 t} = -c_1 e^{r_1 t} \Leftrightarrow$

$c_2 = -c_1 \dfrac{e^{r_1 t}}{e^{r_2 t}} = -c_1 e^{(r_1 - r_2)t}$. But $r_1 > r_2 \Rightarrow r_1 - r_2 > 0$ and since $t > 0$, $e^{(r_1 - r_2)t} > 1$. Thus

$|c_2| = |c_1| e^{(r_1 - r_2)t} > |c_1|$, and the graph of x can cross the t-axis only if $|c_2| > |c_1|$.

13. Here the initial-value problem for the charge is $Q'' + 20Q' + 500Q = 12$, $Q(0) = Q'(0) = 0$. Then

$Q_c(t) = e^{-10t}(c_1 \cos 20t + c_2 \sin 20t)$ and try $Q_p(t) = A \Rightarrow 500A = 12$ or $A = \frac{3}{125}$.

The general solution is $Q(t) = e^{-10t}(c_1 \cos 20t + c_2 \sin 20t) + \frac{3}{125}$. But $0 = Q(0) = c_1 + \frac{3}{125}$ and

$Q'(t) = I(t) = e^{-10t}[(-10c_1 + 20c_2)\cos 20t + (-10c_2 - 20c_1)\sin 20t]$ but $0 = Q'(0) = -10c_1 + 20c_2$. Thus the charge

is $Q(t) = -\frac{1}{250}e^{-10t}(6\cos 20t + 3\sin 20t) + \frac{3}{125}$ and the current is $I(t) = e^{-10t}\left(\frac{3}{5}\right)\sin 20t$.

14. (a) Here the initial-value problem for the charge is $2Q'' + 24Q' + 200Q = 12$ with $Q(0) = 0.001$ and $Q'(0) = 0$.

Then $Q_c(t) = e^{-6t}(c_1 \cos 8t + c_2 \sin 8t)$ and try $Q_p(t) = A \Rightarrow A = \frac{3}{50}$ and the general solution is

$Q(t) = e^{-6t}(c_1 \cos 8t + c_2 \sin 8t) + \frac{3}{50}$. But $0.001 = Q(0) = c + \frac{3}{50}$ so $c_1 = -0.059$. Also

$Q'(t) = I(t) = e^{-6t}[(-6c_1 + 8c_2)\cos 8t + (-6c_2 - 8c_1)\sin 8t]$ and $0 = Q'(0) = -6c_1 + 8c_2$ so

$c_2 = -0.04425$. Hence the charge is $Q(t) = -e^{-6t}(0.059\cos 8t + 0.04425\sin 8t) + \frac{3}{50}$ and the current is

$I(t) = e^{-6t}(0.7375)\sin 8t$.

(b)

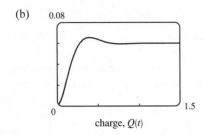

charge, $Q(t)$

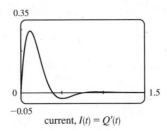

current, $I(t) = Q'(t)$

15. As in Exercise 13, $Q_c(t) = e^{-10t}(c_1 \cos 20t + c_2 \sin 20t)$ but $E(t) = 12\sin 10t$ so try

$Q_p(t) = A\cos 10t + B\sin 10t$. Substituting into the differential equation gives

$(-100A + 200B + 500A)\cos 10t + (-100B - 200A + 500B)\sin 10t = 12\sin 10t \Rightarrow$

$400A + 200B = 0$ and $400B - 200A = 12$. Thus $A = -\frac{3}{250}$, $B = \frac{3}{125}$ and the general solution is

$Q(t) = e^{-10t}(c_1 \cos 20t + c_2 \sin 20t) - \frac{3}{250}\cos 10t + \frac{3}{125}\sin 10t$. But $0 = Q(0) = c_1 - \frac{3}{250}$ so $c_1 = \frac{3}{250}$.

Also $Q'(t) = \frac{3}{25}\sin 10t + \frac{6}{25}\cos 10t + e^{-10t}[(-10c_1 + 20c_2)\cos 20t + (-10c_2 - 20c_1)\sin 20t]$ and

$0 = Q'(0) = \frac{6}{25} - 10c_1 + 20c_2$ so $c_2 = -\frac{3}{500}$. Hence the charge is given by

$Q(t) = e^{-10t}\left[\frac{3}{250}\cos 20t - \frac{3}{500}\sin 20t\right] - \frac{3}{250}\cos 10t + \frac{3}{125}\sin 10t$.

16. (a) As in Exercise 14, $Q_c(t) = e^{-6t}(c_1 \cos 8t + c_2 \sin 8t)$ but try $Q_p(t) = A\cos 10t + B\sin 10t$. Substituting into the

differential equation gives $(-200A + 240B + 200A)\cos 10t + (-200B - 240A + 200B)\sin 10t = 12\sin 10t$, so $B = 0$ and $A = -\frac{1}{20}$. Hence, the general solution is $Q(t) = e^{-6t}(c_1 \cos 8t + c_2 \sin 8t) - \frac{1}{20}\cos 10t$. But

$0.001 = Q(0) = c_1 - \frac{1}{20}$, $Q'(t) = e^{-6t}[(-6c_1 + 8c_2)\cos 8t + (-6c_2 - 8c_1)\sin 8t] - \frac{1}{2}\sin 10t$ and

$0 = Q'(0) = -6c_1 + 8c_2$, so $c_1 = 0.051$ and $c_2 = 0.03825$. Thus the charge is given by

$Q(t) = e^{-6t}(0.051\cos 8t + 0.03825\sin 8t) - \frac{1}{20}\cos 10t$.

(b)

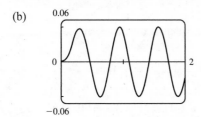

17. $x(t) = A\cos(\omega t + \delta)$ $\iff$ $x(t) = A[\cos\omega t\cos\delta - \sin\omega t\sin\delta]$ $\iff$ $x(t) = A\left(\frac{c_1}{A}\cos\omega t + \frac{c_2}{A}\sin\omega t\right)$ where

$\cos\delta = c_1/A$ and $\sin\delta = -c_2/A$ $\iff$ $x(t) = c_1\cos\omega t + c_2\sin\omega t$. [Note that $\cos^2\delta + \sin^2\delta = 1$ $\Rightarrow$ $c_1^2 + c_2^2 = A^2$.]

18. (a) We approximate $\sin\theta$ by θ and, with $L = 1$ and $g = 9.8$, the differential equation becomes $\dfrac{d^2\theta}{dt^2} + 9.8\theta = 0$. The auxiliary

equation is $r^2 + 9.8 = 0$ $\Rightarrow$ $r = \pm\sqrt{9.8}\,i$, so the general solution is $\theta(t) = c_1\cos(\sqrt{9.8}\,t) + c_2\sin(\sqrt{9.8}\,t)$.

Then $0.2 = \theta(0) = c_1$ and $1 = \theta'(0) = \sqrt{9.8}\,c_2$ $\Rightarrow$ $c_2 = \frac{1}{\sqrt{9.8}}$, so the equation is

$\theta(t) = 0.2\cos(\sqrt{9.8}\,t) + \frac{1}{\sqrt{9.8}}\sin(\sqrt{9.8}\,t)$.

(b) $\theta'(t) = -0.2\sqrt{9.8}\sin(\sqrt{9.8}\,t) + \cos(\sqrt{9.8}\,t) = 0$ or $\tan(\sqrt{9.8}\,t) = \frac{5}{\sqrt{9.8}}$, so the critical numbers are

$t = \frac{1}{\sqrt{9.8}}\tan^{-1}\left(\frac{5}{\sqrt{9.8}}\right) + \frac{n}{\sqrt{9.8}}\pi$ (n any integer). The maximum angle from the vertical is

$\theta\left(\frac{1}{\sqrt{9.8}}\tan^{-1}\left(\frac{5}{\sqrt{9.8}}\right)\right) \approx 0.377$ radians (or about $21.7°$).

(c) From part (b), the critical numbers of $\theta(t)$ are spaced $\frac{\pi}{\sqrt{9.8}}$ apart, and the time between successive maximum values

is $2\left(\frac{\pi}{\sqrt{9.8}}\right)$. Thus the period of the pendulum is $\frac{2\pi}{\sqrt{9.8}} \approx 2.007$ seconds.

(d) $\theta(t) = 0$ $\Rightarrow$ $0.2\cos(\sqrt{9.8}\,t) + \frac{1}{\sqrt{9.8}}\sin(\sqrt{9.8}\,t) = 0$ $\Rightarrow$ $\tan(\sqrt{9.8}\,t) = -0.2\sqrt{9.8}$ $\Rightarrow$

$t = \frac{1}{\sqrt{9.8}}\left[\tan^{-1}(-0.2\sqrt{9.8}) + \pi\right] \approx 0.825$ seconds.

(e) $\theta'(0.825) \approx -1.180$ rad/s.

18.4   Series Solutions   ET 17.4

1. Let $y(x) = \sum\limits_{n=0}^{\infty} c_n x^n$. Then $y'(x) = \sum\limits_{n=1}^{\infty} nc_n x^{n-1}$ and the given equation, $y' - y = 0$, becomes

$\sum\limits_{n=1}^{\infty} nc_n x^{n-1} - \sum\limits_{n=0}^{\infty} c_n x^n = 0$. Replacing n by $n+1$ in the first sum gives $\sum\limits_{n=0}^{\infty}(n+1)c_{n+1}x^n - \sum\limits_{n=0}^{\infty} c_n x^n = 0$, so

$\sum\limits_{n=0}^{\infty}[(n+1)c_{n+1} - c_n]x^n = 0$. Equating coefficients gives $(n+1)c_{n+1} - c_n = 0$, so the recursion relation is

$c_{n+1} = \dfrac{c_n}{n+1}$, $n = 0, 1, 2, \ldots$. Then $c_1 = c_0$, $c_2 = \dfrac{1}{2}c_1 = \dfrac{c_0}{2}$, $c_3 = \dfrac{1}{3}c_2 = \dfrac{1}{3} \cdot \dfrac{1}{2}c_0 = \dfrac{c_0}{3!}$, $c_4 = \dfrac{1}{4}c_3 = \dfrac{c_0}{4!}$, and

in general, $c_n = \dfrac{c_0}{n!}$. Thus, the solution is $y(x) = \sum\limits_{n=0}^{\infty} c_n x^n = \sum\limits_{n=0}^{\infty} \dfrac{c_0}{n!} x^n = c_0 \sum\limits_{n=0}^{\infty} \dfrac{x^n}{n!} = c_0 e^x$.

2. Let $y(x) = \sum\limits_{n=0}^{\infty} c_n x^n$. Then $y' = xy \;\Rightarrow\; y' - xy = 0 \;\Rightarrow\; \sum\limits_{n=1}^{\infty} nc_n x^{n-1} - x\sum\limits_{n=0}^{\infty} c_n x^n = 0$ or

$\sum\limits_{n=1}^{\infty} nc_n x^{n-1} - \sum\limits_{n=0}^{\infty} c_n x^{n+1} = 0$. Replacing n with $n+1$ in the first sum and n with $n-1$ in the second

gives $\sum\limits_{n=0}^{\infty} (n+1)c_{n+1} x^n - \sum\limits_{n=1}^{\infty} c_{n-1} x^n = 0$ or $c_1 + \sum\limits_{n=1}^{\infty} (n+1)c_{n+1} x^n - \sum\limits_{n=1}^{\infty} c_{n-1} x^n = 0$. Thus,

$c_1 + \sum\limits_{n=1}^{\infty} [(n+1)c_{n+1} - c_{n-1}] x^n = 0$. Equating coefficients gives $c_1 = 0$ and $(n+1)c_{n+1} - c_{n-1} = 0$. Thus, the

recursion relation is $c_{n+1} = \dfrac{c_{n-1}}{n+1}$, $n = 1, 2, \ldots$. But $c_1 = 0$, so $c_3 = 0$ and $c_5 = 0$ and in general $c_{2n+1} = 0$. Also,

$c_2 = \dfrac{c_0}{2}$, $c_4 = \dfrac{c_2}{4} = \dfrac{c_0}{4 \cdot 2} = \dfrac{c_0}{2^2 \cdot 2!}$, $c_6 = \dfrac{c_4}{6} = \dfrac{c_0}{6 \cdot 4 \cdot 2} = \dfrac{c_0}{2^3 \cdot 3!}$ and in general $c_{2n} = \dfrac{c_0}{2^n \cdot n!}$. Thus, the solution

is $y(x) = \sum\limits_{n=0}^{\infty} c_n x^n = \sum\limits_{n=0}^{\infty} c_{2n} x^{2n} = \sum\limits_{n=0}^{\infty} \dfrac{c_0}{2^n \cdot n!} x^{2n} = c_0 \sum\limits_{n=0}^{\infty} \dfrac{\left(x^2/2\right)^n}{n!} = c_0 e^{x^2/2}$.

3. Assuming $y(x) = \sum\limits_{n=0}^{\infty} c_n x^n$, we have $y'(x) = \sum\limits_{n=1}^{\infty} nc_n x^{n-1} = \sum\limits_{n=0}^{\infty} (n+1)c_{n+1} x^n$ and

$-x^2 y = -\sum\limits_{n=0}^{\infty} c_n x^{n+2} = -\sum\limits_{n=2}^{\infty} c_{n-2} x^n$. Hence, the equation $y' = x^2 y$ becomes $\sum\limits_{n=0}^{\infty} (n+1)c_{n+1} x^n - \sum\limits_{n=2}^{\infty} c_{n-2} x^n = 0$

or $c_1 + 2c_2 x + \sum\limits_{n=2}^{\infty} [(n+1)c_{n+1} - c_{n-2}] x^n = 0$. Equating coefficients gives $c_1 = c_2 = 0$ and $c_{n+1} = \dfrac{c_{n-2}}{n+1}$

for $n = 2, 3, \ldots$. But $c_1 = 0$, so $c_4 = 0$ and $c_7 = 0$ and in general $c_{3n+1} = 0$. Similarly $c_2 = 0$ so $c_{3n+2} = 0$. Finally

$c_3 = \dfrac{c_0}{3}$, $c_6 = \dfrac{c_3}{6} = \dfrac{c_0}{6 \cdot 3} = \dfrac{c_0}{3^2 \cdot 2!}$, $c_9 = \dfrac{c_6}{9} = \dfrac{c_0}{9 \cdot 6 \cdot 3} = \dfrac{c_0}{3^3 \cdot 3!}, \ldots$, and $c_{3n} = \dfrac{c_0}{3^n \cdot n!}$. Thus, the solution

is $y(x) = \sum\limits_{n=0}^{\infty} c_n x^n = \sum\limits_{n=0}^{\infty} c_{3n} x^{3n} = \sum\limits_{n=0}^{\infty} \dfrac{c_0}{3^n \cdot n!} x^{3n} = c_0 \sum\limits_{n=0}^{\infty} \dfrac{x^{3n}}{3^n n!} = c_0 \sum\limits_{n=0}^{\infty} \dfrac{\left(x^3/3\right)^n}{n!} = c_0 e^{x^3/3}$.

4. Let $y(x) = \sum\limits_{n=0}^{\infty} c_n x^n \;\Rightarrow\; y'(x) = \sum\limits_{n=1}^{\infty} nc_n x^{n-1} = \sum\limits_{n=0}^{\infty} (n+1)c_{n+1} x^n$. Then the differential equation becomes

$(x-3) \sum\limits_{n=0}^{\infty} (n+1)c_{n+1} x^n + 2\sum\limits_{n=0}^{\infty} c_n x^n = 0 \;\Rightarrow\; \sum\limits_{n=0}^{\infty} (n+1)c_{n+1} x^{n+1} - 3\sum\limits_{n=0}^{\infty} (n+1)c_{n+1} x^n + 2\sum\limits_{n=0}^{\infty} c_n x^n = 0 \;\Rightarrow$

$\sum\limits_{n=1}^{\infty} nc_n x^n - \sum\limits_{n=0}^{\infty} 3(n+1)c_{n+1} x^n + \sum\limits_{n=0}^{\infty} 2c_n x^n = 0 \;\Rightarrow\; \sum\limits_{n=0}^{\infty} [(n+2)c_n - 3(n+1)c_{n+1}] x^n = 0$

$\left[\text{since } \sum\limits_{n=1}^{\infty} nc_n x^n = \sum\limits_{n=0}^{\infty} nc_n x^n\right]$. Equating coefficients gives $(n+2)c_n - 3(n+1)c_{n+1} = 0$, thus the recursion relation is

$c_{n+1} = \dfrac{(n+2)c_n}{3(n+1)}$, $n = 0, 1, 2, \ldots$. Then $c_1 = \dfrac{2c_0}{3}$, $c_2 = \dfrac{3c_1}{3(2)} = \dfrac{3c_0}{3^2}$, $c_3 = \dfrac{4c_2}{3(3)} = \dfrac{4c_0}{3^3}$, $c_4 = \dfrac{5c_3}{3(4)} = \dfrac{5c_0}{3^4}$, and

in general, $c_n = \dfrac{(n+1)c_0}{3^n}$. Thus the solution is $y(x) = \sum\limits_{n=0}^{\infty} c_n x^n = c_0 \sum\limits_{n=0}^{\infty} \dfrac{n+1}{3^n} x^n$.

$\left[\text{Note that } c_0 \sum\limits_{n=0}^{\infty} \dfrac{n+1}{3^n} x^n = \dfrac{9c_0}{(3-x)^2} \text{ for } |x| < 3.\right]$

5. Let $y(x) = \sum\limits_{n=0}^{\infty} c_n x^n$ $\Rightarrow$ $y'(x) = \sum\limits_{n=1}^{\infty} nc_n x^{n-1}$ and $y''(x) = \sum\limits_{n=0}^{\infty} (n+2)(n+1)c_{n+2}x^n$. The differential equation

becomes $\sum\limits_{n=0}^{\infty} (n+2)(n+1)c_{n+2}x^n + x\sum\limits_{n=1}^{\infty} nc_n x^{n-1} + \sum\limits_{n=0}^{\infty} c_n x^n = 0$ or $\sum\limits_{n=0}^{\infty} [(n+2)(n+1)c_{n+2} + nc_n + c_n]x^n = 0$

$\left[\text{since } \sum\limits_{n=1}^{\infty} nc_n x^n = \sum\limits_{n=0}^{\infty} nc_n x^n\right]$. Equating coefficients gives $(n+2)(n+1)c_{n+2} + (n+1)c_n = 0$, thus the

recursion relation is $c_{n+2} = \dfrac{-(n+1)c_n}{(n+2)(n+1)} = -\dfrac{c_n}{n+2}$, $n = 0, 1, 2, \ldots$. Then the even

coefficients are given by $c_2 = -\dfrac{c_0}{2}$, $c_4 = -\dfrac{c_2}{4} = \dfrac{c_0}{2 \cdot 4}$, $c_6 = -\dfrac{c_4}{6} = -\dfrac{c_0}{2 \cdot 4 \cdot 6}$, and in general,

$c_{2n} = (-1)^n \dfrac{c_0}{2 \cdot 4 \cdot \cdots \cdot 2n} = \dfrac{(-1)^n c_0}{2^n n!}$. The odd coefficients are $c_3 = -\dfrac{c_1}{3}$, $c_5 = -\dfrac{c_3}{5} = \dfrac{c_1}{3 \cdot 5}$, $c_7 = -\dfrac{c_5}{7} = -\dfrac{c_1}{3 \cdot 5 \cdot 7}$,

and in general, $c_{2n+1} = (-1)^n \dfrac{c_1}{3 \cdot 5 \cdot 7 \cdot \cdots \cdot (2n+1)} = \dfrac{(-2)^n n! c_1}{(2n+1)!}$. The solution is

$$y(x) = c_0 \sum\limits_{n=0}^{\infty} \dfrac{(-1)^n}{2^n n!} x^{2n} + c_1 \sum\limits_{n=0}^{\infty} \dfrac{(-2)^n n!}{(2n+1)!} x^{2n+1}.$$

6. Let $y(x) = \sum\limits_{n=0}^{\infty} c_n x^n$. Then $y''(x) = \sum\limits_{n=2}^{\infty} n(n-1)c_n x^{n-2} = \sum\limits_{n=0}^{\infty} (n+2)(n+1)c_{n+2}x^n$. Hence, the equation $y'' = y$

becomes $\sum\limits_{n=0}^{\infty} (n+2)(n+1)c_{n+2}x^n - \sum\limits_{n=0}^{\infty} c_n x^n = 0$ or $\sum\limits_{n=0}^{\infty} [(n+2)(n+1)c_{n+2} - c_n]x^n = 0$. So the recursion relation

is $c_{n+2} = \dfrac{c_n}{(n+2)(n+1)}$, $n = 0, 1, \ldots$. Given c_0 and c_1, $c_2 = \dfrac{c_0}{2 \cdot 1}$, $c_4 = \dfrac{c_2}{4 \cdot 3} = \dfrac{c_0}{4!}$, $c_6 = \dfrac{c_4}{6 \cdot 5} = \dfrac{c_0}{6!}, \ldots$,

$c_{2n} = \dfrac{c_0}{(2n)!}$ and $c_3 = \dfrac{c_1}{3 \cdot 2}$, $c_5 = \dfrac{c_3}{5 \cdot 4} = \dfrac{c_1}{5 \cdot 4 \cdot 3 \cdot 2} = \dfrac{c_1}{5!}$, $c_7 = \dfrac{c_5}{7 \cdot 6} = \dfrac{c_1}{7!}, \ldots$, $c_{2n+1} = \dfrac{c_1}{(2n+1)!}$. Thus, the solution

is $y(x) = \sum\limits_{n=0}^{\infty} c_n x^n = \sum\limits_{n=0}^{\infty} c_{2n}x^{2n} + \sum\limits_{n=0}^{\infty} c_{2n+1}x^{2n+1} = c_0 \sum\limits_{n=0}^{\infty} \dfrac{x^{2n}}{(2n)!} + c_1 \sum\limits_{n=0}^{\infty} \dfrac{x^{2n+1}}{(2n+1)!}$. The solution can be written

as $y(x) = c_0 \cosh x + c_1 \sinh x$ $\left[\text{or } y(x) = c_0 \dfrac{e^x + e^{-x}}{2} + c_1 \dfrac{e^x - e^{-x}}{2} = \dfrac{c_0 + c_1}{2} e^x + \dfrac{c_0 - c_1}{2} e^{-x}\right]$.

7. Let $y(x) = \sum\limits_{n=0}^{\infty} c_n x^n$ $\Rightarrow$ $y'(x) = \sum\limits_{n=1}^{\infty} nc_n x^{n-1} = \sum\limits_{n=0}^{\infty} (n+1)c_{n+1}x^n$ and $y''(x) = \sum\limits_{n=0}^{\infty} (n+2)(n+1)c_{n+2}x^n$. Then

$(x-1)y''(x) = \sum\limits_{n=0}^{\infty} (n+2)(n+1)c_{n+2}x^{n+1} - \sum\limits_{n=0}^{\infty} (n+2)(n+1)c_{n+2}x^n = \sum\limits_{n=1}^{\infty} n(n+1)c_{n+1}x^n - \sum\limits_{n=0}^{\infty} (n+2)(n+1)c_{n+2}x^n$.

Since $\sum\limits_{n=1}^{\infty} n(n+1)c_{n+1}x^n = \sum\limits_{n=0}^{\infty} n(n+1)c_{n+1}x^n$, the differential equation becomes

$\sum\limits_{n=0}^{\infty} n(n+1)c_{n+1}x^n - \sum\limits_{n=0}^{\infty} (n+2)(n+1)c_{n+2}x^n + \sum\limits_{n=0}^{\infty} (n+1)c_{n+1}x^n = 0$ $\Rightarrow$

$\sum\limits_{n=0}^{\infty} [n(n+1)c_{n+1} - (n+2)(n+1)c_{n+2} + (n+1)c_{n+1}]x^n = 0$ or $\sum\limits_{n=0}^{\infty} [(n+1)^2 c_{n+1} - (n+2)(n+1)c_{n+2}]x^n = 0$.

Equating coefficients gives $(n+1)^2 c_{n+1} - (n+2)(n+1)c_{n+2} = 0$ for $n = 0, 1, 2, \ldots$. Then the recursion relation is

$c_{n+2} = \dfrac{(n+1)^2}{(n+2)(n+1)} c_{n+1} = \dfrac{n+1}{n+2} c_{n+1}$, so given c_0 and c_1, we have $c_2 = \frac{1}{2}c_1$, $c_3 = \frac{2}{3}c_2 = \frac{1}{3}c_1$, $c_4 = \frac{3}{4}c_3 = \frac{1}{4}c_1$, and in general $c_n = \dfrac{c_1}{n}$, $n = 1, 2, 3, \ldots$. Thus the solution is $y(x) = c_0 + c_1 \sum\limits_{n=1}^{\infty} \dfrac{x^n}{n}$. Note that the solution can be expressed as $c_0 - c_1 \ln(1-x)$ for $|x| < 1$.

8. Assuming $y(x) = \sum\limits_{n=0}^{\infty} c_n x^n$, $y''(x) = \sum\limits_{n=2}^{\infty} n(n-1)c_n x^{n-2} = \sum\limits_{n=0}^{\infty} (n+2)(n+1)c_{n+2}x^n$ and

$-xy(x) = -\sum\limits_{n=0}^{\infty} c_n x^{n+1} = -\sum\limits_{n=1}^{\infty} c_{n-1}x^n$. The equation $y'' = xy$ becomes

$\sum\limits_{n=0}^{\infty} (n+2)(n+1)c_{n+2}x^n - \sum\limits_{n=1}^{\infty} c_{n-1}x^n = 0$ or $2c_2 + \sum\limits_{n=1}^{\infty} [(n+2)(n+1)c_{n+2} - c_{n-1}]x^n = 0$. Equating coefficients

gives $c_2 = 0$ and $c_{n+2} = \dfrac{c_{n-1}}{(n+2)(n+1)}$ for $n = 1, 2, \ldots$. Since $c_2 = 0$, $c_{3n+2} = 0$ for $n = 0, 1, 2, \ldots$. Given c_0,

$c_3 = \dfrac{c_0}{3 \cdot 2}$, $c_6 = \dfrac{c_3}{6 \cdot 5} = \dfrac{c_0}{6 \cdot 5 \cdot 3 \cdot 2}, \ldots, c_{3n} = \dfrac{c_0}{3n(3n-1)(3n-3)(3n-4)\cdots 6 \cdot 5 \cdot 3 \cdot 2}$. Given c_1, $c_4 = \dfrac{c_1}{4 \cdot 3}$,

$c_7 = \dfrac{c_4}{7 \cdot 6} = \dfrac{c_1}{7 \cdot 6 \cdot 4 \cdot 3}, \ldots, c_{3n+1} = \dfrac{c_1}{(3n+1)3n(3n-2)(3n-3)\ldots 7 \cdot 6 \cdot 4 \cdot 3}$. The solution can be written

as $y(x) = c_0 \sum\limits_{n=0}^{\infty} \dfrac{(3n-2)(3n-5)\cdots 7 \cdot 4 \cdot 1}{(3n)!}x^{3n} + c_1 \sum\limits_{n=0}^{\infty} \dfrac{(3n-1)(3n-4)\cdots 8 \cdot 5 \cdot 2}{(3n+1)!}x^{3n+1}$.

9. Let $y(x) = \sum\limits_{n=0}^{\infty} c_n x^n$. Then $-xy'(x) = -x \sum\limits_{n=1}^{\infty} nc_n x^{n-1} = -\sum\limits_{n=1}^{\infty} nc_n x^n = -\sum\limits_{n=0}^{\infty} nc_n x^n$,

$y''(x) = \sum\limits_{n=0}^{\infty} (n+2)(n+1)c_{n+2}x^n$, and the equation $y'' - xy' - y = 0$

becomes $\sum\limits_{n=0}^{\infty} [(n+2)(n+1)c_{n+2} - nc_n - c_n]x^n = 0$. Thus, the recursion relation is

$c_{n+2} = \dfrac{nc_n + c_n}{(n+2)(n+1)} = \dfrac{c_n(n+1)}{(n+2)(n+1)} = \dfrac{c_n}{n+2}$ for $n = 0, 1, 2, \ldots$. One of the given conditions is $y(0) = 1$. But

$y(0) = \sum\limits_{n=0}^{\infty} c_n(0)^n = c_0 + 0 + 0 + \cdots = c_0$, so $c_0 = 1$. Hence, $c_2 = \dfrac{c_0}{2} = \dfrac{1}{2}$, $c_4 = \dfrac{c_2}{4} = \dfrac{1}{2 \cdot 4}$, $c_6 = \dfrac{c_4}{6} = \dfrac{1}{2 \cdot 4 \cdot 6}, \ldots,$

$c_{2n} = \dfrac{1}{2^n n!}$. The other given condition is $y'(0) = 0$. But $y'(0) = \sum\limits_{n=1}^{\infty} nc_n(0)^{n-1} = c_1 + 0 + 0 + \cdots = c_1$, so $c_1 = 0$.

By the recursion relation, $c_3 = \dfrac{c_1}{3} = 0$, $c_5 = 0, \ldots, c_{2n+1} = 0$ for $n = 0, 1, 2, \ldots$. Thus, the solution to the initial-value

problem is $y(x) = \sum\limits_{n=0}^{\infty} c_n x^n = \sum\limits_{n=0}^{\infty} c_{2n}x^{2n} = \sum\limits_{n=0}^{\infty} \dfrac{x^{2n}}{2^n n!} = \sum\limits_{n=0}^{\infty} \dfrac{(x^2/2)^n}{n!} = e^{x^2/2}$.

10. Assuming that $y(x) = \sum\limits_{n=0}^{\infty} c_n x^n$, we have $x^2 y = \sum\limits_{n=0}^{\infty} c_n x^{n+2}$ and

$y''(x) = \sum\limits_{n=2}^{\infty} n(n-1)c_n x^{n-2} = \sum\limits_{n=-2}^{\infty} (n+4)(n+3)c_{n+4}x^{n+2} = 2c_2 + 6c_3 x + \sum\limits_{n=0}^{\infty} (n+4)(n+3)c_{n+4}x^{n+2}$.

Thus, the equation $y'' + x^2 y = 0$ becomes $2c_2 + 6c_3 x + \sum\limits_{n=0}^{\infty} [(n+4)(n+3)c_{n+4} + c_n]x^{n+2} = 0$. So $c_2 = c_3 = 0$ and

the recursion relation is $c_{n+4} = -\dfrac{c_n}{(n+4)(n+3)}$, $n = 0, 1, 2, \ldots$. But $c_1 = y'(0) = 0 = c_2 = c_3$ and by the recursion relation, $c_{4n+1} = c_{4n+2} = c_{4n+3} = 0$ for $n = 0, 1, 2, \ldots$. Also, $c_0 = y(0) = 1$, so $c_4 = -\dfrac{c_0}{4 \cdot 3} = -\dfrac{1}{4 \cdot 3}$,

$c_8 = -\dfrac{c_4}{8 \cdot 7} = \dfrac{(-1)^2}{8 \cdot 7 \cdot 4 \cdot 3}, \ldots, c_{4n} = \dfrac{(-1)^n}{4n(4n-1)(4n-4)(4n-5) \cdots \cdot 4 \cdot 3}$. Thus, the solution to the initial-value

problem is $y(x) = \sum\limits_{n=0}^{\infty} c_n x^n = c_0 + \sum\limits_{n=0}^{\infty} c_{4n} x^{4n} = 1 + \sum\limits_{n=1}^{\infty} (-1)^n \dfrac{x^{4n}}{4n(4n-1)(4n-4)(4n-5) \cdots \cdot 4 \cdot 3}$.

11. Assuming that $y(x) = \sum\limits_{n=0}^{\infty} c_n x^n$, we have $xy = x \sum\limits_{n=0}^{\infty} c_n x^n = \sum\limits_{n=0}^{\infty} c_n x^{n+1}$, $x^2 y' = x^2 \sum\limits_{n=1}^{\infty} nc_n x^{n-1} = \sum\limits_{n=0}^{\infty} nc_n x^{n+1}$,

$$y''(x) = \sum\limits_{n=2}^{\infty} n(n-1)c_n x^{n-2} = \sum\limits_{n=-1}^{\infty} (n+3)(n+2)c_{n+3}x^{n+1} \qquad \text{[replace } n \text{ with } n+3\text{]}$$

$$= 2c_2 + \sum\limits_{n=0}^{\infty} (n+3)(n+2)c_{n+3}x^{n+1},$$

and the equation $y'' + x^2 y' + xy = 0$ becomes $2c_2 + \sum\limits_{n=0}^{\infty} [(n+3)(n+2)c_{n+3} + nc_n + c_n]x^{n+1} = 0$. So $c_2 = 0$ and the

recursion relation is $c_{n+3} = \dfrac{-nc_n - c_n}{(n+3)(n+2)} = -\dfrac{(n+1)c_n}{(n+3)(n+2)}$, $n = 0, 1, 2, \ldots$. But $c_0 = y(0) = 0 = c_2$ and by the

recursion relation, $c_{3n} = c_{3n+2} = 0$ for $n = 0, 1, 2, \ldots$. Also, $c_1 = y'(0) = 1$, so $c_4 = -\dfrac{2c_1}{4 \cdot 3} = -\dfrac{2}{4 \cdot 3}$,

$c_7 = -\dfrac{5c_4}{7 \cdot 6} = (-1)^2 \dfrac{2 \cdot 5}{7 \cdot 6 \cdot 4 \cdot 3} = (-1)^2 \dfrac{2^2 5^2}{7!}, \ldots, c_{3n+1} = (-1)^n \dfrac{2^2 5^2 \cdots \cdot (3n-1)^2}{(3n+1)!}$. Thus, the solution is

$$y(x) = \sum\limits_{n=0}^{\infty} c_n x^n = x + \sum\limits_{n=1}^{\infty} \left[(-1)^n \dfrac{2^2 5^2 \cdots \cdot (3n-1)^2 x^{3n+1}}{(3n+1)!} \right].$$

12. (a) Let $y(x) = \sum\limits_{n=0}^{\infty} c_n x^n$. Then $x^2 y''(x) = \sum\limits_{n=2}^{\infty} n(n-1)c_n x^n = \sum\limits_{n=0}^{\infty} (n+2)(n+1)c_{n+2}x^{n+2}$,

$xy'(x) = \sum\limits_{n=1}^{\infty} nc_n x^n = \sum\limits_{n=-1}^{\infty} (n+2)c_{n+2}x^{n+2} = c_1 x + \sum\limits_{n=0}^{\infty} (n+2)c_{n+2}x^{n+2}$, and the equation

$x^2 y'' + xy' + x^2 y = 0$ becomes $c_1 x + \sum\limits_{n=0}^{\infty} \{[(n+2)(n+1) + (n+2)]c_{n+2} + c_n\}x^{n+2} = 0$. So $c_1 = 0$ and the

recursion relation is $c_{n+2} = -\dfrac{c_n}{(n+2)^2}$, $n = 0, 1, 2, \ldots$. But $c_1 = y'(0) = 0$ so $c_{2n+1} = 0$ for $n = 0, 1, 2, \ldots$.

Also, $c_0 = y(0) = 1$, so $c_2 = -\dfrac{1}{2^2}$, $c_4 = -\dfrac{c_2}{4^2} = (-1)^2 \dfrac{1}{4^2 2^2} = (-1)^2 \dfrac{1}{2^4 (2!)^2}$, $c_6 = -\dfrac{c_4}{6^2} = (-1)^3 \dfrac{1}{2^6 (3!)^2}, \ldots$,

$c_{2n} = (-1)^n \dfrac{1}{2^{2n} (n!)^2}$. The solution is $y(x) = \sum\limits_{n=0}^{\infty} c_n x^n = \sum\limits_{n=0}^{\infty} (-1)^n \dfrac{x^{2n}}{2^{2n} (n!)^2}$.

(b) The Taylor polynomials T_0 to T_{12} are shown in the graph.

Because T_{10} and T_{12} are close together throughout the

interval $[-5, 5]$, it is reasonable to assume that T_{12} is a good

approximation to the Bessel function on that interval.

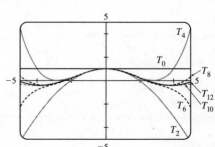

18 Review

CONCEPT CHECK

1. (a) $ay'' + by' + cy = 0$ where a, b, and c are constants.

(b) $ar^2 + br + c = 0$

(c) If the auxiliary equation has two distinct real roots r_1 and r_2, the solution is $y = c_1 e^{r_1 x} + c_2 e^{r_2 x}$. If the roots are real and equal, the solution is $y = c_1 e^{rx} + c_2 x e^{rx}$ where r is the common root. If the roots are complex, we can write $r_1 = \alpha + i\beta$ and $r_2 = \alpha - i\beta$, and the solution is $y = e^{\alpha x}(c_1 \cos \beta x + c_2 \sin \beta x)$.

2. (a) An initial-value problem consists of finding a solution y of a second-order differential equation that also satisfies given conditions $y(x_0) = y_0$ and $y'(x_0) = y_1$, where y_0 and y_1 are constants.

(b) A boundary-value problem consists of finding a solution y of a second-order differential equation that also satisfies given boundary conditions $y(x_0) = y_0$ and $y(x_1) = y_1$.

3. (a) $ay'' + by' + cy = G(x)$ where a, b, and c are constants and G is a continuous function.

(b) The complementary equation is the related homogeneous equation $ay'' + by' + cy = 0$. If we find the general solution y_c of the complementary equation and y_p is any particular solution of the original differential equation, then the general solution of the original differential equation is $y(x) = y_p(x) + y_c(x)$.

(c) See Examples 1–5 and the associated discussion in Section 18.2 [ET 17.2].

(d) See the discussion on pages 1158–1160 [ET 1122–1124].

4. Second-order linear differential equations can be used to describe the motion of a vibrating spring or to analyze an electric circuit; see the discussion in Section 18.3 [ET 17.3].

5. See Example 1 and the preceding discussion in Section 18.4 [ET 17.4].

TRUE-FALSE QUIZ

1. True. See Theorem 18.1.3 [ET 17.1.3].

2. False. The differential equation is not homogeneous.

3. True. $\cosh x$ and $\sinh x$ are linearly independent solutions of this linear homogeneous equation.

4. False. $y = Ae^x$ is a solution of the complementary equation, so we have to take $y_p(x) = Axe^x$.

EXERCISES

1. The auxiliary equation is $r^2 - 2r - 15 = 0 \Rightarrow (r - 5)(r + 3) = 0 \Rightarrow r = 5, r = -3$. Then the general solution is $y = c_1 e^{5x} + c_2 e^{-3x}$.

2. The auxiliary equation is $r^2 + 4r + 13 = 0 \Rightarrow r = -2 \pm 3i$, so $y = e^{-2x}(c_1 \cos 3x + c_2 \sin 3x)$.

3. The auxiliary equation is $r^2 + 3 = 0 \Rightarrow r = \pm\sqrt{3}\,i$. Then the general solution is $y = c_1\cos(\sqrt{3}\,x) + c_2\sin(\sqrt{3}\,x)$.

4. The auxiliary equation is $4r^2 + 4r + 1 = 0 \Rightarrow (2r+1)^2 = 0 \Rightarrow r = -\frac{1}{2}$, so the general solution is
$y = c_1 e^{-x/2} + c_2 x e^{-x/2}$.

5. $r^2 - 4r + 5 = 0 \Rightarrow r = 2 \pm i$, so $y_c(x) = e^{2x}(c_1\cos x + c_2\sin x)$. Try $y_p(x) = Ae^{2x} \Rightarrow y_p' = 2Ae^{2x}$
and $y_p'' = 4Ae^{2x}$. Substitution into the differential equation gives $4Ae^{2x} - 8Ae^{2x} + 5Ae^{2x} = e^{2x} \Rightarrow A = 1$ and
the general solution is $y(x) = e^{2x}(c_1\cos x + c_2\sin x) + e^{2x}$.

6. $r^2 + r - 2 = 0 \Rightarrow r = 1, r = -2$ and $y_c(x) = c_1 e^x + c_2 e^{-2x}$. Try $y_p(x) = Ax^2 + Bx + C \Rightarrow y_p' = 2Ax + B$
and $y_p'' = 2A$. Substitution gives $2A + 2Ax + B - 2Ax^2 - 2Bx - 2C = x^2 \Rightarrow A = B = -\frac{1}{2}, C = -\frac{3}{4}$ so the
general solution is $y(x) = c_1 e^x + c_2 e^{-2x} - \frac{1}{2}x^2 - \frac{1}{2}x - \frac{3}{4}$.

7. $r^2 - 2r + 1 = 0 \Rightarrow r = 1$ and $y_c(x) = c_1 e^x + c_2 x e^x$. Try $y_p(x) = (Ax + B)\cos x + (Cx + D)\sin x \Rightarrow$
$y_p' = (C - Ax - B)\sin x + (A + Cx + D)\cos x$ and $y_p'' = (2C - B - Ax)\cos x + (-2A - D - Cx)\sin x$. Substitution
gives $(-2Cx + 2C - 2A - 2D)\cos x + (2Ax - 2A + 2B - 2C)\sin x = x\cos x \Rightarrow A = 0, B = C = D = -\frac{1}{2}$.
The general solution is $y(x) = c_1 e^x + c_2 x e^x - \frac{1}{2}\cos x - \frac{1}{2}(x+1)\sin x$.

8. $r^2 + 4 = 0 \Rightarrow r = \pm 2i$ and $y_c(x) = c_1\cos 2x + c_2\sin 2x$. Try $y_p(x) = Ax\cos 2x + Bx\sin 2x$ so that no term
of y_p is a solution of the complementary equation. Then $y_p' = (A + 2Bx)\cos 2x + (B - 2Ax)\sin 2x$ and
$y_p'' = (4B - 4Ax)\cos 2x + (-4A - 4Bx)\sin 2x$. Substitution gives $4B\cos 2x - 4A\sin 2x = \sin 2x \Rightarrow$
$A = -\frac{1}{4}$ and $B = 0$. The general solution is $y(x) = c_1\cos 2x + c_2\sin 2x - \frac{1}{4}x\cos 2x$.

9. $r^2 - r - 6 = 0 \Rightarrow r = -2, r = 3$ and $y_c(x) = c_1 e^{-2x} + c_2 e^{3x}$. For $y'' - y' - 6y = 1$, try $y_{p_1}(x) = A$. Then
$y_{p_1}'(x) = y_{p_1}''(x) = 0$ and substitution into the differential equation gives $A = -\frac{1}{6}$. For $y'' - y' - 6y = e^{-2x}$ try
$y_{p_2}(x) = Bxe^{-2x}$ [since $y = Be^{-2x}$ satisfies the complementary equation]. Then $y_{p_2}'(x) = (B - 2Bx)e^{-2x}$ and
$y_{p_2}''(x) = (4Bx - 4B)e^{-2x}$, and substitution gives $-5Be^{-2x} = e^{-2x} \Rightarrow B = -\frac{1}{5}$. The general solution then is
$y(x) = c_1 e^{-2x} + c_2 e^{3x} + y_{p_1}(x) + y_{p_2}(x) = c_1 e^{-2x} + c_2 e^{3x} - \frac{1}{6} - \frac{1}{5}xe^{-2x}$.

10. Using variation of parameters, $y_c(x) = c_1\cos x + c_2\sin x$, $u_1'(x) = -\csc x\sin x = -1 \Rightarrow u_1(x) = -x$, and
$u_2'(x) = \dfrac{\csc x\cos x}{x} = \cot x \Rightarrow u_2(x) = \ln|\sin x| \Rightarrow y_p = -x\cos x + \sin x\ln|\sin x|$. The solution is
$y(x) = (c_1 - x)\cos x + (c_2 + \ln|\sin x|)\sin x$.

11. The auxiliary equation is $r^2 + 6r = 0$ and the general solution is $y(x) = c_1 + c_2 e^{-6x} = k_1 + k_2 e^{-6(x-1)}$. But
$3 = y(1) = k_1 + k_2$ and $12 = y'(1) = -6k_2$. Thus $k_2 = -2, k_1 = 5$ and the solution is $y(x) = 5 - 2e^{-6(x-1)}$.

12. The auxiliary equation is $r^2 - 6r + 25 = 0$ and the general solution is $y(x) = e^{3x}(c_1\cos 4x + c_2\sin 4x)$. But
$2 = y(0) = c_1$ and $1 = y'(0) = 3c_1 + 4c_2$. Thus the solution is $y(x) = e^{3x}\left(2\cos 4x - \frac{5}{4}\sin 4x\right)$.

13. The auxiliary equation is $r^2 - 5r + 4 = 0$ and the general solution is $y(x) = c_1 e^x + c_2 e^{4x}$. But $0 = y(0) = c_1 + c_2$ and $1 = y'(0) = c_1 + 4c_2$, so the solution is $y(x) = \frac{1}{3}(e^{4x} - e^x)$.

14. $y_c(x) = c_1 \cos(x/3) + c_2 \sin(x/3)$. For $9y'' + y = 3x$, try $y_{p_1}(x) = Ax + B$. Then $y_{p_1}(x) = 3x$. For $9y'' + y = e^{-x}$, try $y_{p_2}(x) = Ae^{-x}$. Then $9Ae^{-x} + Ae^{-x} = e^{-x}$ or $y_{p_2}(x) = \frac{1}{10}e^{-x}$. Thus the general solution is

$y(x) = c_1 \cos(x/3) + c_2 \sin(x/3) + 3x + \frac{1}{10}e^{-x}$. But $1 = y(0) = c_1 + \frac{1}{10}$ and $2 = y'(0) = \frac{1}{3}c_2 + 3 - \frac{1}{10}$, so

$c_1 = \frac{9}{10}$ and $c_2 = -\frac{27}{10}$. Hence the solution is $y(x) = \frac{1}{10}[9\cos(x/3) - 27\sin(x/3)] + 3x + \frac{1}{10}e^{-x}$.

15. Let $y(x) = \sum\limits_{n=0}^{\infty} c_n x^n$. Then $y''(x) = \sum\limits_{n=0}^{\infty} n(n-1)c_n x^{n-2} = \sum\limits_{n=0}^{\infty} (n+2)(n+1)c_{n+2} x^n$ and the differential equation

becomes $\sum\limits_{n=0}^{\infty} [(n+2)(n+1)c_{n+2} + (n+1)c_n]x^n = 0$. Thus the recursion relation is $c_{n+2} = -c_n/(n+2)$

for $n = 0, 1, 2, \ldots$. But $c_0 = y(0) = 0$, so $c_{2n} = 0$ for $n = 0, 1, 2, \ldots$. Also $c_1 = y'(0) = 1$, so $c_3 = -\frac{1}{3}$, $c_5 = \frac{(-1)^2}{3\cdot 5}$,

$c_7 = \frac{(-1)^3}{3\cdot 5\cdot 7} = \frac{(-1)^3 2^3 3!}{7!}, \ldots, c_{2n+1} = \frac{(-1)^n 2^n n!}{(2n+1)!}$ for $n = 0, 1, 2, \ldots$. Thus the solution to the initial-value problem

is $y(x) = \sum\limits_{n=0}^{\infty} c_n x^n = \sum\limits_{n=0}^{\infty} \frac{(-1)^n 2^n n!}{(2n+1)!} x^{2n+1}$.

16. Let $y(x) = \sum\limits_{n=0}^{\infty} c_n x^n$. Then $y''(x) = \sum\limits_{n=0}^{\infty} n(n-1)c_n x^{n-2} = \sum\limits_{n=0}^{\infty} (n+2)(n+1)c_{n+2}x^n$ and the differential equation

becomes $\sum\limits_{n=0}^{\infty} [(n+2)(n+1)c_{n+2} - (n+2)c_n]x^n = 0$. Thus the recursion relation is $c_{n+2} = \frac{c_n}{n+1}$ for

$n = 0, 1, 2, \ldots$. Given c_0 and c_1, we have $c_2 = \frac{c_0}{1}$, $c_4 = \frac{c_2}{3} = \frac{c_0}{1\cdot 3}$, $c_6 = \frac{c_4}{5} = \frac{c_0}{1\cdot 3\cdot 5}, \ldots,$

$c_{2n} = \frac{c_0}{1\cdot 3\cdot 5\cdots (2n-1)} = c_0 \frac{2^{n-1}(n-1)!}{(2n-1)!}$. Similarly $c_3 = \frac{c_1}{2}$, $c_5 = \frac{c_3}{4} = \frac{c_1}{2\cdot 4}$,

$c_7 = \frac{c_5}{6} = \frac{c_1}{2\cdot 4\cdot 6}, \ldots, c_{2n+1} = \frac{c_1}{2\cdot 4\cdot 6\cdots 2n} = \frac{c_1}{2^n n!}$. Thus the general solution is

$y(x) = \sum\limits_{n=0}^{\infty} c_n x^n = c_0 + c_0 \sum\limits_{n=1}^{\infty} \frac{2^{n-1}(n-1)! \, x^{2n}}{(2n-1)!} + c \sum\limits_{n=0}^{\infty} \frac{x^{2n+1}}{2^n n!}$. But $\sum\limits_{n=0}^{\infty} \frac{x^{2n+1}}{2^n n!} = x \sum\limits_{n=0}^{\infty} \frac{(\frac{1}{2}x^2)^n}{n!} = xe^{x^2/2}$,

so $y(x) = c_1 xe^{x^2/2} + c_0 + c_0 \sum\limits_{n=1}^{\infty} \frac{2^{n-1}(n-1)! \, x^{2n}}{(2n-1)!}$.

17. Here the initial-value problem is $2Q'' + 40Q' + 400Q = 12$, $Q(0) = 0.01$, $Q'(0) = 0$. Then

$Q_c(t) = e^{-10t}(c_1 \cos 10t + c_2 \sin 10t)$ and we try $Q_p(t) = A$. Thus the general solution is

$Q(t) = e^{-10t}(c_1 \cos 10t + c_2 \sin 10t) + \frac{3}{100}$. But $0.01 = Q'(0) = c_1 + 0.03$ and $0 = Q''(0) = -10c_1 + 10c_2$,

so $c_1 = -0.02 = c_2$. Hence the charge is given by $Q(t) = -0.02e^{-10t}(\cos 10t + \sin 10t) + 0.03$.

18. By Hooke's Law the spring constant is $k = 64$ and the initial-value problem is $2x'' + 16x' + 64x = 0$, $x(0) = 0$,

$x'(0) = 2.4$. Thus the general solution is $x(t) = e^{-4t}(c_1 \cos 4t + c_2 \sin 4t)$. But $0 = x(0) = c_1$ and

$2.4 = x'(0) = -4c_1 + 4c_2 \Rightarrow c_1 = 0, c_2 = 0.6$. Thus the position of the mass is given by $x(t) = 0.6e^{-4t}\sin 4t$.

19. (a) Since we are assuming that the earth is a solid sphere of uniform density, we can calculate the density ρ as follows:

$\rho = \dfrac{\text{mass of earth}}{\text{volume of earth}} = \dfrac{M}{\frac{4}{3}\pi R^3}$. If V_r is the volume of the portion of the earth which lies within a distance r of the

center, then $V_r = \frac{4}{3}\pi r^3$ and $M_r = \rho V_r = \dfrac{Mr^3}{R^3}$. Thus $F_r = -\dfrac{GM_r m}{r^2} = -\dfrac{GMm}{R^3}r$.

(b) The particle is acted upon by a varying gravitational force during its motion. By Newton's Second Law of Motion,

$m\dfrac{d^2y}{dt^2} = F_y = -\dfrac{GMm}{R^3}\,y$, so $y''(t) = -k^2 y(t)$ where $k^2 = \dfrac{GM}{R^3}$. At the surface, $-mg = F_R = -\dfrac{GMm}{R^2}$, so

$g = \dfrac{GM}{R^2}$. Therefore $k^2 = \dfrac{g}{R}$.

(c) The differential equation $y'' + k^2 y = 0$ has auxiliary equation $r^2 + k^2 = 0$. (This is the r of Section 18.1 [ET 17.1],

not the r measuring distance from the earth's center.) The roots of the auxiliary equation are $\pm ik$, so by (11) in

Section 18.1 [ET 17.1], the general solution of our differential equation for t is $y(t) = c_1 \cos kt + c_2 \sin kt$. It follows that

$y'(t) = -c_1 k \sin kt + c_2 k \cos kt$. Now $y(0) = R$ and $y'(0) = 0$, so $c_1 = R$ and $c_2 k = 0$. Thus $y(t) = R \cos kt$ and

$y'(t) = -kR \sin kt$. This is simple harmonic motion (see Section 18.3 [ET 17.3]) with amplitude R, frequency k, and

phase angle 0. The period is $T = 2\pi/k$. $R \approx 3960$ mi $= 3960 \cdot 5280$ ft and $g = 32$ ft/s^2, so

$k = \sqrt{g/R} \approx 1.24 \times 10^{-3}$ s^{-1} and $T = 2\pi/k \approx 5079$ s ≈ 85 min.

(d) $y(t) = 0 \iff \cos kt = 0 \iff kt = \frac{\pi}{2} + \pi n$ for some integer $n \implies y'(t) = -kR\sin\left(\frac{\pi}{2} + \pi n\right) = \pm kR$. Thus the

particle passes through the center of the earth with speed $kR \approx 4.899$ mi/s $\approx 17{,}600$ mi/h.

☐ APPENDIX

1. $(5 - 6i) + (3 + 2i) = (5 + 3) + (-6 + 2)i = 8 + (-4)i = 8 - 4i$

2. $\left(4 - \frac{1}{2}i\right) - \left(9 + \frac{5}{2}i\right) = (4 - 9) + \left(-\frac{1}{2} - \frac{5}{2}\right)i = -5 + (-3)i = -5 - 3i$

3. $(2 + 5i)(4 - i) = 2(4) + 2(-i) + (5i)(4) + (5i)(-i) = 8 - 2i + 20i - 5i^2 = 8 + 18i - 5(-1)$

$$= 8 + 18i + 5 = 13 + 18i$$

4. $(1 - 2i)(8 - 3i) = 8 - 3i - 16i + 6(-1) = 2 - 19i$

5. $\overline{12 + 7i} = 12 - 7i$

6. $2i\left(\frac{1}{2} - i\right) = i - 2(-1) = 2 + i \quad \Rightarrow \quad \overline{2i\left(\frac{1}{2} - i\right)} = \overline{2 + i} = 2 - i$

7. $\dfrac{1 + 4i}{3 + 2i} = \dfrac{1 + 4i}{3 + 2i} \cdot \dfrac{3 - 2i}{3 - 2i} = \dfrac{3 - 2i + 12i - 8(-1)}{3^2 + 2^2} = \dfrac{11 + 10i}{13} = \dfrac{11}{13} + \dfrac{10}{13}i$

8. $\dfrac{3 + 2i}{1 - 4i} = \dfrac{3 + 2i}{1 - 4i} \cdot \dfrac{1 + 4i}{1 + 4i} = \dfrac{3 + 12i + 2i + 8(-1)}{1^2 + 4^2} = \dfrac{-5 + 14i}{17} = -\dfrac{5}{17} + \dfrac{14}{17}i$

9. $\dfrac{1}{1 + i} = \dfrac{1}{1 + i} \cdot \dfrac{1 - i}{1 - i} = \dfrac{1 - i}{1 - (-1)} = \dfrac{1 - i}{2} = \dfrac{1}{2} - \dfrac{1}{2}i$

10. $\dfrac{3}{4 - 3i} = \dfrac{3}{4 - 3i} \cdot \dfrac{4 + 3i}{4 + 3i} = \dfrac{12 + 9i}{16 - 9(-1)} = \dfrac{12}{25} + \dfrac{9}{25}i$

11. $i^3 = i^2 \cdot i = (-1)i = -i$

12. $i^{100} = (i^2)^{50} = (-1)^{50} = 1$

13. $\sqrt{-25} = \sqrt{25}\, i = 5i$

14. $\sqrt{-3}\,\sqrt{-12} = \sqrt{3}\, i\, \sqrt{12}\, i = \sqrt{3 \cdot 12}\, i^2 = \sqrt{36}\,(-1) = -6$

15. $\overline{12 - 5i} = 12 + 15i$ and $|12 - 15i| = \sqrt{12^2 + (-5)^2} = \sqrt{144 + 25} = \sqrt{169} = 13$

16. $\overline{-1 + 2\sqrt{2}\, i} = -1 - 2\sqrt{2}\, i$ and $|-1 + 2\sqrt{2}\, i| = \sqrt{(-1)^2 + \left(2\sqrt{2}\right)^2} = \sqrt{1 + 8} = \sqrt{9} = 3$

17. $\overline{-4i} = \overline{0 - 4i} = 0 + 4i = 4i$ and $|-4i| = \sqrt{0^2 + (-4)^2} = \sqrt{16} = 4$

18. Let $z = a + bi$ and $w = c + di$.

(a) $\overline{z + w} = \overline{(a + bi) + (c + di)} = \overline{(a + c) + (b + d)i} = (a + c) - (b + d)i = (a - bi) + (c - di) = \overline{z} + \overline{w}$

(b) $\overline{zw} = \overline{(a + bi)(c + di)} = \overline{(ac - bd) + (ad + bc)i} = (ac - bd) - (ad + bc)i.$

On the other hand, $\overline{z}\,\overline{w} = (a - bi)(c - di) = (ac - bd) - (ad + bc)i = \overline{zw}.$

(c) Use mathematical induction and part (b): Let S_n be the statement that $\overline{z^n} = \overline{z}^{\,n}$. S_1 is true because $\overline{z^1} = \overline{z} = \overline{z}^{\,1}$.

Assume S_k is true, that is $\overline{z^k} = \overline{z}^{\,k}$. Then $\overline{z^{k+1}} = \overline{z^{1+k}} = \overline{z z^k} = \overline{z}\,\overline{z^k}$ [part (b) with $w = z^k$] $= \overline{z}^{\,1}\overline{z}^{\,k} = \overline{z}^{\,1+k} = \overline{z}^{\,k+1}$,

which shows that S_{k+1} is true. Therefore, by mathematical induction, $\overline{z^n} = \overline{z}^{\,n}$ for every positive integer n.

Another proof: Use part (b) with $w = z$, and mathematical induction.

19. $4x^2 + 9 = 0 \;\Leftrightarrow\; 4x^2 = -9 \;\Leftrightarrow\; x^2 = -\frac{9}{4} \;\Leftrightarrow\; x = \pm\sqrt{-\frac{9}{4}} = \pm\sqrt{\frac{9}{4}}\,i = \pm\frac{3}{2}i.$

20. $x^4 = 1 \;\Leftrightarrow\; x^4 - 1 = 0 \;\Leftrightarrow\; (x^2 - 1)(x^2 + 1) = 0 \;\Leftrightarrow\; x^2 - 1 = 0 \text{ or } x^2 + 1 = 0 \;\Leftrightarrow\; x = \pm 1 \text{ or } x = \pm i.$

21. By the quadratic formula, $x^2 + 2x + 5 = 0 \;\Leftrightarrow\; x = \dfrac{-2 \pm \sqrt{2^2 - 4(1)(5)}}{2(1)} = \dfrac{-2 \pm \sqrt{-16}}{2} = \dfrac{-2 \pm 4i}{2} = -1 \pm 2i.$

22. $2x^2 - 2x + 1 = 0 \;\Leftrightarrow\; x = \dfrac{-(-2) \pm \sqrt{(-2)^2 - 4(2)(1)}}{2(2)} = \dfrac{2 \pm \sqrt{-4}}{4} = \dfrac{2 \pm 2i}{4} = \dfrac{1}{2} \pm \dfrac{1}{2}i$

23. By the quadratic formula, $z^2 + z + 2 = 0 \;\Leftrightarrow\; z = \dfrac{-1 \pm \sqrt{1^2 - 4(1)(2)}}{2(1)} = \dfrac{-1 \pm \sqrt{-7}}{2} = -\dfrac{1}{2} \pm \dfrac{\sqrt{7}}{2}i.$

24. $z^2 + \frac{1}{2}z + \frac{1}{4} = 0 \;\Leftrightarrow\; 4z^2 + 2z + 1 = 0 \;\Leftrightarrow\;$

$z = \dfrac{-2 \pm \sqrt{2^2 - 4(4)(1)}}{2(4)} = \dfrac{-2 \pm \sqrt{-12}}{8} = \dfrac{-2 \pm 2\sqrt{3}\,i}{8} = -\dfrac{1}{4} \pm \dfrac{\sqrt{3}}{4}i$

25. For $z = -3 + 3i$, $r = \sqrt{(-3)^2 + 3^2} = 3\sqrt{2}$ and $\tan\theta = \frac{3}{-3} = -1 \;\Rightarrow\; \theta = \frac{3\pi}{4}$ (since z lies in the second quadrant).

Therefore, $-3 + 3i = 3\sqrt{2}\left(\cos\frac{3\pi}{4} + i\sin\frac{3\pi}{4}\right).$

26. For $z = 1 - \sqrt{3}\,i$, $r = \sqrt{1^2 + \left(-\sqrt{3}\right)^2} = 2$ and $\tan\theta = \frac{-\sqrt{3}}{1} = -\sqrt{3} \;\Rightarrow\; \theta = \frac{5\pi}{3}$ (since z lies in the fourth quadrant).

Therefore, $1 - \sqrt{3}\,i = 2\left(\cos\frac{5\pi}{3} + i\sin\frac{5\pi}{3}\right).$

27. For $z = 3 + 4i$, $r = \sqrt{3^2 + 4^2} = 5$ and $\tan\theta = \frac{4}{3} \;\Rightarrow\; \theta = \tan^{-1}\left(\frac{4}{3}\right)$ (since z lies in the first quadrant). Therefore,

$3 + 4i = 5\left[\cos\left(\tan^{-1}\frac{4}{3}\right) + i\sin\left(\tan^{-1}\frac{4}{3}\right)\right].$

28. For $z = 8i$, $r = \sqrt{0^2 + 8^2} = 8$ and $\tan\theta = \frac{8}{0}$ is undefined, so $\theta = \frac{\pi}{2}$ (since z lies on the positive imaginary axis). Therefore,

$8i = 8\left(\cos\frac{\pi}{2} + i\sin\frac{\pi}{2}\right).$

29. For $z = \sqrt{3} + i$, $r = \sqrt{\left(\sqrt{3}\right)^2 + 1^2} = 2$ and $\tan\theta = \frac{1}{\sqrt{3}} \;\Rightarrow\; \theta = \frac{\pi}{6} \;\Rightarrow\; z = 2\left(\cos\frac{\pi}{6} + i\sin\frac{\pi}{6}\right).$

For $w = 1 + \sqrt{3}\,i$, $r = 2$ and $\tan\theta = \sqrt{3} \;\Rightarrow\; \theta = \frac{\pi}{3} \;\Rightarrow\; w = 2\left(\cos\frac{\pi}{3} + i\sin\frac{\pi}{3}\right).$

Therefore, $zw = 2 \cdot 2\left[\cos\left(\frac{\pi}{6} + \frac{\pi}{3}\right) + i\sin\left(\frac{\pi}{6} + \frac{\pi}{3}\right)\right] = 4\left(\cos\frac{\pi}{2} + i\sin\frac{\pi}{2}\right),$

$z/w = \frac{2}{2}\left[\cos\left(\frac{\pi}{6} - \frac{\pi}{3}\right) + i\sin\left(\frac{\pi}{6} - \frac{\pi}{3}\right)\right] = \cos\left(-\frac{\pi}{6}\right) + i\sin\left(-\frac{\pi}{6}\right)$, and $1 = 1 + 0i = 1(\cos 0 + i\sin 0) \;\Rightarrow\;$

$1/z = \frac{1}{2}\left[\cos\left(0 - \frac{\pi}{6}\right) + i\sin\left(0 - \frac{\pi}{6}\right)\right] = \frac{1}{2}\left[\cos\left(-\frac{\pi}{6}\right) + i\sin\left(-\frac{\pi}{6}\right)\right].$ For $1/z$, we could also use the formula that precedes

Example 5 to obtain $1/z = \frac{1}{2}\left(\cos\frac{\pi}{6} - i\sin\frac{\pi}{6}\right).$

30. For $z = 4\sqrt{3} - 4i$, $r = \sqrt{\left(4\sqrt{3}\right)^2 + (-4)^2} = \sqrt{64} = 8$ and $\tan\theta = \frac{-4}{4\sqrt{3}} = -\frac{1}{\sqrt{3}}$ $\Rightarrow$ $\theta = \frac{11\pi}{6}$ $\Rightarrow$

$z = 8\left(\cos\frac{11\pi}{6} + i\sin\frac{11\pi}{6}\right)$. For $w = 8i$, $r = \sqrt{0^2 + 8^2} = 8$ and $\tan\theta = \frac{8}{0}$ is undefined, so $\theta = \frac{\pi}{2}$ $\Rightarrow$

$w = 8\left(\cos\frac{\pi}{2} + i\sin\frac{\pi}{2}\right)$. Therefore, $zw = 8 \cdot 8\left[\cos\left(\frac{11\pi}{6} + \frac{\pi}{2}\right) + i\sin\left(\frac{11\pi}{6} + \frac{\pi}{2}\right)\right] = 64\left(\cos\frac{\pi}{3} + i\sin\frac{\pi}{3}\right)$,

$z/w = \frac{8}{8}\left[\cos\left(\frac{11\pi}{6} - \frac{\pi}{2}\right) + i\sin\left(\frac{11\pi}{6} - \frac{\pi}{2}\right)\right] = \cos\frac{4\pi}{3} + i\sin\frac{4\pi}{3}$, and

$1 = 1 + 0i = 1(\cos 0 + i\sin 0)$ $\Rightarrow$ $1/z = \frac{1}{8}\left[\cos\left(0 - \frac{11\pi}{6}\right) + i\sin\left(0 - \frac{11\pi}{6}\right)\right] = \frac{1}{8}\left[\cos\left(\frac{\pi}{6}\right) + i\sin\left(\frac{\pi}{6}\right)\right]$.

For $1/z$, we could also use the formula that precedes Example 5 to obtain $1/z = \frac{1}{8}\left(\cos\frac{11\pi}{6} - i\sin\frac{11\pi}{6}\right)$.

31. For $z = 2\sqrt{3} - 2i$, $r = \sqrt{\left(2\sqrt{3}\right)^2 + (-2)^2} = 4$ and $\tan\theta = \frac{-2}{2\sqrt{3}} = -\frac{1}{\sqrt{3}}$ $\Rightarrow$ $\theta = -\frac{\pi}{6}$ $\Rightarrow$

$z = 4\left[\cos\left(-\frac{\pi}{6}\right) + i\sin\left(-\frac{\pi}{6}\right)\right]$. For $w = -1 + i$, $r = \sqrt{2}$, $\tan\theta = \frac{1}{-1} = -1$ $\Rightarrow$ $\theta = \frac{3\pi}{4}$ $\Rightarrow$

$w = \sqrt{2}\left(\cos\frac{3\pi}{4} + i\sin\frac{3\pi}{4}\right)$. Therefore, $zw = 4\sqrt{2}\left[\cos\left(-\frac{\pi}{6} + \frac{3\pi}{4}\right) + i\sin\left(-\frac{\pi}{6} + \frac{3\pi}{4}\right)\right] = 4\sqrt{2}\left(\cos\frac{7\pi}{12} + i\sin\frac{7\pi}{12}\right)$,

$z/w = \frac{4}{\sqrt{2}}\left[\cos\left(-\frac{\pi}{6} - \frac{3\pi}{4}\right) + i\sin\left(-\frac{\pi}{6} - \frac{3\pi}{4}\right)\right] = \frac{4}{\sqrt{2}}\left[\cos\left(-\frac{11\pi}{12}\right) + i\sin\left(-\frac{11\pi}{12}\right)\right] = 2\sqrt{2}\left(\cos\frac{13\pi}{12} + i\sin\frac{13\pi}{12}\right)$, and

$1/z = \frac{1}{4}\left[\cos\left(-\frac{\pi}{6}\right) - i\sin\left(-\frac{\pi}{6}\right)\right] = \frac{1}{4}\left(\cos\frac{\pi}{6} + i\sin\frac{\pi}{6}\right)$.

32. For $z = 4(\sqrt{3} + i) = 4\sqrt{3} + 4i$, $r = \sqrt{\left(4\sqrt{3}\right)^2 + 4^2} = \sqrt{64} = 8$ and $\tan\theta = \frac{4}{4\sqrt{3}} = \frac{1}{\sqrt{3}}$ $\Rightarrow$ $\theta = \frac{\pi}{6}$ $\Rightarrow$

$z = 8\left(\cos\frac{\pi}{6} + i\sin\frac{\pi}{6}\right)$. For $w = -3 - 3i$, $r = \sqrt{(-3)^2 + (-3)^2} = \sqrt{18} = 3\sqrt{2}$ and $\tan\theta = \frac{-3}{-3} = 1$ $\Rightarrow$ $\theta = \frac{5\pi}{4}$ $\Rightarrow$

$w = 3\sqrt{2}\left(\cos\frac{5\pi}{4} + i\sin\frac{5\pi}{4}\right)$. Therefore, $zw = 8 \cdot 3\sqrt{2}\left[\cos\left(\frac{\pi}{6} + \frac{5\pi}{4}\right) + i\sin\left(\frac{\pi}{6} + \frac{5\pi}{4}\right)\right] = 24\sqrt{2}\left(\cos\frac{17\pi}{12} + i\sin\frac{17\pi}{12}\right)$,

$z/w = \frac{8}{3\sqrt{2}}\left[\cos\left(\frac{\pi}{6} - \frac{5\pi}{4}\right) + i\sin\left(\frac{\pi}{6} - \frac{5\pi}{4}\right)\right] = \frac{4\sqrt{2}}{3}\left[\cos\left(-\frac{13\pi}{12}\right) + i\sin\left(-\frac{13\pi}{12}\right)\right]$, and $1/z = \frac{1}{8}\left(\cos\frac{\pi}{6} - i\sin\frac{\pi}{6}\right)$.

33. For $z = 1 + i$, $r = \sqrt{2}$ and $\tan\theta = \frac{1}{1} = 1$ $\Rightarrow$ $\theta = \frac{\pi}{4}$ $\Rightarrow$ $z = \sqrt{2}\left(\cos\frac{\pi}{4} + i\sin\frac{\pi}{4}\right)$. So by De Moivre's Theorem,

$$(1 + i)^{20} = \left[\sqrt{2}\left(\cos\frac{\pi}{4} + i\sin\frac{\pi}{4}\right)\right]^{20} = (2^{1/2})^{20}\left(\cos\frac{20\cdot\pi}{4} + i\sin\frac{20\cdot\pi}{4}\right) = 2^{10}(\cos 5\pi + i\sin 5\pi)$$
$$= 2^{10}[-1 + i(0)] = -2^{10} = -1024$$

34. For $z = 1 - \sqrt{3}i$, $r = \sqrt{1^2 + \left(-\sqrt{3}\right)^2} = 2$ and $\tan\theta = \frac{-\sqrt{3}}{1} = -\sqrt{3}$ $\Rightarrow$ $\theta = \frac{5\pi}{3}$ $\Rightarrow$ $z = 2\left(\cos\frac{5\pi}{3} + i\sin\frac{5\pi}{3}\right)$.

So by De Moivre's Theorem,

$$\left(1 - \sqrt{3}i\right)^5 = \left[2\left(\cos\frac{5\pi}{3} + i\sin\frac{5\pi}{3}\right)\right]^5 = 2^5\left(\cos\frac{5\cdot 5\pi}{3} + i\sin\frac{5\cdot 5\pi}{3}\right) = 2^5\left(\cos\frac{\pi}{3} + i\sin\frac{\pi}{3}\right)$$
$$= 32\left(\frac{1}{2} + \frac{\sqrt{3}}{2}i\right) = 16 + 16\sqrt{3}i$$

35. For $z = 2\sqrt{3} + 2i$, $r = \sqrt{\left(2\sqrt{3}\right)^2 + 2^2} = \sqrt{16} = 4$ and $\tan\theta = \frac{2}{2\sqrt{3}} = \frac{1}{\sqrt{3}}$ $\Rightarrow$ $\theta = \frac{\pi}{6}$ $\Rightarrow$ $z = 4\left(\cos\frac{\pi}{6} + i\sin\frac{\pi}{6}\right)$.

So by De Moivre's Theorem,

$$\left(2\sqrt{3} + 2i\right)^5 = \left[4\left(\cos\frac{\pi}{6} + i\sin\frac{\pi}{6}\right)\right]^5 = 4^5\left(\cos\frac{5\pi}{6} + i\sin\frac{5\pi}{6}\right) = 1024\left[-\frac{\sqrt{3}}{2} + \frac{1}{2}i\right] = -512\sqrt{3} + 512i.$$

36. For $z = 1 - i$, $r = \sqrt{2}$ and $\tan\theta = \frac{-1}{1} = -1$ $\Rightarrow$ $\theta = \frac{7\pi}{4}$ $\Rightarrow$ $z = \sqrt{2}\left(\cos\frac{7\pi}{4} + i\sin\frac{7\pi}{4}\right)$ $\Rightarrow$

$(1 - i)^8 = \left[\sqrt{2}\left(\cos\frac{7\pi}{4} + i\sin\frac{7\pi}{4}\right)\right]^8 = 2^4\left(\cos\frac{8\cdot 7\pi}{4} + i\sin\frac{8\cdot 7\pi}{4}\right) = 16(\cos 14\pi + i\sin 14\pi) = 16(1 + 0i) = 16.$

37. $1 = 1 + 0i = 1\,(\cos 0 + i \sin 0)$. Using Equation 3 with $r = 1$, $n = 8$, and $\theta = 0$, we have

$$w_k = 1^{1/8}\left[\cos\left(\frac{0 + 2k\pi}{8}\right) + i\sin\left(\frac{0 + 2k\pi}{8}\right)\right] = \cos\frac{k\pi}{4} + i\sin\frac{k\pi}{4}, \text{ where } k = 0, 1, 2, \dots, 7.$$

$w_0 = 1(\cos 0 + i \sin 0) = 1$, $w_1 = 1\left(\cos\frac{\pi}{4} + i\sin\frac{\pi}{4}\right) = \frac{1}{\sqrt{2}} + \frac{1}{\sqrt{2}}i$,

$w_2 = 1\left(\cos\frac{\pi}{2} + i\sin\frac{\pi}{2}\right) = i$, $w_3 = 1\left(\cos\frac{3\pi}{4} + i\sin\frac{3\pi}{4}\right) = -\frac{1}{\sqrt{2}} + \frac{1}{\sqrt{2}}i$,

$w_4 = 1(\cos\pi + i\sin\pi) = -1$, $w_5 = 1\left(\cos\frac{5\pi}{4} + i\sin\frac{5\pi}{4}\right) = -\frac{1}{\sqrt{2}} - \frac{1}{\sqrt{2}}i$,

$w_6 = 1\left(\cos\frac{3\pi}{2} + i\sin\frac{3\pi}{2}\right) = -i$, $w_7 = 1\left(\cos\frac{7\pi}{4} + i\sin\frac{7\pi}{4}\right) = \frac{1}{\sqrt{2}} - \frac{1}{\sqrt{2}}i$

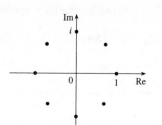

38. $32 = 32 + 0i = 32(\cos 0 + i \sin 0)$. Using Equation 3 with $r = 32$, $n = 5$, and $\theta = 0$, we have

$$w_k = 32^{1/5}\left[\cos\left(\frac{0 + 2k\pi}{5}\right) + i\sin\left(\frac{0 + 2k\pi}{5}\right)\right] = 2\left(\cos\frac{2}{5}\pi k + i\sin\frac{2}{5}\pi k\right), \text{ where } k = 0, 1, 2, 3, 4.$$

$w_0 = 2(\cos 0 + i \sin 0) = 2$

$w_1 = 2\left(\cos\frac{2\pi}{5} + i\sin\frac{2\pi}{5}\right)$

$w_2 = 2\left(\cos\frac{4\pi}{5} + i\sin\frac{4\pi}{5}\right)$

$w_3 = 2\left(\cos\frac{6\pi}{5} + i\sin\frac{6\pi}{5}\right)$

$w_4 = 2\left(\cos\frac{8\pi}{5} + i\sin\frac{8\pi}{5}\right)$

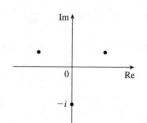

39. $i = 0 + i = 1\left(\cos\frac{\pi}{2} + i\sin\frac{\pi}{2}\right)$. Using Equation 3 with $r = 1$, $n = 3$, and $\theta = \frac{\pi}{2}$, we have

$$w_k = 1^{1/3}\left[\cos\left(\frac{\frac{\pi}{2} + 2k\pi}{3}\right) + i\sin\left(\frac{\frac{\pi}{2} + 2k\pi}{3}\right)\right], \text{ where } k = 0, 1, 2.$$

$w_0 = \left(\cos\frac{\pi}{6} + i\sin\frac{\pi}{6}\right) = \frac{\sqrt{3}}{2} + \frac{1}{2}i$

$w_1 = \left(\cos\frac{5\pi}{6} + i\sin\frac{5\pi}{6}\right) = -\frac{\sqrt{3}}{2} + \frac{1}{2}i$

$w_2 = \left(\cos\frac{9\pi}{6} + i\sin\frac{9\pi}{6}\right) = -i$

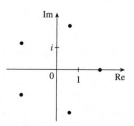

40. $1 + i = \sqrt{2}\left(\cos\frac{\pi}{4} + i\sin\frac{\pi}{4}\right)$. Using Equation 3 with $r = \sqrt{2}$, $n = 3$, and $\theta = \frac{\pi}{4}$, we have

$$w_k = \left(\sqrt{2}\right)^{1/3}\left[\cos\left(\frac{\frac{\pi}{4} + 2k\pi}{3}\right) + i\sin\left(\frac{\frac{\pi}{4} + 2k\pi}{3}\right)\right], \text{ where } k = 0, 1, 2.$$

$w_0 = 2^{1/6}\left(\cos\frac{\pi}{12} + i\sin\frac{\pi}{12}\right)$

$w_1 = 2^{1/6}\left(\cos\frac{3\pi}{4} + i\sin\frac{3\pi}{4}\right) = 2^{1/6}\left(-\frac{1}{\sqrt{2}} + \frac{1}{\sqrt{2}}i\right) = -2^{-1/3} + 2^{-1/3}i$

$w_2 = 2^{1/6}\left(\cos\frac{17\pi}{12} + i\sin\frac{17\pi}{12}\right)$

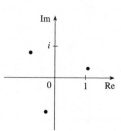

41. Using Euler's formula (6) with $y = \frac{\pi}{2}$, we have $e^{i\pi/2} = \cos\frac{\pi}{2} + i\sin\frac{\pi}{2} = 0 + 1i = i$.

42. Using Euler's formula (6) with $y = 2\pi$, we have $e^{2\pi i} = \cos 2\pi + i\sin 2\pi = 1$.

43. Using Euler's formula (6) with $y = \frac{\pi}{3}$, we have $e^{i\pi/3} = \cos\frac{\pi}{3} + i\sin\frac{\pi}{3} = \frac{1}{2} + \frac{\sqrt{3}}{2}i$.

44. Using Euler's formula (6) with $y = -\pi$, we have $e^{-i\pi} = \cos(-\pi) + i\sin(-\pi) = -1$.

45. Using Equation 7 with $x = 2$ and $y = \pi$, we have $e^{2+i\pi} = e^2 e^{i\pi} = e^2(\cos\pi + i\sin\pi) = e^2(-1+0) = -e^2$.

46. Using Equation 7 with $x = \pi$ and $y = 1$, we have $e^{\pi+i} = e^\pi \cdot e^{1i} = e^\pi(\cos 1 + i\sin 1) = e^\pi \cos 1 + (e^\pi \sin 1)i$.

47. Take $r = 1$ and $n = 3$ in De Moivre's Theorem to get

$$[1(\cos\theta + i\sin\theta)]^3 = 1^3(\cos 3\theta + i\sin 3\theta)$$

$$(\cos\theta + i\sin\theta)^3 = \cos 3\theta + i\sin 3\theta$$

$$\cos^3\theta + 3(\cos^2\theta)(i\sin\theta) + 3(\cos\theta)(i\sin\theta)^2 + (i\sin\theta)^3 = \cos 3\theta + i\sin 3\theta$$

$$\cos^3\theta + (3\cos^2\theta\sin\theta)i - 3\cos\theta\sin^2\theta - (\sin^3\theta)i = \cos 3\theta + i\sin 3\theta$$

$$(\cos^3\theta - 3\sin^2\theta\cos\theta) + (3\sin\theta\cos^2\theta - \sin^3\theta)i = \cos 3\theta + i\sin 3\theta$$

Equating real and imaginary parts gives $\cos 3\theta = \cos^3\theta - 3\sin^2\theta\cos\theta$ and $\sin 3\theta = 3\sin\theta\cos^2\theta - \sin^3\theta$.

48. Using Formula 6,

$$e^{ix} + e^{-ix} = (\cos x + i\sin x) + [\cos(-x) + i\sin(-x)] = \cos x + i\sin x + \cos x - i\sin x = 2\cos x$$

Thus, $\cos x = \dfrac{e^{ix} + e^{-ix}}{2}$. Similarly,

$$e^{ix} - e^{-ix} = (\cos x + i\sin x) - [\cos(-x) + i\sin(-x)] = \cos x + i\sin x - \cos x - (-i\sin x) = 2i\sin x$$

Therefore, $\sin x = \dfrac{e^{ix} - e^{-ix}}{2i}$.

49. $F(x) = e^{rx} = e^{(a+bi)x} = e^{ax+bxi} = e^{ax}(\cos bx + i\sin bx) = e^{ax}\cos bx + i(e^{ax}\sin bx) \Rightarrow$

$$F'(x) = (e^{ax}\cos bx)' + i(e^{ax}\sin bx)'$$

$$= (ae^{ax}\cos bx - be^{ax}\sin bx) + i(ae^{ax}\sin bx + be^{ax}\cos bx)$$

$$= a[e^{ax}(\cos bx + i\sin bx)] + b[e^{ax}(-\sin bx + i\cos bx)]$$

$$= ae^{rx} + b[e^{ax}(i^2\sin bx + i\cos bx)]$$

$$= ae^{rx} + bi[e^{ax}(\cos bx + i\sin bx)] = ae^{rx} + bie^{rx} = (a+bi)e^{rx} = re^{rx}$$

50. (a) From Exercise 49, $F(x) = e^{(1+i)x} \Rightarrow F'(x) = (1+i)e^{(1+i)x}$. So

$$\int e^{(1+i)x}\, dx = \frac{1}{1+i}\int F'(x)\, dx = \frac{1}{1+i}F(x) + C = \frac{1-i}{2}F(x) + C = \frac{1-i}{2}e^{(1+i)x} + C$$

(b) $\int e^{(1+i)x}\, dx = \int e^x e^{ix}\, dx = \int e^x(\cos x + i\sin x)\, dx = \int e^x\cos x\, dx + i\int e^x\sin x$ **(1)**.

Also,

$$\frac{1-i}{2}e^{(1+i)x} = \tfrac{1}{2}e^{(1+i)x} - \tfrac{1}{2}ie^{(1+i)x} = \tfrac{1}{2}e^{x+ix} - \tfrac{1}{2}ie^{x+ix}$$

$$= \tfrac{1}{2}e^x(\cos x + i\sin x) - \tfrac{1}{2}ie^x(\cos x + i\sin x)$$

$$= \tfrac{1}{2}e^x\cos x + \tfrac{1}{2}e^x\sin x + \tfrac{1}{2}ie^x\sin x - \tfrac{1}{2}ie^x\cos x$$

$$= \tfrac{1}{2}e^x(\cos x + \sin x) + i\left[\tfrac{1}{2}e^x(\sin x - \cos x)\right]$$ **(2)**

Equating the real and imaginary parts in **(1)** and **(2)**, we see that $\int e^x\cos x\, dx = \tfrac{1}{2}e^x(\cos x + \sin x) + C$ and $\int e^x\sin x\, dx = \tfrac{1}{2}e^x(\sin x - \cos x) + C$.